Proceedings of the
International Congress of Mathematicians
August 21–29, 1990, Kyoto, Japan

I

Proceedings of the
International Congress of Mathematicians

August 21–29, 1990
Kyoto, Japan

Volume I

The Mathematical Society of Japan

Springer-Verlag

Tokyo Berlin Heidelberg New York
London Paris Hong Kong Barcelona
Budapest

International Congress of Mathematicians
August 21–29, 1990, Kyoto, Japan

Editor:

Ichiro Satake
Mathematical Institute
Faculty of Science
Tohoku University
Sendai 980, Japan

The logo for the ICM-90, designed by Kazuyoshi Aoki and Yuji Komai, symbolizes a Japanese stone lantern, the first character for Kyoto, as well as the character for 10^{16}.

With 96 figures, including 11 halftone illustrations

ISBN 4-431-70047-1 Set (2 volumes)
Springer-Verlag Tokyo Berlin Heidelberg New York

ISBN 3-540-70047-1 in 2 Bänden
Springer-Verlag Berlin Heidelberg New York Tokyo

ISBN 0-387-70047-1 Set (2 volumes)
Springer-Verlag New York Berlin Heidelberg Tokyo

Library of Congress Cataloging-in-Publication Data
International Congress of Mathematicians (1990: Kyoto, Japan)
Proceedings of the International Congress of Mathematicians, August 21-29, 1990, Kyoto/ edited by Ichiro Satake.
p. cm. Includes bibliographical references. ISBN 0-387-70047-1
1. Mathematics – Congresses. I. Satake, Ichiro. 1927-. II. Title QA1.I82 1990 510–dc20
91-4972 CIP

Typesetting: Asco Typesetting Ltd., Hong Kong; Springer T_EX in-house system and typesetting output by Universitätsdruckerei H. Stürtz AG, Würzburg, Fed. Rep. of Germany
Printing and binding: Permanent Typesetting & Printing Co., Ltd., Hong Kong

543210

Preface

The *Proceedings* of the International Congress of Mathematicians 1990, held in Kyoto from August 21 through 29, 1990, are printed in two volumes.

Volume I contains the official record of the Congress, the list of ordinary members, the reports on the work of the Fields Medalists and the Nevanlinna Prize Winner, the plenary addresses, as well as the invited addresses in Sections 1–6. Volume II contains the invited addresses in Sections 7–18. A complete author index is included in both volumes.

The short communications made by members at the Congress are not included in the *Proceedings*, but the names of the communicators are marked in the list of ordinary members. Summaries of these contributions, if received in time, were printed in the *Abstracts*, issued to members during the Congress.

Sendai, April, 1991   The editor

Contents

Volume II

Contents v
Scientific Program vii

Past Congresses

Kyoto International Conference Hall

Past Fields Medalists
and Rolf Nevanlinna Prize Winners

Recipients of Fields Medals

1936 Lars V. Ahlfors
     Jesse Douglas

1950 Laurent Schwartz
     Atle Selberg

1954 Kunihiko Kodaira
     Jean-Pierre Serre

1958 Klaus F. Roth
     René Thom

1962 Lars Hörmander
     John W. Milnor

1966 Michael F. Atiyah
     Paul J. Cohen
     Alexander Grothendieck
     Stephen Smale

1970 Alan Baker
     Heisuke Hironaka
     Sergei P. Novikov
     John G. Thompson

1974 Enrico Bombieri
     David B. Mumford

1978 Pierre R. Deligne
     Charles L. Fefferman
     Grigorii A. Margulis
     Daniel G. Quillen

1982 Alain Connes
     William P. Thurston
     Shing-Tung Yau

1986 Simon K. Donaldson
     Gerd Faltings
     Michael H. Freedman

Rolf Nevanlinna Prize Winners

1982 Robert E. Tarjan

1986 Leslie G. Valiant

Audience in the Main Hall of the Kyoto International Conference Hall

Organization of the Congress

The International Congress of Mathematicians 1990 was held in Kyoto at the invitation of the Science Council of Japan (JSC) and the Mathematical Society of Japan (MSJ) under the cosponsorship of the Japan Society of Mathematical Education, the History of Science Society of Japan, the Institute of Actuaries of Japan, the Japan Society for Software Science and Technology, the Japan Statistical Society, and the Operations Research Society of Japan. The Congress was also supported by the Ministry of Education, Science and Culture, by Kyoto Prefecture, Kyoto City, Kyoto University, and the Information Processing Society of Japan.

The members of the Organizing Committee and its subcommittees are listed on the following pages. However, until the Organizing Committee was established on August 14, 1989, all preparations were carried out by the Committee of ICM-90, which was set up in the MSJ in December, 1986.

The scientific program of the Congress was in the hands of the Program Committee, which was appointed by the International Mathematical Union (IMU) in May, 1987. Its members were Nicolaas H. Kuiper (Chairman), Vladimir I. Arnold, Alain Connes, Ronald L. Graham, Heisuke Hironaka, Masaki Kashiwara, Robert P. Langlands, Sigeru Mizohata, and Daniel G. Quillen.

Recipients of the Fields Medals and the Rolf Nevanlinna Prize were selected by the respective committees appointed by the IMU in April, 1988. The Fields Medal Committee consisted of Ludwig D. Faddeev (Chairman ex officio), Michael F. Atiyah, Jean-Michel Bismut, Enrico Bombieri, Charles L. Fefferman, Kenkichi Iwasawa, Peter D. Lax, and Igor Shafarevich. The Rolf Nevanlinna Prize Committee consisted of László Lovász (Chairman), Alexandre J. Chorin, Michael Rabin and Volker Strassen.

The Organizing Committee was responsible for the other activities of the Congress. The Japan Travel Bureau (JTB) handled accommodations for the participants as the official travel agent.

All sessions of the Congress took place inside the Kyoto International Conference Hall. There were 15 one-hour Plenary Addresses and 138 forty-five-minute Section Lectures in 18 Sections. In addition to these invited lectures, 5 forty-five-minute lectures were invited by the International Commission on Mathematical

Instruction. Participants gave 620 Short Communications in total and conducted more than 40 informal seminars and other meetings.

Various social events were arranged for all participants. Ancient court music and dance accompanied the Opening Ceremony. A reception followed the Presentation of the Works of the Fields Medalists and the Rolf Nevanlinna Prize Winner. A cello recital was held in the evening of August 23, and traditional music performances in the afternoon of August 25. Sunday, August 26 was a holiday, on which several guided tours were arranged by JTB. The Closing Ceremony was followed by a banquet with an entertainment of folk music and dance.

In the evening of August 22, Kyoto Prefecture, Kyoto City, and Kyoto University were hosts to a reception for the Fields Medalists, the Rolf Nevanlinna Prize Winner, invited speakers, and officials of the IMU.

The Organizing Committee of the International Congress of Mathematicians 1990

Honorary Members

Shôkichi Iyanaga
  Member of the Japan Academy, University of Tokyo, Prof. Emer.
Kôsaku Yosida†
  Member of the Japan Academy, University of Tokyo and Kyoto University, Prof. Emer.
Mitio Nagumo
  Osaka University, Prof. Emer.
Saburo Kawai
  Former Chairman of The Institute of Actuaries of Japan, Honorary Chairman of The Kyoei Life Insurance Company, Limited
Masanori Yamanouchi
  Former Chairman of The Institute of Actuaries of Japan

Executive Committee

President
  Kunihiko Kodaira, University of Tokyo, Prof. Emer.
Vice Presidents
  Kiyosi Itô, Kyoto University, Prof. Emer.
  Hikosaburo Komatsu, University of Tokyo
General Secretary
  Huzihiro Araki, RIMS, Kyoto University
Secretary Treasurer
  Hirosi Toda, Kyoto University
Secretaries
  Heisuke Hironaka, Member of the Japan Academy, Harvard University
  Akio Hattori, Meiji University, President of the Mathematical Society of Japan
  Shigeru Iitaka, Gakushuin University
  Sigeru Mizohata, Osaka Electro-Communication University
  Masayoshi Nagata, Okayama University of Science

Kodaira, 1954 Fields Medalist, President of the Organizing Committee

Itô, Honorary President of the Congress

Members
   Hiroshi Fujita, Meiji University
   Ichiro Satake, Tohoku University
   Itiro Tamura, Tokyo Denki University

Organizing Committee

Members representing the Science Council of Japan (JSC)
   Hirotugu Akaike, Institute of Statistical Mathematics, Member of JSC, 4th Division
   Huzihiro Araki, RIMS, Kyoto University, Liaison Committee for Physics
   Hiroshi Fujita, Meiji University, Member of JSC, 4th Division
   Nobuyuki Ikeda, Osaka University, Liaison Committee for Mathematics
   Seizô Itô, Kyorin University, Liaison Committee for Mathematics
   Hikosaburo Komatsu, University of Tokyo, Liaison Committee for Mathematics
   Yoshitomo Matsuo, Science University of Tokyo, Liaison Committee for Mathematics
   Sigeru Mizohata, Osaka Electro-Communication University, Liaison Committee for Mathematics
   Hisasi Morikawa, Nagoya University, Liaison Committee for Mathematics
   Masayoshi Nagata, Okayama University of Science, Member of JSC, 4th Division
   Sadao Nakajima, Department of Physics, Tokai University, Member of JSC, 4th Division
   Ichiro Satake, Tohoku University, Liaison Committee for Mathematics
   Kei Takeuchi, University of Tokyo, Member of JSC, 3rd Division
   Tadashi Ueno, College of Arts and Sciences, University of Tokyo, Member of JSC, 4th Division
   Masaya Yamaguti, Ryukoku University, Liaison Committee for Mathematics
Members representing the Sponsoring Societies
   Akio Hattori, Meiji University, President of the Mathematical Society of Japan
   Heisuke Hironaka, Harvard University
   Shigeru Iitaka, Gakushuin University
   Kiyosi Itô, Kyoto University, Prof. Emer.
   Tosihusa Kimura, Science University of Tokyo
   Kunihiko Kodaira, University of Tokyo, Prof. Emer.
   Isamu Mogi, President of Bunkyo University, President of the Japan Society of Mathematical Education
   Shingo Murakami, Osaka University
   Shuji Naito, Managing Director of Mitsui Mutual Life Insurance Company, President of The Institute of Actuaries of Japan
   Takushiro Ochiai, University of Tokyo

Tadao Oda, Tohoku University
Itiro Tamura, Tokyo Denki University
Hirosi Toda, Kyoto University
Toshio Yamazaki, President of The History of Science Society of Japan
Nobuo Yoneda, Department of Information Science, University of Tokyo,
  President of the Japan Society for Software Science and Technology

Subcommittees of the Organizing Committee

General Committee

Chairman
  Hiroshi Fujita, Meiji University
Members
  Shōzo Koshi, Hokkaido University
  Mutsuhide Matsumura, University of Tsukuba
  Yukiyosi Kawada, University of Tokyo, Prof. Emer.
  Takuo Fukuda, Tokyo Institute of Technology
  Koichi Ogiue, Tokyo Metropolitan University
  Mitsuo Morimoto, Sophia University
  Yuji Ito, Keio University
  Tokitake Kusama, Waseda University
  Takeyuki Hida, Nagoya University
  Nobuo Shimada, Okayama University of Science
  Kenji Ueno, Kyoto University
  Shinzo Watanabe, Kyoto University
  Mikio Sato, RIMS, Kyoto University
  Masaki Kashiwara, RIMS, Kyoto University
  Akira Mizuhara, Kyoto Sangyo University
  Yuzo Hosono, Kyoto Sangyo University
  Shôrô Araki, Osaka City University
  Reiko Sakamoto, Nara Women's University
  Fujitsugu Hosokawa, Kobe University
  Kiyosato Okamoto, Hiroshima University
  Akio Kudō, Kyushu University
  Shigeko Miyajima

Finance Committee

Chairman
  Itiro Tamura, Tokyo Denki University
Members
  Takeyuki Hida, Nagoya University
  Nobuyuki Ikeda, Osaka University
  Hirosi Toda, Kyoto University

Science Committee

Chairman

Sigeru Mizohata, Osaka Electro-Communication University

Members

Huzihiro Araki, RIMS, Kyoto University

Hiroshi Fujita, Meiji University

Heisuke Hironaka, Harvard University

Toshihide Ibaraki, Faculty of Engineering, Kyoto University, Appointed by The Operations Research Society of Japan

Masao Iri, Faculty of Engineering, University of Tokyo

Masaki Kashiwara, RIMS, Kyoto University

Hikosaburo Komatsu, University of Tokyo

Yukio Matsumoto, University of Tokyo

Yoshitomo Matsuo, Science University of Tokyo, Appointed by the Japan Society of Mathematical Education

Tamotsu Murata, Momoyama-gakuin University, Appointed by The History of Science Society of Japan

Akihiro Nozaki, International Christian University, Appointed by the Japan Society for Software Science and Technology

Takushiro Ochiai, University of Tokyo

Tadao Oda, Tohoku University

Yujiro Ohya, Faculty of Engineering, Kyoto University

Kiyosato Okamoto, Hiroshima University

Ichiro Satake, Tohoku University

Ryoichi Shimizu, Institute of Statistical Mathematics, Appointed by The Japan Statistical Society

Shinzo Watanabe, Kyoto University

Local Committee

Chairman

Masayoshi Nagata, Okayama University of Science

Members

Huzihiro Araki, RIMS, Kyoto University

Hirosi Toda, Kyoto University

Masayoshi Miyanishi, Osaka University

Fujitsugu Hosokawa, Kobe University

Haruo Murakami, Faculty of Engineering, Kobe University

Akira Kono, Kyoto University

Yuzo Hosono, Kyoto Sangyo University

Masaki Kashiwara, RIMS, Kyoto University

Masaki Maruyama, Kyoto University

Shinzo Watanabe, Kyoto University

Kyoji Saito, RIMS, Kyoto University

Kenji Ueno, Kyoto University

Tetsuji Miwa, RIMS, Kyoto University
Chiaki Tsukamoto, Kyoto Institute of Technology
Takahiro Kawai, RIMS, Kyoto University
Yasuo Yamazaki, RIMS, Kyoto University

Publications Committee

Chairman
Ichiro Satake, Tohoku University
Members
Tadao Oda, Tohoku University
Shige Toshi Kuroda, Gakushuin University
Takushiro Ochiai, University of Tokyo
Kazuo Okamoto, College of Arts and Sciences, University of Tokyo

Foreign Affairs Committee

Chairman
Akio Hattori, Meiji University
Members
Shingo Murakami, Osaka University
Reiji Takahashi, Sophia University
Takushiro Ochiai, University of Tokyo
Shinichi Kotani, University of Tokyo
Shigeko Miyajima

Publicity Committee

Chairman
Shigeru Iitaka, Gakushuin University
Member
Mitsuo Morimoto, Sophia University

Committee of ICM-90

The members of the Organizing Committee and the members of the General
Committee except Prof. Nakajima of the former constitute the Committee of
ICM-90, which was responsible for the preparations of ICM-90 until August
14, 1989, when the Organizing Committee started. Officers of the Committee of
ICM-90 are the following:

President
Kunihiko Kodaira, University of Tokyo, Prof. Emer.
Vice Presidents
Kiyosi Itô, Kyoto University, Prof. Emer.
Heisuke Hironaka, Harvard University

Chairman
  Hikosaburo Komatsu, University of Tokyo
Secretary
  Huzihiro Araki, Kyoto University
Treasurer
  Hirosi Toda, Kyoto University
Executive Members
  Shigeru Iitaka, Gakushuin University
  Sigeru Mizohata, Osaka Electro-Communication University
  Masayoshi Nagata, Okayama University of Science

The Committee of ICM-90 has the same subcommittees as the Organizing Committee and the following Fund Raising Committee:

Fund Raising Committee

Chairman
  Kiyosi Itô, Kyoto University, Prof. Emer.
Vice Chairmen
  Akio Hattori, Meiji University
  Takeyuki Hida, Nagoya University
  Hiroaki Hijikata, Kyoto University
  Ken Hirose, Waseda University
  Tosihusa Kimura, Science University of Tokyo
  Haruo Murakami, Faculty of Engineering, Kobe University
  Shingo Murakami, Osaka University
  Minoru Nakaoka, Okayama University of Science
  Masahiko Saito, College of Arts and Sciences, University of Tokyo
  Tosiya Saito, Kawaijuku
Members
  Shôrô Araki, Osaka City University
  Koji Doi, Ritsumeikan University
  Yasutaka Ihara, RIMS, Kyoto University
  Shigeru Iitaka, Gakushuin University
  Mitsuru Ikawa, Osaka University
  Hideyuki Matsumura, Nagoya University
  Shigeko Miyajima
  Masatake Mori, Faculty of Engineering, University of Tokyo
  Hisasi Morikawa, Nagoya University
  Mitsuo Morimoto, Sophia University
  Takushiro Ochiai, University of Tokyo
  Akihiro Okuyama, Kobe University
  Kayo Otsuka, Faculty of Economics, Teikyo University
  Tatuo Simizu, Polytechnic Consultants Inc.
  Fumiyuki Terada, Waseda University
  Nobuo Yoneda, Department of Information Science, University of Tokyo

List of Donors

International Mathematical Union
Science Council of Japan
Mathematical Society of Japan
Members of the Mathematical Society of Japan

The Life Insurance Association of Japan
The Taniguchi Foundation
Association of Tokyo Stock Exchange Regular Members
The Federation of Bankers Associations of Japan
The Marine & Fire Insurance Association of Japan, Inc.
The Trust Company Association of Japan

Japan Federation of Construction Contractors, Inc.
The Federation of Electric Power Companies
Fujitsu Limited
Hitachi, Ltd.
The Inamori Foundation
Mitsubishi Electric Corporation
NEC Corporation
Incorporated Foundation Oriental Life Insurance Cultural Development Center
Toshiba Corporation
Commemorative Association for the Japan World Exposition (1970)
Japan Automobile Manufacturers Association, Inc.
Japan Association for Mathematical Sciences

IBM Japan, Ltd.
Nippon Telegraph and Telephone Corporation
The Kajima Foundation
Matsushita Electric Industrial Co., Ltd.
Oki Electric Industry Co., Ltd.
Nihon Unisys, Ltd.
The Portopia 81 Foundation

Nippon Steel Corporation
Hattenkai Volunteers
Kyocera Corporation
Murata Machinery, Ltd.
Murata Manufacturing Co., Ltd.
NTT Data Communications Systems Corporation
Omron Corporation
Sanyo Electric Co., Ltd.
Sharp Corporation
Shimadzu Corporation
Suntory Limited
Wacoal Corporation

NKK Corporation
Kawasaki Steel Corporation
Fuji Country Co., Ltd.
Mitsuru Sangyo Co., Ltd.
Tokyo Shoseki Co., Ltd.
Sumitomo Metal Industries, Ltd.

Kobe Steel, Ltd.
Asahi Glass Co., Ltd.
The Norinchukin Bank
Shinko-Shuppansha Keirinkan
    Publishing Co., Ltd.
Toyobo Co., Ltd.
Kubota Corporation
Dainippon Ink & Chemicals, Inc.
Daiwa House Industry Co., Ltd.
The Ito Foundation
Japan Meat Conference
Japan Tobacco Inc.
Kake Educational Institution,
    Okayama University of Science
Kao Corporation
Meiji Milk Products Co., Ltd.
Nippon Meat Packers, Inc.
Renown Incorporated
Suken Shuppan Co., Ltd.
Nisshin Steel Co., Ltd.
Anonymous

Ricoh Company, Ltd.
Nakayama Steel Works, Ltd.
Godo Steel, Ltd.
Bridgestone Corporation
Dentsu Inc.
Ebara Corporation
Hakuhodo Inc.
Hitachi Hi Soft, Ltd.
Isetan Company Ltd.
Ito-Yokado Co., Ltd.
Japan Information Processing Service
    Co., Ltd.
The Japan Research Institute, Limited
K. I. Systems Inc.
Kumon Institute of Education
Mitsukoshi Ltd.
Obunsha Co., Ltd.
Osaka Shoseki Co., Ltd. Publishers
Toppan Printing Co., Ltd.

Daido Steel Co., Ltd.
Topy Industries, Limited
Gendaisugakusha Co., Ltd.
Kyouritsu Shuppan Co., Ltd.
Takaoka Electric Mfg. Co., Ltd.
Yodogawa Steel Works, Ltd.
Mitsubishi Steel Mfg. Co., Ltd.
The Japan Steel Works, Ltd.

Faddeev giving the opening speech to the Congress

(From left to right: *Araki, Komatsu, Itô, Hattori, Kondo, Lovász, Hori, Aramaki, Tanabe, Nishijima, Onozawa*)

Opening Ceremonies

The opening ceremonies of the Congress were held in the Event Hall of the Kyoto International Conference Hall, starting at 9:30 a.m. on August 21, 1990. The Gagaku Club of Tenri University entertained the participants with a *Gagaku* (court music) recital entitled *Etenraku*. Academician Ludwig D. Faddeev, President of the International Mathematical Union (IMU), opened the Congress with the following speech:

After this symbolic opening let me formally declare the International Congress of Mathematicians (ICM-90) in Kyoto open.

This is the first Congress in the history of the International Mathematical Union to take place outside of Europe and North America. This is consonant with the main goal of the Union – the promotion of mathematical research throughout the world. For this reason, the proposal of the Japanese Committee of Mathematicians to hold the Congress in Kyoto was enthusiastically accepted by the General Assembly of IMU four years ago. I believe that I can already express admiration for the efforts being made by the Organizing Committee to make the Congress work effectively. It appears that the attendance is the highest in the history of ICM.

The scientific program of ICM is traditionally in the hands of the Program Committee appointed by the Executive Committee of IMU. Let me disclose to you the list of its members. Professor Nicolaas H. Kuiper was appointed as chairman. Professors Vladimir I. Arnold, Alain Connes, Ronald L. Graham, Heisuke Hironaka, Masaki Kashiwara, Robert P. Langlands, Sigeru Mizohata, and Daniel G. Quillen served as members. We shall witness the effectiveness of their choice of speakers in the coming days.

Now I come to the important duty of designating the President of the Congress. Let me nominate Professor Hikosaburo Komatsu for this post.

Acclamation.

I take this as confirmation of the proposal. Thank you.

Komatsu, President of the
Congress, showing the first sheet
of the ICM-90 commemorative
stamps

Professor Komatsu, as President of the Congress, made the following address:

Thank you Mr. President. I am honored to serve as President of the Congress.

Ladies and Gentlemen,

On behalf of the Organizing Committee I would like to welcome you to the
International Congress of Mathematicians 1990 (ICM-90). I very much hope
that your stay in the historical city of Kyoto will be a pleasant one.

First of all, let me explain how the Congress has been prepared. It was
more than seven years ago that the Liaison Committee for Mathematics in
the Science Council of Japan (JSC), which is the Committee for Mathematics
in Japan, and the Mathematical Society of Japan (MSJ) started planning to
host ICM-90. After the Feasibility Committee chaired by Professor Kiyosi
Itô made a careful initial investigation of this issue, MSJ decided in 1985 to
invite ICM-90 to Kyoto. Fortunately this invitation was accepted by the Site
Committee of IMU and was approved by the General Assembly of IMU at
Oakland in 1986.

The actual preparations started when MSJ formed the Committee of ICM-
90 in December, 1986. Its Executive Committee consists of the following:

President	Kunihiko Kodaira
Vice Presidents	Kiyosi Itô
	Heisuke Hironaka
Chairman	Hikosaburo Komatsu
Secretary	Huzihiro Araki
Treasurer	Hirosi Toda
Members	Shigeru Iitaka
	Sigeru Mizohata
	Masayoshi Nagata

The full Committee of ICM-90 had an additional 29 members.

By the end of 1987, the Japan Society of Mathematical Education, the History of Science Society of Japan, the Institute of Actuaries of Japan, the Japan Society for Software Science and Technology, the Japan Statistical Society, and the Operations Research Society of Japan decided to sponsor the Congress and elected their representatives to the Committee of ICM-90 and its Science Committee.

In June, 1989, JSC made the final decision of sponsoring ICM-90 with the consent of the Government. Then, according to the general rules of JSC, the Committee of ICM-90 was reorganized into the Organizing Committee in August, 1989.

Second, as the President of IMU reported, the scientific program of the Congress and, in particular, the list of invited speakers were prepared by the Program Committee, appointed by the Executive Committee of IMU. However, a few additional speakers working in Japan, and also substitutes for speakers who declined were invited by the Organizing Committee on its own initiative, in which the latter were selected from the list of alternatives made by the Program Committee.

Finally 15 speakers were invited to give one-hour Plenary Addresses. Among them two declined and were replaced by other choices. Also 144 speakers were invited to give 45-minute Section Lectures, 19 were substituted, and 6 lectures were cancelled. In addition, the International Commission on Mathematical Instruction invited five 45-minute speakers.

We have benefited very much by the recent reconciliation in the world politics and the prosperity of the Japanese economy. As far as I know, no invited speakers declined for political or economic reasons.

Adopting Professor Mary Rudin's proposal, the last General Assembly of IMU in 1986 recommended that Subfields of Mathematics, Women, and Mathematicians in Small Countries should not be overlooked when speakers to the ICM are selected. I believe both the Program Committee and the Organizing Committee have respected this recommendation.

The selection of additional speakers and the scheduling of scientific program were carried out by the Science Committee in the Organizing Committee, chaired by Sigeru Mizohata.

Thirdly, the budget of this Congress amounted to approximately 300,000,000 Yen (approximately US $ 2,000,000). One third of the revenue is the registration fees, one third the donations from private corporations, and the rest consists of subventions from IMU, JSC and MSJ, the donations by individual members of MSJ, and the miscellaneous income (here the donations by 1138 members of MSJ are actually more than the total of all the other subventions). We are very sorry that we had to set a high registration fee of 30,000 Yen but we were forced to do so because the Japanese tax regulations do not allow us to receive tax-exempted donations exceeding the total amount of registration fees.

In addition to the budget of the Organizing Committee, the Japan Association for Mathematical Sciences, MSJ, the Oriental Life Insurance Cultural Development Center, the Commemorative Association for the Japan World Exposition (1970), and The Kajima Foundation allocated a total of 60,000,000 Yen to support 269 foreign participants mostly from developing countries or countries with foreign currency restrictions. This number includes 47 IMU Scholars whose travel expenses are supported by the Special Development Fund of IMU.

Most funds from these foundations also originated from donations solicited from enterprises by the Organizing Committee.

The biggest contributors are insurance and electronic companies. On behalf of all the participants, I would like to thank all these bodies for their generous donations. A list of donors is in the Program and will be published in the Proceedings.

It is not an easy task to raise so much money. But I must confess that it was a pleasant one, too, because every executive I met for this purpose showed a liking for mathematics and appreciated that mathematics had played an important role in the development of the Japanese economy.

The Fund Raising Committee is chaired by Kiyosi Itô and its office was set up in Gakushuin University with Shigeru Iitaka in charge.

Assistance with visa applications was carried out by Shinichi Kotani at the office located in the University of Tokyo.

The other preparations have been planned and implemented by General Secretary Huzihiro Araki, Treasurer Hirosi Toda and the Local Committee chaired by Masayoshi Nagata. The Secretariat was set up in the Research Institute for Mathematical Sciences, Kyoto University.

The Proceedings of the Congress will be edited by the Publications Committee with Ichiro Satake as the chairman. Its office is in Tohoku University.

I would like to thank these institutions and, in particular, Kyoto University for their kind cooperation.

Today, 4,000 mathematicians from 83 countries have assembled here to review our scientific achievements over the last few years and to set goals for the future in all fields of mathematics ranging from pure mathematics through applied mathematics to mathematical education. This seems to be an almost megalomaniac dream at this time of specialization. I do not know any other discipline which attempts to hold this kind of congresses regularly.

I have often wondered why mathematicians do have Congresses and what Congresses mean to them. My answer is that Congresses are to mathematicians what *Bon* and New Year Festivities are to Japanese, in which they abandon their daily life completely.

Japanese are believed to work continuously without vacations, but that is not true. Even in the Edo Period there were two one-week long holidays. One is the New Year Festivities and the other is the *Bon Festivities* which take place a week earlier than this time of the year.

On these holidays people are relieved from labor and go back to their native home. People are not allowed to cook on the first days of the Festivities, so that they have a busy time preparing all meals before the holidays start.

New Year Festivities are associated with the future. We renew everything we can and start again. *Bon Festivities* are for the past. We receive ancestors' ghosts, make conversations with them, and then send them back. In cities like Kyoto people decorate their entrance halls with their treasures and keep the doors open. The whole city becomes a big museum. In the countryside people gather in the village square and dance. There is no distinction between performers and spectators; all dance. That is the way Japanese refresh themselves, inherit their traditions and unite. *Bon* and New Year Festivities also give young people the opportunity to meet together to make a new family.

I would like to ask all speakers at the Congress to make their lectures accessible to a wide audience, and not just to the specialists, at least in the first part. This is certainly not the time to work in a daily manner.

I hope you all enjoy this big event.

Thank you.

Professor Akio Hattori, President of the Mathematical Society of Japan, spoke as follows:

Mr. Chairman, Distinguished Guests, Ladies and Gentlemen,

I should like to extend a warm welcome to you all for having gathered from all parts of the world to attend this Congress. The Japanese mathematical community is proud of a pioneer; I am referring to Professor Kiyosi Itô, a renowned probabilist. He has been instrumental in bringing this Congress to Japan and in its organization. I propose that Professor Itô be elected Honorary President of this Congress.

Professor Itô was elected by acclamation and spoke as follows:

Mr. Chairman, Distinguished Guests, Ladies and Gentlemen,

It is my great honor to speak to you as Honorary President of the Congress.

Before now, some Japanese mathematicians have tried to bring the Congress to Japan. As early as the 1960s, Professors S. Iyanaga and Y. Kawada began campaigning for the Congress to be held in Japan. Now that this Congress is starting, I feel a certain nostalgia for their pioneering efforts.

When we formed the Organizing Committee chaired by Professor Kuni-hiko Kodaira, the first Japanese Fields medalist, our big problem was how to raise funds. In this respect, I would like to give credit to Professor Kodaira. Impressed by his fame and his fine personality Mr. H. Tanimura, Professor Kodaira's friend, together with Messrs. R. Ishikawa and M. Kitoku set up a fund-raising program for us and appealed to leading companies. We thank them for their tremendous efforts, which enabled us to make concrete plans for the Congress. To our great regret, Professor Kodaira is unable to come here today. I would like to express our sincere gratitude to him on this occasion.

The next problem was that we anticipated there would be less participants than usual because of Japan's geographic location and the worldwide economic upheaval. Professor H. Komatsu proposed to make a grant available to help participants from abroad. This idea was welcomed by several foundations whose support contributed to the increase in the number of participants.

In Japan, people have been more concerned with science and technology than with mathematics. Despite this, we are proud of two exceptional projects which have helped mathematicians: the International Symposia supported by Mr. Toyosaburo Taniguchi and secondly, the Exchange Program of the Japan Association for Mathematical Sciences established by Professor H. Hironaka, the second Japanese Fields medalist. Through these projects many Japanese mathematicians have acquired a sense of international cultural exchange, which is perhaps one of the main aims of the Congress.

A large number of members of the Mathematical Society of Japan helped the Organizing Committee in many respects, academically or nonacademically. At the final stage of preparation a tremendous amount of business had to be carried out within a limited time. We are very grateful to Professors H. Komatsu, H. Araki, H. Toda, S. Iitaka, S. Kotani, and their secretaries for all their hard work.

We are very happy to welcome you here in Kyoto. Please enjoy the lectures and participate as much as possible in the informal discussions.

Thank you.

Professor Jiro Kondo, President of the Science Council of Japan, gave the following welcome address:

Mr. Chairman, Distinguished Guests, Ladies and Gentlemen,

It is a great pleasure to have this opportunity, on behalf of the Science Council of Japan, to speak to all of you who have gathered here from as many as 80 countries, at the opening of the International Congress of Mathematicians 1990.

The Science Council of Japan was established in 1949 as government organization representing qualified Japanese scientists both internally and internationally, covering all scientific fields consisting of Cultural, Social and Natural Sciences. The aim of the Council is to promote scientific development and to improve administration, industry and living standards through science.

Since that time, we have been working to contribute to the progress of science in cooperation with the academic organizations of the world by sponsoring many international congresses here in Japan, and by sending Japanese delegations to international congresses held overseas. We do this because we believe that the promotion of international scientific exchanges is one of our most important duties.

We have opened today the International Congress of Mathematicians in cooperation with the Mathematical Society of Japan, the Japan Society of Mathematical Education, the History of Science Society of Japan, the Institute of Actuaries of Japan, the Japan Society of Software Science and Technology, the Japan Statistical Society, and the Operations Research Society of Japan. It is an extraordinary pleasure for me to have this opportunity to be with so many distinguished scientists from around the world, and to listen to your lectures and presentations on the recent achievement.

The first Meeting of the International Congress of Mathematicians was held in Zürich in 1897. Since then, the Congress has been held nearly every four years somewhere in the world, and this is the first Congress to be held outside Europe and North America. I offer my heartfelt congratulations on the remarkable growth of the Congress in terms of both quality and number of participants and at the same time I express my great pleasure in being able to welcome the Congress to Japan for the first time. Since I graduated from the Department of Mathematics, Imperial University of Kyoto in 1940, I am very happy to hold this Congress here in Kyoto. After my graduation I worked mostly in the field of mathematical applications, such as theoretical aerodynamics, operations research, systems engineering and environmental sciences.

Recent international trends clearly indicate the importance of basic research in science and technology. In particular, mathematical sciences are entrusted with the crucial mission of providing the backbone for all sciences and the theoretical basis for the new era of information science in the twenty-first century, and they must also live up to people's high expectations. Moreover, because of the rapid progress in computer technology, many scientists in other fields, as well as engineers and businessmen, are coming to expect a greater contribution from mathematicians.

As researchers of all branches of mathematics gather from all over the world at this International Congress, it is expected that much interaction among the different branches will be made and that this will result in a new development of mathematics as a unified discipline. While we see recently a strong tendency of specialization in all branches of science, I have learned that essential progress in mathematics has often arisen out of the unexpected connection of different areas. The interaction of researchers with different cultural backgrounds is also expected to produce such an effect. I hear that this Congress is organized with an emphasis on interdisciplinary and international interactions and efforts have been made to provide various opportunities for contact among participants. I hope that through the interaction with neighbouring fields of science which will inject new blood into mathematics and through friendship among participants crossing the boundaries of different branches of mathematics and different countries, this Congress serves as a stepping stone towards active worldwide research cooperation in the future.

I earnestly hope that everyone who participates in this Congress, will gain worthwhile experience in both fundamental and applied aspects of mathematics.

In closing, I sincerely hope for the great success of the Congress. I also hope that all of you from abroad will enjoy their stay in Japan. I believe that this Congress will become truly memorable for you through contact with fellow scientists, and I wish this to be a chance for you to learn more about Kyoto, the ancient capital of Japan, and about Japanese culture.

Thank you.

The following congratulatory telegram from Prime Minister Toshiki Kaifu was read both in Japanese and in English by Professor Huzihiro Araki:

PLEASED TO EXTEND A HEARTY WELCOME TO ALL DELEGATES FROM AROUND THE WORLD AT THE OPENING OF THE 11TH GENERAL ASSEMBLY OF THE INTERNATIONAL MATHEMATICAL UNION AND THE INTERNATIONAL CONGRESS OF MATHEMATICIANS 1990 HELD IN JAPAN UNDER THE COSPONSORSHIP OF THE SCIENCE COUNCIL OF JAPAN, MATHEMATICAL SOCIETY OF JAPAN, JAPAN SOCIETY OF MATHEMATICAL EDUCATION, THE OPERATIONS RESEARCH SOCIETY OF JAPAN, THE HISTORY OF SCIENCE SOCIETY, JAPAN SOCIETY FOR SOFTWARE SCIENCE AND TECHNOLOGY, THE JAPAN STATISTICAL SOCIETY AND THE INSTITUTE OF ACTUARIES OF JAPAN.

I WISH A GREAT SUCCESS IN THIS INTERNATIONAL CONGRESS.

TOSHIKI KAIFU
PRIME MINISTER

Mr. Kosuke Hori, Minister of Education, Science and Culture gave a congratulatory address as follows:

Distinguished Delegates, Ladies and Gentlemen of the Audience,

It is a real pleasure and privilege for me to be afforded this opportunity to address congratulatory greetings to this distinguished audience at the opening of the Twenty-first International Congress of Mathematicians.

First of all, I would like to extend my sincerest welcome to you all, especially to those who have travelled far in order to come to Japan. At the same time, I am deeply grateful and wish to pay respect to all the people who have made great efforts in mathematical research.

Recently, pure mathematics, applied mathematics, mathematical education, and other fields of mathematics have achieved remarkable progress in research. This International Congress has been held every four years since

1897 for the purpose of bringing together mathematicians in all fields from all over the world in order to share the significant progress achieved in each area of mathematics and to help further new progress in mathematics. In these respects, the International Congress of Mathematicians has attained excellent achievements.

I am told that this International Congress is the first to be held in Asia, and distinguished mathematicians from Japan and abroad who participate in it will present many new discoveries in mathematics. I am confident that this Congress will further enrich the knowledge that is common property to all human beings.

I hope that those participants who have travelled from overseas will have the opportunity to acquaint themselves with some aspects of Japanese culture and society and that they will have a pleasant stay in Japan.

Finally, I would like to pay our deepest respects to those people who have already worked so hard in organizing this Congress and I anticipate that it will be highly successful.

Thank you very much for your attention.

Mr. Teiichi Aramaki, Governor of Kyoto Prefecture, gave a congratulatory address as follows:

Distinguished Delegates, Ladies and Gentlemen of the Audience,

As Governor of Kyoto Prefecture, I would like to extend my sincerest congratulations for the Twenty-first International Congress of Mathematicians to be held here in Kyoto from today, with distinguished mathematicians assembled from various countries of the world.

The First International Congress of Mathematicians was held in Zürich, Switzerland in 1897 and since then, the congress has produced excellent results, establishing a long history and tradition. The fact that the Fields Medal, which is said to be the Nobel Prize of mathematics, and the Nevanlinna Prize have been awarded to superb mathematicians has attracted the attention of not only the people concerned but also many others.

It is a great pleasure and privilege for us living in Kyoto Prefecture that this authoritative International Congress is being held in Kyoto which itself is recognized as an academic and scientific center. On behalf of the people of Kyoto Prefecture, I would like to extend a cordial welcome to all of you.

The International Congress of Mathematicians has been held only in western cities so far and so it is a great pleasure and honor that this is the first International Congress of Mathematicians to be held outside Europe and North America. I am sure that this fact will further encourage many Japanese mathematicians.

Finally, I would like to pay our deepest respects to the scientists of the Research Institute for Mathematical Sciences, Kyoto University and to those who have gone to great pains in organizing this International Congress. I anticipate that each of the participating mathematicians will take an active role in the discussions and that the Congress will be highly successful.

Thank you for your attention.

Dr. Tomoyuki Tanabe, Mayor of Kyoto City, gave the following congratulatory address:

Ladies and Gentlemen,

Firstly, please let me welcome you all to Kyoto.

I am delighted that the Twenty-first International Congress of Mathematicians is starting today and on behalf of the citizens of Kyoto, I would like to extend a cordial welcome to you all.

As you may know, Kyoto is an ancient city with a history that looks back on 1,200 years of culture and tradition. This is why Kyoto is said to be the spiritual root of Japan.

Kyoto is rich in cultural heritage and a variety of time-honored customs based on Japanese tradition still exist as part of our everyday way of life. Traditional industries which embody the quintessence of Japanese folk art and technology can also be found.

On the other hand, Kyoto is a scientific city which has produced several superb Nobel Prize winners and Fields Medalists. At the same time, Kyoto is an industrial city which has given birth to world leaders in high-tech industries.

About ten years ago, we declared Kyoto as a City Open to the Free Exchange of World Cultures. In this connection, it is a profound feeling of pleasure and privilege that the International Congress of Mathematicians is being held here in Kyoto.

I am confident that everyone who has travelled from afar to participate in this International Congress, will renew old friendships as well as begin new acqaintances. I also hope that you will enjoy the mixture of traditional Japanese culture and the modern atmosphere of our city.

Let me close my greetings by wishing the International Congress of Mathematicians success and combine this with a wish for your good health and happiness.

Thank you for your attention.

Dr. Yasunori Nishijima, President of Kyoto University, delivered this congratulatory address:

Ladies and Gentlemen,

It is my honor and great pleasure to greet this distinguished gathering at the opening of the International Congress of Mathematicians.

I have learned that this Congress had its first meeting in 1897. That is also the year that Kyoto University was founded as the Kyoto Imperial University and this marked the establishment of the modern academe in the old capital. Ninety-four years have passed since then and today Kyoto welcomes this gathering of the world's notable mathematicians to Japan for the first time. Kyoto is most honored to be the chosen site.

Last week, the General Conference of the International Association of Universities took place in Helsinki, Finland. It was very cool and dry, unlike the hot and wet weather here. The Chancellor, Professor Olli Lehto, and all the members of the University of Helsinki, including the most delightful of students, made the meeting a great success.

The theme of the Conference was "The Mission of the University: Universality, Diversity, and Interdependence," and I had the opportunity to chair the whole-day commission on "Diversity within Universality." In this rapidly and dramatically changing world, Diversity tends to be emphasized. However, at the same time, Universality is the human aspiration to create a higher degree of wisdom for the future and this is the mission of the world's academic community.

After the exciting and fruitful meeting in Helsinki, I returned to Japan a few days ago. In the airport transit lounge, I met Chancellor Lehto, who was on his way to this Congress. He was holding a small bag on his lap with great care. I thought it was the medal for the Rolf Nevanlinna Prize which will be presented at today's Ceremony. I said, "Is that top secret?" Chancellor Lehto simply smiled and nodded. We boarded the same airplane for the last stretch of our long journey from Helsinki to Kyoto. This morning, following the performance of a classic Japanese court dance, *Bugaku*, the presentation of the Fields Medals and the Rolf Nevanlinna Prize will reveal all the secrets.

I really think mathematics is the core of the universality of human wisdom. Mathematics has since ancient times conveyed wisdom throughout the history of human endeavor. At the same time, mathematics covers all branches of knowledge and leads us along the road to Truth and Reason.

David Hilbert said, "Mathematics is an organism for whose vital strength the indissoluble union of the parts is a necessary condition."

I would like to congratulate all those who have worked so hard to prepare and organize this most important meeting. I anticipate it will be a great success.

I wish all participants a most fruitful and pleasant meeting in Kyoto.

On behalf of Takashi Fukaya, Minister of Posts and Telecommunications, Mr. Tomoyuki Onozawa, Director General of Posts, presented a sheet of commemorative stamps with the following letter:

To Professor Hikosaburo Komatsu, President of the International Congress of Mathematicians.

It is of high significance and a matter for congratulations that the Twenty-first International Congress of Mathematicians is being held here in Kyoto, for the first time in Asia.

The Ministry of Posts and Telecommunications has issued a specially designed postage stamp to commemorate this event, and the first sheet of stamps is now presented to you.

> From Takashi Fukaya, Minister of Posts
> and Telecommunications
> On August 21, 1990.

Kyoto Gagaku-Kai then performed a *Bugaku* (court dance) entitled *Gosechi no Mai* accompanied by the Gagaku Club of Tenri University which played a *Gagaku* (court music).

Academician Ludwig D. Faddeev, Chairman of the Fields Medal Committee, announced the recipients of the Fields Medals as follows:

The Fields Medal and Prize Committee is appointed by the Executive Committee of the IMU. For this term the following people were appointed and worked for the Committee: Professors M. Atiyah, J. M. Bismut, E. Bombieri, C. Fefferman, K. Iwasawa, P. D. Lax and I. Shafarevich. Since it is the duty of the President of the IMU to chair the work of this Committee, I served as chairman.

After thorough consideration of the material at our disposal, we decided to award four Medals. The recipients are

Vladimir G. Drinfeld
Vaughan F. R. Jones
Shigefumi Mori
Edward Witten

I believe that these names are well known to the mathematical community throughout the world. Their scientific contributions will be described during the afternoon session.

Let me proceed to the pleasant task of presenting the awards. I ask the Honorable Minister Hori to do this.

The winners came forward and received their medals and prize checks from Mr. Kosuke Hori, Minister of Education, Science and Culture.

Professor László Lovász announced the recipient of the Rolf Nevanlinna Prize as follows:

The Nevanlinna Prize Committee consisted of Alexandre Chorin from Berkeley, Michael Rabin from Jerusalem, Volker Strassen from Konstanz, and Lázló Lovász from Budapest as chairman. After considering a number of outstanding candidates, the Committee decided to award the Rolf Nevanlinna

Prize to Alexander A. Razborov from the Steklov Institute in Moscow, for his groundbreaking work on lower bounds for circuit complexity.

The winner came forward and received his medal and prize check from Mr. Kosuke Hori, Minister of Education, Science and Culture.

The opening ceremonies adjourned at 11:00 a.m.

Closing Ceremonies

The closing ceremonies were held in the Main Hall of the Kyoto International Conference Hall, starting at 1:15 p.m. on August 29, 1990.

Academician Ludwig D. Faddeev, President of the International Mathematical Union began the closing ceremonies with the following words:

Time runs very fast and we have come to the end of our Congress. I believe that we can now judge very highly the results of the scientific program and congratulate the Program Committee on their success. Personally, I was glad to observe how prominently Mathematical Physics was represented in its connections with other domains of Mathematics.

The work of the Congress has been so smooth that some of you did not notice the efforts of the Organizing Committee to achieve this. We are all highly indebted to our Japanese colleagues for their excellent job.

Let me also add that the general atmosphere of the Congress has been very friendly and this has allowed us to concentrate on purely mathematical problems.

It is my pleasure to inform you that the Emperor and Empress of Japan invited the winners of the Fields Medals and the Nevanlinna Prize to visit them in Tokyo. The Honorary President of the Congress, Professor Itô, as well as Professor Hironaka and I were also present. The interest expressed by the Imperial Family in the work of our Congress is a great honor to us all.

Let me now inform you of the results of the General Assembly which took place in Kobe just before the Congress. It represented 52 members of the IMU most of whom attended the meeting.

During the last few years Saudi Arabia has been accepted as a new member. Spain and Israel have upgraded their membership to Group III.

The General Assembly elected new Committees and Commissions.

The new Executive Committee is composed as follows:

President   J. L. Lions
Vice-Presidents J. Coates
       D. Mumford

Secretary   J. Palis
Members   J. Arthur
             A. Dold
             H. Komatsu
             L. Lovász
             E. Zehnder

I shall remain on the Executive Committee as the past President.

Half of the Executive Committee has changed. So it is appropriate to cite one of the resolutions adopted at the General Assembly: "The General Assembly gives special thanks to Professor Olli Lehto for his excellent work as Secretary to the IMU during the last eight years, ably assisted by Mrs. Mäkeläinen. It also thanks the University of Helsinki and the Finnish Ministry of Education for their generous support of the IMU secretariat over this period."

The General Assembly also appointed the Commissions.

The ICMI Executive Committee:

President   M. de Guzmán
Vice-Presidents   J. Kilpatrick
                  Lee Peng-Yee
Secretary   M. Niss
Members   Yu. L. Ershov
                  E. Luna
                  A. Sierpinska

The ex-officio members are: the past President of ICMI, the President of IMU, the Secretary of IMU and the IMU representative at CTS (ICSU). The next ICME will take place in 1992 in Quebec.

Commission on Developments and Exchange:

Chairman   M. S. Narasimhan
Members   P. Bérard
            C. Camacho
            A. Grunbaum
            A. O. Kuku
            J. Mawhin
            T. Ochiai
            P. L. Papini
            Wu Wen Tsun

The ex-officio members are: the past Chairman of CDE, the President of IMU, the Secretary of IMU and the IMU representative at COSTED (ICSU).

Besides the administrative business, there was lively and important discussion at the General Assembly on the role of Applied Mathematics and its balance in the program of the ICM, the increasing relevance of Mathematics in Industry, and related problems of mathematical education.

Finally the Site Committee made its proposal to the General Assembly on the time and location of the next Congress, which was adopted after some

Chatterji inviting the
Congress to Zürich

© 1990 H. Kono

discussion. The next Congress, ICM-94 will be held in Zürich, Switzerland. I
shall now step down and give the floor to Professor Chatterji.

Professor S. D. Chatterji of the Ecole Polytechnique Fédérale de Lausanne
invited the audience to the next International Congress of Mathematicians with
these words:

Ladies and Gentlemen,

On behalf of the Swiss Mathematical Society and the entire Swiss mathematical community, it is my great privilege to invite you all to the 1994
International Congress of Mathematicians in Zürich. As you know, the honor
and responsibility of organizing the Congress have fallen on Zürich twice
before in the past: in 1897 and 1932. Zürich is an important financial, com-
mercial and cultural centre of Switzerland; it is also the seat of two of our
important institutions of higher education—the Swiss Federal Institute of
Technology (ETH) and the University of Zürich. Situated in beautiful natural
surroundings in the heart of Europe, Zürich is easily accessible by rail, road
and air.
    Our Japanese hosts have set such high levels of hospitality and efficiency
at this Congress that it would be difficult to match them. However, we shall
do our best to make your participation at the 1994 Congress agreeable and
fruitful.
    Rendezvous then for August 1994 in Zürich.

Dr. Jose Felipe Voloch of the Instituto de Matemática Pura e Aplicada,
Brazil, and Professor Carol Wood of Wesleyan University, USA, President-Elect

of the Association of Women in Mathematics, were invited to give comments on ICM-90.

Professor Hikosaburo Komatsu, President of the Congress, closed the meeting and the Congress with these words:

Thank you for your kind remarks. The organizers of the Congress are really rewarded to hear them. However, all praise should go to Professors Kenji Ueno, Masaki Maruyama, and Chiaki Tsukamoto, as well as to Miss Tanaka, Mrs. Ichiki, Miss Ishii to name just a few of the people who have carried out all the plans of the Congress so perfectly.

Our thanks are also due to Professors Gleason and Mesirov for sending us the material they prepared for the last Congess in Berkeley. This was of great help to us.

This Congress has been admired for its smooth organization but this was true only in appearance. In spite of all the devoted works, I often had to remain in the Secretariat to solve many minor problems. Even when I was attending a lecture, the buzzer, you might have heard, called me back there. Therefore, I must admit that my impression of the Congress is very partial. Nevertheless, I felt that we are at another turning point in the history of mathematics. The previous one was marked at the Second Congress in 1900 when Hilbert gave his famous lecture. Since then we have obtained an enormous number of general results by axiomatization and abstract formulation, often at the hands of mathematical giants. This time it is a transition from abstract simplification to more concrete synthesis. We are now in a fortunate time when we can solve many problems which remained open for many years in spite of all the efforts of past generations of mathematicians. We no longer have a single genius but many people work together developing new strong streams. It was only many brooks last time at the Congress in Berkeley. They meet together and now we see a big river or a sea or even an ocean.

On behalf of all participants, I would like to praise Professor Kuiper and the other members of the Program Committee for their outstanding work in selecting the invited speakers. I would also like to thank the speakers for their admirable efforts.

At our time of democracy, international cooperation is indispensable and so is the unity of mathematics. We are proud of the fact that this Congress has helped these goals.

As reported in the Daily Bulletin, the Congress has been attended by 3,954 ordinary members, 452 accompanying members, and 92 child members from 76 countries.

Unfortunately, all the preregistered members from 7 countries were unable to attend because of the crisis in the Persian Gulf. World politics has once again cast its shadow on the Congress. I do hope that the next Congress in 1994 will be a truly universal one.

I declare the Congress closed.

The closing ceremonies adjourned at 1:45 p.m.

Scientific Program

Invited One-Hour Addresses at the Plenary Sessions

Invited Forty-Five Minute Addresses at the Section Meetings

Section 1: Mathematical Logic and Foundations

There were 15 Short Communications in this section.

Section 2: Algebra

There were 59 Short Communications in this section.

Section 3: Number Theory

There were 29 Short Communications in this section.

Section 4: Geometry

There were 38 Short Communications in this section.

Section 5: Topology

There were 43 Short Communications in this section.

Section 6: Algebraic Geometry

There were 25 Short Communications in this section.

Section 7: Lie Groups and Representations

There were 23 Short Communications in this section.

Section 8: Real and Complex Analysis

There were 73 Short Communications in this section.

Section 9: Operator Algebras and Functional Analysis

There were 52 Short Communications in this section.

Section 10: Probability and Mathematical Statistics

Section 11: Partial Differential Equations

Section 12: Ordinary Differential Equations and Dynamical Systems

Section 13: Mathematical Physics

There were 39 Short Communications in this section.

Section 14: Combinatorics

There were 25 Short Communications in this section.

Section 15: Mathematical Aspects of Computer Science

There were 5 Short Communications in this section.

Section 16: Computational Methods

There were 26 Short Communications in this section.

Section 17: Applications of Mathematics to the Sciences

There were 18 Short Communications in this section.

Section 18: History, Teaching and the Nature of Mathematics

There were 22 Short Communications in this section.

In addition there were 2 Short Communications in the Post Deadline Session.

List of Participants

The 4102 participants are listed alphabetically.
The asterisks indicate those 651 participants who gave or coauthored short communications.
Those 151 with § signs were absent, although they were either invited participants or have paid the registration fee.

ABD-ALLA, Abo-el-nour N., EGYPT
ABE, Eiichi, JAPAN
ABE, Hitoshi, JAPAN
ABE, Kojun, JAPAN
ABE, Masanori, JAPAN
ABE, Naoto, JAPAN
ABE, Takatsugu, JAPAN
ABE, Takehisa, JAPAN
ABE, Yukitaka, JAPAN
ABERER, Karl, SWITZERLAND
ABI KHZAM, Farouk F.*, LEBANON
A'CAMPO, Norbert, SWITZERLAND
ACHUTHAN, Paninjukunnath*, INDIA
ADACHI, Kenzo, JAPAN
ADACHI, Masahisa*, JAPAN
ADACHI, Toshiaki, JAPAN
ADACHI, Yuko, JAPAN
ADAMS, David R., U.S.A.
ADAMYAN, Vadim M.*, U.S.S.R.
AGAOKA, Yoshio, JAPAN
AGEMI, Rentaro, JAPAN
AGOU, Simon J.*, FRANCE
AGUILAR, Marcelo A., MEXICO
AHARA, Kazushi*, JAPAN
AHERN, Patrick R., U.S.A.
AHMED, Naseer*, PAKISTAN
AHMEDOU, Haouba, MAURITANIA
AHUJA, Mangho*, U.S.A.
AIBA, Akira, JAPAN
AIDA, Haruko, JAPAN
AIDA, Shigeki, JAPAN
AIKAWA, Hiroaki*, JAPAN
AIKOU, Tadashi, JAPAN
AITCHISON, Iain Roderick*, AUSTRALIA
AKAGI, Misao, JAPAN
AKAHIRA, Masafumi*, JAPAN

AKAHORI, Takao, JAPAN
AKAMATSU, Toyohiro, JAPAN
AKAO, Kazuo, JAPAN
AKASHI, Shigeo§, JAPAN
AKIBA, Shigeo, JAPAN
AKIMOTO, Yoshihisa, JAPAN
AKITA, Yoshihiko, JAPAN
AKIYAMA, Hiroshi, JAPAN
AKIYAMA, Keiichi, JAPAN
AKIYAMA, Kenzi, JAPAN
AKIYAMA, Koichiro, JAPAN
AKUTAGAWA, Kazuo, JAPAN
AKUTSU, Takashi, JAPAN
AKUTSU, Tatsuya, JAPAN
ALAMATSAZ, M. H.§, IRAN
ALAMOLHODAEI, Hassan, IRAN
ALBAR, Muhammad Alawi§, SAUDI ARABIA
ALBER, Solomon J.*, U.S.A.
ALBERTONI, Sergio, ITALY
ALBERTSON, Michael O., U.S.A.
AL-DHAHIR, M. Wassel§, KUWAIT
ALEFELD, Goetz E., F.R.G.
ALEKSEEV, Anatoly Semenovich, U.S.S.R.
ALEXANDROV, Alexander D., U.S.S.R.
ALEXOPOULOS, George, FRANCE
ALIKHANI-KOOPAEI, Aliasghar*, IRAN
ALLAN, Graham R., U.K.
ALMEIDA, Jorge M.G., PORTUGAL
ALON, Noga M., ISRAEL
AL-SHAKHS, Adnan A.§, SAUDI ARABIA
ALSINA, Claudi, SPAIN
ALTMAN, Allen B., U.S.A.
AMANO, Kazuo, JAPAN
AMASAKI, Mutsumi, JAPAN
AMEMIYA, Ichiro, JAPAN
AMFILOKHIEV, Alexandr A., U.S.S.R.

AMIR, Dan, ISRAEL
AMITSUR, Shimshon A., ISRAEL
ANAND, Kailash K.*, CANADA
ANANTHARAMAN, Claire, FRANCE
ANASTASIEI, Mihai*, ROMANIA
ANCONA, Vincenzo, ITALY
ANDERSON, Jack M.*, U.S.A.
ANDERSSON, Karl G., SWEDEN
ANDO, Hideo, JAPAN
ANDO, Shiro*, JAPAN
ANDO, Shoichi, JAPAN
ANDO, Tetsuo, JAPAN
ANDO, Tsuyoshi, JAPAN
ANDO, Yutaka, JAPAN
ANDOU, Yuuki, JAPAN
ANDREATTA, Marco, ITALY
ANDRONIKOF, Emmanuel, FRANCE
ANDROUAIS, Anne M., FRANCE
ANDRUSKIEWITSCH, Nicolas, ARGENTINA
ANICHINI, Giuseppe*, ITALY
ANTOINE, Jean-Pierre, BELGIUM
ANZAI, Kazuo, JAPAN
AOKI, Kazuyoshi, JAPAN
AOKI, Kenji, JAPAN
AOKI, Noboru, JAPAN
AOKI, Nobuo, JAPAN
AOKI, Norihiro, JAPAN
AOKI, Shigeru, JAPAN
AOKI, Takashi, JAPAN
AOKI, Takatoshi, JAPAN
AOMOTO, Kazuhiko, JAPAN
AOYAMA, Hiroshi, JAPAN
AOYAMA, Yoichi, JAPAN
ARAI, Hiroshi, JAPAN
ARAI, Hitoshi*, JAPAN
ARAI, Masaharu, JAPAN
ARAKAWA, Tsuneo*, JAPAN
ARAKELYAN, Norair, U.S.S.R.
ARAKI, Huzihiro, JAPAN
ARAKI, Shoro, JAPAN
ARAKI, Testu, JAPAN
ARAZY, Jonathan§, ISRAEL
ARENS, Richard F., U.S.A.
ARIKI, Susumu, JAPAN
ARIMA, Satoshi*, JAPAN
ARISAKA, Nakaaki, JAPAN
ARISAWA, Mariko, JAPAN
ARISUMI, Tadashi, JAPAN
ARKOWITZ, Martin*, U.S.A.
ARMENDARIZ, Efraim P., U.S.A.
AROCA, José-Manuel*, SPAIN
ARROWSMITH, David K.*, U.K.
ARTEMIADIS, Nicolas K., GREECE
ARUMUGAM, Gurusamy*, INDIA
ARVANITOYEORGOS, Andreas*, U.S.A.
ARYA, Shashi Prabha*, INDIA
ASADA, Akira*, JAPAN
ASADA, Mamoru, JAPAN

ASADA, Teruko, JAPAN
ASAEDA, You*, JAPAN
ASAHI, Masahiro, JAPAN
ASAI, Kazuto, JAPAN
ASAI, Keisui, JAPAN
ASAI, Teruaki, JAPAN
ASAKAWA, Hidekazu, JAPAN
ASAKURA, Fumioki*, JAPAN
ASAKURA, Soichi, JAPAN
ASAMOTO, Noriko, JAPAN
ASANO, Hiroshi, JAPAN
ASANO, Kazuo, JAPAN
ASANO, Kiyoshi, JAPAN
ASANO, Kouhei, JAPAN
ASANO, Shigemoto, JAPAN
ASANO, Tetsuo, JAPAN
ASH, J.Marshall*, U.S.A.
ASHIKAGA, Tadashi, JAPAN
ASHINO, Ryuichi, JAPAN
ASHOUR, Attia Abdel-Salam, EGYPT
ASKEY, Richard A., U.S.A.
ASOH, Shin-ichi, JAPAN
ASOH, Yasuhiro, JAPAN
ASTALA, Kari, FINLAND
ATSUJI, Atsushi, JAPAN
ATSUMI, Tsuyoshi, JAPAN
ATTALLAH, Sabria M. K.*, EGYPT
AUPETIT, Bernard H., CANADA
AUSLANDER, Maurice, U.S.A.
AVDISPAHIĆ, Muharem, YUGOSLAVIA
AWERBUCH, T. E., U.S.A.
AXELSSON, Axel O., NETHERLANDS
AXLER, Sheldon, U.S.A.
AYUPOV, Shavkat A., U.S.S.R.
AZARPANAH, Fariborz, IRAN
AZIMI, Parviz*, IRAN
AZUHATA, Takashi, JAPAN
AZUKAWA, Kazuo, JAPAN
AZUMAYA, Goro, U.S.A.
BAAK, Eric, NETHERLANDS
BAAYEN, P. C., NETHERLANDS
BABAI, László, HUNGARY
BACHELIS, Gregory F.*, U.S.A.
BACHELOT, Agnès, FRANCE
BACHELOT, Alain*, FRANCE
BADE, William George, U.S.A.
BADIOZZAMAN, Abdol J.§, IRAN
BAGCHI, Sitadri N.*, INDIA
BAILEY, Rosemary A., U.K.
BAILY, Walter L. Jr., U.S.A.
BAK, Anthony, F.R.G.
BAKER, Roger C.*, U.K.
BALAKRISHNAN, V. K., U.S.A.
BALOG, Antal, HUNGARY
BANCHOFF, Thomas F., U.S.A.
BANDO, Shigetoshi, JAPAN
BANNAI, Eiichi, JAPAN
BANNAI, Etsuko, JAPAN

BĀNULESCU, Martha C.*, ROMANIA
BARBASCH, Dan M., U.S.A.
BARBOSA, Joao L. M., BRAZIL
BARDOS, Claude W., FRANCE
BARLOTTI, Adriano, ITALY
BARLOW, Martin T., U.K.
BARNES, Frank William*, KENYA
BARRETT, David E.*, U.S.A.
BARTH, Karl F., U.S.A.
BARTLE, Robert G., U.S.A.
BASS, Hyman, U.S.A.
BATTY, Charles J. K., U.K.
BAUER, Heinz, F.R.G.
BAUTISTA-RAMOS, Raymundo, MEXICO
BAXTER, Rodney J., AUSTRALIA
BAYOD, Jose M.*, SPAIN
BEAULIEU, Liliane*, CANADA
BECKER, Helmut, F.R.G.
BECKER, Jerry P., U.S.A.
BECKER, Ronald I., SOUTH AFRICA
BEDFORD, Eric, U.S.A.
BEHBOODIAN, Javad*, IRAN
BEHFOROOZ, G. G.*, U.S.A.
BEILINSON, Alexander, U.S.S.R.
BELAGE, Abel, FRANCE
BELL, Howard E.*, CANADA
BELTRAMETTI, Mauro C.*, ITALY
BENABDALLAH, Khalid, CANADA
BENSON, Clark T., U.S.A.
BENSON, David J., U.K.
BEN-YELLES, Choukri-Bey, ALGERIA
BÉRARD, Pierre H., FRANCE
BERG, Christian*, DENMARK
BERGER, Marcel, FRANCE
BERGER, Ruth I.*, U.S.A.
BERGERON, François*, CANADA
BERNALDEZ, Jose M., PHILIPPINES
BERTHELOT, Pierre R., FRANCE
BERTIN, Emile M. Y., NETHERLANDS
BESCHLER, Edwin F., U.S.A.
BESSIS, Jean-Pierre, FRANCE
BEZARD, Max Jack, FRANCE
BEZDEK, Andras, HUNGARY
BHANDARI, Ashwani K., INDIA
BHATTARAI, Hom Nath, NEPAL
BICHARA, Alessandro, ITALY
BIERSTEDT, Klaus D.*, F.R.G.
BION-NADAL, Jocelyne*, FRANCE
BIRMAN, Joan S., U.S.A.
BIRNIR, Björn*, ICELAND
BISHOP, Christopher J., U.S.A.
BJÖRCK, Göran*, SWEDEN
BJÖRNER, Anders, SWEDEN
BLACKADAR, Bruce E., U.S.A.
BLACKBURN, Norman, U.K.
BLASIUS, Don M., U.S.A.
BLEILER, Steven A.*, U.S.A.
BLOCH, Spencer J., U.S.A.

BLUM, Lenore C., U.S.A.
BODNARESCU, M. V., F.R.G.
BOECHERER, Siegfried, F.R.G.
BOGOLUBOV, N. N. Jr.*, U.S.S.R.
BOGOMOLOV, Fedor A., U.S.S.R.
BOILEAU, Michel§, FRANCE
BOIVIN, André, CANADA
BOJARSKI, Bogdan*, POLAND
BÖKSTEDT, Marcel Anders, F.R.G.
BONAHON, Francis, FRANCE
BONGIORNO, Benedetto, ITALY
BOOTH, Geoffrey L.*, SOUTH AFRICA
BORWEIN, Peter B., CANADA
BOUKRICHA, Abderrahmen, TUNISIA
BOURGAIN, Jean§, FRANCE
BOURGUIGNON, Jean-Pierre, FRANCE
BOUZAR, Chikh, ALGERIA
BOYLE, Ann K., U.S.A.
BOZORGNIA, Abolghasem*, IRAN
BRAMBILA-PAZ, Leticia G.*, MEXICO
BRATTELI, Ola, NORWAY
BRATTSTRÖM, G. B., SWEDEN
BRAVO, Rafael*, SPAIN
BRECKENRIDGE, John Wylie, AUSTRALIA
BREEN, Lawrence S., FRANCE
BREMNER, Andrew, U.S.A.
BRENNAN, Joseph P.*, U.S.A.
BRENNER, Sheila, U.K.
BRENTI, Francesco, U.S.A.
BRIDGES, Douglas S.*, NEW ZEALAND
BRODMANN, Markus Peter*, SWITZERLAND
BRODSKY, Mikhail*, U.S.A.
BROWDER, Felix E., U.S.A.
BROWN, Leon*, U.S.A.
BROWN, Robert F., U.S.A.
BRUCE, James William, U.K.
BRUCK, Ronald E. Jr.§, U.S.A.
BRUNDU, Michela, ITALY
BRUNING, Erwin A. K.*, SOUTH AFRICA
BRÜNING, Jochen Wulf, F.R.G.
BRUNS, Winfried*, F.R.G.
BRUYÈRE, Véronique, BELGIUM
BRZEZINSKI, Juliusz, SWEDEN
BSHOUTY, Daoud H.*, ISRAEL
BUCHSBAUM, David A., U.S.A.
BUCHWEITZ, Ragnar-Olaf, CANADA
BUCKI, Andrzej J.§, U.S.A.
BUI, Huy-Qui, NEW ZEALAND
BUJALANCE, Emilio*, SPAIN
BUJALANCE, José A.*, SPAIN
BULGARELLI, Ulderico*, ITALY
BULTER, Michael C. R., U.K.
BÜNING, Herbert*, F.R.G.
BURCHARD, Paul*, U.S.A.
BURENKOV, Victor Ivanovich*, U.S.S.R.
BURSTALL, Francis E., U.K.
BUTLER, Lynne M.*, U.S.A.
BUTZ, Jeffrey R., U.S.A.

BYLEEN, Karl E., U.S.A.
BYRNE, Catriona M., F.R.G.
BYUN, Chang Ho, KOREA, R. of
CAMACHO, César, BRAZIL
CAMERON, Peter J., U.K.
CANFELL, Michael John, AUSTRALIA
CANO, Felipe*, SPAIN
CAPPELL, Sylvia E., U.S.A.
CARAYOL, Henri L., FRANCE
CARLESON, Lennart, SWEDEN
CARLSON, James Andrew, U.S.A.
CARLSON, Jon F., U.S.A.
CARNEIRO, Mario J. D., BRAZIL
CARROLL, Robert W.*, U.S.A.
CARTER, Jack A., U.S.A.
CARTER, Roger W., U.K.
CASADIO TARABUSI, Enrico, ITALY
CASSELMAN, William A., CANADA
CASSON, Andrew J., U.S.A.
CATANESE, Fabrizio M.E., ITALY
CATERINA, Maniscalco, ITALY
CATTANEO, Uberto*, SWITZERLAND
CATTANI, Eduardo H., U.S.A.
CAYFORD, Afton H., U.S.A.
CECCHERINI, Pier Vittorio, ITALY
CERAMI, Giovanna M., ITALY
CERCIGNANI, Carlo*, ITALY
CHADEMAN, Arsalan*, IRAN
CHALEYAT-MAUREL, Mireille M., FRANCE
CHAN, Clara S., U.S.A.
CHAN, Raymond H., HONG KONG
CHAN, Shih-Ping, SINGAPORE
CHANDNA, Om P.*, CANADA
CHANG, Der-Chen E., U.S.A.
CHANG, Joo Sup, KOREA, R. of
CHANG, Kun Soo*, KOREA, R. of
CHANG, Mei-Chu, CHINA-TAIWAN
CHANG, Sun-Yung A., U.S.A.
CHANSAENWILAI, Petcharat, THAILAND
CHARI, Vyjayanthi, INDIA
CHATTERJI, Srishti D., SWITZERLAND
CHAVENT, Guy*, FRANCE
CHEE, P. S.*, MALAYSIA
CHEMIN, Jean-Yves*, FRANCE
CHEN, Chao-Nien, CHINA-TAIWAN
CHEN, Larry Lung-Kee§, U.S.A.
CHEN, Louis H. Y.*, SINGAPORE
CHEN, Yun-Gang*, CHINA, P. R. of
CHENG, Chong-Qing*, CHINA, P. R. of
CHENG, Jih-Hsin*, CHINA-TAIWAN
CHEREDNIK, Ivan V., U.S.S.R.
CHERUBINI, Alessandra, ITALY
CHEW, Tuan Seng, SINGAPORE
CHIBA, Keiko, JAPAN
CHIBA, Toru, JAPAN
CHIBA, Toshiyuki§, JAPAN
CHIEN, C.-S.*, CHINA-TAIWAN
CHIKATA, Hiroshi, JAPAN

CHIN, Chou-Hsieng Joe*, CHINA-TAIWAN
CHIN, William, U.S.A.
CHISTOV, Alexander L., U.S.S.R.
CHIYONOBU, Taizo, JAPAN
CHO, Su Yon, KOREA, R. of
CHO, Tae Geun, KOREA, R. of
CHODA, Hisashi, JAPAN
CHODA, Marie*, JAPAN
CHOE, Boo Rim*, KOREA, R. of
CHOE, Geon H.*, KOREA, R. of
CHOU, Kai Seng*, HONG KONG
CHOUDHARY, Ram Chandra*, INDIA
CHOW, Bennett, U.S.A.
CHRIST, Carol S., U.S.A.
CHRIST, Lily E., U.S.A.
CHRIST, Michael, U.S.A.
CHRISTODOULOU, Demetrios, U.S.A.
CHU, Sydney C. K., HONG KONG
CHU, Wenchang*, CHINA, P. R. of
CILIBERTO, Carlo, ITALY
CILIBERTO, Ciro, ITALY
CLAESSON, Tomas, SWEDEN
CLARK, W. Edwin§, U.S.A.
CLEMENS, Laura E., U.S.A.
CLOZEL, Laurent Y., FRANCE
CNOP, Ivan J., BELGIUM
COATES, John Henry, U.K.
CODDINGTON, Earl A.§, U.S.A.
COGDELL, James W., U.S.A.
COHEN, Donald S., U.S.A.
COHEN, Maurice, CANADA
COHN, Harvey§, U.S.A.
COHN, Paul M.*, U.K.
COHN, Richard M., U.S.A.
COIFMAN, Ronald R., U.S.A.
COLLINS, Peter J.*, U.K.
COLMEZ, Pierre, FRANCE
CONREY, John Brian, U.S.A.
CONTE, Alberto, ITALY
CONTESSA, Maria, ITALY
CONTRERAS-BARANDIARÁN, Gonzalo,
    PERU
COOK, Stephen A., CANADA
COOKE, Kenneth L., U.S.A.
COPPEL, William Andrew, AUSTRALIA
CORDUNEANU, Constantin C.*, U.S.A.
CORNEA, Aurel, F.R.G.
CORON, Jean-Michel, FRANCE
COSTA, David G., BRAZIL
COSTE, Michel, FRANCE
COSTES, Constantine N., U.S.A.
CRAPO, Henry*, FRANCE
CRAWLEY-BOEVEY, William W., F.R.G.
CREE, George C., CANADA
CROFT, Hallard Thomas, U.K.
CUBEDDU, Carmen, ITALY
CUEVAS, Félix Humberto, PANAMA
CUI, Chengri, CHINA, P. R. of

CUNTZ, Joachim, F.R.G.
CURTIS, Philip C. Jr., U.S.A.
CURTO, Raul E., U.S.A.
DĂDĂRLAT, Marius Dumitru, ROMANIA
DAHMARDAH, Habib-Olah*, IRAN
DAJCZER, Marcos, BRAZIL
D'AMBROSIO, Ubiratan, BRAZIL
DAMON, James Norman, U.S.A.
DARAFSHEH, Mohammad-Reza*, IRAN
DARMAWIJAYA, Soeparna, INDONESIA
DARMON, Henri René*, CANADA
DATE, Etsuro, JAPAN
DATE, Tamotsu, JAPAN
DAUBEN, Joseph W., U.S.A.
DAVERMAN, Robert J., U.S.A.
DAVID, Sinnou, FRANCE
DAVIES, Roy Osborne§, U.K.
DAVIS, Chandler, CANADA
DEAKIN, Michael A. B.*, AUSTRALIA
DEBNATH, Lokenath*, U.S.A.
DE GRAAF, Jan, NETHERLANDS
DE GUZMÁN, Miguel, SPAIN
DEHGHAN, Mohammad Ali*, IRAN
DEKSTER, Boris V.*, CANADA
DE LA PEÑA, José Antonio M., MEXICO
DE LA PRADELLE, Arnaud, FRANCE
DEL RIEGO, Lilia*, MEXICO
DEMUTH, Michael*, G.D.R.
DENG, Dong-gao*, CHINA, P. R. of
DEWAN, Kum Kum*, INDIA
DIACONIS, Persi, U.S.A.
DIAMOND, Fred I., U.S.A.
DIAS-AGUDO, Fernando R., PORTUGAL
DICKENSTEIN, Alicia M., ARGENTINA
DIEKERT, Volker, F.R.G.
DIMITRIC, Radoslav M.*, YUGOSLAVIA
DINAR, Nathan, ISRAEL
DING, Shisun, CHINA, P. R. of
DING, Tongren*, CHINA, P. R. of
DINH, Dung*, VIETNAM
DINH, Quang Luu, VIETNAM
DIONNE, Philippe-A., FRANCE
DI PIAZZA, Luisa, ITALY
DLAB, Vlastimil*, CANADA
DLOUSSKY, Georges, FRANCE
DOBRUSHIN, Roland L., U.S.S.R.
DO CARMO, Manfredo P., BRAZIL
DODSON, C. T. J.*, CANADA
DOGUWA, Sani Ibrahim, NIGERIA
DOHI, Yutaka, JAPAN
DOI, Masahiro§, JAPAN
DOI, Shinichi, JAPAN
DOI, Yukio, JAPAN
DOKU, Isamu*, JAPAN
DOLCI, Paolo V., ITALY
DÔMAE, Takanobu, JAPAN
DONALDSON, James A., U.S.A.
DOPLICHER, Sergio, ITALY

DOUADY, Adrien, FRANCE
DOUGLAS, Ronald George, U.S.A.
DRANISHNIKOV, Alexander N., U.S.S.R.
DRINFELD, Vladimir G., U.S.S.R.
DUGGAL, Krishan Lal*, CANADA
DUKE, William D.§, U.S.A.
DUMORTIER, Freddy E., BELGIUM
DUOANDIKOETXEA, Javier, SPAIN
DUPONT, Johan L.*, DENMARK
DURFEE, Alan H.*, U.S.A.
DURRETT, Richard Timothy, U.S.A.
DVORETZKY, Aryeh, ISRAEL
DWIVEDI, Brij Narayan*, INDIA
DYMACEK, Wayne M.*, U.S.A.
DYN, Nira*, ISRAEL
DZHUMADIL'DAEV, A. S.*, U.S.S.R.
DZINOTYIWEYI, Heneri A. M.*, ZIMBABWE
DZRBASHIAN, Mhitar M., U.S.S.R.
DZURAEV, Abduhamid D.*, U.S.S.R.
EBERLEIN, Patrick Barry, U.S.A.
EBIHARA, Madoka*, JAPAN
ECALLE, Jean Pascal, FRANCE
EDA, Katsuya, JAPAN
EDAMATSU, Takasi, JAPAN
EDWARDS, Robert D., U.S.A.
EGASHIRA, Shinji, JAPAN
EGGHE, Leo C.*, BELGIUM
EGUCHI, Kazuo, JAPAN
EGUCHI, Masaaki, JAPAN
EGUCHI, Tohru, JAPAN
EHR, Carolyn K., U.S.A.
EHRENPREIS, Leon, U.S.A.
EI, Shinichiro, JAPAN
EIDA, Atsuhiko, JAPAN
EIN, Lawrence, U.S.A.
EJIRI, Norio, JAPAN
EKHAGUERE, Godwin O. S., NIGERIA
ELLERS, Erich W.*, CANADA
ELLERS, Harald E., CANADA
ELLIOTT, George A.*, DENMARK
ELLIS, Alan J., HONG KONG
ELWORTHY, Kenneth David, U.K.
EMARA, Salah Abbas Ahmed*, EGYPT
ENDO, Akira, JAPAN
ENDO, Mikihiko, JAPAN
ENDO, Shizuo, JAPAN
ENGL, Heinz W.*, AUSTRIA
ENOCK, Michel, FRANCE
ENOKI, Ichiro, JAPAN
ENOMOTO, Fumihiko, JAPAN
ENOMOTO, Kazuyuki*, JAPAN
ENOMOTO, Masatoshi, JAPAN
ENRIQUEZ, Benjamin David, FRANCE
ENTA, Yoichi, JAPAN
EPIFANIO, Giuseppina, ITALY
ERBLAND, John Paul, U.S.A.
ERDMANN, Karin, U.K.
ERIKSSON, Folke S.*, SWEDEN

ERIKSSON-BIQUE, Sirkka-L. A.*, FINLAND
ERNÉ, Marcel T.*, F.R.G.
ESCULTURA, Edgar E.*, PHILIPPINES
ESNAULT, Hélène I., FRANCE
ESSÉN, Matts R.*, SWEDEN
ESTRADA, Ricardo*, COSTA RICA
EVANS, David E., U.K.
EVANS, Michael John, CANADA
EZAWA, Hiroshi, JAPAN
EZEILO, James O.C., NIGERIA
FACCHINI, Alberto, ITALY
FADDEEV, Ludwig D., U.S.S.R.
FAGHIH, Nezameddin*, IRAN
FALK, Michael J., U.S.A.
FAN, Peng, CHINA-TAIWAN
FANIA, M.Lucia, ITALY
FARADZEV, Igor A., U.S.S.R.
FARAHI, Mohammad H.*, IRAN
FARMANESH, Mounesali, IRAN
FATHI, Albert*, U.S.A.
FAVIILI, Franco, ITALY
FEE, Greg J., CANADA
FEIG, Ephraim§, U.S.A.
FEIGIN, Boris L., U.S.S.R.
FEIT, Walter, U.S.A.
FELDMAN, Joel S., CANADA
FENG, Keqin*, CHINA, P. R. of
FERENCZI, Sebastien*, FRANCE
FERNANDEZ, Arturo A.*, SPAIN
FIGÀ-TALAMANCA, Alessandro, ITALY
FILLMORE, Peter A., CANADA
FLACH, Matthias, F.R.G.
FLATH, Daniel E.*, U.S.A.
FLATO, Moshé, FRANCE
FLATTO, Leopold, U.S.A.
FLEISCHER, Alexander*, U.S.S.R.
FLICKER, Yuval Z., ISRAEL
FLOER, Andreas, U.S.A.
FOLLAND, Gerald B., U.S.A.
FOROUZANFAR, Abdol M.*, IRAN
FÖRSTER, Karl-Heinz, F.R.G.
FOSSUM, Robert M., U.S.A.
FOSTER, B. L.*, U.S.A.
FOXBY, Hans B., DENMARK
FRANGOS, Nicholas E., U.S.A.
FRANKS, John M., U.S.A.
FREI, Guenther H., SWITZERLAND
FRIBERG, Jöran, SWEDEN
FRIED, David, U.S.A.
FRIEDLAND, Shmuel*, U.S.A.
FRIEDLANDER, John B., CANADA
FRÖBERG, B. Ralf, SWEDEN
FU, Hong Chen*, CHINA, P. R. of
FUCHINO, Sakaé, JAPAN
FUGLEDE, Bent*, DENMARK
FUJII, Akio, JAPAN
FUJII, Junichi, JAPAN
FUJII, Kazuyuki, JAPAN

FUJII, Masatoshi, JAPAN
FUJII, Michihiko, JAPAN
FUJII, Michikazu, JAPAN
FUJII, Seiji, JAPAN
FUJII, Shinichi, JAPAN
FUJII, Toshiakira, JAPAN
FUJIIE, Setsuro, JAPAN
FUJI'I'E, Tatsuo, JAPAN
FUJIKI, Akira, JAPAN
FUJIKOSHI, Yasunori*, JAPAN
FUJIMAGARI, Tetsuo, JAPAN
FUJIMOTO, Hirotaka, JAPAN
FUJIMOTO, Yoshio, JAPAN
FUJIMURA, Shigeyoshi, JAPAN
FUJISAKI, Genjiro, JAPAN
FUJISAKI, Rieko, JAPAN
FUJISAWA, Taro, JAPAN
FUJITA, Hiroshi, JAPAN
FUJITA, Hiroyuki, JAPAN
FUJITA, Hisaaki, JAPAN
FUJITA, Kazunori, JAPAN
FUJITA, Kenji, JAPAN
FUJITA, Kumiko, JAPAN
FUJITA, Osamu, JAPAN
FUJITA, Takahiko, JAPAN
FUJITA, Takao§, JAPAN
FUJIWARA, Daisuke, JAPAN
FUJIWARA, Hidenori, JAPAN
FUJIWARA, Kazuhiro, JAPAN
FUJIWARA, Koji, JAPAN
FUJIWARA, Masahiko, JAPAN
FUJIWARA, Shigeki, JAPAN
FUJIWARA, Tsukasa, JAPAN
FUJIWARA, Tsuyoshi, JAPAN
FUKAGAI, Eiji, JAPAN
FUKAGAI, Nobuyoshi, JAPAN
FUKAISHI, Hiroo, JAPAN
FUKAMIYA, Masanori§, JAPAN
FUKAYA, Kenji, JAPAN
FUKUCHI, Minoru, JAPAN
FUKUDA, Hiromichi, JAPAN
FUKUDA, Nozomu, JAPAN
FUKUDA, Takashi, JAPAN
FUKUDA, Takuo, JAPAN
FUKUHARA, Kenzo, JAPAN
FUKUHARA, Shinji§, JAPAN
FUKUI, Kazuhiko, JAPAN
FUKUI, Seiichi, JAPAN
FUKUI, Toshizumi, JAPAN
FUKUMOTO, Yoshihiro, JAPAN
FUKUSHI, Takeo, JAPAN
FUKUSHIMA, Masatoshi, JAPAN
FUKUTAKE, Takayoshi, JAPAN
FUKUYAMA, Katusi, JAPAN
FUNAHASHI, Ken-ichi, JAPAN
FUNAKI, Tadahisa, JAPAN
FUNAKOSI, Shunsuke, JAPAN
FUNAKURA, Takeo, JAPAN

FUNAYA, Bokurou, JAPAN
FURSTENBERG, Hillel Harry, U.S.A.
FURUKAWA, Yasukuni, JAPAN
FURUSHIMA, Mikio, JAPAN
FURUSHO, Yasuhiro, JAPAN
FURUTA, Kohji, JAPAN
FURUTA, Mikio, JAPAN
FURUTA, Tomonori, JAPAN
FURUTA, Yoshiomi, JAPAN
FURUTO, David Masaru, U.S.A.
FURUTSU, Hirotoshi, JAPAN
FURUYA, Kiyoko, JAPAN
FURUYA, Masako, JAPAN
FUTAKI, Akito§, JAPAN
GABAI, David, U.S.A.
GAMBAUDO, Jean-Marc*, FRANCE
GAMKRELIDZE, Revaz V., U.S.S.R.
GARCÍA-CUERVA, José, SPAIN
GARDNER, Robert B., U.S.A.
GARLING, David J. H., U.K.
GARNER, Cyril W.L., CANADA
GARZON, Max H., U.S.A.
GAUDRY, Garth Ian, AUSTRALIA
GAULD, David B.§, NEW ZEALAND
GEBAUER, Rüdiger, U.S.A.
GEDDES, Keith, CANADA
GELLER, Daryl N., U.S.A.
GEOGHEGAN, Ross, U.S.A.
GERAMI, Nasrollah*, IRAN
GETIMANE, Mário Frengue, MOZAMBIQUE
GETZLER, Ezra, AUSTRALIA
GHOUSSOUB, Nassif, CANADA
GHYS, Etienne, FRANCE
GIACARDI, Livia*, ITALY
GIGA, Mariko, JAPAN
GIGA, Yoshikazu*, JAPAN
GILES, John Robilliard*, AUSTRALIA
GILLARD, Roland D., FRANCE
GILLET, Henri A., U.S.A.
GILPIN, Michael J.*, U.S.A.
GINGOLD, Harry*, U.S.A.
GIORDANO, Thierry*, SWITZERLAND
GIRKO, Vyacheslav L.*, U.S.S.R.
GIUCULESCU, Alexandru*, ROMANIA
GIUSTI, Mark François, FRANCE
GŁAZEK, Kazimierz*, POLAND
GLEASON, Andrew M., U.S.A.
GLOWINSKI, Roland, U.S.A.
GOCHO, Toru, JAPAN
GODA, Hiroshi, JAPAN
GOKA, Yosihiro, JAPAN
GOLAN, Jonathan Samuel, ISRAEL
GOLDBACH, Ronald W., NETHERLANDS
GOLDBLATT, Robert*, NEW ZEALAND
GOLDSTEIN, Jerome A., U.S.A.
GOLDWASSER, Shafi, U.S.A.
GOLODETS, Valentin Y.*, U.S.S.R.
GOMEZ-MONT, Xavier A., MEXICO

GOMI, Kensaku, JAPAN
GOMIKAWA, Taturo, JAPAN
GONCHAR, Andrei A., U.S.S.R.
GONSHOR, Harry*, U.S.A.
GONZALEZ-ACUÑA, Francisco, MEXICO
GONZALEZ-SPRINBERG, G., FRANCE
GOODEARL, Kenneth R., U.S.A.
GOODMAN, Fredrick M., U.S.A.
GOODMAN, Jacob E., U.S.A.
GOODWILLIE, Thomas G., U.S.A.
GORDON, Cameron M., U.S.A.
GOSSEZ, Jean-Pierre*, BELGIUM
GOTO, Midori, JAPAN
GOTO, Morikuni, JAPAN
GOTO, Ryushi, JAPAN
GOTO, Shiro, JAPAN
GOTO, Yukinori, JAPAN
GOWRISANKARAN, Kohur, CANADA
GRAFAKOS, Loukas*, GREECE
GRAHAM, John J., AUSTRALIA
GRAHAM, R. L., U.S.A.
GRANDSARD, Francine L., BELGIUM
GRANIRER, Edmond E., CANADA
GRANVILLE, Andrew J., U.K.
GRATZER, George A., CANADA
GRAY, Jeremy John, U.K.
GREENE, Curtis, U.S.A.
GREENSPOON, Arthur, U.S.A.
GREGUŠ, Michal§, CZECHOSLOVAKIA
GREUEL, Gert-Martin, F.R.G.
GRIGGS, Jerrold R., U.S.A.
GRIGIS, Alain, FRANCE
GRIGORCHUK, Rostislav Ivanovich, U.S.S.R.
GRIGORIEFF, Rolf Dieter, F.R.G.
GRIMALDI-PIRO, Anna, ITALY
GROBBELAAR, Marié*, SOUTH AFRICA
GROENEWALD, Nico J.*, SOUTH AFRICA
GROETSCH, Charles W.*, U.S.A.
GROSS, Leonard, U.S.A.
GROVE, Karsten, DENMARK
GRUNDMAN, Helen G., U.S.A.
GRÜTER, Michael, F.R.G.
GUARALDO, Rosalind, U.S.A.
GUBELADZE, Iosif D., U.S.S.R.
GUEST, Martin A.*, U.S.A.
GUIBAS, Leonidas J., U.S.A.
GUICHARDET, Alain, FRANCE
GUILLOPE, Laurent, FRANCE
GULLIVER, Robert D., U.S.A.
GUNDLACH, Karl-Bernhard, F.R.G.
GÜNTHER, Matthias, G.D.R.
GUO, Mao Zheng, CHINA, P. R. of
GURJAR, Rajendra Vasant*, INDIA
GYOJA, Akihiko, JAPAN
GYÖRI, Ervin, HUNGARY
GYŐRY, Kálmán*, HUNGARY
HA, Huy Bang*, VIETNAM
HABGGER, Nathan Bernard, FRANCE

HADANO, Toshihiro, JAPAN
HADIAN DEHKORDI, Masoud*, IRAN
HAG, Kari Jorun, NORWAY
HAG, Per, NORWAY
HAHN, Saul G., U.S.A.
HAIDA, Minoru, JAPAN
HAJARNAVIS, Charudatta R., U.K.
HAKEDA, Josuke, JAPAN
HAKEDA, Nobuhiro§, JAPAN
HAKOSALO, Milja Riitta, U.S.A.
HALISTE, Kersti A., SWEDEN
HAMACHI, Toshihiro, JAPAN
HAMADA, Hozumi*, JAPAN
HAMADA, Yusaku, JAPAN
HAMANA, Masamichi*, JAPAN
HAMANA, Yuji, JAPAN
HAMANAKA, Yojiro, JAPAN
HAMASAKI, Atsumi, JAPAN
HANDA, Kenji, JAPAN
HANO, Jun-ichi, JAPAN
HANOUZET, Bernard, FRANCE
HANSEN, Frank*, DENMARK
HANSEN, Vagn L.*, DENMARK
HANSEN, Wolfhard, F.R.G.
HANS-GILL, Rajinder Jeet, INDIA
HAPPEL, Dieter, F.R.G.
HARA, Masao, JAPAN
HARA, Shin-ichiro, JAPAN
HARA, Tadayuki, JAPAN
HARA, Takuya, JAPAN
HARA, Tamio, JAPAN
HARA, Yoshihito, JAPAN
HARA, Yuko, JAPAN
HARADA, Masana, JAPAN
HARADA, Mikio, JAPAN
HARAGUCHI, Yuri R., CHILE
HARDER, Günter, F.R.G.
HARDY, Kenneth, CANADA
HARIKAE, Toshio, JAPAN
HARRIS, Michael H., U.S.A.
HART, R. Neal, U.S.A.
HARTEN, Amiram, ISRAEL/U.S.A.
HARZHEIM, Egbert R.*, F.R.G.
HASEGAWA, Fumio, JAPAN
HASEGAWA, Izumi, JAPAN
HASEGAWA, Keizo, JAPAN
HASEGAWA, Kenji, JAPAN
HASEGAWA, Kohji, JAPAN
HASEGAWA, Ryo, JAPAN
HASEGAWA, Ryu, JAPAN
HASEGAWA, Takayuki, JAPAN
HASEGAWA, Youjirou, JAPAN
HASHIGUCHI, Kosaburo, JAPAN
HASHIGUCHI, Masao, JAPAN
HASHIGUCHI, Norikazu, JAPAN
HASHIMOTO, Hideya§, JAPAN
HASHIMOTO, Mitsuyasu, JAPAN
HASHIMOTO, Takahiro, JAPAN

HASHIMOTO, Takashi, JAPAN
HASHIMOTO, Yoshiaki, JAPAN
HASHIMOTO, Yoshihiko, JAPAN
HASHIMOTO, Yoshitake, JAPAN
HASHIMOTO, Yuji, JAPAN
HASSANI, Akbar*, IRAN
HASUI, Satoshi, JAPAN
HASUMI, Morisuke, JAPAN
HATA, Masayoshi, JAPAN
HATADA, Kazuyuki, JAPAN
HATORI, Asako, JAPAN
HATORI, Osamu*, JAPAN
HATTORI, Akio, JAPAN
HATTORI, Tetsuya§, JAPAN
HATTORI, Toshiaki, JAPAN
HATTORI, Yasunao, JAPAN
HATTORI, Yoshinori, JAPAN
HATVANI, László*, HUNGARY
HAWKES, John, U.K.
HAWKES, Trevor Ongley, U.K.
HAYAKATA, Takayuki, JAPAN
HAYAKAWA, Kantaro, JAPAN
HAYAKAWA, Keizou, JAPAN
HAYASAKA, Satoshi, JAPAN
HAYASHI, Etsuo, JAPAN
HAYASHI, Heima, JAPAN
HAYASHI, Hiroki, JAPAN
HAYASHI, Hiroshi§, JAPAN
HAYASHI, Kazumichi, JAPAN
HAYASHI, Kiyoshi, JAPAN
HAYASHI, Makoto, JAPAN
HAYASHI, Makoto, JAPAN
HAYASHI, Mikihiro, JAPAN
HAYASHI, Nakao*, JAPAN
HAYASHI, Takahiro, JAPAN
HAYASHI, Yoshiki, F.R.G.
HAYASHIDA, Tsuyoshi§, JAPAN
HAYASIDA, Kazuya, JAPAN
HAYMAN, Walter Kurt, U.K.
HEARST, III, William R., U.S.A.
HEBISCH, Waldemar, POLAND
HECKMAN, Gerrit, NETHERLANDS
HEDBERG, Lars Inge, SWEDEN
HEILIGMAN, Mark Isaac, U.S.A.
HEINONEN, Juha M., U.S.A.
HEINTZ, Joos U., ARGENTINA
HEINZE, Joachim, F.R.G.
HELLSTRÖM, Lennart, SWEDEN
HELMINCK, Aloysius Gerarous,
    NETHERLANDS
HELMINCK, Gerard G.F., NETHERLANDS
HENRICHS, Rolf Wim*, F.R.G.
HERMAN, Richard H.§, U.S.A.
HERRERA, Rutilio A.*, COSTA RICA
HERRMANN, Manfred H., F.R.G.
HERZ, Prof., CANADA
HERZOG, Jürgen R. G.§, F.R.G.
HIAI, Fumio, JAPAN

HIBI, Takayuki, JAPAN
HIDA, Takeyuki, JAPAN
HIDAKA, Fumio*, JAPAN
HIGA, Tatsuo, JAPAN
HIGASHIYAMA, Teiko, JAPAN
HIGUCHI, Yasunari, JAPAN
HIGUCHI, Yoshiki, JAPAN
HIJIKATA, Hiroaki, JAPAN
HIKIDA, Mizuho, JAPAN
HIKITA, Teruo, JAPAN
HILLEL, Joel*, CANADA
HILSUM, Michel*, FRANCE
HINO, Yosuke, JAPAN
HINOKUMA, Takanori, JAPAN
HINZ, Andreas M.*, F.R.G.
HIRABAYASHI, Mikihito, JAPAN
HIRACHI, Kengo, JAPAN
HIRAGA, Kaoru, JAPAN
HIRAHATA, Hirotoshi*, JAPAN
HIRAI, Etsuko, JAPAN
HIRAI, Takeshi, JAPAN
HIRAI, Yasuhisa, JAPAN
HIRAIDE, Koichi, JAPAN
HIRAKAWA, Fumiko§, JAPAN
HIRAKAWA, Kosaburo, JAPAN
HIRAKAWA, Masataka, JAPAN
HIRAMATSU, Toyokazu, JAPAN
HIRANO, Masaki, JAPAN
HIRASAWA, Yoshikazu, JAPAN
HIRATA, Kazuhiko, JAPAN
HIRATA, Koichi, JAPAN
HIRATA, Masaki, JAPAN
HIRAYAMA, Minoru, JAPAN
HIROHASHI, Masanori, JAPAN
HIROMI, Genkõ, JAPAN
HIROMORI, Katsuhisa, JAPAN
HIRONAKA, Eriko, JAPAN
HIRONAKA, Heisuke, JAPAN
HIRONAKA, Yumiko*, JAPAN
HIROSE, Ken, JAPAN
HIROSE, Susumu, JAPAN
HIROSHIMA, Tsutomu, JAPAN
HISANO, Tamao, JAPAN
HISHIDA, Toshiaki, JAPAN
HITOTUMATU, Sin, JAPAN
HITSUDA, Masuyuki, JAPAN
HOANG, Le Minh, VIETNAM
HODGSON, Bernard R., CANADA
HOFER, Helmut H., F.R.G.
HÖFER, Thomas, F.R.G.
HOFMANN, Bernd, G.D.R.
HOHTI, Aarno*, FINLAND
HOJO, Shun-ichi, JAPAN
HOLLAND, Samuel S. Jr., U.S.A.
HOLME, Audun, NORWAY
HOLMES, Philip J., U.S.A.
HOLOPAINEN, Ilkka O., FINLAND
HOLT, Fred B., U.S.A.

HOLTE, John M.*, U.S.A.
HOMMA, Katsumi, JAPAN
HOMMA, Kiyomi, JAPAN
HOMMA, Masaaki, JAPAN
HOMMA, Tatsuo, JAPAN
HONARY, Bahman K.*, IRAN
HONARY, Taher Ghasemi*, IRAN
HONDA, Ikuji, JAPAN
HONDA, Kin-ya, JAPAN
HONDA, Motohiro, JAPAN
HONDA, Naofumi, JAPAN
HONG, Imsik, JAPAN
HONG, Sungpyo, KOREA, R. of
HONG, Yi, CHINA, P. R. of
HOOPER, Robert Clark, U.S.A.
HORA, Akihito, JAPAN
HORAI, Masako, JAPAN
HORI, Kenta, JAPAN
HORI, Teturô, JAPAN
HORIE, Kuniaki*, JAPAN
HORIE, Mitsuko, JAPAN
HORIGUCHI, Hiroshi, JAPAN
HORIGUCHI, Syunzi, JAPAN
HORIKAWA, Eiji, JAPAN
HORIUCHI, Annick, FRANCE
HORIUCHI, Kiyomitsu, JAPAN
HORIUCHI, Ryutaro, JAPAN
HORIUCHI, Toshio*, JAPAN
HÖRMANDER, Lars V.§, SWEDEN
HOSHINA, Takao, JAPAN
HOSHIRO, Toshihiko, JAPAN
HOSOBUCHI, Masami, JAPAN
HOSOH, Toshio, JAPAN
HOSOKAWA, Fujitsugu, JAPAN
HOSONO, Yuzo, JAPAN
HOSOYA, Naoto, JAPAN
HOSTE, Jim, U.S.A.
HOTTA, Kôsaku, JAPAN
HOTTA, Ryoshi§, JAPAN
HOU, Jinchuan*, CHINA, P. R. of
HOU, Shui-Hung, HONG KONG
HOU, Zixin*, CHINA, P. R. of
HOUH, Chorng-Shi, U.S.A.
HOUSTON, Johnny, U.S.A.
HOUZEL, Christian, FRANCE
HRUSHOVSKI, Ehud, U.S.A.
HSIEH, Po-Fang*, U.S.A.
HSIEH, Ying-Hen*, CHINA-TAIWAN
HU, Bizhong, CHINA, P. R. of
HU, Hesheng, CHINA, P. R. of
HU, Thakyin, CHINA-TAIWAN
HUANG, Cheng-Gui§, CHINA, P. R. of
HUANG, Jing-Song, U.S.A.
HUANG, Qichang*, CHINA, P. R. of
HUANG, Wei-Zhang*, CHINA, P. R. of
HUANG, Xun-Cheng, U.S.A.
HUANG, Zhiyuan, CHINA, P. R. of
HUBBARD, John Hamal, U.S.A.

HUBBUCK, John Reginald, U.K.
HUGHES, Anne C.*, U.S.A.
HUGHES, Kenneth Robert, SOUTH AFRICA
HUKUM, Singh*, INDIA
HULEK, Klaus W.*, F.R.G.
HUNEKE, Craig L.*, U.S.A.
HURUYA, Tadasi, JAPAN
HUXLEY, Martin Neil, U.K.
HUZII, Mituaki, JAPAN
HWANG, Jun Shung, CHINA-TAIWAN
IARROBINO, A. A. Jr.*, U.S.A.
IBARAKI, Toshihide, JAPAN
IBISCH, Horst D., FRANCE
IBUKIYAMA, Tomoyoshi, JAPAN
ICHIDA, Ryosuke, JAPAN
ICHIHARA, Kanji, JAPAN
ICHIJI, Hiroshi, JAPAN
ICHIJYO, Yoshihiro, JAPAN
ICHIKAWA, Akira, JAPAN
ICHIKAWA, Kaoru*, JAPAN
ICHIKAWA, Takashi, JAPAN
ICHIMURA, Humio, JAPAN
ICHINOSE, Takashi, JAPAN
ICHIRAKU, Shigeo, JAPAN
IDA, Tomoya, JAPAN
IGARASHI, Kan, JAPAN
IGARASHI, Masayuki§, JAPAN
IGARI, Katsuju§, JAPAN
IGARI, Satoru, JAPAN
IGUSA, Kiyoshi, U.S.A.
IHARA, Shin-ichiro, JAPAN
IHARA, Shunsuke, JAPAN
IHARA, Yasutaka, JAPAN
II, Kiyotaka, JAPAN
IIDA, Hirokazu, JAPAN
IIDA, Masato, JAPAN
IITAKA, Shigeru*, JAPAN
IKAWA, Mitsuru, JAPAN
IKAWA, Osamu, JAPAN
IKAWA, Toshihiko, JAPAN
IKEBE, Nobunori, JAPAN
IKEDA, Akira, JAPAN
IKEDA, Fumio, JAPAN
IKEDA, Hideo, JAPAN
IKEDA, Kaoru, JAPAN
IKEDA, Kazumasa, JAPAN
IKEDA, Koichiro, JAPAN
IKEDA, Nobuyuki, JAPAN
IKEDA, Noriaki, JAPAN
IKEDA, Shin, JAPAN
IKEDA, Tamotsu, JAPAN
IKEDA, Toru, JAPAN
IKEDA, Toshiharu, JAPAN
IKEDA, Toshio, JAPAN
IKEDA, Tsutomu, JAPAN
IKEDA, Yoshito, JAPAN
IKEGAMI, Giko, JAPAN
IKEGAMI, Teruo, JAPAN

IKEHATA, Masaru, JAPAN
IKESHOJI, Kiyoshi, JAPAN
IKEYAMA, Tamotsu, JAPAN
IKOMA, Yoshio, JAPAN
ILIADIS, Stavros*, GREECE
ILLMAN, Sören A., FINLAND
IL'YASHENKO, Julij S., U.S.S.R.
IM, John J., U.S.A.
IMADA, Naotaka, JAPAN
IMAFUKU, Kentaro, JAPAN
IMAI, Hideo, JAPAN
IMAI, Hideo, JAPAN
IMAI, Hiroshi, JAPAN
IMAI, Keiko, JAPAN
IMAI, Masataka, JAPAN
IMAI, Toshihiro, JAPAN
IMANISHI, Hideki, JAPAN
IMAYOSHI, Yoichi, JAPAN
IMBRIE, John Z., U.S.A.
INABA, Takashi, JAPAN
INAGAKI, Nobuo§, JAPAN
INATOMI, Akira, JAPAN
INNAMI, Nobuhiro, JAPAN
INOUE, Akihiko, JAPAN
INOUE, Junko, JAPAN
INOUE, Jyunji, JAPAN
INOUE, Kazuyuki, JAPAN
INOUE, Masahisa, JAPAN
INOUE, Takayuki, JAPAN
INOUE, Tetsuo*, JAPAN
INOUE, Toru, JAPAN
INOUE, Yoshinari, JAPAN
INOUE, Yoshiyuki, JAPAN
ION, Patrick D. F., U.S.A.
IONESCU, Paltin, ROMANIA
IONESCU, Paul-Cristodor H., ROMANIA
IORIO Jr., Rafael José, BRAZIL
IOZZI, Alessandra, ITALY/U.S.A.
IRAQI MOGHADDAM, G. H., IRAN
IRI, Masao§, JAPAN
IRIE, Shoji, JAPAN
IRITE, Takao, JAPAN
IROKAWA, Susumu, JAPAN
ISBELL, John R., U.S.A.
ISEKI, Kiyoshi, JAPAN
ISHIBASHI, Hiroyuki, JAPAN
ISHIBASHI, Makoto§, JAPAN
ISHIBASHI, Yasunori, JAPAN
ISHIDA, Hisashi, JAPAN
ISHIDA, Masanori, JAPAN
ISHIDA, Mitsuko, JAPAN
ISHIGAMI, Shigeo, JAPAN
ISHIGAMI, Yoshiyasu*, JAPAN
ISHIHARA, Hajime*, JAPAN
ISHIHARA, Ikuo, JAPAN
ISHIHARA, Junichi, JAPAN
ISHIHARA, Kazuo, JAPAN
ISHIHARA, Muneichi, JAPAN

ISHIHARA, Toru, JAPAN
ISHII, Akira, JAPAN
ISHII, Hidenori, JAPAN
ISHII, Hitoshi, JAPAN
ISHII, Ippei, JAPAN
ISHII, Katsuyuki, JAPAN
ISHII, Kazuhiko, JAPAN
ISHII, Noburo, JAPAN
ISHII, Shihoko, JAPAN
ISHII, Yutaka, JAPAN
ISHIKAWA, Goo, JAPAN
ISHIKAWA, Hirofumi, JAPAN
ISHIKAWA, Masao, JAPAN
ISHIKAWA, Masato, JAPAN
ISHIKAWA, Michio, JAPAN
ISHIKAWA, Mituo, JAPAN
ISHIKAWA, Nobuhiro, JAPAN
ISHIKAWA, Saneaki, JAPAN
ISHIKAWA, Shiro, JAPAN
ISHIKAWA, Takeshi, JAPAN
ISHIKAWA, Takeshi, JAPAN
ISHIKAWA, Tsuneo, JAPAN
ISHIKAWA, Yasushi*, JAPAN
ISHIKAWA, Yoichiro, JAPAN
ISHIMOTO, Hiroyasu, JAPAN
ISHIMURA, Naoyuki, JAPAN
ISHIMURA, Ryuichi, JAPAN
ISHITOYA, Kiminao, JAPAN
ISHIZAKI, Katsuya, JAPAN
ISIDRO, José M., SPAIN
ISKOVSKIH, Vasilii A., U.S.S.R.
ISOGAI, Eiichi, JAPAN
ISOZAKI, Hiroshi, JAPAN
ISU, Minoru, JAPAN
ITAI, Masanori*, U.S.A.
ITANO, Mitsuyuki, JAPAN
ITATSU, Seiichi, JAPAN
ITO, Hideji, JAPAN
ITO, Hidekazu, JAPAN
ITO, Hiroshi, JAPAN
ITO, Hiroya, JAPAN
ITO, Hiroyuki, JAPAN
ITO, Ichiro, JAPAN
ITO, Keiichi*, JAPAN
ITO, Kiyosi, JAPAN
ITO, Masami, JAPAN
ITO, Masayuki, JAPAN
ITO, Masayuki, JAPAN
ITO, Ryuichi, JAPAN
ITO, Seizo, JAPAN
ITO, Takashi, JAPAN
ITO, Tatsuro, JAPAN
ITO, Toshikazu, JAPAN
ITO, Yoshifumi*, JAPAN
ITO, Yoshifusa, JAPAN
ITO, Yoshihiko, JAPAN
ITO, Yoshihiko, JAPAN
ITO, Yuji, JAPAN

ITOH, Jin-ichi, JAPAN
ITOH, Masa-aki, JAPAN
ITOH, Shiroh, JAPAN
ITOH, Takashi, JAPAN
ITOH, Tatsuo*, JAPAN
ITOH, Yoshiaki*, JAPAN
ITOKAWA, Yoe, JAPAN
IVANOV, Aleksander A., U.S.S.R.
IVANOV, Andrei V., U.S.S.R.
IVANOV, Kamen Ganchev, BULGARIA
IVANŠIĆ, Ivan*, YUGOSLAVIA
IWAHORI, Nagayoshi, JAPAN
IWAHORI, Nobuko, JAPAN
IWAI, Akira, JAPAN
IWAI, Mieko§, JAPAN
IWAI, Toshihiro, JAPAN
IWAKAMI, Tatsuo, JAPAN
IWAMOTO, Masayuki, JAPAN
IWAMOTO, Takashi, JAPAN
IWANAGA, Yasuo, JAPAN
IWANOWSKI, Peter, F.R.G.
IWASAKI, Chisato, JAPAN
IWASAKI, Ichiro, JAPAN
IWASAKI, Nobuhisa, JAPAN
IWASAKI, Shiro, JAPAN
IWASAWA, Kenkichi§, JAPAN
IWATA, Kazuo, JAPAN
IWATA, Koichi, JAPAN
IWATANI, Teruo, JAPAN
IWATSUKA, Akira, JAPAN
IWAYAMA, Eiko, JAPAN
IYANAGA, Shokichi, JAPAN
IZÉ, Antonio F.*, BRAZIL
IZEKI, Hiroyasu, JAPAN
IZUCHI, Keiji*, JAPAN
IZUMI, Hideo*, JAPAN
IZUMI, Masaki, JAPAN
IZUMI, Shuzo, JAPAN
IZUMISAWA, Masataka, JAPAN
IZUMIYA, Shyuichi, JAPAN
JACO, William H., U.S.A.
JACOB, Niels, F.R.G.
JAEGER, Arno*, F.R.G.
JAROSZ, Krzysztof M., U.S.A.
JAVAME GHAZVINI, M. J., IRAN
JENKINS, James A., U.S.A.
JENSEN, Norbert, F.R.G.
JI, Xinhua*, CHINA, P. R. of
JIJTCHENKO, Aleksei B., U.S.S.R.
JIMBO, Michio, JAPAN
JIMBO, Shuichi, JAPAN
JIMBO, Toshiya, JAPAN
JIN, Gyo Taek, KOREA, R. of
JIN, Naondo, JAPAN
JOHNSON, Dave, U.K.
JOHNSON, Gerald W., U.S.A.
JOHNSON, Roy A., U.S.A.
JOLLY, Michael Summerfield*, U.S.A.

JONES, Lowell Edwin, U.S.A.
JONES, Mark C. W.*, U.K.
JONES, Peter W., U.S.A.
JONES, Vaughan F. R., NEW ZEALAND
JORGENSEN, Palle E. T.*, U.S.A.
JUANG, Jonq*, CHINA-TAIWAN
JURKIEWICZ, Jerzy, POLAND
JUTILA, Matti Ilmari*, FINLAND
KABEYA, Yoshitsugu, JAPAN
KACHI, Hideyuki, JAPAN
KACHI, Yasuyuki, JAPAN
KADISON, Richard V.§, U.S.A.
KADO, Jiro, JAPAN
KADOYAMA, Makoto, JAPAN
KAGA, Toshihiro, JAPAN
KAGESAWA, Masataka, JAPAN
KÅGESTEN, Owe K., SWEDEN
KAGEYAMA, Sanpei, JAPAN
KAHAN, W. M., U.S.A.
KAHANE, Jean-Pierre, FRANCE
KAHN, Bruno*, FRANCE
KAIMANOVICH, Vadim A., U.S.S.R.
KAIZU, Satoshi*, JAPAN
KAIZUKA, Tetsu§, JAPAN
KAJI, Hajime, JAPAN
KAJIKAWA, Yuji*, JAPAN
KAJIKIYA, Ryuji, JAPAN
KAJIMOTO, Hiroshi, JAPAN
KAJITANI, Kunihiko, JAPAN
KAJITANI, Yoji*, JAPAN
KAJIWARA, Joji*, JAPAN
KAJIWARA, Takeshi, JAPAN
KAJIWARA, Tsuyoshi, JAPAN
KAKEHI, Katsuhiko, JAPAN
KAKEHI, Tomoyuki, JAPAN
KAKIE, Kunio, JAPAN
KAKIMIZU, Osamu, JAPAN
KAKITA, Takao, JAPAN
KAKIUCHI, Hideaki, JAPAN
KAKO, Takashi, JAPAN
KAKUBA, Stevens J. K., UGANDA
KAKUDA, Yuzuru, JAPAN
KALLIANPUR, Gopinath, U.S.A.
KALONI, Purna N.§, CANADA
KALYABIN, Gennadiy A.*, U.S.S.R.
KAMADA, Seiichi, JAPAN
KAMAE, Teturo, JAPAN
KAMAL, Ahmed A. M.*, EGYPT
KAMATA, Masayoshi, JAPAN
KAMBAYASHI, Tatsuji, JAPAN
KAMEDA, Masumi, JAPAN
KAMEI, Eizaburo, JAPAN
KAMETAKA, Yoshinori*, JAPAN
KAMETANI, Makoto, JAPAN
KAMETANI, Yukio, JAPAN
KAMEYAMA, Atsushi, JAPAN
KAMIMURA, Yutaka, JAPAN
KAMINKER, Jerry, U.S.A.

KAMIYA, Hisao, JAPAN
KAMIYA, Hisashi, JAPAN
KAMIYA, Noriaki, JAPAN
KAMIYA, Shigeyasu, JAPAN
KAMIYA, Tadashi, JAPAN
KAMO, Shizuo, JAPAN
KAMYAD, A. Vahidian*, IRAN
KANAI, Masahiko, JAPAN
KANAI, Norio, JAPAN
KANAYAMA, Hiroshi, JAPAN
KANDA, Mamoru, JAPAN
KANDA, Shigeo, JAPAN
KANEDA, Eiji, JAPAN
KANEDA, Masaharu, JAPAN
KANEKO, Akira, JAPAN
KANEKO, Hiroshi, JAPAN
KANEKO, Jyoichi, JAPAN
KANEKO, Makoto, JAPAN
KANEKO, Masanobu, JAPAN
KANEMAKI, Shoji, JAPAN
KANEMARU, Tadayoshi, JAPAN
KANEMITSU, Mitsuo, JAPAN
KANEMITSU, Shigeru, JAPAN
KANENOBU, Taizo*, JAPAN
KANETA, Hitoshi, JAPAN
KANETO, Takeshi, JAPAN
KANEYUKI, Soji, JAPAN
KANG, Ming-Chang*, CHINA-TAIWAN
KANG, Seok-Jin, KOREA, R. of
KANG, Tae Ho, KOREA, R. of
KANIUTH, Eberhard*, F.R.G.
KANJIN, Yuichi, JAPAN
KANNO, Kenji*, JAPAN
KANNO, Tsuneo, JAPAN
KANO, Masako, JAPAN
KANO, Tadayoshi, JAPAN
KANTOR, Jean-Michel, FRANCE
KANTOWSKI, Eleanore L., U.S.A.
KANZAKI, Teruo§, JAPAN
KAPLAN, Hadassah, ISRAEL
KAPLANSKY, Irving, U.S.A.
KAPOVICH, Michael Erikovich, U.S.S.R.
KAPUR, Aruna*, INDIA
KARAMATSU, Yoshikazu, JAPAN
KARAMZADEH, Omid*, IRAN
KARANDIKAR, Rajeeva L., INDIA
KARBE, Manfred, F.R.G.
KARHUMÄKI, Juhani E. U.*, FINLAND
KARLSSON, Thomas Nils§, SWEDEN
KARZANOV, Alexander V., U.S.S.R.
KASAHARA, Koji, JAPAN
KASAHARA, Yasushi, JAPAN
KASAHARA, Yuji, JAPAN
KASAI, Shin-ichi, JAPAN
KASAJIMA, Tomomi, JAPAN
KASHANI, S. M. B.*, IRAN
KASHIHARA, Kenji, JAPAN
KASHIMURA, Ryohei, JAPAN

KASHIWABARA, Takuji, JAPAN
KASHIWADA, Toyoko, JAPAN
KASHIWAGI, Toshio§, JAPAN
KASHIWAGI, Yoshimi, JAPAN
KASHIWARA, Hiroko, JAPAN
KASHIWARA, Masaki, JAPAN
KASONGA, Raphael Abel, TANZANIA
KASUE, Atsushi, JAPAN
KATAOKA, Kiyoomi, JAPAN
KATASE, Kiyoshi, JAPAN
KATAYAMA, Shigeru, JAPAN
KATAYAMA, Shin-ichi, JAPAN
KATAYAMA, Yoshikazu, JAPAN
KATO, Akio*, JAPAN
KATO, Fumio*, JAPAN
KATO, Hiroshi, JAPAN
KATO, Hisako, JAPAN
KATO, Hisao§, JAPAN
KATO, Junji, JAPAN
KATO, Kazuhisa, JAPAN
KATÔ, Kazuko, JAPAN
KATO, Kazuya, JAPAN
KATO, Keiichi, JAPAN
KATO, Masahide, JAPAN
KATO, Masakimi, JAPAN
KATO, Minoru, JAPAN
KATO, Mitsuyoshi, JAPAN
KATO, Sadao, JAPAN
KATO, Shin, JAPAN
KATO, Shin-ichi, JAPAN
KATO, Suehiro, JAPAN
KATO, Syohei, JAPAN
KATO, Toyonori, JAPAN
KATO, Yoshifumi, JAPAN
KATO, Yoshio, JAPAN
KATORI, Makoto*, JAPAN
KATSUDA, Atsushi, JAPAN
KATSUMATA, Yasuo*, JAPAN
KATSURA, Shinji, JAPAN
KATSURA, Toshiyuki, JAPAN
KATSURADA, Hidenori, JAPAN
KATSURADA, Masanori, JAPAN
KATSURADA, Masashi, JAPAN
KATSUURA, Hidefumi*, U.S.A./JAPAN
KAUFFMAN, Louis Hirsch, U.S.A.
KAUL, Saroop K., CANADA
KAWABE, Jun, JAPAN
KAWADA, Takayuki, JAPAN
KAWADA, Yukiyosi, JAPAN
KAWADA, Yutaka, JAPAN
KAWAGUCHI, Hiroaki*, JAPAN
KAWAGUCHI, Shinji, JAPAN
KAWAGUCHI, Shokei§, JAPAN
KAWAGUCHI, Tomoaki*, JAPAN
KAWAHARA, Yusaku, JAPAN
KAWAHIGASHI, Yasuyuki, JAPAN
KAWAI, Ei-ichiro, JAPAN
KAWAI, Shigeo, JAPAN

KAWAI, Takahiro, JAPAN
KAWAI, Toru, JAPAN
KAWAI, Tsuneyuki, JAPAN
KAWAKAMI, Masato, JAPAN
KAWAKAMI, Satoshi*, JAPAN
KAWAKAMI, Tomohiro, JAPAN
KAWAKUBO, Katsuo, JAPAN
KAWAMATA, Yujiro, JAPAN
KAWAMOTO, Fuminori, JAPAN
KAWAMOTO, Naoki, JAPAN
KAWAMOTO, Shin-ichi, JAPAN
KAWAMURA, Kazuhiro, JAPAN
KAWAMURA, Michihiko, JAPAN
KAWAMURA, Shinzo, JAPAN
KAWANAGO, Tadashi, JAPAN
KAWANAKA, Noriaki, JAPAN
KAWANO, Haruaki§, JAPAN
KAWANO, Yasuhito, JAPAN
KAWASAKI, Hidefumi, JAPAN
KAWASAKI, Tetsuro, JAPAN
KAWASHIMA, Seiji, JAPAN
KAWASHIMA, Toshio, JAPAN
KAWASHITA, Mishio, JAPAN
KAWASUMI, Shogo, JAPAN
KAWATA, Shigeto, JAPAN
KAWAZOE, Takeshi, JAPAN
KAWAZU, Kiyoshi, JAPAN
KAWAZUMI, Nariya, JAPAN
KAZAMA, Hideaki, JAPAN
KAZAMA, Ken-ichiro, JAPAN
KAZAMAKI, Norihiko, JAPAN
KAZUMI, Tetuya, JAPAN
KEARSLEY, Mary J., U.K.
KEARTON, Cherry, U.K.
KEATING, Kevin Patrick, U.S.A.
KEEN, Linda, U.S.A.
KEGEL, Otto H., F.R.G.
KELLOGG, Franklin R.§, U.S.A.
KEMER, Alexander R., U.S.S.R.
KENDEROV, Peter Stoyanov, BULGARIA
KENIG, Carlos E., U.S.A.
KENMOTSU, Katsuei, JAPAN
KERA, Kazuo, JAPAN
KERAYECHIAN, Asghar*, IRAN
KERMAN, Ron A., CANADA
KERNER, Otto, F.R.G.
KESTENBAND, Barbu C.*, U.S.A.
KEUM, Jonghae, U.S.A./KOREA, R. of
KEUNE, Frans J., NETHERLANDS
KHALIFAH, M.-F. A.-S.*, EGYPT
KHANDUJA, Sudesh K.*, INDIA
KHARAGHANI, Hadi, CANADA
KHELLADI, Abdelkader, ALGERIA
KHRUSHCEV, Sergei V., U.S.S.R.
KIDA, Masanari, JAPAN
KIDA, Yuji, JAPAN
KIGAMI, Jun, JAPAN
KIGURADZE, Ivan T., U.S.S.R.

KIHARA, Hiroshi, JAPAN
KIJIMA, Yoichi, JAPAN
KIKKAWA, Michihiko, JAPAN
KIKUCHI, Akira§, JAPAN
KIKUCHI, Fumio*, JAPAN
KIKUCHI, Hiroko, JAPAN
KIKUCHI, Isamu, JAPAN
KIKUCHI, Kazunori, JAPAN
KIKUCHI, Keiichi, JAPAN
KIKUCHI, Keisuke, JAPAN
KIKUCHI, Koji, JAPAN
KIKUCHI, Masato, JAPAN
KIKUCHI, Norio, JAPAN
KIKUCHI, Shigeki, JAPAN
KIKUCHI, Teppei, JAPAN
KIKUMASA, Isao, JAPAN
KILPELÄINEN, Tero A. V., FINLAND
KIM, Sung Sook, KOREA, R. of
KIM, TongHo, JAPAN
KIM, Yong Woon, KOREA, R. of
KIM, Young Ho, KOREA, R. of
KIMN, Ha-Jine, KOREA, R. of
KIMURA, Hironobu, JAPAN
KIMURA, Hiroshi, JAPAN
KIMURA, Ikuo, JAPAN
KIMURA, Isamu, JAPAN
KIMURA, Kazuhiro, JAPAN
KIMURA, Kyoko, U.S.A.
KIMURA, Makoto, JAPAN
KIMURA, Minako, JAPAN
KIMURA, Morishige, JAPAN
KIMURA, Noriaki, JAPAN
KIMURA, Takahisa, JAPAN
KIMURA, Takashi, JAPAN
KIMURA, Tatsuo, JAPAN
KIMURA, Tetsuzo, JAPAN
KIMURA, Tosihusa, JAPAN
KINOSHITA, Shin'ichi, JAPAN
KINOSHITA, Takafumi, JAPAN
KINOSHITA, Yasushi, JAPAN
KINOSITA, Kiyosi§, JAPAN
KIRILLOV, A. N., U.S.S.R.
KIRK, Paul A., U.S.A.
KISAKA, Masashi, JAPAN
KISHI, Masanori, JAPAN
KISHIMOTO, Kazuo, JAPAN
KISO, Kazuhiro, JAPAN
KISTER, James M., U.S.A.
KISTER, Jane E., U.S.A.
KITA, Hiroo*, JAPAN
KITADA, Akihiko*, JAPAN
KITADA, Toshiyuki, JAPAN
KITADA, Yasuhiko, JAPAN
KITAGAWA, Keiichiro, JAPAN
KITAGAWA, Takashi, JAPAN
KITAGAWA, Tosio§, JAPAN
KITAGAWA, Yoshihisa, JAPAN
KITAHARA, Haruo, JAPAN

KITAHARA, Kiyoshi, JAPAN
KITAKUBO, Shigeru, JAPAN
KITAMI, Tsutomu§, JAPAN
KITAMURA, Ichiziro, JAPAN
KITAMURA, Kazue, JAPAN
KITAMURA, Shin-ichi, JAPAN
KITAZAKI, Kuniaki, JAPAN
KITAZUME, Masaaki, JAPAN
KIYOHARA, Kazuyoshi, JAPAN
KIYOHARA, Mineo, JAPAN
KIYOTA, Masao, JAPAN
KIYOTA, Shoichiroh, JAPAN
KIZUKA, Takashi, JAPAN
KLAWE, Maria M., CANADA
KLEIN, Abraham A.*, ISRAEL
KLEISLI, Heinrich*, SWITZERLAND
KLIN, Mihail Haimovich, U.S.S.R.
KNAPP, Anthony W., U.S.A.
KNOBLOCH, Eberhard H., F.R.G.
KNOPFMACHER, Arnold*, SOUTH AFRICA
KO, Ki Hyoung, KOREA, R. of
KOBAYASHI, Kazuaki, JAPAN
KOBAYASHI, Keijiro§, JAPAN
KOBAYASHI, Keiko, JAPAN
KOBAYASHI, Kusuo, JAPAN
KOBAYASHI, Masako, JAPAN
KOBAYASHI, Masanori*, JAPAN
KOBAYASHI, Mei*, U.S.A.
KOBAYASHI, Minoru, JAPAN
KOBAYASHI, Mitsuko, JAPAN
KOBAYASHI, Osamu, JAPAN
KOBAYASHI, Satoshi§, JAPAN
KOBAYASHI, Shigeru*, JAPAN
KOBAYASHI, Shoji, JAPAN
KOBAYASHI, Shoshichi, U.S.A.
KOBAYASHI, Takao, JAPAN
KOBAYASHI, Teiichi, JAPAN
KOBAYASHI, Toshimasa, JAPAN
KOBAYASHI, Toshiyuki, JAPAN
KOBAYASHI, Tsuyoshi, JAPAN
KOBAYASHI, Yoshikazu, JAPAN
KOBAYASHI, Yuji*, JAPAN
KOBAYASHI, Zenji, JAPAN
KOBAYASI, Kazuo§, JAPAN
KOBORI, Akira§, JAPAN
KOÇAK, Cevdet*, TURKEY
KOCH, Helmut, G.D.R.
KOCHERLAKOTA, Rama Rao*, U.S.A.
KODA, Junji, JAPAN
KODA, Takashi, JAPAN
KODAIRA, Kunihiko§, JAPAN
KODAKA, Kazunori, JAPAN
KODAMA, Akio, JAPAN
KODAMA, Tetsuo, JAPAN
KODAMA, Yukihiro, JAPAN
KODERA, Heiji, JAPAN
KOGISO, Takeyoshi, JAPAN
KOH, Jee Heub*, U.S.A.

KOHDA, Atsuhito, JAPAN
KOHMOTO, Susumu, JAPAN
KOHNO, Mitsuhiko, JAPAN
KOHNO, Toshitake, JAPAN
KOIKE, Kazuhiko, JAPAN
KOIKE, Masao, JAPAN
KOIKE, Minoru§, JAPAN
KOIKE, Naoyuki, JAPAN
KOIKE, Satoshi, JAPAN
KOIKE, Shigeaki, JAPAN
KOISO, Miyuki, JAPAN
KOISO, Norihito, JAPAN
KOITABASHI, Masanori, JAPAN
KOIZUMI, Shoji, JAPAN
KOIZUMI, Sumiyuki, JAPAN
KOJIMA, Hisashi, JAPAN
KOJIMA, Isao, JAPAN
KOJIMA, Jun, JAPAN
KOJIMA, Kiyofumi, JAPAN
KOJIMA, Makoto, JAPAN
KOJIMA, Sadayoshi, JAPAN
KOJYO, Hidemaro§, JAPAN
KOKETSU, Akio, JAPAN
KOKUBU, Hiroshi, JAPAN
KOLLÁR, János, U.S.A.
KOLOSKOV, Valentin Yurievich*, U.S.S.R.
KOLYVAGIN, Victor A., U.S.S.R.
KOMATSU, Gen, JAPAN
KOMATSU, Hikosaburo, JAPAN
KOMATSU, Kazushi, JAPAN
KOMATSU, Keiichi*, JAPAN
KOMATSU, Michiharu, JAPAN
KOMATU, Yusaku, JAPAN
KOMEDA, Jiryo, JAPAN
KOMIYA, Kaname, JAPAN
KOMIYA, Katsuhiro, JAPAN
KOMORI, Tetsushi, JAPAN
KOMORI, Youhei, JAPAN
KOMURA, Mutsumi, JAPAN
KOMURA, Takako, JAPAN
KOMURA, Yukio, JAPAN
KOMURO, Motomasa, JAPAN
KOMURO, Naoto, JAPAN
KONDO, Jiro, JAPAN
KONDO, Seizo, JAPAN
KONDO, Shigeyuki, JAPAN
KONDO, Shoichi, JAPAN
KONDO, Takeshi, JAPAN
KONGSAKORN, Kannika, THAILAND
KONNO, Hiroshi, JAPAN
KONNO, Kazuhiro, JAPAN
KONNO, Norio*, JAPAN
KONNO, Reiji, JAPAN
KONNO, Yasuko, JAPAN
KONNO, Yoshihiko, JAPAN
KONO, Akira, JAPAN
KONO, Hiroaki, JAPAN
KONO, Norio, JAPAN

KONO, Shigeo, JAPAN
KOOHARA, Akira, JAPAN
KOPFERMANN, Klaus, F.R.G.
KOPPELBERG, Sabine, F.R.G.
KORBAŠ, Július*, CZECHOSLOVAKIA
KORENBLUM, Boris*, U.S.A.
KOREVAAR, Jacob*, NETHERLANDS
KORI, Tosiaki, JAPAN
KORIYAMA, Akira§, JAPAN
KOSACHEVSKYA, Elena Aleksandrovna,
    U.S.S.R.
KOSAKI, Hideki, JAPAN
KOSCHORKE, Ulrich*, F.R.G.
KOSHI, Shozo*, JAPAN
KOSHIBA, Shunichi, JAPAN
KOSHIBA, Yoichi, JAPAN
KOSHIBA, Zenichiro, JAPAN
KOSHIKAWA, Hiroaki, JAPAN
KOSHITANI, Shigeo, JAPAN
KOSUDA, Masashi, JAPAN
KOTA, Osamu, JAPAN
KOTAKE, Takeshi, JAPAN
KOTANI, Motoko, JAPAN
KOTANI, Shinichi, JAPAN
KOTO, Toshiyuki, JAPAN
KOTUS, Janina, POLAND
KOUNO, Akira, JAPAN
KOUNO, Masaharu, JAPAN
KOYAMA, Akira, JAPAN
KOYAMA, Shin-ya, JAPAN
KOYAMA, Toshiko, JAPAN
KOYAMA, Yoshinori§, JAPAN
KOYANAGI, Tsunehira, JAPAN
KOZUKA, Kazuhito, JAPAN
KRANZ, Przemo T.*, U.S.A.
KRASNOSCHEKOV, Pavel, U.S.S.R.
KRASNY, Robert, U.S.A.
KRESS, Rainer, F.R.G.
KRGOVIĆ, Dragica N.*, YUGOSLAVIA
KRICHEVER, Igor M., U.S.S.R.
KRISZTIN, Tibor*, HUNGARY
KRONHEIMER, Peter B., U.K.
KUBO, Fumio*, JAPAN
KUBO, Izumi, JAPAN
KUBO, Kyoko*, JAPAN
KUBO, Masahiro*, JAPAN
KUBO, Yoshiko, JAPAN
KUBO, Yoshiyuki, JAPAN
KUBOTA, Koji, JAPAN
KUBOTA, Mitsuru, JAPAN
KUDO, Aichi, JAPAN
KUDO, Akio*, JAPAN
KUDO, Osamu, JAPAN
KUGA, Kenichi, JAPAN
KÜHNEL, Wolfgang, F.R.G.
KUIPER, Nicolaas H., FRANCE
KUKU, Aderemi Oluyomi*, NIGERIA
KUMAGAI, Donna*, U.S.A.

KUMAGAI, Takashi, JAPAN
KUMAHARA, Keisaku, JAPAN
KUMAKURA, Gakuji, JAPAN
KUMAZAWA, Masaaki, JAPAN
KUNITA, Hiroshi, JAPAN
KUNZ, Ernst A., F.R.G.
KUO, H.-H.*, U.S.A.
KUPERBERG, Gregory J.*, U.S.A.
KUPERBERG, Krystyna M.*, U.S.A.
KUPIAINEN, Antti J., FINLAND
KURAMITSU, Hirofumi, JAPAN
KURAMOTO, Yoshiyuki, JAPAN
KURANO, Kazuhiko, JAPAN
KURATA, Kazuhiro, JAPAN
KURATA, Masahiro, JAPAN
KURATA, Yoshiki, JAPAN
KURATSUBO, Shigehiko, JAPAN
KUREPA, Duro R.*, YUGOSLAVIA
KURIBAYASHI, Akikazu*, JAPAN
KURIBAYASHI, Izumi, JAPAN
KURIBAYASHI, Katsuhiko, JAPAN
KURIBAYASHI, Yukio, JAPAN
KURIHARA, Akira, JAPAN
KURIHARA, Masato, JAPAN
KURIHARA, Mitsunobu, JAPAN
KURIYAMA, Ken, JAPAN
KURKE, Herbert, G.D.R.
KURODA, Koji, JAPAN
KURODA, Shigetoshi, JAPAN
KURODA, Sige-Nobu, JAPAN
KUROGI, Tetsunori, JAPAN
KUROKAWA, Nobushige, JAPAN
KUROKAWA, Takahide, JAPAN
KUROKAWA, Takashi, JAPAN
KUROKAWA, Toshiko, JAPAN
KUROKI, Gen, JAPAN
KUROSAKI, Yasuhiro, JAPAN
KUROSE, Akinari, JAPAN
KUROSE, Hideki, JAPAN
KUROSE, Takashi, JAPAN
KURPITA, Bohdan I., U.K.
KURSHAN, Robert P., U.S.A.
KUSABA, Toshikuni, JAPAN
KUSAMA, Tokitake, JAPAN
KUSANAGI, Hiroyuki, JAPAN
KUSUDA, Masaharu, JAPAN
KUSUNOKI, Yukio, JAPAN
KUSUOKA, Shigeo, JAPAN
KUWABARA, Ruishi, JAPAN
KUWADA, Masahide, JAPAN
KUWAE, Kazuhiro, JAPAN
KUZNETSOV, Nicolai Vasilievich, U.S.S.R.
KWAK, Jin Ho, KOREA, R. of
KWON, Yonghoon, KOREA, R. of
KYE, Seung-Hyeok, KOREA, R. of
KYUNO, Shoji, JAPAN
LABAHN, George, CANADA
LABARCA, Rafael E., CHILE

LACEY, H. Elton, U.S.A.
LACOMBLEZ, Chantal, FRANCE
LADAS, Gerasimos*, U.S.A.
LADDE, Gangaram S.*, U.S.A.
LADYZENSKAYA, Olga A., U.S.S.R.
LAFFEY, Thomas Joseph, IRELAND
LAHTINEN, Aatos O.*, FINLAND
LAI, Hang-Chin, CHINA-TAIWAN
LAINE, Ilpo E., FINLAND
LAKSHMIBAI, V.*, U.S.A.
LAM, Kee Yuen, CANADA
LAMB, David A., U.S.A.
LAMBERT, William M., COSTA RICA
LAMPRECHT, Erich H. K., F.R.G.
LANDMAN, Alan§, U.S.A.
LANE, Mary C., U.S.A.
LANGLAIS, Michel R.*, FRANCE
LANIN, Anatolii I., U.S.S.R.
LAPIDUS, Michel L.*, U.S.A.
LARIJANI, Mohammad-Javad A.§, IRAN
LAU, Anthony To-Ming, CANADA
LAU, Kee-Wai, HONG KONG
LAUMON, Gerard H., FRANCE
LAURITZEN, Niels, DENMARK
LAVRENTIEV, Mihail M., U.S.S.R.
LAWNICZAK, Anna T.*, CANADA/U.S.A.
LAWRENCE, Harry Robert, U.S.A.
LAWRENCE, Ruth Jayne, U.S.A.
ŁAWRYNOWICZ, Julian, POLAND
LAZARSFELD, Robert K., U.S.A.
LAZARUS, Andrew J.*, U.S.A.
LAZEBNIK, Felix G.*, U.S.A.
LÊ, Dung Trang, VIETNAM
LE, Hai Khoi*, VIETNAM
LE, Hong-Van*, VIETNAM
LEANDRE, Remi, FRANCE
LEBEAU, Gilles, FRANCE
LEBLANC, Emile A.*, CANADA
LEBORGNE, Daniel§, FRANCE
LE CAM, Lucien M.§, FRANCE
LEE, Chung-Nim, U.S.A.
LEE, Jong-Eao John§, CHINA-TAIWAN
LEE, Jyh-Hao*, CHINA-TAIWAN
LEE, Peter Martin, U.K.
LEE, Sa-Ge*, KOREA, R. of
LEE, Yuh-Jia, CHINA-TAIWAN
LEHTO, Olli, FINLAND
LEIBOWITZ, Daniela, ISRAEL
LEICHTWEISS, Kurt*, F.R.G.
LEMAIRE, Jean-Michel, FRANCE
LEMARIÉ, Pierre Gilles*, FRANCE
LEMPP, Steffen, U.S.A.
LEUNG, Ka Hin*, SINGAPORE
LEUNG, Pui-Fai Fred, SINGAPORE
LEVESQUE, Claude, CANADA
LEVIATAN, Dany, ISRAEL
LEVINE, Jerome P., U.S.A.
LEVITIN, Michael R., U.S.S.R.

LEWIS, Adrian S., CANADA
LEWIS, Keith Allen, U.S.A.
LEWIS, Wayne, U.S.A.
LI, Bang-He*, CHINA, P. R. of
LI, Chun Wah*, HONG KONG
LI, Shangzhi, CHINA, P. R. of
LIANG, Wen-Hai, CHINA, P. R. of
LIANG, Xue-Zhang*, CHINA, P. R. of
LIAO, Keren*, CHINA, P. R. of
LICKORISH, W. B. Raymond, U.K.
LIH, Ko-Wei*, CHINA-TAIWAN
LIM, Chong-Keang, MALAYSIA
LIN, En-Bing*, U.S.A.
LIN, Fang Hua, U.S.A.
LIN, Qing, CHINA, P. R. of
LIN, Song-Sun*, CHINA-TAIWAN
LIND, Douglas A., U.S.A.
LINDENSTRAUSS, Joram, ISRAEL
LINDQVIST, Peter L.*, FINLAND
LINTON, Fred E.J., U.S.A.
LIONS, Pierre Louis, FRANCE
LIPMAN, Joseph, U.S.A.
LIPSHUTZ, Seymour, U.S.A.
LIPSCOMB, Stephen Leon, U.S.A.
LIPSMAN, Ronald L., U.S.A.
LIRON, Nadav, ISRAEL
LIU, De-Fu*, CHINA, P. R. of
LIU, Fon-Che*, CHINA-TAIWAN
LIU, Ming-Chit, HONG KONG
LIULEVICIUS, Arunas L.§, U.S.A.
LIVORNI, Elvira Laura, ITALY
LOGAN, George D., U.S.A.
LOI, Phan Hung, U.S.A.
LONG, Yiming*, CHINA, P. R. of
LOPEZ, Luis-Miguel, FRANCE
LORENZ, Dan H., ISRAEL
LORENZINI, Dino J.*, U.S.A.
LORING, Terry A., U.S.A.
LOVÁSZ, László, HUNGARY
LOVE, Eric Russell*, AUSTRALIA
LOZANO, Maria-Teresa*, SPAIN
LOZI, René P.*, FRANCE
LU, Qi-Keng*, CHINA, P. R. of
LU, Shanzhen*, CHINA, P. R. of
LU, Sylvia Chin-Pi*, U.S.A.
LUCKE, John Edwin, U.S.A.
LUDWIG, Garry, CANADA
LUFT, Erhard*, CANADA
LUK, Hing-Sun, HONG KONG
LUKEŠ, Jaroslav*, CZECHOSLOVAKIA
LUNA, Domingo, FRANCE
LUO, Dingjun*, CHINA, P. R. of
LUO, Tie, U.S.A.
LUSE, Dzidra*, U.S.S.R.
LUSZTIG, George, U.S.A.
LÜTZEN, Jesper, DENMARK
LYUBEZNIK, Gennady, U.S.A.
MA, Lawrence Kwan Ho, U.S.A.

MA, Zhiming, CHINA, P. R. of
MABUCHI, Toshiki, JAPAN
MACHIDA, Hajime, JAPAN
MACHIDA, Yoshinori, JAPAN
MACKENZIE, Kirill C. H.*, U.K.
MACKEY, George W., U.S.A.
MADER, Wolfgang K.W., F.R.G.
MAEBASHI, Toshiyuki, JAPAN
MAEDA, Fumi-Yuki, JAPAN
MAEDA, Hidetoshi*, JAPAN
MAEDA, Hironobu, JAPAN
MAEDA, Masao, JAPAN
MAEDA, Michie, JAPAN
MAEDA, Shigeru, JAPAN
MAEDA, Shuichiro, JAPAN
MAEDA, Takashi, JAPAN
MAEDA, Yoshiaki, JAPAN
MAEDA, Yoshitaka, JAPAN
MAEHARA, Kazuhisa*, JAPAN
MAEHARA, Shôji, JAPAN
MAEJIMA, Makoto, JAPAN
MAESONO, Yoshihiko, JAPAN
MAGID, Andy R., U.S.A.
MAHMOODIAN, Ebadollah Sayed*, IRAN
MINAMOTO, Ryuichi, JAPAN
MAITANI, Fumio*, JAPAN
MAJDA, Andrew J., U.S.A.
MAJIMA, Hideyuki, JAPAN
MAKAROV, Vitalii S.*, U.S.S.R.
MÄKELÄINEN, Tuulikki, FINLAND
MAKI, Haruo, JAPAN
MAKIHARA, Hiroshi, JAPAN
MAKINO, Isao, JAPAN
MAKINO, Tetu*, JAPAN
MAKS, Johannes G.*, NETHERLANDS
MAKSOUDOV, Faramaz, U.S.S.R.
MALLIAVIN, Marie-Paule, FRANCE
MALLIAVIN, Paul, FRANCE
MALTSEV, Arkadii A., U.S.S.R.
MANABE, Hiroki, JAPAN
MANABE, Shojiro, JAPAN
MANDAI, Takeshi, JAPAN
MANFREDI, Juan J., U.S.A.
MANIN, Yuri Ivanovich§, U.S.S.R.
MANN, Larry N., U.S.A.
MANSFIELD, Kevin Graham, NEW ZEALAND
MANTINI, Lisa A.*, U.S.A.
MARATHE, Kishore B.*, U.S.A.
MARCELO, Reginaldo M., PHILIPPINES
MARĈENKO, Vladimir Alexandroviĉ, U.S.S.R.
MARGERIN, Christophe M., FRANCE
MARGULIS, Grigorii A., U.S.S.R.
MARINO, Mario, ITALY
MARITZ, Pieter*, SOUTH AFRICA
MARKUSHEVICH, Dmitri G.*, U.S.S.R.
MARSH, Marcus M., U.S.A.
MARSHALL, Donald E., U.S.A.
MARTENS, Henrik H., NORWAY

MARTIN, Christiane C.M., FRANCE
MARTINELLI, Enzo, ITALY
MARTINEZ, André, FRANCE
MARTINEZ, Ernesto*, SPAIN
MARTINEZ-VILLA, Roberto, MEXICO
MARTIO, Olli Tapani, FINLAND
MARTUCCI, Giovanni*, ITALY
MARUBAYASHI, Hidetoshi, JAPAN
MARUMOTO, Yoshihiko, JAPAN
MARUO, Kenji, JAPAN
MARUO, Osamu, JAPAN
MARUYAMA, Akira, JAPAN
MARUYAMA, Fumitsuna, JAPAN
MARUYAMA, Ken-ichi, JAPAN
MARUYAMA, Masaki, JAPAN
MARUYAMA, Naomasa, JAPAN
MARUYAMA, Noriko, JAPAN
MARUYAMA, Toru, JAPAN
MASCARELLO, Maria, ITALY
MASCIONI, Vania*, SWITZERLAND
MASE, Shigeru, JAPAN
MASHIMO, Katsuya, JAPAN
MASHINCHI, Mashaalah*, IRAN
MASTERSON, John T.*, U.S.A.
MASUDA, Kayo, JAPAN
MASUDA, Kazuo, JAPAN
MASUDA, Kazuyoshi, JAPAN
MASUDA, Kyuya, JAPAN
MASUDA, Masao, JAPAN
MASUDA, Takashi, JAPAN
MASUDA, Teruo, JAPAN
MASUDA, Tetsuya, JAPAN
MASUI, Takahiro, JAPAN
MASUOKA, Akira, JAPAN
MATAGA, Yoshiharu, JAPAN
MATANO, Hiroshi, JAPAN
MATET, Pierre*, FRANCE
MATHIEU, Olivier, FRANCE
MATHIEU, Philippe§, BELGIUM
MATHIEU, Yves E.§, FRANCE
MATROSOV, Vladimir M.*, U.S.S.R.
MATSUBARA, Yoshie, JAPAN
MATSUDA, Hiroo, JAPAN
MATSUDA, Shigeki, JAPAN
MATSUDA, Toshimitsu, JAPAN
MATSUE, Hirofumi, JAPAN
MATSUE, Yumiko, JAPAN
MATSUGU, Yasuo, JAPAN
MATSUI, Kiyosi, JAPAN
MATSUI, Taku, JAPAN
MATSUKI, Mihoko*, JAPAN
MATSUKI, Toshihiko, JAPAN
MATSUMOTO, Hiroyuki, JAPAN
MATSUMOTO, Kazuko, JAPAN
MATSUMOTO, Kazuo, JAPAN
MATSUMOTO, Keiji, JAPAN
MATSUMOTO, Kengo, JAPAN
MATSUMOTO, Kohji, JAPAN

MATSUMOTO, Makoto, JAPAN
MATSUMOTO, Makoto, JAPAN
MATSUMOTO, Shigenori, JAPAN
MATSUMOTO, Takanori, JAPAN
MATSUMOTO, Waichiro*, JAPAN
MATSUMOTO, Yukio, JAPAN
MATSUMURA, Akitaka, JAPAN
MATSUMURA, Hideyuki, JAPAN
MATSUMURA, Mutsuhide, JAPAN
MATSUMURA, Toru, JAPAN
MATSUO, Yoshitomo, JAPAN
MATSUOKA, Choichiro, JAPAN
MATSUOKA, Katsuo, JAPAN
MATSUOKA, Sachiko, JAPAN
MATSUOKA, Takashi, JAPAN
MATSUSHITA, Osamu, JAPAN
MATSUSHITA, Yasuo, JAPAN
MATSUURA, Shigetake, JAPAN
MATSUURA, Shozo, JAPAN
MATSUYAMA, Hiroshi, JAPAN
MATSUYAMA, Yoshio*, JAPAN
MATSUZAKI, Katsuhiko*, JAPAN
MATSUZAWA, Jun-ichi, JAPAN
MATSUZAWA, Tadato, JAPAN
MATTILA, Pertti E. J.*, FINLAND
MATUDA, Tizuko, JAPAN
MATUI, Akinori, JAPAN
MATUMOTO, Hisayoshi, JAPAN
MATUMOTO, Takao, JAPAN
MATUNAGA, Saburô, JAPAN
MATUSZEWSKI, Roman, BELGIUM
MATUURA, Takahide*, JAPAN
MAUDE, Ronald, U.K.
MAUDUIT, Christian*, FRANCE
MAULDIN, Richard D., U.S.A.
MAUMARY, Serge, SWITZERLAND
MAWATA, Christopher P., U.S.A.
MAZUR, Barry C., U.S.A.
McCARTHY, Randy*, U.S.A.
McCOY, Barry M., U.S.A.
McCOY, Peter A.*, U.S.A.
MCCRORY, Clint G., U.S.A.
McDONALD, James Wade, U.S.A.
McDUFF, Dusa, U.S.A./U.K.
McKAY, James H.*, U.S.A.
McKEAN, Henry P., U.S.A.
McMINN, Trevor J., U.S.A.
McMULLEN, Curt, U.S.A.
MEGGINSON, Robert E., U.S.A.
MEHRI, Bahman*, IRAN
MEHTA, Madan Lal*, FRANCE
MEHTA, Vikram B., INDIA
MEIER, David, SWITZERLAND
MELROSE, Richard B., AUSTRALIA
MENDOZA, Leonardo, VENEZUELA
MERKLEN, Hector A., BRAZIL
MESHKANI, Ali*, IRAN
MESTRANO, Nicole, FRANCE

MEURMAN, Arne Erik, SWEDEN
MEYER, David A., U.S.A.
MEYER, Jean-Pierre G., U.S.A.
MEYER, Yves F., FRANCE
MIATELLO, Roberto Jorge, ARGENTINA
MICHAEL, Ernest A., U.S.A.
MICHAELIS, Walter J.*, U.S.A.
MICKELSSON, Jouko A., FINLAND
MIDORIKAWA, Hisaichi, JAPAN
MIKAMI, Kentaro, JAPAN
MIKAMI, Shunsuke, JAPAN
MIKAMI, Toshio, JAPAN
MIKI, Hidefumi, JAPAN
MIKI, Hiroo, JAPAN
MIKI, Yoshikazu, JAPAN
MILICIC, Dragan, U.S.A.
MILLER, Arnold W.*, U.S.A.
MILLER, Sanford S.*, U.S.A.
MILLETT, Kenneth C.*, U.S.A.
MILLSON, John J., CANADA
MILMAN, Vitali D., ISRAEL
MILNER, Eric C., CANADA
MILNES, Paul, CANADA
MIMACHI, Katsuhisa, JAPAN
MIMAR, Arman, U.S.A.
MIMURA, Mamoru, JAPAN
MIMURA, Masayasu, JAPAN
MIMURA, Yoshio, JAPAN
MIN, Kyong Jin, KOREA, R. of
MIN, Kyung Chan, KOREA, R. of
MINAMI, Haruo, JAPAN
MINAMI, Nariyuki, JAPAN
MINAMI, Toshiro, JAPAN
MINDA, Carl David*, U.S.A.
MINEMURA, Katsuhiro, JAPAN
MINGO, James A., CANADA
MINN, Jooha, KOREA, R. of
MIRON, Radu*, ROMANIA
MISAGHIAN, Manochehr§, IRAN
MISAKI, Norihiro, JAPAN
MISAWA, Tetsuya, JAPAN
MISCENKO, Evguenii F., U.S.S.R.
MITAMURA, Takashi, JAPAN
MITCHELL, Josephine M., U.S.A.
MITOMA, Itaru, JAPAN
MITROPOLSKY, Yurii A., U.S.S.R.
MITSUI, Takayoshi, JAPAN
MITSUI, Taketomo, JAPAN
MITSUMATSU, Yoshihiko, JAPAN
MIWA, Megumu, JAPAN
MIWA, Tatsuro*, JAPAN
MIWA, Tetsuji, JAPAN
MIYACHI, Akihiko, JAPAN
MIYADA, Itiro, JAPAN
MIYADERA, Isao§, JAPAN
MIYAGAWA, Masami, JAPAN
MIYAGAWA, Yukitaka, JAPAN
MIYAHARA, Yoshio, JAPAN

MIYAJIMA, Kimio, JAPAN
MIYAJIMA, Shizuo, JAPAN
MIYAKE, Masatake, JAPAN
MIYAKE, Toshitsune, JAPAN
MIYAMOTO, Haruo, JAPAN
MIYAMOTO, Masahiko, JAPAN
MIYAMOTO, Munemi, JAPAN
MIYANISHI, Masayoshi, JAPAN
MIYANO, Hiroshi*, JAPAN
MIYANO, Satoru, JAPAN
MIYAOKA, Etsuo, JAPAN
MIYAOKA, Reiko, JAPAN
MIYAOKA, Yoichi, JAPAN
MIYASAKA, Masanori, JAPAN
MIYATA, Yoichiro, JAPAN
MIYATAKE, Sadao, JAPAN
MIYAUCHI, Mutsuo, JAPAN
MIYAWAKI, Isao, JAPAN
MIYAZAKI, Chikashi, JAPAN
MIYAZAKI, Hitoshi, JAPAN
MIYAZAKI, Isao, JAPAN
MIYAZAKI, Ken-ichi, JAPAN
MIYAZAKI, Yoichi*, JAPAN
MIYAZAWA, Yasuyuki, JAPAN
MIYOSHI, Shigeaki, JAPAN
MIZOGUCHI, Minoru, JAPAN
MIZOHATA, Kiyoshi, JAPAN
MIZOHATA, Sigeru, JAPAN
MIZUHARA, Akira, JAPAN
MIZUHARA, Takahiro, JAPAN
MIZUMACHI, Hitoshi, JAPAN
MIZUMACHI, Ryuichi§, JAPAN
MIZUMOTO, Hisao, JAPAN
MIZUNO, Hirobumi, JAPAN
MIZUNO, Toru, JAPAN
MIZUNOYA, Takeshi, JAPAN
MIZUTA, Yoshihiro, JAPAN
MIZUTANI, Akira, JAPAN
MIZUTANI, Tadayoshi, JAPAN
MOCHIZUKI, Kiyoshi, JAPAN
MOEANADDIN, Rahim*, IRAN
MOEGLIN, Colette, FRANCE
MOGHADDAM, M. R. R.*, IRAN
MOGI, Isamu, JAPAN
MOHAMMADI, Hassanabadi A.*, IRAN
MOHEBI, Hossein, IRAN
MOHRI, Masayuki, JAPAN
MOHSENI, Mahmoud M.*, IRAN
MOKOBODZKI, Gabriel, FRANCE
MOLCHANOV, Stanislav A., U.S.S.R.
MOLCHANOV, Vladimir F.*, U.S.S.R.
MØLLER, Jesper M.*, DENMARK
MOMOSE, Fumiyuki, JAPAN
MONTAKHAB, Mohammad Sadegh§, IRAN
MOODY, Robert Vaughan§, CANADA
MOON, John W., CANADA
MOORE, Charles N., U.S.A.
MOORE, John D., U.S.A.

MORA, Leonardo Enrique, BRAZIL
MORAN, Gadi W.*, ISRAEL
MORDUKHOVICH, Boris S.*, U.S.A.
MORENO, Guillermo, MEXICO
MORENO SOCÍAS, Guillermo,
   SPAIN/FRANCE
MORGENSTERN, Rudolf, F.R.G.
MORI, Eon, JAPAN
MORI, Hiroshi, JAPAN
MORI, Hiroyuki, JAPAN
MORI, Makoto, JAPAN
MORI, Masao*, JAPAN
MORI, Masatake, JAPAN
MORI, Seiki, JAPAN
MORI, Shigefumi, JAPAN
MORI, Shosuke, JAPAN
MORI, Takakazu, JAPAN
MORI, Toshio, JAPAN
MORIKAWA, Hisashi, JAPAN
MORIKAWA, Naoto, JAPAN
MORIKAWA, Ryozo, JAPAN
MORIMOTO, Akira, JAPAN
MORIMOTO, Hiroaki, JAPAN
MORIMOTO, Hiroko, JAPAN
MORIMOTO, Hiroshi, JAPAN
MORIMOTO, Hiroshi, JAPAN
MORIMOTO, Kanji, JAPAN
MORIMOTO, Masaharu, JAPAN
MORIMOTO, Masayosi, JAPAN
MORIMOTO, Mayumi, JAPAN
MORIMOTO, Mitsuo, JAPAN
MORIMOTO, Shoji*, JAPAN
MORIMOTO, Tohru, JAPAN
MORIMOTO, Yoshinori, JAPAN
MORIOKA, Tatsushi, JAPAN
MORISHITA, Kazuhiko, JAPAN
MORISHITA, Mitsuji, JAPAN
MORITA, Jun, JAPAN
MORITA, Shigeyuki, JAPAN
MORITA, Yasuo, JAPAN
MORITA, Yoshihisa, JAPAN
MORITA, Yoshiyuki, JAPAN
MORIWAKI, Atsushi, JAPAN
MOROSAWA, Shunsuke, JAPAN
MOROZ, B. Z., ISRAEL
MORRISON, David R.§, U.S.A.
MORTON, Hugh Reynolds*, U.K.
MOSCOVICI, Henri, U.S.A.
MOSER, Jürgen, SWITZERLAND
MOSER, Lucy, U.S.A.
MOTEGI, Kimihiko*, JAPAN
MOTEGI, Masanori, JAPAN
MOTOHASHI, Nobuyoshi, JAPAN
MOTOSE, Kaoru, JAPAN
MOUNTFORD, Tom*, U.S.A.
MOURAD, Karim Joseph, U.S.A.
MOURRAIN, Bernard, FRANCE
MOUSSAVI, Ahmad*, IRAN

MUKAI, Juno, JAPAN
MUKAI, Shigeru, JAPAN
MUKAWA, Takayuki, JAPAN
MUKOYAMA, Kazuo, JAPAN
MULASE, Motohico, U.S.A.
MULCAHY, Marjorie, U.S.A.
MURA, Toshio§, U.S.A.
MURAI, Jousin, JAPAN
MURAI, Takafumi, JAPAN
MURAKAMI, Haruo, JAPAN
MURAKAMI, Jun, JAPAN
MURAKAMI, Kouichi, JAPAN
MURAKAMI, Maki, JAPAN
MURAKAMI, Shingo, JAPAN
MURAMATU, Tosinobu, JAPAN
MURAMORI, Takao, JAPAN
MURAMOTO, Katsushi, JAPAN
MURASE, Nobuyuki, JAPAN
MURASUGI, Kunio, CANADA
MURATA, Minoru*, JAPAN
MURATA, Tamotsu, JAPAN
MURATA, Yoshihiro, JAPAN
MURAYAMA, Mitutaka, JAPAN
MURAZAWA, Tadashi, JAPAN
MURO, Masakazu*, JAPAN
MURRE, Jacob P., NETHERLANDS
MURTY, Vijayakumar, CANADA
MUTA, Masanori, JAPAN
MUTANGADURA, Simba Arthur*,
   ZIMBABWE
MUTI, Nobol*, JAPAN
MUTO, Yosio§, JAPAN
MUTOU, Hideo, JAPAN
NAEENI, S. M. K., IRAN
NAGAHARA, Takasi, JAPAN
NAGAI, Hideo, JAPAN
NAGAI, Osamu, JAPAN
NAGAI, Sinpei, JAPAN
NAGAI, Tamao, JAPAN
NAGAMATI, Sigeaki*, JAPAN
NAGAMOCHI, Hiroshi, JAPAN
NAGANO, Azuma§, JAPAN
NAGANUMA, Hidehisa, JAPAN
NAGAO, Hirosi, JAPAN
NAGAO, Hisao, JAPAN
NAGAO, Koichi, JAPAN
NAGAOKA, Katsufumi§, JAPAN
NAGAOKA, Ryosuke, JAPAN
NAGASAKA, Kenji*, JAPAN
NAGASAKA, Yukio, JAPAN
NAGASAKI, Ikumitsu, JAPAN
NAGASAWA, Takeyuki, JAPAN
NAGASE, Haruo, JAPAN
NAGASE, Masayoshi, JAPAN
NAGASE, Michihiro, JAPAN
NAGASE, Noriaki, JAPAN
NAGASE, Teruko, JAPAN
NAGASHIMA, Takashi, JAPAN

NAGATA, Chigusa, JAPAN
NAGATA, Chihiro, JAPAN
NAGATA, Masatsugu, JAPAN
NAGATA, Masayoshi, JAPAN
NAGATOMO, Yasuyuki, JAPAN
NAGAYA, Hitoshi, JAPAN
NAGAYAMA, Haruya*, JAPAN
NAGISA, Masaru, JAPAN
NAGURA, Toshinobu, JAPAN
NÄKKI, Raimo T., FINLAND
NAIM, Linda, FRANCE
NAITO, Hirotada, JAPAN
NAITO, Hisashi, JAPAN
NAITO, Manabu, JAPAN
NAITO, Satoshi, JAPAN
NAITO, Shuji, JAPAN
NAITO, Toshiki, JAPAN
NAKABO, Shigekazu, JAPAN
NAKADA, Hitoshi, JAPAN
NAKADA, Masami, JAPAN
NAKAGAMI, Keiko, JAPAN
NAKAGAMI, Yoshiomi, JAPAN
NAKAGAWA, Hisao, JAPAN
NAKAGAWA, Jin, JAPAN
NAKAGAWA, Kiyokazu, JAPAN
NAKAGAWA, Noriaki, JAPAN
NAKAGAWA, Ryosuke, JAPAN
NAKAGAWA, Yasuhiro, JAPAN
NAKAGAWA, Yoko, JAPAN
NAKAGAWA, Yoko, JAPAN
NAKAGOSHI, Norikata, JAPAN
NAKAHARA, Hayao, JAPAN
NAKAHARA, Toru*, JAPAN
NAKAHATA, Koji, JAPAN
NAKAI, Isao, JAPAN
NAKAI, Mitsuru, JAPAN
NAKAI, Yoshinobu, JAPAN
NAKAJIMA, Atsushi, JAPAN
NAKAJIMA, Fumio, JAPAN
NAKAJIMA, Hiraku, JAPAN
NAKAJIMA, Iwao, JAPAN
NAKAJIMA, Katsuhiko, JAPAN
NAKAJIMA, Kazufumi, JAPAN
NAKAJIMA, Shoichi, JAPAN
NAKAJIMA, Tsugukazu, JAPAN
NAKAKI, Tatsuyuki, JAPAN
NAKAMA, Yutaka, JAPAN
NAKAMITSU, Kuniaki, JAPAN
NAKAMULA, Ken, JAPAN
NAKAMURA, Akira, JAPAN
NAKAMURA, Gen, JAPAN
NAKAMURA, Hiroaki, JAPAN
NAKAMURA, Hiroyuki, JAPAN
NAKAMURA, Iku§, JAPAN
NAKAMURA, Kirio, JAPAN
NAKAMURA, Masataka, JAPAN
NAKAMURA, Masato, JAPAN
NAKAMURA, Masatoshi, JAPAN

NAKAMURA, Riichiro§, JAPAN
NAKAMURA, Tetsuo, JAPAN
NAKAMURA, Tokushi, JAPAN
NAKAMURA, Yoshihiro, JAPAN
NAKAMURA, Yoshimasa, JAPAN
NAKAMURA, Yoshio, JAPAN
NAKAMURA, Yukio, JAPAN
NAKANE, Michiyo*, JAPAN
NAKANE, Shizuo, JAPAN
NAKANE, Takashi, JAPAN
NAKANISHI, Kōki, JAPAN
NAKANISHI, Shizu, JAPAN
NAKANISHI, Tomoki, JAPAN
NAKANISHI, Toshihiro, JAPAN
NAKANISHI, Yasutaka*, JAPAN
NAKANO, Minoru, JAPAN
NAKANO, Minoru, JAPAN
NAKANO, Shigeo, JAPAN
NAKANO, Shin, JAPAN
NAKANO, Tetsuo, JAPAN
NAKANO, Yoshihiro, JAPAN
NAKAO, Masahiro, JAPAN
NAKAO, Mitsuhiro, JAPAN
NAKAO, Mitsuhiro T.*, JAPAN
NAKAO, Shintaro, JAPAN
NAKAOKA, Akira, JAPAN
NAKAOKA, Minoru, JAPAN
NAKASHIMA, Katsuya, JAPAN
NAKASHIMA, Kiichi, JAPAN
NAKASHIMA, Masaharu, JAPAN
NAKASHIMA, Toru, JAPAN
NAKASHIMA, Toshiki, JAPAN
NAKATA, Tomoichi, JAPAN
NAKATA, Yoshimoto, JAPAN
NAKATSUKA, Harunori, JAPAN
NAKATSUKASA, Takuma, JAPAN
NAKAUCHI, Nobumitsu*, JAPAN
NAKAYAMA, Chikara, JAPAN
NAKAYAMA, Hiromichi, JAPAN
NAKAYAMA, Maki*, SOUTH AFRICA
NAKAYAMA, Noboru, JAPAN
NAKAYAMA, Takashi, JAPAN
NAKAYASHIKI, Atsushi, JAPAN
NAKAZATO, Hiroshi, JAPAN
NAKAZAWA, Hideaki, JAPAN
NAKAZAWA, Noriyuki§, JAPAN
NAKAZI, Takahiko, JAPAN
NAMBA, Kanji, JAPAN
NAMBA, Makoto*, JAPAN
NAMIKAWA, Yoshinori, JAPAN
NAMIKAWA, Yukihiko, JAPAN
NAMIOKA, Isaac, U.S.A.
NANBU, Tokumori, JAPAN
NAPALKOV, Valeusive, U.S.S.R.
NAPPI, Chiara R., U.S.A.
NARA, Chie, JAPAN
NARASIMHAN, Madumbai Seshachalu,
INDIA

NARAYANASWAMI, Pallasena P., CANADA
NARAZAKI, Takashi, JAPAN
NARITA, Junichiro, JAPAN
NARITA, Masahiro, JAPAN
NARITA, Ryoichi, JAPAN
NARUKAWA, Kimiaki, JAPAN
NARUSE, Hiroshi, JAPAN
NARUSHIMA, Hiroshi, JAPAN
NASATYR, Emile Ben*, U.K.
NASHED, M. ZUHAIR, U.S.A.
NATSUME, Toshikazu, U.S.A.
NAUDÉ, Cornelia G., SOUTH AFRICA
NAUDÉ, Gert, SOUTH AFRICA
NAVARRO, Milagros P., PHILIPPINES
NAVARRO, Vicente, SPAIN
NAWA, Hayato§, JAPAN
NAWATA, Masako, JAPAN
NAYATANI, Shin, JAPAN
NEDELEC, Jean Claude, FRANCE
NEGAMI, Seiya, JAPAN
NEGISHI, Aiko, JAPAN
NEISHTADT, Anatoli I., U.S.S.R.
NEMENZO, Fidel R., PHILIPPINES
NEMOTO, Hiroaki, JAPAN
NEMOTO, Seiji, JAPAN
NESETRIL, Jaroslav, CZECHOSLOVAKIA
NESTERENKO, Yuri, U.S.S.R.
NETO, Orlando M. B., PORTUGAL
NEUBERGER, Barbara O., U.S.A.
NEUBERGER, John W., U.S.A.
NEUEN, Annette I., SWITZERLAND
NEUENSCHWANDER, Daniel,
  SWITZERLAND
NEVANLINNA, Olavi, FINLAND
NEWHOUSE, Sheldon E., U.S.A.
NEWMAN, Michael F., AUSTRALIA
NGHIEM, Xuan Hai*, FRANCE
NGO, Dac Tan, VIETNAM
NGUIFFO BOYOM, Michel B.*, FRANCE
NGUYEN, Dai Tien, VIETNAM
NGUYEN, Dinh Cong, VIETNAM
NGUYEN, Dinh Tri, VIETNAM
NGUYEN, Duc Tuan*, VIETNAM
NGUYEN, Khoa Son, VIETNAM
NGUYEN, Minh Tri*, VIETNAM
NGUYEN, Quang Do Thong*, FRANCE
NGUYEN, Van Chan*, VIETNAM
NGUYEN, Viet Dung*, VIETNAM
NICOLAS, Jean-Louis*, FRANCE
NIELAND, Hendrik Maarten, NETHERLANDS
NII, Shunsaku, JAPAN
NIIKURA, Yasuo, JAPAN
NIINO, Kiyoshi, JAPAN
NIITSUMA, Hiroshi, JAPAN
NIJENHUIS, Albert*, U.S.A.
NIKNAM, Assadollah*, IRAN
NIKOLSKII, Nikolai K.*, U.S.S.R.
NIKOLTJEVA-HEDBERG, M., SWEDEN

NIKULIN, Vjacheslav V., U.S.S.R.
NINNEMANN, Olaf, F.R.G.
NINOMIYA, Hirokazu, JAPAN
NINOMIYA, Nobuyuki*, JAPAN
NINOMIYA, Syoiti, JAPAN
NINOMIYA, Yasushi, JAPAN
NISHI, Haruko, JAPAN
NISHIDA, Goro, JAPAN
NISHIDA, Hiroshi, JAPAN
NISHIDA, Koji, JAPAN
NISHIDA, Takaaki, JAPAN
NISHIGAKI, Sei-ichi, JAPAN
NISHIHARA, Kenji, JAPAN
NISHIHARA, Masaru*, JAPAN
NISHII, Ryuei, JAPAN
NISHIKAWA, Akira, JAPAN
NISHIKAWA, Masayuki, JAPAN
NISHIKAWA, Nobutaka, JAPAN
NISHIKAWA, Seiki, JAPAN
NISHIMI, Jisho, JAPAN
NISHIMORI, Toshiyuki, JAPAN
NISHIMORI, Yasunori, JAPAN
NISHIMOTO, Katsuyuki*, JAPAN
NISHIMURA, Jun-ichi, JAPAN
NISHIMURA, Takashi, JAPAN
NISHIMURA, Takeshi, JAPAN
NISHIMURA, Toshio, JAPAN
NISHIMURA, Yasuichiro, JAPAN
NISHINO, Tetsuro*, JAPAN
NISHIO, Kazuhiro, JAPAN
NISHIO, Masaharu, JAPAN
NISHIO, Masaru, JAPAN
NISHIOKA, Michio, JAPAN
NISHITANI, Tatsuo, JAPAN
NISHIURA, Yasumasa, JAPAN
NISHIYAMA, Akishige, JAPAN
NISHIYAMA, Kiminao, JAPAN
NISHIYAMA, Kyo, JAPAN
NISHIZAWA, Kiyoko, JAPAN
NISIO, Makiko, JAPAN
NISTOR, Victor, ROMANIA
NITTA, Takashi, JAPAN
NIWA, Masahiko, JAPAN
NIWA, Shinji, JAPAN
NIWANO, Eikazu, JAPAN
NIWASAKI, Takashi, JAPAN
NIZAMUDDIN, Kazi G.*, U.S.A.
NODA, Akio, JAPAN
NODA, Kazunari, JAPAN
NOGAMI, Akihiro, JAPAN
NOGAMI, Yoshiko, JAPAN
NOGI, Tatsuo, JAPAN
NOGUCHI, Junjiro, JAPAN
NOGUCHI, Mitsunori, JAPAN
NOHDA, Nobuhiko*, JAPAN
NOKURA, Tsugunori, JAPAN
NOMA, Atsushi, JAPAN
NOMAKUCHI, Kentaro, JAPAN

NOMOTO, Hisao, JAPAN
NOMURA, Kazumasa*, JAPAN
NOMURA, Yasutoshi, JAPAN
NOMURA, Yuji, JAPAN
NONO, Kiyoharu, JAPAN
NONOGUCHI, Norimichi, JAPAN
NOUMI, Masatoshi, JAPAN
NOZAKI, Akihiro, JAPAN
NOZAWA, Ryohei, JAPAN
NUNZIANTE, Diana§, ITALY
OAKU, Toshinori, JAPAN
OBATA, Morio, JAPAN
OBATA, Nobuaki*, JAPAN
OBAYASHI, Tadao, JAPAN
OBERSCHELP, Arnold*, F.R.G.
OBERSTE-VORTH, Ralph W., U.S.A.
OCHIAI, Hiroyuki, JAPAN
OCHIAI, Shoji, JAPAN
OCHIAI, Takushiro, JAPAN
OCNEANU, Adrian, U.S.A.
ODA, Tadao, JAPAN
ODA, Takayuki*, JAPAN
ODAI, Yoshitaka*, JAPAN
ODANI, Kenzi, JAPAN
ODLYZKO, Andrew M., U.S.A.
OEDA, Kazuo, JAPAN
OEHMKE, Robert H., U.S.A.
OEHMKE, Theresa M., U.S.A.
OELLERMANN, Ortrud Ruth*,
    SOUTH AFRICA
OETTLI, Werner, F.R.G.
O'FARRELL, Anthony G., IRELAND
OGATA, Shoetsu, JAPAN
OGAWA, Akihiko, JAPAN
OGAWA, Hiroyuki, JAPAN
OGAWA, Shigeyoshi*, JAPAN
OGAWA, Toshiyuki, JAPAN
OGAWA, Tsukane, JAPAN
OGAWA, Yosuke, JAPAN
OGIUE, Koichi, JAPAN
OGOMA, Tetsushi, JAPAN
OGUCHI, Kunio, JAPAN
OGUISO, Keiji, JAPAN
OGURA, Kazunori, JAPAN
OGURA, Shohei, JAPAN
OGURA, Yukio, JAPAN
OH, Jae-Pill, KOREA, R. of
OH, Junghwan, KOREA, R. of
OHARA, Jun*, JAPAN
OHASHI, Masakazu, JAPAN
OHATA, Koichi, JAPAN
OHBUCHI, Akira, JAPAN
OHHAMA, Minako, JAPAN
OHI, Takeo, JAPAN
OHKUBO, Toshio, JAPAN
OHKURO, Shigeru, JAPAN
OHMIYA, Mayumi, JAPAN
OHMORI, Zyuzyu, JAPAN

OHNISHI, Isamu, JAPAN
OHNITA, Yoshihiro, JAPAN
OHNO, Hiloshi*, JAPAN
OHNO, Koji, JAPAN
OHNO, Masahiro, JAPAN
OHNO, Mayumi, JAPAN
OHNO, Shûichi, JAPAN
OHORI, Masayuki, JAPAN
OHSITA, Akihiro, JAPAN
OHTA, Akira, JAPAN
OHTA, Haruto, JAPAN
OHTA, Hiroshi, JAPAN
OHTA, Masami, JAPAN
OHTA, Minolu, JAPAN
OHTA, Takuya§, JAPAN
OHTAKE, Hiromi, JAPAN
OHTSUKA, Fumiko, JAPAN
OHTSUKA, Makoto, JAPAN
OHTSUKI, Tomotada, JAPAN
OHYA, Yujiro, JAPAN
OHYAMA, Yoshiyuki, JAPAN
OHYAMA, Yousuke, JAPAN
OIKAWA, Kotaro, JAPAN
OISHI, Ryuzo, JAPAN
OKA, Hiroe, JAPAN
OKA, Hirokazu, JAPAN
OKA, Masaaki, JAPAN
OKA, Masatoshi, JAPAN
OKA, Mutsuo, JAPAN
OKA, Shinpei§, JAPAN
OKA, Yukimasa, JAPAN
OKABAYASHI, Shigeyoshi, JAPAN
OKABE, Hiroshi§, JAPAN
OKABE, Tsuneharu, JAPAN
OKABE, Yasunori, JAPAN
OKADA, Kanae, JAPAN
OKADA, Mari, JAPAN
OKADA, Masae, JAPAN
OKADA, Masami, JAPAN
OKADA, Masazumi, JAPAN
OKADA, Nolio, JAPAN
OKADA, Norio, JAPAN
OKADA, Soichi, JAPAN
OKADA, Tatsuya, JAPAN
OKADA, Toshinao, JAPAN
OKADA, Tsutomu, JAPAN
OKADA, Tsutomu, JAPAN
OKADA, Yasunori*, JAPAN
OKADO, Mosato, JAPAN
OKAI, Takayuki, JAPAN
OKAJI, Takashi, JAPAN
OKAMOTO, Hisashi, JAPAN
OKAMOTO, Hisashi, JAPAN
OKAMOTO, Kazuo, JAPAN
OKAMOTO, Kiyosato, JAPAN
OKAMOTO, Shigemi, JAPAN
OKAMURA, Hideaki, JAPAN
OKANO, Takashi, JAPAN

OKANO, Takeshi, JAPAN
OKAYASU, Takashi, JAPAN
OKAYASU, Takateru, JAPAN
OKAZAKI, Kimiko, JAPAN
OKAZAKI, Ryotaro, JAPAN
OKIYAMA, Satoru, JAPAN
OKOCHI, Hiroko, JAPAN
OKSENDAL, Bernt*, NORWAY
OKUBO, Katsumi, JAPAN
OKUBO, Kazuyoshi, JAPAN
OKUGAWA, Kotaro, JAPAN
OKUHARA, Hiroshi, JAPAN
OKUMURA, Hirozo, JAPAN
OKUMURA, Masafumi, JAPAN
OKUMURA, Yoshihide, JAPAN
OKUNO, Toshinao, JAPAN
OKURA, Hiroyuki, JAPAN
OKUTSU, Kosaku*, JAPAN
OKUYAMA, Akihiro, JAPAN
OKUYAMA, Tetsuro, JAPAN
OKUYAMA, Yasuo, JAPAN
OKUYAMA, Yukihiko, JAPAN
OLECH, Czesław, POLAND
OLECHE, Paul Odhiambo, KENYA
OLLIVIER, François, FRANCE
OLSEN, Catherine L., U.S.A.
OMACHI, Eriko, JAPAN
OMORI, Hideki, JAPAN
OMOTO, Chikaharu, JAPAN
ONO, Akira, JAPAN
ONO, Hiroakira, JAPAN
ONO, Kaoru, JAPAN
ONO, Kosuke, JAPAN
ONO, Takashi, JAPAN
ONO, Tomoaki, JAPAN
ONO, Yukio§, JAPAN
ONODA, Nobuharu, JAPAN
ONYANGO-OTIENO, Vitalis P., KENYA
OOHASHI, Tsunemichi, JAPAN
OOISHI, Akira, JAPAN
ORAZOV, Mered B., U.S.S.R.
ORIHARA, Akio, JAPAN
ORLOVA, Luidila, U.S.S.R.
O'ROURKE, Francesca S.*, U.K.
ORRO, Patrice, FRANCE
ORSTED, Bent, DENMARK
OSADA, Hirofumi, JAPAN
OSADA, Masayuki, JAPAN
OSAKA, Hiroyuki, JAPAN
OSAWA, Ikuko, JAPAN
OSAWA, Takeo, JAPAN
O'SHEA, Donal B., U.S.A.
OSHIKAWA, Keiichi, JAPAN
OSHIMA, Kazuyuki, JAPAN
OSHIMA, Toshio, JAPAN
OSHIMA, Yoichi, JAPAN
OSHIRO, Kiyoichi, JAPAN
OSIKAWA, Motosige, JAPAN

OSIPOV, Yuri S., U.S.S.R.
OTA, Kaori Imai, JAPAN
OTA, Minoru, JAPAN
OTA, Schoichi, JAPAN
OTAL, Javier*, SPAIN
OTOFUJI, Takashi, JAPAN
OTSUKA, Kayo, JAPAN
OTSUKA, Kenichi, JAPAN
OTSUKI, Nobukazu, JAPAN
OTSUKI, Tominosuke, JAPAN
OTWAY, Thomas H., U.S.A.
OUCHI, Moto, JAPAN
OUCHI, Osamu, JAPAN
OUCHI, Sunao, JAPAN
OWA, Shigeyoshi*, JAPAN
OWADA, Hiromoto, JAPAN
OWUSU-ANSAH, Twum*, GHANA
OYABU, Takashi, JAPAN
OZAWA, Kazuo, JAPAN
OZAWA, Masanao, JAPAN
OZAWA, Tetsuya, JAPAN
OZEKI, Hideki, JAPAN
OZEKI, Ikuzo*, JAPAN
OZEKI, Kiyota, JAPAN
OZEKI, Tomoko, JAPAN
OZONE, Jun, JAPAN
PACELLA, Filomena, ITALY
PACKER, Judith A.*, U.S.A.
PAGANI, Carlo Domenico, ITALY
PAHK, Dae Hyeon*, KOREA, R. of
PAK, Hong Kyung, KOREA, R. of
PAK, Jim Suk, KOREA, R. of
PALEV, Tchavdar Dimitrov, BULGARIA
PALIS, Jacob Jr., BRAZIL
PANDEY, Jagdish N.*, CANADA
PARANJAPE, Kapil H., INDIA
PARDINI, Rita M., ITALY
PARK, Bong-Kyu, KOREA, R. of
PARK, Chan-Young, KOREA, R. of
PARK, Hye Sook, KOREA, R. of
PARK, Kyewon K., U.S.A.
PARK, Sang Ro§, KOREA, R. of
PARK, Yong Moon, KOREA, R. of
PARTANEN, Juha Veikko*, FINLAND
PARTIS, Michael T., AUSTRALIA
PARVATHAM, Rajagopalan*, INDIA
PĀSĀRESCU, Ovidiu-Florin A., ROMANIA
PASCU, Eugen D.*, ROMANIA
PASEMAN, Gerhard Raymond, U.S.A.
PASK, David A., AUSTRALIA
PASSARE, Mikael, SWEDEN
PATHAK, Ram Shankar*, INDIA
PAWŁUCKI, Wiesław J.*, POLAND
PEARS, Alan R., U.K.
PEDERSEN, Gert K., DENMARK
PEDICCHIO, M. Cristina*, ITALY
PEKONEN, Osmo E. T., FINLAND
PELCZYŃSKI, Aleksander, POLAND

PELLER, Vladimir V.*, U.S.S.R.
PELTONEN, Kirsi Hannele, FINLAND
PENA, Juan Manuel*, SPAIN
PERCUS, Ora Engelberg*, U.S.A.
PERSSON, Jan, SWEDEN
PERSSON, Ulf A., SWEDEN
PETROVIĆ, Ljiljana M.*, YUGOSLAVIA
PHAM, Huy Dien*, VIETNAM
PHAM, Loi Vu, VIETNAM
PHAN, Thien Thach*, VIETNAM
PHILIPPIN, Gérard Auguste, CANADA
PHILLIPS, N. C., U.S.A.
PIARD, Alain A.P., FRANCE
PICKERING, Douglas A.§, CANADA
PIER, Jean-Paul, LUXEMBOURG
PIMSNER, Michael Viktor, F.R.G.
PINCHON, Didier M., FRANCE
PIPPENGER, Nicholas J., CANADA
PISIER, Gilles J., FRANCE
PLATONOV, Vladimir P., U.S.S.R.
PLOTKIN, Eugene*, U.S.S.R.
POGORELOV, Aleksei V.*, U.S.S.R.
POINT, Nelly, FRANCE
POKHARIYAL, Ganesh Prasad*, KENYA
POPA, Sorin Teodor, U.S.A.
PORRU, Giovanni*, ITALY
POTYAGAILO, Leonid D.*, U.S.S.R.
POURABDOLLAH, M. A.*, IRAN
POURCIN, Genevieve, FRANCE
POWELL, Wayne B., U.S.A.
PRASAD, Dipendra, INDIA
PRASAD, Gopal, INDIA
PREISS, David, U.K.
PRESSLEY, Andrew N., U.K.
PRICE, David T., U.S.A.
PRIETO, Carlos, MEXICO
PROPPE, Harold§, CANADA
PTÁK, Vlastimil, CZECHOSLOVAKIA
PURZITSKY, Norman, CANADA
PUTINAR, Mihai I., ROMANIA
PYBER, La'szló, HUNGARY
QIAN, Tao*, CHINA, P. R. of
QIU, Peizhang*, CHINA, P. R. of
QIU, Shuxi, CHINA, P. R. of
QUINN, Declan P.F., U.S.A.
QUINN, Frank S., U.S.A.
QURAISHI, Rehana*, INDIA
RABINDRANATHAN, M., U.S.A.
RADHA, Srinivasa Raghavan*, INDIA
RADJABALIPOUR, Mehdi*, IRAN
RĂDULESCU, Florin G., ROMANIA
RAGNEDDA, Francesco§, ITALY
RAGUSA, Alfio, ITALY
RAIMONDO, Mario, ITALY
RAJVANSHI, Niti§, INDIA
RALLIS, Stephen, U.S.A.
RAMANAN, Sundararaman, INDIA
RAMM, Alexander G.*, U.S.A.

RAMSEY, Helen Mary, U.K.
RAN, Ziv, U.S.A.
RANDELL, Richard, U.S.A.
RANSFORD, Thomas J., U.K.
RASOULIAN, Amid§, IRAN
RASSIAS, Themistocles M.§, GREECE
RATANAPRASERT, Chaweewan, THAILAND
RAUBENHEIMER, Heinrich*,
    SOUTH AFRICA
RAUGEL, Geneviève, FRANCE
RAUTMANN, Reimund K. A.*, F.R.G.
RAWLINGS, Don Paul*, U.S.A.
RAYNER, Francis J.*, U.K.
RAZAVI, Asdollah, IRAN
RAZBOROV, Alexandre A., U.S.S.R.
RECH, Mathias, SWITZERLAND
RECKNAGEL, Winfried, F.R.G.
RECSKI, András*, HUNGARY
REES, S. Mary, U.K.
REID, Alan W., U.K.
REID, Miles, U.K.
REITEN, Idun, NORWAY
REJALI, Ali*, IRAN
REMENYI, Maria, F.R.G.
REN, Jiagang, CHINA, P. R. of
RENAULT, Jean Nicolas, FRANCE
RENEGAR, James M., U.S.A.
REPOVŠ, Dušan*, YUGOSLAVIA
RESHETIKHIN, Nicolai, U.S.A.
REUTENAUER, Christophe*, CANADA
REYES, Bory Juan*, CUBA
RHAI, Tong-Shieng, CHINA-TAIWAN
RHEE, Choon J.*, U.S.A.
RIBES, Luis, CANADA
RICHARDSON, Mike Gerard, U.K.
RICKMAN, Seppo U., FINLAND
RIEDER, Gisèle Ruiz, U.S.A.
RIEMENSCHNEIDER, Oswald, F.R.G.
RIERA, Gonzalo, CHILE
RIGBY, John F.*, U.K.
RINGEL, Claus Michael*, F.R.G.
RIPPON, Philip Jonathan*, U.K.
RISHEL, Thomas W., U.S.A.
ROAN, Shi-Shyr, CHINA-TAIWAN
ROBBIANO, Lorenzo, ITALY
ROBBINS, Neville, U.S.A.
ROBERTS, Anne D., U.S.A.
ROBERTS, Paul C., U.S.A.
ROBINSON, Alan C., U.K.
RODINO, Luigi, ITALY
RÖDL, Vojtech, CZECHOSLOVAKIA/U.S.A.
ROERO, Clara Silvia*, ITALY
ROGGENKAMP, Klaus W., F.R.G.
ROMANOV, Vladimir G., U.S.S.R.
ROMERO M., Guillermo, MEXICO
ROONEY, Paul G., CANADA
ROOS, Guy, FRANCE
ROSATI, Mario, ITALY

ROSENBERGER, Gerhard*, F.R.G.
ROSENKNOP, John Z.*, F.R.G.
ROSENTHAL, Haskell P., U.S.A.
ROTHBERGER, Fritz§, CANADA
ROWE, David E., U.S.A.
ROWLEY, Christopher A., U.K.
ROY, Marie Francoise, FRANCE
ROZANOV, Yurii A., U.S.S.R.
RUAN, Zhong-Jin, U.S.A.
RUBINSTEIN, Joachim H.*, AUSTRALIA
RUDIN, Mary E., U.S.A.
RUDIN, Walter, U.S.A.
RUITENBURG, Wim B.*, U.S.A.
RUIZ, Mari-Jo P., PHILIPPINES
RUMELY, Robert S.*, U.S.A.
RUSHING, Thomas Benny, U.S.A.
RUTTER, John W.*, U.K.
SABURI, Yutaka, JAPAN
SAEKI, Osamu, JAPAN
SAGAN, Bruce E.*, U.S.A.
SAID, Hassan Bin*, MALAYSIA
SAIKAWA, Kazuhiko, JAPAN
SAIKI, Kazuyuki, JAPAN
SAITO, Haruhito, JAPAN
SAITO, Hiroshi, JAPAN
SAITO, Hiroshi, JAPAN
SAITO, Kazuyuki, JAPAN
SAITO, Kichisuke, JAPAN
SAITÔ, Kimiaki, JAPAN
SAITO, Kyoji, JAPAN
SAITO, Masa-Hiko, JAPAN
SAITO, Masahiko, JAPAN
SAITO, Morihiko, JAPAN
SAITO, Mutsumi, JAPAN
SAITO, Satoru, JAPAN
SAITO, Seiji, JAPAN
SAITO, Shiro, JAPAN
SAITO, Shiroshi, JAPAN
SAITO, Shuji, JAPAN
SAITO, Takeshi, JAPAN
SAITO, Tatsuhiko, JAPAN
SAITOH, Hitoshi*, JAPAN
SAITOH, Saburou*, JAPAN
SAITOH, Toshifumi, JAPAN
SAITOU, Noboru, JAPAN
SAKA, Koichi, JAPAN
SAKAGAWA, Hideko, JAPAN
SAKAGUCHI, Koji, JAPAN
SAKAGUCHI, Masanori, JAPAN
SAKAGUCHI, Michinori, JAPAN
SAKAGUCHI, Shigeru§, JAPAN
SAKAI, Akira, JAPAN
SAKAI, Fumio, JAPAN
SAKAI, Koukichi, JAPAN
SAKAI, Makoto, JAPAN
SAKAI, Masami, JAPAN
SAKAI, Sakuko, JAPAN
SAKAI, Shoichiro, JAPAN

SAKAI, Takashi, JAPAN
SAKAI, Yuji, JAPAN
SAKAKI, Makoto, JAPAN
SAKAKIBARA, Kenichi, JAPAN
SAKAKIBARA, Nobuhisa, JAPAN
SAKAMAKI, Kazuhiro, JAPAN
SAKAMOTO, Koichi§, JAPAN
SAKAMOTO, Kunio, JAPAN
SAKAMOTO, Reiko, JAPAN
SAKAMOTO, Takanori, JAPAN
SAKAMOTO, Takeshi, JAPAN
SAKAMOTO, Yoshiteru§, JAPAN
SAKANAGA, Toshiyuki, JAPAN
SAKANE, Yusuke, JAPAN
SAKANO, Kazunori, JAPAN
SAKAROVITCH, Jacques, FRANCE
SAKATA, Hiroshi, JAPAN
SAKATA, Toshio, JAPAN
SAKUMA, Makoto, JAPAN
SAKUMA, Motoyoshi, JAPAN
SAKURADA, Kuninori, JAPAN
SAKURAI, Susumu, JAPAN
SAKURAI, Takatoshi, JAPAN
SAKURAI, Yoshitane, JAPAN
SAKURAMOTO, Atsushi, JAPAN
SALAHITDINOV, Mahmud S., U.S.S.R.
SALBERGER, Per E., FRANCE
SALVATORE, Addolorata, ITALY
SALZER, Herbert E.*, U.S.A.
SAMARSKII, Alexander, U.S.S.R.
SANADA, Katsunori, JAPAN
SANCHEZ-RUIZ, Luis M.*, SPAIN
SANEKATA, Nobuhiro, JAPAN
SANKARAN, Gregory K., U.K.
SANNAMI, Atsuro, JAPAN
SANO, Kimiro, JAPAN
SANO, Shigeru*, JAPAN
SANO, Takashi, JAPAN
SANTHA, Miklos, FRANCE
SANUGI, Bahrom Bin*, MALAYSIA
SAPER, Leslie David, U.S.A.
SARNAK, Peter, U.S.A.
SASABUCHI, Syoichi, JAPAN
SASAJIMA, Yoshiaki, JAPAN
SASAKI, Chikara, JAPAN
SASAKI, Hiroki, JAPAN
SASAKI, Takehiko, JAPAN
SASAKI, Takeshi, JAPAN
SASAKI, Tetsuo, JAPAN
SASAKI, Toru, JAPAN
SASAKURA, Nobuo, JAPAN
SASANO, Kazuhiro, JAPAN
SASAO, Akira, JAPAN
SASAYAMA, Hiroyoshi*, JAPAN
SATAKE, Ichiro, JAPAN
SATAKE, Ikuo, JAPAN
SATO, Atsushi, JAPAN
SATO, Atsushi, JAPAN

SATO, Daihachiro*, JAPAN
SATO, Eiichi, JAPAN
SATO, Enji, JAPAN
SATO, Fumihiro, JAPAN
SATO, Hajime, JAPAN
SATO, Hiroki, JAPAN
SATO, Hiroshi, JAPAN
SATO, Hisashi, JAPAN
SATO, Humitaka, JAPAN
SATO, Iwao, JAPAN
SATO, Kei, JAPAN
SATO, Ken-iti*, JAPAN
SATO, Kenkichi§, JAPAN
SATO, Kunio, JAPAN
SATO, Masahisa, JAPAN
SATO, Masako, JAPAN
SATO, Mikio, JAPAN
SATO, Nobuaki, JAPAN
SATO, Shin, JAPAN
SATO, Shizuko, JAPAN
SATO, Shuichi, JAPAN
SATO, Shun, JAPAN
SATO, Takeshi, JAPAN
SATO, Takeyoshi, JAPAN
SATO, Takuji, JAPAN
SATO, Yoshihisa, JAPAN
SATO, Yoshitaka, JAPAN
SATO, Yumiko, JAPAN
SATOH, Takakazu, JAPAN
SATOH, Yukiti, JAPAN
SAVIN, Gordan, U.S.A.
SAWA, Tatuo, JAPAN
SAWADA, Ken, JAPAN
SAWADA, Toshio, JAPAN
SAWAE, Ryuichi, JAPAN
SAWAHATA, Michimasa, JAPAN
SAWAI, Hitoshi, JAPAN
SAWAKI, Sumio, JAPAN
SAWASHIMA, Ikuko, JAPAN
SAXL, Jan, U.K.
SAYEKI, Hidemitsu§, CANADA
SCEPIN, Evgenii Vitalievich*, U.S.S.R.
SCHAAL, Werner G.H., F.R.G.
SCHAFLITZEL, Reinhard*, F.R.G.
SHAPIRA, Pierre B., FRANCE
SCHEIN, Boris M., U.S.A.
SCIFFMANN, Gerard M., FRANCE
SCHIRESON, Max L.*, U.S.A.
SCHIRMER, Helga H.§, CANADA
SCHMEELK, John*, U.S.A.
SCHMICKLER-HIRZEBRUCH, Ulrike, U.S.A.
SCHMID, Wilfried, U.S.A.
SCHNEIDER, Hans, U.S.A.
SCHNIZER, Ansgar Wolfgang, AUSTRIA
SCHOENFELD, Lowell, U.S.A.
SCHREIBER, Bert M., U.S.A.
SCHWARTZ, Jean-Marie M., FRANCE
SCHWARZ, Albert, U.S.S.R.

SCOPPOLA, Carlo M., ITALY
SCULLY, Catherine N.*, U.K.
SEDDIGHI, Karim§, IRAN
SEGAL, Graeme Bryce, U.K.
SEGAL, Jack, U.S.A.
SEGAWA, Shigeo, JAPAN
SEGERT, Jan*, U.S.A.
SEI, Fumihiro§, JAPAN
SEIBERT, Peter*, MEXICO
SEITOH, Akira, JAPAN
SEKI, Go, JAPAN
SEKIGAWA, Hisao, JAPAN
SEKIGAWA, Kouei, JAPAN
SEKIGUCHI, Hideko, JAPAN
SEKIGUCHI, Katsusuke, JAPAN
SEKIGUCHI, Takeshi, JAPAN
SEKIGUCHI, Tsutomu, JAPAN
SEKINE, Masayuki, JAPAN
SEKINE, Mituhiro, JAPAN
SEKINE, Tadayuki, JAPAN
SEKINO, Kaoru, JAPAN
SEKIZAWA, Masami, JAPAN
SEKO, Yoshiki, JAPAN
SELFRIDGE, John L., U.S.A.
SELIGMAN, George B., U.S.A.
SEMENOV, Evgeny*, U.S.S.R.
SENBA, Takasi, JAPAN
SENDA, Akio, JAPAN
SENDOV, Blagovest Hristov, BULGARIA
SENN, Walter M., SWITZERLAND
SEO, Yuki, JAPAN
SERIZAWA, Hisamitsu, JAPAN
SESAY, Mohamed W.I.§, U.S.A.
SESHADRI, Conjeeveram S., INDIA
SETO, Hisayoshi, JAPAN
SEYDI, Hamet, SENEGAL
SHARFUDDIN, Shaik M.*, BANGLADESH
SHARIF, Habib, IRAN
SHARMA, Banwari Lal, INDIA
SHARP, Rodney Y., U.K.
SHASTRI, Anant R.*, INDIA
SHATZ, Stephen S.§, U.S.A.
SHAW, Sen-Yen*, CHINA-TAIWAN
SHEN, Junpei, CHINA, P. R. of
SHEU, Albert Jeu-Liang*, U.S.A.
SHEU, Shey Shiung*, CHINA-TAIWAN
SHI, Ning-Zhong*, CHINA, P. R. of
SHI, Zhong-Ci*, CHINA, P. R. of
SHIBA, Masakazu*, JAPAN
SHIBAGAKI, Wasao, JAPAN
SHIBANO, Hiroki, JAPAN
SHIBAOKA, Yasumitsu§, JAPAN
SHIBATA, Katsuyuki*, JAPAN
SHIBATA, Keiichi, JAPAN
SHIBATA, Tetsutaro, JAPAN
SHIBATA, Toshio, JAPAN
SHIBAYAMA, Kenshin, JAPAN
SHIBUKAWA, Youichi, JAPAN

SHIBUTANI, Norimasa, JAPAN
SHIEH, Narn R., CHINA-TAIWAN
SHIGA, Hironori, JAPAN
SHIGA, Hiroo, JAPAN
SHIGA, Hiroshige, JAPAN
SHIGA, Kiyoshi, JAPAN
SHIGA, Koji, JAPAN
SHIGA, Tokuzo, JAPAN
SHIGEKAWA, Ichiro, JAPAN
SHIGEMATSU, Keiichi, JAPAN
SHIM, Jae-Ung, KOREA, R. of
SHIMA, Hirohiko, JAPAN
SHIMA, Tadashi, JAPAN
SHIMADA, Hideo, JAPAN
SHIMADA, Ichiro, JAPAN
SHIMADA, Nobuo, JAPAN
SHIMADA, Tsutomu, JAPAN
SHIMAKAWA, Kazuhisa, JAPAN
SHIMAKURA, Norio, JAPAN
SHIMATANI, Kenichiro, JAPAN
SHIMENO, Nobukazu, JAPAN
SHIMIZU, Akinobu, JAPAN
SHIMIZU, Atutoshi, JAPAN
SHIMIZU, Hideo, JAPAN
SHIMIZU, Kenichi, JAPAN
SHIMIZU, Ryoichi*, JAPAN
SHIMIZU, Satoru, JAPAN
SHIMIZU, Yoshiyuki, JAPAN
SHIMIZU, Yuji, JAPAN
SHIMODA, Norio, JAPAN
SHIMODA, Yasuhiro, JAPAN
SHIMOMURA, Hiroaki, JAPAN
SHIMOMURA, Katsunori, JAPAN
SHIMOMURA, Shun, JAPAN
SHIMOMURA, Takashi, JAPAN
SHIMURA, Hiroyuki, JAPAN
SHIMURA, Michio, JAPAN
SHIMURA, Takaaki, JAPAN
SHIN, Kilho, JAPAN
SHINAGAWA, Mitsuo, JAPAN
SHINKAI, Kenzo*, JAPAN
SHINODA, Juichi, JAPAN
SHINODA, Ken-ichi, JAPAN
SHINOHARA, Masahiko, JAPAN
SHINOHARA, Yaichi, JAPAN
SHINOHARA, Yoshitane*, JAPAN
SHINOMIYA, Yoichi§, JAPAN
SHINOZUKA, Shigeo§, JAPAN
SHIODA, Tetsuji, JAPAN
SHIOHAMA, Katsuhiro, JAPAN
SHIOJI, Naoki, JAPAN
SHIOMI, Tatsuyuki, JAPAN
SHIOTA, Jun, JAPAN
SHIOTA, Ken-ichi, JAPAN
SHIOTA, Masahiro, JAPAN
SHIOTA, Takahiro, U.S.A.
SHIOTA, Yasunobu, JAPAN
SHIOYA, Masahiro, JAPAN

SHIOZAKI, Yasutoshi, JAPAN
SHIRAIWA, Kenichi, JAPAN
SHIRAKI, Mitsunobu, JAPAN
SHIRASOU, Takeo, JAPAN
SHIRATANI, Katsumi, JAPAN
SHIROSAKI, Manabu, JAPAN
SHISHIKURA, Mitsuhiro, JAPAN
SHIUE, Wei Kei*, U.S.A.
SHIZUKAWA, Akira, JAPAN
SHIZUTA, Yasushi, JAPAN
SHOHOJI, Takao§, JAPAN
SHOJI, Mayumi, JAPAN
SHOJI, Toshiaki, JAPAN
SHOKUROV, Vyacheslav V., U.S.S.R.
SHPEKTOROV, Sergey Victorovich, U.S.S.R.
SHPILRAIN, Vladimir E.*, U.S.S.R.
SHUKLA, Pradeep K.*, U.S.A.
SHUM, Kar-Ping*, HONG KONG
SHUSTIN, Eugenii I., U.S.S.R.
SI, Si, MYANMAR
SIBONY, Nessim, FRANCE
SIBUYA, Masaaki, JAPAN
SIDARTO, Kuntjoro Adji, INDONESIA
SIDDIQI, Jamil A.*, CANADA
SIDDIQI, Rafat Nabi*, KUWAIT
SIELKIN, Michail V.*, U.S.S.R.
SIEMONS, Johannes, U.K.
SIERSMA, Dirk, NETHERLANDS
SIGAL, Islael Michael, CANADA/U.S.A.
SIKORAV, Jean-Claude, FRANCE
SIM, Chiaw Hock, MALAYSIA
SIMIS, Aron, BRAZIL
SIMIZU, Tatuo*, JAPAN
SIMMS, John C.*, U.S.A.
SIMON, Imre, BRAZIL
SIMON, Jacques C. H., FRANCE
SIMONYI, Gábor*, HUNGARY
SIMPSON, Carlos Tschuodi, U.S.A.
SINAI, Yakov G., U.S.S.R.
SINGH, Neeta*, INDIA
SIRAO, Tunekiti, JAPAN
SIU, Man-Keung, HONG KONG
SJÖGREN, Peter, SWEDEN
SJÖLIN, Per, SWEDEN
SKANDALIS, Georges, FRANCE
SKAU, Christian F., NORWAY
SKORA, Richard K.*, U.S.A.
SKUBACHEVSKII, Alexander L.*, U.S.S.R.
SLAMAN, Theodore A., U.S.A.
SLEMROD, Marshall*, U.S.A.
SLODOWY, Peter J., F.R.G.
SLOVIC, Harold Geoffrey, U.S.A.
SMALØ, Sverre Olaf, NORWAY
SMITH, Martha K., U.S.A.
SMITH, Perry B., U.S.A.
SMITH, Roy Campbell, U.S.A.
SMITH, Stuart H., CANADA
SMITH, Stuart P., U.S.A.

SMITH, Wayne S., U.S.A.
SMOGOR, Louis E., U.S.A.
SOGO, Hideyo, JAPAN
SOLARIN, Adewale R. T.*, NIGERIA
SOLBERG, Øyvind, NORWAY
SOLOMON, David R.*, U.K.
SOLOMON, Ronald M., U.S.A.
SOMA, Sumihiko, JAPAN
SOMEKAWA, Mutsuro, JAPAN
SOMER, Lawrence Eric*, U.S.A.
SOMMER, Manfred, F.R.G.
SOMOLINOS, Alfredo*, SPAIN
SORIA, Fernando, SPAIN
SOTO-ANDRADE, Jorge*, CHILE
SOUČEK, Jiři, CZECHOSLOVAKIA
SOUKUP, Lajos, HUNGARY
SOULÉ, Christophe J., FRANCE
SPAGNOLO, Sergio A., ITALY
SPANIER, Fred, U.S.A.
SPINADEL, Vera W. de*, ARGENTINA
SPINELLI, Giancarlo, ITALY
SRINIVAS, Vasudevan, INDIA
STAFFANS, Olof J., FINLAND
STALLINGS, John R.*, U.S.A.
STAN, Ioan I.*, ROMANIA
STANILOV, Grosio*, BULGARIA
STANLEY, Lee J.*, U.S.A.
STANLEY, Richard P., U.S.A.
STANTON, Robert J., U.S.A.
STEEB, Willi Hans*, SOUTH AFRICA
STEEL, John R., U.S.A.
STEENBRINK, Joseph H. M., NETHERLANDS
STEFĂNESCU, Mirela I.*, ROMANIA
STEFANOV, Plamen D., BULGARIA
STEFÁNSSON, Jón R., ICELAND
STEGEMAN, Jan D., NETHERLANDS
STEGENGA, David A., U.S.A.
STEGER, Tim J.§, U.S.A.
STEINER, Richard J., U.K.
STEPHENSON, Ken, U.S.A.
STERN, Ronald J.§, U.S.A.
STERNHEIMER, Daniel H., FRANCE
STETKAER, Henrik, DENMARK
STETTER, Hans J., AUSTRIA
STEVENS, Glenn H., U.S.A.
STEWART, Ian Nicholas, U.K.
STIEGLER, Karl Drago§, F.R.G.
STOB, Michael J., U.S.A.
STORCH, Uwe, F.R.G.
STORMER, Erling, NORWAY
STORVICK, David Arne*, U.S.A.
STRANO, Rosario, ITALY
STRAY, Arne*, NORWAY
STRÖMBERG, Jan-Olov, SWEDEN
STROOKER, Jan R., NETHERLANDS
STROTH, Gernot*, F.R.G.
STRUWE, Michael, SWITZERLAND
STURM, Karl Theodor*, SWITZERLAND

STURMFELS, Bernd, U.S.A.
SUCCI, Francesco, ITALY
SUCCI CRUCIANI, Rosanna, ITALY
SUDO, Masahiro, JAPAN
SUDO, Masaki, JAPAN
SUEHIRO, Naoki*, JAPAN
SUEYOSHI, Yutaka, JAPAN
SUGA, Shuichi, JAPAN
SUGAHARA, Kunio, JAPAN
SUGANO, Takashi, JAPAN
SUGANO, Tosiaki, JAPAN
SUGAWA, Toshiyuki, JAPAN
SUGAWARA, Masahiro, JAPAN
SUGAWARA, Takeshi, JAPAN
SUGAWARA, Tamio, JAPAN
SUGIE, Jitsuro, JAPAN
SUGIE, Toru, JAPAN
SUGIMOTO, Mitsuru, JAPAN
SUGIMOTO, Shin, JAPAN
SUGINO, Ken, JAPAN
SUGITA, Hiroshi, JAPAN
SUGITA, Kimio, JAPAN
SUGITANI, Sadao, JAPAN
SUGIURA, Mitsuo, JAPAN
SUGIYAMA, Ken-Ichi, JAPAN
SUITA, Nobuyuki, JAPAN
SUKETA, Masaki, JAPAN
SUKLA, Indulata*, INDIA
SULLIVAN, John M., U.S.A.
SULTANGAZIN, Umirzak M., U.S.S.R.
SUMA, Yousuke, JAPAN
SUMI, Makiko, JAPAN
SUMI, Toshio*, JAPAN
SUMIHIRO, Hideyasu, JAPAN
SUMIOKA, Takeshi, JAPAN
SUMITOMO, Takeshi*, JAPAN
SUMMERS, Danny*, CANADA
SUMNERS, De Witt L., U.S.A.
SUN, Wen Xiang*, CHINA, P. R. of
SUN, Ziqi, U.S.A.
SUNADA, Toshikazu, JAPAN
SUNAGA, Junichi, JAPAN
SUNDARAM, Sheila, U.S.A.
SUNDBERG, Gösta, SWEDEN
SUNDER, Viakalathur Shankar, INDIA
SUNLEY, Judith S., U.S.A.
SUNOUCHI, Choichiro, JAPAN
SUNOUCHI, Haruo, JAPAN
SURI, Subhash, U.S.A.
SUTHERLAND, Colin Eric, AUSTRALIA
SUTO, Kiyokazu, JAPAN
SUTTI, Carla Nodari§, ITALY
SUWA, Noriyuki§, JAPAN
SUWA, Tatsuo, JAPAN
SUYAMA, Yoshihiko, JAPAN
SUZUKI, Atsushi, JAPAN
SUZUKI, Chisato, JAPAN
SUZUKI, Haruo, JAPAN

SUZUKI, Hideaki, JAPAN
SUZUKI, Humio, JAPAN
SUZUKI, Kazumasa*, JAPAN
SUZUKI, Kazuo, JAPAN
SUZUKI, Kazuya, JAPAN
SUZUKI, Komei*, JAPAN
SUZUKI, Kunie, JAPAN
SUZUKI, Kyoko, JAPAN
SUZUKI, Makoto, JAPAN
SUZUKI, Masaaki, JAPAN
SUZUKI, Masakazu, JAPAN
SUZUKI, Masashi, JAPAN
SUZUKI, Michio, U.S.A.
SUZUKI, Naoyoshi, JAPAN
SUZUKU, Norio, JAPAN
SUZUKI, Osamu, JAPAN
SUZUKI, Satoshi, JAPAN
SUZUKI, Shinichi, JAPAN
SUZUKI, Takashi, JAPAN
SUZUKI, Takeru, JAPAN
SUZUKI, Toshio, JAPAN
SUZUKI, Toshio, JAPAN
SUZUKI, Toshiyuki, JAPAN
SUZUKI, Yoshiya, JAPAN
SUZUKI, Yuji§, JAPAN
SUZUKI, Yuki, JAPAN
SY, Polly Wee, PHILIPPINES
SYMONDS, Peter, CANADA
SZABO, Zoltan I.§, BOTSWANA
SZŐNYI, Tamás*, HUNGARY
TABATA, Masahisa*, JAPAN
TABATA, Minoru, JAPAN
TACHIKAWA, Atsushi, JAPAN
TACHIKAWA, Hiroyuki, JAPAN
TACHIZAWA, Kazuya, JAPAN
TADA, Minoru, F.R.G.
TADA, Toshimasa, JAPAN
TAFLIN, Erik, FRANCE
TAGUCHI, Yoshiko, JAPAN
TAGUCHI, Yuichiro, JAPAN
TAHARA, Ken-ichi, JAPAN
TAHATA, Rataka, JAPAN
TAHERIZADEH, Abdol-Javad*, IRAN
TAINAKA, Akihiro, JAPAN
TAJIMA, Shinichi, JAPAN
TAKADA, Ichiro, JAPAN
TAKADA, Yoshikazu, JAPAN
TAKAGI, Hiroyuki, JAPAN
TAKAGI, Izumi, JAPAN
TAKAHASHI, Hideo, JAPAN
TAKAHASHI, Hideyuki, JAPAN
TAKAHASHI, Hiroaki, JAPAN
TAKAHASHI, Katsuaki, JAPAN
TAKAHASHI, Katsuo, JAPAN
TAKAHASHI, Katsutoshi, JAPAN
TAKAHASHI, Ken-ichi, JAPAN
TAKAHASHI, Makoto, JAPAN
TAKAHASHI, Masahiro, JAPAN

TAKAHASHI, Moto-o, JAPAN
TAKAHASHI, Reiji, JAPAN
TAKAHASHI, Satoshi, JAPAN
TAKAHASHI, Sechiko, JAPAN
TAKAHASHI, Shuichi, JAPAN
TAKAHASHI, Tadashi, JAPAN
TAKAHASHI, Takeshi, JAPAN
TAKAHASHI, Tetsuya, JAPAN
TAKAHASHI, Toyofumi, JAPAN
TAKAHASHI, Tsunero, JAPAN
TAKAHASHI, Yasuji, JAPAN
TAKAHASHI, Yoichiro, JAPAN
TAKAHASI, Mitiko, JAPAN
TAKAHASI, Sin-Ei, JAPAN
TAKAI, Hiroshi, JAPAN
TAKAKU, Akira, JAPAN
TAKAKURA, Tatsuru, JAPAN
TAKAKUWA, Shoichiro, JAPAN
TAKAMUKI, Takashi, JAPAN
TAKAMURA, Shigeru, JAPAN
TAKANE, Martha Imay, MEXICO
TAKANO, Katsuo, JAPAN
TAKANO, Kazuhiko, JAPAN
TAKANO, Kyoichi, JAPAN
TAKANO, Mitio§, JAPAN
TAKANO, Yoshiko*, JAPAN
TAKANOBU, Satoshi, JAPAN
TAKAO, Hiroshi, JAPAN
TAKAOKA, Masaki§, JAPAN
TAKARAJIMA, Itaru, JAPAN
TAKASAKI, Kanehisa, JAPAN
TAKASE, Koichi, JAPAN
TAKASE, Mitsuo, JAPAN
TAKASHIMA, Keizo, JAPAN
TAKASU, Satoru, JAPAN
TAKATA, Toshie, JAPAN
TAKAYAMA, Nobuki, JAPAN
TAKAYANAGI, Hideshi, JAPAN
TAKEBAYASHI, Tadayoshi, JAPAN
TAKEBE, Takashi, JAPAN
TAKEDA, Masayoshi, JAPAN
TAKEDA, Yoshifumi, JAPAN
TAKEDA, Yuichi, JAPAN
TAKEDA, Yuichiro, JAPAN
TAKEDA, Ziro§, JAPAN
TAKEGOSHI, Kensho, JAPAN
TAKEHANA, Hiroaki, JAPAN
TAKEI, Yoshitsugu, JAPAN
TAKEMOTO, Fumio, JAPAN
TAKENAKA, Shigeo, JAPAN
TAKENOUCHI, Osamu*, JAPAN
TAKEO, Fukiko, JAPAN
TAKESAKI, Masamichi, U.S.A.
TAKESHITA, Akira, JAPAN
TAKEUCHI, Akira, JAPAN
TAKEUCHI, Atsuko, JAPAN
TAKEUCHI, Hiroshi, JAPAN
TAKEUCHI, Jiro, JAPAN

TAKEUCHI, Junji, JAPAN
TAKEUCHI, Kei, JAPAN
TAKEUCHI, Kensuke, JAPAN
TAKEUCHI, Kentaro, JAPAN
TAKEUCHI, Kisao, JAPAN
TAKEUCHI, Kiyohiko, JAPAN
TAKEUCHI, Mamoru, JAPAN
TAKEUCHI, Masahito, JAPAN
TAKEUCHI, Masaru, JAPAN
TAKEUCHI, Nobuko, JAPAN
TAKEUCHI, Shigeru, JAPAN
TAKEUCHI, Teruo, JAPAN
TAKEUCHI, Yasuhiro§, JAPAN
TAKEUCHI, Yasuji, JAPAN
TAKEUCHI, Yoshihiro, JAPAN
TAKEUTI, Gaisi, U.S.A.
TAKEUTI, Yosie, JAPAN
TAKEWAKA, Yoshie, JAPAN
TAKIGAWA, Shinya, JAPAN
TAKIZAWA, Osamu*, JAPAN
TALAGRAND, Michel, FRANCE
TALL, Franklin D.*, CANADA
TAMAKI, Dai, JAPAN
TAMAKI, Kazuhiro, JAPAN
TAMAMURA, Akie, JAPAN
TAMANO, Ken-ichi, JAPAN
TAMANOI, Hirotaka, JAPAN
TAMARI, Fumikazu, JAPAN
TAMBARA, Daisuke, JAPAN
TAMBOUR, Torbjörn, SWEDEN
TAMBURINI, Chiara M., ITALY
TAMURA, Hiroshi, JAPAN
TAMURA, Itiro§, JAPAN
TAMURA, Saburo, JAPAN
TAMURA, Yozo, JAPAN
TAN, Henry K., U.S.A.
TAN, Ki-Seng*, U.S.A.
TANABE, Hiroki, JAPAN
TANABE, Kunio, JAPAN
TANABE, Susumu, JAPAN
TANABÉ, Susumu, JAPAN
TANAHASHI, Katsumi, JAPAN
TANAHASHI, Kotaro, JAPAN
TANAKA, Giichi, JAPAN
TANAKA, Hiroshi, JAPAN
TANAKA, Hiroshi, JAPAN
TANAKA, Junzo, JAPAN
TANAKA, Katsumi, JAPAN
TANAKA, Kazunaga, JAPAN
TANAKA, Kazuyuki, JAPAN
TANAKA, Mahito, JAPAN
TANAKA, Minoru, JAPAN
TANAKA, Naoki, JAPAN
TANAKA, Ryuichi, JAPAN
TANAKA, Shohei, JAPAN
TANAKA, Tadashi, JAPAN
TANAKA, Tamaki*, JAPAN
TANAKA, Tsutomu, JAPAN

TANAKA, Yohei, JAPAN
TANAKA, Yukiyasu, JAPAN
TANDAI, Kwoichi*, JAPAN
TANEMURA, Hideki, JAPAN
TANG, Francis C. Y., CANADA
TANGO, Hiroshi, JAPAN
TANI, Atusi§, JAPAN
TANI, Seiichi, JAPAN
TANI, Tsugio, JAPAN
TANIGAWA, Harumi, JAPAN
TANIGAWA, Yoshio, JAPAN
TANIGUCHI, Hajime, JAPAN
TANIGUCHI, Kazuo*, JAPAN
TANIGUCHI, Masaru, JAPAN
TANIGUCHI, Setsuo, JAPAN
TANIHIRA, Yumiko, JAPAN
TANIKAWA, Akio, JAPAN
TANIKAWA, Masao, JAPAN
TANISAKI, Toshiyuki, JAPAN
TANIYAMA, Kouki, JAPAN
TANNO, Shukichi, JAPAN
TANUMA, Kazumi, JAPAN
TARAMA, Shigeo, JAPAN
TARDOS, Eva, HUNGARY/U.S.A.
TARKALANOV, Krassimir D.§, BULGARIA
TARTAR, Luc C., U.S.A.
TARUI, Jun§, U.S.A.
TARUMI, Tomoyuki, JAPAN
TASAKI, Hiroyuki, JAPAN
TASHIRO, Yoshihiro, JAPAN
TATSUUMA, Nobuhiko, JAPAN
TATUZAWA, Tikao§, JAPAN
TAWARA, Seisui, JAPAN
TAYA, Hisao, JAPAN
TAYLOR, John, U.K.
TAYLOR, John Christopher, CANADA
TAYLOR, Laurence R., U.S.A.
TAYLOR, Michael E., U.S.A.
TAYOSHI, Takao, JAPAN
TAZAWA, Yoshihiko, JAPAN
TELLO, Angela, PANAMA
TENENBLAT, Keti, BRAZIL
TENGSTRAND, Anders Sven, SWEDEN
TERADA, Itaru, JAPAN
TERADA, Junko, JAPAN
TERADA, Toshiaki, JAPAN
TERADA, Toshiji, JAPAN
TERAGAITO, Masakazu, JAPAN
TERAI, Naoki, JAPAN
TERAI, Nobuhiro, JAPAN
TERAO, Taro, JAPAN
TERAOKA, Yoshinobu, JAPAN
TERASAWA, Jun, JAPAN
TERUYA, Tamotsu, JAPAN
THAKUR, Dinesh S., INDIA
THOMASON, Robert W., U.S.A.
THOMASSEN, Carsten, DENMARK
THOMEE, Vidar C., SWEDEN

THOMPSON, Robert J., U.S.A.
THOMSEN, Momme Johs*, F.R.G.
THORNLEY, Gillian M., NEW ZEALAND
TIAN, Gang, CHINA, P. R. of
TIMOURIAN, James G., U.S.A.
TITANI, Satoko, JAPAN
TITI, Edriss S.*, U.S.A.
TOBIN, Sean J., IRELAND
TODA, Hirosi, JAPAN
TODA, Nobushige, JAPAN
TODOROV, Andrey N., BULGARIA
TODOROV, Gordana Glisa, U.S.A.
TOGARI, Yoshio, JAPAN
TOGNOLI, Alberto, ITALY
TOKI, Keisuke, JAPAN
TOKITA, Akira, JAPAN
TOKITA, Takesi, JAPAN
TOKIZAWA, Masamichi, JAPAN
TOKUNAGA, Hideya, JAPAN
TOKUNAGA, Hiroo, JAPAN
TOMABECHI, Satosi, JAPAN
TOMARI, Masataka, JAPAN
TOMARU, Tadashi, JAPAN
TOMEI, Carlos, BRAZIL
TOMIDA, Masamichi, JAPAN
TOMINAGA, Hiroyuki, JAPAN
TOMINAGA, Hisao, JAPAN
TOMINAGA, Tadashi, JAPAN
TOMISAKI, Matsuyo, JAPAN
TOMITA, Koichi, JAPAN
TOMITA, Minoru, JAPAN
TOMITA, Yoshihito, JAPAN
TOMIYAMA, Jun, JAPAN
TOMIZAWA, Akira, JAPAN
TONEGAWA, Takayuki, JAPAN
TON-THAT, Tuong*, U.S.A.
TOPIWALA, Pankaj N., U.S.A.
TORIUMI, Satoshi, JAPAN
TORRES, Rodolfo H., U.S.A./ARGENTINA
TOSE, Nobuyuki*, JAPAN
TOTOKI, Haruo, JAPAN
TOURÉ, Saliou, IVORY COAST
TOUTOUNIAN, Faezeh B.-V.*, IRAN
TOWA, Yuji, JAPAN
TOYODA, Masanori, JAPAN
TOYONARI, Toshitaka, JAPAN
TRAPANI, Camillo, ITALY
TRAUTMANN, Günther*, F.R.G.
TRICERRI, Franco, ITALY
TROTMAN, David, FRANCE
TRUE, Hans*, DENMARK
TSANG, Kai-Man, HONG KONG
TSUBOI, Akito, JAPAN
TSUBOI, Shoji, JAPAN
TSUBOI, Takashi, JAPAN
TSUCHIDA, Kensei, JAPAN
TSUCHIHASHI, Hiroyasu, JAPAN
TSUCHIKAWA, Masao, JAPAN

TSUCHIKURA, Tamotsu, JAPAN
TSUCHIMOTO, Yoshifumi, JAPAN
TSUCHIYA, Akihiro, JAPAN
TSUCHIYA, Masaaki, JAPAN
TSUCHIYA, Morimasa§, JAPAN
TSUCHIYA, Nobuo, JAPAN
TSUCHIYA, Takuya, JAPAN
TSUDA, Ei*, JAPAN
TSUDA, Koichi, JAPAN
TSUDA, Teruko, JAPAN
TSUGAWA, Osamu§, JAPAN
TSUGE, Masayoshi, JAPAN
TSUJI, Hajime, JAPAN
TSUJI, Mikio, JAPAN
TSUJI, Naomitsu, JAPAN
TSUJI, Tadashi, JAPAN
TSUJII, Masato*, JAPAN
TSUJII, Yoshiki, JAPAN
TSUJIOKA, Kunio, JAPAN
TSUJISHITA, Toru, JAPAN
TSUKADA, Kazumi, JAPAN
TSUKADA, Makoto, JAPAN
TSUKAHARA, Kumiko, JAPAN
TSUKAHARA, Shigeo, JAPAN
TSUKAMOTO, Chiaki, JAPAN
TSUKIMOTO, Hiroshi, JAPAN
TSUKUI, Yasuyuki, JAPAN
TSUMURA, Hirofumi, JAPAN
TSUNODA, Shuichiro, JAPAN
TSURUMI, Kazuyuki, JAPAN
TSUSHIMA, Ryuji, JAPAN
TSUSHIMA, Yukio, JAPAN
TSUTSUI, Toru, JAPAN
TSUTSUMI, Akira, JAPAN
TSUTSUMI, Yoshio, JAPAN
TSUYUMINE, Shigeaki, JAPAN
TSUZUKI, Nobuo, JAPAN
TSUZUKU, Tosiro, JAPAN
TUGUÉ, Tosiyuki, JAPAN
TUKADA, Kunio, JAPAN
TUNG, Shih Ping*, CHINA-TAIWAN
TURAEV, Vladimir G., U.S.S.R.
TURQUETTE, Atwell R.§, U.S.A.
TYLLI, Hans-Olav J., FINLAND
TYMCHATYN, Edward Dmytro, CANADA
TYURIN, Andrei N., U.S.S.R.
TZANNES, Vassilis*, GREECE
UCHIBORI, Tomio, JAPAN
UCHIDA, Daiyu, JAPAN
UCHIDA, Koji, JAPAN
UCHIDA, Motoo*, JAPAN
UCHIDA, Yoshiaki, JAPAN
UCHIKOSHI, Keisuke, JAPAN
UCHIMURA, Keisuke, JAPAN
UCHIUMI, Masauki, JAPAN
UCHIYAMA, Jun, JAPAN
UCHIYAMA, Kohei, JAPAN
UCHIYAMA, Koichi, JAPAN

UCHIYAMA, Mitsuru, JAPAN
UDA, Toshio, JAPAN
UDRIŞTE, Constantin N.*, ROMANIA
UE, Masaaki, JAPAN
UEDA, Hideharu, JAPAN
UEDA, Masaru, JAPAN
UEDA, Tetsuo, JAPAN
UEDA, Yoshika, JAPAN
UEHARA, Tsuyoshi, JAPAN
UEKI, Naomasa, JAPAN
UEMURA, Hideaki, JAPAN
UEMURA, Yoshiaki, JAPAN
UENO, Kazuo, JAPAN
UENO, Kazuyoshi, JAPAN
UENO, Kenji, JAPAN
UENO, Kimio, JAPAN
UENO, Machi, JAPAN
UENO, Tadashi, JAPAN
UENO, Yoshiaki, JAPAN
UESAKA, Hiroshi, JAPAN
UESU, Kagumi, JAPAN
UESU, Tadahiro, JAPAN
UETAKE, Tsuneo, JAPAN
UHLENBECK, Karen K., U.S.A.
UJIIE, Katsumi, JAPAN
UKAI, Seiji, JAPAN
UKONN, Michihisa, JAPAN
ULECIA, Teresa*, SPAIN
ULMER, Douglas L., U.S.A.
ULRICH, Bernd, U.S.A.
ULUÇAY, Cengiz§, TURKEY
UMAYA, Noriaki, JAPAN
UMEDA, Tomio, JAPAN
UMEDA, Toru, JAPAN
UMEDA, Yoshio, JAPAN
UMEGAKI, Hisaharu, JAPAN
UMEHARA, Masaaki, JAPAN
UMEMURA, Hiroshi§, JAPAN
UMENO, Takashi, JAPAN
UMEZAWA, Toshio, JAPAN
UMEZU, Yumiko, JAPAN
UNGER, Luise Susanne*, F.R.G.
UNO, Katsuhiro, JAPAN
UNSURANGSIE, Sumalee, THAILAND
URABE, Hironobu, JAPAN
URABE, Tohsuke, JAPAN
URAKAWA, Hajime, JAPAN
URATA, Toshio, JAPAN
USA, Takeshi, JAPAN
USAMI, Hiroyuki, JAPAN
USAMI, Yoko, JAPAN
USHIJIMA, Takeo, JAPAN
USHIJIMA, Teruo*, JAPAN
USHIKI, Shigehiro, JAPAN
USHITAKI, Fumihiro, JAPAN
USUI, Sampei, JAPAN
UTAMURA, Motoaki, JAPAN
UTSUDE, Hisaki, JAPAN

UWANO, Yoshio, JAPAN
UZAWA, Masakatsu, JAPAN
VÄISÄLÄ, Jussi*, FINLAND
VALIBOUZE, Annick Angèle, FRANCE
VALLA, Giuseppe, ITALY
VALLEJO, Ernesto, MEXICO
VALLI, Giorgio, ITALY
VAN DAELE, Alfons, BELGIUM
VAN DEN DRIES, Lou, U.S.A.
VAN DER GEER, Gerard§, NETHERLANDS
VAN DER KALLEN, Wilberd L.,
   NETHERLANDS
VAN DER PUT, Marius, NETHERLANDS
VAN DER WALT, A. P. J., SOUTH AFRICA
VAN ENTER, Aernout C.D.*, NETHERLANDS
VAN GEEL, Jan M. H.*, BELGIUM
VANHECKE, Lieven N.A., BELGIUM
VAN LINT, Jacobus H., NETHERLANDS
VAN WYK, Leon*, SOUTH AFRICA
VARCHENKO, Alexandre, U.S.S.R.
VAROPOULOS, Nicolas Thodore, FRANCE
VASIU, Adrian, ROMANIA
VAUGHAN, Jerry E.*, U.S.A.
VAVILOV, Nikolai A., U.S.S.R.
VEGA, Luis, U.S.A.
VELDSMAN, Stefan*, SOUTH AFRICA
VELOSO, José M. M.§, BRAZIL
VELU, Jacques, FRANCE
VENEMA, Gerard A.*, U.S.A.
VERNIER-PIRO, Stella*, ITALY
VIANA, Marcelo, BRAZIL
VIEHWEG, Eckart E., F.R.G.
VINCENT, Georges A., SWITZERLAND
VIRO, Oleg Ya., U.S.S.R.
VLADIMIROV, Vasilii S., U.S.S.R.
VOEVODIN, Valentin V., U.S.S.R.
VOGEL, Wolfgang*, G.D.R.
VOJTA, Paul A., U.S.A.
VO KHAC, Khoan*, FRANCE
VOLBERG, Alexander L., U.S.S.R.
VOLOCH, José Felipe, BRAZIL
VOLOVICH, Igor V., U.S.S.R.
VOROS, André, FRANCE
VU, Dinh Hoa, G.D.R.
VU, Kim Tuan*, VIETNAM
WADA, Hidekazu, JAPAN
WADA, Hideo, JAPAN
WADA, Junzo, JAPAN
WADA, Kouichi, JAPAN
WADA, Masaaki, U.S.A.
WADA, Ryoko, JAPAN
WADA, Takakazu, JAPAN
WADA, Tomoyuki, JAPAN
WAJNRYB, Bronislaw, ISRAEL
WAKABAYASHI, Isao, JAPAN
WAKABAYASHI, Seiichiro, JAPAN
WAKAE, Masami, JAPAN
WAKATSUKI, Taketo, JAPAN

WAKAYAMA, Masato, JAPAN
WAKIMOTO, Minoru, JAPAN
WALDSPURGER, Jean-Loup, FRANCE
WALKER, Elbert A., U.S.A.
WALL, Charles T. C., U.K.
WALLACE, David A.R.*, U.K.
WALLACH, Nolan R., U.S.A.
WALTER, Wolfgang*, F.R.G.
WAN, Yieh-Hei, U.S.A.
WAN, Zhe-xian, SWEDEN
WANG, Cun-Zheng, CHINA, P. R. of
WANG, Shicheng*, CHINA, P. R. of
WANG, Shikun*, CHINA, P. R. of
WASHIHARA, Masako, JAPAN
WASHINGTON, Lawrence C., U.S.A.
WASHIO, Tadashi, JAPAN
WASSERMANN, Antony J., U.K.
WATABE, Mutsuo, JAPAN
WATABE, Tsuyoshi, JAPAN
WATAMORI, Yoko, JAPAN
WATANABE, Atumi, JAPAN
WATANABE, Hiroshi, JAPAN
WATANABE, Hisako, JAPAN
WATANABE, Hisao, JAPAN
WATANABE, Jiro, JAPAN
WATANABE, Junsei, JAPAN
WATANABE, Kazuhisa, JAPAN
WATANABE, Keiichi, JAPAN
WATANABE, Keiichi, JAPAN
WATANABE, Kimio, JAPAN
WATANABE, Kinji, JAPAN
WATANABE, Kiyoshi, JAPAN
WATANABE, Masayuki, JAPAN
WATANABE, Michiaki§, JAPAN
WATANABE, Nobuya, JAPAN
WATANABE, Osamu, JAPAN
WATANABE, Seiji, JAPAN
WATANABE, Shinzo, JAPAN
WATANABE, Tadashi, JAPAN
WATANABE, Takao, JAPAN
WATANABE, Takesi, JAPAN
WATANABE, Tetsuo, JAPAN
WATANABE, Toshihiro, JAPAN
WATANABE, Toshikazu, JAPAN
WATANABE, Yoshikazu, JAPAN
WATANABE, Yoshiyuki, JAPAN
WATARI, Chinami, JAPAN
WATARI, Satoru, JAPAN
WATATANI, Yasuo, JAPAN
WATKINS, John J., U.S.A.
WEBB, Jeffrey R. L., U.K.
WEBBER, David Brian, U.K.
WEDER, Ricardo*, MEXICO
WEE, In-Suk*, KOREA, R. of
WEHRHAHN, Karl H.*, AUSTRALIA
WEIBEL, Charles A., U.S.A.
WEISFELD, Morris, U.S.A.
WEISS, Asia Ivic, CANADA

WEISS, William, CANADA
WELCH, Philip D.*, U.K.
WELK, Reiner, F.R.G.
WENZEL, Christian*, U.S.A./F.R.G.
WEST, Alan, U.K.
WESTON, Kenneth W.*, U.S.A.
WHITMAN, Nancy C.*, U.S.A.
WIDMAN, Kjell-Owe H., SWITZERLAND
WIEGAND, Sylvia M., U.S.A.
WIGDERSON, Avi, ISRAEL
WIGLEY, Neil M.*, CANADA
WILLIAMS, Daniel A., U.S.A.
WILLIAMS, Robert F., U.S.A.
WILSON, James M., U.S.A.
WILSON, John Stuart, U.K.
WILSON, Pelham M.H., U.K.
WINKLER, Jörg, F.R.G.
WISNER, Robert J., U.S.A.
WITTE, Dave, U.S.A.
WITTEN, Edward, U.S.A.
WITTEN, Louis, U.S.A.
WOGEN, Warren R.*, U.S.A.
WOLPER, James Samuel*, U.S.A.
WONG, Philip Pit Wang, HONG KONG
WONG, Raymond Y., U.S.A.
WONG, Roderick S.-C., CANADA
WONG, Yau-Chuen*, HONG KONG
WOOD, Carol S., U.S.A.
WORONOWICZ, Stanislaw Lech, POLAND
WORSLEY, Keith John, CANADA
WRIGHT, David James*, U.S.A.
WU, Liang Sen*, CHINA, P. R. of
WU, Pei Yuan*, CHINA-TAIWAN
WU, Yihren*, U.S.A.
WULFSOHN, Aubrey*, ISRAEL
XIONG, Jincheng*, CHINA, P. R. of
XU, Dao Yi*, CHINA, P. R. of
XU, Yichao*, CHINA, P. R. of
YABU, Yasuhiko, JAPAN
YABUTA, Kozo, JAPAN
YAGASAKI, Tatsuhiko, JAPAN
YAGI, Akiko, JAPAN
YAGI, Atsushi, JAPAN
YAGI, Hirotomo, JAPAN
YAGI, Katsumi, JAPAN
YAGISHITA, Kimio, JAPAN
YAGITA, Nobuaki, JAPAN
YAGUCHI, Teruo*, JAPAN
YAJIMA, Yukinobu, JAPAN
YAKU, Takeo, JAPAN
YAMADA, Akira, JAPAN
YAMADA, Hiromichi, JAPAN
YAMADA, Kotaro, JAPAN
YAMADA, Naoki, JAPAN
YAMADA, Osanobu, JAPAN
YAMADA, Shuji§, JAPAN
YAMADA, Sumio, JAPAN
YAMADA, Toshihiko, JAPAN

YAMADA, Toshio, JAPAN
YAMADA, Yoshio, JAPAN
YAMADA, Yuji, JAPAN
YAMAGAMI, Shigeru, JAPAN
YAMAGATA, Hideo*, JAPAN
YAMAGATA, Kunio, JAPAN
YAMAGISHI, Kikumichi, JAPAN
YAMAGISHI, Yoshikazu, JAPAN
YAMAGUCHI, Hakuki, JAPAN
YAMAGUCHI, Hiroshi, JAPAN
YAMAGUCHI, Keizo, JAPAN
YAMAGUCHI, Kohhei, JAPAN
YAMAGUCHI, Masaru, JAPAN
YAMAGUCHI, Michinari, JAPAN
YAMAGUCHI, Seiichi, JAPAN
YAMAGUCHI, Tadashi, JAPAN
YAMAGUCHI, Takao, JAPAN
YAMAGUCHI, Toshihiro, JAPAN
YAMAGUTI, Kiyosi*, JAPAN
YAMAGUTI, Masaya, JAPAN
YAMAHARA, Hideo*, JAPAN
YAMAKAWA, Aiko, JAPAN
YAMAKI, Hiroyoshi, JAPAN
YAMAMOTO, Hiro-o, JAPAN
YAMAMOTO, Kazuhiro, JAPAN
YAMAMOTO, Kensho, JAPAN
YAMAMOTO, Makoto, JAPAN
YAMAMOTO, Masahiro, JAPAN
YAMAMOTO, Minoru, JAPAN
YAMAMOTO, Shuichi, JAPAN
YAMAMOTO, Sunao, JAPAN
YAMAMOTO, Takanori, JAPAN
YAMAMOTO, Tetsuro, JAPAN
YAMAMOTO, Tetsuya, JAPAN
YAMAMOTO, Yoshihiko, JAPAN
YAMAMOTO, Yutaka, JAPAN
YAMAMURA, Ken, JAPAN
YAMANAKA, Hajime, JAPAN
YAMANAKA, Takesi, JAPAN
YAMANE, Hiroyuki, JAPAN
YAMANO, Gosuke, JAPAN
YAMANO, Takehisa, JAPAN
YAMANOSHITA, Tsuneyo, JAPAN
YAMANOUCHI, Takehiko, JAPAN
YAMASAKI, Aiichi, JAPAN
YAMASAKI, Hideki, JAPAN
YAMASAKI, Masayuki, JAPAN
YAMASAKI, Motohiro, JAPAN
YAMASAKI, Toshiharu, JAPAN
YAMASAKI, Yasuo, JAPAN
YAMASHIMA, Shigeho, JAPAN
YAMASHITA, Hajime*, JAPAN
YAMASHITA, Satoshi, JAPAN
YAMATO, Hajime, JAPAN
YAMATO, Kazuo, JAPAN
YAMATO, Kenji, JAPAN
YAMATO, Yuiti, JAPAN
YAMAUCHI, Kazunari, JAPAN

YAMAUCHI, Masatoshi, JAPAN
YAMAUCHI, Norio, JAPAN
YAMAWAKI, Noriaki*, JAPAN
YAMAYA, Atsushi, JAPAN
YAMAZAKI, Chikao§, JAPAN
YAMAZAKI, Keijiro, JAPAN
YAMAZAKI, Masao, JAPAN
YAMAZAKI, Mitsuru, JAPAN
YAMAZAKI, Tadashi, JAPAN
YAMAZAKI, Taeko, JAPAN
YAN, Jia-An*, CHINA, P. R. of
YANAGAWA, Minoru*, JAPAN
YANAGAWA, Takaaki, JAPAN
YANAGI, Kenjiro, JAPAN
YANAGIHARA, Hiroshi, JAPAN
YANAGIHARA, Niro, JAPAN
YANAGIMOTO, Shigekazu*, JAPAN
YANAGISAWA, Takashi, JAPAN
YANAGISAWA, Taku, JAPAN
YANAI, Hiromichi, JAPAN
YANG, Jae-Hyun, KOREA, R. of
YANG, Paul C., U.S.A.
YANO, Koichi, JAPAN
YANO, Tamaki*, JAPAN
YAP, Leonard Y. H.*, SINGAPORE
YASUDA, Kumi, JAPAN
YASUGI, Mariko, JAPAN
YASUI, Tsutomu, JAPAN
YASUKURA, Osami, JAPAN
YASUMOTO, Masahiro, JAPAN
YASUNAGA, Hisatoshi, JAPAN
YASUO, Minato, JAPAN
YASURAOKA, Yuzo, JAPAN
YASUTOMI, Sinichi, JAPAN
YASUURA, Hiroto, JAPAN
YATES, Samuel*, U.S.A.
YE, Rugang, U.S.A.
YEH, R. Z., U.S.A.
YEH, Yeong-Nan*, CHINA-TAIWAN
YEUNG, Kit Ming, HONG KONG
YLINEN, Kari E., FINLAND
YOCCOZ, Jean-Christophe, FRANCE
YOKOGAWA, Kouji, JAPAN
YOKOI, Hideo, JAPAN
YOKOI, Yoshitaka, JAPAN
YOKONUMA, Takeo, JAPAN
YOKOTA, Hisashi, JAPAN
YOKOTA, Ichiro*, JAPAN
YOKOTA, Toshiro, JAPAN
YOKOTA, Yoshiyuki, JAPAN
YOKOTE, Ichiro, JAPAN
YOKOYAMA, Kazunori, JAPAN
YOKOYAMA, Kazuo, JAPAN
YOKOYAMA, Misako, JAPAN
YOKOYAMA, Mutsumi, JAPAN
YOKOYAMA, Ryozo, JAPAN
YOKOYAMA, Shin, JAPAN
YOKOYAMA, Takuji, JAPAN

YOKURA, Shoji*, JAPAN
YONEDA, Kaoru, JAPAN
YONEMOTO, Hiroshi, JAPAN
YONEZAWA, Yoshitaka, JAPAN
YOO, Il, KOREA, R. of
YOO, Ki-Jo, KOREA, R. of
YOON, Goog-Joong, KOREA, R. of
YOR, Marc J., FRANCE
YORINAGA, Masataka, JAPAN
YOROZU, Shinsuke§, JAPAN
YOSHIDA, Eiji, JAPAN
YOSHIDA, Haruyo*, JAPAN
YOSHIDA, Hidenobu, JAPAN
YOSHIDA, Hiroaki, JAPAN
YOSHIDA, Hiroyuki, JAPAN
YOSHIDA, Jun-ichi, JAPAN
YOSHIDA, Kenichi, JAPAN
YOSHIDA, Kiyoshi, JAPAN
YOSHIDA, Mamoru, JAPAN
YOSHIDA, Masaaki, JAPAN
YOSHIDA, Masaharu, JAPAN
YOSHIDA, Minoru, JAPAN
YOSHIDA, Nakahiro, JAPAN
YOSHIDA, Nobuo, JAPAN
YOSHIDA, Norio, JAPAN
YOSHIDA, Toshio, JAPAN
YOSHIHARA, Hisao*, JAPAN
YOSHIHARA, Ken-ichi, JAPAN
YOSHIKAWA, Atsushi, JAPAN
YOSHIKAWA, Kazuo, JAPAN
YOSHIKAWA, Kenichi, JAPAN
YOSHIKAWA, Masaki, JAPAN
YOSHIKAWA, Masayuki, JAPAN
YOSHIMATSU, Yumiko, JAPAN
YOSHIMOTO, Akinori, JAPAN
YOSHIMURA, Mitsuhiko, JAPAN
YOSHIMURA, Zenichi, JAPAN
YOSHINAGA, Etsuo, JAPAN
YOSHINAGA, Ken-ichi, JAPAN
YOSHINO, Ken-ichi, JAPAN
YOSHINO, Masafumi, JAPAN
YOSHINO, Takashi, JAPAN
YOSHINO, Yuji, JAPAN
YOSHINOBU, Yasuo, JAPAN
YOSHIOKA, Akiko, JAPAN
YOSHIOKA, Akira, JAPAN
YOSHIOKA, Tsuneo, JAPAN
YOSHITOMI, Kentaro, JAPAN

YOSHIZAWA, Hisaaki, JAPAN
YOSHIZAWA, Taro, JAPAN
YOSIDA, Setuzo, JAPAN
YOSIOKA, Masanori, JAPAN
YOTSUTANI, Shoji, JAPAN
YOZAWA, Takashi, JAPAN
YUAN, Ya-Xiang*, CHINA, P. R. of
YUI, Noriko, CANADA
YUNG, Tin Gun, HONG KONG
YURI, Michiko, JAPAN
ZABCZYK, Jerzy, POLAND
ZAFARANI, Jafar*, IRAN
ZAGIER, Don B., F.R.G.
ZAHAROPOL, Radu*, U.S.A.
ZAHEDANI, Heydar Z.*, IRAN
ZAHEDI, Mohammad Mehdi*, IRAN
ZAICEV, Gennadi Ju.§, U.S.S.R.
ZAINODIN, Haji Jubok, MALAYSIA
ZALCMAN, Lawrence A., ISRAEL
ZARETTI, Anna, ITALY
ZARHIN, Yuri G., U.S.S.R.
ZASLAVSKY, Boris G.*, U.S.S.R.
ZAYED, Ahmed I.*, U.S.A.
ZDRAVKOVSKA, Smilka, U.S.A.
ZEEMAN, E. Christopher, U.K.
ZEINAL-HAMADANI, Ali§, IRAN
ŻELAZKO, Wiesław T.*, POLAND
ZELDITCH, Steven, U.S.A.
ZELEVINSKY, Andrey V., U.S.S.R.
ZELMANOV, Efim I., U.S.S.R.
ZHANG, De-Qi, CHINA, P. R. of
ZHANG, Jin-Hao, CHINA, P. R. of
ZHANG, Li Hua, CHINA, P. R. of
ZHANG, Shunian*, CHINA, P. R. of
ZHANG, Zezeng, CHINA, P. R. of
ZHANG, Zhi-Fen*, CHINA, P. R. of
ZHARN, P. Joseph§, U.S.A.
ZHOU, Xueguang*, CHINA, P. R. of
ZHOU, Zixiang*, CHINA, P. R. of
ZHOU, Zuo Ling*, CHINA, P. R. of
ZHU, Xiao-Wei, U.S.A.
ZIEGLER, Zvi, ISRAEL
ZIEMIAN, Bogdan*, POLAND
ZIMMER, Horst Günter*, F.R.G.
ZIMMERMAN, Grenith J., U.S.A.
ZIMMERMANN, Benno, SWITZERLAND
ZIMMERMANN-HUISGEN, Birge K., U.S.A.
ZOKAEI, Mohammad§, IRAN

Membership by Nationality

Algeria	3	Malaysia	7
Argentina	10	Mauritania	1
Australia	21	Mexico	14
Austria	4	Mozambique	1
Bangladesh	2	Myanmar	1
Belgium	12	Nepal	1
Brazil	15	Netherlands	25
Bulgaria	8	New Zealand	7
Canada	77	Nigeria	5
Chile	4	Norway	11
China, People's Republic of	67	Pakistan	1
China-Taiwan	29	Panama	2
Colombia	1	Peru	2
Costa Rica	2	Philippines	7
Cuba	1	Poland	17
Czechoslovakia	8	Portugal	4
Denmark	15	Romania	18
Egypt	6	Saudi Arabia	2
El Salvador	1	Senegal	1
Finland	29	Sierra Leone	1
France	123	Singapore	3
German Democratic Republic	6	South Africa	13
Germany, Federal Republic of	98	Spain	24
Ghana	1	Sweden	32
Greece	8	Switzerland	14
Hong Kong	14	Tanzania	1
Hungary	15	Thailand	4
Iceland	2	Tunisia	1
India	40	Turkey	2
Indonesia	2	Uganda	2
Iran	57	United Kingdom	94
Iraq	1	Uruguay	1
Ireland	4	USA	396
Israel	30	USSR	110
Italy	69	Venezuela	2
Ivory Coast	1	Vietnam	21
Japan	2409	Yugoslavia	9
Kenya	2	Zimbabwe	2
Korea, Republic of	41	Stateless	3
Lebanon	1		
Luxembourg	1	Total	4102

The Work of the Fields Medalists
and the Rolf Nevanlinna Prize Winner

The works of the Fields Medalists and the Rolf Nevanlinna Prize Winner were presented as follows:

Fields Medalists

The Rolf Nevanlinna Prize Winner

© 1990 H. Kono

Fields Medalists: *Witten, Mori, Jones, Drinfeld*

Photo by A. Mizutani

Razborov (left), the Rolf Nevanlinna Prize winner, *Hori* and *Lovász*

On the Mathematical Work of Vladimir Drinfeld

*Yuri Ivanovich Manin**

Steklov Mathematical Institute, 42 Vavilova, 117966 GSP-1, Moscow, USSR

I

Drinfeld has written his first published paper when he was a schoolboy. He proved there a nice result in the style of Hardy's classic treatise "Inequalities" and solved a problem to which R. A. Rankin devoted two notes. This paper still makes an interesting reading. It starts a series of Drinfeld's works which can be considered somewhat isolated in the general context of his mathematical production but which contains such worthy results as:

- a proof that the cuspidal degree zero cycles on a modular curve generate a torsion subgroup of the Jacobian;
- classification of instantons (ADHM construction, jointly with Atiyah, Hitchin and Manin);
- reduction theory of completely integrable systems of KdV-type (jointly with Sokolov));
- a sharp asymptotic upper bound for the number of points of a curve defined over a finite field of order p^{2n} (jointly with Vladut);
- a proof that an $SO(n+1)$-invariant finitely additive measure on S^n is Lebesgue for $n = 2$ and 3 (the cases $n = 1$ and $n \geq 4$ were treated earlier by Banach, Margulis and Sullivan).

The limitations of space and time make it impossible for me to review Drinfeld's contributions referred to above, and I shall concentrate upon the two subjects that were Drinfeld's main preoccupation in the last decade. These are Langlands' program and quantum groups. In both domains, Drinfeld's work constituted a decisive breakthrough and prompted a wealth of research.

II

Langlands' program is a series of conjectures, theorems and insights aimed to an understanding of the Galois groups of local and global fields of dimension one, that is $\mathbb{Q}$, $\mathbb{F}_p(t)$, their finite extensions and completions.

---

* Delivered by Michio Jimbo.

Proceedings of the International Congress
of Mathematicians, Kyoto, Japan, 1990
© The Mathematical Society of Japan, 1991

One can convincingly argue that these Galois groups constitute a primary object of number theory, more fundamental than the integers themselves. Anyway, most of the classical themes of number theory, like prime numbers, L-functions, and modular forms reveal a lot of hidden structure when viewed from the Galois-theoretic angle.

The highest achievement of the classical period was the class-field theory which gave a description of $\mathrm{Gal}(k^{\mathrm{ab}}/k)$ in terms of the arithmetics of k. If k is local, $\mathrm{Gal}(k^{\mathrm{ab}}/k)$ is (canonically isomorphic to) the profinite completion of k^*; if k is global, $\mathrm{Gal}(k^{\mathrm{ab}}/k)$ is the profinite completion of the group A_k^*/k^* where A_k denotes the adèle ring of k. Of course, to describe $\mathrm{Gal}(k^{\mathrm{ab}}/k)$ is the same as to describe one-dimensional representations of $\mathrm{Gal}(\bar{k}/k)$. Langlands suggested that the next step should be a description of certain n-dimensional representations of $\mathrm{Gal}(\bar{k}/k)$. The main content of his conjectures, in a very simplified and imprecise form, can be summarized in two statements. First, n-dimensional representations of $\mathrm{Gal}(\bar{k}/k)$ are in a natural bijection with automorphic irreducible infinite-dimensional representations of $GL(n, k)$ (resp. $GL(n, A_k)$) in the local (resp. global) case. Second, this correspondence (the non-commutative reciprocity law) is described via the identification of the L-functions that can be constructed in a natural way for both Galois representations and the adèlic general linear group representations. In the classical (global) cases when the structure of these L-functions is known, they are Mellin's transforms of modular forms which represent, roughly speaking, the de Rham aspect of the cohomology of certain modular spaces, while Galois representations embody their étale cohomology. Thus, Langlands' program is related to Grothendieck's motives, a largely conjectural universal cohomology theory of algebraic varieties.

Drinfeld proved Langlands' conjectures for $GL(2)$ over global fields of finite characteristic. His decisive contribution was the discovery of a new class of modular spaces and their detailed investigation. While the classical modular spaces parametrize elliptic curves, abelian varieties or Hodge structures, Drinfeld had an astonishing idea that in order to treat the finite characteristic case one should parametrize objects of a new kind. The first approximation is now called Drinfeld modules (earlier examples of which were introduced by Carlitz). A more general notion indispensable for the complete theory is known under the equally uninspiring names "shtuka" (meaning approximately "a piece of something" in English) and "F-sheaf" (which Drinfeld uses for the lack of something better; as a mild reproach I must say that in the domain of terminology his imagination produces less brilliant solutions than in the theorem-proving).

For completeness, I shall give below a formal definition of an F-sheaf and the statement of the main theorem of Drinfeld's theory.

Let X be a smooth complete model of the ground field $k = \mathbb{F}_q(X)$. A (left) F-sheaf of rank d over a scheme S is a diagram

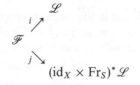

in which $\mathscr{L}$ and $\mathscr{F}$ are locally free sheaves on $X \times S$ of rank d, i and j are injective, $\mathrm{Coker}(i)$ is an invertible sheaf on the graph Γ_α of a certain S-point $\alpha : S \to X$, $\mathrm{Coker}(j)$ is an invertible sheaf on a similar graph Γ_β.

F-sheaves can be rigidified by imposing a level structure. F-sheaves with a non-trivial level structure admit coarse moduli schemes, which can be conveniently compactified. For $d = 2$, $GL(2, A_k)$ acts upon compactified moduli spaces, much in the same way as $GL(2, A_{\mathbb{Q}, f})$ acts upon moduli spaces of elliptic curves via Hecke operators. A Galois-theoretic analysis of this action upon the cohomology furnishes the following results.

Let $W_k \subset \mathrm{Gal}(\overline{k}^s / k)$ be the group of all automorphisms inducing upon $\overline{\mathbb{F}}_q$ an integer power of the Frobenius morphism. Fix a prime $l \neq \mathrm{char}(k)$. Denote by Σ_1^l the set of continuous representations $\varrho : W_k \to GL(n, \overline{\mathbb{Q}}_l)$ with the following properties: i) $\mathrm{Im}(\varrho) \subset GL(n, F)$ for a finite extension F of $\mathbb{Q}_l$; ii) inertia subgroups of almost all points of k are contained in $\mathrm{Ker}(\varrho)$.

Let now E be an algebraically closed field of characteristic 0. A cuspidal representation of $GL(2, A_k)$ is, by definition, an irreducible representation over E belonging to the space of E-valued cusp forms. Denote by Σ_2^l the set of cuspidal representations over $\mathbb{Q}_l$.

The principal theorem of Drinfeld states that Σ_1^l and Σ_2^l are bijective with respect to a map conserving L-functions.

Drinfeld's proof is long and involved, both technically and logically. Although it is contained in a series of journal papers, the most comprehensive treatment of the whole theory is given in his two still unpublished manuscripts (totalling up to about 800 type-written pages). It is to be hoped that after a final brushing up they will be given to a book publisher. The reason it has not been done to this moment seems to be Drinfeld's preoccupation with another new and fascinating subject – quantum groups.

III

Formally speaking, quantum groups constitute a vaguely defined subclass of Hopf algebras. Their first examples were discovered by the mathematical physicists of Leningrad school, students and collaborators of L. D. Faddeev. Drinfeld first summarized the basic definitions and results of this theory, largely conceived or systematized in his own work, in his talk at the Berkeley ICM four years ago. This report and several articles of M. Jimbo played a decisive role in the crystallization of this new domain and drew to it attention of many mathematicians.

Somewhat schematically, one can describe several principal themes of this theory in the following way:

a) For many decades, a common wisdom was that simple Lie groups and Lie algebras are rigid objects: they are classified by discrete data (Dynkin diagrams), they do not vary in a continuous family. A remarkable discovery, due in this generality to Drinfeld and Jimbo, is that this ceases to be true if one considers the deformations in the class of non-commutative and cocommutative Hopf algebras.

More precisely, for any simple Lie algebra $\mathfrak{g}$ (and more generally, for algebras defined by Cartan root data) there exists a one-parametric deformation $U_q(\mathfrak{g})$ of the universal enveloping algebra $U(\mathfrak{g})$, with comultiplication and antipode. More than that, the whole structure and representation theory deforms in this way, with many remarkable new twists.

b) The properties of quantum groups are closely related to certain remarkable non-linear algebraic equations defining the so-called Yang-Baxter operators. The simplest Yang-Baxter equation is a condition on a linear operator $R \in \text{End}(V \otimes V)$ where V is a linear space. The condition is that R satisfies the braid relation: $R_{12}R_{23}R_{12} = R_{23}R_{12}R_{23}$ as operators on $V^{\otimes 3}$. A more general Yang-Baxter equation concerns operators depending on spectral parameters. Many solutions of Yang-Baxter equations were discovered by the specialists working in the two-dimensional statistical physics because such solutions give rise to exactly solvable lattice models generalizing the famous Ising model.

Drinfeld introduced the notion of a universal Yang-Baxter operator which is an invertible element in the (completed) tensor square of a Hopf algebra. He has proved its existence in $U_q(\mathfrak{g})^{\otimes 2}$ and has given a general "double" construction allowing one to generate Yang-Baxter operators in the representation categories of various Hopf algebras.

c) Until recently, the quantum group theory lacked a classification theorem describing precisely what class of objects we want to consider and what is the structure of this class. Such a theorem was recently formulated and proved in the two important papers by Drinfeld. It can be compared with the first theorems of Lie establishing the relations between Lie algebras and local Lie groups.

Indeed, this theorem is local, even formal, in nature, because Drinfeld considers formal deformations of formal groups (or, in the dual language, of their universal enveloping algebras). He also imposes from the very beginning a Yang-Baxter type condition ("quasi-triangularity"). He proves then that the whole deformation is defined by the lowest order non-trivial data.

In the course of proof, he introduces a new notion of quasi-Hopf algebra, weakening in an appropriate way the coassociativity condition on the comultiplication. He connects quasi-Hopf algebras with the Knizhnik-Zamolodchikov differential equation whose monodromy is related to the Drinfeld-Jimbo R-operators as was shown by Kohno. Finally, he shows that in a formal situation the quasi-Hopf algebras can be reduced to the usual Hopf algebras by a kind of gauge transformation. The proof involves killing a lot of cohomological obstructions arising in the complex perturbative scheme.

d) One should mention also the notions of a Poisson-Lie group and a Poisson-Lie action introduced by Drinfeld at an earlier stage of his work on Yang-Baxter equations. They form one of the most basic differential geometric structures connected with Hamiltonian mechanics, and their role in future will certainly grow.

IV

I hope that I conveyed to you some sense of broadness, conceptual richness, technical strength and beauty of Drinfeld's work for which we are now honoring him with the Fields medal. For me, it was a pleasure and a privilege to observe at a close distance the rapid development of this brilliant mind which taught me so much.

The Work of Vaughan F. R. Jones

Joan S. Birman

Department of Mathematics, Columbia University, New York, NY 10027, USA

It gives me great pleasure that I have been asked to describe to you some of the very beautiful mathematics which resulted in the awarding of the Fields Medal to Vaughan F. R. Jones at ICM '90.

In 1984 Jones descovered an astonishing relationship between von Neumann algebras and geometric topology. As a result, he found a new polynomial invariant for knots and links in 3-space. His invariant had been missed completely by topologists, in spite of intense activity in closely related areas during the preceding 60 years, and it was a complete surprise. As time went on, it became clear that his discovery had to do in a bewildering variety of ways with widely separated areas of mathematics and physics, some of which are indicated in Figure 1. These included (in addition to knots and links) that part of statistical mechanics having to do with exactly solvable models, the very new area of quantum groups, and also Dynkin diagrams and the representation theory of simple Lie algebras. The central connecting link in all this mathematics was a tower of nested algebras which Jones had discovered some years earlier in the course of proving a theorem which is known as the "Index Theorem".

My plan is to begin by discussing the Index Theorem, and the tower of algebras which Jones constructed in the course of his proof. After that, I plan to return to the chart in Figure 1 in order to indicate how this tower of algebras served as a bridge between the diverse areas of mathematics which are shown on the chart. I will restrict my attention throughout to one very special example of the tower construction, and so also to one special example of the associated link invariants, in order to make it possible to survey a great deal of mathematics in a very short time. Even with the restriction to a single example, this is a very ambitious plan. On the other hand, it only begins to touch on Vaughan Jones' scholarly contributions.

1. The Index Theorem

Let $\mathbf{M}$ denote a von Neumann algebra. Thus $\mathbf{M}$ is an algebra of bounded operators acting on a Hilbert space $\mathscr{H}$. The algebra $\mathbf{M}$ is called a *factor* if its center consists only of scalar multiples of the identity. The factor is *type* II_1 if it admits a linear functional, called a trace, $\mathrm{tr} : \mathbf{M} \to \mathbb{C}$, which satisfies the following three conditions:

Proceedings of the International Congress
of Mathematicians, Kyoto, Japan, 1990
© The Mathematical Society of Japan, 1991

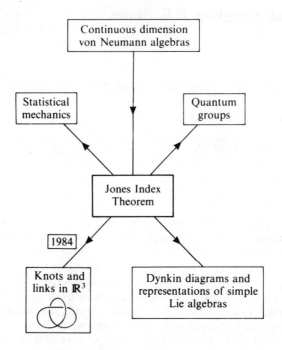

Fig. 1. The Jones Index Theorem

$$\mathrm{tr}(\mathbf{xy}) = \mathrm{tr}(\mathbf{yx}) \text{ for all } \mathbf{x}, \mathbf{y} \in \mathbf{M}$$
$$\mathrm{tr}(\mathbf{1}) = 1.$$
$$\mathrm{tr}(\mathbf{xx}^*) > 0 \text{ for all } \mathbf{x} \in \mathbf{M}, \text{ where } \mathbf{x}^* \text{ is the adjoint of } \mathbf{x}.$$

In this situation it is known that the trace is unique, in the sense that it is the only linear form satisfying the first two conditions. An old discovery of Murray and von Neumann was that factors of type II_1 provide a type of "scale" by which one can measure the dimension $\dim_{\mathbf{M}}(\mathcal{H})$ of $\mathcal{H}$. The notion of dimension which occurs here generalizes the familiar notion of integer-valued dimensions, because for appropriate $\mathbf{M}$ and $\mathcal{H}$ it can be any non-negative real number or ∞.

The starting point of Jones' work was the following question: if $\mathbf{M}_1$ is a type II_1 factor and if $\mathbf{M}_0 \subset \mathbf{M}_1$ is a *subfactor*, is there any restriction on the real numbers which occur as the ratio

$$\lambda = \dim_{\mathbf{M}_0}(\mathcal{H})/\dim_{\mathbf{M}_1}(\mathcal{H}) \ ?$$

The question has the flavor of questions one studies in Galois theory. On the face of it, there was no reason to think that λ could not take on any value in $[1, \infty]$, so Jones' answer came as a complete surprise. He called λ the *index* $[\mathbf{M}_1 : \mathbf{M}_0]$ of $\mathbf{M}_0$ in $\mathbf{M}_1$, and proved a type of rigidity theorem about type II_1 factors and their subfactors.

The Jones Index Theorem. *If* $\mathbf{M}_1$ *is a* II_1 *factor and* $\mathbf{M}_0$ *a subfactor, then the possible values of the index* $[\mathbf{M}_1 : \mathbf{M}_0]$ *are restricted to:*

$$[4, \infty] \cup \{4\cos^2(\pi/p), \quad \text{where } p \in \mathbb{N}, p \geq 3\}.$$

Moreover, each real number in the continuous part of the spectrum $[4, \infty]$ *and also in the discrete part* $\{4\cos^2(\pi/p), p \in \mathbb{N}, p \geq 3\}$ *is realized.*

We now sketch the idea of the proof, which is to be found in [Jo1]. Jones begins with the type II_1 factor $\mathbf{M}_1$ and the subfactor $\mathbf{M}_0$. There is also a tiny bit of additional structure: In this setting there exists a map $\mathbf{e}_1 : \mathbf{M}_1 \to \mathbf{M}_0$, known as the *conditional expectation* of $\mathbf{M}_1$ on $\mathbf{M}_0$. The map $\mathbf{e}_1$ is a *projection*, i.e. $(\mathbf{e}_1)^2 = \mathbf{e}_1$.

His first step is to prove that the ratio λ is independent of the choice of the Hilbert space $\mathscr{H}$. This allows him to choose an appropriate $\mathscr{H}$ so that the algebra $\mathbf{M}_2 = \langle \mathbf{M}_1, \mathbf{e}_1 \rangle$ generated by $\mathbf{M}_1$ and $\mathbf{e}_1$ makes sense. He then investigates $\mathbf{M}_2$ and proves that it is another type II_1 factor, which contains $\mathbf{M}_1$ as a subfactor, moreover the index $|\mathbf{M}_2 : \mathbf{M}_1|$ is equal to the index $|\mathbf{M}_1 : \mathbf{M}_0|$, i.e. to λ. Having in hand another II_1 factor $\mathbf{M}_2$ and its subfactor $\mathbf{M}_1$, there is also a trace on $\mathbf{M}_2$ which (by the uniqueness of the trace) coincides with the trace on $\mathbf{M}_1$ when it is restricted to $\mathbf{M}_1$, and another conditional expectation $\mathbf{e}_2 : \mathbf{M}_2 \to \mathbf{M}_1$. This allows Jones to iterate the construction, to build algebras $\mathbf{M}_1, \mathbf{M}_2 \dots$ and from them a family of algebras:

$$\mathbf{J}_n = \{\mathbf{1}, \mathbf{e}_1, \dots, \mathbf{e}_{n-1}\} \subset \mathbf{M}_n, \quad n = 1, 2, 3, \dots .$$

Rewriting history a little bit in order to make the subsequent connection with knots a little more transparent, we now replace the $\mathbf{e}_k$'s by a new set of generators which are units, defining:

$$\mathbf{g}_k = q\mathbf{e}_k - (1 - \mathbf{e}_k),$$

where

$$(1 - q)(1 - q^{-1}) = 1/\lambda.$$

The $\mathbf{g}_k$'s generate $\mathbf{J}_n$, because the $\mathbf{e}_k$'s do, and we can solve for the $\mathbf{e}_k$'s in terms of the $\mathbf{g}_k$'s. So

$$\mathbf{J}_n = \mathbf{J}_n(q) = \{\mathbf{1}, \mathbf{g}_1, \dots, \mathbf{g}_{n-1}\},$$

and we have a *tower of algebras*, ordered by inclusion:

$$\mathbf{J}_1(q) \subset \mathbf{J}_2(q) \subset \mathbf{J}_3(q) \subset \dots .$$

The parameter q, which replaces the index λ, is the quantity now under investigation.

The parameter q is woven into the construction of the tower. First, defining relations in $\mathbf{J}_n(q)$ depend upon q:

(1)
$$\mathbf{g}_i\mathbf{g}_k = \mathbf{g}_k\mathbf{g}_i \quad \text{if } |i - k| \geq 2,$$

(2)
$$\mathbf{g}_i\mathbf{g}_{i+1}\mathbf{g}_i = \mathbf{g}_{i+1}\mathbf{g}_i\mathbf{g}_{i+1},$$

(3_q)
$$\mathbf{g}_i^2 = (q - 1)\mathbf{g}_i + q,$$

(4)
$$\mathbf{1} + \mathbf{g}_i + \mathbf{g}_{i+1} + \mathbf{g}_i\mathbf{g}_{i+1} + \mathbf{g}_{i+1}\mathbf{g}_i + \quad + \mathbf{g}_i\mathbf{g}_{i+1}\mathbf{g}_i = 0.$$

A second way in which q enters into the structure involves the trace. Recall that since $\mathbf{M}_n$ is type II_1 it supports a unique trace, and since $\mathbf{J}_n$ is a subalgebra it does too, by restriction. This trace is known as a *Markov trace*, i.e. it satisfies the important property:

$$(5_q) \qquad\qquad \mathrm{tr}(\mathbf{w}\mathbf{g}_n) = \tau(q)\,\mathrm{tr}(\mathbf{w}) \quad \text{if } \mathbf{w} \in \mathbf{J}_n,$$

where $\tau(q)$ is a fixed function of q. Thus, for each fixed value of q the trace is multiplied by a fixed scalar when one passes from one stage of the tower to the next, by multiplying an arbitrary element of $\mathbf{J}_n$ by the new generator $\mathbf{g}_n$ of $\mathbf{J}_{n+1}$.

Relations (1) and (2) above have an interesting geometric meaning, familiar to topologists. They are defining relations for the *n-string braid group*, $\mathbf{B}_n$, discovered by Emil Artin [Ar] in a foundational paper written in 1923. We pause to discuss braids.

An n-braid may be visualized by a weaving pattern of strings in 3-space which join n points on a horizontal plane to n corresponding points on a parallel plane, as illustrated in the example in Figure 2, where $n = 4$. In the case $n = 3$, the familiar braid in a person's hair gives another example. The strings are allowed to be stretched and deformed, the key features being that strings cannot pass through one-another and always proceed directly downward in their travels from the upper plane to the lower one. The equivalence class of weaving patterns under such deformations is an *n-braid*. One multiplies braids by concatenation and erasure of the middle plane. This multiplication makes them into a group, the *n*-string braid group $\mathbf{B}_n$. The identity is a braid which, when pulled taut, goes over to n straight lines. Generators are the $n - 1$ elementary braids which (by an abuse of notation) we continue to call $\mathbf{g}_1, \ldots, \mathbf{g}_{n-1}$. The pictures in Figure 3 show that relations (1) and (2) hold between the generators of $\mathbf{B}_n$. In fact, Artin proved they are *defining relations* for $\mathbf{B}_n$. Thus for each n there is a homomorphism from the n-string braid group $\mathbf{B}_n$ into the Jones algebra $\mathbf{J}_n(q)$, and from the group algebra $\mathbb{C}\mathbf{B}_n$ onto $\mathbf{J}_n(q)$.

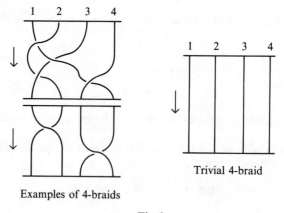

Examples of 4-braids

Trivial 4-braid

Fig. 2

Returning to the business at hand, i.e. the proof of the Index Theorem, Jones next shows that properties (1), (2), (3_q) and (5_q) suffice for the calculation of the

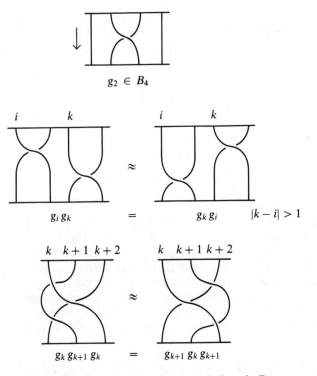

$g_2 \in B_4$

$g_i\, g_k$ $=$ $g_k\, g_i$ $\quad |k - i| > 1$

$g_k\, g_{k+1}\, g_k$ $=$ $g_{k+1}\, g_k\, g_{k+1}$

Fig. 3. Generators and defining relations in $\mathbf{B}_n$

trace of an arbitrary element $\mathbf{x} \in \mathbf{J}_n(q)$. It turns out that trace($\mathbf{x}$) is an *integer polynomial* in $(\sqrt{q})^{\pm 1}$. (We will meet it again in a few moments as the Jones polynomial associated to $\mathbf{x}$.) Jones proof of the Index Theorem is concluded when he shows that the infinite sequence of algebras $\mathbf{J}_n(q)$, with the given trace, could not exist if q did not satisfy the restrictions of the Index Theorem.

2. Knots and Links

We have already seen hints of topological meaning in $\mathbf{J}_n(q)$ via braids. There is more to come. Knots and links are obtained from braids by identifying the initial points and end points of a braid in a circle, as illustrated in Figure 4. It was proved by J. W. Alexander in 1928 that every knot or link arises in this way. Earlier we described an equivalence relation on weaving patterns which yields braids, and there is a similar (but less restrictive) equivalence relation on knots, i.e. a knot or link *type* is its equivalence class under isotopy in 3-space. Note that isotopy in 3-space which takes one closed braid representative of a link to another closed braid representative will pass through a sequence of representatives which are not closed braids in an obvious way. For example see the 2-component link which is illustrated in Figure 4. The left picture is an obvious closed braid representative, whereas the right is not.

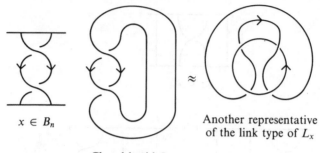

$x \in \mathbf{B}_n$

Closed braid L_x

Another representative
of the link type of L_x

Fig. 4. Braids determine links

Let $\mathbf{B}_\infty$ denote the disjoint union of all of the braid groups $\mathbf{B}_n$, $n = 1, 2, 3, \ldots$. In 1935 the mathematician A. A. Markov proposed the equivalence relation on $\mathbf{B}_\infty$ which corresponds to link equivalence [M]. Remarkably, the properties of the trace, or more particularly the facts that $\text{tr}(\mathbf{xy}) = \text{tr}(\mathbf{yx})$ together with property (5_q), were exactly what was needed to make the trace polynomial into an invariant on Markov's equivalence classes! Using Markov's proposed equivalence relation (which was proved to be the correct one in 1972 [Bi]), Jones proved, with almost no additional work beyond results already established in [Jo1], the following theorem:

Theorem [Jo3]. *If* $\mathbf{w} \in \mathbf{B}_\infty$, *then (after multiplication by an appropriate scalar, which depends upon the braid index n) the trace of the image of $\mathbf{w}$ in $\mathbf{J}_n(q)$ is a polynomial in $(\sqrt{q})^{\pm 1}$ which is an invariant of the link type defined by the closed braid $\mathbf{L_w}$.*

The invariant of Jones' theorem is the one-variable *Jones polynomial* $\mathbf{V}_x(q)$. Notice that the independent "variable" in this polynomial is essentially the index of a type II_1 subfactor in a type II_1 factor! It's discovery opened a new chapter in knot and link theory.

3. Statistical Mechanics

We promised to discuss other ways in which the work of Jones was related to yet other areas of mathematics and physics, and begin to do so now. As it turned out, when Jones did his work the family of algebras $\mathbf{J}_n(q)$ were already known to physicists who were concerned with *exactly solvable models* in Statistical Mechanics. (For an excellent introduction to this topic, see R. Baxter's article in these Proceedings.) One of the simplest examples in this area is known as the *Potts model*. In that model one considers an array of "atoms" arranged at the vertices of a planar lattice with m rows and n columns as in Figure 5. Each "atom" in the system has various possible spins associated to it, and in the simplest case, known as the Ising model, there are two choices, "+" for spin up and "−" for spin down. We have indicated one of the 2^{nm} choices in Figure 5, determining a

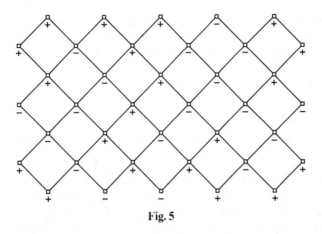

Fig. 5

state of the system. The goal is to compute the free energy of the system, averaged over all possible states.

Letting σ_i denote the spin at site i, we note that an edge e contributes an *energy* $E_e(\sigma_i, \sigma_j)$, where σ_i and σ_j are the states of the endpoints of e. Let E be the collection of lattice edges. Let β be a parameter which depends upon the temperature. Then the Gibbs *partition function* **Z** is given by the formula:

$$\mathbf{Z} = \sum_{\sigma_1,\ldots,\sigma_{mn}} \prod_{e \in E} \exp(-\beta E_e(\sigma_i, \sigma_j)).$$

All of this is microscopic, nevertheless the major macroscopic thermodynamic quantities are functions of the partition function. In particular, the free energy **F**, the object of interest to us at this time, is given by $\mathbf{Z} = \exp(-\beta \mathbf{F})$.

To compute the manner in which the atoms in one row of the lattice interact with atoms in the next, physicists set up the *transfer matrix* **T**, which expresses the row-to-row interactions. It turned out that, in order to be able to calculate the free energy, the transfer matrices must satisfy conditions known as the *Yang-Baxter equations* and (to the great surprise of everyone) they turned out to be the braid relations (1) and (2) in disguise! (Remark: before Jones' work, to the best of our knowledge, it was not known that the Yang-Baxter equation was related to braids or knots.) Even more, the algebra which the transfer matrices generate in the Ising model, known to physicists as the *Temperley-Lieb algebra*, is our algebra $\mathbf{J}_n(q)$. The partition function **Z** is related in a very simple way to the transfer matrix:

$$\mathbf{Z} = \text{trace}(\mathbf{T})^m.$$

In fact, it is closely related to the Jones trace.

The initial discovery of a relationship between the Potts model and links was reported on in [Jo3]. It opened a new chapter in the flow of ideas between mathematics and physics. We give an explicit example of a way in which the relationship of Jones' work to physics led to new insight into mathematics. Learning that the partition function was a sum over states of the system, Louis Kauffman was led to seek a decomposition of the Jones polynomial into a related sum over "states" of knot diagram, and arrived in [K1] at an elegant "states model" for the Jones polynomial. The Jones polynomial, and Kauffman's states model for it, were later seen to generalize to other polynomial invariants, with associated states models, for links in S^3 and eventually into invariants for 3-manifolds M^3 and links in 3-manifolds. The full story is not known at this writing, however we refer the reader to V. Turaev's article in these *Proceedings* for an excellent account of it, as of August 1990.

4. Quantum Groups and Representations of Lie Algebras

We begin by explaining the structure of the algebra $\mathbf{J}_n(q)$. It will be convenient to begin with another algebra $\mathbf{H}_n(q)$, which is generated by symbols $\mathbf{g}_1, \ldots, \mathbf{g}_{n-1}$ (which now have a third meaning), with defining relations (1), (2) and (3_q). The algebra $\mathbf{H}_n(q)$ is very well-known to mathematicians. It's the *Iwahori-Hecke algebra*, also known as the *Hecke algebra of the symmetric group* [Bo]. Its relationship with the symmetric group is simple to describe and beautiful. Notice that when $q = 1$, relation (3_q) simplifies to $(\mathbf{g}_k)^2 = 1$. One recognizes (1), (2) and (3_1) as defining relations for the group algebra $\mathbb{C}S_n$ of the symmetric group S_n. Here $\mathbf{g}_k$ is to be re-interpreted as a transposition which exchanges the symbols $\mathbf{k}$ and $\mathbf{k+1}$. In this way we may view $\mathbf{H}_n(q)$ as a "q-deformation" of the complex group algebra $\mathbb{C}S_n = \mathbf{H}_n(1)$.

The algebra $\mathbb{C}S_n$ is *rigid*, that is if one deforms it in this way its irreducible summands continue to be irreducible summands of the same dimension, in fact $\mathbf{H}_n(q)$ is actually algebra-isomorphic to $\mathbb{C}S_n$ for generic q. Thus $\mathbf{H}_n(q)$ is a direct sum of finite dimensional matrix algebras, its irreducible summands being in one-to-one correspondence with the irreducible representations of the symmetric group S_n. In this setting, Jones showed in [Jo2] that for generic q the algebra $\mathbf{J}_n(q)$ may be interpreted as the algebra associated to the q-deformations of those irreducible representations of S_n which have Young diagrams with at most two rows.

We now explain how $\mathbf{H}_n(q)$ is related to quantum groups. It will be helpful to recall the classical picture. The fundamental representation of the Lie group GL_n acts on $\mathbb{C}^n$, and so its k-fold tensor product acts naturally on $(\mathbb{C}^n)^{\otimes k}$. The symmetric group S_k also acts naturally on $(\mathbb{C}^n)^{\otimes k}$, permuting factors. (Remark: In this latter action, the representations of S_k which are relevant are those whose Young diagrams have $\leq n$ rows.) As is well known, the actions of GL_n and S_k are each other's commutants in the full group of linear transformations of $(\mathbb{C}^n)^{\otimes k}$. If one now replaces GL_n and $\mathbb{C}S_k$ by the quantum group $U_q(GL_n)$ and the Hecke algebra $\mathbf{H}_k(q)$ respectively, then the remarkable fact is that $U_q(GL_n)$ and $\mathbf{H}_k(q)$ are still each other's commutants [Ji]. The corresponding picture for $\mathbf{J}_n(q)$ is obtained by restricting to GL_2 and to representations of S_k having Young diagrams with at most 2 rows.

We remark that these are not isolated instances of algebraic accidents, but rather special cases of a phenomenon which relates a large part of the mathematics of quantum groups to finite dimensional matrix representations of the group algebra $\mathbb{C}B_n$ which support a Markov trace (e.g. see [BW]).

5. Dynkin Diagrams

Dynkin diagrams arise in the tower construction which we described in §1 via the inclusions of the algebras $J_n(q)$ in the Jones tower. The inclusions for the Jones tower are very simple, and correspond to the Dynkin diagram of type A_n. However, other, more complicated towers may be obtained by replacing the II_1 factor M_1 in the tower construction of §1 above by $M_1 \cap (M_0)'$, where $(M_0)'$ is the commutant of M_0 in M_1. We refer the reader to [GHJ] for an introduction to this topic and a discussion of the "derived tower" and the Dynkin diagrams which occur. The connections with the representations of simple Lie algebras can be guessed at from our discussion in §4 above.

6. Concluding Remarks

I hope I have succeeded in showing you some of the ways in which Jones' work created bridges between the areas of mathematics which were illustrated in Figure 1. To conclude, I want to indicate very briefly some of the ways in which those bridges have changed the mathematics which many of us are doing.

There is another link polynomial in the picture, the famous *Alexander polynomial*. It was discovered in 1928, and was of fundamental importance to knot theory, both in the classical case of knots in S^3 and in higher dimensional knotting. Shortly after Jones' 1984 discovery, it was learned that in fact both the Alexander and Jones polynomial were specializations of the 2-variable Jones polynomial. That discovery was made simultaneously by five separate groups of authors: Freyd and Yetter, Hoste, Lickorish and Millett, Ocneanu, and Przytycki and Traczyk, a simple version of the proof of Ocneanu being given in [Jo3]. One of the techniques used in the proof was the combinatorics of link diagrams, and that technique led to the discovery of yet another polynomial, by Louis Kauffman [K2].

From the point of view of algebra, the Jones polynomial comes from a trace function on $J_n(q)$, and the 2-variable Jones polynomial from a similar trace on the full Hecke algebra $H_n(q)$. Beyond that, there is another algebra, the so-called Birman-Wenzl algebra [BW], and Kauffman's polynomial is a trace on it. Even more, physicists who had worked with solutions to the Yang Baxter equation, realized that they knew of still other Markov traces, so they began to grind out still other polynomials, in initially bewildering confusion. That picture is fairly well understood at this moment, however the work of Witten [W] indicates there are still other, related, link invariants. The generalizations are vast, with much work to be done.

There is also a different and very direct way in which Jones has had equal influence. His style of working is informal, and one which encourages the free and open interchange of ideas. During the past few years Jones wrote letters to various people which described his important new discoveries at an early stage, when he did not yet feel ready to submit them for journal publication because he had much more work to do. He nevertheless asked that his letters be shared, and so they were widely circulated. It was not surprising that they then served as a rich source of ideas for the work of others. As it has turned out, there has been more than enough credit to go around. His openness and generosity in this regard have been in the best tradition and spirit of mathematics.

References

[Ar]   Artin, E.: Theorie der Zöpfe. Hamburg Abh. **4** (1925) 47–72
[Bi]   Birman, J.: Braids, links and mapping class groups. Ann. Math. Stud. Princeton
       Univ. Press, Princeton 1974, pp. 37–69
[Bo]   Bourbaki, N.: Groupes et algebres de Lie. Hermann, Paris 1968, Chapitre IV
[BW]   Birman, J., Wenzl, H.: Braids, link polynomials and a new algebra. Trans. AMS
       313 (1989) 249–273
[GHJ] Goodman, F., de la Harpe, P., Jones, V.: Coxeter graphs and towers of algebras.
       MSRI Publications, vol. 15. Springer, Berlin Heidelberg New York 1989
[Jo1] Jones, V.F.R.: Index for subfactors. Invent. math. **72** (1983) 1–25
[Jo2] Jones, V.F.R.: Braid groups, Hecke algebras and subfactors. In: Geometric methods
       in operator algebras. Pitman Research Notes in Mathematics, vol. 123, 1986,
       pp. 242–273
[Jo3] Jones, V.F.R.: Hecke algebra representations of braid groups and link polynomials.
       Ann. Math. **126** (1987) 335–388
[Ji]   Jimbo, M.: A q-analogue of $U(gl(N+1))$, Hecke algebra, and the Yang Baxter
       equation. Letters in Math. Physics **II** (1986) 247–252
[K1]   Kauffman, L.: A states model for the Jones polynomial. Topology **26** (1987)
       385–407
[K2]   Kauffman, L.: An invariant of regular isotopy. Trans. AMS (to appear)
[M]   Markov, A.A.: Über die freie Äquivalenz geschlossener Zöpfe. Recueil Mathema-
       tique Moscou **1** (1935) 73–78
[W]   Witten, E.: Quantum field theory and the Jones polynomial. In: Braid group, knot
       theory and statistical mechanics, eds. Yang and Ge. World Scientific Press, 1989

On the Work of Shigefumi Mori

Heisuke Hironaka

Department of Mathematics, Harvard University, Cambridge, MA 02138, USA

The most profound and exciting development in algebraic geometry during the last decade or so was the *Minimal Model Program* or *Mori's Program* in connection with the classification problems of algebraic varieties of dimension three. Shigefumi Mori initiated the program with a decisively new and powerful technique, guided the general research direction with some good collaborators along the way, and finally finished up the program by himself overcoming the last difficulty. The program was constructive and the end result was more than an existence theorem of minimal models. Even just the existence theorem by itself was the most fundamental result toward the classification of general algebraic varieties in dimension 3 up to birational transformations. The constructive nature of the program, moreover, provided a way of factoring a general birational transformation of threefolds into elementary transformations (divisorial contractions, flips and flops) that could be explicitly describable in principle. Mori's theorems on algebraic threefolds were stunning and beautiful by the totally new features unimaginable by those algebraic geometers who had been working, probably very hard too, only in the traditional world of algebraic or complex-analytic surfaces. Three in dimension was in fact a quantum jump from two in algebraic geometry.

Historically, to classify algebraic varieties has always been a fundamental problem of algebraic geometry and even an ultimate dream of algebraic geometers. During the early decades of this century, many new discoveries were made on the new features in classifying algebraic surfaces, unimaginable from the case of curves. They were mainly done by the so-called Italian school of algebraic geometers, such as Guido Castelnuovo (1865–1952), Federigo Enriques (1871–1946), Francesco Severi (1989–1961) and many others. Since then, there have been several important modernization, precision, reconstruction with rigor, extensions, in the theory of surfaces. The most notable among those were the works by Oscar Zariski during the late 1950s (especially, Castelnuovo's criterion for rational surfaces and minimal models of surfaces) and those by Kunihiko Kodaira during the early 1960s (especially, detailed study on elliptic surfaces and complex-analytic extensions, especially non-algebraic). In particular, Kodaira and his younger colleague S. Iitaka have produced many talented followers and collaborators. Up to and after the "Algebraic Surfaces" by I. R. Shafarevich, Yu. I. Manin, B. G. Moishezon, et al. (1965), the Russian school of algebraic geometers

Proceedings of the International Congress
of Mathematicians, Kyoto, Japan, 1990
© The Mathematical Society of Japan, 1991

advanced the study of some important algebraic surfaces and their deformations, especially the study of Torelli problems. As for the extensions to positive characteristics, outstanding and remarkable were the works of D. Mumford and E. Bombieri, and their followers during the late 1960s and 1970s. There have been so many other important contributors to the theory of surfaces that I would not try here to list all the important names and works in the theory of surfaces. There were also several unique works on classification problems of non-complete surfaces or complete surfaces with specified divisors, whose studies were instigated as being "2.5 dimensional" by S. Iitaka around 1977.

As for the higher dimensional algebraic varieties, the decade of 1970's saw three lines of new progress, just to name a few. One was the discovery, in the early 1970's, of the gap between the *rationality* and *unirationality*. The second was an attempt to classify *Fano 3-folds*, of which the birational classification was begun by Yu. I. Manin (1971), V. A. Iskovskih (1971, 1977–78), and followed by V. V. Shokurov (1979). However the biregular classification of Fano 3-folds in case $B_2 > 1$, with existence and number of moduli, had to wait a few more years to be achieved by Mori, jointly with S. Mukai, after a new powerful technique was invented by Mori in 1980. (The biregular classification in case $B_2 = 1$ was recently completed by S. Mukai using moduli of vector bundles on $K3$ surfaces.) The third line of progress was an early version of serious attempts toward *higher-dimensional classification* problems, largely inspired by S. Iitaka's bold conjectures proposed around 1970, and especially pressed hard during the latter half of the 1970's, by the Tokyo school of complex-algebraic geometers. The essence of their results was reported by H. Esnault in her Bourbaki talk, exp. 568 (1981).

Notably, K. Ueno produced some structure theorems in higher dimensions during 1977–79, which appeared at that time to boost the program of Iitaka et al. However, the scope of their achievements were limited toward the original goal of classifying general algebraic 3-folds in a style of extending the beautiful old theory of classifying surfaces. Their crucial limitation was the lack of a good higher-dimensional analog to Zariski's theory of minimal models. Technically, as it later became clear, the drawback was the absence of Mori's stunning success in analyzing birational transformations from the point of view of extremal contractions in dimensions higher than 2. In short, a new insight had to be injected into the classification program and it came from the technique of extremal rays inaugurated by Mori which inspired M. Reid, J. Kollár, Y. Kawamata et al. to begin working in earnest in search of minimal models around the turn of the decade.

Early in 1979, Mori brought to algebraic geometry a completely new excitement, that was his proof of *Hartshorne's conjecture*, proposed in 1970, which said that the projective spaces are the only smooth complete algebraic varieties with ample tangent bundles. It was also exciting news to differential geometers, such as Y.-T. Siu and S.-T. Yau who subsequently found an independent proof for Frankel's conjecture of 1961, which was implied by Hartshorne's. It is not clear that Hartshorne's conjecture can actually be proven by the differential-geometric method of Siu and Yau. In all approaches, the most important step was to show the existence of *rational curves* in the manifold in question. Mori's idea was simple and natural as good ones were always so, while the proof was not. The idea was that, under some numerical conditions, rational curves should be obtained by deforming a given curve inside the manifold and degenerating it into a bunch of curves of lower genera. But the difficulty was proving such plentifulness of deformations. There, Mori's ingenuity was to overcome this difficulty first in the cases of *positive characteristics*, where the Frobenius maps did a miracle, and then deduce the complex case from them.

Mori extended and reformulated his new and powerful technique of finding rational curves, which was referred to as *extremal rays* in the *cone of curves* by himself in another monumental paper of his on "Threefolds whose canonical bundles are not nef" (*nef* = numerically effective). The cone of curves in a projective 3-fold was defined to be the closure of the real convex hull generated by the numerical equivalence classes (or classes in the second homology group over the real numbers) of irreducible algebraic curves, and an extremal ray was to be an edge of the cone within the region where the canonical divisor takes negative intersection number. Mori's discovery was that the cone was locally finitely generated in the canonically negative region and that each extremal ray was generated by a *rational curve*. In the above "not nef" paper, moreover, Mori established an outstanding theorem which, in dimension three, completely classified the geometric structures of the family of rational curves corresponding to an extremal ray. The union of such curves was either an irreducible divisor (either a smooth projective line bundle, or a projective plane, or a quadric in a projective 3-space) or the entire 3-fold. The former led to a birational blowing down while the latter to a fibration map (either conic bundle, or Del Pezzo fiber space or Fano 3-fold). In each case, the description was precise and the target variety (blown-down birational model or the base of fibration) was again *projective*. This result was absolutely stunning to anybody who have had experiences with non-projective or even non-algebraic examples of birational transformations, and it was so beautiful as to encourage many algebraic geometers once again to look into the birational problems in dimension 3. Subsequently, Mori completed the *classification of Fano 3-folds*, jointly with S. Mukai in 1981, and he established a *criterion for uniruledness*, jointly with Y. Miyaoka in 1985. The Fano 3-folds and uniruled 3-folds were those to be investigated separately in all details and there are still left open some important problems about those special 3-folds. Excluding these, the most important general problem was to prove the *existence of minimal models* in which the canonical bundles were nef.

Mori's "not nef" paper showed, for the first time in the history of algebraic geometry, that a general birational transformation between smooth algebraic 3-folds was not completely untouchable and in fact "finitely manageable" in some sense. Naturally the dimension 3 was far more complicated than the dimension 2, and moreover some sort of singularities (seemingly finitely classifiable after Mori) were created in the process of factoring it into elementary transformations. Excited with Mori's discoveries, several algebraic geometers began clustering to him, this time with much clearer vision and better hope than ever, to work on factorization of birational transformations, and more importantly to work on defining what *minimal models* should be and how to obtain them, notably M.

Reid, V. V. Shokurov, Y. Kawamata, J. Kollár and of course S. Mori himself. *Mori's Program*, named by J. Kollár and meant to imply the process of obtaining minimal models, was as follows: Start from any projective smooth variety X, which is not birationally uniruled, and find a finite succession of "elementary" birational transformations by which X is transformed to a "minimal model". Of course, the central questions here were the meanings of "elementary transformation" and "minimal model". Firstly, the inverse of a blowing-up with a smooth center in a smooth 3-fold (even in a 3-fold with a "mildly" singular point) was definitely elementary. It was called a *divisorial contraction*. This certaily decreased the Néron-Severi number and hence it should make the variety closer to its minimal model. Suppose we could not make a divisorial contraction any more and the Néron-Severi number reached its minimum. Is the variety then good enough to be called a minimal model? The answer was clearly *no*, although the singularities created by those divisorial contractions were quite acceptable. Unlike the 2-dimensional cases, the resulting 3-fold can still have *extremal rays* which cause unpleasant behaviors of the canonical and pluricanonical bundles. The importance of these bundles had been clearly recognized even from the time of Castelnuovo and Enriques in the theory of classification problems as well as in the theory of deformations. Thus, to *eliminate an extremal ray*, a completely new type of elementary transformation was needed. This was the one later called *flip*, which was, roughly speaking, to take out extremal (canonically negative, or with negative intersection number with the canonical divisor) rational curves and put some rational curves back in with a new imbedding type so that new curves are canonically positive. Such a surgery type operation is unique if it exists. The *existence*, however, turned out to be extremely delicate and hard to prove.

As early as in 1981, encouraged by Mori's "not nef" paper, Miles Reid made fairly clear the idea of what minimal models should be and what elementary transformations would be, by publishing a paper on "Minimal models of canonical 3-folds", which was an expansion of his lecture "Canonical 3-folds" in the Journées de Géométrie Algébrique d'Angers 1980. It then became absolutely clear (mildly suggested by Mori's work before) that some special kind of singularities must be permitted in the good notion of minimal models. M. Reid introduced the notions of *canonical singularity* and *terminal singularity*, which turned out to be very useful in the minimal model program of which he was the first to conjecture in literature. (See Miles Reid "Decomposition of Toric morphisms" 1983.) The former was, as the name suggests, the kind of singularity that appeared in the canonical models and could be quite horrible algebraically and geometrically. The latter was the kind that behaved much better than the former and which people hoped to be the only kind to appear in the minimal models. In any case, Reid's definition was simple and clear and some basic theorems were proved by himself, in which he showed that terminal singularities were closely related to the deformations of the classic singularities of Du Val.

Historically, P. Du Val in 1934 systematically studied the singularities that did not affect the condition of adjunction, that is, in the language of Reid, the canonical singularities in dimension 2. In 1966, M. Artin extended and modernized the classification of what he called *rational singularities*. Du Val singularities were exactly rational singularities of multiplicity 2, or *rational double points*. After M. Reid's introduction of canonical and terminal singularities, rational singularities and their deformations were studied once again. Extensive and direct studies on terminal singularities followed M. Reid's works and by 1987 a complete classification of terminal singularities in dimension 3 was achieved with technically useful lemmas by a combination of works by several mathematicians, V. I. Danilov, D. Morrison, G. Stevens, S. Mori, J. Kollár, N. Shepherd-Barron, and others. Some experimental works were done about higher dimensional terminal singularities, such as the one in dimension 4 by S. Mori, D. Morrison and I. Morrison which seemed to indicate far more complexity than the case of dimension 3.

Back to the minimal model problem, Mori's technique of *extremal ray contraction* had to be generalized to singular varieties, at least to the varieties with only terminal singularities, which were even "rationally factorial". It does not look completely hopeless that eventually Mori's all characteristics method can be modified and extended to such singular cases. (See J. Kollár's work that is to appear in the proceedings of the Algebraic Geometry Satellite Conference at Tokyo Metropolitan University, 1990.) However it then was not done in Mori's way. Instead, the generalization was obtained first by generalizing Kodaira's Vanishing Theorem and then by making an ingenious use of this generalization. Here, and subsequently, the contributions of Y. Kawamata were big to the minimal model program.

The celebrated *Vanishing Theorem* was proven by K. Kodaira in 1953. Numerous generalizations, in special cases or in general, were done especially during the 1970's and 80's mostly in an effort to investigate the nicety of the structure of the canonical ring of a projective variety. Here the *canonical ring* means the graded algebra generated by sections of pluricanonical bundles (say, on a desingularized model). As for the nicety, they looked for the property of *being finitely generated* of the canonical algebra or "rationality or better" properties of the singularities of the canonical model. Y. Kawamata had a very clear objective in generalizing and using the Vanishing Theorem, that was to generalize Mori's theory of extremal ray contractions and then verify the minimal model program.

The operation called *flip* is clearly directed, i.e., it changes "canonically negative" to "canonically positive". There is a similar operation for a rational curve with zero intersection number with the canonical divisor, which is called a *flop*. Unlike flips, the inverse of flops are again flops. A flop is symmetric in this sense. A flip changes a variety into another, birational to and better than the original, better in singularities, better in terms of pluricanonical bundles, and so on. Historically, there had been known many examples of flops. The simplest flop, between smooth 3-folds, was known and used even by earlier algebraic geometers. Some flops were shown to be useful in studying degeneration of $K3$ surfaces by V. Kulikov in 1977. In contrast, examples of flips are not so easy to find because 3-folds had to be singular in order to have singularities improved. P. Francia gave an explicit example in his paper published in 1980. Other examples were seen in a paper of M. Reid published in 1983, here with prototypes of minimal models. In 1983, V. V. Shokurov published the *Non-vanishing Theorem*, which was proven by using the Vanishing Theorem. His theorem implied that flips cannot be done infinitely many times.

For Mori's Program, therefore, not only the divisorial contractions are finite but also the flips are finite in any sequence. The very final problem was hence to prove the *existence* of flips for given extremal rational curves, after all the previous works by S. Mori, M. Reid, J. Kollár, V. V. Shokurov and Y. Kawamata, just to name some of the most important contributors to the Program. At any rate, Y. Kawamata reduced the problem of the flip to the existence of a "nice" doubly anticanonical divisor *globally* in a neighborhood of a given extremal rational curve, while the existence had been only proven by M. Reid *locally* about each point of the curve. This seemingly small gap between *global* and *local* was actually enormous. Mori finally overcame this gap by checking cases after cases with very delicate and intricate investigations and established the final existence theorem of algebraic flips by reducing them to sequences of simpler analytic flips and smaller contractions. This monumental paper of Mori was published in the very first issue of the new Journal of the American Mathematical Society, establishing a constructive existence theorem of minimal models which had been shown to have many important consequences.

We need much more of Mori's originality to break the stubborn prejudice that it is a Herculean task to extend the classical classification theorems to all dimensions.

References

1. *Algebraic Manifolds with Ample Tangent Bundles*

[1-1] (with H. Sumihiro) On Hartshorne's conjecture. J. Math. Kyoto Univ. **18** (1978) 223–238
[1-2] Projective manifolds with ample tangent bundles. Ann. Math. **110** (1979) 593–606
[1-3] Hartshorne's conjecture and extremal ray (in Japanese). Sugaku **35**, no. 3 (1983) 193–209 [English translation: Hartshorne conjecture and extremal ray. Sugaku Expositions. Amer. Math. Soc. **1**, no. 1 (1988) 15–37]

2. *Threefolds with Non-nef Canonical Bundles*

[2-1] Threefolds whose canonical bundles are not numerically effective. Proc. Nat. Acad. Sci. USA **77** (1980) 3125–3126
[2-2] Threefolds whose canonical bundles are not numerically effective Ann. Math. **116** (1982) 133–176
[2-3] Threefolds whose canonical bundles are not numerically effective. In: Algebraic Threefolds. Proc. of C. I. M. E. Session, Varenna, 1981 (A. Conte, ed.). Lecture Notes in Mathematics, vol. 947. Springer, Berlin Heidelberg New York 1982, pp. 155–189

3. *Classification of Fano 3-folds*

[3-1] (with S. Mukai) Classification of Fano 3-folds with $B_2 \geq 2$. Manuscripta Math. **36** (1981) 147–162
[3-2] (with S. Mukai) On Fano 3-folds with $B_2 \geq 2$. In: Algebraic Varieties and Analytic Varieties (S. Iitaka, ed.). Advanced Stud. in Pure Math., vol. 1. Kinokuniya and North-Holland, 1983, pp. 101–129

[3-3] Cone of curves and Fano 3-folds. Proc. of ICM'82 in Warsaw 1983, pp. 127–132

[3-4] (with S. Mukai) Classification of Fano 3-folds with $B_2 \geq 2$, I. In: Algebraic and Topological Theories – to the memory of Dr. Takehiko Miyata (M. Nagata et al., eds.). Kinokuniya, Tokyo, 1985, pp. 496–545

4. Characterization of Uniruledness

[4-1] (with Y. Miyaoka) A numerical criterion of uniruledness. Ann. Math. **124** (1986) 65–69

5. Completion of the Minimal Model Program, or Mori's Program

[5-1] On 3-dimensional terminal singularities. Nagoya Math. J. **98** (1985) 43–66

[5-2] Classification of higher dimensional varieties. In: Algebraic Geometry Bowdoin 1985 (S. J. Bloch, ed.), Proc. Symp. in Pure Math., vol. 46, part 1, 1987, pp. 269–331

[5-3] Flip theorem and the existence of minimal models for 3-folds. J. Amer. Math. Soc. **1** (1988) 117–253

[5-4] (with C. Clemens and J. Kollár) Higher Dimensional Complex Geometry, A Summer Seminar at the Univ. of Utah, Salt Lake City, 1987, Astérisque **166**, Soc. Math. France (1988)

[5-5] (announcement of joint work with J. Kollár) Birational classification of algebraic 3-folds. In: Algebraic Analysis, Geometry, and Number Theory, Proc. of JAMI Inaugural Conf. (J.-I. Igusa, ed.), Supplement to Amer. J. Math., Johns Hopkins Univ. Press, 1989, pp. 307–311

On the Work of E. Witten

Ludwig D. Faddeev

Steklov Mathematical Institute, Leningrad 191011, USSR

It is a duty of the chairman of the Fields Medal Committee to appoint the speakers, who describe the work of the winners at this session. Professor M. Atiyah was asked by me to speak about Witten. He told me that he would not be able to come but was ready to prepare a written address. So it was decided that I shall make an exposition of his address adding my own comments. The full text of Atiyah's address is published separately.

Let me begin by the statement that Witten's award is in the field of Mathematical Physics.

Physics was always a source of stimulus and inspiration for Mathematics so that Mathematical Physics is a legitimate part of Mathematics. In classical time its connection with Pure Mathematics was mostly via Analysis, in particular through Partial Differential Equations. However quantum era gradually brought a new life. Now Algebra, Geometry and Topology, Complex Analysis and Algebraic Geometry enter naturally into Mathematical Physics and get new insights from it. And[1]

In all this large and exciting field, which involves many of the leading physicists and mathematicians in the world, Edward Witten stands out clearly as the most influential and dominating figure. Although he is definitely a physicist (as his list of publications clearly shows) his command of mathematics is rivalled by few mathematicians, and his ability to interpret physical ideas in mathematical form is quite unique. Time and again he has surprised the mathematical community by a brilliant application of physical insight leading to new and deep mathematical theorems.

Now I come to description of the main achievements of Witten. In Atiyah's text many references are given to Feynman Integral, so that I begin with a short and rather schematic reminding of this object.

In quantum physics the exact answers for dynamical problems can be expressed in a formal way as follows:

$$Z = \int e^{iA} \prod_x d\mu$$

where A is an action functional of local fields – functions of time and space variables x, running through some manifold M. The integration measure is a

---

[1] Small print type here and after refers to Atiyah's text.

Proceedings of the International Congress
of Mathematicians, Kyoto, Japan, 1990
© The Mathematical Society of Japan, 1991

product of local measures for values of fields in a point x over all M. The result of integration Z could be a number or function of parameters defining the problem – coupling constants, boundary or asymptotical conditions, etc.

In spite of being an ill-defined object from the point of view of rigorous mathematics, Feynman functional integral proved to be a powerful tool in quantum physics. It was gradually realized that it is also a very convenient mathematical means. Indeed the geometrical objects such as loops, connections, metrics are natural candidates for local fields and geometry produces for them interesting action functionals. The Feynman integral then leads to important geometrical or topological invariants.

Although this point of view was expressed and exemplified by several people (e.g., A. Schvarz used 1-forms ω on a three dimensional manifold with action $A = \int \omega d\omega$ to describe the Ray-Singer torsion) it was Witten who elaborated this idea to a full extent and showed the flexibility and universality of Feynman integral.

Now let me follow Atiyah in description of the main achievements of Witten in this direction.

1. Morse Theory

His paper [2] on supersymmetry and Morse theory is obligatory reading for geometers interested in understanding modern quantum field theory. It also contains a brilliant proof of the classic Morse inequalities, relating critical points to homology. The main point is that homology is defined via Hodge's harmonic forms and critical points enter via stationary phase approximation to quantum mechanics. Witten explains that "supersymmetric quantum mechanis" is just Hodge-de Rham theory. The real aim of the paper is however to prepare the ground for supersymmetric quantum field theory as the Hodge-de Rham theory of infinite-dimensional manifolds. It is a measure of Witten's mastery of the field that he has been able to make intelligent and skilful use of this difficult point of view in much of his subsequent work.

Even the purely classical part of this paper has been very influential and has led to new results in parallel fields, such as complex analysis and number theory.

2. Index Theorem

One of Witten's best known ideas is that the index theorem for the Dirac operator on compact manifolds should emerge by a formally exact functional integral on the loop space. This idea (very much in the spirit of his Morse theory paper) stimulated an extensive development by Alvarez-Gaumé, Getzler, Bismut and others which amply justified Witten's view-point.

3. Rigidity Theorems

Witten [7] produced an infinite sequence of such equations which arise naturally in the physics of string theories, for which the Feynman path integral provides a heuristic explanation of rigidity. As usual Witten's work, which was very precise and detailed in its formal aspects, stimulated great activity in this area, culminating in rigorous proofs of these new rigidity theorems by Bott and Taubes [1]. A noteworthy aspect of these proofs is that they involve elliptic function theory and deal with the infinite sequence of operators simultaneously rather than term by term. This is entirely natural from Witten's view-point, based on the Feynman integral.

4. Knots

Witten has shown that the Jones invariants of knots can be interpreted as Feynman integrals for a 3-dimensional gauge theory [11]. As Lagrangian, Witten uses the Chern-Simons function, which is well-known in this subject but had previously been used as an addition to the standard Yang-Mills Lagrangian. Witten's theory is a major breakthrough, since it is the only intrinsically 3-dimensional interpretation of the Jones invariants: all previous definitions employ a presentation of a knot by a plane diagram or by a braid.

Although the Feynman integral is at present only a heuristic tool it does lead, in this case, to a rigorous development from the Hamiltonian point of view. Moreover, Witten's approach immediately shows how to extend the Jones theory from knots in the 3-sphere to knots in arbitrary 3-manifolds. This generalization (which includes as a specially interesting case the empty knot) had previously eluded all other efforts, and Witten's formulas have now been taken as a basis for a rigorous algorithmic definition, on general 3-manifolds, by Reshetikin and Turaev.

Now I turn to another beautiful result of Witten – proof of positivity of energy in Einstein's Theory of Gravitation.

Hamiltonian approach to this theory proposed by Dirac in the beginning of the fifties and developed further by many people has led to a natural definition of energy. In this approach a metric γ and external curvature h on a space-like initial surface $S^{(3)}$ embedded in space-time $M^{(4)}$ are used as parameters in the corresponding phase space. These data are not independent. They satisfy Gauss-Codazzi constraints – highly nonlinear PDE. The energy H in the asymptotically flat case is given as an integral of indefinite quadratic form of $\nabla\gamma$ and h. Thus it is not manifestly positive. The important statement that it is nevertheless positive may be proved only by taking into account the constraints – a formidable problem solved by Yau and Schoen in the late seventies and as Atiyah mentions, "leading in part to Yau's Fields Medal at the Warsaw Congress".

Witten proposed an alternative expression for energy in terms of solution of a linear PDE with the coefficients expressed through γ and h. This equation is

$$\mathscr{D}^{(3)}\psi = 0$$

where $\mathscr{D}^{(3)}$ is the Dirac operator induced on $S^{(3)}$ by the full Dirac operator on $M^{(4)}$. Witten's formula somewhat schematically can be written as follows:

$$H(\psi_0, \psi_0) = \int (|\nabla\psi|^2 + \psi^* G\psi) dS$$

where ψ_0 is the asymptotic boundary value for ψ and G is proportional to the Einstein tensor $R_{ik} - \frac{1}{2}g_{ik}R$. Due to the equation of motion $G = T$, where T is the energy-momentum tensor of matter and thus manifestly positive. So the positivity of H follows.

This unexpected and simple proof shows another ability of Witten – to solve a concrete difficult problem by specific elegant means.

On the Work of Edward Witten

Michael Atiyah

Trinity College, Cambridge CB2 1TQ, England

1. General

The past decade has seen a remarkable renaissance in the interaction between mathematics and physics. This has been mainly due to the increasingly sophisticated mathematical models employed by elementary particle physicists, and the consequent need to use the appropriate mathematical machinery. In particular, because of the strongly non-linear nature of the theories involved, topological ideas and methods have played a prominent part.

The mathematical community has benefited from this interaction in two ways. First, and more conventionally, mathematicians have been spurred into learning some of the relevant physics and collaborating with colleagues in theoretical physics. Second, and more surprisingly, many of the ideas emanating from physics have led to significant new insights in purely mathematical problems, and remarkable discoveries have been made in consequence. The main input from physics has come from quantum field theory. While the analytical foundations of quantum field theory have been intensively studied by mathematicians for many years the new stimulus has involved the more formal (algebraic, geometric, topological) aspects.

In all this large and exciting field, which involves many of the leading physicists and mathematicians in the world, Edward Witten stands out clearly as the most influential and dominating figure. Although he is definitely a physicist (as his list of publications clearly shows) his command of mathematics is rivalled by few mathematicians, and his ability to interpret physical ideas in mathematical form is quite unique. Time and again he has surprised the mathematical community by a brilliant application of physical insight leading to new and deep mathematical theorems.

Witten's output is remarkable both for its quantity and quality. His list of over 120 publications indicates the scope of his research and it should be noted that many of these papers are substantial works indeed.

In what follows I shall ignore the bulk of his publications, which deal with specifically physical topics. This will give a very one-sided view of his contribution, but it is the side which is relevant for the Fields Medal. Witten's standing as a physicist is for others to assess.

Proceedings of the International Congress
of Mathematicians, Kyoto, Japan, 1990
© The Mathematical Society of Japan, 1991

Let me begin by trying to describe some of Witten's more influential ideas and papers before moving on to describe three specific mathematical achievements.

2. Influential Papers

His paper [2] on supersymmetry and Morse theory is obligatory reading for geometers interested in understanding modern quantum field theory. It also contains a brilliant proof of the classic Morse inequalities, relating critical points to homology. The main point is that homology is defined via Hodge's harmonic forms and critical points enter via stationary phase approximation to quantum mechanics. Witten explains that "supersymmetric quantum mechanics" is just Hodge-de Rham theory. The real aim of the paper is however to prepare the ground for supersymmetric quantum field theory as the Hodge-de Rham theory of infinite-dimensional manifolds. It is a measure of Witten's mastery of the field that he has been able to make intelligent and skilful use of this difficult point of view in much of his subsequent work.

Even the purely classical part of this paper has been very influential and has led to new results in parallel fields, such as complex analysis and number theory.

Many of Witten's papers deal with the topic of "Anomalies". This refers to classical symmetries or conservation laws which are violated at the quantum level. Their investigation is of fundamental importance for physical models and the mathematical aspects are also extremely interesting. The topic has been extensively written about (mainly by physicists) but Witten's contributions have been deep and incisive. For example, he pointed out and investigated "global" anomalies [3], which cannot be studied in the traditional perturbative manner. He also made the important observation that the η-invariant of Dirac operators (introduced by Atiyah, Patodi and Singer) is related to the adiabatic limit of a certain anomaly [4]. This was subsequently given a rigorous proof by Bismut and Freed.

One of Witten's best known ideas is that the index theorem for the Dirac operator on compact manifolds should emerge by a formally exact functional integral on the loop space. This idea (very much in the spirit of his Morse theory paper) stimulated an extensive development by Alvarez-Gaumé, Getzler, Bismut and others which amply justified Witten's view-point.

Also concerned with the Direc operator is a beautiful joint paper with Vafa [5] which is remarkable for the fact that it produces sharp uniform bounds for eigenvalues by an essentially topological argument. For the Dirac operator on an odd-dimensional compact manifold, coupled to a background gauge potential, Witten and Vafa prove that there is a constant C (depending on the metric, but independent of the potential) such that *every interval of length C contains an eigenvalue*. This is not true for Laplace operators or in even dimensions, and is a very refined and unusual result.

3. The Positive Mass Conjecture

In General Relativity the positive mass conjecture asserts that (under appropriate hypotheses) the total energy of a gravitating system is positive and can only be zero for flat Minkowski space. It implies that Minkowski space is a stable ground state. The conjecture has attracted much attention over the years and was established in various special cases before being finally proved by Schoen and Yau in 1979. The proof involved non-linear P. D. E. through the use of minimal surfaces and was a major achievement (leading in part to Yau's Fields Medal at the Warsaw Congress). It was therefore a considerable surprise when Witten outlined in [6] a much simpler proof of the positive mass conjecture based on linear P. D. E. Specifically Witten introduced spinors and studied the Dirac operator. His approach had its origin in some earlier ideas of supergravity and it is typical of Witten's insight and technical skill that he eventually emerged with a simple and quite classical proof. Witten's paper stimulated both mathematicians and physicists in various directions, demonstrating the fruitfulness of his ideas.

4. Rigidity Theorems

The space of solutions of an elliptic differential equation on a compact manifold is naturally acted on by any group of symmetries of the equation. All representations of compact connected Lie groups occur this way. However, for very special equations, these representations are trivial. Notably this happens for the spaces of harmonic forms, since these represent cohomology (which is homotopy invariant). A less obvious case arises from harmonic spinors (solutions of the Dirac equation), although the relevant space here is the "index" (virtual difference of solutions of D and D^*). This was proved by Atiyah and Hirzebruch in 1970. Witten raised the question whether such "rigidity theorems" might be true for other equations of interest in mathematical physics, notably the Rarita-Schwinger equation. This stimulated Landweber and Stong to investigate the question topologically and eventually Witten [7] produced an infinite sequence of such equations which arise naturally in the physics of string theories, for which the Feynman path integral provides a heuristic explanation of rigidity. As usual Witten's work, which was very precise and detailed in its formal aspects, stimulated great activity in this area, culminating in rigorous proofs of these new rigidity theorems by Bott and Taubes [1]. A noteworthy aspect of these proofs is that they involve elliptic function theory and deal with the infinite sequence of operators simultaneously rather than term by term. This is entirely natural from Witten's view-point, based on the Feynman integral.

5. Topological Quantum Field Theories

One of the remarkable aspects of the Geometry/Physics interaction of recent years has been the impact of quantum field theory on low-dimensional geometry (of 2, 3 and 4 dimensions). Witten has systematized this whole area by showing that there are, in these dimensions, interesting *topological* quantum field theories [8], [9], [10]. These theories have all the formal structure of quantum field theories but they are purely topological and have no dynamics (i.e. the Hamiltonian is zero). Typically the Hilbert spaces are finite-dimensional and various traces give well-defined invariants. For example, the Donaldson theory in 4 dimensions fits into this framework, showing how rich such structures can be.

A more recent example, and in some ways a more surprising one, is the theory of Vaughan Jones related to knot invariants, which has just been reported on by Joan Birman. Witten has shown that the Jones invariants of knots can be interpreted as Feynman integrals for a 3-dimensional gauge theory [11]. As Lagrangian, Witten uses the Chern-Simons function, which is well-known in this subject but had previously been used as an addition to the standard Yang-Mills Lagrangian. Witten's theory is a major breakthrough, since it is the only intrinsically 3-dimensional interpretation of the Jones invariants: all previous definitions employ a presentation of a knot by a plane diagram or by a braid.

Although the Feynman integral is at present only a heuristic tool it does lead, in this case, to a rigorous development from the Hamiltonian point of view. Moreover, Witten's approach immediately shows how to extend the Jones theory from knots in the 3-sphere to knots in arbitrary 3-manifolds. This generalization (which includes as a specially interesting case the empty knot) had previously eluded all other efforts, and Witten's formulas have now been taken as a basis for a rigorous algorithmic definition, on general 3-manifolds, by Reshetikin and Turaev.

Moreover, Witten's approach is extremely powerful and flexible, suggesting a number of important generalizations of the theory which are currently being studied and may prove to be important.

One of the most exciting recent developments in theoretical physics in the past year has been the theory of 2-dimensional quantum gravity. Remarkably this theory appears to have close relations with the topological quantum field theories that have been developed by Witten [12]. Detailed reports on these recent ideas will probably be presented by various speakers at this congress.

6. Conclusion

From this very brief summary of Witten's achievements it should be clear that he has made a profound impact on contemporary mathematics. In his hands physics is once again providing a rich source of inspiration and insight in mathematics. Of course physical insight does not always lead to immediately rigorous mathematical proofs but it frequently leads one in the right direction, and technically correct proofs can then hopefully be found. This is the case with Witten's work. So far his insight has never let him down and rigorous proofs, of the standard we mathematicians rightly expect, have always been forthcoming. There is therefore no doubt that contributions to mathematics of this order are fully worthy of a Fields Medal.

References

1. R. Bott and C. H. Taubes: On the rigidity theorems of Witten. J. Amer. Math. Soc. **2** (1989) 137
2. E. Witten: Supersymmetry and Morse theory. J. Diff. Geom. **17** (1984) 661
3. E. Witten: An $SU(2)$ anomaly. Phys. Lett. **117 B** (1982) 324
4. E. Witten: Global anomalies in string theory. Proc. Argonne-Chicago Symposium on Geometry, Anomalies and Topology (1985)
5. E. Witten and C. Vafa: Eigenvalue inequalities for Fermions in gauge theories. Comm. Math. Phys. **95** (1984) 257
6. E. Witten: A new proof of the positive energy theorem. Comm. Math. Phys. **80** (1981) 381
7. E. Witten: Elliptic genera and quantum field theory. Comm. Math. Phys. **109** (1987) 525
8. E. Witten: Topological quantum field theory. Comm. Math. Phys. **117** (1988) 353
9. E. Witten: Topological gravity. Phys. Lett. B **206** (1988) 601
10. E. Witten: Topological sigma models. Comm. Math. Phys. **118** (1988) 411
11. E. Witten: Quantum field theory and the Jones polynomial. Comm. Math. Phys. **121** (1989) 351
12. E. Witten: On the structure of the topological phase of two dimensional gravity. Nuclear Phys. B **340** (1990) 281

The Work of A. A. Razborov

László Lovász

Department of Computer Science, Eötvös University, H-1088 Budapest, Hungary
and
Department of Computer Science, Princeton University, Princeton, NJ 08544, USA

Perhaps the most difficult and deepest field in computer science is the derivation of lower bounds for the computational complexity of various problems (i.e., proving that no algorithm whatsoever can solve the given problem within a certain time or space). While these questions are very natural extensions of classical problems on undecidability, they turn out much more difficult. Up to some years ago, very few results were known, and these gave very poor bounds and concerned very restricted models of computation. At the same time, some of the most exciting unsolved problems in theoretical computer science (like $\mathscr{P} = \mathscr{N}\mathscr{P}$) are in this area; to prove those, one would need lower bounds on the running time which is exponential in the size of the input.

One conclusion drawn from the situation was that the methods of mathematical logic used in decidability theory are not powerful enough to yield negative results in complexity theory, and that combinatorial methods would be needed to obtain such results.

The idea of using combinatorial methods is suggested by the *Boolean circuit* model of computation. In this model, the computation is described by an acyclic directed graph, whose nodes (also called *gates*) correspond to elementary steps in the computation. Nodes with no edge entering them are called *input gates*, and the input of the computation is an assignment of a Boolean value (0 or 1) to each input gate. Every other gate computes some simple (Boolean) function of the values computed by the the tails of edges entering it; this recursively defines a value of each gate. For simplicity, we shall assume that there is only one gate with no edge leaving; this serves as the *output gate*.

Every algorithm in other possible models (Turing machine, RAM machine etc.) can be transformed into a Boolean circuit (more exactly, into a series of Boolean circuits, one for each input size). The number of gates in the circuit corresponds to the running time of the algorithm. Another important parameter of the circuit is its *depth*, i.e., the maximum length of a path from an input node to an output node. This corresponds to the running time of the algorithm if parallel processing is allowed; in this case, the *width* (the maximum number of gates mutually inaccessible by paths) corresponds to the number of processors needed for this degree of parallelization.

The two simplest kinds of gates are AND gates and OR gates, computing the logical conjunction and disjunction of Boolean values, respectively. A Boolean circuit consisting of such gates can compute only monotone functions. If we

Proceedings of the International Congress
of Mathematicians, Kyoto, Japan, 1990
© The Mathematical Society of Japan, 1991

allow, however, that each input node receives either a variable or its negation, then every Boolean function can be computed using only AND and OR gates.

The famous $\mathcal{P} \neq \mathcal{NP}$ problem would follow if one could show the following. For every $n \geq 2$, let D_n be a Boolean circuit with AND and OR gates, whose input gates correspond to pairs (i, j) with $1 \leq i < j \leq n$. Every input to such a circuit, i.e., every assignment of values 0 or 1 to the input gates, corresponds to a graph on $\{1, \ldots, n\}$. Assume that D_n computes the value 1 if and only if this graph has a clique larger than $\sqrt{n}$ (or has a Hamiltonian circuit, or is 3-colorable; we could use here any $\mathcal{NP}$-complete property). Conjecture: *the number of gates in D_n is necessarily superpolynomial (perhaps exponential) in n.*

While we are still far from being able to prove this, in the last few years combinatorial methods for proving good (exponential) lower bounds on the circuit complexity of various Boolean functions have been developed. These methods still impose restrictions on the computation model, but these restrictions are quite natural and much weaker than those considered previously. One class of results is for bounded-depth (or, more generally, small-depth) circuits. This development is the result of contributions from many authors (Furst–Saxe–Sipser 1981, Ajtai (1983), Yao (1985), Hastad (1986), Razborov (1987)). The other (mathematically perhaps even more difficult) class of results concerns monotone circuits; this restriction of the model was introduced by Valiant, and the breakthrough is due to Razborov (1985a, 1986). We start with describing this fundamental result.

While every monotone Boolean function can be computed by a monotone Boolean circuit, it can be expected that monotone circuits are less powerful than general circuits. So (as suggested by Valiant), exponential (or at least superpolynomial) lower bounds on the monotone circuit complexity of monotone problems in $\mathcal{NP}$ might be easier to prove, and at the same time they may eventually lead to exponential lower bounds for general circuits.

Razborov (1985a) justified part of this expectation by proving the following fundamental result.

Theorem. *For deciding the existence of a clique with given size in a graph, no polynomial size monotone circuits exist.*

This basic result and its powerful proof method have immediately inspired much further work. Andreev (1985) used similar techniques to obtain an exponential lower bound on a less natural $\mathcal{NP}$-complete problem. Alon and Boppana (1987), by strengthening the combinatorial arguments of Razborov, proved an exponential (not merely superpolynomial) lower bound on the monotone circuit complexity of the clique problem.

Razborov (1985b) proved that even the bipartite matching problem (decide whether a bipartite graph has a perfect matching, which is a special case of the clique problem) needs superpolynomial-size monotone circuits. This problem is solvable in polynomial time if we do not require monotonicity of the circuit; therefore this result separates monotone and non-monotone polynomial time. Such separation results are extremely rare. Tardos [T] used Razborov's methods to prove the existence of a polynomial time computable monotone Boolean function which takes an exponential number of gates if we want to compute it by a monotone Boolean circuit.

The proof technique of Razborov is also very important: he introduces a very powerful new method, namely approximation. The method applies in many situations, but to fill in the details, difficult special considerations are needed. We shall sketch the method below; it is, however, best described through another of its applications to the complexity of bounded-depth circuits.

Every Boolean function can be computed by a circuit with depth 2 (this follows from the existence of the "conjunctive normal form"), but circuits with small depth will typically have exponential size even for quite simple Boolean functions. The parity function (or XOR function) of n bits is defined as the sum of these bits modulo 2. It was proved by Hastad (1986) that *a circuit with* AND *and* OR *gates, with n input bits and depth d, computing the parity function of these bits, has at least* $\exp\left(0.1n^{1/(d-1)}\right)$ *gates.*

Razborov (1985b) showed that the approximation method can also be applied to prove lower bounds on the size of non-monotone but small-depth circuits. Among the really simple functions of n bits, besides AND, OR and XOR, the majority function comes to mind: the value of this function is 1 if at least half of the input bits are 1, and 0 otherwise. Razborov proved an exponential lower bound for the size of small-depth circuits computing the majority function. An interesting feature of the result is that he can even allow XOR gates, computing the mod 2 sum of arbitrarily many bits in a single step. (This shows the complementary nature of this result and Hastad's.) To be exact, Razborov's important result is the following:

Theorem. *A circuit with* AND, OR *and* XOR *gates, with n input bits and depth d, computing the majority function of these bits, has at least* $\exp\left(0.1n^{(1/2d)}\right)$ *gates.*

This application of the approximation method is in fact much simpler, and it is a "textbook" example of a lower-bound proof combining algebraic, probabilistic, and combinatorial ideas.

To sketch the general idea, assume that we want to prove an exponential lower bound on the number of gates in a Boolean circuit computing some Boolean function f. One introduces a notion of "distance" between Boolean functions and specifies a class of "simple" Boolean functions so that the distance of the function f from any "simple" function is exponentially large. Furthermore, one defines gates which "approximate" the original gates in the sense that if we replace an original gate by the approximating gate, the new circuit computes a Boolean function which is at a distance at most 1 from the original. These approximating gates must have the property that if all the original gates are replaced by their approximations, the new circuit will compute a "simple" function. So this new function is exponentially far from the original function f, and hence, we must have changed exponentially many gates. In particular, the circuit must have exponentially many gates.

In the proof of the lower bound for small-depth circuits, "simple" functions will be those expressible as polynomials over GF(2) with small degree. We measure the distance of two functions in terms of the number of inputs on which they differ. An elegant argument shows that the majority function is "far" from every polynomial with low degree. Approximations are only needed for the AND gates; this is achieved by a nice randomized construction.

For proving his lower bound for the monotone complexity of the k-clique problem, Razborov has to use much more involved constructions, and the proofs are based on difficult combinatorial methods, in particular on the powerful method of "sunflowers" from hypergraph theory.

In a recent paper (1989), he analyses the approximation method and in a rather novel manner, is able to establish its boundaries. (This to me shows that this is indeed a "method" and not just a language to present various arguments.) He shows among others that for Boolean circuits without any restriction, this method can yield at most quadratic lower bounds (while, as we have seen, both for monotone and bounded-depth circuits it yields exponential lower bounds).

Razborov's work (of which here we discussed only one cluster of results) was received with great interest and discussed enthusiastically at seminars all around the world. In an area where any step forward seemed almost hopeless (but which was at the same time a central area of theoretical computer science) his results meant that deep methods could be developed and to obtain strong lower bounds for algorithms was not impossible.

References

Ajtai, M.: Σ_1^1-formulae on finite structure. Ann. Pure Appl. Logic **24** (1983) 1–48

Alon, N., Boppana, R.B.: The monotone circuit complexity of Boolean functions. Combinatorica **7** (1987) 1–22

Andreev, A.E.: On a method for obtaining lower bounds for the complexity of individual monotone functions. Dokl. Akad. Nauk SSSR **282** (1985) 1033–1037 (Russian) [English transl.: Soviet Math. Dokl. **31**, 530–534]

Furst, M., Saxe, J., Sipser, M.: Parity, circuits, and the polynomial time hierarchy. Proc. 22nd IEEE Symp. on Found. of Comp. Sci., 1981, pp. 260–270

Hastad, J.: Almost optimal lower bounds for small depth circuits. Proc. 18th ACM Symp. on Theory of Comp., 1986, pp. 6–20

Razborov, A.A.: Lower bounds on the monotone circuit complexity of some Boolean functions. Dokl. Akad. Nauk SSSR **281** (1985a) 798–801 (Russian) [English transl.: Soviet Math. Dokl. **31**, 354–357]

Razborov, A.A.: A lower bound on the monotone network complexity of the logical permanent. Math. Zametki **37**, (1985b) 887–900 (Russian) [English transl.: Math. Notes of the Acad. Sci. USSR **37**(6) 485–493]

Razborov, A.A.: Lower bounds for the monotone complexity of Boolean functions. Proc. of the Internatl. Congress of Math., Berkeley, Calif., 1986, pp. 1478–1487

Razborov, A.A.: Lower bounds on the size of bounded depth network over a complete basis with logical addition. Math. Zametki **41**, (1987) 598–607 (Russian) [English transl.: Math. Notes of the Acad. Sci. USSR **41**(4) 333–338]

Razborov, A.A.: On the method of approximations. Proc. 21st Annual ACM Symp. on Theory of Computing, 1989, pp. 167–176

Tardos, É.: The gap between monotone and non-monotone circuit complexity is exponential. Combinatorica **8** (1988) 141–142

Yao, A.C.-C.: Separating the polynomial-time hierarchy by oracles. Proc. 26th IEEE Symp. on Found. of Comp. Sci., 1985, pp. 1–10

Invited One-Hour Addresses
at the Plenary Sessions

Algebraic K-Theory, Motives, and Algebraic Cycles

Spencer Bloch

Department of Mathematics, University of Chicago, Chicago, IL 60637, USA

The concept of motive was introduced by A. Grothendieck. Roughly, one wants to think of smooth projective varieties over a field k as objects in an additive category $\mathscr{V}(k)$ with morphisms $X \to Y$ being correspondences (i.e., suitable equivalence classes of algebraic cycles on $X \times Y$.) The category of motives, $\mathscr{M}(k)$, is a $\mathbb{Q}$-linear abelian (or weaker, Karoubian) category equipped with a tensor product, an invertible Tate twist functor $? \otimes \mathbb{Q}(1) : \mathscr{M}(k) \to \mathscr{M}(k)$, and a cohomology functor $H^* : \mathscr{V}(k) \to \mathscr{M}(k)$ which is universal for cohomology functors like Betti, Hodge, or étale cohomology. Standard conjectures about algebraic cycles imply $\mathscr{M}(\bar{k})$ is semi-simple, but these conjectures have proved quite intractable.

Inspired by work of Deligne on mixed Hodge structures, Beilinson started to study a more general abelian category of mixed motives $\mathscr{M}\mathscr{M}(k)$. The smallest full abelian subcategory of $\mathscr{M}\mathscr{M}(k)$ containing the Tate motives $\mathbb{Q}(n)$, $n \in \mathbb{Z}$, and closed under extensions, would be the category $\mathscr{M}\mathscr{T}\mathscr{M}(k)$ of mixed Tate motives. Objects in this category would have a canonical weight filtration with $\mathbb{Q}(n)$ of weight $-2n$, and the functor associating to an object M the corresponding $\bigoplus(\mathrm{gr}^W_{-2n} M)(-n)$ should be a fibre functor making $\mathscr{M}\mathscr{T}\mathscr{M}(k)$ a neutral Tannakian category. In other words, automorphisms of the fibre functor compatible with tensor product form an algebraic group $\mathrm{Gal}(\mathscr{M}\mathscr{T}\mathscr{M}(k))$ which, because it must preserve the weight filtration, is of the form $\mathscr{N} \rtimes \mathbb{G}_m$ with $\mathscr{N}$ nilpotent. The category $\mathscr{M}\mathscr{T}\mathscr{M}(k)$ is equivalent to the category of representations over $\mathbb{Q}$ of $\mathscr{N} \times \mathbb{G}_m$. The graded Lie algebra

$$\mathscr{L}(k) \overset{\mathrm{def}}{=} \mathrm{Lie}(\mathscr{N}), \quad \mathscr{L}(k) = \mathscr{L}_{-1} \oplus \mathscr{L}_{-2} \oplus \cdots \tag{0.1}$$

is a fundamental invariant of k. Beilinson further stipulated that $\mathscr{M}\mathscr{T}\mathscr{M}(k)$ should be related to the Quillen K-theory of the field k via the basic formula

$$\mathrm{Ext}^i_{\mathscr{M}\mathscr{T}\mathscr{M}(k)}(\mathbb{Q}, \mathbb{Q}(n)) \cong \mathrm{gr}^n_\gamma K_{2n-i}(k) \otimes \mathbb{Q} \tag{0.2}$$

where the grading is with respect to the γ-filtration on $K_*(k)$.

It seems likely this expected relation between motives and K-theory has a natural interpretation in terms of algebraic cycles. Although technical problems involving moving lemmas currently block any definitive results, I want to show that cycles can be used to define reasonable candidates for motivic cohomology and also for the Lie algebra associated to $\mathscr{M}\mathscr{T}\mathscr{M}(k)$. The basic cycle-theoretic objects, the higher Chow groups $\mathrm{CH}^r(X, n)$ associated to a variety X and non-negative integers

Proceedings of the International Congress
of Mathematicians, Kyoto, Japan, 1990
© The Mathematical Society of Japan, 1991

n and r, are integrally defined (i.e. not $\mathbb{Q}$-vector spaces). In §3 I exploit this to show how, in an arithmetic situation, these groups give natural generalizations of the global points of an abelian variety over a number field and of the corresponding group III. One is led to a general Birch-Swinnerton-Dyer conjecture for motives. This conjecture has been proved by K. Kato [BK] for the case of the Riemann zeta function.

Everything I know about mixed motives I learned from Beilinson and Deligne. I am grateful to M. Hanamura for many helpful conversations about these ideas. A. Suslin pointed out an error in the proof of the moving lemma in [Bl1]. I have another proof, but it is too complicated to inspire confidence at this point, so I prefer to formulate the relation between K-theory and higher Chow groups conditionally. Results about a spectral sequence relating higher Chow groups and K-theory are joint work with S. Lichtenbaum. Finally, the arithmetic conjectures in the last section were inspired by joint work with K. Kato on Tamagawa numbers of motives [BK]. Over the years, he has graciously shared credit with me for any number of his ideas, including a number of those discussed in §3.

1. Higher Chow Groups

Let X be a variety over a field k. A cosimplicial scheme is built from affine spaces $\Delta_n = \operatorname{Spec}(k[t_0, \ldots, t_n]/(\sum t_i - 1))$ with face maps $\Delta_{n-1} \hookrightarrow \Delta_n$ defined by $t_i = 0$ and degeneracies $t_i \mapsto t_i + t_{i+1}$. For any $r \geq 0$ and any algebraic k-scheme X, a simplicial abelian group $\mathscr{Z}^r(X, \cdot)$ is defined by taking $\mathscr{Z}^r(X, n)$ to be the free abelian group generated by irreducible subvarieties $V \subset X \times \Delta_n$ of codimension r meeting all faces of codimension ≥ 1 properly. The boundary maps $\partial_i : \mathscr{Z}^r(X, n) \to \mathscr{Z}^r(X, n-1)$ are given by restriction to the i-th face. One defines the higher Chow groups

$$\operatorname{CH}^r(X, n) \overset{\text{def}}{=} \pi_n(\mathscr{Z}^r(X, \cdot)) \cong H_n(\mathscr{Z}^r(X, \cdot)). \tag{1.1}$$

In particular, one checks easily $\operatorname{CH}^r(X, 0)$ is the classical Chow group, and $\operatorname{CH}^1(X, 1) \cong \Gamma(X, \mathcal{O}_X^*)$ for X regular.

Conjecture (1.2) (Moving Lemma for Cycles). *Let $U \subset X$ be Zariski open. Write $\mathscr{Z}^r(U, \cdot)' = \operatorname{Image}[\mathscr{Z}^r(X, \cdot) \to \mathscr{Z}^r(U, \cdot)]$. Then for all r, X, and U, the inclusion $\mathscr{Z}^r(U, \cdot)' \hookrightarrow \mathscr{Z}^r(U, \cdot)$ is a quasi-isomorphism.*

Conjecture (1.2) would immediately yield a long exact localization sequence for higher Chow groups. With more work, one gets:

Theorem (1.3) [Bl1]. *Assume the moving lemma for cycles holds. Then there are isomorphisms*

$$\operatorname{CH}^r(X, n) \otimes \mathbb{Q} \cong \operatorname{gr}^r_\gamma K'_n(X) \otimes \mathbb{Q}, \tag{1.3.1}$$

where K' is the K-theory of coherent sheaves, and gr_γ is the γ-grading.

Let T be a smooth k-variety. Let $D_1, \ldots, D_m$ be smooth divisors on T, and assume all intersections of the D_i are transversal. The multi-relative K-theory $K_*(T; D_1, \ldots, D_m)$ is defined via a homotopy fibre construction. One has long exact sequences

$$K_*(T; D_1, \ldots, D_m) \to K_*(T; D_1, \ldots, D_{m-1}) \to K_*(D_m; D_1 \cap D_m, \ldots, D_{m-1} \cap D_m)$$

$$\to K_{*-1}(T; D_1, \ldots, D_m) \to \cdots$$

$$\to K_0(D_m; D_1 \cap D_m, \ldots, D_{m-1} \cap D_m).$$

For $W \subset T$ closed, one defines $K_{*W}(T; D_1, \ldots, D_m)$ again by a homotopy fibre so there are long exact sequences

$$K_{*W}(T; D_1, \ldots, D_m) \to K_*(T; D_1, \ldots, D_m) \to K_*(T - W; D_1, \ldots, D_m)$$

$$\to K_{*-1W}(T; D_1, \ldots, D_m) \to \cdots \to K_0(T - W; D_1, \ldots, D_m).$$

Conjecture (1.4) (Moving Lemma for Relative K_0). *Let X be a smooth k-variety. Write $\delta_j = X \times \Delta_{n-1} \hookrightarrow X \times \Delta_n$ for the codimension 1 faces. Let $\mathscr{V}_i \subset X \times \Delta_n$ be the union of all algebraic subsets of dimension i meeting all faces $\bigcap \delta_j$ of codimension ≥ 1 properly. Let $W \subset \mathscr{V}_i$ be an algebraic set of dimension $\leq i - 1$. Then the map*

$$K_{0, W - \mathscr{V}_{i-1} \cap W}(X \times \Delta_n - \mathscr{V}_{i-1}; \delta_0, \ldots, \delta_n) \to K_{0, \mathscr{V}_i - \mathscr{V}_{i-1}}(X \times \Delta_n - \mathscr{V}_{i-1}; \delta_0, \ldots, \delta_n)$$

is zero.

Intuitively, this conjecture says a class in K_0 supported on $W - W \cap \mathscr{V}_{i-1}$ can be moved on $\mathscr{V}_i$ to a class supported on $\mathscr{V}_{i-1}$.

Theorem 1.5 (with S. Lichtenbaum). *Let X be a smooth k-variety, and assume the moving lemma for relative K_0 holds for X and all n, $i \geq 0$. Then there is a spectral sequence*

$$E_2^{p,q} = \mathrm{CH}^{-q}(X, -p - q) \Rightarrow K_{-p-q}(X). \tag{1.5.1}$$

This should be compared with the Atiyah-Hirzebruch spectral sequence in topological K-theory. Unlike the localization spectral sequence in algebraic K-theory, (1.5.1) may be non-trivial for $X = \mathrm{Spec}(k)$. Viewing $\mathscr{L}^r(X, \cdot)$ as a complex in negative degrees, one may conjecture the complex

$$\Gamma(r)^{\cdot} \stackrel{\mathrm{def}}{=} \mathscr{L}^r(X, *)[-2r] \tag{1.6}$$

satisfies the axioms formulated by Beilinson and Lichtenbaum. The spectral sequence can be rewritten

$$E_2^{p, -q} = H^{p+q}(\Gamma(q)) \Rightarrow K_{q-p}(X). \tag{1.7}$$

2. Motives

The philosophy is something like this. Associated to a smooth variety X there should be a Tannakian category $\mathscr{MTM}(X)$ of mixed Tate motives, with a fibre functor $M \mapsto H^0(\bigoplus \mathrm{gr}_{2n}^W R_{\mathrm{Betti}}(M)(n))$, where $R_{\mathrm{Betti}}(M)$ is a Tate variation of Hodge structure on X, the Betti realization of M. There should be a space $\mathrm{Mot}(X)$ with $\pi_1(\mathrm{Mot}(X))^\wedge \cong \mathrm{Gal}(\mathscr{MTM}(X))$ (where "$\wedge$" means nilpotent completion) so mixed

Tate motives (e.g. the $\mathbb{Q}(n)$) can be thought of as $\mathbb{Q}$-local systems on $\mathrm{Mot}(X)$. Indeed, one might hope for natural $\mathbb{Z}$-local systems $\mathbb{Z}(n)$. Moreover one expects isomorphisms in the derived category

$$\Gamma(r)^{\cdot} \cong R\Gamma(\mathrm{Mot}(X), \mathbb{Z}(r)), \tag{2.1}$$

and hence

$$\mathrm{CH}^r(X, p) \cong H^{2r-p}(\mathrm{Mot}(X), \mathbb{Z}(r)). \tag{2.2}$$

Let $\mathrm{Grass}_{X,n+p,n}$ denote the Grassmannian of rank n quotients of $\mathcal{O}_X^{\oplus n+p}$. By analogy with the work of Dwyer and Friedlander [DF], one might hope to identify the K-theory $K_*(X)$ with the homotopy of the space of sections

$$\pi_* \left[\varinjlim_n \varinjlim_p \mathrm{Hom}(\mathrm{Mot}(X), \mathrm{Mot}(\mathrm{Grass}_{X,n+p,n})) \right].$$

In [op. cit.] the l-adic étale homotopy groups of the fibre are calculated:

$$\pi_i \left[\varinjlim \varinjlim \mathrm{fibre}(\mathrm{Grass}_{X,n+p,n,\text{ét}} \to X_{\text{ét}}) \right] \cong \begin{cases} 0 & i = 2j+1 \\ \mathbb{Z}_l(j) & i = 2j. \end{cases} \tag{2.3}$$

The homotopy groups of the fibre $\varinjlim \varinjlim \mathrm{Mot}(\mathrm{Grass}_{X,n+p,n}) \to \mathrm{Mot}(X)$, viewed as local systems on the base should, by the same token, be 0 in odd degrees and $\mathbb{Z}(j)$ in degree $2j$. The usual spectral sequence in topology converging to the homotopy groups of the space of sections of a fibration

$$E_2^{p,-q} = H^p(\mathrm{Base}, \pi_q(\mathrm{Fibre})) \Rightarrow \pi_{q-p}(\mathrm{Space\ of\ Sections}) \tag{2.4}$$

would give

$$E_2^{p,-q} = H^p(\mathrm{Mot}(X), \mathbb{Z}(q/2))) \Rightarrow K_{q-p}(X) \tag{2.5}$$

where by convention $\mathbb{Z}(q/2) = (0)$ for q odd. Note that all even differentials d_{2n} are necessarily 0. Using this it is straightforward to check that the spectral sequences (1.7) and (2.5) are at least potentially in agreement up to renumbering (décalage).

Of course, all this is completely speculative. One can, however, make a start at understanding the space $\mathrm{Mot}(X)$. For $X = \mathrm{Spec}(F)$ a field, one expects $\mathrm{Mot}(X) = \mathrm{Mot}(F)$ to be a $K(\pi, 1)$ with $\pi^\wedge = \mathrm{Gal}(\mathcal{MTM}(F))$. (Note this should not be the general case. Indeed, for X a flag variety over F, $\mathcal{MTM}(X)$ should equal $\mathcal{MTM}(F)$, but the motivic cohomology groups

$$\mathrm{gr}_\gamma^n K_{2n-i}(X) \otimes \mathbb{Q} \neq \mathrm{gr}_\gamma^n K_{2n-i}(F) \otimes \mathbb{Q}.) \tag{2.6}$$

Beilinson and Deligne have made progress in writing down what should be the Lie algebra $\mathcal{L}(F)$ (0.1). I want to propose a candidate $\mathcal{L}_{\text{cycle}}(F)$ for $\mathcal{L}(F)$ constructed using algebraic cycles.

Let F be an algebraically closed field. Write $X = \mathbb{P}_F^1 - \{1\}$, $X^s = X \times \cdots \times X$ (s times). Faces on X^s are obtained by setting various coordinate functions $= 0$ or ∞. The symmetric group $\mathcal{S}_s$ acts on X^s. Let $\mathcal{V}^r(s)$ denote the group of codimension r algebraic cycles with $\mathbb{Q}$-coefficients on X^s which meet all faces properly and which are alternating with respect to the action of $\mathcal{S}_s$. Define

$$\partial = \sum (-1)^{i-1}(\partial_i^0 - \partial_i^\infty), \tag{2.7}$$

where ∂_i^j is the restriction map on cycles obtained by setting the i-th coordinate function $= j$. The map ∂ preserves alternating representations so we have $\partial : \mathscr{V}^r(s) \to \mathscr{V}^r(s-1)$.

Define a graded group $\mathscr{N}^i = \bigoplus_r \mathscr{V}^r(2r-i)$ and a complex $\mathscr{N}^{\bullet}$ which is $\mathscr{N}^i$ in degree i with boundary $\partial : \mathscr{N}^i \to \mathscr{N}^{i+1}$ induced by ∂ on $\mathscr{V}^r(s)$. One has a product structure $\mathscr{V}^r(s) \times \mathscr{V}^t(u) \to \mathscr{V}^{r+t}(s+u)$ given by obvious (external) product of cycles composed with projection on the alternating representation. This induces a graded product structure $\mathscr{N}^i \times \mathscr{N}^j \to \mathscr{N}^{i+j}$. With this product, $\mathscr{N}^{\bullet}$ is a DGA (i.e. associative and graded commutative with the usual identities for ∂. Note $\mathscr{N}^i \neq (0)$ for $i < 0$.)

Define $\mathscr{N}^{0+} = \mathrm{Ker}(\mathscr{N}^0 \to \mathscr{V}^0(0))$. Let $J = [\bigoplus_{i \leq -1} \mathscr{N}^i \oplus \mathscr{N}^{0+}] \cdot \mathscr{N}^{\bullet}$. The ideal J is graded but not stable under ∂. Let $\mathscr{J}$ be the differential graded ideal generated by J. $\mathscr{J}$ is generated as an ideal by $\bigoplus_{i \leq -1} \mathscr{N}^i \oplus \mathscr{N}^{0+} \oplus \partial \mathscr{N}^{0+}$ because $\mathscr{V}^0(p) = (0)$ for $p \neq 0$ so $\partial \mathscr{N}^{-1} \subset \mathscr{N}^{0+}$. One shows for F algebraically closed and $i > 0$,

$$\bigwedge^i (\mathscr{N}/\mathscr{J})^1 \subset (\mathscr{N}/\mathscr{J})^i. \tag{2.8}$$

Writing

$$\mathscr{R} = \{x \in (\mathscr{N}/\mathscr{J})^1 \mid \partial x \in \bigwedge^2 (\mathscr{N}/\mathscr{J})^1\}, \qquad \mathscr{R} = \mathscr{R}_1 \oplus \mathscr{R}_2 \oplus \cdots \tag{2.9}$$

one proves

$$\partial \mathscr{R} \subset \bigwedge^2 \mathscr{R} \subset \bigwedge^2 (\mathscr{N}/\mathscr{J})^1. \tag{2.10}$$

As a consequence, one has a complex

$$\mathscr{R} \xrightarrow{\delta} \bigwedge^2 \mathscr{R} \xrightarrow{\delta} \bigwedge^3 \mathscr{R} \to \cdots. \tag{2.11}$$

Define $\mathscr{L}_{\mathrm{cycle}}(F)$ to be the pro-vector space dual to $\mathscr{R}$, i.e. write $\mathscr{R} = \varinjlim \mathscr{R}_\alpha$ with $\mathscr{R}_\alpha$ finite dimensional and take $\mathscr{L}_{\mathrm{cycle},\alpha} = \mathrm{Hom}(\mathscr{R}_{\mathrm{cycle},\alpha}, \mathbb{Q})$, $\mathscr{L}_{\mathrm{cycle}}(F) = \varprojlim \mathscr{L}_{\mathrm{cycle},\alpha}$. The graded pro-Lie algebra $\mathscr{L}_{\mathrm{cycle}}(F)$ is zero in non-negative degrees.

A (mixed Tate) motive on $\mathrm{Spec}(F)$ is a finite dimensional graded $\mathbb{Q}$-vector space V with a representation (compatible with gradings) $\mathscr{L}(F) \to \mathrm{End}(V)$. This leads to

Conjecture (2.12). *A (mixed Tate) motive on $\mathrm{Spec}(F)$ is an algebraic cycle with coefficients in $\mathrm{End}(V)$, $Z \in \mathscr{R} \otimes \mathrm{End}(V)$ which is of degree 0 for the tensor product grading, and which satisfies $\delta Z = [Z, Z]$.*

By way of example, I want to describe the polylogarithm motives as cycles. (This discussion is modeled on Deligne's interpretation of a conjecture of Zagier on the higher K-theory of number fields [D1].) Write $\mathscr{R} = \bigoplus \mathscr{R}_r$ with $\mathscr{R}_r \subset \mathscr{V}^r(2r-1)/$ (relations). One checks that

$$\mathscr{R}_1 = \mathscr{V}^1(1)/(\text{relations}) \cong F^* \otimes \mathbb{Q}. \tag{2.13}$$

Define a map

$$\varrho_r : \mathbb{Z}[F - \{0, 1\}] \to \mathscr{R}_r \tag{2.14}$$

as follows: take

$$\varrho_1(a) = (1 - a) \in \mathscr{R}_1 \cong F^* \otimes \mathbb{Q}. \tag{2.15}$$

For $r \geq 2$, let $T_0, \ldots, T_{r-1}$ be homogeneous coordinates on $\mathbb{P}^{r-1}$. For $a \in F - \{0, 1\}$, let $V_a \subset X^{2r-1}$ be the subvariety described parametrically by

$$V_a = \{(aT_{r-2}/T_{r-1}, \ldots, aT_0/T_{r-1}, (T_0 - T_1)/T_0, (T_1 - T_2)/T_1, \ldots,$$
$$(T_{r-2} - T_{r-1})/T_{r-2}, (T_{r-1} - aT_0)/T_{r-1})\}. \tag{2.16}$$

One checks that V_a meets faces properly. Define

$$\varrho_r(a) = \text{alt}(V_a) \tag{2.17}$$

where alt is the projection on the alternating representation of $\mathscr{S}_{2r-1}$. (The cycle $\varrho_2(a)$ was found by Totaro [T].) One verifies for $r \geq 2$

$$\partial \varrho_r(a) = -(a) \cdot \varrho_{r-1}(a) = \varrho_{r-1}(a) \cdot (a). \tag{2.18}$$

Define $\mathscr{R}_{\log, r} = \text{Image}(\varrho_r)$. It follows that $\mathscr{R}_{\log} \subset \mathscr{R}$ is a graded sub-co-Lie algebra, with dual graded Lie algebra $\mathscr{L}_{\log}$ a quotient of $\mathscr{L}_{\text{cycle}}$.

Let $N \geq 2$ be an integer. The N-polylog motive, M_N, should be viewed as an object in $\mathscr{M}\mathscr{T}\mathscr{M}(\mathbb{P}^1 - \{0, 1, \infty\})$. We consider its specialization $M_{N,z}$ at some F-point $z \in \mathbb{P}^1 - \{0, 1, \infty\}$. Define (following Deligne [op. cit.]) $(N + 1) \times (N + 1)$ matrices (viewed as acting on $\mathbb{Q}^{[0,N]}$):

$$e_0 = \begin{bmatrix} 0 & & & & 0 \\ 0 & 0 & & & \\ 0 & 1 & 0 & & \\ 0 & 0 & 1 & 0 & \\ & & & \ddots & \\ 0 & & & 1 & 0 \end{bmatrix} \qquad e_1 = \begin{bmatrix} 0 & & & 0 \\ 1 & 0 & & \\ & & \ddots & \\ & 0 & & 0 \end{bmatrix}$$

Then $M_{N,z} \in \mathscr{R}_{\log} \otimes \text{End}(\mathbb{Q}^{[0,N]})$ should be associated to an algebraic cycle Z_z. Taking

$$Z_z = -(z) \otimes e_0 - \sum_{r=1}^{N} \varrho_r(z) \otimes \text{ad}(e_0)^{r-1}(e_1) \tag{2.19}$$

yields

$$\partial Z_z = [Z_z, Z_z] \tag{2.20}$$

so the resulting map $\mathscr{L}_{\text{cycle}} \twoheadrightarrow \mathscr{L}_{\log} \to \text{End}(\mathbb{Q}^{[0,N]})$ is a representation.

Of course much work remains to be done, for example, to spread out the construction to define $\mathscr{M}\mathscr{T}\mathscr{M}(T)$ for a variety T over a field. It seems reasonable to expect a construction analogous to (2.12), applied to sums of cycles of codimension r_i on $T \times X^{2r_i-1}$ which are equidimensional over T, should yield objects in $\mathscr{M}\mathscr{T}\mathscr{M}(T)$.

3. Arithmetic

As suggested above, assuming the moving Conjecture (1.2), the higher Chow groups of a variety over a field can be thought of as giving an integral structure for motivic cohomology. In [BK], we define a Tamagawa number for motives, and we use this to give an integral version of the Beilinson conjectures relating values of a regulator map on rational motivic cohomology with values of *L*-functions. Our formulation only involved motivic cohomology $\otimes \mathbb{Q}$, but availability of integral motivic cohomology suggests a natural interpretation for the group III appearing in the Tamagawa volume formula. I will outline these ideas briefly. In [op. cit.], the local constructions are completely proven and do not depend on any conjectures. This forces us to work with an integral model of our variety. In what follows I will indicate (following the brief discussion at the end of §5 of op. cit.) what is needed to formulate a conjecture depending only on the variety over $\mathbb{Q}$. For a general variety *X*, this formulation involves a number of other "standard" conjectures about étale cohomology, *L*-functions, and *K*-theory. However, for varieties like curves and rational surfaces the necessary conditions can be verified directly.

Let *X* be a smooth projective variety over $\mathbb{Q}$. We think of $\mathrm{CH}^r(X, 2r - n)$ "=" $H^n(\mathrm{Mot}(X), \mathbb{Z}(r))$. It is suggestive (although I do not know how precise) to think of the motive $H^n(X, \mathbb{Z}(r))$ as the "local system" on $\mathrm{Mot}(\mathrm{Spec}(\mathbb{Q}))$ given by $H^n(\mathrm{Fibre}[\mathrm{Mot}(X) \to \mathrm{Mot}(\mathrm{Spec}(\mathbb{Q}))], \mathbb{Z})(r)$. Assuming $2r > n$, this local system would have "weights" < 0 and hence no non-zero sections, so one would get

$$\mathrm{CH}^r(X, 2r - n) \to \mathrm{Ext}^1_{\mathrm{Mot}(\mathrm{Spec}(\mathbb{Q}))}(\mathbb{Z}, H^{n-1}(X, \mathbb{Z}(r))). \tag{3.1}$$

In the case $2r = n$,

$$\mathrm{CH}^r(X)_0 \overset{\mathrm{def}}{=} \mathrm{Ker}[\mathrm{CH}^r(X) \to \mathrm{Hom}_{\mathrm{Mot}(\mathrm{Spec}(\mathbb{Q}))}(\mathbb{Z}, H^{2r}(X, \mathbb{Z}(r)))]$$

$$\to \mathrm{Ext}^1_{\mathrm{Mot}(\mathrm{Spec}(\mathbb{Q}))}(\mathbb{Z}, H^{2r-1}(X, \mathbb{Z}(r))). \tag{3.2}$$

One hopes that (3.1) and (3.2) are isomorphisms.

Next one wants to understand $\mathrm{Ext}^1_{\mathrm{Mot}(\mathrm{Spec}(\mathbb{Q}_p))}(\mathbb{Z}, H^{n-1}(X, \mathbb{Z}(r)))$ for $p \leq \infty$. We assume henceforth that $n - 2r \leq -1$, so the motive

$$M \overset{\mathrm{def}}{=} H^{n-1}(X, \mathbb{Z}(r))$$

has weight $n - 1 - 2r \leq -2$.

Let MHS/$\mathbb{R}$ denote the category of mixed Hodge structures defined over $\mathbb{R}$, i.e. endowed with an $\mathbb{R}$-linear conjugation F_∞. Following [D3], one expects

$$\mathrm{Ext}^1_{\mathrm{Mot}(\mathrm{Spec}(\mathbb{R}))}(\mathbb{Z}, M) \cong \mathrm{Ext}^1_{\mathrm{MHS}/\mathbb{R}}(\mathbb{Z}, H^{n-1}(X, \mathbb{Z}(r)))$$

$$\cong [H_B^{n-1}(X_{\mathbb{C}}, \mathbb{C})/\{H_B^{n-1}(X_{\mathbb{C}}, \mathbb{Z}(r)) + F^r H_B^{n-1}(X_{\mathbb{C}}, \mathbb{C})\}]^+. \tag{3.3}$$

Note that this is a Lie group.

The situation at the non-archimedean primes is more delicate. One wants

$$\mathrm{Ext}^1_{\mathrm{Mot}(\mathrm{Spec}(\mathbb{Q}_p))}(\mathbb{Z}, M) \subset H^1(\mathrm{Gal}(\bar{\mathbb{Q}}_p/\mathbb{Q}_p), M(\hat{\mathbb{Z}})). \tag{3.4}$$

where $M(\hat{\mathbb{Z}}) = H_{\mathrm{ét}}^{n-1}(X_{\bar{\mathbb{Q}}}, \hat{\mathbb{Z}}(r))$. There are defined [op. cit.] subgroups

$$H_e^1 \subset H_f^1 \subset H_g^1 \subset H^1(\mathrm{Gal}(\bar{\mathbb{Q}}_p/\mathbb{Q}_p), M(\hat{\mathbb{Z}})). \tag{3.5}$$

If elements in H^1 are viewed as extensions, the $\mathbb{Z}_l$ part ($l \neq p$) of H_f^1 (resp. H_g^1) coincides with extensions which split after tensoring with $\mathbb{Q}_l$ and restricting to the maximal unramified extension of $\mathbb{Q}_p$ (resp. all extensions). The $\mathbb{Z}_p$ part of H_f^1 (resp. H_g^1) are crystalline (resp. de Rham) extensions, i.e. those which split when tensored with the Fontaine rings B_{cris} (resp. B_{DR}). In [op. cit.] the main object of study is H_f^1, which should be thought of as $\mathrm{Ext}^1_{\mathrm{Mot(Spec}(\mathbb{Z}_p))}$. One knows the $\mathbb{Z}_l$ part of H_g^1/H_f^1 is torsion free of rank equal to the rank of the space of coinvariants for frobenius acting on $\mathrm{Hom}_I(M(\mathbb{Q}_l), \mathbb{Q}_l(1))$, where $I \subset \mathrm{Gal}(\bar{\mathbb{Q}}_p/\mathbb{Q}_p)$ is the inertia group (and there is a similar interpretation of the p-part in terms of $\mathrm{Crys}(M^*(1))/(1 - f)$, cf. op. cit.) Standard conjectures say that these ranks are all equal and coincide with the order of vanishing of the local factor of the dual L-function $L_p(M^*(1), s)^{-1}$ at $s = 0$, i.e. with the rank of $H_{2d-n+1,\mathrm{\acute{e}t}}(X_{\mathbb{F}_p}, \mathbb{Q}_l(r - 1 - d))$, where $d = \dim X_{\mathbb{F}_p}$. What one expects is that the cycle class map [Bl2]

$$\mathrm{CH}^r(X_{\mathbb{Q}_p}, 2r - n) \to H^1(\mathrm{Gal}(\bar{\mathbb{Q}}_p/\mathbb{Q}_p), M(\hat{\mathbb{Z}})) \tag{3.6}$$

has image $\subset H_g^1$ and maps $\mathrm{CH}^r(X_{\mathbb{Z}_p}, 2r - n)$ into H_f^1. Moreover, one has a chern class map $\mathrm{gr}_\gamma^{r-1} K'_{2r-n-1}(X_{\mathbb{F}_p}) \otimes \mathbb{Q}_l \to H_{2d-n+1,\mathrm{\acute{e}t}}(X_{\mathbb{F}_p}, \mathbb{Q}_l(r - 1 - d))$ which by Jannsen's generalization of Tate's conjecture [J] should be an isomorphism. Putting these ideas together with localization yields (conjecturally)

$$\mathbb{Q} \otimes H_g^1/H_f^1 \overset{i}{\hookrightarrow} [\mathrm{CH}^r(X_{\mathbb{Q}_p}, 2r - n)/\mathrm{CH}^r(X_{\mathbb{Z}_p}, 2r - n)] \otimes \mathbb{Q}$$

$$\cong [\mathrm{gr}_\gamma^r K_{2r-n}(X_{\mathbb{Q}_p})/\mathrm{gr}_\gamma^r K_{2r-n}(X_{\mathbb{Z}_p})] \otimes \mathbb{Q} \cong \mathrm{gr}_\gamma^{r-1} K'_{2r-n-1}(X_{\mathbb{F}_p}) \otimes \mathbb{Q}, \tag{3.7}$$

the inclusion i being an isomorphism after tensoring with $\hat{\mathbb{Q}}$. Let $N = (H_g^1/H_f^1) \cap$ image(i), and "define" $\mathrm{Ext}^1_{\mathrm{Mot(Spec}(\mathbb{Q}_p))}(\mathbb{Z}, M) \subset H_g^1$ to be the pre-image of N. One has

$$0 \to \mathrm{Ext}^1_{\mathrm{Mot(Spec}(\mathbb{Z}_p))}(\mathbb{Z}, M) \to \mathrm{Ext}^1_{\mathrm{Mot(Spec}(\mathbb{Q}_p))}(\mathbb{Z}, M) \to N \to 0 \tag{3.8}$$

so $\mathrm{Ext}^1_{\mathrm{Mot(Spec}(\mathbb{Q}_p))}(\mathbb{Z}, M)$ can be thought of as a topological group containing $\mathrm{Ext}^1_{\mathrm{Mot(Spec}(\mathbb{Z}_p))}(\mathbb{Z}, M)$ as an open compact subgroup. In the special case $X = \mathrm{Spec}(\mathbb{Q})$, $n = r = 1$, one sees from op. cit. Example 3.9 that $H_f^1 \cong \mathbb{Z}_p^\times$, $H_g^1 \cong \mathbb{Q}_p^\times \otimes \hat{\mathbb{Q}}$, and $N = \mathbb{Z}$ embedded in H_g^1/H_f^1 in the natural way, so $\mathrm{Ext}^1_{\mathrm{Mot(Spec}(\mathbb{Q}_p))}(\mathbb{Z}, \mathbb{Z}(1)) \cong \mathbb{Q}_p^\times$.

It will be convenient to write

$$B(\mathbb{Z}_p) = H_f^1 = \mathrm{Ext}^1_{\mathrm{Mot(Spec}(\mathbb{Z}_p))}(\mathbb{Z}, M); \quad B(\mathbb{Q}_p) = \mathrm{Ext}^1_{\mathrm{Mot(Spec}(\mathbb{Q}_p))}(\mathbb{Z}, M);$$

$$B(\mathbb{R}) = \mathrm{Ext}^1_{\mathrm{Mot(Spec}(\mathbb{R}))}(\mathbb{Z}, M). \tag{3.9}$$

One of the main results in op. cit. was to endow the groups (compact for $p < \infty$) $B(\mathbb{Z}_p)$ and $B(\mathbb{R})$ with Haar measures ω_p (depending on a choice of trivialization of $\det(H_{\mathrm{DR}}^n(X/\mathbb{Q})/F^r))$. We extend the ω_p to measures on the locally compact groups $B(\mathbb{Q}_p)$, $p \leq \infty$. Writing $L(M, s) = \prod L_p(M, s)$ for the L-function associated to M, one has for almost all p,

$$\omega_p(B(\mathbb{Z}_p)) = L_p(M, 0)^{-1}. \tag{3.10}$$

The resulting measure μ on

$$B(\mathbb{A}_{\mathbb{Q}}) = \prod_{p \leq \infty} B(\mathbb{Q}_p) \text{ (restricted direct product with respect to } B(\mathbb{Z}_p)) \tag{3.11}$$

is independent of the choice of trivialization. Note that this measure converges (i.e. gives finite non-zero measure to the compact set $\prod B(\mathbb{Z}_p)$) when $n - 2r \leq -2$. In general, it is necessary to assume meromorphic continuation of the L-function, define $\omega'_p = \omega_p \cdot L_p(M, 0)$ for all p outside a finite set S of bad primes ($\omega'_p = \omega_p$ for $p \in S$), and then define

$$\mu = \mu'/L_S(M, 0) \tag{3.12}$$

where L_S is the partial L-function with $p \in S$ omitted.

Assume $n - 1 - 2r \leq -3$. The Tamagawa number conjecture, as formulated in op. cit., reads

$$\mu(B(\mathbb{A}_\mathbb{Q})/B(\mathbb{Q})) = \#H^0(\mathrm{Gal}(\bar{\mathbb{Q}}/\mathbb{Q}), M^*(1)(\mathbb{Q}/\mathbb{Z}))/\# \tilde{\text{Ш}}(M) \tag{3.13}$$

where

$$B(\mathbb{Q}) = \{x \in H^1(\mathrm{Gal}(\bar{\mathbb{Q}}/\mathbb{Q}), M(\hat{\mathbb{Z}})) | x \otimes \mathbb{Q} \in \mathrm{image}[\mathrm{gr}_\gamma^r K_{2r-n}(X) \otimes \mathbb{Q}$$
$$\to H^1(\mathrm{Gal}(\bar{\mathbb{Q}}/\mathbb{Q}), M(\hat{\mathbb{Q}}))]\};$$

$$\tilde{\text{Ш}} = \mathrm{Ker}\left[\frac{H^1(\mathrm{Gal}(\bar{\mathbb{Q}}/\mathbb{Q}), M(\mathbb{Q}/\mathbb{Z}))}{B(\mathbb{Q}) \otimes \mathbb{Q}/\mathbb{Z}} \to \bigoplus_{p \leq \infty} \frac{H^1(\mathrm{Gal}(\bar{\mathbb{Q}}_p/\mathbb{Q}_p), M(\mathbb{Q}/\mathbb{Z}))}{B(\mathbb{Q}_p) \otimes \mathbb{Q}/\mathbb{Z}} \right]. \tag{3.14}$$

Construction (3.15). Assume $n - 1 - 2r \leq -2$.
(i) There is a natural map

$$\mathrm{CH}^r(X_\mathbb{Q}, 2r - n) \to B(\mathbb{Q}). \tag{3.15.1}$$

(ii) Assume moreover $H^{n-1}_{\text{ét}}(X_{\bar{\mathbb{Q}}}, \hat{\mathbb{Z}})$ has no torsion. Then there is a natural map

$$\text{Ш}(\mathrm{CH}^r(X_\mathbb{Q}, 2r - n - 1)) \to \tilde{\text{Ш}}(M). \tag{3.15.2}$$

Here, of course,

$$\text{Ш}(\mathrm{CH}^r(X_\mathbb{Q}, 2r - n - 1)) \overset{\text{def}}{=} \mathrm{Ker}(\mathrm{CH}^r(X_\mathbb{Q}, 2r - n - 1)$$
$$\to \prod_{p \leq \infty} \mathrm{CH}^r(X_{\mathbb{Q}_p}, 2r - n - 1)). \tag{3.15.3}$$

The arrow in (3.15.1) is the cycle class [Bl2]. The first step in constructing the map (3.15.2) is to consider the exact sequence (with $\mathscr{Z}^r(X, m)$ as in §1)

$$0 \to Z\mathscr{Z}^r(X_{\bar{\mathbb{Q}}}, 2r - n) \to \mathscr{Z}^r(X_{\bar{\mathbb{Q}}}, 2r - n) \overset{\partial}{\to} B\mathscr{Z}^r(X_{\bar{\mathbb{Q}}}, 2r - n - 1) \to 0. \tag{3.15.4}$$

Taking galois cohomology yields a map

$$\mathrm{Ker}[\mathrm{CH}^r(X_\mathbb{Q}, 2r - n - 1) \to \mathrm{CH}^r(X_{\bar{\mathbb{Q}}}, 2r - n - 1)]$$
$$\to H^1(\mathrm{Gal}(\bar{\mathbb{Q}}/\mathbb{Q}), Z\mathscr{Z}^r(X_{\bar{\mathbb{Q}}}, 2r - n))$$
$$\to H^1(\mathrm{Gal}(\bar{\mathbb{Q}}/\mathbb{Q}), \mathrm{CH}^r(X_{\bar{\mathbb{Q}}}, 2r - n)) \tag{3.15.5}$$

and hence a map

$$\text{Ш}(\mathrm{CH}^r(X_\mathbb{Q}, 2r - n - 1)) \to \text{Ш}(H^1(\mathrm{Gal}(\bar{\mathbb{Q}}/\mathbb{Q}), \mathrm{CH}^r(X_{\bar{\mathbb{Q}}}, 2r - n))). \tag{3.15.6}$$

The rest of the argument is more complicated, and will not be given in detail. Note, however, that in many cases of interest (e.g. $r = n = 2$ and X a curve; or $r = 2, n = 3$, and X a surface over $\mathbb{Q}$ whch is $\overline{\mathbb{Q}}$-rational; or $r = 2, n = 1$, and $X = \mathrm{Spec}(K)$ for K a number field) one has an exact sequence (3.15.7)

$$0 \to H^{n-1}(X_{\overline{\mathbb{Q}}}, \mathbb{Q}/\mathbb{Z}(r)) \to \mathrm{CH}^r(X_{\overline{\mathbb{Q}}}, 2r - n) \to \mathrm{CH}^r(X_{\overline{\mathbb{Q}}}, 2r - n) \otimes \mathbb{Q} \to 0. \quad (3.15.7)$$

In these cases, a straightforward diagram chase gives

$$\text{Ш}(H^1(\mathrm{Gal}(\overline{\mathbb{Q}}/\mathbb{Q}), \mathrm{CH}^r(X_{\overline{\mathbb{Q}}}, 2r - n))) \to \widetilde{\text{Ш}}(M). \qquad (3.15.8) \quad \square$$

Conjecture (3.16). *The map* (3.15.1) *is surjective, and identifies* $B(\mathbb{Q})$ *with the image of* $\mathrm{CH}^r(X_{\mathbb{Q}}, 2r - n) \to \mathrm{CH}^r(X_{\overline{\mathbb{Q}}}, 2r - n)$. *The map* (3.15.2) *is an isomorphism* $\text{Ш}\mathrm{CH}^r(X_{\mathbb{Q}}, 2r - n - 1) \cong \widetilde{\text{Ш}}(M)$.

The author attaches particular importance to the possibility of giving a geometric interpretation of $\widetilde{\text{Ш}}(M)$ via the higher Chow groups.

Example (3.17). Using an important result of Suslin [Su]

$$\mathrm{CH}^r(K, r) \cong K_r^{\mathrm{Milnor}}(K) \qquad (3.17.1)$$

for any field K, and assuming the moving Conjecture (1.2), one gets a localization sequence for any curve

$$K_2(k(C)) \to \coprod_{P \in C} k(P)^* \to \mathrm{CH}^2(C, 1) \to 0, \qquad (3.17.2)$$

whence $\mathrm{CH}^2(C, 1) \cong SK_1'(C)$. Let X be a smooth surface, and let $\mathscr{S} \subset X$ be the union of all curves in X. Repeating the above localization argument yields

$$K_2(k(X)) \to SK_1'(\mathscr{S}) \to \mathrm{CH}^2(X, 1) \to 0, \qquad (3.17.3)$$

so $\mathrm{CH}^2(X, 1) \cong H^1(X, \mathscr{K}_2)$ (Zariski cohomology of the sheaf $K_2[Q]$).

Assume now that X is a surface over $\mathbb{Q}$ which is $\overline{\mathbb{Q}}$-rational, it follows $[CT]$ that for any number field K,

$$\mathrm{CH}^2(X_K, 1) \cong T(K), \qquad (3.17.4)$$

where T is the algebraic torus with character group the $\mathrm{Gal}(\overline{\mathbb{Q}}/\mathbb{Q})$-module $\mathrm{NS}(X_{\overline{\mathbb{Q}}})$ ($=$ Néron-Severi group of divisors on $X_{\overline{\mathbb{Q}}}$.) In this case

$$M(\widehat{\mathbb{Z}}) = H^2_{\text{ét}}(X_{\overline{\mathbb{Q}}}, \widehat{\mathbb{Z}}(2)) \cong NS(X_{\overline{\mathbb{Q}}}) \otimes \mathbb{Z}(1) \qquad (3.17.4)$$

and it is not hard to show

$$\widetilde{\text{Ш}}(M) \cong \text{Ш}(T) \overset{\text{def}}{=} \mathrm{Ker}[H^1(\mathrm{Gal}(\overline{\mathbb{Q}}/\mathbb{Q}), T(\overline{\mathbb{Q}})) \to \prod H^1(\mathrm{Gal}(\overline{\mathbb{Q}}_p/\mathbb{Q}_p), T(\overline{\mathbb{Q}}_p))].$$
$$(3.17.5)$$

The Tamagawa number conjecture (3.13) (suitably modified because here $n - 1 - 2r = -2$) follows in this case from work of Ono [O].

One knows in general for a rational surface that there is an inclusion

$$\mathrm{Ker}(\mathrm{CH}^2(X_{\mathbb{Q}}) \to \mathrm{CH}^2(X_{\overline{\mathbb{Q}}})) \hookrightarrow H^1(\mathrm{Gal}(\overline{\mathbb{Q}}/\mathbb{Q}), T(\overline{\mathbb{Q}})) \quad [\mathrm{CT}]. \quad (3.17.6)$$

If we assume in addition that X admits a rational pencil of genus 0 curves, a beautiful theorem of Salberger [Sa] says that this inclusion induces an isomorphism

$$\mathrm{III CH}^2(X_{\mathbb{Q}}, 0) = \mathrm{III CH}^2(X_{\mathbb{Q}}) \cong \mathrm{III}(T). \quad (3.17.7)$$

Exercise (3.18). Show that for F a number field, $\mathrm{III CH}^2(\mathrm{Spec}(F), 2)$ coincides with the "wild kernel"

$$\mathrm{Ker}(K_2(F) \to \prod K_2(F_v)), \quad (3.18.1)$$

where v runs through all places of F, and F_v denotes the completion of F at v. Show the Tamagawa number conjecture for $\zeta_F(2)$ is compatible with Lichtenbaum's conjecture.

Bibliography

[Be1] Beilinson, A.A.: Polylogarithm and cyclotomic elements. Preprint
[Be2] Beilinson, A.A.: Notes on absolute Hodge cohomology. Contemporary Mathematics, vol. 55, part 1 (1986)
[Be3] Beilinson, A.A., Goncharov, A.B., Schechtman, V.V., Varchenko, A.N.: Aomoto dilogarithms, mixed Hodge structures, and motivic cohomology of pairs of triangles on the plane. In: The Grothendieck Festschrift, vol. 1. (Progress in Mathematics Series, vol. 86) Birkhäuser, Boston 1990
[Be4] Beilinson, A.A., MacPherson, R., Schechtman, V.V.: Notes on motivic cohomology. Duke Math. J. **54**, no. 2 (1987) 679–710
[Bl1] Bloch, S.: Algebraic cycles and higher K-theory. Adv. Math. (1986)
[Bl2] Bloch, S.: Algebraic cycles and the Beilinson conjectures. Contemporary Mathematics, vol. 58, part 1 (1986)
[BK] Bloch, S., Kato, K.: L-functions and Tamagawa numbers of motives. In: The Grothendieck Festschrift, vol. 1 (Progress in Mathematics Series, vol. 86.) Birkhäuser, Boston 1990
[CT] Colliot-Thélène, J.-L.: Hilbert's theorem 90 for K_2, with application to the Chow groups of rational surfaces. Invent. math. **71** (1983) 1–20
[D1] Deligne, P., Interprétation motivique de la conjecture de Zagier reliant polylogarithmes et régulateurs. Preprint
[D2] Deligne, P.: Le groupe fondamental de la droite projective moins trois points. In: Galois groups over $\mathbb{Q}$. Math. Sci. Res. Inst. Publ. 16. Springer, Berlin Heidelberg New York 1989
[D3] Deligne, P.: Valeurs de fonctions L et périodes d'intégrales. In: AMS Proc. Symp. Pure Math., vol. 33, part 2, pp. 313–346 (1979).
[DM] Deligne, P., Milne, J.S.: Tannakian categories, in Hodge cycles, motives, and Shimura varieties. (Lecture Notes in Mathematics, vol. 900.) Springer, Berlin Heidelberg New York 1982
[DF] Dwyer, W., Friedlander, E., Algebraic and étale K-theory. Preprint
[J] Jannsen, U.: Mixed motives and algebraic K-theory. (Lecture Notes in Mathematics, vol. 1400.) Springer, Berlin Heidelberg New York 1990
[Kl] Kleiman, S.: Motives. In: Proc. 5th Nordic Summer School, Oslo 1970. Wolters-Noordhoff, Holland 1972

[L1]   Lichtenbaum, S.: New results on weight two motivic cohomology. In: The Grothen-
       dieck Festschrift, vol. III. (Progress in Mathematics Series, vol. 86.) Birkhäuser,
       Boston 1990
[L2]   Lichtenbaum, S.: The construction of weight two arithmetic cohomology. Invent.
       math. **88** (1987) 183–215
[M]   Manin, Yu.: Correspondences, motives, and monoidal transformations. Mat. Sbornik
       119 (1968) 475–507
[O]   Ono, T.: On the Tamagawa number of algebraic tori. Ann. Math. **78**, (1963) 47–73
[Q]   Quillen, D.: Higher algebraic K-theory I. In: Algebraic K-theory I. (Lecture Notes in
       Mathematics, vol. 341.) Springer, Berlin Heidelberg New York 1973
[Sa]   Salberger, P.: Zero cycles on rational surfaces over number fields. Invent. math. **91**, no.
       3, (1988) 505–524
[Su]   Suslin, A.: Algebraic K-theory of fields. ICM 1986, pp. 222–244
[T]   Totaro, B.: Thesis, Berkeley 1990
[Z1]   Zagier, D.: Polylogarithms, Dedekind zeta functions, and the algebraic K-theory of
       fields. Preprint
[Z2]   Zagier, D.: The Bloch-Wigner-Ramakrishnan polylogarithm function. Math. Ann.
       286 (1990) 613–624

Computational Complexity of Higher Type Functions

Stephen A. Cook [*]

University of Toronto, Toronto, Ontario M5S 1A4, Canada

Abstract. The customary identification of feasible with polytime is discussed. Next, higher type functions are presented as a way of giving computational meaning to theorems. In particular, if a theorem has a "feasibly constructive" proof, the associated functions should be polytime. However examples are given to illustrate the difficulty of capturing the notion of polytime for higher type functions. Finally, König's Lemma is used to illustrate a theorem whose computational meaning is naturally expressed by functions of type level 3.

1. Traditional Complexity and Polytime

The title of Alan Cobham's 1964 lecture [Cob] "The intrinsic computational difficulty of functions" expresses well the subject matter of complexity theory. A major subgoal of the theory is to identify and study the so-called *feasible* functions. It has become customary to identify the notion of feasible with that of polynomial time computable (polytime, for short). In this section we will examine this identification.

In defining polytime, it is important to emphasize that the parameter used is always the length $|x|$ of the input number x, where $|x| = \lceil \log_2(x + 1) \rceil =$ the number of bits in the binary notation for x.

Definition 1. $f : \mathbb{N} \to \mathbb{N}$ is *polytime* iff there is a Turing machine M and a number k such that for all $x \in \mathbb{N}$, M on input x computes $f(x)$ within $O(|x|^k)$ steps.

The class of polytime functions is very robust. Thus, for example, "Turing machine" in the above definition can be replaced by any reasonable abstraction of a computer, and the class remains unchanged.

The equation feasible = polytime needs interpretation before it can be justified. It is not literally true, since the function $f(x) = 2^{|x|^{1000}}$ is polytime but is not computationally feasible in any sense. (We assume that radix notation is used for

---

[*] Supported by the Natural Sciences and Engineering Research Council of Canada.

numbers.) Thus we must restrict attention to functions arising from "natural" computational problems.

Convention 2. NAT is the class of functions arising from natural computational problems.

A problem whose description requires mention of a large exponent such as 1000 is considered contrived, and not natural.

For the sake of argument here, we identify "feasible" with "practically solvable" in the following sense:

Convention 3. *PracSolv* is the class of functions f which can be computed by some actual computer on all inputs of 10,000 bits or less.

Note that a practical program for computing f need not be known. We only require that one exists. The Convention 3 is not as technology dependent as it may appear. There are physical limits on the power of any conceivable actual computer, as illustrated by the examples below.

The equation feasible = polytime can now be replaced by a somewhat more precise statement.

Thesis 4. PracSolv $\cap$ NAT = polytime $\cap$ NAT.

There are countless examples arising from computing theory and practice which tend to support this thesis. To give just a few, the functions $f(x) = x^2$, $f(x, y) = \gcd(x, y)$, and $f(x) = \lceil \sqrt{x} \rceil$ all lie clearly in both the left side and right side of 4. On the other hand, the function $f(x) = 2^x$ lies clearly outside both the left side and right side of 4. This function is not in PracSolv, because when x has length 1000 bits, $f(x)$ has length about 2^{1000} bits, which cannot be physically written using all the atoms on Earth to express bits.

There are also 0-1 valued functions in NAT which are clearly outside both sides of Thesis 4. For instance, let f represent truth in WS1S (the weak second order theory of one successor). That is, $f(x) = 1$ if x codes (say using ASCII conventions) a true formula in WS1S, and $f(x) = 0$ otherwise. Then f is computable, but provably not polytime [Mey]. Furthermore Stockmeyer [Sto] proved that any Boolean circuit that correctly decides the truth of an arbitrary WS1S formula of length 616 symbols must have more than 10^{123} gates. Thus any conceivable actual computer that is correctly programmed to solve an arbitrary such instance would take until the sun burns out to solve at least one of them.

There are many natural problems whose status with respect to both sides of Thesis 4 is questionable. Among these are the 300 or so "*NP* complete" problems listed in the book by Garey and Johnson [GJ]. None of these is known to be polytime and none is known to be in PracSolv. If any one of these problems is polytime, then provably they all are (i.e. $P = NP$). It is conceivable that one (and therefore all) are polytime, but nevertheless the satisfiability problem has no practical algorithm. This would refute Thesis 4. Unless and until this unlikely situation is demonstrated, however, the assumption $P \neq NP$ together with the thesis 4 provides a very useful orientation for programmers in the field. Anyone who is trying to write a program for a problem and discovers the problem is NP complete would be well advised to change goals and work on a more feasible version of the problem.

There are other potential candidates for refuting Thesis 4, and it is instructive to consider some of these.

CANDIDATES FOR (polytime ∩ NAT) − (PracSolv ∩ NAT)

A) Factoring in $\mathbb{Q}[x]$

Lenstra, Lenstra and Lovasz [LLL] were the first to prove that factoring univariate polynomials over the rationals can be performed in polytime. The original runtime proved for their algorithm was $O(n^{12})$, and it did not appear to be practical. However more recent versions [Scho] have smaller runtimes and do appear to work on large inputs in a reasonable amount of time. Before [LLL] there were programs which effectively solved the problem in most cases, although these programs were known to "blow up" on a small set of inputs. These "bad"inputs can arise in algebraic number field computations.

B) Minor-Closed Graph Families

Robertson and Seymour [RS] proved the remarkable result that every minor-closed family of finite graphs has a finite obstruction set. Here an undirected graph H is a *minor* of G iff H can be obtained (up to isomorphism) from G by contracting certain edges and deleting certain edges and vertices. A family $\mathcal{T}$ of graphs is *minor-closed* iff whenever $G \in \mathcal{T}$ and H is a minor of G, then $H \in \mathcal{T}$. A graph G_0 is an *obstruction* of $\mathcal{T}$ iff $G_0 \notin \mathcal{T}$, but if H is any proper minor of G_0, then $H \in \mathcal{T}$. Thus to determine whether a given graph G is in $\mathcal{T}$ it suffices to check, for each obstruction G_0 of $\mathcal{T}$, whether G_0 is a minor of G.

Robertson and Seymour also proved that for each fixed H there is a polytime algorithm which takes an arbitrary graph G as input and determines whether H is a minor of G. It follows that every minor-closed graph family has a polytime recognition algorithm. Later the minor checking algorithm was improved to run in time $O(n^3)$, but in either case the runtime increases by an extremely large constant factor with increasing H. None of these algorithms is practical when H has four or more vertices.

The most famous minor-closed family is the class of planar graphs, which have the Kuratowski graphs K_5 and $K_{3,3}$ as the obstruction set. More generally, for any fixed surface, the set of graphs embeddable on that surface forms a minor-closed family and hence is in polytime. When the surface is a torus, there are some 800 known obstructions, and there may be more. Another interesting minor-closed family consists of graphs embeddable in $\mathbb{R}^3$ in such a way that no two cycles are linked. See [FL] for other examples.

Although the work of Robertson and Seymour does not yield practical algorithms for the recognition of any of these minor-closed families, such algorithms are known in some cases (e.g. planar graphs). Also the knowledge springing from [RS] that these problems are polytime has spurred research to find practical algorithms, with success in some cases [FL]. Time will tell whether some natural minor-closed family resists long and concerted efforts to find a practical recognition algorithm.

CANDIDATES FOR (PracSolv ∩ NAT) − (polytime ∩ NAT)

A) Linear Programming

The simplex method has long been an extremely useful practical method for solving linear programming problems, even though the method, or at least some versions of it, are known to be exponential in the worst case. Thus linear programming used to be a candidate for the above difference set. However, in 1979 Khachian [Kha] proved that the problem is in polytime, using a very different algorithm.

B) Probabilistic Polytime (Verified Answer)

There are some natural problems not known to have (deterministic) polytime algorithms but which do have verified probabilistic polytime algorithms. These "Las Vegas" algorithms require a random source of input bits (e.g. coin tosses). For every problem instance the algorithms terminate in polytime with high probability, and when they do terminate the answer given is always correct. An early example of such a problem is factoring univariate polynomials over large finite fields [Ber]. A more recent example is the set of prime numbers. Probabilistic algorithms for compositeness have been known for some time [SS], [Rab], but these have one-sided error, as in C) below. Recently Goldwasser and Kilian [GK] found a probabilistic polytime algorithm for prime testing which is always correct when it gives an output, but it fails to give an output for certain prime inputs. This was improved [AH] to yield a provably correct but impractical Las Vegas algorithm for prime recognition. A practical version has also been developed [AM], but this algorithm lacks a proof that the runtime is polynomial.

It turns out that both the polynomial factoring algorithm [Ber] and the practical prime recognition algorithm [AM], although probabilistic in principle, work perfectly well in practice when the random source of bits is replaced by some simple deterministically generated sequence of bits. Thus there exist plausible deterministic polytime algorithms for these problems, and it may be only a question of time until someone proves them correct. In fact, Miller [Mil] proved that prime recognition is deterministic polytime, assuming the Extended Riemann Hypothesis.

C) Probabilistic Polytime (One-Sided Error)

Some probabilistic set recognition algorithms always terminate in polytime; if they accept the input then that answer is always correct, but if they reject the input then there is a small error probability. This error probability can be made arbitrarily small simply by repeating the computation a number of times, assuming that the random input bits are truely random and independent.

One example [Schw] of a problem with this kind of probabilistic algorithm, but not known to be in (deterministic) polytime, is the set of nonsingular $n \times n$ matrices whose entries are linear polynomials over $\mathbb{Z}$ in n variables. The probabilistic algorithm consists of choosing small random integer values for the variables and evaluating the resulting determinant. If the value is nonzero, then the original matrix is certainly nonsingular; if the value is 0, that is evidence that the original matrix was singular. The experiment can be repeated any number of times. Assuming we have access to a truely random source of bits, this is a practical algorithm. However in practice truely random sources are scarce or non-existent, so pseudo random number generators with weakly random seeds are used. These *seem* to work well, but if the algorithm repeatedly outputs "singular" for a given input matrix it is hard to justify having total confidence in the result. Hence on the one hand the problem may not really be in PracSolv, while on the other hand it may be deterministic polytime for the reasons outlined in B) above.

A second example in this category is testing irreducibility of polynomials in many variables over $\mathbb{Q}$, using "concise" representation [vzG]. Here a "yes" answer from the probabilistic algorithm means the input polynomial is certainly irreducible, while a "no" leaves some room for doubt. The algorithm has been extended [KalT] to find likely factorization in case of a "no" answer. For each input polynomial of size n (in concise representation), the probability of the factorization algorithm producing a false factor can be made much less than 2^{-n}. Nevertheless, the possibility of error persists, because there is no known deterministic polytime algorithm for testing divisibility of polynomials in many variables.

In sum, there are no very convincing candidate counterexamples to thesis 4. However some natural polytime problems, especially certain minor-closed graph families, certainly do not have known practical algorithms. But the tendency is for practical algorithms for a problem to be discovered, once it is known that the problem is polytime and a concerted effort is made.

2. Realizing Constructive Proofs

A constructive proof of a theorem provides computational information. For example, if the theorem has the form $\forall x \exists y A(x, y)$ the proof should (at least implicitly) provide an algorithm for computing a function f such that $\forall x A(x, f(x))$ holds. We say that f *realizes* the assertion $\forall x \exists y A(x, y)$. We are interested in studying the complexity of functions f needed to realize such assertions; for example, determining whether there is a polytime such f. The notion of a *feasibly constructive* proof is discussed in [CU]. A necessary condition for a proof of a theorem to be feasibly constructive is that it provide polytime algorithms for any functions needed to realize the theorem.

Here are some theorems with feasibly constructive proofs:

Kuratowski's Theorem. Every finite graph G can either be embedded into the plane, or one of the graphs K_5 and $K_{3,3}$ can be homeomorphically embedded in G. If f is a realizing function, then $f(G)$ is either a planar embedding of G or identifies an embedding of either K_5 or $K_{3,3}$ in G. One can in fact infer a polytime such f from the usual proof of Kuratowski's theorem.

Many theorems of graph theory and finite combinatorics can be feasibly realized in a similar manner, including Hall's theorem and Menger's theorem.

The Extended Euclidean Algorithm. For all integers a and b there exist integers x, y, and d such that $d = ax + by$ and $d|a$ and $d|b$. Again Euclid's proof shows how x, y, and d can be computed from a and b in time at worst $O((\log a + \log b)^2)$.

An interesting example whose feasibility status is unknown is Fermat's "Little Theorem", which we state in contrapositive form: For all integers n and a there exists an integer d such that if $0 < a < n$ and $a^{n-1} \not\equiv 1 \pmod{n}$ then $1 < d < n$ and d divides n. A feasibly constructive proof of this theorem would provide a polytime f which produces a divisor $d = f(a, n)$ for n whenever a and n satisfy the hypotheses above. Such an f would provide a major breakthrough in the quest for practical algorithms for large integer factorization. This is because if $n > 1$ is composite and not one of the rare "Carmichael numbers" [HW] then at least half of the integers a such that $0 < a < n$ satisfy $a^{n-1} \not\equiv 1 \pmod{n}$. (The a's in the multiplicative group $\mathbb{Z}_n^*$ which satisfy $a^{n-1} \equiv 1 \pmod{n}$ form a proper subgroup of $\mathbb{Z}_n^*$, assuming n is not a Carmichael number.)

3. Type 2 Functions

Higher type functions take functions (as well as numbers) as arguments. We may as well suppose that such functions take numbers (as opposed to functions) as values, because of the following device. If $F : \mathbb{N}^{\mathbb{N}} \to \mathbb{N}^{\mathbb{N}}$, then we define $F_1 : \mathbb{N}^{\mathbb{N}} \times \mathbb{N} \to \mathbb{N}$ by

$$F_1(g, x) = F(g)(x).$$

Thus F_1 is for practical purposes the same as F.

Higher type functions may be needed to realize theorems for two reasons. First, the theorem may make an assertion about functions, as in $\forall g \forall x \exists y A(g, x, y)$. Second, even if the subject matter is numbers rather than functions, the usual way of realizing an implication $A \to B$ is by a function F such that for all objects g realizing A, $F(g)$ realizes B.

In order to discuss the complexity of such functions F we must explain how to compute them. For now, we restrict attention to "type 2" functions $F(g, x)$ which take as arguments a numerical function $g : \mathbb{N} \to \mathbb{N}$ and a number $x \in \mathbb{N}$. An algorithm for computing F takes x as input in the ordinary way, but it accesses the function g via an "oracle". Thus a machine executing the algorithm has a special query register (or tape). When the machine assumes a query state with a number y in this register, the "oracle" replaces y by $g(y)$.

This is the only mechanism by which the machine can acquire knowledge of the input g. It follows that for fixed g and x, the value $F(g, x)$ can only depend on $g(y)$ for finitely many values y.

In measuring the runtime of the machine which computes $F(g, x)$, we shall make the convention that each oracle call $g(y)$ takes just $|g(y)|$ steps. This pins down the definition of runtime, but does not solve the problem of defining a complexity class, such as polytime, for type 2 functions. For type 1 functions, polytime means time polynomial in the length of the input. To generalize this to type 2, we must decide what is the "length" of an input function $g : \mathbb{N} \rightarrow \mathbb{N}$. The definition below is a first step in that direction.

Definition 5. A type 2 function F is in the class OPT (oracle polytime) iff there is an algorithm which computes F whose runtime on input (g, x) is bounded by a polynomial in m and $|x|$, where m is the length of the maximum query answer $g(y)$ returned during the computation.

If we restrict attention to inputs g which are 0-1 valued, then OPT provides a perfectly reasonable way to formalize polytime for type 2 functions. This was the original method used to define a polytime version of Turing reducibility between sets [Coo1].

However, for arbitrary input functions g, OPT is too broad to be identified with polytime. For example, if

$$F(g, x) = g^{|x|}(x)$$

(g composed with itself $|x|$ times and evaluated at x) then F is in OPT. But if we define g_0 by $g_0(y) = y^2$, and let $f(x) = F(g_0, x)$, then $f(x) = x^{2^{|x|}}$, which is not polytime. Thus there are type 2 functions in OPT which, when interpreted as operators on the set of type 1 functions, do not preserve the set of polytime functions.

Nevertheless, we will take $F \in$ OPT to be a necessary condition for F to be intuitively polytime. This is enough to show that the pigeon hole principle below has no feasibly constructive proof.

Proposition 6 (Pigeon Hole Principle). $\forall g \forall x \exists a \exists b [a, b \leq x$ and $(g(a) \geq x$ or $(a \neq b$ and $g(a) = g(b)))]$

A functional F realizing the above proposition would satisfy $F(g, x) = < a, b >$, where $< a, b >$ is a natural number coding the pair (a, b) using a standard pairing function, and a and b satisfy the condition in Proposition 6.

Theorem 7. *No function F in OPT realizes the Pigeon Hole Principle.*

The proof is an easy consequence of the definition of OPT. Consider an algorithm which computes an F which realizes the Pigeon Hole Principle and fix some input $x > 0$. Each time the algorithm poses a distinct new query $y \leq x$ we choose the answer $g(y)$ to be the least natural number not yet given as a query answer for this computation. In this way we can force the algorithm to query $g(y)$ for each y in $\{0, 1, \ldots, x\}$ before it finds a and b satisfying Proposition 6.

Thus the algorithm must make at least $x + 1$ different queries, a number which grows exponentially in $|x|$.

4. The Basic Polytime Functions

As mentioned above, it seems that some definition of length for functions must be given in order to give a proper generalization for polytime to type 2 functions. One possibility is to define the length $|g|$ of g to be a function, as follows:

Definition 8. If $g : \mathbb{N} \to \mathbb{N}$, then the length $|g|$ of g is the type one function defined by

$$|g|(n) = \max_{|y| \le n} |g(y)|.$$

This definition and the development below appears in [Kap] and [KC].

We define a type 2 polynomial to be any type 2 function $P(L, n)$ represented by a "polynomial expression" in the type 1 variable L and the numerical variable n. More formally, we give a recursive definition of *polynomial expression* as follows:

a) The variable n and the constants 0 and 1 are each polynomial expressions.
b) If P and Q are polynomial expressions, so are $P + Q$, $P \cdot Q$, and $L(P)$.

Each polynomial expression represents a type 2 function (which by definition is a type 2 polynomial) in the obvious way. An example of a type 2 polynomial is

$$P(L, n) = L(L(n + 3) \cdot L(n)).$$

Definition 9. The type 2 functional $F(g, x)$ is *basic polytime* iff for some type 2 polynomial $P(L, n)$ some oracle algorithm computes $F(g, x)$ within $P(|g|, |x|)$ steps, for all g and x.

It can be shown that every basic polytime function is in OPT. Thus, for example, the bounded maximum function

$$M(g, x) = |g|(|x|) = \max_{|y| \le |x|} |g(y)|$$

is not in OPT and hence not basic polytime, since $2^{|x|}$ oracle calls are necessary to evaluate $M(g, x)$ in case g is identically 0. This shows that the permissible runtime bounds for basic polytime functions are not necessarily themselves basic polytime, because $M(g, x) = P(L, n) = L(n)$ is a type 2 polynomial, where $L = |g|$ and $n = |x|$. On the other hand the length maximum function

$$LM(g, x) = \max_{y \le |x|} |g(y)|$$

is basic polytime, with runtime bound

$$P(L, n) = O(n \cdot L(n)),$$

where $n = |x|$ and $L = |g|$.

The basic polytime functions have nice closure properties, and (unlike OPT functions) they preserve the set of type 1 polytime functions. What is more remarkable they turn out essentially to coincide with the polynomial operators defined by Mehlhorn [Meh] in 1976. The latter functions were defined using Cobham's [Cob] limited recursion on notation, and the definition does not mention machines or runtimes. Polynomial operators were introduced to generalize the notion of polytime reducibility from sets to that of functions.

Theorem 10 [Kap, KC]. *A type 2 function $F(g, x)$ is basic polytime iff it is a polynomial operator.*

It is straightforward to prove the "if" direction in the above result. The converse requires considerable effort, however, because of the necessity of pre-bounding the growth rate each time the operation of limited recursion on notation is applied to a function.

Theorem 10 reinforces the conclusion that the basic polytime functionals form a natural robust class. However we will give an example in the next section of a type 2 function outside the class that still seems to merit the label "polytime".

5. Realizing Well-Quasi-Orders

Definition 11. Let Q be a set and $\leq$ be a binary relation in Q. Then $(Q, \leq)$ is a *quasi-order* iff $\leq$ is reflexive and transitive. $(Q, \leq)$ is a well-quasi-order (wqo) if in addition, for every infinite sequence $q_0, q_1, \ldots,$ from Q, there exists indices i, j with $i < j$ and $q_i \leq q_j$.

Note that $(\mathbb{N}, \leq)$ is a wqo; and in fact every well-order is a wqo. A more interesting example is $(\mathcal{G}, \leq_m)$, where $\mathcal{G}$ is the set of finite graphs and $H \leq_m G$ iff H is a minor of G. Robertson and Seymour's monumental work [RS], which will cover more than 1500 pages, shows that $(\mathcal{G}, \leq_m)$ is a wqo. An immediate corollary is the fact that each minor-closed graph family has a finite obstruction set, as mentioned in Sect. 1.

Robertson and Seymour's proof is not only long, it is nonconstructive. Yet the assertion that some pair $(Q, \leq)$ is a wqo should have computational significance, as suggested by the following definition.

Definition 12. A function F *realizes* a wqo $(Q, \leq)$ iff for all $g : \mathbb{N} \to Q$, $F(g) = \langle i, j \rangle$, where $i < j$ and $g(i) \leq g(j)$.

We are interested in the complexity of the functions F required to realize a given wqo $(Q, \leq)$. This gives an indication of what techniques are required to prove that $(Q, \leq)$ is a wqo. In the case of $(\mathbb{N}, \leq)$, it turns out that there is a realizing basic feasible functional. We first note however that the "brute force" algorithm is not polytime. By brute force we mean: query $g(0), g(1), g(2), \ldots g(j)$ until $g(j-1) \leq g(j)$. This algorithm could require up to $g(0) + 1$ queries, and is not in OPT.

A basic polytime realizing function $F(g)$ for $(\mathbb{N}, \le)$ can be computed using binary search. We initialize $a = 0$ and $b = g(0) + 1$. First compare $g(a)$ and $g(b)$; if $g(a) \le g(b)$ then output $< a, b >$ and halt. Otherwise we note that

(*) $0 < g(a) - g(b) < b - a$.

Let $c = \lfloor \frac{1}{2}(a + b) \rfloor$. Assign to the pair (a, b) either (a, c) or (c, b) as follows. If after one of these assignments $a < b$ and $g(a) \le g(b)$ then make that assignment, output $< a, b >$, and halt. Otherwise choose the assignment which maintains the invariant (*) and repeat the above starting with the computation of c.

This algorithm terminates after at most $|g(0)|$ iterations. The largest query argument is $g(0) + 1$, so the time charged for each query is at most $|g|(|g(0) + 1|)$. Hence the runtime is bounded by $P(L) = O(L(0) \cdot (L(L(0) + 1) + L(0)))$, where $L = |g|$. Hence the functional it computes is basic polytime.

As a second example, consider the lexicographical order $(\mathbb{N} \times \mathbb{N}, \le)$, where $(a, b) \le (a', b')$ iff either $a < a'$ or $(a = a'$ and $b \le b')$. Then no basic polytime functional realizes this wqo. In fact no polytime function realizes this wqo under any reasonable definition of polytime. To see this, we define for each parameter $c \in \mathbb{N}$ a sequence $g_c(0), g_c(1), g_c(2), \ldots$ of pairs as follows:

$$\langle c, c \rangle, \langle c, c - 1 \rangle, \ldots, \langle c, 0 \rangle$$
$$\langle c - 1, c^{|c|} \rangle, \langle c - 1, c^{|c|} - 1 \rangle, \ldots, \langle c - 1, 0 \rangle$$
$$\langle c - 2, c^{|c|^2} \rangle, \ldots, \langle c - 2, 0 \rangle$$
$$\vdots$$
$$\langle 0, c^{|c|^c} \rangle, \ldots, \langle 0, 0 \rangle$$
$$\langle 0, 0 \rangle, \ldots \ .$$

We argue below that the function f defined by $f(c, i) = g_c(i)$ is polytime computable. Suppose however that we assume that F realizes the wqo $(\mathbb{N} \times \mathbb{N}, \le)$ and define h by

(**) $h(c) = F(\lambda i f(c, i))$.

Since the sequence $g_c(0), g_c(1), g_c(2), \ldots$ is strictly decreasing with respect to $\le$ for at least the first $c^{|c|^c}$ values, it follows that $j > c^{|c|^c}$, where $h(c) = \langle i, j \rangle$. Hence h is not polytime. Thus (**) indicates a sense in which F does not preserve polytime functions.

To see that f is polytime computable, note that if $g_c(i) = \langle c - k, c^{|c|^k} \rangle$ for some k, $1 \le k \le c$, then $i > c^{|c|^{k-1}}$. Hence $|c|^k = O(|i|)$, so in general $|g_c(i)| = O(|i|^2 + |c|)$.

As a third example, consider the length lexicographical order $(\mathbb{N} \times \mathbb{N}, \le')$, defined by $(a, b) \le' (a', b')$ iff either $|a| < |a'|$ or $(|a| = |a'|$ and $|b| \le |b'|)$. We then have

Theorem 13 [Coo2]. *No basic polytime F realizes $(\mathbb{N} \times \mathbb{N}, \le')$.*

Nevertheless, the brute force algorithm defines a realizing function F which seems intuitively to be polytime. In particular, F is in OPT, and F preserves polytime functions in the sense of (**). This suggests that the class of basic polytime functions should be enlarged if it is to include all intuitively polytime type 2 functions (see [Coo2]).

As a fourth and final example we return to graph minors.

Theorem 14 [FRS]. *Suppose F realizes $(\mathcal{G}, \leq_m)$. Then there is $f : \mathbb{N} \times \mathbb{N} \to \mathcal{G}$ such that $f(c, i)$ has at most $c + i$ nodes, and if*

$$h(c) = F(\lambda i f(c, i)),$$

then h grows faster than any function provably recursive in Peano Arithmetic.

Peano Arithmetic refers to the standard first order theory of $\mathbb{N}$ under $+$ and $\times$. Roughly speaking, it can formalize any argument not involving infinite sets. In particular, Ackermann's function is provably recursive in Peano Arithmetic. Thus any function realizing $(\mathcal{G}, \leq_m)$ is infeasible in a very strong sense.

6. Type 3 Functions

The most natural way to realize the Weak König's Lemma (König's Lemma for binary trees) is via a type 3 function. We state the lemma in contrapositive form, to give it computational content.

Proposition 15 (Weak König's Lemma (WKL)). *If every path is finite in a binary tree, then there is a uniform bound on path length in the tree.*

A realizing functional B should take a function K, which bounds paths, to a uniform bound $B(K)$. To make sense of the path-bounding function K, its argument g should code a potentially infinite branch. This branch can be specified by a function $g : \mathbb{N} \to \{0, 1\}$, where the binary sequence $g(0), g(1), g(2), \ldots$ indicates successive choices (left or right) of children of nodes in the tree, starting with the root. Then $K(g)$ is an upper bound on the longest path in the tree which is specified by an initial segment of the sequence $g(0), g(1), \ldots$. Thus two different functions g and g' might both be extensions of the same finite maximal path in the tree, and we do not require that the two upper bounds $K(g)$ and $K(g')$ for this path be the same. Any total function $K \in \mathbb{N}^{\{0,1\}^{\mathbb{N}}}$ provides enough information to specify a uniform bound $B(K)$ on path length, even though K does not uniquely determine the tree.

Definition 16. The type 3 function B *realizes* WKL iff for all $K \in \mathbb{N}^{\{0,1\}^{\mathbb{N}}}$ and all $g \in \{0, 1\}^{\mathbb{N}}$ there exists $g' \in \{0, 1\}^{\mathbb{N}}$ such that $g(i) = g'(i)$ for all $i \leq K(g')$, and $K(g') \leq B(K)$.

Notice that we do not require $K(g) \leq B(K)$, but only $K(g') \leq B(K)$ for some g' which extends the same finite maximal path as g. (For an alternative definition of realizing WKL see [Fef].)

It turns out that not just the complexity of B is in question, but there is even doubt about whether a computable B exists which realizes WKL in the above sense. In fact computability theory for type 3 functions suffers because there are at least two different plausible but inconsistent definitions of computable function in the literature. Fortunately these definitions are essentially equivalent for type 2 (and type 1) functions; namely *computable* means computable by a Turing machine with oracles. The two definitions we have in mind are Kleene's general recursive functionals and the so-called recursively countable functionals (see [GH] for an excellent presentation of both).

It turns out that no Kleene general recursive functional B realizes WKL. Roughly speaking, this is because an algorithm for computing B can access its input function K only by posing query arguments g to an oracle, which returns $K(g)$. Further, the only possible arguments g that can be posed are recursive. But it turns out (see [GH]) that there exists an infinite binary tree with no infinite recursive branch. Of course this tree has arbitrarily long finite branches. Thus one can define K such that $K(g)$ is an upper bound for the longest path which g extends, when g is recursive, and $K(g)$ is arbitrary if g is not recursive. When presented with such an input K, no Kleene algorithm can compute a sensible value for $B(K)$.

On the other hand, there is a recursively countable function B which realizes WKL. This is partly because recursively countable type 3 functions have as their domain only the set of "continuous" type 2 functions K. Here K is *continuous* iff for each $g \in \{0, 1\}^{\mathbb{N}}$, there exists $m \in \mathbb{N}$ such that $K(g)$ is determined by the restriction of g to $\{0, 1, \ldots, m - 1\}$. (In other words, $K(g') = K(g)$ for all g' such that $g'(i) = g(i)$, $0 \leq i < m$.) A recursively countable oracle only queries its input at finite restrictions $g_0 = g \mid \{0, 1, \ldots, m - 1\}$ of a function g. The oracle may return $K(g)$, or the symbol $\perp$, denoting "unknown". But it can only return $\perp$ for a finite number of restriction sizes m. By systematically querying K at all possible finite 0-1 sequences g_0 of length m, where $m = 1, 2, \ldots$ the algorithm will (by König's Lemma) eventually reach a value of m for which the oracle can give a value $K(g)$ for each g_0 of length m. The algorithm then outputs the maximum of all these values of $K(g)$ as the value of $B(K)$.

7. Conclusion

One general phenomenon we have seen is that the higher the type level, the harder it is to pin down suitable definitions for complexity classes. Thus for type 3 functions we are not sure what computable means. For type 2 functions we are not sure what polytime means. We might add that for type 1 functions we are not sure what linear time means.

We have talked about the notion of realizing theorems informally, but logicians have given various general formal definitions of realizability [Tro]. This development includes specification of a formal system whose theorems are to be realized. Buss's system IS_2^1 [Bus] is especially interesting here, since in some sense it captures first order polynomial time reasoning. In particular, the functions provably recursive in IS_2^1 are precisely the polytime functions. The underlying logic is the intuitionistic predicate calculus, and hence the proofs are "feasibly constructive" (see [CU]).

In [CU] a class of functionals of all finite types is introduced using the typed lambda calculus and a type 2 "recursor" for limited recursion on notation. This class, called the *basic feasible functionals* in [CK], turns out at type level 2 to coincide with the basic polytime functionals discussed in Sect. 4. In [CU] it is shown that the basic feasible functionals provide realizability interpretations for the system IS_2^1 in two different senses: Kreisel's "modified realizability" and Gödel's "*Dialectica* interpretation". These interpretations are used to study the proving power of IS_2^1 and related systems.

A major goal of studying formal systems such as IS_2^1 is proving independence results and conservative extension results. An example of the latter, proved in [Fer] and [Fef], is that when a version of Weak König's Lemma is added as an axiom to IS_2^1, there are no new resulting theorems of the form $\forall x \exists y A(x, y)$, where $A(x, y)$ is a polytime relation.

A nice (but so far elusive) result would be to show that a standard theorem, such as Fermat's Little Theorem, is independent of IS_2^1. This would not show that the factor finding function $f(a, n)$, discussed in Sect. 2, is not polytime computable. But it would show that no algorithm for such a function has a feasibly constructive correctness proof. The significance of this lies in the following general observation: Natural problems solvable by polytime functions can be solved by polytime functions with algorithms which have correctness proofs that involve only polytime notions.

Acknowledgements. I am grateful to Erich Kaltofen, Stephen Bellantoni, Toni Pitassi, and other colleagues for pointing out errors and suggesting improvements in the original manuscript.

References

[AH]   Adleman, L.M., Huang, M.-D.A.: Recognizing primes in random polynomial time. Proc. 19th ACM Symp. on Theory of Computing 1987, pp. 462–469

[AM]   Atkin, O.L., Morain, F.: Elliptic curves and primality proving. Math. Comput. (September 1990) (submitted)

[Ber]   Berlekamp, E.R.: Factoring polynomials over large finite fields. Math. Comp. **24** (1970) 713–735

[Bus]   Samuel R. Buss: The polynomial hierarchy and intuitionistic bounded arithmetic. In: Selman, A.L. (ed) Structure in Complexity Theory. Lecture Notes in Computer Science, vol. 223. Springer, Berlin Heidelberg New York 1986, pp. 77–103

[Cob]   Cobham, A.: The intrinsic computational difficulty of functions. Proc. of the 1964 International Congress for Logic, Methodology, and the Philosophy of Science (Y. Bar-Hillel, ed.). North-Holland, Amsterdam 1964, pp. 24–30

[Coo1]   Cook, S.A.: The complexity of theorem-proving procedures. Proc. 3rd ACM Symp. on Theory of Computing 1971, pp. 151–158

[Coo2]   Cook, S.A.: Computability and complexity of higher type functionals. Proc. MSRI Workshop on Logic from Computer Science, Y. Moschovakis, ed. (1990) (to appear)

[CK]   Cook, S.A., Kapron, B.M.: Characterizations of the basic feasible functionals of finite type. Proc. MSI Workshop on Feasible Mathematics, S. Buss and P. Scott, eds. Birkhäuser 1990, pp. 71–96

[CU]   Cook, S., Urquhart, A.: Functional interpretations of feasibly constructive arith-
         metic. Technical Report 210/88, University of Toronto (1988). Extended Abstract
         in Proc. 21st ACM Symp. on Theory of Computing 1989, pp. 107–112 (to be
         submitted for publication)

[Edm]   Edmonds, J.: Paths, trees, flowers. Canad. J. Math. **17** (1965) 449–467

[Fef]   Feferman, S.: Milking the "Dialectica" interpretation. Manuscript (1990)

[FL]   Fellows, M., Langston, M.: Nonconstructive tools for proving polynomial time
         decidability. JACM **35**(3) (1988) 727–739

[Fer]   Ferreira, F.: Polynomial time computable arithmetic and conservative extensions.
         Ph.D. thesis, Dept. of Mathematics, Pennsylvania State University 1988

[FRS]   Friedman, H., Robertson, N., Seymour, P.: The metamathematics of the graph
         minor theorem. Contemp. Math. **65** (1987) 229–261

[GH]   Gandy, R.O., Hyland, J.M.E.: Computable and recursively countable functions of
         higher type. Logic Colloquium 76, North-Holland 1977, pp. 407–438

[GJ]   Garey, M.R., Johnson, D.S.: Computers and intractability: A guide to the theory
         of NP-completensss. Freeman, San Francisco 1979

[vzG]   von zur Gathen, J.: Irreducibility of multivariate polynomials. J. Comp. System
         Sci. **31** (1985) 225-264

[GK]   Goldwasser, S., Kilian, J.: A provable correct and probably fast primality test.
         Proc. 18th ACM Symp. Theory Comp. 1986, pp. 316–329

[HW]   Hardy, G.H., Wright, E.M.: An introduction to the theory of numbers, fifth edn.
         Oxford 1979

[KalT]   Kaltofen, E., Trager, B.: Computing with polynomials given by black boxes for
         their evaluations: Greatest common divisors, factorization, separation of numera-
         tors and denominators J. Symb. Comp. **9**(3) (1990) 301–320

[Kap]   Kapron B.: Feasible computation in higher types. Ph.D. thesis, University of
         Toronto 1990

[KC]   Kapron, B., Cook, S.: A new characterization of Mehlhorn's polynomial time
         functionals. Manuscript (1990) (submitted)

[Kha]   Khachian, L.G.: A polynomial time algorithm for linear programming. Dokl.
         Akad. Nauk SSSR **244**(5) (1979) 1093–1096. Translated in Sov. Math. Dokl. **20**,
         191–194

[LLL]   Lenstra, A.K., Lenstra, H.W., Lovász, L.: Factoring polynomials with rational
         coefficients. Math. Ann. **261** (1982) 515–534

[Meh]   Mehlhorn, K.: Polynomial and abstract subrecursive classes. JCSS **12** (1976) 147–
         148

[Mey]   Meyer, A.R.: Weak monadic second-order theory of successor is not elementary-
         recursive. Lecture Notes in Mathematics, vol. 453. Springer, Berlin Heidelberg
         New York 1975, pp. 132–154

[Mil]   Miller, G.L.: Riemann's hypothesis and tests for primality. J. Comp. System Sci.
         13 (1976) 300–317

[Rab]   Rabin, M.O.: Probabilistic algorithms for testing primality. J. Number Theory **12**
         (1980) 128–138

[RS]   Robertson, N., Seymour, P.: Graph minors $I, \ldots, XV$. J. Combinatorial Theory
         (Ser. B), beginning 1983. (Not all papers are completed.)

[Scho]   Schönhage, A.: Factorization of univariate integer polynomials by diophantine
         approximation and an improved basis reduction algorithm. Proc. ICALP '84.
         Lecture Notes in Computer Science, vol. 172. Springer, Berlin Heidelberg New
         York 1984, pp. 436–447

[Schw]   Schwartz, J.T.: Fast probabilistic algorithms for verification of polynomial identi-
         ties. J. ACM **27** (1980) 701–717

[Sto]   Stockmeyer, L.J.: The complexity of decision problems in automata theory and
         logic. Ph.D. thesis, MIT (1974). Report TR-133, MIT Laboratory for Computer
         Science
[SS]   Solovay, R.M., Strassen, V.: A fast Monte-Carlo test for primality. SIAM J. Comp.
         6 (1977) 84–85. Correction: **7** (1978) 118
[Tro]   Troelstra, A.S.: Metamathematical investigation of intuitionistic arithmetic and
         analysis. Lecture Notes in Mathematics, vol. 344. Springer, Berlin Heidelberg New
         York 1973

Conformal Field Theory and Cohomologies of the Lie Algebra of Holomorphic Vector Fields on a Complex Curve

Boris L. Feigin

Institute of Solid State Physics, USSR Academy of Sciences, Chernogolovka
142432 Moscow Region, USSR

Introduction

In this text we shall deal with only one aspect of the conformal field theory – the concept of the modular functor. Recall the main structure of the conformal field theory. Let M be a complex curve. First of all the theory associates to M a finite dimensional vector space $H(M)$. The construction of $H(M)$ depends on the choice of a complex structure on M, but really the correspondence $M \to H(M)$ is topological by nature. Our approach to the definition of $H(M)$ is the following. Let $\mathrm{Dif}(M)$ be the group of all C^∞-diffeomorphisms of M. The group $\mathrm{Dif}(M)$ acts on the space of all complex structures on M.

If a group G acts on a manifold S and n is a point of S we can define G_n – the stability group of n. If G acts transitively on S, then $S \cong G/G_n$. It is possible to define the object "G_n" for a non-transitive action so that we have always $S \cong G/"G_n"$. Assume that a group H acts transitively on S and H_n – the stability group of n; define G_n as the following diagram in the category of groups: $G \to H - H_n$. In other words, G_n is the "intersection" G and H_n in H.

So we can assign to a complex structure "com" on M the "subgroup" $\mathrm{Dif}_{\mathrm{com}}(M)$ $\subset \mathrm{Dif}(M)$. It can be shown, that for different complex structures com_1 and com_2 the "subgroups" $\mathrm{Dif}_{\mathrm{com}_1}(M)$ and $\mathrm{Dif}_{\mathrm{com}_2}(M)$ are "conjugate" in some "derived" sense.

We want to think about the space $H(M)$ as the space of cohomologies of the group $\mathrm{Dif}_{\mathrm{com}}(M)$ with coefficients in some one-dimensional representation. This representation is related to the value of the central charge of the theory. In this text we shall give some arguments in favour of this fact. Really, we shall construct the Lie algebra of the group $\mathrm{Dif}_{\mathrm{com}}(M)$, and realize $H(M)$ as the space of "integrable" homologies of this Lie algebra with coefficients in some distinguished one-dimensional representation.

So, from this point of view, a minimal conformal field theory is the investigation of the group of symmetries of a complex structure on a surface. By the same method we can explore the symmetries of other geometrical structures. For example, the Wess-Zumino theory is attached to the group of holomorphic transformations of a vector bundle on M.

Proceedings of the International Congress
of Mathematicians, Kyoto, Japan, 1990
© The Mathematical Society of Japan, 1991

In more general cases we have a sheaf of associative algebras on M (e.g. conformal field theory associated with a W-algebra). The modular functor is the homologies of this object.

The contents of the text are clear from the titles of the sections. The main result is a theorem of the Section VI.

Now a few words about the references. Information about continuous cohomologies of the Lie algebras of vector fields on smooth manifolds is contained in the book: D.B. Fuchs, *Cohomologies of infinite dimensional Lie algebras*, Plenum Publishing Corporation, New York 1986.

Conformal field theories, associated with Kac-Moody algebras are discussed by Akihiro Tsuchiya, Kenji Ueno, and Yasuhiko Yamada in the article *Conformal field theory on universal family of stable curves with gauge symmetries* (Advanced Studies in Pure Math. **19** (1989) 459–565).

The Segal's article in these Proceedings is devoted to the notion of modular functor. Determinant bundles on the moduli space of curves are explored in: A.A. Beilinson and V.A. Schechtman, *Determinant bundles and Virasoro algebras*, Comm. Math. Phys. **118** (1988) 651.

Minimal conformal field theories were constructed in: A.A. Belavin, A.M. Polyakov and A.B. Zamolodchikov, Nucl. Phys. **B241** (1984) 333.

The theory of singular support of representations of a Lie algebra is developed in O. Gabber, *The integrability of the characteristic variety*, Amer. J. Math. **103** (1981) 445.

In the article: B.L. Feigin and B.L. Tsygan, *Riemann-Roch theorem and Lie algebra cohomology I*, Proceedings of the winter school on geometry and physics, Srni, 9–16 January, 1988, pp. 15–52, you can find the calculations that may be useful for generalization of the results of this text on the manifolds of dimension greater than one.

Ideas of deformation theory are in the works of Gerstenhaber, Quillen, Halperin, Stasheff, and others.

The construction of the modular functor in the Virasoro case is contained in: B.L. Feigin and D.B. Fuchs, *Cohomology of some nilpotent subalgebras of the Virasoro and Kac-Moody Lie algebras*, IGP, vol. 5, no. 2 (1988).

I. Definition of the Lie Algebra of Vector Fields on a Complex Manifold

Let M be a complex manifold of dimension n. What is the Lie algebra of holomorphic vector fields on M?

Let T be the tangent bundle of M, $T = T^{0,-1} \oplus T^{-1,0}$ – the Hodge decomposition of the complexification of T. The space of C^∞ sections $\Gamma(T^{-1,0})$ of the bundle $T^{-1,0}$ is closed with respect to the bracket. Locally, an element of $\Gamma(T^{-1,0})$ has the form

$$\sum_i f_i(z_1, \ldots, z_n, \bar{z}_1, \ldots, \bar{z}_n) \frac{\partial}{\partial z_i}.$$

Denote by $T^{i,j}$ the bundle $(\bigwedge^{-i} T^{-1,0}) \otimes (\bigwedge^{-j} T^{0,-1})$, where i, j are integers. In particular $T^{1,1}$ is the bundle of volume forms. Consider the Dolbeault resolution of the sheaf of holomorphic vector fields on M:

$$0 \to T^{-1,0} \xrightarrow{\bar{\partial}} T^{-1,1} \xrightarrow{\bar{\partial}} T^{-1,2} \to \cdots T^{-1,n} \to 0.$$

The bracket on the space $\Gamma(T^{-1,0})$ is extended in a natural way on the space $\Gamma(T^{-1,0}) \oplus \Gamma(T^{-1,1}) \oplus \cdots \oplus \Gamma(T^{-1,n})$. (Since elements of $\Gamma(T^{-1,i})$ are $(0, 1)$-forms with values in the bundle $T^{-1,0}$). It is easy to see that this extended bracket defines a structure of differential. Lie superalgebra on the following complex

$$0 \to \Gamma(T^{-1,0}) \xrightarrow{\bar{\partial}} \Gamma(T^{-1,1}) \xrightarrow{\bar{\partial}} \Gamma(T^{-1,2}) \to \cdots \Gamma(T^{-1,n}) \to 0.$$

We shall call this differential superalgebra the Lie algebra of holomorphic vector fields on M since it is the space of global sections of the resolution of the sheaf of holomorphic vector fields on M. We shall denote this complex by $\Gamma(\text{Lie}(M))$. It is clear that for affine M the algebra $\Gamma(\text{Lie}(M))$ is equivalent to $\text{Lie}(M)$, where $\text{Lie}(M)$ is the Lie algebra of holomorphic vector fields on M. Another approach to the definition of the Lie algebra of vector fields on a manifold is the following. Choose a covering $\{U_i, i \in I\}$ of M, where each U_i is affine. Let us associate to any subset $i_1 < i_2 < \cdots < i_l$ of I the Lie algebra $\text{Lie}(U_{i_1} \cap U_{i_2} \cap \cdots \cap U_{i_l})$. Using this set of data we can define by standard method a cosimplicial object in the category of Lie algebras. We omit details of the construction, but it is possible to show that such cosimplicial Lie algebras for different coverings are equivalent in some sense. In particular, we can use the covering consisting of all open sets. So, we shall call this cosimplicial object the Lie algebra of vector fields on M and denote it by $\text{Lie}_\Delta(M)$. The differential Lie superalgebra $\Gamma(\text{Lie}(M))$ defines in the standard way a cosimplicial Lie algebra which is equivalent to $\text{Lie}_\Delta(M)$.

Remark. Let A be a group, ξ an A-bundle on M, $\text{End}(\xi)$ the bundle of endomorphisms of ξ. Denote by $\text{End}(\xi) \otimes T^{0,i}$ the bundle of $(0, i)$-forms with values in $\text{End}(\xi)$. The Dolbeault complex of $\Gamma \text{End}(\xi)$:

$$0 \to \Gamma(\text{End}(\xi) \otimes T^{0,0}) \xrightarrow{\bar{\partial}} \Gamma(\text{End}(\xi) \otimes T^{0,1}) \xrightarrow{\bar{\partial}} \cdots \to \Gamma(\text{End}(\xi) \otimes T^{0,n}) \to 0$$

has an obvious structure of differential Lie superalgebra. This object is the natural candidate for the Lie algebra of the endomorphisms of the bundle. We can also give the cosimplicial version of this definition, using coverings of M. Note that all statements in this text have their counterpart for the algebra $\Gamma \text{End}(\xi)$, or for the cosimplicial version $\text{End}_\Delta(\xi)$.

II. Continuous Cohomologies of Lie(M)

Using the methods of the Gel'fand-Fuchs theory of the continuous cohomologies of the Lie algebra of vector fields on a smooth manifold it is possible to define and to calculate the continuous cohomologies of the differential Lie superalgebra $\Gamma \text{Lie}(M)$. First we recall main definitions and results from the Gel'fand-Fuchs theory.

Let N be a C^∞-manifold and $L(N)$ be the Lie algebra of C^∞-vector fields on N; $L(N)$ is a topological Lie algebra with respect to the C^∞-topology. Continuous cohomologies of $L(N)$ are defined as the cohomologies of the subcomplex of the standard cohomological complex of $L(N)$ which consists of continuous cochains. The procedure of calculation of the continuous cohomologies $H_c^*(L(N))$ is divided into two parts – "local" and "global". The aim of the "local" part is the calculation of the continuous cohomologies of the Lie algebra W_n of formal vector fields on $\mathbb{R}^n$, where $n = \dim N$. An element of W_n is a vector field on the formal neighbourhood of the origin in $\mathbb{R}^n$. Gel'fand and Fuchs proved, that $H_c^*(W_n)$ is isomorphic to the algebra of cohomologies of some topological space X_n. The space X_n may be described as a preimage of $2n$-skeleton of the base space of the universal $GL(n, \mathbb{C})$-bundle in the total space of this bundle. It is important for us that X_n is a $GL(n, \mathbb{C})$-space and therefore a $GL(n, \mathbb{R})$-space, since $GL(n, \mathbb{R}) \subset GL(n, \mathbb{C})$.

The aim of the "global" part of the calculation is to "sew" the local calculations at all points of N. To do this let us associate with the tangent bundle of N the bundle $\varrho : U \to N$, of which each fiber is isomorphic to X_n. The main theorem on continuous cohomologies asserts that the algebra $H_c^*(L(N))$ is isomorphic to the cohomologies of the topological space of all sections of the bundle ϱ. Note that in the most important case (for us) where $n = 1$ the space X_n is homotopic to the three-dimensional sphere S^3.

The differential Lie superalgebra $\Gamma(\mathrm{Lie}(M))$ is also equipped with C^∞-topology. So we can define the continuous cohomologies and calculate them using the same technique. Let us formulate the result in the case $\dim M = 1$.

Proposition. *In the case* $\dim M = 1$, *the algebra* $H_c^* \Gamma(\mathrm{Lie}(M))$ *is isomorphic to the algebra of cohomologies of the space of all continuous maps* $\mathrm{Hom}(M, S^3)$ *from M (as a topological space) into S^3.*

It is not hard to find out the structure of the algebra $H^*(\mathrm{Hom}(M, S^3), \mathbb{C})$. Consider the natural map $\theta : M \times \mathrm{Hom}(M, S^3) \to S^3; \theta(m, f) = f(m)$, where $m \in M$, $f \in \mathrm{Hom}(M, S^3)$. The corresponding map between cohomologies gives us a map $\mathbb{C} \cong H^3(S^3, \mathbb{C}) \to H^*(M, \mathbb{C}) \otimes H^*(\mathrm{Hom}(M, S^3), \mathbb{C})$ which we can rewrite as a set of three maps $\theta_i : H_i(M, \mathbb{C}) \to H^{3-i}(\mathrm{Hom}(M, S^3), \mathbb{C})$, $i = 0, 1, 2$. It can be proved that the algebra $H^*(\mathrm{Hom}(M, S^3), \mathbb{C})$ is a free skew-commutative algebra and the images of θ_1, θ_2, θ_3 are the space of generators. Now we shall write down the explicit formulas for the cochains in the standard complex of $\Gamma(\mathrm{Lie}(M))$ which represent the classes corresponding to the images of $\{\theta_i\}$.

We need some preparations. First we recall the definition of holomorphic projective connections on a one-dimensional manifold. A projective connection defines the following holomorphic differential operator of order three:

$$D : \Gamma(T^{-1,0}) \to \Gamma(T^{2,0})$$

such that in a neighbourhood of an arbitrary point three independent solutions of the equation $D(\varphi) = 0$ constitute the Lie algebra which is isomorphic to $sl_2(\mathbb{C})$. This operator is defined by the projective connection up to a constant. We can fix this constant if we suppose that in the local coordinate Z the highest symbol of D is equal to $\left(\dfrac{\partial}{\partial Z}\right)^3$. An operator D with these properties defines a projective connection.

Remark. Fix a closed curve $\gamma \in M$ without self-intersections. After restriction on γ the operator D defines the operator of the second hamiltonian structure in the KdV theory. Note that a projective connection provides the family of operators:

$$D_i : \Gamma(T^{-1,i}) \to \Gamma(T^{2,i}).$$

Locally, the kernel of the operator D_i consists of antiholomorphic tensors of type $(0, 1)$ with values in the kernel of $D_0 = D$. We suppose that all operators D_i have the same symbol. It is easy to see that D_i is adjoint to D_{1-i}. (Recall that for any differential operator $U : E \to F$, where E, F are vector bundles on a manifold, the adjoint operator acts from the bundle $F^* \otimes \mathrm{Vol}$ into $E^* \otimes \mathrm{Vol}$, where Vol is the bundle of volume forms, and $*$ is a sign for the dual bundle.)

Let us consider the following maps:

$$\Psi_0 : \Lambda^3 \Gamma(T^{-1,0}) \to \Gamma(T^{0,0}),$$
$$\Psi_1 : \Lambda^2 \Gamma(T^{-1,0}) \to \Gamma(T^{1,0}).$$

Here the operator Ψ_1 is given by the formula: $\Psi_1(a, b) = (Da)b - a(Db)$, a, $b \in \Gamma(T^{-1,0})$. The map Ψ_0 is characterized up to a constant by the property: Ψ_0 is a $\Gamma(T^{-1,0})$-homomorphism. (Note that the spaces $\Gamma(T^{i,j})$ are $\Gamma(T^{-1,0})$-modules.) The explicit formula for Ψ_0:

$$\Psi_0(U_1, U_2, U_3) = \mathrm{Det} \begin{vmatrix} f_1 & f_2 & f_3 \\ f_1' & f_2' & f_3' \\ f_1'' & f_2'' & f_3'' \end{vmatrix},$$

$$U_i \in \Gamma(T^{-1,0}), \qquad U_i = f_i(z, \bar{z}) \frac{\partial}{\partial z},$$

where z is a local coordinate on M, and "$'$" means the derivation $\dfrac{\partial}{\partial z}$. The value of the determinant does not depend on the choice of a coordinate. We need also the maps Ψ_2, Ψ_3.

$$\Psi_2 : \Lambda^2 \Gamma(T^{-1,0}) \otimes \Gamma(T^{-1,1}) \to \Gamma(T^{0,1}),$$
$$\Psi_3 : \Gamma(T^{-1,0}) \otimes \Gamma(T^{-1,1}) \to \Gamma(T^{1,1}).$$

The formula for Ψ_3 is similar to the formula for Ψ_1, namely $\Psi_3(a, b) = D_0(a) \cdot b - a \cdot D_1(b)$, $a \in \Gamma(T^{-1,0})$, $b \in \Gamma(T^{-1,1})$. The map Ψ_2 is a $\Gamma(T^{-1,0})$-homomorphism and is given by the same determinant as Ψ_0, where $U_1 = f_1 \dfrac{\partial}{\partial z}$, $U_2 = f_2 \dfrac{\partial}{\partial z}$ and $U_3 = f_3 \dfrac{\partial}{\partial z} d\bar{z}$.

Proposition. *There is a map* φ *from the standard homological complex of the Lie algebra* $\Gamma(\mathrm{Lie}(M))$ *into the de Rham complex of* M. *The construction of* φ *is clear from the following diagram. In the top of the diagram is a fragment of the homological complex and the de Rham complex is on the bottom.*

$$
\begin{array}{c|c|c|c}
\Lambda^0\Gamma(T^{-1,0}) & \Gamma(T^{-1,0}) & \Lambda^2\Gamma(T^{-1,0})\text{-------} & \Lambda^3\Gamma(T^{-1,0}) \\
 & \oplus & \oplus & \\
 & \Gamma(T^{-1,0})\otimes\Gamma(T^{-1,1}) & \Lambda^2\Gamma(T^{-1,0})\otimes\Gamma(T^{-1,1}) & \Psi_0 \\
\varphi & \Psi_3 & \Psi_2 & \\
 & & \Gamma(T^{1,0})\leftarrow\text{--------} & \\
 & & \oplus & \Psi_1 \\
 & \Gamma(T^{1,1}) & \Gamma(T^{0,1}) & \Gamma(T^{0,0})
\end{array}
$$

Note that the de Rham complex here is reflected and shifted.

Now let β be an element of $H_i(M)$. We can assign a continuous cochain of the Lie algebra $\Gamma(\mathrm{Lie}(M))$ by the formula $u \to \langle \beta, \varphi(u)\rangle$, where u is a chain, and $\langle\,,\,\rangle$ is the integration of $\varphi(u)$ over the cycle β. It is easy to see that this construction gives us the desired map $H_i(M) \to H_c^{3-i}(\Gamma(\mathrm{Lie}(M)))$.

Remark 1. The following part of our diagram

$$
\begin{array}{ccccc}
\leftarrow & \Lambda^2\Gamma(T^{-1,0}) & \leftarrow & \Lambda^3\Gamma(T^{-1,0}) & \leftarrow \\
 & \downarrow & & \downarrow & \\
0 \leftarrow & \Gamma(T^{1,0}) & \leftarrow & \Gamma(T^{0,0}) & \leftarrow 0
\end{array}
$$

is a morphism from the standard complex of the Lie algebra $\Gamma(T^{-1,0})$ into the Dolbeault complex of M. For non-compact M the continuous cohomologies of $\Gamma(T^{-1,0})$ is infinite-dimensional at least in dimension three. This map of complexes induces the map $V^* \to H_c^3(\Gamma(T^{-1,0}))$ where V is the space of holomorphic functions on M. But for compact M we have dim $H_c^i(\Gamma(T^{-1,0})) < \infty$ for arbitrary i. Let us formulate the important corollary from the results of this section.

Corollary. *The dimension of the space* $H_c^1(\Gamma(\mathrm{Lie}(M)))$ *is equal to one if* M *is compact. It means that the algebra* $\Gamma(\mathrm{Lie}(M))$ *has a non-trivial one-dimensional representation.*

Remark 2. If $\mathfrak{U}$ is a finite dimensional Lie algebra and $v : \mathfrak{U} \to \mathrm{End}(V)$ is a finite dimensional representation of $\mathfrak{U}$, then the map $a \mapsto \mathrm{tr}\, v(a)$, $a \in \mathfrak{U}$ is a 1-cocycle of $\mathfrak{U}$. We can interpret this 1-cocycle as an element of the space $H^0(\mathfrak{U}, \mathfrak{U}^*)$. This construction can be easily generalized. The function $a^n \mapsto \mathrm{tr}(v(a^n))/n!$ is an element $\mathrm{tr}_v^{(n)}$ of the space $H^0(\mathfrak{U}, S^n(\mathfrak{U}^*))$. In other word, the representation v defines a functional $\mathrm{tr}_v \in H^0(\mathfrak{U}, S^*(\mathfrak{U}^*))$, $\mathrm{tr}_v = \sum \mathrm{tr}_v^{(n)}$. Let us consider the Lie superalgebra $\mathfrak{U} \oplus \xi\mathfrak{U} = \mathfrak{U} \otimes \mathbb{C}[\xi]$, where $\mathbb{C}[\xi]$ is a Grassmann algebra in one variable; $\mathfrak{U} \oplus \xi\mathfrak{U}$ is a graded

Lie superalgebra with $\deg(\mathfrak{U}) = 0$, $\deg(\xi\mathfrak{U}) = -1$. It is evident that $H^*(\mathfrak{U} \oplus \xi\mathfrak{U}) \cong H^*(\mathfrak{U}, S^*(\mathfrak{U}^*))$. So we have $\mathrm{tr}_v \in H^0(\mathfrak{U} \oplus \xi\mathfrak{U})$. This idea can be applied to the Lie algebra $\Gamma \,\mathrm{Lie}(M)$. Namely, using Gel'fand-Fuchs technique it is possible to find out the continuous cohomologies of the algebra $\Gamma \,\mathrm{Lie}(M) \oplus \xi\Gamma \,\mathrm{Lie}(M)$. This problem is also divided into two parts: "local" and "global". The local part consists of the calculation of the algebra $H_c^*(W_1 \oplus \xi W_1)$ where W_1 is the algebra of formal vector fields on a line. The answer is the following:

$$H_c^*(W_1 \oplus \xi W_1) \cong H_c^*(W_1) \oplus H_c^*(W_1, W_1^*) \oplus H_c^*(W_1, S^2 W_1^*) \oplus \cdots;$$

$$H_c^i(W_1) \cong \mathbb{C} \quad \text{if } i = 0, 3 \quad \text{and} \quad H_c^i(W_1) \cong 0 \quad \text{if } i \neq 0, 3;$$

$$H_c^2(W_1, S^i W_1^*) \cong H_c^3(W_1, S^i W_1^*) \cong \mathbb{C}$$

and in other dimensions the cohomologies are equal to zero ($i \neq 0$). The multiplication in the space $H_c^*(W_1 \oplus \xi W_1)$ is trivial. The cohomologies of the algebra $\Gamma \,\mathrm{Lie}(M) \oplus \xi\Gamma \,\mathrm{Lie}(M)$ are the same as those of the space of all continuous maps of M (as a topological space) into $\mathbb{Y}$, where $\mathbb{Y}$ is the space satisfying $H^*(\mathbb{Y}, \mathbb{C}) \cong H_c^*(W_1 \oplus \xi W_1)$. In particular, we can conclude that $H_c^0(\Gamma \,\mathrm{Lie}(M) \oplus \xi\Gamma \,\mathrm{Lie}(M))$ is an algebra of polynomials in an infinite set of variables $\sigma_1, \sigma_2, \ldots, \in H_c^0(\Gamma \,\mathrm{Lie}(M) \oplus \xi\Gamma \,\mathrm{Lie}(M))$. Therefore we can associate with a representation v of $\Gamma \,\mathrm{Lie}(M)$ an analog of trace – an element of the space $\mathbb{C}[\sigma_1, \sigma_2, \ldots]$.

III. Description of a 1-Cocycle of the Cosimplicial Lie Algebra $\mathrm{Lie}_{\Delta}(M)$

In this section we shall suppose that M is compact and one-dimensional.

Choose the following simple Zarisky covering of M. Let p be a point of M, $U_1 = M\{p\}$, U_2 is a small neighborhood of p; $\{U_1, U_2\}$ is a covering and let $\mathrm{Lie}_0(M)$ be the following diagram of Lie algebras:

$$
\begin{array}{ccc}
\mathrm{Lie}(U_1) & & \mathrm{Lie}(U_2) \\
& \searrow{\scriptstyle \pi_1} \quad \swarrow{\scriptstyle \pi_2} & \\
& \mathrm{Lie}(U_1 \cap U_2) &
\end{array}
\tag{*}
$$

We shall work with this diagram in the category of Lie algebras instead of the full object $\mathrm{Lie}_{\Delta}(M)$. A 1-cocycle of $\mathrm{Lie}_0(M)$ is the following set of data: $\{v_1, v_2, \varrho\}$, here ϱ is a 2-cocycle of the Lie algebra $\mathrm{Lie}(U_1 \cap U_2)$, v_i is a 1-cochain of $\mathrm{Lie}(U_i)$ and $\pi_i^* \varrho = dv_i$, where d is the differential in the standard cohomological complexe. Let z be a local coordinate at p, and ϱ is a standard Virasoro-type cocycle:

$$\varrho\left(f_1 \frac{\partial}{\partial z}, f_2 \frac{\partial}{\partial z}\right) = \mathrm{Res}_{z=0}\left(\frac{\partial^2 f_1}{\partial^2 z}\frac{\partial f_2}{\partial z} - \frac{\partial^2 f_2}{\partial^2 z}\frac{\partial f_1}{\partial z}\right) dz.$$

It can be proved that the restrictions of ϱ on $\mathrm{Lie}(U_1)$ and $\mathrm{Lie}(U_2)$ are homologous to zero. It means that we can find v_1 and v_2 such that $dv_i = \varrho$; these equations

$(i = 1, 2)$ have only one solution since $H^1(\mathrm{Lie}(U_i)) = 0$. Let us write down the formula for ϱ without using a local coordinate.

Let us fix a projective connection on M and the corresponding operator D. The expression for the cochain: $\tilde{\varrho}(u_1, u_2) = \mathrm{Res}_p(Du_1 \cdot u_2 - u_1 \cdot Du_2)$, $u_i \in \mathrm{Lie}(U_1 \cap U_2)$. It is evident that $\tilde{\varrho}$ is homologous to ϱ. The restriction of $\tilde{\varrho}$ onto $\mathrm{Lie}(U_i)$ is zero since in this case $(Du_1)u_2 - u_1(Du_2)$ is holomorphic on U_i, if $u_1, u_2 \in \mathrm{Lie}(U_i)$.

So, we have constructed the class in the space $H^1(\mathrm{Lie}_0(M))$. Really this is the same class as we constructed earlier for $\Gamma(\mathrm{Lie}(M))$. Note that our construction of the cochain of the algebra $\Gamma(\mathrm{Lie}(M))$ depends on the choice of projective connections, but the corresponding cohomology classes are equal. The same is true for the algebras $\mathrm{Lie}_0(M)$ and $\mathrm{Lie}_\Delta(M)$. In this section we have constructed a map $H^2(\mathrm{Lie}(U_1 \cap U_2)) \to H^1(\mathrm{Lie}_0(M)) \cong H^1(\mathrm{Lie}_\Delta(M))$. It enables us to associate a one-dimensional representation of the algebra $\mathrm{Lie}_\Delta(M)$ to any value of the central charge of the Virasoro algebra.

It is not hard to calculate the whole space of cohomologies of $\mathrm{Lie}_\Delta(M)$. The result is: $H^*(\mathrm{Lie}_0(M)) \cong H^*(\mathrm{Lie}_\Delta(M)) \cong H_c^*(\Gamma(\mathrm{Lie}(M)))$. This is a consequence of the following fact. Let N be an affine one-dimensional manifold and $\mathrm{Lie}(N)$ is a Lie algebra of algebraic vector fields on N.

Proposition. *The natural pairing between the spaces $H_i(\mathrm{Lie}(N))$ and $H_c^i(\Gamma(\mathrm{Lie}(N)))$ is non-degenerate for arbitrary i. The algebra $H_c^*(\Gamma(\mathrm{Lie}(N)))$ is free and skew-commutative with one generator in degree three and r generators in degree 2, where $r = \dim H^1(N, \mathbb{C})$.*

IV. Homologies of the Lie Algebra of Holomorphic Vector Fields on a Compact One-dimensional Manifold M

As in the previous section we choose the diagram $(*)$ as a model for the Lie algebra of holomorphic vector fields $\mathrm{Lie}_\Delta(M)$. Let us fix a one-dimensional representation of the diagram $\mathrm{Lie}_0(M)$ which we denote by $\square_c$, $c \in \mathbb{C}$. It means that the corresponding element $H^1(\mathrm{Lie}_0(M))$ is the image of the class $a \in H^2(\mathrm{Lie}(U_1 \cap U_2))$ which is represented by the cocycle:

$$\left(f\frac{\partial}{\partial z}, g\frac{\partial}{\partial z}\right) \mapsto \frac{c}{24}\mathrm{Res}_0(f'' \cdot g' - f' \cdot g'').$$

We shall denote by $\mathrm{Vir}(U_1 \cap U_2) = \mathrm{Lie}(U_1 \cap U_2) \oplus \mathbb{C}C$ the central extension of $\mathrm{Lie}(U_1 \cap U_2)$, where C is the central element, and the cocycle a corresponds to the value $c = 1$.

The diagram $(*)$ is the same as the set of mappings:

$$
\begin{array}{ccc}
 & \mathrm{Lie}(U_1) & \\
 \nearrow & & \searrow \\
\mathrm{Lie}_0(M) & & \mathrm{Lie}(U_1 \cap U_2) \qquad\qquad (**)\\
 \searrow & & \nearrow \\
 & \mathrm{Lie}(U_2) &
\end{array}
$$

Informally speaking, $\mathrm{Lie}_0(M)$ is an intersection of the subalgebras $\mathrm{Lie}(U_1)$ and $\mathrm{Lie}(U_2)$ in $\mathrm{Lie}(U_1 \cap U_2)$. By the standard arguments in the spirit of the Frobenius reciprocity law the homologies $H_i(\mathrm{Lie}_0(M), \square_c)$ are isomorphic to

$$H_i(\mathrm{Lie}(U_1), \mathrm{Ind}(\mathrm{Lie}_0(M), \square_c; \mathrm{Lie}(U_1))).$$

Here we denote by the symbol $\mathrm{Ind}(A, M; B)$ the induced representation of the Lie algebra B from the representation M of subalgebra $A \subset B$. It is evident from the diagram $(**)$ that the representation $\mathrm{Ind}(\mathrm{Lie}_0(M), \square_c; \mathrm{Lie}(U_1))$ is isomorphic to the restriction onto $\mathrm{Lie}(U_1)$ of the $\mathrm{Vir}(U_1 \cap U_2)$-representation $\mathrm{Ind}(\mathrm{Lie}(U_2) \oplus \mathbb{C}C, 1_c;$ $\mathrm{Vir}(U_1 \cap U_2))$. Here 1_c is the one-dimensional module of the algebra $\mathrm{Lie}(U_2) \oplus \mathbb{C}C$, where $\mathrm{Lie}(U_2)$ acts in trivially on 1_c and C acts by multiplication with $c \in \mathbb{C}$. It is easy to see that if the genus of M is greater than one, then the representation $\mathrm{Ind}(\mathrm{Lie}_0(M), \square_c; \mathrm{Lie}(U_1))$ is free over the universal enveloping algebra of $\mathrm{Lie}(U_1)$. Now we are ready to formulate the result about the homologies but first we recall the standard notations about the representations of the Virasoro algebra. The algebra Vir is the central extension of the Lie algebra of vector fields on a punctured disk around the origin in $\mathbb{C}$. Let C be the central element of Vir. There are two important subalgebras Vir^+ and $\overline{\mathrm{Vir}}^+$ in $\mathrm{Vir} : \mathrm{Vir}^+ = \left\{ f(z)\dfrac{\partial}{\partial z} \oplus \mathbb{C}C \mid f(0) = 0 \right\}$ and $\overline{\mathrm{Vir}}^+ = \left\{ f(z)\dfrac{\partial}{\partial z} \oplus \mathbb{C}C \right\}$, where f is regular at $z = 0$. Denote by 1_c the one-dimensional $\overline{\mathrm{Vir}}^+$-module, where C acts by multiplication with $c \in \mathbb{C}$; $1_{h,c}$ denotes the one-dimensional Vir^+-module, whose corresponding character is $f(z)\dfrac{\partial}{\partial z} \mapsto hf'(0)$, $C \mapsto c$, for $h, c \in \mathbb{C}$. We also put $M_c = \mathrm{Ind}(\overline{\mathrm{Vir}}^+, 1_c; \mathrm{Vir})$ and $M_{h,c} = \mathrm{Ind}(\overline{\mathrm{Vir}}^+, 1_{h,c};$ $\mathrm{Vir})$. Usually $M_{h,c}$ and M_c are called a Verma module and a parabolic Verma module respectively.

We can attach to an arbitrary point $p \in M$ the Virasoro algebra $\mathrm{Vir}(p)$ which is the central extension of the Lie algebra of holomorphic vector fields in a punctured neighborhood $U(p)$ of p. The algebras $\overline{\mathrm{Vir}}^+(p)$ and $\mathrm{Vir}^+(p)$ consist of regular vector fields and regular vector fields which are zero at p respectively. So, for any $p \in M$ we have the $\mathrm{Vir}(p)$-modules $M_{h,c}(p)$ and $M_c(p)$. It is easy to see that the representation $M_c(p)$ is isomorphic to $\mathrm{Ind}(\mathrm{Lie}(U_2) \oplus \mathbb{C}C, 1_c; \mathrm{Vir}(U_1 \cap U_2))$.

Proposition. *Suppose that the genus of $M > 1$. Then for any $i \neq 0$ $H_i(\mathrm{Lie}_0(M), \square_c) \cong$*
$H_i(\mathrm{Lie}_\Delta(M), \square_c) = 0$; $H_0(\mathrm{Lie}_0(M), \square_c) \cong H_0(\mathrm{Lie}_\Delta(M), \square_c) \cong M_c(p)/(\mathrm{Lie}(U_1)M_c(p))$; the last term is the space of $\mathrm{Lie}(U_1)$ – coinvariants in the module $M_c(p)$.

Now we fix a set of distinct points $p_1, p_2, \ldots, p_n$ of M and let $W(p_1, p_2, \ldots, p_n) = M \setminus \{p_1, p_2, \ldots, p_n\}$. We want to describe the induced representation $\mathrm{Ind}(\mathrm{Lie}_\Delta(M),$ $\square_c; \mathrm{Lie}\, W(p_1, p_2, \ldots, p_n))$. Let Z_n be the sum $\mathrm{Lie}\, U(p_1) \oplus \mathrm{Lie}\, U(p_2) \oplus \cdots \mathrm{Lie}\, U(p_n)$ and let $\hat{Z}_n$ be the central extension of the Lie algebra Z_n which corresponds to the cocycle:

$$\left(f_1(z_1)\frac{\partial}{\partial z_1} + f_2(z_2)\frac{\partial}{\partial z_2} + \cdots + f_n(z_n)\frac{\partial}{\partial z_n}, \right.$$

$$\left. g_1(z_1)\frac{\partial}{\partial z_1} + g_2(z_2)\frac{\partial}{\partial z_2} + \cdots + g_n(z_n)\frac{\partial}{\partial z_n} \right)$$

$$\mapsto \frac{c}{24}\sum (f_i'' \cdot g_i' - f_i' \cdot g_i''), \quad c \in \mathbb{C},$$

where z_i is a local coordinate at p_i. It is clear that the algebra $\hat{Z}_n$ acts on the space $M_c(p_1) \otimes M_c(p_2) \otimes \cdots \otimes M_c(p_n)$. The Lie algebra Lie $W(p_1, p_2, \ldots, p_n)$ is a subalgebra of Z_n, and the imbedding $\theta : \mathrm{Lie}\, W(p_1, p_2, \ldots, p_n) \to Z_n$ sends a vector field $a \in \mathrm{Lie}(W(p_1, p_2, \ldots, p_n))$ to the set of the Laurent expansions of a at the points p_1, $p_2, \ldots, p_n$. It may be shown that the imbedding θ has a lift to an imbedding $\hat{\theta} : \mathrm{Lie}(W(p_1, p_2, \ldots, p_n)) \to \hat{Z}_n$. Using the same arguments as in the beginning of this section we can see that the module $\mathrm{Ind}(\mathrm{Lie}_\Delta(M), \square_c; \mathrm{Lie}(W(p_1, p_2, \ldots, p_n)))$ is isomorphic to the restriction of the $\hat{Z}_n$-module $M_c(p_1) \otimes M_c(p_2) \otimes \cdots \otimes M_c(p_n)$ on $\mathrm{Lie}(W(p_1, p_2, \ldots, p_n))$.

It is easy to define the (cosimplicial) Lie subalgebra $\mathrm{Lie}_\Delta(M, p_1, p_2, \ldots, p_n) \subset \mathrm{Lie}_\Delta(M)$ which consists of holomorphic vector fields on M with zeros at the points $p_1, p_2, \ldots, p_n$.

Each point p_i defines the homomorphism $\varphi(h_i) : \mathrm{Lie}_\Delta(M, p_1, p_2, \ldots, p_n) \to \mathbb{C}$, $h_i \in \mathbb{C}$ which assigns to a vector field in a neighbourhood of p the number by the rule: $\varphi(h_i)\left(f_1(z_1)\frac{\partial}{\partial z_1} \right) = h_i f(0)$. We also have a map $\mu : \mathrm{Lie}_\Delta(M, p_1, p_2, \ldots, p_n) \to \mathrm{Lie}_\Delta(M) \to \mathbb{C}$, where the last arrow corresponds to the representation $\square_c$. The tensor product of all these 1-dimensional modules gives us the representation $\square_{h_1, \ldots, h_n; c}$ of the algebra $\mathrm{Lie}_\Delta(M, p_1, p_2, \ldots, p_n)$. Consider the imbedding

$$\mathrm{Lie}_\Delta(M, p_1, p_2, \ldots, p_n) \to \mathrm{Lie}(W(p_1, p_2, \ldots, p_n))$$

and the corresponding induced representation.

Proposition. 1) *The module* $\mathrm{Ind}(\mathrm{Lie}_\Delta(M, p_1, p_2, \ldots, p_n), \square_{h_1, \ldots, h_n; c}; \mathrm{Lie}(W(p_1, p_2, \ldots, p_n)))$ *is isomorphic to the restriction of the representation of the Lie algebra* $\hat{Z}_n$ *in the space* $M_{h_1, c}(p_1) \otimes M_{h_2, c}(p_2) \otimes \cdots \otimes M_{h_n, c}(p_n)$ *on* $\mathrm{Lie}(W(p_1, p_2, \ldots, p_n))$.

2) For $i \neq 0$, $H_i(\mathrm{Lie}_\Delta(M, p_1, p_2, \ldots, p_n), \square_{h_1, \ldots, h_n; c}) = 0$ *if the genus* $g(M) > 1$ *or* $g(M) = 1$ *and* $n \geq 1$ *or* $g(M) = 0$ *and* $n \geq 3$.

3) $H_0(\mathrm{Lie}_\Delta(M, p_1, p_2, \ldots, p_n), \square_{h_1, \ldots, h_n; c}) \cong M_{h_1, c}(p_1) \otimes M_{h_2, c}(p_2) \otimes \cdots \otimes M_{h_n, c}(p_n)/\mathrm{Lie}(W(p_1, p_2, \ldots, p_n))(M_{h_1, c}(p_1) \otimes M_{h_2, c}(p_2) \otimes \cdots \otimes M_{h_n, c}(p_n))$.

Remark. Let γ_i be a small circle around the point $p_i \in M$ and $D(\gamma_i)$ is the group of diffeomorphisms of γ_i with itself; $D(\gamma_i)$ is a real infinite-dimensional Lie group, and let $\mathrm{Lie}^{\mathbb{R}}(\gamma_i)$ be the Lie algebra of the group $D(\gamma_i)$. Denote by $\mathrm{Lie}(\gamma_i) = \mathbb{C} \otimes_{\mathbb{R}} \mathrm{Lie}^{\mathbb{R}}(\gamma_i)$

the complexification of Lie$^{\mathbb{R}}(\gamma_i)$. It is impossible to construct the complexification of the group $D(\gamma_i)$, but it is possible to define a local complex group with the Lie algebra Lie(γ_i). Fix two subalgebras L^+ and L^- in Lie$(\gamma_1) \oplus$ Lie$(\gamma_2) \oplus \cdots \oplus$ Lie(γ_n). The algebra Lie$^+$ consists of vector fields which admit extensions on the small disks with the bounderies $\gamma_1, \ldots, \gamma_n$. The Lie algebra L^- is the image of Lie$(W) \subset \oplus$Lie(γ_i), where W is the manifold obtained by deleting the disks from M. Let D, D^+, D^- be the local groups with the Lie algebras $\oplus$Lie(γ_i), L^+, L^- respectively. Define also the local group $D_0^+ \subset D^+$, where an element of the Lie algebra of D_0^+ is a set of vector fields on the disks $(f_1, \ldots, f_n)$ such that $f_i(p_i) = 0$.

Denote by $M(g, n)$ the moduli space of curves of genus g with n marked points. Our object $(M, p_1, \ldots, p_n)$ is a point of $M(g, n)$. It is easy to see that the double coset space $D_0^+ \backslash D / D^-$ is isomorphic to the formal neighbourhood of $(M, p_1, \ldots, p_n)$ in the $M(g, n)$. The double coset space $D^+ \backslash D / D^-$ is the formal neighbourhood of M in the moduli space $M(g) = M(g, 0)$. As in the Section III a value $c \in \mathbb{C}$ of the central charge defines a central extension of the Lie algebra $\oplus$ Lie(γ_i) and therefore a central extension of the local group D^∞. This extension is the map $\hat{D}_c \overset{\lambda}{\to} D$, whose the kernel is isomorphic to $\mathbb{C}$. Let us fix liftings $\mu_0 : D_0^+ \to \hat{D}_c$, $\mu_1 : D^- \to \hat{D}_c$. The local group D_0^+ has the family of one-dimensional representations $\theta(h_1, \ldots, h_n) : (\exp(u_1), \ldots, \exp(u_n)) \mapsto \exp(\sum h_i u_i'(P_i))$, where $h_i \in \mathbb{C}$, and u_i is a little vector field on the disk with boundary γ_i.

Using these data we can construct a line bundle $\xi_{h_1, \ldots, h_n; c}$ over the formal neighbourhood of $(M, p_1, \ldots, p_n) \in M(g, n)$. First define a line bundle ξ_c by the map $\xi_c : \mu_0(D_0^+) \backslash \hat{D}_c / \mu_1(D^-) \to D_0^+ \backslash D / D^-$. The representation $\theta(h_1, \ldots, h_n)$ defines a homogeneous line bundle on $D_0^+ \backslash D$, hence a line bundle $\xi_{h_1, \ldots, h_n}$ over $D_0^+ \backslash D / D^-$. Now put $\xi_{h_1, \ldots, h_n; c} = \xi_{h_1, \ldots, h_n} \otimes \xi_c$.

Denote by ξ_i the following line bundle on $M(g, n)$; the fiber of ξ_i is the tangent vector space to M at p_i. Let Δ be the pull-back on $M(g, n)$ of the bundle of volume forms on $M(g, 0)$ (it is given by the natural map from $M(g, n)$ onto $M(g, 0)$). Then, the bundle $\xi_{h_1, \ldots, h_n; c}$ is isomorphic to the tensor product $\xi_1^{h_1} \otimes \xi_2^{h_2} \otimes \cdots \otimes \xi_n^{h_n} \otimes \Delta^{c/26}$. (If the numbers $h_1, \ldots, h_n$ and $c/26$ are integers, then $\xi_{h_1, \ldots, h_n; c}$ is a bundle defined over the whole $M(g, n)$, otherwise $\xi_{h_1, \ldots, h_n; c}$ exists only in a small neighbourhood.)

The space of coinvariants $M_{h_1, c}(p_1) \otimes M_{h_2, c}(p_2) \otimes \cdots \otimes M_{h_n, c}(p_n) / \text{Lie}(W(p_1, p_2, \ldots, p_n))(M_{h_1, c}(p_1) \otimes M_{h_2, c}(p_2) \otimes \cdots \otimes M_{h_n, c}(p_n))$ can be identified with the space of generalized sections of the bundle $\xi_{h_1, \ldots, h_n; c}$ with support in the point $(M, p_1, \ldots, p_n)$. More precisely; the space of coinvariants is:

$$H_{(M, p_1, \ldots, p_n)}^N(\xi_{h_1, \ldots, h_n; c}),$$

where $H_X^N(F)$ is the local cohomology of a sheaf F with support in a submanifold X, in our case $N = n + 3g - 3$.

V. An Analytic Version of the Homology of the Lie Algebra of Holomorphic Vector Fields on M with Coefficients in the One-dimensional Representation $\square_c$

In this section we shall calculate the homology of the differential Lie superalgebra $\Gamma(\text{Lie}(M))$ with coefficients in a one-dimensional representation. Let us write down the standard homological complex:

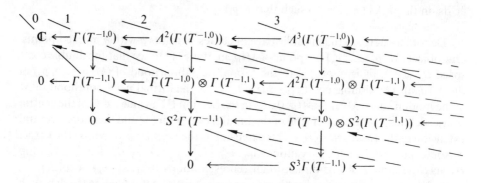

Here the horizontal and vertical arrows are derived from the bracket in $\Gamma(\text{Lie}(M))$. They constitute the differential in the standard complex of $\Gamma(\text{Lie}(M))$ with trivial coefficients. The dashed arrows are the following maps:

$$\Lambda^i \Gamma(T^{-1,0}) \otimes S^j \Gamma(T^{-1,1}) \to \Lambda^{i-1} \Gamma(T^{-1,0}) \otimes S^{j-1} \Gamma(T^{-1,1}).$$

They arise as a contraction with the multiple of the tensor corresponding to the map $\Gamma(T^{-1,0}) \otimes \Gamma(T^{-1,1}) \xrightarrow{\Psi_3} \Gamma(T^{1,1}) \to \mathbb{C}$, where the last map is the integration of forms over the manifold. (Recall that the map Ψ_3 is related to the operator D with the highest symbol $(\partial/\partial z)^3$. If we multiply Ψ_3 with $c/24$ we get the one-dimensional representation of $\Gamma(\text{Lie}(M))$ corresponding to the value c of the central charge.)

Our complex has a natural filtration $\{K_n\}$, the subcomplex K_n consisting of the first n columns. First let us calculate the first term of the spectral sequence, i.e., the homologies of the columns. If the genus of M is greater than one, then the first term is concentrated in degree zero and is equal to the sum $\bigoplus S^i V$, where V is the space of homologies of the complex $\Gamma(T^{-1,0}) \to \Gamma(T^{-1,1})$, of $\dim V = 3g - 3$. The first term of the spectral sequence coincides with the limit term.

Proposition. $H_*(\Gamma \text{Lie}(M), \square_c) \cong H_*(\text{Lie}_\Delta(M), \square_c).$

The previous calculation illustrates this fact. We saw in the Section III, that $H_i(\text{Lie}_\Delta(M), \square_c) = 0$, for $i \neq 0$; and $H_0(\text{Lie}_\Delta(M), \square_c)$ is the space of distributions on the space of complex structures on M.

Remark 1. Recall the idea of the deformation theory (in a very rough form). Let A be an algebraic object (algebra, Lie algebra, algebraic manifold), RA be a resolution of A. For example, if A is a ring, then RA is a simplicial free ring; if A is a commutative algebra, then RA is the Tyurin resolution. Let G be the group of transformations of RA and $H^*(G)$ be the algebra of cohomologies of G. Then, the spectrum of the skew-commutative algebra $H^*(G)$ is the space of versal deformations of A. The model for deformation is the standard complex of G with coefficients in RA. The results of the Sections V and IV are realizations of this idea in the case where A is a one-dimensional manifold M. We associated to M the cosimplicial Lie algebra of transformations of M (or the differential Lie algebra $\Gamma(\mathrm{Lie}(M))$). In terms of this cosimplicial Lie algebra we calculated the small neighbourhood of M in the moduli space. A representation π of the group (or Lie algebra) of transformations defines a sheaf over the moduli space (versal deformations), of which the space of sections is $H^*(G, \pi)$.

Remark 2. There is another (very informal) way of thinking about all these facts. Let M be a compact complex manifold and $\mathrm{Dif}(M)$ be a group of all C^∞-diffeomorphisms of M and $\mathrm{Dif}_c(M) \subset\subset \mathrm{Dif}(M)$ be the subgroup of diffeomorphisms preserving the complex structure. If $\dim M = 1$ and the genus of M is greater than one, then $\mathrm{Dif}_c(M)$ is very little (finite). But we want to think about the group $\mathrm{Dif}_c(M)$ in a "derived" sense. The Lie algebra of $\mathrm{Dif}_c(M)$ is the differential Lie algebra $\Gamma(\mathrm{Lie}(M))$ and the Euler characteristic of the complex $\Gamma(\mathrm{Lie}(M))$ is the formal dimension of the group $\mathrm{Dif}_c(M)$. So, if M is a curve, then $\dim(\mathrm{Dif}_c(M))$ is negative and equal to $3 - 3g$, where g is the genus of M. The double coset space $\mathrm{Dif}_c(M)\backslash \mathrm{Dif}(M)/\mathrm{Dif}(M)$ is exactly the space of all complex structures on M. In this remark we suppose $\dim M = 1$. The group of $\mathrm{Dif}_c(M)$ has a negative dimension, so the dimension of the double coset space is equal to $3g - 3 = \dim M(g, 0)$. A one-dimensional representation of the group $\mathrm{Dif}_c(M)$ defines a homogeneous line bundle over the factor space $\mathrm{Dif}_c(M)\backslash \mathrm{Dif}(M)$ and, therefore, over the moduli space. By this construction we get powers of the determinant bundle. From this point of view it is very interesting to investigate all representations of the group $\mathrm{Dif}_c(M)$.

Example. Consider the complex $L_n = \{\Gamma(T^{n,0}) \to \Gamma(T^{n,1})\}$, $n \in \mathbb{Z}$. It is clear that L_n is a $\Gamma(\mathrm{Lie}(M))$-module. It means that we have a representation $v_n : \Gamma(\mathrm{Lie}(M)) \to \mathrm{End}(L_n)$. The trace of this representation is an element of $H^1(\Gamma(\mathrm{Lie}(M)))$. The corresponding value of the central charge is equal to $-2(6n^2 - 6n + 1)$.

From the presentation of $M(g, 0)$ as a double coset space $\mathrm{Dif}_c(M)\backslash \mathrm{Dif}(M)/\mathrm{Dif}(M)$ we see that there is a $\mathrm{Dif}_c(M)$-bundle $* = \mathrm{Dif}(M)/\mathrm{Dif}(M) \to M(g)$.

This is a principle $\mathrm{Dif}_c(M)$-bundle. We can apply the Weil construction of the characteristic classes to this bundle. Let G be a Lie group and $\mathscr{G}$ be the Lie algebra of G. The usual Weil homomorphism associates with a principal G-bundle $E \to S$ and an invariant polynomial p on the space $\mathscr{G}$, $p \in S^n(\mathscr{G}^*)$ a cohomology class of the space S of dimension $2n$. In our case we have also a map: $H_c^0(\Gamma(\mathrm{Lie}(M))$, $S^j(\Gamma(\mathrm{Lie}(M)^*)) \to H^{2j}(M(g, 0), \mathbb{C})$.

We saw that the algebra $H_c^0(\Gamma(\mathrm{Lie}(M)), S^j(\Gamma(\mathrm{Lie}(M)^*))$ is an algebra of polynomials in an infinite number of variables $\mathbb{C}[\sigma_1, \sigma_2, \dots]$. So we get the (well-known) map $\mathbb{C}[\sigma_1, \sigma_2, \ldots] \to H^*(M(g, 0), \mathbb{C})$. The calculation of this map gives us the Riemann-Roch theorem for families of curves.

VI. Integrable Homologies of the Lie Algebra $\text{Lie}_{\Delta}(M)$

For any open subset (in the Zarisky topology) $U \subset M$ we have an imbedding $\text{Lie}_{\Delta}(M) \subset \text{Lie}(U)$. A representation v of the algebra $\text{Lie}_{\Delta}(M)$ is actually the set of Lie(U)-modules $H(U) \cong \text{Ind}(\text{Lie}_{\Delta}(M), v; \text{Lie}(U))$ and isomorphisms $\varphi(U_1, U_2)$: $\text{Ind}(\text{Lie}(U_2), H(U_2); \text{Lie}(U_1)) \to H(U_1)$, where U_1 is a subset of U_2. Such morphisms $\varphi(U_1, U_2)$ should satisfy the evident compatibility condition: for a couple of imbeddings $U_3 \to U_2 \to U_1$ the composition of the maps

$$\text{Ind}(\text{Lie}(U_2), \text{Ind}(\text{Lie}(U_1), H(U_1); \text{Lie}(U_2)); \text{Lie}(U_3))$$

$$\xrightarrow{\text{Ind }\varphi(U_2, U_1)} \text{Ind}(\text{Lie}(U_2), H(U_2); \text{Lie}(U_3)) \xrightarrow{\varphi(U_3, U_2)} H(U_3)$$

is equal to $\varphi(U_3, U_1)$.

In a more general context it is reasonable to call such a set of data as a representation of the algebra $\text{Lie}_{\Delta}(M)$ also in the case when the maps $\varphi(U_1, U_2)$ are not necessarily isomorphisms. It is easy to define the homologies of the algebra $\text{Lie}_{\Delta}(M)$ with coefficients in such a representation.

Let $\mathfrak{U}$ be a Lie algebra and π be a representation of $\mathfrak{U}$. Let us denote by $ss(\pi) \subset \mathfrak{U}^*$ the singular support of π. We shall call π integrable (or lisse) if $ss(\pi) = 0$. Note that for finite dimensional $\mathfrak{U}$ the class of integrable representations coincides with the class of finite dimensional ones. We shall call a representation of the algebra $\text{Lie}_{\Delta}(M)$ lisse if for any open set $U \subset M$ the singular support of the corresponding $\text{Lie}(U)$-module $H(U)$ is zero. (We also suppose that $H(U)$ has a finite set of generators.)

Proposition. *Let π be a representation of the Virasoro algebra from the category of representations with highest weight. We suppose that the central element $C \in \text{Vir}$ acts on π by multiplication with a constant $c \in \mathbb{C}$. The following conditions are equivalent:*

(1) π is a lisse representation of Vir.

$$(2) \quad c = c_{p,q} = 1 - \frac{6 \cdot (p - q)^2}{p \cdot q},$$ *where p, q are two natural relatively prime numbers. The representation π decomposes as $\pi_1 \oplus \pi_2 \oplus \cdots \oplus \pi_N$, where π_i is an irreducible representation of the corresponding minimal conformal field theory.*

Let us denote by $L_{h,c}$ the irreducible quotient of the Verma representation $M_{h,c}$. Fix two natural relatively prime numbers (p, q). The irreducible representations of the minimal (p, q)-theory are $L_{h,c_{p,q}}$, where $h = h_{r,s}$ is equal to $\dfrac{(rp - sq)^2 - 1}{4pq}$, $1 \le r \le p - 1$, $1 \le s \le r$; $L_{h_{1,1}}$ is called the vacuum representation.

Now we attach to a minimal (p, q)-theory a lisse representation of the algebra $\text{Lie}_{\Delta}(M)$. A $\text{Lie}_{\Delta}(M)$-module $\square_{c_{p,q}}$ is defined in the following way. Consier the open set $W(p_1, \ldots, p_n) = M \setminus \{p_1, \ldots, p_n\}$. We previously attached to each point p_i a copy of the Virasoro algebra and let $L(p_i)$ be its vacuum representation. The

Lie algebra Lie $W(p_1, \ldots, p_n)$ acts on the space $L(p_1) \otimes L(p_2) \otimes \cdots \otimes L(p_n) = \square_{c_{p,q}}(W(p_1, \ldots, p_n))$. The induced representation from the Lie $W(p_1, \ldots, p_n)$-module $\square_{c_{p,q}}(W(p_1, \ldots, p_n))$ to the algebra Lie $W(p_1, \ldots, p_{n+1})$ is isomorphic to $L(p_1) \otimes L(p_2) \otimes \cdots \otimes L(p_n) \otimes M_{0,c_{p,q}}(p_{n+1})$. So it is clear that there is a natural map

$$\text{Ind}(\text{Lie } W(p_1, \ldots, p_n); \overline{\square}_{c_{p,q}}; \text{Lie } W(p_1, \ldots, p_{n+1})) \to \overline{\square}_{c_{p,q}}(W(p_1, \ldots, p_{n+1})).$$

We see that $\overline{\square}_{c_{p,q}}$ is the lisse quotient of the Lie$_\Delta(M)$-representation $\square_{c_{p,q}}$. If $c \neq c_{p,q}$ for any p, q, then the module $\square_c$ has no lisse quotient.

Fix a set of distinct points $\{a_1, \ldots, a_l\}$ of M. We can define in a similar way the notion of lisse representations of Lie$_\Delta(M, a_1, \ldots, a_l)$. An example of a lisse representation of Lie$_\Delta(M, a_1, \ldots, a_l)$ is the following. Let us attach to each point a_i the number $h(a_i) = h_{r_i, s_i}$. If a point a is not in $\{a_1, \ldots, a_l\}$ put $h(a) = 0$. The representation $\overline{\square}(h(a_1), \ldots, h(a_l); c_{p,q})$ associates to the open set $W(p_1, \ldots, p_n)$ the space $L_{h(p_1);c_{p,q}} \otimes L_{h(p_2);c_{p,q}} \otimes \cdots \otimes L_{h(p_n);c_{p,q}}$.

The conformal field theory defines the so-called modular functor. This functor associates with a Riemann surface a finite dimensional vector space (another name of this space is the conformal block for the holomorphic statistical sum). The definition of the modular functor for the minimal conformal field theory is the following: $H(M, p, q) = L_{0,c_{p,q}}/\mathfrak{U} \cdot L_{0,c_{p,q}}$. Here $L_{0,c_{p,q}}$ is the vacuum representation of Vir attached to an arbitrary point of M, and $\mathfrak{U}$ is the Lie algebra of vector fields on M-{point}.

Theorem. 1) *The algebra* $H_c^*(\Gamma \text{ Lie}(M))$ *acts on the space* $H_*(\text{Lie}_\Delta(M), \overline{\square}_{c_{p,g}})$ *and the dual space is a free* $H_c^*(\Gamma \text{ Lie}(M))$-*module with generators in degree zero.*

2) $H_0(\text{Lie}_\Delta(M), \overline{\square}_{c_{p,q}}) \cong H(M, p, q)$.

A similar statement is true for the algebra Lie$_\Delta(M, a_1, a_2, \ldots, a_n)$.

Elliptic Methods in Variational Problems

Andreas Floer [*]

Ruhr-Universität Bochum, Postbox 102148, W-4630 Bochum, Fed. Rep. of Germany

1. Introduction

Morse theory, invented in the 1930s, has had remarkable applications both in Topology and Analysis. Applied to infinite dimensional manifolds, it leads to existence proofs for solutions of certain differential equations which entirely circumvent the problem of calculation. One of the earliest examples is Schnirelman's existence proof for closed geodesics on spheres obtained by considering the energy function on loop space. Conversely, Morse theory on the same function led in 1956 to Bott's periodicity theorem [5] for classical Lie groups.

In the 70ies the hypotheses for infinite dimensional Morse theory were formalized into a condition of "essential positivity" of the function [1], and an analytic compactness condition [23], the much discussed Palais–Smale condition. In the following decade, attention focussed on certain essentially positive functions which just fail to satisfy this second condition, called "borderline cases". The most prominent examples are the minimal-surface problem [25] and Yang-Mills theory in dimension 4. These problems were not only these very "natural " problems from the geometric and physical point of view, but also revealed similar special properties.

In this contribution, we consider functions which are also of "Palais Smale borderline" type, but which are not "essentially positive". The point is that while the usual approach to Morse theory does not apply to such cases, a different approach yields applications to existence of solutions to differential equations [2], as well as to topology. Instead of developing an abstract setting for this method, we present the two main applications. First, however, we discuss the two different approaches to Morse theory in the finite dimensional case.

2. Morse Theory in Finite Dimensions

Consider a smooth real valued function $f : M \to \mathbb{R}$ on a smooth (finite dimensional) manifold M. Morse theory is concerned with the relation between the

---

[*] † May 15, 1991.

[1] Meaning boundedness from below and finiteness of the Morse indices.

[2] i.e. Hamilton's equation.

Proceedings of the International Congress
of Mathematicians, Kyoto, Japan, 1990
© The Mathematical Society of Japan, 1991

critical set

$$C(f) = \{x \in M \mid df(x) = 0\} \tag{1}$$

of f and the topology of M. Here, we assume that $C(f)$ is "nondegenerate", i.e. that for every $x \in C(f)$, the second derivative of f (called the Hessian of f at x) is a nondegenerate quadratic form. That such relation should exist is seen most easily in the "waterlevel picture": Consider as a standard example of a manifold the two-torus T^2, embedded in 3-space $\mathbb{R}^3$ in the usual way as the surface of a tire. Let f be given by the height [3], such that $C(f)$ consists of four points. Now, imagine $\mathbb{R}^3$ to be flooded, and consider the portion

$$M_a = \{x \in M \mid f(x) \le a\} \tag{2}$$

of M below the "waterlevel" $a \in \mathbb{R}$. The crucial observation is that the topology of M_a changes precisely at those levels which are the values of a critical point of f. Moreover, up to homotopy, this change is described by gluing to $M_{a-\varepsilon}$ the "unstable manifold" of x, which is a cell whose dimension is called the Morse index $\mu(x)$, and equals the dimension of the negative "eigenspace" of the Hessian at x. This successive attaching of cells implies the well-known "Morse inequalities" in homology, which can be stated as follows:

If M is closed, then $H_*(M)$ is obtained from a chain complex, which is given by a boundary operator ∂ on the free abelian group over the critical set.

We will denote this "chain complex" by

$$C_*(f) \simeq \mathbb{Z}^d \qquad \text{with} \qquad d = \mid C(f) \mid \;.$$

To make this procedure more precise, we have to introduce an auxiliary metric γ on M, and consider the corresponding "gradient flow trajectories" of f, i.e. smooth paths $u : \mathbb{R} \to M$ satisfying the equation

$$\frac{du(t)}{dt} + \nabla_\gamma f(u(t)) = 0 \;. \tag{3}$$

With the exception of the constant solutions $u(t) \equiv x$ for $x \in \mathscr{C}(f)$, the function decreases along gradient flow trajectories. The stable and unstable manifolds of $x \in \mathscr{C}(f)$ are the sets $W_\gamma^\pm(x)$ of all trajectories u with $\lim_{t \to \pm\infty} = x$.[4] The metric on $TM(x)$ also converts the Hessian into a symmetric operator on $TM(x)$ whose negative eigenspace $E_\gamma^-(x)$ is the tangent space at x of $W_\gamma(x)$.[5]

There is a different, in some sense dual approach to Morse theory, which instead of the unstable manifolds considers only the spaces

$$\mathscr{M}_\gamma(x,y) = \{u : \mathbb{R} \to M \mid \quad u \text{ satisfies (5).} \quad\} = W_\gamma^-(x) \cap W_\gamma^+(y) \;. \tag{4}$$

---

[3] The tire assumed to be in upright position.
[4] We can always identify trajectories and points in M through the correspondence $u \hat{=} u(0)$.
[5] In fact, it is globally diffeomorphic to $W_\gamma^-(x)$.

with

$$
\begin{cases}
\dfrac{du}{dt} = -\nabla_\gamma f(u(t)) \ , \\[2mm]
\lim_{t \to -\infty} u(t) = \quad x \ , \\[2mm]
\lim_{t \to \infty} u(t) = \quad y \ .
\end{cases}
\tag{5}
$$

These are spaces of gradient flow lines connecting two given critical points $x, y \in C(f)$.

For "generic" γ, these intersections are transversal, resulting in smooth manifolds $\mathcal{M}_\gamma(x, y)$ of dimension $\mu(x) - \mu(y)$. Such gradient flows are called "Morse Smale flows". Note that if we assume an orientation on $E_\gamma^-(x)$ at all critical points $x \in C(f)$, then $\mathcal{M}_\gamma(x, y)$ is naturally oriented as the intersections of the oriented $W_\gamma^-(x)$ with the co-oriented $W_\gamma^+(y)$. Finally the translational symmetry implies a splitting

$$
\mathcal{M}_\gamma(x, y) = \hat{\mathcal{M}}_\gamma(x, y) \times \mathbb{R} \cup \{ \text{ constant trajectories } \} \ .
\tag{6}
$$

where $\mathcal{M}_\gamma(x, y)$ will be called the reduced trajectory space. To formulate the Morse inequalities, we consider for $x, y \in \mathscr{C}(f)$ the integer

$$
< \partial x \mid y > = \#\hat{\mathcal{M}}_\gamma(x, y)
\tag{7}
$$

where $\#$ applied to an oriented manifold denotes its algebraic number, counted with sign, if it is discrete, and is zero otherwise. In other words, $< \partial x \mid y >$ is the number of "isolated trajectories" of the $\nabla_\gamma f$-flow, i.e. of components of $\mathcal{M}_\gamma(x, y)$ of type $\mathbb{R}$, between x and y, counted positively, if its "intrinsic" orientation points in the flow direction and negatively otherwise.

We interpret (7) as the matrix elements of an endomorphism on $C_*(f)$. That is, if $x \in \mathscr{C}(f)$, then ∂x is the formal sum over the endpoints of all isolated trajectories originating at x, counted by sign. Then we have the following

Theorem 1 (Morse Inequalities). *If M is closed and (f, γ) defines a Morse Smale flow on M, then*
1. *the discrete parts of the sets $\hat{\mathcal{M}}_\gamma(x, y)$ are finite, and $\partial\partial = 0$*
2. *$\ker /\operatorname{im} \partial \cong H_*(M)$*

In the example of the torus, we may have to perturb the metric given by standard embedding slightly to obtain a Morse Smale flow. For example, there should be no trajectories connecting the two critical points of Morse index 1. For any Morse Smale flow, however, we will find that isolated trajectories come in pairs of opposite sign: the boundary operator is zero, since $H_*(M) = C_*(f) = \mathbb{Z}^4$. (This is called a perfect Morse function.) An example with nontrivial ∂ is easily constructed e.g. by creating an additional local maximum of f.

Let us have just a glimpse of the proof of the boundary property of ∂. The matrix element $< \partial\partial x \mid y >$ of the homomorphism $\partial\partial$ for $x, y \in \mathscr{C}(f)$ is given by the algebraic number of "double trajectories", i.e. pairs of isolated flow lines connecting x and z with a stopover at a third critical point y. In this situation, it can be shown that if we change the outgoing direction at x slightly in a particular way, we generate a 1-parameter family in $\hat{\mathscr{M}}_\gamma(x, z)$ for any Morse Smale flow. In fact, there exist local diffeomorphisms

$$\hat{\mathscr{M}}_\gamma(x, y) \times (0, 1] \times \hat{\mathscr{M}}_\gamma(y, z) \hookrightarrow \hat{\mathscr{M}}_\gamma(x, z). \tag{8}$$

Now, every non-compact component of a 1-dimensional manifold must have two ends, and compactness of M implies that such ends always correspond to "double trajectories". Analysing the orientation involved, we find that each of these pairs of double trajectories cancel in $< \partial \partial x \mid y >$. We see that compactness properties of the spaces $\mathscr{M}_\gamma(x, y)$ are crucial in the definition of ∂ as well as in the proof of its boundary property. These compactness properties can be derived from compactness of M, but they would follow in the same way from compactness of

$$S = \{ u(0) \mid u \in \mathscr{M}_\gamma(x, y) \quad \text{for some} \quad x, y \in \mathscr{C}(f) \} \tag{9}$$

as subset of M. Hence, if this set, the union of all bounded flow trajectories, is compact in M, then we can associate to it the homology group

$$I_*(S, f, \gamma) = \ker / \operatorname{im} \partial. \tag{10}$$

If M is noncompact, then this group will in general differ from $H_*(M)$; it is a new invariant of the flow and of its compact invariant set S. One can show that $I_*(S, f, \gamma)$ is invariant under the deformation of f and γ (through non–Morse–Smale flows) as long as S remains compact throughout the deformation. In fact, $I_*(S, f, g)$ can be shown to be the homology of the "Conley Index" of S; which is the homology of a neighbourhood U of S relative to a suitably defined "exit set" $A \subset U$. For details, see [10] and [17].

3. The Symplectic Action

We want to extend the idea of the Conley index to infinite dimensional variational problems. Consider the "phase space" $\mathbb{R}^{2n}$, of positions and moments of a linear mechanical system, parameterized by $Z = \{p_i, q_i\}_{i=1}^n$. For every loop $z : \mathbf{S}^1 \to \mathbb{R}^{2n}$, we define the "action function"

$$a_h(z) = \int p \, dq + \int h_s(z(s)) ds \tag{11}$$

where $h : \mathbf{S}^1 \times \mathbb{R}^{2n} \to \mathbb{R}$ is a smooth function, considered as a "time dependent Hamiltonian". In the simplest case, it is the sum of the kinetic term $\frac{m}{2} p^2$ and a potential term $V_s(q)$. Hamilton's principle states that the critical points of a_h in $\Omega(\mathbb{R}^{2n}) = C^\infty(\mathbf{S}^1, \mathbb{R}^{2n})$ are the solutions of Hamilton's equation. If $\mathbb{R}^{2n}$ is identified with $\mathbb{C}^n$ by means of a complex structure J and a metric g, then these equations take the form[6]

$$J \frac{dz}{ds} + \nabla_g h_s(z(s)) = 0. \tag{12}$$

---

[6] Here, $\omega := g(J \cdot, \cdot)$ is assumed to be skew–symmetric, and the equation (12) is essentially determined by this "symplectic form" alone.

The left hand side of (12) can be considered as the gradient of a_h on $\Omega(\mathbb{R}^{2n})$ with respect to the "L^2–metric" on this infinite dimensional space defined by g.

To apply Morse theory to Hamilton's principle, one might extend $\Omega(\mathbb{R}^{2n})$ to some Banach space $\bar{\Omega}$, and modify the gradient to a smooth vector field on $\bar{\Omega}$. Then the gradient flow would be available as a homotopy on $\bar{\Omega}$ decreasing the action. Note, however, that the Hessian of a_h at a critical point is a zero–order perturbation of the selfadjoint operator $J\frac{d}{dz}$, which has infinite dimensional positive and negative eigenspaces. Hence, the unstable manifolds are infinite dimensional cells, the addition of which does not change the homotopy type. Therefore, one cannot relate critical points of a_h to the topology of subsets of $\Omega(\mathbb{R}^{2n})$. Nevertheless, the problem was attacked in 1978 by Rabinowitz [24]: He restricted the function to finite dimensional Fourier-subspaces of loops, and used a method of Morse theory which was "stabilized" with an increase of dimension. Thus the topological invariants used by Rabinowitz were trivial in the total loopspace, but existed nevertheless in some sort of a "stable limit". In the following decade, one came to understand these invariants in the framework of a special homology theory in $\Omega(\mathbb{R}^{2n})$ which pays special respect to the decomposition into positive and negative Fourier components of loops, and in which the unstable and stable manifolds would define cycles and cocycles, respectively. Though this idea was never made precise, it inspired various existence proofs for closed solutions of Hamiltonian systems in $\mathbb{R}^{2n}$.

The dual approach to Morse theory can be used to define such "half–infinite" dimensional homology groups without having to consider infinite dimensional subsets. Since we are interested in individual trajectories rather than in the entire gradient flow, we do not have to "regularize" the gradient, but can consider maps $u : \mathbb{R} \times \mathbb{S}^1 \to \mathbb{R}^{2n}$ satisfying the equation

$$\frac{\partial u(s,t)}{\partial t} + J\frac{\partial u(s,t)}{\partial s} + (\nabla_g h_s)\,(u\,(s,t)) = \left(\bar{\partial}_J u + D_g h(u)\right)(s,t) = 0. \qquad (13)$$

Here, we have combined the derivative in the flow time and the differential part of the gradient of a_h to the Cauchy Riemann operator, whose zero set is the set of holomorphic maps from $\mathbb{R} \times \mathbb{S}^1 \cong \mathbb{C}/i\mathbb{Z}$ to $\mathbb{C}^n$. Thus the "gradient flow equation" for a_h is a zero order perturbation of the equation for "holomorphic curves" in $\mathbb{C}^n$, which Gromov [21] introduced to symplectic geometry in 1985. Ellipticity of $\bar{\partial}_J$ and nondegeneracy of $\mathscr{C}(a_h)$ implies that the spaces

$$\mathscr{M}_h(x) = \left\{ u : \mathbb{R} \times \mathbb{S}^1 \to \mathbb{R}^{2n} \mid u \text{ satisfies (15) } \right\} \qquad (14)$$

$$\begin{cases} \bar{\partial}_J u + D_g h(u) = 0 \;, \\[2mm] \lim_{t \to -\infty} u(t,s) = x(s) \;, \\[2mm] \lim_{t \to \infty} u(t,s) = y(s) \end{cases} \qquad (15)$$

are, for generic h and J, finite dimensional manifolds, with an orientation defined through a "determinant bundle". Moreover, Gromov's compactness theory of (pseudo)-holomorphic curves implies that the integers (7) are still defined for suitable h.

The trajectory spaces (14) can be defined equally well if $\mathbb{R}^{2n}$ is replaced by any symplectic manifold (P, ω) with symplectic (i.e. nondegenerate closed) form $\omega \in \Omega^2(P)$. In fact, let us assume that P is compact. In this case, Gromov's compactness theory predicts that the spaces $\mathcal{M}_\gamma(x, y)$ are compact (in the topology of local convergence) up to the phenomenon of "bubbles": A sequence u_k in $\mathcal{M}_\gamma(x, y)$ may converge to u outside a point $(s, t) \in \mathbb{S}^1 \times \mathbb{R}$, but may "blow up" to form a holomorphic map $(S^2, \infty) \longrightarrow (P, u(s, t))$ in arbitrarily small neighbourhoods of (s, t). Such "holomorphic spheres" always define nontrivial elements in $\pi_2(P)$. This leads to the following chain complex:

Theorem 2. *Let (P, ω) be a compact symplectic manifold with $\pi_2(P) = 0$. Then for generic h and J, the trajectory spaces (14) define a boundary operator $\partial = \partial_{h,J}$ on $C_*(a_h)$.*

Its homology is independent of the choice of h and J. For the proof of the boundary property $\partial\partial = 0$, the "gluing maps" (8) are now defined by a version of Taubes' grafting constructions, which was first defined for instantons, but applies to other elliptic equations as well. The crucial point is invariance property, which implies that the "new homology group" depends only on P, and that it may be calculated e. g. for "small" Hamiltonians. Since $\mathcal{C}(a_0)$ is the set of constant loops, it turns out that it coincides with $H_*(P)$. The result [18] is a "Morse inequality" between the set of closed solutions of (12) and the homology of P; in particular, the number of such solutions must be greater than or equal to the dimension of $H_*(P)$. Recall that all solutions were assumed to be non-degenerated. In the general case, the method can be extended to yield "cuplenght estimates". This fact, conjectured by Arnold in [3,2], had been established by Conley and Zehnder in [8] for the torus T^{2n}, applying the Conley index to a "reduced" finite dimensional problem. The construction of the complex in Theorem 2 was in fact an attempt to define the Conley index directly in an infinte dimensional situation. It can be defined in various other situations involving the symplectic action, e.g. to Lagrangian intersections [15] and symplectic manifolds with contact boundaries[7]. It is not always as readily calculable, in fact, it may yield new invariants in symplectic geometry which are yet to be explored.

4. Gauge Theory

The method described in the preceeding section uses strongly the special geometric and analytic features of the symplectic action function. There exists, however, another natural variational problem with remarkably similar properties. To describe it, we have to replace the loop space $\Omega(P)$ by the space of SU_2-connections on a closed oriented three-manifold M. Recall that all SU_2-bundles of M are trivial, and that with respect to such a trivialization, the space of connections on $M \times SU_2$ can be identified with

$$\mathcal{A}_M := \Omega^1(M) \otimes su_2 \tag{16}$$

---

[7] Work in progress.

where su$_2$ is the Lie algebra of SU$_2$. A connection a defines a covariant derivative $\nabla_a \xi = \nabla \xi + a\xi$ of "sections" ξ. $M \to \mathbb{C}^2$, where ∇ is the usual derivative, and su$_2$ acts on $\mathbb{C}^2$ in the usual way. The gauge group

$$\mathscr{G}_M = C^\infty(M, \mathrm{SU}_2) \qquad (17)$$

of bundle transformations of $M \times \mathbb{C}^2$ also acts on such sections, and hence on $\mathscr{A}_M$ through

$$g^{-1} \nabla_a g = \nabla_{g*a} \; ; \; g*a = g^{-1} \nabla g + g^{-1} ag. \qquad (18)$$

The curvature-2-form F_a of $a \in \mathscr{A}_M$ is $\mathscr{G}_M$-equivariant, where $\mathscr{G}_M$ acts linearly on $\Omega^2(M) \otimes$ su$_2$, i.e. we have $F_{g*a} = g^{-1}F_a g^{-1}$. Such a $\mathscr{G}_M$-equivariant map can also be considered as a section of an infinite dimensional vector bundle[8] over the quotient space $\mathscr{B}_M = \mathscr{A}_M / \mathscr{G}_M$. The latter is known to contain as an open subset the infinite dimensional manifold $\mathscr{B}_M^*$ of irreducible connection classes.

So far, we have not made use of the dimension of M. However, the particular position of $\Omega^2(M)$ in the De Rham complex of M lends a special aspect to gauge theory in dimensions $2, 3$ and 4. In the present case, note that curvature defines a 1-form on $\mathscr{A}_M$, which turns out to be exact. More precisely, it defines up to an additive constant a function $S : \mathscr{A}_M \to \mathbb{R}$ by

$$\int_M \mathrm{tr}(F_a \wedge b) = dS(a)b. \qquad (19)$$

This function is almost gauge invariant: It defines a function $S : \mathscr{B}_M \to \mathbb{R}/\mathbb{Z}$ first considered by Chern and Simons. Thus the critical point set $\mathscr{C}(S)$ of this Chern Simons functions is the set of flat connections, which can be described algebraically through their holonomy:

$$\mathscr{C}(S) = \{a \in \mathscr{A}_M \mid F_a = 0\} / \mathscr{G}_M \qquad (20)$$
$$= \mathrm{Hom}\,(\pi_1(M), \mathrm{SU}_2) / \mathrm{ad}(\mathrm{SU}_2).$$

Though the actual description of (17) may be a rather complicated problem, we cannot really expect infinite dimensional Morse theory to be of much help. In fact, the objective in this case is entirely reversed: If R_M is related by such a Morse theory to some homology group, then this group could define a new invariant of M. Since the Hessian of s at $\alpha \in R_M$, which is given by the symmetric bilinear form

$$D^2 S(\alpha)(\xi, \zeta) = \int_M \mathrm{tr}(\xi \wedge d_\alpha \zeta) \qquad (21)$$

on $\Omega^1(M) \otimes$ su$_2$, has again an infinite dimensional negative and positive subspaces, we cannot expect the topology of $\mathscr{B}_M$ itself to come in. Just as solutions of the Cauchy-Riemann equation on $\mathbb{R} \times S^1$ are the gradient flow trajectories for the symplectic action, the gradient flow trajectories for S are given by instantons on $\mathbb{R} \times M$. To see this, note that every connection on $\mathbb{R} \times M$ is gauge equivalent

---

[8] With fibre $\Omega^2(M) \otimes$ su$_2$.

to a connection a with component in $\mathbb{R}$-direction[9] vanishing. Such a connection describes a family in $\mathscr{B}_M$ by $a(s) = a_{\{s \times M\}}$. If γ_M is a metric on M, and a metric on $\mathbb{R} \times M$ is defined by $(ds)^2 + \gamma_M$, then the antiself duality equation on $\mathbb{R} \times M$ takes on the form

$$0 = \frac{d}{ds} a(s) + *_M F_{a(s)} \tag{22}$$

where $*_M$ denotes Hodge duality in $\Omega^*(M)$ with respect to γ_M. By (17), we can consider $*F_a$ as the gradient of the Chern-Simons function with respect to the L^2-metric on $\mathscr{A}_M$ defined by γ_M. We arrive at the conclusion that instantons on $\mathbb{R} \times M$ represent "trajectories of the L^2-gradient flow" of the Chern-Simons function S, even though such a flow is not defined everywhere as a homotopy on $\mathscr{B}_M$. The analogue of (4) and (14) is now the family of instanton spaces

$$\mathscr{M}(\alpha \mid \mathbb{R} \times M \mid \beta) = \{a \in \mathscr{A}_{\mathbb{R} \times M} \mid \quad A \text{ satisfies (24)} \quad \} / \mathscr{G}_{\mathbb{R} \times M} \tag{23}$$

with

$$\begin{cases} F_a^+ = 0 \,, \\ \lim_{t \to -\infty} a(t) = \alpha \,, \\ \lim_{t \to \infty} a(t) = \beta \,. \end{cases} \tag{24}$$

The connection $\alpha \in \mathscr{C}(S)$ is nondegenerate as a critical point of S in $\mathscr{B}_M$ iff the first cohomology group $H^1_\alpha(M, \mathrm{su}_2)$ with coefficients twisted by α vanishes. Under this condition, it follows from Taubes' analysis that "generically" the spaces (18) are smooth finite dimensional manifolds. (Here, we may need perturbations not only of the metric but also of S itself, and such perturbations may also be used to make $\mathscr{C}(S)$ nondegenerate whenever this is not a priori the case. However, for the sake of brevity we will have to ignore this point here.) Orientations can be derived from Donaldson's work [11, 12] and a "weak compactness theory" of instanton spaces was developed by Uhlenbeck [28, 29]. With these three ingredients, we can again set up the dual "Morse inequalities". Assume first that $H_1(M) = 0$, to avoid "abelian" elements of $C(S)$. Denote by $C_*(M)$ the free abelian group over $\mathscr{C}(S) \setminus \{\theta\}$ (which is finite when nondegenerate), and define

$$\partial : C_*(M) \to C_*(M) \qquad < \partial \alpha, \beta > = \# \hat{\mathscr{M}}(\alpha \mid \mathbb{R} \times M \mid \beta) \tag{25}$$

with $\#$ and $\hat{\mathscr{M}}$ defined as before. The proof of the boundary property $\partial \partial = 0$ now relies on an analogue of (8) which is almost the original Taubes grafting procedure. The result is the following new topological invariant:

Theorem 3. *We have $\partial \partial = 0$, and the homology groups*

$$F_*(M) = \ker \partial / \mathrm{im} \partial$$

---

[9] Which we will consider as the zero–component a_0.

do not depend on the choice of the metric γ_M. Moreover, for every oriented homology 3-sphere there are perturbations of S whose critical set is nondegenerate, and which give rise to a chain complex $(C_*(M, h), \partial_{\gamma,h})$, whose homology depends only on M.

The groups $F_*(M)$ were calculated by Fintushel and Stern for Brieskorn spheres $M(p, q, r)$. Here S truns out to be "perfect", and $F_*(M) = C_*(M)$. For more complex Brieskorn spheres $M(p_1, \ldots, p_n)$, S is still perfect in the sense that although $C(S)\backslash\{\theta\}$ is not discrete, its homology coincides with $F_*(M)$, see $[4, 22]$[10]. There is, however, no reason to believe that S will be perfect for any large class of 3-manifolds, and other methods of calculation are needed. These rely very much on the functorial property of F_* with respect to (oriented) cobordism, and on certain exact sequences which define a system of "axioms" for the "homology theory" F_*. To explain functoriality, consider a smooth compact oriented cobordism W between M and N, which we denote as $W : M \rightarrow N$. Add half cylinders $\mathbb{R}_- \times M$ and $\mathbb{R}_+ \times N$ to W to obtain an open 4-manifold W_α, and consider a metric on W_α which is cylindrical on the ends. Then we can define finite dimensional, oriented, and weakly compact manifolds $\mathcal{M}_\gamma(\alpha \mid W \mid \beta)$ for $\alpha \in R_M$ and $\beta \in R_N$ in much the same way as in the case $W = \mathbb{R} \times M$. (Of course, the translational symmetry is not defined anymore.) It turns out that, counting the discrete part of the (unreduced) moduli spaces, we obtain a chain homomorphism

$$W_* : C_*(M) \rightarrow C_*(N) \qquad < \alpha \mid W_* \mid \beta >= \#\mathcal{M}(\alpha \mid W \mid \beta). \qquad (26)$$

The proof of the chain property is similar to the proof of $\partial\partial = 0$. Hence, cobordisms $W : M \rightarrow N$ of homology 3-spheres induce homomorphisms $W_* : F_*(M) \rightarrow F_*(W)$. This construction turns out to be functorial with respect to the obvious composition law for cobordism.

In the category of 3-dimensional cobordisms between surfaces, such a functor (taking values in finite dimensional Hilbert spaces) has been recently discovered, the Witten–Jones "topological quantum field theory". Its initial definition uses the path integral of the Quantum–Field–theory on 3-dimensions, in which the Chern–Simons function S plays the role of the Lagrangian.[11] The path integral itself does not have an entirely rigorous mathematical meaning, but physical reasoning implies certain axiomatic properties the resulting functor which allow its construction by combinatorial techniques. To a certain degree the functor F_* may also be considered as a topological quantum field theory, though its "axioms", besides functoriality itself, differ strongly from those of Witten–Jones

---

[10] This is similar to the determination of $I_*(P)$ in the symplectic case.

[11] It is quite remarkable that this function thus leads to two apparently quite independent sets of topological invariants for three manifolds, but no direct relation can be derived from the definition.

theory: the central ingredient seems to be an exact sequence

$$F_*(M)$$

$$F_*(M_0) \longrightarrow F_*(M_1) \tag{27}$$

where M_0 and M_1 are given by two different kinds of surgery on the same knot in M, see [19]. Moreover, even though such exact sequences can be shown to define a "complete" axiomatic system, i.e. to "characterize" the functor F_* in the same way as the Eilenberg–Steenrod axioms characterize usual homology, they do not yield a straight forward calculation for the groups F_* due to the ambiguities inherent in exact sequences. Exact sequences are nevertheless the most useful tools for calculations in ordinary homology, and we expect the same to be true for F_*. The functorial structure also suggests that F_* should find applications to special (e.g. integral) cobordism groups of 3-manifolds. For example, parts of the work of Fintushel and Stern on the integral cobordism classes of Brieskorn spheres can be put into a "homological" framework, and can thus be extended to more general classes of manifolds. Moreover, F_* comes up in connection with Donaldson's applications of instantons on 4-manifolds. For a manifold W with boundary $\partial W = M$, the Donaldson polynomials can be extended to $\mathbb{Z}$–linear maps

$$S_* H_2(W) \to F_*(M). \tag{28}$$

This allows us to formulate "obstructions" to closing a 4-manifold. (In particular, we have $F_*(M) \neq 0$ whenever a homology 3-sphere M bounds a 4-manifold whose intersection form is "forbidden" by one of Donaldson's theorems.) Moreover, with luck, one may be able to analyse the change of the Donaldson polynomial (22) under handle addition. The exact triangle for surgery on M may be considered as one step in this direction.

References

1. S. Akbulut, J. McCarthy: Casson's invariant for oriented homolgy 3-spheres – an exposition. Mathematical Notes. Princeton University Press, 1987
2. V.I. Arnold: Mathematical methods of classical mechanics. Chapter Appendix 9. Springer, 1978
3. V.I. Arnold: Sur une propriété topologique des applications globalement canoniques de la méchanique classique. C. R. Acad. Paris (1965) **261**, 3719–3722
4. S. Bauer, C. Okonek: The algebraic geometry of representation spaces associated to Seifert fibred homology 3-spheres. Preprint.
5. R. Bott: An application on the Morse theory to the topology of Lie-groups. Bull. Soc. Math. France **84** (1956) 251–281
6. R. Bott: Lectures on Morse theory, old and new. Bull. AMS **7** (1982) 331–358
7. A. Casson: An invariant for homology 3-spheres. 1985. Lectures at MSRI Berkeley

8. C.C. Conley, E. Zehnder: The Birkhoff-Lewis fixed point theorem and a conjecture of V. I. Arnold. Invent. math. **73** (1983) 33–49

9. S.S Chern, J. Simons: Characteristic forms and geometric invariants. Ann. Math. **99** (1974) 48–69

10. C.C. Conley: Isolated invariant sets and Morse index. CBMS Reg. Conf. Series in Math. **38**. AMS, 1978

11. S.K. Donaldson: An application of gauge–theory to the topology of 4-manifolds. J. Diff. Geom. **18** (1983) 269–316

12. S.K Donaldson: The orientation of Yang–Mills moduli spaces and 4-manifold topology. J. Diff. Geom. **26** (1987) 397–428

13. R. Fintushel, R.J Stern: Instanton homology groups of Seifert fibred homology 3-spheres. Preprint

14. R.Fintushel, R.J. Stern: SO(3)-connections and the topology of 4-manifolds. J. Diff. Geom. **20** (1984) 523–539

15. A. Floer: Morse thoery for Lagrangian intersection theory. J. Diff. Geom. **18** (1988) 513–517

16. A. Floer: Symplectic fixed points and holomorphic spheres. Comm. Math. Physics**120** (1989) 576–611

17. A. Floer: The unregularised gradient flow of the symplectic action. Comm. Pure Appl. Math. **41** (1988) 775–813

18. A. Floer: Witten's complex and infinite dimensional Morse theory. J. Diff. Geom. **30** (1989) 207–221

19. A. Floer: Instanton homology and dehn surgery. In preparation.

20. W.M. Goldman: Invariant functions on lie groups and Hamiltonian flows of surface group representation. MSRI preprint, 1985

21. M. Gromov: Pseuodoholomorphic curves in symplectic manifolds. Inv. math. **82** (1985) 307–347

22. P.A. Kirk, P.E. Klassen: Representation spaces of Seifert fibred homology spheres. To appear in Topology

23. R.S. Palais: Morse theory on Hilbert manifolds. Topology **2** (1963) 229–340

24. P. Rabinowitz: Periodic solutions of Hamiltonian systems. Comm. Pure Appl. Math. **31** (1978) 157–184

25. J. Sachs, K.K. Uhlenbeck: The existence of minimal 2-spheres. Ann. Math. **113** (1981) 1–24

26. C.H. Taubes: A framework for Morse theory for the Yang-Mills functionals. preprint Havard University, 1986

27. C.H. Taubes: Self-dual Yang-Mills connections on non-self-dual 4-manifolds. J. Diff. Geom. **17** (1982) 139–170

28. K.K. Uhlenbeck: Connections with L_p-bounds on curvature. Commun. Math. Phys. (1982) 32–32

29. K.K. Uhlenbeck: Removable singularities in Yang-Mills fields. Commun. Math. Phys. (1982) 11–29

30. E. Witten: Supersymmetry and Morse theory. J. Diff. Geom. **17** (1982) 661–692

31. E. Witten: Topological quantum field theory. Preprint IAS, Feb. 1988

Braids, Galois Groups, and Some Arithmetic Functions

Yasutaka Ihara

Research Institute for Mathematical Sciences, Kyoto University, Kyoto 606, Japan

This lecture is about some new relations among the classical objects of the title. The study of such relations was started by [B_1, G, De, Ih_1] from independent motivations, and was developed in [A_3, C_3, A-I, IKY, Dr_2, O, N], etc. It is still a very young subject, and there are several different approaches, each partly blocked by its own fundamental conjectures! But it is already allowing one to glimpse some new features of the classical "monster" $\mathrm{Gal}(\bar{\mathbb{Q}}/\mathbb{Q})$, and providing a bridge connecting $\mathrm{Gal}(\bar{\mathbb{Q}}/\mathbb{Q})$ even with such "modern" objects as the quantum groups [Dr_2]. I will not try to "explain" any general philosophies that are still in the air, but to draw a few lines sketching the concretely visible features of the subject.

§1. Introduction

The absolute Galois group over the rational number field $\mathbb{Q}$, denoted by $G_{\mathbb{Q}} = \mathrm{Gal}(\bar{\mathbb{Q}}/\mathbb{Q})$, is one of the classical mathematical objects which we really need to understand better. It is primarily a (huge) topological group; the automorphism group of the field $\bar{\mathbb{Q}}$ of all algebraic numbers in $\mathbb{C}$, equipped with the Krull topology. It is moreover equipped with arithmetic structure, i.e., the system of conjugacy classes of embeddings $G_{\mathbb{Q}_p} \to G_{\mathbb{Q}}$ of all local absolute Galois groups into $G_{\mathbb{Q}}$. Through the arithmetic structure of $G_{\mathbb{Q}}$, each "natural" representation of $G_{\mathbb{Q}}$ provides the set of all prime numbers with a "natural" additional structure. For example, the character $\chi_N : G_{\mathbb{Q}} \to (\mathbb{Z}/N)^\times$ defining the action of $G_{\mathbb{Q}}$ on the group of N-th roots of unity gives rise to the classification of prime numbers modulo N. Study of the group $G_{\mathbb{Q}}$ and its natural representations has an *ultimate* goal to understand the "total structure" of the set of all prime numbers.

Now if X is an algebraic variety over $\mathbb{Q}$, the fundamental group $\pi_1(X(\mathbb{C}), b)$, when suitably completed, is equipped with a natural action of $G_{\mathbb{Q}}$. Here, b is a $\mathbb{Q}$-rational base point of X. Already in some basic cases, this action gives rise to a very big and interesting representation of $G_{\mathbb{Q}}$. What this action amounts to is the following. (We may, and will, assume X to be geometrically connected.) As is well known, all finite (topological) coverings of $X(\mathbb{C})$ and morphisms between two such coverings are algebraic and defined over $\bar{\mathbb{Q}}$. The above action contains all information related to the "field of definition" and the $G_{\mathbb{Q}}$-conjugations of

Proceedings of the International Congress
of Mathematicians, Kyoto, Japan, 1990
© The Mathematical Society of Japan, 1991

these coverings, morphisms and the points above b. In particular, if they are all defined over some Galois extension $\Omega/\mathbb{Q}$ ($\Omega \subset \bar{\mathbb{Q}}$), then $G_{\mathbb{Q}}$ acts via its factor group $\text{Gal}(\Omega/\mathbb{Q})$. (The converse is also valid if the completion of π_1 has trivial center.) Choice of completions of π_1 depends on the family of finite coverings one wants to consider. If this is all the finite coverings, the choice should be the profinite completion $\hat{\pi}_1 = \varprojlim(\pi_1/N)$, the projective limit of all finite factor groups of π_1. It is also important to consider the pronilpotent completion π_1^{nil}, where one allows only those π_1/N (the covering groups) that are nilpotent.

If there is no family of finite coverings of X having a remarkable common field of definition, then the representation of $G_{\mathbb{Q}}$ on $\hat{\pi}_1(X(\mathbb{C}), b)$ would not be so interesting. But already when $X = \mathbb{P}^1 - \{0, 1, \infty\}$, there are several natural families with various remarkable arithmetic features in their fields of definition (see (4) below). One of the reasons why this type of representations is interesting lies in that these different arithmetic features can be viewed simultaneously from a certain height, the representation of $G_{\mathbb{Q}}$ on $\hat{\pi}_1$. Other reasons and motivations will also be explained below, with names of the main contributors. Before this we shall give a sequence of basic examples of X starting with $\mathbb{P}^1 - \{0, 1, \infty\}$. For $n \geq 4$, X_n is the moduli space of ordered n-tuples $(x_1, \ldots, x_n)$ of distinct points on the projective line $\mathbb{P}^1$. The corresponding fundamental group $P_n = \pi_1(X_n(\mathbb{C}), b)$ is the quotient modulo center ($\cong \mathbb{Z}/2$) of the pure sphere braid group on n strings. Note that $X_4 \simeq \mathbb{P}^1 - \{0, 1, \infty\}$, $P_4 \simeq F_2$ (free, rank 2).

Recently, several substantial works have been done in connection with these "big Galois representations." They include the following (1)~(6):

(1) Already for more than a decade, this type of Galois representations has been *used* effectively to construct finite Galois extensions over $\mathbb{Q}$ with given Galois groups, for various cases of finite simple groups (G. V. Belyĭ, M. Fried, B. Matzat, K. Y. Shih, J. G. Thompson, ...). By the Hilbert irreducibility theorem, it suffices to construct a Galois extension over some rational function field $\mathbb{Q}(t_1, \ldots, t_m)$ ($m \geq 1$) having a given Galois group. Many such extensions have been detected among the (function fields of) coverings of X_{m+3} (etc.) by combining (i) only the basic knowledge on how $G_{\mathbb{Q}}$ acts on $\hat{\pi}_1(X(\mathbb{C}), b)$, with (ii) a deep knowledge on specific finite groups and their characters. As there is a distinguished report on this topic in the last ICM [B$_2$] (see also [Ma] for the later development), we shall only recall and stress the following:

Belyĭ proved, among other things in [B$_1$] that the canonical representation

$$\varphi_X : G_{\mathbb{Q}} \to \text{Out}\, \hat{\pi}_1(X(\mathbb{C}))$$

for $X = \mathbb{P}^1 - \{0, 1, \infty\}$ is *injective*. This sets in evidence the importance of the problem to characterize the image. Here, $\text{Out}\, \hat{\pi}_1$ is the outer automorphism group, and φ_X is induced from the $G_{\mathbb{Q}}$-action on $\hat{\pi}_1(X(\mathbb{C}), b)$. φ_X is "independent" of b in the obvious sense.

(2) A. Grothendieck made some basic proposals related to the study of the $G_{\mathbb{Q}}$-action on $\hat{\pi}_1(X(\mathbb{C}), b)$ [G]. One of them is as follows. Take $X = X_4 = \mathbb{P}^1 - \{0, 1, \infty\}$. Then it has the obvious S_3-symmetry but there is no S_3-invariant choice of a

base point b. This already indicates that the use of the fundamental *groupoid* on a suitable S_3-stable set B of base points would be better. (The natural action of S_4 on X_4 factors through its quotient $\simeq S_3$.) He also suggests using all possible combinatorial relations among the $X_n(\mathbb{C})$ $(n = 4, 5, \ldots)$, and that, in a certain sense, the two cases of n with $\dim X_n = 1, 2$ (i.e., $n = 4, 5$) would be basic. In particular, understanding and using full relationship between $X_4(\mathbb{C})$ and $X_5(\mathbb{C})$ would give, presumably, all the crucial non-obvious information on the image of φ_X for $X = X_4$.

We shall describe more about these in the main text; the action of $G_\mathbb{Q}$ on the completed fundamental groupoids (§2), Deligne's tangential base points (which will serve as B) (§2.3), Drinfeld's new information on the image of φ_X (§3), and the Lie version of the study of the $G_\mathbb{Q}$-action on $\{\pi_1^{\mathrm{nil}}(X_n(\mathbb{C}))\}_n$ (§5).

(3) π_1^{nil} *as a new test case for motivic philosophy.*

In [De], P. Deligne develops a motivic theory of nilpotent quotients of fundamental groups in algebraic geometry. The first main point is that π_1^{nil} is not just a topological group with a $G_\mathbb{Q}$-action but is a limit of objects with more structures $-$: To each $\pi_1 = \pi_1(X(\mathbb{C}), b)$, one can associate, via Malcev, some projective system $\{U^m \pi_1\}_{m \geq 1}$ of linear unipotent algebraic groups $U^m \pi_1$ over $\mathbb{Q}$. One may assume that the group of $\mathbb{Z}$-valued points $(U^m \pi_1)(\mathbb{Z})$ is the quotient modulo torsion of $\pi_1 / \pi_1(m+1)$, where $\pi_1(m+1)$ is the $(m+1)$-th member of the lower central series of π_1. The Galois group $G_\mathbb{Q}$ acts on the profinite groups $(U^m \pi_1)(\widehat{\mathbb{Z}})$ $(m \geq 1)$, from which the $G_\mathbb{Q}$-action on π_1^{nil} can be almost recovered. Under some assumptions on X, the underlying vector space over $\mathbb{Q}$ of the Lie algebra $\mathrm{Lie}(U^m \pi_1)$ has mixed Hodge structure (J. Morgan, D. Sullivan, K. T. Chen, R. Hain, $\ldots$). Deligne adds more structures to $\mathrm{Lie}(U^m \pi_1)$ (and accordingly, on $U^m \pi_1$), the mixed motif structure, in terms of various *realizations*.

When $X = \mathbb{P}^1 - \{0, 1, \infty\}$, the motivic $\mathrm{Lie}(U^m \pi_1)$ is a successive extension of Tate's motives "$\mathbb{Q}(k)$" $(1 \leq k \leq m)$. Knowledge on the motivic extension of "$\mathbb{Q}$" by "$\mathbb{Q}(k)$" turns out to be crucial in order to understand a certain portion of the Galois action on $(U^m \pi_1)(\widehat{\mathbb{Z}})$. Deligne constructs a basic extension of "$\mathbb{Q}$" by "$\mathbb{Q}(k)$" $(k \geq 3, \text{odd})$, describes a certain portion of the Galois action on the double commutator quotient of π_1^{nil}, and under the hypothesis "$\mathrm{Ext}^1(\mathbb{Q}, \mathbb{Q}(k))$ *for such k is one-dimensional*", derives further desirable consequences related to the size and the "$\mathbb{Q}$-rationality" of the Galois image. (See §5.4 below.)

(4) Some new objects in number theory, such as adelic analogues of beta and gamma functions, and higher circular l-units, have been constructed and used to give explicit comparisons of the $G_\mathbb{Q}$-actions, on π_1^{nil} of $\mathbb{P}^1 - \{0, 1, \infty\}$, on torsion points of Fermat Jacobians, and on higher circular l-units (G. Anderson, R. Coleman, the author, $\ldots$). This series of work was started in [Ih$_1$] and was developed in [A$_3$, C$_3$, IKY, Ih$_i$, A-I$_{1,2}$], etc., by combining with other ideas [A$_1$, C$_1$, C$_2$], $\ldots$.

The following special towers of coverings of $\mathbb{P}^1 - \{0, 1, \infty\}$ are relevant; (i) *the meta-abelian tower,* (ii) *the nilpotent tower,* and (iii) *the genus 0 tower.* Here, (i) corresponds to the double commutator quotient $\hat{\pi}_1 / \hat{\pi}_1''$ of $\hat{\pi}_1$, (ii) to π_1^{nil}, and (iii) is also quite big $-$ what it generates is bigger than (ii). As for the common field of definition, (i) is related to abelian extensions over the cyclotomic field $\mathbb{Q}(\mu_\infty)$, (ii) to a very natural sequence of Galois extensions over $\mathbb{Q}$ that are nilpotent over $\mathbb{Q}(\mu_\infty)$, and (iii) to (the field generated by) higher circular l-units.

As for (i), the conclusion is that the Galois actions on $\hat{\pi}_1/\hat{\pi}_1''$, on torsion points of Fermat Jacobians, and on roots of circular units, can be compared with each other in terms of *explicit universal formulas*. The size of the Galois image is also measured explicitly. The theory developed to that of Anderson's hyperadelic gamma function, which also plays the role of a bridge connecting Gauss sums with circular units.

As for (ii) and (iii), the main conclusion is that the action of $\sigma \in G_\mathbb{Q}$ on π_1^{nil} can be expressed explicitly   mod l^n, for any $n \geq 1$, in terms of its action on the group of higher circular l-units.

For more of these, see §6.

(5) V. G. Drinfeld more recently discovered a striking connection between (what is expected to be) the automorphism group of the tower $\{\pi_1(X_n(\mathbb{C}))\}_{n \geq 4}$ and a "universal braid transformation group"

$$(A, \Delta, \varepsilon, \phi, R) \rightarrow (A, \Delta, \varepsilon, \phi', R')$$

acting on the structures of quasi-triangular quasi-Hopf algebras [Dr$_2$]. These groups become non-trivial after suitable *completions* of π_1, etc. The Galois image in Aut $\hat{\pi}_1(X_4(\mathbb{C}))$ is contained in these (essentially the same) groups. (See §4.)

(6) There are several other important works on this subject including the following.

(i) Grothendieck raises a question as to whether the intertwiners of two Galois representations in $\hat{\pi}_1$ arising from two algebraic varieties always correspond to algebraic morphisms in the "non-abelian" situation. In the special case of $\mathbb{P}^1$ minus finitely many points, H. Nakamura [N] gives an affirmative answer.

(ii) T. Oda and Y. Matsumoto each gives, from different viewpoints, a non-abelian analogue of the Néron-Ogg-Šafarevič criterion for good reduction of curves, using π_1^{nil} instead of H_1^{et}; cf. [O].

Due to the lack of time, these will not be included in this lecture.

We shall start by defining the Galois action on the completed fundamental groupoids.

§2. The Basic Definitions

2.1 Let X be an algebraic variety defined over $\mathbb{Q}$, and $X(\mathbb{C})$ be the set of $\mathbb{C}$-rational points of X equipped with the usual (complex analytic) topology. We assume X to be geometrically connected, which is equivalent with $X(\mathbb{C})$ being connected. For $a, b \in X(\mathbb{C})$, call $\pi_1(X(\mathbb{C}); a, b)$ the set of homotopy classes (rel. to a, b) of paths from a to b on $X(\mathbb{C})$. When $a = b$, it is denoted by $\pi_1(X(\mathbb{C}), b)$. The composition rule

$$\pi_1(X(\mathbb{C}); b, c) \times \pi_1(X(\mathbb{C}); a, b) \longrightarrow \pi_1(X(\mathbb{C}); a, c) \qquad (2.1.1)$$

gives the system $\{\pi_1(X(\mathbb{C}); a, b)\}_{a,b \in B}$ structure of groupoid, the fundamental groupoid of $X(\mathbb{C})$ with base point set B ($B \subset X(\mathbb{C})$). When $B = \{b\}$, $\pi_1(X(\mathbb{C}), b)$ is the fundamental group.

We shall recall the definitions of the profinite completion $\hat{\pi}_1(X(\mathbb{C}); a, b)$ and the $G_{\mathbb{Q}}$-action on $\hat{\pi}_1(X(\mathbb{C}); a, b)$ when $a, b \in X(\mathbb{Q})$ (the $\mathbb{Q}$-rational points of X).

First, $\hat{\pi}_1(X(\mathbb{C}); a, b)$ is the completion of $\pi_1(X(\mathbb{C}); a, b)$ with respect to the following topology. Two elements p, p' of the latter set are sufficiently close to each other if the associated "round trip" $p'^{-1} \cdot p$ belongs to a sufficiently small subgroup of $\pi_1(X(\mathbb{C}), a)$ with finite index. The completed set $\hat{\pi}_1(X(\mathbb{C}); a, b)$ is then profinite, i.e., compact and totally disconnected topological space. The composition (2.1.1) is continuous and carries over to the completion. Each path class $p \in \pi_1(X(\mathbb{C}); a, b)$ gives rise to a "compatible" system of fiber bijections $p_f : f^{-1}(a) \xrightarrow{\sim} f^{-1}(b)$, where f runs over all finite coverings of $X(\mathbb{C})$, and p_f is induced by tracing above a path representing p. Here, "compatible" means that for any f, f', p_f and $p_{f'}$ are compatible with the fiber projections induced from each element of $\mathrm{Hom}(f, f')$. This procedure $p \to \{p_f\}_f$ induces the bijection

$$\hat{\pi}_1(X(\mathbb{C}); a, b) \approx \left\{ \begin{array}{l} \text{compatible systems of fiber} \\ \text{bijections} \quad \hat{p}_f : f^{-1}(a) \xrightarrow{\sim} f^{-1}(b) \end{array} \right\}_f .$$

This is because $X(\mathbb{C})$ is locally arcwise connected and locally simply connected, being an underlying space of an analytic space.

Since all finite coverings of $X(\mathbb{C})$ are algebraic and defined over $\overline{\mathbb{Q}}$ (the generalized Riemann existence theorem), we may assume that f runs over all finite etale coverings of $X \otimes \overline{\mathbb{Q}}$. If $a, b \in X(\mathbb{Q})$, each $\sigma \in G_{\mathbb{Q}}$ induces the fiber bijections $f^{-1}(a) \xrightarrow{\sim} (\sigma f)^{-1}(a)$, $f^{-1}(b) \xrightarrow{\sim} (\sigma f)^{-1}(b)$; hence σ acts on $\hat{\pi}_1(X(\mathbb{C}); a, b)$ by

$$\{\hat{p}_f\} \longrightarrow \{\sigma \circ \hat{p}_{\sigma^{-1}f} \circ \sigma^{-1}\} .$$

It is easy to see that this action is compatible with the composition induced from (2.1.1) by completion. Therefore, $G_{\mathbb{Q}}$ acts as an automorphism group of the completed fundamental groupoid of $X(\mathbb{C})$ with base point set $B \subset X(\mathbb{Q})$; in particular, as that of the completed fundamental group with base point $b \in X(\mathbb{Q})$. These actions are compatible with the groupoid homomorphisms induced from any algebraic morphisms $X' \to X$ over $\mathbb{Q}$.

Example. Take $X = \mathbb{P}^1 - \{0, \infty\}$ and $b = 1$. Then $\pi_1(X(\mathbb{C}), 1) \simeq \mathbb{Z}$, being generated by the class p of the loop $\tau \to \exp(2\pi i \tau)$ ($0 \le \tau \le 1$), and $\hat{\pi}_1(X(\mathbb{C}), 1) = \hat{\mathbb{Z}} = \varprojlim(\mathbb{Z}/N)$. If f_N is the cyclic covering $t \to t^N$ of X ($N = 1, 2, \ldots$), then

$$f_N^{-1}(1) = \{1, \zeta_N, \ldots, \zeta_N^{N-1}\}, \quad \zeta_N = \exp(2\pi i/N) .$$

Write $\sigma(\zeta_N) = \zeta_N^{\chi(\sigma)_N}$ ($\sigma \in G_{\mathbb{Q}}, \chi(\sigma)_N \in \mathbb{Z}/N$). Then p_{f_N} acts on $f_N^{-1}(1)$ by $\theta \longrightarrow \zeta_N \theta$, and σ acts on $f_N^{-1}(1)$ by $\theta \longrightarrow \theta^{\chi(\sigma)_N}$ ($\theta \in f_N^{-1}(1)$). Therefore,

$$\sigma \circ p_{f_N} \circ \sigma^{-1} = p_{f_N}^{\chi(\sigma)_N} . \tag{2.1.2}$$

Call $\chi(\sigma) = \varprojlim \chi(\sigma)_N \in \widehat{\mathbb{Z}}^\times$. Then χ is a (continuous) homomorphism

$$\chi : G_{\mathbb{Q}} \longrightarrow \widehat{\mathbb{Z}}^\times , \qquad\qquad (2.1.3)$$

called the cyclotomic character. By (2.1.2), σ acts on $\hat{\pi}_1 = \widehat{\mathbb{Z}}$ via $\chi(\sigma)$-multiplication. (Here and in the following, for any associative ring A with unit element, $A^\times$ denotes the group of invertible elements of A.)

2.2 We return to the general case. The Galois group $G_{\mathbb{Q}}$ thus acts on the completed fundamental groupoid $\{\hat{\pi}_1(X(\mathbb{C}); a, b)\}_{a,b \in B}$ ($B \subset X(\mathbb{Q})$) and in particular on $\hat{\pi}_1(X(\mathbb{C}), b)$ ($b \in X(\mathbb{Q})$) as an automorphism group:

$$\varphi_{X,b} : G_{\mathbb{Q}} \longrightarrow \operatorname{Aut} \hat{\pi}_1(X(\mathbb{C}), b) .$$

The induced homomorphism obtained by forgetting the role of the base point is also of interest:

$$\varphi_X : G_{\mathbb{Q}} \longrightarrow \operatorname{Out} \hat{\pi}_1(X(\mathbb{C}))$$

$(\operatorname{Out}(\hat{\pi}_1) = \operatorname{Aut}(\hat{\pi}_1)/\operatorname{Int}(\hat{\pi}_1)$, where $\operatorname{Int}(\hat{\pi}_1)$ is the inner automorphism group). Replacing $\hat{\pi}_1$ by the pronilpotent completion π_1^{nil}, we obtain $\varphi_{X,b}^{\mathrm{nil}}$, etc.

As for X, we shall mainly consider the varieties

$$X = X_n = \{(x_1, \ldots, x_n) \in (\mathbb{P}^1)^n; \quad x_i \neq x_j \quad \text{for} \quad i \neq j\}/PGL(2)$$

$(n \geq 4)$, where $PGL(2) = \operatorname{Aut} \mathbb{P}^1$ acts on X diagonally. Note that

$$X_4 \simeq \mathbb{P}^1 - \{0, 1, \infty\},$$
$$X_5 \simeq (X_4)^2 - \Delta \quad (\Delta : \text{the diagonal}),$$

$P_n = \pi_1(X_n(\mathbb{C}), *)$ is the quotient modulo center $(= \pi_1(PGL(2, \mathbb{C})) \simeq \mathbb{Z}/2)$ of the pure sphere braid group on n strings, and that

$$P_4 \simeq F_2 \quad \text{(free, rank 2)}$$
$$P_5 = F_2 \ltimes F_3 \quad \text{(semi-direct, } F_2 \text{ acts on } F_3) .$$

Now let $X = X_4 = \mathbb{P}^1 - \{0, 1, \infty\}$, so that $\pi_1(X(\mathbb{C})) \simeq F_2$. As reviewed in §1, Belyǐ proved that φ_X is then injective. This focuses light on the following

Question 1. What can one say about the images of $\varphi_X, \varphi_X^{\mathrm{nil}}$, etc.?

If one can characterize the image of φ_X explicitly, then one is led to a completely different description of $G_{\mathbb{Q}}$.

Question 2. What can one say about the "elementwise description" of $\varphi_X, \varphi_X^{\mathrm{nil}}$, etc.?

Actually, this $Q 2$ is not well posed at this stage of the development of mathematics, as we know no good system of NAMES for elements of $G_{\mathbb{Q}}$. As far as the author knows, no element of $G_{\mathbb{Q}}$ other than the identity element and the complex conjugation (remember that $\overline{\mathbb{Q}} \subset \mathbb{C}$ in our formulation) has an explicit name to identify itself. Nor is it so for $\widehat{F}_2$. Before asking to give an explicit description of the homomorphism φ_X, we must ask ourselves the possibility of giving good names to elements of $G_{\mathbb{Q}}, \widehat{F}_r (r \geq 2)$. If, however, we replace $\widehat{F}_r$ by the nilpotent completion F_r^{nil}, each element of F_r^{nil} then has a good name to identify itself which is obtained from the non-commutative Taylor expansion in r variables with coefficients in $\widehat{\mathbb{Z}}(\cong \prod_l \mathbb{Z}_l)$. Thus, at present, the sense of $Q\,2$ is to ask to *give a new system of names for elements of $G_{\mathbb{Q}}$ using φ_X^{nil} and describe* any interplay in terms of these names. (φ_X^{nil} is no longer injective; hence it concerns with a certain quotient of $G_{\mathbb{Q}}$ which is "big and small".)

We shall mainly discuss:

1) Works of Deligne, Drinfeld, and the author on $Q\,1$ (§§3, 4, 5),

and

2) Works of Anderson, Coleman, and the author on $Q\,2$ (§6).

Among them, 1) is closely related to $\varphi_{X,B}$ for $X = X_5$.

2.3 Let $X = \mathbb{P}^1 - \{0, 1, \infty\}$, and F_2 be the free group of rank 2 on two letters x, y. Our first goal is to show that for each $\sigma \in G_{\mathbb{Q}}$, $\varphi_X(\sigma)$ is determined by two "coordinates", $\chi(\sigma)$ and f_σ, where $\chi : G_{\mathbb{Q}} \to \widehat{\mathbb{Z}}^\times$ is the cyclotomic character and f_σ is an element of $\widehat{F}_2' = (\widehat{F}_2, \widehat{F}_2)$, the commutator subgroup of $\widehat{F}_2$. This can also be done relying more on group-theoretic normalization as in [B$_1$], [Ih$_1$], but we proceed more "conceptually" using Deligne's tangential base points [De].

Let $\mathbb{B}$ be the set of "arrows" $\overrightarrow{ij}$ with $i, j \in \{0, 1, \infty\}, i \neq j$. Thus, $\mathbb{B}$ has six elements and the symmetric group S_3 acts simply transitively on $\mathbb{B}$. For $a, b \in \mathbb{B}$, Deligne defines $\pi_1 = \pi_1(X(\mathbb{C}); a, b)$ and the $G_{\mathbb{Q}}$-action on its profinite completion. Topologically, it is clear what π_1 should mean when, in general, a, b are *simply connected subspaces* of $X(\mathbb{C})$. The base point $\overrightarrow{ij}$ plays the same role as the open interval I_{ij} on $\mathbb{R}$ bounded by i, j *and* not containing the third point k from $\{0, 1, \infty\}$. For a finite etale covering $Y \to X$ over $\overline{\mathbb{Q}}$, the fiber above $\overrightarrow{ij}$ consists of points $P \in \overline{Y}(\overline{\mathbb{Q}})$ above i given together with a "topological branch" (i.e., a lifting of I_{ij}) at each P. Here $\bar{f} : \overline{Y} \to \mathbb{P}^1$ is the compactification of f. In order to define the $G_{\mathbb{Q}}$-action on $\hat{\pi}_1(X(\mathbb{C}); a, b)$ for $a, b \in \mathbb{B}$, it suffices to give an *algebraic* interpretation of the branches at P. One way to put it is as follows. (This device proved to be useful [A–I$_1$].) Let t_{ij} be the linear fractional function $\mathbb{P}^1 \to \mathbb{P}^1$ which maps i, j, k to $0, 1, \infty$, respectively. Then a branch at P is a local embedding of the local ring of $\overline{Y}$ at P into the ring of *Puiseux series* in t_{ij} which extends: (i) the obvious embedding of the local ring of $\mathbb{P}^1$ at i into the ring of power series in t_{ij}, and (ii) the residue field embedding determined by the geometric point P. The corresponding topological branch is obtained by the principle to choose "the positive real root for $t_{ij}^{1/e}$, on I_{ij}". The group $G_{\mathbb{Q}}$ acts on the fibers above $\overrightarrow{ij}$ via its action on the Puiseux coefficients $\in \overline{\mathbb{Q}}$. One may prefer to reinterpret this in terms of the normalization of the fiber product of f with $\operatorname{Spec} \mathbb{C}[t_{ij}^{1/e}]$ (e: the ramification index).

Now we consider $\pi_1(X(\mathbb{C}); \overrightarrow{01}, \overrightarrow{10})$, $\pi_1(X(\mathbb{C}), \overrightarrow{01})$ and the Galois action on their completions. The first set contains an obvious element defined from the interval $(0,1)$. Call it p. The second group contains a small positive loop around 0, called x, and $y = p^{-1} \circ x' \circ p$, where x' is the transform of x by $t \to 1 - t$ $(t = t_{01})$. This group is free on x, y.

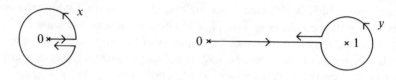

Now for each $\sigma \in G_{\mathbb{Q}}$, put

$$f_\sigma = p^{-1} \circ \sigma(p) \in \hat{\pi}_1(X(\mathbb{C}), \overrightarrow{01}) . \tag{2.3.1}$$

Then σ acts on the generators of $\hat{\pi}_1(X(\mathbb{C}), \overrightarrow{01}) = \widehat{F}_2$ as

$$x \longrightarrow x^{\chi(\sigma)}, \quad y \longrightarrow f_\sigma^{-1} \cdot y^{\chi(\sigma)} \cdot f_\sigma . \tag{2.3.2}$$

It follows easily that $f_\sigma \in \widehat{F}_2'$, and that (2.3.2) with this requirement characterizes f_σ. When σ is the complex conjugation, $\chi(\sigma) = -1$, $f_\sigma = 1$.

Remark. By (3.1.1)(I),(II) below, it follows also that $z = (xy)^{-1}$ (a loop around ∞) is mapped to $g_\sigma^{-1} z^{\chi(\sigma)} g_\sigma$, where $g_\sigma = f_\sigma(x, z) x^{\frac{1}{2}(1-\chi(\sigma))}$.

Although $\widehat{F}_2$ contains much more than the free words on x, y, we shall express an element of this group conveniently as $f(x, y)$, because it will then make sense to speak of $f(\xi, \eta)$ for any elements $\xi, \eta \in G$ of any profinite group G; the image of f under the unique homomorphism $\widehat{F}_2 \to G$ mapping x, y to ξ, η respectively.

§3. The Galois Action (Profinite)

3.1   So what is the image of the mapping $G_{\mathbb{Q}} \to \widehat{\mathbb{Z}}^\times \times \widehat{F}_2'$ defined by $\sigma \mapsto (\chi(\sigma), f_\sigma)$? The known equations satisfied by $\lambda = \chi(\sigma)$, $f = f_\sigma$ are as follows.

(I) $f(x, y)f(y, x) = 1;$

(II) $f(z, x)z^m f(y, z)y^m f(x, y)x^m = 1,$
             $\text{if} \quad xyz = 1, \quad m = \frac{1}{2}(\lambda - 1);$ (3.1.1)

(III) (Drinfeld)   Let $P_5 = \pi_1(X_5(\mathbb{C}), \mathscr{B}_5)$ and $x_{ij} \in P_5$

$(1 \le i, j \le 5)$ be as defined below. Then in $\widehat{P}_5$,

$$f(x_{12}, x_{23})f(x_{34}, x_{45})f(x_{51}, x_{12})f(x_{23}, x_{34})f(x_{45}, x_{51}) = 1 .$$

Remark. Drinfeld's formula given in [Dr₂] is in terms of *plane* braid group on 4 strings and is non-cyclic. The above formula is equivalent to his. The author previously wrote down more complicated formulas as 4 transposition relations (w.r.t. (1 2),(2 3),(3 4),(4 5) in S_5 instead of (12345)) in [Ih₆]. Drinfeld thinks this type of formula may very well be known to Grothendieck.

As for the more obvious question: "use of X_n $(n \geq 6)$...?", see §3.3.

The definitions of $\mathscr{B}_n$ and x_{ij}. Let $\tilde{\mathscr{B}}_n$ be the space of all n-tuples $(b_1, \ldots, b_n)$ of distinct points of $\mathbb{R}^{\cup}(\infty)$ satisfying the condition: b_{i+1} *is next to b_i in the positive direction for all i $(1 \leq i \leq n-1)$ including the case of passing through* ∞. Then $PGL_2^+(\mathbb{R})$, the real projective linear group of degree two with positive determinant, acts on $\tilde{\mathscr{B}}_n$ diagonally, and the quotient space $\mathscr{B}_n = \tilde{\mathscr{B}}_n / PGL_2^+(\mathbb{R})$ is naturally embedded into $X_n(\mathbb{C})$. The space $\mathscr{B}_n$ is simply connected and hence it makes sense to speak of the fundamental group $P_n = \pi_1(X_n(\mathbb{C}), \mathscr{B}_n)$. It is generated by the elements x_{ij} $(1 \leq i, j \leq n)$ shown below:

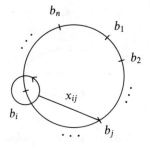

3.2 *Sketch of proof of (3.1.1).* (I) Apply the automorphism $\theta : t \to 1 - t$ to both sides of (2.3.1).

(II) Let r be the element of $\pi_1(X(\mathbb{C}); \overrightarrow{10}, \overrightarrow{1\infty})$ corresponding to the rotation of argument π at the point 1:

Then it is easy to see that

$$r^{-1} \cdot \sigma(r) = \theta(x)^{\frac{1}{2}(\chi(\sigma)-1)} \in \hat{\pi}_1(X(\mathbb{C}), \overrightarrow{10}) .$$

Therefore, if $q = r \circ p$, we have $\sigma(q) = qy^{\frac{1}{2}(\chi(\sigma)-1)} f_\sigma(x, y)$. Let ω be the automorphism $t \to (1-t)^{-1}$ of X. Then $\omega^2(q)\omega(q)q = 1$. Apply σ on this to obtain (II).

(III) The symmetric group S_5 acts on X_5 from the left by substitution of coordinates, as $(x_i) \to (x_{s^{-1}i})$ $(s \in S_5)$. Let $s = (13524)$. Then for a certain "tangential base point"β and a path class $q \in \pi_1(X_5(\mathbb{C}); \beta, s(\beta))$, one can naturally identify $\pi_1(X_5(\mathbb{C}), \beta)$ with $\pi_1(X_5(\mathbb{C}), \mathscr{B}_5) = P_5$ and show that

$$\sigma(q) = q \cdot f_\sigma(x_{45}, x_{34}) \ , \qquad (3.2.1)$$

$$s^4(q) \cdot s^3(q) \cdot s^2(q) \cdot s(q) \cdot q = 1 \ . \qquad (3.2.2)$$

The equation (III) follows directly from these. A topological illustration:

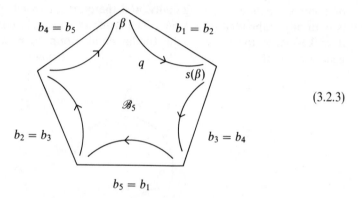

$$(3.2.3)$$

The author used a double Puiseux series expansion to define the algebraic interpretation of β.

3.3 As in [Dr$_2$], define $\widehat{GT}$ (the "Grothendieck-Teichmüller group") to be the subgroup of Aut $\widehat{F}_2$ consisting of all automorphisms of the form

$$x \longrightarrow x^\lambda, \quad y \longrightarrow f^{-1} \cdot y^\lambda \cdot f \qquad (\lambda \in \widehat{\mathbb{Z}}^\times, f \in \widehat{F}_2') \ ,$$

where λ and f satisfy (3.1.1)(I),(II),(III). (It forms a subgroup!) Then by definition, the image of $\varphi_{X,b}$ ($b = \vec{01}$) is contained in $\widehat{GT}$. As φ_X is injective (Belyĭ), so is $\varphi_{X,b}$. Therefore, $\varphi_{X,b}$ induces an inclusion

$$\varphi_{X,b} : G_\mathbb{Q} \hookrightarrow \widehat{GT} \ . \qquad (3.3.1)$$

Two questions arise:

Question 3.3.2 Can one still obtain new relations using X_n for higher n's ?

Question 3.3.3 Can one characterize the image of $G_\mathbb{Q}$ using these types of relations (or to ask more strongly, is (3.3.1) already a bijection) ?

If one believes a strict analogy with conformal field theory, then no more essentially new relations would be obtained by using X_n for higher n ([M-S, G,

T-K, Dr$_2$]). Drinfeld's new quantum group theoretic interpretation of $\widehat{GT}$ (see §4 below) gives a very impressive version of this philosophical reasoning.

See §5.3 for the Lie version of these questions.

§4. Connection with Transformation of Structure of "Complete" Quasi-Triangular Quasi-Hopf Algebras

4.1 Drinfeld introduced the concept of quasi-triangular quasi-Hopf algebra (abbrev. *qtqH* algebra) [Dr$_1$, Dr$_2$]. It is more general than Hopf algebra in that the coassociativity and the cocommutativity assumptions are weakened (equalities replaced by conjugations). Thus, a *qtqH* algebra over a field k is a quintuple $(A, \Delta, \varepsilon, \phi, R)$, where (i) A is an associative k-algebra with unit element 1, (ii) $\Delta : A \to A \otimes A$ (comultiplication) and $\varepsilon : A \to k$ (counit) are k-algebra homomorphisms mapping 1 to 1 ($\otimes$: the tensor product over k), such that

$$(\varepsilon \otimes \mathrm{id}) \circ \Delta = (\mathrm{id} \otimes \varepsilon) \circ \Delta = \mathrm{id} \; , \tag{4.1.1}$$

(iii) $\phi \in (A \otimes A \otimes A)^\times$ and $R \in (A \otimes A)^\times$ satisfy:

(A) The basic conjugacy relations:

$$(\mathrm{id} \otimes \Delta)(\Delta(a)) = \phi \cdot (\Delta \otimes \mathrm{id})(\Delta(a)) \cdot \phi^{-1} \; , \tag{4.1.2}$$

$$^t\Delta(a) = R \cdot \Delta(a) \cdot R^{-1} \; , \tag{4.1.3}$$

for all $a \in A$, where $b \to {}^t b$ ($b \in A \otimes A$) denotes the transposition of two factors. It is *not* assumed that $^t R \cdot R = 1$.

(B) The compatibility equalities:

$$(\mathrm{id} \otimes \varepsilon \otimes \mathrm{id})\phi = 1 \; , \tag{4.1.4}$$

(the pentagon relation) $\tag{4.1.5}$

$$(\mathrm{id} \otimes \mathrm{id} \otimes \Delta)\phi \cdot (\Delta \otimes \mathrm{id} \otimes \mathrm{id})\phi = (1 \otimes \phi) \cdot (\mathrm{id} \otimes \Delta \otimes \mathrm{id})\phi \cdot (\phi \otimes 1) \; ,$$

(two hexagon relations) omitted (cf. [Dr$_2$](I·6, ab)). $\tag{4.1.6}$
Finally, the existence of an antipode is assumed.

As is explained in [Dr$_2$] the assumptions in (iii) can be understood more conceptually in terms of isomorphisms between tensor products of A-modules. If V, W are (left) A-modules (w.r.t. the k-algebra structure of A), then their tensor product $V \otimes W$ over k is natually an $(A \otimes A)$-module, and via Δ, again an A-module. If V_1, V_2, V_3 are three A-modules, then the $(A \otimes A \otimes A)$-module $V_1 \otimes V_2 \otimes V_3$ can be regarded as an A-module in two different ways, and (4.1.2) imposes that the ϕ-multiplication induce an A-isomorphism

$$(V_1 \otimes V_2) \otimes V_3 \xrightarrow{\sim} V_1 \otimes (V_2 \otimes V_3) \; .$$

The pentagon relation requires the diagram

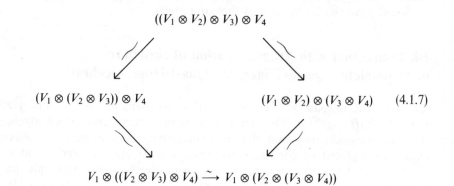

$$((V_1 \otimes V_2) \otimes V_3) \otimes V_4$$

$$(V_1 \otimes (V_2 \otimes V_3)) \otimes V_4 \qquad\qquad (V_1 \otimes V_2) \otimes (V_3 \otimes V_4) \qquad (4.1.7)$$

$$V_1 \otimes ((V_2 \otimes V_3) \otimes V_4) \xrightarrow{\sim} V_1 \otimes (V_2 \otimes (V_3 \otimes V_4))$$

to be commutative. In the diagrams for hexagon relations, the choice of directions of arrows related to "R" is delicate. The two diagrams (3.2.3),(4.1.7) are related to each other via Knizhnik–Zamolodchikov equations ([Dr₂]§2).

4.2   Now take $m \in \mathbb{Z}$ and $f(x, y) \in F_2$. If we replace ϕ, R by

$$\begin{cases} \phi' & = \phi \cdot f(R^{21}R^{12},\ \phi^{-1}R^{32}R^{23}\phi)\ , \\ R' & = R \cdot ({}^t RR)^m\ , \end{cases} \qquad (4.2.1)$$

respectively, then the 2 equalities in (A) are still satisfied. Here, if $R = \sum a_i \otimes b_i$, $R^{23} = \sum 1 \otimes a_i \otimes b_i$, $R^{32} = \sum 1 \otimes b_i \otimes a_i$, etc. Drinfeld shows that $(A, \Delta, \varepsilon, \phi', R')$ satisfies (B) *if and only if* (λ, f) satisfies the 3 equalities of (3.1.1), where $m = \frac{1}{2}(\lambda - 1)$. Actually, the only pairs of λ and f satisfying (3.1.1) are $(\lambda, f) = (\pm 1, 1)$. But if $\mathbb{Z}$ and F_2 are replaced by suitable *completions*, then there are many solutions. It suffices to recall that for $\lambda \in \widehat{\mathbb{Z}}^\times$ and $f \in \widehat{F}'_2$, (3.1.1) was the system of defining equations for the group $\widehat{GT}$, and that $\widehat{GT}$ contains $G_{\mathbb{Q}}(!)$. The above two explicit solutions correspond to the identity element and the complex conjugation of $G_{\mathbb{Q}}$. Drinfeld considers some types of *complete qtqH algebras* over the ring of formal power series $k[[h]]$ over a field k of characteristic 0, called *quantized universal enveloping algebras* (*abbrev.* QUE-algebras). In this case, $\otimes$ is replaced by the completed tensor product $\widehat{\otimes}$. For such *qtqH* algebras, the transformation (4.2.1) will make sense if λ, f are elements of $k^\times$, $F_2^{\mathrm{nil}}(k)$ (the "k-nilpotent completion" of F_2) respectively, and if R is sufficiently close to 1. Thus, one may define the group $GT(k)$ which is still "big" and has a meaning as the group of transformations of structures of QUE-algebras over $k[[h]]$. (The group law for $\widehat{GT}$ corresponds to the reverse of the composition of transformations (4.2.1).) Roughly speaking, one may consider $\widehat{GT}$ as the universal group of transformations of structures of *qtqH* algebras.

Remark. There is another type of transformations of structures of *qtqH* algebras called *twists* [Dr₂]. The two types of transformations commute with each other, and the twists do not change the structure (as quasi-tensor category) of the category of A-modules.

Question 4.2.2. *Is there any "profinite" $(A, \Delta, \varepsilon, \phi, R)$ over some ring related to $\overline{\mathbb{Q}}$ on which $G_{\mathbb{Q}}$ acts "naturally" and for which the natural action of σ^{-1} coincides with the transformation (4.2.1) obtained by $(\chi(\sigma), f_\sigma)$?*

§5. The Galois Action (Pronilpotent)

5.1 For $\pi_1 = \pi_1(X(\mathbb{C}), *)$, instead of $\hat{\pi}_1$ we may also consider the pronilpotent completion π_1^{nil}, the projective limit of all finite nilpotent factor groups of π_1. It is the direct product of its l-Sylow subgroups $\pi_1^{(l)}$, the pro-l completion of π_1, where l runs over all prime numbers. For a topological group G, we denote by

$$G = G(1) \supseteq G(2) \supseteq \cdots \supseteq G(m+1) \supseteq \cdots$$

its lower central series defined by $G(m+1) = (G, G(m))$ $(m \geq 1)$. Here, $(\,,\,)$ is the closure of the algebraic commutator. Note that

$$\bigcap_m G(m) = \{1\}, \quad \text{for} \quad G = \pi_1^{\text{nil}}, \ \pi_1^{(l)} \ .$$

The representations $\varphi_X, \varphi_{X,b}$ of $G_{\mathbb{Q}}$ induce the quotient representations $\varphi_X^{\text{nil}}, \varphi_X^{(l)}$, etc. There are various advantages of passage to these quotients: (i) For a "canonical" choice of X such as $\mathbb{P}^1 - \{0, 1, \infty\}$, φ_X^{nil} gives rise to a natural filtration of some quotient of $G_{\mathbb{Q}}$ which is of arithmetic interest. (ii) Each element of π_1^{nil}, for $\mathbb{P}^1 - \{0, 1, \infty\}$ etc., allows an explicit presentation as formal non-commutative power series over $\widehat{\mathbb{Z}}$. So, we may ask more explicit questions about the projection of $f_\sigma(x, y)$ on π_1^{nil} (than about $f_\sigma(x, y)$ itself). (iii) For π_1^{nil}, we may use Lie algebra techniques. (iv) The nilpotent quotients of the fundamental group π_1 are, in some sense, close to cohomology groups and are known to have additional structures. Under some assumptions on X, the Lie algebra of $\pi_1/\pi_1(m+1)$ has *mixed Hodge structure* (J. Morgan, D. Sullivan, K. T. Chen, R. Hain), and further, *mixed motif structure* (Deligne).

We shall start by explaining (i).

5.2 For each prime number l, there is a canonical sequence

$$\mathbb{Q} \subseteq \mathbb{Q}^{(l)}(1) \subseteq \cdots \subseteq \mathbb{Q}^{(l)}(m) \subseteq \cdots \subseteq \mathbb{Q}^{(l)}(\infty) = \bigcup \mathbb{Q}^{(l)}(m) \qquad (5.2.1)$$

of (infinite) Galois extensions over $\mathbb{Q}$, starting with $\mathbb{Q}^{(l)}(1) = \mathbb{Q}(\mu_{l^\infty})$ (μ_{l^∞} : the group of roots of unity of l-power order), and an associated graded Lie algebra $\mathfrak{g}^{(l)}$, defined as follows. For each $m \geq 1$, $\mathbb{Q}^{(l)}(m)$ is the field corresponding to the kernel of the representation

$$G_{\mathbb{Q}} \longrightarrow \text{Out}(F_2^{(l)}/F_2^{(l)}(m+1))$$

induced from $\varphi_X^{(l)}$ for $X = X_4 = \mathbb{P}^1 - \{0, 1, \infty\}$. This kernel will not change if X_4 is replaced by X_n $(n \geq 5)$ [Ih$_5$]. The union field $\mathbb{Q}^{(l)}(\infty)$ corresponds to the kernel of $\varphi_X^{(l)}$ for $X = X_n$, for any $n \geq 4$. It is a pro-l (non-abelian) extension over $\mathbb{Q}(\mu_{l^\infty})$ unramified outside l. For each $m \geq 1$, the Galois group $\mathrm{Gal}(\mathbb{Q}^{(l)}(m+1)/\mathbb{Q}^{(l)}(m))$ is a free $\mathbb{Z}_l$-module of finite rank (call it $r^{(l)}(m)$). It is centralized by $\mathrm{Gal}(\mathbb{Q}^{(l)}(m+1)/\mathbb{Q}^{(l)}(1))$, and as $\mathrm{Gal}(\mathbb{Q}^{(l)}(1)/\mathbb{Q})$-module, has the Tate twist m. The graded Lie algebra $\mathfrak{g}^{(l)}$ is the direct sum of its m-th graded pieces

$$\mathrm{Gal}(\mathbb{Q}^{(l)}(m+1)/\mathbb{Q}^{(l)}(m)) \otimes_{\mathbb{Z}_l} \mathbb{Q}_l \quad (m = 1, 2, 3, \dots),$$

each of which is a $\mathbb{Q}_l$-module ($\mathbb{Q}_l$: the l-adic number field); cf. [Ih$_4$]. This graded Lie algebra $\mathfrak{g}^{(l)}$ over $\mathbb{Q}_l$ is a standard "approximation" of the filtered Galois group $\mathrm{Gal}(\mathbb{Q}^{(l)}(\infty)/\mathbb{Q})$.

Here is a set of very basic open questions.

Question 5.2.2 (i) What is the structure of $\mathfrak{g}^{(l)}$? (ii) Does there exist a natural graded Lie algebra $\mathfrak{g}$ over $\mathbb{Q}$ such that $\mathfrak{g}^{(l)} \simeq \mathfrak{g} \otimes \mathbb{Q}_l$ for all l ? (iii) In particular, is the rank $r^{(l)}(m)$ for each m independent of l ?

We shall discuss two approaches to Q 5.2.2 (ii).

5.3 We shall define a candidate for the Lie algebra $\mathfrak{g}$ in question (cf. [Ih$_5$]). As before, let $P_n = \pi_1(X_n(\mathbb{C}), \mathscr{B}_n)$ $(n \geq 4)$ (see §3.1), and let $\mathfrak{P}_n$ be the graded Lie algebra over $\mathbb{Q}$ associated with the lower central series of P_n. Thus, the m-th graded piece $\mathrm{gr}^m \mathfrak{P}_n$ of $\mathfrak{P}_n$ is the $\mathbb{Q}$-module

$$(P_n(m)/P_n(m+1)) \otimes \mathbb{Q} \quad (m \geq 1) .$$

If X_{ij} $(1 \leq i, j \leq n)$ denotes the element of $\mathfrak{P}_n$ of degree 1 represented by x_{ij} (§3.1), then $\mathfrak{P}_n$ is the graded Lie algebra over $\mathbb{Q}$ generated by the X_{ij}'s which satisfy the fundamental relations

$$X_{ii} = 0, \quad X_{ij} = X_{ji}, \quad \sum_{k=1}^n X_{ik} = 0 \qquad (1 \leq i, j \leq n)$$

$$[X_{ij}, X_{kl}] = 0 \qquad \text{if } \{i, j\} \cap \{k, l\} = \phi$$

([K, Ih$_5$]). A *special derivation* of $\mathfrak{P}_n$ of degree m (≥ 1) is a derivation D of $\mathfrak{P}_n$ into itself such that $D(X_{ij}) = [T_{ij}, X_{ij}]$ with some $T_{ij} \in \mathrm{gr}^m \mathfrak{P}_n$ $(1 \leq i, j \leq n)$. Let $\mathscr{D}_n$ be the graded Lie algebra over $\mathbb{Q}$ whose m-th graded piece is the $\mathbb{Q}$-module of all *symmetric special outer* derivations of $\mathfrak{P}_n$ of degree m $(m \geq 1)$. Here, "symmetric" refers to the invariance with respect to the obvious S_n-action on $\mathfrak{P}_n$, and "outer" refers to considering modulo inner derivations.

The sense of considering such a Lie algebra $\mathscr{D}_n$ is that the $G_{\mathbb{Q}}$-action on $P_n^{(l)}$ gives rise to a degree-preserving Lie algebra embedding

$$\mathfrak{g}^{(l)} \hookrightarrow \mathscr{D}_n \otimes \mathbb{Q}_l \quad (n \geq 4) \tag{5.3.1}$$

for each prime number l. Now, when n increases, $\mathscr{D}_n$ gets *smaller*. More precisely, the projection $\mathfrak{P}_n \to \mathfrak{P}_{n-1}$, defined by letting $X_{ij} \to 0$ for i or $j = n$, induces a Lie homomorphism $\mathscr{D}_n \to \mathscr{D}_{n-1}$. We know that this is *injective* for $n \geq 5$. Thus, these embeddings give rise to an infinite chain

$$\mathscr{D}_\infty = \bigcap_{n \geq 4} \mathscr{D}_n \subseteq \cdots \subseteq \mathscr{D}_n \subseteq \cdots \subseteq \mathscr{D}_5 \subseteq \mathscr{D}_4 , \qquad (5.3.2)$$

of graded Lie algebras over $\mathbb{Q}$. By (5.3.1) and some compatibility,

$$\mathfrak{g}^{(l)} \subseteq \mathscr{D}_\infty \otimes \mathbb{Q}_l . \qquad (5.3.3)$$

Now we ask the following Lie versions of $Q\,3.3.2$, $Q\,3.3.3$:

Question 5.3.4 (i) $\mathscr{D}_\infty = \mathscr{D}_5$? (ii) $\mathfrak{g}^{(l)} = \mathscr{D}_\infty \otimes \mathbb{Q}_l$ *for all* l ?

Remark. From (5.3.2) (5.3.3), it follows that

$$r^{(l)}(m) \leq \dim \mathrm{gr}^m \mathscr{D}_\infty \leq \cdots \leq \dim \mathrm{gr}^m \mathscr{D}_5 \leq \dim \mathrm{gr}^m \mathscr{D}_4$$

for each $m \geq 1$. It is interesting to see which of the $\leq$ are equalities. For $m < 7$, $r^{(l)}(m) = \dim \mathrm{gr}^m \mathscr{D}_4$. But $r^{(l)}(7) = 1$ and $\dim \mathrm{gr}^7 \mathscr{D}_4 = 2$. Terada-Ihara and Drinfeld independently verified that $\dim \mathrm{gr}^7 \mathscr{D}_5 = 1$ ([Ih7, Dr2]), which is in favor of the affirmative aspect of $Q\,5.3.4$. See [Ih3, Ih4] for more about these ranks.

Problem 5.3.5. *Construct elements of $\mathscr{D}_\infty$ by algebraic or topological means.*

5.4 Deligne constructs a basic motivic extension of the Tate motives, "$\mathbb{Q}$" by "$\mathbb{Q}(m)$" ($m \geq 3$, odd) [De]. Moreover, assuming his conjecture (*loc. cit.* (8.1)) which asserts that such extensions form a 1-dimensional space generated by his extension, and using [So$_{1,2}$], he proves the following.

The Galois representation in $\pi_1^{(l)}/\pi_1^{(l)}(m+1)$ is "$\mathbb{Q}$-rational" in the following sense (for $X = \mathbb{P}^1 - \{0, 1, \infty\}$, $\pi_1 = \pi_1(X(\mathbb{C}), \overrightarrow{01})$): There exists a linear algebraic group Del_m over $\mathbb{Q}$ and a short exact sequence

$$1 \longrightarrow U\mathrm{Del}_m \longrightarrow \mathrm{Del}_m \longrightarrow GL(1) \longrightarrow 1 ,$$

with $U\mathrm{Del}_m$ unipotent, all independent of l, such that for each l the Galois representation in $\pi_1^{(l)}/\pi_1^{(l)}(m+1)$ factors through a representation $G_\mathbb{Q} \to \mathrm{Del}_m(\mathbb{Q}_l)$ which has an open image at least if $l > 2$. Moreover, the abelianization of $U\mathrm{Del}_m$, with the $GL(1)$-action, decomposes as

$$(U\mathrm{Del}_m)^{\mathrm{ab}} \simeq \bigoplus_{3 \leq k \leq m, \, odd} \mathbb{A}^1(k)$$

($\mathbb{A}^1$: *the affine line, k : the Tate twist*). *Finally, $\{\mathrm{Del}_m\}_{m \geq 1}$ forms a $G_\mathbb{Q}$-compatible projective system.*

In particular, if one assumes the above conjecture on the extension of "$\mathbb{Q}$" by "$\mathbb{Q}(m)$", then $Q\,5.2.2\,(ii)$ will have an affirmative answer with an additional information: the graded Lie algebra $\mathfrak{g}$ is generated by some subset of the form $\{s_m\}_m$, where m runs over *odd* integers ≥ 3 and s_m is of degree m.

§6. Arithmetic Aspects

6.1 In §6, we shall review some works of Anderson, Coleman, the author, ... , on the "elementwise description" and the arithmetic study of the representation φ_X^{nil} for $X = \mathbb{P}^1 - \{0, 1, \infty\}$.

Let $X = \mathbb{P}^1 - \{0, 1, \infty\}$ and, as in §2.3, identify the fundamental group $\pi_1(X(\mathbb{C}), \overrightarrow{01})$ with the free group F_2 on x, y. The Galois group $G_\mathbb{Q}$ acts on $\hat{\pi}_1(X(\mathbb{C}), \overrightarrow{01})$, and hence on $\widehat{F}_2$ and $F_2^{\mathrm{nil}} = \prod_l F_2^{(l)}$. Take any $\sigma \in G_\mathbb{Q}$. Then the action of σ on $\widehat{F}_2$ can be expressed by two coordinates $\chi(\sigma)$ and f_σ ($\chi(\sigma) \in \widehat{\mathbb{Z}}^\times, f_\sigma \in \widehat{F}_2'$) (see §2.3). Therefore, its action on F_2^{nil} can be expressed by $\chi(\sigma)$ and the projection f_σ^{nil} of f_σ on F_2^{nil}. Call $\mathscr{A} = \widehat{\mathbb{Z}} \ll \xi, \eta \gg$ the *non-commutative* power series algebra in two variables over $\widehat{\mathbb{Z}}$. Then F_2^{nil} can be embedded into $\mathscr{A}^\times$ via $x \to 1 + \xi$, $y \to 1 + \eta$, and f_σ^{nil} can be regarded as an element of $\mathscr{A}^\times$. Each coefficient of the power series f_σ^{nil} may be regarded as *a new invariant* of σ. One is motivated to compare it with other invariants of σ, the old ones (§§6.2, 6.3) or the newly constructed ones (§6.5).

Now as a $\widehat{\mathbb{Z}}$-module, $\mathscr{A}$ is the direct sum

$$\mathscr{A} = \widehat{\mathbb{Z}} \cdot 1 \oplus \mathscr{A} \cdot \xi \oplus \mathscr{A} \cdot \eta \,, \tag{6.1.1}$$

and $(f_\sigma^{\mathrm{nil}})^{-1}$ decomposes as

$$(f_\sigma^{\mathrm{nil}})^{-1} = 1 + A_1 \xi + A_2 \eta \quad (A_1, A_2 \in \mathscr{A}) \,.$$

Put $\psi_\sigma(\xi, \eta) = 1 + A_1 \xi$. Then it follows easily from (3.1.1)(I) that

$$f_\sigma^{\mathrm{nil}} = \psi_\sigma(\eta, \xi) \cdot \psi_\sigma(\xi, \eta)^{-1} \,.$$

Thus, *knowing* ψ_σ *is equivalent to knowing* f_σ^{nil}. Moreover, ψ_σ is an anti 1-cocycle

$$\psi_{\sigma\tau} = \sigma(\psi_\tau) \cdot \psi_\sigma \quad (\sigma, \tau \in G_\mathbb{Q})$$

with respect to the action of $G_\mathbb{Q}$ on $\mathscr{A}$ extending that on F_2^{nil}, and is more convenient for describing the σ-action on abelian subquotients of $\hat{\pi}_1$ ([Ih₂]; cf. also [A-I₂§2]).[1]

Our first subject now is an explicit formula for the *commutative* power series $\psi_\sigma^{\mathrm{ab}}$ obtained from ψ_σ by letting ξ and η commute. Let $\mathscr{A}^{\mathrm{ab}} = \widehat{\mathbb{Z}}[\![\xi, \eta]\!]$ be the *commutative* formal power series algebra, with the induced $G_\mathbb{Q}$-action $1 + \xi \to (1 + \xi)^{\chi(\sigma)}$, $1 + \eta \to (1 + \eta)^{\chi(\sigma)}$ ($\sigma \in G_\mathbb{Q}$). Then $G_\mathbb{Q} \to (\mathscr{A}^{\mathrm{ab}})^\times$ ($\sigma \mapsto \psi_\sigma^{\mathrm{ab}}$) is a

---

[1] In these papers, the base point is $\overrightarrow{\infty 1}$ and x, y are loops around $0, 1$, respectively. So, the definitions are slightly different. See *Remark* at the end of §6.5.

1-cocycle, and it turns out that each coefficient of ψ_σ^{ab} can be expressed in terms of "old invariants" of σ which we now recall.

6.2 The Old Invariants. Fix a prime number l.

(i) *The l-adic cyclotomic character* $\chi^{(l)}$ is the l-component of χ, i.e., $\chi(\sigma) = (\chi^{(l)}(\sigma))$, $\chi^{(l)}(\sigma) \in \mathbf{Z}_l^\times$.

(ii) *The cyclotomic elements* (Soulé, Deligne). These are certain continuous mappings

$$\kappa_m^{(l)} : G_\mathbf{Q} \longrightarrow \mathbf{Z}_l \quad (m \geq 1, odd) ,$$

satisfying the 1-cocycle relation

$$\kappa_m^{(l)}(\sigma\tau) = \kappa_m^{(l)}(\sigma) + \chi^{(l)}(\sigma)^m \kappa_m^{(l)}(\tau) \quad (\sigma, \tau \in G_\mathbf{Q}) . \tag{6.2.1}$$

[Construction] Let $n \geq 1$ (but $n \geq 2$ if $l = 2$). Put $\zeta_n = \exp\left(\frac{2\pi i}{l^n}\right)$ and

$$\varepsilon_{m,n} = \prod_a (\zeta_n^a - 1)^{<a^{m-1}>} ,$$

where the product is over all integers a such that $0 < a < l^n$ and $(a, l) = 1$; $\langle a^{m-1} \rangle$ is the smallest positive integer congruent to a^{m-1} mod l^n. Note that $\varepsilon_{m,n}$ is totally real and totally positive (because m is *odd*). It is easy to see that each of $\varepsilon_{m,n+1}/\varepsilon_{m,n}$ and $\sigma(\varepsilon_{m,n})/\varepsilon_{m,n}^b$ is an l^n-th power of a totally positive element of $\mathbf{Q}(\mu_{l^\infty})$, where $\sigma \in G_\mathbf{Q}$, $b \in \mathbf{Z}$, $b \equiv \chi^{(l)}(\sigma)^{1-m}(\mathrm{mod}\, l^n)$. Hence there is a unique $\kappa_m^{(l)}(\sigma) \in \mathbf{Z}_l$ such that

$$\sigma((\varepsilon_{m,n})^{1/l^n}) = (\sigma(\varepsilon_{m,n}))^{1/l^n} \cdot \zeta_n^{\chi^{(l)}(\sigma)^{1-m} \cdot \kappa_m^{(l)}(\sigma)}$$

holds for all $n \geq 2$. Moreover, $\kappa_m^{(l)}$ satisfies (6.2.1). Here, for any positive real number c, c^{1/l^n} denotes its *positive real* root. By (6.2.1), $\kappa_m^{(l)}$ factors through $\mathrm{Gal}(\mathbf{Q}(\mu_{l^\infty})^{ab}/\mathbf{Q})$, $\mathbf{Q}(\mu_{l^\infty})^{ab}$ being the maximal *abelian* extension of $\mathbf{Q}(\mu_{l^\infty})$. Moreover, by Soulé [So$_{1,2}$], these 1-cocycles $\kappa_m^{(l)}$ do not vanish at least if $l > 2$.

6.3 The following *explicit formula* for the coefficients of ψ_σ^{ab} is due to the contributions of Anderson, Coleman, Deligne, the author, Kaneko and Yukinari; cf. [A$_3$, C$_3$, IKY]. (See also [Ich] for a simplification of [IKY].)

For each $\sigma \in G_\mathbf{Q}$, define $\kappa_m^*(\sigma) \in \widehat{\mathbf{Z}}^\times = \prod_l \mathbf{Z}_l^\times$ by

$$\kappa_m^*(\sigma) = ((l^{m-1} - 1)^{-1} \kappa_m^{(l)}(\sigma))_l .$$

Theorem [A$_3$, C$_3$, IKY]. *The commutative power series* $\psi_\sigma^{ab}(\xi, \eta)$ *can be expressed explicitly as follows.*

$$\psi_\sigma^{ab}(\xi, \eta) = \exp\left\{ \sum_{m \geq 3, odd} \frac{\kappa_m^*(\sigma)}{m!}((X+Y)^m - X^m - Y^m) \right\}$$

$$\times \exp\left\{ -\frac{1}{2} \sum_{m \geq 2, even} \frac{b_m(1 - \chi(\sigma)^m)}{m!}((X+Y)^m - X^m - Y^m) \right\} ,$$

where $X = \log(1 + \xi)$, $Y = \log(1 + \eta)$, and the constants b_m are defined by

$$\log\left(\frac{1 - e^{-t}}{t}\right) = \sum_{m \geq 1} \frac{b_m}{m!} t^m \ .$$

(*mb_m is the m-th Bernoulli number.*)

Note that $\psi_\sigma^{ab}(\xi, \eta)$ is of the form

$$\gamma_\sigma(\xi)\gamma_\sigma(\eta)\gamma_\sigma((1 + \xi)(1 + \eta) - 1)^{-1} \ .$$

This power series γ_σ is closely related to Anderson's hyperadelic gamma function Γ_σ.

The image of $\sigma \to \psi_\sigma^{ab}(\xi, \eta)$ was studied closely by Coleman [C3], Ichimura and Kaneko [IK]. The expected image can be figured out via Coleman theory [C1, C2], and the difference from the expected image can be measured in terms of the "Vandiver gap".

6.4   We now explain another aspect of ψ_σ^{ab}. It is a connection with the action of σ on the *double* commutator quotient $\widehat{F}_2/\widehat{F}_2''$ of $\widehat{F}_2$, or equivalently, on torsion points of Fermat Jacobians. Put $\mathscr{F} = \widehat{F}_2$, and consider the abelianizations $\mathscr{F}^{ab} = \mathscr{F}/\mathscr{F}'$ and $\mathscr{F}'^{ab} = \mathscr{F}'/\mathscr{F}''$ of $\mathscr{F}$ and $\mathscr{F}'$ first as (additive) $\widehat{\mathbb{Z}}$-modules. Then $\mathscr{F}^{ab} = \widehat{\mathbb{Z}}\underline{x} \oplus \widehat{\mathbb{Z}}\underline{y}$ on which σ acts via $\chi(\sigma)$-multiplication (because of (2.3.2)). Here $\underline{x}, \underline{y}$ are the classes of x, y. Now $\mathscr{F}^{ab}$ acts on $\mathscr{F}'^{ab}$ by conjugation. Therefore, $\mathscr{F}'^{ab}$ may be regarded as a module over the completed group algebra $\widehat{\mathbb{Z}}[\![\mathscr{F}^{ab}]\!]$. But one can show that *this module is free of rank* 1 *generated by the class* θ' *of* $(x, y) = xyx^{-1}y^{-1}$. As σ acts semi-linearly on $\mathscr{F}'^{ab}$, this action is presented by the unique element B_σ' of $\widehat{\mathbb{Z}}[\![\mathscr{F}^{ab}]\!]$ such that

$$\sigma(\theta') = B_\sigma' \cdot \theta' \ . \tag{6.4.1}$$

Now define $B_\sigma \in \widehat{\mathbb{Z}}[\![\mathscr{F}^{ab}]\!]$ by the formula

$$B_\sigma' = \left(\frac{\underline{x}^{\chi(\sigma)} - 1}{\underline{x} - 1} \cdot \frac{\underline{y}^{\chi(\sigma)} - 1}{\underline{y} - 1}\right) B_\sigma \ . \tag{6.4.2}$$

This B_σ is connected with ψ_σ^{ab} as follows. Consider the projection

$$\text{pr} : \widehat{\mathbb{Z}}[\![\mathscr{F}^{ab}]\!] \longrightarrow \widehat{\mathbb{Z}}[\![\xi, \eta]\!] = \mathscr{A}^{ab} \quad \begin{cases} \underline{x} \mapsto 1 + \xi, \\ \underline{y} \mapsto 1 + \eta. \end{cases} \tag{6.4.3}$$

Then

$$\text{pr}(B_\sigma) = \psi_\sigma^{ab} \ . \tag{6.4.4}$$

The projection pr has a big kernel $\mathscr{K}$. In fact, $\mathscr{K} \cdot \theta'$ $(\subset \mathscr{F}'/\mathscr{F}'')$ is the kernel of $\mathscr{F}'/\mathscr{F}'' \to F_2^{nil}/(F_2^{nil})''$. Thus, ψ_σ^{ab} is the power series which describes the σ-action on $(F_2^{nil})'/(F_2^{nil})''$ universally.

This power series was first treated in [Ih₁], including its connections with the l-power torsion points of Fermat Jacobians of l-power degree and Jacobi sums. We shall explain this briefly keeping in sight its generalized and refined version due to Anderson [A₃]. First, we make the following identification

$$\mathscr{F}^{ab} = \mathrm{Hom}((\mathbb{Q}/\mathbb{Z})^2, \mu_\infty) \tag{6.4.5}$$

(μ_∞ : the group of roots of unity in $\mathbb{C}$), with $\underline{x}$ (resp. $\underline{y}$) corresponding to $(s,t) \to \exp(2\pi i s)$ (resp. $\exp(2\pi i t)$); $s, t \in \mathbb{Q}/\mathbb{Z}$. Note that the $G_\mathbb{Q}$-action on $\mathscr{F}^{ab}$ is recovered from its action on μ_∞ via (6.4.5). Through (6.4.5) we may regard each element of $\widehat{\mathbb{Z}}[\![\mathscr{F}^{ab}]\!]$ as a function

$$(\mathbb{Q}/\mathbb{Z})^2 \mapsto \widehat{\mathbb{Z}} \otimes \mathbb{Q}(\mu_\infty) \ . \tag{6.4.6}$$

The above B_σ, considered as a function (6.4.6), is the *adelic beta function* $B_\sigma(s,t)$ $(s,t \in \mathbb{Q}/\mathbb{Z})$. It is strikingly analogous to the classical beta function [A₃]. Very roughly speaking, B_σ plays the role of (the classical beta)$^{\sigma-1}$.

Now, the abelian covering of $X = \mathbb{P}^1 - \{0,1,\infty\}$ corresponding to $\mathscr{F}^{ab}/(N)$ $(N = 1,2,\dots)$ is the Fermat curve

$$Y_N : \{u^N + v^N = 1; \quad uv \neq 0\} \ .$$

The covering map is given by $(u,v) \to u^N$. The abelian coverings over Y_N are controlled by the group of torsion points of the Jacobian of $\overline{Y}_N$ (the compactification of Y_N). Thus the action of σ on $\mathscr{F}/\mathscr{F}''$ is directly tied to that on the group of these torsion points ([Ih₁, A₂,₃]). As the Frobenius elements act on the latter group by multiplication of Jacobi sums (etc.), these collected together in terms of ψ_σ^{ab} (or B_σ) give a universal expression of Jacobi sums. For ψ_σ^{ab}, it is:

Theorem [Ih₁]. *Let l be a prime number, $n \geq 1$, $\mathfrak{p}$ be a prime ideal of $\mathbb{Q}(\mu_{l^n})$ not lying above l, and $\sigma = \sigma_\mathfrak{p}$ be a Frobenius element of $\mathfrak{p}$. Then for any l^n-torsion points s, t of $\mathbb{Q}/\mathbb{Z}$ with s, t, $s+t \neq 0$, the special value of the l-component of ψ_σ^{ab} $(\in \mathbb{Z}_l[\![\xi,\eta]\!])$ at $\xi = \exp(2\pi i s) - 1$, $\eta = \exp(2\pi i t) - 1$, is essentially the Jacobi sum (w.r.t. $\mathfrak{p}, l^n, s, t$).*

More generally, the values of $B_\sigma(s,t)$ $(s,t \in \mathbb{Q}/\mathbb{Z})$ are related to Jacobi sums and also Gauss sums [A₃]. Note that the two theorems of §6.3, §6.4, combined, give a direct connection between the circular units and the Jacobi sums.

Anderson [A₃, A₄] defined the hyperadelic gamma function

$$\Gamma_\sigma : \mathbb{Q}/\mathbb{Z} \longrightarrow \text{(some arithmetic ring)}$$

which factors B_σ just as γ_σ factors ψ_σ^{ab}. It interpolates Gauss sums, and its "logarithmic derivative" can be given explicitly in terms of circular units so that it forms a bridge connecting Gauss sums and circular units. The last connection was partly established independently by a different method by Miki [Mi]. See Coleman [C₃] for connections with and applications to other aspects of cyclotomy.

6.5 $\psi_\sigma(\xi, \eta)$ **and Higher Circular** l**-Units (Anderson-Ihara [AI$_{1,2}$]).** This connection arises from a comparison of the tower of nilpotent coverings and of genus 0 coverings.

Call a finite subset $S \subset \mathbb{P}^1(\mathbb{C})$ l-*elementary*, if S is obtained from $S_0 = \{0, 1, \infty\}$ by finite number of operations of the form $S \to S^{1/l}$ (all l-th roots), $S \to T_{a,b,c}(S)$. Here, $a, b, c \in S$ (distinct), and $T_{a,b,c}$ is the projective linear transformation of $\mathbb{P}^1$ that maps a, b, c to $0, 1, \infty$ respectively.

Definition 6.5.1. $E^{(l)}$ is the subgroup of $\mathbb{C}^\times$ generated by the constituents of $S - \{0, \infty\}$, where S runs over all l-elementary subsets of $\mathbb{P}^1(\mathbb{C})$.

It is easy to see that elements of $E^{(l)}$ are l-units, i.e., element of $\overline{\mathbb{Q}}$ which, together with its reciprocal, is integral over $\mathbb{Z}[1/l]$. They are called *higher circular* l-*units*. The group $E^{(l)}$ contains such elements as

$$1 - \zeta_n, \quad (1 - (1 - \zeta_n)^{1/l^n})^{1/l^n}, \dots,$$

where $\zeta_n = \exp(2\pi i/l^n)$ $(n = 1, 2, \dots)$.

Theorem [A-I$_2$]. *Each coefficient of* $\psi_\sigma(\xi, \eta)$ *(mod* l^n*) can be expressed explicitly in terms of the* σ-*action on* $E^{(l)}$.

A key dialogue : "How to distinguish different l-th roots in $E^{(l)}$ intelligibly?" : "In terms of a natural structure of "forest" with vertices in $E^{(l)}$."

Corollary [A-I$_1$]

$$\mathbb{Q}^{(l)}(\infty) = \mathbb{Q}(E^{(l)}) \; .$$

We shall conclude this lecture with two additional open questions.

Question 6.5.2. i) *Is* $\mathbb{Q}^{(l)}(\infty)$ *the maximal pro-*l *extension over* $\mathbb{Q}(\mu_{l^\infty})$ *unramified outside* l ?

ii) *How big is* $E^{(l)}$ *and* $E^{(l)} \cap \mathbb{Q}(\mu_{l^\infty})$?

Remark. The non-commutative (resp. commutative) l-adic power series $\psi(\sigma) \in \mathbb{Z}_l \ll u, v \gg$ of [Ih$_2$] §1 (D) Ex. 1 (resp. $F_\sigma \in \mathbb{Z}_l[\![u, v]\!]$ of [Ih$_1$] §2) are related to the above ψ_σ (resp. ψ_σ^{ab}) as follows. Write ξ, η instead of u, v respectively. Then $\psi(\sigma)$ and F_σ are the l-components of

$$f_\sigma(y, z) y^{\frac{1}{2}(\chi(\sigma) - 1)} \cdot \psi_\sigma(\eta, \xi)$$

and

$$\frac{x^{\chi(\sigma)} - 1}{x - 1} \cdot \frac{y^{\chi(\sigma)} - 1}{y - 1} \cdot y^{\frac{1}{2}(\chi(\sigma) - 1)} \cdot \psi_\sigma^{ab}(\xi, \eta) \; ,$$

respectively, where $z = (xy)^{-1}$. Note that $f_\sigma^{ab} = 1$ and that ψ_σ^{ab} is symmetric in ξ and η.

Added in proof. The author found later that Question 5.3.4 (i) has an affirmative answer. The proof is based on [Dr$_2$, Ih$_5$].

References

[A$_1$]   Anderson, G.: Cyclotomy and an extension of the Taniyama group. Compositio Math. **57** (1986) 153–217

[A$_2$]   Anderson, G.: Torsion points on Fermat Jacobians, roots of circular units and relative singular homology. Duke Math. J. **54** (1987) 501–561

[A$_3$]   Anderson, G.: (a) The hyperadelic gamma function. Invent. math. **95** (1989) 63–131. (b) ibid (a précis). Adv. Stud. Pure Math. **12** (1987) 1–19

[A$_4$]   Anderson, G.: Normalization of the hyperadelic gamma function. In: Galois groups over $\mathbb{Q}$. Publ. MSRI, no. 16 (1989) 1–31. Springer, Berlin Heidelberg New York

[A-I$_1$]   Anderson, G., Ihara, Y.: Pro-l branched covering of $\mathbb{P}^1$ and higher circular l-units. Ann. Math. **128** (1988) 271–293

[A-I$_2$]   Anderson, G., Ihara, Y.: ibid Part 2. Int'l J. Math. **1** (1990) 119–148

[B$_1$]   Belyi, G. V.: On Galois extensions of a maximal cyclotomic field. Izv. Akad. Nauk USSR **43** (1979) 267–276; transl. Math. USSR Izv. **14** (1980) 247–256

[B$_2$]   Belyi, G. V.: On the commutator of the absolute Galois group. Proc. ICM 1986. American Mathematical Society, Berkeley 1987, pp. 346–349 (in Russian)

[C$_1$]   Coleman, R.: The dilogarithm and the norm residue symbol. Bull. Soc. Math. France **109** (1981) 373–402

[C$_2$]   Coleman, R.: Local units modulo circular units. Proc. Amer. Math. Soc. **89** (1983) 1–7

[C$_3$]   Coleman, R.: Anderson-Ihara theory: Gauss sums and circular units. Adv. Stud. Pure Math. **17** (1989) 55–72

[De]   Deligne, P.: Le groupe fondamental de la droite projective moins trois points. In: Galois groups over $\mathbb{Q}$. Publ. MSRI, no. 16 (1989) 79–298. Springer, Berlin Heidelberg New York

[Dr$_1$]   Drinfeld, V.G.: Quasi-Hopf algebras (Russian). Algebra and Analysis **1**, no. 6 (1989) 114–148.

[Dr$_2$]   Drinfeld, V.G.: On quasi-triangular quasi-Hopf algebras and some group closely associated with Gal($\overline{\mathbb{Q}}/\mathbb{Q}$). Preprint 1990 (in Russian)

[G]   Grothendieck, A.: Esquisse d'un programme. Mimeographed note (1984)

[Ich]   Ichimura, H.: On the coefficients of the universal power series for Jacobi sums. J. Fac. Sci. Univ. Tokyo IA **36** (1989) 1–7

[IK]   Ichimura, H., Kaneko, M.: On the universal power series for Jacobi sums and the Vandiver conjecture. J. Number Theory **31** (1989) 312–334

[Ih$_1$]   Ihara, Y.: Profinite braid groups. Galois representations and complex multiplications. Ann. Math. **123** (1986) 43–106

[Ih$_2$]   Ihara, Y.: On Galois representations arising from towers of coverings of $\mathbb{P}^1 \backslash \{0, 1, \infty\}$. Invent. math. **86** (1986) 427–459

[Ih$_3$]   Ihara, Y.: Some problems on three point ramifications and associated large Galois representations. Adv. Stud. Pure Math. **12** (1987) 173–188

[Ih$_4$]   Ihara, Y.: The Galois representation arising from $\mathbb{P}^1 - \{0, 1, \infty\}$ and Tate twists of even degree. In: Galois groups over $\mathbb{Q}$. Publ. MSRI, no. 16 (1989) 299–313. Springer, Berlin Heidelberg New York

[Ih$_5$]   Ihara, Y.: Automorphisms of pure sphere braid groups and Galois representations. In: The Grothendieck Festschrift, vol. 2. Progress in Mathematics, vol. 87. Birkhäuser, Basel 1991, pp. 353–373

[Ih₆]   Ihara, Y.: Galois groups over $\mathbb{Q}$ and monodromy (in Japanese). In: Prospects of Algebraic Analysis, RIMS report **675** (1988) 23–34

[Ih₇]   Ihara, Y.: Derivations of the Lie algebra associated with $F_{0,n}\mathbb{P}^1$ and the image of Gal($\overline{\mathbb{Q}}/\mathbb{Q}$) (in Japanese). In: Algebraic number theory, RIMS report **721** (1990) 1–8

[IKY]   Ihara, Y., Kaneko, M., Yukinari, A.: On some properties of the universal power series for Jacobi sums. Adv. Stud. Pure Math. **12** (1987) 65–86

[K]   Kohno, T.: On the holonomy Lie Algebra and the nilpotent completion of the fundamental group of the complement of hypersurfaces. Nagoya Math. J. **92** (1983) 21–37

[Ma]   Matzat, B. H.: Braids and Galois groups. Doğa Mat. **14** (1990) no. 2, 57–69

[M-S]   Moore, G., Seiberg, N.: Classical and quantum conformal field theory. Commun. Math. Phys. **123** (1989) 177–254

[Mi]   Miki, H.: On the l-adic expansion of certain Gauss sums and its applications. Adv. Stud. Pure Math. **12** (1987) 87–118

[N]   Nakamura, H.: Galois rigidity of the etale fundamental groups of punctured projective lines. J. Reine Angew. Math. **411** (1990) 205–216

[O]   Oda, T.: A note on ramification of the Galois representation on the fundamental group of an algebraic curve. J. Number Theory **34** (1990) 225–228; Part 2 in preparation

[So₁]   Soulé, C.: On higher p-adic regulators. Lecture Notes in Mathematics, vol. 854. Springer, Berlin Heidelberg New York 1981, pp. 372–401

[So₂]   Soulé, C.: Éléments cyclotomiques en K-théorie. Astérisque **147–148** (1987) 225–257

[T-K]   Tsuchiya, A., Kanie, Y.: Vertex operators in conformal field theory on $\mathbb{P}^1$ and monodromy representations of braid group. Adv. Stud. Pure Math. **16** (1987) 297–372

Von Neumann Algebras in Mathematics and Physics

Vaughan F. R. Jones

Department of Mathematics, University of California, Berkeley, CA 94720, USA

Motivation

The last ten years or so have seen a considerable synthesis in mathematics and mathematical physics. In this talk I will be concerned only with those topics appearing in Fig. 1, all connected by having something to do with braids.

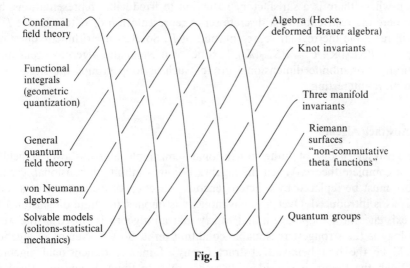

Conformal field theory

Functional integrals (geometric quantization)

General quantum field theory

von Neumann algebras

Solvable models (solitons-statistical mechanics)

Algebra (Hecke, deformed Brauer algebra)

Knot invariants

Three manifold invariants

Riemann surfaces "non-commutative theta functions"

Quantum groups

Fig. 1

Many very different themes could be used for a talk such as this one, but I have chosen von Neumann algebras because they are what led me into this circle of ideas. Thus the presentation will be historical rather than logical.

A *von Neumann algebra* M is a $*$-algebra of bounded operators on a Hilbert space $(\mathcal{H}, \langle \, , \, \rangle)$ which contains the identity and is closed in the weak operator topology, i.e. if a_n is a net of operators in M and $\langle a_n\xi, \eta \rangle \to \langle a\xi, \eta \rangle$ for some a and all ξ and η in $\mathcal{H}$, then a is in M. Most of the interest is when M is infinite dimensional so it should be pointed out at the outset that a finite dimensional von Neumann algebra is just a direct sum of matrix algebras, each acting with a certain multiplicity on $\mathcal{H}$.

Proceedings of the International Congress of Mathematicians, Kyoto, Japan, 1990

These algebras were first introduced by Murray and von Neumann in [MvN1]. Their motivations were many, not the least being a beautiful result of von Neumann in [vN] which showed that von Neumann algebras could equally be defined as *commutants* of self-adjoint sets (notation: if S is a set of bounded operators, S', the commutant of S, is by definition the set of all bounded operators x such that $xs = sx$ for all s in S). This means that von Neumann algebras are the algebras of *symmetries* for any structure belonging to Hilbert space, such as geometric configurations of subspaces or unitary group representations. Let me discuss some other motivations for looking at von Neumann algebras.

a) Unitary Group Representations

The abstract theory of representations of a compact group is very complete but as soon as the group is not compact, many different phenomena occur. A unitary representation no longer decomposes simply as a direct sum of irreducibles, which is reflected in a possibly exotic structure of the commutant of the group. This already happens for, say, the regular representation of the free group on two generators. Of course a semisimple Lie group G is an example of a "type I" group where there is a satisfactory reduction to irreducible representations, but even here the restriction of a discrete series representation of G to a lattice Γ in G will resemble the regular representation of Γ. So even in arithmetic questions (say $G = PSL_2(\mathbb{R})$, $\Gamma = PSL_2(\mathbb{Z})$), "exotic" von Neumann algebras occur quite naturally. For infinite dimensional groups such as loop groups the situation is even more interesting.

b) Abstract Algebra

The abstract theory of finite dimensional semisimple algebras over a field is also a complete theory. When the algebra becomes infinite dimensional, general theory must be replaced by a heterogeneous collection of examples unless some analysis is introduced. Over $\mathbb{C}$ semisimplicity is implied (in finite dimensions) by the existence of a * operation and to say that an algebra is closed in the weak topology is the strongest reasonable condition on it. So von Neumann algebras should be the best behaved abstract family of infinite dimensional algebras. Although the theory is considerably richer than in finite dimensions there do exist non-trivial general results and there is a significant class of von Neumann algebras (those approximable by finite dimensional algebras) for which a complete classification exists.

c) Unbounded Operators

At first sight the important operator d/dx appears unnatural on Hilbert space since it is not defined on all vectors (in $L^2(\mathbb{R})$). However the opposite is true. Provided d/dx is given the right *domain* its graph is a closed subspace and hence natural to Hilbert space. In general an operator is called pre-closed if it is densely defined and the closure of its graph is the graph of an operator. All interesting linear operators seem to be pre-closed.[1]

The trouble with domains comes when one tries to add and multiply unbounded operators. For this reason one would hope to replace unbounded operators by bounded ones. Thus the relations $[P, Q] = \mathrm{id}$ can be handled in the Weyl form with two unitary groups $U(t)$ and $V(t)$ with $U(t)V(s) = e^{2\pi i s t}V(s)U(t)$. In general one could use the von Neumann algebra of all bounded operators having the same symmetries as the unbounded ones.

d) Quantum Theory

The very language of quantum mechanics suggests von Neumann algebras. States are vectors in a Hilbert space. Observables are self-adjoint operators and numerical information about observables for systems in states is given by scalar products. Thus it is quite natural to consider a von Neumann algebra of observables associated with any subsystem of a quantum system. Certainly von Neumann was thinking along these lines. Such an approach was indeed adopted as an "algebraic" approach to quantum field theory by Haag and Kastler [HK] who postulated von Neumann algebras associated with regions of space time and satisfying certain causality, positivity and Lorentz covariance conditions. Although it is very abstract and difficult to produce mathematical examples, some things can be deduced from such a general theory and new results in von Neumann algebras have recently added to the possibilities.

1. Factors and Their Types

The spectral theorem shows that abelian von Neumann algebras are abstractly of the form $L^\infty(X, \mu)$ acting with some multiplicity on a Hilbert space. Attention turns immediately to *factors* which are by definition von Neumann algebras with trivial centre. The simplest example of such a factor is $\mathscr{B}(\mathscr{H})$, the algebra of all bounded operators. One of the fundamental discoveries of Murray and von Neumann was that of factors not abstractly isomorphic to any $\mathscr{B}(\mathscr{H})$.

The first example was that of the commutant M of the left regular representation of any discrete group Γ all of whose (non-identity) conjugacy classes are infinite (such as free groups F_n or $PSL_n(\mathbb{Z})$ for $n \geq 2$). One can show that this von Neumann algebra is generated by the right regular representation and that it is a factor. Also if $\xi \in l^2(\Gamma)$ is the characteristic function of the identity then the linear functional $\mathrm{tr}(x) = \langle x\xi, \xi \rangle$ defines a *trace* on M (i.e. $\mathrm{tr}(xy) = \mathrm{tr}(yx)$). There is no such functional on $\mathscr{B}(\mathscr{H})$ if $\dim \mathscr{H} = \infty$.

An infinite dimensional factor admitting such a trace is called a *type* II_1 *factor* and a factor of the form $\mathscr{B}(\mathscr{H}) \otimes$ (a type II_1 factor) is called a type II_∞ factor if $\dim \mathscr{H} = \infty$. A factor which is neither of type I ($\cong \mathscr{B}(\mathscr{H})$) nor of type II is called a type III factor.

---

[1] I can only think of one exception – the "derivations" in Fox's free differential calculus are not pre-closed on l^2 of the free group.

A II_1 factor shares some of the nice features of a finite dimensional matrix algebra (e.g. it is simple). Its most seductive feature is continuous dimensionality. If one looks at the numbers trace(p) (trace normalized so that trace(1) = 1), where $p \in M_n(\mathbb{C})$ are projections, one obtains the numbers m/n for $m = 0, 1, \ldots, n$, where m is of course the rank of p. In a II_1 factor one obtains the whole unit interval $[0, 1]$.

2. GNS Construction

An important construction is the Gelfand-Naimark-Segal construction. One begins with a ∗-algebra A and a linear functional (state) $\varphi : A \to \mathbb{C}$ with $\varphi(a^*a) \geq 0$. Define $\langle \, , \, \rangle$ on A by $\langle a, b \rangle = \varphi(b^*a)$. Quotienting if necessary A becomes a pre-Hilbert space and its completion is written $\mathscr{H}_\varphi$. Under mild conditions A will act on $\mathscr{H}_\varphi$ by left multiplication and the von Neumann algebra it generates is said to result from the GNS construction from φ. It should be thought of as the completion of A with respect to φ. Examples of all types of factors can now be obtained, using $A = \otimes_{i=1}^\infty M_2(\mathbb{C})$. If $h_i \in M_2(\mathbb{C})$ are positive matrices of trace 1 then the formula $\varphi(\otimes_{i=1}^\infty x_i) = \prod_{i=1}^\infty \text{trace}(h_i x_i)$ defines a state, said to be a product state. The result of the GNS construction is then always a factor. It is of type I if $h_i = \begin{pmatrix} 1 & 0 \\ 0 & 0 \end{pmatrix}$, of type II_1 if $h_i = (1/2) \begin{pmatrix} 1 & 0 \\ 0 & 1 \end{pmatrix}$ and of type III if $h_i = (1 + \lambda)^{-1} \begin{pmatrix} 1 & 0 \\ 0 & \lambda \end{pmatrix}$ for $0 < \lambda < 1$. In the last case the state is known as the Powers state after R. Powers [Pow] who proved that the factors for different λ are mutually non-isomorphic.

3. Modular Theory

The Tomita-Takesaki theory shows that to every weakly continuous state φ on a von Neumann algebra M there is a one parameter group σ_t of automorphisms of M characterized by the KMS condition $\varphi(xy) = \varphi(\sigma_i(y)x)$. It allows one to subdivide the type III factors into type III_λ, $\lambda \in [0, 1]$ where the Powers factors are of type III_λ, $\lambda \in (0, 1)$. Types III_0 and III_1 can be obtained from product states by suitable choices of the h_i's. Generically a factor is of type III_1. For details we refer to Connes' Helsinki congress talk [Co1], or [Ta].

4. Hyperfiniteness

A von Neumann algebra M is *hyperfinite* if there is an increasing sequence of finite dimensional ∗-subalgebras whose union is weakly dense in M. Thus our infinite tensor product factors are hyperfinite, by construction, but Murray and von Neumann showed that the free group II_1 factor is not hyperfinite, nor are the examples coming from lattices in semisimple Lie groups.

All hyperfinite factors are known. Here is a table of them.

Table 1. Hyperfinite factors

Type I_n $n = 1, 2, \ldots, \infty$	One for each n Proof: elementary
Type II_1	A unique factor, denoted R. Uniqueness proved by Murray and von Neumann in [MvN2]
Type II_∞	A unique factor. Uniqueness proved by Connes in [Co2]
Type III_0	One for each ergodic non-transitive flow. Proved by Krieger [Kr] and Connes [Co2]
Type III_λ $0 < \lambda < 1$	Powers factors are the unique examples. Uniqueness proved by Connes [Co2]
Type III_1	A unique example, first analysed by Araki and Woods in [AW]. Uniqueness proved by Haagerup [Ha] and Connes [Co3]

In quantum field theory it is expected that the von Neumann algebra of observables localized in a nice region is a hyperfinite type III_1 factor.

The classification of hyperfinite (also called "injective") factors is a great achievement. Among other things it paves the way for the study of subfactors, to which I now turn.

5. Index for Subfactors

The representation theory of a II_1 factor is very simple. There is a single parameter – a positive real number (or ∞), $\dim_M(\mathscr{H})$, which measures the size of $\mathscr{H}$ compared to the Hilbert space $L^2(M)$ obtained from M by the GNS construction using the trace (for which $\dim_M(L^2(M)) = 1$). Thus if $N \subset M$ are II_1 factors we define the index of N in M to be the real number (≥ 1):

$$[M : N] = \dim_N(L^2(M)).$$

Examples:

(i) $[N \otimes M_n(\mathbb{C}) : N] = n^2$
(ii) $[M : M^G] = |G|$ if G is a group of outer automorphisms of M.

The next theorem shows that this is an interesting notion.

Theorem [Jo1]. a) *If* $[M : N] < 4$ *then there is an* $n \in \mathbb{Z}$, $n \geq 3$ *for which* $[M : N] = 4 \cos^2 \pi/n$.

b) *All values of the index of part* a) *are realized, as is any real number* ≥ 4, *by subfactors of the hyperfinite* II_1 *factor.*

This theorem may be proved by iterating a certain *basic construction* which associates to $N \subset M$ an extension $\langle M, e_N \rangle$ of M where e_N projects from $L^2(M)$ to $L^2(N)$. One obtains a tower M_i with $N = M_0$, $M = M_1$ and $M_n = \langle M, e_1, \ldots, e_{n-1} \rangle$, e_j being orthogonal projection from $L^2(M_j)$ to $L^2(M_{j-1})$.

The e_i's satisfy the following properties

1) $e_i^2 = e_i = e_i^*$
2) $e_i e_j = e_j e_i$ if $|i - j| \geq 2$
3) $e_i e_{i \pm 1} e_i = \tau e_i$ $\qquad (\tau = [M : N]^{-1})$
4) $\mathrm{tr}(w e_{n+1}) = \tau \, \mathrm{tr}(w)$ if w is a word on $1, e_1, \ldots, e_n$.

The hermitian form $\mathrm{tr}(x^* y)$ on the algebra generated by the e_i's must be positive definite and this forces $[M : N]$ to be $4 \cos^2 \pi/n$. We will see later how the same form arises in connection with surgery on three-manifolds and this result about degeneracies of the form is precisely what allows a simple explicit formula for some new three-manifold invariants!

The construction of examples of subfactors proceeds by the explicit construction of a sequence of e_i's and a trace satisfying 1)–4) above. The II_1 factor is then obtained by the GNS construction and the subfactor is that generated by $e_2, e_3, \ldots$. In the case of index > 4 these subfactors have non-trivial centralizer. For more details on what happens in index > 4 see Popa's talk in this volume.

6. Commuting Squares

The tower M_i arising from $N \subset M$ defines two towers of finite dimensional algebras which we will call the centralizer towers. They are the commutants $A_i = M' \cap M_i$ and $B_i = N' \cap M_i$. Clearly $A_i \subset B_i$ and they satisfy the "commuting square" condition (first used by Popa in a different context) that the orthogonal projections from B_{i+1} to A_{i+1} and B_i commute and $B_i \cap A_{i+1} = A_i$.

This condition allows one to control the von Neumann algebra inclusion of the GNS closures of $\cup A_i$ and $\cup B_i$, which would be impossible without the condition.

In fact commuting squares give subfactors automatically by iterating the basic construction and the study of commuting squares constitutes a new and intriguing problem in finite dimensional linear algebra. A machine for producing examples from quantum groups has been developed by Wenzl. Other examples abound – see [Su], [HJ] and [HS], but the general structure of commuting squares is quite unclear.

7. Finite Depth, Classification Results

Actions of finite groups are completely classified on the hyperfinite type II_1 factor ([Jo7]) and one might hope for an extension of these results to finite index subfactors. This is unlikely since one may take any finite set of automorphisms $\alpha_1, \alpha_2, \ldots, \alpha_n$ and form the subfactor

$$\left\{ \begin{pmatrix} x & & \\ & \alpha_1(x) & \quad O \\ & & \ddots \\ O & & \alpha_n(x) \end{pmatrix} \middle| x \in R \right\}$$

of $R \otimes M_n(\mathbb{C})$. This subfactor remembers too much about the group generated by $\alpha_1, \alpha_2, \ldots, \alpha_n$ and although if this group is amenable we are in a good situation ([O1]), if it is not there will be no nice classification.

On the other hand there is a class of subfactors, first stressed by Ocneanu, of R for which classification is a well-posed problem. A subfactor $N \subset M$ is said to be of *finite depth* if the dimensions of the centres of the centralizer towers A_i and B_i are bounded. It is then easy to see that the centralizer towers exhibit periodicity beyond a certain point. The index of a finite depth subfactor is given by the square of the norm of the matrix, with integer entries, describing the stabilized inclusion for $A_i \subset A_{i+1}$. See [GHJ].

Popa has shown in [Po] that the stabilized commuting square of centralizer towers is a complete invariant for subfactors of R of finite depth. Another version of the result is claimed by Ocneanu who has a more elaborate and computable version of the invariant.

If the index is less than 4 a complete classification is possible. Coxeter-Dynkin diagrams arise out of the combinatorics of the centralizer towers and according to Ocneanu subfactors of index < 4 are classified by Dynkin diagrams of types A_n, D_{2n}, E_6 and E_8, there being one for each A_n and D_{2n} and two for each of E_6 and E_8. The index of the subfactor is $4 \cos^2 \pi/n$, n being the Coxeter number (see [O2, GHJ]). Popa has extended this to index 4, where infinite depth and extended Dynkin diagrams occur (see [GHJ] and Popa's talk in these proceedings).

Popa has given a deep generalization of his result to cases of infinite depth provided certain asymptotic behavior of the combinatorics can be controlled. Combined with ideas of Wasserman in [Was] this gives new results about compact group actions. See Popa's paper in these proceedings.

8. Statistical Mechanical Models

The abstract algebra presented by the e_i relations 1), 2) and 3) was used by Temperley and Lieb [TL, Ba] to show the equivalence of the ice-type and self-dual Potts models of statistical mechanics. The relations are satisfied by certain matrices which combine to give the row-to-row transfer matrices of the models. The same abstract algebra occurs in the models of Andrews, Baxter and Forrester [ABF] but the parameter is now in the discrete series $4 \cos^2 \pi/n$. Pasquier used the ADE Coxeter graphs to get more models ([Pa]).

The ice-type model is a vertex model where the interactions between the elementary components of a system take place at the vertices of a graph. The Potts model is a spin model where interactions are on the edges and the ABF and Pasquier models are IRF models with many-spin interactions around faces of a planar graph.

Many more elaborate models can be obtained from quantum groups and there are corresponding algebraic relations generalizing the e_i ones. One may use these models to construct subfactors including the Wenzl ones but apparently others as well ([Jo2, D+]).

The relation of subfactors to the solvability of the model, if any, remains unclear.

9. Bimodules, Hypergroups, Paragroups, Quantized Groups ...

The combinatorics of the centralizer towers are very rich and attempts are being made to extract the data in them. The most ambitious project is Ocneanu's. He uses bimodules and intertwiners and obtains a structure with many properties of an IRF model of statistical mechanics, where the "faces" become closed paths around induction-restriction diagrams. See his notes in these proceedings.

Sunder has used a less detailed structure called a hypergroup which is only supposed to contain the combinatorial structure of tensor products of bimodules ([Su]).

The idea of using bimodules was first stressed by Connes. See [Co4, Jo3].

10. Braid Groups

The braid group B_n on n strings may be presented on $n-1$ generators $\sigma_1, \sigma_2, \ldots, \sigma_{n-1}$ with relations $\sigma_i \sigma_{i+1} \sigma_i = \sigma_{i+1} \sigma_i \sigma_{i+1}$ and $\sigma_i \sigma_j = \sigma_j \sigma_i$ if $|i - j| \geq 2$ (see [Bi]). These relations bear a resemblance to the e_i ones and one may represent the braid group by sending σ_i to $te_i - (1 - e_i)$ if $2 + t + t^{-1} = \tau^{-1}$. The discrete series corresponds to $t = e^{2\pi i/n}$, σ_i unitary, and the continuous part to $t > 0$, σ_i self-adjoint.

The presence of such braid group representations is a pervasive feature of all the generalizations to do with quantum groups and solvable models. The e_i representations correspond to the 2-dimensional representation of $U_q(sl_2)$. It is not known whether these representations, either collectively or individually, are faithful for $n > 3$. Lawrence has found these representations as natural actions of braid groups on homology groups – see [La]. See also Varchenko's talk in these proceedings.

11. Hecke Algebras

If $H < G$ are groups, the Hecke algebra is the commutant of the representation of G on the vector space of functions on G/H. In the case where G is $GL_n(\mathbb{F}_q)$ and H is the upper triangular matrices, this algebra admits a presentation on generators $g_1, g_2, \ldots, g_{n-1}$ with relations

$$g_i^2 = (q - 1)g_i + q, \quad g_i g_{i+1} g_i = g_{i+1} g_i g_{i+1}, \quad g_i g_j = g_j g_i \text{ if } |i - j| \geq 2.$$

For $|q| \neq 1$ or 0 this algebra is isomorphic to the group algebra of the symmetric group S_n (see [Bo]).

Clearly the braid group can be represented in the Hecke algebra in the obvious way. In fact these representations contain the e_i ones (with $q = t$) as direct summands. Under the isomorphism with $\mathbb{C}S_n$, the e_i representations correspond to Young diagrams with at most 2 rows. (See [Jo4].)

12. Knot Polynomials

Braids may be closed by tying the top to the bottom to form links (see below).

The braid $\sigma_1\sigma_2^{-1}\sigma_1\sigma_2^{-1}$:   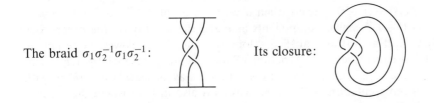   Its closure:

It is clear that the closure of a braid $\alpha \in B_n$ does not change if α is conjugated in B_n, nor if α is embedded in B_{n+1} and multiplied by $\sigma_n^{\pm 1}$. These two "Markov moves" generate the equivalence relation of having the same closure. It follows from this and relation 4) in the e_i algebra that a normalized version of the trace of the algebra element representing a braid will be an invariant of the closure of the braid. For a link L this invariant turns out to be a polynomial $V_L(t)$ in the variable t (or $\sqrt{t}$) and can be calculated, though not always rapidly, from the following "skein relation".

Skein Relation ([Cnw]). If L_+, L_- and L_0 are links identical except near one crossing where they are as below

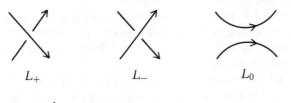

$$L_+ \qquad\qquad L_- \qquad\qquad L_0$$

then
$$\frac{1}{t}V_{L_+} - tV_{L_-} = (\sqrt{t} - 1/\sqrt{t})V_{L_0}.$$

The right-handed trefoil has $V(t) = t + t^3 - t^4$ and the left-handed one has $V(t) = 1/t + 1/t^3 - 1/t^4$. The Alexander polynomial has a similar relation and this prompted many people to develop a 2-variable polynomial called the HOMFLY polynomial $P_L(l,m)$ having arbitrary coefficients in its skein relation ([Jo5, F+, Li1]).

Kauffman found an explicit formula for $V_L(t)$ in [Ka1] from an arbitrary knot diagram which was used by him, Murasugi and Thistlethwaite to prove some old conjectures of Tait about alternating knots ([Ka1, Mu, Th]). He also found another two-variable polynomial generalization of $V(t)$, called the Kauffman polynomial $F(a,x)$. It does not contain the Alexander polynomial.

The polynomials are very useful in calculating the minimal number of crossings for a diagram of a knot and the minimal number of strings for a closed braid representation of a knot ([LT, Mo, FW]).

13. The R-Matrix, Powers State Picture

A special representation of the relations 1)–4) of §5 was discovered by Pimsner and Popa in [PP]. If $e \in \text{End}(\mathbb{C}^2 \otimes \mathbb{C}^2)$ is defined by $e = \tau e_{11} \otimes e_{22} + \sqrt{\tau(1-\tau)}(e_{12} \otimes e_{21} + e_{21} \otimes e_{12}) + (1-\tau)e_{22} \otimes e_{11}$, then if we let e_i on $(\mathbb{C}^2)^{\otimes n}$ be defined between the ith and $(i+1)^{th}$ tensor components as e and the identity on the others, one finds that 1), 2) and 3) are satisfied. Moreover if φ_λ is the Powers state (giving a type III_λ factor), then its restriction to the algebra generated by $1, e_1, \ldots, e_n$ defines a trace satisfying 4). This gives a useful way of calculating $V_L(t)$, with $t = \lambda$. This representation, without φ_λ, was used also in [TL] where the e_i's give the transfer matrix for an ice-type model.

At this point a remarkably rapid development in the understanding of the polynomials was made possible by the existence of the theory of quantum groups. Fadeev's Leningrad group, particularly Sklyanin [Sk] and Kulish and Reshetikhin [KR] had uncovered a new structure related to the ice-type model for which the relevant Lie algebra was $sl_2(\mathbb{C})$. For background see [Fa]. This had been generalized by Jimbo [Ji1] and Drinfeld [Dr] to produce analogues of the braiding matrices obtained from the e_i representation I have just described, one for each finite dimensional representation of every simple Lie algebra. The analogue of the Powers state was soon found and the following picture was established by Reshetikhin [Re] and Rosso [Ro]:

Let $\mathscr{G}$ be a simple finite dimensional complex Lie algebra and let $\mathscr{L}$ be the class of links with distinguished components $C_1, C_2, \ldots, C_n$. Then to each way of assigning finite dimensional representations of $\mathscr{G}$ to $C_1, \ldots, C_n$ there is a polynomial invariant of isotopy for links in $\mathscr{L}$.

The polynomial $V_L(t)$ corresponds to $\mathscr{G} = sl_2$ and the assignment of its 2-dimensional representation to all components. The HOMFLY and Kauffman polynomials are obtained in a similar way from sl_n and the symplectic (or orthogonal) algebras respectively.

In this picture the geometric operations of cabling (and satellites in general) can be understood in terms of tensor products of representations ([MSt]).

Explicit formulae on any (not necessarily braided) picture of a link may be given. This was first done for HOMFLY in [Jo6] and generalized to the Kauffman polynomial in [Tu1].

14. Positivity of the Markov Trace

Ocneanu's approach to the HOMFLY polynomial was a direct generalization of my construction of $V_L(t)$, by defining a trace on the Hecke algebra by the property $\text{tr}(wg_{n+1}) = z\,\text{tr}(w)$ if w is a word on $g_1, g_2, \ldots, g_n$, and where z is a new variable. Subfactors occur for the values of (q, z) allowing a *-algebra structure on the Hecke algebra for which the trace is positive. This set of values was determined by Ocneanu, and Wenzl constructed the subfactors and calculated their indices ([F+, We1]). It is convenient to use the variables $\tau = q/(1+q)^2$ and $\eta = (1+z)/(1+q)$. Then the "positivity spectrum" (Fig. 2) is as follows:

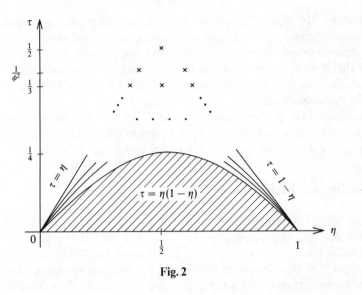

Fig. 2

The discrete points are the intersections in $[0, 1] \times [0, 1]$ of the curves $\eta = P_k(\tau)/P_{k-1}(\tau)$ with the curves $\eta = \tau P_{k'-2}(\tau)/P_{k'-1}(\tau)$ as indicated, where $P_k(\eta)$ are the polynomials given by $P_{k+1} = P_k - \eta P_{k-1}$, $P_0 = 1$, $P_1 = 1$, and $k + k' = l + 3$, $k = 2, 3, \ldots, l + 1$, l indicating the horizontal row on which the point is situated. The index of the subfactor at the point labelled (l, k) is $\sin^2 k\pi/(l+3)/\sin^2 \pi/(l+3)$ ([We1]).

15. Brauer and BMW Algebras

A new algebra was developed in [BW] and [Muk] to play the role for the Kauffman polynomial that the Hecke algebra plays for the HOMFLY polynomial. The idea, inspired by Kauffman, is to add objects like $|\ldots| \asymp |\ldots|$ to the braid group generators. Calling these generators E_i and the usual braid group generators G_i, the BMW algebra with parameters a, x, has presentation:

$$G_i G_{i+1} G_i = G_{i+1} G_i G_{i+1}, \quad G_i G_j = G_j G_i \text{ if } |i - j| \geq 2,$$

$$G_i + G_i^{-1} = x(1 + E_i), \quad E_i G_i = G_i E_i = a E_i, \quad E_i^2 = (a + a^{-1} - x) x^{-1} E_i,$$

$$E_i G_{i\pm 1}^{\pm 1} E_i = a^{\mp 1} E_i, \quad E_i G_{i\pm 1} G_i = E_i E_{i\pm 1}.$$

These relations are best understood in terms of diagrams. The third is used because of the Kauffman polynomial. Using Markov traces and the basic construction of subfactor analysis Wenzl determined the structure of the BMW algebra for generic values of the parameters and a lot about it for special values. Its dimension is $1 \cdot 3 \cdots (2n + 1)$, n being the number of G_i's as above. It has the Hecke algebra as a natural quotient and the e_i algebra as a natural subalgebra.

Brauer in [Br] defined algebras of the same dimension as a model for the commutant of the symplectic and orthogonal groups in tensor powers of the vector representation. Wenzl in [Wez] solved the problem of the generic structure of the Brauer algebra essentially by viewing it as a specialization of the BMW algebra. Wenzl has also obtained subfactors from the BMW algebra using positivity considerations.

The commutant of the quantum group $U_q(sl_n)$ on the tensor powers of the n-dimensional representation is the Hecke algebra ([Ji2]). The BMW algebra does the same job for quantum groups of types B and C (see [We3]).

16. Conformal Field Theory

There are many similarities between subfactor theory and conformal field theory. I cite the discrete and continuous values of the central charge c ([FQS]), the ADE classification for $c < 1$ ([CIZ]), the "fusion rules" corresponding to the combinatorics of the centralizer towers ([GW]) and the existence of a tractable subclass called rational CFT's corresponding to finite depth subfactors. But the most compelling evidence for a connection is that the braid group representations which occur canonically in the subfactor picture for index < 4 occur also for the $SU(2)$ case of one of the main models of CFT – the Wess-Zumino-Witten model. Following [KZ], Tsuchiya and Kanie calculated the monodromy of the n-point functions in the holomorphic sector of this model and found the e_i-braid group representations (for spin $1/2$). Thus the subfactors themselves could be constructed out of the WZW theory but in a somewhat indirect way.

Moore and Seiberg showed in [MSe] that the braid group situation was just the genus zero case of a theory that works in arbitrary genus for any conformal field theory. To do this they checked that the defining relations ([Waj]) for the mapping class group do indeed follow from their axioms for CFT.

Conformal field theories are often obtained as continuum limits of critical 2-dimensional classical statistical mechanical systems. It is not clear that there is any deep relationship between the appearances of the braid group in solvable models and in the CFT of the continuum limit.

17. Algebraic Quantum Field Theory, Superselection Sectors

Several people ([Fr, FRS, Lo]) have noticed that the braiding and Markov trace structure is inherent in any (not necessarily conformal) low dimensional QFT. In the framework of algebraic QFT ([HK]), Doplicher, Haag and Roberts in [DHR] introduced superselection sectors as representations of the observable algebra which are equivalent to the vacuum representation when restricted to the algebra of the causal complement of some bounded region. Haag duality is the property that the algebras of such bounded regions and their causal complements are each other's commutants. This duality is supposed to hold in the vacuum sector. A simple calculation shows that superselection sectors give rise to endomorphisms of the local observable algebra. Using geometric properties of space-time, implementers of the endomorphism give rise to unitary braid group representations where the endomorphism takes σ_i to σ_{i+1}. The Markov trace may be obtained as a simple weak limit.

The square root of the index of the subfactor which is the image of the endomorphism was called the "statistical dimension" of the sector in [DHR], but because they were considering 4 dimensions this dimension was necessarily an integer in their theory.

One of the problems with all this work is that there do not seem to be any concrete examples where the subfactors and braid group representations have been calcualted to the satisfaction of an expert in von Neumann algebras.

18. Loop Groups and Subfactors

The loop group $LSU(2)$ has a discrete series of "positive energy" projective unitary reprseentations labelled by a level l and a spin j, $0 \le 2j \le l$. Inspired by [TK] and in an attempt to understand and implement the ideas of superselection sectors, and to provide naturally occuring examples of subfactors, A. Wassermann and I have been working on the best known example, WZW theory, especially for $LSU(2)$. We interpret the superselection sectors as being the discrete series of the loop group for a fixed level. Given an interval I in the circle S^1 (I^c will denote the complementary interval), let $L_I G$ be the group of loops supported in I. The von Neumann algebra $(L_I G)''$ corresponds to the local algebra. Haag duality (which we have proved) says that for the vacuum sector (spin $= 0$), $(L_I G)' = (L_{I^c} G)''$. These von Neumann algebras are type III_1 factors in any sector and we have shown that, if the level is fixed, the representations of $L_I G$ for any spin are unitarily equivalent.

We conjecture that $[(L_I G)' : (L_{I^c} G)''] = \sin^2\{(2j+1)\pi/(l+2)\}/\sin^2(\pi/(l+2))$ and in general that the subfactor is the tensor product of a Wenzl subfactor and the hyperfinite type III_1 factor. The vertex operators of Tsuchiya and Kanie, if made to become unbounded operator valued distributions on S^1, should provide explicit intertwiners between the representations of $L_I G$ for fixed level and different spin.

19. Witten's Interpretation of $V_L(t)$ and Its Generalizations

Motivated by his own work, ideas of Atiyah and Segal and the relations between subfactors, knots and CFT, Witten proposed the following formula for a link L with components $L_1, L_2, \ldots, L_n$ ([Wi1]).

$$V_L(e^{2\pi i/k+2}) = \int_A [\mathscr{D}A] e^{ik \int_{S^3} \mathrm{tr}(A \wedge dA + \frac{2}{3} A \wedge A \wedge A)} \prod_{i=1}^{n} W(L_i)$$

where A runs over all $su(2)$-valued 1-forms on S^3 (modulo the gauge group) and $W(L_i)$ is the trace of the parallel transport using A (in the 2-dimensional representation) along the closed curve L_i. A measure $[\mathscr{D}A]$ with the appropriate properties has not been shown to exist so this formula must be taken in the context of Witten's topological QFT ([A]), which predicts enough formal properties of such an integral for it to be calculated (i.e. identified with $V_L(t)$). This can be thought of as the solvability of this particular topological QFT.

The generalizations are now apparent. $SU(2)$ can be replaced by any compact Lie group G and one may choose any finite dimensional representation of G per component of L. This reproduces the ingredients of the picture that emerged from quantum groups in §13.

More significantly S^3 may be replaced by any closed 3-manifold so that Witten's theory predicts the existence of invariants for links in 3-manifolds and gives explicit formulae for calculating them from a surgery description of the link in the 3-manifold. These formulae have been checked using the Kirby calculus ([FR]) at least for $SU(2)$ in [RT] and [KM] using quantum groups. In [KM] certain explicit evaluations are given in terms of classical invariants. In [Li2] an elementary formula for these invariants occurs using no more than cabling and $V_L(t)$. The key ingredient in the proof is the degeneracy of the trace on the e_i algebra of §5. One wonders whether the 3-manifold invariants may be obtained directly from subfactors.

20. Topological Quantum Field Theory

Witten has developed a formalism in which a quantum field theory in $d + 1$ dimensions assigns a "Hilbert space" to any d-dimensional manifold Σ with extra structure and every time this manifold is the boundary of a $d + 1$ dimensional manifold M with compatible extra structure there is a vector in the Hilbert space of the boundary. In the simplest case $M = \Sigma \times [0, 1]$ the vector is supposed to define an operator giving the time evolution of a system from time 0 to time 1. The Hilbert spaces and vectors are supposed to satisfy certain important and powerful axioms. If the "extra structure" is little more or less than an orientation one talks of *topological* quantum field theory. See [A] where examples are given connected with Donaldson and Floer theory.

In the case of the theory given by the Chern-Simons action $\mathrm{tr}(\mathrm{Ad}\, A + (2/3)A^3)$, $d = 2$ and Witten identifies the Hilbert space corresponding to a surface Σ as being the (finite dimensional) vector space of conformal blocks for the corresponding Wess-Zumino-Witten theory with the same gauge group. This is crucial for his calculations as it allows him to assert that, for the $SU(n)$ theory the vector space corresponding to the sphere with 4 marked points is 2-dimensional. This, together with the "gluing" axioms of topological QFT allows the formal calculation of the functional integral via skein theory.

The Hilbert space for a surface Σ is also deduced via a "geometric quantization" approach using a complex structure on Σ. That the Hilbert space should be independent of the complex structure is interpreted as implying the existence of a flat connection on certain natural bundles over Teichmüller space. See the talk by Tsuchiya in these proceedings.

Witten's surgery formula comes from identifying a basis for the Hilbert space of a torus with the vectors obtained by realizing the torus as the boundary of a solid torus containing one simple homotopically non-trivial closed curve to which an irreducible representation of the compact group is assigned. The action of the diffeomorphisms in this basis (all that is required for surgery formulae given the gluing axioms) is precisely that of $SL(2, \mathbb{Z})$ on the characters of the relevant affine Lie algebra at the given level.

References

[AW]   Araki, H., Woods, E.: A classification of factors. Publ. RIMS, Kyoto University (1968) 51–130

[A]   Atiyah, M.: Topological quantum field theory. Publ. Math. I.H.E.S. **68** (1989) 175–186

[Ba]   Baxter, R.: Exactly solved models in statistical mechanics. Academic Press, New York 1982

[BW]   Birman, J., Wenzl, H.: Braids, link polynomials and a new algebra. Trans. AMS **313** (1989) 269–273

[Bi]   Birman, J.: Braids, links and mapping class groups. Ann. Math. Studies **82** (1974)

[Bo]   Bourbaki, N.: Groupes et algèbres de Lie, IV, V, VI. Masson, Paris 1981

[Br]   Brauer, R.: On algebras which are connected with the semisimple continuous groups. Ann. Math. **38** (1937) 856–872

[CIZ]   Cappelli, A., Itzykson, C., Zuber, J. B.: Nucl. Phys. B **280** (1987) 445–

[Co1]   Connes, A.: von Neumann algebras. Proceedings ICM Helsinki, 1978, vol. 1, pp. 97–109

[Co2]   Connes, A.: Classification of injective factors. Ann. Math. **104** (1976) 73–115

[Co3]   Connes, A.: Factors of type III$_1$, property L'_λ and the closure of inner automorphisms. J. Operator Theory **14** (1985) 189–211

[Co4]   Connes, A.: Classification des facteurs. In: Operator Algebras and Applications, Proc. Symp. Pure Appl. Math. **38** (1982) part II, 43–109

[Cnw]   Conway, J.: An enumeration of knots and links. Computational Problems in Abstract Algebra (ed. J. Leech). Pergamon Press 1969, pp. 329–358

[D+]   Date, E., Jimbo, M., Miki, K., Miwa, T.: Cyclic representations of $U_q(sl(n+1, \mathbb{C}))$ at $q^N = 1$. RIMS Preprint 1990

[DHR]   Doplicher, S., Haag, R., Roberts, J.: Local observables and particle statistics I, II. Comm. Math. Phys. **23** (1971) 199–230 and **35** (1974) 49–85

[Dr]   Drinfeld, V.: Quantum groups. Proc. ICM 1986, vol. 1, pp. 798–820

[Fa]   Faddeev, L.: Integrable models in $(1 + 1)$-dimensional quantum field theory. (Lectures in les Houches, 1982) Elsevier Science Publishers, 1984, pp. 563–608

[FR]   Fenn, R., Rourke, C.: On Kirby's calculus of links. Topology **18** (1979) 1–15

[FW]   Franks, J., Williams, R.: Braids and the Jones-Conway polynomial. Trans. AMS **303** (1987) 97–108

[FRS]   Fredenhagen, K., Rehren, Schroer, B.: Superselection sectors with braid group statistics and exchange algebras. Comm. Math. Phys. **125** (1989) 201–226

[F+]   Freyd, P., Yetter, D., Hoste, J., Lickorish, W., Millett, K., Ocneanu, A.: A new polynomial invariant of knots and links. Bull. AMS **12** (1985) 183–190

[FQS]   Friedan, D., Qiu, Z., Shenker, S.: Conformal invariance, unitarity and critical exponents in two dimensions. Phys. Rev. Lett. **52** (1984) 1575–1578

[Fr]   Fröhlich, J.: Statistics of fields, the Yang-Baxter equation and the theory of knots and links. Proceedings Cargise, ed. G. 't Hooft et al. (1987)

[GHJ]   Goodman, F., delaHarpe, P., Jones, V.: Coxeter graphs and towers of algebras. MSRI Publications (Springer) **14** (1989)

[GW]   Goodman, F., Wenzl, H.: Littlewood Richardson coefficients for Hecke algebras
       at roots of unity. Adv. Math.

[HK]   Haag, R., Kastler, D.: An algebraic approach to quantum field theory. J. Math.
       Phys. **5** (1964) 848–861

[Ha]   Haagerup, U.: Connes' bicentralizer problem and the uniqueness of the injective
       factor of type III_1. Acta. Math. **158** (1987) 95–148

[HS]   Haagerup, U., Schou, J.: Some new subfactors of the hyperfinite II_1 factor.
       Preprint 1989

[HJ]   delaHarpe, P., Jones, V.: Paires de sous-algèbres semisimples et graphes fortement
       réguliers. C.R. Acad. Sci. Paris **311** (1990) 147–150

[Ji1] Jimbo, M.: A q-difference analogue of $U(\mathfrak{g})$ and the Yang-Baxter equation. Lett.
       Math. Phys. **102** (1986) 537–567

[Ji2] Jimbo, M.: A q-analogue of $U(sl(N + 1))$, Hecke algebra and the Yang-Baxter
       equation. Lett. Math. Phys. **11** (1986) 247–252

[Jo1] Jones, V.: Index for subfactors. Invent. math. **72** (1983) 1–25

[Jo2] Jones, V.: Notes on subfactors and statistical mechanics. In: Braid Group, Knot
       Theory and Statistical Mechanics (ed. Yang and Ge). World Scientific, 1989, pp.
       1–25

[Jo3] Jones, V.: Index for subrings of rings. Contemp. Math. **43** (1985) 181–190

[Jo4] Jones, V.: Braid groups, Hecke algeras and type II_1 factors. In: Geometric
       Methods in Operator Algebras (ed. Araki and Effros). Pitman Res. Notes in
       Math. (1983) 242–273

[Jo5] Jones, V.: A polynomial invariant for knots via von Neumann algebras. Bull.
       AMS **12** (1985) 103–111

[Jo6] Jones, V.: On knot invariants related to some statistical mechanical models.
       Pacific J. Math. **137** (1989) 311–334

[Jo7] Jones, V.: Actions of finite groups on the hyperfinite type II_1 factor. Memoirs
       AMS **237** (1980)

[Ka]   Kauffman, L.: State models and the Jones polynomial. Topology **26** (1987)
       395–401

[KM]   Kirby, R., Melvin, P.: The three manifold invariants of Witten and Reshetikhin-
       Turaev. Preprint 1990

[KZ]   Knizhnik, V., Zamolodchikov, A.: Current algebra and Wess-Zumino models in
       two dimensions. Nuc. Phys. B **247** (1984) 83–103

[Kr]   Krieger, W.: On ergodic flows and the isomorphism of factors. Math. Ann. **223**
       (1976) 19–70

[KR]   Kulish, P., Reshetikhin, N.: Quantum linear problem for the sine-Gordon equation
       and higher representations. J. Sov. Math. **23** (1983) 2435–2441

[La]   Lawrence, R.: Homological representations of the Hecke algebra. Comm. Math.
       Phys. (to appear)

[Li1] Lickorish, W.: Polynomials for links. Bull. LMS **20** (1988) 558–588

[Li2] Lickorish, W.: Three-manifold invariants from the combinatorics of the Jones
       polynomial. Pacific Math. J. (to appear)

[LT]   Lickorish, W., Thistlethwaite, M.: Some links with non-trivial polynomials and
       their crossing-numbers. Comment. Math. Helv. **63** (1988) 527–539

[Lo]   Longo, R.: Index of subfactors and statistics of quantum fields I. Comm. Math.
       Phys. **126** (1989) 217–247

[MSe] Moore, G., Seiberg, N.: Classical and quantum conformal field theory. Comm.
       Math. Phys. **123** (1989) 177–254

[Mo]   Morton, H.: Closed braid representations for a link and its 2-variable polynomial.
       Preprint Liverpool, 1985

[MSt]   Morton, H., Strickland, P.: Jones polynomial invariants for knots and satellites. Univ. of Liverpool Preprint 1989

[Mu]   Murasugi, K.: Jones polynomials and classical conjectures in knot theory. Topology **26** (1987) 187–194

[Muk]   Murakami, J.: The Kauffman polynomial of links and representation theory. Osaka J. Math. **24** (1987) 745–758

[MvN1]   Murray, F., von Neumann, J.: On rings of operators. Ann. Math. **37** (1936) 116–229

[MvN2]   Murray, F., von Neumann, J.: On rings of operators IV. Ann. Math. **44** (1943) 716–808

[O1]   Ocneanu, A.: Actions of discrete amenable groups on von Neumann algebras. Lecture Notes in Mathematics, vol. 1138. Spinger, Berlin Heidelberg New York 1985

[O2]   Ocneanu, A.: Quantized groups, string algebras and Galois theory for algebras. In: Operator Algebras and Applications (eds. Evans and Takesaki), 1988, pp. 119–172

[Pa]   Pasquier, V.: Two-dimensional critical systems labelled by Dynkin diagrams. Nucl. Phys. B **285** (1987) 162–172

[Po]   Popa, S.: Classification of subfactors: the reduction to community squares. Invent. math. **101** (1990) 19–43

[PP]   Pimsner, M., Popa, S.: Entropy and index for subfactors. Ann. Sci. Ec. Norm. Sup. **19** (1986) 57–106

[Pow]   Powers, R.: Representations of uniformly hyperfinite algebras and their associated von Neumann rings. Ann. Math. **86** (1967) 138–171

[Re]   Reshetikhin, N.: Quantized universal enveloping algebras, the Yang-Baxter equation and invariants of links I, II. LOMI Preprints 1988

[RT]   Reshetikhin, N., Turaev, V.: Invariants of 3-manifolds via link polynomials and quantum groups. Invent. math. (to appear)

[Ro]   Rosso, M.: Groupes quantiques et modèles à vertex de V. Jones en théorie des noeuds. C. R. Acad. Sci. Paris **307** (1988) 207–210

[Sk]   Sklyanin, E.: Some algebraic structures connected with the Yang-Baxter equation. J. Sov. Math. **19** (1982) 1546–1596

[Su]   Sunder, V. S.: Π_1-factors, their bimodules and hypergroups. Preprint 1989

[Ta]   Takesaki, M.: Tomita's theory of modular Hilbert algebras and its applications. (Lecture Notes in Mathematics, vol. 128.) Springer, Berlin Heidelberg New York 1970

[TL]   Temperley, H., Lieb, E.: Relations between the 'percolation' and... . Proc. Roy. Soc. Ser A **322** (1971) 251–280

[Th]   Thistlethwaite, M.: A spanning tree expansion of the Jones polynomial. Topology **26** (1987) 297–309

[TK]   Tsuchiya, A., Kanie, Y.: Vertex operators in conformal field theory on $\mathbb{P}^1$ and monodromy representations of braid group. Adv. Stud. Pure Math. **16** (1988) 297–372

[Tu]   Turaev, V.: The Yang-Baxter equation and invariants of links. Invent. math. **92** (1988) 527–553

[vN]   von Neumann, J.: Zur Algebra der Funktionaloperatoren. Math. Ann. **102** (1929) 370–427

[Waj]   Wajnryb, B.: A simple presentation for the mapping class group of an orientable surface. Israel J. Math. **45** (1983) 157–174

[Wass]   Wassermann, A.: Coactions and Yang-Baxter equations for ergodic actions and subfactors. In: Operator Algebras and Applications (ed. Evans and Takesaki). LMS Lecture Notes **136**, vol. 2 (1988) 203–236

[We1]   Wenzl, H.: Hecke algebras of type A_n and subfactors. Invent. math. **92** (1988) 349–383

[We2]   Wenzl, H.: On the structure of Brauer's centralizer algebras. Ann. Math. **128** (1988) 173–193

[We3]   Wenzl, H.: Quantum groups and subfactors of type B, C and D. Comm. Math. Phys. **133** (1990) 383–432

[Wi1]   Witten, E.: Quantum field theory and the Jones polynomial. Comm. Math. Phys. **121** (1989) 351–399

Geometric Algorithms and Algorithmic Geometry

László Lovász

Department of Computer Science, Eötvös University, H-1088 Budapest, Hungary and
Department of Computer Science, Princeton University, Princeton, NJ 08544, USA

0. Introduction

In this paper I would like to illustrate two facts. First, that ideas borrowed from convex, projective, and other classical branches of geometry play an important role in the design of algorithms for problems that do not seem to have anything to do with geometry: for problems in optimization, combinatorics, algebra, and number theory. Second, that such applications of geometry suggest some very elementary algorithmic questions concerning geometric notions, whose solution is far from complete. How to compute the volume? How to decide whether or not a convex body contains the other? How to present a convex body as an input to an algorithm? Even partial answers to these problems require a wealth of mathematical ideas, often (again) quite unrelated to the original question. On the other hand, answers to these questions have a very wide range of applicability.

In the first chapter some typical constructions are surveyed which lead from non-geometric problems to geometric ones. Such is, of course, the classical field of the "Geometry of Numbers". This is the application of lattice geometry and convexity to number theory, and was initiated by Minkowski around the turn of the century. We only touch upon this area to point out that the recent shift of interest from structural problems to algorithmic ones has induced a lot of activity, and this new approach has even fertilized classical types of investigations.

We also give a brief introduction to polyhedral combinatorics, developed in the 60s by Ford, Fulkerson, Hoffman, Edmonds, and others. Here polyhedral theory and linear programming are applied to combinatorial optimization problems. This approach yields surprisingly successful algorithms both from a theoretical and practical point of view. We also show how some enumeration problems are related to the computation of the volume of certain polytopes.

The second chapter treats verious forms of presentation of a convex body to an algorithm. This seemingly technical issue leads to a powerful equivalence principle between different ways of presentation. A combination of this principle with polyhedral combinatorics provides (at least theoretically) efficient algorithms for most combinatorial optimization problems that can be solved efficiently by any other means.

Chapter 3 describes a basic construction in algorithmic geometry, namely the Löwner-John ellipsoids. These are used in the ellipsoid method of Shor, Yudin, Nemirovskii and Khachiyan (in particular in establishing the equivalence principle formulated in the previous chapter) and in many other geometric

Proceedings of the International Congress
of Mathematicians, Kyoto, Japan, 1990
© The Mathematical Society of Japan, 1991

algorithms. We also sketch the geometric background of Karmarkar's celebrated linear programming algorithm.

In the last chapter we discuss recent developments concerning the problem of computing the volume of a convex body. After some very discouraging negative results, Dyer, Frieze and Kannan designed a polynomial time randomized algorithm which approximates the volume with an arbitrarily small error. It turns out that the crucial issue is to generate a random point uniformly distributed over a general convex body. The solution of this problem leads to Markov chains, eigenvalues of matrices, differential geometry, and even to some algebraic topology.

There is an extremely important branch of algorithmic geometry which is not treated here; this is usually called computational geometry. Polyhedral combinatorics leads to high-dimensional problems; other – more immediate – applications in image processing and robotics lead to two- and three-dimensional questions. In such cases, a different notion of efficiency is the crucial one. Our concern will be polynomial time; in computational geometry, usually linear or almost-linear time is the target. The interested reader may consult the monographs by Preparata and Shamos (1985) and by Edelsbrunner (1987).

1. Number Theory, Combinatorics, and Convex Sets

In this chapter we illustrate some of the most important constructions which transform algorithmic problems in various branches of mathematics into geometric questions, and in fact into very simple ones. These examples will also serve to help us put the corresponding general algorithmic problems into the right framework.

1.1 Geometry of Numbers

This is the classical area where the application of geometric ideas to non-geometric problems has been very successful ever since Minkowski's work. In the last decade, algorithmic questions arising from numerical methods, primality testing, computational algebra, cryptography, and other areas have revigorated the field; it turns out that to solve these algorithmic problems, often new structural insight is needed.

The classical problem in number theory which lead Minkowski to the "Geometry of Numbers" is the problem of *simultaneous diophantine approximation*: given n real numbers $\alpha_1, \ldots, \alpha_n$, and an "error bound" $\varepsilon > 0$, find integers $p_1, \ldots, p_n$ and q such that $q > 0$ and

$$\left| \alpha_i - \frac{p_i}{q} \right| \le \frac{\varepsilon}{q} \qquad (i = 1, \ldots, n).$$

The answer is trivial for $\varepsilon = 1/2$ (we can choose q arbitrarily); it was proved by Dirichlet that such integers exist for every $\varepsilon > 0$ and in fact we can require that $q \le \varepsilon^{-n}$.

While the proof of Dirichlet's theorem is quite easy (using "Dirichlet's Principle"), no efficient algorithm is known to *find* such an approximation. A geometric translation is useful, among others, in finding a solution with a worse bound on q.

Consider the following vectors in $\mathbb{R}^{n+1}$: $e_1, \ldots, e_n$ (the first n basis vectors) and $(\alpha_1, \ldots, \alpha_n, 1)^T$. The linear combinations of these vectors with integer coefficients form a lattice L, whose typical point looks like $(q\alpha_1 - p_1, \ldots, q\alpha_n - p_n, q)^T$ with integral $p_1, \ldots, p_n$ and q. Dirichlet's Theorem is equivalent to asserting that the lattice L has a non-zero lattice point in the brick

$$K = [-\varepsilon, \varepsilon] \times \ldots \times [-\varepsilon, \varepsilon] \times [-\varepsilon^{-n}, \varepsilon^{-n}].$$

Minkowski's famous "First Theorem" shows that the fact that K is a brick is irrelevant: all that matters is that K is a convex body centrally symmetric with respect to the origin with volume at least 2^{n+1} times the determinant of the lattice.

The convex body K can be viewed as the unit ball of a norm, and so Minkowski's theorem guarantees the existence of a "short" non-zero lattice vector in every lattice, measured in an arbitrary norm.

Many other problems turn out equivalent to the existence of a short non-zero lattice vector in appropriately defined lattices. For example, let $f(x) = \sum_{i=1}^{n} a_i x^i$ be a polynomial with integral coefficients; we want to know whether f is irreducible. Let α be a root of f; for simplicity, assume that α is real. Let K be a sufficiently large real number (computable from f) and let $L \subseteq \mathbb{R}^{n+1}$ be the lattice generated by the vectors $e_i + K\alpha^{i-1} e_{n+1}$, $1 \leq i \leq n$. If f is reducible over the rational field, then L contains a non-zero vector with (euclidean) length at most $2^n \sqrt{\sum_i a_i^2}$. On the other hand, if f is irreducible then every non-zero vector in the lattice is at least 2^n times this length. This fact is the basis of the efficient (polynomial time) algorithm for factoring polynomials (Lenstra, Lenstra and Lovász 1982).

We shall not go into geometric algorithms involving lattices in this survey; instead, we refer to (Lovász 1989) for a survey.

1.2 Polyhedral Combinatorics

Polyhedral combinatorics provides perhaps the most successful general approach to various combinatorial problems. To illustrate the idea, consider the following simple graph-theoretic problem. Let G be a (finite) graph with node set V and edge set E. A set of nodes of G is called *stable* if no two elements of it are connected by an edge. Let $\alpha(G)$ denote the maximum cardinality of a stable set in G. To determine $\alpha(G)$ is difficult (NP-hard) in general; but the approach of polyhedral combinatorics suggests efficiently solvable special cases, as well as efficiently computable estimates of $\alpha(G)$.

Let us construct the following convex polytope: for every stable set S of nodes, let χ^S denote the incidence vector of S (in the space $\mathbb{R}^V$ of vectors indexed by the nodes of G), and let STAB(G) be the convex hull of such incidence vectors. Then

$$\alpha(G) = \max\{|S| : S \text{ stable}\} = \max\{1 \cdot \chi^S : S \text{ stable}\}$$
$$= \max\{\textstyle\sum_i x_i : x \in \text{STAB}(G)\}.$$

(since the maximum of a linear objective function over a polytope is automatically assumed at a vertex).

The hope is to apply the powerful methods of linear programming to find this maximum. To this end, however, we have to find a representation of STAB(G) as the solution set of a system of linear inequalities. Such a representation exists of course; but how to find it? Note that the arguments presented so far would work for the problem of finding a maximum cardinality member in any collection of subsets of a finite set V; where we have to be problem-specific is in finding the linear representation of the corresponding polytope.

There is a very large number of results presenting linear descriptions of various combinatorial polyhedra; we shall restrict ourselves to the stable set polytope and be content with giving a couple of illustrations. A natural starting point is the following set of linear inequalities:

$$0 \le x_i \le 1 \qquad (i \in V), \tag{1}$$

$$x_i + x_j \le 1 \qquad (ij \in E). \tag{2}$$

It is clear that the incidence vector of any stable set, and therefore every vector in STAB(G), satisfies inequalities (1) and (2). The solution set of this system is, however, a larger polytope than STAB(G) in general. In fact, *inequalities (1) and (2) suffice to describe* STAB(G) *if and only if G is a bipartite graph*. So at least for bipartite graphs, $\alpha(G)$ can be determined using linear programming algorithms. Or, applying the Duality Theorem of linear programming, one can obtain a min-max formula for $\alpha(G)$ (which, after some transformations, turns out to be equivalent to the König Theorem).

If the graph is non-bipartite then it contains a circuit of odd length and we can use such a circuit to add further constraints to (1) and (2): for every odd circuit C we write up the inequality

$$\sum_{i \in V(C)} x_i \le \frac{|C| - 1}{2}. \tag{3}$$

Graphs for which (1), (2) and (3) suffice to describe STAB(G) are called *t-perfect*; these graphs are less well understood than bipartite graphs, but several important classes of them are known. We mention *series-parallel graphs*, i.e., graphs which can be obtained by the repeated application of series and parallel extensions (Chvátal 1975, Boulala and Uhry 1979)

Assume that G is *t*-perfect; then $\alpha(G)$ can be expressed as the optimum value of a linear program with constraints (1)–(2)–(3). Note, however, that (3) includes possibly exponentially many constraints (in $n = |V|$), and so writing up this program and calling a linear program solver would be inefficient. We shall see that general geometric considerations provide efficient solution methods for such systems which do not need the whole system explicitly.

Further natural inequalities valid for STAB(G), but not implied by the previous ones, can be written up. Let B be a clique (a set of mutually adjacent nodes in G). Then every stable set meets B in at most one node and hence the inequality

$$\sum_{i \in B} x_i \le 1 \tag{4}$$

is valid for STAB(G). Those graphs for which (1) and (4) suffice to describe STAB(G) are called *perfect*. This rich class has been defined by Berge (1961) before these polyhedral methods were introduced, motivated by many classes of examples. Let us mention one: Let $(V, \le)$ be a (finite) partially ordered set.

Define a graph G by connecting two elements of V iff they are comparable. *Comparability graphs* obtained this way are perfect.

It is beyond the scope of this paper to treat perfect graphs; we refer to Berge and Chvátal (1984) and Grötschel, Lovász and Schrijver (1988). It should be mentioned, however, that polyhedral methods play a central role in their study, even in proving seemingly elementary properties.

Polyhedral combinatorics is closely related to *integer linear programming*. The basic problem here is to solve a system of linear inequalities *in integers*. For example, the integer solutions of the system (1)–(2) are exactly the incidence vectors of stable sets. The problem of simultaneous diophantine approximation can also be viewed as an integer programming problem: Given α_i, ε, and Q, find a solution of

$$-\varepsilon \le q\alpha_i - p_i \le \varepsilon \qquad (i = 1, \ldots, n),$$
$$1 \le q \le Q.$$

For a long while, integer programming and lattice geometry have developed independently, and used rather different methods. A first substantial connection was established by Lenstra (1983), who designed a polynomial time algorithm to solve integer linear programs with a bounded number of variables, using methods borrowed from the geometry of numbers. This approach to integer programming seems to gain further momentum in recent years.

1.3 Enumeration and Volume

Some enumeration problems also have useful translations into geometry. This connection is not so well understood and we only give one example. Let $P = (E, \le)$ be a partially ordered set. A linear order of E which is compatible with the given partial order is called a *linear extension* of the partial order. The number of linear extensions is a measure of how incomplete the partial order is, and it plays an important role in several algorithmic and other questions. There is no reasonable formula or efficient algorithm known to find this number. In fact, no such algorithm can be expected by the recent important result of Brightwell and Winkler (1990), which asserts that to determine the number of linear extensions of a poset is #P-complete. But an efficient approximation algorithm can be obtained, which is based on the following construction (Stanley 1986).

Consider the linear space $\mathbb{R}^E$ and the incidence vectors of filters in P. The convex hull of these incidence vectors is a polytope FILT(P). By the methods of polyhedral combinatorics mentioned in the previous section, it can be shown that FILT(P) is defined by the inequalities

$$0 \le x_i \le 1 \qquad (i \in E),$$
$$x_i \le x_j \qquad (i \le j).$$

This fact can be used, as sketched above, to solve optimization problems involving filters. Right now, however, the following fact is important: *the number of linear extensions of P equals $n!$ times the volume of* FILT(P). This observation reduces the problem of enumerating linear extensions to the problem of determining the volume of a polytope, which is described as the solution set of a small number of simple linear inequalities (in n-space we have $O(n^2)$ inequalities). We shall return to this general geometric problem in Chapter 4.

2. What is a Convex Body?

We have seen that various number theoretic, algebraic, and combinatorial questions can be reduced to quite fundamental problems in geometry, such as computing the volume or finding the maximum of a linear objective function over a convex body. Before treating geometric algorithms to solve such problems, we have to introduce the right framework.

A convex body is a closed, bounded, full-dimensional convex set in $\mathbb{R}^n$. This simple definition becomes insufficient, however, if we are interested in algorithmic questions. In this chapter we discuss our aspects for algorithms and the algorithmic notion of a convex body.

2.1 Basic Algorithmic Problems for a Convex Body

Our condition for the "efficiency" of an algorithm is worst-case polynomial time: this means that there exists a constant $c > 0$ such that for every input of length n, the algorithm makes $O(n^c)$ bit-operations. Here the length of the input is the total number of bits needed to describe the input. We assume that input numbers are always rational, and their contribution to the input length is the total number of digits in the binary representation of the numerator and of the denominator.

For the theory of polynomial time algorithms, and for related notions in complexity theory like NP and NP-hard, see e.g. (Garey and Johnson 1979). To follow this paper, it should suffice to translate "NP-hard" or "#P-complete" as "it is hopeless to find a polynomial-time algorithm for this problem".

We want to study algorithms whose input is a convex body, and with this we run into trouble. How to present a convex body? Various forms occur in various problems. In linear programming, one always considers convex polyhedra, presented as the solution set of a system of linear inequalities. An equally natural form in which convex bodies come up is a convex polytope, presented as the convex hull of an explicitly given set of vectors. In Banach space theory, convex bodies arise as unit balls of norms, where the norm may be given by some formula; e.g. $\sum_i |x_i|^3 \leq 1$ defines the unit ball of the ℓ_3-norm. In geometry, a convex body $K \subseteq \mathbb{R}^n$ is sometimes described by its support function: this is essentially the function $\varrho_K : \mathbb{R}^n \to \mathbb{R}$, defined by $\varrho_K(u) = \max\{u \cdot x : x \in K\}$.

The polytope STAB(G) is defined as the convex hull of a set of vectors (incidence vectors of stable sets), but to describe it, it would be very inefficient to list these vectors; it suffices to specify the graph G. So STAB(G) is presented by *implicitly* specifying its vertices.

We are interested in algorithms that are as independent of the specifics of the presentation of the body as possible. It turns out that most geometric algorithms depend on the possibility to carry out one or more of the following tasks:

- *Membership Test*: Given a (rational) vector x, decide whether or not $x \in K$.
- *Separation*: Given a (rational) vector x, decide whether or not $x \in K$, and if not, find a hyperplane separating x from K.
- *Validity Test*: Given a linear inequality $a \cdot x \leq \alpha$ (with rational coefficients), decide whether or not the inequality is valid for all $x \in K$.
- *Violation*: Given a linear inequality $a \cdot x \leq \alpha$ (with rational coefficients), decide whether or not the inequality is valid for all $x \in K$, and if not, find an $x \in K$ violating it.

It is clear that finding a separating hyperplane is a more difficult task than testing for membership, and finding a violating point is more difficult than testing for validity. Scanning through the examples above, we see that depending on the presentation of K, one or the other of these tasks can be carried out easily while others appear non-trivial. For example, if the body is given as the solution set of linear inequalities, then it is trivial to test membership by substituting x in each of the defining inequalities; and if it violates one, this also yields a separating hyperplane. On the other hand, there is no obvious way to test the validity of an inequality. Similarly, if the body is presented as the convex hull of vectors, then it is trivial to test validity of a linear inequality, but testing membership is non-trivial.

For STAB(G), neither one of the above tasks is easy, at least if we allow only polynomial time in n (note that the input size in $O(n^2)$). In fact, each of the above tasks is NP-hard.

To get presentation-independent results, we define a *membership oracle* as a black box which works as follows: if we plug in a vector $x \in \mathbb{R}^n$, it returns "YES" or "NO". Its answers must be consistent with the interpretation that "YES" means $x \in K$ for some convex body K. We can call such an oracle in an algorithm; one call counts as a single step. Of course, if we have an algorithm to test membership in polynomial time, then we can "put this inside the black box", and this increases the running time by a polynomial factor only.

One can introduce separation, validity, and violation oracles in a similar way.

The following is a surprising and powerful principle (not a theorem!):

Equivalence Principle. *The four oracles above are equivalent from the point of view of polynomial time algorithms.*

In other words, if for a class of convex bodies we have a polynomial time algorithm to solve either one of them, then all the others can be solved in polynomial time. This principle is true only under some technical assumptions and restrictions which, however, seldom cause any problem in its applications. There are various ways to formulate technical conditions making the Equivalence Principle valid; we only sketch some, and refer for a complete discussion to the monograph (Grötschel, Lovász and Schrijver 1988).

One set of these technicalities consists of making the "boundedness" and "full-dimensionality" properties of convex bodies effective: we need to know a number $R > 0$ such that K is contained in the ball with radius R about the origin, and another number $r > 0$ such that K contains a ball with radius r. These numbers must be considered part of the input to any algorithm, so the number of bits needed to write them down must be included in the input size. If the body is given by a membership oracle, and we want to solve the other tasks, then in addition the center of an inscribed ball with radius r must be given in advance. It is easy to argue that without this kind of information, the Equivalence Principle would not be valid.

Another set of limitations comes from numerical errors. We have to re-define the oracles so that an "error bound" $\varepsilon > 0$ is also part of the input, and then allow a small error in the answer. For example, in a membership oracle we should allow either a "YES" or a "NO" answer if the distance of x from the boundary of K is less than ε. One can formulate "weak" versions of all the above oracles in an analogous way. A precise form of the Equivalence Principle holds for these weak versions. (If the convex body K is a polytope with (say) 0-1 vertices, then we do not have to restrict ourselves to weak versions.)

The main ingredient in the proof of the Equivalence Principle is the *Ellipsoid Method*, to be sketched in the next chapter.

2.2 Applications of the Equivalence Principle

Some consequences of the Equivalence Principle are immediate. Since testing membership in a polyhedron presented by an (explicit) system of linear inequalities is trivial, we can test validity of a linear inequality for such polyhedra in polynomial time. This result implies a polynomial time test for the solvability of a system of linear inequalities, which in turn implies a polynomial time algorithm to solve linear programs.

In a similar way it follows that quadratic programs with positive definite constraints can be solved in polynomial time.

Several basic algorithmic issues concerning convex bodies can be solved using just the Equivalence Principle; let us mention two of these. The issue of validity of a linear constraint can be viewed as a special case of either of the following questions:

- Given two convex bodies K_1 and K_2, are they disjoint?
- Given two convex bodies K_1 and K_2, is $K_1 \subseteq K_2$?

Due to the necessary uncertainty around the boundary, in both cases we can only expect an approximate answer; we specify an $\varepsilon > 0$ and if $\mathrm{vol}(K_1 \cap K_2) < \varepsilon$, or if $\mathrm{vol}(K_1 \setminus K_2) < \varepsilon$, then the "YES" answer is acceptable.

Having separation oracles for K_1 and K_2, we can design a separation algorithm for $K_1 \cap K_2$ trivially. Suppressing technical details (the number r), the Equivalence Principle yields a polynomial time solution for the violation problem for $K_1 \cap K_2$. Let $c \cdot x \leq \gamma$ be any inequality which is invalid for all points in K_1, and solve the violation problem for $K_1 \cap K_2$. If the inequality is valid for $K_1 \cap K_2$ then this intersection must be empty; else, the violation algorithm returns a common point.

The second problem is more difficult; in fact, to solve it (even with the ε-tolerance) would yield polynomial-time solutions for NP-complete problems. But using the randomized methods discussed in Chapter 4 below, we can test $K_1 \subseteq K_2$ (with the ε-tolerance) in time polynomial in n and $1/\varepsilon$. (Note that a "truely" polynomial algorithm ought to be polynomial in $\log(1/\varepsilon)$.)

The most interesting applications of the Equivalence Principle are in the field of combinatorial optimization. We illustrate this by two results due to Grötschel, Lovász and Schrijver (see 1988). Consider the problem of determining $\alpha(G)$ for a t-perfect graph G. As discussed earlier, this is equivalent to maximizing the linear objective function $\sum_i x_i$ over STAB(G), which could be easily solved by binary search if we could test validity for STAB(G). By the Equivalence Principle, it would suffice to find a polynomial time membership test for STAB(G). In other words, given a vector $x \in \mathbb{R}^V$, is $x \in$ STAB(G)?

At this point we want to use the linear description of STAB(G), which we have determined for t-perfect graphs: STAB(G) is the solution set of inequalities (1)–(2)–(3) above. So all we have to do is to test whether x satisfies these inequalities.

For (1) and (2), this is trivially checked by simply substituting x in them. But for (3) this does not work: there are (typically) exponentially many inequalities in (3), and we do not want to generate them all.

However, there is a way to check in polynomial time whether or not x satisfies all these inequalities. We may assume that we have checked (1) and (2) and that x satisfies these. For every edge $ij \in E$, let $y_{ij} = 1 - x_i - x_j$. Since x satisfies (2), these numbers are non-negative, and we may consider them as "lengths" of the edges. Let C be an odd circuit in G, then

$$\sum_{i \in E(C)} y_{ij} = |C| - \sum_{i \in V(C)} x_i,$$

and hence x satisfies the constraint of (3) belonging to this circuit C if and only if the "length" of C is at least 1. So x satisfies (3) iff the length of every odd circuit is at least 1.

Now there is a rather simple version of breadth-first-search that finds the shortest odd circuit in any graph with non-negative edge-lengths (the details of this do not belong here). Using this, all we have to do is to check whether this minimum length is less than 1 or not.

The stability number of a perfect graph can also be determined in polynomial time. This algorithm is also based on the Equivalence Principle, but it is substantially more complicated. If we want to copy the above argument, we run into the following difficulty: the membership test for STAB(G) is equivalent to the validity test for the stable set polytope of the complementary graph, and so the equivalence principle reduces the problem of determining a maximum stable set to (esssentially) the same problem for the complementary graph. One has to go up to the $|V|^2$ dimensional space to apply the Equivalence Principle successfully.

This type of application of the Equivalence Principle is not rare; in fact, it is shown by Grötschel, Lovász and Schrijver (see 1988) that most of those combinatorial optimization problems whose polynomial time solvability was known by (often quite involved) ad hoc algorithms, can be solved in polynomial time using a combination of the Equivalence Principle with very elementary algorithmic ideas. There are also several combinatorial optimization problems which can be solved in polynomial time, but for which no polynomial time algorithm avoiding the use of the Equivalence Principle is known (two very important such problems are finding the independence number of a perfect graph and minimizing a submodular setfunction).

Unfortunately, algorithms derived from the Equivalence Principle are polynomial, but very slow (their running time is a polynomial with a very high degree), and therefore practically useless. This is quite natural, considering how general this method is. These results should be interpreted as "existence proofs" for polynomial time solvability of the problem; having established this much, one can try to design problem-specific algorithms which are more efficient. The remarks above show that this is often very difficult.

3. Convex Bodies and Ellipsoids

3.1 The Löwner-John Ellipsoid

It was proved by Löwner that *for each convex body K, there exists a unique ellipsoid E with minimum volume containing it.* John proved that *if we shrink this ellipsoid of a convex body K from its center by a factor of n, we obtain an ellipsoid that is contained in K.* We call this ellipsoid the *Löwner-John ellipsoid* of the body (see (Grötschel, Lovász and Schrijver 1988) for details). Analogues of the theorems above hold if we consider an inscribed ellipsoid with largest volume.

If we restrict ourselves to centrally symmetric convex bodies, then of course the center will be the center of the Löwner-John ellipsoid as well, and in John's theorem it suffices to shrink by a factor of $\sqrt{n}$.

The Löwner-John ellipsoid itself may be difficult to compute. However, an ellipsoid E with somewhat weaker properties can be computed in polynomial time, using a version of the shallow cut ellipsoid method due to Yudin and Nemirovskii (1976). We call an ellipsoid E a *weak Löwner-John ellipsoid* for K, if E contains K and if we shrink E from its center by a factor of $2n^{3/2}$, we obtain an ellipsoid that is contained in K.

Theorem 3.1 *Given a separation oracle for a convex body K, a weak Löwner-John ellipsoid for K can be computed in polynomial time.*

We remark that if K is given in a more explicit manner (e.g., as the solution set of a system of linear inequalities or the convex hull of a set of vertices) then the factor $n^{3/2}$ can be improved to $2n$.

The algorithm proving Theorem 3.1 is based on a natural proof of John's theorem. Let E be the ellipsoid with smallest volume containing K. Applying an affine transformation, we may assume that E is the unit ball about the origin. Let E' denote the ball with radius $1/n$ about the origin, and assume that E' is not contained in K. Let $x \in E' \setminus K$, and let H be a hyperplane separating x from K. H cuts E into two parts, one of which, say E_1, includes K. Now E_1 is smaller or only slightly larger than a half-ball, since $x \notin E_1$. Therefore it is a routine computation in linear algebra to verify that E_1 can be included in an ellipsoid which has smaller volume than E.

To turn this proof into an algorithm, two ideas have to be added. We want to start out with some ellipsoid containing K (say, with the ball with radius R about the origin), and replace it as in the above proof with ellipsoids of smaller and smaller volume until a weak Löwner-John ellipsoid is obtained. In order to guarantee a good running time, we have to reduce the volume by a substantial factor like $1 - (1/n^2)$; this will be achieved if the radius of E' is chosen, say, $1/2n$ instead of $1/n$. The other problem is to find the vector $x \in E' \setminus K$. This is easy if K is the solution set of a system of linear inequalities, but hard in general. What we can do is to test the $2n$ intersection points of the axes with the surface of E' for membership in K; if one of them is not in K, then the separation algorithm yields the hyperplane H. If each of them is in K then K contains the smaller ball obtained from E' by shrinking by a factor of $\sqrt{n}$. Thus E is a weak Löwner-John ellipsoid.

Recently some interesting new results concerning the computation of the "true" Löwner-John ellipsoid have been obtained. If K is a convex polytope presented by an explicit list of its vertices, then for any $\varepsilon > 0$ an approximation of the Löwner-John ellipsoid with error at most ε can be found in polynomial time. The "polar" problem of finding an inscribed ellipsoid with (approximately) maximum volume can also be solved in polynomial time, provided the polytope is presented as the solution set of an explicit system of linear inequalities. Such an algorithm was given by Nesterov and Nemirovskii (1989), and improved by Khachiyan and Todd (1990). These results are very presentation-dependent; Khachiyan and Todd conjecture that to compute an approximate Löwner-John ellipsoid for a polytope presented by linear inequalities is NP-hard.

The diameter and width of the weak Löwner-John ellipsoid can be used to estimate the diameter and width of K with relative error at most $n^{3/2}$. This estimate is not as bad as it looks at a first glance; it can be shown (Bárány and Füredi 1986) that if K is given by any of the oracles above, then any estimate on the width of a body K computable in polynomial time is off by a factor which grows as a power of n.

3.2 The Ellipsoid Method

The algorithm sketched above is the heart of various versions of the algorithm called the Ellipsoid Method, developed by Shor (1970) and Yudin and Nemirovkii (1976). The method became widely known when Khachiyan (1979) applied it to obtain the first polynomial time linear programming algorithm. To show the idea, assume that we want to find the optimum of a linear objective function $c \cdot x$ over a convex body $K \subseteq \mathbb{R}^n$ for which we have a separation oracle. Design a black box which does the following: we plug in a vector $x_0 \in \mathbb{R}^n$. If $x_0 \notin K$, it returns a hyperplane separating x from K. If $x_0 \in K$, it returns the hyperplane $c \cdot x = c \cdot x_0$. This is essentially a separation oracle for the set of vectors in K optimizing the linear objective function $c \cdot x$. Even though this convex set is not full-dimensional, an appropriate modification of the argument above gives that we can use this black box as a separation oracle in the ellipsoid method and obtain an "approximately optimal" point in K.

Once we can solve the optimization problem over K, the validity and violation problems are easily settled.

It is more difficult to handle these problems when K is given by a membership test rather than by a separation oracle. It takes a nice but difficult argument by Yudin and Nemirovskii (1976) to settle this case.

3.3 Karmarkar's Method

The ellipsoid method yields a polynomial time algorithm to solve linear programs, but it is too slow in practice to be useful in actual computations. The classical method to solve linear programs, namely Dantzig's simplex method, takes exponential time in the worst case but works very efficiently for the majority of real-life problems. A linear programming algorithm which is both theoretically efficient (polynomial-time) and competitive with the simplex method in practice was given by Karmarkar (1984). Since this method also has a very geometric background, it is worth sketching here.

Assume that we want to maximize a linear objective function $c \cdot x$ over a convex body K. Also assume that we have already found an interior point x_0 and we want to find a sequence of interior points $x_1, x_2, \ldots$ converging to the optimum point. When stepping from x_i to x_{i+1}, it would be natural to move in the direction of c, or, in other words, orthogonal to the hyperplane $c \cdot x = \text{const}$. This leads us, however, to points close to the boundary and (possibly) far from the optimum, and then we have to move in some other direction.

However, it is not really natural to move orthogonal to c; this is not invariant under affine transformations while the hole task is. The idea of many optimization algorithms is to change the metric of the space so that the notion of "orthogonality" should be tailored to the particular body (such are the variable metric methods, but the ellipsoid method can also be viewed this way).

There is a very natural metric inside every convex body, introduced by Hilbert. The distance of any two points $u, v \in \text{int}(K)$ is defined as the logarithm of the cross ratio of u, v, and the two points of intersection of the line through u and v with the boundary of K. (This construction may also be familiar from the Caley-Klein model of hyperbolic geometries.) It is not difficult to see that if we move in the direction of steepest descent with respect to this metric to, say, half way to the boundary, and then repeat this, then we get very close to the optimum very fast. For example, if K is centrally symmetric with respect to x_0 then we approach the optimum on a straight line and the distance from the optimum is halved at each step.

Unfortunately, there is no easy way to compute the Hilbert metric in a convex body, but we can approximate it in the neighborhood of an interior point v by finding a projective transformation which maps K onto a convex body K' so that the unit ball includes K, K includes the ball with a reasonably large radius ϱ about the origin, and v is mapped onto the origin. Then the euclidean metric will approximate the projective metric. Except for the condition on v, this sounds like the ellipsoid method, but we have more freedom because we are allowed to use projective transformations. (The optimization problem itself is not projective invariant, but one can transform it into projective invariant problems easily; for example, the Violation Problem above is projective invariant.)

The crucial part of Karmarkar's method is to construct such a projective transformation very fast in the case of polytopes presented by an explicit system of m linear inequalities. The inner radius ϱ he achieves is $1/m$; this shows that (unlike the ellipsoid method) this method is sensitive to the number of constraints. This is of course only a very rough sketch; many ingenious details must be added to make this procedure work for really large linear programs. For details, extensions, and related algorithms see the survey of Todd (1989).

4. The Volume of a Convex Body

Now we turn to the fundamental problem of determining, or at least estimating, the volume of a convex body. This question has recently brough exciting developments. For a while, a number of negative results were obtained (Section 4.1), which showed that even to compute an estimate in polynomial time with decent relative error is hopeless. But recently Dyer, Frieze and Kannan (1989) designed a *randomized* polynomial time algorithm (i.e., an algorithm making use of a random number generator) which computes an estimate of the volume such that the probability that the relative error is larger than a prescribed $\varepsilon > 0$ is arbitrarily small. This outstanding result uses a number of tools from probability and geometry, and we sketch it in Section 4.2.

4.1 The Difficulty of Computing the Volume

One thing we can do is to find a weak Löwner-John ellipsoid of the body. Since it is easy to follow how an affine transformation modifies the volume (it multiplies by the determinant of the corresponding matrix), we may assume that the body is contained in the unit ball and contains the ball with radius $1/2n^{3/2}$ about the origin. In this case, we have the trivial bounds on the volume:

$$\frac{\pi^{n/2}}{2^n n^{3n/2} \Gamma(1+n/2)} \le \mathrm{vol}(K) \le \frac{\pi^{n/2}}{\Gamma(1+n/2)}.$$

So for an arbitrary convex body (given by, say, a separation oracle), we can compute an upper bound on its volume with relative error at most $2^n n^{3n/2}$.

The following surprising result of Bárány and Füredi (1986), improving a somewhat weaker bound given by Elekes (1986), shows that no substantially better estimate can be given on the volume, at least if the convex body is given by one of the four equivalent oracles discussed in Chapter 2.

Theorem 4.1. *Consider any polynomial time algorithm which assigns to every convex body K, given by (say) a membership oracle, an upper bound $w(K)$ on $\mathrm{vol}(K)$. Then there is a constant $c > 0$ such that in every dimension n there exists a body K for which $w(K) > n^{cn} \mathrm{vol}(K)$.*

Let us sketch the proof of this result. It depends on the following geometric lemma:

Lemma 4.2. *There exists a constant $c > 0$ such that the volume of the convex hull of any $p > 0$ points in the unit ball is less than $p \cdot n^{-cn}$.*

To prove Theorem 4.1, first apply the algorithm to the unit ball B. The algorithm runs and asks the membership of a polynomial number p of points $v_1, \ldots, v_p$ from the oracle, and computes an upper bound $w(B)$ of the volume. Now apply the algorithm to the convex hull K of $\{e_1, \ldots, e_n, v_1, \ldots, v_p\} \cap B$. It follows that the algorithm runs exactly as in the previous case: it asks the same questions from the oracle and therefore it gets the same answers. So the algorithm finds the same estimate $w(K) = w(B)$. But since the volume of K is much smaller than the volume of B, this estimate must have a large relative error.

This result does not say anything about computing the volume of convex bodies given in any specific way, say as the solution set of a system of linear inequalities. However, in both cases the exact computation of the volume is NP-hard (Dyer and Frieze 1988, Khachiyan 1988) and even #P-hard (Khachiyan 1989). (This latter fact also follows from the result of Brightwell and Winkler (1990) mentioned in Section 1.3.)

The approximate computation of the volume of explicitly described polyhedra is an open problem.

4.2 Markov Chains, Isoperimetric Inequalities, and Approximating the Volume

In this section we sketch the randomized algorithm of Dyer, Frieze and Kannan (and variants) for estimating the volume of a convex body $K \subseteq \mathbb{R}^n$. About this body, we only need to assume that it is given by a separation oracle.

At this point, many readers may ask: what's wrong with the straightforward Monte-Carlo algorithm? We have already made the assumption that K is included in the unit ball B. Let us generate many random points in B, and count how often we hit K. This gives us an estimate on the ratio of the volumes of K and B.

The problem is that the volume of K may be smaller than the volume of B by an exponential factor (in n). Hence the first exponentially many random points will miss the body K. This method can be applied to estimate the ratio of the volumes of two convex bodies (one including the other) only if this ratio is not too small.

This suggests the first trick: "connect" K and B by a sequence of convex bodies $K = K_0 \subseteq K_1 \subseteq \ldots \subseteq K_m = B$, so that $\mathrm{vol}(K_i)/\mathrm{vol}(K_{i+1}) \geq 1/2$ (and m is polynomial in n). Then these ratios can be estimated by the Monte-Carlo method, and their product gives an estimate on the ratio $\mathrm{vol}(K)/\mathrm{vol}(B)$. Such a sequence is easily constructed: we can take e.g. $K_i = B \cap (1 + \frac{1}{2n})^i K$. Since K contains the ball $(1/(2n^{3/2}))K$, it follows that $K_i = B$ for $i \geq 4n \log n$.

However, estimating $\mathrm{vol}(K_i)/\mathrm{vol}(K_{i+1})$ by the Monte-Carlo method is not so easy; the key question in this algorithm (and in all versions of it) is:

How to generate a random point (with uniform distribution) in a convex body?

The solution by Dyer, Frieze and Kannan is the following. Consider the lattice of vectors whose coordinates are integer multiples of a sufficiently small number δ. Call two lattice points *adjacent* if their distance is exactly δ. Starting from the origin, take a random walk on the lattice points in K. If we are at a lattice point v, select an adjacent lattice point w at random. If $w \in K$, then move to w; else, stay at v. (For technical reasons, we flip a coin before the move and if it falls on head, we stay where we are anyway.) After an appropriate number of steps, we stop; our current position can be considered as a random point in K.

It is easy to see, using the theory of Markov chains, that if w_t denotes the (random) lattice point obtained after t steps, then the distribution of w_t tends to the uniform distribution over the set V of lattice points accessible from the origin along legal walks. (Note that V is essentially $L \cap K$, except possibly for some lattice points near the boundary of K.)

The problem is to find a good bound on the rate of this convergence (on the *mixing rate* of the Markov chain $(w_0, w_1, \ldots)$). By general results on the mixing rate of Markov chains (Sinclair and Jerrum 1989), we know that this depends on the *conductance* of the Markov chain. This (in our case) can be defined as the largest $\Phi > 0$ such that for every $S \subseteq V$, the number of pairs (u, v) of adjacent lattice points with $u \in S$ and $v \in V \setminus S$ is at least $\Phi \cdot (2n) \cdot \min\{|S|, |V \setminus S|\}$. The least t for which the distribution w_t is essentially uniformly distributed is about $1/\Phi^2$.

So the question is to find a lower bound on the conductance of this Markov chain. Let $S \subseteq V$ and let K_1 be the set of points in K nearer to S than to $V \setminus S$. Then we expect that

(a) the volume of K_1 is about $\delta^n |S|$,

(b) the volume of $K \setminus K_1$ is about $\delta^n |V \setminus S|$,

(c) the surface area of the K_1 inside K is about δ^{n-1} times the number of adjacent pairs of lattice points (u, v) with $u \in S$ and $v \in V \setminus S$.

If we accept these approximations (which is a technically quite difficult part of the argument), then the problem reduces to the following isoperimetric inequality:

Theorem 4.3. *Let K be a convex body in $\mathbb{R}^n$ with diameter d. Let F be a surface with $(n-1)$-dimensional measure f, cutting K into two parts with volumes v_1 and v_2. Then*

$$f \geq \frac{1}{d} \min\{v_1, v_2\}.$$

A weaker inequality (but still sufficient to prove the polynomiality of the volume algorithm) was proved by Dyer, Frieze and Kannan. This form is due to Karzanov and Khachiyan (1990) (who use methods from differential geometry) and to Lovász and Simonovits (1990) (who use the so-called Ham-Sandwich Theorem; to justify the remark that algebraic topology also plays some role, let us mention that the Ham-Sandwich Theorem is derived from the Brouwer Fixed Point Theorem).

Unfortunately, even with these improvements the algorithm is practically useless; its running time grows with the 16th power of n. The main problem is that random walks are slow in getting to the distant part of K. It is easy to come up with other Markov chains of points in K which "jump around" faster (and also have uniform limit distribution). Unfortunately, to prove the corresponding analogues of the isoperimetric inequality in Theorem 4.3 seems to take new methods. It is clear the algorithm of Dyer, Frieze and Kannan, representing a theoretical breakthrough, is not the last word in the area of volume algorithms.

An algorithm to efficiently generate random points in a convex body has many further applications. Numerical integration by Monte-Carlo methods is an obvious one. We have also mentioned in Section 2.2 the problem to test whether a convex body K_1 is included in another convex body K_2. This can be done by generating many random points in K_1 and test whether these are contained in K_2. (To be convinced that at most a fraction of ε of the volume of K_1 is not contained in K_2, we have to generate about $1/\varepsilon$ points. Thus this algorithm is not "truely" polynomial if ε is part of the input.)

References

Bárány, I., Füredi, Z. (1986): Computing the volume is difficult. Proc. of the 18th Annual ACM Symp. on Theory of Computing. Assoc. Comput. Machin., New York, pp. 442–447

Berge, C. (1961): Färbung von Graphen, deren sämtliche bzw. deren ungerade Kreise starr sind. Wiss. Zeitung, Martin Luther Univ. Halle-Wittenberg, 1961, p. 114

Boulala, M., Uhry, P. (1979): Polytope des indépendants dans un graph série-parallèle. Discrete Math. **27**, 225–243

Brightwell, G., Winkler, P. (1990): Counting linear extensions is #P-complete. Bellcore preprint

Chvátal, V. (1975): On certain polytopes associated with graphs. J. Combin. Theory B **18**, 138–154

Dyer, M., Frieze, A. (1988): On the complexity of computing the volume of a polytope. SIAM J. Comp. **17**, 967–974

Dyer, M., Frieze, A., Kannan, R. (1989): A random polynomial time algorithm for approximating the volume of convex bodies. Proc. of the 21st Annual ACM Symposium on Theory of Computing. Assoc. Comput. Machin., New York, pp. 375–381

Edelsbrunner, H. (1987): Algorithms in combinatorial geometry. Springer, Berlin Heidelberg New York

Elekes, G. (1986): A geometric inequality and the complexity of computing volume. Discrete and Computational Geometry **1**, 289–292

Garey, M.R., Johnson, D.S. (1979): Computers and intractability: A guide to the theory of NP-completeness. Freeman, San Francisco

Grötschel, M., Lovász, L., Schrijver, A. (1986): Relaxations of vertex packing. J. Comb. Theory B **40**, 330–343

Grötschel, M., Lovász, L., Schrijver, A. (1988): Geometric algorithms and combinatorial optimization. Springer, Berlin Heidelberg New York

Khachiyan, L.G. (1988): On the complexity of computing the volume of a polytope. Izv. Akad. Nauk SSSR, Engineering Cypernetics **3**, 216–217

Khachiyan, L.G. (1989): The problem of computing the volume of polytopes is NP-hard. Uspekhi Mat. Nauk **44**, 199–200

Khachiyan, L.G., Karzanov, A. (1990): On the conductance of order Markov chains. Rutgers Univ. Tech. Report DCS TR 268

Khachiyan, L.G., Todd, M.J. (1990): On the complexity of estimating the maximal inscribed ellipsoid for a polytope. Cornell Univ. SORIE Tech. Report 893

Lenstra, H.W., Jr. (1983): Integer programming with a fixed number of variables. Oper. Res. **8**, 538–548

Lenstra, A.K., Lenstra, H.W., Jr., Lovász, L. (1982): Factoring polynomials with integral coefficients. Math. Ann. **261**, 515–534

Lovász, L. (1989): Geometry of numbers and integer programming. In: Mathematical programming, recent developments and applications (eds. M. Iri, K. Tanabe). Kluwer Academic Publishers, Dordrecht Boston London, pp. 177–201

Lovász, L., Simonovits, M. (1990): Mixing rate of Markov chains, an isoperimetric inequality, and computing the volume. Proc. 31st Annual Symp. on Found. of Computer Science. IEEE Computer Soc., New York (to appear)

Preparata, F.P., Shamos, M.I. (1985): Computational geometry, an introduction. Springer, New York Berlin Heidelberg

Todd, M.J. (1989): Recent developments and new directions in linear programming. In: Mathematical programming, recent developments and applications (eds. M. Iri, K. Tanabe). Kluwer Academic Publishers, Dordrecht Boston London, pp. 109–157.

Yudin, B.D., Nemirovskii, A.S. (1976): Informational complexity and efficient methods for the solution of convex extremal problems. Ekonomika i Mat. Metody **12**, 357–369 (Russian) [English transl.: Matekon **13**, 25–45]

Intersection Cohomology Methods in Representation Theory[*]

George Lusztig

Department of Mathematics, Room 2-276
Massachusetts Institute of Technology, Cambridge, MA 02139, USA

1. Introduction

In recent years, the theory of group representations has greatly benefited from a new approach provided by the topology of singular spaces, namely intersection cohomology (IC) theory.

Let G be a connected reductive algebraic group over an algebraically closed field (which will be assumed to be $\mathbf{C}$ unless otherwise specified); although G itself is clearly a non-singular variety, the study of the representations of G (or its Lie algebra, or its forms over various fields) leads to singular varieties connected with G.

The main theme of this article is that the invariants of IC theory, applied to the singular varieties arising from G appear again and again in a broad range of problems of representation theory: construction of representations, computation of their character, construction of nice bases for representations.

Chronologically, both the earliest [22, 23] and the most recent [45] such problem was that of constructing canonical bases in certain representations of Iwahori-Hecke algebras (resp. in irreducible, finite dimensional representations of G). The relevant singularities were Schubert singularities (resp. quiver singularities); their local IC provided the necessary ingredients.

Certain misterious and apparently intractable quantities in representation theory, like the multiplicities in the Jordan-Hölder series of Verma modules, were recognized to be expressible in terms of certain local IC spaces and this has led to understanding and computing them.

IC methods are needed for the classification of complex irreducible representations of a reductive group over a finite field; furthermore, the character values of such representations appear to be intimately related to certain IC complexes, the character sheaves.

IC methods have been used to give a new construction of the Springer representations of Weyl groups; an elaboration of this idea has been used to construct representations of affine Iwahori-Hecke algebras and hence of p-adic groups.

[*] Supported in part by National Science Foundation grant DMS 8702842.

Proceedings of the International Congress
of Mathematicians, Kyoto, Japan, 1990
© The Mathematical Society of Japan, 1991

Unfortunately, several important topics had to be omitted from this account. Among them are the use of $\mathcal{D}$-modules, the role of the Fourier-Deligne transform and the geometric theory of automorphic forms (see Laumon [26]).

2. *IC* Theory

This theory has been originally introduced by Goresky and MacPherson [15]. (See also MacPherson's survey [48].) Subsequently, it has been developed into a very powerful tool by Beilinson, Bernstein, Deligne and Gabber [4], who blended *IC* theory into Deligne's theory of weights.

Consider an irreducible complex algebraic variety Y. An *IC datum* for Y consists of an open dense smooth subvariety Y_0 of Y together with a local system $\mathcal{L}$ on Y_0. ($\mathcal{L}$ corresponds to a finite dimensional complex representation of the fundamental group of Y_0.)

In general, it is not possible to extend $\mathcal{L}$ to a local system on the whole of Y; even if such extension existed, local Poincaré duality with coefficients in it would not necessarily hold, when Y has singularities. Now *IC* theory provides a canonical extension $\mathcal{L}^\sharp$ of $\mathcal{L}$ from Y_0 to Y, not necessarily as a local system, but as a complex of sheaves (well defined up to quasi-isomorphism) which has constructible cohomology sheaves $\mathcal{H}^i\mathcal{L}^\sharp$, does satisfy local Poincaré duality and is the most economical possible with these properties.

We say that $\mathcal{L}^\sharp$ is the *IC extension* of $\mathcal{L}$ to Y or that $\mathcal{L}^\sharp$ is an *IC complex with support* Y. The stalks $\mathcal{H}^i_x\mathcal{L}^\sharp$ at points $x \in Y$ are the *local IC spaces* of Y with coefficients in $\mathcal{L}$. The *global IC* of Y with coefficients in $\mathcal{L}$ is by definition the cohomology of Y with coefficients in $\mathcal{L}^\sharp$.

These concepts are also well defined in the case of algebraic varieties over an algebraic closure $\bar{F}_q$ of a finite field F_q; one uses l-adic local systems and l-adic (étale) cohomology.

We shall give three examples, which will be amplified in the following sections.

In the first example (the prototype of a Schubert variety singularity, see §3), we take Y to be the variety of all flags of subspaces $V_1 \subset V_2 \subset V_3$ in $\mathbf{C}^4$ (dim $V_i = i$) such that V_2 has non-zero intersection with a fixed two dimensional subspace $E_2 \subset \mathbf{C}^4$. We take Y_0 to be the open subset of Y defined by the condition $V_2 \neq E_2$ and we take $\mathcal{L} = \mathbf{C}$. The stalks $\mathcal{H}^i_x\mathcal{L}^\sharp$ of the *IC* extension of $\mathcal{L}$ to Y can be described as follows. If $x \in Y_0$ we have dim $\mathcal{H}^0_x\mathcal{L}^\sharp = 1$ and $\mathcal{H}^i_x\mathcal{L}^\sharp = 0$ for $i \neq 0$. If $x \in Y - Y_0$ we have dim $\mathcal{H}^0_x\mathcal{L}^\sharp = 1$, dim $\mathcal{H}^2_x\mathcal{L}^\sharp = 1$ and $\mathcal{H}^i_x\mathcal{L}^\sharp = 0$ for $i \neq 0, 2$.

In the second example (the prototype of a character sheaf, see §14) we take

$$Y = \left\{ \begin{pmatrix} a & b \\ c & d \end{pmatrix} \mid a, b, c, d \in \mathbf{C}, ad - bc = 1 \right\}$$

and we take Y_0 to be the open set of Y defined by the condition $a + d \neq \pm 2$ (matrices with distinct eigenvalues). Let $\pi : \tilde{Y}_0 \to Y_0$ be the two-fold covering defined by the equation $z^2 = (a + d)^2 - 4$. We define $\mathcal{L}$ to be the local system on Y_0 whose stalk $\mathcal{L}_y$ at $y \in Y_0$ is the one dimensional vector space of functions

$\pi^{-1}(y) \to \mathbf{C}$ with sum of values equal to 0. The stalks $\mathscr{H}_x^i \mathscr{L}^\sharp$ of the IC extension of $\mathscr{L}$ to Y can be described as follows. If $x \in Y_0$ we have $\mathscr{H}_x^0 \mathscr{L}^\sharp = \mathscr{L}_x$ and $\mathscr{H}_x^i \mathscr{L}^\sharp = 0$ for $i \neq 0$. If x is $\pm$ the identity matrix, we have $\dim \mathscr{H}_x^2 \mathscr{L}^\sharp = 1$ and $\mathscr{H}_x^i \mathscr{L}^\sharp = 0$ for $i \neq 2$. If x has eigenvalues ± 1 but it is not $\pm$ the identity matrix, then $\mathscr{H}_x^i \mathscr{L}^\sharp = 0$ for all i.

In the third example (the prototype of a cuspidal character sheaf, see §§13, 14) we take

$$Y = \left\{ \begin{pmatrix} a & b \\ c & d \end{pmatrix} \mid a, b, c, d \in \mathbf{C}, ad - bc = 1, a + d = 2 \right\}$$

and we take Y_0 to be the open set defined by the condition $(b, c) \neq (0, 0)$. We define a two-fold covering $\pi : \tilde{Y}_0 \to Y_0$ by

$$\tilde{Y}_0 = \{(z_1, z_2) \in \mathbf{C}^2 \mid (z_1, z_2) \neq (0, 0)\}, \quad \pi(z_1, z_2) = \begin{pmatrix} z_1 z_2 + 1 & -z_1^2 \\ z_2^2 & -z_1 z_2 + 1 \end{pmatrix}$$

and we associate to it a local system $\mathscr{L}$ on Y_0 as in the second example. The stalks $\mathscr{H}_x^i \mathscr{L}^\sharp$ of the IC extension of $\mathscr{L}$ to Y can be described as follows. If $x \in Y_0$ we have $\mathscr{H}_x^0 \mathscr{L}^\sharp = \mathscr{L}_x$ and $\mathscr{H}_x^i \mathscr{L}^\sharp = 0$ for $i \neq 0$. If x is the identity matrix, we have $\mathscr{H}_x^i \mathscr{L}^\sharp = 0$ for all i.

3. Left Cells in Weyl Groups and Schubert Varieties

In this section we describe the first instance when IC methods made their appearance in representation theory.

The theory of primitive ideals in enveloping algebras leads naturally to a remarkable partition of the Weyl group W of G into subsets called *left cells* (see Joseph [18]); the left cells can be used to parametrize primitive ideals. (This generalizes a partition of the symmetric group S_n which was found earlier by combinatorists.) It is known that each left cell carries a natural representation of W with a canonical basis (in 1-1 correspondence with the elements in the left cell). These representations provide a decomposition of the regular representation of W into representations which are close to being irreducible.

Let H be the Iwahori-Hecke algebra over $\mathbf{Z}[v, v^{-1}]$ corresponding to W; here, v is an indeterminate. H has a standard basis $\tilde{T}_w$ (as a $\mathbf{Z}[v, v^{-1}]$-module) indexed by elements $w \in W$. Its multiplication rule is a deformation of the multiplication rule in the group algebra of W:

$$\tilde{T}_w \tilde{T}_{w'} = \tilde{T}_{ww'} \quad \text{if } l(ww') = l(w) + l(w'),$$

$$(\tilde{T}_s + v^{-1})(\tilde{T}_s - v) = 0 \quad \text{if } l(s) = 1;$$

here, $l : W \to \mathbf{N}$ is the length function.

In [22], Kazhdan and the author came across a new basis C_w ($w \in W$) of H which had some very remarkable properties with respect to multiplication. It can be defined as follows.

Let $^-: H \to H$ be the ring homomorphism which takes $\tilde{T}_w$ to $\tilde{T}_{w^{-1}}^{-1}$ for all w and v to v^{-1}. Let $\mathscr{L}$ be the $\mathbf{Z}[v^{-1}]$-submodule of H generated by the standard basis and let $\pi : \mathscr{L} \to \mathscr{L}/v^{-1}\mathscr{L}$ be the canonical projection. It turns out that for each $w \in W$, there is a unique element $C_w \in \mathscr{L}$ such that $\bar{C}_w = C_w$ and $\pi(C_w) = \pi(\tilde{T}_w)$.

This definition is equivalent to (but simpler than) the one given in [22]; unlike the definition in [22], it does not use the partial order on W.

It turns out that

$$C_w = \sum_{y \in W} (-1)^{l(w)-l(y)} v^{l(w)-l(y)} P_{y,w}(v^{-2}) \tilde{T}_y$$

where the $P_{y,w}$ are polynomials with integral coefficients. Moreover this basis is, in principle, computable.

In terms of this basis, we found a new definition of left cells (which was later shown to be the same as the one of Joseph). Our definition had the advantage of being elementary, of carrying representations of H (not only of W) endowed with canonical bases, and of making sense for an arbitrary Coxeter group.

Let $\mathscr{F}$ be the collection of all subsets $K \subset W$ with the following property: the submodule of H spanned by the elements C_w ($w \in K$) is a left ideal. The definition [22] of left cells ean be given as follows. Two elements $w, w' \in W$ are in the same left cell if the following condition is satisfied: for any $K \in \mathscr{F}$ we have $w \in K$ if and only if $w' \in K$.

In [23], Kazhdan and the author found a geometric interpretation of the polynomials $P_{y,w}$ above. To explain the result, we need some notation.

Assume that we are given a Borel subgroup B of G, with unipotent radical U_B and a maximal torus T of B. The orbits of B acting by conjugation on the variety X of all Borel subgroups (the *flag manifold*) of G are naturally parametrized by the elements of W. Let $\mathscr{O}_w$ be the orbit corresponding to $w \in W$ and let $\bar{\mathscr{O}}_w$ be its closure in X (a *Schubert variety*). Let $\mathscr{L}$ be the local system $\mathbf{C}$ on $\mathscr{O}_w$ and let $\mathscr{L}^\sharp$ be its *IC* extension to $\bar{\mathscr{O}}_w$. If $y, w \in W$, we say that $y \leq w$ if $\mathscr{O}_y \subset \bar{\mathscr{O}}_w$. We then showed that, for such y, w, we have $P_{y,w}(v^2) = \sum_i n_{y,w,i} v^i$, where $n_{y,w,i} = \dim \mathscr{H}_x^i \mathscr{L}^\sharp$ (notation of §2), for any $x \in \mathscr{O}_y$. (If $y \not\leq w$, we have $P_{y,w} = 0$.) Thus, the polynomials $P_{y,w}$ contain encoded all information about the local *IC* of Schubert varieties.

Apart from their relevance to the theory of primitive ideals, the theory of left cells in W has been crucial in the study [33] of representations of groups over F_q (and also in [35]); in the opposite direction, the results of [33] gave new information on the structure of left cells, in particular they have made possible the explicit determination [37] of all representations of W which are carried by left cells. (For example, if G is a classical group, these representations of W are multiplicity free, with a number of irreducible components equal to a power of 2.)

Another definition of left cells has been found in [36] in terms of leading coefficients of the structure constants $h_{x,y,z} \in \mathbf{Z}[v, v^{-1}]$ ($x, y, z \in W$) of the algebra H with respect to the basis C_w. Namely, for each $z \in W$, there is a well defined integer $a(z) \geq 0$ such that $v^{a(z)} h_{x,y,z} \in \mathbf{Z}[v]$ for all x, y and $v^{a(z)-1} h_{x,y,z} \notin \mathbf{Z}[v]$ for some x, y. Let $\gamma_{x,y,z^{-1}}$ be the constant term of $(-v)^{a(z)} h_{x,y,z^{-1}}$. Quite remarkably, it turns out that the rule $t_x t_y = \sum_{z \in W} \gamma_{x,y,z^{-1}} t_z$ defines a structure of associative ring with 1 on the free abelian group J with basis $(t_x)_{x \in W}$. Then the left cells of W can be characterized as follows: we have $t_x t_y \neq 0$ if and only if x, y^{-1} are in the same left cell.

The ring J has several interesting features. First, there is a explicit homomorphism of algebras $H \to J \otimes \mathbf{Z}[v, v^{-1}]$ (see [36]) which becomes an isomorphism whenever v is specialized to any non-zero complex number which is either 1 or is not a root of 1. Second, the multiplication in J can be (conjecturally, see [39]) described in very simple terms, in terms of convolution of equivariant vector bundles on some finite sets with some finite group action. (The conjecture of [39] gives a precise description of the apparently complicated pattern in which left cells meet right cells, i.e. images of left cells under $w \to w^{-1}$.)

4. Highest Weight Modules

Let $\mathbf{g}, \mathbf{b}, \mathbf{n}, \mathbf{t}$ be the Lie algebras of G, B, U_B, T. In [22], Kazhdan and the author formulated a conjecture relating the polynomials $P_{y,w}$ (see §3) to multiplicities in the Verma modules of $\mathbf{g}$. More precisely, assuming that λ is a linear form on $\mathbf{t}$, we denote by M_λ the Verma module of $\mathbf{g}$ with highest weight λ and by L_λ its unique irreducible quotient. Let $\varrho : \mathbf{t} \to \mathbf{C}$ be half the sum of the positive roots. Our conjecture was that for all $w \in W$ we have

$$L_{-w(\varrho)-\varrho} = \sum_{y \leq w} (-1)^{l(w)-l(y)} P_{y,w}(1) M_{-y(\varrho)-\varrho} \qquad \text{(a)}$$

in the appropriate Grothendieck group. By an inversion formula for the $P_{y,w}$ proved in [22], one sees that (a) admits the following equivalent formulation

$$M_{y(\varrho)-\varrho} = \sum_{w \geq y} P_{y,w}(1) L_{w(\varrho)-\varrho}. \qquad \text{(b)}$$

In view of [23], this relates multiplicities in Verma modules to local IC of Schubert varieties. Our conjecture has been proved by Beilinson, Bernstein [2] and by Brylinski, Kashiwara [8] using the theory of $\mathcal{D}$-modules.

Not only the value of $P_{y,w}$ at 1 but also the coefficients of the various powers of v^2 in it have representation theoretic meaning (actually two different ones): they can be interpreted as dimensions of higher Ext groups between a Verma module and a simple module (Vogan [59]) or as multiplicities of a simple module in the successive quotients of the Jantzen filtration of a Verma module (Beilinson, Bernstein [3]).

5. Symmetric Spaces

Let $\theta : G \to G$ be an involution and let K be the identity component of the fixed point set G^θ. Let Y_0 be a K-orbit on X and let $\mathscr{L}$ be a one dimensional K-equivariant local system on Y_0.

By [2, 59], the local IC of the closure $\bar{Y}_0$ with coefficients in $\mathscr{L}$ enters in an essential way in the representation theory of the real reductive group attached to G, θ.

An algorithm for computing this local IC (generalizing that in [22]) has been proposed by Vogan and has been established in [47].

6. Groups over F_q

In this section we take G, X as before, but we assume that the ground field is $\bar{F}_q$ rather than $\mathbf{C}$ and that we are given an F_q-rational structure on G with corresponding Frobenius map $F : G \to G$. This induces a Frobenius map $F : X \to X$. Following [12], we consider for any $w \in W$ the set X_w of all Borel subgroups B' of G such that $B', F(B')$ are in relative position w. The finite group $G(F_q)$ acts on X_w by conjugation; this induces an action on the l-adic cohomology groups of X_w with compact support. Although one has from [12] good information about the alternating sum of these cohomology groups (as a virtual representation) one knows very little about the representations on the individual cohomology groups, due to the fact that X_w is non-proper. The situation is much better if we replace X_w by its closure $\bar{X}_w$ and consider the natural representations of $G(F_q)$ in the global IC groups of $\bar{X}_w$ with constant coefficients. Although $\bar{X}_w$ is singular in general, it is locally isomorphic to a Schubert variety hence its local IC is as in §3. Hence we still have good information on the alternating sum of these IC spaces (as a virtual representation); but this time we can also get good information on the individual IC spaces using especially the Weil conjectures in IC (see [4]). We can in this way compute explicitly [33] the way in which individual IC spaces decompose in irreducible representations and at the same time classify the irreducible representations occuring there (these are the *unipotent representations* of $G(F_q)$).

This method extends to non-unipotent representations (see [33]).

7. Affine Schubert Varieties

In this section we take G to be simply connected, almost simple.

Let $\check{Q}$ be the the lattice of coroots of G and let $\tilde{W}$ be the the semidirect product of W with $\check{Q}$; thus, $\tilde{W}$ is the *affine Weyl group*.

Let $G((\varepsilon))$ (resp. $G[[\varepsilon]]$) be the group of points of G over $\mathbf{C}((\varepsilon))$ (resp. $\mathbf{C}[[\varepsilon]]$) where ε is an indeterminate. The inverse image of B under the canonical homomorphism $G[[\varepsilon]] \to G$ induced by $\varepsilon \to 0$ is denoted by I. Let $\tilde{X}$ be the set of all subgroups of $G((\varepsilon))$ which are conjugate to I. It is called the *affine flag manifold*; a closely related space has been first studied in Bott's work [7]. Now $G((\varepsilon))$ acts transitively on $\tilde{X}$ and the stabilizer of I is known to be I itself. The restriction of this action to I has countably many orbits on $\tilde{X}$ which may be naturally put in 1-1 correspondence (as in Iwahori and Matsumoto [17]) with the elements of $\tilde{W}$. Let $\mathcal{O}_w$ be the I-orbit corresponding to $w \in \tilde{W}$. Now $\mathcal{O}_w$ may be naturally regarded as an affine space over $\mathbf{C}$ of dimension $l(w)$ where $l : \tilde{W} \to \mathbf{N}$ is the length function (see [17]).

One can regard naturally $\tilde{X}$ as an increasing union of projective algebraic varieties over $\mathbf{C}$, each of the varieties in this union being I-stable. (See [23, 31, 58].) Then the closure $\bar{\mathcal{O}}_w$ of $\mathcal{O}_w$ in $\tilde{X}$ is a well defined projective variety over $\mathbf{C}$, called an *affine Schubert variety*. We can then define the relation $y \le w$ on $\tilde{W}$, as in §3. Now the polynomials $P_{y,w}$ in §3 are well defined for $y, w \in \tilde{W}$ (see [22]). They again record the dimensions of the local IC of $\bar{\mathcal{O}}_w$ at a point in $\mathcal{O}_y \subset \bar{\mathcal{O}}_w$, for $y \le w$, as in §3 (see [23]).

(The elementary definition of the polynomials $P_{y,w}$ given in [22] makes sense for arbitrary Coxeter groups; it is expected [23] that they always have ≥ 0 coefficients, although it is not clear what the geometrical interpretation of these coefficients should be in the non-crystallographic case. This positivity property has been verified by Alvis, for the finite Coxeter group of type H_4, using a computer.)

8. Highest Weight Modules for Affine Lie Algebras

In this section we assume that G is almost simple, simply connected and of simply laced type. We shall recall some basic facts on affine Lie algebras. Consider the Lie algebras obtained by extending scalars: $\mathbf{g}[\varepsilon, \varepsilon^{-1}] = \mathbf{g} \otimes \mathbf{C}[\varepsilon, \varepsilon^{-1}]$, $\mathbf{g}[\varepsilon] = \mathbf{g} \otimes \mathbf{C}[\varepsilon]$. We regard $\mathbf{g}$ as a subalgebra of $\mathbf{g}[\varepsilon]$ by $x \to x \otimes 1$. It is known that there is a non-trivial central extension $\pi : \tilde{\mathbf{g}} \to \mathbf{g}[\varepsilon, \varepsilon^{-1}]$ of Lie algebras over $\mathbf{C}$ with one dimensional kernel D; moreover, it is unique up to isomorphism. The Lie algebra $\tilde{\mathbf{g}}$ is said to be an *affine Lie algebra*. Let $\tilde{U}$ be its enveloping algebra.

Let $\tau \in \mathbf{t}$ be a coroot and let $\tau', \tau'' \in \tilde{\mathbf{g}}$ be such that $\pi(\tau') = \varepsilon\tau$, $\pi(\tau'') = \varepsilon^{-1}\tau$. Then $[\tau', \tau'']$ is a non-zero element of D, independent of the choice of τ, τ', τ''; we denote it c.

Let $k \in \mathbf{C}$. A $\tilde{U}$-module is said to have *central charge* k if c acts on it as k times identity.

Let $\tilde{\mathbf{p}}$ be the inverse image of $\mathbf{g}[\varepsilon]$ under π. It is known that $\pi : \tilde{\mathbf{p}} \to \mathbf{g}[\varepsilon]$ has a unique cross-section $\mathbf{g}[\varepsilon] \to \tilde{\mathbf{p}}$ which is a homomorphism of Lie algebras (over $\mathbf{C}$); using this we may identify $\tilde{\mathbf{p}} = D \oplus \mathbf{g}[\varepsilon]$ as Lie algebras.

Let $\mathbf{i}$ (resp. $\mathbf{i}_0$) be the inverse image of $\mathbf{b}$ (resp. of $\mathbf{n}$) under the canonical homomorphism $\mathbf{g}[\varepsilon] \to \mathbf{g}$ induced by $\varepsilon \to 0$. We have a direct sum decomposition $\mathbf{i} = \mathbf{t} \oplus \mathbf{i}_0$ as vector spaces. Let $\tilde{\mathbf{t}} = D \oplus \mathbf{t}$, $\tilde{\mathbf{i}} = D \oplus \mathbf{i} = \tilde{\mathbf{t}} \oplus \mathbf{i}_0$ and let $\tilde{\lambda} : \tilde{\mathbf{t}} \to \mathbf{C}$ be a linear form. Consider the $\mathbf{C}$-linear map $\tilde{\mathbf{i}} \to \mathbf{C}$ given by $\tilde{\lambda}$ on $\tilde{\mathbf{t}}$ and by 0 on $\mathbf{i}_0$. This is a Lie algebra homomorphism hence it extends to a homomorphism of the enveloping algebra $\tilde{U}_1$ of $\tilde{\mathbf{i}}$ to $\mathbf{C}$. We induce this to a $\tilde{U}$-module; the resulting $\tilde{U}$-module $M_{\tilde{\lambda}} = \tilde{U} \otimes_{\tilde{U}_1} \mathbf{C}$ is called a Verma module; it has central charge $\tilde{\lambda}(c)$.

Let h be the Coxeter number of G. The existence of Segal-Sugawara operators implies that $M_{\tilde{\lambda}}$ has a unique simple quotient module (denoted $L_{\tilde{\lambda}}$) provided that $\tilde{\lambda}(c)$ is not equal to $-h$.

Let $\tilde{\varrho} : \tilde{\mathbf{t}} \to \mathbf{C}$ be the linear form which is equal to ϱ on $\mathbf{t}$ (see §4) and takes the value h on c. Let Q be the lattice of roots of $\mathbf{g}$ (regarded as a group of linear forms $\mathbf{t} \to \mathbf{C}$). Let $\tilde{W}$ be the semidirect product of W and Q (the affine Weyl group); this may be identified with the affine Weyl group in §7 since we are in the simply laced case. We have an action of $\tilde{W}$ on the space of linear forms on $\tilde{\mathbf{t}}$ as follows: an element $\xi \in Q$ acts by $\tilde{\lambda} \to \tilde{\lambda} + \tilde{\lambda}(c)\xi$; the action of W is the contragredient of the action on $\tilde{\mathbf{t}}$ which is the usual one on $\mathbf{t}$ and the identity on D.

The statements (a) and (b) below (which are entirely analogous to §4(a),(b)) have been recently proved by Casian [11] and by Casian [10], Kashiwara[19]:

$$L_{-w(\tilde{\varrho})-\tilde{\varrho}} = \sum_{y \leq w} (-1)^{l(w)-l(y)} P_{y,w}(1) M_{-y(\tilde{\varrho})-\tilde{\varrho}} \qquad (a)$$

$$M_{y(\tilde{\varrho})-\tilde{\varrho}} = \sum_{w \geq y} P_{y,w}(1) L_{w(\tilde{\varrho})-\tilde{\varrho}}. \qquad (b)$$

Here $w, y \in \tilde{W}$; the sums are in the appropriate Grothendieck groups (but the sum in (b) is infinite) and $P_{y,w}$ are the polynomials in §7 describing the local IC of affine Schubert varieties. Note that $L_{-w(\tilde{\varrho})-\tilde{\varrho}}$ and $L_{w(\tilde{\varrho})-\tilde{\varrho}}$ are well defined since $-w(\tilde{\varrho}) - \tilde{\varrho}$ and $w(\tilde{\varrho}) - \tilde{\varrho}$ take at c the values $-2h$ and 0, which are $\neq -h$. While the identities §4(a),(b) are equivalent to each other, the identities (a),(b) above are quite different: (a) is typical for integral highest weights with central charge $< -h$, while (b) is typical for integral highest weights with central charge $> -h$. Note that in [19] the geometry used is not the one in §7, but that of "affine Schubert varieties of finite codimension" (see [23, 20]). Note also that, in [10, 19], a formula like (b) is proved for any symmetrizable Kac-Moody Lie algebra.

9. Modular Representations

Assume that G is as in §8. Let $\tilde{W}^0$ be the set of all elements in the affine Weyl group $\tilde{W}$ which have maximal length in their left W-coset.

Let p be a (not too small) prime number. Let $\mathscr{C}$ be the category of finite dimensional rational representations (over $\bar{F}_p$) of the algebraic group over $\bar{F}_p$ of the same type as G.

The simple objects of $\mathscr{C}$ are indexed by the set of dominant integral weights $\mathbf{t} \to \mathbf{C}$ but their structure is unknown. They appear as the unique simple quotients of certain "Weyl modules" whose weight structure is well understood (it is given by Weyl's character formula). By Steinberg's tensor product theorem, the structure of simple objects of $\mathscr{C}$ would be known in general if it were known for simple modules with *restricted* highest weight (i.e. with coordinates in $[0, p-1]$). In [29] a conjecture was formulated expressing the character of certain simple modules in $\mathscr{C}$ (including those with restricted highest weight) as linear combinations of Weyl modules with explicit coefficients involving the values at 1 of the polynomials $P_{y,w}(y, w \in \tilde{W}^0)$ as in §7. This conjecture has been tested in a variety of ways and seems to be extremely plausible. However, the connection between local IC of affine Schubert varieties and modular representations is quite misterious. It seems likely that to establish this connection one should use two intermediate steps, which we now explain.

Besides $\mathscr{C}$ we consider two other categories $\mathscr{C}', \mathscr{C}''$ of representations.

$\mathscr{C}'$ has as objects the finite dimensional representations of the quantized enveloping algebra corresponding to G with parameter $\exp(2\pi i/p)$ (in the sense of [43]) in which the p-th powers of the canonical generators K_i (see [43]) act as identity.

$\mathscr{C}''$ has as objects the $\tilde{U}$-modules with central charge $-p-h$ (see §8), of finite length, with all composition factors of form $L_{\tilde{\lambda}}$ (as in §8) where the restriction of $\tilde{\lambda}$ to $\mathbf{t}$ is a dominant integral weight.

In both these categories, the simple objects are naturally indexed by the same set: the set of dominant integral weights $\mathbf{t} \to \mathbf{C}$. The simple modules appear again as the unique simple quotients of certain "Weyl modules" (also contained in our category). For $\mathscr{C}'$, the Weyl modules have the same weight structure as the corresponding simple module of G (over $\mathbf{C}$) hence are described by Weyl's character formula; for $\mathscr{C}''$ a Weyl module is defined by inducing from the enveloping algebra of $\tilde{\mathbf{p}}$ (see §8) to $\tilde{U}$ the representation of $\tilde{\mathbf{p}}$ on which c acts as $-p-h$ times identity and $\mathbf{g}[\varepsilon]$ acts through a finite dimensional irreducible representation of its quotient $\mathbf{g}$.

In [44] it is conjectured that

(a) *the categories $\mathscr{C}'$, $\mathscr{C}''$ are equivalent; the equivalence should preserve the indexing of simple modules and takes Weyl modules to Weyl modules.*

(For this, p need not be a prime.)

Although $\mathscr{C}, \mathscr{C}'$ are not equivalent, one can expect a very close connection between them; in particular, one can expect [43] that

(b) *two simple modules in $\mathscr{C}, \mathscr{C}'$ which have the same restricted highest weight should have the same weight structure.*

Now, in $\mathscr{C}''$, the simple modules can be expressed in terms of Weyl modules with coefficients involving the values at 1 of the polynomials $P_{y,w}(y, w \in \tilde{W}^0)$ as in §7. (This can be deduced by a "translation principle" from §8(a), see [25].) In view of (a), (b), the same should hold for $\mathscr{C}'$ and, in the restricted case, for $\mathscr{C}$. In this way, our conjecture on modular representations would follow from (a) and (b).

In the non-simply laced case one should use twisted affine Lie algebras.

10. Affine Schubert Varieties and Weight Multiplicities

A number of features of the theory of finite dimensional representations of G can be recovered in a purely geometric way from the IC of affine Schubert varieties. We preserve the assumptions of §7, 8. If $\lambda : \mathbf{t} \to \mathbf{C}$ is a dominant weight contained in Q, we define $n_\lambda \in \tilde{W}$ to be the unique element of maximal length in the $W - W$ double coset of $\tilde{W}$ containing λ. Consider the finite dimensional representation L_λ of G with highest weight λ and consider another dominant weight $\lambda' \in Q$. The following result is proved in [31].

In the weight decomposition of L_λ, the weight λ' appears with multiplicity equal to $P_{n_{\lambda'},n_\lambda}(1)$ if $n_{\lambda'} \leq n_\lambda$ and zero, otherwise.

Taking the polynomial itself instead of its value at 1 we get a q-analog of weight multiplicities. An interpretation of this in representation theoretic terms has been found by R. K. Brylinski [9] (modulo some restrictions which were removed by Ginsburg [16]).

11. Periodic Affine Schubert Varieties

We preserve the setup of §7. We want to discuss (in a somewhat non-rigorous way) some variants of affine Schubert varieties. We need some further notation.

Let $B((\varepsilon))$ (resp. $U_B((\varepsilon))$) be the group of $\mathbf{C}((\varepsilon))$-points of B (resp. U_B). Let $T[[\varepsilon]]$ be the group of $\mathbf{C}[[\varepsilon]]$-points of T. Let $R = T[[\varepsilon]]U_B((\varepsilon))$ be the "identity component" of $B((\varepsilon))$, so that $B((\varepsilon))/R \cong \check{Q}$.

Now I acts by left translation on $G((\varepsilon))/B((\varepsilon))$ with finitely many orbits, which may be put naturally in 1-1 correspondence with the elements of W.

We have a "principal fibration" $\pi : G((\varepsilon))/R \to G((\varepsilon))/B((\varepsilon))$ with group $\check{Q}$ and any I-orbit on $G((\varepsilon))/R$ (for the left translation action) is mapped by π bijectively onto an I-orbit on $G((\varepsilon))/B((\varepsilon))$. Thus, $\check{Q}$ acts freely on the set of I-orbits on $G((\varepsilon))/R$ with quotient space W. More precisely, we see that the set of I-orbits on $G((\varepsilon))/R$ may be put in 1-1 correspondence with $\tilde{W}$; however, this is canonical only up to composition with a translation in $\check{Q}$.

Let F be the fixed point set of T acting on $G((\varepsilon))/R$ by left translation. Each I-orbit $\mathcal{O}$ on $G((\varepsilon))/R$ contains a unique point $x_{\mathcal{O}} \in F$.

The orbits of I on $G((\varepsilon))/R$ are seen to be both of infinite dimension and of infinite codimension. The "closure" $\bar{\mathcal{O}}$ of such an orbit $\mathcal{O}$ can be defined as follows.

Let $G[\varepsilon^{-1}]$ be the group of points of G over $\mathbf{C}[\varepsilon^{-1}]$; let B^- be the Borel subgroup of G such that $B \cap B^- = T$ and let I^- be the inverse image of B^- under the canonical homomorphism $G[\varepsilon^{-1}] \to G$ induced by $\varepsilon^{-1} \to 0$. By definition, $\bar{\mathcal{O}}$ is the union of all I-orbits $\mathcal{O}'$ such that $\mathcal{O}$ has non-empty intersection with the the I^--orbit $I^- x_{\mathcal{O}'}$ of $x_{\mathcal{O}'}$. For such $\mathcal{O}'$, the intersection $I^- x_{\mathcal{O}'} \cap \bar{\mathcal{O}}$ is a (finite dimensional) irreducible variety and $I^- x_{\mathcal{O}'} \cap \mathcal{O}$ is an open dense smooth subvariety of it.

Then $\bar{\mathcal{O}}$ is a union of countably many I-orbits of larger and larger (finite) codimension; the orbits and their closure relation form a periodic pattern, with $\check{Q}$ as the group of periods. The orbit closures $\bar{\mathcal{O}}$ may be called *periodic affine Schubert varieties*.

We can define the local IC of the closure $\bar{\mathcal{O}}$ of one I-orbit along another I-orbit $\mathcal{O}' \subset \bar{\mathcal{O}}$ to be the local IC of $I^- x_{\mathcal{O}'} \cap \bar{\mathcal{O}}$ at $x_{\mathcal{O}'}$ with respect to the local system $\mathbf{C}$ on $I^- x_{\mathcal{O}'} \cap \mathcal{O}$.

I was led to consider these periodic affine Schubert varieties about ten years ago while trying to find support for the conjecture on modular representations mentioned in §9. This has led to a periodicity property of the polynomials $P_{y,w}$, for the affine Weyl group, with $y, w \in \tilde{W}^0$ as in §9, which I could prove directly [28]. The periodicity property means that these polynomials (in the stable range) may be regarded as being indexed by a pair of I-orbits on $G((\varepsilon))/R$; it turns out that they can be interpreted geometrically as the polynomials which record the local IC spaces of the periodic affine Schubert varieties.

The previous discussion suggests that the singularities of periodic affine Schubert varieties might be isomorphic to singularities of (ordinary) affine Schubert varieties. This is false; there is equality only for local IC spaces, which are not sensitive to torsion phenomena. In fact, periodic affine Schubert varieties are less singular than the corresponding ordinary ones; for example, if $G = SL_2$, the periodic affine Schubert varieties are non-singular, while the ordinary ones are rational homology manifolds with singularities.

The periodic affine Schubert varieties have been considered independently by B. Feigin and E. Frenkel in their work [14] on Wakimoto modules. It would be very interesting to find a connection between their representation theoretic considerations and the local IC spaces considered above.

12. Local IC of Closures of Unipotent Classes

Let $\mathscr{U}$ be the variety of unipotent elements of G. We denote by $\mathscr{I}$ the set of all pairs $(C, \mathscr{L})$ where C is a unipotent class in G and $\mathscr{L}$ is an irreducible G-equivariant local system on C defined up to isomorphism. (Thus, $\mathscr{L}$ corresponds to an irreducible representation of the group of components of the centralizer of an element of C; this group is naturally a quotient of the fundamental group of C.) Let $(C, \mathscr{L}) \in \mathscr{I}$. Then the cohomology sheaf $\mathscr{H}^i \mathscr{L}^\sharp$ of the IC complex on the closure $\bar{C}$ of C can be restricted to a unipotent class $C' \subset \bar{C}$ and is a G-equivariant local system there; a given irreducible G-equivariant local system $\mathscr{L}'$ on C' appears with a certain multiplicity, say $m_{C', \mathscr{L}'; C, \mathscr{L}; i}$ in this restriction. The polynomials

$$\Pi_{C', \mathscr{L}'; C, \mathscr{L}} = \sum_i m_{C', \mathscr{L}'; C, \mathscr{L}; i} v^i \tag{a}$$

describe the local IC of $\bar{C}$ with coefficients in $\mathscr{L}$. We want to describe a natural partition of $\mathscr{I}$ into *blocks*. We define a relation $\preceq$ on $\mathscr{I}$ as follows: $(C', \mathscr{L}') \preceq (C, \mathscr{L})$ means that $C' \subset \bar{C}$ and $\Pi_{C', \mathscr{L}'; C, \mathscr{L}} \neq 0$. A pair $(C, \mathscr{L}) \in \mathscr{I}$ is said to be *minimal* if $(C', \mathscr{L}') \preceq (C, \mathscr{L})$ implies $(C', \mathscr{L}') = (C, \mathscr{L})$. For a minimal pair $(C, \mathscr{L})$, we define the block of $(C, \mathscr{L})$ to be the set of all $(C'', \mathscr{L}'') \in \mathscr{I}$ such that $(C, \mathscr{L}) \preceq (C'', \mathscr{L}'')$. This block contains a unique minimal pair: $(C, \mathscr{L})$. Any element of $\mathscr{I}$ is contained in a unique block, i.e. the blocks form a partition of $\mathscr{I}$.

Let us fix a minimal pair $(C, \mathscr{L})$ and choose $u \in C$. Let $\mathscr{T}$ be a maximal torus of the centralizer of u in G, let L be the centralizer of $\mathscr{T}$ in G and let $\mathscr{W}$ be the normalizer of $\mathscr{T}$ modulo L; one can identify naturally $\mathscr{W}$ with the Weyl group of a maximal reductive connected subgroup of the centralizer of u.

(b) *The pairs $(C', \mathscr{L}')$ in the block of $(C, \mathscr{L})$ are in natural 1-1 correspondence with the irreducible representations of $\mathscr{W}$.*

Originally, a statement of this kind has been established by Springer [57] in the case where $C = \{1\}, \mathscr{L} = \mathbf{C}$; in that case, we have $\mathscr{W} = W$. Springer's construction has been reformulated in IC terms in [30] (see also [6]) and, in [34], the "generalized Springer correspondence" (b) has been established.

The construction of [30, 34] is based on the following idea. We can start with a very simple minded action of $\mathscr{W}$ on a local system on an open dense smooth subset of some irreducible variety (the action being the identity on the base). By functoriality of the IC complex, this extends automatically to an action of $\mathscr{W}$ on the corresponding local IC at any point of our variety and this provides many interesting representations of $\mathscr{W}$.

The Springer correspondence (resp. the generalized one) has been explicitly described in each case, see [1, 51, 52, 55] (resp. [34, 46, 56]).

The polynomials (a) have been described in all cases in the form of tables or algorithms. In the first case which has been considered, type A (see [30]), the computation of these polynomials was based on a connection between unipotent classes and affine Schubert varieties. The case of the block containing $(\{1\}, \mathbf{C})$ for types $\neq A$, was done in [5, 53, 54], using results of [6, 21]. The general case is considered in [35] in the framework of the theory of character sheaves.

These results (in l-adic version) remain valid in characteristic p, at least if p is a good prime for G. In this form they are of direct significance for the representation theory of reductive groups over F_q; indeed, the polynomials (a) for the block of $(\{1\}, \bar{\mathbf{Q}}_l)$ are very closely related with Green functions (character values of generic representations at unipotent elements), while in the case of the other blocks, they are closely related with character values of more degenerate representations at unipotent elements. (See [42].)

13. Cuspidal Pairs

A pair $(C, \mathscr{L})$ is said to be *cuspidal* (cf. [34, 35]) if for any proper parabolic subgroup P of G with unipotent radical U and any unipotent element $g \in P$, the cohomology with compact support of $gU \cap C$ with coefficients in the restriction of $\mathscr{L}$ is zero. This implies that $(C, \mathscr{L})$ forms a block by itself; conversely, if a block of $\mathscr{I}$ consists of a single element, then that element is cuspidal. The cuspidal pairs are quite rare. For example, if G is of type E_8, there is a single cuspidal pair; its C is the unique unipotent class whose fundamental group is the symmetric group S_5 and $\mathscr{L}$ corresponds to the sign representations of S_5. If $G = Sp_{2n}$, then G has exactly one cuspidal pair if n is a triangular number and none, otherwise.

14. Character Sheaves

It is possible to immitate very closely the representation theory of a reductive group over F_q, using certain IC complexes (called *character sheaves*) on G or subvarieties of G as substitutes for the irreducible representations. One obtains in this way a purely geometric character theory which makes sense over any algebraically closed field (in particular, over $\mathbf{C}$).

We refer to [34, 35] for the definition of character sheaves. Here we give two (extreme) examples. Let Y_0 be the set of regular semisimple elements in G. The fundamental group of Y_0 has as a canonical quotient the semidirect product of W with the lattice of coweights of G; consider an irreducible representation of the fundamental group which factors through a finite quotient of this semidirect product and let $\mathscr{L}$ be the corresponding local system on Y_0. The *IC* complex on G corresponding to $(Y_0, \mathscr{L})$ is an example of a character sheaf. If $G = GL_n$ this gives all character sheaves; for other types, there may be in addition character sheaves supported by a proper closed subset of G; in particular, if G is semisimple, the *IC* complex on $\bar{C}$ corresponding to a cuspidal pair $(C, \mathscr{L})$ (see §13) is a character sheaf (said to be a cuspidal character sheaf). The existence of these character sheaves with thin support is the main reason why representation theory for groups other than GL_n is so much more complicated than for GL_n itself.

The character sheaves have been classified in [34, 35] in two different ways: in terms of their support and also in terms of objects connected with the dual group. This last classification is extremely similar to that of the irreducible representations of a reductive group over a finite field [33]. In both cases the essential ingredients are: a semisimple element in the dual group, a "special" unipotent element in its centralizer, and a certain (small) finite group.

There are at least two, quite different, applications of character sheaves to representation theory. The first one is to characters of groups over a finite field (see §15). The second one is to the construction of representations of p-adic groups (see §§16, 17).

15. Character Sheaves and Frobenius

In this section we assume that G is as in §6.

The theory of character sheaves remains valid (in the l-adic version) for G, at least in good characteristic (and probably in bad characteristic as well). Consider a character sheaf of A of G such that A and its inverse image F^*A under the Frobenius map $F : G \to G$ have the same support and are isomorphic (as objects in the derived category of that support). Choose a specific isomorphism and, for x a fixed point of F in the support, take the alternating sum of traces of that isomorphism on the stalks of the cohomology sheaves of A at x. We obtain a class function $\chi_A : G^F \to \bar{\mathbf{Q}}_l$ on the finite group $G^F = G(F_q)$ (for $x \in G^F$, not in the support of A, we set $\chi_A(x) = 0$). It is independent, up to a non-zero factor, of the choice of isomorphism. One of the main results of [35] is that, when A runs over the character sheaves as above, the functions χ_A (suitably normalized) have values in the cyclotomic integers and they form an orthonormal basis of the space of all class functions on G^F. Thus, this space has two natural bases; one, of geometric origin, is provided by the functions χ_A; the other one is formed by the characters of the irreducible representations of G^F. The elements in the first basis are essentially computable, hence the character table of G^F would be known if the transition matrix between these two bases would be determined. For GL_n, this transition matrix is the identity, but in general it is not so. However, there is considerable evidence that this transition matrix is quite simple, almost diagonal; more precisely this matrix should consist of small diagonal blocks, and each such diagonal block should be the matrix of a *non-abelian Fourier transform* (as in [33, 39]) over a small finite group. (This has been verified in a number of cases [38].)

16. Some Algebras Attached to Blocks

We fix $r_0 \in \mathbf{C}^*$; let

$$\mathbf{V} = \{(N, \sigma) \in \mathbf{g} \times \mathbf{g} | N \text{ nilpotent, } \sigma \text{ semisimple, } [\sigma, N] = 2r_0 N\}.$$

We have a natural action of G on $\mathbf{V}$ by conjugation on both factors. Let $\tilde{\mathcal{I}}$ be the set of pairs $(\mathcal{O}, \mathcal{F})$ where $\mathcal{O}$ is a G-orbit on $\mathbf{V}$ and $\mathcal{F}$ is an irreducible G-equivariant local system on the variety $\mathcal{O}$. We consider some $(\mathcal{O}, \mathcal{F}) \in \tilde{\mathcal{I}}$. The first projection defines a map $\mathcal{O} \to \mathbf{c}$ for a well defined nilpotent G-orbit $\mathbf{c}$ on $\mathbf{g}$. We identify $\mathbf{c}$ with a unipotent class C in G using the exponential map; we obtain a map $\mathcal{O} \to C$. We can find a G-equivariant local system $\mathcal{L}$ on C whose inverse image under the last map contains $\mathcal{F}$ as a direct summand. Let $\mathcal{B}$ be the block of $\mathcal{I}$ containing $(C, \mathcal{L})$. One can show that $\mathcal{B}$ is independent of the choice of $\mathcal{L}$; it depends only on $(\mathcal{O}, \mathcal{F})$. We thus have a map $(\mathcal{O}, \mathcal{F}) \to \mathcal{B}$ from $\tilde{\mathcal{I}}$ to the set of blocks of $\mathcal{I}$; its fibres define a partition of $\tilde{\mathcal{I}}$ into subsets in 1-1 correspondence with the blocks of $\mathcal{I}$. These subsets are called the *blocks of* $\tilde{\mathcal{I}}$.

It turns out that the elements of $\tilde{\mathcal{I}}$ in a fixed block (corresponding, say, to $\mathcal{B}$) index in a natural way the isomorphism classes of simple modules over a certain algebra $\mathbf{H}$.

To define this algebra, we need some notation. Let $(C, \mathcal{L})$ be the unique minimal pair in $\mathcal{B}$; choose $u \in C$ and associate to it $\mathcal{T}, L, \mathcal{W}$ as in §11. We choose a parabolic subgroup P of G having L as Levi subgroup; let U_P be its unipotent radical. Let $P_1, \ldots, P_m$ be the parabolic subgroups which contain strictly P and are minimal with this property. Let L_i be the Levi subgroup of P_i which contains L. Let $\mathbf{l}, \mathbf{l}_i, \mathbf{n}'$ be the Lie algebras of L, L_i, U_P. Let $\mathbf{h}, \mathbf{h}_i$ be the centres of $\mathbf{l}, \mathbf{l}_i$. For any linear form $\alpha : \mathbf{h} \to \mathbf{C}$ we set $\mathbf{g}_\alpha = \{x \in \mathbf{g} | [y, x] = \alpha(y)x, \forall y \in \mathbf{h}\}$. For each i, there is a unique non-zero linear form $\alpha_i : \mathbf{h} \to \mathbf{C}$ such that $0 \neq \mathbf{g}_{\alpha_i} \subset \mathbf{l}_i \cap \mathbf{n}'$ and $\mathbf{g}_{\alpha_i/2} = 0$. Let $\mathbf{S}$ be the symmetric algebra of $\mathbf{h}^*$.

Let s_i be the unique non-trivial element of $\mathcal{W}$ which can be represented by an element of L_i. Then $s_1, \ldots, s_m$ are Coxeter generators of $\mathcal{W}$. We have a natural action $\xi \to^w \xi$ of $\mathcal{W}$ on $\mathbf{h}$ hence on $\mathbf{S}$. Let $x_0 \in \mathbf{l}$ be the nilpotent element such that $\exp(x_0) = u$. For each i, let c_i be the largest integer ≥ 2 such that $\operatorname{ad}(x_0)^{c_i-2} : \mathbf{l}_i \cap \mathbf{n}' \to \mathbf{l}_i \cap \mathbf{n}'$ is $\neq 0$. We define a $\mathbf{C}$-vector space $\mathbf{H} = \mathbf{S} \otimes \mathbf{C}[\mathcal{W}]$. There is a unique structure of associative $\mathbf{C}$-algebra with unit $1 \otimes e$ on $\mathbf{H}$ such that the following properties hold. First, the maps $\mathbf{S} \to \mathbf{H}$ ($\xi \to \xi \otimes e$) and $\mathbf{C}[\mathcal{W}] \to \mathbf{H}$ ($w \to 1 \otimes w$) are algebra homomorphisms. Second, we have $(\xi \otimes e)(1 \otimes w) = \xi \otimes w$ for any $\xi \in \mathbf{S}, w \in \mathcal{W}$. Finally, we have

(a) $(1 \otimes s_i)(\xi \otimes e) - (^{s_i}\xi \otimes e)(1 \otimes s_i) = c_i r_0 (\xi -^{s_i} \xi)\alpha_i^{-1} \otimes e, \quad (\xi \in \mathbf{S}, 1 \leq i \leq m).$

One can show that

(b) *the isomorphism classes of simple $\mathbf{H}$-modules are in natural 1-1 correspondence with the elements in the given block of $\tilde{\mathcal{I}}$.*

The connection between these kind of objects is given in [40]. In that paper an **H**-module is constructed for any element of $\tilde{\mathscr{I}}$. The idea is to apply the IC construction [34] of generalized Springer representations using fully its symmetry group (i.e. in equivariant homology); thus we do that construction with parameters in the classifying space of that symmetry group, or a smooth, finite dimensional approximation of it.

In the case where $\mathscr{B}$ is the block of $(\{1\}, \mathbf{C})$, the algebra **H** is closely related to an affine Iwahori-Hecke algebra whose simple modules were classified in [24]. In fact, the proof of (b) shares a number of features with the proof of the main result in [24]; one important difference is that in [24] the main technique is equivariant K-theory where IC methods are not available.

In the case where $G = SL_n$, $\mathscr{B}$ is the block of $(\{1\}, \mathbf{C})$ and $r_0 = 1$, the algebra **H** appears also in Drinfeld's work [13] on Yangians of type A. Can one construct finite dimensional representations of Yangians of general type using IC methods parallel to those just described ?

17. Unramified Representations of Simple p-Adic Groups

In this section we take G as in §7. Let $\mathscr{G}$ be the group of rational points of a simple split adjoint algebraic group of type dual to G over a nonarchimedean local field with residue field F_q. An irreducible admissible representation V of $\mathscr{G}$ is said to be *unramified* if there exists a parahoric subgroup $\mathscr{P}$ of $\mathscr{G}$ with "unipotent radical" $U_{\mathscr{P}}$ and "reductive part" $\bar{\mathscr{P}} = \mathscr{P}/U_{\mathscr{P}}$ (a reductive group over F_q) such that the $U_{\mathscr{P}}$-invariant part of V contains some unipotent cuspidal representation Ψ of the finite group $\bar{\mathscr{P}}$. (We get the same concept if we omit the word "cuspidal".) The following refinement of the Deligne-Langlands conjecture has been formulated in [32].

There should be a natural 1-1 correspondence between the set of isomorphism classes of unramified representations of $\mathscr{G}$ and the set of triples (N, s, ϕ) up to G-conjugacy, where N is a nilpotent element of $\mathbf{g}$, s is a semisimple element of G such that $Ad(s)N = qN$ and ϕ is an irreducible representation of the (finite) group of components of the simultaneous centralizer $Z(N, s)$ of N and s on which the centre of G acts trivially.

This is rather similar to the classification of unipotent representations for a reductive group over F_q given in [33].

(One can allow $\mathscr{G}$ to be not necessarily split, but an inner form of a split group; this gives rise to a character χ of the centre of G and one should take the same set of parameters except that the centre should be required to act on ϕ according to χ.)

We would like to suggest a way prove this. Let us fix $\mathscr{P}, \Psi$ as above. To these one can associate an Iwahori-Hecke algebra $\mathscr{H}$ corresponding to a certain (extended) Dynkin diagram Δ and some parameters $q^{m_i}, (m_i \geq 1)$, one for each vertex t_i of Δ. Let $\tilde{\Gamma}$ be the (extended) Dynkin diagram corresponding to $\mathscr{G}$ and let Γ be the its subdiagram (of finite type) corresponding to $\mathscr{P}$. By definition, the vertices of Δ are the vertices of $\tilde{\Gamma}$ which are not in Γ, assuming that there are at least two such vertices; otherwise, Δ is defined to be empty. Assume that Δ is non-empty and let t_i be one of its vertices. Let $\mathscr{P}_i \supset \mathscr{P}$ be the parahoric subgroup corresponding to the subgraph whose vertices are those of Γ together with t_i. The representation of $\bar{\mathscr{P}}_i$ induced from $\bar{\mathscr{P}}$ by Ψ decomposes into a sum of two irreducible representations; the quotient of their dimension is known to be of form q^{m_i} for some integer $m_i \geq 1$. (This integer is explicitly known by [27, p.35].) Now let t_i, t_j be two different vertices of Δ; we want to define the multiplicity of the bond joining them in Δ. If Δ has exactly two vertices, this multiplicity is defined to be infinite. If Δ has at least three vertices, we denote by $\mathscr{P}_{i,j} \supset \mathscr{P}$ the parahoric subgroup corresponding to the subgraph whose vertices are those of Γ together with t_i, t_j. The representation of $\bar{\mathscr{P}}_{i,j}$ induced from $\bar{\mathscr{P}}$ by Ψ decomposes according to the representations of an Iwahori-Hecke algebra (with possibly unequal parameters) of type $A_1 \times A_1, A_2, B_2$ or G_2; accordingly, the sought for multiplicity is by definition 0,1,2 or 3. (See again [27, p.35].) One can check that the Δ is an (extended) Dynkin diagram. This completes the definition of $\mathscr{H}$ (except that there is an indeterminacy due to isogeny, but we will skip over this difficulty).

It seems likely (and probably not difficult to prove) that the unramified representations of $\mathscr{G}$ which correspond to our $\mathscr{P}, \Psi$ are in natural 1-1 correspondence with the simple $\mathscr{H}$-modules.

This is well known in the case where $\mathscr{P}$ is an Iwahori subgroup. In that case, $\mathscr{H}$ is an Iwahori-Hecke algebras with equal parameters and one can use the classification [24] of its simple modules.

But it seems to be difficult to extend the K-theoretic method of [24] to the general case. Instead, in the general case we can argue as follows. We can replace $\mathscr{H}$ by its generic version (over $\mathbf{C}[v, v^{-1}]$) and take (as in [41]) certain associated graded algebras (over $\mathbf{C}[r]$) of it and then specialize r to $r_0 = \log(q)/2$. The resulting algebras turn out to be of the type considered in §16 (hence their simple modules can be classified as in §16(b)); on the other hand, from the representation theory of these algebras one can completely recover the representation theory of $\mathscr{H}$, by the results of [41]. This should give the desired result.

18. Cells in Affine Weyl Groups

We preserve the assumptions of §7. The definition of left cells and right cells of W given in [22] (see also §3) extends without change to the case of $\tilde{W}$. (But in this case, this is the only known definition; it is not known how to extend the definition from the theory of primitive ideals).

One can also define a partition of $\tilde{W}$ into *two-sided cells*: the two sided cells are the smallest subsets of $\tilde{W}$ which are unions of left cells and also unions of right cells.

It is known (see [36]) that there are only finitely many left cells in $\tilde{W}$ but their structure is not well understood except in type A (see Shi [50]); however, the two-sided cells of $\tilde{W}$ have been shown in [36] to be in natural 1-1 correspondence with the unipotent classes in the corresponding group over $\mathbf{C}$. The proof in [36] is quite complicated. One ingredient is the use of the function $a : \tilde{W} \to \mathbf{N}$ defined as in §3; another one is the use of the results of [24] on representations of affine Iwahori-Hecke algebras; in addition, the theory of character sheaves also plays a role in the proof.

It is interesting that many invariants of a unipotent class can be reconstructed from the corresponding two-sided cell, in purely combinatorial terms. For example, if a two-sided cell corresponds to the class of the unipotent element u, then the variety of Borel subgroups containing u has dimension $a(z)$ where z is any element in the two-sided cell.

19. Quiver Singularities and Canonical Bases

In this section we assume that G is as in §8. In [45], methods of IC have been used to construct some particularly nice bases of the finite dimensional irreducible representations of $\mathbf{g}$. We will explain how this is done. (Another approach to these questions has been recently found by Kashiwara; see his lecture at this Congress.)

We assume that we are given a basis element in each root space in $\mathbf{n}$ corresponding to a simple root; we denote these elements by $e_1, \ldots, e_n$. (They are in 1-1 correspondence with the vertices of the Dynkin graph, denoted $1, \ldots, n$.)

Let $\mathbf{U}$ (resp. $\mathbf{U}^+$) be the enveloping algebra of $\mathbf{g}$ (resp. of $\mathbf{n}$). Recently, Ringel [49] gave a very elegant description of $\mathbf{U}^+$ in terms of representations of a *quiver* (the Dynkin graph of G with a fixed orientation). All representations of the quiver are assumed to be finite dimensional over $\mathbf{C}$. The dimension of a representation is naturally a vector $\mathbf{d} = (d_1, \ldots, d_n) \in \mathbf{N}^n$.

Given three representations V, V', V'' of the quiver, we can consider the set of all subrepresentations of V'' which are isomorphic to V and such that the corresponding quotient is isomorphic to V'. This set is naturally an algebraic variety and we denote by $c_{V,V',V''}$ its Euler characteristic in cohomology with compact support. Let $\mathscr{R}$ be the $\mathbf{C}$-vector space with basis elements V indexed by the isomorphism classes of representations of the quiver. Then the $c_{V,V',V''}$ can be regarded as the structure constants of an algebra structure on $\mathscr{R}$ and Ringel shows that there is a unique algebra isomorphism $\mathbf{U}^+ \cong \mathscr{R}$ which takes each $e_i \in \mathbf{U}^+$ to the irreducible representation of our quiver corresponding to the vertex i of the Dynkin graph. Via this isomorphism, $\mathbf{U}^+$ aquires from $\mathscr{R}$ a $\mathbf{C}$-basis. Note that this basis depends on the chosen orientation of the Dynkin graph.

We want to look at ways in which a representation of the quiver can be deformed to a more degenerate one. It is known that the isomorphism classes of representations of our quiver correpond to certain orbits of algebraic group actions and we are lead to consider closures of such orbits. One of the ideas of [45] was to exploit the fact that these closures of orbits may be singular hence have invariants coming from IC theory. (These singularities may be called *quiver singularities*.)

We shall define such invariants $j_{V,V'} \in \mathbf{N}$ for any two representations V, V' of the quiver. In the case where V, V' have different dimensions we set $j_{V,V'} = 0$. Assume now that they have the same dimension $\mathbf{d} = (d_1, \ldots, d_n)$. Let

$$\mathbf{E} = \oplus_{i \to j} \mathrm{Hom}(\mathbf{C}^{d_i}, \mathbf{C}^{d_j})$$

sum over all arrows $i \to j$ in the given orientation, and let $G_{\mathbf{d}} = \prod_i \mathrm{GL}_{d_i}(\mathbf{C})$. The group $G_{\mathbf{d}}$ acts naturally on the $\mathbf{C}$-vector space $\mathbf{E}$ by $(g_i) : (f_{ij}) \to (g_j f_{ij} g_i^{-1})$. Any point in $\mathbf{E}_{\mathbf{d}}$ may be regarded as a representation of dimension $\mathbf{d}$ of our quiver and all representations of dimension $\mathbf{d}$ arise in this way. Moreover two points in $\mathbf{E}$ define isomorphic representations if and only if they are in the same $G_{\mathbf{d}}$-orbit. Thus, V, V' correspond to two $G_{\mathbf{d}}$-orbits $\mathcal{O}, \mathcal{O}'$ on $\mathbf{E}$. If $\mathcal{O}$ is contained in the closure of $\mathcal{O}'$, we define $j_{V,V'}$ to be the Euler characteristic of the local IC of that closure (with coefficients $\mathbf{C}$) at a point in $\mathcal{O}$. Otherwise, we set $j_{V,V'} = 0$.

The elements $\sum_V j_{V,V'} V$ (for various representations V' of the quiver) form a new basis of $\mathscr{R}$; upon transfering it to $\mathbf{U}^+$, we obtain a basis of $\mathbf{U}^+$ which, by one of the main results of [45], is independent of the chosen orientation. For this reason, we call it the *canonical basis*.

Assume that we are given an irreducible, finite dimensional representation of $\mathbf{U}$ with a specified lowest weight vector. If we apply the elements of the canonical basis to this lowest weight vector, the resulting elements less 0 will miraculously form a basis for the representation space. This is a canonical basis of the given representation of $\mathbf{U}$. It is a basis with extremely favorable properties (see [45]).

All these bases have natural q-analogs in the framework of quantized enveloping algebras.

References

1. D. Alvis and G. Lusztig: On Springer's correspondence for simple groups of type E_n, ($n = 6, 7, 8$). Math. Proc. Camb. Phil. Soc. **92** (1982) 65–72
2. A. A. Beilinson and J. Bernstein: Localisation des **g**-modules. C.R. Acad. Sci. Paris **292** (1981) 15–18
3. A. A. Beilinson and J. Bernstein: A proof of Jantzen conjecture. Preprint
4. A. A. Beilinson, J. Bernstein and P. Deligne: Faisceaux pervers. Astérisque **100** (1982)
5. W. Beynon and N. Spaltenstein: Green functions of finite Chevalley groups of type E_n ($n = 6, 7, 8$). J. Algebra **88** (1984) 584–614
6. W. Borho and R. MacPherson: Représentations des groupes de Weyl et homologie d'intersection pour les variétés nilpotentes. C. R. Acad. Sci. Paris **292** (1981) 707–710
7. R. Bott: An application of the Morse theory to the topology of Lie groups. Bull. Soc. Math. France **84** (1956) 251–282
8. J.-L. Brylinski and M. Kashiwara: Kazhdan-Lusztig conjecture and holonomic systems. Invent. math. **64** (1981) 387–410
9. R. K. Brylinski: Limits of weight spaces, Lusztig's q-analogs and fiberings of adjoint orbits. J. Amer. Math. Soc. **2** (1989) 517–533
10. L. Casian: Kazhdan-Lusztig multiplicity formulas for Kac-Moody algebras. C. R. Acad. Sci. Paris **310** (1990) 333–337
11. L. Casian: Kazhdan-Lusztig conjecture in the negative level case (Kac-Moody algebras of affine type). Preprint
12. P. Deligne and G. Lusztig: Representations of reductive groups over a finite field. Ann. Math. **103** (1976) 103–161
13. V. G. Drinfeld: Degenerate affine Hecke algebras and Yangians. Funkt. Anal. Prilozh. **20** (1986) 69–70 (Russian)

14. B. L. Feigin and E. V. Frenkel: Affine Kac-Moody algebras and semi-infinite flag manifolds. Commun. Math. Phys. **128** (1990) 161–189
15. M. Goresky and R. MacPherson: Intersection homology theory. Topology **19** (1980) 135–162
16. V. Ginsburg: Perverse sheaves on loop groups and Langlands duality. Preprint
17. N. Iwahori and H. Matsumoto: On some Bruhat decomposition and the structure of the Hecke ring of p-adic Chevalley groups. Publ. Math. IHES **25** (1965) 237–280
18. A. Joseph: W-module structure in the primitive spectrum of the enveloping algebra of a semisimple Lie algebra. (Lecture Notes in Mathematics, vol. 728.) Springer, Berlin Heidelberg New York 1979, pp. 116–135
19. M. Kashiwara: Kazhdan-Lusztig conjecture for symmetrizable Kac-Moody Lie algebra. Preprint
20. M. Kashiwara and T. Tanisaki: Kazhdan-Lusztig conjecture for symmetrizable Kac-Moody Lie algebra; II, Intersection cohomologies of Schubert varieties. Preprint
21. D. Kazhdan: Proof of Springer's hypothesis. Isr. J. Math. **28** (1977) 272–286
22. D. Kazhdan and G. Lusztig: Representations of Coxeter groups and Hecke algebras. Invent. math. **53** (1979) 165–184
23. D. Kazhdan and G. Lusztig: Schubert varieties and Poincaré duality. Proc. Symp. Pure Math. **36** (1980) 185–203, Amer. Math. Soc.
24. D. Kazhdan and G. Lusztig: Proof of the Deligne-Langlands conjecture for Hecke algebras. Invent. math. **87** (1987) 153–215
25. S. Kumar: Proof of Lusztig's conjecture concerning negative level representations of affine Lie algebras. preprint
26. G. Laumon: Faisceaux automorphes liés aux séries d'Eisenstein. Automorphic forms, Shimura varieties and L-functions, vol. I (L. Clozel, J. S. Milne, ed.). Academic Press, New York 1990, pp. 227–281
27. G. Lusztig: Representations of finite Chevalley groups. Regional Conf. Series in Math. 39. Amer. Math. Soc., 1978
28. G. Lusztig: Hecke algebras and Jantzen's generic decomposition patterns. Adv. Math. **37** (1980) 121–164
29. G. Lusztig: Some problems in the representation theory of finite Chevalley groups. Proc. Symp. Pure Math., vol. 37. Amer. Math. Soc. 1981, pp. 313–317
30. G. Lusztig: Green polynomials and singularities of unipotent classes. Adv. Math. **42** (1981) 169–178
31. G. Lusztig: Singularities, character formulas and a q-analog of weight multiplicities. Astérisque **101–102** (1983) 208–229
32. G. Lusztig: Some examples of square integrable representations of semisimple p-adic groups. Trans. Amer. Math. Soc. **227** (1983) 623–653
33. G. Lusztig: Characters of reductive groups over a finite field. Ann. Math. Studies, vol. 107. Princeton University Press, 1984
34. G. Lusztig: Intersection cohomology complexes on a reductive group. Invent. math. **75** (1984) 205–272
35. G. Lusztig: Character sheaves. Adv. Math. **56** (1985) 193–237; II, **57** (1985) 226–265; III, **57** (1985) 266–315; IV, **59** (1986) 1–63; V, **61** (1986) 103–155
36. G. Lusztig: Cells in affine Weyl groups. Algebraic groups and related topics. Advanced Studies in Pure Math., vol. 6. Kinokuniya and North-Holland, Tokyo and Amsterdam, 1985; II, J. Alg. **109** (1987) 536–548; III, J. Fac. Sci. Tokyo University, IA **34** (1987) 223–243; IV, J. Fac. Sci. Tokyo University, IA **36** (1989) 297–328
37. G. Lusztig: Sur les cellules gauches des groupes de Weyl. C. R. Acad. Sci. Paris **302** (1986) 5–8
38. G. Lusztig: On the character values of finite Chevalley groups at unipotent elements. J. Algebra **104** (1986) 146–194

39. G. Lusztig: Leading coefficients of character values of Hecke algebras. Proc. Symp. Pure Math. **47** (1987) 253–262. Amer. Math. Soc.
40. G. Lusztig: Cuspidal local systems and graded Hecke algebras. *I*, Publ. Math. I.H.E.S. **67** (1988) 145–202; *II*, in preparation
41. G. Lusztig: Affine Hecke algebras and their graded version. J. Amer. Math. Soc. **2** (1989) 599–635
42. G. Lusztig: Green functions and character sheaves. Ann. Math. **131** (1990) 355–408
43. G. Lusztig: Finite dimensional Hopf algebras arising from quantized universal enveloping algebras. J. Amer. Math. Soc. **3** (1990)
44. G. Lusztig: On quantum groups. J. Algebra **131** (1990) 466–475
45. G. Lusztig: Canonical bases arising from quantized enveloping algebras. J. Amer. Math. Soc. **3** (1990) 447–498; *II*, Progress of Theor. Physics (to appear)
46. G. Lusztig and N. Spaltenstein: On the generalized Springer correspondence for classical groups. Algebraic groups and related topics. Advanced Studies in Pure Math., vol. 6. Kinokuniya and North-Holland, Tokyo and Amsterdam 1985, pp. 289–316
47. G. Lusztig and D. Vogan: Singularities of closures of K-orbits on a flag manifold. Invent. math. **71** (1983) 365–379
48. R. MacPherson: Global questions in the topology of singular spaces. Proc. I.C.M., Warszawa 1983, North-Holland, 1984, pp. 213–235
49. C. M. Ringel: Hall algebras and quantum groups. Invent. math. (1990)
50. J.-Y. Shi: The Kazhdan-Lusztig cells in certain affine Weyl groups. (Lecture Notes in Mathematics, vol. 1179.) Springer, Berlin Heidelberg New York 1986
51. T. Shoji: On the Springer representations of the Weyl groups of classical algebraic groups. Comm. in Alg. **7** (1979) 1713–1745; 2027–2033
52. T. Shoji: On the Springer representations of the Chevalley groups of type $F4$. Comm. Alg. **8** (1980) 409–440
53. T. Shoji: On the Green polynomials of Chevalley groups of type $F4$. Comm. Alg. **10** (1982) 505–543
54. T. Shoji: On the Green polynomials of classical groups. Invent. math. **74** (1983) 239–267
55. N. Spaltenstein: Appendix. Math. Proc. Camb. Phil. Soc. **92** (1982) 73–78
56. T. Shoji: On the generalized Springer correspondence for exceptional groups. Algebraic groups and related topics. Advanced Studies in Pure Math., vol. 6. Kinokuniya and North-Holland, Tokyo and Amsterdam 1985, pp. 317–338
57. T. A. Springer: Trigonometric sums, Green functions of finite groups and representations of Weyl groups. Invent. math. **36** (1976) 173–207
58. J. Tits: Résumé du cours. Annuaire du Collège de France (1981–82)
59. D. Vogan: Irreducible characters of semisimple Lie groups. *I*, Duke Math. J. **46** (1979) 61–108; *II*, The Kazhdan-Lusztig conjectures. Duke Math. J. **46** (1979) 805–859; *III*, Proof of the Kazhdan-Lusztig conjecture in the integral case. Invent. math. **71** (1983) 381–418; *IV*, Character-multiplicity duality. Duke Math. J. **49** (1982) 943–1073

The Interaction of Nonlinear Analysis and Modern Applied Mathematics

Andrew J. Majda [*]

Department of Mathematics and Program in Applied and Computational Mathematics
Fine Hall, Washington Road, Princeton University, Princeton, NJ 08544, USA

Dedicated to Ron Diperna (1947–1989)

Introduction

Many of the physical phenomena that are currently important research topics in applied mathematics involve strong, irregular, and typically unstable fluctuations with intense activity occurring on extremely small length scales. Examples include fully developed turbulence and the instability of thin layers in fluid flows, phase transitions and homogenization in materials, dendrite formation in crystal growth, the focusing of laser beams, and various problems in classical statistical physics such as the kinetic theory of gases. For a discussion of the mathematical progress in this last topic, see the report of P.L. Lions at this Congress. The quantitative physical phenomena are usually described by solutions of nonlinear partial differential equations in various singular limits involving either small spatial/temporal scales or coefficients of the equation. A prototypical example of such a process is the high Reynolds number limit of the Navier-Stokes equations,

$$\frac{\partial v^\varepsilon}{\partial t} + \text{div} \, (v^\varepsilon \otimes v^\varepsilon) = -\nabla p^\varepsilon + \varepsilon \Delta v^\varepsilon \, ,$$

$$\text{div} \, v^\varepsilon = 0 \, . \tag{0.1}$$

Here $v = {}^t(v_1, v_2, v_3)$ is the fluid velocity, p is the scalar pressure, $v \otimes v = (v_i v_j)$, and ε is the reciprocal of the Reynolds number. Fully developed turbulence is characterized by the fact that the parameter ε in (0.1) is extremely small, i.e. $\varepsilon \ll 1$; the formal limiting equations with $\varepsilon = 0$ in (0.1) are called the Euler equations.

The recent development of powerful, user friendly, supercomputers and work-stations combined with high resolution graphics and sophisticated numerical algorithms has enabled researchers to discover interesting and unexpected new phenomena in the solutions of nonlinear P.D.E.'s with a resolution often exceeding the capabilities of contemporary experiments. Understanding these phenomena has created the need for a highly interdisciplinary interaction involving ideas from nonlinear analysis, numerical computation, and sophisticated asymptotic methods. The reports by Mimura [51] and Krasny [33] at this Congress illustrate aspects of this type of interaction in modern applied mathematics; some recent

[*] Partially supported by research grants NSF DMS-9001805, ARO DAAL03-89-K-0013, ONR N00014-89-J-1044.

papers of the author [46, 36, 37] describe this modern research mode in the context of various problems in fluid dynamics. Except for brief remarks and some references, it will not be possible due to limitations on length to describe examples of such exciting interdisciplinary developments in any detail in this article and the main emphasis will involve the nonlinear analysis of the new phenomena.

The problems which I discuss below have a common theme. There is a sequence of solutions of a nonlinear P.D.E. which satisfies the bound,

$$\int_\Omega |v^\varepsilon|^2 dy \leq C \tag{0.2}$$

as $\varepsilon \downarrow 0$ so that there is a function v and a subsequence with

$$v^\varepsilon \rightharpoonup v \text{ i.e. weakly in } L^2(\Omega). \tag{0.3}$$

In applications, the bound in (0.2) is usually guaranteed by a physical principle such as an energy estimate and the small parameter $\varepsilon \downarrow 0$ represents the limit of increasingly smaller scale fluctuations. The use of the L^2-norm in (0.2) is for simplicity in exposition. Various kinds of defects can happen in such a limiting process. These are illustrated by the heuristic decomposition

$$v^\varepsilon = v^\varepsilon_{\text{osc}} + v^\varepsilon_{\text{conc}} + \bar{v}^\varepsilon \tag{0.4}$$

of v^ε. Here $\bar{v}^\varepsilon$ converges strongly in L^2 so that there is a $\bar{v}$ with

$$\int_\Omega |\bar{v}^\varepsilon - \bar{v}|^2 dy \to 0 \text{ as } \varepsilon \downarrow 0. \tag{0.5}$$

The functions $v^\varepsilon_{\text{osc}}$ measure the small scale oscillations in the limit in the sense that

$$|v^\varepsilon_{\text{osc}}(y)|_{L^\infty} \leq C, \tag{0.6A}$$

and there is a $v_{\text{osc}} \in L^\infty$ so that

$$\int_\Omega \phi v^\varepsilon_{\text{osc}} dy \to \int_\Omega \phi v_{\text{osc}} dy \qquad \text{for all functions } \phi \in C_0(\Omega), \tag{0.6B}$$

but $v^\varepsilon_{\text{osc}}$ does not converge strongly in L^2 to v_{osc}. A prototypical example of a sequence with small scale oscillations in R^1 is given by $v^\varepsilon_{\text{osc}} = v\left(\frac{y}{\varepsilon}\right)$ where $v(y)$ is a fixed smooth periodic function. The functions $v^\varepsilon_{\text{conc}}$ in the heuristic decomposition in (0.4) measure the development of intense small scale concentrations in the limit in the sense that $v^\varepsilon_{\text{conc}}$ satisfies the uniform L^2-bound in (0.2) and there is a function v_{conc} so that

$$v^\varepsilon_{\text{conc}} \to v_{\text{conc}} \text{ a.e. in } \Omega, \tag{0.7}$$

but nevertheless, $v^\varepsilon_{\text{conc}}$ does not converge strongly to v_{conc} in L^2. A prototypical example illustrating concentration in $L^2(R^N)$ is given by the sequence $v^\varepsilon_{\text{conc}} = \varepsilon^{\frac{-N}{2}} \varrho\left(\frac{x}{\varepsilon}\right)$ with $\varrho \in C_0^\infty$, $\varrho \geq 0$, $\int \varrho = 1$.

Most of the research in nonlinear P.D.E. prior to the late 1970's involved problems where suitable apriori estimates automatically guaranteed strong convergence so that $v^\varepsilon_{\text{osc}}$ and $v^\varepsilon_{\text{conc}}$ in (0.4) were zero and the relevant nonlinear maps $g(v^\varepsilon)$ are continuous in the limit. With the strong motivation from understanding many of the new phenomena described earlier, an important current research direction in nonlinear P.D.E.'s involves situations where both fine scale oscillations and/or concentrations occur in various physical problems so that v_{osc}^{ε} and/or v_{conc}^{ε} are non-zero in the heuristic decomposition in (0.4). In the next section some of the mathematical tools and physical problems that have been studied recently with oscillations and/or concentrations are described briefly. Of course, this article is not intended to be a detailed survey and the discussion is limited by both the taste and expertise of the author. The final sections of the article contain a more detailed report on problems with oscillations and concentrations in incompressible fluid flow – this is a rich and important class of problems where both effects occur simultaneously and many interesting problems remain unsolved.

1. The Development of Oscillations and Concentrations in Problems from Applied Mathematics

1A. Some Mathematical Tools

An effective tool for nonlinear analysis associated with a sequence v_{osc}^{ε} satisfying the L^{∞}-bound in (0.6A) is the Young measure introduced by Tartar in this context. If the vector field v is a mapping from $\Omega \subseteq R^N$ to R^M, the Young measure theorem asserts that there is a family of probability measures on R^M, $\{v_y\}$ indexed by $y \in \Omega$ so that after passing to subsequence, weak limits of composite non-linear maps $g(v^{\varepsilon})$ are given by

$$\lim_{\varepsilon \to 0} \int_{\Omega} \phi(y) g(v_{osc}^{\varepsilon}) dy = \int_{\Omega} \phi(y) \int_{R^M} g(\lambda) dv_y(\lambda) dy. \tag{1.1}$$

There is strong convergence of v_{osc}^{ε} to v_{osc} on Ω if and only if the probability measure v_y is given by the Dirac mass, $\delta_{v_{osc}(y)}$, for each $y \in \Omega$. An elementary probabilistic interpretation of the Young measure utilizing the Lebesque differentiation theorem has been developed recently by Ball [6]. Diperna [18, 19] introduced the concept of measure-valued solution of a system of conservation laws associated with an L^{∞} bounded sequence so that only oscillations occur in the limit. The effectiveness of this tool for interpreting the limit of fine scale oscillations in model problems such as the zero dispersion limit of solutions of the KdV equation has been surveyed earlier by Diperna (see [18, 19]). A very general extension of the Young measure theorem for general sequences with only the uniform $L^2(L^p)$ bound in (0.2) has been developed by Diperna and the author [20] for the specific applications of constructing suitable measure-valued solutions of problems from mathematical physics where both fine scale oscillations and concentrations develop in the limit. Strong motivation for this work was provided by the problems involving the high Reynolds number limit of solutions of the Navier-Stokes equations with finite kinetic energy described in (0.1); the limiting solution is a suitable measure-valued solution of the Euler equations with both oscillations and concentrations (see [20]). Unlike (1.1), the generalized Young measures in this context are often singular continuous with respect to Lebesque measure, reflecting the development of concentrations – more will be said about this topic later in this paper.

In physical problems where the phenomena of concentration from (0.7) occurs without the development of oscillations (so that $v_{\text{osc}}^{\varepsilon} = 0$ in (0.4)) it is very natural to measure the size of the set where concentration occurs; two types of defect measures have been introduced recently for this purpose, the weak* defect measure [35] and the reduced defect measure [22]. With the uniform L^2-bound in (0.2) and $v^{\varepsilon} \rightharpoonup v$, there is a non-negative Radon measure σ so that

$$\int \varphi d\sigma = \lim_{\varepsilon \to 0} \int_{\Omega} \varphi |v^{\varepsilon} - v|^2 dy \tag{1.2}$$

for all $\varphi \in C_0(\Omega)$. The measure σ is the weak* defect measure; if σ is zero on an open set, there is strong convergence on that open set. Motivated by problems in the calculus of variations with critical Sobolev exponents, P.L. Lions [35] has proved a beautiful simple theorem guaranteeing that the weak* defect measure is always concentrated on a rather small set in this context. An illustration of this theorem is provided by the following example: if v^{ε} is a sequence of functions in R^2 satisfying the bound

$$\int_{R^2} |v^{\varepsilon}|^2 + \int_{R^2} |\nabla v^{\varepsilon}| \le C, \tag{1.3A}$$

then there is a countable number of points $\{y_j\}_{j=1}^{\infty}$ and weights $\alpha_j \ge 0$ with $\sum \alpha_j^{1/2} < \infty$, so that necessarily

$$\sigma = \sum_j \alpha_j \delta_{y_j}, \tag{1.3B}$$

with δ_{y_j} the Dirac mass at y_j.

The reduced defect measure [22] is the finitely sub-additive outer measure θ defined by

$$\theta(E) = \limsup \int_E |v^{\varepsilon} - v|^2 dx. \tag{1.4}$$

Obviously, the reduced defect measure has the property that $\theta(E) = 0$ for any set E if and only if there is strong convergence on E. Furthermore, for any closed set F, it is easy to see that

$$\theta(F) \le \sigma(F), \tag{1.5}$$

so that $\theta(F)$ can be very small even though $\sigma(F)$ is large. Examples of the use of estimates for the size of the reduced defect measure occur in the important physical problem involving vortex sheets for incompressible fluid flow and will be discussed in the later sections of this paper [22]. In these examples the support of σ can include all of R^3 while θ concentrates on a set of small measure with Hausdorff dimension one and this is crucial for the applications.

The discussion just presented of some tools for measuring phenomena in weak convergence is necessarily sketchy and incomplete. The interested reader should consult the recent lecture notes of L.C. Evans [25] for a more complete discussion of these matters. There is a need to develop other theoretical tools to measure additional structure in weak convergence with defects; one interesting new concept with potential applications is the H-measure introduces independently by Tartar [57] and Gerard [26]. Tartar's lecture at this Congress [58] provides an introduction to these topics.

1B. The Development of Fine Scale "Twinning" and Defects in Materials

Very recently an elegant mathematical theory has been developed to explain at the macroscopic level certain phase transitions in solids as the temperature, θ, varies [7, 8, 13, 16, 32]. In such austenitic-martensitic phase transitions there is a decrease in crystal symmetry at a transition temperature; for example, from a high temperature solid phase with cubic symmetry (austenite) to a low temperature solid phase with tetragonal symmetry (martensite). At the transition temperature, it is observed in experiments that the martensitic phase develops small scale regular fluctuations and is often "finely twinned" along planes related to the crystal lattice and also there is an austensite/finely twinned martensite interface. The key point in the mathematical analysis is the fact that the symmetries in a macroscopic continuum theory of crystals for both the austensite and martensite phases at the critical temperature θ_0 leads to a variational problem to minimize the free energy over deformations $y(x)$: $\Omega \supseteq R^3 \rightarrow R^3$ with a structure so that the finite infimum *is not achieved*. In other words, for the appropriate free energy density, $W(A, \theta)$ with $A \in M_+^{3 \times 3}$, the set of 3×3 matrices with positive determinant, I is finite with

$$I = \min_{y} \int_{\Omega} W\left(\frac{dy}{dx}, \theta_0\right) dx, \tag{1.6}$$

but there is no weak deformation y so that the minimum, I, is achieved. In particular, the appropriate free energy functionals $W(A, \theta)$ at the critical temperature, θ_0, are necessarily not rank-one convex and thus are not quasi-convex [5]. The rigorous theory predicts that in the process of attempting to achieve the minimum, the sequence of deformation gradients, $\frac{dy^\varepsilon}{dx}$, necessarily develops apriori fine scale but regular oscillations in the martensitic phase which admit an elegant characterization via classical Young measures (see (1.1)) as combinations of weighted Dirac masses located on an appropriate set of matrices. Numerical computations [16] confirm the continuum theory and specific tests for detailed experiments based on the mathematical theory have recently been proposed [8]. An extremely readable mathematical account of this interdisciplinary research is presented in the paper by James and Kinderlehrer [32].

In the previous paragraph, an application in materials science with fine scale oscillations was described. Next the solution of a model problem motivated by the issue of prescribing strengths and locations of defects in the theory of liquid crystals [24] will be described where concentrations develop in a certain fashion so that the weak defect measure from (1.2) is nontrivial. The model problem analyzed by Brezis, Coron, and Lieb [10] concerns harmonic maps with values in S^2 and defects at prescribed locations a_i with specified (Brouwer) degree, deg (ϕ, a_i), at the location a_i for $1 \leq i \leq N$. Thus, the admissible maps are given by

$$\mathscr{E} = \left\{ \phi \in C^1(R^3 \sim \cup_{i=1}^N \{a_i\}; S^2) \big| \deg(\phi, a_i) = d_i \text{ and } \int_{R^3} |\nabla \phi|^2 < \infty \right\}$$

The model problem consists of finding

$$E = \inf_{\phi \in \mathscr{E}} \int |\nabla \phi|^2. \tag{1.7}$$

The number E is finite and there is a beautiful explicit formula for this value [10]; as in (1.6), the infimum in (1.7) is not achieved but concentrations develop in the minimizing deformation gradients rather than the fine scale oscillations from (1.6). Specifically, in attempting to achieve the minimum, $v^\varepsilon = \nabla \phi^\varepsilon$ concentrates so that $v^\varepsilon \rightharpoonup 0$ but $|v^\varepsilon|^2 \rightharpoonup \sigma$ in the sense of Radon measures where the weak* defect measure is given by $\sigma = 8\pi\delta_C$; the set C consists of a finite union of straight line segments in R^3 defined by some minimal connection, where δ_C is the uniform one-dimensional Lebesque measure on these lines (see [10]).

1C. The Development of Concentrations in Focusing Solutions of Nonlinear Schrödinger Equations

This is a very rich class of problems providing simplified models for focusing laser beams where concentrations develop spontaneously in time in a problem with critical Sobolev exponents. These equations have the form

$$iv_t = -\Delta v - |v|^{4/N} v, \qquad v(x, 0) = v_0(x), \tag{1.8}$$

for $x \in R^N$. The equations in (1.8) have a Hamiltonian structure and a special conformal invariance as a consequence of the special exponent in (1.8); furthermore, the L^2-norm is conserved in time on intervals of existence. Weinstein [59] has proved that solutions of (1.8) exist for all time provided that $||v_0||_{L^2} < ||Q||_{L^2}$ where $Q(x)$ is a suitable ground state. For suitable initial data with critical mass so that $||v_0||_{L^2} = ||Q||_{L^2}$, he exploited the conformal invariance to construct explicit focusing solutions which develop dynamic concentrations as $t \nearrow T_* < \infty$ and studied the stability of this process through methods of concentration compactness [60]. In very recent work, Merle [50] has attacked the intriguing question of continuation of these special solutions past the blow up time. He considers sequences of smooth initial data v_0^ε regularizing the data $v_0(x)$ which develops singularities so that the regularized data v_0^ε satisfy $||v_0^\varepsilon|| < ||Q||$ and $||v_0^\varepsilon - v_0|| \to 0$. As mentioned earlier, the corresponding solutions $v^\varepsilon(x, t)$ are defined globally in time and the limiting behavior as $\varepsilon \downarrow 0$ is analyzed in [50] for $t > T_*$; the singular solution is unstable in time in the sense that the singularity disappears in this limit for $t > T_*$ and suitable special subsequences of v^ε converge to explicit defocusing solutions of (1.8) characterized by arbitrary phase shift parameters. There is "loss of information" in time once the dynamic concentration develops because these phase shifts are not uniquely determined and depend on the special subsequence. This author conjectures that appropriate measure-valued solutions [20] of (1.8) arise for $t > T_*$ in the general case in order to reflect the loss of information. The nature of singularity formation for solutions of (1.8) with $||v_0||_{L^2} > ||Q||_{L^2}$ is an active research topic in applied mathematics [34] with rich additional phenomena in the structure of singular solutions beyond those discussed briefly here.

1D. Nonlinear Geometric Optics and Oscillations for Hyperbolic Systems of Conservation Laws

Recently, there has been a great deal of activity in developing the theory of nonlinear geometric optics for general hyperbolic systems of conservation laws in multi-D [29, 42, 30, 28, 49]. As in linear geometric optics, these formal asymptotic theories utilize short wave length approximations so they involve the propagation of oscillations in hyperbolic systems. In nonlinear geometric optics, approximate solutions of general quasi-linear hyperbolic systems are constructed with the general form,

$$v^\varepsilon = \bar{v}(x,t) + \varepsilon v_1\left(\frac{\phi}{\varepsilon}, x, t\right) + 0(\varepsilon), \qquad \varepsilon \ll 1 \qquad (1.9)$$

where $\phi = (\phi_1(x,t), ..., \phi_d(x,t))$ is a vector of phase functions. Through self-consistent asymptotic approximations, simplified equations for the phase functions and amplitudes in v_1 are developed and in many instances these approximations can be solved exactly; thus, a great deal of genuine insight can be gained into complex physical problems. One of the theoretical achievements is a general theory for resonantly interacting small amplitude periodic wave trains with many new phenomena that are completely absent in geometric optics for linear problems [42, 30, 49, 12, 43]. In applications, the general theory of nonlinear geometric optics has been used recently to explain a variety of complex phenomena in physical problems involving compressible fluid flow [43, 3, 4], wave propagation in reacting materials [41, 2, 45] and dynamic nonlinear elasticity [49]. For example, a completely new theory of nonlinear instability for supersonic vortex sheets has been developed recently by the author and his recent Ph.D. student, Artola which combines the theory of hyperbolic mixed problems with nonlinear geometric optics [3, 4, 48] in a sophisticated fashion; this theory has been motivated and confirmed by large scale computations of Woodward (see the discussion in [36]. Another important use of nonlinear geometric optics is to extract mathematically tractable model equations which retain some of the essential features of much more complex physical phenomena [52, 44, 45].

This general topic has been mentioned briefly here because it involves interdisciplinary interactions in applied mathematics with the propagation of oscillations, and there is a definite need for more work on the rigorous theory of nonlinear geometric optics beyond the contribution of Diperna and the author [23]. Some important and accessible open theoretical problems are discussed in two review articles [37, 40]. Unlike the linear case, the effects of diffraction are not understood even at the formal level for small amplitude discontinuities. Current surveys of the formal aspects of nonlinear geometric optics including a discussion of problems with diffraction have been given by Hunter [31] and Rosales [53]. In interesting recent work, D. Serre [55, 54] has used the classical Young measure to describe the propagation of large amplitude oscillations in special quasi-linear hyperbolic systems with linearly degenerate wave fields.

1E. Oscillations and Concentrations for Incompressible Fluid Flow

Ingenious numerical computations [15, 14] and physical experiments reveal bewildering complexity at finite times in solutions of the incompressible fluid equations from (0.1) both at high Reynold's number for the Navier-Stokes equations with $\varepsilon \ll 1$ in (0.1) and as various numerical regularization parameters tend to zero for the incompressible Euler equations (the equations in (0.1) with $\varepsilon = 0$). The work of Diperna and the author [20, 22, 21] involving measure-valued solutions of the three-dimensional Euler equations is a first attempt to quantify such observed catastrophic instabilities in a mathematical fashion. Due to the rapid amplification of vorticity in three-dimensional fluid flows [36], very little apriori information is available and the only generally bounded positive physical quantity of an approximating sequence, v^ε, is the local kinetic energy. Thus, it is reasonable to assume the uniform bounds

$$\max_{0 \le t \le T} \int_{|x| \le R} |v^\varepsilon|^2(x,t)dx \le C_{R,T} \qquad (1.10)$$

for any $R > 0, T > 0$. In [20], measure-valued solutions of the three-dimensional Euler equations are constructed which arise from the weak limit of suitable approximate solution sequences satisfying the uniform bound in (1.10); a prototypical example from [20] is the fact that in the limit as $\varepsilon \downarrow 0$, a family of Leray-Hopf weak solutions of (0.1) converges to a measure-valued solution of the 3-D Euler equations. In general, these measure-valued solutions allow for both oscillations and concentrations and several examples of both of these phenomena are discussed in [20, 21].

The simplest way to generate examples of approximate solution sequences and measure-valued solutions for 3-D Euler is to take a sequence of smooth exact solutions and examine the generalized Young measure in the limit. It is an easy exercise for the reader to check that any sequence of smooth velocity fields

$$v^\varepsilon = (0, 0, v_3^\varepsilon(x_1, x_2)), \qquad (1.11A)$$

with the kinetic energy bound in (1.10), defines a sequence of exact smooth solutions of the 3-D Euler equations converging to a measure-valued solution of 3-D Euler; clearly all of the phenomena of development of simultaneous oscillations and concentrations in solutions of the fluid equations can be displayed in such solution sequences by mimicking the constructions below (0.6) and (0.7). If $v_3^\varepsilon \rightharpoonup v_3(x_1, x_2)$ in L^2, then

$$v = (0, 0, v_3(x_1, x_2)) \qquad (1.11B)$$

defines a weak solution of the 3-D Euler equations despite the fact that typically $(v_3^\varepsilon)^2$ does not converge weakly to $(v_3)^2$. A more interesting example of an exact solution sequence for the 3-D Euler equations displaying the development of fine scale oscillations [20] is given by

$$v^\varepsilon = \left(v\left(\frac{x_2}{\varepsilon}, x_2\right), 0, w\left(x_1 - v\left(\frac{x_2}{\varepsilon}, x_2\right)t, x_2, \frac{x_2}{\varepsilon}\right) \right), \qquad (1.12)$$

where v and w are fixed smooth bounded functions with $v(y, x_2)$ periodic in y; in particular, this sequence satisfies the uniform L^∞-bound, $|v^\varepsilon|_{L^\infty} \le C$. Unlike the example in (1.11), *the weak limit of* (1.12) *is an explicit smooth function but* this smooth function is *no longer an exact solution of the 3-D Euler equations* –

new phenomena occur through the persistence of oscillations and the limiting process is an explicit example of a nontrivial measure-valued solution. It is worth remarking that both of the exact solution sequences for the Euler equations in (1.11) and (1.12) inherently involve three space dimensions; furthermore, these examples are constructed in a systematic fashion beginning with suitable exact two-dimensional solutions, $(0,0)$ and $\left(v\left(\frac{x_2}{\varepsilon},x_2\right),0\right)$ respectively (see p. 445 of [17] for a description of the general recipe). Steady exact solutions of the 2-D Euler equations (see [47]) are given by the radial eddies,

$$v = \left(\frac{-x_2}{r^2}, \frac{x_1}{r^2}\right) \int_0^r s\omega(s)\,ds,\qquad(1.13)$$

with $r = (x_1^2 + x_2^2)^{1/2}$. Elementary examples of exact solution sequences for the 2-D Euler equations exhibiting concentrations can be generated in two different ways:

A)   Pick a *positive* function $\omega(r) \geq 0$ with bounded support and define the v^ε by

$$v^\varepsilon = \left(\log \frac{1}{\varepsilon}\right)^{-1/2} \varepsilon^{-1} v\left(\frac{x}{\varepsilon}\right)\qquad(1.14A)$$

with v given from ω by (1.13).

B)   Pick a function $\omega(r)$ with compact support but $\int_0^\infty s\omega(s)\,ds = 0$ so that there is zero circulation and define v^ε by

$$v^\varepsilon = \varepsilon^{-1} v\left(\frac{x}{\varepsilon}\right)\qquad(1.14B)$$

with v given from ω by (1.13).

The scalings in both examples have been chosen to guarantee the uniform local kinetic energy bounds in (1.10). In both examples, $v^\varepsilon \to 0$ but the nonlinear terms in the Euler equations, $v \otimes v$ from (0.1), satisfy

$$v^\varepsilon \otimes v^\varepsilon \to C \begin{pmatrix} 1 & 0 \\ 0 & 1 \end{pmatrix} \delta(x)\qquad(1.15)$$

with weak convergence in the sense of measures. Here $\delta(x)$ is the Dirac mass at the origin and the constant C varies in the two examples and depends on the core (see [20] and [21] for the details). Such examples illustrate the phenomena of concentration in solutions of the 2-D Euler equations; these two exact solution sequences generate measure-valued solutions of the 2-D Euler equations where the associated Young measure is singular continuous with respect to Lebesque measure due to the appearance of concentrations. It is useful to note here that in example (1.14A)

$$\int_{R^2} |\text{curl } v^\varepsilon| \leq C, \quad \text{but} \quad \overline{\lim}_{\varepsilon \downarrow 0} \int_{R^2} |\nabla v^\varepsilon| = +\infty\qquad(1.16)$$

while $\|\nabla v^\varepsilon\|_{L^1} \leq C$ in example (1.14B). Other amusing examples of measure-valued solutions for the 3-D Euler equations with fine scale oscillations have been developed in interesting recent work by Brenier [9].

The following general problem is of interest in the present context:

Problem. Characterize the closure in the L^2-weak topology of smooth exact solutions of the 3-D Euler equations.

The above explicit examples all provide insight into this question. There is a need to find other interesting examples like the one in (1.12) which illustrate that the weak closure is larger than merely the weak solutions of 3-D Euler; such solutions involving interacting oscillations and concentrations would be particularly useful.

2. Vortex Sheets, Potential Theory, and Concentration-Cancellation for 2-D Incompressible Flow

Incompressible fluid flow in two-space dimensions is generally much simpler than that in three space dimensions because given the velocity $v = (v_1, v_2)$, the scalar vorticity, $\omega = (v_2)_{x_1} - (v_1)_{x_2}$, is conserved along fluid particle paths for the 2-D Euler equations. Thus, it is a classical fact that smooth solutions of the 2-D Euler equations exist for all time while the analogous question for 3-D Euler is an outstanding and very important open problem due to vorticity amplification [36, 47]. Nevertheless, there are important practical problems involving weak solutions of the 2-D Euler equations, and here recent progress [22, 21, 27, 1, 61] is described in understanding the new mathematical phenomena and structure that occurs for vortex sheet initial data.

A two dimensional incompressible velocity field, v_0, defines *vortex sheet initial data* provided that curl $v_0 = \omega_0$ is a finite Radon measure and v_0 has locally finite kinetic energy. In applications, ω_0 is typically a surface Dirac measure supported on a smooth curve. Recent ingenious large scale numerical computations reveal tremendous complexity in the solutions of the 2-D fluid equations with vortex sheet initial data as various regularization parameters converge to zero; there is also much more complexity in the computed solutions when ω_0 changes sign than if ω_0 has a distinguished sign i.e. $\omega_0 \geq 0$ (see Krasny's report at this Congress [33] and also [46] for a detailed description and references for the numerical results). Furthermore, classical linearized stability analysis of the simplest vortex sheets reveals that they define a classical Hadamard ill-posed initial value problem like that for the Laplace equation, so both the computational and theoretical issues are extremely subtle. All of this complexity combines with the fact that vortex sheets are ubiquitous as approximations to velocity fields in engineering applications [11] to make the study of conceivable weak solutions for the 2-D Euler equations a fascinating contemporary research topic. One large research effort (see [11] for references) restricts the study of vortex sheets to analytic initial curves and attempts to use complex analysis to study the first relatively faint singularity formation in such vortex sheets. In typical computational and practical applications, vortex sheet initial data are usually not defined by real analytic curves [46, 38]. The methods and results described here [22, 21, 27, 1, 61] are completely different and address the global complexity of approximations for vortex sheets for all times with general (non-analytic) vortex sheet initial data through ideas naturally involving the weak topology.

Vortex sheets are regularized [21] by a sequence of globally smooth solutions of the 2-D Euler equations, $v^\varepsilon(x,t)$, satisfying the bounds,

$$\max_{0 \le t \le T} \left(\int_{|x| \le R} |v^\varepsilon(x,t)|^2 dx + \int_{R^2} |\omega^\varepsilon(x,t)| dx \right) \le C_{R,T}, \tag{2.1}$$

for any $R, T > 0$ with $\omega^\varepsilon = $ curl v^ε; a very general mathematical framework for these and other important regularizations satisfying (2.1) such as the high Reynolds number limit of 2-D Navier-Stokes and computational vortex methods is developed in [21]. If $v^\varepsilon(x,0) \rightharpoonup v_0$ where v_0 defines vortex sheet initial data, does $v^\varepsilon(x,t)$ converge to a weak solution of the 2-D Euler equations with this vortex sheet initial data? Are there new phenomena in the limiting process? There has been partial progress on the first question which is described here along with the fact that striking new mathematical phenomena do occur. In particular, one result in [21] is the existence of a measure-valued solution of 2-D Euler defined for all time with a special structure so that there are only concentrations and not oscillations (such as in (1.12)) in the limit as $\varepsilon \downarrow 0$ with the bounds in (2.1). The key fact proved in [21] is that with (2.1) there is a subsequence v^ε so that $v^\varepsilon \rightharpoonup v$ in L^2_{loc} and

$$\int_0^T \int_{|x| \le R} |v^\varepsilon - v| dx\, dt \to 0 \tag{2.2}$$

for any $T > 0, R$ so that in particular $v^\varepsilon \rightharpoonup v$ a.e. and only concentrations can occur (see (0.7)). The elementary examples for 2-D Euler described in (1.14)–(1.16) above satisfy the bounds in (2.1) and illustrate that the phenomena of concentration can occur with the natural bounds in (2.1) for vortex sheets. An important issue in deciding whether the function v defined in (2.2) is a suitable weak solution of 2-D Euler involves various estimates on the size of the concentration set [22, 21, 27, 1, 61]. These topics are discussed next.

2A. Size of the Weak Star and Reduced Defect Measures for the Problem of Vortex Sheets

The problem of existence for vortex sheet initial data with the bounds in (2.1) has some of the flavor of a problem involving critical Sobolev exponents. This fact is supported by the result proved in [21] that if the approximating sequence v^ε satisfies the L^p-bound on the initial vorticity, $\|\omega_0^\varepsilon\|_{L^p} \le C$ for some $p > 1$, in addition to (2.1), then v^ε converges strongly to v in L^2 and v defines a classical weak solution for 2-D Euler. Also recall that the bounds already presented in (1.3) in discussing concentration compactness resemble those in (2.1) except that in (2.1) only div $v^\varepsilon = 0$ and curl v^ε are bounded in L^1 rather than the complete gradient. The elementary example of concentration with positive vorticity in (1.14A) and (1.16) directly illustrates that the bounds in (2.1) for exact solutions of 2-D Euler are not sufficient to guarantee the stronger bound in (1.3A). Thus, one suspects that sequences of smooth solutions of 2-D Euler satisfying (2.1) will not have a weak * defect measure with the simple structure from (1.3B) involving a countable number of Dirac masses. C. Greengard and Thomann [27] have constructed a beautiful family of explicit examples which illustrate this point in dramatic fashion.

Theorem [27]. *Consider an arbitrary positive finite Radon measure σ on R^2, then there is a sequence of exact steady solutions of the 2-D Euler equations, v^ε satisfying the bounds in (2.1) with $v^\varepsilon \rightharpoonup 0$ in L^2 so that the weak*defect measure associated with this sequence is σ.*

Actually, the example in [27] produces uniform Lebesque measure for the unit square as weak* defect but it is an elementary exercise for the reader involving simple functions to deduce the more general result in the theorem from the construction in [27]; the construction superimposes "clouds of vortices" like the ones in (1.14B) but utilizes two scales in such elementary solutions in a clever fashion. In agreement with some of the trends of numerical solutions when the vorticity changes sign [46], the theorem indicates that almost arbitrary complexity can occur in the weak topology in concentrations when the bound in (2.1) is satisfied.

One of the main results in [22] is that despite this almost arbitrary complexity as regards the weak* defect measure for sequences v^ε with the bounds in (2.1), the reduced defect measure, θ, defined in (1.4) nevertheless concentrates on small sets. The following operational definition is useful and implicit in the constructions in [22].

Definition. The reduced defect measure, θ, concentrates on a *set with Hausdorff dimension p* provided that there is a constant C depending only on δ and closed sets F_r so that $\theta(F_r) = 0$ with the complement of F_r, F_r^c, satisfying $H_r^{p+\delta}(F_r^c) \leq C$ for any $\delta > 0$ and $r \leq r_0$. Here H_r^d denotes Hausdorff premeasure of dimensional d and level r.

In particular, $H_r^{p+2\delta}(F_r^c)$ vanishes as $r \downarrow 0$ for any $\delta > 0$. Recall from (1.5) that θ satisfies $\theta(F) \leq \sigma(F)$ for closed sets; thus, it is possible that even though σ has very large support and charges a closed set F_r, θ satisfies $\theta(F_r) = 0$. In fact, the following is true:

Theorem [22]. *A) Assume that v^ε is a sequence of steady time-independent velocity fields with div $v^\varepsilon = 0$ and satisfying the bounds in (2.1), then the reduced defect measure associated with a subsequence concentrates on a set with Hausdorff dimension zero.*

B) Assume that v^ε is a time-dependent approximate solution sequence for 2-D Euler (see [21, 22]) satisfying the bounds in (2.1), then the reduced defect measure concentrates on a set with space-time cylindrical Hausdorff dimension one.

The examples in (1.14) show that the theorem is sharp in general. Here are some brief comments on the proof of the theorem. There is a stream function ψ^ε so that $\Delta\psi^\varepsilon = \omega^\varepsilon$ and $v^\varepsilon = \nabla^\perp \psi^\varepsilon = \left(-\psi^\varepsilon_{x_2}, \psi^\varepsilon_{x_1}\right)$ and thus

$$v^\varepsilon(x,t) = \int_{R^2} K(x-y)\omega^\varepsilon(y,t)dy, \tag{2.3}$$

with kernel $K(x) = (2\pi|x|)^{-2}(-x_2, x_1)$. There are two main steps in the proof from [22] for time independent velocity fields. In the first step, (see Theorem 4.1 in [22], the radial distribution function of the vorticity at an arbitrary point is introduced,

$$\omega^\varepsilon(B(s,x)) = \int_{B(s,x)} |\omega^\varepsilon(y)| dy, \tag{2.4}$$

where $B(s,x)$ denotes the open ball of radius s centered at x. The key point is to prove a uniform decay estimate for the vorticity distribution function for an appropriate subsequence, i.e. for any $\delta > 0$ there exists a closed set F_R so that

$$F_R \subset \{x | \omega^\varepsilon(B(s,x)) \le K s^\delta, \quad 0 \le s \le 1\}, \tag{2.5}$$

and $H_R^{3\delta}(F_R^c) \le C$. Here the constants K and C depend on δ. For a fixed measure with finite total mass, the conclusion in (2.5) follows from standard covering arguments (see Stein [56]); however, it is *important* here *that the set F_R in (2.5) is uniform for the entire sequence* so a different proof is needed – Riesz potentials and a refined Chebyshev inequality for Hausdorff premeasure are utilized to achieve this. In the second step (see thm 6.1 in [22], the uniform decay estimate in (2.5) is used with the convolution representation in (2.3) yielding the improved velocity estimate

$$\int_{F_R} |v^\varepsilon|^{p'} \le C(R) \qquad \text{for some } p' > 2. \tag{2.6}$$

The facts in (2.6) and (2.2) are then combined with simple interpolation to establish that $\theta(F_R) = 0$ and this completes the proof of the theorem for steady velocity fields. For the time-dependent case of the theorem, these estimates are combined with uniform Hölder estimates in time

$$\left(\int_{B(R,0)} |v^\varepsilon(x,t_1) - v^\varepsilon(x,t_2)|^2 \right)^{1/2} \le C_{q,\alpha} |t_1 - t_2|^\alpha \tag{2.7}$$

for $1 < q < 2$, $0 < \alpha < 1$ which are valid for approximate solution sequences for 2-D Euler. One subtlety is that the estimates in (2.7) are also critical in the sense that $C_{q,\alpha} \nearrow \infty$ as $q \nearrow 2$ or $\alpha \nearrow 1$. Recently, L.C. Evans [25] has found another proof of part A) of the theorem for steady velocity fields by utilizing capacity theory.

2B. Concentration-Cancellation

The terminology "concentration-cancellation" was introduced in [21, 22] and describes the phenomena that even though $v^\varepsilon \otimes v^\varepsilon$ does not converge weakly to $v \otimes v$ due to the development of concentrations while $\|v^\varepsilon - v\|_{L^1} \to 0$, the velocity field v nevertheless defines a weak solution for 2-D Euler – in simple exact solutions such as those in (1.14), the concentrations in (1.15) actually cancel against test functions in the weak form in the limit (see [21]). At the present time, it is an important open problem to determine whether concentration-cancellation always occurs in the limit for solution sequences satisfying the bounds in (2.1) which are appropriate for vortex sheet initial data. Nevertheless, there are some interesting partial results [22, 27, 1, 61]. The first results are given in the following

Theorem [22]. *A) For steady solution sequences for 2-D Euler satisfying the bounds in (2.1), concentration-cancellation always occurs.*

B) For time-dependent solution sequences for 2-D Euler satisfying the bounds in (2.1) assume additionally that the reduced defect measure, θ, concentrates on a

set with Hausdorff dimension strictly less than one, then concentration-cancellation occurs.

Lack of space prevents a detailed discussion of the proof of the theorem but a few comments are appropriate. The proofs crucially exploit the special nonlinear covariant structure in the weak form of the Euler equations which is a direct consequence of the rotational invariance of solutions of (0.1), i.e. if $v^\varepsilon(x, t)$ satisfies 2-D Euler, so does ${}^t\sigma v^\varepsilon(\sigma x, t)$ for any rotation matrix σ. The proofs also crucially use the fact that the reduced defect measure, θ, concentrates on a small set (see the theorem in the previous subsection) to build a "shielding sequence" of test functions which shields the limit process from the concentration defects (see Section 3 of [22]). A simpler proof of part B) of the theorem was presented in [27]. Recently, in interesting work, Alinhac [1] and Zheng [61] have proved that concentration-cancellation occurs provided it is assumed that the weak star defect measure (not the reduced defect measure!) has support which is a set of roughly one-dimensional Hausdorff measure. These proofs use the "shielding sequence" strategy from [22] in an essential fashion. All of these partial results still leave unanswered the following basic issue:

Problem. Does concentration-cancellation always occur for time-dependent solution sequences with the bound in (2.1) when θ concentrates on a set with space-time cylindrical Hausdorff measure exactly one?

There is probably an affirmative answer to this question when the vorticity has a distinguished sign, i.e. $\omega^\varepsilon(x, t) \geq 0$. It should be mentioned here that for smooth solutions of 2-D Euler a non-negative vorticity at later times is automatically guaranteed by a non-negative initial vorticity, $\omega_0^\varepsilon(x)$ because the vorticity is transported along fluid particle trajectories. In particular, Theorem 3.1 from [21] guarantees that for $\omega^\varepsilon(x, t) \geq 0$ the solution sequence v^ε converges strongly to v in L^2 and v defines a weak solution of 2-D Euler provided that the vorticity maximal function using (2.4) decays at the extremely slow rate,

$$\max_{\substack{x \in R^2 \\ 0 \leq t \leq T}} \omega^\varepsilon(B(r, x), t) \leq C \left(\log \left(\frac{1}{r} \right) \right)^{-(1+\beta)}, \quad r \leq 1 \tag{2.8}$$

for some $\beta > 0$ and any $T > 0$. On the other hand, it is not difficult to prove [38] the following apriori decay estimate for this maximal function assuming positive vorticity distributions, $\omega^\varepsilon(x, t) \geq 0$ with bounded support:

$$\max_{\substack{x \in R^2 \\ 0 \leq t < +\infty}} \omega^\varepsilon(B(r, x), t) \leq C \left(\log \left(\frac{1}{r} \right) \right)^{-1/2}, \quad r \leq 1. \tag{2.9}$$

Thus, for positive signed vorticity, (2.8) and (2.9) indicate that there is an extremely small discrepancy between conditions guaranteeing strong convergence to a weak solution and the general case. Of course, the elementary examples in (1.14A) can be used to show that concentration does occur with positive vorticity distributions and also that the estimate in (2.9) is sharp so that some assessment of concentration-cancellation is still needed. On the other hand, the elementary examples in (1.14B) illustrate that there is no apriori decay rate for the vorticity maximal function when the vorticity

changes sign. The interested reader can consult [46] for a report on the striking computational evidence indicating vastly increased complexity in numerical
solutions when the vorticity changes sign.

References

[1]   Alinhac, S.: Un phenomene de concentration evanescente pour des flots non-
      stationnaires incompressibles en dimension deux. Comm. Math. Phys. **127** (1990)
      pp. 585–596
[2]   Almgren, R., Majda, A., Rosales R.: Rapid initiation through high frequency resonant
      nonlinear acoustics. Phys. Fluids A **2** (1990) 1014–1029
[3]   Artola, M., Majda, A.: Nonlinear development of instabilities in supersonic vortex
      sheets I: the basic kink modes. Physica **28D** (1988) 253–281
[4]   Artola, M., Majda, A.: Nonlinear development of instabilities in supersonic vortex
      sheets II: resonant interaction among kink modes. S.I.A.M. J. Appl. Math. **49** (1989)
      1310–1349
[5]   Ball, J.: Energy-minimizing configurations in nonlinear elasticity. Proceedings of
      I.C.M. Warsaw, 1983, pp. 1309–1314
[6]   Ball, J.: A version of the fundamental theorem for Young measures. (Lecture Notes in
      Physics, vol. 344, eds. M. Rascle, D. Serre, M. Slemrod). Springer, Berlin Heidelberg
      New York 1981, pp. 207–215
[7]   Ball, J., James, R.D.: Fine phase mixtures as minimizers of energy. Arch. Rat. Mech.
      Anal. **100** (1987) 13–52
[8]   Ball, J., James, R.D.: Proposed experimental tests of a theory of fine microstructure
      and the two-well problem. Preprint 1990
[9]   Brenier, Y.: The least action principle and the related concept of generalized flows
      for incompressible perfect fluids. J. Am. Math. Soc. **2** (1989) 225–255
[10] Brezis, H., Coron, J.M., Lieb, E.: Harmonic maps with defects. Comm. Math. Phys.
      107 (1987) 649–705
[11] Caflisch, R.: Mathematical aspects of vortex dynamics. S.I.A.M. Publications,
      Philadelphia 1989
[12] Cehelsky, P., Rosales, R.: Resonantly interacting weakly nonlinear hyperbolic waves
      in the presence of shocks: A single space variable in a homogeneous time independent
      medium. Stud. Appl. Math. **74** (1986) 117–138
[13] Chipot, M., Kinderlehrer, D.: Equilibrium configurations of crystals. Arch. Rat.
      Mech. Anal. **103** (1988) 237–277
[14] Chorin, A.J.: Estimates of intermittency, spectra and blow-up in developed turbulence.
      Comm. Pure Appl. Math. **34** (1981) 853–866
[15] Chorin, A.J.: The evolution of a turbulent vortex. Comm. Math. Phys. **83** (1982)
      517–535
[16] Collins, C., Luskin, M.: The computation of the austenitic-martensitic phase tran-
      sition. Lecture Notes in Physics, vol. 344 (eds. M. Rascle, D. Serre, M. Slemrod).
      Springer, Berlin Heidelberg New York 1989, pp. 34–50
[17] Constantin, P., Majda, A.: The Beltrami spectrum for incompressible fluid flows.
      Comm. Math. Phys. **115** (1988) 435–456
[18] Diperna, R.: Measure-valued solutions to conservation laws. Arch. Rat. Mech. Anal.
      88 (1985) 223–270
[19] Diperna, R.: Compactness of solutions to nonlinear P.D.E. Proceedings of I.C.M.
      Berkeley, 1986, pp. 1057–1063
[20] Diperna, R., Majda, A.: Oscillations and concentrations in weak solutions of the
      incompresible fluid equations. Comm. Math. Phys. **108** (1987) 667–689

[21] Diperna, R., Majda, A.: Concentrations in regularizations for 2-D incompressible flow. Comm. Pure Appl. Math. **40** (1987) 301–345
[22] Diperna, R., Majda, A.: Reduced Hausdorff dimension and concentration-cancellation for 2-D incompressible flow. J. Am. Math. Soc. **1** (1988) 59–95
[23] Diperna, R., Majda, A.: The validity of nonlinear geometric optics for weak solutions of conservation laws. Comm. Math. Phys. **98** (1985) 313–347
[24] Ericksen, J.L., Kinderlehrer, D. (eds.): Theory and application of liquid crystals, vol. 5. I.M.A. Volumes in Math. and Appl. Springer, New York
[25] Evans, L.C.: Weak Convergence Methods for Nonlinear Partial Differential Equations. Regional Conf. Series in Mathematics. Am. Math. Soc., Providence, RI, 1990
[26] Gerard, P.: Compacite par compensation et regularite 2-microlocale. Seminaire d'EDP (Ecole Polytechnique, Paris), et Article a paraitre 1988–1989
[27] Greengard, C., Thomann, E.: On Diperna-Majda concentration sets for two-dimensional incompressible flow. Comm. Pure Appl. Math. **41** (1988) 295–303
[28] Hunter, J., Keller, J.B.: Weak shock diffraction. Wave Motion **6** (1984) 79–89
[29] Hunter, J.K., Keller, J.B.: Weakly nonlinear high frequency waves. Comm. Pure Appl. Math. **36** (1983) 543–569
[30] Hunter, J.K., Majda, A., Rosales, R.: Resonantly interacting weakly nonlinear hyperbolic waves, II: several space variables. Stud. Appl. Math. **75** (1986) 187–226
[31] Hunter, J.: Nonlinear geometrical optics. To appear in I.M.A. volume in Math. and Appl. titled on *Multi-dimensional Hyperbolic Waves* edited by J. Glimm and A. Majda
[32] James, R.D., Kinderlehrer, D.: Theory of diffusionless phase transformations. (Lecture Notes in Physics, vol. 344, eds. M. Rascle, D. Serre, M. Slemrod). Springer, Berlin Heidelberg New York 1989, pp. 51–84
[33] Krasny, R.: Proceedings of this International Congress, p. 1573
[34] Landman, M., Papanicolaou, G.C., Sulem, C. Sulem, P.L.: Rate of blow-up for solutions of the nonlinear Schrödinger equation in critical dimension. Phys. Rev. **A 38** (1988) 3834–3843
[35] Lions, P.L.: The concentration-compactness principle in the calculus of variations: The limit case, Part I and Part II. Rev. Mat. Iberoamericana, vol. 1, no. 1, pp. 145–201 et no. 2, pp. 45–121 (1985)
[36] Majda, A.: Vorticity, turbulence, and acoustics in fluid flow. The 1990 John von Neumann lecture of the Society for Industrial and Applied Mathematics. To appear in S.I.A.M. Review in 1991
[37] Majda, A.: One perspective on open problems in multi-dimensional conservation laws. To appear in I.M.A. volume in Math. and Appl. titled *Multi-dimensional Hyperbolic Waves* edited by J. Glimm and A. Majda
[38] Majda, A.: Lecture notes on oscillations and concentrations for incompressible fluid flow. Princeton University (1988–1989 academic year)
[39] Majda, A.: Compressible fluid flow and systems of conservation laws in several space variables. (Applied Mathematical Sciences, vol. 53.) Springer, Berlin Heidelberg New York 1984
[40] Majda, A.: Nonlinear geometric optics for hyperbolic systems of conservation laws. In: Oscillation theory, computation, and methods of compensated compactness, IMA vol. 2. Springer, New York 1986 pp. 115–165
[41] Majda, A., Rosales, R.: Nonlinear mean field-high frequency wave interactions in the induction zone. S.I.A.M.J. Appl. Math. **47** (1987) 1017–1039
[42] Majda, A., Rosales, R.: Resonantly interacting weakly nonlinear hyperbolic waves, I: A single space variable. Stud. Appl. Math. **71** (1984) 149–179
[43] Majda, A., Rosales, R., Schonbek, M.: A canonical system of integro-differential equations arising in resonant nonlinear acoustics. Stud. Appl. Math. **79** (1988) 205–261

[44] Majda, A.: High mach number combustion. In: Reacting flows: Combustion and chemical reactors. AMS Lectures in Applied Mathematics **24** (1986) 109–184

[45] Majda, A., Rosales, R.: A Theory for spontaneous mach stem formation in reacting shock fronts, I: The basic perturbation analysis. S.I.A.M. J. Appl. Math **43** (1983) pp. 1310–1334

[46] Majda, A.: Vortex dynamics: Numerical analysis, scientific computing, and mathematical theory. Proceedings First International Congress for Industrial and Applied Mathematics. S.I.A.M. Publications, 1988, pp. 153–182

[47] Majda, A.: Vorticity and the mathematical theory of incompressible fluid flow. Comm. Pure Appl. Math., vol 39, 1986, pp. S187–S220

[48] Majda, A., Artola, M.: Nonlinear geometric optics for hyperbolic mixed problems. In: Analyse mathematique et applications. In honor of J.L. Lions, Gauthier-Villars, Paris 1988, pp. 319–356

[49] Maslov, V.P.: Mathematical aspects of integral optics. Moscow VINITI 1983

[50] Merle, F.: On uniqueness and continuation properties after blow-up time of self-similar solutions of nonlinear Schrödinger equation with critical exponent and critical mass. Preprint 1990

[51] Mimura, M.: Proceedings of this International Congress, p. 1627

[52] Rosales, R., Majda, A.: Weakly nonlinear detonation waves. S.I.A.M. J. Appl. Math. **43** (1983) 1086–1118

[53] Rosales, R.: A introduction to weakly nonlinear geometrical optics. To appear in I.M.A. volume on *Multi-dimensional Hyperbolic Waves* edited by J. Glimm and A. Majda

[54] Serre, D.: Richness and the classification of quasilinear hyperbolic systems. Preprint 1989

[55] Serre, D.: Oscillations non lineaires des systemes hyperboliques: Methods et resultats qualitatifs. Preprint 1989

[56] Stein, E.M.: Singular integrals and the differentiability properties of functions. Princeton University Press, NJ, 1970

[57] Tartar, L.: H-measures, a new approach for studying homogenization, oscillations and concentration effects in partial differential equations. To appear in Proc. Roy. Soc. Edinburgh

[58] Tartar, L.: Proceedings of this International Congress

[59] Weinstein, M.I.: Nonlinear Schrödinger equations and sharp interpolation estimates. Comm. Math. Phys. **87** (1983) 567–576

[60] Weinstein, M.I.: On the structure and formation of singularities in solutions to the nonlinear dispersive evolution equations. Comm. Partial Differential Equations **11** (1986) 545–565

[61] Zheng, Y.: Concentration-cancellation phenomena for weak solutions to certain nonlinear partial differential equations. Ph.D. thesis, Univ. of Calif. at Berkeley (April 1990)

Dynamical and Ergodic Properties
of Subgroup Actions on Homogeneous Spaces
with Applications to Number Theory

Grigorii A. Margulis

Institute for Problems in Information Transmission, Academy of Sciences of USSR
ul. Ermolovoi 19, Moscow 101447, USSR

A subgroup action on a homogeneous space is a classical object of study in ergodic theory. Some old and new results about ergodic properties of this section will be stated. We shall also formulate results on orbit closures and give applications of these results to number theory and, in particular, to the study of sets of values of indefinite quadratic forms at integral points. In the last section, formulations of some new results about actions on general manifolds and measure spaces will be given.

1. Ergodicity Theorems

Let G be a Lie group, F a subgroup of G and D a closed subgroup of G. Let us consider the natural action of F by left translations on the homogeneous space G/D. This action is called *ergodic* if, for every F-invariant Borel subset A of G/D, either $\mu(A) = 0$ or $\mu((G/D) - A) = 0$ where μ is a G-quasiinvariant measure on G/D. One can easily prove the following two assertions.

Proposition 1. *The action of F on G/D is ergodic if and only if the closure of F acts ergodically on G/D.*

Proposition 2 (Duality Principle). *Let us assume that F is closed. Then the action of F on G/D is ergodic if and only if the action of D on G/F is ergodic.*

We say that the homogeneous space G/D has finite volume if there exists a finite G-invariant measure on G/D. If G/D *has finite volume* and D is discrete then D is called a *lattice*.

1.1 Mautner Property

If G/D carries a G-invariant measure μ then one can define a continuous unitary representation ϱ of the group G on the Hilbert space $L_2(G/D, \mu)$ of square μ-integrable functions on G/D by the formula $(\varrho(g)f)(x) = f(g^{-1}x)$ where $f \in L_2(G/D, \mu)$, $g \in G$ and $x \in G/D$. It turns out that many ergodic properties of the action of F on G/D

Proceedings of the International Congress
of Mathematicians, Kyoto, Japan, 1990
© The Mathematical Society of Japan, 1991

can be reformulated in terms of the representation ϱ. This observation gives an approach to the study of ergodic properties based on representation theory; this approach was first used by I.M. Gelfand and S.V. Fomin (see [28]).

One can easily see that, in the case where G/D has finite volume, the action of F on G/D is ergodic if and only if each $\varrho(F)$-invariant function $f \in L_2(G/D, \mu)$ is constant or, in other words, if f is $\varrho(G)$-invariant. Therefore Lemma 1 stated below is true. Before the formulation of this lemma we give

Definition 1. We say that a subgroup H of G has *Mautner property* with respect to F if, for any continuous unitary representation φ of G on a Hilbert space W and any $w \in W$ such that $\varphi(F)w = w$, we have that $\varphi(H)w = w$.

Lemma 1. *Let the homogeneous space G/D have finite volume, and let H be a subgroup of G. Let us assume that H has Mautner property with respect to F and that the closure of the subgroup generated by F and H acts ergodically on G/D. Then the action of F on G/D is ergodic.*

Mautner property was first used by F.J. Mautner to prove the ergodicity of geodesic flows on symmetric Riemannian spaces (see [43]). C.C. Moore proved Theorem 1 formulated below about Mautner property of Ad-compact normal subgroups. Let us give at first the definition of these subgroups.

Definition 2. The Ad-*compact normal subgroup* for F is the smallest subgroup in the class of all connected normal subgroups H of G satisfying the following condition

(A) for each g from the closure of F, the linear transformation $\mathrm{Ad}(\pi(g))$ is diagonalizable and its eigenvalues have absolute value 1 where Ad is the adjoint representation of G/H and $\pi: G \to G/H$ is the canonical epimorphism.

Remark. It is not difficult to prove that the condition (A) is equivalent to the condition

(A') the subgroup $\mathrm{Ad}(\pi(F))$ is relatively compact in the group $\mathbf{GL}(\mathfrak{G}/\mathfrak{H})$ of linear transformations of the Lie algebra $\mathfrak{G}/\mathfrak{H}$ of the group G/H.

Let us denote the Ad-compact normal subgroup for F by H_F. If G is connected and semisimple then H_F is the product of all almost simple factors G_i of G such that the restriction of $\mathrm{Ad}(F)$ to the Lie algebra $\mathfrak{G}_i$ of G_i is not relatively compact in $\mathbf{GL}(\mathfrak{G}_i)$.

Theorem 1 (see [45]). *The subgroup H_F has Mautner property with respect to F.*

Remark 1. It is not difficult to show that the closure of the subgroup generated by F and H_F is the maximal subgroup of G having Mautner property with respect to F.

Remark 2. An analogue of Theorem 1 for semisimple groups over locally compact fields of characteristic zero was proved by S.G. Dani in [10].

Roughly speaking, the proof of Theorem 1 is based on the following simple

Lemma 2. *Let G be a topological group, F a subgroup of G, φ a continuous unitary representation of G on a Hilbert space W, and let $g_0 \in G$ and $w \in W$. Assume that, for any neighbourhood $V \subset G$ of the identity $e \in G$, g_0 is contained in he closure of the subset FVF. Then $\varphi(g_0)w = w$ whenever $\varphi(F)w = w$.*

To prove Lemma 2, it is enough to consider the function $f(g) = \langle \varphi(g)w, w \rangle$, $g \in G$, and remark that (a) since φ is unitary and $\varphi(F)w = w$, the function f is constant on a double coset modulo F; (b) the function f is continuous; (c) $f(e) = \|w\|^2$ and $f(g) = \|w\|^2$ if and only if $\varphi(g)w = w$. A special case of Lemma 2 is the following

Lemma 3 (Generalized Mautner Lemma). *Let G be a topological group and let x, $y \in G$ be the elements such that the sequence $\{x^n y x^{-n}\}$ converges to e as $n \to +\infty$. If φ is a continuous unitary representation of the group G on a Hilbert space W, $w \in W$ and $\varphi(x)w = w$ then $\varphi(y)w = w$.*

Example 1. Let $G = \mathbf{SL}_2(\mathbb{R})$ be the group of real unimodular 2×2 matrices and let $g = \begin{pmatrix} a & 0 \\ 0 & a^{-1} \end{pmatrix} \in \mathbf{SL}_2(\mathbb{R})$ where $|a| \neq 1$. Let F denote the cyclic subgroup generated by g, and let

$$U = \left\{ \begin{pmatrix} 1 & t \\ 0 & 1 \end{pmatrix} \middle| t \in \mathbb{R} \right\} \quad \text{and} \quad U^- = \left\{ \begin{pmatrix} 1 & 0 \\ t & 1 \end{pmatrix} \middle| t \in \mathbb{R} \right\}.$$

One can easily check that if $y \in U \cup U^-$ then one of the sequences $\{g^n y g^{-n}\}$ and $\{g^{-n} y g^n\}$ converges to e as $n \to +\infty$. In view of Lemma 3, it implies that both U and U^- have Mautner property with respect to F. But these subgroups generate G. Hence G has Mautner property with respect to F.

Example 2. Let G, g, a, F and U denote the same as in Example 1. If $c \neq 0$ we have that

$$\begin{pmatrix} 1 & c^{-1}(1-a) \\ 0 & 1 \end{pmatrix} \begin{pmatrix} a & 0 \\ c & a^{-1} \end{pmatrix} \begin{pmatrix} 1 & c^{-1}(1-a^{-1}) \\ 0 & 1 \end{pmatrix} = \begin{pmatrix} 1 & 0 \\ c & 1 \end{pmatrix}$$

and hence $\begin{pmatrix} a & 0 \\ c & a^{-1} \end{pmatrix} \in U \begin{pmatrix} 1 & 0 \\ c & 1 \end{pmatrix} U$. It implies that, for any neighbourhood $V \subset G$ of e, g is contained in the closure of the subset UVU. Now applying Lemma 2, we get that F has Mautner property with respect to U. On the other hand according to Example 1, G has Mautner property with respect to F. Hence G has Mautner property with respect to U.

1.2 Criteria of Ergodicity for Semisimple G

Theorem 1 and Lemma 1 imply

Theorem 2 (see [45]). *Let G be a Lie group, F a subgroup of G, and D a closed subgroup of G such that G/D has finite volume. Let H_F denote the* Ad-*compact normal subgroup for F. Then the action of F on G/D is ergodic if and only if the closure of the subgroup FH_F acts ergodically on G/D.*

Using Theorem 2 and Propositions 1 and 2, it is not difficult to prove

Theorem 3 (see [44], [45] and [46]). *Let G, F, D and H_F be the same as in Theorem 2.*

(a) *If the subgroup DH_F is dense in G then the action of F on G/D is ergodic.*
(b) *Let us assume that G is semisimple and that the action of F on G/D is ergodic. Then DH_F is dense in G.*

Definition 2. A lattice $\Gamma \subset G$ in a connected semisimple Lie group without compact factors is *reducible* if G admits connected normal subgroups H, H' such that $HH' = G$, $H \cap H'$ is discrete and $\Gamma/(\Gamma \cap H)(\Gamma \cap H')$ is finite. A lattice is *irreducible* if it is not reducible.

It is known (see [54, Corollary 5.21]) that if a lattice $\Gamma \subset G$ in a connected semisimple Lie group without compact factors is irreducible then, for any nontrivial connected normal subgroup H of G, the subgroup ΓH is dense in G. Therefore Theorem 3 implies

Theorem 4 (see [44] or [80, Theorem 2.2.6]). *Let G be a connected semisimple Lie group without compact factors, Γ an irreducible lattice in G, and $F \subset G$ a subgroup which is not relatively compact in G. Then the action of F on G/Γ is ergodic.*

Let us note that Theorem 4 is used in an essential way in the proofs of the strong rigidity theorem and the superrigidity theorem (see [40] and [47]).

Example 3. Let G, g, F and U denote the same as in Example 1 and let Γ be a lattice in G. Then in view of Theorem 4, both F and U act ergodically on G/Γ. This assertion follows also from Lemma 1 and Examples 1 and 2. Let us note that the ergodicity of actions of F and U on G/Γ implies (in fact, is equivalent to) the classical results of E. Hopf and G.A. Hedlund on ergodicity of geodesic and horocycle flows on a surface of constant negative curvature and finite area (see [29]).

1.3 Criteria of Ergodicity for Nilpotent and Solvable G

For the case of nilpotent G, there is a good ergodicity criterion due to L. Green. Before we formulate this criterion, let us recall that (a) G/D is compact if and only if G/D has finite volume; (b) whenever G/D is compact, the subgroup $D \cdot [G, G]$ is closed and, consequently, $G/D \cdot [G, G]$ is a torus where $[G, G]$ denotes the commutator subgroup of G (see [54, Chap. II]).

Theorem 5 (see [1, Chap. V]). *Let G be a connected nilpotent Lie group, F a subgroup of G, and D a closed subgroup of G such that G/D is compact. The action of F on G/D is ergodic if and only if $\pi(F)$ is dense in the torus $G/D \cdot [G, G]$ where $\pi : G \to G/D \cdot [G, G]$ denotes the canonical epimorphism.*

Remark 1. In [1], Theorem 5 is formulated only in the case of one-parameter F. But the proof given there is applicable in general case.

Remark 2. It is possible to deduce Theorem 5 from Theorem 3(a).

If G is solvable but not nilpotent then the ergodicity criterion is rather complicated (see Appendix 1 to the Russian translation of [1], [5] or [67]). This criterion is based on the reduction to the case of nilpotent G. The reduction consists of two steps: (1) the reduction to the case of solvable groups of type (R) (by definition, a group G is of type (R) if, for each $g \in G$, all eigenvalues of Ad g have absolute value 1); (2) the reduction from the case of solvable groups of type (R) to the case of nilpotent groups. The first reduction is based on Mautner lemma (Lemma 3), and the second one is based on Malcev's construction of semisimple splitting.

1.4 Ergodicity Criterion for Arbitrary G

Let G be a connected Lie group and D a closed subgroup of G. By a quotient of G/D we mean a homogeneous space G/B where $B \subset G$ is a closed subgroup which contains D. If B contains a closed normal subgroup L of G such that the factor group G/L is Euclidean (resp. semisimple) then the quotient G/B is called *Euclidean* (resp. *semisimple*); a connected solvable Lie group is called *Euclidean* if it is locally isomorphic to an extension of a vector group by a compact commutative Lie group. It is clear that there exists the maximal Euclidean (resp. semisimple) quotient G/B of G/D, i.e. if G/B' is another Euclidean (resp. semisimple) quotient of G/D then $B' \supset B$ or, in other words, G/B' is a quotient of G/B. Let us note that if G/D has finite volume then arbitrary (resp. arbitrary Euclidean) quotient of G/D has finite volume (resp. is compact).

Theorem 6 (see [5] and [12]). *Let G be a connected Lie group, $F \subset G$ a subgroup which is either one-parameter or cyclic, and $D \subset G$ a closed subgroup such that G/D has finite volume. The action of F on G/D is ergodic if and only if the actions of F on the maximal Euclidean and semisimple quotients of G/D are both ergodic.*

Remark. It is assumed in [5] and [12] that G/D is so called admissible homogeneous space. But according to [66], this condition is satisfied whenever G/D has finite volume.

1.5 Ergodic Decomposition of Nonergodic Flows

A.N. Starkov obtained a number of results on the ergodic decomposition of nonergodic flows (i.e. nonergodic actions of one-parameter subgroups). Let us formulate some of these results in the form of the following

Theorem 7 (see [65], [67] and [69]). *Let G be a connected Lie group, $F = \{f_t\}$ a one-parameter subgroup of G, and D a closed subgroup of G such that G/D has finite volume. Then there exists two partitions E and $\widetilde{E}$ of G into closed F-invariant submanifolds $E(x)$ and $\widetilde{E}(x)$ with the following properties*

(i) $x \in E(x) \subset \widetilde{E}(x)$ *for each* $x \in G/D$;

(ii) $E(x) = \widetilde{E}(x)$ *for almost all $x \in G/D$ (i.e. for all $x \in (G/D) - A$ where $\mu(A) = 0$ and μ is a G-invariant measure on G/D);*

(iii) *for each $x \in G/D$, there exist smooth finite F-invariant measures μ_x on $E(x)$ and $\widetilde{\mu}_x$ on $\widetilde{E}(x)$;*

(iv) *for each $x \in G/D$, the action of F on $E(x)$ is ergodic (relative to the measure μ_x);*

(v) $\widetilde{E}(x) = \overline{FH_F x}$ *and* $E(x) = \overline{FI_F x}$ *for each $x \in G/D$ where $\overline{A}$ denotes the closure of a subset A, H_F denotes the Ad-compact normal subgroup for F, and $I_F \subset H_F$ is a normal subgroup of G. (The subgroup I_F is called the Dani subgroup and is characterized as the smallest connected normal subgroup of G such that, for each $g \in F$, the operator Ad g have on $\mathfrak{G}/\mathfrak{I}$ only eigenvalues with absolute value 1 and this operator is diagonalizable on $\mathfrak{G}/\mathfrak{R} + \mathfrak{I}$ where $\mathfrak{G}$ (resp. $\mathfrak{I}$) is the Lie algebra of G (resp. I_F) and $\mathfrak{R}$ is the radical of $\mathfrak{G}$).*

(vi) *for each $x \in G/D$, there exist a connected Lie group G_x, a one-parameter subgroup $F_x = \{f_{x,t}\} \subset G_x$, a closed subgroup $D_x \subset G_x$, and a finite covering $\pi: G_x/D_x \to E(x)$ such that G_x/D_x has finite volume, the action of F_x on G_x/D_x is ergodic, and $\pi(f_{x,t}y) = f_t\pi(y)$ for each $y \in G_x/D_x$.*

2. Spectrum and Mixing

We refer to [9] and [63] about the notions of pure point spectrum, Lebesgue spectrum, K-flow, Bernoulli shift etc. and about results from ergodic theory.

2.1 Spectrum of Flows on Homogeneous Spaces

Let G be a connected Lie group, $F = \{f_t\}$ a one-parameter subgroup of G, and D a closed subgroup of G such that G/D has finite volume. Let μ denote the normalized $(\mu(G/D) = 1)$ Borel G-invariant measure on G/D. As in Sect. 1, let us define the unitary representation ϱ on $L_2(G/D, \mu)$. For any quotient $E = G/B$ of G/D, let

$$L_2(E) = \{f \in L_2(G/D, \mu) | f(g_1 D) = f(g_2 D) \text{ if } g_1 \in g_2 B\}.$$

It is clear that $L_2(E)$ is $\varrho(F)$-invariant.

Theorem 8 (see [5]). *Let E denote the maximal Euclidean quotient of $M = G/D$. Suppose that the action of F on M is ergodic. Then the restriction of $\varrho(F)$ to $L_2(E)$ has pure point spectrum (the set of eigenvalues being a finitely generated subgroup of the real line of rank equal to the dimension of E). If $L_2(E) \neq L_2(M)$ i.e. if M is not Euclidean, the restriction of $\varrho(F)$ to the orthogonal complement $L_2(M) \ominus L_2(E)$ has Lebesgue spectrum of infinite multiplicity.*

Remark 1. It follows from Theorems 7 and 8 that, also in the case where the action of F on G/D is not ergodic, the space $L_2(G/D, \mu)$ can be decomposed into the direct sum of two $\varrho(F)$-invariant subspaces H_1 and H_2 such that the restriction of $\varrho(F)$ to H_1 has pure point spectrum and the restriction of $\varrho(F)$ to H_2 has Lebesgue spectrum of infinite multiplicity.

Remark 2. The proof of Theorem 8 is based on some results about unitary representations of semi-direct products and about the spectrum of K-flows.

Remark 3. Theorem 8 generalizes some earlier results about the spectrum of flows on homogeneous spaces (see [1, 10, 12, 28, 43, 44, 45, 53, 61, 70, 71]).

2.2 Mixing Properties

Theorem 8 implies the following

Theorem 9 (see [5]). *Suppose that the action of F on G/D is ergodic and that there are no nontrivial Euclidean quotients of G/D. Then the section of $F = \{f_t\}$ on G/D is mixing, i.e.*

$$\lim_{t \to \infty} \mu(f_t A \cap B) = \mu(A)\mu(B)$$

for any measurable susets $A, B \subset G/D$.

Analogues of Theorems 8 and 9 for cyclic F are proved in [12]. In [10, 11] and [12], S.G. Dani obtained criteria when the translation $T_g : G/D \to G/D$, $T_g(x) = gx$, $x \in G/D$, is a K-automorphism, when T_g is a Bernoulli shift, and when T_g has positive entropy. In particular, he proved that T_g has positive entropy if and only if $|\lambda| \neq 1$ for at least one eigenvalue λ of Ad g (the same is true for algebraic groups over nonarchimedean locally compact fields).

B. Marcus proved that, in many cases, the action of F on G/D is mixing of all degrees. In particular, he proved

Theorem 10 (see [34]). *Suppose that the group G is semisimple and has no compact factors and that the action of $F = \{f_t\}$ on G/D is ergodic. Then this action is mixing of all degrees, i.e.*

$$\lim_{t_1,\ldots,t \to +\infty} \mu(A_0 \cap f_{t_1} A_1 \cap f_{t_1+t_2} A_2 \cap \cdots \cap f_{t_1+t_2+\cdots+t_r} A_r) = \mu(A_0) \cdot \ldots \cdot \mu(A_r)$$

for any integer $r \geq 1$ and any measurable subsets $A_0, A_1, \ldots, A_r \subset G/D$.

Combining Theorem 4 and Theorem 10, we get

Theorem 11. *Let G be a connected semisimple Lie group without compact factors, Γ an irreducible lattice in G, and $F \subset G$ a one-parameter subgroup which is not relatively compact in G. Then the action of F on G/Γ is mixing of all degrees.*

A special case of Theorem 11 is

Corollary (see [34]). *The horocycle flow on a surface of constant negative curvature and finite area is mixing of all degrees.*

This corollary proves the conjecture which was posed by Ya.G. Sinai on the International congress of mathematicians in Stockholm (1962). In [34], B. Marcus stated the conjecture that "mixing $\Rightarrow$ mixing of all degrees" for the action of F on G/D.

3. Invariant Measures for Actions of Unipotent Subgroups

Let G be a Lie group. An element u of G will be called *unipotent* if the transformation Ad u of the Lie algebra of G is unipotent, i.e. if all eigenvalues of Ad u are equal to 1. A subgroup U of G will be called *unipotent* if it consists of unipotent elements.

3.1 Description of Finite Ergodic Measures

Let F be a subgroup of G and D a closed subgroup of G. A F-invariant Borel measure μ on G/D is called F-*ergodic* if the action of F on G/D is ergodic relative to μ, i.e., if for every F-invariant Borel subset A of G/D, either $\mu(A) = 0$ or $\mu((G/D) - A) = 0$. Let us note that each F-invariant Borel measure on G/D can be decomposed into a (continuous) sum of F-ergodic measures. It seems impossible to describe in the general case all F-invariant Borel probability measures on G/D. But in view of the following recent fundamental theorem of M. Ratner, this is possible in the case where F is unipotent and D is discrete.

Theorem 12 (see [58, 59] and [60]). *Let G be a Lie group, Γ a discrete subgroup of G (not necessarily a lattice) and U a unipotent subgroup of G. Then for any U-invariant U-ergodic Borel probability measure σ on G/Γ, there exist $x \in G/\Gamma$ and a closed subgroup $H \subset G$ containing U such that the set Hx is closed in G/Γ and σ is a finite H-invariant measure the support of which is Hx. (In other words, H_x is a lattice in H and the measure σ is the image of a H-invariant measure on H/H_x under the canonical map $H/H_x \to Hx$ where $H_x = \{h \in H | hx = x\}$ is the stabilizer of x in H.)*

Roughly speaking, the difference between the case of unipotent subgroups and the general case is explained by the fact that, for an action of unipotent subgroups, the divergence of trajectories is "polynomial" in contrast to an action of general F where the divergence of trajectories is usually exponential.

Theorem 12 proves a conjecture which M. Ratner suggests to call Raghunathan's measure conjecture. It was first posed in [13]. The conjecture was suggested by the results of H. Furstenberg and W. Parry for nilpotent G and of S.G. Dani for the case where G is reductive and U is horosprerical (see [13], if G/Γ is compact see also [72]; a subgroup $U \to G$ is called *horospherical* if there exists $g \in G$ such that

$$U = \{u \in G | g^j u g^{-j} \to e \text{ as } j \to +\infty\}.$$

Generalizing a result of H. Furstenberg from [27], S.G. Dani proved the conjecture for the case where $G = \mathbf{SL}_2(\mathbb{R})$ (see [14]).

Theorem 12 provides many important ergodic theoretic consequences. In particular, it implies the joinings theorem and the factor theorem (see [59]). The joinings theorem describes invariant ergodic measures for a product of two actions of cyclic unipotent groups, and the factor theorem describes U-invariant measurable partitions of G/Γ where U is a cyclic unipotent subgroup which acts ergodically on G/Γ. These theorems generalize analogous results for $G = \mathbf{SL}_2(\mathbb{R})$ obtained by M. Ratner (see [56] and [57]). Let us formulate a corollary of the joinings theorem.

Theorem 13 (The Rigidity Theorem; see [59]). *Let G_i be a connected semisimple Lie group, Γ_i a lattice in G_i containing no non-trivial normal subgroups of G_i and u_i a unipotent element of G_i, $i = 1, 2$. Let U_i denote the cyclic subgroup generated by u_i. Suppose that the action of U_1 on G_1/Γ_1 is ergodic and there is a measure preserving map $\psi : G_1/\Gamma_1 \to G_2/\Gamma_2$ such that $\psi(xu_1) = \psi(x)u_2$ for μ_1-almost every $x \in G_1/\Gamma_1$, where μ_1 denotes a G_1-invariant measure on G_1/Γ_1. Then one can find $h \in G_1$ and a surjective homomorphism $\alpha : G_1 \to G_2$ such that $\alpha(\Gamma_1) \subset h\Gamma_2 h^{-1}$ and $\psi(g\Gamma_1) = \alpha(g)h\Gamma_2$ for μ_1-almost every $g\Gamma_1 \in G_1/\Gamma_1$. Also α is a local isomorphism whenever ψ is finite to one or G_1 is simple and it is an isomorphism whenever ψ is one-to-one or G_1 is simple with trivial center.*

Theorem 13 generalizes Ratner's rigidity theorem for $G_i = \mathbf{SL}_2(\mathbb{R})$ (see [55]). This theorem had been previously obtained by D. Witte (see [78] and [79]) with the help of methods from [55] and [57].

3.2 Finiteness of Ergodic Measures

Theorem 12 gives the description of finite ergodic measures for actions of unipotent groups. Now let us formulate a theorem about the finiteness of ergodic measures for these actions.

Theorem 14 (see [15]). *Let G be a connected Lie group, Γ a lattice in G, and U a unipotent subgroup of G. Let ν be a locally finite U-invariant measure on G/Γ. Then there exist Borel U-invariant subsets X_i, $i \in \mathbb{N}$, such that $\nu(X_i) < \infty$ for all i and $G/\Gamma = \bigcup_{i=1}^{\infty} X_i$. In particular every locally finite U-ergodic U-invariant measure is finite.*

For the case where $G = \mathbf{SL}_n(\mathbb{R})$ and $\Gamma = \mathbf{SL}_n(\mathbb{Z})$, Theorem 14 is essentially equivalent to the following assertion: for any lattice Λ in R^n and any unipotent one-parameter subgroup $\{u_t\}$ of $\mathbf{SL}_n(\mathbb{R})$, one can find $\delta > 0$ such that the set

$$\Omega(\Lambda, \delta) = \{t \geq 0 \mid \|u_t z\| > \delta \text{ for all } z \in \Lambda - \{0\}\}$$

has positive density. A strengthened version of this assertion is proved in [17] and, in a weak form, it is proved in [35] (namely it is proved in [35] that $\Omega(\Lambda, \delta)$ is unbounded for some $\delta = \delta(\Lambda) > 0$).

Using Theorem 1, one can deduce from Theorem 14

Theorem 15. *Let G be a connected Lie group, Γ a lattice in G, and H a connected subgroup of G such that the quotient of H by its unipotent radical is semisimple (by the unipotent radical of H we mean the maximal connected normal unipotent subgroup of H). Let v be a locally finite H-invariant measure on G/Γ. Then there exist Borel H-invariant subsets X_i, $i \in \mathbb{N}$, such that $v(X_i) < \infty$ for all i and $G/\Gamma = \bigcup_{i=1}^{\infty} X_i$. In particular every locally finite H-ergodic H-invariant measure is finite.*

One can see that Theorem 15 differs from Theorem 14 only in the replacement of a unipotent subgroup U by a connected subgroup H such that the quotient of H by its unipotent radical is semisimple.

The following is a special case of Theorem 15.

Theorem 16. *Let G, Γ and H be the same as in Theorem 15, and let $x \in G/\Gamma$. Suppose that the orbit Hx is closed in G/Γ. Then $H \cap G_x$ is a lattice in G where $G_x = \{g \in G | gx = x\}$ is the stabilizer of x.*

Applying Theorem 16 in the case where $G = \mathbf{SL}_n(\mathbb{R})$, $\Gamma = \mathbf{SL}_n(\mathbb{Z})$, $x = e\Gamma$, and $H \subset G$ is the set of $\mathbb{R}$-rational points in a $\mathbb{Q}$-subgroup of G, we obtain the following theorem of Borel and Harish-Chandra.

Theorem 17. *Let H be a connected $\mathbb{Q}$-group. Suppose that the quotient of H by its unipotent radical is semisimple. Then $H(\mathbb{Z})$ is a lattice in $H(\mathbb{R})$.*

Remark. In the most general form of Borel-Harish-Chandra theorem, the condition "the quotient of H by its unipotent radical is semisimple" is replaced by a weaker condition "H has no non-trivial $\mathbb{Q}$-rational characters".

4. Orbit Closures

4.1 Raghunathan's Conjecture

Roughly speaking, ergodicity theorems of Sect. 1 claim that "almost all" orbits of subgroup actions on a homogeneous space are dense. But these theorems say nothing about the behaviour of "individual" orbits. M.S. Raghunathan stated a conecture about closures of orbits of unipotent groups and also noted the connection of his conjecture with Oppenheim-Davenport conjecture.

Conjecture 1 (Raghunathan's Conjecture). *Let G be a connected Lie group, Γ a lattice in G and U a unipotent subgroup of G. Then for any $x \in G/\Gamma$, there exists a closed subgroup $L = L(x) \subset G$ containing U such that the closure of the orbit Ux coincides with Lx.*

Conjecture 1 can be generalized in the following way:

Conjecture 2 (Generalized Raghunathan's Conjecture). *Let G be a connected Lie group, Γ a lattice in G and H a subgroup of G. Suppose that H is generated by unipotent* elements. Then for any $x \in G/\Gamma$ there exists a closed subgroup $L = L(x) \subset G$ containing H such that the closure of the orbit Hx coincides with Lx.

Raghunathan's conjecture has been proved in the following cases: (a) G is reductive and U is horospherical (see [18]); (b) $G = \mathbf{SL}_3(\mathbb{R})$ and $U = \{u(t)\}$ is a one-parameter unipotent subgroup of G such that $u(t) - I$ has rank 2 for all $t \neq 0$ where I is the identity matrix (see [21]); (c) G is solvable (see [68] and [69]).

Let G, Γ and U be the same as in Conjecture 1. A closed U-invariant subset $Y \subset G/\Gamma$ is called U-*minimal* if it does not contain any proper closed U-invariant subset or, equivalently, if Uy is dense in Y for any $y \in Y$. It is proved in [38] that any closed U-invariant U-minimal subset of G/Γ is compact. On the other hand since the unipotent subgroup U is nilpotent and, consequently, amenable, we have that, for any compact U-invaraint subset $Y \subset G/\Gamma$, there exists a U-invariant Borel probability measure on Y. In view of these observations, Theorem 12 implies that any closed U-invariant U-minimal subset of G/Γ is an orbit of a subgroup $L \subset G$. This proves Conjecture 1 in the case where the closure of Ux is U-minimal.

Remark 1. It is not possible to assume in Conjecture 2 that H is an arbitrary subgroup of G. For example, if

$$G = \mathbf{SL}_2(\mathbb{R}), \qquad \Gamma = \mathbf{SL}_2(\mathbb{Z}), \qquad D = \left\{ \begin{pmatrix} \lambda & 0 \\ 0 & \lambda^{-1} \end{pmatrix} \middle| \lambda \in \mathbb{R}, \lambda \neq 0 \right\}.$$

then there exists $x \in G/\Gamma$ such that the closure of Dx is not a manifold. On the other hand in Conjecture 2, the condition "H is generated by unipotent elements" is not the most general of all possible conditions. Let $G = \mathbf{SL}_2(\mathbb{R})$, let Γ be a lattice in G, and let $P = \left\{ \begin{pmatrix} a & b \\ 0 & a^{-1} \end{pmatrix} \middle| a, b \in \mathbb{R}, a \neq 0 \right\}$. Then for any point $x \in G/\Gamma$, the orbit Px is dense in G/Γ. Also it seems plausible that if G is a semisimple Lie group without compact factors, $D \subset G$ is a maximal $\mathbb{R}$-split torus in G, $\Gamma \subset G$ is an irreducible lattice, $x \in G/\Gamma$, Dx is relatively compact in G/Γ and $\mathbb{R}$-rank of G is greater than 1, then the closure of Dx is an orbit of a subgroup $L \subset G$.

Remark 2. Let $L \subset G$ be a closed subgroup, $x \in G/\Gamma$ and $G_x = \{g \in G | gx = x\}$. Then the orbit Lx is closed in G/Γ if and only if the canonical map $L/L \cap G_x \to G/\Gamma$ is proper. Using this observation and Theorem 15, it is not difficult to prove that, in Conjectures 1 and 2, Lx admits a finite L-invariant measure or, equivalently, $L \cap G_x$ is a lattice in L.

Remark 3. Using Remark 2 and Borel's density theorem (see [3]), it is not difficult to prove that if $G = \mathbf{SL}_n(\mathbb{R})$, $\Gamma = \mathbf{SL}_n(\mathbb{Z})$ and $x = g\Gamma \in G/\Gamma$ then, in Conjectures 1 and 2, the subgroup $g^{-1}L(x)g$ is algebraic and defined over $\mathbb{Q}$ (more precisely, $g^{-1}L(x)g$ is commensurable with the set of $\mathbb{R}$-rational points of a $\mathbb{Q}$-subgroup of $\mathbf{SL}_n$).

Remark 4. It seems that Conjecture 2 should follows from Conjecture 1. This is so if Γ is an arithmetic subgroup of G. In that case as it was noted by M.S. Raghunathan, the mentioned reduction follows from Remark 3 and the countability of the set of $\mathbb{Q}$-subgroups.

Let us say that an element g of a Lie group is *quasiunipotent* if all eigenvalues of the transformation Ad g have absolute value 1. If Conjecture 2 is proved then it will be possible to prove the following conjecture; for one-parameter F, this conjecture was stated by A.N. Starkov in [69].

(∗) *Let G be a connected Lie group, F a subgroup of G, and D a closed subgroup of G such that G/D has finite volume. Suppose that F is generated by quasiunipotent elements. Then for any $x \in G/D$, the closure of the orbit Fx is a manifold.*

4.2 Closures of Orbits of Non-Quasiunipotent One-Parameter Subgroups

It is not difficult to prove (see for example [68]), that whenever D is a lattice in G, F is one-parameter and F does not consist of quasiunipotent elements, there exists $x \in G/D$ such the closure of Fx is not a manifold.

Conjecture (A) stated below has been proved by S.G. Dani in the following cases: (i) $G = \mathbf{SL}_n(\mathbb{R})$ and $g_t = \mathrm{diag}(e^{-t}, \ldots, e^{-t}, e^{\lambda t}, \ldots, e^{\lambda t})$ where λ is such that the determinant is 1 (see [16] where some results of W.M. Schmidt on badly approximable systems of linear forms are used); (ii) G is a connected semisimple Lie group of $\mathbb{R}$-rank 1 (see [19]).

Conjecture (A). *Let G be a Lie group and Γ a lattice in G. Let $\{g_t\}$ be a one-parameter subgroup of G such that g_1 is not quasiunipotent. Then for any nonempty open subset Ω of G/Γ*

$$\{x \in \Omega | the \ \{g_t\}\text{-oribt of } x \text{ is bounded (relatively compact)}\}$$

is of Hausdorff dimension equal to the dimension of G.

It is plausible that the following strengthening of Conjecture (A) is true.

Conjecture (B). *Let G, Γ and $\{g_t\}$ be the same as in Conjecture (A), and let Y be a finite subset of G/Γ. Then for any nonempty open subset Ω of G/Γ*

$$\{x \in \Omega | the \ \{g_t\}\text{-orbit of } x \text{ is bounded, the closure of this orbit is of Hausdorff}$$
dimension less than the dimension of G, and the intersection of this closure with Y is empty}

is of Hausdorff dimension equal to the dimension of G.

5. Values of Indefinite Quadratic Forms at Integral Points

In this section, we give some applications of results on orbit closures to the study of sets of values of indefinite quadratic forms at integral points.

5.1 Proof of Oppenheim-Davenport Conjecture

According to Meyer s theorem (see [6] or [62]), if B is a rational nondegenerate indefinite quadratic form in n variables and $n \geq 5$ then B represents zero over $\mathbb{Z}$ nontrivially i.e. there exist integers $x_1, \ldots, x_n$ not all equal to 0 such that $B(x_1, \ldots, x_n) = 0$. Theorem 18 stated below can be considered as an analogue of this assertion in the case where B is not a multiple of a rational form. Note that in Theorem 18, the condition "$n \geq 5$" is replaced by a weaker condition "$n \geq 3$".

Theorem 18 (see [36] and [37]; a simplified proof is given in [22] and [39]). *Let B be a real nondegenerate indefinite quadratic form in n variables. Suppose that $n \geq 3$ and that B is not a multiple of a rational form. Then for any $\varepsilon > 0$, there exists $x \in \mathbb{Z}^n - \{0\}$ such that $|B(x)| < \varepsilon$.*

Let us formulate a stronger version of Theorem 18.

Theorem 18' (see [37]). *Let B and n be the same as in Theorem 18. Then for any $\varepsilon > 0$, there exists $x \in \mathbb{Z}^n - \{0\}$ such that $0 < |B(x)| < \varepsilon$.*

Theorems 18 and 18' are of course equivalent for forms that do not represent zero over $\mathbb{Z}$ nontrivially. One can easily see that if Theorems 18 and 18' are proved for some n_0 then they are proved for all $n \geq n_0$. So it is enough to prove these theorems for $n = 3$. We note that, in Theorems 18 and 18', the condition "$n = 3$" can not be replaced by the condition "$n = 2$"; to see this, consider the form $x_1^2 - \lambda x_2^2$ where λ is an irrational positive number such that $\sqrt{\lambda}$ has a continued fraction development with bounded partial quotients.

Theorem 18 was conjectured by A. Oppenheim in 1929 for $n \geq 5$ (see [48] and [49]) and by H. Davenport in 1946 for all $n \geq 3$ (see [25]). Theorem 18' was conjectured in 1953 by A. Oppenheim (see [51]).

Theorem 18' had been proved earlier by B.J. Birch, H. Davenport and H. Ridout for $n \geq 21$ (see [26] or [31, Sect. 42.4]) and by H. Davenport and H. Heilbronn for diagonal forms in 5 variables i.e. for forms B of the type $B(x_1, \ldots, x_5) = \lambda_1 x_1^2 + \cdots + \lambda_5 x_5^2$ (see [25]). G.L. Watson [77] extended the result of Davenport and Heilbronn to forms which include a single cross-product term (say $\lambda_6 x_4 x_5$). R.C. Baker and H.P. Schlickewey [2] proved Theorem 18' for n equal 20, 19 or 18 but under some restrictions on the signature of B. Theorem 18' was proved by A. Oppenheim for forms in $n \geq 4$ variables representing zero over Z nontrivially (see [51] and [52]) and by H. Iwaniec [30] and G.L. Watson [76] for some types of forms in 3 and 4 variables. The proofs mentioned above are in the context of analytic number theory and reduction theory.

Theorems 18 and 18' are deduced from the following

Theorem 19. *Let $G = \mathbf{SL}_3(\mathbb{R})$ and $\Gamma = \mathbf{SL}_3(\mathbb{Z})$. Let B be a real nondegenerate indefinite quadratic form in 3 variables. Let us denoted by H_B the group of elements of G preserving the form B and by Ω the space of lattices in $\mathbb{R}^3$ having determinant 1. (The quotient space G/Γ can be canonically identified with Ω. Under this identification*

$g\Gamma$ goes to $g\mathbb{Z}^3$.) Let G_y denote the stabilizer $\{g \in |gy = y\}$ of $y \in \Omega$. If $z \in \Omega = G/\Gamma$ and the orbit $H_B z$ is relatively compact in Ω then the quotient space $H_B/H_B \cap G_z$ is compact.

Theorem 19 proves generalized Raghunathan's conjecture in a very special case. Now we explain how Theorem 18 is deduced from Theorem 19. As mentioned above, it is enough to prove Theorem 18 for $n = 3$. Suppose that the assertion of Theorem 18 is not true. Then one can easily show using Mahler's compactness criterion that the set $H_B \mathbb{Z}^3$ is relatively compact in Ω. Now we apply Theorem 19 for $z = \mathbb{Z}^3$ and get that the quotient space $H_B/H_B \cap \Gamma$ is compact. Then in view of Borel's density theorem, $H_B \cap \Gamma$ is Zariski dense in H_B. But $\Gamma = \mathbf{SL}_3(\mathbb{Z})$. Therefore the algebraic subgroup $H_B \subset G$ is defined over $\mathbb{Q}$, and hence B is a multiple of a rational form. Contradiction.

5.2 Values of Quadratic Forms at Primitive Integral Points

Let B be a real nondegenerate indefinite quadratic form in n variables. In [51], it is shown for $n \geq 3$ that if the set $(0, \varepsilon) \cap B(\mathbb{Z}^n)$ is not empty for any $\varepsilon > 0$, then the same is true for the form $-B$. On the other hand, it is clear that $B(\mathbb{Z}^n)$ is invariant under multiplication by square of integers. Thus Theorem 18' implies that if $n \geq 3$ and B is not a multiple of a rational form then $B(\mathbb{Z}^n)$ is dense in $\mathbb{R}$. Let us consider now the set of values of B at primitive integral points, namely on $P(\mathbb{Z}^n) = \{x \in \mathbb{Z}^n | x \neq ky$ for any $y \in \mathbb{Z}^n$ and $k \in \mathbb{Z}$ with $|k| \geq 2\}$. Then Theorem 18' doesn't imply, under the condition of that theorem, that $B(\mathfrak{P}(\mathbb{Z}^n))$ is dense in $\mathbb{R}$. Nevertheless it is possible to prove the following

Theorem 20 (see [20]). *Let B be a real nondegenerate indefinite quadratic form on $\mathbb{R}^n$, where $n \geq 3$, which is not a multiple of a rational form. Let B_2 be the corresponding bilinear form defined by $4B_2(v, w) = B(v + w) - B(v - w)$ for all $v, w \in \mathbb{R}^n$. Let a, b, c be such that there exist $v, w \in \mathbb{R}^n$ for which $B(v) = a$, $B(w) = b$, $B_2(v, w) = c$. Then for any $\varepsilon > 0$ there exist $x, y \in \mathfrak{P}(\mathbb{Z}^n)$ such that $|B(x) - a| < \varepsilon$, $|B(y) - b| < \varepsilon$ and $|B_2(x, y) - c| < \varepsilon$. In particular $B(\mathfrak{P}(\mathbb{Z}^n))$ is dense in $\mathbb{R}$.*

Theorem 20 is easily deduced from Theorem 16 and the following strengthening of Theorem 19.

Theorem 21 (see [20]). *Let G, Γ, Ω and H_B be the same as in Theorem 19. Let $x \in \Omega = G/\Gamma$. Then the orbit $H_B x$ is either closed or dense in G/Γ.*

Remark. Generalized Raghunathan's conjecture entails that if $n \geq 3$, B is a nondegenerate indefinite quadratic form on $\mathbb{R}^n$ and $H_B = \{g \in \mathbf{SL}_n(\mathbb{R}) | gB = B\}$ then, for any $x \in \mathbf{SL}_n(\mathbb{R})/\mathbf{SL}_n(\mathbb{Z})$, the orbit $H_B x$ is either closed or dense in $\mathbf{SL}_n(\mathbb{R})/\mathbf{SL}_n(\mathbb{Z})$. This and Theorem 16 easily imply that if B as above is not a multiple of a rational form and B_2 is the corresponding bilinear form then for any $\{a_{ij} | i, j = 1, 2, \ldots, n - 1\} \subset \mathbb{R}$ for which there exist $v_1, \ldots, v_{n-1} \in \mathbb{R}^n$ such that $B_2(v_i, v_j) = a_{ij}$ and for any $\varepsilon > 0$ there also exist $x_1, \ldots, x_{n-1} \in \mathfrak{P}(\mathbb{Z}^n)$ such that $|B_2(x_i, x_j) - a_{ij}| < \varepsilon$ for all $i, j = 1, \ldots, n - 1$.

As mentioned in Sect. 4, Raghunathan's conjecture is proved in [21] in the case where $G = \mathbf{SL}_3(\mathbb{R})$ and $U = \{u(t)\}$ is a one-parameter unipotent subgroup of G such that $u(t) - I$ has rank 2 for all $t \neq 0$. This and Theorem 16 easily imply the following

Theorem 22 (see [21]). *Let B_1 and B_2 be two real quadratic forms on $\mathbb{R}^3$ for which the following conditions are satisfied:*

(i) *there exists a basis of $\mathbb{R}^3$ such that*

$$B_1(x) = 2x_1 x_3 - x_2^2 \qquad and \qquad B_2(x) = x_3^2$$

where x_1, x_2, x_3 are coordinates of x with respect to the basis;

(ii) *no nonzero linear combination of B_1 and B_2 is a multiple of a rational form.*

Then for any $a, b \in \mathbb{R}, b > 0$, and any $\varepsilon > 0$ there exists a primitive integral vector p such that

$$|B_1(p) - a| < \qquad and \qquad |B_2(p) - b| < \varepsilon.$$

Digressing from quadratic forms, let us note at the end of the section that there a connection between results on closures of orbits of horospherical subgroups and some results on orbits of frames under discrete linear groups (see [13], [17] and [23]). These results on orbits of frames have the flavour of results on Diophantine approximation with matrix argument.

6. Markov Spectrum

Let $B(x) = \sum_{1 \leq i, j \leq n} b_{ij} x_i x_j$, $b_{ij} = b_{ji}$, be a real nondegenerate indefinite quadratic form in n variables and let us denote by Φ_n the set of all such forms. Let $d(B)$ denote the determinant of the matrix (b_{ij}). Let us set

$$m(B) = \inf_{x \in \mathbb{Z}^n - \{0\}} |B(x)| \qquad and \qquad \mu(B) = \frac{m(B)^n}{|d(B)|}.$$

It is clear that $\mu(\lambda B) = \mu(B)$ for every $\lambda \in \mathbb{R}, \lambda \neq 0$. Let M_n denote $\mu(\Phi_n)$. The set M_n is called *Markov spectrum*. It easily follows from Mahler's compactness criterion that M_n is bounded and closed.

Two forms $B, B' \in \Phi_n$ will be called *equivalent* if there exist $g \in \mathbf{SL}_n(\mathbb{Z})$ and $\lambda \in \mathbb{R}$ such that $gB = \lambda B'$. One can easily see that $\mu(B) = \mu(B')$ if B and B' are equivalent.

6.1 Discreteness of Markov Spectrum for $n \geq 3$

In 1880, A.A. Markov described the intersection of M_2 with the segment $(4/9, \infty)$ and described corresponding quadratic forms (see [7] and [41]). It follows from this description that $M_2 \cap (4/9, \infty)$ is a discrete countable subset of $(4/9, 4/5)$ and that, for any $a > 4/9$, there are only finitely many equivalence classes of forms $B \in \Phi_2$ with $\mu(B) > a$. On the other hand, the intersection $M_2 \cap [0, 4/9]$ is not countable and moreover has quite a complicated topological structure (see [7]).

It follows from Meyer's theorem and Theorem 18 that $M_n = \{0\}$ whenever $n \geq 5$. In 1955 (see [8]), J.W.S. Cassels and H.F.P. Swinnerton-Dyer proved the conditional result that Theorem 18 implies the following

Theorem 23. *Let $n \geq 3$ and $\varepsilon > 0$. Then the set $M_n \cap (\varepsilon, \infty)$ is finite and moreover there are only finitely many equivalence classes of forms $B \in \Phi_n$ with $\mu(B) > \varepsilon$.*

For rational forms B, Theorem 23 had been proved earlier by L.Ya. Vulakh (see [75]). Really he proved a stronger assertion that there are only finitely many equivalence classes of forms B with $\mu_0(B) > \varepsilon$ where

$$\mu_0(B) = \frac{m_0(B)^n}{|d(B)|} \quad \text{and} \quad m_0(B) = \inf_{x \in \mathbb{Z}^n - \{0\}, B(x) \neq 0} |B(x)|.$$

As noted in [74], Vulakh has obtained the complete description of spectra of nonzero minima of rational Hermitian forms.

Both the proof of Theorem 23 based on Theorem 18 and the proof given in [75] are non-effective. In particular these proofs do not give an explicit upper bound for the cardinality of $M_n \cap (\varepsilon, \infty)$ for $n = 3, 4$. It would be very interesting to get such a bound.

For $n = 3$, A.A. Markov determined first 4 values of $\mu(B)^{-1}$ and B.A. Venkov determined next 7 values (see [42] and [73]). These values are

$$3/2,\ 5/2,\ 3,\ 25/8,\ 1200/7^3 = 3{,}498\ldots,\ 7/2,\ 15/4,\ 3 \cdot 13^2/5^3 = 4{,}056,$$

$$112/3^3 = 4{,}148\ldots,\ 135/2^5 = 4{,}218\ldots,\ 9/2.$$

For $n = 4$ and forms of signature 0, A. Oppenheim determined first 7 values of $\mu(B)^{-1}$ (see [50]). These values are

$$9/4,\ 17/4,\ 25/4,\ 117/16,\ 33/4,\ 9,\ 625/64.$$

It should be noted that there are two inequivalent forms which correspond to the value 9. For $n = 4$ and forms of signature -2 or 2, A. Oppenheim determined also the first value of $\mu(B)^{-1}$ which is 7/4.

6.2 A Strengthened Version of the Discreteness Theorem

Let us formulate now a strengthened version of Theorem 23.

Theorem 24 (see [39]). *Let $a, b, c \in \mathbb{R}$ and let*

$$\Phi_n(a, b, c) = \left\{ B \in \Phi_n \, \middle|\, \text{there exist } v, w \in \mathbb{R}^n \text{ such that} \right.$$

$$\frac{B(v)^n}{d(B)} = a, \quad \frac{B(w)^n}{d(B)} = b \text{ and } \frac{B_2(v, w)^n}{d(B)} = c \text{ where } B_2 \text{ is the}$$

$$\left. \text{bilinear form corresponding to } B \right\}.$$

Suppose that $n \geq 3$. Then for any $\varepsilon > 0$, there are only finitely many equivalence classes of forms $B \in \Phi_n(a, b, c)$ for which it is not possible to find primitive integral vectors $x, y \in \mathfrak{P}(\mathbb{Z}^n)$ such that

$$\left| \frac{B(x)^n}{d(B)} - a \right| < \varepsilon, \qquad \left| \frac{B(y)^n}{d(B)} - b \right| < \varepsilon \qquad and \qquad \left| \frac{B_2(x, y)^n}{d(B)} - c \right| < \varepsilon.$$

The proof of Theorem 24 is based on the following slight generalization of Theorem 21.

Theorem 21' (see [39]). *Let G, Γ, Ω and H_B be the same as in Theorem 21. Let A be a closed H_B-invariant subset of $\Omega = G/\Gamma$. Suppose that $A \neq \Omega$. Then A is the union of finite number of closed H_B-orbits.*

A. Borel and G. Prasad proved in [4] a generalization of Theorem 18' for a family $\{B_s\}$ where $s \in S$, S is a finite set of places of a number field k containing the set S_∞ of archimedean places, B_s is a quadratic form on k_s^n, and k_s is the completion of k at s. It would be interesting to prove analogues of Theorems 20, 22, 23 and 24 for families $\{B_s\}$.

7. Generic Points

Let X be a locally compact σ-compact topological space and φ a homeomorphism of X. We say that a point $x \in X$ is *generic* (with respect to φ) if there exists a Borel probability measure μ_x on X such that for any bounded continuous function f on X

$$\lim_{n \to \infty} \frac{1}{n} \sum_{i=0}^{n-1} f(\varphi^i x) = \int_X f \, d\mu_x.$$

According to Birkhoff individual ergodic theorem for any φ-invariant Borel probability measure μ on X, one can find a subset $A \subset X$ such that $\mu(A) = 0$ and each point $x \in X - A$ is generic.

The following conjecture was stated in [14] (for the case where $G = \mathbf{SL}_n(\mathbb{R})$ and $\Gamma = \mathbf{SL}_n(\mathbb{Z})$).

Conjecture 3. *Let G be a connected Lie group, Γ a lattice in G and u a unipotent element of G. Then every point $x \in G/\Gamma$ is generic with respect to the homeomorphism $y \to uy$, $y \in G/\Gamma$.*

Conjecture 3 can be strengthened in the following way:

Conjecture 4. *Let G, Γ and u be the same as in Conjecture 3. Then for any $x \in G/\Gamma$, there exist a closed subgroup $L \subset G$ containing u and a L-invariant Borel probability measure μ on G/Γ such that the orbit Lx is closed, the support of the measure μ is Lx, and for any bounded continuous function f on G/Γ*

$$\lim_{n \to \infty} \frac{1}{n} \sum_{i=0}^{n-1} f(u^i x) = \int_{G/\Gamma} f \, d\mu.$$

Remark. Considering appropriate increasing sequences $\{S_n\} \subset U$ and replacing $\frac{1}{n} \sum_{i=0}^{n-1} f(u^i x)$ by $\frac{1}{\sigma(S_n)} \int_{S_n} f(ux) \, d\sigma(u)$ where σ is Haar measure on U, one can formulate an analogue of Conjecture 4 for an arbitrary closed unipotent subgroup $U \subset G$. Let us also note the connection of Conjecture 4 with Theorems 12 and 14.

It is clear that for a cyclic U, Raghunathan's conjecture follows from Conjecture 4. Conjecture 4 was proved for nilpotent G by E. Lesigne (see [32] and [33]) and for $G = \mathbf{SL}_2(\mathbb{R})$ by S.G. Dani and J. Smillie (see [24]; for $\Gamma = \mathbf{SL}_2(\mathbb{Z})$ see also [14]). For $G = \mathbf{SL}_2(\mathbb{R})$, we have the following formulation

Theorem 25 (see [24]). *Let $G = \mathbf{SL}_2(\mathbb{R})$, let Γ be a lattice in G, let $x \in G/\Gamma$, let $\{u_t\}$ be a one-parameter unipotent subgroup of G, and let u denote u_1. Suppose that the orbit $\{u_t x | t \in \mathbb{R}\}$ is not periodic i.e. that $u_t x \neq x$ for every $t \neq 0$. Then the orbit $\{u_t x | t \in \mathbb{R}\}$ and the sequence $\{u^n x | n \in \mathbb{Z}\}$ are uniformly distributed with respect to the G-invariant Borel probability measure μ on G/Γ; i.e. for any bounded continuous function f on G/Γ*

$$\lim_{T \to \infty} \frac{1}{T} \int_0^T f(u_t x) \, dt = \lim_{N \to \infty} \frac{1}{N} \sum_{n=0}^{N-1} f(u^n x) = \int_{G/\Gamma} f \, d\mu.$$

The following number theoretic assertion is deduced from Theorem 25.

Theorem 26 (see [14]). *For $t \in \mathbb{R}$, let $[t]$ denote the largest integer x with $x \leq t$ and let $\{t\} = t - [t]$. For $m, n \in \mathbb{N}^+$, let (m, n) denote the g.c.d. of m and n. Then for any irrational θ*

$$\lim_{T \to \infty} \frac{1}{T} \sum_{\substack{0 \leq m \leq T\{m\theta\} \\ (m, [m\theta]) = 1}} \{m\theta\}^{-1} = \frac{1}{\zeta(2)} = \frac{6}{\pi^2}$$

where ζ denotes Riemann zeta-function.

Conjecture 4 entails quantative versions of results from Sect. 5. In particular, if this conjecture is proved it will be possible to prove the following assertions.

(i) *Let n, B, B_2, a, b and c be the same as in Theorem 20. Then for any $\varepsilon > 0$, one can find $d = d(\varepsilon, a, b, c, B) > 0$ such that for all sufficiently large S there exist at least dS pairs of primitive integral vectors $x, y \in \mathfrak{P}(\mathbb{Z}^n)$ for which*

$$\|x\| < S, \quad \|y\| < S, \quad |B(x) - a| < \varepsilon, |B(y) - b| < \varepsilon \quad and \quad |B_2(x, y) - c| < \varepsilon$$

where $\| \ \|$ denotes the Euclidean norm on $\mathbb{R}^n$. Moreover we can choose $d(\varepsilon, a, b, c, B)$, so that it is a rational function of ε, a, b, c and coefficients of B.

(ii) *Let B_1 and B_2 be the same as in Theorem 22. Then for any $a, b \in \mathbb{R}$, $b > 0$, and any $\varepsilon > 0$, one can find $c = c(\varepsilon, a, b, B_1, B_2)$ such that for all sufficiently large S there exist at least $cS^{1/2}$ primitive integral vectors p for which*

$$\|p\| < S, \quad |B_1(p) - a| < \varepsilon \quad and \quad |B_2(p) - b| < \varepsilon.$$

Moreover of $c(\varepsilon, a, b, B_1, B_2)$, can be chosen to be a rational function of ε, a, b and coefficients of B_1 and B_2.

8. Actions of Semisimple Groups and Discrete Subgroups on Manifolds and Measure Spaces

Let us give formulations of some recent results about actions on general manifolds and measure spaces. We refer to [80] and [81] for the definition of notions which are used in these formulations.

Theorem 27 (see [64]). *Let G be a connected simple Lie group with finite center, finite fundamental group, and $\mathbb{R}$-rank(G) $\geq$ 2. Let M be a compact manifold and suppose there is a nontrivial real analytic action of G on M preserving a (real analytic) connection and a finite measure. Then $\pi_1(M)$ is not isomorphic to the fundamental group of any complete Riemannian manifold N with negative curvature bounded away from 0 and $-\infty$.*

Theorem 28 (see [82]). *Let a locally compact group G act continuously on a compact metrizable space M. Suppose G preserves a finite measure on M with respect to which the G-action is ergodic. Suppose further that G is a semisimple group of higher rank (i.e. G is a finite product $\prod G_i$ where each G_i is the set of k_i-points of a k_i-simple connected k_i-group of k_i-rank at least 2). If $P \to M$ is a principal H-bundle on which G acts by principle bundle automorphisms, where H is a real algebraic group, then the algebraic hull of the action is a reductive group with finite center.*

Theorem 29 (see [83]). *Let G be a semisimple group of higher rank and $\Gamma \subset G$ a lattice. Suppose Γ acts smoothly on a compact manifold M preserving a volume density. Then the following are equivalent:*

(a) *The Γ-action has zero entropy, i.e. the entropy $h(\gamma)$ equals 0 for all $\gamma \in \Gamma$.*
(b) *The Γ-action is measurably isometric, i.e. it is measurably conjugate to an action which obtained by a homomorphism of Γ into a compact group K which acts continuously on M.*
(c) *There is a Γ-invariant measurable Riemannian metric on M.*

Concluding Remarks Made During the Congress

(i) I learnt at the beginning of the Congress that very recently (about two weeks before the Congress) M. Ratner proved Conjectures 1, 2, 3 and 4. Also recently, N.A. Shah proved some partial results in the direction of the proof of Conjecture 4 (even these results are enough to prove assertions (i) and (ii) from Sect. 7 and the assertion mentioned in the remark after Theorem 21). Proofs of M. Ratner and N.A. Shah are based on the reduction to Theorems 12 and 14.

(ii) Let $G = \mathbf{SL}_2(\mathbb{R})$, let $\Gamma = \mathbf{SL}_2(\mathbb{Z})$ and let $U = \{u_t\}$ be a one-parameter unipotent subgroup of G. For each $t > 0$ there is a unique closed orbit of U on G/Γ of period t, call it C_t ($u_t x = x$ and $u_{t'} x \neq x$ for $x \in C_t$ and $0 < t' < t$). P. Sarnak informed me about the following observation which was essentially due to D. Zagier. The famous Riemann hypothesis about zeroes of Riemann zeta-function is equivalent to the following statement. For any function $f \in C_0^\infty(G/\Gamma)$ (i.e. f is smooth and has a compact support) and any $\varepsilon > 0$

$$\frac{1}{t}\int_{C_t} f\,dl = \int_{G/\Gamma} f\,d\mu + 0_{\varepsilon,f}(t^{-(3/4)+\varepsilon})$$

as $t \to \infty$ where l is the Lebesgue measure on C_t and μ is the G-invariant probability Borel measure on G/Γ. It should be noted that (a) the power $-\dfrac{3}{4} + \varepsilon$ is different from the usual power $-\frac{1}{2} + \varepsilon$ which appears in ergodic theory; (b) the equivalence remains true if one considers only K-invariant functions f where K is a maximal compact subgroup of G.

References

1. Auslander, L., Green, L., Hahn, F.: Flows on homogeneous spaces (Annals of Mathematics Studies, no 53). Princeton University Press, 1963
2. Baker, R.C., Schlickewey, H.P.: Indefinite quadratic forms. Proc. London Math. Soc. **54** (1987) 383–411
3. Borel, A.: Density properties of certain subgroups of semisimple groups. Ann. Math. **72** (1960) 179–188
4. Borel, A., Prasad, G.: valeurs de formes quadratiques aux points entiers. C. R. Acad. Sci. Paris **307**, Serie 1 (1988) 217–220
5. Brezin, J., Moore, C.C.: Flows on homogeneous spaces: a new look. Amer. J. Math **103** (1981) 571–613
6. Cassels, J.W.S.: Rational quadratic forms. Academic Press, London New York, 1978
7. Cassels, J.W.S.: An introduction to Diophantine approximation. Cambridge Univ. Press, Cambridge, 1957
8. Cassels, J.W.S., Swinnerton-Dyer, H.P.F.: On the product of three homogeneous linear forms and indefinite ternary quadratic forms. Philos. Trans. Roy. Soc. London **248**, Ser. A (1955) 73–96
9. Cornfeld, I.P., Sinai, Ya.G., Fomin, S.V.: Ergodic theory. Gosud. Izd. Fiz. Mat. Lit., Moscow, 1980 (Russian) [English transl.: Springer, Berlin Heidelberg New York 1982].
10. Dani, S.G.: Kolmogorov automorphisms on homogeneous spaces. Amer. J. Math. **98** (1976) 119–163
11. Dani, S.G.: Bernoullian translations and minimal horospheres on homogeneous spaces. J. Ind. Math. Soc. **39** (1976) 245–294
12. Dani, S.G.: Spectrum of an affine transformation. Duke Math. J. **44** (1977) 129–155
13. Dani, S.G.: Invariant measures and minimal sets of horospherical flows. Invent. math. **64** (1981) 357–385
14. Dani, S.G.: On uniformly distributed orbits of certain horocycle flows. Ergod. Theor. Dynam. Syst. **2** (1982) 139–158
15. Dani, S.G.: On orbits of unipotent flows on homogeneous spaces. Ergod. Theor. Dynam. Syst. **4** (1984) 25–34
16. Dani, S.G.: Divergent trajectories of flows on homogeneous spaces and Diophantine approximation. J. Reine Angew. Math. **359** (1985) 55–89
17. Dani, S.G.: On orbits of unipotent flows on homogeneous spaces II. Ergod. Theor. Dynam. Syst. **6** (1986) 167–182
18. Dani, S.G.: Orbits of horospherical flows. Duke Math. J. **53** (1986) 177–188
19. Dani, S.G.; Bounded orbits of flows on homogeneous spaces. Comment. Math. Helv. **61** (1986) 636–660
20. Dani, S.G., Margulis. G.A.: Values of quadratic forms at primitive integral points. Invent. math. **98** (1989) 405–424

21. Dani, S.G., Margulis, G.A.: Orbit closures of generic unipotent flows on homogeneous spaces of SL(3, ℝ). Math. Ann. **286** (1990) 101–128
22. Dani, S.G., Margulis, G.A.: Values of quadratic forms at primitive integral points. l'Enseignement Math. (to appear)
23. Dani, S.G., Raghavan, S.: Orbits of Euclidean frames under discrete linear groups. Isr. J. Math. **36** (1980) 300–320
24. Dani, S.G., Smillie, J.: Uniform distribution of horocycle flows for Fuchsian groups. Duke Math. J. **51** (1984) 185–194
25. Davenport, H., Heilbronn, H.: On indefinite quadratic forms in five variables. J. London Math. Soc. **21** (1946) 185–193
26. Davenport, H., Ridout, H.: Undefinite quadratic forms. Proc. London Math. Soc. **9** (1959) 544–555
27. Furstenberg, H.: The unqiue ergodicity of the horocycle flow. In: Recent Advances in Topological Dynamics (ed. A. Beck). Springer, Berlin Heidelberg New York 1972, pp. 95–115
28. Gelfand, I.M., Fomin, S.V.: Geodesic flows on manifolds of constant negative curvature. Usp. Mat. Nauk **7** (1952) 118–137 (Russian)
29. Hopf, E.: Statistik der geodätischen Linienin Mannigfaltigkeiten negativer Krümmung. Ber. Ver. Sächs. Akad. Wiss. Leipzig **91** (1939) 261–304
30. Iwaniec, H.: On indefinite quadratic forms in four variables. Acta Arith. **33**, 209–229 (1977)
31. Lekkerkerker, C.G.: Geometry of numbers. Walters-Noordhoff, Groningen; North-Holland, Amsterdam London, 1969
32. Lesigne, E.: Théorèmes ergodiques pour une translation sur nilvariété. Ergod. Theor. Dynam. Syst. **9** (1989) 115–126
33. Lesigne, E.: Sur une nilvariété, les parties minimales associées à une translation sont uniquement ergodiques. Preprint.
34. Marcus, B.: The horocycle flow is mixing of all degrees. Invent. math. **46** (1978) 201–209
35. Margulis, G.A.: On the action of unipotent groups in the space of lattices. In: Proc. of the Summer school on group representations (bolyai Janos Math. Soc., Budapest, 1971). Akadémiai Kiado, Budapest 1975, pp. 365–370
36. Margulis, G.A.: Formes quadratiques indéfinies et flots unipotents sur les espaces homogènes. C. R. Acad. Sci. Paris **304**, Serie 1 (1987) 249–253
37. Margulis, G.A.: Discrete subgroups and ergodic theory. In: Proc. of the conference "Number theory, trace formulas and discrete groups" in honour of A. Selberg (Oslo, 1987). Academic Press, London New York, 1988, pp. 377–398
38. Margulis, G.A.: Compactness of minimal closed invariant sets of actions of unipotent groups. Geometriae Dedicata (to appear)
39. Margulis, G.A.: Orbits of group actions and values of quadratic forms at integral points. Preprint IHES, 1990
40. Margulis, G.A.: Discrete subgroups of semisimple groups. Springer, Berlin Heidelberg New York 1990
41. Markov, A.A.: On binary quadratic forms of positive determinant. SPb, 1880; Usp. Mat. Nauk **3**(5) (1948) 7–51 (Russian)
42. Markov, A.A.: Sur les formes quadratiques ternaires indéfinies. Math. Ann. **56** (1903) 233–251
43. Mautner, F.J.: Geodesic flows on symmetric Riemannian spaces. Ann. Math. **65** (1957) 416–431
44. Moore, C.C.: Ergodicity of flows on homogeneous spaces. Amer. J. Math. **88** (1966) 154–178
45. Moore, C.C.: The Mautner phenomenon for general unitary representations. Pacific J. Math. **86** (1980) 155–169

46. Mostow, G.D.: Intersections of discrete subgroups with Cartan subgroups. Indian J. Math. **34** (1970) 203–214
47. Mostow, G.D.: Strong rigidity of locally symmetric spaces. (Annals of Mathematics Studies, no 78). Princeton University Press, 1973
48. Oppenheim, A.: The minima of indefinite quaternary quadratic forms. Proc. Acad. Sci. USA **15** (1929) 724–727
49. Oppenheim, A.: The minima of indefinite quaternary quadratic forms. Ann. Math. **32** (1931) 271–298
50. Oppenheim, A.: Minima of quaternary quadratic forms of signature O. Proc. London Math. Soc. **37** (1934) 63–81
51. Oppenheim, A.: Values of quadratic forms I, II. Quart. J. Math. Oxford Ser. (2), **4** (1953) 54–59, 60–66
52. Oppenheim, A.: Values of quadratic forms III. Monatsh. Math. Phys. **57** (1953) 97–101
53. Parasyuk, O.S.: Horocycle flows on surfaces of constant negative curvature. Usp. Mat. Nauk **8**(3) (1953) 125–126 (Russian)
54. Raghunathan, M.S.: Discrete subgroups of Lie groups. Springer, Berlin Heidelberg New York, 1972
55. Ratner, M.: Rigidity of horocycle flows. Ann. Math. **115** (1982) 597–614
56. Ratner, M.: Factors of horocycle flows. Ergod. Theor. Dynam. Syst. **2** (1982) 465–489
57. Ratner, M.: Horocycle flows: joinings and rigidity of products. Ann. Math. **118** (1983) 277–313
58. Ratner, M.: Strict measure rigidity for unipotent subgroups of solvable groups. Invent. math. **101** (1990) 449–482
59. Ratner, M.: On measure rigidity for unipotent subgroups of semisimple groups. Acta Math. (to appear)
60. Ratner, M.: On Raghunathan's measure conjecture. Preprint.
61. Safonov, A.V.: on the spectral type of ergodic G-induced flows. Funkt. Anal. i Priložen **14**(4) (1980) 81–82 (Russian)
62. Serre, J.P.: A course in arithmetic. Springer, Berlin Heidelberg New York 1973
63. Sinai, Ya.G.: Dynamical systems with countably-multiple Lebesgue spectrum II. Izv. Akad. Nauk SSSR Ser. Mat. **30**, 15–68 (Russian) [English transl.: Amer. Math. Soc. Transl. (2) **68** (1988) 34–88]
64. Spatzier, R.J., Zimmer, R.J.: Fundamental groups of negatively curved manifolds and actions of semisimple groups. Topology (to appear)
65. Starkov, A.N.: The ergodic behaviour of flows on homogeneous spaces. Dokl. Akad. Nauk SSSR **273**, 538–540 (Russian) [English transl.: Soviet Math. Dokl. **28** (1983) 675–676]
66. Starkov, A.N.: On spaces of finite volume. Vestnik MGU **5** (1986) 64–65 (Russian)
67. Starkov, A.N.: The ergodic decomposition for homogeneous flows. Izv. Akad. Nauk SSSR **51** (1987) 1191–1213 (Russian)
68. Starkov, A.N.: Solvable homogeneous flows. Mat. Sbornik **176** (1987) 242–259 (Russian)
69. Szarkov, A.N.: The ergodic decomposition of flows on homogeneous spaces of finite volume. Mat. Sbornik **180** (1989) 1614–1633 (Russian)
70. Stepin, A.M.: Flows on solvable manifolds. Usp. Mat. Nauk **24**(5) (1989) 241–242 (Russian)
71. Stepin, A.M.: Dynamical systems on homogeneous spaces of semisimple groups. Izv. Akad. Nauk SSSR Ser. Mat. **37**, 1091–1107 (Russian) [English transl. Math. USSR Izv. **7** (1973) 1089–1104]
72. Veech, W.A.: Unique ergodicity of horospherical flows. Amer. J. Math. **99** (1977) 827–859
73. Venkov, B.A.: On Markov's extremal problem for ternary quadratic forms. Izv. Akad. Nauk SSSR Ser. Mat. **9** (1945) 429–434 (Russian)
74. Vulakh, L.Ya.: On minima of rational indefinite Hermitian forms. Ann. N.Y. Acad. Sci. **70** (1983) 99–106

75. Vulakh, L.Ya.: On minima of rational indefinite quadratic forms. J. Number Theory **21** (1985) 275–285
76. Watson, G.L.: On indefinite quadratic forms in three or four variables. J. London Math. Soc. **28** (1953) 239–242
77. Watson, G.L.: On indefinite forms in five variables. Proc. London Math. Soc. (3) **3** (1953) 170–181
78. Witte, D.: Rigidity of some translations on homogeneous spaces. Invent. math. **85** (1985) 1–27
79. Witte, D.: Zero entropy affine maps on homogeneous spaces. Amer. J. Math. **109** (1987) 927–961
80. Zimmer, R.J.: Ergodic theory and semisimple groups. Birkhäuser, Boston, 1984
81. Zimmer, R.J.: Actions of semisimple groups and discrete subgroups. Proc. Int. Math. Cong., Berkeley, California, USA, 1986, pp. 1247–1258
82. Zimmer, R.J.: On the algebraic hull of an automorphism group of a principal bundle. Comment. Math. Helv. (to appear)
83. Zimmer, R.J.: Spectrum, entropy, and geometric structures for smooth actions of Kazhdan groups. Preprint

Pseudodifferential Operators, Corners and Singular Limits

Richard B. Melrose

Department of Mathematics, Room 2-243, Massachusetts Institute of Technology
Cambridge, MA 02139, USA

In the first part of my talk I shall describe some of the properties one should expect of a calculus of pseudodifferential operators which corresponds to the microlocalization of a Lie algebra of vector fields. This is not intended to be a formal axiomatic program but it leads one to consider conditions on the Lie algebra for such microlocalization to be possible. The symbolic structure of the calculus also shows how it can be applied in the solution of analytic questions related to the Lie algebra, especially to the inversion of elliptic elements of the enveloping algebra.

In the second part I shall consider several such analytic question which arise in various differential-geometric settings and are, or appear to be, amenable to the application of these pseudodifferential techniques. For those examples which have already been analyzed the Lie algebra is identified and then a specific question is discussed using the calculus of pseudodifferential operators which arises from it.

Finally in the third part of the talk I shall briefly outline a general strategy for the construction of 'the' calculus of pseudodifferential operators which microlocalizes a given Lie algebra satisfying conditions which make it a boundary-fibration structure. This construction applies to most of the situations described in the second part of the talk and, I conjecture, can be extended to apply to the others.

I. Pseudodifferential Operators

Even though many of the examples of interest here arise on singular or non-compact spaces it is very convenient from an analytic point of view to work always in a fixed category of spaces. The simplest class of spaces which seems to be sufficiently wide to allow many different types of problems to be attacked is that of compact manifolds with corners; the morphisms are the b-maps, discussed below. Such a space is a compact topological manifold with boundary, X, with a $\mathscr{C}^\infty$ structure locally modeled on the spaces

$$\mathbb{R}^n_k = [0, \infty)^k \times \mathbb{R}^{n-k}. \tag{1}$$

This fixes the space $\mathscr{C}^\infty(X)$ of smooth functions on X; locally these are just the functions on $\mathbb{R}^n_k$ which are the restrictions of $\mathscr{C}^\infty$ functions on $\mathbb{R}^n$. For

Proceedings of the International Congress
of Mathematicians, Kyoto, Japan, 1990
© The Mathematical Society of Japan, 1991

simplicity we also insist that all the boundary hypersurfaces of X are embedded. If we let $M_1(X)$ be the set of boundary hypersurfaces then this means that each $H \in M_1(X)$ has a defining function

$$\varrho_H \in \mathscr{C}^\infty(X) \text{ s.t. } H = \{\varrho_H = 0\}, \ \varrho_H \geq 0, \ d\varrho_H \neq 0 \text{ at } H. \tag{2}$$

We denote by $M_{(1)}(X)$ the set of true boundary faces, the connected closed submanifolds locally given by intersection of the boundary hypersurfaces and set $M(X) = M_{(1)}(X) \cup \{X\}$; all these submanifolds are embedded. Limiting the analysis to compact manifolds with corners implies that in many cases a preliminary geometric step involving a compactification or blow-up procedure is required.

It is always possible to embed a compact manifold with corners as a subset of a compact manifold, $\tilde{X}$, without boundary:

$$\iota_X : X \hookrightarrow \tilde{X}, \ \mathscr{C}^\infty(X) = \iota_X^* \left(\mathscr{C}^\infty(\tilde{X}) \right). \tag{3}$$

This allows one to readily discuss any of the usual structures on a $\mathscr{C}^\infty$ manifold, by restriction from $\tilde{X}$, provided only that such structures are invariant under diffeomorphisms of $\tilde{X}$ which fix each point of ∂X. In particular $\mathscr{C}^\infty$ vector fields, the tangent bundle, cotangent bundle, form bundles etc. can all be defined in this way; of course they also have more intrinsic definitions.

In most cases the structure we are seeking is most readily expressed in terms of a space of real $\mathscr{C}^\infty$ vector fields:

$$\mathscr{V}(X) \subset \mathscr{C}^\infty(X; TX). \tag{4}$$

The two most basic properties we insist upon are:

$$\mathscr{V}(X) \text{ is a Lie algebra} \tag{5}$$

under the commutation bracket and

$$\mathscr{V}(X) \text{ is a } \mathscr{C}^\infty(X)\text{-module}. \tag{6}$$

Of course the space of all smooth vector fields, $\mathscr{C}^\infty(X; TX)$, has these two properties. However, in order that the vector fields in $\mathscr{V}(X)$ correspond to infinitesmal diffeomorphisms of X we also require that

$$\text{each } V \in \mathscr{V}(X) \text{ is tangent to each } H \in M_1(X). \tag{7}$$

The simplest example is just the space $\mathscr{V}_b(X)$ of all $\mathscr{C}^\infty$ vector fields on X satisfying (7). Thus we can summarize these properties as requiring

$$\mathscr{V}(X) \subset \mathscr{V}_b(X) \text{ is a Lie subalgebra and } \mathscr{C}^\infty(X)\text{-submodule}. \tag{8}$$

Near any point $\bar{x} \in X$ there are local coordinates $x_1, \ldots, x_n$ in which $\bar{x}$ is mapped to the origin of $\mathbb{R}^n_k$, for some k. Then $\mathscr{V}_b(X)$ is locally the span, as a $\mathscr{C}^\infty$-module, of the vector fields

$$x_1 \partial_{x_1}, \ldots, x_k \partial_{x_k}, \partial_{x_{k+1}}, \ldots, \partial_{x_n}. \qquad (9)$$

There are no relations between these vector fields so there is a natural vector bundle, which we denote ${}^b TX$, of which $\mathscr{V}_b(X)$ forms the space of all sections. Since these sections are vector fields there is a natural vector bundle map

$$ {}^b TX \longrightarrow TX \qquad (10)$$

which is an isomorphism over the interior, but has corank k at a boundary face of codimension k, since the first k elements in (9) are mapped to zero. In general we insist that the Lie algebra $\mathscr{V}(X)$ have a similar structure. Thus we require that there be a vector bundle ${}^{\mathscr{V}} TX$ and vector bundle maps

$$ {}^{\mathscr{V}} TX \longrightarrow {}^b TX \longrightarrow TX \qquad (11)$$

where the overall map is $\iota_{\mathscr{V}}$ and

$$\mathscr{V}(X) = \iota_{\mathscr{V}} \circ \mathscr{C}^\infty(X; {}^{\mathscr{V}} TX). \qquad (12)$$

We denote by ${}^{\mathscr{V}} T^* X$ the dual bundle to ${}^{\mathscr{V}} TX$, and by ${}^{\mathscr{V}} \Lambda^k X$ the associated form bundles.

Since $\mathscr{V}$ is a Lie algebra of operators acting on $\mathscr{C}^\infty(X)$ the enveloping algebra is naturally a filtered ring of operators, denoted $\mathrm{Diff}_{\mathscr{V}}^m(X)$, acting on $\mathscr{C}^\infty(X)$. In fact, since the elements of $\mathscr{V}$ are local diffeomorphisms, twisted versions, $\mathrm{Diff}_{\mathscr{V}}^m(X; E, F)$, of these spaces of operators are defined between sections of any two vector bundles, E and F, over X.

One easy way to see why we might be interested in these structures is the following elementary result.

Proposition (13). *Exterior differentiation defines a complex*

$$\mathscr{C}^\infty(X) \xrightarrow{d} \mathscr{C}^\infty(X; {}^{\mathscr{V}} \Lambda^1) \xrightarrow{d} \ldots \xrightarrow{d} \mathscr{C}^\infty(X; {}^{\mathscr{V}} \Lambda^q) \qquad (14)$$

where q is the fibre dimension of ${}^{\mathscr{V}} TX$. If g is a fibre metric on ${}^{\mathscr{V}} TX$ then the associated Laplacian is an element of $\mathrm{Diff}_{\mathscr{V}}^2(X; {}^{\mathscr{V}} \Lambda^k)$ for each k, $0 \le k \le q$.

The assumption (12) means that the filtration of the ring $\mathrm{Diff}_{\mathscr{V}}^*(X)$ is given by symbol maps, $\sigma_{k,\mathscr{V}}$, giving a short exact sequence

$$0 \longrightarrow \mathrm{Diff}_{\mathscr{V}}^{k-1}(X; E, F) \hookrightarrow \mathrm{Diff}_{\mathscr{V}}^k(X; E, F) \xrightarrow{\sigma_{k,\mathscr{V}}} P^k({}^{\mathscr{V}} T^* X; E, F) \longrightarrow 0, \qquad (15)$$

where $P^k({}^{\mathscr{V}} T^* X; E, F)$ is the space of $\mathscr{C}^\infty$ homomorphisms of the lifts to ${}^{\mathscr{V}} T^* X$ of the vector bundles E, F which are homogeneous polynomials of degree k on the fibres. An operator is ($\mathscr{V}$-) elliptic if its symbol in this sense is invertible off the zero section of ${}^{\mathscr{V}} T^* X$. The symbol of the Laplacian is always given by multiplication by the square of the length, so it is always elliptic as an element of $\mathrm{Diff}_{\mathscr{V}}^k(X; {}^{\mathscr{V}} \Lambda^k)$.

Now by a *small* calculus of $\mathscr{V}$-pseudodifferential operators we mean linear spaces of operators

$$\Psi^m_{\mathscr{V}}(X;E,F) \ni A : \mathscr{C}^\infty(X;E) \longrightarrow \mathscr{C}^\infty(X;F), \; m \in \mathbb{R}, \tag{16}$$

with (at least) the following additional properties. First these spaces should extend the ring of $\mathscr{V}$-differential operators:

$$\mathrm{Diff}^k \mathscr{V}(X;E,E') \subset \Psi^k_{\mathscr{V}}(X;E,E') \tag{17}$$

$$\Psi^m_{\mathscr{V}}(X;E,E') \circ \Psi^{m'}_{\mathscr{V}}(X;E',E'') \subset \Psi^{m+m'}_{\mathscr{V}}(X;E,E''). \tag{18}$$

Moreover we require that the filtration be fixed by symbol maps giving short exact sequences analogous to (15):

$$0 \longrightarrow \Psi^{m-1}_{\mathscr{V}}(X;E,F) \longhookrightarrow \Psi^m_{\mathscr{V}}(X;E,F) \xrightarrow{\sigma_{m,\mathscr{V}}} S^{[m]}(^{\mathscr{V}}T^*X;E,F) \longrightarrow 0, \tag{19}$$

where now $S^{[m]}(^{\mathscr{V}}T^*X;E,F)$ is the quotient of the space of homomorphisms, as before, but now only required to be symbols of order m, (see (79)) by the subspace of symbols of order $m-1$. In fact we require that (19) should split (continuously), i.e. there should be a $\mathscr{V}$-quantization map:

$$q_{\mathscr{V}} : S^m(^{\mathscr{V}}T^*X;E,F) \longrightarrow \Psi^m_{\mathscr{V}}(X;E,F) \tag{20}$$

which splits (19) and which is surjective modulo the residual algebra

$$\Psi^{-\infty}_{\mathscr{V}}(X;E,F) = \bigcap_m \Psi^m_{\mathscr{V}}(X;E,F). \tag{21}$$

Naturally the map (19) is to be consistent with (15) as is the product formula:

$$\sigma_{m+m',\mathscr{V}}(A \circ B) = \sigma_{m,\mathscr{V}}(A) \circ \sigma_{m',\mathscr{V}}(B). \tag{22}$$

Finally we require the algebra to be asymptotically complete with respect to the filtration. Thus if $A_j \in \Psi^{m-j}_{\mathscr{V}}(X;E,F)$ then there must exist $A \in \Psi^m_{\mathscr{V}}(X;E,F)$ such that

$$A \sim \sum_j A_j, \text{ i.e. } A - \sum_{j<N} A_j \in \Psi^{m-N}_{\mathscr{V}}(X;E,F) \quad \forall \, N \in \mathbb{N}. \tag{23}$$

Then A is determined up to the addition of an element of the residual algebra.

These conditions hold for the calculus of pseudodifferential operators on a compact manifold without boundary. Various of the standard theorems in that case can be extended to any algebra with these properties. For example

Proposition (24). *If $A \in \Psi^m_{\mathscr{V}}(X;E,F)$ is elliptic, in the sense that its symbol has a representative which is invertible outside a compact set, and the conditions (16)-(21) hold then there exists $B \in \Psi^{-m}_{\mathscr{V}}(X;F,E)$ such that*

$$A \circ B - \mathrm{Id} \in \Psi^{-\infty}_{\mathscr{V}}(X;F,F), \; B \circ A - \mathrm{Id} \in \Psi^{-\infty}_{\mathscr{V}}(X;E,E). \tag{25}$$

In the familiar case, that is $\partial X = \emptyset$ and $\mathscr{V} = \mathscr{C}^\infty(X;TX)$, the residual calculus consists of smoothing operators, which are compact operators on any Sobolev space. Then (25) is a rather strong conclusion. In other cases, such as those discussed below, the residual calculus does not consist of compact operators. Then for most applications (25) must be improved. The proof of (25) itself is straightforward. Using the surjectivity of the symbol map in (19), the composition formula (22) and the exactness in (19) one can make an initial choice B_1 for B which satisfies the weaker form of (25):

$$A \circ B_1 - \mathrm{Id} = R_1 \in \Psi_{\mathscr{V}}^{-1}(X;F). \tag{26}$$

The space $\Psi_{\mathscr{V}}^{-1}(X;E)$ is an ideal in $\Psi_{\mathscr{V}}^0(X;E)$ so, again using (18) the Neumann series for $(\mathrm{Id}+R_1)^{-1}$ converges asymptotically in the sense of (23), allowing $B \sim B_1(\mathrm{Id}+R_1)^{-1}$ to be constructed. To improve (25) we should therefore look for other ideals which allow finer approximations to an inverse to be constructed.

To find such ideals in the calculus we need to assume more about the algebra $\mathscr{V}$. Through any point $x \in X$ let $Q = Q(x)$ be the closure of the set of endpoints of integral curves of $\mathscr{V}$ starting at x. We shall demand that this be always a p-submanifold. This is the strongest condition on a submanifold in the category of manifolds with corners, it just means that near each point of Q there is a coordinate system reducing the point to the origin in $\mathbb{R}_k^n$, and such that Q is locally given by the intersection of coordinate planes; the first k of these being the local boundary hypersurfaces. Restriction to Q of the elements of $\mathscr{V}$ defines a Lie algebra of vector fields on Q:

$$\mathscr{V}(X) \longrightarrow \mathscr{W}_{\mathscr{V}}(Q) \subset \mathscr{V}_b(Q). \tag{27}$$

We demand that

$$\mathscr{W}_{\mathscr{V}}(Q) = \mathscr{W}(Q) \text{ span } \mathscr{V}_b(Q) \text{ over the interior of } Q. \tag{28}$$

It follows that the null space of the restriction (27), $\mathscr{N}(\mathscr{V};Q)$, defines a subbundle

$$^{\mathscr{V}}NQ \subset {}^{\mathscr{V}}T_Q X. \tag{29}$$

The fibres of this bundle are finite dimensional Lie algebras and

$$\mathscr{W}(Q) = \mathscr{C}^\infty(Q; {}^{\mathscr{V}}T_Q X / {}^{\mathscr{V}}NQ)$$

i.e. the structure bundle, $^{\mathscr{W}}TQ$, of $\mathscr{W}$ is just the quotient bundle $^{\mathscr{V}}T_Q X / {}^{\mathscr{V}}NQ$.

We actually require that these conditions on the integral leaves of $\mathscr{V}$ hold locally uniformly in an appropriate sense. To explain this we first consider the notion of a b-map between compact $\mathscr{C}^\infty$ manifolds with corners. A $\mathscr{C}^\infty$ map

$$f : X \longrightarrow X' \tag{30}$$

is a b-map if, for each boundary hypersurface $H' \in M_1(X')$ of X' the lift of the ideal of functions, $\mathscr{I}(H') \subset \mathscr{C}^\infty(X')$, vanishing at H' is the product of such ideals in X :

$$f^* \mathscr{I}(H') = \prod_{H \in M_1(X)} \mathscr{I}(H)^{e(H,H')}. \tag{31}$$

In terms of defining functions of the boundary hypersurfaces this just means

$$f^* \varrho_{H'} = a_{H'} \prod_{H \in M_1(X)} \varrho_H^{e(H,H')}, \ 0 < a_{H'} \in \mathscr{C}^\infty(X). \tag{32}$$

The powers $e(H, H')$ are non-negative integers. One important consequence of (32) is that the differential of f extends to the b-tangent space ${}^b TX$, introduced above:

$${}^b f_* : {}^b T_x X \longrightarrow {}^b T_{f(x)} X' \quad \forall \ x \in X. \tag{33}$$

We introduce three refinements of the notion of a b-map related to the properties if this b-differential. Namely a b-map is said to be a b-*submersion* if the b-differential, (33), is everywhere surjective. It is said to be a b-*normal* map if

$$\forall \ H \in M_1(X) \ e(H, H') \neq 0 \text{ for at most one } H' \in M_1(X'). \tag{34}$$

Finally a b-map is said to be a b-*fibration* if it is both b-normal and a b-submersion. A fibration of compact manifolds with corners is easily seen to be a b-fibration. The converse statement is not true. For any b-fibration $\phi : X \longrightarrow X'$ the subspace $\mathscr{V}_\phi(X) \subset \mathscr{V}_b(X)$, just the sections of the null space of ${}^b\phi_*$, is a Lie algebra of vector fields satisfying all the conditions imposed so far, as is the weighted version

$$\varrho^k \mathscr{V}_\phi(X) = \{ V \in \mathscr{V}_b(X); V = \varrho^k W, \ W \in \mathscr{V}_\phi(X) \}$$
$$\varrho^k = \prod_{h \in M_1(X)} \varrho_H^{k(h)} \tag{35}$$

for any map $k : M_1(X) \longrightarrow \mathbb{N}_0 = \{0, 1, \dots\}$.

The final condition we wish to impose on $\mathscr{V}$ is that each boundary face $F \in M(X)$ (including X itself) should have a b-fibration

$$\phi_F : F \longrightarrow L(F) \tag{36}$$

such that

$$Q(x) = \phi_F^{-1}(\phi_F(x)) \ \forall \ x \text{ in the interior of } F. \tag{37}$$

Moreover we require that the Lie algebra is not too far from being that of the b-fibration corresponding to the interior:

$$\mathscr{V}(X) \supset \varrho^k \mathscr{V}_\phi(X) \text{ for some } k : M_1(X) \longrightarrow \mathbb{N}_0 \tag{38}$$

where $\phi = \phi_X$. Although the integral leaves through the interior points of F are just the leaves of ϕ_F through those points this still allows considerable freedom for the Q to change, in dimension, from face to face.

Definition (39). A boundary-fibration structure on a compact manifold with corners is a subspace $\mathscr{V}(X) \subset \mathscr{V}_b(X)$ satisfying (8), (11), (12), (28), (37) and (38).

With these additional conditions it follows that for each integral submanifold $Q = Q(x)$

$$\text{each fibre } {}^\mathscr{V} N_p Q, \ p \in Q, \text{ is a solvable Lie algebra}. \tag{40}$$

Notice that we are immediately forced to consider a more general notion than that of a boundary-fibration structure, namely the structure induced by a boundary-fibration structure on each of the leaves. In this extended type of structure there are hidden variables, namely those in the bundle, $^{\mathcal{V}}NQ$ of solvable Lie algebras. A small calculus of pseudodifferential operators, still denoted $\Psi_{\mathcal{V}}^{m}(X)$, corresponding to such a structure should behave as convolution operators on the fibres of the associated bundle of Lie groups.

The ideals needed to obtain finer approximate inverses of elliptic operators are associated with the quotients

$$0 \longrightarrow \mathscr{I}_{Q}(X) \cdot \mathscr{V}(X) \longrightarrow \mathscr{V}(X) \longrightarrow \mathscr{U}(Q) = \mathscr{C}^{\infty}(Q; {}^{\mathcal{V}}T_{Q}X) \longrightarrow 0. \quad (41)$$

Since $\mathscr{I}(Q) \cdot \mathscr{V}$ is an ideal this gives rise to a homomorphism of the enveloping algebras, which we call the normal operator at Q:

$$N_{Q} : \operatorname{Diff}_{\mathscr{V}}^{k}(X; E, F) \longrightarrow \operatorname{Diff}_{\mathscr{U}}^{k}(Q; E, F). \quad (42)$$

We require (as a condition on the $\mathscr{V}$-pseudodifferential operators) that there be an extension of this map

$$N_{Q} : \Psi_{\mathscr{V}}^{m}(X; E, F) \longrightarrow \Psi_{\mathscr{U}}^{m}(Q; E, F). \quad (43)$$

The ideal

$$\Psi_{\mathscr{V}}^{m,1}(X; E, F) = \{A \in \Psi_{\mathscr{V}}^{m}(X; E, F); N_{Q}(A) = 0 \ \forall \ Q \subset \partial X\} \quad (44)$$

and its powers

$$\Psi_{\mathscr{V}}^{m,k}(X; E, F) = \Psi_{\mathscr{V}}^{m,1}(X; E, F) \cdot \Psi_{\mathscr{V}}^{0,k-1}(X; E) \quad (45)$$

are then of considerable interest and in particular we would expect that

$$\Psi_{\mathscr{V}}^{-\infty,\infty}(X; E, F) \ni A : \mathscr{C}^{\infty}(X; E) \longrightarrow \dot{\mathscr{C}}^{\infty}(X; F), \quad (46)$$

the latter space being the subspace of $\mathscr{C}^{\infty}(X; F)$ consisting of the sections vanishing to all order at ∂X.

Error terms of the type in (46) would be extremely satisfactory, but in most practical cases it is not possible to obtain them directly. This can be understood from the discussion of (26). In order to get the error in $\Psi_{\mathscr{V}}^{-1,1}(X; F)$, which could be improved by iteration, as in the proof of (25), we need to be able to solve the *model problem*

$$N_{Q}(A) \cdot N_{Q}(B) = \operatorname{Id} \quad (47)$$

exactly, for all integral submanifolds $Q \subset \partial X$. This can often be done, and this is precisely the method used in the problems discussed below. However the inverse, $N_{Q}(B)$, is seldom in the small calculus associated to $\mathscr{U}$. In practice it is therefore necessary to consider *full* calculi which have more off-diagonal singularities. This is discussed in III below.

II. Applications and Examples

Next I shall consider various examples of Lie algebras of vector fields, $\mathscr{V}$, and give some representative results concerning $\mathscr{V}$-differential operators which can be obtained by use of the calculus of $\mathscr{V}$-pseudodifferential operators.

(i) b-Structure

The first example is the one already mentioned, namely $\mathscr{V}_b(X)$. This is naturally defined on any compact manifold with corners. For a compact manifold with boundary the associated b-calculus was described in [22], see also [15; Chapter 18], and called the totally characteristic calculus. For the general case of a compact manifold with corners the small calculus is described in [25] and the full calculus in [23]. Certainly $\mathscr{V}_b(X)$ is a boundary-fibration structure and the calculus has all the properties described above. In the case of a compact manifold with boundary the only integral submanifolds are X and the components of the boundary $H \in M_1(X)$. The b-fibrations required in (37) are all the single-leaf fibrations. Consider the normal operator for the calculus, as indicated in (43). Since this is such a natural calculus the normal operator can be realized as an operator on an associated geometric space. Namely the normal bundle to a component of the boundary, $N\{X; H\}$, $H \in M_1(X)$, can be compactified to a bundle of intervals, denoted X_H over H. This has a natural action of $[0, \infty)$. The normal operator takes the form

$$N_H : \Psi_b^m(X) \longrightarrow \Psi_{b,I}^m(X_H) \tag{48}$$

where the suffix I denotes the $[0, \infty)$-invariant elements. The invertibility properties of these operators can be discussed using the Mellin transform. In fact the Mellin transform of the normal operator gives an entire family of pseudodifferential operators on H, called the indicial family:

$$\mathbb{C} \ni \lambda \longmapsto I(A, \lambda) \in \Psi^m(H) . \tag{49}$$

If A is elliptic as a b-pseudodifferential operators this family is elliptic for each λ (in fact the principal symbol is independent of λ) and $I(A, \lambda)$ is invertible except for a discrete set of points:

$$\mathrm{spec}_b(A) = \{\lambda \in \mathbb{C} : I(A, \lambda) \text{ is not invertible}\} . \tag{50}$$

Theorem (51) (Melrose and Mendoza [24]). *An elliptic element of* $\Psi_b^m(X)$ *is Fredholm as a map on the space* $\varrho^r \mathscr{C}^\infty(X)$ *if and only if*

$$\mathrm{Im}\,\lambda \neq -r \quad \forall\, \lambda \in \mathrm{spec}_b(A) . \tag{52}$$

The formula for the index of such an elliptic b-pseudodifferential operators involves extensions of the work of Atiyah, Patodi and Singer [1] on index theory and the η-invariant.

Metrics on the bundle ${}^b T X$ are related to conic geometry. If X is a compact manifold with boundary then near ∂X it decomposes as a product, since ∂X has a (global) defining function $x \in \mathscr{C}^\infty(X)$. By a *conic metric* (of weight s) on X we mean a Riemann metric on the interior which, near ∂X, takes the form

$$g_b \sim x^{2s} \left[\left(\frac{dx}{x} \right)^2 + h(x, y, dx, dy) \right] \tag{53}$$

for some defining function x, and with $h(0, y, 0, dy)$ a metric on ∂X. This is slightly more special than simply a fibre metric on ${}^b T X$. Cheeger considered such questions as the identity of the L^2 cohomology groups of X with respect to g_b (see [6, 7, 8]). These questions can be treated rather directly by constructing a good parametrix for the Laplacian using the b-calculus, see [23].

Another differential-geometric question involving the b-calculus is suggested by the work of Baum, Douglas and Taylor [3]). Namely to find an analytic construction of Poincaré duality for K-theory (Kasparov [16, 17]) for compact manifolds with boundary, extending the original motivation for K-homology of Atiyah.

Theorem (54) (Melrose and Piazza [25]). *A b-quantization map induces isomorphisms realizing Poincaré duality:*

$$K^i(T^* X) \longleftrightarrow K_i(X, \partial X), \ K^i(T^* X, \partial T^* X) \longleftrightarrow K_i(X), \ i = 0, 1.$$

This means that the K_0-homology groups can be identified with equivalence classes of elliptic b-pseudodifferential operators on X. The K_1 groups can be identified with such classes on $X \times [0, 1]$, so the calculus on manifolds with corners is useful even if X is a manifold with boundary.

The b-calculus also has other more analytic applications, particularly as related to spaces of conormal distributions.

(ii) 0-Structure

A second natural boundary-fibration structure on any compact manifold with boundary is the 0-structure. The Lie algebra of vector fields is

$$\mathcal{V}_0(X) = \{ V \in \mathcal{V}_b(X) ; V = 0 \text{ at } \partial X \} . \tag{55}$$

The associated calculus described in [19] has all the properties introduced above. The integral submanifolds are simply X and the individual points of the boundary. The b-fibrations are therefore again trivial, the single-fibre fibration of X and the point-fibre fibration of the boundary. Since each boundary point is a leaf the fibre of the structure bundle ${}^0 T_p X$ is a solvable Lie algebra for each $p \in \partial X$; it is just the homogeneous extension of an abelian algebra. The normal operators at each boundary point are convolution operators on the associated solvable Lie group. Again the naturality of the 0-structure means that the normal operators actually act on a geometric space, in this case on the inward-pointing half of the tangent space to X at p.

Fibre metrics on the structure bundle ${}^0TX$ are always of the form

$$g_0 = \frac{h}{x^2} \tag{56}$$

where h is a metric in the ordinary sense on X and x is a defining function for the boundary. The invariant metric on hyperbolic n-space, represented as the ball $\{|x| < 1\}$ in $\mathbb{R}^n$,

$$g_H = \frac{|dx|^2}{(1 - |x|^2)}$$

is clearly an example of such a metric. In fact the normal operator of the Laplacian of a metric (56) is always reducible to the hyperbolic Laplacian. This can be used to show that the spectral theory (indeed the scattering theory) of a 0-metric is similar to that of the hyperbolic Laplacian. For example

Theorem (57) (Mazzeo and Melrose [19]). *If* $0 < \kappa \in \mathscr{C}^\infty(X)$ *and* $\kappa|dx|^2_h = 1$ *at* ∂X *then the kernel of the normalized resolvent of the metric* (56),

$$[\varDelta - \kappa s(n - s)]^{-1}, \tag{58}$$

extends to a meromorphic family of 0-pseudodifferential operators, $s \in \mathbb{C}$.

Mazzeo in [18] used the construction of a parametrix for the Laplacian to find the Hodge cohomology of a 0-metric as in (56). Metrics of this form arise on cuspless, infinite volume, quotients of hyperbolic space by a geometrically finite discrete group (e.g. Schottky groups).

Other problems can be tackled using these operators, in particular the solution operators to elliptic boundary problems lie in this calculus.

(iii) Bergman Geometry

Any $\mathscr{C}^\infty$ bounded strictly pseudoconvex domain $\Omega \subset \mathbb{C}^n$ carries Kähler metrics with Kähler form

$$\omega_\varrho = -i\partial\bar{\partial}\log\varrho \tag{59}$$

where $-\varrho$ is a plurisubharmonic defining function for $\partial\Omega$. The Laplacian of this metric corresponds to a boundary-fibration structure on $X = \Omega_{\frac{1}{2}}$, the manifold obtained from (and diffeomorphic to) Ω by adjoining $\sqrt{\varrho}$ to $\mathscr{C}^\infty(\Omega)$, to give $\mathscr{C}^\infty(\Omega_{\frac{1}{2}})$. Thus there is a $\mathscr{C}^\infty$ map (with fold singularity at the boundary) $\iota_{\frac{1}{2}} : X \longrightarrow \Omega$. On X the boundary-fibration structure is determined by the contact form θ on $\partial\Omega$. This form fixes a class of 1-forms on X such that for any representative, Θ,

$$\mathscr{V}_\Theta(X) = \{V \in \mathscr{V}_0(X); \Theta(V) = O(r^2)\} \tag{60}$$

where $r = (\iota_{\frac{1}{2}})^* \sqrt{\varrho} \in \mathscr{C}^\infty(X)$.

Theorem (61) (Epstein, Melrose and Mendoza [11]). *The resolvent of the Bergman-type Laplacian, of the metric corresponding to* (59), *has an analytic extension similar to that of the Laplacian on the ball:*

$$[\Delta - s(1 - s)]^{-1}$$

extends from $\operatorname{Re} s > \frac{1}{2}$ *to be a meromorphic family of* Θ*-pseudodifferential operators,* $s \in \mathbb{C}$.

Using this same approach, i.e. by studying the Bergman-type Laplacian, but now on $(n, 0)$ and $(n, 1)$-forms one can provide an alternate solution of Kohn's $\bar{\partial}$-Neumann problem:

$$\bar{\partial}u = f, \; f \in \mathscr{C}^\infty(\Omega; \Lambda^{0,1}), \; \bar{\partial}f = 0$$
$$u \perp \mathscr{H}(\Omega) = \{u \in L^2(\Omega), \; \bar{\partial}u = 0\} . \tag{62}$$

Corollary (63). *On any* $\mathscr{C}^\infty$ *strictly pseudoconvex domain Kohn's* $\bar{\partial}$*-Neumann problem* (62) *is solved,* $u = Nf$, *by a* Θ*-pseudodifferential operator of order* -1.

The standard regularity properties, see [13], then follow readily.

(iv) Adiabatic Limit of a Fibration

Consider a fibration of compact manifolds

$$F \underline{\hspace{2cm}} M$$
$$\downarrow \phi$$
$$Y .$$

On the total space, M, consider a 1-parameter family of metrics which becomes singular as the parameter, $x \downarrow 0$:

$$g_x = g_\infty + \frac{\phi^* h}{x^2} . \tag{64}$$

Here g_∞ is a metric on M (or just a symmetric cotensor which is non-degenerate on the fibres) and h is a metric on Y.

This is an example of a singular limit. It was considered by Witten ([28]) in a special case to study the η-invariant. In general the behaviour of the adiabatic limit of the η-invariant was discussed by Bismut and Cheeger [2] and by Dai [9]. The analytic torsion or Ray and Singer can be examined by similar methods (see Dai, Epstein and Melrose [10]).

The associated boundary-fibration structure provides a simple way to describe the adiabatic limit of the Hodge cohomology. The manifold with corners on which we work is $X = M \times [0, 1]$ where the extra factor is the parameter x. Then set

$$\mathscr{V}_a(X) = \{V \in \mathscr{V}_b(X); Vx \equiv 0 \text{ and}$$
$$V \text{ is tangent to the fibres of } \phi \text{ at } x = 0\} . \tag{65}$$

If local coordinates are taken for the fibration, y in the base and z in the fibres then locally

$$\mathscr{V}_a(X) = \mathrm{sp}\,\{x\partial y, \partial z\} = \mathscr{C}^\infty(X;{}^a T X). \tag{66}$$

The b-fibrations required, in (37), to show that $\mathscr{V}_a(X)$ is a boundary-fibration structure are given by the projection $X \longrightarrow M$ over the interior and $\{x = 1\}$ and by the fibration ϕ over $\{x = 0\}$. The metrics (64) fix a fibre metric on ${}^a T X$ and so, from (13) the Laplacian $\Delta_a \in \mathrm{Diff}_a^2(X;{}^a\Lambda^k)$. Then

$$\mathscr{H}^k = \left\{u \in \mathscr{C}^\infty(X;{}^a\Lambda^k); \Delta_a u = 0\right\}$$

is the space of all $\mathscr{C}^\infty$ sections of a vector bundle over $[0,1]$ with fibre $H_x^k \subset \mathscr{C}^\infty(X_x;{}^a\Lambda^k)$. The behaviour of these fibres at 0 brings out the Leray spectral sequence for the fibration. Namely

Proposition (67) (Mazzeo and Melrose [20]). *For each k the spaces, defined for each $\ell \in \mathbb{N}_0$,*

$$E^{k,\ell} = \left\{u_0 \in \mathscr{C}^\infty(X_0;{}^a\Lambda^k); \exists\ \tilde{u} \in \mathscr{C}^\infty(X;{}^a\Lambda^k) \right.$$
$$\left. \text{with } \tilde{u}\big|_{x=0} = u_0,\ \Delta_a\tilde{u} = 0(x^{2\ell})\right\} \tag{68}$$

form a decreasing sequence of vector spaces stabilizing to H_0^k, which has the dimension of the cohomology of M:

$$E^{k,0} \supset E^{k,1} \supset E^{k,2} \supset \cdots \supset E^{k,N} = \cdots = E^{k,\infty} = H_0^k. \tag{69}$$

Here $E^{k,0} = \mathscr{C}^\infty(X_0;{}^a\Lambda^k)$, $E^{1,k}$ is the space of sections of a vector bundle over Y and $E^{k,2}$ is finite dimensional.

(v) Töplitz Correspondence

Any compact $\mathscr{C}^\infty$ manifold, Y, can be embedded as a totally real submanifold of a complex manifold, Ω, with $\dim_{\mathbb{C}} \Omega = \dim_{\mathbb{R}} Y$, (Bruhat and Whitney [5]). There exists $\varrho \in \mathscr{C}^\infty(\Omega)$ with

$$\varrho \geq 0, \quad Y = \{\varrho = 0\}$$

and a non-degenerate minimum at, i.e. non-degenerate Hessian normal to, Y. Then the tubes around Y

$$\Omega_\varepsilon = \left\{z \in \Omega; \varrho(z) \leq \varepsilon^2\right\}$$

are, for $\varepsilon > 0$ small enough, strictly pseudoconvex neighbourhoods of Y. A fibration of Ω near Y and transversal to it can be chosen so that $\mathrm{Im}\,\partial\varrho$ vanishes on the fibres. The Töplitz correspondence is the map obtained by fibre-integration of holomorphic $(n,0)$ forms:

$$T_\varepsilon : \left\{u \in \mathscr{C}^\infty(\Omega_\varepsilon; \Lambda^{n,0}); \ \bar\partial u = 0\right\} \overset{\overset{\int}{\text{fibre}}}{\longrightarrow} \mathscr{C}^\infty(Y). \tag{70}$$

Boutet de Monvel and Guillemin [4] showed that this map is Fredholm and conjectured the following

Theorem (71) (Epstein and Melrose [12]). *There exists $\varepsilon_0 > 0$ such that for $0 < \varepsilon < \varepsilon_0$ the Töplitz correspondence* (70) *is an isomorphism.*

The main step in the proof is the uniform solution of the $\bar{\partial}$-Neumann problem in Ω_ε as $\varepsilon \downarrow 0$. For this we start with the manifold $M \simeq \Omega_{\varepsilon_0} \times [0, \varepsilon_0]$, obtained by blowing up the singular set in the cone formed by the shrinking tubes,

$$Y \times \{0\} \subset \left\{(z, \varepsilon); \varrho(z) \le \varepsilon^2, \, 0 \le \varepsilon \le \varepsilon_0\right\}.$$

Thus M is a manifold with corners, it has three boundary hypersurfaces. The blow-up procedure introduces an 'adiabatic' boundary H_a, and this has a fibration, $\phi_a : H_a \longrightarrow Y$. There is a trivial boundary at $\{\varepsilon = \varepsilon_0\}$ and the θ-boundary, H_θ. The θ-boundary corresponds to the boundary of Ω and has a contact bundle defined on it. Let X be the manifold with the square-root $\mathscr{C}^\infty$ structure introduced at H_θ, there is a conformal class of 1-forms, with representative Θ, defined at the new H_Θ. On X the α-structure is obtained by combining the Θ-structure of (iii) with the adiabatic structure of (iv):

$$\mathscr{V}_\alpha(X) = \{V \in \mathscr{V}_b(X); Vx \equiv 0,$$
$$V \text{ is tangent to the fibres of } \phi_a \text{ and} \tag{72}$$
$$V \text{ vanishes at } H_\Theta \text{ with } \Theta(V) = O(\varrho_\Theta^2)\}.$$

Here ϱ_Θ is a defining function for H_Θ; $\mathscr{V}_\alpha(X)$ is a boundary-fibration structure. The adiabatic Bergman-type metric with Kähler form

$$-i\partial\bar{\partial} \log\left(\frac{\varrho}{\varepsilon^2} - 1\right)$$

is a fibre metric on the structure bundle ${}^\alpha TX$ of $\mathscr{V}_\alpha$. Thus, by Proposition (13) the Laplacian is an elliptic element of $\mathrm{Diff}_\alpha^2(X; {}^\alpha \Lambda^{p,q})$. One of the crucial steps in proving Theorem (71) is to show that the Bergman projection, onto holomorphic functions on the tube of radius ε, is an α-pseudodifferential operator.

(vi) Analytic Surgery

Suppose M is a compact manifold without boundary and $H \subset M$ is an embedded, oriented closed hypersurface. If ϱ is a defining function for H and x is a parameter consider the 1-parameter family of metrics

$$g_x = a \, d\varrho^2 + (\varrho^2 + x^2)h.$$

Here $a > 0$ is a $\mathscr{C}^\infty$ function in (ϱ, x)-polar coordinates, i.e. on the manifold, $X = [M \times [0, 1]; H \times \{0\}]$, obtained by blowing up $H \times \{0\}$ in $M \times [0, 1]$ and h is a metric on M. As $x \downarrow 0$ the metric g_x degenerates to a conic metric (incomplete) on the manifold with boundary obtained by closing $M \backslash H$ with two copies of H. The boundary-fibration structure on X is

$$\mathscr{V}_v = \{V \in \mathscr{V}_b(X); Vx \equiv 0\} \, . \tag{73}$$

McDonald in his PHD thesis [21] has constructed the v-calculus which contains the inverse of the Laplacian of g_x. This allows one to describe, somewhat as in (iv), the behaviour of the Hodge cohomology as $x \downarrow 0$. Partial results of this type have been obtained by Seeley and Singer [27] and more recently extended by Seeley [26].

(vii) Projective Algebraic Varieties

One rather intriguing question is whether the notion of boundary-fibration in Definition (39) applies to a singular projective algebraic variety M. More precisely let X be a compact manifold with corners obtained from M in two steps. First let $F : \widetilde{M} \longrightarrow M$ be a resolution of M, so the preimage of the singular locus of M is a finite union of hypersurfaces in $\widetilde{M}$ with only normal intersections. From the real point of view these are embedded submanifolds of codimension two, let X be obtained by real blow-up of these submanifolds. Now we ask:

Question (74). Is there are boundary-fibration structure on X (for an appropriate choice of $\widetilde{M}$) such that the induced (Fubini-Study) metric on the regular part of M is a weighted fibre metric on the structure bundle?

A weighted metric is as in (53), i.e. a fibre metric with a conformal factor which is a product of powers of defining functions for boundary hypersurfaces. If such a boundary-fibration structure exists then the construction of the corresponding $\mathscr{V}$-pseudodifferential calculus, as outlined next, should allow a rather direct treatment of the Hodge cohomology of the Laplacian.

III. Microlocalization

In the final part of my talk I wish to give a brief description of a general procedure to construct a calculus of $\mathscr{V}$-pseudodifferential operators. For simplicity I shall only consider the cases involving no parameters, i.e. when $\mathscr{V}$ spans the tangent space over the interior of X. This description therefore only applies to cases (i)–(iii) although the general case is quite similar, simply more complicated. Also for simplicity of presentation I shall suppress various line bundles, typically weighted density bundles, which occur during the construction by describing their sections as 'distributional densities'.

Schwartz' kernel theorem asserts that the space of continuous linear maps

$$A : \dot{\mathscr{C}}^{\infty}(X) \longrightarrow \mathscr{C}^{-\infty}(X), \tag{75}$$

where the range space is the space of extendible distributions on a compact manifold with corners, can be identified with a space distributional densities over the product $X^2 = X \times X$; namely the sections of the 'kernel density bundle' KD. Thus k_A is the kernel if A if

$$A\psi(x) = \int_X k_A(x, x')\psi(x'). \tag{76}$$

If $\partial X = \emptyset$ then under this identification

$$\Psi^m(X) \longleftrightarrow I^m(X^2, \Delta; KD). \tag{77}$$

The notation for the space of conormal distributions on the right, on X^2 associated to the diagonal, $\Delta = \{(x, x') \in X^2; x = x'\}$, is that of Hörmander ([14]). The differential operators on X are precisely the local pseudodifferential operators so:

$$\text{Diff}^k(X) \longleftrightarrow \{k \in I^k(X^2, \Delta; KD); \text{supp}(k) \subset \Delta\}. \tag{78}$$

This means that the kernels of differential operators are Dirac sections of KD.

The conormal distributions in (77) are those distributions which are singular only at Δ and in any local coordinates have Fourier transform in directions transversal to Δ a symbol:

$$a(x, \xi) = \int k(x, x')e^{i(x-x')\cdot\xi}dx' \in S^m \text{ i.e. satisfies}$$

$$|\partial_x^\alpha \partial_\xi^\beta a(x, \xi)| \le C_{\alpha,\beta}(1 + |\xi|)^{m-|\beta|} \quad \forall \, \alpha, \beta. \tag{79}$$

A function which is smooth in x and a polynomial in ξ certainly satisfies these estimates. The differential operators are pseudodifferential operators for which a is a polynomial. Thus if we think of the algebra extension

$$\text{Diff}^*(X) \longleftrightarrow \Psi^*(X) \tag{80}$$

as microlocalization of $\mathscr{C}^\infty(X; TX)$ it consists in the replacement of all polynomials by all symbols in (79), or all Dirac sections smooth along Δ in (76) by all conormal sections. Then one can think of pseudodifferential operators as non-commutative symbolic functions of vector fields, just as differential operators are non-commutative polynomials in vector fields.

Algebraically it is clear that this construction should be extendible to boundary-fibration structures. Indeed the assumption (12) means that the elements of $\mathscr{V}$ are just linear functions on the fibres of the bundle $^{\mathscr{V}}T^*X$. Thus one can think of the elements of $\text{Diff}^*_{\mathscr{V}}(X)$ as corresponding to all polynomials on $^{\mathscr{V}}T^*X$. However we need to realize this in a more geometric fashion to generalize (77). Note that $\text{Diff}^k_{\mathscr{V}}(X)$ cannot correspond to all the Dirac sections at the diagonal, since the full algebra of differential operators does!

So we look for a new manifold with corners, the $\mathscr{V}$-stretched product $X^2_{\mathscr{V}}$, which resolves the diagonal from this point of view. More precisely there should be a b-map

$$\beta_{\mathscr{V}} : X^2_{\mathscr{V}} \longrightarrow X^2 \tag{81}$$

which is a diffeomorphism of the interior of $X^2_{\mathscr{V}}$ to the interior of X^2 and a p-submanifold $\Delta_{\mathscr{V}} \subset X^2_{\mathscr{V}}$ such that

$$\beta_{\mathscr{V}} : \Delta_{\mathscr{V}} \longleftrightarrow \Delta. \tag{82}$$

Moreover $X^2_\mathcal{V}$ should be constructed in such a way that each element of $\mathcal{V}$, lifted to X^2 acting on the left factor of X, is $\beta_\mathcal{V}$-related to a vector field on $X^2_\mathcal{V}$, i.e. the Lie algebra lifts from the left factor, and so that the lifted algebra is transversal to $\varDelta_\mathcal{V}$, meaning it spans the normal bundle. It is also natural to demand symmetry of the construction, so that the reflection giving interchange of the factors,

$$\tau : X^2 \ni (x, x') \longmapsto (x', x) \in X^2$$

should lift to diffeomorphism $\tau_\mathcal{V}$ of $X^2_\mathcal{V}$.

Once we find such a resolution we can define a corresponding class of operators. The fact the $\beta_\mathcal{V}$ is a b-map which is an isomorphism over the interior means that

$$\beta^*_\mathcal{V} : \mathscr{C}^{-\infty}(X^2) \longleftrightarrow \mathscr{C}^{-\infty}(X^2_\mathcal{V}). \tag{83}$$

Thus the kernel of an operator can just as well be considered as a distributional density on $X^2_\mathcal{V}$ as on X^2. Next we need to find a replacement for the space on the right in (77). This is also easy since the assumption that $\varDelta_\mathcal{V}$, the lifted diagonal, is a p-submanifold means that it is natural just to take $I^m(X^2_\mathcal{V}, \varDelta_\mathcal{V})$ to be the set of restrictions to $X^2_\mathcal{V}$ of the conormal distributions with respect to an extension of $\varDelta_\mathcal{V}$ in an extension of X^2 as in (3). Away from $\varDelta_\mathcal{V}$ these distributions are $\mathscr{C}^\infty$ so we can impose additional conditions to get

$$I^m_{sm}(X^2_\mathcal{V}, \varDelta_\mathcal{V}) = \big\{ \kappa \in I^m(X^2_\mathcal{V}, \varDelta_\mathcal{V}); \kappa \equiv 0 \text{ at all}$$
$$H \in M_1(X^2_\mathcal{V}) \text{ s.t. } H \cap \varDelta_\mathcal{V} = \emptyset \big\}. \tag{84}$$

This allows us to define the small calculus by analogy with (77):

$$\Psi^m_\mathcal{V}(X) \longleftrightarrow I^m_{sm}(X^2_\mathcal{V}, \varDelta_\mathcal{V}; KD). \tag{85}$$

In all the cases (i)–(vi) (and various others) which have been analyzed the stretched product can be defined by a process of blowing up the integral submanifolds Q of $\mathcal{V}$ in ∂X as submanifolds of the boundary of the diagonal in X^2.

This definition automatically gives the inclusion (17) and a symbol map as in (19). The product formula (18) is not so immediate. Before discussing this briefly let me note where the 'full' calculus, alluded to above, comes from. The kernels in (85) are singular just at the diagonal. However it is rather natural to expect similar conormal singularities to arise at the other naturally defined submanifolds of $X^2_\mathcal{V}$, namely the boundary hypersurfaces. Thus we can consider more general spaces of conormal distributions, denoted $I^{m,\mathfrak{m}}(X^2_\mathcal{V}, \varDelta)$, where $\mathfrak{m} : M_1(X) \longrightarrow \mathbb{R}$ associates an order to each boundary face. Then

$$\Psi^{m,\mathfrak{m}}_\mathcal{V}(X) \longleftrightarrow I^{m,\mathfrak{m}}(X^2_\mathcal{V}, \varDelta_\mathcal{V}; KD). \tag{86}$$

In practice one should consider polyhomogeneous spaces, with elements having expansions in terms of powers of the defining functions at the boundary hypersurfaces and the diagonal. The small calculus then just corresponds to kernels of order $+\infty$ (meaning rapidly vanishing) at hypersurfaces not meeting $\varDelta_\mathcal{V}$ and smooth up to the other hypersurfaces (having expansions with non-negative integral powers only).

The most important outstanding point is the proof of a product formula, (18), or more generally for the full calculus (although this is not usually an algebra so the orders must satisfy appropriate bounds for operators to compose). This proof is also carried out geometrically in these cases. The crucial construction is that of the $\mathscr{V}$-triple product, $X_{\mathscr{V}}^3$. This is related to $X_{\mathscr{V}}^2$ in essentially the same way that X^3 is related to X^2. Namely there are three maps

$$\pi_{o,\mathscr{V}}^3 : X_{\mathscr{V}}^3 \longrightarrow\!\!\!\!\!\rightarrow X_{\mathscr{V}}^2 \tag{87}$$

for $o = f, c, s$ corresponding to projection off the left, the central and the right factor of X. The notation corresponds to the fact that in the product formula

$$C = A \cdot B \tag{88}$$

the kernel of C, the composite operator, can be computed from those of A, the second operator, and B, the first operator, by pull-back, product and push-forward operations:

$$\kappa_C = (\pi_{c,\mathscr{V}}^3)_* \left[(\pi_{s,\mathscr{V}}^3)^* \kappa_A \cdot (\pi_{f,\mathscr{V}}^3)^* \kappa_B \right] . \tag{89}$$

Here of course one needs to interpret the operations appropriately in terms of distributional densities. Thus the functorial properties of conormal distributions are of primary importance. In particular it is crucial that the stretched projections (87) be b-fibrations since the push-forward and pull-back of polyhomogeneous conormal distributions under b-fibrations are polyhomogeneous conormal, see [23].

References

1. Atiyah, M.F., Patodi, V.K., Singer, I.M: Spectral asymmetry and Riemannian geometry. Math. Proc. Camb. Phil. Soc. **77** (1975) 43–69
2. Bismut, J.-M., Cheeger, J.: Invariants êta et indices des familles pour les variétés à bord. C.R. Acad. Sci. Paris **305** (1987) 127–130
3. Baum, P., Douglas, R.G., Taylor, M.E.: Cycles and relative cycles in analytic K-theory. J. Diff. Geom. **30** (1989) 761–804
4. Boutet de Monvel, L., Guillemin, V.W.: The spectral theory of Toeplitz operators. Princeton Univ. Press, Princeton 1981
5. Bruhat, F., Whitney, H.: Quelques propriétés fondamentales des ensembles analytiques-réels. Comm. Math. Helv. **33** (1959) 132–160
6. Cheeger, J.: On the Hodge theory of Riemannian pseudomanifolds. Proc. Symp. Pure Math. **36** (1980) 91–146
7. Cheeger, J.: Spectral geometry of singular Riemannian spaces. J. Diff. Geom. **18** (1983) 575–657
8. Cheeger, J.: Eta invariants, the adiabatic approximation and conical singularities. J. Diff. Geom. **26** (1987) 175–221
9. Dai, X.: Adiabatic limits, Non-multiplicativity of the signature and the Leray spectral sequence. PhD Thesis, S.U.N.Y. Stony Brook, 1989
10. Dai, X., Epstein, C.L., Melrose, R.B.: Adiabatic limit of the Ray-Singer analytic torsion. (In preparation)

11. Epstein, C.L., Melrose, R.B., Mendoza, G.: Resolvent of the Laplacian on strictly pseudoconvex domains. Acta math. (To appear)
12. Epstein, C.L., Melrose, R.B.: Shrinking tubes and the $\bar{\partial}$-Neumann problem. Preprint (1990)
13. Folland, G.B., Kohn, J.J.: The Neumann problem for the Cauchy-Riemann complex. Ann. Math. Studies, No. 75, Princeton Univ. Press, Princeton 1972
14. Hörmander, L.: Fourier integral operators I. Acta Math. **127** (1971) 79–183
15. Hörmander, L.: The analysis of linear partial differential operators III. Springer, Berlin Heidelberg New York 1985
16. Kasparov, G.G.: Topological invariants of elliptic operators, I: K-homology. Math. USSR Izv. **9** (1975) 751–792
17. Kasparov, G.G.: Equivariant KK-theory and the Novikov conjecture. Invent. math. **91** (1988) 147–201
18. Mazzeo, R.: Hodge cohomology of a conformally compact metric. J. Diff. Geom. **28** (1988) 309–339
19. Mazzeo, R., Melrose, R.B.: Meromorphic extension of the resolvent on complete spaces with asymptotically negative curvature. J. Funct. Anal. **75** (1987) 260–310
20. Mazzeo, R., Melrose, R.B.: The adiabatic limit, Hodge cohomology and Leray's spectral sequence for a fibration. J. Diff. Geom. **31** (1990) 185–213
21. McDonald, P.T.: The Laplacian for spaces with cone-like singularities. PhD Thesis, M.I.T. (1990)
22. Melrose, R.B.: Transformation of boundary problems. Acta Math. **147** (1981) 149–236
23. Melrose, R.B.: Differential analysis on manifolds with corners. (In preparation)
24. Melrose, R.B., Mendoza, G.: Elliptic operators of totally characteristic type. MSRI Preprint (1983)
25. Melrose, R.B., Piazza, P.: Analytic K-theory on manifolds with corners. Adv. Math. (To appear)
26. Seeley, R.: Conic degeneration of the Dirac operator. Preprint
27. Seeley, R., Singer, I.M.: Extending $\bar{\partial}$ to singular Riemann surfaces. J. Geom. Phys. **5** (1988) 121–136
28. Witten, E.: Global gravitational anomalies. Comm. Math. Phys. **100** (1985) 197–229

Birational Classification of Algebraic Threefolds

Shigefumi Mori

Research Institute for Mathematical Sciences, Kyoto University, Kyoto 606, Japan

§1. Introduction

Let us begin by explaining the background of the birational classification. We will work over the field $\mathbb{C}$ of complex numbers unless otherwise mentioned.

Let X be a non-singular projective variety of dimension r. The canonical divisor class K_X is the only divisor class (up to multiples) naturally defined on an arbitrary X. Its sheaf $\mathcal{O}_X(K_X)$ is the sheaf of holomorphic r-forms. An alternative description is $K_X = -c_1(X)$, where $c_1(X)$ is the first Chern class of X. Therefore it is natural to expect some role of K_X in the classification of algebraic varieties.

The classification of non-singular projective curves C is classical, and summarized in the following table, where $g(C)$ is the genus (the number of holes) of C, $H = \{z \in \mathbb{C} \mid \mathrm{Im} z > 0\}$ and Γ is a subgroup of $SL_2(\mathbb{R})$:

(1.1)

$g(C)$	0	1	≥ 2
$\deg K_C$	-2	0	$2g(C) - 2$
C	$\mathbb{P}^1$	$\mathbb{C}/(\text{lattice})$	H/Γ

Here we see three different situations. For instance, everything is explicit if $g(C) = 0$; the moduli (to parametrize curves) is the main interest if $g(C) \geq 2$.

Our interest is in generalizing this to higher dimensions. The first difficulty which arises in the surface case is that there are too many varieties for genuine classification (*biregular classification*).

(1.2) For a non-singular projective surface X and an arbitrary point $x \in X$, there is a birational morphism $\pi : B_x X \to X$ from a non-singular projective surface $B_x X$ such that $E = \pi^{-1}(x)$ is isomorphic to $\mathbb{P}^1$ (E is called a (-1)-*curve*) and π induces an isomorphism $B_x X - E \simeq X - x$.

In view of (1.2), it is impractical to distinguish X from $B_x X, B_y B_x X, \ldots$ if we want a reasonable classification list. More generally, we say that two algebraic varieties X and Y are *birationally equivalent* and we write $X \sim Y$ if there is a birational mapping $X \cdots \to Y$ or equivalently if their rational function fields $\mathbb{C}(X)$ and $\mathbb{C}(Y)$ are isomorphic function fields over $\mathbb{C}$. We did not face this phenomenon in the curve case, since $X \simeq Y$ iff $X \sim Y$ for curves X and Y.

Proceedings of the International Congress
of Mathematicians, Kyoto, Japan, 1990
© The Mathematical Society of Japan, 1991

In view of the list (1.1) for curves, we need to divide the varieties into several classes to formulate more precise problems. This is why the *Kodaira dimension* $\kappa(X)$ of a non-singular projective variety X was introduced by [Iitaka1] and [Moishezon].

(1.3)   Let $H^0(X, \mathcal{O}(vK_X))$ be the space of global v-ple holomorphic r-forms $(v \geq 0, r = \dim X)$, and $\phi_0, \ldots, \phi_N$ be its basis. If $N \geq 0$, then

$$\Phi_{vK_X} : X \cdots \to \mathbb{P}^N \text{ given by } \Phi_{vK}(x) = (\phi_0(x) : \cdots : \phi_N(x))$$

is a rational map. We set $P_v(X) = N + 1$. It is important that $H^0(X, \mathcal{O}(vK_X))$ and Φ_{vK} are birational invariants, that is $X \sim Y$ induces $H^0(X, \mathcal{O}(vK_X)) = H^0(Y, \mathcal{O}(vK_Y))$ for $v > 0$. We set $\kappa(X) = -\infty$ if $P_v(X) = 0$ for all $v > 0$. If $P_e(X) > 0$ for some $e > 0$, then

$$\kappa(X) := \text{Max}\{\dim \Phi_{vK_X}(X) \mid v > 0\}.$$

In particular, $P_v(X)$ and $\kappa(X)$ are birational invariants of X.

We remark that $\kappa(X) \in \{-\infty, 0, 1, \ldots, \dim X\}$, and that X with $\kappa(X) = \dim X$ is said to be *of general type*. We have the following table for curves.

(1.4)

$g(C)$	0	1	≥ 2
$\kappa(C)$	$-\infty$	0	1

To have some idea on higher dimensions, we can use the easy result $\kappa(X \times Y) = \kappa(X) + \kappa(Y)$. In particular,

(1.5)   case $(\kappa(X) = -\infty)$ $\kappa(\mathbb{P}^1 \times Y) = -\infty$,

(1.6)   case $(0 < \kappa(X) < \dim X)$

$$\kappa(\underbrace{E \times \cdots \times E}_{a \text{ times}} \times \underbrace{C \times \cdots \times C}_{b \text{ times}}) = b \text{ if } g(E) = 1 \text{ and } g(C) \geq 2.$$

The case $0 < \kappa(X) < \dim X$ is studied by the Iitaka fibration.

(1.7)   **Iitaka Fibering Theorem** [Iitaka2]. *Let X be a non-singular projective variety with $0 < \kappa(X) < \dim X$. Then there is a morphism $f : X' \to Y'$ of non-singular projective varieties with connected fibers such that $X' \sim X$, $\dim Y' = \kappa(X)$ and $\kappa(f^{-1}(y)) = 0$ for a sufficiently general point $y \in Y'$.*

In (1.7), we cannot expect $\kappa(Y') = \dim Y'$ or even $\kappa(Y') \geq 0$. Therefore X' is not so simple as (1.6). Nevertheless (1.7) reduces the case $0 < \kappa(X) < \dim X$ to the cases $\kappa(X) = -\infty, 0, \dim X$. Thus we can explain the birational classification as in (1.1) for higher dimensions.

§2. Birational Classification

For a non-singular projective variety X, we define a graded ring (called the *canonical ring*)

$$R(X) = \oplus_{v \geq 0} H^0(X, \mathcal{O}(vK_X)).$$

If $\kappa(X) \geq 0$, the v-canonical image $\Phi_{vK}(X)$ is a birational invariant of X. The existence of stable canonical image is interpreted in terms of $R(X)$ by the following easy proposition.

(2.1) **Proposition.** *Let X be a non-singular projective variety of $\kappa(X) \geq 0$. Then $\Phi_{vK}(X)$ for sufficiently divisible $v > 0$ are all naturally isomorphic iff $R(X)$ is a finitely generated $\mathbb{C}$-algebra.*

(2.2) For X of general type, constructing moduli spaces is one of our main interests. One standard way is to try to find a uniform v such that $\Phi_{vK} : X \cdots \to \Phi_{vK}(X)$ is birational for all X and classify the image. One can expect nice properties of the image (canonical model to be explained later) if there is a stable canonical image. Therefore we would like to ask whether the canonical ring is finitely generated for X of general type (2.1).

(2.3) The case $\kappa(X) = \dim X$ suggests to reduce the birational classification of all varieties to the biregular classification of standard models (like $\Phi_{vK}(X)$ for sufficiently divisible v). However, when $\kappa(X) < \dim X$, there are no obvious candidates for the standard models. For $0 \leq \kappa(X)$, we can ask to find some "standard" models.

We only say the following for $\kappa(X) \leq 0$ at this point.

(2.4) For X with $\kappa(X) = 0$, we would like to find some "standard" model $Y \sim X$ and to classify all such Y.

(2.5) For many X with $\kappa(X) = -\infty$, there exist infinitely many "standard" models $\sim X$. To study the relation among these models is a role of birational geometry. We would like to have a structure theorem of such models. One general problem is to see if all such X are *uniruled*, i.e. there exists a rational curve through an arbitrary point of X, or equivalently there is a dominating rational map $\mathbb{P}^1 \times Y \cdots \to X$ for some Y of dimension $n - 1$. (It is easy to see that uniruled varieties have $\kappa = -\infty$ as in (1.5).)

Since we use the formulation by Iitaka and Moishezon, one basic problem will be the deformation invariance of κ.

(2.6) **Conjecture** [Iitaka1, Moishezon]. *Let $f : X \to Y$ be a smooth projective morphism with connected fibers and connected Y. Then $\kappa(f^{-1}(y))$ and $P_v(f^{-1}(y))$ $(v \geq 1)$ are independent of $y \in Y$.*

§3. Surface Case

We review a few classical results on surfaces which may help the reader to understand the results for 3-folds.

The basic result is the inverse process of (1.2).

(3.1) **Castelnuovo-Enriques.** *Let E be a curve on a non-singular projective surface X'. Then E is a (-1)-curve (i.e. $X' = B_x X$ and E is the inverse image of x for some non-singular projective surface X and $x \in X$) iff $E \simeq \mathbb{P}^1$ and $(E \cdot K_{X'}) = -1$. We write $\mathrm{cont}_E : X' \to X$ and call it the contraction of the (-1)-curve E.*

Finding a (-1)-curve in every exceptional set, we have the following:

(3.2) **Factorization of Birational Morphisms.** *Let $f : X \to Y$ be a birational morphism of non-singular projective surfaces. Then f is a composition of a finite number of contractions of (-1)-curves.*

Starting with a non-singular projective surface X, we can keep contracting (-1)-curves if there are any. After a finite number of contractions, we get a non-singular projective surface $Y (\sim X)$ with no (-1)-curves. Depending on whether K_Y is nef $((K_Y \cdot C) \geq 0$ for all curves $C)$, $\kappa(X)$ takes different values.

(3.3) Case where K_Y is nef. Then Y is the only non-singular projective surface $\sim X$ with no (-1)-curves. To be precise, if Y' is a such surface, then the composite $Y \cdots \to X \cdots \to Y'$ is an isomorphism. This Y is called the *minimal model* of X and denoted by $X_{\min}$. In this case, we have $\kappa(X) \geq 0$.

(3.4) Case where K_Y is not nef. Then an arbitrary Y' (including Y) which is birational to X and has no (-1)-curves is isomorphic to either $\mathbb{P}^2$ or a $\mathbb{P}^1$-bundle over some non-singular curve. In this case, X has no minimal models and we have $\kappa(X) = -\infty$ by (1.5).

The above (3.3) together with (3.4) says that the birational classification of X with $\kappa \geq 0$ is equivalent to the biregular classification of minimal models.

Based on (3.3) and (3.4), the canonical model is defined.

(3.5) Let X be a non-singular projective surface of general type. Then there exists exactly one normal projective surface $Z (\sim X)$ such that Z has only Du Val (rational double) points and K_Z is ample, where Du Val points are defined by one of the following list.

$$A_n : xy + z^{n+1} = 0 \ (n \geq 0),$$
$$D_n : x^2 + y^2 z + z^{n-1} = 0 \ (n \geq 4),$$
$$E_6 : x^2 + y^3 + z^4 = 0,$$
$$E_7 : x^2 + y^3 + yz^3 = 0,$$
$$E_8 : x^2 + y^3 + z^5 = 0.$$

Such Z is called the *canonical model* of X and denoted by X_{can}. The natural map $X_{min} \cdots \to X_{can}$ is a morphism which contracts all the rational curves C with $(C \cdot K_{X_{min}}) = 0$ into Du Val points and is isomorphic elsewhere.

(3.5.1) **Remark.** This X_{can} can also be obtained as $\Phi_{vK}(X_{min}) = \Phi_{vK}(X)$ for an arbitrary $v \geq 5$ (Bombieri).

(3.6) Let X be a non-singular projective minimal surface with $\kappa = 0$. Thus X has torsion K_X, i.e. some non-zero multiple of it is trivial. There is a precise classification of all such X.

(3.7) The deformation invariance of $\kappa(x)$ and $P_v(X)$ was done by [Iitaka3] using the classification of surfaces. [Levine] gave a simple proof without using classification.

§4. The Extremal Ray Theory (The Minimal Model Theory)

The first problem in generalizing the results in §3 to higher dimensions is to find some class of varieties in which there is a reasonable contraction theorem because there is no immediate generalization of (3.1) to 3-folds, since the contraction process inevitably introduces singularities [Mori1]. To define the necessary class of singularities, the first important step was taken by Reid [Reid1,3].

(4.1) **Definition** [Reid3]. Let (X, P) be a normal germ of an algebraic variety (or an analytic space) which is normal. We say that (X, P) has *terminal singularities* (resp. *canonical singularities*) iff

(i) K_X is a $\mathbb{Q}$-*Cartier* divisor, i.e. rK_X is Cartier for some positive integer r (minimal such r is called the *index* of (X, P)), and

(ii) for some (or equivalently, every) resolution $\pi : Y \to (X, P)$, we have $a_i > 0$ (resp. $a_i \geq 0$) for all i in the expression:

$$rK_Y = \pi^*(rK_Y) + \sum a_i E_i,$$

where E_i are all the exceptional divisors and $a_i \in \mathbb{Z}$.

For surfaces, a terminal (resp. canonical) singularity is smooth (resp. a Du Val point). We note that, for projective varieties X with only canonical singularities, the same definitions of $P_v(X)$, Φ_{vK} and $\kappa(X)$ work and these are still birational invariants. We can also talk about the ampleness of K_X and the intersection number $(K_X \cdot C) \in \mathbb{Q}$ for such X.

The idea of the cone of curves which is the core of the extremal ray theory was first introduced in Hironaka's thesis [Hironaka].

(4.2) **Definition.** Let X be a projective n-fold. A 1-cycle $\sum a_C C$ is a formal finite sum of irreducible curves C on X with coefficients $a_C \in \mathbb{Z}$. For a 1-cycle Z and a $\mathbb{Q}$-Cartier divisor D, the intersection number $(Z \cdot D) \in \mathbb{Q}$ is defined. Then

$$N_1(X)_{\mathbb{Z}} = \{1\text{-cycles}\}/\{1\text{-cycles } Z \mid (Z \cdot D) = 0 \text{ for all } D\}$$

is a free abelian group of finite rank $\varrho(X) < \infty$. Thus $N_1(X) = N_1(X)_{\mathbb{Z}} \otimes_{\mathbb{Z}} \mathbb{R}$ is a finite dimensional Euclidean space. The classes $[C]$ of all the irreducible curves C span a convex cone $NE(X)$ in $N_1(X)$. Taking the closure for the metric topology, we have a closed convex cone $\overline{NE}(X)$. Then

(4.3) **Cone Theorem.** *If X has only canonical singularities, then there exist countably many half lines $R_i \subset NE(X)$ such that*
   (i) $\overline{NE}(X) = \sum_i R_i + \{z \in \overline{NE}(X) \mid (z \cdot K_X) \geq 0\}$,
   (ii) for an arbitrary ample divisor H of X and arbitrary $\varepsilon > 0$, there are only finitely many R_i's contained in

$$\{z \in \overline{NE}(X) \mid (z \cdot K_X) \leq -\varepsilon(z \cdot H)\}.$$

Such an R_i is called an *extremal ray* of X if it cannot be omitted in (i) of (4.3). We note that an extremal ray exists on X iff K_X is not nef. Each extremal ray R_i defines a contraction of X.

(4.4) **Contraction Theorem.** *Let R be an extremal ray of a projective n-fold X with only canonical singularities. Then there exists a morphism $f : X \to Y$ to a projective variety Y (unique up to isomorphism) such that $f_*\mathcal{O}_X = \mathcal{O}_Y$ and an irreducible curve $C \subset X$ is sent to a point by f iff $[C] \in R$. Furthermore $\operatorname{Pic} Y = \operatorname{Ker}[(C \cdot) : \operatorname{Pic} X \to \mathbb{Z}]$ for such a contracted curve C. This f is called the contraction of R and denoted by cont_R.*

The contraction of an extremal ray is not always birational.

(4.5) *Let X be a smooth projective surface with an extremal ray R. Then cont_R is one of the following.*
   (i) the contraction of a (-1)-curve,
   (ii) a $\mathbb{P}^1$-bundle structure $X \to C$ over a non-singular curve,
   (iii) a morphism to one point, when $X \simeq \mathbb{P}^2$.

The description of all the possible contractions for a nonsingular projective 3-fold X is given in [Mori1]. Here we only remark that $\operatorname{cont}_R X$ can have a terminal singularity $\mathbb{C}^3/<\sigma>$ of index 2, where σ is the involution $\sigma(x, y, z) = (-x, -y, -z)$.

(4.6) The category of varieties in which we play the game of the minimal model program is the category $\mathscr{C}$ of projective varieties with only terminal singularities which are $\mathbb{Q}$-*factorial* (i.e. every Weil divisor is $\mathbb{Q}$-Cartier). The goal of the game is to get a *minimal* (resp. *canonical*) model, i.e. a projective n-fold X with only terminal (resp. canonical) singularities such that K_X is nef (resp. ample). Let us first state the minimal model program which involves two conjectures.

(4.7) Let X be an n-fold $\in \mathscr{C}$. If K_X is nef, then X is a minimal model and we are done. Otherwise, X has an extremal ray R. Then $\operatorname{cont}_R : X \to X'$ satisfies one of the following.

(4.7.1) Case where dim $X' <$ dim X. Then cont$_R$ is a surjective morphism with connected fibers of dimension > 0 and relatively ample $-K_X$ (like $\mathbb{P}^1$-bundle), and X is uniruled ([Miyaoka-Mori]). This is the case where we can never get a minimal model, and we stop the game since we have the global structure of X, cont$_R : X \to X'$.

(4.7.2) Case where cont$_R : X \to X'$ is birational and contracts a divisor. This cont$_R$ is called a *divisorial contraction*. In this case $X' \in \mathscr{C}$ and $\varrho(X') < \varrho(X)$. Therefore we can work on X' instead of X.

(4.7.3) Case where cont$_R : X \to X'$ is birational and contracts no divisors. In this case, $K_{X'}$ is not $\mathbb{Q}$-Cartier and $X' \notin \mathscr{C}$. So we cannot continue the game with X'. This is the new phenomenon in dimension ≥ 3.

To get around the trouble in (4.7.3) and to continue the game, Reid proposed the following.

(4.8) **Conjecture (Existence of Flips).** *In the situation of (4.7.3), there is an n-fold $X^+ \in \mathscr{C}$ with a birational morphism $f^+ : X^+ \to X'$ which contracts no divisors and such that K_{X^+} is f^+-ample. The map $X \cdots \to X^+$ is called a flip.*

Since $\varrho(X^+) = \varrho(X)$ in (4.8), the divisorial contraction will not occur for infinitely many times. Therefore the following will guarantee that the game will be over after finitely many steps.

(4.9) **Conjecture (Termination of Flips).** *There does not exist an infinite sequence of flips $X_1 \cdots \to X_2 \cdots \to \cdots$.*

Therefore the minimal model program is completed only when the conjectures (4.8) and (4.9) are settled affirmatively.

The conjecture (4.9) was settled affirmatively by [Shokurov1] for 3-folds and by Kawamata-Matsuda-Matsuki [KMM] for 4-folds. (4.8) was first done by [Tsunoda], [Shokurov2], [Mori3] and [Kawamata6] in a special but important case. Finally (4.8) was done for 3-folds by [Mori5] using the work of [Kawamata6] mentioned above.

(4.10) Thus for 3-folds, we can operate divisorial contractions and flips for a finite number of times and get either a minimal model $\in \mathscr{C}$ or an $X \in \mathscr{C}$ which has an extremal ray R of type (4.7.1). Thus we can get 3-fold analogues of results in §3.

(4.11) For simplicity of the exposition, we did not state the results in the strongest form and we even omitted various results. Therefore we would like to mention names and give a quick review.

After the prototype of the extremal ray theory was given in [Mori1], the theory has been generalized to the relative setting with a larger class of singularities (toward the conjectures of Reid [Reid3,4]) by Kawamata, Benveniste, Reid, Shokurov and Kollár (in the historical order) and perhaps some others.

First through the works of [Benveniste] and [Kawamata2], Kawamata introduced a technique [Kawamata3] which was an ingenious application of the
Kawamata-Viehweg vanishing ([Kawamata1] and [Viehweg2]). Based on the
works by [Shokurov1] (Non-vanishing theorem) and [Reid2] (Rationality theorem), [Kawamata4] developped the technique to prove the Base point freeness
theorem (and others) in arbitrary dimensions. The discreteness of the extremal
rays was later done by [Kollár1]. As for this section, we refer the reader to the
talk of Kawamata.

§5. Applications of the Minimal Model Program (MMP) to 3-Folds

Considering MMP in relative setting, one has the factorization generalizing (3.2):

(5.1) **Theorem.** *Let $f : X \to Y$ be a birational morphism of projective 3-folds
with only $\mathbb{Q}$-factorial terminal singularities. Then f is a composition of divisorial
contractions and flips.*

Since minimal 3-folds have $\kappa \geq 0$ by the hard result of Miyaoka [Miyaoka1–
3], one has the following (cf. (3.3) and (3.4)).

(5.2) **Theorem.** *A 3-fold X has a minimal model iff $\kappa(X) \geq 0$.*

Unlike the surface case, the minimal model of a 3-fold X is not unique; it is
unique only in codimension 1. If we are given a $\mathbb{Q}$-factorial minimal model $X_{\min}$,
every other $\mathbb{Q}$-factorial minimal model of X is obtained from $X_{\min}$ by operating
a simple operation called a *flop* for a finite number of times ([Kawamata6],
[Kollár4]). Many important invariants computed by minimal models do not
depend on the choice of the minimal model. We refer the reader to the talk of
Kollár.

(5.3) **Theorem.** *For a 3-fold X, the following are equivalent.*
   (i) $\kappa(X) = -\infty$,
   (ii) X is uniruled,
   *(iii) X is birational to a projective 3-fold Y with only $\mathbb{Q}$-factorial terminal
sigularities which has an extremal ray of type (4.7.1).*

It will be an important but difficult problem to classify all the possible Y
in (iii) of (5.3). There are only finitely many families of such Y with $\varrho(Y) = 1$
([Kawamata7]).
Since a canonical model exists if a minimal model does ([Benveniste] and
[Kawamata2]), one has the following (cf. (3.5)).

(5.4) **Theorem.** *If X is a 3-fold of general type, then X has a canonical model and
the canonical ring $R(X)$ is a finitely generated $\mathbb{C}$-algebra.*

The argument for (5.4) can be considered as a generalization of the argument for (3.5.1). However the effective part "$v \geq 5$" of (3.5.1) has not yet been generalized to dimension ≥ 3.

To study varieties X with $\kappa \geq 0$, [Kawamata4] posed the following.

(5.5) **Conjecture (Abundance Conjecture).** *If X is a minimal variety, then rK_X is base point free for some $r > 0$.*

For 3-folds, there are works by [Kawamata4] and [Miyaoka4] (cf. [KMM]). However the torsionness of K for minimal 3-folds with $\kappa = 0$ is unsolved, and it remains to prove:

(5.6) **Problem.** *Let X be a minimal 3-fold with H an ample divisor such that $(K_X^3) = 0$ and $(K_X^2 \cdot H) > 0$. Then prove that $\kappa(X) = 2$.*

(5.7) **Remark** ($\kappa = 0$). The 3-folds X with $\kappa(X) = 0$ and $H^1(X, \mathcal{O}_X) \neq 0$ were classified by [Viehweg1] and (5.5) holds for these. This was based on Viehweg's solution of the addition conjecture for 3-folds, and we refer the reader to [Iitaka4]. However not much is known about the 3-folds X with $\kappa(X) = 0$ (or even K_X torsion) and $H^1(X, \mathcal{O}_X) = 0$: so far many examples have been constructed and it is not known if there are only finitely many families. There is a conjecture of [Reid6] in this direction.

By studying the flips more closely, [Kollár-Mori] proved the deformation invariance of κ and P_v (cf. (3.7)):

(5.8) **Theorem.** *Let $f : X \to \Delta$ (unit disk) be a projective morphism whose fibers are connected 3-folds with only $\mathbb{Q}$-factorial terminal singularities. Then*
   (i) $\kappa(X_t)$ is independent of $t \in \Delta$, where $X_t = f^{-1}(t)$,
   (ii) $P_v(X_t)$ is independent of $t \in \Delta$ for all $v \geq 0$ if $\kappa(X_0) \neq 0$.

Indeed for such a family X/Δ, the simultaneous minimal model program is proved and the (modified) work of [Levine] is used to prove (5.8). We cannot drop the condition "$\kappa(X_0) \neq 0$" at present since the abundance conjecture is not completely solved for 3-folds.

As for other applications (e.g. addition conjecture, deformation space of quotient surface singularities, birational moduli), we refer the reder to [KMM] and [Kollár6].

§6. Comments on the Proofs for 3-Folds

Many results on 3-folds are proved by using only the formal definitions of terminal singularities. However some results on 3-folds rely on the classification of 3-fold terminal singularities [Reid3], [Danilov], [Morrison-Stevens], [Mori2] and [KSB] (cf. Reid's survey [Reid5] and [Stevens].) The existence of flips and flops heavily rely on it. Thus generalizing their proofs to higher dimension seems hopeless. At present, there is no evidence for the existence of flips in higher dimensions except that they fit in the MMP beautifully. I myself would accept them as working hypotheses. A more practical problem will be to complete the log-version of the minimal model program for 3-folds [KMM]. This is related to the birational classification of open 3-folds and n-folds with $\kappa = 3$. Since log-terminal singularities have no explicit classification, this might be a good place to get some idea on higher dimension. Shokurov made some progress in this direction [Shokurov3].

There are two other results relying on the classification.

(6.1)   **Theorem** [Mori4]. *Every 3-dimensional termal singularity deforms to a finite sum of cyclic quotient terminal singularities (i.e. points of the form* $\mathbb{C}^3/\mathbb{Z}_r(1,-1,a)$ *for some relatively prime positive integers a and r).*

This was used in the Barlow-Fletcher-Reid plurigenus formula for 3-folds [Fletcher] and [Reid5] (cf. also [Kawamata5]). Given a 3-fold X with only terminal singularities, each singularity of X can be deformed to a sum of cyclic quotient singularities $\mathbb{C}^3/\mathbb{Z}_r(1,-1,a)$. Let $S(X)$ be the set of all such (counted with multiplicity). For each $P = \mathbb{C}^3/\mathbb{Z}_r(1,-1,a) \in S(X)$, we let

$$\phi_P(m) = (m - \{m\}_r)\frac{r^2-1}{12r} + \sum_{j=0}^{\{m\}_r-1} \frac{\{aj\}_r(r-\{aj\}_r)}{2r},$$

where $\{m\}_r$ is the integer $s \in [0, r-1]$ such that $s \equiv m(\mathrm{mod}\ r)$. For a line bundle L on X, let $\chi(L) = \sum_j(-1)^j \dim H^j(X,L)$ and let $c_2(X)$ be the second Chern class of X, which is well-defined since X has only isolated singularities. Then the formula is stated as the following.

(6.2)   **The Barlow-Fletcher-Reid Plurigenus Formula.**

$$\chi(\mathcal{O}_X(mK_X)) = \frac{m(m-1)(2m-1)}{12}(K_X^3) + (1-2m)\chi(\mathcal{O}_X) + \sum_{P\in S(X)} \phi_P(m),$$

$$\chi(\mathcal{O}_X) = -\frac{1}{24}(K_X \cdot c_2(X)) + \sum_{P\in S(X)} \frac{r^2-1}{24r}.$$

This is important for effective results on 3-folds (cf. §7).

(6.3)   **Theorem** ([KSB]). *A small deformation of a 3-dimensional terminal singularity is terminal.*

This is indispensable in the construction of birational moduli. An open problem in this direction is

(6.4)   **Problem.** *Is every small deformation of a 3-dimensional canonical singularity canonical?*

Since this remains unsolved, we cannot put an algebraic structure on

$$\{\text{canonical 3-folds}\}/\text{isomorphisms}.$$

§7. Related Results

I would like to list some of the directions, which I could not mention in the previous sections. This is by no means exhaustive. For instance, I could not mention the birational automorphism groups (cf. [Iskovskih] for the works before 1983) due to the lack of my knowledge.

(7.1)   *Effective Classification.* The Kodaira dimension κ is not a simple invariant. For instance, we know that $\kappa(X) = -\infty$ iff $P_v(X) = 0 (\forall v > 0)$. Therefore $P_{12}(X) = 0$ was an effecitve criterion for a surface X to be ruled, while $\kappa(X) = -\infty$ was not. The 3-dimensional analogue is not known yet.

There are results by Kollár [Kollár2] in the case $\dim H^1(X, \mathcal{O}_X) \geq 3$ (cf. [Mori4]). The Barlow-Fletcher-Reid plurigenus formula (6.2) is applied for instance to get $aK_X \sim 0$ with some effectively given $a > 0$ for 3-folds X with numerically trivial K_X by [Kawamata5] and [Morrison], and to get $P_{12}(X) > 0$ for canonical 3-folds X with $\chi(\mathcal{O}_X) \leq 1$ by [Fletcher].

(7.2)   *Differential Geometry.* As shown by [Yau], there are differential geometric results (especially when K is positive) which seem out of reach of algebraic geometry. Therefore we welcome differential geometric approaches. In this direction is Tsuji's construction of Kähler-Einstein metrics on canonical 3-folds [Tsuji].

(7.3)   *Characteristic p.* [Kollár5] generalized [Mori1] (extremal rays of smooth projective 3-folds over $\mathbb{C}$) to char p. This suggests the possibility of little use of vanishing theorems in MMP for 3-folds. A goal will be the MMP for 3-folds in char p. However even the classification of terminal singularities is open.

(7.4)   *Mixed Characteristic Case.* One can ask about the extremal rays (and so on) for arithmetic 3-folds X/S. The methods of [Shokurov2] and [Tsunoda] might work, if X/S is semistable. In the general case, I do not know any results in this direction.

(7.5)   *Analytic or Non-projective 3-Folds.* Studying analytic or non-projective 3-folds will require a substitute for the cone of curves modulo numerical equivalence. However analytic or non-projective minimal 3-folds can be handled by the flop [Kollár4]. There is a work of [Kollár6].

References

[Benveniste] X.Benveniste: Sur l'anneau canonique de certaines variétés de dimension 3. Invent. math. **73** (1983) 157–164

[Danilov] V.I.Danilov: Birational geometry of toric 3-folds. Math. USSR Izv. **21** (1983) 269–279

[Fletcher] A.Fletcher: Contributions to Riemann-Roch on projective 3-folds with only canonical singularities and applications, In: Algebraic Geometry, Bowdoin 1985. Proc. Symp. Pure Math. **46** (1987) 221–232

[Hironaka] H.Hironaka: On the theory of birational blowing-up. Thesis, Harvard University 1960

[Iitaka1] S.Iitaka: Genera and classification of algebraic varieties. 1 (in Japanese). Sugaku **24** (1972) 14–27

[Iitaka2] S.Iitaka: On D-dimensions of algebraic varieties. J. Math. Soc. Japan **23** (1971) 356–373

[Iitaka3] S.Iitaka: Deformations of complex surfaces. II. J. Math. Soc. Japan **22** (1971) 274–261

[Iitaka4] S.Iitaka: Birational geometry of algebraic varieties. Proc. ICM83, Warsaw 1984, pp. 727–732

[Iskovskih] V.A.Iskovskih: Algebraic threefolds with special regard to the problem of rationality. Proc. ICM83, Warsaw 1984, pp. 733–746

[Kawamata1] Y.Kawamata: A generalisation of Kodaira-Ramanujam's vanishing theorem. Math. Ann. **261** (1982) 43–46

[Kawamata2] Y.Kawamata: On the finiteness of generators of the pluri-canonical ring for a threefold of general type. Amer. J. Math. **106** (1984) 1503–1512

[Kawamata3] Y.Kawamata: Elementary contractions of algebraic 3-folds. Ann. Math. **119** (1984) 95–110

[Kawamata4] Y.Kawamata: The cone of curves of algebraic varieties. Ann. Math. **119** (1984) 603–633

[Kawamata5] Y.Kawamata: On the plurigenera of minimal algebraic 3-folds with $K \approx 0$. Math. Ann. **275** (1986) 539–546

[Kawamata6] Y.Kawamata: The crepant blowing-up of 3-dimensional canonical singularities and its applications to the degeneration of surfaces. Ann. Math. **127** (1988) 93–163

[Kawamata7] Y.Kawamata: Boundedness of Ⅺ-Fano threefolds. Preprint (1990)

[KMM] Y.Kawamata, K.Matsuda, K.Matsuki: Introduction to the minimal model problem. In: Algebraic Geometry, Sendai 1985. Adv. Stud. Pure Math. **10** (1987) 283–360

[Kollár1] J.Kollár: The cone theorem. Ann. Math. **120** (1984) 1–5

[Kollár2] J.Kollár: Higher direct images of dualizing sheaves. Ann. Math. (2) **123** (1986) 11–42

[Kollár3] J.Kollár: Higher direct images of dualizing sheaves. II. Ann. Math. (2) **124** (1986) 171–202

[Kollár4] J.Kollár: Flops. Nagoya Math. J. **113** (1989) 14–36

[Kollár5] J.Kollár: Extremal rays on smooth treefolds. Ann. Sci. ENS (to appear)

[Kollár6] J.Kollár: Flips, flops, minimal models etc. Proc. Diff. Geom. Symposium at Harvard, May 1990

[Kollár-Mori] J.Kollár, S.Mori: To be written up. (1991)

[KSB] J.Kollár, N.Shepherd-Barron: Threefolds and deformations of surface singularities. Invent. math. **91** (1988) 299–338

[Levine] M.Levine: Pluri-canonical divisors on Kähler manifolds. Invent. math. **74** (1983) 293–303

[Miyaoka1] Y.Miyaoka: Deformations of a morphism along a foliation. In: Algebraic Geometry, Bowdoin 1985. Proc. Symp. Pure Math. **46** (1987) 245–268

[Miyaoka2] Y.Miyaoka: The Chern classes and Kodaira dimension of a minimal variety. In: Algebraic Geometry, Sendai 1985. Adv. Stud. Pure Math. **10** (1987) 449–476

[Miyaoka3] Y.Miyaoka: On the Kodaira dimension of minimal threefolds. Math. Ann. **281** (1988) 325–332

[Miyaoka4] Y.Miyaoka: Abundance conjecture for threefolds: $v = 1$ case. Comp. Math. **68** (1988) 203–220

[Miyaoka-Mori] A numerical criterion for uniruledness. Ann. Math. **124** (1986) 65–69

[Moishezon] B.G.Moishezon: Algebraic varieties and compact complex spaces. ICM70, Nice 1971, pp. 643–648

[Mori1] S.Mori: Threefolds whose canonical bundles are not numerically effective. Ann. Math. **116** (1982) 133–176

[Mori2] S.Mori: On 3-dimensional terminal singualrities. Nagoya Math. J. **98** (1985) 43–66

[Mori3] S.Mori: Minimal models for semistable degenerations of surfaces. Lectures at Columbia University 1985, unpublished

[Mori4] S.Mori: Classification of higher-dimensional varieties. In: Algebraic Geometry, Bowdoin 1985. Proc. Symp. Pure Math. **46** (1987) 269–332

[Mori5] S.Mori: Flip theorem and the existence of minimal models for 3-folds. J. Amer. Math. Soc. **1** (1988) 117–253

[Morrison] D.Morrison: A Remark on Kawamata's paper "On the plurigenera of minimal algebraic 3-folds with $K \approx 0$". Math. Ann. **275** (1986) 547–553

[Morrison-Stevens] D.Morrison, G.Stevens: Terminal quotient singularities in dimensions three and four. Proc. AMS **90** (1984) 15–20

[Nakamura-Ueno] I.Nakamura, K.Ueno: An addition formula for Kodaira dimensions of analytic fibre bundles whose fibres are Moishezon manifolds. J. Math. Soc. Japan **25** (1973) 363–371

[Reid1] M.Reid: Canonical threefolds. In: Géometrie Algébrique Angers. Sijthoff & Noordhoff (1980) 273–310

[Reid2] M.Reid: Projective morphisms according to Kawamata. Preprint, Univ. of Warwick (1983)

[Reid3] M.Reid: Minimal models of canonical threefolds. In: Algebraic and Analytic Varieties. Adv. Stud. Pure Math. **1** (1983) 131–180

[Reid4] M.Reid: Decomposition of toric morphisms. Progress in Math. **36** (1983) 395–418

[Reid5] M.Reid: Young person's guide to canonical singularities. In: Algebraic Geometry, Bowdoin 1985. Proc. Symp. Pure Math. **46** (1987) 345–416

[Reid6] M.Reid: The moduli space of 3-folds with $K = 0$ may nevertheless be irreducible. Math. Ann. **278** (1987) 329–334

[Shokurov1] V.Shokurov: The nonvanishing theorem. Izv. Akad. Nauk SSSR Ser. Mat. **49** (1985) 635–651

[Shokurov2] V.Shokurov: Letter to M.Reid, 1985

[Shokurov3] V.Shokurov: Special 3-dimensional flips. Preprint, MPI 1989

[Stevens] J.Stevens: On canonical singularities as total spaces of deformations. Abh. Math. Sem. Univ. Hamburg **58** (1988) 275–283

[Tsuji] H.Tsuji: Existence and degeneration of Kähler-Einstein metric for minimal algebraic varieties of general type. Math. Ann. **281** (1988) 123–133

[Tsunoda] S.Tsunoda: Degenerations of Surfaces. In: Algebraic Geometry, Sendai 1985. Adv. Stud. Pure Math. **10** (1987) 755–764

[Viehweg1] E.Viehweg: Klassifikationstheorie algebraischer Varietäten der Dimension drei. Comp. Math. **41** (1981) 361–400

[Viehweg2] E.Viehweg: Vanishing theorems. J. Reine Angew. Math. **335** (1982) 1–8

[Yau] S.-T.Yau: On the Ricci curvature of a compact Kähler manifold and the complex Monge-Ampère equation I. Comm. Pure Appl. Math. **31** (1978) 339–411

Hyperbolic Billiards

Yakov G. Sinai

Landau Institute of Theoretical Physics, Academy of Sciences of USSR
Moscow, 117334 USSR

1. Definition of Billiards

Billiards are dynamical systems which correspond to the uniform motion of a material point inside a domain on a Riemannian manifold with elastic reflections off the boundary. Thus billiards are geodesic flows on manifolds with boundaries. Specific features of billiards arise when the role of the boundary is much more important than the role of the underlying manifold. Ergodic properties of billiards have been discussed already in the works of Hadamard and Birkhoff. An important contribution was made by the Soviet physicists N. S. Krylov [K1]. During the last two decades the theory of billiards has been developed enormously in various directions and some part of this development will be the main content of this lecture. Billiards constitute an important class of dynamical systems due to the following reasons:

1. some classes of billiards demonstrate a strong chaotic behavior and can be considered among the best examples of deterministic chaos; the theory of such billiards is discussed below in detail;
2. many interesting examples of dynamical systems of physical or mechanical origin can be reduced to billiards; especially it is true for systems where the interaction involves elastic collisions between the particles;
3. a deep analysis of properties of billiards is essential for some problems in the theory of quantum chaos and asymptotical problems in partial differential equations;
4. the theory of billiards suggests many beautiful geometrical problems.

A rich theory of billiards relates to cases where the table of the billiard is a closed compact subset of Euclidean space R^d or d-dimensional flat torus Tor^d, d is the dimension. The tables which we shall consider here can be described as follows. Let there be given r C^∞-functions $f_i(q), 1 \le i \le r$, defined on R^d or Tor^d and such that $f_i^{-1}(0)$ do not contain critical points of f_i. Then $Q = \{q \in R^d | f_i(q) \ge 0, 1 \le i \le r\}$ or $Q = \{q \in Tor^d | f_i(q) \ge 0, 1 \le i \le r\}$. In order to avoid some trivial complications it is natural to assume also that if $q \in Q$ and $f_{i_1}(q) = f_{i_2}(q) = 0$ then grad $f_{i_1}(q)$, grad $f_{i_2}(q)$ are non-collinear. The boundary ∂Q is the union $\partial Q = \bigcup_{i=1}^r \partial Q_i$ where $\partial Q_i = \{q \in \partial Q | f_i(q) = 0\}$. The points of a

Proceedings of the International Congress
of Mathematicians, Kyoto, Japan, 1990
© The Mathematical Society of Japan, 1991

component ∂Q_i which do not belong to other components ∂Q_j, $j \neq i$, are called regular points of the boundary. At regular points $q \in \partial Q$ the unit normal vector $n(q)$ directed inside Q as well as the operator $K(q)$ of the second fundamental form of the boundary are defined. Recall that $n(q + dq) = n(q) + K(q)dq$ where $dq \in \mathcal{T}_q(\partial Q)$ and $\mathcal{T}_q(\partial Q)$ is the tangent space to ∂Q at q. The geometrical properties of the boundary which we shall use below will be expressed in terms of $K(q)$.

As was already mentioned billiards correspond to the uniform motion of a point (billiard ball) inside Q with elastic reflections off ∂Q. This means that at the reflection the normal component of the velocity changes its sign while the tangential component remains unaltered. The phase space M of the billiard is the unit tangent bundle over Q. The points of M are denoted as $x = (q, v)$ where $q \in Q$, $v \in \mathcal{T}_q(Q)$, $\|v\| = 1$. The canonical projection $x \to q$ is denoted by π. The space M is also a manifold with a piece-wise smooth boundary, the boundary $\partial M = \bigcup_{i=1}^r \partial M_i, \partial M_i = \pi^{-1}(\partial Q_i)$. Regular points of ∂M are such points $x \in \partial M_i$ that $\pi(x)$ are regular points of ∂Q.

Introduce the set of incoming vectors

$$\partial M^- = \{x = (q, v) \in \partial M \mid q \in \partial Q \text{ is regular and } (v, n(q)) \leq 0\}$$

and the set of outgoing vectors

$$\partial M^+ = \{x = (q, v) \in \partial M \mid q \in \partial Q \text{ is regular and } (v, n(q)) \geq 0\}.$$

The law of reflection corresponds to the transformation $R : \partial M^- \to \partial M^+$ where

$$Rx = R(q, v) = (q, v - 2(n(q), v)n(q)) \in \partial M^+ \quad \text{if} \quad (q, v) \in \partial M^-.$$

A point moves inside M and when it reaches the boundary ∂M^- it jumps instantly to a new point of ∂M^+ under the action of R. In what follows we consider only a subset $M^1 \subset M$ consisting of such x whose trajectories pass only through regular points of the boundary and have finitely many reflections during any finite interval of time. If $d\mu = \text{const}\, dq d\omega_q(v)$ where $d\omega_q(v)$ is the Lebesgue measure on the unit sphere $S_q^{d-1} \subset \mathcal{T}_q(Q)$ is the normed Liouville measure on M then it is easy to show that $\mu(M^1) = 1$. Thus we may define the one-parameter group $\{S^t\}$ of transformations of M^1 onto itself which corresponds to the billiard motion and preserves μ (see [CFS1]). The flow $\{S^t\}$ is discontinuous because the dynamics is discontinuous. It is an often situation in ergodic theory that discontinuous flows are easier for an analysis than their smooth approximations.

Sometimes it is useful also to consider the induced map or the so-called billiard ball map $T : \partial M^+ \cap M^1 \to \partial M^+ \cap M^1$ which transforms a point $x \in \partial M^+ \cap M^1$ to the point $x^1 \in \partial M^+ \cap M^1$ corresponding to the next reflection. The map T preserves the measure $d\mu_1 = \text{const}\, (n(q), v) \cdot d\sigma(q)\omega_q(v)$ where $d\sigma(q)$ is the Riemannian volume on ∂Q.

The first examples of billiards are billiards inside polygons or polyhedra. Some of them appear in connection with the dynamics of one-dimensional particles on a segment interacting through elastic collisions. The billiards in triangles correspond to the dynamics of two such particles. It turns out that even this simple case is very difficult for an analysis and still many problems remain open. Deep results here were obtained by Kerchkoff, Mazur and Smillie (see [KMS1]) using the theory of quadratic differentials and some methods of algebraic geometry. In particular they showed that for billiards inside polygons with commensurate angles the system has only obvious first integrals. The first applications of these methods appeared in the papers by Veech (see [V1]).

2. Hyperbolic Billiards

The main content of this lecture will concern the theory of hyperbolic billiards, i.e., billiards where the behavior of trajectories has many common features with the behavior of geodesics in manifolds of negative curvature. Take a point $x = (q, v) \in M^1$ and a $(d-1)$-dimensional open C^∞-submanifold $\tilde{\gamma} \subset Q$ passing through q, orthogonal to v at q and homeomorphic to the $(d-1)$-dimensional disk. Denote by γ its continuous framing by unit normal vectors such that $x \in \gamma$.

Definition 1. A submanifold γ is called a *local stable manifold (lsm)* of x if during the dynamics $S^t x$ and $S^t y$ for any $y \in \gamma$ have reflections from the same components of the boundary ∂M for all $t > 0$ and one can find positive numbers $C(x), \lambda(x)$ such that for all $t > 0$

$$\text{dist}\,(\pi(S^t x), \pi(S^t y)) \leq C(x) \exp\{-\lambda(x)|t|\}\,.$$

If in this definition we take $t < 0$ we shall get the definition of a *local unstable manifold (lum)* of x. Lsm and lum will be denoted by $\gamma^{(s)}(x)$ and $\gamma^{(u)}(x)$, respectively.

Definition 2. A global stable (unstable) manifold of $x \in M$ consists of all y for which one can find $\tau \geq 0\,(\tau \leq 0)$ such that $S^\tau y$ belongs to $\gamma^{(s)}(S^\tau x)\,(\gamma^{(u)}(S^\tau x))$.

We shall denote them by $\Gamma^{(s)}(x)$ and $\Gamma^{(u)}(x)$ and use the abbreviations gsm and gum. Stable and unstable manifolds are the main objects in the theory of smooth hyperbolic dynamical systems like Axiom-A systems of Smale [Sm1] and Anosov systems [An1]. Contrary to the case of smooth hyperbolic systems gsm and gum in billiards have singularities (see examples below) which arise from trajectories tangent to the boundary. Definitions 1 and 2 can be introduced in the same way for the induced map T.

We shall now introduce the main definition.

Definition 3. Billiard in a domain Q is called hyperbolic if μ-almost every point has gsm and gum.

We shall discuss now various conditions under which billiard is hyperbolic. As it will be seen the corresponding problems are particular cases of a more general problem of finding sufficient conditions for a dynamical system to display some kind of a chaotic behavior.

Definition 4. Billiard is called dispersing if at all regular points of the boundary the operator $K(q) > 0$.

This definition implies that at any regular point the boundary is locally strictly concave.

Theorem 1. *Every dispersing billiard is hyperbolic.*

This theorem is in fact some version of the well-known Hadamard-Perron theorem (see [An1, G1, S2] and [KS1]), which gives sufficient conditions for the existence of lsm and lum in the case of smooth dynamical systems. Theorem 1 gives an easy way to construct examples of hyperbolic billiards.

Dispersing billiards are in many respects similar to geodesic flows on compact manifolds of negative curvature. For many applications it is important to consider billiards which correspond to geodesic flows on manifolds of non-positive curvature.

Definition 5. Billiard is called semi-dispersing if at all regular points of the boundary $K(q) \geq 0$.

Semi-dispersing billiards arise naturally when we study the dynamics of hard disks or spheres interacting via elastic collisions (see for example [CFS1]). In those cases the boundary is the union of subsets of cylinders having many flat directions. It is a more difficult problem to find conditions under which a semi-dispersing billiard is hyperbolic. A natural approach is the following (see [CS1]). Let us write down the sequence of positive times of reflections from the boundary,

$0 < t_1 < t_2 < \cdots < t_n, \ldots$, and let $\tau_j = t_j - t_{j-1}, \tau_1 = t_1$. Introduce also the points of the boundary where these reflections take place: $q_1, q_2 \ldots$ and $K_j = K(q_j)$. We shall need also the operators U_i mapping $\mathcal{T}_i^+$ onto $\mathcal{T}_i^-$ in parallel to $n(q_i)$ where the subspace $\mathcal{T}_i^+(\mathcal{T}_i^-)$ is orthogonal to $v_i^+(v_i^-), S^{t_i \pm 0} x = (q, v_i^{\pm})$, and the operators V_i mapping $\mathcal{T}_i^+$ onto $\mathcal{T}_{q_i}(\partial Q)$ in parallel to v_i^+. The adjoint operator V_i^* maps $\mathcal{T}_{q_i}(\partial Q)$ onto $\mathcal{T}_i^+$ in parallel to $n(q_i)$. Now write down the following operator-valued continued fraction

$$B^+(x) = \cfrac{I}{\tau_1 I + U_1 \cfrac{I}{2\cos\phi_1 V_1^* K_1 V_1 + U_2 \cfrac{I}{\tau_2 I + \ldots\ldots} U_2^{-1}} U_1^{-1}}. \tag{1}$$

Here $\cos\phi_j = (v_i^+, n(q_i))$, ϕ_j is the reflection angle. For semi-dispersing billiards all operators $V_j^* K_j V_j$ are non-negative. It is easy to show that this continued fraction is converging as soon as $t_n \to \infty$ and $B^+(x) \geq 0$. The following theorem is a particular case of Chernov's theorem (see [C1]).

Theorem 2. *If $B^+(x) > 0$ a.e. then the semi-dispersing billiard is hyperbolic.*

This theorem can be derived also from the results in the book by Katok and Strelcyn (see [KS1]).

The operator $B^+(x)$ for $x = (q, v)$ is a self-adjoint operator acting on the $(d-1)$-dimensional space passing through q orthogonal to v. It has a simple geometrical meaning. Namely, if $\gamma^{(u)}(x)$ is a lum of x, $\widetilde{\gamma}^{(u)}(x) = \pi(\gamma^{(u)}(x))$ then $\widetilde{\gamma}^{(u)}(x)$ is a $(d-1)$-dimensional submanifold lying inside Q and $B^+(x)$ is the operator of the second fundamental form of $\widetilde{\gamma}^{(u)}(x)$ at $q = \pi(x)$. The condition $B^+(x) > 0$ means that the submanifold $\widetilde{\gamma}^{(u)}(x)$ is strictly concave. The appearance of continued fractions in (1) is quite natural. If we write down the Jacobi equations corresponding to our flow then they will be second order linear differential equations with piecewise constant coefficients having discontinuites at the moments of reflections. It is well-known that the solutions to such equations can be written as continued fractions. The operators $B^+(x)$ are exactly these solutions.

An interesting non-trivial example of the semi-dispersing billiard where one can already see the main difficulties of the theory was proposed by Kramli, Simanyi and Szasz (see [KSS1]). The table consists of the three-dimensional torus from which two non-parallel cylinders are removed. As was already mentioned the system of r hard disks or balls moving uniformly inside a cube and interacting via elastic collisions can be reduced to semi-dispersing billiards. Apparently these billiards are hyperbolic but now it is known only for small values of r.

It is very surprising that hyperbolic billiards are encountered also among two-dimensional convex domains. The first example was proposed in the famous paper by Bunimovich [B1]. The table is the Bunimovich's stadium where the boundary consists of two straight segments and two circular arcs. The boundary is a C^1-curve and it is essential that the curvature is a discontinuous function on the boundary. If it were sufficiently smooth then the corresponding billiard cannot be hyperbolic because a theorem of Lazutkin (see [L1]) says that in such cases there exist so many caustics that the set of their tangent vectors has positive measure. Recently the result by Bunimovich was extended by Bunimovich (see [B2]) and Donnay (see [D1]). Namely the form of the stadium can be perturbed in such a way that two segments remain but the circular arcs can be replaced by arbitrary C^2-arcs whose curvature is close to a constant.

In connection with [B1, B2, D1] a general problem emerges of describing two-dimensional tables for which the billiard is hyperbolic. Wojtkowski in [W1] proposed a general approach to this problem based on the existence of invariant families of cones (see also [S3]) and found several new examples of hyperbolic billiards. Some other examples were constructed in the paper by Markarian (see [M1]).

3. Fundamental Theorem in the Theory of Hyperbolic Billiards

In this section we discuss the problems related to ergodicity and mixing of hyperbolic billiards. We remark first that the existence of lsm and lum for the flow $\{S^t\}$ implies the existence of these manifolds for the induced map T. Their dimensions are equal to $d-1$ while the dimension of ∂M is $2(d-1)$. It is an easy part of the theory of piecewise smooth hyperbolic dynamical systems which says that ergodic components of T and $\{S^t\}$ have positive measures (see, for example, [P1] and [KS1]). The proof is based upon an idea of Hopf [H1] according to which the ergodic component of a.e. x contains also $\Gamma^{(u)}(x)$ and $\Gamma^{(s)}(x)$. Thus we can take a lum $\gamma^{(u)}(x)$ and construct lsm $\gamma^{(s)}(y)$ for every $y \in \gamma^{(u)}(x)$. It needs some efforts to show that starting with an a.e. x we get a set of positive measure. The main ingredient of the proof is the so-called property of absolute continuity of foliations generated by $\Gamma^{(s)}, \Gamma^{(u)}$ (see details in [KS1]). However it is a much more difficult problem to prove that the ergodic component of T or $\{S^t\}$ is unique. The difficulty is due to the fact that because of singularities the sizes of lsm or lum are everywhere discontinuous and can be arbitrarily small. Therefore apriori it is not clear whether one can connect two arbitrary points by a chain of sets of the described form.

In [S2] a theorem was proven which showed how to overcome the difficulty in the case of two-dimensional dispersing billiards. Later it was called the fundamental theorem in the theory of dispersing billiards. The idea was to show that in a neighborhood of a typical point for a given lum $\gamma^{(u)}$ the probability to find $\gamma^{(s)}(y)$, $y \in \gamma^{(u)}$, whose length is greater than $Cl(\gamma^{(u)})$, l is the length, C is an arbitrary constant, tends to 1 as the radius of the neighborhood tends to zero. It follows from this statement that a small neighborhood of this point lies within one ergodic component. The ergodicity follows from an easy statement that the set of typical points is connected. There are two other publications [BS1] and [S4] which contain some modifications of the original proof and a proof of ergodicity based on a slightly different idea.

In our joint paper with Chernov [CS1] the ideas of the fundamental theorem were extended further. Let $R \subset \partial M$ be the union of singular points of ∂M, of tangent points of $\partial M^+ \cup \partial M^-$, i.e., points $x = (q,v)$ where $(n(q),v) = 0$ and the points of discontinuity of T. We imposed the following condition: for a.e. point of R the image under dynamics of a plane framed by unit normals locally extends in all directions. Under this condition we proved that typical points have neighborhoods belonging to one ergodic component (local ergodicity). An immediate consequence of this theorem is

Theorem 3. *Every dispersing billiard is ergodic.*

Kramli, Simanyi and Szasz in [KSS2] made an important improvement in the conditions of the fundamental theorem which is essential for semi-dispersing billiards and gave a very detailed proof of their version of the fundamental theorem which they called "A transversal fundamental theorem".

Using their improvement they proved the following remarkable result (see [KSS3]).

Theorem 4. *Consider the system of three balls on $Tor^d, d \geq 2$. Then it is ergodic on the subspace of the phase space where the total momentum is zero and the centrum of mass is fixed.*

In the paper by Bunimovich [B3] it was shown that the fundamental theorem is applied to the billiards in stadium and some other convex domains. Thus these billiards are also ergodic.

In systems with hyperbolic behavior ergodicity usually implies much stronger statistical properties. Namely it is easy to show under the conditions of Theorems 3 and 4 that the billiard is a K-flow. It means in particular that the adjoint group of unitary operators has countable Lebesgue spectrum and strong properties of mixing. It is a more difficult problem to prove the Bernoulli property which says that the induced map T is isomorphic to a Bernoulli shift. This was done by Gallavotti and Ornstein in [GO1] for two-dimensional dispersing billiards.

4. Markov Partitions and Strong Statistical Properties of Two-Dimensional Hyperbolic Billiards

If we intend to study more deeply statistical properties of hyperbolic billiards like central limit theorem and decay of time-correlation functions we must use some special symbolic representations of these systems. The most convenient one is the representation which can be obtained with the help of a Markov partition. Let us recall the corresponding definitions (see [Bo1, BoR1, R1, R2, S5, S7]). We shall deal with the induced map T.

Take a point $x \in \partial M^+$ and lum $\gamma^{(u)}(x)$, lsm $\gamma^{(s)}(x)$, choose subsets $\Pi^{(u)} \subset \gamma^{(u)}(x), \Pi^{(s)} \subset \gamma^{(s)}(x)$.

Definition 6. A subset $\Pi \subset \partial M^+$ is called a parallelogram with basic sets $\Pi^{(u)}, \Pi^{(s)}$ if for any $y_1 \in \Pi^{(u)}, y_2 \in \Pi^{(s)}$ the lsm $\gamma^{(s)}(y_1)$ and the lum $\gamma^{(u)}(y_2)$ intersect each other and the intersection $\gamma^{(s)}(y_1) \cap \gamma^{(u)}(y_2)$ consists of one point also belonging to Π.

If Π is a parallelogram then for any $x' \in \Pi$ the sets $\gamma^{(u)}(x') \cap \Pi, \gamma^{(s)}(x') \cap \Pi$ can be taken as basic sets. If Π', Π'' are parallelograms then $\Pi' \cap \Pi''$ is also a parallelogram. Each parallelogram has two natural partitions $\xi^{(u)}(\Pi), \xi^{(s)}(\Pi)$. The elements of these partitions are $C_{\xi^{(u)}(\Pi)}(y) = \Pi \cap \gamma^{(u)}(y)$, $C_{\xi^{(s)}(\Pi)}(y) = \Pi \cap \gamma^{(s)}(y)$ respectively. Assume that we are given a partition η whose elements are parallelograms $\Pi_1, \Pi_2, \ldots, \Pi_r, \ldots$. This means that $\mu(\Pi_{i_1} \cap \Pi_{i_2}) = 0$ for $i_1 \neq i_2$ and $\mu(\cup \Pi_j) = 1$. We shall denote by $\xi^{(u)}, \xi^{(s)}$ the partitions of ∂M^+ where for a.e. x we have $C_{\xi^{(u)}}(x) = C_{\xi^{(u)}(\Pi_j)}(x), C_{\xi^{(s)}}(x) = C_{\xi^{(s)}(\Pi_j)}(x)$ when $x \in \Pi_j$.

Definition 7 (see [Bo1, S5, S7]). The partition η is called Markov partition if for a.e. x

$$T^{-1}C_{\xi^{(u)}}(x) \subseteq C_{\xi^{(u)}}(T^{-1}x), \quad TC_{\xi^{(s)}}(x) \subseteq C_{\xi^{(s)}}(T_x).$$

In the case of smooth hyperbolic systems natural Markov partitions are usually finite. For billiards and other hyperbolic discontinuous systems such partitions are by necessity countable. If $\mu(\Pi) > 0$ then the basic sets $\Pi^{(u)}, \Pi^{(s)}$

are subsets of $\gamma^{(u)}(x), \gamma^{(s)}(x)$ which do not contain open components and usually they are Cantor-like sets of positive measure. The reason is again due to the fact that in any neighborhood of a typical point lum and lsm can be arbitrarily small. Apparently similar Markov partitions can appear in other dynamical systems with a non-uniform hyperbolic behavior like some classes of one-dimensional maps, Henon maps etc.

Returning to billiards suppose that we have an arbitrary partition η of ∂M^+ which is the phase space for T. Having a partition η we construct the symbolic representation for T as follows. Assume that all Π_j are closed and write down the inclusions $T^n x \in \Pi_{j_n}, -\infty < n < \infty$.

Definition 8. The sequence $j = \{j_n\}$ is called the symbolic representation of x.

It is clear that if $j \Leftrightarrow x$ then $T \Pi_{j_{n-1}} \cap \Pi_{j_n} \neq \phi$. The usefulness of Markov partitions is due to the fact that the inverse statement is in a sense also true: if j is any sequence for which $\mu(T \Pi_{j_{n-1}} \cap \Pi_{j_n}) > 0$ for all $n, -\infty < n < \infty$, then there exists x for which j is a symbolic representation. In other words the phase space ∂M^+ of T is coded by realizations of a topological Markov chain up to a subset of measure zero. In the paper [BS2] we constructed Markov partitions for two-dimensional dispersing billiards. In the recent paper [BCS1] using the same ideas we proved under some technical assumptions the following theorem.

Theorem 5. *For two-dimensional hyperbolic billiards there exist countable Markov partitions. Moreover, the diameters of parallelograms can be made less than any given $\varepsilon > 0$.*

In particular such Markov partitions exist for dispersing billiards, Bunimovich's stadium and its perturbations and some other cases.

Describe now some corollaries which follow directly from Theorem 5.

Corollary 1. *The set of periodic orbits of a hyperbolic billiard is everywhere dense.*

It was shown to us by Pesin how this statement can be proven by "standard arguments from hyperbolic theory". L. Stojanov in [St1] proved that for any multi-dimensional semi-dispersing billiard in R^d the number $N(R)$ of periodic orbits whose period is not more than R is not bigger than $C_1 \exp\{C_1 R\}$ for some $C_1 > 0$. Theorem 5 gives a possibility to obtain an estimation of $N(R)$ from below.

Corollary 2. *The number $N(R)$ is bigger than $\exp\{C_2 R\}$ for some $C_2 > 0$ and all sufficiently large R.*

Apparently as in the case of smooth hyperbolic system there exists the asymptotics $\lim_{R \to \infty} \frac{\ln N(R)}{R}$ connected with the topological entropy of billiard but it still remains as an open problem.

In our paper [BS3] such statistical properties as the central limit theorem, time decay of correlation functions for two-dimensional dispersing billiards were studied. These results can be extended to a wider class of two-dimensional hyperbolic billiards (Bunimovich, Chernov, Sinai – in preparation). Using Markov partitions one can prove for these billiards some version of H-theorem in statistical mechanics.

5. Some Other Results and Concluding Remarks

As one can see from what was said above the theory of dispersing billiards is similar to some extent to the theory of hyperbolic dynamical systems. The theory of semi-dispersing billiards is in the same sense similar to the theory of partially hyperbolic dynamical system. Unfortunately it is not still developed in a sufficiently general manner. Only some particular results have been obtained. Chernov in [Ch1] constructed lsm and lum under rather general conditions. Their dimensions may be less than $d - 1$. He gave also (see [Ch2]) a detailed proof of the formula for the measure-theoretic entropy

$$h(\{S^t\}) = \int tr(B^+(x))d\mu(x)$$

which is positive unless $K(q) \equiv 0$ like in the case of flat boundaries. Wojtkowski in [W2] studied the asymptotical behavior of entropy when the mean free path tends to zero.

Some examples of multi-dimensional hyperbolic billiards analogous in sense to billiard in stadium are described in [B3] (see, however, the remarks in [W3]).

Several Hamiltonian systems behave in many respects as hyperbolic billiards. In the papers by Kubo [Ku1] (see also [S6]) the motion of a particle on the two-dimensional torus under the action of some repelling forces was studied. It was shown that it is hyperbolic (in the same sense as in the definition 3). A surprising result was obtained in a recent paper by Donnay and Liverani [DL1]. The authors found examples of attractive potentials for which the dynamics is also hyperbolic. In the paper by Wojtkowski [W4] (see there some other references) a system of particles with different masses moving along a semi-line with a constant acceleration and undergoing elastic collisions was studied. It was shown that it has several positive Lyapunov exponents which implies through the Pesin's formula [P1] that this system has positive entropy and a partially hyperbolic behavior.

Arnold in [Ar1] gave an heuristic explanation why two-dimensional dispersing billiards should behave like geodesic flows on surfaces of negative curvature. He proposed to consider doubles of the corresponding tables and represent them as limits of smooth surfaces of negative (more precisely, non-positive) curvature. However, this idea being very beautiful has a defect. The flow which appears under this limiting transition contains "extra" trajectories which can have common parts with the boundary. In this connection some questions may be asked. For example, one can take the double of the billiard table, its universal covering and ask the following:

1. will the lift of a billiard trajectory to the universal covering go to infinity;
2. if so, what is the structure of points at infinity which are the end-points of the lifted trajectories, going out of some points;
3. will a lifted trajectory of the billiard lie to a finite distance from some Lobachevsky geodesics.

Some answers to these questions were obtained recently by Babenko [Ba1].

In conclusion I thank the Organizational Committee of ICM-90 for the honorable invitation to present this talk. I thank V. I. Arnold and J. Lebowitz who read the text and made many useful remarks. Also I am very glad to express my sincere gratitude to Bunimovich, Chernov, Kramli, Szasz with whom we worked together for many years on various problems concerning billiards.

References

[An1]   Anosov D.V.: Geodesic flows on closed Riemannian manifolds of negative curvature. Proc. Steklov Institute **90** (1967) 1–235
[Ar1]   Arnold V.I.: Small denominators and problems of stability of motion in classical and celestical mechanics. Uspehi Math. Sci. **18:6** (1963) 85–191
[Ba1]   Babenko I.K.: Behavior of trajectories of dispersing billiards on the flat torus. Math. Sbornik. (In press)
[Bo1]   Bowen R.: Equilibrium states and ergodic theory of Anosov diffeomorphisms. (Lecture Notes in Mathematics, vol. 470.) Springer, Berlin Heidelberg New York 1975
[BR1]   Bowen R., Ruelle D.: The ergodic theory of Axiom A flows. Invent. math. **29** (1975) 181–202
[B1]   Bunimovich L.A.: On the ergodic properties of certain billiards. Funct. Anal. Appl. **8** (1974) 254–255
[B2]   Bunimovich L.A.: On stochastic dynamics of rays in resonators. Radiofizika **28** (1985) 1601–1602
[B3]   Bunimovich L.A.: A theorem on ergodicity of two-dimensional hyperbolic billiards. Comm. Math. Phys. **130** (1990) 599–621
[B4]   Bunimovich L.A.: Many-dimensional nowhere dispersing billiards with chaotic behavior. Physica **D33** (1988) 58–63
[BS1]   Bunimovich L.A., Sinai Ya.G.: On a fundamental theorem in the theory of dispersing billiards. Math. Sbornik **19** (1973) 407–423
[BS2]   Bunimovich L.A., Sinai Ya.G.: Construction of Markov partitions for dispersing billiards. Comm. Math. Phys. **73** (1980) 247–280
[BS3]   Bunimovich L.A., Sinai Ya.G.: Statistical properties of Lorentz gas with periodic configuration of scatterers. Comm. Math. Phys. **78** (1981) 479–497
[BCS1] Bunimovich L.A., Chernov N.I., Sinai Ya.G.: Markov partitions for two-dimensional hyperbolic billiards. Uspehi Math. Sci. **45:3** (1990) 97–134
[C1]   Chernov N.I.: Construction of transversal fibers for multi-dimensional semi-dispersing billiards. Funct. Anal. Appl. **16** (1982) 35–46
[C2]   Chernov N.I.: Entropy of semidispersing billiards. (In preparation)
[CS2]   Chernov N.I., Sinai Ya.G.: Ergodic properties of some systems of 2-D disks and 3-D spheres. Uspehi Math. Sci. **42:3** (1987) 153–174
[CFS1] Cornfeld I.P., Fomin S.V., Sinai Ya. G.: Ergodic theory. Springer, Berlin Heidelberg New York 1981
[D1]   Donnay V.J.: Convex billiards with positive entropy. (In preparation)

[DL1]   Donnay V.J., Liverani C.: Ergodic properties of particle motion in potential fields. Comm. Math. Phys. (In press)

[G1]   Gallavotti G.: Lectures on the billiard. (Lecture Notes in Physics, vol. 38. Moser, Ju., ed.) Springer, Berlin Heidelberg New York 1975, pp. 236–295

[GO1]   Gallavotti G., Ornstein D.: Billiards and Bernoulli schemes. Comm. Math. Phys. **38** (1974) 83–101

[H1]   Hopf E.: Statistik der geodätischen Linien in Mannigfaltigkeiten negativer Krümmung. Ber. Verhandl. Sächs. Akad. Wiss. **91** (1939) 261–304

[K1]   Krylov N.S.: Works on the foundations of statistical physics. Princeton University Press, Princeton 1979

[Ku1]   Kubo I.: Perturbed billiard systems. Nagoya Math. J. **61** (1976) 1–57

[KMS1]   Kerchhoff S., Masur H., Smillie J.: Ergodicity of Billiard flows and quadratic differentials. Ann. Math. **124** (1986) 293–311

[KS1]   Katok A.B., Strelcyn J.-M.: Invariant manifolds, entropy and billiards, smooth maps with singularities. (Lecture Notes in Mathematics, vol. 1222.) Springer, Berlin Heidelberg New York 1986

[KSS1]   Kramli A., Simany N., Szasz D.: Ergodic properties of semi-dispersing billiards. Two cylindric scatterers in the 3-D torus. Nonlinearity (1989) 311–326

[KSS2]   Kramli A., Simanyi N., Szasz D.: A "transversal" fundamental theorem for semi-dispersing billiards. Comm. Math. Phys. **129** (1990) 535–560

[KSS3]   Kramli A., Simanyi N., Szasz D.: Three billiard balls on the v-dimensional torus is a K-flow. Ann. Math. (In press)

[L1]   Lazutkin V.F.: The existence of caustics for the billiard problem in a convex domain. Izv. Acad. of Sci. Math. Series **37** (1973) 188–223

[M1]   Markarian R.: Billiards with Pesin region of measure one. Comm. Math. Phys. **118** (1988) 87–97

[P1]   Pesin Ya.B.: Lyapunov characteristic exponents and smooth ergodic theory. Uspehi Math. Sci. **32:4** (1977) 55–112

[R1]   Ruelle D.: A measure associated with Axiom A attractors. Amer. J. Math. **98** (1976) 619–654

[R2]   Ruelle D.: Thermodynamic formalism. (Encyclopedia of Math. and its Appl. 5) Addison-Wesley, 1978

[S1]   Sinai Ya.G.: Introduction to Ergodic Theory. Princeton University Press, Princeton 1977

[S2]   Sinai Ya.G.: Dynamical systems with elastic collisions. Uspehi Math. Sci. **25:2** (1970) 141–192

[S3]   Sinai Ya.G.: Appendix to the first edition of G.M.Zaslavsky's book "Statistical irreversibility in non-linear systems" ("Nauka" Press, Moscow) 114–139, 1970

[S4]   Sinai Ya.G.: Ergodic properties of Lorentz gas. Funct. Anal. Appl. **13** (1979) 46–59

[S5]   Sinai Ya.G.: Markov partitions and U-diffeomorphisms. Funct. Anal. Appl. **2** (1968) 61–82

[S6]   Sinai Ya.G.: A "physical" system with positive entropy. Vestnik Moscow State University **5** (1963) 6–12

[S7]   Sinai Ya.G.: Gibbs measures in ergodic theory. Uspehi Math. Sci. **27:4** (1972) 21–64

[Sm1]   Smale S.: Differentiable dynamical systems. Bull. Amer. Math. Soc. **73** (1967) 747–817

[St1]   Stojanov L.: An estimate from above of the number of periodic orbits for semi-dispersed billiards. Comm. Math. Phys. **124** (1989) 217–227

[V1]   Veech W.A.: Interval exchange transformations. J. Anal. Math. **33** (1978) 222–272

[W1]   Wojtkowski M.P.: Principles for the design of billiards with nonvanishing Lya-
        punov exponents. Comm. Math. Phys. **105** (1986) 391–414
[W2]   Wojtkowski M.P.: Measure theoretic entropy of the system of hard spheres. Erg.
        Theor. Dyn. Syst. **8** (1988) 133–153
[W3]   Wojtkowski M.P.: Linearly stable orbits in 3 dimensional billiards. Comm. Math.
        Phys. **129** (1990) 319–328
[W4]   Wojtkowski M.P.: A system of one dimensional balls with gravity. Comm. Math.
        Phys. **126** (1990) 507–534

Applications of Non-Linear Analysis in Topology[*]

Karen Uhlenbeck

Sid Richardson Foundation Regents' Chair in Mathematics, Department of Mathematics, The University of Texas at Austin, Austin, TX 78712, USA

If there is any single characterizing feature of the mathematics of the last few years, it is the interactions among subdisciplines. This activity is well-documented by the talks at ICM90. This lecture contains a small part of the background for applications of non-linear analysis in the field of topology. My discussion covers roughly the last twelve years, with emphasis on the earlier period of this time. There are many other articles in this volume which describe in more detail current areas of research. We refer particularly to Floer's Plenary address, and section lectures by McDuff, Simpson, Tian, Kronheimer, and many talks related to topological quantum field theory.

The specific mathematical tools I am considering in this paper are those of "hard" analysis, by which we mean two things. Hard analysis refers in graduate student slang to the use of estimates. But here it refers as well to what Gromov [Gr] calls "hard" — the realization of "soft" or "flabby" topological concepts via the solution of specific rigid partial differential equations. These techniques are naturally not to every mathematician's taste. Topologists have spent a great deal of effort reproving theorems such as the Bott periodicity theorem in ways more related to the general constructive methods of algebraic topology. For mathematicians like myself these geometric methods provide a concrete geometric realization of what is otherwise very much algebraic abstraction.

1. Background Discussion

Our first step is to illustrate the content of this talk with an example which uses only advanced calculus. We define a real-valued function

$$f : \mathbb{C}^{n+1} - \{0\} \to \mathbb{R}$$

by the formula

$$f(\mathbf{v}) = \frac{(A\mathbf{v} \cdot \bar{\mathbf{v}})}{(\mathbf{v} \cdot \bar{\mathbf{v}})}$$

for $\mathbf{v} = (z^1, \ldots, z^n) \in \mathbb{C}^{n+1} - \{0\}$. Here A is any Hermitian complex $(n+1) \times (n+1)$ matrix. It is easy to see that if $\alpha \in \mathbb{C} - \{0\}$, $f(\alpha \mathbf{v}) = f(\mathbf{v})$. So f induces a map

---

[*] Partially supported by grants from the American Mathematical Society and the National Science Foundation.

Proceedings of the International Congress
of Mathematicians, Kyoto, Japan, 1990
© The Mathematical Society of Japan, 1991

$$[f] = (\mathbb{C}^{n+1} - \{0\})/(\mathbb{C} - \{0\}) = \mathbb{C}P^n \to \mathbb{R}$$

by $[f]([\mathbf{v}]) = f(\mathbf{v})$. The critical points of f (and $[f]$) are points where the derivative vanishes. It is a standard exercise in advanced calculus to show that the equations for critical points of $[f]$ are the lines $[\mathbf{v}]$ for

$$A\mathbf{v} - \lambda\mathbf{v} = 0$$

where $\lambda = f(\mathbf{v})$.

Now the connection with topology comes from the relation between the critical points of $[f]$ and the topology of $\mathbb{C}P^n$. If the eigenvalues of A are distinct, one can compute that the smallest is the minimum, the largest the maximum, and the $n - 1$ other critical points occur with index (number of negative directions in the second derivative) $2, 4, \ldots, 2(n - 1)$. Standard Morse theory tells us that $\mathbb{C}P^n$ is built from one handle in each even dimension. The Betti numbers $b_{2i} = 1$, $0 \le i \le n$ and $b_{2i+1} = 0$. Of course, this is not the usual way of computing the Betti numbers of $\mathbb{C}P^n$.

However, much closer in spirit to many of our examples is the case when A is a Hermitian projection operator of rank $n - k$. Then the minimum of $[f]$ on $\mathbb{C}P^n$ is zero and it occurs on a $\mathbb{C}P^k$ sitting in $\mathbb{C}P^n$. The other critical points consist of a $\mathbb{C}P^{n-k-1}$ on which the maximum occurs. We obtain a very simple topological result by using the gradient flow for $[f]$ to retract $\mathbb{C}P^n - \mathbb{C}P^{n-k-1}$ into $\mathbb{C}P^k$. The result is that the embedding $\mathbb{C}P^k \subseteq \mathbb{C}P^n$ is a homotopy equivalence up to dimension k.

The prototype theorem for this talk is the Bott Periodicity theorem for the unitary group, as originally proved by Bott in 1959 [B]. Here the theory of ordinary differential equations replaces the advanced calculus of our first example. The space which replaces $\mathbb{C}P^n$ is $C_{-I,I}^\infty([0, 1], SU(2m))$, or the parametrized smooth curves between $-I$ and $+I$ in the special unitary group $SU(2m)$. The function $[f]$ is replaced by energy

$$E(s) = \frac{1}{2} \int_0^1 |\dot{s}(t)|^2 \, dt.$$

The equation for critical curves is the equation for geodesics

$$D_t \dot{s}(t) = 0.$$

The minimum for this functional occurs on the set of great circles between $(-I, +I)$. This can be checked to be the complex Grassmannian $G(m, 2m)$ of m planes in $2m$ space. All the other critical points have index at least $2m + 2$. The gradient flow provides a homotopy equivalence between the loop space and the Grassmannian up to dimension $2m$. Namely for $0 \le i \le 2m$, we have the result that

$$\pi_{i+1} SU(2m) = \pi_i(\Omega SU(2)) = \pi_i G(m, 2m).$$

There are methods from algebraic topology which show that

$$\pi_{i-1} SU(m) = \pi_{i-1} SU(k)$$

for $i \le 2m \le 2k$ and

$$\pi_i G(m, 2m) = \pi_{i-1} SU(m) \quad \text{for } i \le 2m.$$

Bott Periodicity Theorem [B].

$$\pi_{i-1} SU(k) = \pi_{i+1} SU(k), \qquad i \leq k.$$

Other older results which use the theory of geodesics have more to do with differential geometry.

Theorem (Hadamard). *Suppose M is a compact, connected manifold with negative sectional curvature, then $\pi_{i+1}(M) = 0$ for $i \geq 1$.*

One version of this proof is obtained by showing that all geodesics are of shortest length. As a consequence, every connected component of the loop space is topologically trivial. Hence $\pi_i(\Omega(M)) = \pi_{i+1}(M) = 0$ for $i \geq 1$.

Theorem (Myers). *If M is compact with positive Ricci curvature, then $\pi_1(M)$ is finite.*

This can be proved by showing all the minimizing geodesics are short. A general reference is Milnor's book on Morse theory [Mi].

The infinite dimensional loop spaces $\Omega(M)$ were originally handled by retraction onto finite dimensional spaces using piecewise solutions as approximations. This does not work in more than one variable because of the difficulties involved in gluing small solution pieces together. It may have been that mathematicians hoped that the multivariable problems could be easily handled once the proper tools for treating global problems were developed. This turned out to be not quite true.

The modern developments *do* rest entirely on the foundations of functional analysis and elliptic operator theory. The analytic tools are Hölder spaces, Sobolev spaces, embedding theorems, interpolation theorems and the fundamental estimates for elliptic and parabolic systems.

In the 1960s, an ambitious subject called "global analysis" developed with the explicit goal of solving non-linear problems via methods from infinite dimensional differential topology. During this period, a different set of tools was developed. A short list of these tools includes: the notion of Fredholm operator and the the Atiyah-Singer index theorems (1963) [A-S]; the definition of infinite dimensional manifolds [L]; metric structures and refinements such as layer structures and Fredholm structures; the definition of a non-linear Fredholm operator and Smale's extension of the Sard theorem [Sm] (1965); the Palais-Smale conditions and applications in the calculus of variations (1964) [P-S]; in addition, several variants of infinite dimensional degree theorems, K-theory and transversality. A good sense of the spirit of this development can be obtained by browsing through the three volumes of the proceedings of the Berkeley 1968 AMS conference organized by S.S. Chern and S. Smale [C-S].

The optimism of the era of global analysis has ultimately been justified, but this did not happen immediately. The problem is essentially as follows: In order to discover properties of solutions of ordinary or partial differential equations which have global significance, it is essential to make estimates. Now, it usually happens that certain estimates are natural to the problem. Sometimes it may be an estimate on a maximum of a norm, or more usually an integral estimate on solutions is available. Typically the estimate is on the L^2 norm of the first derivative of a solution. In other words, an estimate in the Banach space L_1^2 is natural. However, in order to obtain the results which have topological meaning, the estimates have to imply something about the space of continuous solutions, or at the very least, some information about continuity. This occurs when the Banach space L_1^2 lies in C^0. However, $L_1^2 \subset C^0$ is a dimensionally dependent inclusion. It holds for $n = 1$, or for the case of ordinary differential equations, but not for $n \geq 2$, the dimensions of partial differential equations.

Hence the explanation for the success of the cited examples on loop space is not exactly what it was expected to be. The problems work at least partly because the Sobolev embedding theorem $L_1^2 \subset C^0$ is true in dimension 1, and naive attempts to apply the theories to partial differential equations do not work except in restrictive cases. Many of the applications which were ultimately found are extremely deep. Especially significant are the dimensional differences. In retrospect, one could not expect one single aspect of non-linear analysis to magically provide for a variety of deep applications.

2. Results

S.T. Yau's proof of the Calabi conjecture, published in 1978, showed for the first time the effect which modern methods of solving partial differential equations could have on other fields. The analytic theorem Yau proved was for an arbitrary complex Kähler manifold with non-positive first Chern class. Yau proved that there is a Kähler metric which solves Einstein's equation.

Theorem [Y-1]. *Let M be a complex Kähler manifold with $c_1(M) \leq 0$. Then there is a Kähler metric g in the same Kähler class as the given one with*

$$\text{Ricci}(g) - Rg = 0.$$

The topological conditions imply that $R \leq 0$, where the constant R is a multiple the constant scalar curvature of the Einstein metric.

This result had been conjectured by Calabi and partially proved by Aubin. However, its importance in algebraic geometry lies in the following application.

Corollary. *If M is a complex Kähler manifold with $c_1(M) \leq 0$, then*

$$(-1)^n c_2(M) c_1(M)^{n-2}[M] \geq \frac{(-1)^n n}{2(n+1)} c_1(M)^n [M].$$

This theorem is cited in every theorem on the classification of higher dimensional algebraic varieties. It is essentially the only topological restriction known for algebraic manifolds. The theorem is not true for positive Chern class. One of the satisfying results of Yau's proof is that the importance of the topological condition $c_1(M) \leq 0$ is apparent. Yau's original theorem (in the case of complex 3-folds with $c_1 = 0$) is fundamental in the current model for fundamental physics (string theory).

In the ensuing years, the applications of partial differential equations have been extensive, and we give a very brief survey of the initial results in each field. It would not be possible to list all the latest results and their fine points in a general survey article.

Minimal Surfaces

Yau and his coworkers obtained a number of interesting results. Many but not all of these results have alternate proofs. The first theorem is in the spirit of Myer's theorem, and was published in 1980 by Schoen and Yau. I call it the topological positive mass theorem, as it is the topological version of the well-known positive mass theorem of general relativity. (Schoen and Yau used the same techniques to prove the theorem in general relativity.) Scalar curvature is local mass. An alternate proof using linear analysis (Dirac operators) now exists. The analytical basis for the Schoen-Yau proof is a theorem on minimal surfaces.

Theorem [S-Y, S-U]. *If Σ_g is a surface of genus $g \geq 0$, and M any compact manifold with $\pi_1(\Sigma) \subseteq \pi_1(M)$, then there exists an area minimizing branched immersion of Σ in M.*

Application (Topological Positive-Mass) [S-Y]. *If $M = M^3$ has non-negative scalar curvature, then the group $\pi_1(M)$ does not contain $\pi_1(\Sigma_g)$ as a subgroup unless $\Sigma_g = S^1 \times S^1$ and $M^3 = S^1 \times S^1 \times S^1/\Gamma$ is a quotient of the flat torus.*

To prove this, look at the second variation of the minimal surface. The positive scalar curvature forces it to have a negative direction.

Meeks and Yau proved a series of results which show that minimal surfaces in 3-manifolds are embedded. In many cases, this provides alternate more rigid proofs of basic theorems in 3-manifold topology, such as Dehn's lemma, the loop theorem, and the sphere theorem. As part of this program they gave the first proof of the equivariant loop theorem.

Equivariant Loop Theorem (Meeks-Yau). *Let M^3 be a handle-body with boundary Σ_g, and $K \subseteq \text{Diff}(M^3, \Sigma_g)$ be a finite group. Assume M^3 is given a metric in which K acts as isometries, and in which Σ_g has positive scalar curvature (outward like $S^2 = \partial D^3$). Then there exists an embedded minimal disk $(D^2, S^1) \subseteq (M^3, \Sigma_g)$ such that the elements $k \in K$ either leave D^2 invariant, or map D^2 to a disk $k(D^2)$ with $k(D^2) \cap D^2 = \emptyset$.*

The first step in the analysis consists in showing that there is indeed a smallest disc in M^3 with boundary on Σ_g (this leads to a boundary value problem which is a combination of Dirichlet and Neumann conditions). If the solution is not embedded, or if it intersects an iterate, it turns out that it cannot really be of smallest area. Meeks and Yau pioneered the use of 3-manifold techniques to show that minimal surfaces in 3-manifolds are often embedded rather than immersed [M-Y].

The proof due to Siu and Yau of the Frankel conjecture dates from the same year as the proof due to Mori.

Frankel Conjecture. *If M^n is a complex Kähler manifold with positive bisectional curvature, then M^n is biholomorphically equivalent to $\mathbb{C}P^n$.*

Siu and Yau use the minimal 2-spheres shown to exist by Sacks and Uhlenbeck [S-U]. Positive curvature tends to place restrictions on what can be minimal (as in Myer's theorem and in Schoen and Yau's proof of the positive mass conjecture). In this case, Siu and Yau show that the minimal sphere is actually a holomorphic curve [Si-Y].

Finally, we mention a much more recent result. The method due to Sacks and myself for finding minimal 2-spheres in manifolds had been used by Meeks and Yau to handle embedding problems for spheres in 3-manifolds, as well as by Siu and Yau in the Frankel conjecture. However, these proofs use the area minimizing spheres, whereas Micallef and Moore [M-M] later found a use for the non-minimizing critical points. Their isotropic curvature condition is satisfied if the Riemannian curvature is pinched between K and $4K$.

Theorem [S-U]. *If M is a compact manifold and $\pi_i(M) \neq 0$ for some $i \neq 1$, then there is a 2-sphere which is a stationary point of the area functional.*

Sphere Theorem of Micallef and Moore. *If M^n is simply connected and has positive curvature on isotropic 2-planes, then M^n has the homotopy type of a n-sphere.*

Proof. Show that $\pi_i(M) = 0$ for $i \leq [n/2]$. If this is true, the Hurewicz homomorphism and Poincaré duality complete the proof. If $\pi_i(M)$ is the first non-zero homotopy group, a difficult minimax argument leads to the construction of a minimal 2-sphere of index at most $(i-2)$. However, the curvature condition forces the existence of at least $[n/2] - 1$ directions in which the second variation is negative. This leads to a contradiction.

Gauge Theory and 4-Manifolds

Donaldson's announcement of the restrictions on the topology of 4-manifolds with differentiable structures is a more recent mathematical event. Donaldson's startling use of gauge theory in four dimensions followed almost immediately the successful use of minimal surfaces, and development of these gauge theory techniques is still an active field.

The program instigated and to a great extent carried out by Donaldson consists of encoding the properties of differential structures on 4-manifolds by studying the self-dual Yang-Mills equations on the manifold. The difficulty is that the mathematician must introduce a metric onto the smooth manifold. The first step is to understand the analysis, then the dependence on the metric must still be analyzed.

The non-linear analysis part of Donaldson's theory consists in construction of the space of self-dual solutions of Yang-Mills equations over a conformal manifold M in a bundle with structure group $SU(2)$ and second Chern class $-k$. (In the more recent literature, the orientation is reversed to study the anti-self-dual equations in a bundle of second Chern class $+k$. This fits in better with complex analysis). The basic ingredients is the list of theorems developed by global analysts. The Atiyah-Singer theorem determines the dimension of the moduli space and the Sard-Smale theorem can be used to show it is generically a manifold. Taubes' implicit function theorem developed to construct solutions by gluing instantons on a manifold was later modified by Donaldson to include the construction of solutions on the connected sum $M_1 \# M_2$ from solutions on M_1 and M_2. The solution spaces are not compact. However, the boundary is well-understood via exactly the arguments developed to understand the convergence of minimal surfaces.

Theorem (Donaldson) [D-2]. *If M^4 is a simply connected 4-manifold with positive definite self-intersection form, then the moduli space of solutions to the self-dual Yang-Mills equations with $k = -1$ and group $SU(2)$ is generically an oriented manifold with isolated singularities whose boundary can be identified with M.*

The isolated singularities correspond to solutions where $E = L \oplus L^{-1}$ splits into line bundles. This is a theorem an analyst might well have proved, although it would certainly not be obvious to include the orientability. However, the topological use of this theorem appears a a corollary.

Corollary. *M^4 is topologically the connected sum of $\mathbb{C}P^2$'s.*

Each singularity looks like a $\mathbb{C}P^2$ with a positive orientation. There isn't anything else with definite form oriented cobordant to this sum of $\mathbb{C}P^2$'s.

The theory has been developed further, and more elaborate properties of the solution space of the Yang-Mills equation are used in later results. We refer the readers to a forthcoming survey by Freed [Fr] and the article by Donaldson [D-1].

Some of the properties of the moduli spaces in gauge theory are quite similar to properties of Gromov's pseudoholomorphic curves. McDuff has used these to distinguish different symplectic forms [Gr, McD].

Complex Moduli Spaces

The results about 4-manifolds obtained from the Yang-Mills equation are obtained by looking at the topology of the "moduli spaces of solutions." Here by moduli space we refer to the actual solutions to Yang-Mills divided out by the natural geometric equivalence. This is in many ways similar to older examples of moduli spaces, such as the Riemann moduli space of conformal structures on surfaces. In a development which is related to the moduli space of self-dual Yang-Mills equations, a whole class of problems in algebraic geometry can be put in a general framework which we might call infinite dimensional geometric invariant theory.

The foundational paper is the paper by Atiyah and Bott on Yang-Mills equations over Riemann surfaces [A-B]. An older result of Narasimhan and Sheshadri [N-S] proves that the moduli space of stable holomorphic bundles M_g over a complex curve Σ_g of genus $g \geq 2$ can be identified with the moduli space of projectively flat connections. These connections can be identified with the minima of the Yang-Mills functional on connections in Σ_g. Atiyah and Bott conjectured that there is a very beautiful analytic picture which fits this functional. The symmetry group of Yang-Mills is the real group of gauge transformations $\mathcal{G}$. However, its complexification $\mathcal{G}_\mathbb{C}$ acts and stratifies the sets of connections $\mathfrak{A}$ into classes of holomorphic structures $\mathfrak{A}/\mathcal{G}_\mathbb{C}$ on the bundle. The Yang-Mills functional is the L^2 norm of moment map for this action. Stable orbits are (essentially) the orbits on which the action is free and the quotient Hausdorff. By a very general principle, the equivariant topology of the stable moduli space can be computed from the topology of $\mathfrak{A}/\mathscr{G}$ by examining the Morse theory.

Atiyah and Bott were not able to carry through the analysis, but obtained their results from algebraic geometry. Frances Kirwan [Ki] carried out this very general program in a finite dimensional setting. This pattern of development is strikingly similar to the original construction of the Morse theory of geodesics, which inspired the very useful finite dimensional development. Donaldson [Do-3,4], Hitchin [Hi], Daskalapoulos [Da], Bradlow [Br], Corlette [Co] and Simpson [Si] have carried out the analysis and extended the picture to cover coupled equations and complex manifolds of higher dimension. Naturally, the topological results are much better in complex dimension 1. This is because of the Sobolev inequalities. We discuss this in more detail in the next section.

We finish this section by stating one of the basic results of these computations. It corresponds to the bundle version of Yau's solution of the Calabi conjecture. The topological consequences are similar, although as with the Calabi and Frankel conjectures, there is also an algebraic proof.

Theorem (Donaldson, Uhlenbeck-Yau) [D-5, U-Y]. *The moduli space of stable bundles on a complex Kähler manifold is isomorphic to the space of irreducible solutions to the holomorphic Yang-Mills equations (an extension of the anti-self dual equations to arbitrary dimension).*

Corollary. *If E is a stable holomorphic bundle of rank r on a complex Kähler manifold of dimension n, then*

$$c_1(E)^2 \wedge \omega^{n-2}[M] \leq \frac{2r}{r+1} c_2(E) \wedge \omega^{n-2}[M].$$

This is a purely topological statement about the cohomology class of the Kähler form and the characteristic classes of the complex tangent bundle. This inequality is easily seen to be true by applying Chern-Weil theory to the holomorphic Yang-Mills connections.

The Poincaré Conjecture

The Poincaré conjecture refers to one of the best known problems in topology. If a manifold M^n has the homotopy type of S^n, is it S^n (differentiably or continuously)? I don't think that analysts have gotten very close to proving or disproving the Poincaré conjecture. It is perhaps not so well-known that this conjecture has been the inspiration for a number of fundamental developments in analysis.

Most of the ideas have focussed on Einstein's equation. This equation is the Euler-Lagrange equation for a critical metric for the variational integral $\int_M K_g \, d\mu_g$. Here vol $M = \int_M d\mu_g$ is kept fixed. Here K_g is the scalar curvature of the metric g. Einstein's equation reads

$$\mathrm{Ricci}(g) - \lambda_n K g = 0.$$

It goes without saying that we know more about a manifold if it has a solution of Einstein's equation on it. In particular, for a 3-manifold Einstein's equation is equivalent to the metric having constant Riemannian curvature. Clearly not every 3-manifold supports a solution to Einstein's equation because there is no metric of constant Riemannian curvature on most 3-manifolds. Nevertheless, this fact may not have been widely recognized by analysts in the past.

Solving Einstein's equation is very difficult. Yamabe [Yam] proposed as the first step to fix the conformal structure and vary only the function describing lengths. This leads to a much studied conformally invariant problem usually called the Yamabe Problem [L-P].

Palais was motivated to construct a very general theory of the calculus of variations [P]. He claims that he hoped to apply it to the Einstein functional. However, he very quickly realized that the critical points of the Einstein functional have infinite index and coindex. Hence they cannot be detected by the topological methods he developed. There is still some hope of using a minimax argument and the known solution to the Yamabe problem. I am not sure how much faith any of us have in this project, though.

Hamilton's results on the Einstein equation (1982) were very surprising and promising. He showed via a heat flow argument that a 3-manifold with positive Ricci curvature supports a solution to Einstein's equation. This makes it the quotient of S^3 by a finite group. This result has been very influential on analysis in general without providing a solution to the Poincaré conjecture.

Finally, I would like to comment that the inspiration of the Poincaré conjecture is still very much with analysts. Thurston's results on 3-manifolds go a long way towards describing the geometry of 3-manifolds. Present thinking is that it may be possible to use some other variant of a curvature integral. An example might be

$$\int_M |\mathrm{Riem}(g)|^p \, d\mu(g)$$

for $p \geq 3/2$ in 3 dimensions. The hope is that, by transposing gauge theoretic techniques over to manifolds, the obstructions to minimizing such integrals can be better understood.

Conjecture (Due to Deane Yang). *If M^3 is aspherical and atoroidal, then the minimum of the integral*

$$\int_M |\mathrm{Riem}(g)|^{3/2} \, d\mu(g)$$

under the constraint vol $M = 1$ *is either zero, or is taken on by a metric of constant negative curvature.*

This result is in fact nearly implied by Thurston's conjectures on 3-manifolds. Yang's conjecture should lead to interesting analysis, even if it doesn't touch the Poincaré conjecture.

3. Analytical Technique

The variety of topological results cited in the previous section is matched by the variety of different analytical methods which were used. There is no neat classification which matches a set of results with a set of techniques. I roughly classify the methods under the subheadings:

> Continuity method,
> Borderline dimension,
> Gauge theory,
> Heat equation methods.

This is not an exhaustive list of methods. A theory such as that constructed by Andreas Floer uses nearly all the ideas mentioned in this section and more. Many of the latest developments involve topological index theory constructions on the solution spaces as are needed for Donaldson's invariants of 4-manifolds and applications of Gromov's pseudo-holomorphic curves.

Continuity Method

The continuity method is on the surface naive. The goal is to solve an equation

$$F(g) = 0$$

where F is a non-linear elliptic system. To use the continuity method, start with a trial g_0, and compute $F(g_0) = F_0$. Put in a parameter $\varepsilon \in [0, 1]$ and solve

$$F(g_\varepsilon) = (1 - \varepsilon)F_0.$$

This is done by showing that $dF(g_\varepsilon)$ is invertible as a map between appropriately introduced Banach spaces, and that the solution g_t stays bounded in the tangent norm. The invertibility of dF implies via the implicit function theorem that the set of $\varepsilon \in [0, 1]$ for which we have a solution to our equation is open. The estimates (and some weak convergence) show this set is closed. The sophistication comes from the usual necessity of dealing with a number of Banach spaces, and from the further necessity of estimating the inverse of dF.

This is the method used by S.T. Yau to solve the Calabi conjecture. While the equation is Einstein's equation

$$\text{Ricci}(g) - \lambda(g) = 0,$$

in the Kähler case an equation can be written for a Kähler potential $\varphi(\varepsilon)$ where $g(t) = g_0 + \partial\bar{\partial}\varphi(\varepsilon)$ is a new metric. The equation becomes one for the potential function φ. The first stage of the estimates follow easily from the maximum principle. However, essentially all derivatives of φ have to be estimated, although the estimates become iterative after the third derivatives. No method of solution has been found which avoids estimates (nor would we expect this to happen). However the estimates themselves have been given a more geometric foundation and have found a wider application in the general study of Monge-Ampere equations.

Uhlenbeck and Yau used the same method to solve the holomorphic Yang-Mills equations on a Kähler manifold. Here, the metric is in a bundle, and there is no potential, but there is a maximum principle. By adding an ε, we obtain an invertible equation

$$F_g + \varepsilon \ln g = 0.$$

For $\varepsilon > 0$, it is easy to solve this equation. As $\varepsilon \to 0$, the solution does blow up unless an extra geometric condition of stability is satisfied. We show that, if blow-up does occur, that the normalized solutions $\frac{g(\varepsilon)}{\mu(\varepsilon)}$ converge to a degenerate metric π which violates the stability condition. The sophisticated analysis occurs in higher order estimates, not in the initial outline.

Finally, Taubes' construction of instantons on four manifolds is via an implicit function theorem proved in the same style [T-1]. His original iterative proof can be rewritten using the continuity method. He glues in the standard instantons on $\mathbb{R}^4$, localized to lie in a ball of radius λ into a ball centered at p on an arbitrary compact four manifold. The resulting approximate solution can be modified by a small amount to give a solution to the instanton equation. Donaldson further needs to obtain a moduli space with the parameters (λ, p). Again, it is the estimates and the invertibility of an operator very close to the derivative operator which are essential [D-2].

The continuity method is best suited to cases where invertibility of the derivative is built into the situation. In all the cases just cited, there is a moduli space of solutions whose dimension and structure can be understood by using the appropriate choice of coordinates (or gauge) when looking for the solution. A maximum principle is available for the Kähler-Einstein and Holomorphic Yang-Mills examples. Invertibility in Taubes' construction comes from topological constraints and knowledge of S^4. It remains a question when Taubes' gluing of point localized solutions applies to other partial differential equations. Schoen has applied a somewhat similar idea in constructing solutions of the Yamabe problem with point singularities [S] and Kapouleas [Ka] has shown the existence of many complicated surfaces of constant curvature in $\mathbb{R}^3$ by a not unrelated technique. These problems all exhibit conformal invariance, which allows the scaling of solutions. The resulting approximate solutions have a number of parameters. In the simple cases there is a projection onto a moduli space; in complicated examples the parameters must be chosen carefully.

Borderline Dimension

The critical ingredient in geometric problems is the Sobolev embedding theorems. For the usual mapping examples, $L_1^2 \subset C^0$ is true in dimension 1, false in higher dimensions. Dimension 2 is the borderline or scale invariant dimension, because the integral $\iint |ds|^2 (dx)^2$ scales the same way as $\max_{x \in M} |s(x)|$. Both are scale invariant. A similar phenomena is observed for the Yamabe problem. In dimension n, $L_1^2 \subset L^p$ for $p < \frac{2n}{n-2}$. The relevant power of p in the Yamabe problem is $p = \frac{2n}{n-2}$ [L-P].

For borderline problems, solutions can easily be found using minimization and weak convergence methods. The difficulty is that the limiting functions may not satisfy the constraint satisfied by the approximating functions. A typical example would be to fix a domain bounded by a simple closed curve Γ in the plane. Consider maps $s : D^2 \to \mathbb{R}^2$ such that $s|S^1 : S^1 \to \Gamma$ has degree one. Now minimize

$$\iint\limits_{D^2} \left(\left| \frac{\partial s}{\partial x} \right|^2 + \left| \frac{\partial s}{\partial y} \right|^2 \right) dx\,dy\,.$$

The minimum occurs on a conformal map, and will provide a solution of the Riemann mapping problem. The difficulty is to keep the weak limit of a minimizing sequence from being trivial (i.e., to preserve the condition of $s|S^1 : S^1 \to \Gamma$ is of degree one). One solution is to use the conformal invariance to fix the map on three points of S^1. C.B. Morrey [Mo] used these ideas (which originate with Douglas' solution of the Plateau problem) to find minimal surfaces in arbitrary Riemannian manifolds.

The fundamental observation is that in conformally invariant problems energy estimates plus scale invariance imply that the estimates and convergence are valid except possibly at an isolated set of points. At these points, small neighborhoods may conformally expand to cover large pieces of the geometric solution. The mathematician can recover the geometric solution by blow-up. For $\varepsilon_i \to 0$, set

$$s_i \left(\frac{x - x_0}{\varepsilon_i} \right) = \hat{s}_i(x)\,.$$

Then $\hat{s}_i(x) \to s : \mathbb{R}^n = S^n - \{p\} \to M$ will be a solution on S^n [S-U1].

This principle allows the construction of a large number of two-dimensional minimal surfaces, and underlines our present understanding of the Yamabe problem. Later it came to be fundamental in the analysis of solutions of the self-dual Yang-Mills equation on 4-manifolds. The borderline dimension for Yang-Mills turns out to be 4 instead of 2, and many results for harmonic maps and minimal surfaces have counterparts in theorems on Yang-Mills in dimension four [Ma, Se]. However, the topological results have come from Donaldson's study of the solution spaces of the self-dual Yang-Mills equations. Here existence theorems are proved by the implicit function theorem mentioned in the previous section. Whatever compactness of the solution space has is shown by the same techniques used in the construction of minimal surfaces [Do-2, F-U]. Gromov later applied these same ideas back in two dimensions to study pseudo-holomorphic 2-spheres in symplectic manifolds [Gr, McD].

It is an interesting question whether mathematicians will discover new and useful scale-invariant geometric partial differential equations. These techniques are just waiting to be used again!

Gauge Theory (New Sobolev-Inequalities)

Until gauge field theory appeared in mathematics, well-posed, natural, topological calculus of variations problems seemed confined to one dimension. It is possible to artificially construct variational problems of all sorts. However, the geometrically natural variational problems seem to be the only ones which are useful in topology. These are (a) first order in derivatives and (b) quadratic. Conditions (a) and (b) imply that the natural estimates are L_1^2. Topological results come from C^0, so we are stuck in the classical case of dimension one with geodesics for good problems and in the case of dimension 2 with minimal surfaces for the borderline case.

However, in gauge field theory on a manifold M, the unknown is the connection A which is a one-form. The natural function is the L^2 norm of curvature

$F_A = dA + [A, A]$, which is the first derivative of A. However, topology comes from the overlap functions, which relate A's via their derivatives. Intuitively speaking, curvature is the second derivative of the structure functions of a bundle. The relevant Sobolev embedding is $L_2^2 \subset C^0$. This embedding is true in dimensions 2 and 3, and borderline in dimension 4.

Some basic estimates need to be obtained before analysis can be done. Gauge theory problems have an infinite dimensional symmetry group, and the curvature $F_A = dA + [A, A]$ contains only part of the full derivative of A. These problems are related and are taken care of in the same analytical lemma. Because of the nonlinearity in the problem, the estimates are stated globally in terms of convergence.

Theorem [U-1]. *Let D_i be a sequence of connections on a compact manifold M of dimension n. If F_i is the curvature of D_i, and $F_i \in L^p$ forms a bounded sequence for $p > n/2$, then there exists a subsequence $D_{i'}$ and a sequence of gauge transformations $s_{i'}$ such that $s_{i'}^* D_{i'} \to D$ in L_1^p.*

A weaker version of this lemma applies in the borderline case $p = n/2$.

Yang-Mills in dimensions 2 and 3 gives us new examples of variational problems in which the topology and analysis match. The 2-dimensional example has been extensively studied [A-B] and gives us the topological results on the moduli space of stable bundles over Riemann surfaces. Recently the analytical details were completed by Daskalapoulos [Da]. The 3-dimensional problem is not very well understood geometrically. We do not as yet know how to use the many solutions to Yang-Mills which exist on a 3-manifold.

Heat Equation Methods

One of the earliest global non-linear results was that of Eells and Sampson [E-S]. Recall that the energy functional is defined on maps between two compact Riemannian manifolds $s : M \to N$

$$E(s) = \int_M |ds|^2 \, d\mu.$$

Critical maps are called harmonic maps. Eells and Sampson constructed a harmonic map in every homotopy class of maps $s \in [M, N]$ when N has non-positive sectional curvature. They did this by following L^2 gradient curves. Thus L^2 flow is a non-linear system of parabolic equations

$$\frac{\partial s}{\partial t} = \nabla_s^* \, ds.$$

In fact, early in his career S.T. Yau wrote a number of papers which apply this existence theory [Y-2].

A potential method of solving a non-linear elliptic equation is to solve the associated non-linear parabolic equations and follow the solutions as $t \to \infty$. This at first seems unduly cumbersome, and at least one mathematician expended some effort in finding "better" ways to find the Eells-Sampson harmonic maps [U-2]. In 1983 Richard Hamilton [H] was able to solve Einstein's equation on some 3-manifolds by solving the associated parabolic equation and following the time

dependent solution to a solution of the elliptic equation. This result has already
been noted in the section on the Poincaré conjecture in section 2. Hamilton's
result has had a remarkable effect on geometric PDE, since this result surely
cannot be obtained by variational methods. In the wake of Hamilton's work,
both the Einstein and the Kähler-Einstein equation have been reexamined and
solved in certain contexts using the heat equation method.

The non-linear heat flow was the geometric gradient used by Atiyah and
Bott to study the Yang-Mills equations. Their analysis was shown to be rigorous
by Donaldson, and was used by Donaldson to solve the holomorphic version of
Yang-Mills in arbitrary complex dimension [Do-3,Do-4]. Carlos Simpson [Si] has
shown that this approach fits in very nicely with geometric invariant theory. The
method can be used to obtain the same results as obtained by Yau and myself
[U-Y] and is more philosophically satisfying than the perturbation method.

Strüwe has gone a long way towards showing that many of the useful
properties of harmonic maps (or minimal surfaces) can be obtained via the
parabolic equation [St]. In many cases, the rigid method of looking for solutions
of a geometric elliptic equation via the parabolic equation will give the most
delicate results. This fits in best with the topology when there are moduli spaces
of solutions, the equation is geometrically natural, and fewer choices are involved
in following a parabolic equation compared to picking a minimizing sequence.

4. Failures

There are a number of theorems and conjectures which seem to be suitable for
attack by the methods discussed in the previous section. Some mathematicians
might regard these as "open problems." This is a matter of perspective. Perhaps,
since I confess to having spent considerable time on them myself, the failure is
personal.

Mostow-Rigidity Theorem. *If $s : M^n \to N^n$, $n > 2$, is a diffeomorphism between
hyperbolic manifolds, then s is homotopic to an isometry.*

The proof of this theorem ought to go somewhat as follows. Let $\hat{s}$ be the
harmonic map homotopic to s. The existence of this map dates back to [E-S] in
1963. Because M^n and N^n have constant curvature -1, the map $\hat{s}$ is an isometry.

Unfortunately, the only way we know that $\hat{s}$ is an isometry is by applying
Mostow rigidity in a circular argument. This is, of course, only the simplest version
of the Mostow rigidity theorem. There are some extensions due to Gromov which
ought to be accessible by these means. Ultimately we would generally like to know
better the relationship between curvature and volume. So far partial differential
equations have not been helpful.

A related theorem by Siu uses harmonic maps to establish rigidity of complex
manifolds with negative bisectional curvature [Si]. Other work has been done
by Corlette, Gromov and Schoen, but a partial differential equation's proof of
Mostow's theorem remains elusive.

Theorem [Se]. *Let $C_d^\infty(S^2,S^2)$ be the space of maps of degree d between S^2 and
S^2. Let $M_d^\infty(S^2,S^2) \subseteq C_d^\infty(S^2,S^2)$ be the meromorphic maps of degree d. Then the
inclusion of M_d in C_d^∞ is a homotopy equivalence up to dimension d.*

The proof ought to proceed as follows: Consider the energy integral

$$E(s) = \frac{1}{2} \iint\limits_{S^2} |ds|^2 \, d\mu$$

on maps $s : S^2 \to S^2$. The meromorphic functions are the set on which E takes on its minimum. After removing a set of codimension $d + 2$, the gradient flow retracts the remainder of the space of C^∞ functions onto the minimum.

This is a conformally invariant problem and we have not learned to handle the finer points of the topology of the flow! Segal's theorem is actually very extensive, applying to surfaces of arbitrary genus as the domain and a whole class of (positively curved) complex manifolds as the image [Se]. It is related to the Atiyah-Jones conjecture on the topology of instantons embedded in the space of connections [A-J].

There is actually an analytic proof of Segal's theorem. It can pieced together by applying a theorem of Donaldson [Do-6] on monopoles and Taubes' proof of a Morse theory for the monopole equation [T-2]. This seems too unwieldy to be a model proof, however.

Thurston's Techniques for 3-Manifolds. As part of his theory of 3-manifold topology, Thurston develops techniques for analyzing hyperbolic 3-manifolds which depend heavily on rigidly embedded surfaces in the manifolds. These surfaces are geodesically embedded with constant Gaussian curvature, and are broken along geodesics. It is more natural from the point of view of analysis to examine the properties of constant mean curvature surfaces in 3-manifolds. (Zero mean curvature characterizes locally minimizing surfaces.) Up until this date, attempts to replace Thurston's broken surfaces with smooth surfaces have been strikingly unsuccessful. While some results in 3-manifolds are obtained by partial differential equations techniques, Thurston's program has been unaffected by these methods.

Jones-Witten Invariants. Two of the four fields medals at this congress were given out at least in part for a theory designed to produce 3-manifold invariants. There are many approaches to this theory, but the unifying approach of Witten is to start with a classical geometrical integral similar to those which have been used already in the applications of analysis to topology. Witten in fact takes the integral used by Floer. The classical Chern-Simons integral is defined for an $su(N)$ valued one-form A on a 3-manifold.

$$CS(A) = \frac{1}{4\pi} \int_{M^3} \text{tr} \left(dA \wedge A + \frac{2}{3} A \wedge A \wedge A \right).$$

The Jones-Witten invariant for the manifold has the form

$$W(k, N) = \oint_{\text{all } A} e^{ikCS(A)} \, dA.$$

The symbol $\oint$ indicates the Feynman path integral used heavily in quantum field theory. Unfortunately, this is not a situation which has been made completely rigorous. However, many tools exist as input into calculations in quantum field theory. We list a few, in the hopes of giving some flavor of the subject.

Perturbation Theory. Calculations in quantum field theory which relate to physical experiment are meant to obtain asymptotic formulas as $1/k \to 0$. They are done via Feynman diagrams, which represent power series expansion around a vacuum (usually $A = 0$). In all physical cases I know of, the domain manifold is $\mathbb{R}^n$, $n \le 4$. These methods would fail here, due to the lack of a proper Green's function.

Stationary Phaze (and Ghosts). There are two separate problems which can be formally dealt with in more geometric problems. One is that of several important classical solutions and the other is the lack of ellipticity. Witten has carried out these asymptotic expansions and obtained the first order asymptotic in k [W-2]. There are clearly severe problems in the theory since further calculations cannot even be guessed at. No one has yet obtained lower order terms for manifolds, although some calculations exist for knots [B-N].

Finite Approximation. The approximation of the infinite dimensional integral $\oint$ by finite integrals is known as lattice gauge theory. A large amount of super-computer time is spent on more down-to-physics calculations than those of Chern-Simons theory. The value of these type of calculations is not clear. They certainly shed no light on arbitrary 3-manifolds.

Axiomatic Approach The quantum field theory involved in the Chern-Simons theory is particularly simple, and the formulation of the correct axioms seems to have been the greatest success of the theory so far. Atiyah has formuated the axioms for topological quamtum field theory (TQFT) [A-1].

Geometric Quantization. This formulation of quantum theory is probably the best understood by mathematicians. Witten and his students have a successful formulation of the Chern-Simons theory in terms of calculations on the moduli space of flat bundles [A-PD-W].

Canonical Quantization. It is usually necessary to use both the path integral and the Hamiltonian approach in computing the ingredients of quantum field theories. In many contexts, the Hamiltonian approach makes contact with group representations. We know that the input from 2-d conformal field theory is very useful in setting up the building blocks for the 3-d Chern-Simons theory. We can hence include on our list of useful mathematics:

> *Group representation theory,*
> *Integrable lattice models of statistical mechanics,*
> *Quantization of completely integrable systems.*

Quantum Groups. Finally, some topologists have found that the direct quantum group (Hopf algebra) approach leads to the most direct construction of the Jones-Witten invariants [Q,Wa].

In classical mechanics, Lagrangian and Hamiltonian formulations are equivalent. In quantum field theory, both are necessary and complementary. What remains in question is the consistency (not equivalence) of the facts gained from the different approaches. To someone like myself, who has worked in topological applications of partial differential equations, the situation is analogous to the plurality of approaches which can be made to understanding the Laplace operator on a Lie group. It is possible to get input from a wide variety of mathematical directions: separation of variables, special functions, finite-element approximation schemes, group representations, asymptotic heat kernel methods and a variety of different types of geometric constructions. However, we know what the Laplace operator itself is. This holds the different ideas together. The study of Jones-Witten invariants is similar, but we are missing the central ingredient which corresponds to the Laplace operator — a proper definition of $\not{L}$.

Recent computer calculations of D. Freed and R. Gompf indicate that two approaches (both due to Witten), from conformal field theory and from stationary phase approximation actually agree for lens spaces and some homology spheres [F-G]. Any mathematical demonstrations of this agreement would of necessity contain the proof tht analytic torsion agrees with combinatorial (Reidemeister) torsion. Perhaps we should look forward to a construction of $\not{L}$ which will pull this theory together.

References

[A-1]   Atiyah, M.: Topological quantum field theories. Publ. Math. Inst. Hautes Etudes Sci. Paris **68** (1989) 175–186

[A-2]   Atiyah, M.: The geometry and physics of knots. Lezioni Lincee, Cambridge University Press, Cambridge, 1990

[A-B]   Atiyah, M.F., Bott, R.: The Yang-Mills equations on Riemann surfaces. Phil. Trans. R. Soc. London A**308** (1982) 523–615

[A-J]   Atiyah, M.F., Jones, J.: Topological aspects of Yang-Mills theory. Comm. Math. Phys. **61** (1978) 97–118

[A-S]   Atiyah, M.F., Singer, I.: The index of elliptic operators on compact manifolds. Bull. Amer. Math. Soc. **69** (1963) 422–433

[A-DP-W]   Axelrod, S., Della Pietra, S., Witten, E.: Geometric quantization of Chern-Simons gauge theory. To appear

[B-N]   Bar-Natan, D.: Perturbative Chern-Simons theory. Preprint. Princeton University

[B]   Bott, R.: The stable topology of the classical groups. Ann. Math. **70** (1965) 313–337

[Br]   Bradlow, S.: Special metrics and stability for holomorphic bundles with global sections. J. Diff. Geom. (to appear)

[C-S]   Chern, S.S., Smale, S. (eds.): Global Analysis. Proc. of Sym. in Pure Math XIV-XVI, 1968

[Co]   Corlette, K.: Flat G-bundles with canonical metrics. J. Diff. Geom. **28** (1988) 361–382

[D-1]   Donaldson, S.K.: The geometry of 4-manifolds. ICM 86, vol. 1 (1987) pp. 43–54

[D-2]   Donaldson, S.K.: A application of gauge theory to 4-dimensional topology. J. Diff. Geom. **18** (1983) 81–83

[D-3]   Donaldson, S.K.: A new proof of a theorem of Narasimhan and Seshadri. J. Diff. Geom. **18** (1983) 269–277

[D-4]   Donaldson, S.K.: Anti-self-dual connections on complex algebraic surfaces and stable vector bundles. Proc. Lond. Math. Soc. **50** (1985) 1–26

[D-5]   Donaldson, S.K.: Infinite determinants, stable bundles and curvature. Proc. London Math. Soc. **55**(3) (1987)

[D-6]   Donaldson, S.K.: Nahm's equation and the classification of monopoles. Comm. Math. Phys. **96** (1984) 387–407

[Da]   Daskalopoulos, G.: The topology of the space of stable bundles on a Riemann surface. Thesis. University of Chicago, 1989

[E-S]   Eells, J., Sampson, J.: Harmonic mapping of Riemannian manifolds. Amer. J. Math. **86** (1964) 109–140

[F]   Freed, D.: A survey of the application of gauge theory methods to 4-manifolds. In progress

[F-G]   Freed, D., Gompf, R.: Computer calculation of Witten's 3-manifold invariant. Preprint. University of Texas

[F-U]   Freed, D., Uhlenbeck, K.: Instantons and 4-manifolds. MSRI Publications #1, (Second edition) 1990

[Gr]   Gromov, M.: Soft and hard symplectic geometry. ICM 86, **1** (1987) 81–89

[H]   Hamilton, R.: Three manifolds with positive Ricci curvature. J. Diff. Geom. **17** (1982) 255–306

[Hi]   Hitchin, N.: The self-duality equations on a Riemann surface. Proc. London Math. Soc. **55** (1987) 59–126

[Ka]   Kapouleas, N.: Thesis. Stanford, 1988

[K]   Kirwan, F.: Cohomology of quotients in symplectic and algebraic geometry. Mathematical Notes #31. Princeton University Press, 1987

[L]   Lang, S.: Introduction to differentiable manifolds. Interscience, New York, 1962

[L-P]   Lee, J., Parker, T.: The Yamabe problem. Bull. Amer. Math. Soc. **17** (1987) 37–81

[Ma]   Marini, A.: Dirichlet and Neumann boundary value problems for Yang-Mills connections. Thesis. University of Chicago, 1990

[McD]   McDuff, D.: Elliptic methods in symplectic geometry. Bull. Amer. Math. Soc. **23** (1990) 311–359

[M-Y]   Meeks, III, W., Yau, S.T.: The equivariant loop theorem. In: Morgan and Bass (eds.), The Smith Conjecture. Academic Press, New York 1984

[M-M]   Micallif, M., Moore, J.: Minimal 2-spheres and the topology of manifolds with positive curvature on totally isotropic 2-planes. Ann. Math. **127** (1988) 196–228

[Mi]   Milnor, J.: Morse theory. Ann. Math. Studies, vol. 51. Princeton University Press, 1968

[Mo]   Morrey, C.B.: The problem of Plateau on a Riemannian manifold. Ann. Math. **49** (1948) 807–851

[N-S]   Narasimhan, M.S., Sheshadri, S.: Stable unitary vector bundles on a compact Riemann surface. Ann. Math. **89** (1965) 540–567

[P]   Palais, R.: Foundations of global non-linear analysis. Benjamin, New York 1968

[P-S]   Palais, R., Smale, S.: A generalized Morse theory. Bull. Amer. Math. Soc. **70** (1964) 165–171

[Q]   Quinn, F.: Private communication

[S-U1]   Sacks, J., Uhlenbeck, K.: The existence of minimal immersions of 2-spheres. Ann. Math. **113** (1981) 1–24

[S-U2]   Sacks, J., Uhlenbeck, K.: Minimal immersions of closed Riemann surfaces. Trans. Amer. Math. Soc. **271** (1982) 639–652

[S]   Schoen, R.: The existence of weak solutions with prescribed singular behavior for a conformally invariant scalar equation. Comm. Pure Appl. Math. **41** (1988) 317–392

[S-Y]   Schoen, R., Yau, S.T.: Existence of incompressible minimal surfaces and the topology of manifolds with non-negative scalar curvature. Ann. Math. **110** (1970) 127–142

[Sed]   Sedlacek, S.: A direct method for minimizing the Yang-Mills functional in 4-manifolds. Comm. Math. Phys. **86** (1982) 515–527

[Se]   Segal, G.: Topology of spaces of rational functions. Acta Math. **143** (1979) 39–51

[Sim]   Simpson, C.: Constructing variations of Hodge structure using Yang-Mills theory and applications to uniformization. J. Amer. Math. Soc. **1** (1988) 867–918

[Si]   Siu, Y.T.: The complex analyticity of harmonic maps and the strong rigidity of Kähler manifolds. Ann. Math. **112** (1980) 73–111

[Si-Y]   Siu, Y.T., Yau. S.T.: Compact Kähler manifolds of positive bisectional curvature. Inventiones Math. **59** (1980) 189–204

[Sm]   Smale, S.: An infinite dimensional version of Sard's theorem. Amer. J. Math. **87** (1965) 861–866

[St]   Struwe, M.: On the evolution of harmonic mappings of Riemann surfaces. Comment. Math. Helv. **60** (1985) 558–581

[T-1]   Taubes, C.: Self-dual connections on non-self-dual 4-manifolds. J. Diff. Geom. **17** (1982) 139–170

[T-2]   Taubes, C.: Min-Max Theory for the Yang-Mills-Higgs Equations. Comm. Math. Phys. **97** (1985) 473–540

[U-1]   Uhlenbeck, K.: Connections with L^p bounds on curvature. Comm. Math. Phys. **88** (1982) 11–30

[U-2]   Uhlenbeck, K.: Morse theory by perturbation methods with applications to harmonic maps. Trans. Amer. Math. Soc. **267** (1981) 569–583

[U-Y]   Uhlenbeck, K., Yau, S.T.: On the existence of Hermitian-Yang-Mills Connections in Stable Vector Bundles. Comm. Pure Appl. Math. (1986) 257–293

[W-1]   Witten, E.: Topological quantum field theory. Comm. Math. Phys. **117** (1988) 353–386

[W-2]   Witten, E.: Quantum field theory and the Jones polynomial. Comm. Math. Phys. **131** (1989) 351–399

[Ya]   Yamabe, H.: On a deformation of Riemannian structures on compact manifolds. Osaka Math. J. **12** (1960) 21–37

[Y-1]   Yau, S.T.: On the Ricci curvature of a compact Kähler manifold and the complex Monge-Ampere equation. Comm. Math. Appl. Math. **31** (1978) 339–411

[Y-2]   Yau, S.T.: On the fundamental group of compact manifolds of non-positive curvature. Ann. Math. **93** (1971) 579–585

Multidimensional Hypergeometric Functions in Conformal Field Theory, Algebraic K-Theory, Algebraic Geometry

Alexandre Varchenko

Moscow Institute of Gas and Oil, Leninski Prospekt 65, 117917 Moscow, USSR

Rudolf Arnheim in the book *Visual Thinking* (L.A. 1969) writes that usually concepts tend to crystallize into simple, well-shaped forms. They are tempted by Platonic rigidity. This creates troubles when the range they are intended to cover includes relevant qualitative differences. The variations can be so different from each other that to see them as belonging to one family of phenomena requires mature understanding. To the young mind, they look as different from each other as did the morning star from the evening star to the ancients.

The notion of a general hypergeometric function was introduced by I.M. Gelfand in the mid 80s. Now it is clear that general hypergeometric functions play a major role in interesting parts of mathematics such as Conformal Field Theory, Representation Theory, Algebraic K-Theory, Algebraic Geometry and provide new connections among them.

The general hypergeometric functions are generalizations of the Euler beta-function. The beta-function is the integral of a product of powers of linear functions over the segment. In the generalization the segment is replaced by a polytope and the integral

$$I(\Delta, f, \alpha) = \int_\Delta f_1^{\alpha_1} \ldots f_N^{\alpha_N} \, dx_1 \ldots dx_n$$

is considered as a function of the polytope $\Delta \subset \mathbb{R}^n$, the linear functions $\{f_j\}$, and the exponents $\{\alpha_j\} \subset \mathbb{C}$. The simplest examples are the classical hypergeometric function, the Euler dilogarithm, the volume of a polytope. The systematic study of the general hypergeometric functions was begun only recently in works of I.M. Gelfand's school and K. Aomoto.

There are three basic reasons for the appearance of general hypergeometric functions: the general hypergeometric functions satisfy remarkable differential equations, the general hypergeometric functions satisfy remarkable functional equations, the general hypergeometric functions, as analytic functions of their arguments, have remarkable monodromy groups.

Proceedings of the International Congress
of Mathematicians, Kyoto, Japan, 1990
© The Mathematical Society of Japan, 1991

1. Functional Equations

1.1 Volume of Polytope

The volume of a convex polytope in $\mathbb{R}^3$ has the following properties. The volume does not change under movements of the polytope. If a polytope is divided into two parts by a plane, then the volume of the polytope is equal to the sum of the volumes of the parts. These properties are functional equations of the volume considered as a function on the space of all convex polytopes. The properties of the volume suggest the following definition. The *group of polytopes* in $\mathbb{R}^3$ is the abelian group generated by the symbols (Δ), where $\Delta \subset \mathbb{R}^3$ is any convex polytope, subject to the relations

(a) $(\Delta) = (g\Delta)$ for any motion g.
(b) $(\Delta) = (\Delta_1) + (\Delta_2)$, if Δ is divided by a plane into parts Δ_1, Δ_2.

Is an element of this group uniquely determined by its volume? Is a regular tetrahedron equivalent to a cube of the same volume? These questions form the content of the third Hilbert problem. The third Hilbert problem was solved in Dehn's articles in 1900–1902, before it was published. It turned out that the regular tetrahedron cannot be composed from a cube because the two have different Dehn invariants.

Consider an edge of a polytope. An edge has two characteristics: the length l and the angle θ at the edge. The angle is defined up to a multiple of π and lies in $\mathbb{R}/\pi\mathbb{Z}$. The *Dehn invariant* of a polytope is the expression $D(\Delta) = \sum l_i \otimes \theta_i$, where the sum is taken over all edges of the polytope. The Dehn invariant is an element of the group $\mathbb{R} \otimes_{\mathbb{Z}} (\mathbb{R}/\pi\mathbb{Z})$. It is obvious that the invariant does not change under movements of the polytope and is additive under cuttings. It is easy to see that the Dehn invariant of a cube is equal to zero (as is that of any prism) and that the Dehn invariant of a regular tetrahedron is not equal to zero. So a regular tetrahedron cannot be composed from a cube. According to Sydler, 1965, the equivalence class of a polytope is uniquely determined by its volume and Dehn invariant, see also [C1, Du, DuPS, DuS, S].

This example demonstrates the scheme leading from a hypergeometric function to interesting algebraic concepts. Given a function on some space satisfying some functional equations one considers the abelian group generated by the points of the space subject to relations given by the functional equations of the initial function. If the initial function satisfies some differential equations then the group has additional structure.

1.2 Example

The *Euler dilogarithm* is the function defined by the power series

$$\mathrm{Li}_2(z) = \sum_{m=1}^{\infty} z^m/m^2 \qquad \text{for } |z| < 1.$$

The dilogarithm is a hypergeometric function. It has the integral representation

$$\mathrm{Li}_2(z) = \int_{\Delta} \frac{dx}{x} \wedge \frac{dy}{y}$$

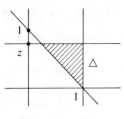

Fig. 1

where the triangle $\triangle$ is shown in Fig. 1. The dilogarithm satisfies the differential equation

$$d\,\mathrm{Li}_2(t) = \ln(1-t)\,d\,\ln(t)$$

and the functional equation

$$\mathrm{Li}_2(x) - \mathrm{Li}_2(y) + \mathrm{Li}_2(y/x) - \mathrm{Li}_2\left(\frac{y(1-x)}{x(1-y)}\right) + \mathrm{Li}_2\left(\frac{1-x}{1-y}\right)$$

$$= n^2/6 - \ln(x)\ln\left(\frac{1-x}{1-y}\right).$$

for $0 < y < x < 1$.

The algebraic construction. The *Bloch group* of a field F is the abelian group B_2, generated by the symbols (t), where $t \in F\backslash\{0, 1\}$, subject to the relations

$$(x) - (y) + (y/x) - \left(\frac{y(1-x)}{x(1-y)}\right) + \left(\frac{1-x}{1-y}\right) = 0 \tag{1}$$

for any $x, y \in F\backslash\{0, 1\}$.

Consider the multiplicative group F^* of a field F and its exterior square $F^* \wedge_{\mathbf{Z}} F^*$. The map

$$t \mapsto (t) \wedge (1-t),$$

sending the symbol $(t), t \in F\backslash\{0, 1\}$, to the element $(t) \wedge (1-t)$ of the group $F^* \wedge F^*$, has a remarkable property. It sends the alternating sum of the elements on the left-hand side of (1) to the zero element of the group $F^* \wedge F^*$. This gives a well-defined homomorphism

$$S : B_2 \to F^* \wedge F^*,$$

called the *Bloch complex*.

The homology groups of the Bloch complex are connected with the Quillen K-groups of the field F. In algebraic K-theory with any field F there is associated the sequence of groups $K_n(F), n \geq 0$. For these groups a multiplication $K_p \otimes K_q \to K_{p+q}$ is defined.

Theorem (Matsumoto, Suslin).

1) $\mathrm{Coker}\,S \simeq K_2(F);$

2) $(\text{Ker } S) \otimes \mathbb{Q} \cong K_3^{\text{ind}}(F) \otimes \mathbb{Q},$

where $K_3^{\text{ind}} = K_3/K_1^3$ is the indecomposable part of K_3.

This theorem gives an elementary definition of K_2 and the indecomposable part of K_3 (modulo torsion). An interesting problem is to find elementary definitions of all the groups $K_n(F)$. Conjecturally such definitions arise from functional and differential equations of polylogarithmic functions.

There are several possible generalizations of the logarithm and the Euler dilogarithm [A3, Le, GM, HaM]. One of them is the Aomoto polylogarithms.

1.3 Aomoto Polylogarithms

A simplex in $P^n(\mathbb{C})$ is an ordered set $L = (L_0, L_1, \ldots, L_n)$ of hyperplanes. A simplex defines the differential n-form

$$\omega_L = d \ln(z_1/z_0) \wedge \cdots \wedge d \ln(z_n/z_0),$$

where $z_i = 0$ is a homogeneous equation of L_i. With a second simplex $M = (M_0, \ldots, M_n)$ n-chain $\varDelta_M$ in $P^n(\mathbb{C})\backslash L$ is connected. $\varDelta_M$ is a curved oriented

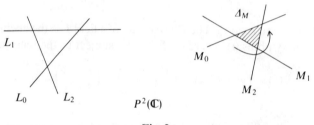

Fig. 2

n-simplex with boundary in M, see Fig. 2. With a pair of simplices an integral

$$a_n(L; M) = \int_{\varDelta_M} \omega_L$$

is associated called the *Aomoto polylogarithm* of order n. The integral depends on the choice of the chain $\varDelta_M$, but does not change under its deformation. The Aomoto polylogarithm has the following properties.

(2) *Antisymmetry.* The integral is antisymmetric with respect to renumbering of the hyperplanes of the first or the second simplices.
(3) *Additivity with respect to a form.* If $L_0, \ldots, L_{n+1}$ are $(n + 2)$ hyperplanes and $L^i := (L_0, \ldots, \hat{L}_i, \ldots, L_{n+1})$, then

$$\sum_{j=0}^{n+1} (-1)^j a_n(L^j; M) = 0.$$

(4) *Additivity with respect to a chain.* If $M_0, \ldots, M_{n+1}$ are $(n + 2)$ hyperplanes, then

$$\sum_{j=0}^{n+1} (-1)^j a_n(L; M^j) = 0$$

for a suitable choice of $\varDelta_{M^j}$.

(5) *Projective invariance.* For every $g \in PGL(n + 1, \mathbb{C})$ if $\varDelta_{gM} = g\varDelta_M$, then

$$a_n(gL; gM) = a_n(L; M).$$

For example, an Aomoto polylogarithm of order 1 is defined by two pairs of points $L = (L_0, L_1)$, $M = (M_0, M_1)$ on $P^1(\mathbb{C})$ and is equal to an integral of the form $\omega_L = d \ln(z_1/z_0)$ over a path going from M_1 to M_0. $a_1(L; M)$ is equal to the logarithm of the cross ratio of the four points (L_0, L_1, M_0, M_1). The Euler dilogarithm is a special case of the Aomoto dilogarithm, see Fig. 1.

There are two connections of polylogarithms of different orders: the multiplication and the differential equation.

(6) The product of Aomoto polylogarithms of orders p, q may be expressed as a sum of Aomoto polylogarithms of order $p + q$, see [BGSV].

(7) The differential of the polylogarithm of order n, considered as a function on the space of all configurations (L, M) may be expressed through suitable polylogarithms of orders $n - 1$ and 1 [A3].

Aomoto polylogarithms describe parameters of the mixed Hodge structure of cohomology groups of a pair $P^n(\mathbb{C}) \backslash L$, $M \backslash M \cap L$, where L, M are configurations of hyperplanes, see [BMS, BGSV]. For example if two points $L_0, L_1 \in P^1(\mathbb{C})$ are removed and two points $M_0, M_1 \in P^1(\mathbb{C})$ are identified then the mixed Hodge structure of the first cohomology group of this space is defined by the number $\exp(a_1(L_0, L_1; M_0, M_1))$.

1.4 Hopf Algebra of Pairs of Simplices

Taking properties of Aomoto polylogarithms as a starting point, it is possible to suggest a definition of a graded Hopf algebra $A(F) = A_0 \oplus A_1 \oplus A_2 \oplus \cdots$ for any field F [BMS, BGSV]. Here $A_0 := \mathbb{Z}$, A_n is the abelian group, generated by the symbols $(L; M)$, where $L = (L_0, \ldots, L_n)$, $M = (M_0, \ldots, M_n)$ are ordered sets of hyperplanes in $P^n(F)$, subject to relations similar to (2)–(5). The multiplication $\mu_{p,q}: A_p \otimes A_q \to A_{p+q}$ and the comultiplication $v_n = \bigoplus_{p=0}^{n} v_{p,n-p}: A_n \to \bigoplus_{p=0}^{n} A_p \otimes A_{n-p}$ of the Hopf algebra are modeled by properties (6), (7).

There is the sequence of complexes $A[n]$, $n \in \mathbb{Z}_+$, associated with any graded Hopf algebra A. The simplest of these complexes are

$$A[1] : 0 \to A_1 \to 0,$$

$$A[2] : 0 \to A_2 \xrightarrow{v_{1,1}} A_1 \otimes A_1 \to 0,$$

$$A[3] : 0 \to A_3 \xrightarrow{v_{2,1} + v_{1,2}} A_2 \otimes A_1 \oplus A_1 \otimes A_2 \xrightarrow{v_{1,1} \otimes 1 - 1 \otimes v_{1,1}} A_1 \otimes A_1 \otimes A_1 \to 0.$$

The complex $A[n]$ is concentrated in degrees from 1 to n. Conjecturally, the cohomology groups of these complexes give the K-groups of the field F modulo torsion.

Conjecture (Beilinson) [BGSV].

$$K_n(F) \otimes \mathbb{Q} \simeq \bigoplus_{j=0}^{[(n-1)/2]} H^{n-2j}(A[n-j] \otimes \mathbb{Q}).$$

Theorem [BGSV]. *The conjecture holds for $n \leq 3$:*

$$K_1 \simeq H^1(A[1]), \quad K_2 \simeq H^2(A[2]), \quad K_{3,\mathbb{Q}} \simeq H^3(A[3]_\mathbb{Q}) \oplus H^1(A[2]_\mathbb{Q}).$$

In the last years efforts have been undertaken to construct a motivic cohomology theory of algebraic manifolds which would be an arithmetical variant of the singular cohomology theory. According to Beilinson a category of motivic sheaves over the spectrum of a field F is a category of graded modules over a suitable algebra and a possible candidate for it could be the algebra $(A(F) \otimes \mathbb{Q})^*$ dual to the algebra $A(F) \otimes \mathbb{Q}$.

1.5 Bloch-Wigner Function

The Euler dilogarithm may be continued to a multivalued analytic function on $\mathbb{C} \setminus \{0, 1\}$. The *Bloch-Wigner* function is its imaginary part

$$D(z) = \text{Im}(\text{Li}_2(z)) + \arg(1 - z) \ln|z|,$$

see [B2, Z]. The Bloch-Wigner function has the following properties.

(8) $D(z)$ is single-valued real analytic on $\mathbb{C}$ except at the points 0 and 1, where it is only continuous.

(9)
$$D(x) - D(y) + D(y/x) - D\left(\frac{y(1-x)}{x(1-y)}\right) + D\left(\frac{1-x}{1-y}\right) = 0$$

for any $x, y \in \mathbb{C} \setminus \{0, 1\}$.

In particular, for any field $F \subset \mathbb{C}$ the Bloch-Wigner function defines a homomorphism $B_2(F) \to \mathbb{R}$ of the Bloch group to real numbers.

1.6 Polylogarithms and Zeta-Function

Let F be an algebraic number field of degree n over $\mathbb{Q}$ with r_1 real and r_2 complex places, $r_1 + 2r_2 = n$. Let $\zeta_F(s)$ be the Dedekind zeta-function of F. The value $\zeta_F(2)$ is expressed in terms of the values of the Bloch-Wigner function at points of the field F.

Theorem [Bo, B, Su, Z]. *In the Bloch complex of F the group* Ker S *is isomorphic (modulo torsion) to* $\mathbb{Z}^{r_2}$. *The co-volume of the image of the map* Ker $S \to \mathbb{R}^{r_2}$, *defined by the composition of complex imbeddings and the Bloch-Wigner function, is a non-zero rational multiple of* $\pi^{-2(r_1+r_2)}|d_F|^{1/2}\zeta_F(2)$, *where d_F is the discriminant of F.*

D. Zagier conjectured that for any natural number m the value $\zeta_F(m)$ has similar expression in terms of values of the classical polylogarithms of order at most m at points of F, [Z]. Recently A. Goncharov proved this conjecture for $m = 3$.

Let $\mathbb{Z}[F\backslash\{0, 1\}]$ be the free abelian group generated by the symbols (t), $t \in F\backslash\{0, 1\}$. $\mathbb{Q}[F\backslash\{0, 1\}] := \mathbb{Z}[F\backslash\{0, 1\}] \otimes \mathbb{Q}$. For any function $D : \mathbb{C}\backslash\{0, 1\} \to \mathbb{R}$ there is a homomorphism $D : \mathbb{Q}[\mathbb{C}\backslash\{0, 1\}] \to \mathbb{R}$, $D : \sum n_i(t_i) \to \sum n_i D(t_i)$.

Let $\{a_j\}$, $j = 1, \ldots, r_1 + 2r_2$, be all possible imbeddings $F \to \mathbb{C}$, $\bar{a}_{r_1+k} = a_{r_1+r_2+k}$.

Theorem [Go]. *There exist* $z_1, \ldots, z_{r_1+r_2} \in \mathbb{Q}[F\backslash\{0, 1\}]$, *such that*

$$\zeta_F(3) = \pi^{3r_2}|d_F|^{-1/2}\det(D_3(\sigma_j(z_i))),$$

where $j = 1, \ldots, r_1 + r_2$,

$$D_3(z) = \mathrm{Re}(\mathrm{Li}_3(z) - \ln|z|\mathrm{Li}_2(z) + \tfrac{1}{3}\ln^2|z|\mathrm{Li}_1(z)),$$

$\mathrm{Li}_n(z)$, $n \geq 1$, *is the classical polylogarithm of order* n, *defined by the power series*

$$\mathrm{Li}_n(z) = \sum_{m=1}^{\infty} z^m/m^n \qquad \textit{for } |z| < 1.$$

2. Hypergeometric Integrals

2.1 Classical Hypergeometric Function

The hypergeometric series

$$F(a, b; c; z) = 1 + \frac{ab}{1c}z + \frac{a(a + 1)b(b + 1)}{1 \cdot 2c(c + 1)}z^2$$

$$+ \frac{a(a + 1)(a + 2)b(b + 1)(b + 2)}{1 \cdot 2 \cdot 3c(c + 1)(c + 2)}z^3 + \cdots$$

satisfies the differential equation

$$z(1 - z)F'' + (c - (a + b + 1)z)F' - abF = 0$$

and has the integral representation

$$\frac{\Gamma(b)\Gamma(c - b)}{\Gamma(c)}F(a, b; c; z) = \int_1^{\infty} t^{a-c}(t - 1)^{c-b-1}(t - z)^{-a}\,dt.$$

Thus the classical hypergeometric function has three definitions: as a power series, as the solution of a differential equation, as an integral depending on a parameter.

These objects are associated with the family of the configurations of the triples of the points $0, 1, z$ of the complex line and are generalized naturally to the case of a family of configurations of hyperplanes in an affine space [A, G, GGZ, GKZ].

2.2 Cohomology of Complement of Configuration of Hyperplanes

Let $\mathscr{C}$ be a finite set of hyperplanes in $\mathbb{C}^n$. Choose a linear equation $f_H = 0$ for any hyperplane. Define the closed differential form df_H/f_H. The *Orlik-Solomon algebra* is the exterior algebra generated by 1 and the forms df_H/f_H, $H \in \mathscr{C}$.

Theorem [Ar1, Bri]. *The Orlik-Solomon algebra is naturally isomorphic to the complex cohomology ring of the complement of a configuration.*

In other words every cohomology class can be represented as a polynomial in the forms df_H/f_H, such a polynomial defines the zero class only if it is equal to zero.

Example [Ar1]. The Orlik-Solomon algebra of the configuration of all diagonal hyperplanes $t_i - t_j = 0$ in $\mathbb{C}^n$ is isomorphic to the exterior algebra generated by 1 and the symbols w_{ij} subject to the relations $w_{ij} = w_{ji}$, $w_{ij} \wedge w_{jk} + w_{jk} \wedge w_{ki} + w_{ki} \wedge w_{ij} = 0$ for pairwise different i, j, k. $P(t) = (1 + t)(1 + 2t)\ldots(1 + (n - 1)t)$ is the Poincaré polynomial of the algebra, see also [Or, OS].

V.I. Arnold computed this example in connection with study of superpositions of algebraic functions and Hilbert's 13th problem; see [Ar2].

A configuration is called weighted if a complex number $\alpha(H)$ is assigned to each hyperplane H. The weights define the function

$$l_\alpha = \prod_{H \in \mathscr{C}} f_H^{\alpha(H)}.$$

This function is a multivalued function on the complement of a configuration. The differential of this function has the form

$$dl_\alpha = l_\alpha \sum_{H \in \mathscr{C}} \alpha(H) \, df_H/f_H.$$

A *hypergeometric differential form* of a weighted configuration is any form $l_\alpha \omega$, where ω is a differential form of the Orlik-Solomon algebra.

Hypergeometric forms form a finite-dimensional complex, as the differential of a hypergeometric form is a hypergeometric form:

$$d(l_\alpha \omega) = l_\alpha \sum_{H \in \mathscr{C}} \alpha(H) \, df_H/f_H \wedge \omega.$$

The *weight local system* $\mathscr{S}(\alpha)$ on the complement of a configuration is the complex one-dimensional local system of coefficients with the monodromy around a hyperplane H equal to the multiplication by $\exp(-2\pi i\alpha(H))$.

The cohomology of the complement of a configuration with coefficients in the weight local system is computed by the complex of hypergeometric forms. More precisely for any number t denote by $t\alpha$ the weights $H \mapsto t\alpha(H)$, $H \in \mathscr{C}$, homothetic to the initial ones.

Theorem [SV3]. *For almost all $t \in \mathbb{C}$ the cohomology of the finite dimensional complex of hypergeometric forms with weights $t\alpha$ is naturally isomorphic to the cohomology of the complement of a configuration with coefficients in $\mathscr{S}(t\alpha)$.*

For example, this is true for $t = 0$ according to the Arnold-Brieskorn theorem. Exceptional values of t form a descrete set. Conjecturally the exceptions form explicitly given arithmetical progressions [Ao1].

2.3 Determinant Formulas

The Euler beta-function is the alternating product of Euler gamma-functions:

$$B(\alpha, \beta) = \Gamma(\alpha)\Gamma(\beta)/\Gamma(\alpha + \beta).$$

There is a generalization of this formula to the case of a configuration of hyperplanes in an affine space [V, Se, A6].

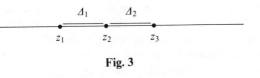

Fig. 3

Example. Consider the configuration of three points z_1, z_2, z_3 on a line, see Fig. 3. The point z_j is the zero of the function $f_j = t - z_j$. Put

$$l_\alpha = (t - z_1)^{\alpha_1}(t - z_2)^{\alpha_2}(t - z_3)^{\alpha_2},$$

$$\omega_1 = \alpha_1 l_\alpha d(t - z_1)/(t - z_1), \qquad \omega_2 = \alpha_2 l_\alpha d(t - z_2)/(t - z_2),$$

then

$$\det\left(\int_{\Delta_i} \omega_j\right) = \frac{\Gamma(\alpha_1 + 1)\Gamma(\alpha_2 + 1)\Gamma(\alpha_3 + 1)}{\Gamma(\alpha_1 + \alpha_2 + \alpha_3 + 1)} \prod_{i \neq j} f_i^{\alpha_i}(z_j).$$

Thus the determinant of integrals of basic hypergeometric forms of a configuration over all bounded components of the complement of the configuration is equal to the product of values of powers of linear functions at the vertices of the configuration up to a multiplicative constant equal to an alternating product of values of the gamma function [V].

The formula has an arithmetical analog. F. Loeser extended it to the case of a configuration of hyperplanes in an affine space over a finite field [Lo]. In this case the gamma-functions are replaced by Gauss sums, the determinant of the hypergeometric integrals is replaced by an alternating product of the determinants of the Frobenius operator in suitable cohomology groups.

3. Hypergeometric Functions and Representation Theory of Lie Algebras

The appearance of hypergeometric functions in the representation theory of Kac-Moody algebras, their quantum deformations, and in the Conformal Field Theory (CFT) is connected with integrals of the form

$$\underbrace{I(t_1, \ldots, t_n)}_{n} = \int \prod_{1 \leq j < k \leq N} (t_j - t_k)^{\lambda_{jk}} \underbrace{dt_{n+1} \wedge \cdots \wedge dt_N}_{m},$$

$$N = m + n.$$

Such integrals correspond to special configurations. The characteristics of these configurations, in particular the homology groups with twisted coefficients of the complement, the complex of hypergeometric forms are interpreted as objects of the representation theory of Kac-Moody algebras. Such integrals satisfy the differential equation which is described in terms of representation theory and is known in CFT as the Knizhnik-Zamolodchikov equation. The branching of such integrals is described in terms of quantum groups corresponding to Kac-Moody algebras.

The appearance of these integrals in physical models isn't surprising. Imagine a model in which points $t_1, \ldots, t_N$ of the line pairwise interact. The interaction is described by the function $(t_j - t_k)^{\lambda_{jk}}$. In this case the average of the interaction over all positions of the last m points is an integral characteristic of the first n points, described by the hypergeometric integral.

In applications the constants of the interaction λ_{jk} have the form

$$\lambda_{jk} = B(v_j, v_k)/\kappa,$$

where $v_1, \ldots, v_N$ are vectors of some complex linear space V, B is a symmetric bilinear form on V, and κ is a complex parameter of the model.

In applications to the representation theory of Kac-Moody algebras, V is the dual space to the Cartan subalgebra of the Kac-Moody algebra, B is the Killing form on it. The vectors $v_{n+1}, \ldots, v_N$ corresponding to the averaged points belong to the set of simple negative roots. The vectors $v_1, \ldots, v_n$ corresponding to the parameters of the integral are the highest weights of the representations of the Kac-Moody algebra.

3.1 Hypergeometric Construction

Assume given

(a) natural numbers $n \le N$, $N = m + n$;
(b) a complex linear space V, symmetric bilinear form B on V and an ordered set of (weight) vectors $v_1, \ldots, v_N \in V$, not necessarily different.

From these data the construction builds a complex linear space W and a differential equation on a W-valued function $\phi(z_1, \ldots, z_n)$:

$$d\phi = \kappa^{-1} \sum_{1 \le j < k \le n} \Omega_{jk} \phi \, d(z_j - z_k)/(z_j - z_k),$$

where $\Omega_{jk} : W \to W$ are suitable linear operators, κ is a complex parameter.

The Construction. Consider the configuration of all the diagonal hyperplanes $t_j = t_k$ in $\mathbb{C}^N$. Define the weights of the diagonals: $\lambda_{jk} = B(v_j, v_k)/\kappa$.

The weights define the complex of hypergeometric forms, the one-dimensional complex local system $\mathscr{S}$ on the complement of the diagonals with the monodromy around the diagonal $t_j = t_k$ equal to the multiplication by $\exp(2\pi i \lambda_{jk})$. Hypergeometric forms have well-defined integrals over chains with coefficients in $\mathscr{S}$.

Let $\mathbb{C}^N \to \mathbb{C}^n$ be the projection on the first n coordinates. A fiber over a point z is the space $\mathbb{C}^m$ in which the diagonals cut a configuration depending on z. Denote it by $\mathscr{C}(z)$.

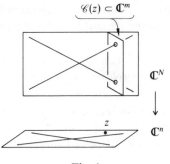

Fig. 4

A point of the base is called *discriminantal* if the configuration in the fiber over it is degenerate. The discriminantal points form the discriminant, the configuration of all diagonal hyperplanes. Over the complement of the discriminant, fibers with distinguished configurations form a locally trivial bundle. The fundamental group of the complement of the discriminant is the pure braid group on n strings.

Consider the restriction $\mathscr{S}(z)$ to the fiber $\mathbb{C}^m \backslash \mathscr{C}(z)$ over a point z of the local system $\mathscr{S}$, and the top homology group $H_m(\mathbb{C}^m \backslash \mathscr{C}(z), \mathscr{S}(z))$ of the fiber with coefficients in this restriction. The *homology bundle* is the complex vector bundle over the complement of the discriminant with fiber $H_m(\mathbb{C}^m \backslash \mathscr{C}(z), \mathscr{S}(z))$ over the point z.

The homology group $H_m(\mathbb{C}^m \backslash \mathscr{C}(z), \mathscr{S}(z))$ depends on z and is uniquely translated along paths in the base. The *Gauss-Manin connection* is this integrable connection on the homology bundle.

The *monodromy representation* of the Gauss-Manin connection is the representation of the pure braid group in the automorphism group of the homology group, induced by translations of the homology group along loops in the complement of the discriminant.

The monodromy representation gives the well-known Burau representation of the braid group in the special case of one-dimensional fiber and equal weights [K2, GiS, Gi]. As will be explained the representations of the braid groups appearing in the theory of quantum groups are closely connected with the monodromy representations of the constructed Gauss-Manin connections.

An integral of a hypergeometric form on $\mathbb{C}^N$ over any cycle in a fiber is equal to zero certainly if this form is the differential of some form or the restriction of this form on any fiber equals zero.

Accordingly we define the *hypergeometric cohomology group* as

$$\mathscr{H}^m = \Omega^m / (Z^m + d\Omega^{m-1}),$$

where Ω^m is the space of hypergeometric m-forms on $\mathbb{C}^N$, $Z^m \subset \Omega^m$ is the subspace of forms with zero restriction on any fiber, $d\Omega^{m-1} \subset \Omega^m$ is the subspace of differentials of hypergeometric $(m-1)$-forms.

The integration defines the homomorphism

$$\iota : H_m(\mathbb{C}^m \backslash \mathscr{C}(z), \mathscr{S}(z)) \to (\mathscr{H}^m)^*$$

of the homology group over a point z to the space independent of z. The homomorphism depends on z and the parameter κ, $\iota = \iota(z, \kappa)$.

Theorem [SV]. *For almost all κ^{-1} with exceptions in a suitable discrete subset of $\mathbb{C}$ the homomorphism $\iota(z, \kappa)$ is an isomorphism for all points z outside the discriminant.*

Theorem [Ao, SV]. *There exist linear operators $\Omega_{jk} : (\mathcal{H}^m)^* \to (\mathcal{H}^m)^*, 1 \leq j < k \leq n$, with the following properties. For any locally constant homology class $\Delta(z) \in H_m(\mathbb{C}^m \backslash \mathscr{C}(z), \mathscr{S}(z))$ the $(\mathcal{H}^m)^*$-valued function $\phi(z) := \iota(z, \kappa)[\Delta(z)]$ satisfies the differential equation*

$$d\phi = \kappa^{-1} \sum_{1 \leq j < k \leq n} \Omega_{jk} \phi \, d(z_j - z_k)/(z_j - z_k). \tag{10}$$

In other words hypergeometric integrals satisfy the differential equation (10) and for a general κ all solutions of the equation are given by the integrals.

This picture has a symmetry group. Any permutation of the coordinates in $\mathbb{C}^N$ accordingly enumerates the weight vectors $v_1, \ldots, v_N \in V$. Any permutation of the last m coordinates preserving the ordered set of the weight vectors preserves the fibers of the projection, the weights of the diagonals, acts on the complex of hypergeometric forms, on homology groups of fibers, on solutions of the differential equation (10).

The *twisted homology group* $H_m(\mathbb{C}^m \backslash \mathscr{C}(z), \mathscr{S}(z))_{\text{ant}}$ is the antisymmetric part of the group $H_n(\mathbb{C}^m \backslash \mathscr{C}(z), \mathscr{S}(z))$ with respect to the action of the permutation group of the last m coordinates preserving the ordered set of weight vectors. The *twisted hypergeometric cohomology group* $\mathcal{H}^m_{\text{ant}}$ is the antisymmetric part of the hypergeometric cohomology group with respect to the action of the same group.

The result of the construction is the complex linear space $(\mathcal{H}^m_{\text{ant}})^*$ and a $(\mathcal{H}^m_{\text{ant}})^*$-valued differential equation (10) on $\mathbb{C}^n$.

3.2 Representations of Lie Algebras and Knizhnik-Zamolodchikov Equation

As an example consider the Lie algebra $\mathfrak{g} = \mathfrak{sl}_2(\mathbb{C})$ of complex 2×2-matrices with the zero trace. $\mathfrak{g}$ is generated by the standard generators e, f, h subject to the relations $[e, f] = h$, $[h, f] = -2f$, $[h, e] = 2e$.

Fix an invariant scalar product on $\mathfrak{g}$ (the Killing form). Let $\Omega \in \mathfrak{g} \otimes \mathfrak{g}$ be the tensor corresponding to the invariant scalar product (the Casimir operator).

Let $L_1, \ldots, L_n$ be representations of $\mathfrak{g}$, $L = L_1 \otimes \cdots \otimes L_n$. Let Ω_{jk} be the linear operator on $L_1 \otimes \cdots \otimes L_n$, acting as the Casimir operator on $L_j \otimes L_k$ and as the identity operator on the other factors. The *Knizhnik-Zamolodchikov* (KZ) *equation* on the L-valued function $\phi(z_1, \ldots, z_n)$ is the system of the differential equations

$$d\phi = \kappa^{-1} \sum_{1 \leq j < k \leq n} \Omega_{jk} \phi \, d(z_j - z_k)/(z_j - z_k),$$

where κ is a complex parameter.

The KZ equation defines the integrable connection on the trivial bundle $L \times \mathbb{C}^n$ with singularities over the diagonals. This connection has a remarkable property:

parallel translations of this connection commute with the action of $\mathfrak{g}$ on fibers. Thus the eigenspaces of the operators e, f, h are invariant under parallel translations.

It turns out that the KZ equation restricted on suitable invariant subspaces coincides with suitable hypergeometric equations constructed above.

More precisely, let $\mathfrak{h} \subset \mathfrak{g}$ be the Cartan subalgebra generated by h. The *Verma module* of $\mathfrak{g}$ with the highest weight $\Lambda \in \mathfrak{h}^*$ is the infinite dimensional representation of $\mathfrak{g}$ generated by one (vacuum) vector v with the properties $ev = 0$, $hv = \langle h, \Lambda \rangle v$. Verma modules are the simplest representations from which all finite-dimensional representations may be constructed. Let $M_1, \ldots, M_n$ be Verma modules with the highest weights $\Lambda_1, \ldots, \Lambda_n \in \mathfrak{h}^*$. Put $M = M_1 \otimes \cdots \otimes M_n$, $\Lambda = \Lambda_1 + \cdots + \Lambda_n$. M is the direct sum of the eigenspaces of the operator h, $M = \bigoplus_{m \geq 0} M_{\Lambda - m\alpha}$, where $M_{\Lambda - m\alpha}$ is the eigenspace with the eigenvalue $\langle h, \Lambda - m\alpha \rangle$, $\alpha \in \mathfrak{h}^*$ is the single positive root of $\mathfrak{sl}_2(\mathbb{C})$. The *vacuum subspace* $\mathrm{Vac}_{\Lambda - m\alpha} \subset M_{\Lambda - m\alpha}$ is the subspace of all vectors annihilated by the operator e.

Theorem [SV, DJMM]. *The vacuum subspace* $\mathrm{Vac}_{\Lambda - m\alpha}$ *and the KZ equation restricted on it are canonically isomorphic, respectively, to the space* $(\mathscr{H}^m_{\mathrm{ant}})^*$ *dual to the twisted hypergeometric cohomology group, and to the differential equation with values in* $(\mathscr{H}^m_{\mathrm{ant}})^*$ *constructed by the hypergeometric construction from the projection* $\mathbb{C}^{m+n} \to \mathbb{C}^n$, *the linear space* $\mathfrak{h}^*$, *the Killing form on* $\mathfrak{h}^*$ *and the ordered set of the vectors* $\Lambda_1, \ldots, \Lambda_n, \underbrace{-\alpha, \ldots, -\alpha}_{m} \in \mathfrak{h}^*$.

Corollary. *For a general* κ *all solutions of the KZ equation with values in a tensor product of Verma modules are given by the hypergeometric functions.*

An analogous picture has a place if the algebra $\mathfrak{sl}_2(\mathbb{C})$ is replaced by any Kac-Moody algebra, see [SV, DJMM, Ma, Ch, CF, L].

This theorem shows that the KZ equation has topological nature, it is just the Gauss-Manin connection of the simple fiber bundle.

3.3 Hypergeometric Functions in the Conformal Field Theory

The KZ equation was invented in the CFT. Its solutions describe $(n + 1)$-point correlation functions on the Riemann sphere in the Wess-Zumino-Witten model [BPZ, KZ]. In the minimal models of the CFT correlation functions on the sphere have also integral representations in terms of the same configurations [DF]. For the first time integral representations of correlation functions in the CFT appeared in the works by Dotsenko and Fateev.

Any model of the CFT have a certain set of primary fields $\{\phi_n(z, \bar{z})\}$ and an operator algebra

$$\phi_n(z, \bar{z})\phi_m(0, 0) = \sum_p C^p_{nm} z^{\Delta_p - \Delta_m - \Delta_n} \bar{z}^{\bar{\Delta}_p - \bar{\Delta}_m - \bar{\Delta}_n} \phi_p(0, 0) + \cdots,$$

where the numbers Δ_n, $\bar{\Delta}_n$ are the conformal dimensions of the field ϕ_n, the numbers C^p_{nm} are the structure constants of the model, in this expression the terms containing non-primary fields are omitted, see [BPZ]. The integral representations of the correlation functions allowed one to calculate the structure constants of the minimal models of the CFT and of the Wess-Zumino-Witten model corresponding to $\mathfrak{sl}_2(\mathbb{C})$ [DF, Do]. It turns out that in these cases the structure constants are equal to certain alternating products of values of the gamma-function similar to that appearing in the determinant formula of Sect. 2.3. In these cases any structure constant was represented by a certain hypergeometric integral. Probably in other models of the CFT the structure constants are connected with suitable determinants of hypergeometric integrals which in their turn may be expressed as alternating products of values of the gamma-function.

3.4 Hypergeometric Functions and Quantum Groups

The hypergeometric integrals as functions of the parameters satisfy the KZ equation. The monodromy of the KZ equation is described in terms of quantum groups.

The *monodromy representation* of the KZ equation is the representation of the pure braid group on n strings in the group $\text{Aut}(L_1 \otimes \cdots \otimes L_n)$ generated by analytic continuation of solutions along loops in the base. Denote it by τ_κ. Algebraically we define another representation of the pure braid group.

According to Faddeev, Kulish, Reshetikhin, Sklynin, Jimbo, Drinfeld [Dr1, J] the universal enveloping algebra $U\mathfrak{g}$ of the algebra $\mathfrak{g} = \mathfrak{sl}_2(\mathbb{C})$ is deformed to the quantum universal enveloping algebra $U_q\mathfrak{g}$ (the quantum group) depending on the complex parameter q.

For the quantum group a comultiplication $\Delta : U_q\mathfrak{g} \to U_q\mathfrak{g} \otimes U_q\mathfrak{g}$ is defined. If V_1, V_2 are representations of the quantum group then the comultiplication induces a representation structure on their tensor product. The representations $V_1 \otimes V_2$ and $V_2 \otimes V_1$ are isomorphic. The isomorphism is defined by the formula

$$V_1 \otimes V_2 \xrightarrow{R} V_1 \otimes V_2 \xrightarrow{P} V_2 \otimes V_1,$$

where P is the transposition of the factors and $R \in U_q\mathfrak{g} \otimes U_q\mathfrak{g}$ is the distinguished element called the *universal R-matrix* of the quantum group.

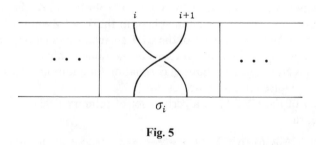

Fig. 5

Let $V_1, \ldots, V_n$ be representations of the quantum group. Then the pure braid group acts on their tensor product. Namely let $\sigma_1, \ldots, \sigma_{n-1}$ be the elementary braids shown in Fig. 5. To any braid σ_i assign the linear operator

$$V_1 \otimes \cdots \otimes V_i \otimes V_{i+1} \otimes \cdots \otimes V_n \to V_1 \otimes \cdots \otimes V_{i+1} \otimes V_i \otimes \cdots \otimes V_n$$

acting as PR on the i-th and $(i + 1)$-th factors and as the identity operator on the other factors. These operators define the representation ϱ_q of the pure braid group on n strings in $\mathrm{Aut}(V_1 \otimes \cdots \otimes V_n)$.

For general values of q the representation theory of the quantum group $U_q\mathfrak{g}$ is the same as the representation theory of the algebra $\mathfrak{g}$, any representation of the algebra $\mathfrak{g}$ is deformed canonically to the representation of the quantum group. Let $L_1, \ldots, L_n$ be representations of $\mathfrak{g}$ and $L_{1_q}, \ldots, L_{n,q}$ be their quantum deformation.

Theorem [K, Dr2]. *The monodromy representation τ_κ of the KZ equation with values in $\mathrm{Aut}(L_1 \otimes \cdots \otimes L_n)$ is equivalent to the R-matrix representation ϱ_q, $q = \exp(2\pi i/\kappa)$, with values in $\mathrm{Aut}(L_{1,q} \otimes \cdots \otimes L_{n,q})$, if κ is not a rational number.*

An analogous statement holds if $\mathfrak{sl}_2(\mathbb{C})$ is replaced by any Kac-Moody algebra.

This theorem is a wonderful statement. Given the differential equation of the CFT described in terms of the representation theory of the Lie algebra one considers its global characteristic, the monodromy representation defined by analytic continuation of solutions and one gets the representation defined by R-matrix of the quantum group which in its turn is defined as the deformation of the universal enveloping algebra of the initial Lie algebra.

It is interesting that new invariants of knots defined by Jones and others are constructed in terms of the same representations of the braid groups, see [C2, Jo, L, RT].

The realization of the KZ equation as the Gauss-Manin connection allows one to make explicit the equivalence of the representations τ_κ and ϱ_q, $q = \exp(2\pi i/\kappa)$. It turns out that for real $z_1 < \cdots < z_n$ there is a canonical isomorphism of the quantum deformation of the vacuum subspace to the twisted homology group:

$$\mathrm{Vac}_{\Lambda - m\alpha, q} \to H_m(\mathbb{C}^m \setminus \mathscr{C}(z), \mathscr{S}(z))_{\mathrm{ant}},$$

which sends the R-matrix representation to the monodromy representation of the Gauss-Manin connection. The equivalence of the Kohno-Drinfeld theorem is the composition

$$\mathrm{Vac}_{\Lambda - n\alpha, q} \to H_m(\mathbb{C}^m \setminus \mathscr{C}(z), \mathscr{S}(z))_{\mathrm{ant}} \xrightarrow{\iota} (\mathscr{H}_{\mathrm{ant}}^m)^* \to \mathrm{Vac}_{\Lambda - n\alpha}.$$

Thus a correspondence between a representation of Lie algebra and its quantum deformation is the correspondence between the twisted homology group and the dual space to the twisted hypergeometric cohomology group given by the integration of the hypergeometric forms.

The details of the proof of this statement and the analogous statement for arbitrary Kac-Moody algebra are being checked in [SV4]. On this subject see also [L].

4. Concluding Remarks

Many interesting works on hypergeometric functions are not mentioned in this talk. Among them there are works by Deligne and Mostow, Beukers and Heckman on monodromy of hypergeometric functions, works by Gelfand, Kapranov, Zelevinsky on combinatorial description of differential equations on hypergeometric functions, works on q-analogs of hypergeometric functions, on difference equations on hypergeometric functions, works by Heckman and Opdam on hypergeometric functions and root systems.

The main problem is to unify numerous and remote investigations on hypergeometric functions into a united theory in which it might be possible to pass from Conformal Field Theory to Algebraic K-Theory by analytic continuation on parameters.

References

[A1]   Aomoto, K.: Un théorème du type de M-M concernant l'integrale des fonctions multiformes. J. Math. Pures Appl. **52** (1973) 1–11

[A2]   Aomoto, K.: On vanishing of cohomology attached to certain many valued meromorphic functions. J. Math. Soc. Japan **27** (1975) 248–255

[A3]   Aomoto, K.: On the structure of integrals of power product of linear functions. Sc. Papers of the College of General Education. Tokyo Univ. **27** (1977) 49–61

[A4]   Aomoto, K.: Addition theorem of Abel type for hyper-logarithms. Nagoya Math. J. **88** (1982) 55–71

[A5]   Aomoto, K.: Configurations and invariant theory of Gauss-Manin systems. Adv. Stud. Pure Math. **4** (1984) 165–179

[A6]   Aomoto, K.: Jacobi polynomials associated with Selberg integrals. SIAM J. Math. Anal. **18** (1987) 545–549

[A7]   Aomoto, K.: On the complex Selberg integral. Quart. J. Math. Oxford **38** (1987) 385–399

[A8]   Aomoto, K.: Special value of the hypergeometric function $_3F_2$ and connection formulae among asymptotic expansions. J. Indian Math. Soc. **51** (1987) 161–221

[A9]   Aomoto, K.: Gauss-Manin connection of integral of difference products. J. Math. Soc. Japan **39** (1987) 191–208

[A10]   Aomoto, K.: Special values of hyperlogarithms and linear difference schemes. Preprint 1988, pp. 1–47

[A11]   Aomoto, K.: Finiteness of a cohomology associated with certain Jackson integrals. Preprint 1989, pp. 1–48

[Ar1]   Arnold, V.I.: The cohomology ring of the colored braid group. Math. Notes **5** (1969) 138–140

[Ar2]   Arnold, V.I.: Topological invariants of algebraic functions. II. Funct. analys i ego pril. **4** (1970) 1–9

[As1]   Askey, R.S.: The q-gamma and q-beta functions. Appl. Anal. **8** (1978) 125–141

[As2]   Askey, R.S.: Some basic hypergeometric extensions of integrals of Selberg and Andrews. SIAM J. Math. **11** (1980) 938–951

[Bei]   Beilinson, A.: Polylogarithms and cyclotomic elements. Preprint MIT 1990, pp. 1–30

[BGSV1] Beilinson, A., Goncharov, A., Schechtman, V., Varchenko, A.: Aomoto dilogarithms, mixed Hodge structures and motivic cohomology of pairs of triangles on the plane. The Grothendieck Festschrift, vol. 1. Birkhäuser, 1990, pp. 135–172

[BGSV2] Beilinson, A., Goncharov, A., Schechtman, V., Varchenko, A.: Projective Geometry and K-theory. Algebra Anal. 2 (1990) n°3, 78–131

[BMS] Beilinson, A., MacPherson, R., Schechtman, V.: Notes on motivic cohomology. Duke Math. J. 54 (1987) 679–710

[BPZ] Belavin, A., Polyakov, A., Zamolodchokov, A.: Infinite conformal symmetry in two-dimensional quantum field theory. Nucl. Phys. B241 (1984) 333

[BeH] Beukers, F., Heckman, G.: Monodromy for hypergeometric functions $_nF_{n-1}$. Invent. math. 95 (1989) 325–354

[B1] Bloch, S.: Higher regulators, algebraic K-theory, and zeta functions of elliptic curves. Lecture Notes. U.C. Irvine, 1977

[B2] Bloch, S.: Applications of the dilogarithm function in algebraic K-theory and algebraic geometry. Proc. of the Intern. Symp. Alg. Geometry. Tokyo, 1978

[Bo] Borel, A.: Commensurability classes and volumes of hyperbolic 3-manifolds. Ann. Sc. Norm. Sup. Pisa 8 (1981) 1–33

[BrGM] Brasselet, J., Goresky, M., MacPherson, R.: Simplicial differential forms with poles. Preprint 1988, pp. 1–40

[Bri] Brieskorn, E.: Sur les groupes de tresses. Séminaire Bourbaki (Lecture Notes in Mathematics, vol. 317.) Springer, Berlin Heidelberg New York, pp. 21–44

[C1] Cartier, P.: Decomposition des polyhèdres: le point sur le troisieme problème de Hilbert. Séminaire Bourbaki, 1985–86, n. 646

[C2] Cartier, P.: Developpements recents sur les groupes de tresses. Applications à la topologie et a l'algèbre. Séminaire Bourbaki, 1989–90, n. 716

[Ch1] Cherednik, V.: Monodromy representations for generalized Knizhnik-Zamolodchikov equation. Preprint ITP-89-74E, Kiev 1990, pp. 1–21

[Ch2] Cherednik, V.: Integral solutions of trigonometric Knizhnik-Zamolodchikov equations and Kac-Moody algebras. Preprint RIMS 1990, pp. 1–31

[CF] Christe, P., Flume, R.: The four-point correlations of all primary operators of the $d = 2$ conformally invariant $SU(2)$ σ-model with Wess-Zumino term. Nucl. Phys. B282 (1987) 466–494

[CW1] Cohen, P., Wolfart, J.: Monodromie et plongement modulaire. Preprint 1989, pp. 1–10

[CW2] Cohen, P., Wolfart, J.: Monodromie des fonctions d'Appell, varietes abéliennes et plongement modulaire. Preprint 1989, pp. 1–6

[DJMM] Date, E., Jimbo, M., Matsuo, A., Miwa, T.: Hypergeometric-type integrals and the $sl(2, \mathbb{C})$ Knizhnik-Zamolodchikov equation. Preprint RIMS (1989) 1–12

[D1] Dehn, M.: Ueber raumgleiche Polyeder. Nachr. Akad. Wiss., Göttingen, Math.-Phys. Kl. 1900, pp. 345–354

[D2] Dehn, M.: Uber den Rauminhalt. Math. Ann. 55 (1902) 465–478

[De1] Deligne, P.: Théorie de Hodge. I. Proc. Int. Cong. Nice 1970, vol. 1, 425–430; II. – Publ. Math. IHES 40 (1971) 5–58; III. – Publ. Math. IHES 44 (1972) 5–77

[De2] Deligne, P.: Equations différentielles à points singuliers réguliers. (Lecture Notes in Mathematics, vol. 163.) Springer, Berlin Heidelberg New York 1970

[DeM] Deligne, P., Mostow, G.D.: Monodromy of hypergeometric functions and non-lattice integral monodromy. Publ. Math. IHES 63 (1986) 5–89

[Do] Dotsenko, V.: Solving the SU(2) conformal field theory with the Wakimoto free field representation. Moscow (1990) 1–31

[DF1] Dotsenko, V., Fateev, V.: Conformal algebra and multipoint correlation functions in 2D statistical models. Nucl. Phys. B240 [FS12] (1984) 312–348

[DF2]   Dotsenko, V., Fateev, V.: Four-point correlation functions and the operator algebra in 2D conformal invariant theories with central charge $C \leq 1$. Nucl. Phys. **B251** [FS13] (1985) 691–734

[Dr1]   Drinfeld, V.: Quantum groups. Proc. Int. Cong. Math. Berkeley 1986, vol. 1, pp. 798–820

[Dr2]   Drinfeld, V.: Quasi-Hopf algebras. Algebra Anal. **1** (1989) 30–46

[Du]   Dupont, J.: Algebra of polytopes and homology of flag complexes. Osaka J. Math. **19** (1982) 599–641

[Du2]   Dupont, J.: On polylogarithms. Preprint (1987) 1–29

[DuPS]   Dupont, J., Parry, W., Sah, C.-H.: Homology of classical Lie groups made discrete, II. J. Algebra **113** (1988) 215–260

[DuS1]   Dupont, J., Sah, C.-H.: Scissors congruences, II. J. Pure and Appl. Alg. **25** (1982) 159–195

[DuS1]   Dupont, J., Sah, C.-H.: Homology of Euclidean Motion groups made discrete and Euclidean scissors congruences. Preprint 1989, pp. 1–39

[FSV]   Feigin, B., Schechtman, V., Varchenko, A.: On algebraic equations satisfied by correlators in WZW model. Letters in Math. Phys. **20**, 291–297 (1990)

[G]   Gelfand, I.M.: General theory of hypergeometric functions. Dokl. Akad. Nauk. SSSR **288** (1986) 573–576

[GG]   Gelfand, I.M., Graev, M.I.: Duality theorem for general hypergeometric functions. Dokl. Akad. Nauk SSSR **289** (1986) 19–23

[GGZ]   Gelfand, I.M., Graev, M.I., Zelevinsky, A.V.: Holonomic systems of equations and series of hypergeometric type. Dokl. Akad. Nauk SSSR **295** (1988) 14–19

[GKZ1]   Gelfand, I.M., Kapranov M.M., Zelevinsky, A.V.: Projective dual varieties and hyperdeterminants. Dokl. Akad. Nauk SSSR **305** (1989) 1294–1298

[GKZ2]   Gelfand, I.M., Kapranov M.M., Zelevinsky, A.V.: A-discriminants and Cayley-Koszul complexes. Dokl. Akad. Nauk SSSR **307** (1989) 1307–1311

[GKZ3]   Gelfand, I.M., Kapranov M.M., Zelevinsky, A.V.: Newton polyhedra of principal A-determinants. Dokl. Akad. Nauk SSSR **308** (1989) 20–23

[GKZ4]   Gelfand, I.M., Kapranov M.M., Zelevinsky, A.V.: Hypergeometric functions and toric varieties. Funct. analys i ego pril. **23** (1989) 12–26

[GM]   Gelfand, I.M., MacPherson, R.: Geometry in grassmanians and a generalization of the dilogarithm. Adv. Math. **44** (1982) 279–312

[GVZ]   Gelfand, I.M., Vasiliev, V.A., Zelevinsky, A.V.: General hypergeometric functions on complex grassmanians. Funct. analys i ego pril. **21** (1987) 23–38

[GV]   Gelfand, I.M., Varchenko, A.N.: On Heaviside functions of configuration of hyperplanes. Funct. analys i ego pril. **21** (1987) 1–18

[GZ]   Gelfand, I.M., Zelevinsky, A.V.: Algebraic and combinatoric aspects of general theory of hypergeometric functions. Funct. analys i ego pril. **20** (1986) 17–34

[Gi]   Givental, A.: Twisted Picard-Lefschetz formulas. Funct. analys i ego pril. **22** (1988), 1, 12–22

[GiS]   Givental, A., Schechtman, V.: Monodromy groups and Hecke algebras. Usp. Mat. Nauk **42** (1987), 6, 138

[Go]   Goncharov, A.: The classical threelogarithm, algebraic theory of fields and Dedekind zeta functions. Moscow 1990, pp. 1–9

[GorM]   Goresky, M., MacPherson, R.: Stratified morse theory. Springer, Berlin Heidelberg New York 1988

[H]   Heckman, G.: Root systems and hypergeometric functions II. Comp. Math. **64** (1987) 353–373

[HO]   Heckman, G., Opdam, E.: Root systems and hypergeometric functions I. Comp. Math. **64** (1987) 329–352

[HaM]   Hain, R., MacPherson, R.: Higher logarithms. Seattle 1990, pp. 1–84

[J]   Jessen, B.: The algebra of polyhedra and the Dehn-Sydler theorem. Math. Scand. **22** (1968) 241–256

[Ji]   Jimbo, M.: Lett. Math. Phys. **10** (1985) 63–69

[Jo]   Jones, V.F.R.: Hecke algebra representations of braid groups and link polynomials. Ann. Math. **126** (1987) 335–388

[KZ]   Knizhnik, V., Zamolodchikov, A.: Current algebras and Wess-Zumino models in two dimensions. Nucl. Phys. **B247** (1984) 83–103

[K1]   Kohno, T.: Monodromy representations of braid groups and Yang-Baxter equations. Ann. Inst. Fourier **37** (1987) 139–160

[K2]   Kohno, T.: Linear representations of braid groups and classical Yang-Baxter equations. Contemp. Math. **78** (1988) 339–363

[K3]   Kohno, T.: Integrable connections related to Manin and Schechtman's higher braid groups. Preprint (1989) 1–13

[L1]   Lawrence, R.: A topological approach to the representations of the Iwahori-Hecke algebra. International J. Modern Physics A **5** (1990) No 16

[L2]   Lawrence, R.: The Homological Approach to Higher Representations. Preprint 1990, pp. 1–9

[L3]   Lawrence, R.: A Functorial Approach to the One-Variable Jones Polynomial. Preprint 1990, pp. 1–15

[Le]   Lewin, L.: Dilogarithms and associated functions. MacDonald, London, 1958

[Lo]   Loeser, F.: Arrangements d'hyperplans et sommes de Gauss Preprint 1989, pp. 1–31

[M]   Manin, Yu. I.: Correspondences, motives and monoidal transformations. Mat. Sb. (N.S.) **119** (1968) 475–507

[MaSY]   Matsumoto, K., Sasaki, T., Yoshida, M.: The period map of a 4-parameter family of K3 surfaces and the Aomoto-Gelfand hypergeometric function of type (3, 6). Proc. Japan Acad. **64** (1988) 307–310

[Ma]   Matsuo, A.: An application of Aomoto-Gelfand hypergeometric functions to the $SU(n)$ Knizhnik-Zamolodchikov equation. Preprint RIMS 1990, pp. 1–18

[Mi1]   Milne, S.: A q-analog of the Gauss summation theorem for hypergeometric series on $U(n)$. Adv. Math. **72** (1988) 59–131

[Mil]   Milne, S.: A q-analog of hypergeometric series well-poised in $SU(n)$ and invariant G-functions. Adv. Math. **58** (1985) 1–60

[Mo]   Mostow, G.D.: Generalized Picard lattices arising from half-integral conditions. Publ. Math. IHES **63** (1986) 91–106

[Mo]   Mostow, G.D.: Braids, hypergeometric functions, and lattices. Bull. Am. Math. Soc. **16** (1987) 2, 225–246

[N]   Nishiyama, S.: Appell's hypergeometric function F_2 and periods of certain elliptic K3 surfaces. Tokyo J. Math. **10** (1987) 33–67

[O1]   Opdam, E.: Root systems and hypergeometric functions III. Comp. Math. **67** (1988) 21–49

[O2]   Opdam, E.: Root systems and hypergeometric functions IV. Comp. Math. **67** (1988) 191–209

[O3]   Opdam, E.: Some applications of hypergeometric shift operators. Invent. math. **98** (1988) 1–18

[Or]   Orlik, P.: Introduction to arrangements. Preprint 1988, pp. 1–127

[OS]   Orlik, P., Solomon, L.: Combinatorics and topology of complements of hyperplanes. Invent. math. **56** (1980) 167–189

[Q]   Quillen, D.: Higher K-theory I. (Lecture Notes in Mathematics, vol. 341) Springer, Berlin Heidelberg New York, pp. 85–147

[RT]   Reshetikhin, N., Turaev, V.: Invariants of 3-manifolds via link polynomial and quantum groups. Invent. math. **103** (1991)

[R]   Rogers, L.: On function sum theorems connected with the series $\sum_1 (x^n/n^2)$. Proc. London Math. Soc. **4** (1907) 169–189

[S1]   Sah, C.-H.: Homology of classical Lie groups made discrete, I. Comment. Math. Helv. **61** (1986) 308–347

[S2]   Sah, C.-H.: Homology of classical Lie groups made discrete, III. J. Pure Appl. Algebra (to appear)

[Se]   Selberg, A.: Bemerkninger om et multiplet integral. Norsk Mat. Tidsskr. **26** (1944) 71–78

[SV1]   Schechtman, V., Varchenko, A.: Integral representations of N-point conformal correlators in the WZW model. Bonn 1989, pp. 1–21

[SV2]   Schechtman, V., Varchenko, A.: Hypergeometric solutions of Knizhnik-Zamolodchikov equations. Letters in Math. Phys. **20** (1990) 279–283

[SV3]   Schechtman, V., Varchenko, A.: Arrangements of hyperplanes and Lie algebras homology. Moscow 1990, pp. 1–130

[SV4]   Schechtman, V., Varchenko, A.: Quantum groups and homology local systems. Preprint 1990, pp. 1–30

[Su]   Suslin, A.: Algebraic K-theory of fields. Proc. of the Int. Congr. Math., Berkeley, 1986, pp. 222–244

[Sy]   Sydler, J.-M.: Conditions necessaires et suffissantes pour l'équivalence des polyèdres de l'espace euclidien à trois dimensions. Comment. Math. Helv. **40** (1965) 43–80

[T]   Terada, T.: Fonctions hypergeometriques F_1 et fonctions automorphes. – I. – J. Math. Soc. Japan **35** (1983) 451–475. II. – Math. Soc. Japan **37** (1985) 173–185

[TK]   Tsuchiya, A., Kanie, Y.: Vertex operators in conformal field theory on $\mathbb{P}^1$ and monodromy representations of braid groups. Adv. Stud. Pure Math. **16** (1988) 297–372

[V1]   Varchenko, A.: Euler beta-function, Vandermond determinant, Legendre equation and critical values of linear functions on configuration of hyperplanes. I. Izv. Akad. Nauk SSSR (1989) 1206–1235. II. Izv. Akad. Nauk SSSR (1990) 146–158

[V2]   Varchenko, A.: Critical values and determinant of periods. Uspehi Mat. Nauk (1989) 235–236

[V3]   Varchenko, A.: Hodge filtration of hypergeometric integrals associated with an affine configuration of general position and a local Torelli theorem. Preprint (1989) 1–13

[V4]   Varchenko, A.: Determinant formula for Selberg type integrals. Preprint (1990) 1–3

[Z1]   Zagier, D.: The remarkable dilogarithm. Bonn 1987, pp. 1–15

[Z1]   Zagier, D.: The Bloch-Wigner-Ramakrishnan polylogarithm function. Bonn 1989, pp. 1–10

Invited Forty-Five Minute Addresses at the Section Meetings

Degree Structures

Theodore A. Slaman

Mathematical Sciences Research Institute
1000 Centennial Drive, Berkeley, CA 94720, USA

The Turing degrees $\mathscr{D}$ were introduced by Kleene and Post ([9], 1954) to isolate and study those properties of the subsets of the natural numbers $\mathbb{N}$ which are expressed purely in terms of relative computability. Intuitively, we form $\mathscr{D}$ by identifying any pair of subsets of $\mathbb{N}$ which are mutually computable and ordering the resulting equivalence classes by relative computability. The natural hierarchies of definability within arithmetic, analysis and higher fragments of set theory all have sharply focused images in $\mathscr{D}$.

We will focus our attention on the second order properties of the Turing degrees. We will phrase our discussion of the known results and especially of the techniques of their proofs within as large a context as possible, so to apply to degree structures based on other forms of relative definability, as well as to $\mathscr{D}$.

Formally, suppose that A and B are subsets of $\mathbb{N}$, henceforth called *reals*. We say that A is *Turing reducible* to B ($A \leq_T B$) if there is a computational procedure which takes an input n from $\mathbb{N}$; over the course of its execution on input n, asks whether various numbers are in B; and, if it receives the correct responses to those questions, after finitely many computational steps returns the answer as to whether n is an element of A. In other words, if we were given B then we would be able to compute A. We say that A and B are Turing equivalent ($A \equiv_T B$) if $A \leq_T B$ and $B \leq_T A$. The Turing degrees D are the $\equiv_T$-equivalence classes. The degree structure $\mathscr{D}$ associated with Turing reducibility is the partial ordering $\langle D, \leq_T \rangle$, the Turing degrees with the ordering inherited from $\leq_T$.

$\mathscr{D}$ is the most intensively studied degree structure but not the only one. Although we cannot introduce them all in detail, we will point out a few other examples. Among the many degree structures on the real numbers obtained by varying the notion of relative definability, we might mention the arithmetic degrees, the hyperarithmetic degrees, the Σ_k-degrees ($A \leq_{\Sigma_k} B$ if A is in the least set which is Σ_k-admissible relative to B), the many-one degrees, the enumeration degrees, the Δ_2^1-degrees and the degrees of constructibility. Alternately, we can retain relative computability and vary the class of reals on which it acts. We can form the Turing degrees of the recursively enumerable sets, the Turing degrees

The author was partially supported by N.S.F. grant DMS-8902437 and Presidential Young Investigator Award DMS-8451748.

of the Δ_2^0-sets (i.e. the structure $\mathcal{D}(\leq_T 0')$ on the degrees below $0'$) or the Turing degrees of the arithmetically definable sets. We will refer to these examples as *local substructures* of $\mathcal{D}$. For any degree x, we can also look at $\mathcal{D}(_T \geq x)$, the partial order of the degrees greater than or equal to x.

Extending our scope to include degree notions on other than sets of integers, we can also mention the Kleene degrees of sets of reals, the degrees of sets of reals modulo ^3E, and the α-degrees of subsets of an admissible ordinal α. Similarly, we can form their local substructures such as α-recursively enumerable α-degrees or the Kleene degrees of the Π_1^1-subsets of the continuum.

To some extent, in our discussion of these structures, we will follow the historical order in which their properties where discovered. We will focus our discussion on the Turing degrees since historically $\mathcal{D}$ has been the proving ground for the analysis of degree structures. To begin with the obvious: there is a least degree consisting of the computable reals; since there are only countably many computational procedures, each Turing degree is countable and has only countably many predecessors, any two degrees x and y have a least upper bound $x \vee y$ (so $\mathcal{D}$ is an upper-semi-lattice with a natural operation of join) and the cardinality of the whole structure is the same as the continuum. To continue beyond these immediate observations we must directly analyze the notion of relative computability.

In the first section, we will discuss the early results about the Turing degrees. For the most part, these were to the effect that $\mathcal{D}$ has a rich existential theory, showing that $\mathcal{D}$ plays various roles as a universal object. In the second section, we will describe the subsequent results which simultaneously limit $\mathcal{D}$'s universal role and reveal its detailed structure. In the final sections, we will discuss some recent results and conjectures of Slaman and Woodin. If true, the conjectured structural theorem would provide a complete logical characterization of $\mathcal{D}$. We prove the conjecture for the hyperarithmetic degrees and for $\mathcal{D}$ with finitely many additional parameters.

§1. The Embedding Theorems of the Late 1950s, 1960s and Early 1970s

Partial Order Embeddings

In their 1954 paper [9], Kleene and Post used a Baire category argument, one involving a primitive form of Cohen forcing, to show that there are sets of incomparable Turing degree. In fact, Kleene and Post proved that every finite partial order can be embedded in $\mathcal{D}$. Sacks ([18], 1963) extended the Kleene-Post embedding theorem to show that every countable partial ordering, and even every one of size $\aleph_1$ which is locally countable, could be embedded in $\mathcal{D}$.

Sacks conjectured:

(1) (Sacks ibid., 1963) **Conjecture:** A partially ordered set P is imbeddable in the (Turing) degrees if and only if P has cardinality at most that of the continuum and each member of P has at most countably many predecessors.

The Kleene-Post and Sacks partial order embedding results indicated a universal quality of $\mathcal{D}$. Sacks expressed a belief in this quality by conjecturing that the strongest possible purely existential property would be satisfied by $\mathcal{D}$. (Surprisingly, this conjecture is still open.) Furthermore, the algebraic properties of $\mathcal{D}$ seemed to be based more in the nature of relative definability than in specific properties of computability. The proofs of the embedding theorems apply to any of the above mentioned degree notions on the reals which are countably based, i.e. excluding only the Δ_2^1-degrees and the degrees of constructibility.

Initial Segments

Say that a subset I of $\mathcal{D}$ is an *ideal* if it is an initial segment and closed under join. I is a *principal ideal* if in addition it has a greatest element. Spector ([23], 1956) answered a question of Kleene-Post [9] by constructing a two element ideal, i.e. a minimal nontrivial degree. Spector's theorem rules out the potential classification of the principal ideals in $\mathcal{D}$ (as the countable model of an $\aleph_0$-categorical theory). For example, there are two nontrivial ideals which are not isomorphic. Spector's method could be extended to build many examples of isomorphism types of initial segments in $\mathcal{D}$. Prompted by Spector's results, Sacks made the following conjecture.

(2) (Sacks [19], 1966) **Conjecture:** S is a finite, initial segment of the degrees if and only if S is order isomorphic to a finite, initial segment of some upper-semi-lattice with a least member.

Sacks's conjecture postulated that another universal property would hold of $\mathcal{D}$: the initial segments of $\mathcal{D}$ would realize all the finite possibilities. Extending work of Thomason and Lachlan, Lerman ([12], 1971) confirmed this conjecture. Lachlan and Lebeuf ([11], 1976) showed that every initial segment of a countable upper-semi-lattice with least element is isomorphic to an initial segment of $\mathcal{D}$. Ultimately, Abraham and Shore ([1], 1986) showed that every initial segment of an upper-semi-lattice which is locally countable and of cardinality $\aleph_1$ is isomorphic to an initial segment of $\mathcal{D}$. Groszek and Slaman ([4], 1983) showed that the Abraham-Shore theorem is best possible; it is independent of $ZFC + 2^\omega > \aleph_1$ whether every initial segment of an upper-semi-lattice which is locally countable and of cardinality $\aleph_2$ can be embedded in $\mathcal{D}$ as an upper-semi-lattice.

The proofs of these theorems, while more technically demanding than the proofs of the embedding theorems, could still be applied to a wide range of countably based degree structures. The enumeration degrees are the only notable exception.

From the progress on the Sacks conjectures, degree theorists speculated whether $\mathcal{D}$ might have an algebraic characterization, or at least occupy a distinguished position among the upper-semi-lattices.

Questions from the 1960s

Workers in the field faced several fundamental questions. We pose them here for $\mathscr{D}$, but they are completely general. Is there a global structure theory for $\mathscr{D}$. Is the theory of $\mathscr{D}$ specific to Turing reducibility or is it applicable to a general class of degree structures? Is the structure of $\mathscr{D}$ tied to the continuum or is it reflected in the local substructures of $\mathscr{D}$?

We recall some specific questions from that time.

(3) (Sacks [19], 1966) Is the theory of $\mathscr{D}$ decidable?

(4) (Sacks ibid., 1966) Are the Turing degrees and the Turing degrees of the arithmetic sets ($\mathscr{A}$) elementarily equivalent?

(5) (Rogers [17], 1967) For a degree a, let $\mathscr{D}(\geq_T a)$ denote the restriction of $\mathscr{D}$ to those degrees above a. Is it the case that for all a and b, $\mathscr{D}(\geq_T a) \overset{\sim}{\to} \mathscr{D}(\geq_T b)$?

(6) (Rogers ibid., 1967) Is there a nontrivial automorphism of $\mathscr{D}$? (If not then we say that $\mathscr{D}$ is *rigid*.)

Rogers defined a relation R on degrees to be *absolutely definable in* $\mathscr{D}$ if R is invariant under all automorphisms of $\mathscr{D}$.

(7) (Rogers ibid., 1967) Are the Turing jump and the relation *recursively enumerable in* absolutely definable in $\mathscr{D}$? In general, which relations are absolutely definable in $\mathscr{D}$?

§2. Coding and Definability Theorems of the 1970s and Early 1980s

The results of the 1970s to the mid 1980s ruled out any reasonable understanding of the Turing degrees in algebraic terms. But then, the exact properties of $\mathscr{D}$ that make it algebraically intractable were used during this period to settle almost all of the Sacks and Rogers questions. Their solutions illustrate the complexity of $\mathscr{D}$: $\mathscr{D}$ is not decidable; the theory of $\mathscr{D}$ is not equal to the theory of the Turing degrees of the arithmetic sets; there are a and b such that $\mathscr{D}(_T \geq a)$ and $\mathscr{D}(_T \geq b)$ are not isomorphic. Rogers's question whether the jump is definable was only recently settled (the jump is definable), but the techniques in its solution were steadily developed through this period. The only questions that remain open are whether $\mathscr{D}$ is rigid and to give a classification of which relations are definable in $\mathscr{D}$.

The trend is for the global properties of degree structures to follow those of the Turing degrees. The notable exception has been the many-one degrees. There, Ershov ([3], 1975) and Paliutin ([16], 1975) succeeded in obtaining an algebraic characterization for the partial ordering of the many-one degrees. The reader might also see Odifreddi ([15], 1989).

A Cone of Minimal Covers

One of the earlier questions to fall was whether $\mathscr{D}$ is elementarily equivalent to $\mathscr{A}$. Jockusch and Soare ([8], 1970) showed that there is no arithmetic degree x such that every y above x is a minimal cover of some z less than y. Jockusch ([5], 1973) showed that there is a nonarithmetic degree with the above property. Thus, the Turing degrees of the arithmetic sets are not elementarily equivalent to $\mathscr{D}$.

Coding in $\mathscr{D}$ and Undecidability

A primary ingredient in the work during this period was the proving and exploiting of *coding theorems*.

Definition. Suppose that $\mathfrak{A}$ is a model of the finite language $\mathscr{L}$ and $\vec{p}$ is a finite sequence of parameters from $\mathscr{D}$. $\mathfrak{A}$ is *coded by $\vec{p}$ in $\mathscr{D}$* if there is an isomorphic image of $\mathfrak{A}$ whose universe, relations, functions, constants and quantifiers are all first order definable in the language of $\mathscr{D}$ with additional symbols for the parameters $\vec{p}$.

A disparate sequence of coding schemes preceded the one which we have isolated below, see Simpson ([20], 1977) or Nerode-Shore ([13], 1979). Say that *a relation R is countable* if there is a countable subset of the degrees such that all of the solutions to R come from that set.

Coding Lemma (Slaman-Woodin [21], 1986). *For any countable relation R on degrees there are parameters $\vec{p}$ such that R is definable in $\mathscr{D}$ from $\vec{p}$.*

The coding lemma is uniform in the following sense. For each n, there is a fixed first order formula φ such that for every countable n-ary relation R on $\mathscr{D}$ there is a sequence $\vec{p}$ so that R is defined from $\vec{p}$ using the formula φ in $\mathscr{D}$. The proof of the coding lemma uses finite conditions to construct $\vec{p}$ from R; consequently, it does not use any machinery that is special to the Turing degrees. It applies or can be modified to apply to a very broad class of degree structures, even to some without minimal degrees such as the enumeration degrees, see Slaman-Woodin ([22], to appear).

We can use the coding lemma to present the solution to Sacks's question whether $\mathscr{D}$ is decidable. By the coding lemma, we can both code the standard model of arithmetic and also define the collection of codes of standard models. In addition, we can interpret second order quantifiers over a coded countable model by quantifiers over sequences in $\mathscr{D}$ which define unary relations. Thus, (Simpson [20], 1977) there is an interpretation of second order arithmetic in the first order theory of $\mathscr{D}$. Lachlan ([10], 1968) gave the original solution to Sacks's question but only showed that the theory of $\mathscr{D}$ is not recursive. Simpson calculated the degree of the first order theory of $\mathscr{D}$.)

Shore's Program

The next developments which we will discuss were initiated by Nerode and Shore ([14], 1980) and pursued extensively by Shore and his collaborators.

Suppose that R is a relation on degrees. We note that the degrees which are produced by the coding lemma to define R in $\mathscr{D}$ are recursion theoretically close to R. By this we mean the following. Let X be a real of degree x. If R is recursively presented relative to X then there is a sequence of degrees $\vec{p}$ which codes R in $\mathscr{D}$ and is arithmetic in x. As a special case, there is a sequence $\vec{P}_X$ of sets which is arithmetically definable from X and whose degrees code (in $\mathscr{D}$) an isomorphic copy of the standard model of arithmetic with a unary predicate for X. Thus, the set X is coded in $\mathscr{D}$ by parameters which are near its degree. Furthermore, if X is sufficiently complicated, say above $0'$, then a sequence of parameters whose degrees code X in $\mathscr{D}$ can be found recursively in X, that is below x in $\mathscr{D}$.

The next step was to find a notion of *neighborhood* which would be first order definable in $\mathscr{D}$ and link an arbitrary degree x to the reals coded in its neighborhood. The Jockusch-Soare theorems pointed in the correct direction. Jockusch and Soare defined a filter in $\mathscr{D}$ and proved that it is disjoint from the degrees of the arithmetic sets. Their theorem suggested that there might be a related definable filter whose complement was exactly the arithmetic degrees or even the degrees below $0'$.

The search for such a filter lead to the development of the theory of REA-operators, see Jockusch-Shore ([7], 1984). Recently, Cooper ([2], 1990) combined the analysis of REA-operators with his study of the degrees of differences of recursively enumerable sets to prove that the Turing jump is definable in the Turing degrees. In fact, Cooper proved the stronger theorem that the relation x *is recursively enumerable in and above* y is definable in $\mathscr{D}$ (solving Roger's question). We remark that this approach is specific to the Turing degrees. The proofs directly exploit the way that Turing reducibility can be recursively approximated.

We list some applications, in the strongest form obtained by these techniques. (Unfortunately, we do not have the space to describe the historical development and give credit where it is due.)

Theorem.

- (Jockusch-Shore [7], 1984) *For all x and y, if $\mathscr{D}(_T \geq x) \stackrel{\sim}{\to} \mathscr{D}(_T \geq y)$ then x and y have the same arithmetic degree.*
- (Cooper [2], 1990) *If π is an automorphism of $\mathscr{D}$ then π is the identity on all degrees greater than or equal to $0'''$.*
- (Cooper ibid.) *If R is a relation on the reals which are Turing above $0'''$, R is invariant under Turing degree and R is definable in second order arithmetic then the relation induced by R is definable in $\mathscr{D}$.*

The last two examples were known to follow from the definability of the jump, see Nerode-Shore ([14], 1980).

A Shortcoming of Method

The three structural properties of $\mathcal{D}$ we isolated above were proven above by following some practical advice: examine the neighborhood of x and recover some information about which reals belong to x. In fact, we will exactly recover the set of representatives of x provided that the set of representatives of x is coded in the neighborhood of x and that the neighborhood of x and the method employed decode the set of its representatives are uniformly recursive in x. Since the statement of the Turing order between reals involves Σ_3^0 and Π_3^0 predicates, x has to provide $0'''$, the information necessary to evaluate these predicates. Consequently, our conclusions for the Turing degrees have all been confined to the realms of sufficiently large degrees or arithmetic equivalence between Turing degrees. Even using Cooper's theorem to get the strongest possible bound on the orbit of $0'$, these methods have (so far) been confined by this three quantifier limit.

We should remark that this shortcoming does not apply to the degrees of constructibility. In that notion of degree, the partial ordering of the degrees of constructibility is below x is just another set constructed from x. In fact, with reasonable set theoretic hypotheses the techniques we have touched upon are sufficient to give a complete analysis of the global theory of the constructibility degrees, in the sense of the next section.

§3. Recent Results

In this section, we will discuss some recent work of Slaman and Woodin ([22], to appear). Unless we specifically indicate otherwise, all of the following results, conjectures and even remarks are drawn from that source.

Assignment of Representatives

Suppose that $\mathscr{E}$ is a degree structure on a set of reals, possibly the set of all reals. Say that $\mathscr{E}$ is determined by the equivalence relation $\equiv_E$ and the ordering $\leq_E$.

Definition. $\mathscr{E}$ has a *first order assignment of representatives* if
(1) for every $\mathscr{E}$-degree x there is a sequence $\bar{p}$ from $\mathscr{E}$ which uniformly codes a representative of x (say using the formulas $\varphi_1, \ldots \varphi_k$);
(2) the relation $\bar{p}$ *codes a representative of* x is an $\mathscr{E}$-definable relation (say by the formula ψ).

We say that $\mathscr{E}$ has a *first order assignment of representatives in parameters* if the same conditions hold as above except that we allow $\varphi_1, \ldots \varphi_k$ and ψ to mention parameters from $\mathscr{E}$.

Suppose that $\mathscr{E}$ has a first order assignment of representatives. If $\mathscr{E}$ is definable in second order arithmetic and is a degree structure on all of the reals then we say that $\mathscr{E}$ is *biinterpretable with second order arithmetic*. If $\mathscr{E}$ is definable in first order arithmetic and is a degree structure on a uniformly arithmetic set of reals then we say that $\mathscr{E}$ is *biinterpretable with first order arithmetic*. Similarly, we can define biinterpretability with parameters.

If $\mathscr{E}$ is biinterpretable with the standard model second order arithmetic then all of the logical properties of $\mathscr{E}$ can be reduced to the reals. We will list some examples. Others will undoubtedly occur to the reader.

Assuming $\mathscr{E}$ is biinterpretable with second order arithmetic:

-   A relation is definable in $\mathscr{E}$ if and only if it is induced by an $\equiv_E$-invariant relation that is definable in second order arithmetic.
-   $\mathscr{E}$ has no nontrivial automorphism, i.e. $\mathscr{E}$ is *rigid*.

Assuming that $\mathscr{E}$ is biinterpretable with second order arithmetic, we sketch an example proof. The standard model of first order arithmetic is rigid because each number is definable. Consequently, the standard model of second order arithmetic is also rigid since each subset of $\mathbb{N}$ is determined by its elements. The fact that no representative of a degree can be moved by an automorphism of the standard model of second order arithmetic implies that an automorphism of $\mathscr{E}$ can only move a degree to another one with the same set of representatives. Of course, this last condition is another way of saying that the only automorphism of $\mathscr{E}$ is the identity.

If we allow parameters we obtain the boldface versions of the above conclusions.

Assuming $\mathscr{E}$ is biinterpretable with second order arithmetic using the parameters $\bar{e}$:

-   A relation is definable from finitely many parameters in $\mathscr{E}$ if and only if it is induced by an $\equiv_E$-invariant relation on reals that definable from finitely many parameters in second order arithmetic, i.e. is induced by a *projective* relation.
-   Any automorphism of $\mathscr{E}$ is determined by its action on $\bar{e}$. That is, $\bar{e}$ is an *automorphism base*.

Just knowing that $\mathscr{E}$ has a finite automorphism base gives us some information about its group of automorphisms. For example, if $\mathscr{E}$ is a degree structure on the reals then the cardinality of the orbit of any finite sequence from $\mathscr{E}$ is bounded by the cardinality of the continuum. Thus, if $\mathscr{E}$ has a finite automorphism base then it has at most continuum many automorphisms. Similarly, if $\mathscr{E}$ is a degree structure on a countable set of reals and has a finite automorphism base then the automorphism group for $\mathscr{E}$ is countable.

A Finite Automorphism Base

We will now sketch a heuristic approach to proving that a degree structure on the reals has a finite automorphism base. The method directly applies to the specific structures of the Turing degrees, the arithmetic degrees, the hyperarithmetic degrees, the Σ_k-admissible degrees, the PTIME Turing degrees and the enumeration degrees.

Let $\mathscr{D}$ denote any of the above degree structures on the reals. In the next few paragraphs we will be measuring definability in terms of a number of jumps. The interpretation of *jump* depends on the interpretation of $\mathscr{D}$. By *arithmetic in,* we mean below some finite number of jumps. The reader may comfortably imagine that we are discussing the Turing degrees with the Turing jump.

For the first step, we prove the coding lemma with its full uniformity. In particular, there is a degree z_0 such that if x is above z_0 then all the representatives of x are coded below x. Some examples for z_0 are $0''$ in the Turing degrees, 0^ω in the arithmetic degrees and the degree of Kleene's $\mathcal{O}$ in the hyperarithmetic degrees. Let z_0 be fixed.

We continue with some countable algebra for $\mathscr{D}$.

Definition. Let I be a countable ideal in $\mathscr{D}$ and let ϱ be an automorphism of I. We say that ϱ is *persistent* if for every x in $\mathscr{D}$ there are an ideal J and an automorphism ϱ^* of J such that $x \in J$, $I \subseteq J$ and ϱ^* agrees with ϱ on I.

Next we prove that every persistent countable automorphism extends to an automorphism of $\mathscr{D}$. The proof employs a generalization of an insight due to Odifreddi and Shore: the coding lemma can be used to show that the restriction of an automorphism of $\mathscr{D}$ to a countable ideal is recursion theoretically close to any uniform upper bound on that ideal. Specifically, we show that if ϱ is a persistent automorphism of the ideal I then ϱ is arithmetic in any upper bound of I.

We now introduce some metamathematical methods. We know that there is a nontrivial automorphism of $\mathscr{D}$ if and only if there is a nontrivial countable automorphism that is persistent. The latter condition is upwards absolute between well-founded models of ZFC.

Applying results of Slaman-Woodin ([21], 1986), we obtain the following theorem.

Theorem. $\mathscr{D}$ *is rigid if and only if* $\mathscr{D}$ *is biinterpretable with second order arithmetic.*

Continuing the metamathematical discussion, let V denote the universe of sets. Let $V[\mathscr{G}]$ be a generic extension of V. Using the absoluteness theorem, we show that if π is an automorphism of $\mathscr{D}^V$, the degrees in V, then π lifts to an automorphism π^* of $\mathscr{D}^{V[\mathscr{G}]}$, the degrees in the generic extension. By moving to a generic extension of V, we can use the definition of forcing to analyze π^*. In particular, if $\mathscr{G}$ is generic with respect to the partial order to add ω_1 Cohen reals to V then π^* is represented as a continuous function on the set of generic reals. In fact, we can use our proof that $\pi(x)$ is close to x to show the following. There is a recursive functional $\{e\}$ and an integer n such that if G is a Cohen generic real over V then the degree of $\{e\}(G \oplus (\pi^{-1}(Z_0))^{(n)})$ is π^* of the degree of G. Here we let Z_0 denote a representative of z_0, $\pi^{-1}(Z_0)$ denote a representative of $\pi^{-1}(z_0)$ and $(\pi^{-1}(Z_0))^{(n)}$ denote the nth jump of $\pi^{-1}(z_0)$.

Our next step is to extract a representation of π on a comeager set of reals in V from the representation of π^* on the comeager set of generic reals in $V[\mathscr{G}]$.

We prove that the same representation of π (using the functional $\{e\}$ relative to $\pi^{-1}(Z_0)^{(n)}$) holds on the set C of reals which are sufficiently generic relative to $\pi^{-1}(z_0)$. The level of genericity required is only finitely many jumps, in the sense of the reducibility determining $\mathscr{D}$. Furthermore, if there is one G in C such that G and $\{e\}(G \oplus (\pi^{-1}(Z_0))^{(n)})$ have the same degree then π is the identity on the degrees represented in C.

In ([6], 1981), Jockusch and Posner show that the Turing degrees represented by any comeager set of reals generate the Turing degrees under the operations of meet and join. Their argument is completely general, so we may conclude that C is an automorphism base.

Now we can make two observations.

First, suppose that $\pi(z_0) = z_0$. Then the set C consists of those reals which are arithmetically generic relative to z_0. Let g_0 be any degree of an element of C. If π also maps g_0 to g_0 then π must be the identity on C and therefore π must be the identity on all of $\mathscr{D}$. Thus, $\{z_0, g_0\}$ is a finite automorphism base for $\mathscr{D}$.

Second, from the arithmetic representation of π on C relative to $\pi^{-1}(Z_0)$ we can find a function arithmetic in $\pi^{-1}(Z_0)$ which represents π on all the reals. Consequently, the fact that the rigidity of $\mathscr{D}$ is absolute may be explained by the fact that every automorphism of $\mathscr{D}$ is arithmetically definable in a real parameter.

Biinterpretability with Parameters

The fact that $\mathscr{D}$ has a finite automorphism base can be combined with the analysis of persistent automorphisms to show that $\mathscr{D}$ is biinterpretable with second order arithmetic in parameters.

Let Z_0 and G_0 denote representatives of the degrees z_0 and g_0 above. Suppose that ψ is a map from the reals onto $\mathscr{D}$ which induces an automorphism on $\mathscr{D}$, i.e. ψ is degree invariant, preserves order in the sense of $\mathscr{D}$-reducibility and has distinct values on reals of distinct degree. If ψ maps Z_0 to z_0 and G_0 to g_0 then ψ must induce the identity automorphism. In other words, ψ must be an assignment of representatives to $\mathscr{D}$. Let the parameters $\vec{p}_{Z_0}$ code Z_0 and $\vec{p}_{G_0}$ code G_0. With some finesse, we can use the coding lemma to express the following condition in D: *There is a persistent countable assignment of representatives sending Z_0 to z_0, G_0 to g_0 and X to x.* By the characterization of persistent countable automorphisms as restrictions of global automorphisms, this statement is equivalent to one saying that there is a map from the reals to $\mathscr{D}$ with the values as above that induces an automorphism. Of course, this automorphism is the identity. Thus, the statement above defines *X is a representative of x* as expressed in the codes for the real X. Consequently, $\mathscr{D}$ is biinterpretable with second order arithmetic using the parameters $\vec{p}_{Z_0}, z_0, \vec{p}_{G_0}$ and g_0.

Theorem. *The structures of the Turing degrees, the arithmetic degrees, the hyperarithmetic degrees, the Σ_k-admissible degrees, the PTIME Turing degrees and the enumeration degrees are all biinterpretable with second order arithmetic in parameters.*

Special Arguments for the Turing, Arithmetic, Hyperarithmetic and Σ_k-Admissible Degrees

In this section, we will discuss results obtained using special properties common to the Turing, Arithmetic, Hyperarithmetic and Σ_k-admissible reducibilities. We should note that these arguments do not intersect with those of Jockusch-Shore and Cooper and provide new proofs of the three structural properties of the Turing degrees mentioned above.

Since the proofs here are more technically involved than the ones that we have discussed so far, we have to resign ourselves to merely stating results. We can only say that the proofs of these results involve a direct analysis how a continuous function must behave to represent the restriction of a degree structure automorphism to a comeager set.

By the remark of the previous section, for all of our degree structures of interest, rigidity is equivalent to biinterpretability with second order arithmetic. Thus, the central problem for any of these structures is whether it has a nontrivial automorphism.

Theorem.
(1) *The hyperdegrees are rigid (and thus are biinterpretable with second order arithmetic.) Similarly, all of the Σ_k-admissible degrees structures are biinterpretable with second order arithmetic.*
(2) *Any automorphism of the arithmetic degrees is the identity above 0^ω.*
(3) *Any automorphism of the Turing degrees is the identity above $0''$.*

Corollary. *There are only countably many automorphisms of the Turing degrees. Any automorphism of the Turing degrees is represented by an arithmetically definable function on reals.*

A similar corollary holds for the arithmetic degrees; any automorphism is hyperarithmetically representable.

Special Arguments for the Turing Degrees

We now restrict $\mathscr{D}$ to denote the Turing degrees. We also drop any restraint against using very special properties of $\mathscr{D}$. In particular, we will make full use of Cooper's theorem that the relation x is recursively enumerable in and above y is definable in $\mathscr{D}$.

Earlier, we sketched a metamathematical proof that any automorphism of $\mathscr{D}$ is arithmetically definable. We can also give more traditional, purely recursion theoretic proof of this fact. Of course, this proof is a more difficult local version of its metamathematical progenitor. With the sharper argument, we can replace the full structure of $\mathscr{D}$ by any ideal in $\mathscr{D}$ which has $0^{(7)}$ as element.

Theorem. *Suppose the I is an ideal in $\mathscr{D}$ and $0^{(7)} \in I$.*

(1) *If $\pi : I \xrightarrow{\sim} I$ then π is represented by an arithmetically definable function on reals.*

(2) *I is biinterpretable in parameters with the fragment of second order arithmetic in which the second order quantifiers range over the reals whose degrees lie in I.*

As a corollary to the theorem, we can demonstrate a connection between the existence of a local automorphism and a global one. By the Kleene basis theorem, if an arithmetic function does not represent an automorphism of $\mathscr{D}$ then there is a counter-example which is recursive in Kleene's $\mathcal{O}$. Thus, we can conclude that any automorphism of $\mathscr{D}(\leq_T \mathcal{O})$ extends to an automorphism of $\mathscr{D}$.

A Concrete Automorphism Base for the Turing Degrees

We already know that there is a finite automorphism base for $\mathscr{D}$. The question arises as to how concrete can we make the base elements. Outside of proving that the Turing degrees are rigid, we can give the best possible result. We show that there is a finite set of recursively enumerable degrees which is an automorphism base for $\mathscr{D}$.

Our approach is to provide a generating family of first order formulas such that the smallest set including the recursively enumerable degrees and closed under definition by these formulas includes the degrees of $0''$ (i.e. z_0) and of a real G which is sufficiently generic for the pair of degrees to form a base. We isolate an result which we prove along the way which is of independent interest.

Theorem.

(1) *$\mathscr{D}(\leq_T 0')$ is biinterpretable with first order arithmetic in parameters. In fact, we may take the parameters to be recursively enumerable degrees.*

(2) *Any automorphism of $\mathscr{D}(\leq_T 0')$ is arithmetically definable.*

Ultimately, we prove that there is a finite set F of recursively enumerable degrees such that $\mathscr{D}$ is biinterpretable with second order arithmetic in the parameters from F. In fact, the analogous theorem holds of any ideal I such that $0^{(7)} \in I$. Thus, we have the following theorem.

Theorem.

(1) *The recursively enumerable degrees are an automorphism base for $\mathscr{D}$. Further, $\mathscr{D}$ is biinterpretable with second order arithmetic using recursively enumerable parameters.*

(2) *If the recursively enumerable degrees are rigid then $\mathscr{D}$ is rigid.*

(3) *If $\mathscr{D}(\leq_T 0')$ is rigid then $\mathscr{D}$ is rigid.*

It is open whether every automorphism of the recursively enumerable degrees extends to one of $\mathscr{D}(\leq_T 0')$ or even to $\mathscr{D}$. Any result along this line would be valuable.

§4. Conjectures

We make the following conjectures jointly with W. H. Woodin.

We say that a real number G is *1-generic* if for every recursively enumerable set S of finite Cohen conditions either there is an element of S which is satisfied by G or there is a neighborhood condition satisfied by G which is incompatible with all the elements of S.

Conjecture I. If I is an ideal in the Turing degrees such that there is the degree of a 1-generic real in I then I has a first order assignment of parameters.

In particular, we believe that the partial ordering of the Turing degrees is biinterpretable with second order arithmetic and that $\mathscr{D}(\leq_T 0')$ is biinterpretable with first order arithmetic.

Conjecture II. The partial ordering of recursively enumerable degrees is biinterpretable with the standard model of first order arithmetic.

We end with a question. Is there a general proof of rigidity and equivalently of biinterpretability with second order arithmetic that is based on simple properties of Turing reducibility and applies to a wide range of degree structures?

References

1. Abraham, U., Shore, R. A.: Initial segments of the degrees of size $\aleph_1$. Israel J. Math. **53** (1986) 1–51
2. Cooper, S. B.: The jump is definable within the structure of the Turing degrees. Bull. A.M.S. **23** (1990) 151–158
3. Ershov, Y. L.: The uppersemilattice of enumerations of a finite set. Alg. Log. **14** (1975) 258–284 ; transl. **14** (1975) 159–175
4. Groszek, M. J., Slaman, T. A.: Independence results on the global structure of the Turing degrees. Trans. Am. Math. Soc. **277** (1983) 579–588
5. Jockusch, C. A., Jr.: An application of Σ_4^0 determinacy to the degrees of unsolvability. J. Sym. Logic **38** (1973) 293–294
6. Jockusch, C. A., Jr., Posner, D.: Automorphism bases for degrees of unsolvability. Israel J. Math. **40** (1981) 150–164
7. Jockusch, C. A., Jr., Shore, R. A.: Pseudo-jump operators II: Transfinite iterations, hierarchies and minimal covers. J. Sym. Logic **49** (1984) 1205–1236
8. Jockusch, C. A., Jr., Soare, R. I.: Minimal covers and arithmetical sets. Proc. Amer. Math. Soc. **25** (1970) 856–859
9. Kleene, S. C., Post, E. L.: The upper semi-lattice of degrees of unsolvability. Ann. Math. **59** (1954) 379–407
10. Lachlan, A. H.: Distributive initial segments of the degrees of unsolvability. Z. Math. Logik Grundlagen Math. **14** (1968) 457–472
11. Lachlan, A. H., Lebeuf, R.: Countable initial segments of the degrees of unsolvability. J. Sym. Logic **41** (1976) 289–300
12. Lerman, M.: Initial segments of the degrees of unsolvability. Ann. Math. **93** (1971) 365–389

13. Nerode, A., Shore, R. A.: Second order logic and first order theories of reducibility. orderings. In: The Kleene Symposium. Barwise, J., Keisler, J., Kunen, K., eds. North-Holland, Amsterdam 1979, pp. 181–200
14. Nerode, A., Shore, R. A.: Reducibility orderings: theories, definability and automorphism. Ann. Math. Logic **18** (1980) 61–89
15. Odifreddi, P. G.: Classical Recursion Theory. North-Holland, Amsterdam 1989
16. Paliutin, E.: Addendum to the paper of Ershov [1975]. Alg. Log. **14** (1975) 284–287; transl. **14** (1975) 176–178
17. Rogers, H., Jr.: Some problems of definability in recursive function theory. In: Sets Models and Recursion Theory, Proc. Summer School in Math. Logic and Tenth Logic Colloq., Lercester, August–September 1965. J.N. Crossley, ed. North-Holland, Amsterdam 1967
18. Sacks, G. E.: Degrees of Unsolvability. Princeton University Press, Princeton, New Jersey 1963
19. Sacks, G. E.: Degrees of Unsolvability, Second Edition. Princeton University Press, Princeton, New Jersey 1966
20. Simpson, S. G.: First order theory of the degrees of recursive unsolvability. Ann. Math. **105** (1977) 121–139
21. Slaman, T. A., Woodin, W. H.: Definability in the Turing degrees. Illinois J. Math. **30** (1986) 320–334
22. Slaman, T. A., Woodin, W. H.: Definability in degree structures. (To appear)
23. Spector, C.: On degrees of recursive unsolvability. Ann. Math. **64** (1956) 581–592

Cohomology and Modules over Group Algebras

Jon F. Carlson

Department of Mathematics, University of Georgia, Athens, GA 30602, USA

Introduction

The last several years have witnessed an increasing interaction between homological algebra and the representation theory of finite groups. The aim of the effort has primarily been the development of a module theory for group rings and algebras. Of course, it is not surprising that homological methods would play a role in such an investigation. After all, cohomology theory really is a theory of extensions and, in small degrees, its relevance to the structure of modules is apparent. However the connection has proven to be deeper. The module theory is directly related to some of the larger features of the cohomology of groups. By this we mean such items as the ring theoretic properties of cohomology – the maximal ideals, Krull dimension, and modules for cohomology rings. In this lecture I will survey some of the developments in this area. For the most part, we will stick to the case in which the group G is finite and the coefficient ring K is a field of characteristic $p > 0$. Nevertheless, many of the results hold true for integral coefficients and for more general groups.

The work which I wish to survey is based on several finiteness conditions. First, we consider only finitely generated modules. But more than that, the foundation rests on two fundamental results of group cohomology.

Theorem (Evens [14]). $H^*(G, K) = \mathrm{Ext}^*_{KG}(K, K)$ is a finitely generated K-algebra. If M is a KG-module, then $H^*(G, M)$ is a finitely generated module over $H^*(G, K)$.

Even's Theorem permits us to consider the maximal ideal spectrum, $V_G(K)$, of $H^*(G, K)$. We need not be concerned here that $H^*(G, K)$ is not commutative. The only noncommutativity occurs among elements of odd degree. These, however, are all in the radical and hence are in every maximal ideal. Note that $V_G(K)$ is a homogeneous affine variety. For a KG-module M, we can define $V_G(M)$ to be the subvariety of $V_G(K)$ associated to the annihilator of $\mathrm{Ext}^*_{KG}(M, M) \cong H^*(G, \mathrm{Hom}_K(M, M))$. It is the set of all maximal ideals containing the annihilator.

The other foundational result is the "Dimension Theorem" of Quillen.

Proceedings of the International Congress
of Mathematicians, Kyoto, Japan, 1990
© The Mathematical Society of Japan, 1991

Theorem [19]. *The components of $V_G(K)$ are in one-to-one correspondence with the conjugacy classes of maximal elementary abelian p-subgroups $E(\cong (\mathbb{Z}/p)^r \hookrightarrow G)$ via the restriction maps.*

For algebraists, the best proof of the theorem is given in [21]. All proofs have required some variation on a theorem of Serre [22] which says that for p-groups which are not elementary abelian, the product of all Bocksteins of elements of degree 1 is nilpotent. One of the inspirations in the development of the subject was the extension of Quillen's theorem to the varieties of modules. The first attempt considered only the complexity or dimension of the varieties [3]. The full result was obtained independently by Alperin and Evens and by George Avrunin [4, 5]. The importance of this step is obvious. It says that in this context we can restrict our attention to the elementary abelian p-groups.

Elementary Abelian Groups

For this section assume that the field K is algebraically closed. Suppose that $E = \langle x_1, \cdots, x_n \rangle \cong (\mathbb{Z}/p)^n$ is an elementary abelian p-group. It is well known that

$$H^*(E, K)/\mathrm{Rad} = K[\zeta_1, \cdots, \zeta_n]$$

is a polynomial ring in $n = \mathrm{rank}(E)$ generators. Here Rad denotes the radical of $H^*(G, K)$. If $p = 2$ then Rad $= 0$ and the generators $\zeta_1, \cdots, \zeta_n$ have degree 1. If $p > 2$ then Rad is generated by the elements of degree 1 and each ζ_i has degree 2. In either case, the maximal ideal spectrum $V_E(K) = K^n$, is affine n-space.

For $\alpha = (\alpha_1, \cdots, \alpha_n) \in K^n$, let

$$u_\alpha = 1 + \sum_{i=1}^{n} \alpha_i(x_i - 1) \in KE.$$

It is easy to see that u_α is a unit of order p in KE and that the inclusion map $K\langle u_\alpha \rangle \hookrightarrow KE$ is an injection of group algebras. Moreover KE is a free $K\langle u_\alpha \rangle$-module. Hence, it makes sense to speak of a restriction map on cohomology from E to $\langle u_\alpha \rangle$. Thus, for a KE-module M, the set of all u_α which will measure nonzero cohomology of M should be special. This leads to the following definition.

Definition. Let M be a KE-module with K, E as above. The set

$$V_E^r(M) = \{\alpha \in K^n | M \text{ is not free over } K\langle u_\alpha \rangle\} \cup \{0\}$$

is a homogenous affine variety called the *rank variety* of M.

The fact that it is a variety is fairly easy to show. The proof shows further that $V_E^r(M)$ can be computed directly from elementary information such as the representation $\varrho : E \to GL(m, K)$ which defines M as a module [9,10]. But more importantly we have

Theorem. $V_E(M) \cong V_E^r(M)$.

This is only an isomorphism of algebraic sets, subsets of K^n. There seem to be no natural, corresponding maps on the coordinate rings. For $p = 2$, it can be viewed as an equality. However for $p > 2$, the restriction map to the subgroups $\langle u_\alpha \rangle$ introduces a Frobenius twist on cohomology. The existence of a relationship between the varieties was clear from the beginning. Avrunin and Leonard Scott[6] proved the more difficult containment by altering the Hopf-algebra structure required for cup products.

For more general groups G, the variety $V_G(M)$ can be pieced together from the $V_E(M)$'s, $E \le G$. This is true in theory. In practice it may be very difficult. Avrunin and Scott also showed that $V_G(M)$ is stratified as in [20].

The variety $V_G(M)$ has several useful properties. Some can be derived easily by considering the rank variety.

Theorem. (a) $V_G(M) = \{0\}$ if and only if M is projective.

(b) $V_G(M \otimes_K N) = V_G(M) \cap V_G(N)$.

(c) If $0 \to A_1 \to A_2 \to A_3 \to 0$ is exact then $V_G(A_i) \subseteq V_G(A_j) \cup V_G(A_k)$ for $\{i, j, k\} = \{1, 2, 3\}$.

(d) If $W \subseteq V_G(K)$ is a (Zariski) closed subvariety, then there exists a module M with $V_G(M) = W$.

(e) If M is indecomposable then the corresponding projective variety $\overline{V}_G(M)$ is connected [11].

Property (d) has been particulary useful. Its proof is completely constructive. Suppose that we are given $\zeta \in H^m(G, K), m \ge 2$. Let $P_0, \cdots, P_{m-1}$ be minimal projective modules so that

$$0 \to \Omega^m(K) \to P_{m-1} \to \cdots \to P_1 \to P_0 \twoheadrightarrow K \to 0$$

is exact and $\Omega^m(K)$ has no projective submodules. Then ζ is represented by a cocycle $\hat{\zeta} : \Omega^m(K) \to K$. Assuming that $\zeta \ne 0$, we have an exact sequence

$$0 \to L_\zeta \to \Omega^m(K) \xrightarrow{\hat{\zeta}} K \to 0.$$

By using the rank variety we may show that $V_G(L_\zeta) = V_G(\zeta)$, the subvariety corresponding to the ideal generated by ζ. Thus (d) can be proved by choosing a set of generators for the ideal of W and using (b).

Two Asides

I want to mention two related areas which have been the subject of some recent study. The first involves the structure of the cohomology rings $E^*(M) = \mathrm{Ext}^*_{KG}(M, M) = H^*(G, \mathrm{Hom}_K(M, M))$ for M a KG-module. The variety of M is certainly associated to this ring but not as its maximal ideal spectrum. The problem is that $E^*(M)$ is not commutative. We do not simply mean that it is graded commutative, but rather, it can be noncommutative in an essential way[12]. There exist examples, which are not exotic, where the ring $E^*(M)$ has simple modules of K-dimension greater than one. That is, there exists a surjective homomorphism of $E^*(M)$ onto the ring of $n \times n$ matrices for $n > 1$. However the news is not all bad. $E^*(M)$ is a PI-ring and hence all of its simple modules have finite dimension. Recently Niwasaki has settled a question concerning the nature of the simple modules.

Theorem [17]. *If $U \subset \text{Ext}^*_{KG}(M, M)$ is a maximal ideal then there exists an elementary abelian p-subgroup $E = \langle x_1, \cdots, x_n \rangle \subset G$ and a unit $u_\alpha = 1 + \sum_{i=1}^{n} \alpha_i(x_i - 1) \in KE$ such that U contains the kernel of the composition of the restrictions of $\text{Ext}^*_{KG}(M, M)$ to E and to $\langle u_\alpha \rangle$.*

In another direction, a very similar sort of theory has been developed for the restricted p-Lie algebras associated to an algebraic group in characteristic p (see [18] for a survey). Suppose that G is a connected linear algebraic group defined over the prime field $\mathbb{F}_p$. Let G_1 denote the kernel of the Frobenius map on G. Here G_1 should be regarded as a group scheme or as a functor, since as a group it has only a single element. The Lie algebra $\underline{g}$ of G may be equipped with a p-power operation $x \to x^{[p]}$, thus making it a restricted p-Lie algebra. The representation theory of rational G-modules is related to that of rational G_1-modules via spectral sequences. Moreover the category of rational G_1-modules is equivalent to the category of restricted $\underline{g}$-modules.

For restricted $\underline{g}$-modules it is possible to define cohomological support varieties and rank varieties as in the finite group case. That is, $H^*(\underline{g}, K)$ is a finitely generated ring and $H^*(\underline{g}, M)$ is a finitely generated module over it, provided M is finitely generated. The analogue of the rank variety is the null cone

$$\mathcal{N} = \{x \in \underline{g} \,|\, x^{[p]} = 0\}$$

which is isomorphic to the maximal ideal spectrum of the cohomology ring [16]. For M, the variety is the set of points $x \in \mathcal{N}$ such that M is not injective as a Kx-module. Many of the same properties for the variety have been shown to hold in this context. A major part of this work was done jointly by Friedlander and Parshall (e.g. see [15]).

It should be noted that our friends in the field of algebraic groups have left us with some sticky problems. What happens with G_2 or G_r for $r \geq 2$? Is $H^*(G_r, K)$ finitely generated? The answers are not known except in some special cases.

Projective Resolutions and Homogeneous Parameters

In recent years, some interesting results have been obtained on systems of parameters for cohomology rings and for the cohomology of a module. Suppose that $\zeta_1, \cdots, \zeta_n$ are homogeneous elements in $H^*(G, K)$ which have the property that $H^*(G, K)$ is a finitely generated module over $K[\zeta_1, \cdots, \zeta_n]$, the polynomial ring in the symbols $\zeta_1, \cdots, \zeta_n$. Another way of stating the condition is that $\cap V_G(\zeta_i) = \{0\}$. For each i, let $m_i = \deg \zeta_i$ and let $\hat{\zeta}_i$ be a cocycle representing ζ_i. Then we have the following commutative diagram.

$$
\begin{array}{ccccccccccc}
\cdots & \longrightarrow & P_{m_i+1} & \longrightarrow & P_{m_i} & \longrightarrow & P_{m_i-1} & \longrightarrow & P_{m_i-2} & \longrightarrow \cdots \longrightarrow & P_0 & \longrightarrow K \longrightarrow 0 \\
& & & & \downarrow & \hat{\zeta}_i \downarrow & & \downarrow & & \downarrow & \| \qquad \| \\
E^{(i)}: & & 0 & \longrightarrow & K & \xrightarrow{j} & L_i & \longrightarrow & P_{m_i-2} & \longrightarrow \cdots \longrightarrow & P_0 \xrightarrow{\varepsilon} K \longrightarrow 0 \\
& & & & & & \| & & \| & & \| \\
C^{(i)}: & & & & 0 & \longrightarrow & C^{(i)}_{m_i-1} & \longrightarrow & C^{(i)}_{m_i-2} & \longrightarrow \cdots \longrightarrow & C^{(i)}_0 \longrightarrow 0
\end{array}
$$

Here (P_*, ε) is a minimal projective resolution of K. The middle row $E^{(i)}$ is the pushout along $\hat{\zeta}_i$ of the upper row and is exact. The complex $C^{(i)}$ has homology $H_j(C^{(i)}) = K$ for $j = m_i-1, 0$ and $H_j(C^{(i)}) = 0$ otherwise. Let $X^{(i)}$ be the complex which is obtained by splicing copies of $E^{(i)}$ as

$$(X^{(i)}) \qquad \cdots \to P_0 \xrightarrow{j\varepsilon} L_i \to \cdots \to P_0 \xrightarrow{j\varepsilon} L_i \to \cdots \to P_0 \to 0.$$

The first result on systems of parameters for cohomology was the following theorem proven in joint work with Dave Benson some 5 years ago.

Theorem [7]. $X = X^{(1)} \otimes_K \cdots \otimes_K X^{(n)}$ *is a projective resolution of* K.

The proof is very easy. The exactness except in degree zero follows from the Künneth formula, while the projectivity comes mostly from the properties of varieties. Actually the theorem can be stated more generally. If $\mathrm{Ext}^*_{KG}(M, M)$ is finitely generated as a module over $K[\zeta_1, \cdots, \zeta_r]$, $\zeta_i \in H^*(G, K)$, then $X = X^{(1)} \otimes_K \cdots \otimes_K X^{(r)} \otimes_K M$ is a projective resolution of M. Hence every module has a projective resolution which is a tensor product of periodic complexes. Moreover the number of such complexes can be taken to be the complexity of M or the dimension of $V_G(M)$.

With appropriate modifications a similar theorem can be proved for a $\mathbb{Z}G$-lattice. Adem [2] has adapted some of these ideas to get a similar result for groups with finite virtual cohomological dimension.

Remark. *Similar techniques will also prove facts about the exponents of cohomology with integral coefficients. For example, if $H^*(G, \mathbb{Z})$ is finitely generated over $\mathbb{Z}[\zeta_1, \cdots, \zeta_n]$ then the order of G divides the product $\prod \exp(\zeta_i)$, where $\exp(\zeta_i)$ is the additive order of $\zeta_i \in H^*(G, \mathbb{Z})$ [13]. The method has also been used to deal with some questions concerning equivariant cohomology of topological spaces with an action of a finite group [1]. The theorem can be extended to exponents of $\mathbb{Z}G$-lattices and to a similar result on the Farrell cohomology of a group with finite virtual cohomological dimension.*

As an example we wish to consider the case in which $n = 3$. That is, we assume that $H^*(G, M)$ is finitely generated over $K[\alpha, \beta, \gamma]$. Then we can form the Kozsul complex:

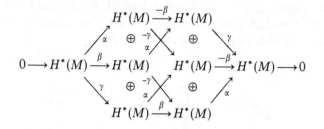

where $H^*(M) = H^*(G, M)$. The maps are cup products with the indicated elements. Let l, m, n be the degree of α, β and γ respectively. We need to assume that the degrees are even if p is odd. The complex has a graded version

$$
\begin{array}{ccccccc}
& & & H^{s+l} & & H^{s+l+m} & \\
& & & \oplus & & \oplus & \\
(*_s) & 0 \longrightarrow & H^s \longrightarrow & H^{s+m} \longrightarrow & H^{s+l+n} \longrightarrow & H^{s+l+m+n} \longrightarrow & 0 \\
& & & \oplus & & \oplus & \\
& & & H^{s+n} & & H^{s+m+n} &
\end{array}
$$

where $H^r = H^r(G, M)$. Of course for any set of parameters $\zeta_1, \cdots, \zeta_t$ we get similar complexes $(*_s)$ which have length t. In any case it is not difficult to prove the following.

Proposition. *If s is sufficiently large then $(*_s)$ is an exact sequence.*

The proof is based on the facts that polynomial rings have finite global dimension and that the Koszul complex with $H^*(G, M)$ replaced by the polynomial ring is a projective resolution of K. The proposition allows us to define a new invariant for modules. Notice first that if we use Tate cohomology then s can be negative. Moreover the sequence $(*_s)$ for M is the same as the complex $(*_{s+1})$ for $\Omega(M)$. Here $\Omega(M)$ is the kernel of the epimorphism $P \to M$ where P is a projective cover for M.

Definition. The *index* of a KG-module M is the least integer r such that $(*_s)$ is exact for all $s \geq r$.

The index of a module is a benchmark for the degree shifting process in cohomology. Clearly $\text{Index}(\Omega(M)) = \text{Index}(M) + 1$. Also we have the following.

Proposition. *The index of M does not depend on the choice of parameters. That is, the index is the same for any set $\zeta_1, \cdots, \zeta_m$ such that $H^*(G, M)$ is finitely generated over $K[\zeta_1, \cdots, \zeta_m]$.*

Some interesting questions are raised by the construction. For example, is the index of trivial module zero? The answer is yes when the p-rank of G is two and in several other situations [8]. What is the index of an irreducible module? Is it zero?

One of the connections with the representation theory is contained in the following. We state this only in the context of the example ($n = 3, \alpha, \beta, \gamma$ as above). The proof follows from results in [8].

Theorem. *If $s \geq \mathrm{Index}(U \otimes_K M^*)$ for all simple modules U, then there is an exact sequence*

$$
\begin{array}{ccccccc}
& & \Omega^{s+l+m}(M) & & \Omega^{s+l}(M) & & \\
& & \oplus & & \oplus & & \\
0 \longrightarrow & \Omega^{s+l+m+n}(M) \longrightarrow & \Omega^{s+l+n}(M) \longrightarrow & \Omega^{s+m}(M) \longrightarrow & \Omega(M) \longrightarrow & 0 \\
& & \oplus & & \oplus & & \\
& & \Omega^{s+m+n}(m) & & \Omega^{s+n}(M) & &
\end{array}
$$

where the maps are induced by (± 1) times α, β, γ as appropriate.

There is another view which we may take of this theory. Consider the complex $C^{(i)}$ associated to ζ_i. Recall that $H_j(C^{(i)})$ is K if $j = 0, m_i - 1$ and zero otherwise. Of course, $\zeta_1, \cdots, \zeta_n$ is a homogeneous set of parameters for $H^*(G, K)$. Let

$$C = C^{(1)} \otimes_K \cdots \otimes_K C^{(n)}.$$

Theorem [8]. *The complex C is a Poincaré duality complex of projective KG-modules. That is, there is a duality $C^* \to C$ which is a homotopy equivalence.*

Among other things the duality implies that when $H^*(G, K)$ is a Cohen-Macaulay ring then its Poincaré series, $f(t)$, satisfies the functional equation $f(\frac{1}{t}) = (-t)^r f(t)$, where $r = p\text{-rank}(G)$. It is conjectured that the condition is equivalent to $H^*(G, K)$ being Cohen-Macaulay.

It can be seen that $H^*(C) = \Lambda(\hat{\zeta}_1, \cdots, \hat{\zeta}_n)$, an exterior algebra, where $\hat{\zeta}_i$ is a cohomology element (in degree $\deg(\zeta_i) - 1 = m_i - 1$) coming from the cohomology of $C^{(i)}$. In some cases the cup product here is G-equivariant. We can form the hypercohomology spectral sequence whose E_0-term is

$$E_0^{r,s} = \mathrm{Hom}_{KG}(P_r \otimes C_s, K)$$

and whose E_2-term is

$$E_2^{r,s} = H^r(G, H^s(C)) = H^r(G, K) \otimes_K H^s(C),$$

and which converges to the hypercohomology

$$H_G^{r+s}(C, K) = H^{r+s}(\mathrm{Hom}_{KG}(C, K)).$$

The early differentials in the spectral sequence are simply cup products with the corresponding group cohomology elements. That is,

$$d_{m_i}(\alpha \otimes \hat{\zeta}_i) = \alpha \zeta_i \otimes 1.$$

The higher differentials are a sort of twisted Massey product – a matric Massey product. The results seem to indicate that some of these Massey products are zero in high enough degrees. The questions about the index of the trivial module K are closely related to the nature of the differentials in this spectral sequence. A smiliar construction can be made for the cohomology of modules.

References

1. Adem, A.: Torsion in equivariant cohomology. Comm. Math. Helv. **64** (1989) 401–411
2. Adem, A.: On the exponent of cohomology of discrete groups. Bull. London Math. Soc. **21** (1989) 585–590
3. Alperin, J. L., Evens, L.: Representations, resolutions and Quillen's dimension theorem. J. Pure Appl. Algebra **22** (1981) 1–9
4. Alperin, J. L. and Evens, L.: Varieties and elementary abelian subgroups. J. Pure Appl. Algebra **26** (1982) 221–227
5. Avrunin, G. S.: Annihilators of cohomology modules. J. Algebra **69** (1981) 150–154
6. Avrunin, G. S., Scott, L.: Quillen stratification for modules. Invent. math. **66** (1982) 277–286
7. Benson, D. J., Carlson, J. F.: Complexity and multiple complexes. Math. Zeit. **195** (1987) 221–238
8. Benson, D. J., Carlson, J. F.: Projective resolutions and Poincaré duality complexes. (To appear)
9. Carlson, J. F.: The complexity and varieties of modules. (Lecture Notes in Mathematics, vol. 882.) Springer, Berlin Heidelberg New York, 1981, pp. 415–422
10. Carlson, J. F.: The varieties and the cohomology ring of a module. J. Algebra **85** (1983) 104–143
11. Carlson, J. F.: The variety of an indecomposable module is connected. Invent. math. **77** (1984) 291–299
12. Carlson, J. F.: Cohomology rings of induced modules. J. Pure Appl. Algebra **44** (1987) 85–97
13. Carlson, J. F.: Exponents of modules and maps. Invent. math. **95** (1989) 13–24
14. Evens, L.: The cohomology ring of a finite group. Trans. Amer. Math. Soc. **101** (1961) 224–239
15. Friedlander, E. M., Parshall, B. J.: Support varieties for restricted Lie algebras. Invent. math. **86** (1986) 553–562
16. Jantzen, J. C.: Kohomologie von p-Lie-Algebren und nilpontente Elemente. Abh. Math. Sem. Univ. Hamburg **76** (1986) 191–219
17. Niwasaki, T.: On Carlson's conjecture for cohomology rings of modules. (To appear)
18. Parshall, B. J.: Cohomology of algebraic groups. Proc. Symp. Pure Math. **47** (1987) 233–248
19. Quillen, D.: The spectrum of an equivariant cohomology ring, I, II. Ann. Math. (2) **94** (1971) 549–602
20. Quillen, D.: A cohomological criterion for p-nilpotency. J. Pure Appl. Algebra **1** (1971) 361–372
21. Quillen, D., Venkov, B. B.: Cohomology of finite groups and elementary abelian subgroups. Topology **11** (1972) 317–318
22. Serre, J. P.: Sur la dimension cohomologique des groupes profinis. Topology **3** (1965) 413–420

On Growth in Group Theory

Rostislav I. Grigorchuk

Moscow Institute of Railway Transportation Engineers
ul. Obraszowa 15, Moscow, USSR

1. General

The concept of growth appeared in group theory in the mid-fifties and now it plays
an increasing role in this theory. Among different notions concerning the idea of
growth in group theory the most important and successful is the notion of growth
of a finitely generated (f.g.) group.

We start with the main definition, then enumerate the results concerning this
definition and at the end we shall touch some other aspects of growth in group
theory.

1.1 The Main Definition

Let G be a finitely generated (f.g.) group with generator system $A = \{a_1, \ldots, a_m\}$,
and let $\partial(g)$ be the length of the element $g \in G$ with respect to the system A, in other
words the minimal number k such that g can be represented in the form

$$g = a_{i_1}^{\varepsilon_1} \ldots a_{i_k}^{\varepsilon_k},$$

$$\varepsilon_j = \pm 1, \qquad j = 1, \ldots, k.$$

The growth function of the group G with respect to the system A is the function

$$\gamma(n) = \operatorname{card}\{g \in G; \partial(g) \le n\}.$$

As $\gamma(n)$ depends on the generator system it is convenient to introduce an equivalence relation on the set of growth functions:

$$\gamma_1(n) \sim \gamma_2(n) \Leftrightarrow \exists C(\gamma_1(n) \le \gamma_2(Cn) \& \gamma_2(n) \le \gamma_1(Cn))$$

and the preordering relation:

$$\gamma_1(n) \preceq \gamma_2(n) \Leftrightarrow \exists C(\gamma_1(n) \le \gamma_2(Cn)).$$

The equivalence class $[\gamma_g^A(n)]$ is an invariant of the group G. We shall call it the
growth degree of the group G.

Obviously the growth degrees of a group and of any of its subgroup of finite
index coincide.

The partially ordered (with respect to $\preceq$) set $\mathfrak{W}$ of the growth degrees of f.g.
groups will be the main object of our consideration.

Proceedings of the International Congress
of Mathematicians, Kyoto, Japan, 1990
© The Mathematical Society of Japan, 1991

1.2 Examples

If $G = \mathbb{Z}^d$ is the free abelian group of rank d, then $\gamma(n) \sim n^d$. If $G = F_m$ is an (absolute) free group of rank $m \geq 2$ then $\gamma(n) \infty e^n$. Thus the growth of a group can be polynomial of any degree $d \in \mathbb{Z}_+$ and can be exponential.

1.3 Connection with Geometrical Growth

The notion of the growth function was introduced by A.S. Schwarzc [1] and independently by J. Milnor [2]. This notion is a combinatorial form of the geometrical growth notion explored by V.A. Efremovich [3].

Let M be a Riemannian manifold. Its growth at infinity is characterized by the growth when $r \to \infty$ of the function

$$v(r) = \text{Vol}(B_x(r)),$$

expressing the volume of the geodesic ball of radius r with center at the point $x \in M$. The examples of the Euclidean space $\mathbb{R}^d$ and hyperbolic space $\mathbb{H}^d$ show that this function in the first case grows as the polynomial r^d and in the second case as the exponential function e^r.

On the other hand the growth function $\gamma_G(n)$ of the f.g. group G expresses the volume of the ball $B_1(n)$ of radius n with center at the unit element $1 \in G$ if the group G is supplied with the metric $d(g, h) = \partial(g^{-1}h)$ and Haar measure.

The connection between geometric and algebraic growths can be also illustrated by A.S. Schwarzc's theorem [1]: if the manifold $\tilde{M}$ is a universal covering manifold of the compact Riemannian manifold M then

$$v_{\tilde{M}}(r) \sim \gamma_{\pi_1(M)}(r).$$

In other words the growth of the covering manifold $\tilde{M}$ is the same as the growth of the fundamental group $\pi_1(M)$ of the manifold M.

2. Milnor's Problems

2.1 Some Results

Intensive investigation of the growth functions of f.g. groups began after J. Milnor's paper [2] was published. A series of important results was obtained within a short time interval. Here are some of them.

2.1.1 A f.g. nilpotent group G has polynomial growth (Wolf [4]) like n^d, where

$$d = \sum_k k \, \text{rank}_Q(G_k/G_{k+1})$$

and $\{G_k\}$ is the lower central series of the group G (Bass [5]).

2.1.2 A f.g. solvable group has exponential growth if it is not virtually nilpotent (Wolf [4], Milnor [6]). (Virtually nilpotent means: contains a nilpotent subgroup of finite index).

2.1.3 A f.g. linear group has exponential growth if it is not virtually nilpotent (Tits [7]).

2.1.4 A free periodic Burnside group

$$B(m, n) = \langle a_1, \ldots, a_m | g^n = 1, g \in G \rangle$$

has exponential growth when $m \geq 2$ and $n \geq 665$ is odd (Adyan [8]).

In 1968 J. Milnor [9] proposed a number of problems concerning growth of groups.

1) *"Is the growth function $\gamma(n)$ necessarily equivalent either to a power of n or to the exponential function 2^n?"* 2) *In particular, is the growth exponent*

$$d = \lim_{n \to \infty} \frac{\log \gamma(n)}{\log n} \tag{1}$$

always either a well-defined integer or infinitely? For which groups is $d < \infty$? (A possible conjecture would be that $d < \infty$ if and only if G contains a nilpotent subgroup of finite index)".

(Division of Milnor's problem into Parts 1) and 2) has been made by me).

As a conjecture Problem 1) was formulated by Wolf [4] and Bass [5].

Conjecture: *"A f.g. group of non-exponential growth is virtually nilpotent".*

2.2 Two Definitions

A group whose growth function is not majorized by any function n^d and is not equivalent to the function e^n is called an intermediate growth group. And a group G is called a group of subexponential growth if $\gamma(n) \prec e^n$ (strong inequality), in other words, if

$$\lim_{n \to \infty} \sqrt[n]{\gamma(n)} = 1$$

(the limit exists due to the semi-multiplicativity of the function $\gamma(n)$).

3. Answer to Milnor's Problem 1)

Milnor's problem 1) was solved negatively in [10, 11, 12].

3.1 On the Construction of Intermediate Growth Groups

The simplest example of such a group is the group Γ from the paper [13] which was constructed as a simple example of a f.g. infinite periodic group. Γ is defined as the group of transformations of the interval $[0, 1]$ from which rational points of the form $k/2^n$, $n = 1, 2, \ldots, 0 \leq k \leq 2^n$ are removed. This group is generated by four transformations a, b, c, d, where a is the permutation of the halves of the interval $[0, 1]$ and b, c, d, are defined with the help of infinite periodic words

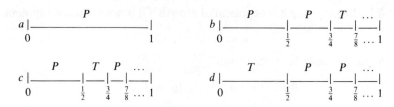

Fig. 1. (P denotes the permutation of the halves of the interval, T is the identity transformation).

$$PPT\ PPT\dots$$
$$PTP\ PTP\dots$$
$$TPP\ TPP\dots$$

as in Figure 1.

We constructed uncountably many intermediate growth groups on the basis of this example (see [10–12]).

3.2 The Main Idea of the Growth Function $\gamma_\Gamma(n)$ Upper Estimation

The group Γ contains a subgroup H of finite index, which allows an embedding Ψ in the direct product of eight copies of the group Γ:

$$H \overset{\Psi}{\cong} \hat{H} \le \Gamma \times \Gamma \times \Gamma \times \Gamma \times \Gamma \times \Gamma \times \Gamma \times \Gamma,$$

and $\hat{H}$ is a subgroup of finite index in Γ^8.

This embedding has the following property: if $g \in H$ and

$$\Psi(g) = (g_1, g_2, \dots, g_8) \in \hat{H}$$

then

$$\sum_{i=1}^{8} \partial(g_i) \le k\partial(g),$$

where k is some constant less than 1. Hence

$$\gamma(n) \le C \sum \gamma(n_1) \dots \gamma(n_8) \tag{2}$$

(C is some constant), where the summation in (2) is taken over those sets of non-negative integers $n_1, \dots, n_8$ for which the inequality

$$\sum_{i=1}^{8} n_i \le kn$$

is valid.

If we assume that

$$\lim_{n \to \infty} \sqrt[n]{\gamma(n)} = \lambda > 1$$

then we shall obtain a contradiction $\lambda < \lambda^k$ from (2). So Γ has sub-exponential growth.

The fact that the function $\gamma(n)$ grows faster than any polynomial results, for example, from the following consideration.

The group Γ is commensurable with its square, in other words the groups Γ and $\Gamma \times \Gamma$ contain isomorphic subgroups of finite index. Hence $\gamma(n)$ satisfies the so-called Δ^2-condition $\gamma(n) \sim \gamma^2(n)$, which results in the lower estimate by some function of the type e^{n^β}, where $\beta > 0$.

However, the lower estimate of the function $\gamma_\Gamma(n)$ can be obtained directly. It is proved more precisely in [11] that

$$e^{\sqrt{n}} \le \gamma_\Gamma(n) \le e^{n^\alpha},$$

where $\alpha = \log_{32} 31$. This obviously gives a negative answer to Milnor's problem 1).

On the basis of a construction which generalizes this example we proved the following theorem (see [11, 14]).

Theorem 1. *The set of growth degrees of f.g. groups has the cardinality of the continuum. It contains a chain of the cardinality of the continuum, and an antichain of the cardinality of the continuum.*

I would like to stress that we are speaking precisely about intermediate degrees of groups in this theorem.

Another example of a group of intermediate growth was constructed by J. Fabrykowski and N. Gupta [15].

3.3 Some Properties of Intermediate Growth Groups

Intermediate growth groups can be constructed both in the class of p-groups (p is any given prime) and in the class of torsion free groups [11, 12].

The intermediate growth groups known up to this moment belong to the class of residually finite groups. Moreover, the groups from the papers [10–12] are residually-p-finite.

The periodic groups constructed in [11, 12] have the property that all their proper factor groups are finite. At the same time in [16] for any prime p, a f.g. p-group of intermediate growth having continuum nonisomorphic factor groups was constructed.

The torsion free group of intermediate growth constructed in [12] admits an invariant linear ordering.

The intermediate growth groups from [11, 12] are not finitely presentable.

There exist recursively presentable (by generating elements and defining relations) intermediate growth groups having solvable word problem. At the same time there are analogous groups with unsolvable word problem [11, 14].

4. On Groups with Polynomial Growth

An exhaustive answer to Part 2) of Milnor's problem was given in the papers [17, 18] and Milnor's conjecture suggested in this part was confirmed. Note that a more general result was obtained in [18], although this paper is mainly devoted to proving Gromov's theorem by methods of nonstandard analysis.

4.1 On Gromov's Theorem

Gromov's theorem proved in [17] gives the description of f.g. groups whose growth functions admit polynomial estimate

$$\gamma(n) \precsim n^d, \tag{3}$$

(d is some constant). This theorem states that a group G is virtually nilpotent if the estimate (3) is valid for its growth function.

Hence the estimate (3) results in an equivalence $\gamma(n) \sim n^d$ for a suitable $d \in \mathbb{Z}_+$.

The proof of this theorem is based on geometric considerations. The left invariant metric $d(g, h) = \partial(g^{-1}h), g, h \in G$ is built on the group G and the sequence

$$X_n = (G, \tfrac{1}{n}d), \qquad n \in \mathbb{N}$$

of metric spaces is considered.

Gromov proved that from this sequence one can extract some subsequence which converges in an exactly defined way to some metric space X_∞ if the condition (3) is valid. A homomorphism $G \to \mathrm{Isom}(X_\infty)$ from the group G into the isometry group of X_∞ arises and $\mathrm{Isom}(X_\infty)$ is a Lie group with a finite number of connected components. Some simple algebraic considerations complete the proof of this theorem.

4.2 On L. van den Dries and A. Wilkie's Theorem

The main result of [18] is: if the inequality

$$\gamma(n) \le Cn^d$$

is valid on some infinite subset $\mathbb{N}_0 \subseteq \mathbb{N}$ of the set of natural numbers then the group G is virtually nilpotent.

From this statement if follows that the limit (1) (finite or infinite) always exists, and also $d < \infty$ if and only if the group G is virtually nilpotent.

5. On the Growth of Cancellation Semigroups

The notion of the growth function of a f.g. semigroup is defined as in the case of a group. The growth of a semigroup can be very strange even in the case when the condition (3) holds true. But if we consider the class of semigroups with left and right cancellation laws then in this case the statement similar to Gromov's theorem is valid.

5.1 The Nilpotency of Semigroups Due to A.I. Malcev

Let x, y, ξ_1, ξ_2, ..., ξ_n, ... be symbols denoting variables running through a semigroup S. Let us denote $X_0 = x$, $Y_0 = y$ and after that by induction

$$X_{n+1} = X_n \xi_n Y_n, \qquad Y_{n+1} = Y_n \xi_n X_n.$$

The semigroup S whose elements satisfy the identity $X_n = Y_n$ for some n is called nilpotent.

A.I. Malcev proved [19] that the classical nilpotency identity for groups is equivalent to the semigroup identity $X_n = Y_n$ for corresponding n.

5.2 On Cancellation Semigroup of Polynomial Growth

Let $S_0 \subseteq S$ be a subsemigroup. We shall say that S_0 has finite index in S if there exists a finite subset $K \subseteq S$ such that for any $s \in S$ there exists $k \in K$ for which $sk \in S_0$.

Theorem 2. *A f.g. cancellation semigroup S has polynomial growth (Condition (3)) if and only if S contains a nilpotent subsemigroup of finite index.*

The proof of this theorem is based on the remark that for a cancellation semigroup of subexponential growth the Ore condition is satisfied and this semigroup possesses a group $G_S = S^{-1}S$ of left quotients.

The estimate (3) on the growth function of the semigroup S makes it possible to give a polynomial type estimate on the growth function of the group G_S after detailed analysis. Then Gromov's theorem can be applied.

5.3 Some Remarks

In [20] it was proved that if a semigroup S with cancellation has polynomial growth, then the group G_S of left quotients also has polynomial growth of the same degree.

It would be interesting to construct examples of semigroups with cancellation of subexponential growth for which there is a jump of growth degree in the diagram $S \to G_S$.

If one could construct a semigroup S with cancellations of subexponential growth such that G_S has exponential growth, then Problem N12 from [21] will be positively solved.

6. On Lacunae in the Set of Growth Degrees of Residually Nilpotent Groups

Recall that a group G is residually-p if for any nonunit element $g \in G$ there exists a finite p-group K and a homomorphism $\varphi : G \to K$ such that $\varphi(g) \neq 1$.

Theorem 3 [22]. *Let p be any prime, let the f.g. group G be residually-p and its growth function satisfy the estimate (3). Then G is virtually nilpotent and so has polynomial growth.*

6.1 Proof of Theorem 3

Let F_p be a finite prime field, $F_p[G]$ the group algebra, $\varDelta \le F_p[G]$ the augmentation ideal, in other words the ideal generated by elements of the form $g - 1$, $g \in G$, $\mathrm{gr}(G)$

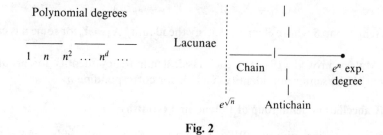

Fig. 2

the associative graded algebra defined by means of the powers of the augmentation ideal

$$gr(G) = \bigoplus_{n=0}^{\infty} A_n = \bigoplus_{n=0}^{\infty} \Delta^n/\Delta^{n+1},$$

let $f_{G,p}(t)$ be the Hilbert-Poincaré series of the algebra $gr(G)$:

$$f_{G,p}(t) = \sum_{n=0}^{\infty} a_n(G)t^n,$$

where $a_n(G) = \dim_{F_p} A_n$.

It follows from Lazard's Theorem 3.11 [23] that either the sequence $a_n(G)$ has polynomial growth and then the p-completion $\hat{G}$ is p-analytic or $a_n(G) \succeq e^{\sqrt{n}}$. Besides if $\gamma(n)$ is the growth function of the group G with respect to any system of generators then the inequality $a_n(G) \leq \gamma(n)$, $n = 1, 2, \ldots$ is valid.

Using Lazard's result and Condition (3) we obtain that the growth of coefficients $a_n(G)$ when $n \to \infty$ is polynomial and so the p-completion $\hat{G}$ is analytic. It follows from Tit's theorem [7] that either G contains a free subgroup with two generators or G is a virtually solvable group. But the first case is impossible due to limitation on the growth, and in the second case, we can conclude that G is virtually nilpotent due to results of [4, 6].

Recently A. Lubotzky and A. Mann [24] have pointed out that in Theorem 3 the assumption that G is residually-p can be changed by the assumption that G is residually nilpotent.

6.2 On the Scale of Growth Degrees of Residually-p groups

Let p be any prime and denote by $\mathfrak{W}_p$ the set of growth degrees of f.g. residually-p groups. Theorem 1 formulated above for the whole class of groups is also true for every set $\mathfrak{W}_p$. So, due to Theorem 3 the structure of the partially ordered set $\mathfrak{W}_p$ can be presented in the form as in Figure 2.

7. On the Generating Series of Growth Function

Many questions of a theoretical and applied character require more detailed investigation of the asymptotic behavior of the growth function $\gamma(n)$ than up to the equivalence $\sim$, defined above. For this it is sometimes useful to connect the generating series

$$\Gamma(t) = \sum_{n=0}^{\infty} \gamma(n) t^n$$

to the function $\gamma(n)$.

7.1 Some Cases When the Function $\Gamma(t)$ is Rational for any System of Generators

i) The group is virtually abelian (Benson [25])
ii) The group is a cocompact group of isometries of the hyperbolic space $\mathbb{H}^d$ (Cannon [26])
iii) The group is hyperbolic (Gromov [27]).

For many nilpotent groups the series the $\Gamma(t)$ also represents a rational function (see for example [28]). At the same time the conjecture that for any f.g. nilpotent group and any of its finite generating set the function $\Gamma(t)$ is rational proved to be false. Namely F. Grunewald proved that for the nilpotent group

$$G = \langle a_1, a_2, b_1, b_2, z | [a_1, b_1] = [a_2, b_2] = z, z \in \mathbb{Z}(G),$$

$$[a_1, a_2] = [b_1, b_2] = [a_1, b_2] = [a_2, b_1] = 1 \rangle$$

($\mathbb{Z}(G)$ is the centre of the group G and the function $\Gamma(t)$ defined using the system of generators $\{a_1, a_2, b_1, b_2, z\}$), the generating series $\Gamma(t)$ is not a rational function.

7.2 Information on the Growth Functions of Nilpotent Groups

Let G be a nilpotent group, $\gamma(n)$ the growth function of G with respect to any system of generators, d the power of polynomial growth of G. Then

a) there exists the limit

$$\lim_{n \to \infty} \frac{\gamma(n)}{n^d} = C$$

(Pancu [29]),
b) the estimate

$$\gamma(n) = Cn^d + O(n^{d-1/2})$$

is valid (Grunewald, unpublished).

It is interesting to find new examples of rationality of the growth function $\Gamma(t)$ and also to obtain another information about the generating series of growth functions.

8. Some Problems and Conjectures

The theory of growth degrees of groups has been developed for more than three decades and has already accumulated a lot of unsolved problems. Let me formulate some of them.

8.1 (A well known problem.) *Is there a f.g. group of intermediate growth with a finite set of defining relations?*

8.2 *Is it true that if the growth function of a f.g. group G grows slower than the function $e^{\sqrt{n}}$, then G is virtually nilpotent?*

8.3 *Does there exist a f.g. group whose growth function is equivalent to the function $e^{\sqrt{n}}$?*

8.4 *Is it true that any group of subexponential growth is residually finite?*

8.5 *To find the asymptotic behaviour of the growth function of the group Γ from [13] (conjecture is that $\gamma(n) \sim e^{\sqrt{n}}$ in this case).*

8.6 *Is it true that for any f.g. nilpotent group the function $\Gamma(t)$ is meromorphic or even algebraic?* (F. Grunewald).

Let

$$G = \langle A | r = 1 \ (r \in R) \rangle$$

be f.g. and $\mathfrak{J}$ be the Cayley graph of G with respect to the system of generators A. A spanning tree $T \subseteq \mathfrak{J}$ is regular if it is defined by a finite automaton. A spanning tree T is minimal if every word in T is the shortest.

8.7 Conjecture. *G has a rational growth function if and only if $\mathfrak{J}$ has a minimal regular spanning tree* (Machi, Schupp).

9. Other Aspects of Growth in Groups

9.1 Cogrowth

Let a f.g. group G be realized as a factor group F_m/H of the free group F_m of rank m, $H \leq F_m$ a normal subgroup, the elements of which will be further considered as reduced words over the basis of F_m. Denote by $h(n)$ the number of words of length $\leq n$ in H. Obviously $h(n) \leq 2m(2m-1)^{n-1}$. Let

$$\alpha_H = \limsup_{n \to \infty} \sqrt{h(n)}, \qquad \mathcal{H}(t) = \sum_{n=0}^{\infty} h(n)t^n.$$

The value α_H which appeared in [30] is called the growth exponent of the normal subgroup H and belongs to the interval $(\sqrt{2m-1}, 2m-1]$.

In [31] the following amenability criterion was proved.

Theorem 4. *The group $G = F_m/H$ is amenable if and only if $\alpha_H = 2m - 1$.*

The function $\mathcal{H}(t)$ and growth exponent α_H can be defined also for any subgroup $H \leq F_m$. In [31] it was proved that if H is a f.g. group then $\mathcal{H}(t)$ is a rational function which can be effectively calculated if the system of generators of H is known.

9.2 Subgroup Growth

Let G be a f.g. group, and let $\varrho_n(G)$ be the (finite) number of its subgroups of index n. The problem of investigation of the asymptotic behavior of the sequence $\varrho(n)$ when $n \to \infty$ has been drawing more and more interest. For example in [32] F. Grunewald, D. Segal and G. Smith investigated in detail the Dirichlet series

$$\xi_G(s) = \sum_{1}^{\infty} \varrho_n(G)n^{-s}$$

of the sequence $\varrho_n(G)$ for the case when G is nilpotent.

On the other hand A. Lubotzky and A. Mann [24] showed that if G is a residually nilpotent f.g. group then $\varrho_n(G)$ has polynomial growth if and only if G is solvable of finite rank.

The general case of a residually finite group is still open. See [24] for other references and comments.

9.3 On the Growth of Graded Algebras Associated to Groups

During investigations of groups different graded algebras associated to these groups arise. For example the associative graded algebra gr G was defined above. Studying the growth of algebras and in particular graded algebras associated to groups is important from different points of view and is connected with different applied aspects of algebra [33–37].

9.4 On the Asymptotic Behavior of Random Walks on Groups

Investigation of random walks on groups was started by H. Kesten in [38]. It is interesting to investigate the asymptotic behavior when $n \to \infty$ of the probability $P_{1,1}^{(n)}$ of returning to the unit element after n steps [38, 39], to calculate the entropy of random walks [40], to describe Martin's and Poisson's boundaries [41, 42] and so on.

10. Applications

10.1 To the Theory of Invariant Means

The definition of an invariant means and the corresponding notion of amenable groups belong to J. von Neumann [43].

Finite and abelian groups are amenable. The class of amenable groups is closed under operations of extension and inductive limit.

Let us denote the class of amenable groups by AG, the class of groups without free subgroups with two generators by NF, and the class of elementary amenable groups constructed by J. von Neumann by EG, i.e. the minimal class of groups containing finite groups, abelian groups and closed under the operations of extension and inductive limit. The following imbeddings EG $\subseteq$ AG $\subseteq$ NF hold true.

In 1957 M. Day [44] proposed the following two problems:

(1) *Is it true that AG = NF?*
(2) *Is it true that AG = EG?*

Some mathematicians ascribe Problem (1) to J. von Neumann but there is no written confirmation of the fact that it was he who suggested this problem.

In 1979 to disprove the conjecture AG = NF I suggested A.Yu. Olshanskii to apply the amenability criterion (Theorem 4) to the groups constructed by him at that moment for the solution of some famous problems of group theory. This suggestion was realized in the short paper [45].

A little later S.I. Adyan [46] also applied the amenability criterion to prove that free periodic Burnside groups $B(m, n)$ are nonamenable when $m \geq 2$ and $n \geq 665$ is odd (the conjecture that these groups are nonamenable was expressed by S.I. Adyan in 1975).

The negative answer to the problem (2) was given in [11].

Theorem 5. *There exist uncountably many amenable groups not belonging to the class EG.*

Hence the class AG of amenable groups is much more than the class EG of elementary amenable groups.

Theorem 5 is an easy corollary to the existence of a continuum of nonisomorphic intermediate growth groups.

Using intermediate growth groups we managed [48] to disprove one of Rosenblatt's conjectures [47] (see also [21], Problem 11) about the so-called superamenable groups.

10.2 To Riemannian Geometry

The results concerning the growth functions of groups are widely used in the classification theory of Riemannian manifolds up to quasi-isometries, in the theory of foliations, in the investigation of Laplace operator on Riemannian manifolds (see [49, 50]), ...

10.3 Other Applications

Information concerning the growth functions of groups is used in many, sometimes unexpected branches of mathematics: in the theory of random walks [38, 39], ergodic theory [50, 52], the theory of finite automata [20] and so on.

In my lecture I did not touch upon many other aspects of applications of the notion of growth in group theory, many of which at this moment are only in their initial stage of development.

References

1. Schwarzc, A.S.: A volume invariant of coverings. Dokl. Ak. Nauk USSR **105** (1955) 32–34
2. Milnor, J.: A note on curvature and fundamental group. J. Diff. Geom. **2** (1968) 1–7

3. Efremovic, V.A.: The proximity geometry of Riemannian manifolds, Uspekhi Math. Nauk **8** (1953) 189 (Russian)
4. Wolf, J.: Growth of finitely generated solvable groups and curvature of Riemannian manifolds. J. Diff. Geom. **2** (1968) 424–446
5. Bass, H.: The degree of polynomial growth of finitely generated nilpotent groups. Proc. London Math. Soc. **25** (1972) 603–614
6. Milnor, J.: Growth in finitely generated solvable groups. J. Diff. Geom. **2** (1968) 447–449
7. Tits, J.: Free subgroups in linear groups. J. Algebra **20** (1972) 250–270
8. Adyan S.I.: The Burnside problem and identities in groups. Nauka, Moscow 1975 (Russian) [English transl.: Proc. Steklov Inst. Math. (1970) 142]
9. Milnor, J.: Problem 5603. Amer. Math. Monthly **75** (1968) 685–686
10. Grigorchuk, R.I.: On Milnor's problem of group growth. Dokl. Ak. Nauk SSSR **271** (1983) 31–33 (Russian) [English transl.: Soviet Math. Dokl. **28** (1983) 23–26]
11. Grigorchuk, R.I.: The growth degrees of finitely generated groups and the theory of invariant means. Izv. Akad. Nauk SSSR. Ser. Math. **48** (1984) 939–985 (Russian) [English transl.: Math. SSSR Izv. **25** (1985)]
12. Grigorchuk, R.I.: On the growth degrees of p-groups and torsion-free groups. Math. Sbornik **126** (1985) 194–214 (Russian) [English transl.: Math. USSR Sbornik **54** (1986) 185–205]
13. Grigorchuk, R.I.: On the Burnside problem for periodic groups. Funct. Anal. Prilozen. **14** (1980) 53–54 [English transl.: Funct. Anal. Appl. **14**, 41–43]
14. Grigorchuk, R.I.: A groups with intermediate growth function and its applications. Second degree doctoral thesis, Moscow 1985
15. Fabrykowski, J., Gupta, N.: On groups with subexponential growth functions. J. Ind. Math. Soc. **49** (1985) 249–256
16. Grigorchuk, R.I.: The construction of intermediate growth groups having continuum factor groups. Algebra and Logic **23** (1984) 383–394
17. Gromov, M.: Groups of polynomial growth and expanding maps. Publ. Math. IHES **53** (1981) 53–73
18. Van den Dries, L., Wilkie, A.J.: Gromov's theorem on groups of polynomial growth and elementary logic. J. Alg. **89** (1984) 349–374
19. Malcev, A.I.: Nilpotent semigroups. Uchen. Zap. Ivanovsk. Ped. In-ta **4** (1953) 107–111 (Russian)
20. Grigorchuk R.I.: On the cancellation semigroup of polynomial growth. Matem. Zametki **43** (1988) 305–319 (Russian)
21. Wagon, S.: The Banach-Tarski Paradox. Encyclopedia of mathematics and its applications 24. Cambridge University Press 1985
22. Grigorchuk, R.I.: On the Hilbert-Pouncare series of the graded algebras associated to groups. Mat. Sbornik **180** (1989) 307–225
23. Lazard, M.: Groupes analytiques p-adiques. Publ. Math. IHES **26** (1965) 389–603
24. Lubotzky, A., Mann, A.: On groups of polynomial subgroup Growth. Preprint 1990, pp. 1–22
25. Benson, M.: Growth series of finite extentions of $\mathbb{Z}^d$ are rational. Invent. math. **73** (1983) 251–269
26. Cannon, J.: The combinatorial structure of cocompact discrete hyperbolic groups. Geometricae Dedicata **16** (1984) 123–148
27. Gromov, M.: Hyperbolic groups. Essays in group theory. MSRI Publications 8 (ed. S.M. Gersten). Springer, Berlin Heidelberg New York 1987, pp. 75–263
28. Shapiro, M.: A geometrical approach to the almost convexity and growth of some nilpotent groups. Preprint 1989
29. Pancu, P.: Croessance des boules et des geodeseques fermees dans les nilvarietes. Ergodic theory and dynamic systems, 1983, pp. 415–445

30. Grigorchuk, R.I.: Symmetrical random walks on discrete groups. Uspekhi Math. Nauk **XXXII** (1977) 217–218 (Russian)
31. Grigorchuk, R.I.: Symmetrical random walks on discrete groups. In: Multicomponent random systems (ed. R.L. Dobrushin, Ya.G. Sinai). Nauka, Moscow 1978, pp. 132–152. [English transl.: Advances in probability and related topics, vol. 6. Marcel Dekker 1980, pp. 285–325]
32. Grunewald, F.J., Segal, D., Smith, G.C.: Subgroups of finite index in nilpotent groups. Invent. math. **93** (1988) 185–233
33. Gelfand, I.M., Kirillov, A.A.: Sur les corps lie's aux algebres enveloppantes des algebres de Lie. Publ. Math. IHES **31** (1966) 5–19
34. Krause, G.R., Lenagan, T.H.: Growth of algebras and Gelfand-Kirillov dimention. Pitman Advanced Publishing Program, Boston London Melbourne 1985
35. Babenco, I.C.: The problems of growth and rationality in Algebra and Topology. Uspekhi Math. Nauk **41** (1989) 95–142
36. Grigorchuk, R.I.: On the topological and metrical types of a regular covering surfaces. Izv. Akad. Nauk SSSR. **54** (1990) 498–536 (Russian)
37. Bereznyi, A.E.: Discrete sub-exponential groups Zap. Nauch Sem. LOMI, **123** (1983) 155–166
38. Kesten, H.: Symmetric random walks on groups. Trans. Amer. Math. Soc. **92** (1959) 336–354
39. Kesten, H.: Full Banach mean values on countable groups. Math. Scand. **7** (1959) 146–156
40. Kaimanovich, V.A., Vershik, A.M.: Random walks on discrete groups: boundary and entropy. Ann. Probab. **11** (1983) 457–490
41. Dynkin, E.B., Maljutov, M.B.: Random walks on groups with a finite number of generators. Soviet Math. Dokl. **2** (1961) 339–402
42. Furstenberg, H.: Random walks and Discrete subgroup of Lie groups. Adv. Probab. Related Topics **1**, 3–63. Dekker, New York (1971)
43. Von Neumann, J.: Zur allgemeinen Theorie des Masses. Fund. Math. **13** (1929) 73–116
44. Day, M.: Amenable semigroups. Ill. J. Math. **1** (1957) 509–544
45. Olshanskii, A.Yu.: On the problem of the existence of an invariant mean on a group. Uspekhi Mat. Nauk **35** (1980) 1165–1166
46. Adyan, S.I.: Random walks on free periodic groups. Izv. Akad. Nauk SSSR, Ser. Mat. **46** (1982) 1139–1149 (Russian) [English transl.: Math. USSR Izv. **21** (1983)]
47. Rosenblatt, J.: Invariant measures and growth conditions. Trans. Amer. Math. Soc. **193** (1974) 33–53
48. Grigorchuk, R.I.: Supramenability and problem of embedding a free semigroups. Funkt. Anal. Priloz. **21** (1987) 74–75 (Russian)
49. Brooks, R.: The fundamental group and the spectrum of the Laplacian. Comm. Math. Helv. **56** (1981) 581–598
50. Varopoulos, N.: Brownian motion and transient groups. Ann. Inst. Fourier (Grenoble) **34** (1984) 243–269

Absolute Integral Closure and Big Cohen-Macaulay Algebras

*Craig Huneke**

Department of Mathematics, Purdue University, West Lafayette, IN 47907, USA

Introduction

Our purpose in this paper is to discuss a recent result by this author jointly with Melvin Hochster which gives the existence of big Cohen-Macaulay (C-M) algebras in equicharacteristic. In general proofs will be omitted, though sometimes sketchy arguments will be given. Before stating the main theorem, we give a review of the basic definitions and concepts used in this paper, particularly the property of a ring's being C-M. Throughout this paper, "ring" means commutative noetherian ring with identity, with one notable exception: the integral closure of a domain R in an algebraic closure of its quotient field will come into play. However, as these rings are never noetherian if $\dim(R) > 0$, no confusion will arise.

Definition 1.1. *Let R be a ring, let M be an R-module, and let $x_1, \ldots, x_n \in R$. The elements $x_1, \ldots, x_n$ are said to be a regular sequence on M (or simply an M-sequence) if* i) *x_i is not a zerodivisor on the module $M/(x_1, \ldots, x_{i-1})M$ for each i between 1 and n, and* ii) *$M \neq (x_1, \ldots, x_n)M$.*

Condition ii) is a nondegeneracy restriction, which is especially relevant when the module M is not finitely generated. For example if R is a domain and M is its quotient field, then i) is trivially satisfied, but ii) fails.

The first condition can be rephrased in several ways of which we list two:

i)′ whenever $\sum_{i=1}^{k} y_i x_i = 0$ with $y_i \in M$, $1 \leq k \leq n$, then $y_k \in (x_1, \ldots, x_{k-1})M$.

If M is finitely generated and the x_i are in the Jacobson radical of R, then i) is equivalent to,

i)″ The first Koszul homology $H_1(\mathbf{x}, M) = 0$, where here and below $\mathbf{x}$ stands for all the x_i. (The Koszul complex $K.(\mathbf{x}, M)$ is the tensor product of the n complexes $0 \to R \xrightarrow{x_i} R \to 0$ with M.)

---

* Partially supported by the NSF.

Proceedings of the International Congress
of Mathematicians, Kyoto, Japan, 1990
© The Mathematical Society of Japan, 1991

In any case, if the **x** form a regular sequence on M, then the first Koszul homology is 0; it is the converse that requires some restrictions.

The (Krull) *dimension* of a ring R, $\dim(R)$, is $\sup\{n \in \mathbf{N} \mid P_0 \subsetneq P_1 \subsetneq \cdots \subsetneq P_n : P_i$ are prime ideals of R.}

A ring $(R, \mathfrak{m})$ is *local* if there is a unique maximal ideal $\mathfrak{m}$ of R. The dimension of a local ring $(R, \mathfrak{m})$ can also be characterized as the least integer n for which there exist n elements $x_1, \ldots, x_n$ with the property that the nilradical of the ideal they generate is exactly $\mathfrak{m}$. Such a system of elements is called a *system of parameters* (s.o.p.). A sequence of elements which can be extended to an s.o.p. will be called *parameters* in this paper.

Definition 1.2. *A local ring R (necessarily noetherian, recall) is* Cohen-Macaulay, *if there is a regular sequence $x_1, \ldots, x_n$ in $\mathfrak{m}$ whose length $n = \dim(R)$. A noetherian ring R is* C-M *if $R_\mathfrak{p}$ is C-M for every prime ideal $\mathfrak{p}$ of R (equivalently, every maximal ideal $\mathfrak{p}$ of R).*

Any sequence as in (1.2) will have to be an s.o.p. Moreover if R is C-M then any s.o.p. will form a regular sequence. In general the length of a maximal regular sequence of elements in the maximal ideal $\mathfrak{m}$ of a local ring $(R, \mathfrak{m})$ is called the *depth of R*, denoted $\operatorname{depth}(R)$. It always holds that $\operatorname{depth}(R) \leq \dim(R)$; R is C-M precisely when equality occurs. Some examples of C-M rings and how they occur are given below.

Example 1.3. Any 0-dimensional ring is C-M vacuously. Any 1-dimensional domain R is C-M, since any non-zero element is a nonzerodivisor.

Example 1.4. Not all 2-dimensional rings, even domains, are C-M (cf. 1.9 below). However, integrally closed 2-dimensional domains are C-M. As integral closure and finite maps play an important role in this paper, we review these concepts here. If $R \subseteq S$, an element $s \in S$ is said to be *integral* over R if s satisfies a monic polynomial equation with coefficients in R. The set of all elements of S integral over R form a ring T called the *integral closure of R in S*. If R is a domain and S is its quotient field, then T is called the integral closure of R. If further $R = T$, R is said to be *integrally closed*. A homomorphism from R into S is said to be *finite* if S is a finite R-module. This is equivalent to saying S is a finitely generated R-algebra in which every element is integral over R.

Example 1.5. Any regular ring is C-M. A local ring $(R, \mathfrak{m})$ is *regular* if the maximal ideal $\mathfrak{m}$ can be generated by an s.o.p. A noetherian ring is regular if all its local rings at prime (equivalently maximal) ideas are regular local rings. Regular local rings are exactly those local rings in which every finitely generated module has finite projective dimension. Examples of regular rings include polynomial rings and power series rings over fields. Regularity is the analogue of the geometric notion of being non-singular.

Example 1.6. If $(R, \mathfrak{m})$ is C-M local and $x_1, \ldots, x_k$ are parameters (hence also a regular sequence) then the quotient ring $R/(x_1, \ldots, x_k)$ is also C-M. Combining this example with (1.5) gives that the so-called complete intersections are C-M. A *complete intersection* is by definition a ring which is a regular ring modulo an ideal generated by a regular sequence. More generally a local Gorenstein ring is C-M. Gorenstein local rings are characterized by the property that they have finite injective dimension over themselves, or alternatively that every s.o.p. generates an irreducible ideal (*irreducible* means not the intersection of two strictly larger ideals).

Example 1.7. If G is a linearly reductive algebraic group over k acting k-linearly on a polynomial ring $S = k[x_1, \ldots, x_n]$, then the ring of invariants $R = S^G$ is C-M. This is due to Hochster and Roberts [H-R], and includes an enormous number of examples. A more general result was recently obtained by myself and Hochster [H-H1]: if S is any regular local ring containing a field, and R is a subring of S such that the inclusion of R into S splits R-linearly, then R is C-M (in positive characteristic this was known [H-R]; see also [Bo]). Our proof uses the theory of tight closure which is discussed below, cf. (2.3).

Example 1.8. If $X \subseteq \mathbb{P}_k^n$ is a projective variety of dimension d which is arithmetically normal (i.e., its homogeneous coordinate ring R is integrally closed) then X is arithmetically C-M (equivalently, R is C-M) iff for all integers t and all i, $1 \le i \le d - 1$, $H^i(X, \mathcal{O}_X(t)) = 0$. X is said to be C-M itself if its local rings are C-M. Geometrically C-M varieties are quite an important and good class; for instance Serre duality holds if X is C-M. Furthermore the computation of the degree of X (equivalently the multiplicity of R) is made easy if X is arithmetically C-M – one only needs to take a general linear space L of complementary dimension such that $X \cap L$ is empty, and then $\deg(X) = \dim_k(S/I(X) + I(L))$, where $I(Y)$ is the homogeneous ideal of functions vanishing on Y and $S = k[X_0, \ldots, X_n]$. In fact, this property characterizes when X is arithmetically C-M.

Example 1.9. To see an explicit ring which is *not* C-M take R to be the subring $k[t^4, t^3s, ts^3, s^4]$ of the polynomial ring $k[t, s]$ in two variables over a field k. It is easy to see that t^4 and s^4 form a homogeneous s.o.p. for R but are not a regular sequence, since $t^4(ts^3)^2 = s^4(t^3s)^2$, but t^4 does not divide t^6s^2 in R, as it would have to do if R is C-M. The element t^2s^2 is missing. If we adjoin this element to R, we actually arrive at the integral closure of R, which is then C-M as it is a 2-dimensional integrally closed domain (see (1.4)).

Example 1.10. While C-M rings can be highly singular, nonetheless they are in a partial sense homologically trivial over regular rings – this is made precise below for two important classes of rings, complete local rings and nonnegatively graded rings over a field. To this end, let $(R, \mathfrak{m}, k)$ denote either a complete local ring with maximal ideal $\mathfrak{m}$ (*complete* means complete in the topology determined by the powers of the maximal ideal) and coefficient field k (a *coefficient field* is a field k contained in R such that the map from k to $R/\mathfrak{m}$ is an isomorphism; if R contains a field and is complete, then such fields always exist), or let $R = \bigoplus_{i \ge 0} R_i$ be a graded ring with $R_0 = k$, and $\mathfrak{m} = \bigoplus_{i \ge 1} R_i$. In this case $R/\mathfrak{m} \simeq R_0 = k$. We will say we are in case 1 if $(R, \mathfrak{m}, k)$ is complete local, and in case 2 if $(R, \mathfrak{m}, k)$ is graded. In both of these fundamental cases, R can be represented as finite over a regular ring A. In case 2, the Noether normalization theorem shows that if $x_1, \ldots, x_n$ is any homogeneous s.o.p. for R, then the subring $k[x_1, \ldots, x_n]$ of R has the properties that R is a finite A-module, and A is isomorphic to a polynomial ring in n-variables over k, i.e., the elements $x_1, \ldots, x_n$ are algebraically independent. In case 1, the Cohen structure theorem for complete local rings shows that if $x_1, \ldots, x_n$ is any s.o.p. for R and we take A to be the complete subring $k[[x_1, \ldots, x_n]]$ of R generated by $x_1, \ldots, x_n$, then R is a finite A-module, and A is isomorphic to a power series ring in n-variables over k, i.e., $x_1, \ldots, x_n$ are analytically independent. We may thus study the properties of R through thinking of it as an A-module in both cases. The homological properties of A are much nicer in general – regular rings are characterized by the fact that modules over them have finite free resolutions.

The property that R is C-M can now be easily expressed in terms of the A-module structure of R. In both cases, R is C-M iff R is a free A-module, i.e., is "homologically trivial" over A. This is a useful point of view.

2. Trivializing Relations on Parameters

Let me return for a moment to (1.9). There we had a 'bad' relation on the parameters t^4 and s^4, namely the relation $t^4(ts^3)^2 = s^4(t^3s)^2$. This relation is nontrivial in the sense that it does not come from the trivial relation on t^4 and s^4, i.e., the relation $t^4(s^4) = s^4(t^4)$. To be precise, if $x_1, \ldots, x_n$ are elements in R, we say a relation $(r_1, \ldots, r_n)$ on these elements (i.e., $\sum r_i x_i = 0$) is *trivial* if it is an R-linear combination of the Koszul relations $(0, \ldots, 0, x_j, 0, \ldots, 0, -x_i, 0, \ldots, 0)$ where the x_j is in the ith spot and the $-x_i$ is in the jth spot. In a C-M local ring all relations on parameters are trivial; this is the fundamental property of C-M rings. However, even if R is not C-M we can ask if a given relation on parameters $x_1, \ldots, x_k$ can be made trivial in some nice extension of R. This phenomenon occurred in (1.9), where by passing to the integral closure of R, we add the element t^2s^2, and then write our nontrivial relation (over R) $(t^2s^6, -t^6s^2)$ as $t^2s^2(s^4, -t^4)$ (over S).

Unfortunately, or perhaps fortunately, else life would be too simple, it is not enough to pass to the integral closure to ensure relations on parameters become trivial, since not all integrally closed rings are C-M. The homogeneous coordinate ring of an elliptic curve cross $\mathbb{P}^1$ is a 3-dimensional integrally closed non C-M ring. Nonetheless one might naively hope that it is possible to trivialize relations on parameters of R in some finite extension S of R. There has been no known natural way to increase depth past 2 (one can obtain depth 2 by taking integral closure). A similar situation prevails for modules over normal rings, where by taking double duals one can pass to modules having depth 2, but no construction is known to 'make' a module have depth bigger than 2 if it does not already possess this property.

In characteristic 0 it is not in general possible to trivialize relations in finite extensions. Indeed, if R is an integrally closed local domain containing the rationals, and $\sum_{i=1}^k r_i x_i = 0$ is a nontrivial relation on parameters $x_1, \ldots, x_k$ of R, then this relation can *never* be made trivial by passing to a finite extension S of R: if such an S existed, then $r_k \in (x_1, \ldots, x_{k-1})S$. Let β denote the map $(1/[L:K])\mathrm{tr}_{L/K}$, where K

is the quotient field of R, L the quotient field of S, and tr( ) is field trace. Then β is an R-linear map from S to R, splitting the inclusion of R into S. In particular, $r_k = \beta(r_k) \in (\beta(x_1), \ldots, \beta(x_{k-1}))R = (x_1, \ldots, x_{k-1})R$, a contradiction. This impossibility in characteristic 0 makes the following main theorem (in characteristic p) somewhat miraculous.

Main Theorem 2.1 [H-H5]. *Let $(R, \mathfrak{m})$ be an excellent local domain of positive characteristic p, and let R^+ denote the integral closure of R in an algebraic closure of the quotient field K of R (called an* absolute integral closure *in [Ar]). If $x_1, \ldots, x_n$ is an s.o.p. of R, then $x_1, \ldots, x_n$ is a regular sequence in R^+, i.e., R^+ is a big C-M algebra for R.*

Here 'big' refers to the fact that R^+ is not a finitely generated R-module. In fact, since the quotient field of R^+ is algebraically closed, R^+ cannot even be noetherian (unless $R = K$ is a field). We will not give an explicit definition of "excellent" except to say all algebras essentially of finite type over a field and all complete local rings are excellent.

The meaning of (2.1) is that given *any* nontrivial relation on the parameters $x_1, \ldots, x_n$ of R, there will be a finite extension domain S of R in which the relation becomes trivial! Of course, passing to S may reap a harvest of new bad relations, but each one can again be trivialized in a further finite extension. While no one finite extension S may suffice to trivialize all relations (in such a case S would be C-M), by passing to the union of all finite extensions, namely R^+, we do obtain such trivialization.

The Frobenius map $F : R \to R$ sending r to r^p plays an important role in the proof of (2.1), as one might expect. To explain how the Frobenius is used I need to digress to talk about the unmixed part of an ideal.

Every ideal in a noetherian ring R has a primary decomposition. An ideal J is *unmixed* if all the primes associated to primary components of J are minimal over J, i.e., there are no embedded components. (One note for experts; one usually also demands that minimal components of J have the same dimension to call J unmixed, but in the context in which we will use this concept, this latter condition will be automatic.) C-M rings got their name from theorems of Macaulay (for polynomial rings) and Cohen (for general regular rings) which state that in a regular local ring R ideals generated by parameters are unmixed. Nagata showed that this unmixedness property characterizes C-M rings.

If J is an ideal of R, we let J^u be the unmixed part of J, i.e., J^u equals the intersection of the minimal primary components of J. If J is generated by parameters in a local ring R, and R is C-M, then $J = J^u$, so it is no surprise that (2.1) is equivalent to the following claim:

(2.2)   If R, R^+, are as in (2.1) and J is an ideal of R generated by parameters $x_1, \ldots, x_k$, then $J^u R^+ = JR^+$. In fact there is a finite extension domain S of R such that $J^u S = JS$. Note however that $(JS)^u$ need not be equal to JS.

The last part of (2.2) follows at once from the first, since J^u is finitely generated, and hence by collecting coefficients in R^+, one can get the extension S.

The first general problem confronting one in attacking this claim, without which one cannot even leave the ground, is this: given two ideals I, J in a ring R, how can one force J to be contained in I in some finite extension S of R? We want this only in the special case where I is generated by parameters and J is I^u, but this case is no easier than the general problem. Of course, one cannot arbitrarily force an ideal J into I in some finite extension; there are obstructions. In fact in characteristic p, if this occurs then J is in the *tight closure* of I. The tight closure of an ideal was introduced in [H-H1], and the study of tight closure was a key in proving (2.1), although the proof is independent of the theory of tight closure. The definition of tight closure is given below.

Definition 2.3. *Let R be a ring of characteristic $p > 0$, let I be an ideal of R, and let $x \in R$. We say that x is in the tight closure of I, denoted by I^*, if there exists an element c of R, not in any minimal prime of R, such that $cx^q \in I^{[q]} = \langle i^q : i \in R \rangle$ for all $q = p^e \gg 0$. Note that $I^{[q]}$ is the image of I under the e-th iteration of the Frobenius. (We also note there is a similar definition for the tight closure of a submodule N of a module M, though we do not give the definition here.)*

While this definition is technical, it has proved to be a powerful tool, and has been used by Hochster and myself (cf. [H-H1-10]) to give strong improvements and new proofs in equicharacteristic (especially simple in characteristic p) of many theorems, which are not apparently related: the intersection conjecture [P-S, Ho1, R1-2], the criterion for acyclicity of Buchsbaum and Eisenbud [B-E], the syzygy theorem [E-G], the monomial conjecture [Ho1], the Briançon-Skoda theorem [B-S], [L-S], and the Hochster-Roberts theorem mentioned in (1.7) [H-R, K, Bo]. Many other results also ensue from the theory.

We return to the general problem of finite extensions. Although it is true that if J is forced into I in a finite extension then $J \subseteq I^*$, the starting point for us was that we knew this to be true in the case where I is generated by parameters and $J = I^u$ [H-H1, 4.7]. Being in the tight closure gives strong equational constraints upon the elements of J in terms of the elements of I. However, as far as we know it is not necessarily true that if I is generated by parameters in a complete local domain R, then $I^*R^+ = IR^+$, although this is true for parameters of length at most three and is probably true. A careful analysis of the equations for tight closure does lead, nonetheless, to the following lemma which gives a sufficient criterion for pushing J into the extension of I in a finite extension of R:

Equational Lemma 2.4. *Let R be a domain of char $p > 0$, and let I, J be ideals of R. Write $I = (x_1, \ldots, x_k)$. If*

$$J^{[p]} \subseteq I^{[p]}R^+ + (x_1 \ldots x_k)^{p-1}JR^+, \tag{#}$$

then there is a finite extension domain S of R such that $JS \subseteq IS$.

We would like to apply (2.4) in the case where I is generated by parameters and where $J = I^u$. Unfortunately, the condition (#) is quite a strong one and certainly will not be true for general parameters $x_1, \ldots, x_k$. The difficulty in proving (2.2) by using (2.4) is finding appropriate parameters (which we call *extraordinary*, cf. (2.7) below) which are special enough to satisfy ($\#$), with $J = I^u$, but general enough so that they exist, and in enough profusion to allow us to pass from arbitrary s.o.p.'s to extraordinary ones in the proof of (2.2). One thing that makes this hard is that whatever the definition of extraordinary might be, it will surely be lost under finite extensions, so one cannot carry through the entire proof using only special types of parameters. There is a delicate balance between the two necessary properties needed by these special parameters. However, *there is* an appropriate definition: we want s.o.p.'s which kill as much cohomology as possible. To describe this precisely, we need a couple more definitions.

Definition 2.5. *A parameter x of R is said to be a* CM multiplier *if for all parameters $x_1, \ldots, x_{k+1}$, x annihilates the quotient module $N = (x_1, \ldots, x_k) : x_{k+1}/(x_1, \ldots, x_k)$, where, in general, $I : J = \{r \in R : rJ \subseteq I\}$.*

The quotient module N of (2.5) can often be interpreted as some cohomology group of the ring. Notice that $N = 0$ if R is C-M. It is not clear that such x exist; the quantification over all parameters makes the existence of such x a problem. (For instance if we set $x_1 = x^n$ for large n then it is immediate that x^n annihilates N, but not at all clear that x will. Unless x is chosen quite carefully, $xN \neq 0$.) The next lemma, found in [H-H1, 11.5] gives the existence of many such elements in excellent rings.

Lemma 2.6. *Let $x \in R$ be such that R_x is C-M. Then there is a fixed power of x which is a CM multiplier.*

In an excellent domain R, the locus of the non Cohen-Macaulay primes of R is closed, and the defining radical ideal has height at least 2, since 1-dimensional domains are C-M. (2.6) guarantees that we can always find two parameters x, y in such a ring, each of which is a CM multiplier. This allows a flexibility which is crucial to the proof.

Definition 2.7. *Parameters $x_1, \ldots, x_k$ in a local ring R are said to be* extraordinary *if x_k is a CM multiplier, and the image of x_i is a CM multiplier in the ring $R_i = R/(x_1, \ldots, x_{i-1})^u$.*

One quick note – in the paper [H-H5], a slightly different requirement is given for x_k but our change here does not affect the proof in any essential way.

Sequences of extraordinary parameters do exist, in any length up to the dimension of the ring, in excellent local domains. Furthermore *any* s.o.p. in such a ring can be moved to an extraordinary s.o.p. by a sequence of the following three types of operations:

   i) Switching two of the parameters.
   ii) Replacing x_i by $x_i + rx_j$ for any $j \neq i$,
   iii) Replacing parameters $x_1, \ldots, x_k$ with $yx_1, x_2, \ldots, x_k$, where $y, x_2, \ldots, x_k$ are also parameters.

The third condition allows enormous change.

Suppose that we know by induction on k, in the context of (2.2), that any $k - 1$ parameters form a regular sequence in R^+. Then it turns out that any extraordinary parameters $x_1, \ldots, x_k$ satisfy the equational lemma (2.4) with $I = (x_1, \ldots, x_k)$ and $J = I^u$. In particular we can trivialize relations on the parameters $x_1, \ldots, x_k$; said otherwise, the map on Koszul homology, $H_1(\mathbf{x}; R) \to H_1(\mathbf{x}; R^+)$ is zero. If this latter property holds for $\mathbf{x}$, which is derived from $\mathbf{y}$ by repeated applications of i)–iii) above, then it also holds for $\mathbf{y}$. This gives us that relations on parameters can be trivialized in R^+, which finishes the sketch of the proof of (2.2) and thus (2.1).

3. Remarks

In this section we give a little background as well as some indications of how the main theorem can be applied. The existence of big C-M modules was shown by Hochster [Ho1] in equicharacteristic around 1975. He went on to show that the existence of such a module, even though not finitely generated, implied many of the so-called "local homological conjectures" in equicharacteristic. Most remain open in mixed characteristic, although the recent work of Paul Roberts [Rol-2], using the intersection theory of Fulton, MacPherson, and Baum, has resulted in the solution of many of the conjectures, even in mixed characteristic.

Hochster's construction of a big C-M module was universal in the sense that he showed if any Cohen-Macaulay module exists for a given local ring, then his construction must also necessarily give one. This very universality makes it difficult to understand the structure of the modules he constructs. Moreover although the module is canonically constructed from the ring the actual structure of the module is mysterious. It is difficult to interpret its structure in terms of the original ring. Hochster also gave a construction of a big C-M algebra, without identity, [Ho2]. Again it is somewhat universal in nature and hard to understand its structure. On the other hand the ring R^+ of (2.1) has a great deal of structure and provides a wealth of insight into why such modules even exist. Moreover its structure is attached to that of R: for instance the going up and going down theorems hold between R and R^+ in the case where R is integrally closed. Another example is provided in the remarks below concerning the structure of R^+: if R, a char. p excellent local domain, maps to S, also an excellent local domain of char. p, then the map extends (not naturally, however) to a map from R^+ to S^+. If the original map is injective (surjective) then the map from R^+ to S^+ can be chosen to be injective (surjective). In other words, the + construction is enough like a functor to be helpful.

Moreover although (2.2) is false in characteristic 0 one can show, by standard methods of reduction to characteristic p, there exist big Cohen-Macaulay algebras (with identity) in equicharacteristic 0. The existence of big C-M algebras (as opposed to big C-M modules) has recently been used by Ian Aberbach [Ab].

We next make some remarks, which we collect in a lemma, concerning some of what we know about the structure of R^+.

Proposition 3.1. *Let R be an excellent local domain, and let R^+ be as in Section 2. Then,*

i) *the sum of any two prime ideals of R^+ is again a prime ideal, or is the whole ring (cf. [Ar]);*

ii) *if $\mathfrak{p} \in \mathrm{Spec}(R)$, and $P \in \mathrm{Spec}(R^+)$ lies over $\mathfrak{p}$, then $(R/\mathfrak{p})^+ \simeq R^+/P$;*

iii) *If W is a multiplicative system in R, then $(W^{-1}R)^+ \simeq W^{-1}(R^+)$.*

Thus R^+ is, both locally and globally, a C-M algebra, even modulo any prime ideal! Part ii) also explains the comments above concerning R^+ mapping onto S^+ when S is a domain which is a homomorphic image of R; simply write $S = R/\mathfrak{p}$, and use ii).

It is interesting to see what (2.1) says when one restricts to case 1 of Example 1.10. In this case R is a complete local domain, finite over a power series subring A generated (as a complete subring of R) by any s.o.p. $x_1, \ldots, x_n$. Since R is finite over A, $R^+ = A^+$. Any given s.o.p. in R can be used to define such an A, so (2.1) says simply that when A is a power series ring over a field of characteristic p, A^+ is a C-M A-algebra. Then it is not difficult to see that Theorem 2.1 is equivalent to the following theorem:

Theorem 3.2. *Let A be a power series ring over a field of characteristic p, and let A^+ be the integral closure of A in an algebraic closure of its quotient field. Then A^+ is flat over A.*

Lest this theorem sound too simple, recall that it is wildly false for dimension ≥ 3 in characteristic 0.

The flatness of A^+ over A has strong consequences concerning the syzygies of certain finitely generated modules M over complete local non C-M domains R. Suppose that the R-module M comes via base change from a module N over a power series ring A contained in R over which R is finite (i.e., $M \cong N \otimes_A R$). The module N has a finite resolution of the form,

$$\mathbb{F} : 0 \to F_d \to F_{d-1} \to \cdots \to F_1 \to F_0 \to N \to 0,$$

where F_i is a finitely generated free A-module. When we tensor with R to obtain M, the sequence is no longer exact; the homology comes exactly from the failure of R to be C-M. (Recall from 1.10 that R is C-M iff R is free as an A-module.) However since $R^+ = A^+$ is flat over A, when we tensor $\mathbb{F} \otimes_A R$ with R^+ the sequence becomes exact. Thus, e.g., the relations on the generators of M over R, while not coming from the original relations of N over A, *are* linear combinations of the original relations over some finite extension of R. This has profound consequences, especially in the graded case, which we discuss below.

4. The Graded Case

We close this paper by making a few remarks on the graded case. Let R be a domain of char. $p > 0$ and as in case 2 of 1.10. We are able to prove a graded version of

Theorem 2.1 by replacing R^+ by a subring R^{+gr} of R^+ which contains R, and is $\mathbb{N}$-graded extending the grading on R. In this case (2.1) becomes the following theorem.

Theorem 4.1. *Any s.o.p. in R consisting of homogeneous elements is a regular sequence in R^{+gr}.*

If A is a graded polynomial subring of R as in case 2 of 1.10 then $A^{+gr} = R^{+gr}$ is free over A, and the closing remarks of Section 3 apply in this case as well for graded modules and resolutions. Theorem 4.1 can be translated into a vanishing theorem on maps of cohomology of projective varieties as follows.

Theorem 4.2. *Let X be a closed, irreducible, reduced, d-dimensional subscheme of $\mathbb{P}^n_k$, where k is a field of characteristic $p > 0$. Then for all i, $1 \leq i \leq d - 1$, and all integers t, there is a finite surjective morphism $f_{i,t} = f : Y \to X$ from a projective scheme Y, also reduced and irreducible, over a finite extension k' of k such that the induced map $H^i(X, O_X(t)) \to H^i(Y, f^*O_X(t))$ is zero.*

A similar result holds without assuming X is irreducible and reduced; one only needs to assume that X is unmixed, and in this case the corresponding projective scheme Y can be chosen to be unmixed.

One very special case of this result, which already has many consequences is that in any reduced graded ring R, any s.o.p. $x_1, \ldots, x_d$ generated by one forms is "C-M" up to degree 2, i.e., $\sum r_i x_i = 0$, with r_i homogeneous, then up to a trivial syzygy, $\deg(r_i) \geq 2$.

For many problems, where a ring R is assumed to be C-M, what is really important is not that R is C-M, but rather that a given finite set of relations can be trivialized in the ring. Moreover proof analysis often shows that it suffices that the relations be trivialized only is some finite extension of the ring; a good example of this is a connectedness theorem of Hartshorne which says that if R is C-M (or even has Serre's property S_2) then R is connected in codimension 1. An analysis of this proof, in light of (2.1), shows that in characteristic p (and hence also equi-characteristic 0 by reduction to equicharacteristic p) that the assumptions on depth are not really needed, and allow one to prove the following generalization: if R is complete local domain of dimension d and $x_1, \ldots, x_k$ are elements of R with $k \leq d - 2$, then the punctured spectrum of $R/(x_1, \ldots, x_k)$ is connected. This was already known and is a result of Faltings. The point is that (2.1) provides considerable insight into what might be true and provides the ammunition to carry proofs through. It seems that significant insight can be gained if one follows the intuition that "all" domains in characteristic p are C-M; while this is false, passing to R^+ will recapture the C-M property.

Acknowledgement. I'd like to thank Mel Hochster and Bill Heinzer for reading a preliminary version of this manuscript and making valuable suggestions.

References

[Ab] Aberbach, I.: *Finite phantom projective dimension and a phantom analogue of the Auslander-Buchsbaum theorem.* Thesis, Univ. of Michigan, 1990

[Ar] Artin, M.: *On the joins of Hensel rings.* Adv. Math. **7** (1971) 282–296

[Bo] Boutot, J.-F.: *Singularitiés rationelles et quotients par les groupes reductifs.* Invent. Math. **88** (1987) 65–68

[B-R] Briançon, J., Skoda, H.: *Sur la cloture integrale d'un ideal de germes de fonctions holomorphes en un point de* **C**ⁿ. C. R. Acad. Sci. Paris Ser. A **278** (1974) pp. 949–951

[E-G] Evans, G., Griffith, P.: *The syzygy problem.* Annals of Math. **114** (1981) 323–333

[F] Faltings, G.: *A contribution to the theory of meromorphic functions.* Nagoya Math. J. **77** (1980) 99–106

[Ha] Hartshorne, R.: *Complete intersections and connectedness.* Amer. J. Math. **84** (1962) 497–508

[Ho1] Hochster, M.: Topics in the homological theory of modules over commutative rings. C.B.M.S. Regional Conf. Ser. in Math. No. 24 A.M.S., Providence, RI 1975

[Ho2] Hochster, M.: *Big Cohen-Macaulay modules and algebras and embeddability in rings of Witt vectors.* Proceedings of the Queen's Univ. Commutative Algebra Conference, Queen's Papers in Pure Appl. Math. **42** (1975) 106–195

[H-H1] Hochster, M., Huneke, C.: *Tight closure, invariant theory, and the Briançon-Skoda theorem.* Journal of the Amer. Math. Soc. **3** (1990) 31–116

[H-H2] Hochster, M., Huneke, C.: *Phantom homology.* Preprint 1990

[H-H3] Hochster, M., Huneke, C.: *F-regularity, test elements and smooth base change.* In preparation

[H-H4] Hochster, M., Huneke, C.: *Tight closure of parameter ideals and splitting in module-finite extensions.* In preparation

[H-H5] Hochster, M., Huneke, C.: *Infinite integral extensions and big Cohen-Macaulay algebras.* Ann. Math. (to appear)

[H-H6] Hochster, M., Huneke, C.: *Applications of the Cohen-Macaulay property for absolute integral closures.* In preparation

[H-H7] Hochster, M., Huneke, C.: *Tightly closed ideals.* Bull. Amer. Math. Soc. **18** (1988) 45–48

[H-H8] Hochster, M., Huneke, C.: *Tight Closure.* Commutative Algebra, Math. Sci. Res. Inst. Publ., no. 15, Springer, New York, Berlin, Heidelberg, London, Paris, Tokyo 1989, pp. 305–324

[H-H9] Hochster, M., Huneke, C.: *Tight closure and strong F-regularity.* Soc. Math. France Memoire **38** (1989) 119–133

[H-H10] Hochster, M., Huneke, C.: *Absolute integral closures are big Cohen-Macaulay algebras in characteristic* p. Bull. Amer. Math. Soc. **24** (1991) 137–143

[H-R] Hochster, M., Roberts, J.: *Rings of invariants of reductive groups acting on regular rings are Cohen-Macaulay.* Adv. Math. **13** (1974) 115–175

[K] Kempf, G., *The Hochster-Roberts theorem of invariant theory.* Michigan Math. J. **26** (1979) 19–32

[L-S] Lipman, J., Sathaye, A.: *Jacobian ideals and a theorem of Briançon-Skoda,* Michigan Math. J. 28 (1981) 199–222

[P-S] Peskine, C., Szpiro, L.: *Dimension projective finie et cohomologie locale.* I.H.E.S. Publ. Math. **42** (1973) 323–395

[Ro1] Roberts, P.: *The vanishing of intersection multiplicities of perfect complexes.* Bull. Amer. Math. Soc. **13** (1985) 127–130

[Ro2] Roberts, P.: *Le théorème d'intersection.* C. R. Acad. Sc. Paris Ser. I, **304** (1987) 177–180

Identities of Associative Algebras

Alexander R. Kemer

Ulyanovsk Branch of Moscow State University, Str. Tolstogo 42, Ulyanovsk 432700, USSR

Preliminaries

Let F be field, $F\langle X\rangle$ the free associative algebra over F generated by a countable set of variables X. One may consider an element from $F\langle X\rangle$ as a polynomial in non-commutative variables from X.

Definition 1. We shall say that an associative algebra A over F satisfies an identity $f(x_1, \ldots, x_n) = 0$, where $f = f(x_1, \ldots, x_n) \in F\langle X\rangle$, if for arbitrary elements $a_i \in A$ the equality $f(a_1, \ldots, a_n) = 0$ is always valid in A.

Definition 2. The set of all polynomials $f \in F\langle X\rangle$ such that A satisfies the identity $f = 0$ is said to be the ideal of identities of the algebra A. We shall denote this ideal by $T[A]$. An ideal of $F\langle X\rangle$ which is an ideal of identities of some algebra is called a T-ideal.

Definition 3. An Algebra A is said to be PI-algebra if it satisfies a non-trivial identity (i.e. $T[A] \neq 0$).

The structure theory for PI-algebras is well developed. Some results of this theory are classic now. One of them is Kaplansky's theorem which asserts that a primitive PI-algebra is finite dimensional over its centre. Another example is the theorem of Nagata-Higman which asserts that any algebra over a field of zero characteristic satisfying identity $x^n = 0$ is nilpotent.

In 1957 A.I. Shirshov proved his famous Height Theorem:

Theorem (A.I. Shirshov [1]). *For any finitely generated PI-algebra A there exist a number h and elements $a_1, \ldots, a_n \in A$ such that elements*

$$a_{i_1}^{\alpha_1} \cdot \ldots \cdot a_{i_k}^{\alpha_k}, \qquad k < h$$

generate A as a space.

The Height Theorem implies the positive solution of Kurosh's problem for PI-algebras: A finitely generated algebraic PI-algebra is finite dimensional.

Proceedings of the International Congress
of Mathematicians, Kyoto, Japan, 1990
© The Mathematical Society of Japan, 1991

Although Kurosh's problem for PI-algebras was solved earlier in 1948 by I. Kaplansky, Shirshov's approach shows clearly the contributions of both conditions (PI and algebraicity) in order that algebra be finite dimensional. The Height Theorem shows also that finitely generated PI-algebras are close to finite dimensional algebras.

Shirshov's Height Theorem was the first result which gave some information about identities of PI-algebras.

This paper is also devoted to the study of the structure of identities of associative algebras.

1. Finitely Generated PI-Algebras

First of all we recall the famous result in PI-theory which was proved by A. Braun in 1982.

Theorem (A. Braun [2]). *The radical of a finitely generated PI-algebra is nilpotent.*

This theorem has the following important corollary:

Corollary. *A finitely generated PI-algebra satisfies for some n the identity*

$$\sum_{\sigma \in S(n)} (-1)^\sigma x_{\sigma(1)} y_1 x_{\sigma(2)} y_2 \cdot \ldots \cdot y_{n-1} x_{\sigma(n)} = 0 \tag{1}$$

where $S(n)$ is the symmetric group of degree n; $(-1)^\sigma = 1$ if σ is an even permutation, $(-1)^\sigma = -1$ if σ is odd.

The identity (1) is called the Capelli identity of n-th order.

It is well known that if char $F = 0$ then the Capelli identities "distinguish" finitely generated PI-algebras from infinitely generated algebras.

Theorem. *Let an associative PI-algebra A over a field of zero characteristic satisfy the Capelli identity of some order. Then there exists a finitely generated PI-algebra B such that $T[A] = T[B]$.*

If char $F \neq 0$ then the conclusion of this theorem is not true. We shall discuss this case later.

The main result about identities of finitely generated PI-algebras is the following theorem which was proved by the author in 1988.

Theorem 1 ([3, 4]). *For any finitely generated PI-algebra A over an infinite field F there exists a finite dimensional F-algebra C such that $T[A] = T[C]$.*

In other words finitely generated PI-algebras cannot be distinguished from finite dimensional algebras in the language of identities.

If F is a finite field then the conclusion of Theorem 1 is not valid, but Theorem 1 gives a full information about identities which are homogeneous with respect to every variables.

One may consider Theorem 1 as a theorem on classification of finitely generated PI-algebras in the language of identities.

Definition 4. Let Γ be a T-ideal in the free algebra $F\langle X\rangle$. The factor algebra $F\langle X\rangle/\Gamma$ is said to be a relatively free algebra generated by the set of variables X.

Definition 5. An algebra over a field F is called representable if it can be embedded into a matrix algebra of finite order over some extension of the basic field F.

Theorem 1 has the following important corollary:

Corollary. *Any relatively free PI-algebra of finite rank is representable* (*the basic field is infinite*).

Since the proof of Theorem 1 has been published we give the main ideas of this proof.

Let A be a finitely generated PI-algebra. Using the theorem of Braun and Levin's theorem it is easy to prove that

$$T[A] \supseteq T[C]$$

for some finite dimensional algebra C. That is the initial situation. Then we want to find an intermediary finite dimensional algebra C_1 such that

$$T[A] \supseteq T[C_1] \supsetneqq T[C].$$

Then we want to find a finite dimensional algebra C_2 such that

$$T[A] \supseteq T[C_2] \supsetneqq T[C_1]$$

and so on.

If we shall follow this way then two problems arise.

1) How can one find those intermediary algebras?
2) Does every ascending chain of ideals of identities of finite dimensional algebras terminate?

We shall not discuss the first problem because it is the most technical part of the proof. We construct those algebras using the so-called trace identities.

The solution of the second problem is also technical. The main idea is to translate structure properties of a finite dimensional algebra into the language of identities.

We assume for simplicity that the basic field is algebraically closed. Consider a finite dimensional algebra C with a unit. The algebra C may be represented in the form

$$C = P + \operatorname{Rad} C$$

where P is the semisimple part of C. The algebra P is a direct sum of full matrix algebras $M_{n_i}(F)$. We define the following parametres of the algebra C:

$$\alpha(C) = \sum n_i$$

$$\beta(C) = \dim_F P$$

$$\gamma(C) = \text{the index of nilpotency of Rad } C.$$

The triple

$$t(C) = (\alpha(C), \beta(C), \gamma(C))$$

is called the type of the algebra C. We order the types lexicographically.

How do we translate the properties of the algebra C into the language of identities? For example, the property $\alpha(C) = n$ is translated into the following identity

$$\sum_{\sigma_1,\ldots,\sigma_k \in S(n+1)} (-1)^{\sigma_1} \cdot \ldots \cdot (-1)^{\sigma_k} y^{\sigma_1(1)+\cdots+\sigma_k(1)-k} x_1 y^{\sigma_1(2)+\cdots+\sigma_k(2)-k}$$

$$\cdots \cdot x_n y^{\sigma_1(n+1)+\cdots+\sigma_k(n+1)-k} = 0 \qquad (2)$$

where $k = \gamma(C)$.

If $T[C'] \supseteq T[C]$ for another finite dimensional algebra C', then it is proved that there exist finite dimensional algebras $C_1, \ldots, C_m$ such that

$$\alpha(C_i) \leq \alpha(C), \qquad T[C'] = T\left[\bigoplus_{i=1}^{m} C_i\right].$$

Other identities translate some other properties. Finally we can prove the following:

Proposition. *Let* C, C' *be finite dimensional algebras over an infinite field,* $T[C'] \supseteq T[C]$. *Then there exist finite dimensional algebras* $C_1, \ldots, C_m$ *such that*

$$T\left[\bigoplus_{i=1}^{m} C_i\right] = T[C'].$$

This proposition is principal for the solution of the second problem.

2. Infinitely Generated PI-Algebras

The methods of studying the identities of infinitely generated algebras highly depend on the characteristic of the basic field.

2.1 Algebras over Field of Zero Characteristic

Definition 6. Let A be an associative algebra, A_0, A_1 subspaces of A such that

$$A = A_0 \oplus A_1$$

$$A_0 A_0, A_1 A_1 \subseteq A_0; \qquad A_0 A_1, A_1 A_0 \subseteq A_1.$$

Then the algebra $A = A_0 \oplus A_1$ is said to be the Z_2-graded algebra or superalgebra graded by A_0, A_1.

Consider the Grassmann algebra G generated by elements $e_1, e_2, e_3, \ldots$ satisfying the relations $e_i e_j = -e_j e_i$ for every i, j. Let G_0 be the subspace of G generated by monomials of even degree, G_1 the subspace generated by monomials of odd degree. Then $G = G_0 \oplus G_1$ is the Grassmann superalgebra.

Definition 7. The subalgebra $G(A) = G_0 \otimes_F A_0 + G_1 \otimes A_1$ of the algebra $G \otimes_F A$ is called the Grassman Hull of the superalgebra $A = A_0 \oplus A_1$.

In 1981 the author proved that any T-ideal equals the ideal of identities of the Grassmann Hull of some finitely generated PI-superalgebra. This theorem has reduced the studying of identities of infinitely generated PI-algebras to the studying of graded identities of finitely generated PI-superalgebras. We recall that the characteristic of the basic field equals zero. Theorem 1 formulated above is valid also in the graded case ([3]). Hence we obtain the following main result about the identities of PI-algebras over the field of zero characteristic:

Theorem 2 [3]. *For any PI-algebra A there exists a finite dimensional superalgebra C such that*

$$T[A] = T[G(C)].$$

Theorem 2 has an interesting corollary:

Corollary. *A relatively free PI-algebra can be embedded into the matrix algebra of finite order over an algebra satisfying the identity $[[x, y], z] = 0$ ($[x, y] = xy - yx$).*

2.2 Specht's Problem

We want to classify associative PI-algebras in the language of identities. Theorem 2 is a satisfactory theorem of classification for algebras of zero characteristic. In the general case we do not have any hypothesis about the classification.

Definition 8. The minimal set of identities of an algebra A which implies all the other identities of A is said to be the base of identities of the algebra A.

In 1950 W. Specht formulated the following problem [5]: Has any associative algebra of zero characteristic a finite base of identities?

The problem of finite base may be formulated not only for algebras of zero characteristic but also for arbitrary algebras, rings, groups and so on. The problem of finite base may be considered as a strict formulation for problem of classification (if there is no more satisfactory hypothesis).

In 1986 the author [6] solved Specht's Problem positively:

Theorem 3. *Any associative algebra over a field of zero characteristic has a finite base of identities.*

Theorem 3 has other formulations:

1) Any T-ideal is finitely generated as a T-ideal.
2) The set of T-ideals satisfies the ascending chain condition.

In the case of non-zero characteristic the problem of finite base is open, but Theorem 1 yields the positive solution of local Specht's problem for algebras over an infinite field.

Theorem 4. *Any* T-*ideal of a finitely generated free algebra is generated by a finite set of polynomials (as a* T-*ideal).*

2.3 Identities of Algebras over a Field of Characteristic p

In the case of characteristic p Theorem 2 is not true. Moreover the author has no classification hypothesis. The structure of identities of algebras over fields of characteristic p is not clear and there are no strong results on this theme.

In 1981 I.B. Volichenko formulated the strange problem: Does any PI-algebra over a field of non-zero characteristic satisfy the standard identity

$$\sum_{\sigma \in S(n)} (-1)^{\sigma} x_{\sigma(1)} \cdot \ldots \cdot x_{\sigma(n)} = 0$$

for some n?

It is not difficult to prove that Volichenko's hypothesis is true if and only if any PI-algebra of non-zero characteristic satisfies the symmetrized standard identity

$$\sum_{\sigma \in S(n)} x_{\sigma(1)} \cdot \ldots \cdot x_{\sigma(n)} = 0$$

for some n.

Indeed, let A be any PI-algebra over a field F and G a Grassmann algebra over F. It is well-known that the algebra $A \otimes_F G$ is a PI-algebra. If $A \otimes_F G$ satisfies the standard identity (symmetrized standard identity) of some degree then it is easy to see that the algebra A satisfies the symmetrized standard identity (standard identity) of the same degree.

Note that the symmetrized standard identity of n-th degree is a full linearization of the identity $x^n = 0$.

The author has solved these problems recently.

Theorem 5. *Any PI-algebra over a field of non-zero characteristic satisfies the standard identity and the symmetrized standard identity of some degree.*

Proof. First of all we remark that it is enough to prove the theorem for algebras with unit over algebraically closed field.

Let Γ be the T-ideal of identities of the given algebra with unit $\Gamma \subseteq F\langle X \rangle$, $X = \{x_1, x_2, \ldots\}$. Consider the finitely generated PI-algebras $F_k / \Gamma \cap F_k$, where $F_k = F\langle x_1, \ldots, x_k \rangle$. By Theorem 1

$$T[F_k / \Gamma \cap F_k] = T[A_k]$$

for some finite dimensional algebra A_k.

The primitive images of the algebra $F\langle X\rangle/\Gamma$ are full matrix algebras over the basic field F. The maximal order of these matrix algebras is called the complexity of the T-ideal Γ.

We represent the algebra A_k in the form

$$A_k = P_k + \mathrm{Rad}\, A_k,$$

where P_k is the semisimple part of A_k. The algebra P_k may be represented in the form

$$P_k = e_0 P_k + e_1 P_k + \cdots + e_s P_k, \qquad s = s(k),$$

where $e_0, e_1, \ldots, e_s$ are orthogonal idempotents, $e_i P_k = M_q(F)$, $i > 0$ (q the complexity of Γ); $e_0 P_k$ a direct sum of other matrix algebras $M_n(F)$, $n < q$.

We prove the theorem by induction. The base of induction ($q = 1$) and the inductive step will be proved simultaneously. Consider the ideal I_k of the algebra A_k generated by all mixed elements $e_i a e_j$, $i \neq j$, $a \in A_k$. Note that it is sufficient to prove that the algebras I_k satisfy the symmetrized standard identity of some degree $n = n(\Gamma)$.

Indeed, if $q = 1$ we put $k = 4n$. The algebra A_k/I_k satisfies the identity

$$\underbrace{[x, y, \ldots, y]}_{p^m} = 0 \tag{3}$$

for some m. Therefore the algebra A_k satisfies the identities

$$S_n^+(z_1[x_1, y_1, \ldots, y_1]t_1, \ldots, z_n[x_n, y_n, \ldots, y_n]t_n) = 0 \tag{4}$$

(S_n^+ is the symmetrized standard polynomial), where z_i, $t_j \in X \cup \{1\}$. Since the identities (4) contain at most k variables, these identities are valid modulo Γ. Hence we obtain that the algebra $F\langle X\rangle/\Gamma$ is an extension of some algebra satisfying (3) by an algebra satisfying the symmetrized standard identity of degree n. It is easy to verify that the full linearization of (3) is the symmetrized standard identity of degree p^m, hence the algebra $F\langle X\rangle/\Gamma$ satisfies the symmetrized standard identity of degree $p^m \cdot n$.

If $q > 1$ then we may assume that the theorem is proved in the case when the complexity is less than q.

Let B be an algebra such that $T[M_q(B)] \supseteq \Gamma$. Since q is the complexity of Γ the complexity of the T-ideal $T[B]$ is equal to 1. Hence by the inductive hypothesis the algebra B must satisfy the symmetrized standard identity of some degree. It is easy to prove that then the algebra $M_q(B)$ must satisfy the symmetrized standard identity of some degree $m = m(\Gamma)$.

Let $h(x_1, \ldots, x_r)$ be a polynomial such that for all s the algebra $M_q(F)$ does not satisfy the identity $h^s = 0$, but the algebra $M_{q-1}(F)$ satisfies the identity $h^s = 0$ for some s.

We put $k = (m + r + 3)n$. Consider the algebra A_k/I_k. This algebra is a direct sum of algebras of the type $M_d(B)$, where B is a local algebra, $d \leq q$. Therefore the algebra A_k/I_k must satisfy the identities

$$f(x_1, \ldots, x_m, y_1, \ldots, y_r, u, v, w) = u S_m^+(x_1, \ldots, x_m) v (h(y_1, \ldots, y_r))^t w = 0 \tag{5}$$

for some m, t, where $u, v, w \in X \cup \{1\}$. Hence we obtain that the algebra A_k satisfies the identities

$$S_n^+(f^{(1)}, \ldots, f^{(n)}) = 0, \tag{6}$$

where $f^{(i)} = f(x_1^{(i)}, \ldots, x_m^{(i)}, y_1^{(i)}, \ldots, y_r^{(i)}, u^{(i)}, v^{(i)}, w^{(i)})$, $u^{(i)}, v^{(i)}, w^{(i)} \in X \cup \{1\}$. Since the identities (6) contain at most k variables these identities are valid modulo Γ. Therefore the algebra $F\langle X \rangle / \Gamma$ is an extension of some algebra satisfying (5) by an algebra satisfying the symmetrized standard identity of degree n. By the inductive hypothesis the identity $h^t = 0$ implies the symmetrized standard identity of some degree n_0. Hence we obtain that the algebra $F\langle X \rangle / \Gamma$ satisfies the symmetrized standard identity of degree $(n_0 + m) \cdot n$.

Let us prove that the algebra I_k satisfies the symmetrized standard identity of some degree $n = n(\Gamma)$.

Let $h(x_1, \ldots, x_r)$ be a central polynomial for the algebra $M_q(F)$. Then the algebra $F\langle X \rangle / \Gamma$ satisfies all identities of the type

$$[h(u_1, \ldots, u_r), v] x^{i_1} [h(u_1, \ldots, u_r), v] x^{i_2} \cdot \ldots \cdot x^{i_{m-1}} [h(u_1, \ldots, u_r), v] = 0. \tag{7}$$

Since $T[A_k] \supseteq \Gamma$, the algebra A_k also satisfies the identities (7). If $m > s = s(k)$ then we substitute into the identity (7) the following elements of A_k: $x = 1$, $v = \sum_{i=1}^{m} e_i a_i e_{i+1}$, $u_i = b_i$, where b_i are such elements that $h(b_1, \ldots, b_r) = \sum \alpha_i e_i$, $\alpha_i \neq \alpha_j$, $\alpha_i \in F$ (these elements exist because h is a central polynomial for $M_q(F)$). As a result we get the equality $v^m = 0$. Hence we obtain

$$0 = e_1 v^m = e_1 a_1 e_2 a_2 \ldots e_m a_m e_{m+1}.$$

It means that we may assume that $s \leq m$.

If we substitute into (7) $x = b \in I_k$, $v = e_i a e_j$, $i \neq j$, $u_t = b_t$, where b_t are such elements that $h(b_1, \ldots, b_r) = e_j$, then we get

$$v a^{i_1} v a^{i_2} \ldots a^{i_{m-1}} v = 0.$$

Linearizing this equality we can see that if we substitute into the polynomial $S_n^+(x_1, \ldots, x_n)$ a lot of elements of mixed type $e_i a e_j$, $i \neq j$, then we annihilate this polynomial. Therefore it is sufficient to prove that for all $i, j, i \neq j$, the algebra $B_k = e_i A_k e_j A_k e_i$ satisfies the symmetrized standard identity of some degree $n = n(\Gamma)$.

Put $i_1 = i_3 = i_5 = \cdots = 0$ in (7). Substituting into (7) $v = y + z$, $y \in e_i A_k e_j$, $z \in e_j A_k e_i$; $x \in e_i A_k e_i$, $h(v_1, \ldots, v_r) = e_i$, we get the equality

$$(z - y)^2 x^{j_1} (z - y)^2 x^{j_2} \ldots x^{j_{t-1}} (z - y)^2 = 0$$

where $j_d = i_{2d}$, $t = [(m + 1)/2]$. Multiplying this equality by e_i from the left side we get the identity

$$x^{j_0} yzx^{j_1} yz \ldots x^{j_{t-1}} yzx^{j_t} = 0 \tag{8}$$

for all $y \in e_i A_k e_j$, $z \in e_j A_k e_i$, $x \in e_i A_k e_i$.

Substituting $y = y + y_1$ into (8) and taking the component of degree 1 by y_1, we get the identity

$$x^{j_0} y_1 zx^{j_1} yz \cdots + x^{j_0} yzx^{j_1} y_1 z \cdots + \cdots = 0. \tag{9}$$

Substitute $y_1 = xy_1$ into (9) and subtract the result from (9) where j_0 is replaced by $j_0 + 1$. We get the identity of the type

$$x^{j_0}yzx^{j_1}(\alpha_1 y_1 zx^{j_2}yz \cdots + \alpha_2 yzx^{j_2}y_1 z \cdots + \cdots) = 0, \tag{10}$$

$\alpha_i \in F$. Then we again substitute $y_1 = xy_1$ into (10) and subtract the result from (10) where j_1 is replaced by $j_1 + 1$. Continue the described process. Finally we get a non-trivial identity of the type

$$\sum_{(j)=(j_0, j_1, \ldots, j_t)} \alpha_{(j)} x^{j_0}yzx^{j_1} \cdot \ldots \cdot yzx^{j_{t-1}}y_1 zx^{j_t} = 0. \tag{11}$$

We repeat this process again t times and then repeat this process with respect to z. Finally we get the identity of the type

$$\sum_{(j)=(j_0, \ldots, j_t)} \beta_{(j)} x^{j_0}y_t z_t x^{j_1} \cdot \ldots \cdot x^{j_{t-1}}y_1 z_1 x^{j_t} = 0$$

for all $x \in e_i A_k e_i$, $y \in e_i A_k e_j$, $z \in e_j A_k e_i$. Hence we obtain that the algebra B_k satisfies the non-trivial identity of the type

$$\sum_{(j)} \beta_{(j)} x^{j_0}x_1 x^{j_1}x_2 \cdot \ldots \cdot x_t x^{j_t} = 0 \tag{12}$$

for all $x, x_i \in B_k$.

So we may assume that the algebra $F\langle X \rangle / \Gamma$ satisfies the identity (12). Therefore the algebra A_k satisfies the identity (12). Substituting $x = a + e_j$, $a \in B_k$, $x_1 = e_i b_1 e_j$, $x_2 = e_j b_2 e_i$, $x_i \in B_k$, $i \geq 3$, we see that the algebra B_k satisfies the identity of the type (12) where t is replaced by $t - 1$. $\qquad\square$

References

1. Shirshov, A.I.: On rings with identical relations. Mat. Sbornik **43** (1957) 277–283
2. Braun, A.: The radical in finitely generated PI-algebra. Bull. Amer. Math. Soc. **7** (1982) 385–386
3. Kemer, A.R.: Representability of relatively free algebras. Algebra i Logika **27** (1988) 274–294
4. Kemer, A.R.: Identities of finitely generated algebras over infinite field. Izv. Acad. Nauk SSSR. Ser. Mat. **54** (1990) 726–753
5. Specht, W.: Gesetze in Ringen. 1. Math. Z. **52** (1951) 557–589
6. Kemer, A.R.: Finite base of identities of associative algebras. Algebra i Logika **26** (1987) 597–641

Intersection Theory and the Homological Conjectures in Commutative Algebra

Paul C. Roberts

Department of Mathematics, University of Utah, Salt Lake City, UT 84112, USA

The subject of this article is a set of conjectures which have been a central topic of investigation in Commutative Algebra for several years. Many of these conjectures are closely related to questions in Intersection Theory, and in some cases they arose directly from attempts to define intersection multiplicities in an algebraic setting. We begin with a discussion of some of these questions in Intersection Theory and their influence in Commutative Algebra. Next we discuss a few of the main conjectures related to this topic. Finally, we show how recent developments in Intersection Theory have made it possible to settle some of these problems.

1. Introduction: Serre's Definition of Intersection Multiplicities

We begin this section with some background on the problem of defining intersection multiplicities. The general question is as follows: given a variety V, a point p of V, and subvarieties X and Y of V such that p is an isolated point of the intersection $X \cap Y$, we wish to to define the multiplicity of the intersection of X and Y at the point p. We are interested here in an algebraic definition (which does not involve topology, for instance), and in one which is defined locally at the point.

As an illustration, we consider the case in which X and Y are curves in the plane. Let A be the local ring at the point p; that is, the ring of rational functions defined in a neighborhood of p. Let the curves X and Y be defined locally near p as the sets of zeros of polynomials $f(x, y)$ and $g(x, y)$. Then the correct definition for the multiplicity is the length of the quotient ring $(A/(f, g))$, where (f, g) is the ideal generated by f and g (or, in the complex case, the dimension of $A/(f, g)$ as a vector space over $\mathbb{C}$). This definition agrees with intuition; if the curves are smooth and not tangent at p it gives multiplicity one, and if they are tangent it is greater than one. In addition, it satisfies Bézout's Theorem, which states that if V is the projective plane, and if X and Y are subvarieties of V of degrees m and n respectively, then the total number of intersections, counted with multiplicities, is mn. This theorem is very important in enumerative geometry.

Before proceeding, we recall some basic definitions and notation. Let A be a commutative Noetherian local ring with maximal ideal $\mathfrak{m}$ and residue field $A/\mathfrak{m} = k$. Let $\dim(A) = d$ denote the Krull dimension of A; this can be defined

Proceedings of the International Congress
of Mathematicians, Kyoto, Japan, 1990
© The Mathematical Society of Japan, 1991

either as the maximum length of a chain of prime ideals or as the minimum number d such that there are elements $x_1, \ldots x_d \in \mathfrak{m}$ which generate an ideal primary to the maximal ideal. Such a sequence of elements is called a *system of parameters*. The ring A is *regular* if the maximal ideal can be generated by d elements; in the geometric situation, the local ring at a point is regular when the variety is smooth at the point. The ring is *Cohen-Macaulay* when the system of parameters form a regular sequence; that is, when $ax_i \in (x_1, \ldots, x_{i-1})$ implies $a \in (x_1, \ldots, x_{i-1})$ for each i. We remark that the above definition of multiplicities for curves works because A is Cohen-Macaulay and (f, g) is a system of parameters in that case.

If A is an integral domain, we say that A is *equicharacteristic* if the characteristic of A is the same as the characteristic of k; otherwise A has *mixed characteristic*.

We now return to the question of defining intersection multiplicities. Let A be a local ring, and let I and J be ideals of A; in the geometric situation, I and J will be the ideals of functions vanishing on subvarieties X and Y. The straightforward generalization of the definition for curves would be to define the multiplicity to be the length of the quotient $A/(I + J)$. However, this will not satisfy Bézout's Theorem (among other things).

The idea of Serre was to add extra correction terms. In addition, it is more convenient to generalize from the quotients A/I and A/J of the ring A to more general finitely generated modules M and N.

Definition (Serre [28]). Let A be a regular local ring and let M and N be finitely generated A-modules such that $M \otimes_A N$ is a module of finite length. Then the intersection multiplicity of M and N is defined to be

$$\chi(M, N) = \sum_{i \geq 0} (-1)^i \operatorname{length}(\operatorname{Tor}_i(M, N)).$$

This definition makes sense since all modules over regular local rings have finite projective dimension. Furthermore, it satisfies Bézout's Theorem, and it agrees with the definition given above in the case of curves. Serre stated three additional properties and proved them in the equicharacteristic case.

1. $\dim(M) + \dim(N) \leq \dim(A)$.

2. (Vanishing) If $\dim(M) + \dim(N) < \dim(A)$, then $\chi(M, N) = 0$.

3. (Positivity) If $\dim(M) + \dim(N) = \dim(A)$, then $\chi(M, N) > 0$.

The first of these Serre proved in general; we note that it can be interpreted as a statement on the topology of intersections in smooth varieties. The second and third were proven in the equicharacteristic case, and were conjectured in the general regular case.

2. The Peskine-Szpiro Intersection Theorem

Attempts to prove Serre's conjectures in the ensuing years included various generalizations of them to non-regular rings. The three statements listed above make sense for any local ring as long as M is assumed to have finite projective dimension (and, of course, $M \otimes_A N$ has finite length). In fact, the first statement makes sense even without the assumption of finite projective dimension, but it is false in this generality; whether it holds with this assumption is not known.

The next major advance was the proof of the "Intersection Theorem" in a quite general situation by Peskine and Szpiro [16, 17]. This theorem states:

Theorem (Intersection Theorem). *Let M be a module of finite projective dimension and let N be a module such that $M \otimes_A N$ has finite length as a module. Then*

$$\dim N \leq \operatorname{proj dim} M.$$

If one replaces projective dimension in this theorem by codimension (i.e. $\dim(A) - \dim(M)$), it becomes the first statement of Serre's conjectures, and, like that statement, it can be considered to describe the topology of the intersection of a variety which is the support of a module of finite projective dimension. The importance of the theorem was demonstrated by the fact that it implies several other conjectures (Peskine-Szpiro [17]):

Corollary (Bass's Conjecture). *If A has a non-zero finitely generated module of finite injective dimension, then A is Cohen-Macaulay.*

Corollary (Zero-Divisor Conjecture). *If M is an A-module of finite projective dimension and x in A is a zero-divisor on A, then x is a zero-divisor on M.*

Shortly thereafter, a stronger version of this theorem was stated (Peskine-Szpiro [18], Roberts [20]):

Theorem (New Intersection Theorem). *Let*

$$0 \to F_k \to \ldots \to F_0 \to 0$$

be a complex of finitely generated free modules with homology of finite length. Then if $F_\bullet$ is not exact, we have $\dim(A) \leq k$.

This theorem implies the original Intersection Theorem.

In the paper cited above, the Intersection Theorem was proven for rings of positive characteristic and for rings whose completion is the completion of a ring of finite type over a ring of characteristic zero. The method, which has since been applied successfully to a wide variety of questions, was to reduce (using the Artin Approximation Theorem) to the case of a ring of positive characteristic p, and then to use iterations of the Frobenius map (the map which sends a to a^p) to prove the theorem in the positive characteristic case. The method of reducing using the Artin Approximation Theorem was extended by Hochster [11] to the general equicharacteristic case. These techniques did not extend to the case of mixed characteristic.

3. The Homological Conjectures

The conjectures of Serre and the Intersection Conjecture in mixed characteristic, together with a number of others, became known as the Homological Conjectures. These conjectures dealt with intersection multiplicities, modules of finite projective dimension, the existence of Cohen-Macaulay modules, and properties of systems of parameters, together with relations between these topics. A summary of these conjectures in 1975 can be found in Hochster [12]; in this monograph Hochster introduced big Cohen-Macaulay modules and proved their existence (thus implying several other conjectures, including the Intersection Conjecture) in the equicharacteristic case. The proof made use of reduction to positive characteristic.

In a somewhat different approach, in Roberts [21] the notion of Cohen-Macaulay complex is defined and proven to exist in the complex algebraic case via the Grauert-Riemenschneider vanishing theorem. The connection between the technique of reduction to positive characteristic and vanishing theorems in cohomology coming from Hodge theory was further shown by a proof of the Kodaira vanishing theorem by reduction to positive characteristic, conjectured by Szpiro and proven recently by Deligne and Illusie [2].

Among these conjectures, as mentioned above, was the question of whether Serre's vanishing and positivity conjectures held for arbitrary rings, when M was assumed to have finite projective dimension. This conjecture was proved for graded modules over a graded ring by Peskine and Szpiro [18] using computations with Hilbert polynomials in a free resolution, and they showed that several other conjectures followed from this one. These results led to specuation that an appropriate theory of Chern classes and Riemann-Roch theorem over general local rings could be used to prove the multiplicity conjectures, and hence many others, in general. The idea was that a bounded complex of free modules $F_\bullet$ would define a sequence of Chern characters $\mathrm{ch}_i(F_\bullet)$, and that these would vanish for i less than the codimension of the support of $F_\bullet$. In codimension one this program was carried out by Foxby [7], using a construction of MacRae [15], leading to proofs of some of the conjectures in low dimension.

However, Dutta, Hochster and McLaughlin [4] constructed an example which showed that this generalization of Serre vanishing was false in dimension three. They construct a module M of finite projective dimension and finite length over the ring $A = k[[X, Y, Z, W]]/(XY - ZW)$, which has dimension three, such that if $N = A/(X, Z)$, then $\chi(M, N) = -1$. Thus the theory of local Chern characters, even if it existed, could not have the hoped for vanishing properties. In the next section we describe such a theory and then show how in spite of these setbacks, it could solve some of the conjectures.

4. Localized Chern Characters

The Chern characters alluded to in the previous section were defined by Baum, Fulton, and MacPherson [1]. There were two major problems in defining such a theory for complexes over a local ring. The first was that there is no cohomology theory. This problem was answered by defining Chern characters as operators on the Chow group. The second problem was that there are no non-trivial vector bundles over a local ring, so that straightforward definitions of Chern characters tend to give zero. The solution here was to give a "localized" definition, with values not in the Chow group of the entire ring, but in the Chow group of the support of the complex. We give a very brief summary of the part of the theory we need; we refer to Fulton [8] for a more complete description.

Let $V = \mathrm{Spec}(A)$, and let Y be a closed subset of V. For each integer i, let $Z_i(Y)$ denote the rational group of cycles of dimension i; that is, $Z_i(Y)$ consists of the free $\mathbb{Q}$-module with generators all reduced and irreducible subschemes of Y of dimension i, or, equivalently, all prime ideals P with $\dim(A/P) = i$. Then the i-th graded piece of the Chow group, denoted $A_i(Y)$, is $Z_i(Y)$ modulo rational equivalence, where rational equivalence is defined by setting the divisor of a rational function on a reduced and irreducible suscheme of dimension $i + 1$ to zero.

Let $F_{\bullet}$ be a bounded complex of free modules with support Z. Then there are Chern characters $\mathrm{ch}_k(F_{\bullet})$ for each $k \geq 0$, which map $A_i(X)$ to $A_{i-k}(X \cap Z)$.

The connection with Euler characteristics was given by the local Riemann-Roch theorem (Fulton [8], Example 18.3.12). We state a very special case of this theorem here. Let $F_{\bullet}$ be a complex of free modules with homology of finite length, or, equivalently, with support at the closed point of $\mathrm{Spec}(A)$. Assume that A is an integral domain, and let $[A]$ denote the class of $\mathrm{Spec}(A)$ in the highest dimensional graded piece of the Chow group $A_d(\mathrm{Spec}(A))$. Then there is a class $\tau(A)$, with $[A]$ as its highest component, which satisfies

$$\chi(F_{\bullet}) = \sum_{i \geq 0} \mathrm{ch}_i(F_{\bullet})(\tau_i(A)).$$

In this equation, the left hand side is an integer, and the right hand side is an element of the Chow group of the support of $F_{\bullet}$. In this case, the support of $F_{\bullet}$ is a point p, so its Chow group reduces to $A_0(p)$, which we identify with $\mathbb{Q}$.

Furthermore, if A is regular (or, more generally, a complete intersection), then $\tau(A) = [A]$.

5. The Proofs of Serre's Vanishing Conjecture and the Intersection Conjecture

Using the theory described in the last section, it is possible to prove Serre's vanishing conjecture without regard to characteristic and to prove the Peskine-Szpiro Intersection Theorem in mixed characteristic. We briefly outline these proofs here and attempt to show how the theory of localized Chern characters is used in their solution.

Let M and N be modules over a regular local ring A such that $M \otimes_A N$ has finite length and $\dim(M) + \dim(N) < \dim(A)$. Let $F_{\bullet}$ and $G_{\bullet}$ be free resolutions of M and N respectively. The problem is to show that $\chi(F_{\bullet} \otimes G_{\bullet}) = 0$, since the homology of $F_{\bullet} \otimes G_{\bullet}$ is precisely $\mathrm{Tor}(M, N)$. By the local Riemann-Roch formula we have

$$\chi(F_{\bullet} \otimes G_{\bullet}) = \mathrm{ch}(F_{\bullet} \otimes G_{\bullet})([A]).$$

We must now use some properties of local Chern characters. First, they are multiplicative, which means that we can write

$$\mathrm{ch}(F_\bullet \otimes G_\bullet)([A]) = \sum_{i+j=d} \mathrm{ch}_i(F_\bullet)(\mathrm{ch}_j(G_\bullet))([A]).$$

Let X and Y denote the supports of M and N respectively; these are also the supports of $F_\bullet$ and $G_\bullet$. Now suppose that $d - j > \dim(Y)$. Then the element $\mathrm{ch}_j(G_\bullet)([A])$, which is in the group $A_{d-j}(Y)$, must be zero, since there can be no subvarieties of Y of dimension larger than the dimension of Y. Thus all terms for which $d - j > \dim(Y)$ are zero. Using the commutativity of Chern characters (proven in Roberts [23]), one shows also that terms with $d - i > \dim(X)$ are zero. The hypothesis that $\dim(X) + \dim(Y) < d$ implies that all terms fall into one of these two sets, and therefore the entire sum is zero, as was to be shown.

This theorem also holds if A is a complete intersection. It was proven independently using Adams operations in K-theory by Gillet and Soulé [9, 10]. It was proven when the singular locus has dimension at most one in Roberts [23]. We note that this argument gives a vanishing theorem for intersection multiplicities defined by local Chern characters for modules of finite projective dimension in general, and this approach appears to work better than the approach based on Euler characteristics for singular points.

We next describe how localized Chern characters are used in the proof of the New Intersection Theorem in mixed characteristic. Let A be a domain of mixed characteristic of dimension d, and let

$$0 \rightarrow F_{d-1} \rightarrow \ldots \rightarrow F_0 \rightarrow 0$$

be a complex of free modules with homology of finite length which is not exact. We wish to prove that such a complex cannot exist. Let $\overline{F}_\bullet$ denote $F_\bullet \otimes A/pA$, where p is the characteristic of the residue field of A. Then $\overline{F}_\bullet$ is a complex of free modules over A/pA.

The main idea of the proof is to compute $\mathrm{ch}_{d-1}([A/pA])$ in two different ways. First, using the properties of localized Chern characters and the fact that $\overline{F}_\bullet$ comes from restriction to a divisor of a complex supported at one point, we deduce that $\mathrm{ch}_{d-1}([A/pA]) = 0$. Second, we use that $\overline{F}_\bullet$ is a non-trivial complex of length $d-1$ over a ring of positive characteristic of dimension $d-1$. By means of an asymptotic Euler characteristic introduced by Dutta [3] and put on a firm foundation by Seibert [27], together with a local cohomology argument, we can then show that it must be positive (details can be found in Roberts [24, 25]). This contradiction proves the theorem.

6. Open Questions

We conclude with some remarks on three questions which are still open.

1. Serre's Positivity Conjecture. The method of local Chern characters provides a geometric approach to the question of positivity, but as of yet it has not led to a solution. A result of Tennison [30] states that if one computes $\chi(A/P, A/Q)$, where $\dim(A/P) + \dim(A/Q) = \dim(A)$, and if the subschemes defined by the prime ideals P and Q are not tangent at the closed point of $\mathrm{Spec}(R)$ (this condition can be stated precisely in terms of associated graded rings), then $\chi(R/P, R/Q)$ is the product of the multiplicities of the A/P and A/Q and is therefore positive. We remark that an example of Roberts [26] shows that positivity is unlikely for two modules of finite projective dimension in the non-regular case, even using Chern characters rather than Euler characteristics, although this question is also open.

2. The Improved New Intersection Conjecture (Evans-Griffith [5, 6]). This conjecture states the following: Let $F_\bullet = 0 \to F_k \to \ldots \to F_0 \to 0$ be a complex of free modules with finite length homology except possibly for $H_0(F_\bullet)$. Assume that there is a minimal generator of $H_0(F_\bullet)$ which is annihilated by a power of the maximal ideal of A. Then $k \geq \dim(A)$.

It seems reasonable to try to prove this theorem using the methods which worked for the "unimproved" version, but this version appears to be more difficult. This conjecture is equivalent to several others, such as Hochster's Monomial Conjecture (see [13]), and is open in dimension three in mixed characteristic.

3. Another Conjecture on Bounded Complexes. The New Intersection Theorem states that the minimum length of a bounded complex of free modules is d. Suppose that $F_\bullet$ has length exactly d. In this case, the question is whether the cycles Z_i are integral over the boundaries B_i in degrees larger than zero. This result would imply the Monomial Conjecture, it can be proven in positive characteristic by the theory of tight closure of Hochster and Huneke [14], and it has been proven by Rees [19] for the Koszul complex in any characteristic.

References

1. Baum, P., Fulton, W., MacPherson, R.: Riemann-Roch for singular varieties. Publ. Math. IHES **45** (1975) 101–145
2. Deligne, P., Illusie, L.: Relèvements modulo p^2 et décomposition du complexe de de Rham. Invent. math. **89** (1987) 247–270
3. Dutta, S.P.: Frobenius and multiplicities. J. Algebra **85** (1983) 424–448
4. Dutta, S.P., Hochster, M., McLaughlin, J.E.: Modules of finite projective dimension with negative intersection multiplicities. Invent. math. **79** (1985) 253–291
5. Evans, E.G., Griffith, P.: The syzygy problem. Ann. Math. **114** (1981) 323–333
6. Evans, E.G., Griffith, P.: The syzygy problem: a new proof and historical perspective. Commutative Algebra (Durham 1981), London Math Soc. Lecture Note Series **72** (1982) 2–11
7. Foxby, H.-B.: The MacRae invariant. Commutative Algebra (Durham 1981), London Math Soc. Lecture Note Series **72** (1982) 121–128
8. Fulton, W.: Intersection theory. Springer, Berlin Heidelberg New York 1984
9. Gillet, H., Soulé, C.: K-théorie et nullité des multiplicités d'intersection. C. R. Acad. Sc. Paris Série I, no. 3, **300** (1985) 71–74
10. Gillet, H., Soulé, C.: Intersection theory using Adams operations. Invent. math. **90** (1987) 243–277
11. Hochster, M.: The equicharacteristic case of some homological conjectures on local rings. Bull. Amer. Math. Soc. **80** (1974) 683–686
12. Hochster, M.: Topics in the homological theory of modules over commutative rings. Regional Conference Series in Mathematics **24**, 1975
13. Hochster, M.: Canonical elements in local cohomology modules and the direct summand conjecture. J. Algebra **84** (1983) 503–553
14. Hochster, M., Huneke, C.: Tight closure, invariant theory, and the Briançon-Skoda Theorem. J. Amer. Math. Soc. **1** (1990) 31–116
15. MacRae, R.E.: On an application of the Fitting invariants. J. Algebra **2** (1965) 153–169

16. Peskine, C., Szpiro, L.: Sur la topologie des sous-schémas fermés d'un schéma localement noethérien, définis comme support d'un faisceau cohérent localement de dimension projective finie. C. R. Acad. Sci. Paris Sér. A **269** (1969) 49–51

17. Peskine, C., Szpiro, L.: Dimension projective finie et cohomologie locale. Publ. Math. IHES **42** (1973) 47–119

18. Peskine, C., Szpiro, L.: Syzygies et Multiplicités. C. R. Acad. Sci. Paris Sér. A **278** (1974) 1421–1424

19. Rees, D.: Reduction of Modules. Math. Proc. Camb. Philos. Soc. **101** (1987) 431–449

20. Roberts, P.: Two applications of dualizing complexes over local rings. Ann. Sci. Éc. Norm. Sup. **9** (1976) 103–106

21. Roberts, P.: Cohen-Macaulay complexes and an analytic proof of the New Intersection Conjecture. J. Algebra **66** (1980) 220–225

22. Roberts, P.: The vanishing of intersection multiplicities of perfect complexes. Bull. Amer. Math. Soc. **13** (1985) 127–130

23. Roberts, P.: Local Chern characters and intersection multiplicities. Proc. Sympos. Pure Math. **46** 2, Amer. Math. Soc., Providence, R.I. 1987, pp. 389-400

24. Roberts, P.: Le théorème d'intersection. C. R. Acad. Sc. Paris Sér. I, no. 7, **304** (1987) 177–180

25. Roberts, P.: Intersection theorems. Commutative Algebra, Proceedings of an MSRI Microprogram. Springer, Berlin Heidelberg New York 1989, pp. 417–436

26. Roberts, P.: Negative intersection multiplicities on singular varieties. To appear in the Proceedings of the Zeuthen Conference, Copenhagen 1989

27. Seibert, G.: Complexes with homology of finite length and Frobenius functors. J. Algebra **125** (1989) 278–287

28. Serre, J.-P.: Algèbre locale – multiplicités. (Lecture Notes in Mathematics, vol. 11.) Springer, Berlin Heidelberg New York 1961

29. Szpiro, L.: Sur la théorie des complexes parfaits. Commutative Algebra (Durham 1981), London Math. Soc. Lecture Note Series **72** (1982) 83–90

30. Tennison, B.R.: Intersection multiplicities and tangent cones. Math. Proc. Camb. Philos. Soc. **85** (1979) 33–42

The Isomorphism Problem
for Integral Group Rings of Finite Groups

Klaus W. Roggenkamp

Report on joint work with L.L. Scott 1980–1990

Mathematisches Institut B, Universität Stuttgart, Pfaffenwaldring 57
W-7000 Stuttgart 80, Fed. Rep. of Germany

1. The History of the Isomorphism Problem

The modern theory of groups originated with the treatments of *Galois* (1811–1832), *Cauchy* (1789–1857) and *Serret* (1819–1885) on finite discontinuous substitution groups. That is permutation groups, or group actions on sets, to use modern language. At that time the theory of abstract groups was only little developed. So it was natural next to study abstract groups by letting them act on sets or linearly on vector spaces. Time has shown that studying linear actions of groups on vector spaces is a much richer theory than letting groups act on sets, since one can invoke the arithmetic of the general linear groups.

In this spirit *W. Burnside, F. G. Frobenius* and later *I. Schur* developed the ordinary (complex) representation theory of finite groups in the years 1896–1910.

In modern language, the idea of Frobenius, Burnside and Schur was to study homomorphisms

$$\varphi \; : \; G \to GL(n, R) \;, \tag{1}$$

of G into $(n \times n)$-matrices over the commutative ring R and use the informations on the matrix groups $\mathrm{Im}(\varphi)$ to obtain informations on the abstract finite group G ; originally R was the ring of complex numbers. (Strictly speaking Burnside and Frobenius, especially, only worked with traces of these matrices.)

An important demonstration of the power of complex representation theory, where one did use the arithmetical properties of $\mathrm{Im}(\varphi) \subset GL(n, \mathbb{C})$, was *Burnside's Theorem* (1911): A finite group, whose order involves at most two different primes is solvable.

In modern times a purely group theoretical proof was found, though a theorem of Frobenius of the same spirit has never been proved without representation theory.

Modular representation theory – i.e. representations in $GL(n, K)$, where K is a field such that $\mathrm{char}(K)$ divides the order of the group – lay dormant until *R. Brauer* in 1935 – at the suggestion of I. Schur – developed the theory further, using ring theory, which enters as follows: Given a representation

$$\varphi \; : \; G \to GL(n, R) \subset \mathrm{Mat}(n, R) \;,$$
$$g \mapsto \varphi(g) \;.$$

Proceedings of the International Congress
of Mathematicians, Kyoto, Japan, 1990
© The Mathematical Society of Japan, 1991

Since Mat(n, R), the ring of $n \times n$ matrices over R, is an R-algebra, we may associate with φ the R-algebra

$$\Lambda_\varphi = \{\Sigma_{g \in G}\ r_g \cdot \varphi(g) | r_g \in R\}\ , \tag{2}$$

generated by Im(φ) over R. Studying the representation φ is tantamount to studying the R-algebra Λ_φ. The universal such R-algebra is the *group ring*, consisting of formal linear combinations,

$$RG = \{\Sigma_{g \in G}\ r_g \cdot g | r_g \in G\}\ ; \tag{3}$$

addition is componentwise, and the R-linear multiplication is induced from the multiplication in G. This group ring maps onto each of the various Λ_φ. The original motivation of Burnside, Frobenius, Schur and Brauer can thus be rephrased as the question:

Which properties of the finite group G are reflected in RG?

Already in early time it was known, that $\mathbb{C}G$ does not reflect all properties of G, since finite abelian groups G and H have isomorphic complex group algebras if and only if $|G| = |H|$.

However, if one considers the group rings KG for all fields K, then the various isomorphism types of abelian groups can be distinguished by their group algebras. That this is not so in general was shown only in 1971 by E. Dade:

Theorem 1 (Dade [D]). *There are two non-isomorphic finite metabelian groups G and H of order $p^6 \cdot q^6$, p and q different primes, such that $KG \simeq KH$ for* **every** *field K.*

It is still an open

Problem 2. Let char(K) $= p > 0$, and let G and H be p-groups, such that $KG \simeq KH$. Is then $G \simeq H$?

Very little is known here, except, that – by computer-analysis – for 2-groups of order $\leq 2^6$, the problem has a positive answer [Pas] and [RS4].

Now the ring of integers $\mathbb{Z}$ is the universal commutative ring, and if for two finite groups the integral group rings $\mathbb{Z}G$ and $\mathbb{Z}H$ are isomorphic, then $RG \simeq RH$ for **every** commutative ring. Thus we come to

Problem 3 (Isomorphism Problem). Assume that for two finite groups the integral group rings $\mathbb{Z}G$ and $\mathbb{Z}H$ are isomorphic. Is then $G \simeq H$?

A positive answer would imply that representation theory – especially, the integral representation theory of a finite group would determine the group up to isomorphism. This issue has a fundamental appeal in and of itself, aside from the original motivation by Burnside and Frobenius. There is a huge literature [Sa].

The rational group algebra, say for a p-group G of odd order, has the form:

$$\mathbb{Q}G = \mathbb{Q}^+ \prod \left(\prod_{i=1}^{v} \mathrm{Mat}(n_i, K_i) \right),$$

where K_i are algebraic number fields, $1 \leq i \leq v$, and $\mathbb{Q}^+$ comes from the trivial representation.

The integral group ring $\mathbb{Z}G$ is then in general a subring of finite index in

$$\mathbb{Z}G \subset \mathbb{Z}^+ \prod \left(\prod_{i=1}^{v} \text{Mat}(n_i, R_i) \right), \tag{4}$$

where R_i are the algebraic integers in the number fields K_i, $1 \leq i \leq v$.

2. Major Results Until 1980

The isomorphism problem was first considered by *Graham Higman*, a student of G. Whitehead, in his thesis in 1939. One of his most spectacular results was:

Theorem 4 (G. Higman [Hi]). *Let u be a unit of finite order in $\mathbb{Z}A$ for a finite abelian group A. Then $u = \pm a$ for some group element $a \in A$.*

The *proof* uses the fact, that the group ring of a finite group G over the commutative ring R is not only an augmented R-algebra, with augmentation

$$\varepsilon_G : \Sigma_{g \in G} \, r_g \cdot g \to \Sigma_{g \in G} \, r_g, \tag{5}$$

but also a Hopf-algebra with antipode

$$*_G : \Sigma_{g \in G} \, r_g \cdot g \to \Sigma_{g \in G} \, r_g \cdot g^{-1}. \tag{6}$$

Let now — more generally — G be a finite group and $u \in \mathbb{Z}G$ a *central* unit of finite order n. (We shall here only deal with the case, where n is odd.) Then $u^* = *_G(u)$ is also a central unit of order n, as well as $v = u \cdot u^*$ (note that u is central), which is fixed under $*_G$, and hence must be equal to 1. Now the coefficient of 1 in v is $1 = \Sigma \, z_g^2$. Since $z_g \in \mathbb{Z}$, we have $u = \pm g_0$ for some $g_0 \in G$.

Though this argument has heavily used the fact that the ring of coefficients is $\mathbb{Z}$, the result is still valid for any integral domain R of characteristic zero, in which no rational prime divisor of $|G|$ is invertible. We shall call such a ring G-*adapted*.

Definition 5. $U(RG)$ stands for the units in RG and

$V(RG) = \{u \in U(RG) \mid \varepsilon_G(u) = 1\}$ are the *units of augmentation* 1.

Then $U(RG) = V(RG) \times U(R)$.

Let us turn to our aim, the isomorphism Problem 4:
If $RG = RH$, then $H \subset U(RG)$, in general however, it will not be contained in $V(RG)$. But this can easily be remedied: Replace H by $H' = \{h \cdot \varepsilon_G(h)^{-1} \mid h \in H\}$. Then $H \simeq H'$ and $RH = RH'$; moreover, $H' \subset V(RG)$.

Thus from now on we will always assume that an equality of group rings $RG = RH$ is always augmented; *i.e.* $H \subset V(RG)$.

Higman's result then says that the abelian group A is unique in $\mathbb{Z}G$ as augmented group of units. This result is a very elegant positive answer to the isomorphism problem.

Actually, Higman, in his 1939 thesis "Units in group rings", was the first to speculate about the isomorphism problem: "Whether it is possible for two non isomorphic groups to have isomorphic group rings I do not know; but the results ... (here) ... suggest that it is unlikely". So, Higman was a bit in favor of a positive answer to the isomorphism problem.

In this connection I should also mention *Richard Brauer*, who in his "Harvard Lecture on Modern Mathematics, Representations of finite groups" in 1963 [Br] listed the isomorphism problem as one of the important open questions in representation theory; however, he did not commit himself in either direction. Contrary to *Hans Zassenhaus*, who firmly believes in a positive answer to the isomorphism problem and even made a much stronger conjecture in 1974 [Za] (cf. below).

The isomorphism problem, $\mathbb{Z}G = \mathbb{Z}H$ as augmented algebras, appears in a new light, if we assume that $*_G = *_H$ (cf. (5)).

Proposition 6 (Banachevski [Ba]). *Let $\mathbb{Z}G = \mathbb{Z}H$ as augmented algebras and assume that both rings have the same antipode i.e. $*_G = *_H$. Then $G = H$ in $\mathbb{Z}G = \mathbb{Z}H$.*

The *proof* is in the same spirit as the proof of Higman's result.

Given an equality of augmented algebras $RG = RH$. Then the elements $\{g \mid g \in G\}$ form a finite subgroup in $V(RH)$, consisting of R-linearly independent elements. S. D. Berman [Ber, Sak] has observed, that the linear independence is often automatic for G-adapted rings. This result allows to phrase the isomorphism problem differently:

Problem 7. Let U be a finite subgroup in $V(\mathbb{Z}G)$, with $|G| = |U|$. Is then $U \simeq G$?

In order to gain more insight into the structure of $V(\mathbb{Z}G)$, one should even ask: How is U embedded in $V(\mathbb{Z}G)$? G. Higman gave the answer for abelian G. However, in general, one can not expect that $U = G$, since $V(\mathbb{Z}G)$ need not be abelian and G need not be normal in $V(\mathbb{Z}G)$. Moreover, even for the dihedral group D of order 8, there exists a unit $u \in \mathbb{Q}D \backslash V(\mathbb{Z}D)$, such that $D \neq u \cdot D \cdot u^{-1} \in \mathbb{Z}D$, and this conjugation is not inner in $\mathbb{Z}D$.

In this connection, *H. Zassenhaus*[Za] made a far reaching conjecture:

Zassenhaus Conjecture. *Let $U \leq V(\mathbb{Z}G)$ be a finite subgroup with $|V| = |G|$. Then there exists a unit $a \in \mathbb{Q}G$ with*

$$a \cdot V \cdot a^{-1} = G. \tag{7}$$

A direct extension of Higman's result for G a finite group and R G-adapted is given by

Theorem 8 (Berman [Ber], Glauberman (Passman [Pa]), Saksonov [Sak]). *Let* $\mathbb{Z}G = \mathbb{Z}H$ *as augmented algebras and let*

$$K_g = \sum_{x \in G/C_G(g)} {}^x g$$

be a class sum in G — $C_G(g)$ *denotes the centralizer of* g *in* G. *Then* $K_g = K_h$ *is a class sum in* RH.

An immediate consequence is:

Corollary 9. *Let* N *be a normal subgroup of* G, *and let* $RG = RH$ *as augmented algebras. Then in* $RG = RH$ *we have* $\sum_{n \in N} n = \sum_{m \in M} m$ *for a normal subgroup* M *in* H.

Using the above result, we can *reformulate the Zassenhaus conjecture:*

Isomorphism Problem over the Class Sums: $\mathbb{Z}G = \mathbb{Z}H$ *implies that there exists an* **isomorphism** $\chi : G \to H$ *inducing the class sum correspondence; i.e. for* $g \in G$,

$$K_g = K_{\chi(g)}. \tag{8}$$

The next result pushes the isomorphism problem further to metabelian groups:

Theorem 10 (Whitcomb [Wh], Jackson [J] 1968). *Assume that* G *is metabelian with a normal abelian subgroup* A *and* G/A *abelian.*
$$\text{If} \quad RG = RH, \quad \text{then} \quad G \simeq H$$
for R *a* G-*adapted ring.*

This follows easily from Higman's result and by considering the "*small group ring*" $RG/(I(A) \cdot I(G) \cdot RG)$. Here we have denoted for a normal subgroup N of G by $I(N) \cdot RG$, the *augmentation ideal of* N, the kernel of the natural map $RG \to RG/N$. We note that the proof does not give any information of how G is embedded into $V(RH)$.

The following result was in 1989 obtained by Kimmerle, Lyons and Sandling, using the classification of finite simple groups:

Theorem 11 [KLS]. *Let* $\mathbb{Z}G = \mathbb{Z}H$ *as augmented algebras, and let*

$$1 = N_0 < \dots < N_t = G$$

be a chief series of G. *Then* H *has a chief series of the same length, and both have isomorphic chief factors.*

This shows in particular that finite simple groups are determined by their integral group rings – most of them though are already determined by their order.

3. Recent Progress

When, in 1980, we started to consider the isomorphism problem, the Zassenhaus conjecture was at first a tempting target for a counterexample. At that time we were not successful, and eventually started to believe, that the Zassenhaus conjecture might be true for certain classes of groups; in fact, it was a guide for our work on p-groups.

Roughly speaking, the Zassenhaus conjecture states, that there are only few automorphisms of the integral group ring $\mathbb{Z}G$ of the finite group G. After having done plenty of calculations, we were looking for some heuristic evidence for the Zassenhaus conjecture. Borrowing ideas from Lie theory we found it in the fact (first noted by us, though the proof is easy), that every derivation $\delta : \mathbb{Z}G \to \mathbb{Z}G$ is inner.

One might even go one step further than Zassenhaus and ask – this problem was briefly touched upon by *Berman and Rossa* [BR] for complete rings, like the p-adic integers $\hat{\mathbb{Z}}_p$:

Problem 12 (Conjugacy Problem). If $V(\hat{\mathbb{Z}}_p G) = V(\hat{\mathbb{Z}}_p H)$, are G and H conjugate in $V(\hat{\mathbb{Z}}_p G)$, provided G is a p-group?

3.1 Progress on the Isomorphism Problem

The next results were obtained with Leonard Scott in 1985 [RS1] and 1987 [RS3]:

Theorem 13. *Let G be a finite group such that there exists an exact sequence*

$$1 \to A \to G \to N \to 1,$$

where A is abelian and N is nilpotent. If $\mathbb{Z}G \simeq \mathbb{Z}H$, then G is isomorphic to H. (More generally, we can replace N by a product of p_i-constrained groups G_i with $O_{p_i'}(G_i) = 1$ of relatively prime order, with slight restrictions on A.)

Remark. For a solvable group G the above implies:

a) For every prime p, the group $G/O_{p'}(G)$ is determined by $\mathbb{Z}G$, where $O_{p'}(G)$ is the largest normal subgroup of G of order prime to p.

b) The Sylow subgroups of G are determined – up to isomorphism – by $\mathbb{Z}G$. It was shown in [KR], that for $\mathbb{Z}G = \mathbb{Z}H$ there exists for each prime p and a Sylow p-subgroup P of G a unit $a(p) \in \mathbb{Q}G$ such that $a(p) \cdot P \cdot a(p)^{-1}$ is a Sylow p-subgroup of H.

3.2 Drawbacks and Progress on the Zassenhaus Conjecture

Let us recall, that the Zassenhaus conjecture (7) implies that the action of every automorphisms α of $\mathbb{Z}G$ on the center of $\mathbb{Z}G$ can be compensated there by a group automorphism.

Theorem 14. *The Zassenhaus conjecture is true for the following classes of groups:*

(i) G is nilpotent. (It has been reported to us, that in this case A.Weiss has shown that for every finite subgroup U of $V(\mathbb{Z}G)$, there exists a unit $a \in \mathbb{Q}G$ such that $a \cdot U \cdot a^{-1} \subset G$ [We].)

(ii) There exists a prime $p \in \mathbb{Z}$, such that G has a normal p-subgroup P with $C_G(P)$ a p-group.

On the other hand, the Zassenhaus conjecture is false [RS2, Ro]:

Theorem 15. *There exists a finite metabelian – even supersolvable – group G, and an automorphism α of $\mathbb{Z}G$, commuting with the augmentation ε_G, such that $\alpha \cdot \varrho$ is not a central automorphism for any group automorphism ϱ of G.*

Note that the groups in Dade's example (Theorem 1) are metabelian!

3.3 Progress on the Conjugacy Problem

In connection with the possibility, that group bases may be conjugate (cf. conjugacy Problem 12), we have obtained (1985, 1986) the following result [RS3]:

Theorem 16. *Let G be a finite group with normal Sylow p-subgroup, such that $O_{p'}(G) = 1$. Let α be an augmented automorphism of $\hat{\mathbb{Z}}_p G$. Then there exists a unit $u \in V(\hat{\mathbb{Z}}_p G)$, such that $u \cdot G \cdot u^{-1} = \alpha(G)$.*

The theorem we have proved is actually *much more general:* Let G be a finite p-constrained group with $O_{p'}(G) = 1$ and let $N = O_p(G)$. If α is an augmented automorphism of $\hat{\mathbb{Z}}_p G$, stabilizing $I_{\hat{Z}_p}(N)G$, then G and $\alpha(G)$ are conjugate in $V(\hat{\mathbb{Z}}_p G)$. This stability always occurs for $N = O_p(G)$, if α arises from an augmented automorphism of $\mathbb{Z}G$.

The result applies in particular to p-groups. For the groups in Theorem 16, the Zassenhaus conjecture is true. Our result is so strong, that it allows to compute the Picard group of the group ring $\mathbb{Z}N$ of a nilpotent group N semilocally [RS, R1]. However, the conjugacy problem has a negative answer both for $V(\mathbb{Z}N)$ and for $V(\hat{\mathbb{Z}}_p N)$. Our results, and the evidence we have gathered up to now, let it appear reasonable to ask whether $V(\hat{\mathbb{Z}}_p G)$ has the

Sylow Property. *Let G be a finite p-group and U a finite subgroup of $V(\hat{\mathbb{Z}}_p G)$.*

$$\text{Is } U \text{ conjugate in } V(\hat{\mathbb{Z}}_p G) \text{ to a subgroup of } G \text{ ?} \tag{9}$$

Even more generally, one might formulate some version of this for the principal p-block, in the spirit of the defect group question of [S3].

Because of the results in [RS] we know, this is true in case $|U| = |G|$. Theorem 16 can be interpreted as a first step to prove Sylow's theorems for finite subgroups of $V(RG)$. In an attempt to prove (9) we recalled that a fancy way of proving Sylow's theorems for the finite group H is to show that the spectrum of the cohomology ring $H^*(H, \mathbb{F}_p)$ is connected; this is also true for the unit group.

More precisely, let V be a profinite p-group, and denote by $H^*(V, \mathbb{F}_p)$ the continuous (even-dimensional for p odd) cohomology ring of V with coefficients in $\mathbb{F}_p$, the field with p elements. If the spectrum of $H^*(V, \mathbb{F}_p)$ is connected, we say that the variety of V is connected. We shall write $VC(V, \mathbb{F}_p)$ for the variety of $H^*(V, \mathbb{F}_p)$.

Though we could not reach our original goal (in the spirit of [Ca]), we were able to prove for R a complete Dedekind domain of characteristic zero with residue field of characteristic p:

Theorem 17. *The following conditions are equivalent for a p-group G:*

(i) Every finite p-subgroup U of $V(RG)$ is conjugate in $V(RG)$ to a subgroup of G.

(ii) For every p-subgroup P of G, the natural inclusion

$$N_G(P)/P \to N_{V(RG)}(P)/P$$

induces a continuous map

$$VC(N_G(P)/P, \mathbb{F}_p) \to VC(N_{V(RG)}(P)/P, \mathbb{F}_p),$$

which is a bijection.

(iii) The variety of $N_{V(RG)}(P)/P$ is connected for every p-subgroup P of G.

(In the statement of the result we have used $N_A(B)$ to denote the normalizer of B in A.)

We want to point out, that the statements (i) and (iii) are not true in general for profinite p-groups; in fact we have examples of unit groups of orders, where (i) is false. It would be interesting to have a *group theoretical criterion* for when (iii) is true for a profinite p-group V and $P = 1$.

Let us return to a discussion of (9). Some years ago we found a proof of (9) for G a 2-group [R1]. Later on we were able to handle groups of order p^3, and we developed a sketch of the proof in the general p-group case in Nov. 1985 [S1]. Since in this proof we had not worked out all the details, we only made at the Arcata-meeting in 1986 the conjecture, that (9) is true for p-groups. (In the meantime, Gary Thompson, a student of Leonard Scott, has worked out the details in [S1, Th].) In October 1986 we learnt that Al Weiss from the University of Alberta [W] had a different proof of:

Theorem 18 (Subgroup Rigidity Theorem). *Let G be a finite p-group, and U a finite subgroup of $V(\hat{\mathbb{Z}}_p G)$. Then U is conjugate in $V(\hat{\mathbb{Z}}_p G)$ to a subgroup of G.*

4. Some Remarks to the Proof of Theorem 16

Questions about isomorphisms of group rings reduce to questions of automorphisms by Kimmerle's trick [Ki, RS]. The main result, Theorem 16, is a consequence of the following statement: Let G be a finite group and R an unramified extension of $\hat{\mathbb{Z}}_p$. Put $N = O_p(G)$.

Theorem 19. *Assume that $C_G(N) \subset N$ and let α be an augmented automorphism of RG with $\alpha(N) \subset I_R(N) \cdot G$ – the induced augmentation ideal of N. Then $\alpha(G)$ is conjugate to G in the units of RG.*

Remark. 1. It turns out, that in the proof of Theorem 19 we do not really require that α stabilizes $I_R(N) \cdot G$, but only that it stabilizes $I_{\mathbb{F}}(N) \cdot G$, where $\mathbb{F} = R/\text{rad } R$.

2. We do not know, whether the stabilization of either ideal in 1.) is automatic for an augmented automorphism α of RG. This is indeed the case when N is a Sylow p-subgroup, since then RG/N is the largest R-algebra homomorphic image of RG, which is torsionfree and whose reduction over the residue field $\mathbb{F}$ of R is semi-simple. The obstruction lies in the fact, that we do not know, whether a class sum correspondence holds for classes of p-power elements in RG.

3. The Theorem 19 is likely to be true for any complete discrete valuation domain of characteristic zero with residue field of characteristic p, assuming appropriate generalizations of the results of Weiss [W], which have been obtained by Gary Thompson [Th] and in [R].

Corollary 20. *Let G be a p-constrained group such that $O_{p'}(G) = 1$ for some prime p. Then the isomorphism problem and the Zassenhaus conjecture have a positive answer for RG. (Any two normalized group bases are even p-adically conjugate.)*

Corollary 21. *Let G be a finite group with a normal Sylow p-subgroup for some prime p. Then the defect group of the principal R-block of G — R is an unramified extension of $\hat{\mathbb{Z}}_p$ — is uniquely determined, up to conjugacy in the block, by the principal block, (i.e., if $\alpha : RG \to RG$ is an augmented automorphism, then $\alpha(P)$ and P are conjugate in the principal block).*

This result was extended by L. L. Scott in [S3], who proved it for a finite group with a cyclic Sylow p-subgroup, which is a T.I. set.

We shall now state the main ingredients of the proof of Theorem 19. There are three main ingredients in the proof. The *first one*, which we already proved in April 1986 is:

Theorem 22. *G is a p-constrained group with P a Sylow p-subgroup. R is a complete Dedekind domain of characteristic zero with $pR \neq R$. Let α be a normalized automorphism of the principal block $B_0 = R[G/O_{p'}(G)]$ of RG – note that B_0 is an augmented algebra, the augmentation being induced from that of RG. Assume that α stabilizes the image of P in B_0; i.e. $\alpha(P) = P$ in B_0. Then $\alpha|_P$ — the restriction of α to P — is induced from an automorphism β of $G/O_{p'}(G)$, such that the automorphism induced from β on B_0 agrees with α up to inner automorphisms centralizing P.*

The proof is based on the following two results:

Lemma 23 (Generalized Coleman-Ward Result [C,Wa]). *Let G be a finite group, P a p-subgroup of G, and S an integral domain, in which p is not invertible. Let $V = V(SG)$. Then we have for the normalizers*

$$N_V(P) = N_G(P) \cdot C_V(P).$$

We shall be using the following *notation*: Given an automorphism α of RG, and an RG-bimodule M, we denote by $_\alpha M_1$ the bimodule, which has the original action on the right, but the left action is twisted by α, i.e. $x \cdot m \cdot y = \alpha(x)my$, $x, y \in RG$, $m \in M$.

Lemma 24. *R is a complete Dedekind domain of characteristic zero with $pR \neq R$, G is a finite group and B a block of RG with defect group D. Let α be an automorphism of B, which stabilizes the image of D in B, b is the Brauer correspondent in $N = N_G(D)$ to B. Then the Green correspondent to the twisted module $_\alpha B_1$ on $G \times G$, is $_\beta b_1$ on $N \times N$ for some automorphism β of RN. Moreover, $\alpha|_D = \beta|_D$, where we regard $D \subset b \subset B$.*

The *second main ingredient* is the following: Let R be an unramified extension of $\hat{\mathbb{Z}}_p$ with residue field $\mathbb{F}$. Let N be a normal p-subgroup of a finite group G, and assume $C_G(N) \subset N$.

Theorem 25. *Let α be an augmented automorphism of RG, with N and G as above, and assume that α stabilizes the ideal $I_{\mathbb{F}}(N) \cdot G$ in $\mathbb{F}G$. Then $\alpha(N)$ is conjugate to N by a unit of RG.*

The *third main ingredient* is an extension of results of Weiss [W] combined with a criterion on the indecomposability of permutation modules, which is the main tool to show that group bases are conjugate. We let R be an unramified extension of $\hat{\mathbb{Z}}_p$, the p-adic integers. $\mathbb{F} = R/pR$ is the residue field of R, and we denote by $x \to \bar{x}$ the natural projection from R to $\mathbb{F}$, and also for any R-module X we put $\overline{X} = X/pX$.

Theorem 26. *Assume that P and Q are p-groups. Let M be an $R(P \times Q)$-module (with the action $(x,y) \cdot m = x \cdot m \cdot y^{-1}$, $x \in P$, $y \in Q$, $m \in M$), which is free of finite rank as RQ-module. If $\overline{M/(M \cdot I_R(Q))}$ is a permutation module for $\mathbb{F}P$, then $\overline{M}$ is a permutation module for $\mathbb{F}(P \oplus Q)$.*

The next result is the place where we use that $C_G(N) \subset N$.

Lemma 27. *Suppose $C_G(N) \subset N$. Then the $\mathbb{F}(N \times G)$-module $\mathbb{F}G$ — with the action $(n, g) \cdot x = n \cdot x \cdot g^{-1}$, $n \in N$, $g \in G$, $x \in \mathbb{F}G$ — is absolutely indecomposable. Moreover, the same is true if the action of N on the left is twisted by an augmented $\mathbb{F}$-algebra automorphism α of $\mathbb{F}G$, or by a group automorphism ϱ of N – i.e., $(n, g) \cdot x = \varrho(n) \cdot x \cdot g^{-1}$ for $n \in N$, $g \in G$ and $x \in \mathbb{F}G$.*

We shall next come to the proof of the "Subgroup Rigidity Theorem" 18 as it was generalized from [W] in [R]. R will be a complete Dedekind domain of finite rank over $\hat{\mathbb{Z}}_p$, the p-adic integers, K is the field of fractions of R and $\text{rad}(R) = \pi \cdot R$. G is a finite p-group. Theorem 18 is an immediate consequence of the following result, which generalizes and was inspired by the work of Al Weiss [W]:

Theorem 28. *Let M be an RG-lattice, and let N be a normal subgroup of G. Assume that*

1. $M\downarrow_N$ is a free RN-module,
2. $M/(I(N) \cdot M)$ is a permutation module for G/N.
Then M is an RG-permutation module.

The next result plays a major role in the proof of Theorem 19, though it is also of interest for its own sake. A generalized permutation module is a direct sum of modules induced from one dimensional representations of subgroups.

Theorem 29 [R]. *1. Assume that K is a splitting field for G and all of its subgroups. If ζ be a primitive p-th root of unity in R, then we require $\pi^t \cdot R \subset (1 - \zeta) \cdot R$. An RG-lattice M is a generalized permutation module if and only if M/π^t is a generalized permutation module.*

2. If $\pi^t \cdot R \nsubseteq (1-\zeta)\cdot R$, then there is an RG-lattice M, which is not a generalized permutation module, but M/π^t is a generalized permutation module.

Let us point out, where the restrictive condition on R/π^t becomes apparent:

Lemma 30. *Let R be a complete Dedekind domain of finite rank over $\hat{\mathbb{Z}}_p$. The induced map*

$$\operatorname{Ext}^1_{RG}(M, N) \to \operatorname{Ext}^1_{R/\pi^t G}(M/\pi^t, N/\pi^t)$$

is injective for all generalized permutation lattices M and N if and only if either R does not contain a primitive p-th root of unity or $\pi^t \cdot R \subset (1 - \zeta) \cdot R$, where ζ is a primitive p-th root of unity.

References

[Ba]   Banachevski, B.: Integral group rings of finite groups. Can. Math. Bull. **10** (1967) 635–642

[Ber]   Berman, S. D.: On a necessary condition for isomorphisms of integral group rings. Dopovidi Akad. Nauk Ukrain. RSR 1953, 313–316

[BR]   Berman, S. D., A.R. Rossa: Integral group rings of finite and periodic groups. Algebra and Math. Logic: Studies in Algebra, pp. 44–53. Izdat., Kiew 1966

[Br]   Brauer, R.: Representations of finite groups. In: Harvard Lectures on Modern Mathematics, 1963

[Ca]   Carlson, J.: The variety of a module. (Lecture Notes in Mathematics. vol. 1142.) Springer, Berlin Heidelberg New York 1985, pp. 88–95

[C]   Coleman, D. B.: On the modular group ring of a p-group. Proc. AMS. **5** (1964) 511–514

[D]   Dade, E.: Deux groupes finis distinctes ayant la même algèbre de groupes sur tout corps. Math. Z. **119** (1971) 345–348

[Hi]   Higman, G.: Units in group rings. D. Phil. Thesis. Oxford University, 1940

[J]   Jackson, D.A.: The groups of units of the integral group rings of finite metabelian and finite nilpotent groups. Quat. J. Math. Oxford Ser. **2**, 20 (1969) 319–331

[Ki]   Kimmerle, W.: Personal communication 1985

[KLS]   Kimmerle, W., R. Lyons, R. Sandling: Composition factors for group rings and Artin's theorem on orders of simple groups. Preprint 1987

[KR]   Kimmerle, W, K.W. Roggenkamp: A Sylow like theorem for integral group rings of finite solvable groups. To appear in Archiv der Mathematik (1990)

[Pa]   Passman, D.S.: Isomorphic groups and group rings. Pacific J. Math. **15** (1965) 561–583

[Pas]   Passman, D.: Group algebras of groups of order p^4 over a modular field. Mich. Math. J. **12** (1965) 405–415

[P]   Puig, L.: Pointed groups and construction of characters. Math. Z. **176** (1981) 265–292

[Rey]   Reynolds, *W. F.*: Blocks and normal subgroups of finite groups. Nagoya Math. J. **22** (1963) 15–32

[RiS]   Ritter, J., S.K. Sehgal: On a conjecture of Zassenhaus on torsion units in integral group rings. Preprint, Univ. Augsburg 1983

[RS]   Roggenkamp, K.W., L.L. Scott: Isomorphisms of p-adic group rings. MS Sept. 85, Ann. Math. **126** (1987) 593–647

[RS1]   Roggenkamp, K.W., L.L. Scott: The isomorphism theorem for integral group rings of nilpotent by abelian groups. MS, May 86, pp. 1–14 (to be published)

[RS2]   Roggenkamp, K.W., L.L. Scott: On a conjecture on group rings by H. Zassenhaus. MS, March 87, pp. 1–39 (to be published)

[RS3]   Roggenkamp, K.W., L.L. Scott: A strong answer to the isomorphism problem for finite p-solvable groups with a normal p-subgroup containing its centralizer. MS, 87 (to be published)

[RS4]   Roggenkamp, K.W., L.L. Scott: Automorphisms and non abelian cohomology. MS, 89

[R1]   Roggenkamp, K.W.: Picard groups of integral group rings of nilpotent groups. Proc. Symp. Pure Math. **47** (1987) 477–485

[R2]   Roggenkamp, K.W.: Units in integral metabelian group rings I. Jackson's unit theorem revisited. Quat. J. Math. **23** (1981) 209–224

[R]   Roggenkamp, K.W.: Subgroup rigidity of p-adic group rings (Weiss arguments revisited). MS 1989

[Ro]   Roggenkamp, K.W.: Observation to a conjecture of H. Zassenhaus. Proceedings of Groups, St. Andrews 1989 (to appear)

[Sak]   Saksonov, A.I.: Group rings of finite groups I. Publ. Math. Debrecen **18** (1971) 187–209

[Sa]   Sandling, R.: The isomorphism problem for group rings: A survey. (Lecture Notes in Mathematics, vol. 1142.) Springer, Berlin Heidelberg New York 1985, pp. 239–255

[S1]   Scott, L.L.: Report on the isomorphism problem. Proc. Symp. Pure Math. **47** (1987) 259–273

[S2]   Scott, L.L.: The modular theory of permutation representations. Representation theory of finite groups and related topics. AMS Meeting at Madison, April 1970, pp. 137–144

[S3]   Scott, L.L.: Defect groups and the Isomorphism Problem. Preprint 1988

[Th]   Thompson, G.: Conjugacy in p-adic group rings. Ph. D. thesis, Univerity of Virginia 1989

[Wa]   Ward, H.N.: Some results on the group algebra of a group over a prime field. Seminar on Finite Groups and Related Topics. Harvard University 1960–1961

[W]   Weiss, A.: p-adic rigidity of p-torsion. Ann. Math. **127** (1987) 317–332

[We]   Weiss, A.: Torsion units in integral group rings. MS, 89

[Wh]   Whitcomb, A.: The group ring problem. Ph. D. thesis, Chicago 1968

[Za]   Zassenhaus, H.: On the torsion units of finite group rings. Studies in Mathematics, Instituto de Alta Cultura, Lisboa 1974, pp. 119–126.

The Local to Global Principle in Algebraic K-Theory

*Robert W. Thomason**

CNRS, U.F.R. de Mathématiques, Université de Paris 7, F-75251 Paris Cedex 05, France

In the early 70s, Quillen [Q] succeeded in finding a good definition of the higher algebraic K-groups, generalizing the usual Grothendieck group K_0. To a noetherian ring or scheme X, his theory associates two distinct series of K-groups: $K_*(X)$, the theory associated to the exact category of finitely generated projective modules or algebraic vector bundles on X; and $G_*(X)$ or $K'_*(X)$, the theory associated to the abelian category of all finitely generated modules or coherent sheaves on X. Quillen's localization theorem for the K-theory of abelian categories applies to $G_*(X)$, and enables him to prove many fundamental results for this theory, including invariance of G_* under replacing the ring R by the ring of polynomials $R[T]$ or the scheme X by $X[T]$, a Mayer-Vietoris exact sequence for a cover of X by two open subschemes, and a Brown-Gersten local-to-global spectral sequence which reduces many G-theory problems to the case of local rings. For regular rings or schemes X, the existence of finite projective resolutions for all finitely generated modules shows that $K_*(X) = G_*(X)$, so all these good results apply to $K_*(X)$ in the regular case. However if X has singularities, Quillen's methods give many fewer results about $K_*(X)$. Unfortunately, this case arises often: for example the integral group ring of a non-zero finite group is always singular. The essential difficulty was the lack of the key tool: a good localization theorem for K-theory.

The author and Thomas F. Trobaugh have remedied this lack by finding the good localization result, using the techniques of Waldhausen's "algebraic K-theory of spaces" [Wa2] and returning to some ideas of Grothendieck about defining K-theory using "perfect complexes" in place of single vector bundles [SGA6]. Using a characterization of perfect complexes, essentially defined as finite complexes of algebraic vector bundles, as being the finitely presented objects in the derived category of modules, we obtain the result that the K_0 class is the only obstruction to extension up to quasi-isomorphism of a perfect complex on an open subscheme of X to à perfect complex on all of X. This extension result and Waldhausen's theory give the localization theorem for K-theory, Theorem 2.1 below. This in turn un-leashes a pack of new fundamental results for K-theory, including the Mayer-

---

* Author partially supported by NSF, The Johns Hopkins University and Université de Paris 7.

Vietoris exact sequence, the Brown-Gersten local-to-global spectral sequences for both the Zariski and Nisnevich topologies, the generalizations to schemes from rings of the results of Weibel, Ogle, and Goodwillie controlling the failure of K_* to be invariant under polynomial or nilpotent extension or to satisfy Mayer-Vietoris for closed covers, and the isomorphism between mod l algebraic K-theory localized by inverting the Bott element and etale topological K-theory. Recently Dongyuan Yao has obtained some similar localization results for the K-theory of non-commutative rings. Hiding in the background of all this is the fact that the Wald-hausen K-theory of a category of complexes depends essentially only on the derived category; this idea has been much advanced by work in progress of Giffen and Neeman, which unfortunately I will not discuss below.

§1. Perfect Complexes and Waldhausen K-Theory

We consider schemes X which are quasi-compact and quasi-separated. *Quasi-compact* means that X is covered by a finite number of open subschemes which are affine, i.e. isomorphic to Spec(R) for some commutative rings R. *Quasi-separated* means that the intersection of any two such affines is again quasi-compact. These conditions are indeed very mild, and are met by all algebraic varieties over a field, by all schemes of finite type over a noetherian ring, and by Spec(R) for any commutative ring R.

Recall that a map between two chain complexes in an abelian category is a *quasi-isomorphism* if it induces an isomorphism on the homology of the complexes. Two complexes E and F are quasi-isomorphic if they are connected by a chain of quasi-isomorphisms

$$E \xleftarrow{\simeq} A \xrightarrow{\simeq} \cdots \xrightarrow{\simeq} F$$

Definition (Grothendieck [SGA6]). *A perfect complex on a scheme X is a complex E of $\mathcal{O}_X$-modules, such that for every affine open $U = $ Spec(R) in X, the restriction $E|U$ is quasi-isomorphic to a finite complex of finitely generated projective R-modules.*

For E to be perfect, it suffices that $E|U$ be quasi-isomorphic on U to a finite complex of such projective modules for all the U in some affine open cover of X. In particular, for $X = $ Spec(R), the perfect complexes are those quasi-isomorphic to such complexes of projective R-modules.

The category of perfect complexes on X is a "category with cofibrations and weak equivalences" in the sense of [Wa2] 1.2, i.e., it is a Waldhausen category. The cofibrations are the maps of complexes which are degree-wise split monomorphisms; the weak equivalences are the quasi-isomorphisms. For Y a closed subspace of X, there is a full Waldhausen subcategory of those perfect complexes which are co-homologically supported on Y; i.e., those which are acyclic on $X - Y$.

The theory of Waldhausen [Wa2] associates higher algebraic K-groups to Waldhausen categories: these are the homotopy groups of a space, or even better, of an infinite loop space or spectrum in the sense of algebraic topology. Let $K_*(X)$ denote the Waldhausen algebraic K-groups associated to the category of perfect complexes on X, and $K(X)$ denote the associated K-theory spectrum. Let $K_*(X$ on $Y)$ and $K(X$ on $Y)$ denote the K-groups and spectrum associated to the Waldhausen subcategory of those perfect complexes on X which are acyclic on $X - Y$.

$K_0(X)$ is the Grothendieck group of [SGA6], generated by quasi-isomorphism classes $[E]$ of perfect complexes on X, modulo an Euler characteristic relation that $[E] = [E'] + [E'']$ whenever one can choose representatives of the quasi-isomorphism classes that fit in a short exact sequence of complexes $0 \to E' \to E \to E'' \to 0$, that is, whenever they are the vertices of an exact triangle in the derived category $D(X)$ of $\mathcal{O}_X$-modules.

Suppose X satisfies the mild hypothesis of having an "ample family of line bundles" in the sense of [SGA6] II.2.2.3. This condition is inherited by all subschemes of X and is satisfied when X is a commutative ring, or is quasi-projective over a commutative ring, or is a separated regular noetherian scheme. In particular, this condition holds for all classical algebraic varieties, since they are quasi-projective over a field. For a scheme X with an ample family of line bundles, a global resolution theorem of Illusie ([SGA6] II.2.2.8) says that every perfect complex on X is globally quasi-isomorphic to a finite complex of algebraic vector bundles defined on all of X. Then the $K_0(X)$ above is isomorphic to the "naive" Grothendieck group of algebraic vector bundles on X, and the Waldhausen $K_*(X)$ are isomorphic to the Quillen K-groups of X ([TT2] 3.10, 3.9).

We have an easy to verify but very useful characterization of perfect complexes in terms of the derived category $D(X)$. Recall that this category is obtained from the category of all chain complexes of $\mathcal{O}_X$-modules by inverting the quasi-isomorphisms (see Verdier [Ve]). $D(X)_{qc}$ denotes the full subcategory of those complexes whose homology sheaves are *quasi-coherent*, i.e., which locally on affine subschemes $\mathrm{Spec}(R)$ are isomorphic to R-modules. $D^+(D)_{qc}$ is the further subcategory of complexes E such that the homology sheaves $H_k(E)$ are 0 for all k sufficiently large.

Proposition 1.1. *Let X be a quasi-compact and quasi-separated scheme. Let E be an object of $D^+(X)_{qc}$. Then the following are equivalent:*

a) *E is a perfect complex*

b) *For any direct system of complexes F_α with quasi-coherent homologies, the canonical map is an isomorphism of morphism sets:*

$$\mathrm{colim}_\alpha \, D(X)(E, F_\alpha) \xrightarrow{\sim} D(X)(E, \mathrm{colim}_\alpha F_\alpha)$$

c) *The functor $D(X)(E, \ )$ sends infinite direct sums in $D^+(X)_{qc}$ to direct sums in the category of abelian groups.*

Proof. [TT2] 2.4.3 yields that a) iff b). That b) implies c) is clear, and that c) implies b) follows from a demonstration dual to that of [T3] 2.6. $\square$

Comparing condition b) with Grothendieck's characterization of finitely presented modules or algebras A over a ring or scheme as those such that $\mathrm{Mor}(A, \ )$

preserves direct colimits ([EGA] IV 8.14), we see this proposition characterizes perfect complexes as finitely presented objects in the derived category. This suggests interesting generalizations. Thus, the "perfect complexes" in the stable homotopy category turn out to be the homotopy finite cell spectra: they have an associated K-theory, which turns out to be Waldhausen's $A(\text{pt})$ ([Wa1] § 1). The proposition also suggests that we could apply the yoga for extending finitely presented objects from an open subscheme to all of X ([EGA] I.6.9). This suggestion leads to:

Key Proposition 1.2. *Let X be a quasi-compact and quasi-separated scheme, $j : U \to X$ the inclusion of a quasi-compact open subscheme, and F a perfect complex on U. Then F is quasi-isomorphic to the restriction j^*E of a perfect complex on X if and only if the class $[F]$ of F in $K_0(U)$ is in the image of $j^* : K_0(X) \to K_0(U)$.*

Proof. [TT2] 5.2.2 The basic idea is to write the complex with quasi-coherent homology $Rj_*(F)$ on X as quasi-isomorphic to a direct colimit of perfect complexes, colim $E_\alpha \approx Rj_*F$. Then on U, F is quasi-isomorphic to colim j^*E_α. As F is perfect, Proposition 1.1 shows that this quasi-isomorphism factors through some j^*E_α. Thus F is quasi-isomorphic to a summand of j^*E_α, the restriction of a perfect complex E_α on X. This is the miracle for general U, although it would be trivial for affine U. The next step, more ordinary, is to show that given a morphism on U between restrictions of perfect complexes, $j^*E_1 \to j^*E_2$, one can extend this to a morphism $E_1 \to E_2$ on X after replacing the old E_1 by another perfect complex whose restriction to U is quasi-isomorphic to that of the old one. An induction and excision reduces this to the case where X has an ample family of line bundles. If $X - U$ is a divisor, one succeeds after replacing E_1 by its tensor product with a power of the line bundle of the divisor. For $X - U$ general, one succeeds after replacing E_1 by its tensor power with the positive degree part of a Koszul complex of line bundles for divisors whose intersection is $X - U$. From this extension result on morphisms, it follows that if 2 out of 3 terms in an exact triangle of perfect complexes in $D(U)$ extend, so does the third. We now adapt an idea of Grayson from cofinality theory: consider the abelian monoid of quasi-isomorphism classes of perfect complexes on U, and take the quotient monoid by setting equal to 0 the classes that are restrictions of perfect complexes on X; using the preceding sentence one sees that the class of a perfect complex on U is zero in this quotient if and only if it extends to a perfect complex on X. The miracle above shows this monoid has inverses, and so is an abelian group. We see from the above that this obstruction group to extension satisfies the Euler characteristic relation for exact triangles, and thus in fact that it is the quotient of $K_0(U)$ by the image of $K_0(X)$. □

In particular, this key Proposition gives a K_0-criterion for an extending up to quasi-isomorphism an algebraic vector bundle on U to a perfect complex on X, that is, to a finite complex of algebraic vector bundles on X in the usual case where X has an ample family. No similar criterion is possible for the question of extension to a single vector bundle on X: it is for this reason that it is necessary to go to the framework of perfect complexes and leave the framework of algebraic vector bundles to prove a good localization theorem for algebraic K-theory. The impossibility of such a criterion was noted by Serre ([Se] 5a), who observed that as the tangent vector bundle to the projective plane $\mathbb{P}^2$ is not a sum of line bundles, its preimage on $\mathbb{A}^3 - 0$ is a vector bundle which is not free, and which does not extend to a vector bundle on $\mathbb{A}^3$, even though $K_0(\mathbb{A}^3) \to K_0(\mathbb{A}^3 - 0) = \mathbb{Z}$ is an isomorphism, and though this vector bundle on $\mathbb{A}^3 - 0$ is the quotient of a monomorphism of free vector bundles on $\mathbb{A}^3 - 0$, which of course do extend. The vast literature on the Serre conjecture contains many negative and a few special positive results on the problem of extending to vector bundles on X for X regular: one may consult the articles of Horrocks, Swan, Murthy, Vasserstein, Bass, M. Kumar, Suslin, and Quillen on this conjecture. Very few have dared to say anything about the case where X is singular.

§ 2. The Localization Theorem and Its Basic Consequences

Theorem 2.1 (Localization Theorem) *Let X be a quasi-compact and quasi-separated scheme, and let Y be a closed subspace of X such that $X - Y$ is quasi-compact. Then there is a natural homotopy fibre sequence, and an associated long exact sequence of homotopy groups*:

$$K(X \text{ on } Y) \to K(X) \to K(X - Y)$$

$$\cdots \to K_n(X \text{ on } Y) \to K_n(X) \to K_n(X - Y) \to K_{n-1}(X \text{ on } Y) \to$$

Proof. [TT2] 7.4. For $n \geq 0$, this results easily from the Key Proposition above and Waldhausen's fibration and approximation theorems [Wa2] 1.6.4 and 1.6.7. As the map $K_0(X) \to K_0(X - Y)$ is not always surjective in the singular case, it is necessary to construct new non-zero negative K-groups K_n for $n < 0$ to continue the exact sequence to the right, and to replace the spectrum $K(X)$ by a new non-connective spectrum of which the Waldhausen K-theory spectrum is a covering space. This is done by following an inductive procedure due to Bass [B], which depends on the part of the localization theorem already proved for higher n, and which uses the exact sequence of 2.2 to inductively define K_{n-1} for $n \leq 0$ given K_n. $\square$

The special case of 2.1 where $X - Y$ is affine and Y is a Cartier divisor is due to Quillen [Gr] (at least in light of [TT2] 5.7 which identifies $K(X \text{ on } Y)$ with Quillen's third term $K(H_Y X)$ in this case). Painful efforts to squeeze a bit more out of Quillen's method have been made by Levine [Le3]. The counterexample of Deligne given in Gersten's paper ([Ge] § 7) shows the limits of an exact category approach to localization.

Corollary 2.2 (Bass Fundamental Theorem). *Let X be a quasi-compact and quasi-separated scheme. Denote by $X[T]$ the scheme $X \times \mathbb{A}^1 = X \otimes_{\mathbb{Z}} \mathbb{Z}[T]$, and similarly by $X[T^{-1}]$ another polynomial extension of X, and by $X[T, T^{-1}]$ the Laurent polynomial extension. Then there is a natural exact sequence for all integers $n \in \mathbb{Z}$:*

$$0 \to K_n(X) \to K_n(X[T]) \oplus K_n(X[T^{-1}]) \to K_n(X[T, T^{-1}]) \to K_{n-1}(X) \to 0$$

Similarly for $K_*(X \text{ on } Y)$.

Proof. [TT2] 7.5. This is proved by descending induction on n using the formula of Quillen ([Q] § 8) that $K_n(\mathbb{P}_X^1) \cong K_n(X) \oplus K_n(X)$, and the Mayer-Vietoris Corollary 2.3 to the localization theorem applied to the cover of $\mathbb{P}_X^1$ by $X[T]$ and $X[T^{-1}]$ with intersection $X[T, T^{-1}]$. For X affine and $n \leq 1$, the result is indeed due to Bass [B]; and for X affine and $n \geq 2$, it is due to Quillen [Gr]. □

Corollary 2.3 (Mayer-Vietoris Theorem). *Let* U, V *be two quasi-compact open subschemes of a quasi-separated scheme* X. *Then there is a homotopy cartesian square:*

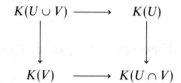

and an associated long exact Mayer-Vietoris sequence:

$$\cdots \to K_n(U \cup V) \to K_n(U) \oplus K_n(V) \to K_n(U \cap V) \to K_{n-1}(U \cup V) \to \cdots$$

Proof. [TT2] 8.1. The localization theorem calculates the homotopy fibres of the horizontal maps in the square, and an easy excision result shows they are equivalent, as U is an open neighborhood of $Y = (U \cup V) - V = U - (U \cap V)$ in $U \cup V$. □

Theorem 2.4 (Brown-Gersten or Local-to-Global Spectral Sequence, Spectral Sequence of Cohomological Descent). *Let* X *be a noetherian scheme of finite dimension. Then there is a strongly converging spectral sequence whose* E^2 *term is the cohomology of* X *for the Zariski topology with coefficients in the sheaf associated to the presheaf of K-groups:*

$$E_2^{p,-q} = H^p(X_{\mathrm{Zar}}; K_q) \Rightarrow K_{q-p}(X)$$

In fact, the augmentation map to the Zariski hypercohomology spectrum ([T1] § 1) *is a homotopy equivalence identifying this spectral sequence to the canonical hypercohomology spectral sequence.*

$$K(X) \xrightarrow{\sim} \mathbb{H}(X_{\mathrm{Zar}}; K)$$

One has the same results after replacing the Zariski topology by the slightly finer Nisnevich topology ([N] § 1, [TT2] Appendix E).

Proof. [TT2] 10.3, 10.8. In the Zariski case, Brown and Gersten showed this would follow from Mayer-Vietoris [BG], and proved the case for regular X or for G-theory. The argument for the Nisnevich topology is similar, but needs a slightly stronger excision result that $K(X \text{ on } Y)$ is unchanged by replacing X by its henselization along Y. Nisnevich [N] proved the Nisnevich case for regular X or for G-theory. Weibel [We4] proved the theorem in the case X has isolated singular points, using ideas of Collino and Pedrini. We note that the Nisnevich topology plays a big role in the work of Kato and Saito [KS], who call it the henselian topology. □

2.5 This spectral sequence reduces many problems to the case of local rings for the appropriate topology, since the stalks of the coefficient sheaves are the values of K_q at these local rings. For the Zariski topology, these are the local rings of X in the usual sense. For the Nisnevich topology, the local rings are the hensel local rings of X, the henselizations of the usual local rings. It is a theorem of Gabber [Ga], completing ideas of Suslin, that for R_m a hensel local ring and n an integer such that $1/n \in R_m$, then the reduction map to the residue field, $R_m \to R/m$, induces an isomorphism on mod n K-theory, at least in positive degrees, $K/n_*(R_m) \cong K/n_*(R/m)$. Thus many problems for mod n K-theory reduce to the case of fields. For example, this reduction plays an essential role in the proof of the étale cohomological descent Theorem 4.1 below.

2.6 In the previously known case where X is a regular scheme of finite type over a field k, this spectral sequence has interesting relations with classical intersection theory because of the formula for $A^p(X)$, the Chow group of codimension p algebraic cycles on X, inspired by Gersten, proved by Quillen ([Q] §7.5), and ascribed to Bloch (who in fact did prove the case $p = 2$, the case $p = 1$ being due to Cartier): $A^p(X) = H^p(X; K_p)$. As Nisnevich observed, this works either for the Zariski or the Nisnevich topology. It would be interesting to have a geometric interpretation of $H^p(X; K_p)$ and a relation to some kind of intersection theory in the singular case. Some work on this has been done by Collino [Co], Gillet [Gi], Levine [Le1, Le2, LeW], Pedrini [PW1, PW2], Weibel, and Barbieri Viale [Vi]. In the singular case, the Zariski and Nisnevich cohomologies are different, and there is some evidence that the Nisnevich cohomology works better.

2.7 The hard-core algebraist may have been wondering if all this scheme business gives anything new for the K-theory of ordinary commutative rings. The answer is that it does. For the proof of spectral sequence 2.4 for the case of a noetherian ring A, $X = \mathrm{Spec}(A)$, and thus the reduction of problems about $K(A)$ to the case of local rings, depends on an induction which requires considering arbitrary open subschemes U of $\mathrm{Spec}(A)$, not all of which will be affine when the dimension of A is greater than one. This is one of the reasons why Quillen's localization result with its hypothesis that $X - Y$ be affine did not suffice to prove Theorem 2.4 for noetherian rings. I also offer the following two new results:

Proposition 2.8. *Let R be a commutative ring, and let $r, s \in R$ be two elements such that the ideal generated by r and s is all R. Then there is a long exact Mayer-Vietoris sequence:*

$$\cdots \to K_n(R) \to K_n(R[1/r]) \oplus K_n(R[1/s]) \to K_n(R[1/rs]) \to K_{n-1}(R) \to \cdots.$$

Proof. This is a special case of 2.3 with $U = \mathrm{Spec}(R[1/r])$ and $V = \mathrm{Spec}(R[1/s])$. Since the ideal (r, s) is all of R, $\mathrm{Spec}(R) = U \cup V$. The result would follow from

Quillen's localization theorem [Gr] under the additional hypothesis that either r or s is not a zero-divisor in R, but this hypothesis is often embarrassing, e.g. in the case of group rings.   □

Proposition 2.9 (Weibel). *Let R be a commutative ring, and $r \in R$. Denote by $R_r^\wedge$ the completion* $\lim_k R/r^k R$. *Then there is an exact Mayer-Vietoris sequence*:

$$\cdots \to K_n(R) \to K_n(R_r^\wedge) \oplus K_n(R[1/r]) \to K_n(R_r^\wedge [1/r]) \to K_{n-1}(R) \to \cdots$$

Proof. This follows from 2.1 and an excision result that $K(R \text{ on } R/r) \approx K(R_r^\wedge$ on $R_r^\wedge/r)$, which Weibel observed could be proved by the method of [TT2] 2.6.3 and 3.19.   □

§ 3. Generalizations to Schemes of Some Results Known in the Affine Case

The Mayer-Vietoris Theorem 2.3 often permits one to generalize results known for affine schemes to general quasi-compact separated schemes X, by inducting on the number of affines in an open cover of X. (A second induction on the number of quasi-affines needed to cover X then relaxes the hypothesis of separation to that of quasi-separation). Work of Weibel [We1, We2, We3, We5], Ogle [Og, OW], Geller, Goodwillie [Go], Soulé [So], Staffeldt [Sta], van der Kallen [vdK], and Vorst [Vo] have produced many results controlling the failure of K-theory of rings to be invariant under polynomial or nilpotent extensions, and the failure of Mayer-Vietoris for closed covers of affine schemes. Their demonstrations depend on the use of projective objects and other ring-theoretic techniques, and do not readily generalize to schemes, where the locally projective algebraic vector bundles are not globally projective. However, our method generalizes their conclusions to the case of schemes. As a sample, I give the following results. The first two concern the mod n K-theory defined by Karoubi and Browder; these groups fit in a short exact universal coefficient sequence:

$$0 \to K_q(X) \otimes \mathbb{Z}/n \to K/n_q(X) \to \mathrm{Tor}_{\mathbb{Z}}(K_{q-1}(X), \mathbb{Z}/n) \to 0$$

Theorem 3.1 (Weibel) *Let X be a quasi-compact and quasi-separated scheme, and let $W \to X$ be the total space of a vector bundle over X (or even a torsor under a vector bundle). Let n be an integer such that $1/n \in \mathcal{O}_X$. Then $W \to X$ induces an isomorphism on* mod n *K-groups:*

$$K/n_*(X) \xrightarrow{\sim} K/n_*(W)$$

In particular, this is true when $W = X[T]$.

Proof. [TT2] 9.5. This is deduced by the method above, using the 5-lemma and the Mayer-Vietoris sequence 2.3, from the affine case due to Weibel [We1] 3.3, [We3]. The critical case is that of the polynomial extension $R[T]$ of a ring R. In general, the induced map $K_*(R) \to K_*(R[T])$ has a non-trivial cokernel, although it is obviously injective, split by the map induced by the ring homomorphism sending T to 0. Stienstra [Sti], following work of Almkvist, Grayson, and Bloch, showed that the cokernel was a module over the ring of Witt vectors of R. Weibel observed this shows the cokernel is uniquely n-divisible if $1/n \in R$, yielding the result in the critical case. $\qquad\qquad\qquad\qquad\qquad\qquad\qquad\qquad\qquad\qquad\qquad\qquad\qquad\quad\Box$

Theorem 3.2 (Weibel). *Let X be a quasi-compact and quasi-separated scheme, with two closed subschemes Y, Z, such that $X = Y \cup Z$ as spaces. Give the intersection $Y \cap Z$ the scheme structure of the fibre product of Y and Z over X. Let n be an integer such that $1/n \in \mathcal{O}_X$. Then there is a homotopy cartesian square:*

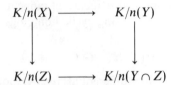

and an associated long exact Mayer-Vietoris sequence:

$$\cdots \to K/n_q(X) \to K/n_q(Y) \oplus K/n_q(Z) \to K/n_q(Y \cap Z) \to K/n_{q-1}(X) \to \cdots.$$

Proof. [TT2] 9.8. Again one reduces to the affine case, which is due to Weibel, essentially [We2] 1.3. The corresponding statement for K_* in place of K/n_* would not be true, but its failure can be analyzed locally using cyclic homology [OW], which result can be globalized by the method of 3.3 below. $\qquad\qquad\qquad\quad\Box$

3.3 Much recent work has focused on cyclic homology HC_*, which serves as a sort of "linear approximation" to algebraic K-theory, and is more readily calculable. Discovered for operator algebras by Connes, cyclic homology has been extended to algebras over a field k by Loday-Quillen and Feigin-Tsygan, and has been extensively developed by J. Block, Brylinksi, Burghelea, Carlsson, Cathelineau, R. Cohen, Goodwillie, J. Jones, Karoubi, Kassel, Ogle, Staffeldt, Vigué-Poirier, Weibel, Wodzicki, and others in a torrent of articles [Lo].

Independently, several people, including J. Block, Loday, and Weibel have proved that HC_* has a Cech-Mayer-Vietoris spectral sequence for a cover of $\mathrm{Spec}(R)$ by affine open subschemes. This allows one to extend the definition of $HC_*(X)$ to schemes X over a field k, by taking homology of the total complex of the Cech complex of an affine cover of X, with coefficients in the cyclic homology complex HC. We deduce, as suggested by Weibel:

Theorem 3.4 (Goodwillie). *Let X' be a quasi-compact and quasi-separated scheme over $\mathrm{Spec}(\mathbb{Q})$. Let $i : X \to X'$ be a closed immersion defined by a sheaf of nil ideals. Then there is a natural isomorphism between the relative cyclic homology over $\mathbb{Q}$ and the relative algebraic K-theory:*

$$K_n(X \to X') \cong HC_{n-1}(X \to X').$$

Proof. Here $K(X \to X')$ is the homotopy fibre of the map $i^* : K(X') \to K(X)$, so that the relative K-groups $K_*(X \to X')$ fit in a long exact sequence with $K_*(X)$ and

$K_*(X')$; and similarly for the relative cyclic homology complex $HC(X \to X')$. For X' affine, the main theorem of [Go] gives a natural homotopy equivalence between $K(X \to X')$ and the complex $HC(X \to X')$ shifted one degree and considered as a generalized Eilenberg-MacLane spectrum. The result for general X then follows by Mayer-Vietoris, cf. [TT2] 9.10. □

Levine has remarked that this result is interesting when X is smooth over a field k of characteristic 0, and X' is the singular infinitesimal thickening of X to $X[\varepsilon]/\varepsilon^2$. Then $K_*(X \to X')$ is by Grothendieck's definition the "tangent space" to K_* at X, and the result shows that it indeed equals the tangent space to the "linear approximation" HC_*. One has the spectral sequence of tangent spaces given by applying 2.4 to the fibre of $K(X') \to K(X)$; using cyclic homology to calculate the E_2 term gives:

$$E_2^{p,-q} = H^p(X; \Omega_X^{q-1} \oplus \Omega_X^{q-3} \oplus \Omega_X^{q-5} \oplus \cdots)$$

where Ω_X are the absolute Kähler differentials over $\mathbb{Q}$, not the relative ones over k. This does suggest somewhat obscure relations between this tangent space, Hodge theory, and infinitesimal deformations of algebraic cycles.

§4. Comparison of Algebraic and Etale Topological K-Theory

4.0 Suppose now that X is a scheme of finite type over a ring k, where k is one of the following: an algebraically closed field, a separably closed field, a number field, a ring of integers in a number field, $\mathbb{Z}_p^\wedge$, $\mathbb{Q}_p^\wedge$, $\mathbb{F}_q[[t]]$, $\mathbb{F}_q((t))$, or $\mathbb{F}_q$. Fix a prime power $l^\nu \geq 3$. Suppose that $1/l \in \mathcal{O}_X$, and that if $l = 2$ then $\mathcal{O}_X$ contains a square root of -1.

If k contains all the l^ν-th roots of unity, this gives a torsion subgroup $\mathbb{Z}/l^\nu$ in the group of units k^*, and hence in $K_1(k)$. By the universal coefficient sequence, this corresponds to an element β in $K/l^\nu_2(k)$. One may then localize the ring $K/l^\nu_*(X)$ by inverting the image of this Bott element β. If k doesn't contain all the l^ν-th roots of unity, one can still form this $K/l^\nu(X)[\beta^{-1}]$: as Dwyer showed, essentially a power of β exists in $K/l^\nu_*(k)$, although β in degree 2 does not ([T1], Appendix A).

Theorem 4.1. *For X as in 4.0, there is a strongly converging spectral sequence from etale cohomolgy:*

$$E_2^{p,-q} \begin{cases} H^p(X_{et}; \mathbb{Z}/l^\nu(i)) & q = 2i \\ 0 & q \text{ odd} \end{cases} \Rightarrow K/l^\nu_{q-p}(X)[\beta^{-1}]$$

The canonical map which forgets the algebraic structure on an algebraic vector bundle and remembers only the underlying topological vector bundle induces an equivalence with the Dwyer-Friedlander etale topological K-theory [DF]:

$$\varrho : K/l^\nu(X)[\beta^{-1}] \xrightarrow{\simeq} K/l^{\nu \text{Top}}(X)$$

Proof. [TT2] 11.5. The method of 2.5 reduces this to the case where X is replaced by its various residue fields, which case was done in [T1], as was the case where X is regular. □

One notes that the results of Suslin-Gabber-Gillet-Thomason [Ga, Su1, Su2, GiT] on the K-theory of strict hensel local rings, the local rings for the etale topology, show that the values of the sheaves K/l^v_q in the etale topology are the i-times Tate-twisted cyclic groups $\mathbb{Z}/l^v(i)$ for $q = 2i$ even and non-negative, and are 0 for q odd and positive. Thus this spectral sequence is the spectral sequence of etale cohomological descent for K/l^v_*, at least in positive degrees. However, it converges to $K/l^v_*(X)[\beta^{-1}]$, and not to $K/l^v_*(X)$, which is in general different. I believe that $K/l^v_q(X)[\beta^{-1}]$ equals $K/l^v_q(X)$ for q sufficiently large (cf. the results of [T2]), but this is definitely false for q small, e.g. for $q = 0$ and X a $K3$ surface over $\mathbb{C}$ or $\mathrm{Spec}(R[1/l])$, where R is a ring of integers in a number field which has more than one prime lying over l ([T1] Example 4.5).

The relation between algebraic K-theory and etale cohomology remains a very active area. Leaving aside the infinity of baseless conjectures, there remain some other actual theorems in special cases: the surjectivity results of Soulé in the case of rings of integers, as perfected by Dwyer-Friedlander [DF]; complete results on K_2 of fields due to Merkurjev and Suslin [MS1]; less complete results about K_3 of fields due to Merkurjev-Suslin ([MS2], [MS3]), Levine [Le4], and Rost [R]; and the descent result of Carlsson for $\pi_* BGL_N^+$ with respect to finite Galois extensions of certain localized rings of integers. The K_2 result of Merkurjev-Suslin has been extensively employed to study algebraic cycles in codimension 2, notably in the hands of Colliot-Thélène, Coombes, M. Gros, Raskind, Salberger, Sansuc, and Soulé. M. Harada has proved a Riemann-Roch theorem for singular varieties without a quasi-projectivity hypothesis by using the homology version of 4.1 for G-theory to reduce Riemann-Roch to the projection formula ([H], [T4] §15).

§5. The Case of Non-commutative Rings

D. Yao has begun the study of the localization theorem for non-commutative rings [Y]. This involves many new technical problems, and the construction of a sort of algebraic geometry of Grothendieck abelian categories. As a sample of his results, one has:

Theorem 5.1 (D. Yao) *Let R be a ring, not necessarily commutative. Suppose that s_1, $s_2, \ldots, s_n$ are elements of R such that the ideal they generate is all of R. Suppose there exist ring automorphisms ϕ_i for $i = 1, 2, \ldots, n$ such that:*

1) *for all $r \in R$, $\phi_i(r)s_i = s_i r$*
2) *$\phi_i(s_j) = s_j$*
3) *$\phi_i \phi_j = \phi_j \phi_i$*

(*For example, the s_i could be central, and the $\phi_i = \mathrm{id}$.*) *Then there is a strongly converging Cech cohomology spectral sequence computing $K_*(R)$ from the K-groups of the localizations:*

$$E_2^{p,-q} = H^p\left(\cdots \to \bigoplus_{i_0 < i_1 < \cdots < i_n} K_q(R[1/s_{i_0} \cdot 1/s_{i_1} \cdot \ldots \cdot 1/s_{i_n}]) \to \cdots\right) \Rightarrow K_{q-p}(R).$$

References

[B]   H. Bass: Algebraic K-theory. Benjamin 1968

[BG]   K. Brown, S. Gersten: Algebraic K-theory as generalized sheaf homology. In:
         Higher K-theories. (Lecture Notes in Mathematics, vol. 341) Springer, Berlin
         Heidelberg New York 1973, pp. 266–292

[Co]   A. Collino: Quillen's K-theory and algebraic cycles on almost nonsingular varieties
         Ill. J. Math. **25** (1981) 654–666

[DF]   W. Dwyer, E. Friedlander: Etale K-theory and arithmetic. Trans. Amer. Math. Soc.
         292 (1985) 247–280

[Ga]   O. Gabber: K-theory of henselian local rings and henselian pairs Preprint 1985

[Ge]   S. Gersten: The localization theorem for projective modules. Comm. Alg. **2** (1974)
         307–350

[Gi]   H. Gillet: On the K-theory of surfaces with multiple curves and a conjecture of
         Bloch. Duke Math. J. **51** (1984) 195–233

[GiT]   H. Gillet, R. Thomason: The K-theory of strict hensel local rings and a theorem of
         Suslin. J. Pure Appl. Alg. **34** (1984) 241–254

[Go]   T. Goodwillie: Relative algebraic K-theory and cyclic homology. Ann. Math. **124**
         (1986) 347–402

[Gr]   D. Grayson: Higher algebraic K-theory II (after D. Quillen). In: Algebraic K-theory:
         Evanston 1976. (Lecture Notes in Mathematics, vol. 551) Springer, Berlin Heidel-
         berg New York 1976, pp. 217–240

[H]   M. Harada: A proof of the Riemann-Roch theorem. Preprint 1988

[KS]   K. Kato, S. Saito: Global class field theory of arithmetic surfaces. In: Applications
         of algebraic K-theory to algebraic geometry and number theory, part I. (Contem-
         porary Mathematics, vol. 55) AMS, 1986, pp. 255–331

[vdK]   W. van der Kallen: Descent for K-theory of polynomial rings. Math. Z. **191** (1986)
         405–415

[Lec]   F. Lecomte: Rigidité des groupes de Chow. Duke J. Math. **53** (1986) 405–426

[Le1]   M. Levine: Bloch's formula for singular surfaces. Topology **24** (1985) 165–174

[Le2]   M. Levine: Zero cycles and K-theory on singular varieties. In: Algebraic geometry:
         Bowdoin 1985. (Proc. Symp. Pure Mathematics, vol. 46, part 2) AMS 1987, pp.
         451–462

[Le3]   M. Levine: Localization on singular varieties. Invent. math. **91** (1988) 423–464

[Le4]   M. Levine: The indecomposable K_3 of fields. Ann. Sci. Ec. Norm. Sup. (4) **22** (1989)
         255–344

[LeW]   M. Levine, C. Weibel: Zero cycles and complete intersections on singular varieties.
         J. Reine Angew. Math. **359** (1985) 106–120

[Lo]   J.-L. Loday: Cyclic homology: a survey. In: Geometric and algebraic topology.
         (Banach Center Publication, vol. 18) PWN 1986, pp. 285–307

[MS1]   А.С. Меркурьев, А.А. Суслин: К-когомологий многооБразий Севери-Брауэра
         и гомоморфизм норенного вычета. In: Изв. Акал. Наук С.С.С.Р. Сер. Мат.
         46 (1982) 1011–1046 [English translation in Math. USSR Izv. **21** (1983) 307–340]

[MS2]   A.S. Merkurjev, A.A. Suslin: On the norm residue homomorphism of degree three.
         LOMI preprint 1986

[MS3]   A.S. Merkurjev, A.A. Suslin: On K_3 of a field. LOMI preprint 1987

[N]   Y. Nisnevich: The completely decomposed topology on schemes and associated
         descent spectral sequences in algebraic K-theory. In: Algebraic K-theory: Connec-
         tions with geometry and topology. Kluwer 1989, pp. 241–342

[Og]   C. Ogle: On the K-theory and cyclic homology of a square-zero ideal. J. Pure Appl.
         Alg. **46** (1987) 233–248

[OW]   C. Ogle, C. Weibel: Relative K-theory and cyclic homology (to appear)

[PW1] C. Perdrini, C. Weibel: K-theory and Chow groups on singular varieties. In: Applications of algebraic K-theory to algebraic geometry and number theory, part I. (Contemporary Mathematics, vol. 55) AMS 1986, pp. 339–370

[PW2] C. Pedrini, C. Weibel: Bloch's formula for varieties with isolated singularities. Comm. Alg. **14** (1986) 1895–1907

[Q] D. Quillen: Higher algebraic K-theory I. In: Higher K-theories. (Lecture Notes in Mathematics, vol. 341) Springer, Berlin Heidelberg New York 1973, pp. 85–147

[R] M. Rost: Hilbert's theorem 90 for $K^M{}_3$. Preprint 1986

[Se] J.-P. Serre: Prolongement des faisceaux analytiques cohérents. Ann. Inst. Fourier **16** (1966) 363–374

[So] C. Soulé: Rational K-theory of the dual numbers of a ring of algebraic integers. In: Algebraic K-theory, Evanston 1980. (Lecture Notes in Mathematics, vol. 854) Springer, Berlin Heidelberg New York 1981, pp. 402–408

[Sta] R. Staffeldt: Rational algebraic K-theory of certain truncated polynomial rings. Proc. Amer. Math. Soc. **95** (1985) 191–198

[Sti] J. Stienstra: Operations in the higher K-theory of endomorphisms. In: Current trends in algebraic topology, part 2. AMS 1982, pp. 59–115

[Su1] A.A. Suslin: On the K-theory of algebraically closed fields. Invent. math. **73** (1983) 241–245

[Su2] A.A. Suslin: On the K-theory of local fields. J. Pure Appl. Alg. **34** (1984) 301–318

[Su3] A.A. Suslin: Algebraic K-theory of fields. Proceedings of the International Congress of Mathematicians; Berkeley 1986, vol. 1. AMS, pp. 222–244

[T1] R.W. Thomason: Algebraic K-theory and etale cohomology. Ann. Sci. Ec. Norm. Sup. (4) **18** (1985) 437–552; erratum (4) **22** (1989) 675–677

[T2] R.W. Thomason: Bott stability in algebraic K-theory. In: Applications of algebraic K-theory to algebraic geometry and number theory, part I. (Contemporary Mathematics, vol. 55) AMS 1986, pp. 389–406

[T3] R.W. Thomason: The finite stable homotopy type of some topoi. J. Pure Appl. Alg. **47** (1987) 89–104

[T4] R.W. Thomason: Survey of algebraic vs. etale topological K-theory. In: Algebraic K-theory and algebraic number theory. (Contemporary Mathematics, vol. 83) AMS 1989, pp. 393–443

[TT1] R.W. Thomason, T.F. Trobaugh: Le théorème de localisation en K-théorie algébrique. C.R. Acad. Sci. Paris Ser. I **307** (1988) 829–831

[TT2] R.W. Thomason, T.F. Trobaugh: Higher algebraic K-theory of schemes and of derived categories. To appear in the Festschrift for Grothendieck (Progress in Math.) Birkhäuser, Basel

[Ve] J.-L. Verdier: Catégories dérivées. In: SGA4&1/2: Cohomologie étale. (Lecture Notes in Mathematics, vol. 569) Springer, Berlin Heidelberg New York 1977, pp. 262–311

[Vi] L. Barbieri Viale: Cohomology theories and algebraic cycles on singular varieties. Preprint 1989

[Vo] T. Vorst: Localization of the K-theory of polynomial extensions. Math. Ann. **244** (1979) 333–353

[Wa1] F. Waldhausen: Algebraic K-theory of spaces, localization, and the chromatic filtration of stable homotopy. In: Algebraic topology, Aarhus 1982. (Lecture Notes in Mathematics, vol. 1051) Springer, Berlin Heidelberg New York 1984, pp. 173–195

[Wa2] F. Waldhausen: Algebraic K-theory of spaces. In: Algebraic and geometric topology. (Lecture Notes in Mathematics, vol. 1126) Springer, Berlin Heidelberg New York 1985, pp. 318–419

[We1] C. Weibel: Mayer-Vietoris sequences and module structures on NK. In: Algebraic

           K-theory, Evanston 1980 (Lecture Notes in Mathematics, vol. 854) Springer, Berlin
           Heidelberg New York 1981, pp. 466–493

[We2]   C. Weibel: Mayer-Vietoris sequences and mod p K-theory. In: Algebraic K-theory,
           Oberwolfach 1980. (Lecture Notes in Mathematics, vol. 966) Springer, Berlin
           Heidelberg New York 1982, pp. 390–407

[We3]   C. Weibel: Module structures on the K-theory of graded rings. J. Alg. **105** (1987)
           465–483

[We4]   C. Weibel: A. Brown-Gersten spectral sequence for the K-theory of varieties with
           isolated singularities. Adv. Math. **73** (1989) 192–203

[We5]   C. Weibel: "Homotopy algebraic K-theory. In: Algebraic K-theory and number
           theory. (Contemporary Mathematics, vol. 83) AMS 1989, pp. 461–488

[Y]   D. Yao: Thesis, The Johns Hopkins University 1990

[SGA6]   P. Berthelot, A. Grothendieck, L. Illusie: SGA6: Théories des intersections et
           théorème de Riemann-Roch. (Lecture Notes in Mathematics, vol. 225) Springer,
           Berlin Heidelberg New York 1971

[EGA]   A. Grothendieck, J. Dieudonné: Eléments de géométrie algébrique. (Publ. Math.
           I.H.E.S. Nos. 8, 11, 17, 20, 24, 28, 32) Presse Univ. France (1961–1967). Chapitre 1
           refondu: (Grundlehren der mathematischen Wissenschaften, Bd. 166) Springer,
           Berlin Heidelberg New York 1971

On the Restricted Burnside Problem

Efim I. Zelmanov

Institute of Mathematics, Universitetskii pr. 4, Novosibirsk-90, 630 090 USSR

In 1902, W. Burnside formulated his famous problems for periodic groups [6]:

The Burnside Problem (also known as the Ordinary Burnside Problem): Is it true that every finitely generated group of bounded exponent is finite?

The General Burnside Problem: Is it true that every finitely generated periodic group is finite?

After many unsuccessful attempts to obtain a proof in the late 30s–early 40s the following weaker version of The Burnside Problem was studied: Is it true that there are only finitely many m-generated finite groups of exponent n? In other words the question is whether there exists a universal finite m-generated group of exponent n having all other finite m-generated groups of exponent n as homomorphic images. Later (thanks to W. Magnus [35]) this question became known as **The Restricted Burnside Problem**.

In 1964 E. S. Golod gave a negative answer to The General Burnside Problem (cf. [9]). Since then a considerable array of infinitely generated periodic groups was constructed by other authors (cf. Alyoshin [2], Suschansky [44], Grigorchuk [11], Gupta-Sidki [54]).

In 1968 P. S. Novikov and S. I. Adian [39] constructed counter-examples to The Burnside Problem for groups of odd exponents $n \geq 4381$ (now for odd exponents $n \geq 115$, cf. I. Lysenok [33]). Olshansky's Monsters (cf. [40]) shows how wildly periodic groups may behave.

At the same time there were two major reasons to believe that The Restricted Burnside Problem would have a positive solution. One of these reasons was the reduction theorem obtained by Ph. Hall and G. Higman [14]. Let $n = p_1^{k_1} \ldots p_r^{k_r}$, where p_i are distinct prime numbers, $k_i \geq 1$, and assume that (a) The Restricted Burnside Problem for groups of exponents $p_i^{k_i}$ has a positive solution, (b) there are only a finite number of finite simple groups of exponent n, (c) the factor group $\text{Out}(G) = \text{Aut}(G)/\text{Inn}(G)$ is solvable for any finite simple group of exponent n. Then The Restricted Burnside Problem for groups of exponent n also has positive solution.

Another reason was the close relation of The Problem to Lie algebras. Suppose $n = p^k$, where p is a prime number. Then the finite group G of exponent p^k is clearly nilpotent. It is easy to see that it is sufficient to find an upper bound

Proceedings of the International Congress
of Mathematicians, Kyoto, Japan, 1990
© The Mathematical Society of Japan, 1991

$f(m, p^k)$ for the class of nilpotency of all m-generated groups G of exponent p^k. Consider the lower central series $G = \gamma_1(G) > \gamma_2(G) > \cdots > \gamma_s(G) = \langle 1 \rangle, \gamma_{i+1}(G) = (\gamma_i(G), G), 1 \leq i < S$, and the direct sum of abelian groups

$$L_\gamma(G) = \overset{s-1}{\underset{i=1}{\oplus}} \gamma_i(G)/\gamma_{i+1}(G).$$

Brackets $[a_i\gamma_{i+1}(G), b_j\gamma_{j+1}(G)] = (a_i, b_j)\gamma_{i+j+1}(G)$, where $a_i \in \gamma_i(G), b_j \in \gamma_j(G), (,)$ is the group commutator, define the structure of a Lie ring on $L_\gamma(G)$ (cf. [4], [12, 34, 48]). It is obvious that the Lie ring $L_\gamma(G)$ has the same class of nilpotency as the group G. If G is generated by elements $x_1, \ldots, x_m$ then $L_\gamma(G)$ is generated by $x_i\gamma_2(G), 1 \leq i \leq m$.

If $n = p$ is a prime number then $L_\gamma(G)$ is an algebra over the field $Z_p, |Z_p| = p$, which satisfies Engel's identity

$$[\ldots[y, \underbrace{x], x], \ldots, x}_{p-1}] = 0 \qquad\qquad E_{p-1}$$

(cf. [35]). Thus The Problem for groups of prime exponent has been reduced to the following problem in Lie algebras: is it true that a Lie algebra over Z_p, which satisfies Engel's identity E_{p-1}, is locally nilpotent?

The last problem was successfully solved by A. P. Kostrikin [27, 28] wo solved in this way The Restricted Burnside Problem for groups of prime exponent.

If G is a finite group of prime power exponent p^k then $L_\gamma(G)$ is no longer an algebra over the field Z_p, it is an algebra over the ring $Z(p^k)$ of residues modulo p^k. That's why along with the lower central series $\gamma_i(G)$ we shall consider the lower central p-series of G (cf. [21], [31, 48, 46]):

$$G = G_1 \supset G_2 \supset \ldots,$$

where G_i is the subgroup of G generated by commutators $((\ldots(x_1, x_2), x_3), \ldots, x_r)$, $r \geq i$, and powers $((\ldots(x_1, x_2), x_3), \ldots, x_r)^{p^\ell}, r \cdot p^\ell \geq i$. It is easy to see that $G_i \supseteq \gamma_i(G) \supseteq G_{i \cdot p^k}$,

$$L(G) = \underset{i \geq 1}{\oplus} G_i/G_{i+1}$$

is an algebra over Z_p. Neither the Lie ring $L_\gamma(G)$ nor $L(G)$ need necessarily satisfy Engel's identity E_{p^k-1} (cf. [13, 16]) but

(1) the Lie algebra $L(G)$ satisfies the linearized Engel's identity E_{p^k-1}, that is, for arbitrary elements $a_1, \ldots, a_{p^k-1} \in L(G)$ we have

$$\Sigma \text{ad}(a_{\sigma(1)}) \ldots \text{ad}(a_{\sigma(p^k-1)}) = 0, \quad \sigma \in S_{p^k-1},$$

(G. Higman, [17]).

(2) for an arbitrary commutator ϱ on the generators $x_i G_2, 1 \leq i \leq r$, we have

$$\text{ad}(\varrho)^{p^k} = 0$$

(I. N. Sanov, [42]).

Now let us turn to what was happening in associative and Lie nil algebras.

A. G. Kurosch [30]* and independently J. Levitzky (cf. [3]) formulated two problems for nil algebras which were similar to Burnside's problems.

The General Kurosh-Levitzky Problem: Is every finitely generated nil algebra nilpotent?

The (Ordinary) Kurosh-Levitzky Problem: Is every finitely generated nil algebra of bounded degree nilpotent?

Actually it was a counterexample to the General Kurosh-Levitzky Problem which was constructed in the paper, [9] of E. S. Golod, then this counterexample was used to construct the first counterexample to the General Burnside Problem. Remark that so far it remains the only counterexample to the General Kurosh-Levitzky Problem.

For the (Ordinary) Kurosh-Levitzky Problem we have a quite different situation. Unlike Group Theory it has only positive solutions in all important classes of algebras. To appreciate the impact this problem had on the Ring Theory let us mention that N. Jacobson's Structure Theory of Algebras was stimulated by the Kurosh Problem and I. Kaplansky introduced the concept of a PI-algebra in search of the most general conditions which ensure the positive solution of the Problem. The following result (in its final form) was due to I. Kaplansky [23].

Theorem. *A finitely generated nil algebra which satisfies a polynomial identity is nilpotent.*

In 1956 A. I. Shirshov suggested another purely combinatorial direct approach to the Kurosh-Levitzky Problem.

Theorem (A. I. Shirshov [43]). *Suppose that an associative algebra A is generated by elements $x_1, \ldots, x_r$ and assume that (1) A satisfies a polynomial identity of degree n, (2) every monomial in $\{x_i\}$ of degree $\leq n$ is nilpotent. Then A is nilpotent.*

It is very important that the nilpotency assumption here is imposed not on every element of A, but only on the monomials in the generators (even on a finite collection of them).

Now let us turn to Lie algebras. It is natural to call an element $a \in L$ nilpotent if the operator ad (a) is nilpotent. With this definition both Kurosh-Levitzky problems become meaningful for Lie algebras. Moreover, by the results of G. Higman and I. N. Sanov (cf. above) the Lie algebra $L(G)$ of a finite group G of exponent p^n satisfies the assumptions of the Kurosh-Levitzky Problem in the form of A. I. Shirshov (the role of monomials is played by commutators).

In [52, 43] we solved this problem for those Lie algebras which satisfy the linearized Engel's identity E_n.

---

* Actually Kurosh's Problem concerned algebraic algebras but we consider only the important case of nil algebras.

Theorem 1. *Suppose that a Lie algebra L is generated by elements $x_1, \ldots, x_r$ and assume that there exist integers $n \geq 1, m \geq 1$ such that (1) L satisfies the linearized Engel's identity E_n, (2) for an arbitrary commutator ϱ on the generators x_i we have $\mathrm{ad}\,(\varrho)^m = 0$. Then L is nilpotent.*

Corollary. *A Lie ring which satisfies Engel's identity is locally nilpotent.*

From Theorem 1 we derive

Theorem 2. *The Restricted Burnside Problem has a positive solution for groups of exponent p^k.*

By the Reduction Theorem of Ph. Hall and G. Higman, the Restricted Burnside Problem has a positive solution in the class of soluble groups, in particular it has a positive solution for groups of odd exponent (by the celebrated Theorem of W. Feit and J. Thompson [8]) and for groups of exponent $n = p^\alpha q^\beta, p, q$ are prime numbers (cf. W. Burnside, [7]).

The announced classification of finite simple groups (cf. [10]) implies that the Restricted Burnside Problem has a positive solution for groups of arbitrary exponent. Now we shall try to explain briefly the idea of the proof of Theorem 1.

In [50] we proved that to prove Theorems 1, 2 it suffices to prove that a restricted (in the sense of N. Jacobson [18]) Lie algebra over an infinite field, which satisfies an Engel's identity, is locally nilpotent. An element a of a Lie algebra L is called sandwich if

$$[[L, a], a] = 0, \quad [[[L, a], L], a] = 0$$

(cf. A. I. Kostrikin, [27]). In case of odd characteristics the second equality easily follows from the first one, however if char $= 2$ then both conditions are necessary. We call a Lie algebra a sandwich algebra if it is generated by a finite collection of sandwiches. The following theorem is due to A. I. Kostrikin and the author.

Theorem About Sandwich Algebras [49]. *A sandwich Lie algebra is nilpotent.*

This theorem suggests the following plan of attack on Theorem 1 (which has been outlined in [50]). Assume that there exists a nonzero Lie algebra L over an infinite field K which satisfies an Engel's identity but isn't locally nilpotent. Then taking the factor-algebra of L modulo its locally nilpotent radical (cf. [26, 41]) we may assume L doesn't contain any nonzero locally nilpotent ideals. Suppose we manage to construct a Lie polynomial $f(x_1, \ldots, x_r)$ such that f is not identically zero on L and for arbitrary elements $a_1, \ldots, a_r \in L$ the value $f(a_1, \ldots, a_r)$ is a sandwich of L. The K-linear span of $f(L) = \{f(a_1, \ldots, a_r) | a_1, \ldots, a_r \in L\}$ is an ideal in L. By the theorem about sandwich algebras the ideal $Kf(L)$ is locally nilpotent which contradicts our assumption.

However one year of effort didn't bring us a desired sandwich-valued polynomial (its existence a posteriori follows from Theorem 1). Instead in November of 1988 we constructed an even sandwich-valued superpolynomial f, which means that for a Lie superalgebra $L = L_0 + L_1$ satisfying the superization of E_n every value of f is a sandwich of L_0. It turned out to be a good substitute of sandwich-valued polynomials. The sketch of this rather complicated construction appeared in [51]. Unfortunately it worked only for characteristics $\neq 2, 3$.

In January of 1989 we constructed another "generalized" nonzero sandwich-valued polynomials (this time involving "divided powers" of ad-operators), their full linearizations being ordinary polynomials. Every value of such a full linearization is a linear combination of sandwiches. This approach worked for any p (cf. [52, 53]).

Some lengthy computations from the proof (which are really hard to read) may be explained within the framework of Jordan Algebra Theory (cf. [19, 20]). We shall demonstrate the idea for the simpler case $p \neq 2, 3$. The first (less computational) part of the construction of a sandwich-valued polynomial is a construction of a polynomial f such that $f(L) \neq 0$ and for an arbitrary element $a \in f(L)$ we have $\mathrm{ad}\,(a)^3 = 0$. Choose arbitrary elements $a, b \in f(L)$ and consider the subspaces $\mathcal{2}^+ = L\,\mathrm{ad}\,(a)^2$, $\mathcal{2}^- = L\,\mathrm{ad}\,(b)^2$. Then for an arbitrary element $c \in \mathcal{2}^-$ the operation $x \circ y = [x, c, y]; x, y \in \mathcal{2}^+$, defines the structure of a Jordan algebra on $\mathcal{2}^+$ (cf. [5, 19, 22, 25]). The pair of subspaces $(\mathcal{2}^-, \mathcal{2}^+)$ is a so-called Jordan pair (cf. [32, 38]).

For $p = 2$ or $p = 3$ we define $\mathcal{2}^-$ and $\mathcal{2}^+$ with the divided powers of adjoint-operators and apply Kevin McCrimmon's Theory of Quadratic Jordan Algebras ([20, 36, 37]).

For odd p we managed to translate Jordan arguments into the language of elementary computations in [52]. For $p = 2$ this substitute didn't work so Jordan Pairs and Algebras played an important role in our paper [53]. However, recently M. Vaughan-Lee succeeded in getting rid of Jordan Algebra Theory even in the case $p = 2$.

Not much is known about the upper bound for classes of nilpotency of r-generated finite groups of exponent p (let alone the exponent $p^k, k > 1$). S. I. Adian and N. N. Repin [1] proved that it grows at least exponentially with respect to p. For comparison let us mention the recent result of A. Belov which asserts that there exists a constant α such that an arbitrary r-generated associative ring which satisfies the identity $x^n = 0$ is nilpotent of degree $\leq r^{\alpha n}$.

Conjecture. *There exists a constant α such that an arbitrary r-generated finite group of exponent p is nilpotent of class $\leq r \cdot \alpha^p$.*

Residually Finite Groups and Compact Groups

The following generalization of Theorem 1 solves the Kurosh-Levitzky Problem (in Shirshov's form) in the class of Lie PI-algebras.

Theorem 3. *Suppose that a Lie algebra L is generated by a finite subset $X \subseteq L$, $|X| = m$ and assume that*
(1) L satisfies a polynomial identity of degree n,

(2) *for an arbitrary commutator ϱ on X of weight $\leq h(m, n)$ the operator $ad\,(\varrho)$ is nilpotent.*

Then L is nilpotent.

This theorem has some applications to the General Burnside Problem in group varieties and to compact groups. It is well known that there are counterexamples to the General Burnside Problem even among residually finite p-groups (such are the counterexamples of E. S. Golod and R. I. Grigorchuk). However Theorem 3 implies

Theorem 4. *A residually finite p-group which satisfies a nontrivial group identity is locally finite.*

Apparently this theorem can be generalized from p-groups to periodic groups in the spirit of the theorem of P. Hall and G. Higman [14].

The assumption of a nontrivial identity can be further weakened to the "infinitesimal" assumption that the adjoint Lie algebra $L(G)$ is PI. The last assertion in its turn is related to the General Burnside Problem for compact groups. V. P. Platonov conjectured that periodic compact (Hausdorff) groups are locally finite. J. S. Wilson [47] proved that (under the assumption that there are finitely many simple sporadic groups) it suffices to prove the conjecture for pro-p-groups. That's what is done in the following theorem.

Theorem 5. *Every periodic pro-p-group is locally finite.*

Indeed, let G be a periodic pro-p-group. Consider the closed subsets $G_{(n)} = \{g \in G | g^{p^n} = 1\}, G = U G_{(n)}$. By Baire's Category theorem one of the subsets $G_{(n)}$ contains some neighborhood, that is $G_{(n)} \supset gH$, where H is a normal subgroup of G of finite index. Then we show that every finitely generated subgroup of H which is invariant under conjugation by g, satisfies an "infinitesimal" identity.

From Theorem 5 combined with [47] and with what is known about locally finite groups [15, 24] there follows

Theorem 6. *Every infinite compact group contains an infinite abelian subgroup.*

Remark as far as Theorem 6 is concerned the reduction to pro-p-groups in [47] didn't use the classification of finite simple groups.

References

1. Adian, S. I., Repin, N. N.: On exponential lower bound for class of nilpotency of Engel Lie algebras. Matem. Zametky **39**, no. 3 (1986) 444–452
2. Alyoshin, S. V.: Finite automata and the Burnside problem on periodic groups. Matem. Zametky **11**, no. 3 (1972) 319–328
3. Amitsur, S. A.: Jacob Levitzki 1904–1956. Israel J. Math. **19** (1974) 1–2

4. Baer, R.: The higher commutator subgroups of a group. Bull. Amer. Math. Soc. **50** (1944) 143–160
5. Benkart, G.: Inner Ideals and the Structure of Lie algebras. Dissertation. Yale University 1974
6. Burnside, W.: On an unsettled question in the theory of discontinuous groups. Quart. J. Pure Appl. Math. **33** (1902) 230–238
7. Burnside, W.: Theory of groups of finite order. Cambridge Univ. Press, 1911; Reprint, Dover, New York 1955
8. Feit, W., Thompson, G.: Solvability of groups of odd order. Pacific J. Math. **13** (1963) 755–1029
9. Golod, E. S.: On nil algebras and residually finite p-groups. Izv. Akad. Nauk SSSR **28**, no. 2 (1964) 273–276
10. Gorenstein, D.: Finite simple groups. New York 1982
11. Grigorchuk, R. I.: On the Burnside Problem for periodic groups. Funct. Anal. Appl. **14**, no. 1 (1980) 53–54
12. Grün, O.: Zusammenhang zwischen Potenzbildung und Kommutatorbildung. J. Reine Angew. Math. **182** (1940) 158–177
13. Grunewald, F. J., Havas, G., Mennicke, J. L., Neumann, M. F.: Groups of exponent eight. Bull. Austral. Math. Soc. **20** (1979) 7–16
14. Hall, P., Higman, G.: On the p-length of p-soluble groups and reduction theorems for Burnside's problem. Proc. London Math. Soc. GN3 (1956) 1–42
15. Hall, P., Kulatilaka, C. R.: A property of locally finite groups. J. London Math. Soc. **39** (1964) 235–239
16. Havas, G., Newmann, M. F.: Application of computers to questions like those of Burnside. (Lecture Notes in Mathematics, vol. 806.) Springer, Berlin Heidelberg New York 1980, pp. 211–230
17. Higman, G.: Lie ring methods in the theory of finite nilpotent groups. Proc. Intern. Congr. Math. Edinburgh 1958, pp. 307–312
18. Jacobson, N.: Restricted Lie algebras of characteristic p. Trans. AMS **50** (1941) 15–25
19. Jacobson, N.: Structure and representations of Jordan algebras. AMS, Providence, RI 1969
20. Jacobson, N.: Lectures on quadratic Jordan algebras. Bombay. Tata Institute of Fundamental Research, 1969
21. Jennings, S. A.: The structure of the group ring of a p-group over a modular field. Trans. AMS **50** (1941) 175–185
22. Kantor, I. L.: Classification of irreducible transitively differential groups. Sov. Math. Dokl. SN SSSR **5** (1964) 1404–1407
23. Kaplansky, I.: Rings with a polynomial identity. Bull. AMS **54** (1948) 575–580
24. Kargapolov, M. I.: On a problem of O. Ju. Smidt. Siber. Math. Ž. **4** (1963) 232–235
25. Koecher, M.: Imbedding of Jordan algebras into Lie algebras. Amer. J. Math. **89** (1967) 787–816
26. Kostrikin, A. I.: On Lie rings with Engel's condition. Dokl. Akad. Nauk SSSR **108**, no. 4 (1956) 580–582
27. Kostrikin, A. I.: On the Burnside Problem. Izv. Akad. Nauk SSSR **23**, no. 1 (1959) 3–34
28. Kostrikin, A. I.: Sandwiches in Lie algebras. Matem. Sb. **110** (1979) 3–12
29. Kostrikin, A. I.: Around Burnside. Nauka, Moscow 1986
30. Kurosh, A. G.: Problems in ring theory which are related to the Burnside Problem for periodic groups. Izv. Akad. Nauk SSSR **5**, no. 3 (1941) 233–240
31. Lazard, M.: Sur les groupes nilpotents et les anneaux de Lie. Ann. Sci. Ecole Norm. Sup. **71**, no. 3 (1954) 101–190
32. Loos, O.: Jordan pairs. Springer, Berlin Heidelberg New York 1975

33. Lysenok, I.: On the Burnside problem for odd exponents $n \geq 115$. Materials of the International Conference in Algebra, Novosibirsk, 1989
34. Magnus, W.: Über Gruppen und zugeordnete Liesche Ringe. J. Reine Angew. Math. **182** (1940) 142–159
35. Magnus, W.: A connection between the Baker-Hausdorff formula and a problem of Burnside. Ann. Math. **52** (1950) 11–26; Errata Ann. Math. **57** (1953) 606
36. McCrimmon, K.: A general theory of Jordan rings. Proc. Nat. Acad. Sci. USA **56** (1966) 1072–1079
37. McCrimmon, K., Zelmanov, E.: The structure of strongly prime quadratic Jordan algebras. Adv. Math. **69** (1988) 113–222
38. Meyberg, K.: Lectures on algebras and triple systems. Lecture Notes, The University of Virginia, Charlottesville, 1972
39. Novikov, P. S., Adían, S. I.: On infinite periodic groups. I, II, III. Izv. Akad. Nauk SSSR **32**, no. 1 (1968) 212–244; no. 2, 251–254; no. 3, 709–731
40. Olshansky, A. Yu.: Geometry of defining relations in groups. Nauka, Moscow 1989
41. Plotkin, B. I.: Algebraic sets of elements in groups and Lie algebras. Uspekhi Mat. Nauk **13**, no. 6 (1958) 133–138
42. Sanov, I. N.: On a certain system of relations in periodic groups of prime power exponent. Izv. Akad. Nauk SSSR **15** (1951) 477–502
43. Shirshov, A. I.: On rings with identical relations. Mat. Sb. **43** (1957) 277–283
44. Suschansky, V. I.: Periodic p-groups of permutations and the General Burnside Problem. Dokl. Akad. Nauk SSSR **247**, no. 3 (1979) 447–461
45. Vaughan-Lee, M.: The Restricted Burnside's Problem. Oxford University Press, 1990
46. Wall, G. E.: On the Lie ring of a group of prime exponent. Proc. 2nd Intern. Conf. Theory of Groups, Canberra, 1973, pp. 667–690
47. Wilson, J. S.: On the structure of compact torsion groups. Monatsh. Math. **96** (1983) 404–410
48. Zassenhaus, H.: Ein Verfahren, jeder endlichen p-Gruppe einem Lie-Ring mit der Charakteristik p zuzuordnen. Abh. Math. Sem. Univ. Hamburg **13** (1940) 200–207
49. Zelmanov, E. I., Kostrikin, A. I.: A theorem on sandwich algebras. Proceedings of the V. A. Steklov Math. Institute of Akad. Nauk SSSR **183** (1988) 142–149
50. Zelmanov, E. I.: On some problems in the theory of groups and Lie algebras. Mat. Sb. **180**, no. 2 (1989) 159–167
51. Zelmanov, E. I.: On the restricted Burnside problem. Siberian Math. J. **30**, no. 6 (1989) 68–74
52. Zelmanov, E. I.: The solution of the restricted Burnside problem for groups of odd exponent. Izv. Akad. Nauk SSSR **54** (1990)
53. Zelmanov, E. I.: The solution of the restricted Burnside problem for 2-groups. To appear in Mat. Sb.
54. Gupta, N., Sidki, S.: On the Burnside problem for periodic groups. Math. Z. **182** (1983) 385–386

A Riemann-Roch Theorem in Arithmetic Geometry

Henri Gillet [*]

Department of Mathematics, Statistics, and Computer Science, (m/c 249)
University of Illinois at Chicago, Box 4348, Chicago, IL 60680, USA

1. Introduction

This talk describes joint work with Christophe Soulé; for full details the reader
can consult the papers [22, 24].

Given an arithmetic variety X, together with a Hermitian metric on its set
of complex points, the arithmetic Riemann-Roch theorem computes, given a
Hermitian vector bundle $\bar{E}$ on X, the degree (in the sense of Arakelov [1,2]) of
the determinant of its cohomology equipped with the Quillen metric. The Quillen
metric is an invariant of the Laplace operator on $X(\mathbb{C})$ with coefficients in E.
The computation is in terms of characteristic classes, which are invariants of the
Hermitian bundle and the arithmetic variety itself, in the 'arithmetic Chow ring'
of the variety. The arithmetic Chow ring is inspired by the ideas of Arakelov
(ibid.) and Deligne [15] in the case of arithmetic surfaces, and the arithmetic
Riemann-Roch theorem itself generalizes the work of Faltings [16] and Deligne
(ibid.) in the relative dimension one case. One application of the theorem, to the
existence of sections with small norm, will be described.

2. Preliminaries

2.1 Basic Definitions

Definition 1. In this note by an *arithmetic variety* over $\mathbb{Z}$, we mean a regular
scheme X, flat and quasi-projective over $\mathbb{Z}$. However in general we need only
assume that the generic fibre $X_{\mathbb{Q}}$ is smooth, and for much of what follows we
can also replace the integers $\mathbb{Z}$ by a more general 'arithmetic ring'; see [18]. We
shall write F_{∞} for the antiholomorphic involution of the complex manifold $X(\mathbb{C})$
induced by complex conjugation.

Definition 2. A *Hermitian bundle* $\bar{E} = (E, h)$ over X consists of a vector bundle
E over X, and a choice of C^{∞} Hermitian metric (i.e. positive definite Hermitian
inner product) h on E over $X(\mathbb{C})$, which is invariant under F_{∞}.

[*] Supported by N.S.F. grant DMS-8901784.

Proceedings of the International Congress
of Mathematicians, Kyoto, Japan, 1990
© The Mathematical Society of Japan, 1991

Example 1. If $X = \mathrm{Spec}(\mathbb{Z})$, then a Hermitian bundle over X is the same thing as a lattice in $\mathbb{R}^n$ for some n.

Example 2. If $X = \mathrm{Spec}(\mathcal{O}_K)$ for $\mathcal{O}_K$ the ring of integers in a number field K, then a Hermitian bundle $\bar{E}$ consists of a finitely generated projective $\mathcal{O}_K$-module E together with a Hermitian inner product on $E \otimes_{\mathcal{O}_K} K_v$ for each archimedian completion K_v of K. The set of isomorphism classes of Hermitian bundles over X is classified by the double coset space

$$GL_n(K) \backslash GL_n(\mathbf{A}_K)/U ,$$

where $\mathbf{A}_K$ is the ring of adeles of K, and U is the maximal compact subgroup of $GL_n(\mathbf{A}_K)$.

   These examples show that the if one views $\mathrm{Spec}(\mathcal{O}_K)$ as being compactified by adding the archimedian places of K as points at infinity, then Hermitian vector bundles over $\mathrm{Spec}(\mathcal{O}_K)$ have a similar adelic description to vector bundles over a smooth projective curve. In this case a Hermitian bundle has a well-defined degree. Let $\det(\bar{E})$ be the top exterior power of E, with the induced Hermitian form, and let $a \in \det(E)$ be a non-zero element. Then set

$$\widehat{\deg}(\bar{E}) = \log(\#(\det(E)/\mathcal{O}_K a)) - \sum_v \log \|a\|_v$$

where the sum is over the complex embeddings $v : K \hookrightarrow \mathbb{C}$, and $\|\ \|_v$ denotes the absolute value corresponding to the choice of metric on $E \otimes_v \mathbb{C}$. Equivalently the degree is equal to minus the logarithm of the covolume (i.e. the volume of a fundamental domain), with respect to the euclidean measure determined by the Hermitian inner product, of the lattice $E \subset E \otimes \mathbb{R}$. See [1,2].

2.2 Riemann-Roch for Curves, and Its Analog for Lattices

Let X be a smooth projective curve over a field k, and let E be a rank n vector bundle over X. Recall that the degree of E equals the degree of $\det E$, which is the degree of the divisor of a meromorphic section of $\det E$. We write g_X for the genus of X. Then the classical Riemann-Roch theorem for curves is:

$$\chi(X, E) = n(1 - g_X) + \deg(E) .$$

If k is a finite field of order q, and V is a finite dimensional k-vector space, then $\dim_k V = \log_q \#(V)$. This leads, given a Hermitian bundle over $\mathrm{Spec}(\mathbb{Z})$, i.e. a lattice $\Lambda \subset \mathbb{R}^n$, to the folowing definition of a numerical analog $h^0(\Lambda)$ of the rank of $H^0(E)$ of a vector bundle E over a curve.

Definition 3. Let $\Lambda \subset \mathbb{R}^n$ be a lattice. We set

$$h^0(\Lambda) = \log \#\{x \in \Lambda \mid \|x\| \leq 1\} .$$

We also write Λ^* for the dual or 'polar' lattice.

It is an immediate consequence of Minkowski's theorem that there exists a constant a such that:

$$h^0(\Lambda) \geq \widehat{\deg}(\Lambda) + an \ . \tag{1}$$

This is analogous to Riemann's theorem:

$$h^0(E) \geq \deg(E) + (1 - g_X)n \ .$$

See [28] for a discussion of this analogy. The following result is more analogous to the Riemann-Roch theorem, in that it estimates the difference between the two sides of (1). See [23] for a proof.

Theorem 4. *With the above notation,*

$$|h^0(\Lambda) - h^0(\Lambda^*) - \widehat{\deg}(\Lambda)| \leq n \log(6/\sqrt{\pi}) + \log \left(2^{-1} \Gamma \left(\frac{n}{2} + 1\right)\right) \ .$$

2.3 Grothendieck-Riemann-Roch for Varieties over a Curve

Let X be a smooth projective variety over a field k, and let E be a vector bundle over X. Suppose that $f : X \to C$ is a map of X onto a smooth projective curve, and let us write $\mathrm{Td}(X/C) = \mathrm{Td}(X)f^*\mathrm{Td}(C)^{-1}$ for the relative Todd class of X over C in any suitable cohomology theory, such as the Chow ring $CH^*(X)_{\mathbb{Q}}$.

On C there is a line bundle $\lambda(E) = \lambda(X/C, E) = \det(Rf_*E)$. The fiber of $\lambda(E)$ at a point $y \in C$ is naturally isomorphic to $\bigotimes_{i \geq 0} \det[H^i(f^{-1}(y), E)]^{(-1)^i}$. (Here, if L is a line bundle, L^{-1} denotes its dual). It follows from the Grothendieck-Riemann-Roch theorem that

$$\deg \lambda(E) = \deg(\mathrm{ch}(E)\mathrm{Td}(X/C)) \ .$$

Here $\deg : CH^*(X)_{\mathbb{Q}} \to CH^0(\mathrm{Spec}(k))_{\mathbb{Q}} \simeq \mathbb{Q}$ is the map which, if X has dimension n, is zero on $CH^i(X)_{\mathbb{Q}}$ for $i \neq n$, and on $CH^n(X)_{\mathbb{Q}}$ takes a dimension zero cycle to its degree. The Riemann-Roch theorem for a curve tells us that:

$$\chi(E) = \chi(f^{-1}(y), E)(1 - g_C) + \deg(\lambda(E)) \ . \tag{2}$$

Here $y \in C$ is arbitrary. Notice that this formula allows one to deduce the Hirzebruch-Riemann-Roch formula for E over X from the corresponding formula for the generic fiber of X over C, together with the computation of $\deg(\lambda(E))$.

3. The Arithmetic Riemann-Roch Theorem

3.1 Motivation

Suppose now that $f : X \to \mathrm{Spec}(\mathbb{Z})$ is a projective arithmetic variety, and we fix a Kähler metric ω on $X(\mathbb{C})$ which is invariant under F_∞. What might a Riemann-Roch theorem for a Hermitian bundle $\bar{E}$ on X look like? If we view f as being analogous to a map to a curve, then (2) suggests that such a theorem combines three ingredients: Theorem 4, the Hirzebruch-Riemann-Roch theorem for E on the generic fibre $X_{\mathbb{Q}}$, and the computation of the degree of a Hermitian line bundle $\lambda(X, \bar{E})$. We need therefore to define a metric on $\lambda(E) = Rf_*(E)$. Note that such a metric on $\lambda(E) = Rf_*(E)$ can be interpreted as measuring the torsion in Rf_*E supported on the 'closed points at infinity.'

3.2 The Quillen Metric

Procedures have been proposed for putting a metric on $\lambda(E) = Rf_*(E)$ by Faltings [16] (for X an arithmetic surface) and Quillen [25] . We shall follow the method of Quillen.

Let $f : M \to B$ be a proper smooth map of complex manifolds. (This of course includes the case of $X(\mathbb{C})$ mapping to a point, for X an arithmetic variety). Suppose that ω is a closed $(1,1)$-form on M which restricts to a Kähler form on each fiber of f. If $\bar{E}$ is a Hermitian holomorphic vector bundle on M, *i.e.* a locally free coherent analytic sheaf E equipped with a C^∞ Hermitian metric, then $\lambda(E) = \lambda(M/B, E) = \det(Rf_*E)$ is a holomorphic line bundle on B. Here Rf_* is the direct image in the derived category of perfect complexes of analytic sheaves. If $b \in B$ is a point, the Hermitian metric on E, together with the induced Kähler metric on $M_b = f^{-1}(b)$ and the $\bar{\partial}$ operator on M_b determine an inner product and a laplace operator Δ_i^b on $A^{0,i}(M_b, E)$. By Hodge theory we may identify canonically the cohomology $H^i(M_b, E)$ of E over M_b, with the subspace $\text{Ker}(\Delta_i^b) \subset A^{0,i}(M_b, E)$. The inner product on $A^{0,i}(M_b, E)$ then induces an inner product, the L^2 inner product h_{L^2}, on $H^i(M_b, E)$ and hence on $\lambda(M_b, E) = \bigotimes_{i \geq 0}[H^i(M_b, E)]^{(-1)^i}$. There is a canonical isomorphism $\lambda(M_b, E) \simeq \lambda(M/B, E)_b$; however in general the L^2 inner product on this complex line will not vary in a C^∞ fashion with $b \in B$. Following Quillen, we modify the L^2 inner product as follows. For each $i \geq 0$ and $b \in B$ we consider the zeta function $\zeta_i^b(s) = \sum 1/\lambda_n^s$, where the sum is over the non zero eigenvalues of Δ_i^b. The sum defining $\zeta_i^b(s)$ converges absolutely for the real part of s large enough, and $\zeta_i^b(s)$ has a meromorphic continuation to the whole complex plane, with a regular point at zero. The Ray-Singer analytic torsion, [26], is then defined as

$$\tau(M_b, \bar{E}) = \sum_{q \geq 0} (-1)^{q+1} q \zeta_q'(0) \ .$$

Finally we define $\lambda(\bar{E}) = (\lambda(E), h_Q = h_{L^2} \exp(\tau(\bar{E})))$. The following theorem was joint work with J.-M. Bismut, and C. Soulé, see [8] for details.

Theorem 5. *Let $f : M \to B$, ω, and $\bar{E}$, be as above. Then:*
1. *The Quillen metric h_Q on the holomorphic line bundle $\lambda(E)$ on B is C^∞.*
2. *There is an equality of $(1,1)$-forms on B:*

$$\text{ch}_1(\lambda(E), h_Q) = \int_f \text{ch}(\bar{E})\text{Td}(\bar{T}_{M/B}) \ .$$

Here $\text{ch}(\bar{E})$ is the representative of the Chern class in deRham cohomology obtained by applying the Chern-Weil homomorphism to the unique unitary connection of type $(0, 1)$ determined by the metric and holomorphic structure on E.

Remark. The Grothendieck Riemann-Roch theorem predicts the equality of cohomology classes which corresponds to the above equation. This result is built on some earlier work of Bismut and Freed on C^∞ determinant bundles for families of Dirac operators.

3.3 Arithmetic Chow Groups

If $f : X \rightarrow Y$ is a proper morphism between arithmetic varieties which restricts to a (proper) smooth map $X(\mathbb{C}) \rightarrow Y(\mathbb{C})$, and $\bar{E}$ is a Hermitian bundle on X, $\lambda(X/Y, \bar{E})$ is a Hermitian line bundle on Y. It is the isomorphism class, up to torsion, of this Hermitian line bundle which we would like to determine. Notice that the "classical" Grothendieck-Riemann-Roch theorem computes the Chern character $\mathrm{ch}_1(\lambda(X/Y, E)) \in CH^1(Y)_{\mathbb{Q}}$ while by Theorem 5 the Quillen metric determines a form representing the de Rham representative of this class.

If X is an arithmetic variety, and $\bar{E} = (E, h)$ is a Hermitian bundle on X, then we have the Chern character $\mathrm{ch}(E) \in CH^*(X)_{\mathbb{Q}}$ in the Chow group of cycles modulo rational equivalence, and forms $\mathrm{ch}(\bar{E}) \in Z^*(X) \overset{\mathrm{def}}{=} \bigoplus_{i \geq 0} Z^{i,i}(X(\mathbb{C}))$ determined by the Chern-Weil homomorphism. (Here $Z^{i,i}(X(\mathbb{C}))$ denotes closed forms of type (i, i)). We shall combine these together into a single theory carrying the characteristic classes we need to compute $\lambda(X/Y, \bar{E})$.

Definition 6. Let X be an arithmetic variety as above. A codimension i *arithmetic cycle* on X is a pair (Z, g_Z) consisting of a codimension i algebraic cycle Z on X, and a current (*i.e.* a form with distribution coefficients) of type $(i-1, i-1)$, such that $dd^c g_Z + \delta_Z$ is C^∞, where δ_Z is the current of type (i, i) given by integration over the complex points of Z. We also require that $F_\infty^* g_Z = (-1)^{i-1} g_Z$. We call g_Z a *Green current* for Z, and we view two Green currents g_Z and g_Z' as equivalent if their difference lies in the image of $\partial + \bar{\partial}$.

If $W \subset X$ is an integral subscheme of codimension $i - 1$, and $f \in k(W)$ is a rational function on W, we define $\widehat{\mathrm{div}}(f) = (\mathrm{div}(f), -\log |f|^2)$. Here $\log |f|^2$ is the current of type $(i-1, i-1)$ obtained by integration against the function $\log |f|^2$ on the complex points of W: by the Poincaré-Lelong lemma $\log |f|^2$ is a Green current for $\mathrm{div}(f)$. We say that an arithmetic cycle is rationally equivalent to zero if it is the sum of cycles of the form $\widehat{\mathrm{div}}(f)$, and we define the codimension i *arithmetic Chow group* $\widehat{CH}^i(X)$ of X to be the quotient of the group of codimension i arithmetic cycles by the subgroup of cycles rationally equivalent to zero.

An element of $\widehat{CH}^i(X)$ determines a rational equivalence class of algebraic cycles on X, a C^∞ form representing their de Rham cohomology class on $X(\mathbb{C})$, together with secondary data. In the paper [18] we make this relationship precise and show how to lift the product and pull back structures from cycles and forms to the arithmetic Chow groups:

Theorem 7. *Suppose that X is an arithmetic variety, with $X_{\mathbb{Q}}$ projective. Then there is an exact sequence*

$$CH^{i-1,i}(X) \overset{\varrho}{\rightarrow} H^{i-1,i-1}(X) \overset{a}{\rightarrow} \widehat{CH}^i(X) \overset{\zeta \oplus \omega}{\rightarrow} CH^i(X) \oplus Z^{i,i}(X) \overset{c-h}{\rightarrow} H^{i,i}(X) .$$

Here $CH^{i-1,i}(X)$ is a group which is up to torsion the weight i part of $K_1(X)$, ϱ is the Beilinson regulator ([4]), $H^{i-1,i-1}(X)$ is the subspace of $H^{i-1,i-1}(X(\mathbb{C}))$ on which F_∞ acts by $(-1)^{i-1}$, a is the map $\alpha \mapsto (0, \alpha)$, ζ is the map forgetting the Green current, $\omega(Z, g_Z) = dd^c g_Z + \delta_Z$, c maps a cycle to its cohomology class and h maps a closed form to its cohomology class. If we had not assumed that $X_{\mathbb{Q}}$ was projective we would have to modify the definition of the Dolbeault cohomology groups in the sequence; cf. [18].

Example 3. Let K be a number field and $X = \text{Spec}(\mathcal{O}_K)$. Then

$$\widehat{CH}^1(X) \simeq K^* \backslash \mathbf{I}_K / U ,$$

where $\mathbf{I}_K$ is the idele group of K, and $U \subset \mathbf{I}_K$ is its maximal compact subgroup. If $\text{Cl}(\mathcal{O}_K)$ is the ideal class group of K, the exact sequence of Theorem 7 becomes:

$$\mathcal{O}_K^* \to \mathbb{R}^{r_1 + r_2} \to \widehat{CH}^1(X) \to \text{Cl}(\mathcal{O}_K) \to 0 .$$

Example 4. For a general arithmetic variety X, let $\widehat{\text{Pic}}(X)$ be the group of isomorphism classes of Hermitian line bundles on X, with the group operation given by tensor product. Then there is an isomorphism

$$\hat{C}_1 : \widehat{\text{Pic}}(X) \to \widehat{CH}^1(X) ,$$

given by $\bar{L} \mapsto (\text{div}(s), -\log \|s\|^2)$, for s an arbitrary rational section of L.

Example 5. On projective space over the integers $\mathbb{P}_{\mathbb{Z}}^n$, viewed as an arithmetic variety, we have a tautological Hermitian line bundle $\widehat{\mathcal{O}(1)}$, with the metric induced by orthogonal projection from the trivial bundle. If $x : \text{Spec}\mathbb{Z} \to \mathbb{P}_{\mathbb{Z}}^n$ is an integral point, then $\widehat{\deg} x^*(\widehat{\mathcal{O}(1)})$ is the Arakelov height of x; cf. [27].

Theorem 8. *1. If $f : X \to Y$ is a map between (regular) arithmetic varieties, there is a pull back map $f^* : \widehat{CH}^*(Y) \to \widehat{CH}^*(X)$. If $f : X \to Y$ and $g : Y \to Z$, then $f^* g^* = (gf)^*$.*

2. $\widehat{CH}^*(X)_{\mathbb{Q}}$ is an associative ring with unit, and the product is compatible, via the maps ζ and ω, with the intersection product on cycles and the wedge product on forms.

3. If $f : X \to Y$ is a proper map of relative dimension d, smooth over $\mathbb{Q}$, there is a direct image map $f_* : \widehat{CH}^*(X) \to \widehat{CH}^{*-d}(Y)$, with the obvious projection formula holding.

The proof of the theorem, and in particular the construction of the product, involves showing that if Y and Z are two cycles on a projective variety over $\mathbb{C}$ which intersect properly, choices of Green currents g_Y and g_Z determine a Green current $g_Y * g_Z$, with good properties, for the product cycle $Y \cdot Z$. This product of Green currents must then be combined with the intersection product of cycles on the regular scheme underlying the arithmetic variety. However, because there is as yet no intersection theory with integral coefficients on a general regular scheme, we use our construction in [17], via Adams operations on K_0, of an intersection theory with rational coefficients. Observe that if $f : X \to \text{Spec}(\mathbb{Z})$ is projective, the pushforward f_* induces a map $\widehat{\deg} : \widehat{CH}^{d+1}(X) \to \mathbb{R}$, where

d is the relative dimension of X over $\mathbb{Z}$; explicitly, $(Z, g_Z) \mapsto \log \# \Gamma(Z, \mathcal{O}_Z) + \frac{1}{2} \int_{X(\mathbb{C})} g_Z$. Composing the product with the degree map gives a pairing $\widehat{CH}^p(X) \otimes \widehat{CH}^{d+1-p}(X) \to \mathbb{R}$. If we restrict this pairing to the subgroup $\widehat{CH}^*(X)_0 = \ker(\omega)$, it factors, at least if X has no bad fibers, through the subgroup of $CH^*(X_{\mathbb{Q}})$ consisting of cycles which are homologically equivalent to zero, where it becomes the height pairing defined by Beilinson and Bloch [5], [13] generalizing the Neron height pairing.

3.4 Characteristic Classes for Hermitian Bundles

In [19] we show that any characteristic class defined by invariant polynomials has a common lifting from the Chow ring and the ring of differential forms to the arithmetic Chow ring. The following theorem summarizes this result in the case of the Chern character, which will be sufficient for our purposes, since the Chern character determines the Todd genus.

Theorem 9. *There is one and only one way of assigning to each Hermitian bundle $\bar{E}$ on an arithmetic variety X a class $\widehat{\mathrm{ch}}(\bar{E}) \in \widehat{CH}^*(X)_{\mathbb{Q}}$, the "arithmetic Chern character", with the following properties:*

1. Under the maps ζ and ω it maps to the classes $\mathrm{ch}(E) \in CH^*(X)_{\mathbb{Q}}$ and $\mathrm{ch}(\bar{E}) \in Z^*(X)$.

2. If $\bar{L}$ is a Hermitian line bundle, then $\widehat{\mathrm{ch}}(\bar{L}) = \exp(\widehat{C}_1(\bar{L}))$.

3. Given two Hermitian bundles $\bar{E}$ and $\bar{F}$ we have:

$$\widehat{\mathrm{ch}}(\bar{E} \otimes \bar{F}) = \widehat{\mathrm{ch}}(\bar{E}) \cdot \widehat{\mathrm{ch}}(\bar{F}) ,$$

and

$$\widehat{\mathrm{ch}}(\bar{E} \oplus \bar{F}) = \widehat{\mathrm{ch}}(\bar{E}) + \widehat{\mathrm{ch}}(\bar{F}) .$$

Here $\bar{E} \oplus \bar{F}$ is the orthogonal direct sum.

4. $\widehat{\mathrm{ch}}$ commutes with pullbacks.

Remark. If $\mathscr{E} : 0 \to E' \to E \to E'' \to 0$ is an exact sequence of vector bundles on an arithmetic variety X, and we choose metrics h', h, h'' on the three bundles, then

$$\widehat{\mathrm{ch}}(E', h') + \widehat{\mathrm{ch}}(E'', h'')) - \widehat{\mathrm{ch}}(E, h) = (0, \widetilde{\mathrm{ch}}(\mathscr{E}, h', h, h'')) ,$$

which is the secondary chern character defined by Bott and Chern [14] satisfying $dd^c \widetilde{\mathrm{ch}}(\mathscr{E}, h', h, h'') = \mathrm{ch}(E', h') + \mathrm{ch}(E'', h'') - \mathrm{ch}(E, h)$, i.e. it measures the failure of the Whitney sum formula to be true at the level of forms.

3.5 The Main Theorem

Let $f : X \to Y$ be a proper map between nonsingular arithmetic varieties, which restricts to a smooth map $X_{\mathbb{Q}} \to Y_{\mathbb{Q}}$; fix a Kähler metric on $X(\mathbb{C})$ which is invariant under F_∞. Let $\bar{E}$ be Hermitian bundle on X, and let $\lambda(\bar{E})$ be the corresponding determinant line bundle on Y, equipped with the Quillen metric. Let $\bar{T}_{X/Y}$ denote the relative tangent bundle with the metric induced by the

Kähler metric on $X(\mathbb{C})$; note that strictly speaking, unless f is smooth, this relative tangent bundle is a virtual object. Consider the difference

$$\widehat{ch}_1(\lambda(\bar{E})) - [f_*(\widehat{ch}(\bar{E})\widehat{Td}(\bar{T}_{X/Y}))]^1 \ ,$$

where $[\ ]^1$ denotes the component in $\widehat{CH}_{\mathbb{Q}}^1$. By the 'classical' Grothendieck-Riemann-Roch theorem, together with Theorem 5 we know that this difference has image zero under both ζ and ω, and so by Theorem 7 must lie in the image of the map $a : H^{0,0}(Y) \to \widehat{CH}^1(Y)_{\mathbb{Q}}$. If one computes this difference for $X = \mathbb{P}_{\mathbb{Z}}^1$ and $Y = \mathrm{Spec}(\mathbb{Z})$, one does not get zero, showing that a naive translation of the Grothendieck-Riemann-Roch theorem is false. By computing, with the help of D. Zagier, the difference for the trivial bundle on $X = \mathbb{P}_{\mathbb{Z}}^n$ for all n, we were led in [20] to define an "arithmetic Todd genus":

Definition 10. Consider the power series

$$R(x) = \sum_{m \text{ odd}} [2\zeta'(-m) + \zeta(-m)(1 + 1/2 + \ldots + 1/m)] \left(\frac{x^m}{m!}\right) \ .$$

Using the splitting principle, we have an additive characteristic class $R(E) \in H^*(X)$ for complex vector bundles E over topological spaces X, such that $R(L) = R(C_1(L))$ for a line bundle L. Now given $f : X \to Y$ as above, define the *arithmetic Todd genus*:

$$Td^A(X/Y) = \widehat{Td}(\bar{T}_{X/Y})(1 + a(R(T_{X(\mathbb{C})/Y(\mathbb{C})}))) \ .$$

Remark. Notice that power series $R(x)$ used to define the correction term in Td^A has a mysterious relationship to the power series used to define the Todd genus itself:

$$Td(x) = 1 - x \sum_{m \geq 0} \zeta(-m)\frac{x^m}{m!} \ .$$

Theorem 11. *Let $f : X \to Y$, $\bar{E}$, be as above. Then we have*

$$\widehat{ch}_1(\lambda(\bar{E})) = [f_*(\widehat{ch}(\bar{E})Td^A(\bar{T}_{X/Y}))]^1 \ .$$

The strategy of the proof is based on the method used to prove the "classical" version of the theorem. That is, to factor the map f as the composition of a closed immersion $i : X \to \mathbb{P}_Y^n$ followed by the projection $g : \mathbb{P}_Y^n \to Y$. The main ingredients are:

1. The computation, with Zagier, of the analytic torsion of the trivial bundle on $\mathbb{P}_Y^n$, [20].

2. Following the series of papers [6] (by J.-M. Bismut), and [9] and [10] (with J.-M. Bismut and C. Soulé), given a resolution $\eta : \bar{\xi}_. \to i_*E$, by a complex of Hermitian bundles, the 'torsion' of $\bar{\xi}_.$ is defined. This is a current on $\mathbb{P}^n(\mathbb{C})$, and in [9] it is shown that it plays for closed immersions between complex manifolds a role similar to that played by the Ray-Singer torsion for submersions. In [10] we then prove a Riemann-Roch theorem for i.

3. The papers [7], by J.-M. Bismut, and [11], by J.-M. Bismut and G. Lebeau, compute the ratio of the Quillen metrics on $\lambda(\bar{E})$ and $\lambda(\bar{\xi}.)$ in terms of the the the torsion of $\bar{\xi}$. This can be viewed as showing that $f_! = g_! i_!$ in the K-theory of Hermitian bundles, up to a correction involving the exotic class $R(N_{X/\mathbb{P}^n_Y})$.

Remark. 1. If $Y = \mathrm{Spec}(\mathbb{Z})$, then the theorem gives:

$$\widehat{\deg}\lambda(\bar{E}) = \widehat{\deg}[\widehat{\mathrm{ch}}(\bar{E})\mathrm{Td}^A(X/\mathbb{Z})] \ ,$$

which is the analogue of the Riemann-Roch theorem for a map to curve that we were looking for.

2. We have proved a more general result, in the style of [3], in which X is allowed to have singularities away from its generic fibre. This requires a careful analysis of the behaviour of complexes under the Grassmannian-graph construction of [3].

3. G. Faltings has recently given a more direct proof of the result, without appealing to the results of [6, 7, 9, 10, 11].

3.6 An Application

Let X be an arithmetic variety of dimension $d + 1$, and fix a Kähler metric on $X(\mathbb{C})$. Let $\bar{E}$ be a Hermitian vector bundle, and $\bar{L}$ a Hermitian line bundle, on X. Suppose that L is relatively ample over X and that $C_1(\bar{L})$ is strictly positive. Write $h^0(X, \bar{E})$ for $\log \#\{s \in H^0(X, E) \,|\, \|s\|_{L^2} \leq 1\}$. The proof of the following theorem uses an earlier, weaker, version of the arithmetic Riemann-Roch theorem, Minkowski's theorem, and a result of J.-M. Bismut and E. Vasserot, [12]. See [21] for further details.

Theorem 12. *If* $\widehat{\deg}(\hat{C}_1(\bar{L})^{d+1}) > 0$*, then:*

$$h^0(\bar{E} \otimes \bar{L}^n) \geq n^{d+1} \frac{r}{(d+1)!} \widehat{\deg}(\hat{C}_1(\bar{L})^{d+1}) + O(n^d \log(n)) \ .$$

In particular, if n *is sufficiently large,* $E \otimes L^n$ *has a non-zero section with* L^2 *norm less than one.*

References

1. Arakelov, S.J.: Intersection theory for divisors on an arithmetic surface. Math. USSR Izv. **8** (1974) 1167–1180
2. Arakelov, S.J.: Theory of intersections on an arithmetic surface. Proc. Int. Cong. of Math., Vancouver, vol. 1, 1978, pp. 405–408
3. Baum, P., Fulton, W., Macpherson, R.: Riemann-Roch for singular varieties. Publ. Math. I.H.E.S. **45** (1975) 101–146
4. Beilinson, A.A.: Higher regulators and values of L-functions. J. Sov. Math. **30** (1985) 2036–2070
5. Beilinson, A.A.: Height pairings between algebraic cycles. Contemp. Math. **67** (1987) 1–24

6. Bismut, J.-M.: Superconnection currents and complex immersions. Invent. math. **99** (1990) 59–113

7. Bismut, J.-M.: Koszul complexes, harmonic oscillators, and the Todd class. Prépublication Orsay 1989, pp. 89–96

8. Bismut, J.-M., Gillet, H., Soulé, C.: Analytic torsion and holomorphic determinant bundles, I, II, III. Comm. Math. Phys. **115** (1988) 49–78, 79–126, 301–351

9. Bismut, J.-M., Gillet, H., Soulé, C.: Bott–Chern currents and complex immersions. Duke Math. J. **60** (1990) 255–284

10. Bismut, J.-M., Gillet, H., Soulé, C.: Complex immersions and Arakelov geometry. In: The Grothendieck Festschrift, vol. I (P. Cartier et al., ed.). Birkhäuser, Boston 1990, pp. 249–231

11. Bismut, J.-M., Lebeau, G.: Immersions complexes et métriques de Quillen (to appear).

12. Bismut, J.-M., Vasserot, E.: Comportement asymptotique de la torsion analytique associée aux puissance d'un fibré en droites positif. C.R. Acad. Sci. Paris **307** (1988) 779–781

13. Bloch, S.: Height pairings for algebraic cycles. In: Proc. Luminy conference on algebraic K-theory. J. Pure App. Alg. **34** (1984) 119–145

14. Bott, R., Chern S.S.: Hermitian vector bundles and the equidisitribution of the zeroes of their holomorphic sections. Acta math. **114** (1968) 71–112

15. Deligne, P.: Le déterminant de la cohomologie. In: Current Trends in Arithmetical Algebraic Geometry (K.Ribet, ed.). Contemp. Math. **67** (1987) 93–177

16. Faltings, G.: Calculus on arithmetic surfaces. Ann. Math. **119** (1984) 387–424

17. Gillet, H., Soulé, C.:Intersection theory using Adams operations. Invent. math. **90** (1987) 243–277

18. Gillet, H., Soulé, C.: Arithmetic intersection theory. Publ. Math. I.H.E.S. **72** (1990) 94–174

19. Gillet, H., Soulé, C.: Characteristic classes for algebraic vector bundles with Hermitian metric. Ann. Math. **131** (1990) 163–238

20. Gillet, H., Soulé, C.: Analytic torsion and the arithmetic Todd genus. Topology **30** (1991) 21–54

21. Gillet, H., Soulé, C.: Amplitude arithmétique. C.R. Acad. Sci. Paris **307** (1988) 887–890

22. Gillet, H., Soulé, C.: Un théorème de Riemann-Roch-Grothendique arithmétique. C.R. Acad. Sci. Paris **309** (1989) 929–932

23. Gillet, H., Soulé, C.: On the number of lattice points in convex symmetric bodies and their duals. Prépublication I.H.E.S. 1990. Israel Math. (to appear)

24. Gillet, H., Soulé, C.: An arithmetic Riemann-Roch theorem. Preprint.

25. Quillen, D.: Determinants of Cauchy-Riemann operators over a Riemann surface. Funct. Anal. Appl. **14** (1985) 31–34

26. Ray, D.B., Singer, I.M.: Analytic torsion for complex manifolds. Ann. Math. **98** (1973) 154–177

27. Silverman, J.H.: A survey of the theory of height functions, In: Current Trends in Arithmetical Algebraic Geometry (K.Ribet, ed.). Contemp. Math. **67** (1987) 269–278

28. Weil, A.: Sur l'analogie entre les corps de nombres algébriques, et les corps de fonctions algébriques. Collected Papers, vol. 1. Springer, Berlin Heidelberg New York 1980, pp. 236–240

Area, Lattice Points and Exponential Sums

Martin N. Huxley

School of Mathematics, University of Wales, College of Cardiff, Senghenydd Road
Cardiff CF2 4AG. Wales, UK

Suppose you have a closed curve. How do you find the area inside? While I was writing my first paper on exponential sums and lattice points, my seven year old daughter came home from school and said. "I know how you find the area of a curve. You count the squares". In other words, copy the curve onto squared paper and count how many squares lie inside the curve. If you want the area to greater accuracy, use paper with smaller squares. If the side of the squares is $1/M$, and you count N squares, then the area A is approximately N/M^2. As a number theorist I prefer to take the squares as unit squares, the curve as enlarged by a factor M, and the relation as being that N is approximately AM^2.

Some squares are inconveniently cut by the curve. When do you count them?

Rule 1. Count a square if its centre lies inside.
Rule 2. Count a square if its lower left corner lies inside.
Rule 3. Count all incomplete squares as half a square.

Rules 1 and 2 are really the same rule: if we shift the squared paper by half a unit in the x and y directions, then the corners of the squares are now where the centres were before. Rule 3 is locally like Rule 2, but it counts extra squares where the curve has maxima and minima, and where x takes its extreme values. Take Rule 2 as the basic rule. The corners of squares are the points in two dimensional space with integer coordinates x, y, the lattice points.

How accurate is the rule? If the curve fits into a rectangular box C high, B broad (in terms of unit squares), then the number of squares cut is at most $2B + 2C + 4$. As M tends to infinity, this discrepancy has order of magnitude M, whilst the area has order M^2.

Lattice points on the curve limit the accuracy. Whether by convention one counts them in or out, changing M to $M + \varepsilon$ makes little change to the area, and counts them all in, whilst changing M to $M - \varepsilon$ counts them all out. If the curve is a polygon, especially a rectangle with sides parallel to the axes, there can be M lattice points on the sides of the rectangle, so that order of magnitude M is then best possible.

So take a smooth curve, convex for convenience. What is the mathematics available? The curve is described in a coordinate-free way by an equation connecting the arc length s with the tangent angle ψ, the radius of curvature being $ds/d\psi$. The

Proceedings of the International Congress
of Mathematicians, Kyoto, Japan, 1990
© The Mathematical Society of Japan, 1991

lattice of integers has an algebraic isomorphism group SL(2, Z), the two by two
integer matrices of determinant one. The lattice is also periodic in two dimensions,
so that you can take Fourier transforms. These tell you that for a smooth curve the
discrepancy $N - AM^2$ has root mean square lying between bounded multiples of
$\sqrt{M}$ (bounded in terms of the shape of the curve). The mean is taken over translations of the curve. As a consequence, for some constant c,

$$|N - AM^2| < c\sqrt{M}$$

for at least half the positions of the curve on the squared paper. The two dimensional
Fourier transform does not help much with pointwise upper bounds, even for a
circle centred on a lattice point, the Gauss circle problem.

The upper bound so far corresponds to replacing the curve by a step function. Voronoi and Sierpinski used Archimedes' idea of replacing the curve by a
polygon. The polygon is chosen so that the gradients of its sides are small rational
numbers. Small is measured by the height norm: the height of a/q in lowest terms
is $\max(|a|, |q|)$. As far as I know, they did not treat the general curve, but only the
hyperbola (for the Dirichlet divisor problem) and the circle. If you assume merely
that the curve has a radius of curvature bounded away from zero and infinity, then
you can show that the discrepancy $|N - AM^2|$ has order of magnitude at most
$(M \log^2 M)^{2/3}$ by using the estimate for the discrepancy of a right-angled triangle
in terms of the continued fraction expansion of the gradient a/q. Sierpinski had this
result in the Gauss circle problem without the logarithmic factor, so being a perfect
circle is a little help. The sides of the polygon have rational gradients, so that they
may contain many lattice points.

The next step is obvious with hindsight: instead of approximating the curve by
a piecewise constant or piecewise linear function, approximate by a piecewise
quadratic function. Here the argument becomes highly analytic. Define $e(t)$ to be
$\exp 2\pi i t$, and $\varrho(t)$ to be the row-of-teeth function, with $\varrho(t) = 1/2 - t$ for $0 \le t < 1$,
and with period one. Then $\varrho(t)$ has the Fourier series

$$\varrho(t) \approx \sum_{\substack{h=-\infty \\ h \ne 0}}^{\infty} \frac{e(ht)}{2\pi i h}.$$

The number of integers n in an interval $a < n \le b$ is $b - a + \varrho(b) - \varrho(a)$. I can now
explain Iwaniec and Mozzochi's attack on the circle problem (1987). It is closely
related to Bombieri and Iwaniec's great paper (1986) on the size of the Riemann
zeta function. Bombieri and Iwaniec's paper turns out to be related to Jutila's work
on the Dirichlet series of modular forms. The connection is that modular forms are
functions of two dimensional lattices.

The first step is to divide the curve into arcs corresponding to the sides of the
Voronoi-Sierpinksi polygon, and to take a new basis for the integer lattice so that
(q, a) is a basis vector. This brings in a 2×2 integer matrix. The curve is then
approximated as a quadratic in a coordinate n in the direction of the vector (q, a)
tangential to the curve.

The sum involving $\varrho(t)$ gives an exponential sum in two summands n and h when the Fourier series for $\varrho(t)$ is truncated. The next two steps are Poisson summation in each variable. There is a serious danger of error terms adding up. It is lucky that the integral to compute is a Bessel function of order one half which has an exact expression of the form

$$e(t)/\sqrt{t}.$$

After approximating, the main terms involve an inner product:

$$e(x^{(k,l)} \cdot y^{(j)}),$$

where $x^{(k,l)}$ is a four dimensional vector

$$(kl, l, l\sqrt{k}, l/\sqrt{k})$$

indexed by two integers k and l, and $y^{(j)}$ is a four dimensional vector constructed from the quadratic approximation. The index j refers to the arc of the curve.

If a/q has small height (a major arc), then k and l take a bounded number of values, and you estimate trivially (in fact one need not go so deep). If a/q has large height (a minor arc), then k and l run through complicated ranges. When these have been simplified, the next step is a form of the large sieve.

The modern large sieve is best described as a sort of Sobolev inequality relating the discrete L_2 norm of a function to the ordinary L_2 norms of the function and its derivatives. The discrete L_2 norm in one dimension is

$$\sup_S \sum_S |f(z_i)|^2$$

taken over sets $S = \{z_1, \ldots, z_R\}$ with $|z_i - z_j| \geq \delta$ for $i \neq j$. In number-theoretic applications we have a fixed set S, usually corresponding to a subset of the rational numbers, and the function $f(z)$ is an exponential sum, which can be regarded as an integral with respect to a discrete measure supported on the integers. In this case there is a duality principle and an underlying bilinear form, in which the vectors x are on the same footing as the vectors y. A further generalisation replaces the condition $|z_i - z_j| \geq \delta$ by a factor in the upper bound which counts the number of pairs of points differing by less than δ (the convolution of the discrete measure supported on the points of S with itself, against a kernel function supported on a δ-neighbourhood of the diagonal).

The next two steps are counting the number of coincidences among sums of pairs of x vectors, which is number theory, and the number of coincidences among pairs of y vectors, which is number theory mixed with analysis. A coincidence between two y vectors means that on two different arcs, the curve weaves between lattice points in the same way. A large set of mutually coincident vectors will add up to a systematic error in the discrepancy $N - AM^2$.

Rather than write down the conditions for entries of the y vectors to coincide, I try to describe them. A y vector corresponds to a minor arc and a two by two integer matrix. Coincidence in the first entries means that the two integer matrices are close in the sense that PQ^{-1} is a matrix with small integer entries, and so PQ^{-1} is a short word in the generators $\begin{pmatrix} 1 & 1 \\ 0 & 1 \end{pmatrix}$ and $\begin{pmatrix} 1 & 0 \\ 1 & 1 \end{pmatrix}$. Coincidence in the second entry involves the matrix, the constant term and the gradient. Coincidence in the third entry is a relation between the denominators q of the gradients a/q and the coefficients of n^2 in the quadratic approximations. Coincidence in the fourth entry involves all the coefficients of the approximating polynomial, but if the other three entries coincide, then the condition simplifies to one involving the constant terms and the denominators q of the gradients a/q.

What arguments can one use in counting the number of pairs of coincident vectors? The first is compactness, or in its discrete version, Dirichlet's box principle. If there are many points in a bounded region, then there is a small set containing many of them. The second principle is approximation, or the mean value theorem. A smooth function on a small set can be approximated by a polynomial, or still better, a linear function. The third principle is the boon and bane of analytic number theory. The only arbitrarily small integer is zero. An inequality which is strengthened too much turns into an equation. When these three arguments are used in turn, we pass from discrete to continuous and back to discrete again.

At present we are overestimating the number of coincidences because we cannot handle the constant term in the quadratic polynomial. So the second and fourth entries of the y vectors are not used. Iwaniec and Mozzochi (1987) could count coincidence only for the rectangular hyperbola. Huxley and Watt (1988) generalised the relevant lemma in Bombieri and Iwaniec (1986), getting bounds for one dimensional exponential sums, and Huxley (1990) adapted the Iwaniec-Mozzochi method to general smooth closed curves to get

$$N = AM^2 + O(M^{7/11}(\log M)^{47/22}).$$

Can this be improved? There is some hope of using the second and fourth entries of the y vectors, which would give a small improvement on the exponent $7/11$. To reach one half one must avoid taking moduli. Bombieri and Iwaniec (1986) suggest using the Fourier theory of the modular group, which acts on the vectors (q, a) corresponding to the sides of the Voronoi-Sierpinski polygon. This Fourier theory was suggested by Selberg (1956) and worked into a usable form by Kuznietsov (1980) and Deshouillers and Iwaniec (1982).

There are many related questions. Bombieri and Pila (1989) have shown by algebraic geometry that if the curve is not algebraic, then the number of integer points on it is $O(M^\varepsilon)$ for any positive ε, improving the earlier result of Swinnerton-Dyer (1974). Huxley (1988, 1989) has considered the number of integer points within a small distance δ of the curve, using a mixture of elementary and exponential sum techniques. The lattice point problem is a special case of rounding error in numerical approximation, and one can give estimates there too (Huxley 1991).

This method seems to be purely two dimensional, because a quadratic approximation to a surface contains too many terms. Even the continued fraction rule does not generalise. But I hope that someone will take up the challenge of finding an arithmetic method for counting lattice points in three or more dimensions.

References

Bombieri, E. (1987): Le grand crible dans le théorie analytique des nombres, 2de edn. Astérisque, Paris

Bombieri, E., Iwaniec, H. (1986): On the order of $\zeta(1/2 + it)$. Ann. Sc. Norm. Sup. Pisa Cl. Sci. (4) **13**, 449–472

Bombieri, E., Pila, J. (1989): The number of integral points on arcs and ovals. Duke Math. J. **59**, 337–357

Deshouillers, J.-M., Iwaniec, H. (1982): Kloosterman sums and Fourier coefficients of cusp forms, Invent. math. **70**, 219–288

Fouvry, E., Iwaniec, H. (1989): Exponential sums for monomials, J. Number Theory **33**, 311–333

Heath-Brown, D.R., Huxley, M.N. (1990): Exponential sums with a difference. Proc. London Math. Soc. (3) **61**, 227–250

Huxley M.N. (1988): The fractional parts of a smooth sequence. Mathematika **35**, 292–296

Huxley M.N. (1989): The integer points close to a curve. Mathematika **36**, 198–215

Huxley M.N. (1990): Exponential sums and lattice points. Proc. London Math. Soc. (3) **60**, 471–502

Huxley M.N. (1991): Exponential sums and rounding error. J. London Math. Soc. (to appear)

Huxley, M.N., Kolesnik, G. (1991): Exponential sums and the Riemann zeta function III. Proc. London Math. Soc. (3) **62** (to appear)

Huxley, M.N., Watt, N. (1988): Exponential sums and the Riemann zeta function. Proc. London Math. Soc. (3) **57**, 1–24

Huxley, M.N., Watt, N. (1989): Exponential sums with a parameter. Proc. London Math. Soc. (3) **59**, 233–252

Iwaniec, H., Mozzochi, C.J. (1988): On the divisor and circle problems. J. Number Theory **29**, 60–93

Jutila, M. (1987): Lectures on a method in the theory of exponential sums. Tata Institute Lectures in Mathematics and Physics, vol. 80. Springer, Bombay

Krätzel. E., (1988): Lattice points. D.V.W., Berlin

Kuznietsov, N.V. (1980): Peterson's conjecture for cusp forms of weight zero and Linnik's conjecture on sums of Kloosterman sums. Mat. Sbornik **111**. 334–383

Selberg, A. (1956): Harmonic analysis and discontinuous groups in weakly symmetric Riemannian spaces with applications to Dirichlet series. J. Indian Math. Soc. **20**, 47–87

Sierpinski, W. (1906): Sur un problème du calcul des fonctions asymptotiques. Prace Mat.-Fiz **17**, 77–118

Swinnerton-Dyer, H.P.F. (1974): The number of lattice points on a convex curve. J. Number Theory **6**, 128–135

Voronoi, G. (1906): Sur un problème du calcul des fonctions asymptotiques. J. Reine Angew. Math. **126**

Watt, N. (1989a): A problem on semicubical powers. Acta Arith. **52**, 119–140

Watt, N. (1989b): Exponential sums and the Riemann zeta function II, J. London Math. Soc. **39**, 385–404

Watt, N. (1990): A problem on square roots of integers. Periodica Math. Hung. **21**, 55–64

Watt, N. (1991): Exponential sums with a character. (To appear)

Generalized Class Field Theory

Kazuya Kato

Department of Mathematics, Faculty of Science
University of Tokyo, Hongo, Bunkyo-ku, Tokyo 113, Japan

This paper is a survey of the K-theoretic generalization of class field theory.

For a field K, let K^{ab} be a maximal abelian extension of K, that is, the union of all finite abelian extensions of K in a fixed algebraic closure of K. The classical local (resp. global) class field theory says that if K is a finite extension of the p-adic (resp. rational) number field $\mathbf{Q}_p$ (resp. $\mathbf{Q}$), the Galois group $\mathrm{Gal}(K^{ab}/K)$ is approximated by the multiplicative group $K^\times$ (resp. the idele class group C_K), and via this approximation, we can obtain knowledge on abelian extensions of K.

In §1 (resp. §2), we give a K-theoretic generalization of the classical local (resp. global) class field theory. There finite extensions of $\mathbf{Q}_p$ (resp. $\mathbf{Q}$) are replaced by "higher dimensional local fields" (resp. finitely generated fields over prime fields), and the group $K^\times$ (resp. C_K) is replaced by Milnor's K-group $K_n^M(K)$ (resp. by the K_n^M-idele class group), where n is the "dimension" of K.

In §3, we discuss some other aspects of generalizations of local class field theory.

In §4, we discuss generalizations of the classical ramification theory to higher dimensional schemes.

1. Local Class Field Theory

An n-dimensional local field is defined inductively as follows. A 0-dimensional local field is a finite field. For $n \geq 1$, an n-dimensional local field is a complete discrete valuation field whose residue field is an $(n-1)$-dimensional local field.

For example, a finite extension of $\mathbf{Q}_p$ is a one dimensional local field.

For a field k, let $K_*^M(k)$ be Milnor's K-group of k defined by

$$K_q^M(k) = ((k^\times)^{\otimes q})/J$$

where J is the subgroup of the q-fold tensor product of $k^\times$ (as a $\mathbf{Z}$-module) generated by elements of the form $a_1 \otimes \dots \otimes a_q$ such that $a_i + a_j = 1$ for some $i \neq j$. (Cf. Milnor [Mi].) The main result of the local class field theory of an n-dimensional local field is the following (Parsin [Pa₂], Kato [Ka₁] II).

Proceedings of the International Congress
of Mathematicians, Kyoto, Japan, 1990
© The Mathematical Society of Japan, 1991

Theorem 1. *Let K be an n-dimensional local field. Then, there exists a canonical homomorphism*

$$\varrho : K_n^M(K) \longrightarrow \mathrm{Gal}(K^{\mathrm{ab}}/K)$$

which induces an isomorphism $K_n^M(K)/N_{L/K}K_n^M(L) \xrightarrow{\cong} \mathrm{Gal}(L/K)$ for each finite abelian extension L of K. Here $N_{L/K} : K_n^M(L) \to K_n^M(K)$ is the norm homomorphism ([Ka$_1$] II 1.7). The correspondence $L \mapsto N_{L/K}K_n^M(L)$ is a bijection from the set of all finite abelian extensions of K in a fixed algebraic closure of K onto the set of all open subgroups of $K_n^M(K)$ of finite indices. (For the definition of the openness of a subgroup of $K_n^M(K)$, see [Ka$_2$]).

2. Global Class Field Theory

Let X be a proper integral scheme over the ring of rational integers $\mathbf{Z}$ and let K be the function field of X. For simplicity we assume here $\mathrm{char}(K) = 0$ and that K has no ordered field structure.

For a non-zero coherent ideal I of $\mathcal{O}_X$ and for $q \geq 1$, define the sheaf of abelian groups $K_q^M(\mathcal{O}_X, I)$ on X_{zar} by

$$K_q^M(\mathcal{O}_X, I) = \mathrm{Ker}(K_q^M(\mathcal{O}_X) \to K_q^M(\mathcal{O}_X/I))$$

where

$$K_q^M(\mathcal{O}_X) = ((\mathcal{O}_X^\times)^{\otimes q})/J$$

with J the subgroup sheaf of the tensor product generated by local sections of the form $a_1 \otimes \dots \otimes a_q$ such that $a_i + a_j = 1$ for some $i \neq j$, and $K_q^M(\mathcal{O}_X/I)$ is defined similarly. Define

$$C_I(X) = H^n(X_{\mathrm{zar}}, K_n^M(\mathcal{O}_X, I)), \quad \text{where} \quad n = \dim(X).$$

If I and I' are non-zero coherent ideals of $\mathcal{O}_X$, the inclusion $I \subset I'$ induces a surjection $C_I(X) \to C_{I'}(X)$. The main result of the class field theory of K is the following.

Theorem 2.1. *(1) $C_I(X)$ is a finite group for any I.*

(2) We have a canonical isomorphism of profinite abelian groups

$$\varprojlim_I C_I(X) \cong \mathrm{Gal}(K^{\mathrm{ab}}/K),$$

where I ranges over all non-zero coherent ideals of $\mathcal{O}_X$.

(3) For a non-empty regular open subscheme U of X, there exists a canonical isomorphism of profinite abelian groups

$$\varprojlim_I C_I(X) \cong \pi_1^{\mathrm{ab}}(U),$$

where I ranges over all non-zero coherent ideals of $\mathcal{O}_X$ such that $U \cap \mathrm{Spec}(\mathcal{O}_X/I) = \phi$, and $\pi_1^{\mathrm{ab}}(U)$ is the quotient of the algebraic fundamental group $\pi_1(U)$ of U modulo the closure of its commutator subgroup.

(4) If X is regular, then $C_{\mathcal{O}_X}(X)$ is isomorphic to the group $CH_0(X)$ of the classes of zero cycles on X modulo rational equivalence, and we have a canonical isomorphism of finite abelian groups

$$CH_0(X) \cong \pi_1^{\mathrm{ab}}(X).$$

An essential part of this theorem was proved by Bloch [Bl₁] in the case $\dim(X) = 2$. Cf. also [Pa₁] for the two dimensional case. The general case was proved in [KS₂] (see also S. Saito [SS₁]) by using the method in Bloch [Bl₁], except that we adopted in [KS₂] another definition of $C_I(X)$ which uses Nisnevich topology [Ni], a Grothendieck topology defined by Nisnevich [Ni] (called the henselian topology in [KS₂]), instead of Zariski topology. ($C_I(X) \overset{\mathrm{def}}{=} H^n(X_{\mathrm{Nis}}, K_n^M(\mathcal{O}_X, I))$ in [KS₂].). It was found later that Nisnevich topology and Zariski topology give the same $C_I(X)$ [KS₃].

The relation of the above theorem with the classical global class field theory is that if $X = \mathrm{Spec}(O_K)$ for a finite extension K of $\mathbf{Q}$, then $C_I(X)$ coincides with the ideal class group of conductor I of O_K.

The positive characteristic version of (4) of the above theorem is that if X is a proper smooth variety over a finite field k, $CH_0(X)^0 = \mathrm{Ker}(\deg : CH_0(X) \to \mathbf{Z})$ is finite and canonically isomorphic to $\mathrm{Ker}(\pi_1^{\mathrm{ab}}(X) \to \mathrm{Gal}(k^{\mathrm{ab}}/k))$. This fact was proved in [KS₁], and another proof was given in Colliot-Thélène, Sansuc and Soulé [CSS] and Gros [Gr]. That the former group surjects onto the latter was proved long ago by Lang whose paper [Lan] is the first work on the higher dimensional class field theory.

As an application of the generalized global class field theory, we have

Theorem 2.2. *For any scheme S of finite type over $\mathbf{Z}$, the abelian group $CH_0(S)$ is of finite type.*

The case $\dim(S) = 2$ of this theorem was proved by Bloch [Bl₁], and the general case was proved in [KS₂].

Some finiteness theorems on $\pi_1^{\mathrm{ab}}(S)$ for arithmetic schemes S were proved in Katz-Lang [KL].

3. Some Aspects of the Generalized Local Class Field Theory

I give rough indications on some aspects of generalizations of local class field theory, which are not included in §1.

3.1 Explicit Reciprocity Law

The explicit reciprocity law for a finite extension of $\mathbf{Q}_p$ has been generalized to higher dimensional local fields of mixed characteristic by Vostokov-Kirrilov [VK] and Vostokov [V]. A generalization to complete discrete valuation fields of mixed characteristic $(0, p)$ with residue field F such that $[F : F^p] < \infty$ is given in [Ka4], as an application of the p-adic cohomology theory of Fontaine-Messing [FM].

Let K be "the most important two dimensional local field"

$$(\varprojlim_{n} (\mathbf{Z}/p^n\mathbf{Z})[[q]][q^{-1}]) \otimes_{\mathbf{Z}} \mathbf{Q}$$

where q is the q-invariant in the theory of moduli of elliptic curves. This field appears as a certain p-adic completion at infinity of a modular curve. As is shown in [Ka5], the explicit reciprocity laws of the two dimensional local fields $K(\zeta_{p^n})$, where ζ_{p^n} is a primitive p^n-th root of 1, are related to the special values of L-functions of elliptic modular forms and to an Iwasawa theory of elliptic modular forms, just as the explicit reciprocity laws of $\mathbf{Q}_p(\zeta_{p^n})$ are related to the special values of Riemann zeta function and to the classical Iwasawa theory.

3.2 Semi-Global Theories

The class field theory of curves over (usual) local fields was studied by Coombes [Co] and S. Saito [SS₂], and that of surfaces over local fields was studied recently by S. Saito and Salberger.

The class field theory of complete discrete valuation fields whose residue fields are (usual) global fields of positive characteristic was sought for first by Ihara [Ih], and studied by [Ka₁] III.

The class field theory of two dimensional complete noetherian local rings with finite residue fields (for example $\mathbf{Z}_p[[T]]$) was studied by S. Saito [SS₃].

3.3 Serre's Local Class Field Theory

Serre's local class field theory of a complete discrete valuation field with algebraically closed residue field [Se₂], and its generalization by Hazewinkel to the perfect residue field case [Ha], are generalized to the imperfect residue field case as a duality theorem of the following form: If K is a complete discrete valuation field with residue field F such that $\mathrm{char}(F) = p > 0$ and $[F : F^p] = p^r < \infty$, then the Galois cohomologies

$$(*) \qquad R\Gamma(K, \mathbf{Z}/p^n\mathbf{Z}(s)) \quad \text{and} \quad R\Gamma(K, \mathbf{Z}/p^n\mathbf{Z}(t)), \qquad s + t = r + 1$$

are in perfect duality via the dualizing functor $R\,\mathrm{Hom}_{\mathbf{Z}/p^n\mathbf{Z}}(\ , W_n\Omega^r_{F,\log})[r + 1]$. Here the objects $(*)$ and the logarithmic part of the de Rham-Witt sheaf $W_n\Omega^r_{F,\log}$ ([Il]) are not regarded just as a complex of abelian groups or as objects on the small etale site on $\mathrm{Spec}(F)_{\mathrm{et}}$, but regarded as objects on a much bigger site. (Precisely speaking, they are regarded as objects on the site of schemes S over

F which are locally isomorphic to "relative perfections" ([Ka$_3$] I, II) of smooth schemes over F, endowed with the etale topology.) The details of this duality theorem are given in [Ka$_3$] III.

For example, if F is an algebraically closed field, this duality and the inclusion

$$U_K/(U_K)^{p^n} \subset K^\times/(K^\times)^{p^n} = H^1(K, \mathbf{Z}/p^n\mathbf{Z}(1))$$

induces $H^1(K, \mathbf{Z}/p^n\mathbf{Z}) \cong \mathrm{Ext}^1(U_K, \mathbf{Z}/p^n\mathbf{Z})$, where U_K is the unit group of K which is not regarded just as a group, but regarded as a pro-algebraic group over F. This reproduces the p-primary part (the essential part) of Serre's local class field theory

$$\pi_1(U_K) \cong \mathrm{Gal}(K^{\mathrm{ab}}/K).$$

4. Ramification Theory

In the classical ramification theory, we mainly consider a finite extension B/A of discrete valuation rings with perfect residue fields. We have the following three kinds of important invariants of ramification: The different $\delta(B/A) \in \mathbf{Z}$; in the case of Galois extension with Galois group G, the Lefschetz numbers

$$i(\sigma) = \mathrm{length}(B/I_\sigma) \in \mathbf{Z} \quad \text{for} \quad \sigma \in G - \{1\},$$

with I_σ the ideal of B generated by $\{\sigma(b) - b \; ; \; b \in B\}$; also in the Galois case, the Artin (or Swan) conductors of representations of G. Is it possible to generalize these invariants and relations between them such as

$$(*) \qquad\qquad \delta(L/K) = \sum_{\sigma \in G - \{1\}} i(\sigma)$$

to higher dimensional schemes?

Concerning $\delta(B/A)$ and $i(\sigma)$, Bloch obtained in [Bl$_2$] a nice ramification theory in higher dimensions, generalizing $\delta(B/A)$ and $i(\sigma)$ to zero cycle classes on the ramification locus (i.e. to elements of $CH_0(E)$ where E is the ramification locus). His projection formula [Bl$_2$] (7.1) is an extension of (*) to two dimensional schemes.

Concerning conductors, there are many different attempts of generalizations (Deligne [De], Laumon [La], S. Saito [SS$_4$], Berthelot [Be$_1$, Be$_2$]). In the following, I only introduce my method on the conductors of one dimensional Galois representations [Ka$_7$] which is closely related to the generalized local class field theory. I generalize, under the influence of Bloch's theory [Bl$_2$], the Swan conductors to zero cycle classes on the wild ramification locus in the two dimensional case (the plan exists even for dimension ≥ 3), and give applications Theorems (4.1)–(4.3).

Let X be an excellent connected normal scheme, U a regular dense open subscheme of X, and let $\chi : \pi_1^{\mathrm{ab}}(U) \to \Lambda^\times$ be a continuous homomorphism where Λ is a discrete (commutative) field. We are interested in the wild ramification of χ

on X. We assume that the wild ramification locus E of χ on X (with the reduced scheme structure) is a disjoint union of schemes of finite type over perfect fields. This condition is satisfied for example if X is of finite type over $\mathbf{Z}$ or over a perfect field.

First, we have a divisor called "Swan conductor divisor" ([Ka$_6$]) sw$(\chi) = \sum_{\mathfrak{p}} \mathrm{sw}_{\mathfrak{p}}(\chi) \overline{\{\mathfrak{p}\}}$ on X ($\overline{\{\mathfrak{p}\}}$ denotes the closure of $\{\mathfrak{p}\}$), where $\mathfrak{p}$ ranges over all points of codimension one. The integer $\mathrm{sw}_{\mathfrak{p}}(\chi)$ has the following properties: (i) If $\dim(X) = 1$, it coincides with the classical Swan conductor of χ at $\mathfrak{p}$. (ii) $\mathrm{sw}_{\mathfrak{p}}(\chi) > 0$ if and only if χ is wildly ramified at $\mathfrak{p}$. Furthermore, if the following condition

(C) X is regular and $D = (X - U)_{\mathrm{red}}$ is a divisor with normal crossings on X,

we have a canonical global section

$$\mathrm{rsw}(\chi) \in \Gamma(E, \mathcal{O}_E \otimes_{\mathcal{O}_X} \Omega^1_X(\log(D)) \otimes_{\mathcal{O}_X} \mathcal{O}_X(\mathrm{sw}(\chi)))$$

called the refined Swan conductor ([Ka$_7$] I). Here $\Omega^1_X(\log(D))$ is the $\mathcal{O}_X$-module on X_{et} defined by generators $\mathrm{dlog}(a)$ ($a \in j_*(\mathcal{O}_U^\times)$ with j the inclusion map $U \to X$) subject to natural relations such as "$a\mathrm{dlog}(a)$ is additive in a", and $\mathcal{O}_X(\mathrm{sw}(\chi))$ denotes the invertible $\mathcal{O}_X$-module corresponding to the divisor $\mathrm{sw}(\chi)$. In this case, E coincides with the support of the divisor $\mathrm{sw}(\chi)$. It can be shown that the $\mathcal{O}_E$-module $\mathcal{O}_E \otimes_{\mathcal{O}_X} \Omega^1_X(\log(D))$ is locally free of rank (locally) $\dim(E)+1$. If X is of finite type over $\mathbf{Z}$, these $\mathrm{sw}(\chi)$ and $\mathrm{rsw}(\chi)$ have explicit descriptions in terms of K-theoretic class field theory ([Ka$_7$] I). We say (X, U, χ) is clean if at any point x of E, the stalk of $\mathrm{rsw}(\chi)$ at x (which is always non-zero) is a part of a basis of $\mathcal{O}_E \otimes_{\mathcal{O}_X} \Omega^1_X(\log(D)) \otimes_{\mathcal{O}_X} \mathcal{O}_X(\mathrm{sw}(\chi))$ at x. If (X, U, χ) is clean, we define a cycle $\mathrm{Char}(X, U, \chi)$ on the vector bundle $\mathcal{O}_E \otimes_{\mathcal{O}_X} \Omega^1_X(\log(D))$ over E by

$$\mathrm{Char}(X, U, \chi) = \sum_{\mathfrak{p}} \mathrm{sw}_{\mathfrak{p}}(\chi)\, \mathrm{Image}(\varphi_{\mathfrak{p}}),$$

where $\varphi_{\mathfrak{p}}$ is the map $\mathcal{O}_E \otimes_{\mathcal{O}_X} \mathcal{O}_X(-\mathrm{sw}(\chi)) \to \mathcal{O}_E \otimes_{\mathcal{O}_X} \Omega^1_X(\log(D))$ induced by $\mathrm{rsw}(\chi)$, and $\mathrm{Image}(\varphi_{\mathfrak{p}})$ is regarded as a subbundle of $\mathcal{O}_E \otimes_{\mathcal{O}_X} \Omega^1_X(\log(D))$. In the clean case, we define our generalization of the Swan conductor

$$c(X, U, \chi) \in CH_0(E)$$

to be $(-1)^{\dim(E)}$ times the intersection class of the two cycles $\mathrm{Char}(X, U, \chi)$ and the zero section of the vector bundle $\mathcal{O}_E \otimes_{\mathcal{O}_X} \Omega^1_X(\log(D))$. This $\mathrm{Char}(X, U, \chi)$ is an analogue of the charateristic cycle in the theory of $\mathscr{D}$-modules.

Now without the condition (C), I conjecture that there exists a proper birational morphism $f : X' \to X$ such that $f^{-1}(U) \overset{\cong}{\to} U$ and $(X', f^{-1}(U), \chi)$ is clean. It can be proved ([Ka$_7$] I) that if $\dim(X) \leq 2$, this conjecture is true and

$$c(X, U, \chi) \overset{\mathrm{def}}{=} f_* c(X', f^{-1}(U), \chi) \in CH_0(E)$$

is independent of the choice of such X'.

Theorem 4.1. *Let X be a proper normal surface over an algebraically closed field k, U a regular dense open subscheme of X, ℓ a prime number different from $\mathrm{char}(k)$, $\chi : \pi_1^{\mathrm{ab}}(U) \to \overline{\mathbf{F}}_\ell^\times$ a continuous homomorphism, and $\mathcal{F}$ the $\overline{\mathbf{F}}_\ell$-sheaf on U_{et} of rank one corresponding to χ. Then,*

$$\chi(U, \mathcal{F}) = \chi(U, \overline{\mathbf{F}}_\ell) - \deg(c(X, U, \chi)).$$

Here $\chi(\ )$ denotes the Euler-Poincaré characteristic $\sum_i (-1)^i \dim_{\overline{\mathbf{F}}_\ell} H_{\mathrm{et}}^i(\ )$.

This theorem was generalized by T. Saito [ST$_2$] to the case $\dim(X)$ is arbitrary under a certain assumption on χ .

The Riemann-Roch formula for the Euler-Poincaré characteristics of ℓ-adic sheaves ($\ell \neq$ characteristic), which should generalize the Grothendieck-Ogg-Shafarevich formula on ℓ-adic sheaves on curves to higher dimensional varieties, was sought for first by Grothendieck. A formula in the characteristic zero case was obtained by MacPherson [Ma]. In the positive characteristic case, results for surfaces, of types different from (4.1), have been obtained by Deligne [De], Laumon [Lau] (using ideas of Deligne), S. Saito [SS$_4$]. Deligne is the first person who had the idea of the characteristic cycle of an ℓ-adic sheaf. Recently, Berthelot obtained a Riemann-Roch formula for $\mathscr{D}$-modules with Frobenius in characteristic p by defining the characteristic cycle of a $\mathscr{D}$-module with Frobenius ([Be$_2$] basing on his theory [Be$_1$]). There should be close relations between his characteristic cycle, the characteristic cycle of Deligne, and the characteristic cycle $\mathrm{Char}(X, U, \chi)$ discussed above.

Theorem 4.2. *Let A be a complete discrete valuation ring with field of fractions k and with perfect residue field F. Let X be a regular connected A-scheme which is proper flat of relative dimension one over A, U a dense open subscheme of $X_k = X \otimes_A k$, let ℓ be a prime different from $\mathrm{char}(F)$, let $\chi : \pi_1^{\mathrm{ab}}(U) \to \overline{\mathbf{F}}_\ell^\times$ be a continuous homomorphism, and let $\mathcal{F}$ be the $\overline{\mathbf{F}}_\ell$-sheaf of rank one on U corresponding to χ . Assume χ is at worst tamely ramified at points in $X_k - U$. (This last condition is satisfied automatically if $\mathrm{char}(k) = 0$). Then*

$$\mathrm{sw}(R\Gamma((U_{\overline{k}})_{\mathrm{et}}, \mathcal{F})) = \mathrm{sw}(R\Gamma((U_{\overline{k}})_{\mathrm{et}}, \overline{\mathbf{F}}_\ell)) + \deg(c(X, U, \chi)).$$

Here $\mathrm{sw}(R\Gamma(\ )) = \sum_i (-1)^i \mathrm{sw}(H^i(\ ))$, with sw the Swan conductor of a representation of $\mathrm{Gal}(k^{\mathrm{sep}}/k)$.

In [Bl$_2$], Bloch obtained a formula which expresses $\mathrm{sw}(R\Gamma((U_{\overline{k}})_{\mathrm{et}}, \overline{\mathbf{F}}_\ell))$ in terms of the differential module $\Omega_{X/A}^1$.

Finally we discuss a conjecture of Serre. Let B be a regular local ring and let G be a finite subgroup of $\mathrm{Aut}(B)$. Assume that the following conditions (i)(ii) are satisfied.

(i) For any $\sigma \in G - \{1\}$, B/I_σ is of finite length where I_σ denotes the ideal of B generated by $\{\sigma(b) - b \, ; \, b \in B\}$.

(ii) Let $A = \{b \in B \, ; \, \sigma(b) = b$ for any $\sigma \in G\}$. Then, A is noetherian and $A/m_A \to B/m_B$ is an isomorphism (m_* denote the maximal ideals).

Define the function $a_G : G \to \mathbf{Z}$ by

$$a_G(\sigma) = -\operatorname{length}(B/I_\sigma) \quad \text{for} \quad \sigma \in G - \{1\},$$

$$a_G(1) = -\sum_{\sigma \in G - \{1\}} a_G(\sigma).$$

If $\dim(B) = 1$, a_G is a character of a representation of G called Artin representation. Serre conjectures that a_G is a character of a representation of G even when $\dim(B) > 1$ ([Se$_1$]).

Theorem 4.3. *The conjecture is true if* $\dim(B) = 2$.

This theorem was proved in Kato, S. Saito, T. Saito [KSS] in the equal characteristic case, and in T. Saito [ST$_1$] in the mixed characteristic case under a certain assumption.

The outline of the proof of (4.3) given in [Ka$_7$] II is very similar to the proof for the one dimensional case. It is sufficient to prove that if χ is the character of a representation of G, then

$$\operatorname{Card}(G)^{-1} \sum_{\sigma \in G} a_G(\sigma)\chi(\sigma) \in \mathbf{Z}.$$

By using the two dimensional version in Bloch [Bl$_2$] of the formula (*) at the bigining of §4 and by the theory of Brauer, just as in the one dimensional case, we are reduced to the case χ is of degree one. We may assume that A is complete, A/m_A is perfect, and χ is wildly ramified. Let $X = \operatorname{Spec}(A)$, $E = \operatorname{Spec}(A/m_A) \subset X$ and $U = X - E$. Then, we can prove

$$\operatorname{Card}(G)^{-1} \sum_{\sigma \in G} a_G(\sigma)\chi(\sigma) = 1 + \deg(c(X, U, \chi)) \in \mathbf{Z}$$

where deg is the canonical isomorphism $CH_0(E) \xrightarrow{\cong} \mathbf{Z}$.

References

[Be$_1$] Berthelot, P.: Cohomologie rigid et théorie des $\mathscr{D}$-modules. Preprint
[Be$_2$] Berthelot, P.: In preparation
[Bl$_1$] Bloch, S.: Algebraic K-theory and class field theory for arithmetic surfaces. Ann. Math. **114** (1981) 229–266
[Bl$_2$] Bloch, S.: Cycles on arithmetic schemes and Euler characteristics of curves. Proc. Symp. Pure Math. AMS **46** Part 2 (1987) 421–450
[CSS] Colliot-Thélène, J.-L., Sansuc, J.-J., Soulé, C.: Torsion dans le groupe de Chow de codimension deux. Duke Math. J. **50** (1983) 763–801
[Co] Coombes, K. R.: Local class field theory for curves. Contemp. Math. **55** I (1983) 117–134
[De] Deligne, P.: Letter to Illusie, Nov. 4, 1976
[FM] Fontaine, J. M., Messing, W.: p-adic periods and p-adic etale cohomology. Contemp. Math. **67** (1987) 179–207

[Fu]   Fulton, W.: Intersection theory. Springer, Berlin Heidelberg New York 1984

[Gr]   Gros, M.: Sur la partie p-primaire du groupe de Chow de codimension deux. Comm. Alg. **13** (1985) 2407–2420

[Ha]   Hazewinkel, M.: Corps de classes local, Appendix to Demazure, M., Gabriel, P.: Groupes algébriques, North-Holland Publ. 1970, pp. 648-674

[Ih]   Ihara, Y.: On a problem on some complete p-adic function fields (in Japanese). Kokuroku 41, RIMS, Kyoto Univ. (1968) 7–17

[Il]   Illusie, L.: Complexe de de Rham-Witt et cohomologie cristalline. Ann. Sci. Ec. Norm. Sup. **12** (1976) 501–661

[Ka$_1$]   Kato, K.: A generalization of local class field theory by using K-groups, I., J. Fac. Sci. Univ. of Tokyo, Sec. IA, **26** (1979) 303–376; II, ibid. **27** (1980) 603–683; III, ibid. **29** (1982) 31–43

[Ka$_2$]   Kato, K.: Class field theory and algebraic K-theory. (Lecture Notes in Mathematics, vol. 1016.) Springer, Berlin Heidelberg New York, pp. 109–127

[Ka$_3$]   Kato, K.: Duality theories for p-primary etale cohomology. Algebraic and topological theories. I, Kinonuniya (1985) 127–148; II, Comp. Math. **63** (1987) 259–270; III. In preparation

[Ka$_4$]   Kato, K.: The explicit reciprocity law and the cohomology of Fontaine-Messing. Preprint

[Ka$_5$]   Kato, K.: p-adic Hodge theory and values of zeta functions of elliptic cusp forms. In preparation

[Ka$_6$]   Kato, K.: Swan conductors for characters of degree one in the imperfect residue field case. Contemp. Math. **83** (1989) 101–132

[Ka$_7$]   Kato, K.: Class field theory, $\mathscr{D}$-modules, and ramification on higher dimensional schemes. I, to appear in Amer. J. Math.; II. In preparation

[KL]   Katz, N., Lang, S.: Finiteness theorems in geometric class field theory. l'Enseignement Mathématique **27** (1981) 285–319

[KS$_1$]   Kato, K., Saito, S.: Unramified class field theory of arithmetical surfaces. Ann. Math. **118** (1983) 241–275

[KS$_2$]   Kato, K., Saito, S.: Global class field theory for arithmetic schemes. Contemp. Math. **5** Part I (1986) 255–330

[KS$_3$]   Kato, K., Saito, S.: In preparation

[KSS]   Kato, K., Saito, S., Saito, T.: Artin characters for algebraic surfaces. Amer. J. Math. **110** (1988) 49–76

[Lan]   Lang, S.: Unramified class field theory over function fields in several variables. Ann. Math. **64** (1956) 286–325

[Lau]   Laumon, G.: Caractéristique d'Euler Poincaré des faisceaux constructibles sur une surface. Asterisque **101–102** (1983) 193–207

[Ma]   MacPherson, R.: Chern classes of singular varieties. Ann. Math **100** (1974) 423–432

[Mi]   Milnor, J.: Algebraic K-theory and quadratic forms. Inv. math. **9** (1970) 318–344

[Ni]   Nisnevich, A.: Arithmetic and cohomology invariants of semi-simple group schemes and compactifications of locally symmetric spaces. Funct. Anal. Appl. **14**, no. 1 (1980) 61–62

[Pa$_1$]   Parsin, A. N.: Abelian coverings of arithmetic schemes. Sov. Math. Dokl. **19**, no. 6 (1978) 1438–1442

[Pa$_2$]   Parsin, A. N.: Local class field theory. Proc. Steklov Inst. Math. **165** (1985) 157–185

[Se$_1$]   Serre, J.-P.: Sur la rationalité des représentations d'Artin. Ann. Math. **72** (1960) 406–420

[Se$_2$]   Serre, J.-P.: Sur les corps locaux à corps résiduel algébriquement clos. Bull. Soc. Math. France **89** (1961) 105–154

[SS$_1$]   Saito, S.: Unramified class field theory for arithmetic schemes. Ann. Math. **121** (1985) 251–281

[SS₂] Saito, S.: Class field theory for curves over local fields. J. Number Theory **21** (1985) 44–80

[SS₃] Saito, S.: Class field theory for two dimensional local rings. Adv. Stud. Pure Math. **12** (1987) 343–373

[SS₄] Saito, S.: General fixed point formula for an algebraic surfaces and the theory of Swan represenations for two dimensional local rings. Amer. J. Math. **109** (1987) 1009–1042

[ST₁] Saito, T.: Bloch's 0-cycles and Artin characters of arithmetic surfaces. Preprint (1986)

[ST₂] Saito, T.: The Euler numbers of ℓ-adic sheaves of rank 1 in positive characteristic. Preprint

[Vo]   Vostokov, S. V.: Explicit construction of class field theory for a multidimensional local field. Math. USSR Izv. **26** (1986) 263–287

[VK]   Vostokov, S. V., Kirillov A. N.: Normed pairing in a two dimensional local field. J. Sov. Math. **30**, no. 1 (1985) 1847–1853

On the Mordell-Weil Group and the Shafarevich-Tate Group of Modular Elliptic Curves

Victor Alecsandrovich Kolyvagin*

Steclov Mathematical Institute, u. Vavilova 42, 117966 Moscow GSP-1, USSR

The main purpose of this paper is to describe some recent results pertaining to the diophantine analysis of elliptic curves. A new element is an extension of the set of explicit cohomology classes, see Section 2.

1. The Conjecture of Birch and Swinnerton-Dyer and the Hypothesis of Finiteness of the Shafarevich-Tate Group

Let E be an elliptic curve defined over the field of rational numbers $\mathbb{Q}$, for example, by its Weierstrass equation $y^2 = 4x^3 - g_2 x - g_3$. Let R be a finite extension of $\mathbb{Q}$. We are interested in the group $E(R)$ called the Mordell-Weil group of E over R and the Shafarevich-Tate group $\text{III}(R, E)$. The group $\text{III}(R, E)$ is, by definition, $\ker(H^1(R, E) \to \prod_v H^1(R(v), E))$, where v runs through the set of all places (equivalence classes of valuations) of R, $R(v)$ is the v-adic completion of R. For an arbitrary extension L of $\mathbb{Q}$, we let $\bar{L}$ denote an algebraic closure of L. If V/L is a Galois extension, then $G(V/L)$ denotes its Galois group, and $H^1(L, E) = H^1(G(\bar{L}/L), E(\bar{L}))$.

Let Y be some set of algebraic curves over R. By definition, the Hasse principle holds for Y, if for all $X \in Y$ one has: $X(R)$ is nonempty $\Leftrightarrow X(R(v))$ is nonempty for each v. The group $\text{III}(R, E)$ is the obstacle to the Hasse principle for the set $Y(R, E)$ of main principal homogeneous spaces over E defined over R. In particular, the Hasse principle holds for $Y(R, E)$ if and only if the group $\text{III}(R, E)$ is trivial.

According to the Mordell-Weil theorem, $E(R) \simeq F \times \mathbb{Z}^{r(R,E)}$, where $F \simeq E(R)_{\text{tor}}$ is a finite group, and $r(R, E)$ is a nonnegative integer called the rank of E over R. Concerning the group $\text{III}(R, E)$, it is conjectured that it is finite. In general, it is known that $\text{III}(R, E)$ is a torsion group (being a subgroup of the torsion group $H^1(R, E)$) and for a natural number M its subgroup $\text{III}(R, E)_M$ is finite. If A is an abelian group, we let A_M denote its subgroup of all elements of exponents M. Only recently in works of Rubin and the author, the finiteness of $\text{III}(R, E)$ was proved for some E and R. We shall discuss these results later.

---

* This work was partly prepared during a visit of the author at the Max-Planck-Institute für Mathematik at Bonn. He wishes to express his gratitude for the support and hospitality provided by this institute.

The elements of $E(R)_{tor}$ can be effectively calculated. For example, let R be $\mathbb{Q}$ and let E be defined by an equation $u^2 = w^3 + \alpha w + \beta$, where α, $\beta \in \mathbb{Z}$, $\delta = 4\alpha^3 + 27\beta^2 \neq 0$ (this is always possible). According to the Nagell-Lutz theorem, if $P \in E(\mathbb{Q})_{tor}$ is nonzero, then $u(P) = 0$ or $u(P)^2|\delta$. Mazur determined all possible types of $E(\mathbb{Q})_{tor}$, in particular, $[E(\mathbb{Q})_{tor}] \leq 16$.

We are interested here in the case $R = \mathbb{Q}$. No algorithm is known in general for calculating $r(\mathbb{Q}, E)$ and generators of $E(\mathbb{Q})/E(\mathbb{Q})_{tor}$. But recently here and in the study of $\text{Ш}(R, E)$ essential progress was made.

More specifically, it is connected to advances towards proving the Birch-Swinnerton-Dyer conjecture (BSD) which predicts a connection between the arithmetic of E and its L-function.

We let $L(E, s)$ denote the L-function of E over $\mathbb{Q}$, defined for $\text{Re}(s) > 3/2$ as

$$\prod_q L_q(E, s) = \sum_{n=1}^{\infty} a_n n^{-s}, \qquad a_n \in \mathbb{Z}.$$

Here q runs through the set of rational primes. Let $N \in \mathbb{N}$ be the conductor of E. If $(q, N) = 1$, then $L_q(E, s) = (1 - a_q q^{-s} + q^{1-2s})^{-1}$, where $a_q = q + 1 - [\tilde{E}(\mathbb{Z}/q\mathbb{Z})]$, $\tilde{E}$ being the reduction of E modulo q (E has the good reduction at q). If $q|N$, then $L_q(E, s) = 1, (1 \pm q^{-s})^{-1}$ depending on the type of bad reduction of E at q.

Assume that E is modular, that is there exists a weak Weil parametrization $\gamma : X_0(N) \to E$ [12]. Here $X_0(N)$ is the modular algebraic curve over $\mathbb{Q}$ parametrizing classes of isogenies of elliptic curves with cyclic kernel of order N. According to the Taniyama-Shimura-Weil conjecture, every elliptic curve over $\mathbb{Q}$ is modular. Then $L(E, s)$ has an analytic continuation to an entire function on the complex plane which satisfies a functional equation

$$Z(E, 2 - s) = \varepsilon Z(E, s) \tag{1}$$

where $Z(E, s) = (2\pi)^{-s} N^{s/2} \Gamma(s) L(E, s)$ and $\varepsilon = \pm 1$ depends on E.

An analogeous L-function $L(R, E, s)$ of E over R can be defined (its definition is essential for us only up to a finite product of Euler factors), having analogous properties. We let $ar(R, E)$ denote the oder of vanishing $L(R, E, s)$ at $s = 1$. According to BSD, one conjectures the identity:

$$r(R, E) = ar(R, E). \tag{2}$$

Moreover BSD connects the first nonzero coefficient of the expansion of $L(R, E, s)$ around $s = 1$ with the order of $\text{Ш}(R, E)$ (using the hypothesis that $\text{Ш}(R, E)$ is finite) and other parameters of E, but we do not go into this here.

In the sequel we will omit the letter $\mathbb{Q}$ in the notations $\text{Ш}(\mathbb{Q}, E), r(\mathbb{Q}, E), ar(\mathbb{Q}, E)$. It follows from (1) that $ar(E)$ is even when $\varepsilon = 1$, $ar(E)$ is odd when $\varepsilon = -1$. E is called even or odd, respectively.

For $R = \mathbb{Q}$ the current state of conjecture (2) and of the hypothesis of finiteness of $\text{Ш}(E)$ is expressed by the result:

Theorem 1. *The equality $r(E) = ar(E)$ holds and $\text{Ш}(E)$ is finite if $ar(E) \leq 1$.*

We remark that empirical material shows that curves with $ar(E) > 1$ compose a relatively small part in the set of all curves. Apparently (taking into account the Taniyama-Shimura-Weil conjecture), Theorem 1 covers a substantial part of all elliptic curves over $\mathbb{Q}$.

Further we discuss a scheme of the proof of Theorem 1, formulate earlier results and give some examples.

Let D be a fundamental discriminant of the imaginary-quadratic field $K = \mathbb{Q}(\sqrt{D})$ such that $D \equiv \square (\bmod\ 4N)$, $D \neq -3, -4$. As E is modular, there exists the Heegner point $P_D \in E(K)$ (which will be defined later), it satisfies the condition:

$$\sigma e P_D = -\varepsilon e P_D \tag{3}$$

where $e =$ exponent of $E(\mathbb{Q})_{\mathrm{tor}}$, σ is the generator of $G(K/\mathbb{Q})$. The author proved [6]–[8]:

Theorem 2. *The equality $r(E) = ar(E)$ holds and $Ш(E)$ is finite if* 1) $ar(E) \leq 1$, 2) $\exists D \mid P_D$ *has infinite order.*

From the Gross and Zagier results [5] it follows

Theorem 3. *If $(D, 2N) = 1$, then $ar(K, E) \geq 1$, $ar(K, E) = 1 \Leftrightarrow P_D$ has infinite order.*

Waldspurger [21] for $ar(E) = 1$ and, independently, Bump, Friedberg, Hoffstein [2] and M. Murty, V.K. Murty [14] for $ar(E) = 0$ proved

Theorem 4. *If $ar(E) \leq 1$, then $(D, 2N) = 1$ and $ar(K, E) = 1$ for an infinite set of values of D.*

So from Theorems 3, 4 it then follows that condition 2) in Theorem 2 follows from condition 1), that is Theorem 2 is equivalent to Theorem 1.

From (1) we have that $ar(E) = 0 \Rightarrow \varepsilon = 1$, $ar(E) = 1 \Rightarrow \varepsilon = -1$. Using (3), we deduce from the conditions: P_D has infinite order, $r(K, E) = 1$, and $ar(E) \leq 1$, that $r(E) = ar(E)$. The kernel of the natural homomorphism $Ш(E) \to Ш(K, E)$ is $Ш(E) \cap H^1(G(K/\mathbb{Q}), E(K)) \subset Ш(E)_2$ which is a finite group.

Thus Theorem 2 is a consequence of the author's result [8]:

Theorem 5. *The equality $r(K, E) = 1$ holds, and $Ш(K, E)$ is finite, if P_D has infinite order.*

We note that Theorems 5, 3 give (1) for $R = K$ when $ar(K, E) = 1$. The inequality $r(E) \geq 1$ when $ar(E) = 1$ follows already from Theorem 3 and Waldspurger's result.

A subclass in the class of modular elliptic curves is formed by elliptic curves with complex multiplication: $\mathrm{End}(E) \neq \mathbb{Z}$ and then $\mathrm{End}(E)$ is an order with class number one of an imaginary-quadratic extension k of $\mathbb{Q}$. We let W' denote this subclass. The modular invariant $j = g_2^3/(g_2^3 - 27g_3^2)$, which runs through all rational numbers on the set of elliptic curves over $\mathbb{Q}$, takes on 13 values on the set W'.

The specific property of a curve from W' is the possibility to use, in studying it, the theory of abelian extensions of k because $E(\mathbb{Q})_{\text{tor}} \subset E(k^{ab})$ for $E \in W'$. In particular, by using so called elliptic units, Coates and Wiles [3] proved (2) for $E \in W'$, $ar(E) \neq 0$. Recently Rubin [17], also using elliptic units (we will come back to this later), proved under the same condition that $\text{III}(E)$ is finite. This gave the first examples of finite groups $\text{III}(E)$. Moreover he proved that, for $E \in W'$, $ar(E) = 1 \Rightarrow r(E) \leq 1$.

2. Explicit Cohomology Classes

Now we discuss briefly the method of proof of Theorem 5.

For an arbitrary extension L of $\mathbb{Q}$ the exact sequence $0 \to E_M \to E(\bar{L}) \to E(\bar{L}) \to 0$ $(E_M = E(\bar{\mathbb{Q}})_M)$ induces the exact sequence

$$0 \to E(L)/ME(L) \to H^1(L, E_M) \to H^1(L, E)_M \to 0. \qquad (4)$$

The Selmer group $S_M(R, E)$, by definition, is the subgroup of $H^1(R, E_M)$ consisting of elements whose image in $H^1(R(v), E_M)$ lies in $E(R(v))/ME(R(v))$ for all places v of R. In particular, (4) induces the exact sequence

$$0 \to E(R)/ME(R) \to S_M(R, E) \to \text{III}(R, E)_M \to 0. \qquad (5)$$

It is known (the weak Mordell-Weil theorem) that $S_M(R, E)$ is a finite M-torsion group. In particular, $\text{III}(R, E)_M$ is a finite group as we remarked before.

Let $R = K$. If $P = P_D$ has infinite order, then we define $C = C_D$ to be the maximal natural number dividing the image of P in $E(K)/E(K)_{\text{tor}} \simeq \mathbb{Z}^{r(K,E)}$. We let $C = 0$ if $P \in E(K)_{\text{tor}}$. Thus P has infinite order $\Leftrightarrow C \neq 0$. We let S'_M denote the factor group of $S_M(K, E)$ modulo the subgroup generated by P. Taking into account (5) and the Mordell-Weil theorem: $E(K) \simeq F \times \mathbb{Z}^{r(K,E)}$, with F finite, Theorem 5 will follow from the existence of $C' \in \mathbb{N}$ such that $C'S'_M = 0 \, \forall \, M \in \mathbb{N}$.

The non-degenerate alternating Weil pairing $[\ ,\ ]_M : E_M \times E_M \to \mu_M = \bar{\mathbb{Q}}^*_M$ induces a pairing

$$\langle\ ,\ \rangle_{M,v} : H^1(K(v), E_M) \times H^1(K(v), E_M) \to H^2(K(v), \mu_M).$$

For $v = \infty$ the field $K(\infty) \simeq \mathbb{C}$ and the corresponding cohomology groups are trivial. For $v \neq \infty$ the group $H^2(K(v), \mu_M)$ is identified canonically with $\mathbb{Z}/M\mathbb{Z}$ by local class field theory. If $a, b \in H^1(K, E_M)$, then $\langle a, b \rangle_{M,v} \overset{\text{def}}{=} \langle a(v), b(v) \rangle_{M,v}$, where $a(v), b(v)$ are the localizations of a, b. According to global class field theory (the reciprocity law) $\langle a, b \rangle_{M,v} \neq 0$ only for a finite set of places v and the following relation holds:

$$\sum_{v \neq 0} \langle a, b \rangle_{M,v} = 0. \qquad (6)$$

Relation (6) can be considered as a condition on a if an element b is fixed. To use (6) for the study of $S_M(K, E)$ it is necessary to find explicit elements b. This was my strategy. Thus I constructed a set T of explicit elements of $H^1(K, E_M)$ by using Heegner points over ring class fields of K. The special properties of these elements allowed to deduce from (6) with $a \in S_M(K, E)$ and $b \in T$ the relation $C'S'_M = 0$ for some $C' \in \mathbb{N}$, the divisor and main component of which is C.

Now we describe the construction of an element from T. First we define the Heegner points. Fix an ideal i in the ring of integers O of K such that $O/i \simeq \mathbb{Z}/N\mathbb{Z}$ (i exists in view of the assumptions on D). If $\lambda \in \mathbb{N}$, then K_λ denotes the ring class field of K of conductor λ. It is a finite abelian extension of K. Let O_λ be $\mathbb{Z} + \lambda O$, $i_\lambda = i \cap O_\lambda$. If $(\lambda, N) = 1$, we define the point $z_\lambda \in X_0(N)(K_\lambda)$ as corresponding to the class of the isogeny $\mathbb{C}/O_\lambda \to \mathbb{C}/i_\lambda^{-1}$, where i_λ^{-1} is the inverse of i_λ in the group of proper O_λ-ideals. We let $y_\lambda = \gamma(z_\lambda) \in E(K_\lambda)$, $P = P_D =$ the norm of y_1 from K_1 to K. The points y_λ, P are called Heegner points (corresponding to the parametrization $\gamma : X_0(N) \to E$, $K = \mathbb{Q}(\sqrt{D})$ and i).

We use the notation p (or p with (a subscript) for rational primes which do not divide N and remain prime in K. We let Λ^r denote the set of all products $p_1 \cdots p_r$ with distinct p_m, $\Lambda = \bigcup_{n=1}^\infty \Lambda^r$.

Let $\lambda \in \Lambda$, $G_\lambda = G(K_\lambda/K_1)$. The group G_λ is the direct product of the subgroups $G_{\lambda, p} = G(K_\lambda/K_{\lambda/p})$ for $p|\lambda$. The natural homomorphism $G_{\lambda, p} \to G_p$ is an isomorphism. The group G_p is isomorphic to the group $\mathbb{Z}/(p + 1)\mathbb{Z}$. For each p, we fix a generator $t_p \in G_p$; $t_p \in G_{\lambda, p}$ denotes the corresponding generator of $G_{\lambda, p}$. We let $\mathrm{Tr}_p = \sum_{j=0}^p t_p^j$. Recall that $\sum_{n=1}^\infty a_n n^{-s} = L(E, s)$ for $\mathrm{Re}(s) > 3/2$. For $p|\lambda$ one finds the relation:

$$\mathrm{Tr}_p y_\lambda = a_p y_{\lambda/p}. \tag{7}$$

The relations (7) are the basis for the definition of explicit cohomology classes.

Let Δ_λ denote the ring $\mathbb{Z}[G_\lambda]$. We define a Δ_λ-module B_λ in the following way. Let F_λ be the direct sum $\sum_{\eta|\lambda} \Delta_\eta$, where G_λ acts on Δ_η by the natural homomorphism $\Delta_\lambda \to \Delta_\eta$. Let 1_η denote the unit of Δ_η, H_λ be the Δ_λ-submodule of F_λ generated by the elements $\mathrm{Tr}_p 1_\eta - a_p 1_{\eta/p}$ for all $p|\eta|\lambda$. Then $B_\lambda = F_\lambda/H_\lambda$.

It is not difficult to prove that $(B_\lambda)_{\mathrm{tor}} = 0$. Let $1'_\eta$ be the image of 1_η in B_λ, then $\{1'_\eta, \eta|\lambda\}$ is a system of generators of B_λ over Δ_λ. By (7) $\exists!$ homomorphism $\varphi : B_\lambda \to E(K_\lambda)$ such that $1'_\eta \to y_\eta$. We let $I_p = -\sum_{j=1}^p j t_p^j \in \Delta_\lambda$, $I_\lambda = \prod_{p|\lambda} I_p$. Let Q_λ be the element $I_\lambda 1'_\lambda$.

For $M \in \mathbb{N}$ we define $\Lambda(M)$ as the subset of Λ consisting of elements λ such that $M|(p + 1)$, $M|a_p \forall p|\lambda$. Further, $\Lambda^r(M) = \Lambda^r \cap \Lambda(M)$. We claim that $(1 - g)Q_\lambda \in MB_\lambda$ for $\lambda \in \Lambda(M)$ and $g \in G_\lambda$. It is enough to verify this for $g = t_p$, where $p|\lambda$. It is clear that

$$(1 - t_p)I_p = \mathrm{Tr}_p - (p + 1). \tag{8}$$

Thus, we have $(1 - t_p)Q_\lambda = I_{\lambda/p}(1 - t_p)I_p 1'_\lambda = I_{\lambda/p}(\mathrm{Tr}_p - (p + 1))1'_\lambda = I_{\lambda/p}(a_p 1'_{\lambda/p} - (p + 1)1'_\lambda) \in MB_\lambda$.

As $(B_\lambda)_{\mathrm{tor}} = 0$, there exists a unique element $((1 - g)Q_\lambda)/M \in B_\lambda$. We define the element $\tau'_\lambda(M) \in H^1(K_1, E_M)$ to be the class of the cocycle:

$$\psi : g \mapsto (g - 1)(\varphi(Q_\lambda)/M) + \varphi(((1 - g)Q_\lambda)/M),$$

where $g \in G(\bar{K}_1/K_1)$. The element $\tau_\lambda(M) \in H^1(K, E_M)$ we define as the corestriction of $\tau'_\lambda(M)$. We call T the set $\{\tau_\lambda(M), M \in \mathbb{N}, \lambda \in \Lambda(M)\}$.

Let (b) denote the image of $b \in H^1(K, E_M)$ in $H^1(K, E)_M$, $c_\lambda(M) = (\tau_\lambda(M))$. That is, $c_\lambda(M)$ is the corestriction of the element of $H^1(K_1, E)_M$ defined by the cocycle, $g \mapsto \varphi((1 - g)Q_\lambda)/M)$. If $\lambda \in \Lambda^r(M)$, then the automorphism $\sigma \in G(K/\mathbb{Q})$ acts on $c_\lambda(M)$ by multiplication by $(-1)^{r+1}\varepsilon$. The symbol $\langle a, b \rangle_{M,v}$ depends only on (b), if $a \in S_M(K, E)$.

The elements $c_p(M)$ were defined first, see [6]. This allowed to prove the relation $C'(\sigma + \varepsilon)S_M(K, E) = 0$, which is equivalent to the finiteness of $E(\mathbb{Q})$ and $\text{III}(E)$ when $\varepsilon = 1$, and to the finiteness of $E_{(D)}(\mathbb{Q})$ and $\text{III}(E_{(D)})$ when $\varepsilon = -1$. Here $E_{(D)}$ is the elliptic curve (the form of E over K) defined by the equation $Dy^2 = 4x^3 - g_2 x - g_3$.

In [8] there were defined elements $\tau_\lambda(M)$ for some subset of the set $\{M \in \mathbb{N}, \lambda \in \Lambda(M)\}$ containing the set $\{M | (M, d) = 1, \lambda \in \Lambda(M)\}$, where $d = $ exponent of $E(\mathbb{K})_{\text{tor}}$, $\mathbb{K}$ is the composite of the $K_{\lambda'}$ for $\lambda' \in \Lambda$. By using here the modules B_λ and the property $(B_\lambda)_{\text{tor}} = 0$ we shake off the additional restrictions on (M, λ) when $(M, d) > 1$. The relation (6) with $(b) = c_\lambda(M)$ when $\lambda \in \Lambda^r(M)$, $r \leq 2$, allowed to prove the relation $C'S'_M = 0$.

We note that an application of the elements $\tau_\lambda(M)$ when $\lambda \in \Lambda^r$ with arbitrary $r \geq 0$ allowed in [8] to pass from a relation of the type $C \, \text{III}(K, E) = 0$ to a relation of the type $[\text{III}(K, E)] | C^2$. Because of the existence on $\text{III}(K, E)$ of a non-degenerate (as $\text{III}(K, E)$ is finite) alternate Cassels pairing with values in $\mathbb{Q}/\mathbb{Z}$, it then follows that the second relation implies the first relation.

In [20] Thaine used the cyclotomic units for a new proof of annihilating relations in the ideal class groups of real abelian extensions of $\mathbb{Q}$. Rubin [16] adapted Thaine's approach, using elliptic units instead of cyclotomic units, for proving annihilating relations in the ideal class groups of abelian extensions of the imaginary-quadratic field $k = \text{End}(E) \otimes \mathbb{Q}$ when $E \in W'$. By using the natural connection between ideal class groups and the Selmer group $S_M(\mathbb{Q}, E)$ Rubin proved an universal annihilating relation for $S_M(\mathbb{Q}, E)$ by the condition that $ar(E) = 0$.

A comparison of the approaches of Thaine [20] and of the author [6] for proving annihilating relations in the ideal class groups and in the Selmer groups, respectively, suggested the possibility in [7] of combining them into a single general framework. A further step was a construction and use in [8] of sets of cohomology classes of the type T, both in the theory of modular elliptic curves and in the theory of ideal class groups of abelian extensions of $\mathbb{Q}$ or an imaginary-quadratic extension of $\mathbb{Q}$. For information on this theory and some further applications we refer to the papers [8, 1, 4, 9, 10, 11, 13, 15, 18, 19].

3. Examples

Example 1 (Rubin [17]). For the curves with complex multiplication $(k = \mathbb{Q}(\sqrt{-1}))$ $y^2 = x^3 - x$, $y^2 = x^3 + 17x$ we have: $r(E) = ar(E) = 0$, $\text{III}(E) = 0$, $\mathbb{Z}/2\mathbb{Z} + \mathbb{Z}/2\mathbb{Z}$, respectively.

Example 2 (Kolyvagin [7]). Let $E : y^2 = 4x^3 - 4x + 1$. It is an odd modular curve without complex multiplication, of conductor $N = 37$. Let $(D, 2N) = 1$. The curves $E_{(D)}$:

$$Dy^2 = 4x^3 - 4x + 1 \tag{9}$$

are even and have no complex multiplication. For computation of $L(E_{(D)}, 1)$ and C_D the following identity can be used:

$$L(E_{(D)}, 1) = 2 \sum_{n=1}^{\infty} \frac{a_n}{n} \left(\frac{D}{n}\right) \exp(-2\pi n/(|D|\sqrt{37})) = (2\Omega_-/\sqrt{D}) C_D^2 \tag{10}$$

where Ω_--the imaginary period of E, $\left(\dfrac{D}{n}\right)$-the Legendre symbol. See [22] for (10); the connection between $L(E_{(D)}, 1)$ and C_D is a consequence of the results of Gross and Zagier [5].

Let $L(E_{(D)}, 1) \neq 0$ or, equivalently, $C_D \neq 0$. Then $E_{(D)}(\mathbb{Q})$ is finite and, moreover, is trivial because always $E_{(D)}(\mathbb{Q})_{tor} = 0$. That is equation (9) has no solutions in rational numbers. Further, $\text{III}(E_{(D)})$ is finite and $C_D \text{III}(E_{(D)}) = 0$. For example, if $D = -7, -11$ then $C_D = 1$, so $\text{III}(E_{(D)}) = 0$. See [7] for further information on this example.

We recall that $C_D \neq 0$ for an infinite set of values of D according to a result of Waldspurger.

It is a classical fact tht $E(\mathbb{Q}) \simeq \mathbb{Z}$ is generated by the point $(y = 1, x = 0)$. Of course, $ar(E) = 1$, see [22], for example. The author proved [8] that $\text{III}(E) = 0$.

References

1. Bertolini, M., Darmon, H.: Kolyvagin's descent and Mordell-Weil groups over ring class fields. J. Reine Angew. Math. **412** (1990) 63–74
2. Bump, D., Friedberg, S., Hoffstein, J.: A non vanishing theorem for derivatives of automorphic L-functions with applications to elliptic curves. Bull. Amer. Math. Soc. **21** (1989) 89–93
3. Coates, J., Wiles, A.: On the conjecture of Birch and Swinnerton-Dyer. Invent. math. **39** (1977) 223–251
4. Gross, B.H.: Kolyvagin's work on modular elliptic curves. Proceedings of Durham Conference on L-functions and Arithmetic, 1989. Cambridge University Press (to appear)
5. Gross, B.H., Zagier, D.B.: Heegner points and derivatives of L-series. Invent. math. **84** (1986) 225–320
6. Kolyvagin, V.A.: Finiteness of $E(\mathbb{Q})$ and $\text{III}(E, \mathbb{Q})$ for a subclass of Weil curves. Izv. Akad. Naùk SSSR, Ser. Mat. **52** (1988) 522–540 [English transl.: Math USSR Izv. **32** (1989) 523–542]
7. Kolyvagin, V.A.: On the Mordell-Weil and Shafarevich-Tate groups for Weil elliptic curves. Izv. SSSR, Ser. Mat. **52** (1988) 1154–1180 [English transl.: Math USSR Izv. **33** (1989) 474–499]
8. Kolyvagin, V.A.: Euler systems. The Grothendieck Festschrift, vol. 2. Progr. in Math., vol. 87. Boston: Birkhäuser 1991, pp. 435–483
9. Kolyvagin, V.A., Logachev, D.Y.: Finiteness of the Shafarevich-Tate group and the group of rational points for some modular abelian varieties. Algebra and Analysis 1, no. 5 (1989)
10. Kolyvagin, V.A.: On the structure of Shafarevich-Tate groups. Proceedings of USA-USSR Symposium on Algebraic Geometry, Chicago, 1989. (Lecture Notes in Mathematics.) Springer, Berlin Heidelberg New York (to appear)
11. Kolyvagin, V.A.: On the structure of Selmer groups. (1990). Math. Ann. (to appear)

12. Mazur, B., Swinnerton-Dyer, H.P.F.: Arithmetic of Weil curves. Invent. math. **25** (1974) 1–61
13. McCallum, W.G.: Kolyvagin's work on Shafarevich-Tate groups. Proceedings of Durham Conference on *L*-functions and Arithmetic, 1989. Cambridge University Press (to appear)
14. Murty, M.R., Murty, V.K.: Mean values of derivatives of modular *L*-series. (1989). Ann. Math. (to appear)
15. Perrin-Riou, B.: Travaux de Kolyvagin et Rubin. Séminaire Bourbaki 717 (1989/1990)
16. Rubin, K.: Global units and ideal class groups. Invent. math. **89** (1987) 511–526
17. Rubin, K.: Tate-Shafarevich group and *L*-functions of elliptic curves with complex multiplications. Invent. math. **89** (1987) 527–560
18. Rubin, K.: The Main Conjecture. Appendix to: Cyclotomic fields I–II by S. Lang. (Graduate Texts in Mathematics, vol. 121.) Springer, New York Berlin Heidelberg 1990, 2nd edn., pp. 397–419
19. Rubin, K.: The "main conjectures" of Iwasawa theory for imaginary quadratic fields. Invent. math. **103** (1991) 25–68
20. Thaine, F.: On the ideal class groups of real abelian extensions of $\mathbb{Q}$. Ann. Math. **128** (1988) 1–18
21. Waldspurger, J.-L.: Sur les valeurs de certaines fonctions *L* automorphes en leur centre de symmetrie. Comp. Math. **54** (1985) 173–242
22. Zagier, D.B.: Modular points, modular curves, modular surfaces and modular forms. (Lecture Notes in Mathematics, vol. 1111.) Springer, Berlin Heidelberg New York 1985, pp. 225–246

La Transformation de Fourier Géométrique et ses Applications

Gérard Laumon

Université Paris-Sud, URA D 0752, Bâtiment 425, F-91405 Orsay Cedex, France

1. La transformation de Fourier géométrique
[Br, De1, Il1, Ka1, Ka-La, La1]

Dans tout cet exposé, on fixe deux nombres premiers distincts p et ℓ, une clôture algébrique $\overline{\mathbb{Q}}_\ell$ du corps $\mathbb{Q}_\ell$ des nombres ℓ-adiques et un caractère additif non trivial $\psi : \mathbb{F}_p \longrightarrow \overline{\mathbb{Q}}_\ell^\times$. On fixe aussi un corps algébriquement clos k de caractéristique p. On désignera par q une puissance de p ($q = p^f$, $f \in \mathbb{N}^\times$), par $\mathbb{F}_q$ l'unique sous-corps de k à q éléments et par $\mathrm{Frob}_q \in \mathrm{Gal}(k/\mathbb{F}_q)$ l'inverse de l'élément de Frobenius ($\mathrm{Frob}_q(\alpha)^q = \alpha$, $\forall \alpha \in k$). On utilisera librement le formalisme des 6 opérations de Grothendieck entre les catégories dérivées $D_c^b(X, \overline{\mathbb{Q}}_\ell)$ (X un k-schéma de type fini) ainsi que le formalisme des t-structures (cf. [SGA4, SGA5, SGA7, B-B-D, De2 et Ek]).

Soient S un k-schéma de type fini, $\pi : E \longrightarrow S$ un fibré vectoriel de rang constant d, $\pi' : E' \longrightarrow S$ le fibré dual, $\mathrm{pr} : E' \times_S E \longrightarrow E$, $\mathrm{pr}' : E' \times_S E \longrightarrow E'$ les projections canoniques et $\langle , \rangle : E' \times_S E \longrightarrow \mathbb{A}_S^1$ l'accouplement de dualité. Sur $\mathbb{A}_S^1 = S[u]$, on a le revêtement d'Artin-Schreier

$$S[v] \longrightarrow S[u], \quad v \longmapsto u = v^p - v$$

qui est étale, galoisien, de groupe de Galois $\mathbb{F}_p$. Si l'on pousse ce $\mathbb{F}_p$-torseur par le caractère ψ^{-1} on obtient un $\overline{\mathbb{Q}}_\ell$-faisceau lisse de rang 1 (un "$\overline{\mathbb{Q}}_\ell^\times$-torseur") sur $\mathbb{A}_S^1$, noté $\mathscr{L}_\psi$. La *transformation de Fourier géométrique*, inventée par Deligne, est l'opération

$$\mathscr{F}_\psi : D_c^b(E, \overline{\mathbb{Q}}_\ell) \longrightarrow D_c^b(E', \overline{\mathbb{Q}}_\ell),$$

$$\mathscr{F}_\psi(-) = R\,\mathrm{pr}'_!(\mathrm{pr}^*(-) \overset{L}{\otimes} \langle , \rangle^* \mathscr{L}_\psi)[d].$$

Théorème 1.1 (Deligne). *(i) $\mathscr{F}_\psi$ est une équivalence de catégories triangulées, de quasi-inverse $\mathscr{F}'_\psi(-)(d)$, où $\mathscr{F}'_\psi$ est la transformation de Fourier géométrique pour $\pi' : E' \longrightarrow S$ (on a identifié E'' à E par $e \longmapsto (e' \longmapsto -\langle e', e \rangle)$).*

(ii) Si $E = E_1 \times_S E_2$ pour deux fibrés vectoriels E_1 et E_2 de rang constant sur S, on a

$$\mathscr{F}_\psi\left(\mathrm{pr}_1^*(-) \overset{L}{\otimes} \mathrm{pr}_2^*(-)\right) \simeq \mathrm{pr}_1'^*\,\mathscr{F}_{\psi,1}(-) \overset{L}{\otimes} \mathrm{pr}_2'^*\,\mathscr{F}_{\psi,2}(-),$$

*où $\mathcal{F}_{\psi,i}(-)$ est la transformation de Fourier géométrique pour E_i et $\mathrm{pr}_i : E_1 \times_S$
$E_2 \longrightarrow E_i$ (resp. $\mathrm{pr}'_i : E'_1 \times_S E'_2 \to E'_i$) la projection canonique.*

*(iii) Si $f : E_1 \longrightarrow E_2$ est un morphisme de fibrés vectoriels de rang constant
sur S, on a*

$$f'^* \mathcal{F}_{\psi,1}(-)[d_2 - d_1] \simeq \mathcal{F}_{\psi,2} Rf_!(-),$$

*où f' est le transposé de f, où d_i est le rang de E_i et où $\mathcal{F}_{\psi,i}$ est la transformation
de Fourier géométrique pour E_i.*

Théorème 1.2 (Verdier). *Pour tout $K \in \mathrm{ob}\, D^b_c(E, \overline{\mathbb{Q}}_\ell)$, la flèche d'oubli des supports*

$$R \mathrm{pr}'_!(\mathrm{pr}^* K \overset{L}{\otimes} \langle,\rangle^* \mathcal{L}_\psi) \longrightarrow R \mathrm{pr}'_*(\mathrm{pr}^* K \overset{L}{\otimes} \langle,\rangle^* \mathcal{L}_\psi)$$

est un isomorphisme dans $D^b_c(E', \overline{\mathbb{Q}}_\ell)$.

Compte-tenu de la dualité de Verdier, de la t-exactitude à droite de f_* pour f
affine (prouvée par Artin) et de la forme forte des conjectures de Weil (prouvée
par Deligne), le Théorème 1.2 admet les corollaires suivants:

Corollaire 1.3. *On a un isomorphisme canonique*

$$R\mathcal{H}om(\mathcal{F}_\psi(K), \pi'^! L) \simeq \mathcal{F}_{\psi^{-1}}(R\mathcal{H}om(K, \pi^! L))(d)$$

bi-fonctoriel en $(K, L) \in \mathrm{ob}(D^b_c(E, \overline{\mathbb{Q}}_\ell)^{\mathrm{opp}} \times D^b_c(S, \overline{\mathbb{Q}}_\ell))$.

Corollaire 1.4. *$\mathcal{F}_\psi$ est t-exact et induit une équivalence de catégories abéliennes*

$$\mathcal{F}_\psi : \mathrm{Perv}(E, \overline{\mathbb{Q}}_\ell) \longrightarrow \mathrm{Perv}(E', \overline{\mathbb{Q}}_\ell)$$

de quasi-inverse $\mathcal{F}_\psi(-)(d)$.

Corollaire 1.5. *Si S et $E \overset{\pi}{\longrightarrow} S$ sont définis sur $\mathbb{F}_q$, pour tout $K \in \mathrm{ob}\, D^b_c(E, \overline{\mathbb{Q}}_\ell)$
défini sur $\mathbb{F}_q$ et pur de poids w, $\mathcal{F}_\psi(K) \in \mathrm{ob}\, D^b_c(E', \overline{\mathbb{Q}}_\ell)$ est défini sur $\mathbb{F}_q$ et pur de
poids $w + d$.*

2. Application aux sommes trigonométriques
[Br, Ka2, Ka3, Ka-La]

Dans ce paragraphe $S = \mathrm{Spec}(k)$ et $E = \mathrm{Spec}(k[x_1, \ldots, x_d])$ (avec sa $\mathbb{F}_q$-structure
naturelle). On identifie alors E' à $\mathrm{Spec}(k[x'_1, \ldots, x'_d])$ de sorte que

$$\langle x', x \rangle = \sum_{\delta=1}^d x'_\delta x_\delta.$$

Pour tout schéma X de type fini sur k, défini sur $\mathbb{F}_q$, pour tout $x \in X(\mathbb{F}_q) = \{x \in
X(k) | \mathrm{Frob}_q(x) = x\}$ et pour tout $M \in \mathrm{ob}\, D^b_c(X, \overline{\mathbb{Q}}_\ell)$, défini sur $\mathbb{F}_q$, Frob_q agit sur
la fibre de M en x et on note $t_M(x)$ la trace de cet endomorphisme de M_x. La
formule des traces de Grothendieck pour $\widehat{\mathrm{Frob}}_q$ entraîne immédiatement:

Proposition 2.1. *Pour tout* $K \in$ ob $D_c^b(E, \overline{\mathbb{Q}}_\ell)$, *défini sur* $\mathbb{F}_q$, *et tout* $x' \in E'(\mathbb{F}_q) = (\mathbb{F}_q)^d$, *on a*

$$t_{\mathscr{F}_\psi(K)}(x') = (-1)^d \sum_{x \in (\mathbb{F}_q)^d} t_K(x)\psi \circ \mathrm{tr}_{\mathbb{F}_q/\mathbb{F}_p}(\langle x', x\rangle)$$

(en d'autres termes, $t_{\mathscr{F}_\psi(K)}$ *est, au signe près, la transformée de Fourier de la fonction* t_K).

Soit R une sous-$\mathbb{Z}$-algèbre de type fini de $\mathbb{C}$, soit X un R-schéma affine, lisse, purement de dimension relative m, et soit $f = (f_1, \ldots, f_d) : X \longrightarrow \mathbb{A}_R^d$ un R-morphisme fini. Pour tout homomorphisme d'anneaux $v : R \longrightarrow k$ avec $v(1) = 1$, on note $(-)_v$ le changement de base par v. Alors, $K_v = (f_v)_* \overline{\mathbb{Q}}_\ell[m]$ est un $\overline{\mathbb{Q}}_\ell$-faisceau pervers sur $\mathbb{A}_k^d = E$, K_v est défini sur $\mathbb{F}_q \subset k$ dès que $\mathbb{F}_q \supset R/\mathrm{Ker}(v)$ et K_v est pur de poids m. Par suite, $\mathscr{F}_\psi(K_v)$ est aussi sur $\overline{\mathbb{Q}}_\ell$-faisceau pervers sur $\mathbb{A}_k^d = E'$, $\mathscr{F}_\psi(K_v)$ est aussi défini sur $\mathbb{F}_q \subset k$ dès que $\mathbb{F}_q \supset R/\mathrm{Ker}(v)$ et $\mathscr{F}_\psi(K_v)$ est aussi pur de poids $m + d$ (cf. (1.4), (1.5)). De plus, d'après (2.1), on a

$$t_{\mathscr{F}_\psi(K_v)}(x') = (-1)^{d+m} \sum_{x \in X_v(\mathbb{F}_q)} \psi \circ \mathrm{tr}_{\mathbb{F}_q/\mathbb{F}_p}(\langle x', f_v(x)\rangle_v)$$

pour tout $\mathbb{F}_q$ avec $k \supset \mathbb{F}_q \supset R/\mathrm{Ker}(v)$.

Comme l'a remarqué le premier Brylinski, il résulte du théorème de structure des $\overline{\mathbb{Q}}_\ell$-faisceaux pervers simples et purs qu'il existe un entier $\chi_v \geq 0$ et un polynôme non nul $\varphi_v(x')$ dans $(R/\mathrm{Ker}(v))[x_1', \ldots, x_d']$ tels que

$$(2.2) \qquad \left| \sum_{x \in X_v(\mathbb{F}_q)} e^{\frac{2\pi i}{p} \mathrm{tr}_{\mathbb{F}_q/\mathbb{F}_p}(\langle x', f_v(x)\rangle_v)} \right| \leq \chi_v q^{m/2}$$

dès que $k \supset \mathbb{F}_q \supset R/\mathrm{Ker}(v)$ et que $x' \in (\mathbb{F}_q)^d$ satisfait $\varphi_v(x') \neq 0$.

Théorème 2.3 (Katz et Laumon). *Il existe* $\chi \in \mathbb{N}$ *et* $\varphi(x') \in R[x_1', \ldots, x_d']$ *non nul tels que, pour tout* k *et tout* v, $\chi_v = \chi$ *et* $\varphi_v = v(\varphi)$ *fassent marcher les estimations* (2.2) *ci-dessus.*

Exemple 2.4 (Adolphson et Sperber [Ad-Sp], Denef et Lœser [De-Lo]). Soient $I \subset \mathbb{Z}^m$ avec $|I| = d$, $X = (\mathbb{G}_{m,R})^m$ et $f = (f_i)_{i \in I} : X \longrightarrow \mathbb{A}_R^I = \mathbb{A}_R^d$ avec $f_i(x) = x^i$. Alors f est fini (et par suite (2.3) s'applique) si et seulement si l'enveloppe convexe de I dans $\mathbb{R}^m$ est de dimension m et contient 0 dans son intérieur. Dans ce cas, χ et φ sont déterminés explicitement dans loc. cit.

3. Localisation de caractéristiques d'Euler-Poincaré et de déterminants [Ka1, Il1, La1]

Soient X un k-schéma de type fini, $A = \mathrm{Spec}(k[x])$, $D = A \cup \{\infty\}$ la droite projective correspondante, $f : X \longrightarrow D$ un k-morphisme propre, $A' = \mathrm{Spec}(k[x'])$ la droite affine duale de A et $D' = A' \cup \{\infty'\}$ la droite projective correspondante. On pose $Y = f^{-1}(\infty)$, $U = X - Y = f^{-1}(A)$. On note $j : A \hookrightarrow D$, $j' : A' \hookrightarrow D'$ les inclusions, $\mathrm{pr}_X : D' \times_k X \longrightarrow X$, $\mathrm{pr}_{D'} : D' \times_k X \longrightarrow D'$ les projections canoniques, $\mathscr{L}_\psi(x'x)$ le $\overline{\mathbb{Q}}_\ell$-faisceau lisse de rang 1 sur $A' \times_k A$ associé au revêtement d'Artin-Schreier $v^p - v = x'x$ et au caractère ψ et prolongé par 0 à $D' \times_k D$ tout entier et $\mathscr{L}_\psi(x'f)$ le $\overline{\mathbb{Q}}_\ell$-faisceau sur $D' \times_k X$ image réciproque de $\mathscr{L}_\psi(x'x)$ par $D' \times_k f$. Enfin, soit $K \in \mathrm{ob}\, D_c^b(X, \overline{\mathbb{Q}}_\ell)$.

Posons
$$K' = j'_! \mathscr{F}_\psi(j^* Rf_* K)[-1] \in \mathrm{ob}\, D_c^b(D', \overline{\mathbb{Q}}_\ell)\,.$$

On a encore
$$K' = R\,\mathrm{pr}_{D'_*}(\mathrm{pr}_X^* K \overset{L}{\otimes} \mathscr{L}_\psi(x'f))\,.$$

Si l'on note o' l'origine de A', $\eta_{o'}$ (resp. $\eta_{\infty'}$) le point générique de l'hensélisé de D' en o' (resp. ∞') et $\overline{\eta}_{o'}$ (resp. $\overline{\eta}_{\infty'}$) un point géométrique localisé en $\eta_{o'}$ (resp. ∞'), on a le triangle distingué

$$(3.1) \qquad R\Gamma_c(U, K) \longrightarrow K'_{\overline{\eta}_{o'}} \longrightarrow R\Gamma(X, R\Phi_{\overline{\eta}_{o'}}) \longrightarrow$$

$(K'_{o'} = R\Gamma_c(U, K))$ et l'isomorphisme

$$(3.2) \qquad\qquad\qquad K'_{\overline{\eta}_{\infty'}} \overset{\sim}{\longrightarrow} R\Gamma(X, R\Phi_{\overline{\eta}_{\infty'}})$$

$(K'_{\infty'} = 0)$, où $R\Phi_{\overline{\eta}_{o'}} \in \mathrm{ob}\, D_c^b(X, \overline{\mathbb{Q}}_\ell[\mathrm{Gal}(\overline{\eta}_{o'}/\eta_{o'})])$ (resp. $R\Phi_{\overline{\eta}_{\infty'}} \in \mathrm{ob}\, D_c^b(X, \overline{\mathbb{Q}}_\ell[\mathrm{Gal}(\overline{\eta}_{\infty'}/\eta_{\infty'})])$) sont les cycles évanescents pour $\mathrm{pr}_{D'}$ relativement à $\mathrm{pr}_X^* K \overset{L}{\otimes} \mathscr{L}_\psi(x'f)$ en o' (resp. ∞').

Le support de $R\Phi_{\overline{\eta}_{o'}}$ est clairement contenu dans Y. Soit $V \subset U$ l'ouvert formé des points $x \in U$ tels que f et les $\overline{\mathbb{Q}}_\ell$-faisceaux de cohomologie $\mathscr{H}^i(K)$ ($i \in \mathbb{Z}$) soient lisses en x et soit $Z = U - V$ le fermé complémentaire.

Théorème 3.3 (Laumon). *Le support de $R\Phi_{\overline{\eta}_{\infty'}}$ est contenu dans $Y \cup Z$.*

Ce théorème est l'analogue en cohomologie ℓ-adique du principe de la phase stationnaire: si U est une variété différentielle $\mathscr{C}^\infty$, si $f : U \longrightarrow \mathbb{R}$ est une fonction $\mathscr{C}^\infty$ et si ω est une densité $\mathscr{C}^\infty$ à support compact sur U telle que

$$\mathrm{Supp}(\omega) \cap \{x \in U | df(x) = 0\} = \emptyset,$$

on a
$$\int_U e^{ix'f} \omega = O(x'^{-n})$$

pour tout entier $n \geq 0$ quand $x' \in \mathbb{R}$ tend vers ∞.

On ne peut pas comparer directement $K_{\overline{\eta}_{o'}}$ et $K_{\overline{\eta}_{\infty'}}$. Par contre, $K_{\overline{\eta}_{o'}}$ et $K_{\overline{\eta}_{\infty'}}$ ont même caractéristique d'Euler-Poincaré. On déduit donc de (3.1), (3.2) et (3.3) que :

$$(3.4) \qquad \chi(X, K) = \chi(Y, K) - \chi(Y, R\Phi_{\overline{\eta}_{o'}}) + \chi(Y \cup Z, R\Phi_{\overline{\eta}_{\infty'}}).$$

Application 3.5. Si X est une variété projective sur k et si $K_1, K_2 \in \text{ob } D_c^b(X, \overline{\mathbb{Q}}_\ell)$ sont localement isomorphes pour la topologie étale sur X, Deligne a démontré que $\chi(X, K_1) = \chi(X, K_2)$ (cf. [Il2]). On peut en donner une autre démonstration par récurrence sur la dimension de X basée sur (3.4).

On peut aussi comparer les déterminants de $K_{\overline{\eta}_{o'}}$ et $K_{\overline{\eta}_{\infty'}}$. En particulier, on peut déduire de (3.1), (3.2) et (3.3) la conjecture de Deligne suivante :

Théorème 3.6 (Laumon). *Soient X une courbe projective, lisse et connexe sur k, ω une 1-forme méromorphe non identiquement nulle sur X et $K \in \text{ob } D_c^b(X, \overline{\mathbb{Q}}_\ell)$. On suppose X, ω et K définis sur $\mathbb{F}_q$. Alors la "constante globale" de Grothendieck*

$$\varepsilon(X, K) = \det(-\text{Frob}_q, R\Gamma(X, K))^{-1}$$

admet la formule du produit

$$\varepsilon(X, K) = q^{(1-g)r(K)} \prod_{x \in |X|} \varepsilon(X_{(x)}, K|X_{(x)}, \omega|X_{(x)})$$

où g est le genre de X, $r(K)$ est le rang générique de K et, pour tout point fermé x de X, $X_{(x)}$ est l'hensélisé de X en x et $\varepsilon(X_{(x)}, K|X_{(x)}, \omega|X_{(x)})$ est la "constante locale" de Deligne, Dwork et Langlands.

Récemment, Loeser a obtenu par la même méthode une formule du produit pour le déterminant de la cohomologie étale du complémentaire d'un arrangement d'hyperplans dans $\mathbb{P}_k^n$ à valeurs dans un $\overline{\mathbb{Q}}_\ell$-faisceau lisse de rang 1 de type Kummer (cf. [Lo]).

4. Transformations de Fourier locales [La1]

Il y en a de trois types $\mathscr{F}_\psi^{(o,\infty')}$, $\mathscr{F}_\psi^{(\infty,o')}$ et $\mathscr{F}_\psi^{(\infty,\infty')}$. On ne s'intéressera ici qu'au premier type. Soient $T = \text{Spec}(k[[\varpi]])$ (resp. $T' = \text{Spec}(k[[\varpi']])$) avec sa $\mathbb{F}_q$-structure naturelle (ϖ (resp. ϖ') est une indéterminée sur k), η (resp. η') le point générique et s (resp. s') le point fermé de T (resp. T'), $\overline{\eta}$ (resp. $\overline{\eta}'$) un point géométrique localisé en η (resp. η') et $\mathscr{G}$ (resp. $\mathscr{G}'$) la catégorie des $\text{Gal}(\overline{\eta}/\eta)$ (resp. $\text{Gal}(\overline{\eta}'/\eta')$)-modules sur $\overline{\mathbb{Q}}_\ell$. Si $V \in \text{ob } \mathscr{G}$, on note encore V le $\overline{\mathbb{Q}}_\ell$-faisceau lisse sur η correspondant et $V_!$ sur prolongement par zéro à T tout entier. Alors par définition

$$\mathscr{F}_\psi^{(o,\infty')}(V) = R^1 \Phi_{\overline{\eta}'}(\text{pr}^*(V_!) \overset{L}{\otimes} \mathscr{L}_\psi(\varpi/\varpi'))_{(s',s)} \in \text{ob } \mathscr{G}'$$

où $\text{pr} : T' \times_k T \longrightarrow T$ et $\text{pr}' : T' \times_k T \longrightarrow T'$ sont les projections canoniques et où $\mathscr{L}_\psi(\varpi/\varpi')$ est le $\overline{\mathbb{Q}}_\ell$-faisceau lisse de rang 1 sur $\eta' \times_k T$ associé au revêtement d'Artin-Schreier $v^p - v = \varpi/\varpi'$ et au caractère ψ et prolongé par zéro à $T' \times_k T$ tout entier.

Théorème 4.1 (Laumon).

(i) $\mathscr{F}_\psi^{(0,\infty')}$ est un foncteur exact de $\mathscr{G}$ vers $\mathscr{G}'$.

(ii) Pour tout $V \in \mathrm{ob}\,\mathscr{G}$, on a

$$r(\mathscr{F}_\psi^{(0,\infty')}(V)) = r(V) + s(V)$$

$$s(\mathscr{F}_\psi^{(0,\infty')}(V)) = s(V)$$

où $r(-)$ est le rang et $s(-)$ le conducteur de Swan, de sorte que $\mathscr{F}_\psi^{(0,\infty')}(\mathscr{G}) \subset \mathscr{G}'_{[0,1[}$, où $\mathscr{G}'_{[0,1[}$ est la sous-catégorie pleine de $\mathscr{G}'$ formée des objets V' dont tous les sous-quotients irréductibles W' vérifient $s(W') < r(W')$.

(iii) Le foncteur $\mathscr{F}_\psi^{(0,\infty')} : \mathscr{G} \longrightarrow \mathscr{G}'_{[0,1[}$ est une équivalence de catégories abéliennes.

(iv) $\mathscr{F}_\psi^{(0,\infty')}(V)^\vee \simeq \mathscr{F}_{\psi^{-1}}^{(0,\infty')}(V^\vee)(1)$.

(v) Si V est défini sur $\mathbb{F}_q$, il en est de même de $\mathscr{F}_\psi^{(0,\infty')}(V)$ et la "constante locale" $\varepsilon_0(T, V, d\varpi)$ de Deligne, Dwork et Langlands est égale à

$$\varepsilon_0(T, V, d\varpi) = (-1)^{r(V')} \det(V')(\varpi')$$

où $V' = \mathscr{F}_\psi^{(0,\infty')}(V)$ (on a identifié $\mathrm{Gal}(\overline{\mathbb{F}_q((\varpi'))}/\mathbb{F}_q((\varpi')))^{\mathrm{ab}}$ à $(\mathbb{F}_q((\varpi'))^\times)^\wedge$ par la théorie du corps de classe abélien).

Cette transformation de Fourier locale donne une construction cohomologique locale de la représentation d'Artin et de la "constante locale". Elle a aussi permis à Henniart de démontrer la conjecture de Langlands locale numérique, i.e. de dénombrer les objets simples de $\mathscr{G}$ de conducteur borné (cf. [He]).

5. Une nouvelle preuve d'une conjecture de Weil [La1]

Si K est un $\overline{\mathbb{Q}}_\ell$-faisceau pervers, irréductible, défini sur $\mathbb{F}_q$ et pur de poids 0, sur une droite projective D définie sur $\mathbb{F}_q$, Deligne a montré que $R\Gamma(D, K)$ est aussi pur de poids 0. La transformation de Fourier géométrique permet d'en donner une nouvelle démonstration, inspirée par la preuve de Witten des inégalités de Morse. Supposons que $D = A \cup \{\infty\}$, où $\infty \in D(\mathbb{F}_q)$ et $A = \mathrm{Spec}(k[x])$ est défini sur $\mathbb{F}_q$, et que les $\overline{\mathbb{Q}}_\ell$-faisceaux de cohomologie $\mathscr{H}^i(K)$ ($i \in \mathbb{Z}$) sont lisses en ∞. Alors la restriction de $\mathscr{F}_\psi(K|A)$ à $A' - \{o'\}$ ($A' = \mathrm{Spec}(k[x'])$) est de la forme $\mathscr{J}'[1]$ pour un $\overline{\mathbb{Q}}_\ell$-faisceau lisse irréductible $\mathscr{J}'$. La remarque essentielle est que $\mathscr{J}'$ est un constituant du $\overline{\mathbb{Q}}_\ell$-faisceau lisse **réel**

$$\mathscr{H}^{-1}(\mathscr{F}_\psi((K \oplus K^\vee)|A) \oplus \mathscr{F}_{\psi^{-1}}((K \oplus K^\vee)|A))|A' - \{0'\}$$

où $K^\vee = R\mathscr{H}om(K, \overline{\mathbb{Q}}_\ell[2](1))$ est le dual de Verdier de K. Par suite, d'après un théorème clé de Deligne, $\mathscr{J}'$ est lui même pur d'un certain poids (on a fixé un isomorphisme $\imath : \overline{\mathbb{Q}}_\ell \xrightarrow{\sim} \mathbb{C}$, de sorte que $\imath \circ \psi^{-1} = \overline{\imath \circ \psi}$).

6. La transformation de Radon géométrique [Br, Il1]

Il s'agit d'une variante homogène de la transformation de Fourier. Soient $\overline{\pi}$: $P = \mathbb{P}(E) \longrightarrow S$ et $\overline{\pi}' : P' = \mathbb{P}(E') \longrightarrow S$ les fibrés projectifs associés aux fibrés vectoriels du paragraphe 1. On note $Z \subset P' \times_S P$ la variété d'incidence entre ces deux fibrés projectifs en dualité et $\varrho : Z \longrightarrow P$, $\varrho' : Z \longrightarrow P'$ les projections canoniques. La *transformation de Radon géométrique*, inventée par Brylinski, est l'opération

$$\mathscr{R} : D^b_c(P, \overline{\mathbb{Q}}_\ell) \longrightarrow D^b_c(P', \overline{\mathbb{Q}}_\ell)$$

$$\mathscr{R}(-) = R\varrho'_* \varrho^*(-)[d-2].$$

On note $\mathscr{S} \subset \mathrm{Perv}(P, \overline{\mathbb{Q}}_\ell)$ (resp. $\mathscr{S}' \subset \mathrm{Perv}(P', \overline{\mathbb{Q}}_\ell)$) la sous-catégorie strictement pleine formée des $\overline{\mathbb{Q}}_\ell$-faisceaux pervers sur P (resp. P') isomorphes à $\overline{\pi}^* L[d-1]$ (resp. $\overline{\pi}'^* L'[d-1]$) pour un $L \in \mathrm{ob}\, \mathrm{Perv}(S, \overline{\mathbb{Q}}_\ell)$ (resp. $L' \in \mathrm{ob}\, \mathrm{Perv}(S, \overline{\mathbb{Q}}_\ell)$) ; c'est une sous-catégorie de Serre.

Théorème 6.1 (Brylinski). *(i) Pour tout $K \in \mathrm{ob}\, \mathrm{Perv}(P, \overline{\mathbb{Q}}_\ell)$ et tout entier $i \neq 0$, ${}^p\mathscr{H}^i(\mathscr{R}(K)) \in \mathrm{ob}\, \mathscr{S}'$.*

(ii) Le foncteur $K \longmapsto {}^p\mathscr{H}^0(\mathscr{R}(K))$ induit une équivalence de catégories abéliennes de la catégorie quotient $\mathrm{Perv}(P, \overline{\mathbb{Q}}_\ell)/\mathscr{S}$ sur la catégorie quotient $\mathrm{Perv}(P', \overline{\mathbb{Q}}_\ell)/\mathscr{S}'$; un quasi-inverse est induit par $K' \longmapsto {}^p\mathscr{H}^0(\mathscr{R}'(K'))(d-2)$, où $\mathscr{R}'$ est la transformation de Radon géométrique pour P'.

Brylinski applique ce théorème à l'étude, dans le cas singulier, de la monodromie des pinceaux de Lefschetz.

Dans [La2], nous utilisons cette transformation de Radon pour l'étude des faisceaux automorphes cuspidaux de Drinfeld.

7. Une application à la théorie des représentations des groupes réductifs sur les corps finis [Kaz-La]

Soient V un espace vectoriel de dimension 2 sur k et $\omega \in \Lambda^2 V - \{0\}$ une forme volume. On suppose que V et ω sont définis sur $\mathbb{F}_q$. On identifie le dual V' de V à V à l'aide de la forme symplectique ω. On a donc une involution

$$\mathscr{F}_\psi(-)(1) : \mathrm{Perv}(V, \overline{\mathbb{Q}}_\ell) \longrightarrow \mathrm{Perv}(V, \overline{\mathbb{Q}}_\ell)$$

si l'on considère V comme un fibré vectoriel sur $\mathrm{Spec}(k)$ de manière évidente. Soit $\mathscr{A}$ la catégorie abélienne des paires (A, φ) où $A \in \mathrm{ob}\, \mathrm{Perv}(V, \overline{\mathbb{Q}}_\ell)$ et $\varphi : \mathscr{F}_\psi(\mathrm{Frob}_q^* A)(1) \xrightarrow{\sim} A$ est un isomorphisme dans $\mathrm{Perv}(V, \overline{\mathbb{Q}}_\ell)$. Soit $K(\mathscr{A})$ le groupe de Grothendieck de $\mathscr{A}$. On considère l'application

$$\mathrm{ob}\, \mathscr{A} \times \mathrm{ob}\, \mathscr{A} \longrightarrow \overline{\mathbb{Q}}_\ell$$

$$((A_1, \varphi_1), (A_2, \varphi_2)) \longmapsto \sum_{i \geq 0} (-1)^i \, \mathrm{tr}(\varphi^i, \mathrm{Ext}^i_{\mathrm{Perv}(V, \overline{\mathbb{Q}}_\ell)}(A_1, DA_2))$$

où $DA_2 = R\mathcal{H}om(A_2, \overline{\mathbb{Q}}_\ell[2](1))$, où les Ext^i ($i \geq 0$) sont les Ext^i de Yoneda dans la catégorie abélienne $\text{Perv}(V, \overline{\mathbb{Q}}_\ell)$ (ce sont des $\overline{\mathbb{Q}}_\ell$-espaces vectoriels de dimension finie, nuls pour $i \gg 0$, d'après [Be]) et où φ^i est induit par φ_1 et φ_2. Elle induit un accouplement

$$(,) : K(\mathcal{A}) \times K(\mathcal{A}) \longrightarrow \overline{\mathbb{Q}}_\ell$$

dont on notera $N(\mathcal{A})$ le noyau.

Le groupe algébrique sur k, $\text{Aut}(V, \omega) \times \mathbb{G}_{m,k}$, agit sur V par $(g, t) \cdot v = g(tv) = tg(v)$ d'une part et par $(g, t) \cdot v = g(t^{-1}v) = t^{-1}g(v)$ d'autre part, d'où deux actions de $\text{Aut}(V, \omega) \times \mathbb{G}_{m,k}$ sur $\text{Perv}(V, \overline{\mathbb{Q}}_\ell)$ échangées par $\mathscr{F}_\psi(-)(1)$. On en déduit une action de $\text{Aut}(V, \omega)(\mathbb{F}_q) \times \{t \in k | t^{q+1} = 1\}$ sur $K(\mathcal{A})$ qui respecte l'accouplement $(,)$ et donc son noyau.

Théorème 7.1 (Kazhdan). *(i) $K(\mathcal{A})/N(\mathcal{A})$ a une structure naturelle de $\overline{\mathbb{Q}}_\ell$-espace vectoriel de dimension finie sur lequel $\text{Aut}(V, \omega)(\mathbb{F}_q) \times \{t \in k | t^{q+1} = 1\}$ agit.*

(ii) Si $\chi : \{t \in k | t^{q+1} = 1\} \longrightarrow \overline{\mathbb{Q}}_\ell^\times$ est un caractère régulier ($\chi^2 \neq 1$) la composante isotypique σ_χ de χ dans $K(\mathcal{A})/N(\mathcal{A})$ est la représentation irréductible de la série discrète de $\text{Aut}(V, \omega)(\mathbb{F}_q) \simeq SL_2(\mathbb{F}_q)$ associée à χ par la construction de Weil.

Dans [Kaz-La], nous construisons un analogue de $K(\mathcal{A})/N(\mathcal{A})$ pour un groupe semi-simple et simplement connexe arbitraire (à la place de $\text{Aut}(V, \omega)$) et conjecturons une généralisation de (7.1).

Remarque 7.2. Pour d'autres applications à la théorie des représentations des groupes réductifs sur les corps finis, voir [Br, De3, Lu].

Remerciement. Je remercie Mme Le Bronnec pour la magnifique composition de ce manuscrit.

Bibliographie

[Ad-Sp]   Adolphson, A., Sperber, S.: Exponential sums and Newton polyhedra, cohomology and estimates. Ann. Math. **130** (1989) 367–406

[Be]   Beilinson, A.: On the derived category of perverse sheaves. (Lecture Notes in Mathematics, vol. 1289.) Springer, Berlin Heidelberg New York 1987, pp. 27–41

[B-B-D]   Beilinson, A., Bernstein, J., Deligne, P.: Faisceaux pervers. Astérisque **100** (1983)

[Br]   Brylinski, J.-L.: Transformations canoniques, Dualité projective, Théorie de Lefschetz, Transformations de Fourier et sommes trigonométriques. Astérisque **140–141** (1986) 3–134

[De1]   Deligne, P.: Lettre à D. Kazhdan du 29 novembre 1976

[De2]   Deligne, P.: La conjecture de Weil II. Publ. Math. IHES **52** (1980) 137–252

[De3]   Deligne, P.: Métaplectique. Communication personnelle

[De-Lo]   Denef, J., Lœser, F.: Weights of exponential sums, intersection cohomology, and Newton polyhedra. Preprint 1990

[Ek]   Ekedahl, T.: On the ℓ-adic formalism in the "Grothendieck Festschrift", vol. 2. Birkhäuser, Basel (to appear)

[He]   Henniart, G.: La conjecture de Langlands locale numérique pour $GL(n)$. Ann. Sci. Ec. Norm. Sup. **21** (1988) 497–544

[Il1]   Illusie, L.: Deligne's ℓ-adic Fourier transform. Proceedings of Symposia in Pure Math., vol. 46, part 2. A.M.S. 1987, pp. 151–163

[Il2]   Illusie, L.: Théorie de Brauer et caractéristique d'Euler-Poincaré d'après P. Deligne. Astérisque **82–83** (1981) 161–172

[Ka1]   Katz, N.M.: Travaux de Laumon, Séminaire Bourbaki. Astérisque **161-162** (1988) 105–132

[Ka2]   Katz, N.M.: Gauss sums, Kloosterman sums, and monodromy. Ann. Math. Stud., vol. 116. Princeton Univ. Press, 1988

[Ka3]   Katz, N.M.: Perversity and exponential sums. Tohoku Math. J. (to appear)

[Ka-La]   Katz, N.M., Laumon, G.: Transformation de Fourier et majoration de sommes exponentielles. Publ. Math. IHES **62** (1985) 361–418

[Kaz-La]   Kazhdan, D., Laumon, G.: Gluing of perverse sheaves and discrete series representations. J. Geom. Phys. **5** (1988) 63–120

[La1]   Laumon, G.: Transformation de Fourier, constantes d'équations fonctionnelles et conjecture de Weil. Publ. Math. IHES **65** (1987) 131–210

[La2]   Laumon, G.: Correspondance de Langlands géométrique pour les corps de fonctions. Duke Math. J. **54** (1987) 309–359

[Lo]   Lœser, F.: Arrangements d'hyperplans et sommes de Gauss. Prépublicaton École Polytechnique, Palaiseau 1989

[Lu]   Lusztig, G.: Fourier transforms on a semi-simple Lie algebra over $\mathbb{F}_q$. (Lecture Notes in Mathematics, vol. 1271.) Springer, Berlin Heidelberg New York 1987, pp. 117–188

[SGA]   Grothendieck, A., et al.: Séminaire de Géométrie Algébrique du Bois-Marie, SGA4, SGA5, SGA7. (Lecture Notes in Mathematics, vols. 269, 305, 589, 288, 340.) Springer, Berlin Heidelberg New York 1972 à 1977

Algebraic Independence of Values of Analytic Functions

Yuri Nesterenko

Department of Mathematics, Moscow State University, 119899, Moscow, USSR

The first result on the algebraic independence of values of analytic functions was obtained by F. Lindemann in 1882:

If $a_1, \ldots, a_m$ are algebraic numbers linearly independent over $\mathbb{Q}$, then numbers

$$e^{a_1}, \ldots, e^{a_m}$$

are algebraically independent over $\mathbb{Q}$.

This means that for any polynomial $P \in \mathbb{Z}[x_1, \ldots, x_m]$, $P \not\equiv 0$, the inequality $|P(e^{a_1}, \ldots, e^{a_m})| > 0$ holds. The algebraic independence of any numbers $\omega_1, \ldots, \omega_m$ over $\mathbb{Q}$ implies, in particular, that all the numbers ω_i are transcendental. Side by side with problems of the algebraic independence of numbers, an object of study in the theory of transcendental numbers is to obtain quantitative characteristics of independence. By a measure of algebraic independence of complex numbers $\omega_1, \ldots, \omega_m$ we mean a function of two variables $f(d, H)$ such that for any polynomial $P \in \mathbb{Z}[X_1, \ldots, X_m]$, $P \not\equiv 0$, whose degree does not exceed d and whose coefficients have modulus not greater than H, the estimate

$$|P(\omega_1, \ldots, \omega_m)| > f(d, H)$$

holds.

It is easy to prove for all complex numbers $\omega_1, \ldots, \omega_m$ the inequality

$$\phi(\bar{\omega}; d, H) = \min |P(\omega_1, \ldots, \omega_m)| < H^{-c_1 d^m},$$

where $c_1 = 1/(2m!)$ and minimum is taken over all polynomials

$$P \in \mathbb{Z}[x_1, \ldots, x_m], \qquad P \not\equiv 0, \qquad \deg P \le d, \qquad H(P) \le H$$

($H(P)$ is the maximum of the moduli of the coefficients of P). But any lower estimate of this minimum $\phi(\bar{\omega}; d, H)$ depends on the individual properties of numbers $\omega_1, \ldots, \omega_m$ and estimates are obtained only for special kinds of numbers.

In 1932 K. Mahler [16] proved that, under the conditions of Lindemann's theorem,

$$\phi(e^{a_1}, \ldots, e^{a_m}; d, H) > H^{-c_2 d^m},$$

Proceedings of the International Congress
of Mathematicians, Kyoto, Japan, 1990
© The Mathematical Society of Japan, 1991

where $c_2 > 0$ depends only on numbers $\alpha_1, \ldots, \alpha_m$, and H exceeds some bound, which depends on $\alpha_1, \ldots, \alpha_m$ and d.

1. *E*-Functions

Function $e^{\alpha z}$ with algebraic number α, $\alpha \neq 0$, belong to a certain class of entire functions, introduced in 1929 by C. Siegel [39] and called by him *E*-functions. This class of functions contains hypergeometric functions

$$pFq\begin{pmatrix} a_1, \ldots, a_p \\ b_1, \ldots, b_q \end{pmatrix} z^{q-p+1} \end{pmatrix}, \qquad p \leq q,$$

with rational parameters a_i, b_j and is a ring, closed with respect to differentiation, integration between 0 and z, and replacing z by λz, where λ is algebraic. After the basic articles of Siegel the complete result was proved in 1955 by A.B. Shidlovskij [38].

Theorem 1. *Suppose that the E-functions $f_1(z), \ldots, f_m(z)$ form a solution of the system of linear differential equations*

$$y_k' = q_{k0}(z) + \sum_{i=1}^m q_{ki}(z)y_i, \qquad k = 1, 2, \ldots, m, \tag{1}$$

$q_{ki}(z) \in \mathbb{C}(z)$; *suppose that α is an algebraic number not equal to zero and the singular points of the system* (1). *Then*

$$\operatorname{tr\,deg}_{\mathbb{Q}} \mathbb{Q}(f_1(\alpha), \ldots, f_m(\alpha)) = \operatorname{tr\,deg}_{\mathbb{C}(z)} \mathbb{C}(z)(f_1(z), \ldots, f_m(z)).$$

Lindemann's theorem follows from the theorem of Shidlovskij, if we take $f_i(z) = e^{\alpha_i z}$ and $\alpha = 1$.

Theorem 1 gave rise to a great number of results concerning specific *E*-functions (see [38]). For example, the function

$$\Psi_k(z) = \sum_{n=1}^\infty \frac{z^{kn}}{(n!)^k},$$

satisfies a linear differential equation of order k with coefficients in $\mathbb{C}(z)$; it can be proved that, for any algebraic number $\alpha \neq 0$, the $r(r+1)/2$ numbers $\Psi_k^{(i)}(\alpha), 0 \leq i < k, 1 \leq k \leq r$, are algebraically independent over $\mathbb{Q}$.

There exists an important problem, connected with applications of Theorem 1: *proving the algebraic independence of solutions of linear differential equations over* $\mathbb{C}(z)$. In different specific cases it was the subject of many papers. We mention here only recent papers by W. Salichov [34, 35] and F. Beukers, D. Brownawell and G. Heckmann [2]. This subject is related to the Picard-Vessiot theory of linear differential equations and the calculation of Galois groups of such equations.

In 1962, using the Shidlovskij results, S. Lang [13] obtained a quantitative variant of the above theorem in a form similar to the Mahler's theorem (see also the article by Galochkin [8]). In the next assertion, which was proved in 1977 [20], the dependence on d of the lower bound for H is effective.

Theorem 2. *Suppose that the E-functions $f_1(z), \ldots, f_m(z)$ and the number α satisfy the conditions of Theorem 1. There exists constant $\tau > 0$, which depends only on the system* (1), *the functions $f_1(z), \ldots, f_m(z)$ and the number α, such that for*

$$\ln \ln H > \tau d^{2m} \ln(d+1)$$

the following inequality holds:

$$\phi(f_1(\alpha), \ldots, f_m(\alpha); d, H) > H^{-c_3 d^m},$$

where $c_3 = 4^m \varkappa^m (m \varkappa^2 + \varkappa + 1)$, and $\varkappa$ is the degree of the algebraic field, which contains α and the coefficients of Taylor expansions of $f_i(z)$.

In many cases the constant τ may be computed effectively, [4] and [25].

2. Mahler's Functions

Here we briefly discuss the values of power series (in general, in several variables) which satisfy certain functional equations. Mahler was the first to study such series in 1929 (see [17]). Besides the works of Mahler, we mention here the results of J. Loxton and A.J. van der Poorten, K.K. Kubota, D.W. Masser. Recently interesting relations between Mahler's functions and automata theory were found [14].

For an integer $n \geq 1$ let $T = (t_{ij})$ be a nonsingular matrix of order n with nonnegative integer entries such that none of the eigenvalues of T is a root of unity. We define a transformation $T: \mathbb{C}^n \to \mathbb{C}^n$ by the rule

$$T(z_1, \ldots, z_n) = \left(\prod_{i=1}^{n} z_i^{t_{1i}}, \ldots, \prod_{i=1}^{n} z_i^{t_{ni}} \right).$$

The next theorem was proved by Loxton and van der Poorten [15].

Theorem 3. *Let $A(\bar{z})$ be an $m \times m$ matrix whose entries are rational functions of the variable $\bar{z}$ with algebraic coefficients such that $\det A(0) \neq 0$. Let $\bar{f}(\bar{z}) = (f_1(\bar{z}), \ldots, f_m(\bar{z}))$ be a vector of power series in $\bar{z} = (z_1, \ldots, z_n)$ with coefficients in an algebraic number field K, $[K : \mathbb{Q}] < \infty$, which converge in some neighbourhood U of the origin in $\mathbb{C}^n$ and satisfy the equations*

$$\bar{f}(\bar{z}) = A(\bar{z}) * \bar{f}(T\bar{z}).$$

If the matrix A and vector $\bar{\alpha}$ with algebraic components satisfy some technical conditions, then

$$\operatorname{tr} \deg_{\mathbb{Q}} \mathbb{Q}(f_1(\bar{\alpha}), \ldots, f_m(\bar{\alpha})) = \operatorname{tr} \deg_{\mathbb{C}(\bar{z})} \mathbb{C}(\bar{z})(f_1(\bar{z}), \ldots, f_m(\bar{z})).$$

Recently K. Nishioka [27], using a method introduced in [23] and based on the general theory of elimination (see § 5), in the case $n = 1$ proved the same assertion under weaker conditions: α is an algebraic number such that $0 < |\alpha| < 1$, $A(\alpha^{d^l})$ is nonsingular and $a(\alpha^{d^l}) \neq 0$ for all $l \geq 0$, where $a(z)$ is a common denominator of entries of $A(z)$.

The first estimate for the measure of algebraic independence of values of Mahler's functions were proved in 1985 in [23]. But this estimate was not effective in the degrees of polynomials. Using the zero estimate for Mahler functions, which was proved by Nishioka, Becker found effective algebraic independence measures of those numbers (see [1, 28]).

Theorem 4. *Let $A(z)$ be an $m \times m$ matrix and $B(z)$ be an m-dimensional vector whose entries are rational functions of one variable z with algebraic coefficients. Let $\bar{f} = (f_1(z), \ldots, f_m(z))$ be a vector of power series with algebraic coefficients which converge in some neighbourhood U of the point $z = 0$ and let $\bar{f}$ satisfy*

$$\bar{f}(z^t) = A(z) * \bar{f}(z) + B(z),$$

where $t \geq 2$ is an integer, and which are algebraically independent over $\mathbb{C}(z)$. Suppose that α is an algebraic number, $\alpha \in U$, $0 < |\alpha| < 1$, and the numbers α^{t^l}, $l \geq 0$, are distinct from the poles of $A(z)$ or $B(z)$. Then

$$\phi(f_1(\alpha), \ldots, f_m(\alpha); d, H) > \exp(-c_4 d^m (\ln H + d^{m+2})).$$

3. Exponential Functions

One of the problems suggested by D. Hilbert in 1900 was the conjectural transcendence of values of the function α^z with algebraic base $\alpha \neq 0, 1$ at algebraic irrational points β. This assertion was proved in 1934 by A.O. Gelfond and independently by Th. Schneider. The natural analog of the Lindemann's theorem is the next

Gelfond's Conjecture. *If $\alpha \neq 0, 1$ is an algebraic number and algebraic numbers $\beta_1, \ldots, \beta_m$ and 1 are linearly independent over $\mathbb{Q}$, then the numbers*

$$\alpha^{\beta_1}, \ldots, \alpha^{\beta_m}$$

are algebraically independent over $\mathbb{Q}$.

This problem is equivalent to *the algebraic independence over $\mathbb{Q}$ of numbers α^β, $\alpha^{\beta^2}, \ldots, \alpha^{\beta^{d-1}}$*, where β is an algebraic number of degree $d \geq 3$. It was proved for $d = 3$ by Gelfond [9] in 1948. Now the best result in general case is the lower estimate tr deg $K \geq \left[\dfrac{d+1}{2}\right]$ for transcendence degree over $\mathbb{Q}$ of the field $K = \mathbb{Q}(\alpha^\beta, \alpha^{\beta^2}, \ldots, \alpha^{\beta^{d-1}})$, where [ ] denotes the integer part.

Let us consider a more general situation. Suppose $a_1, \ldots, a_p$ and $b_1, \ldots, b_q$ be complex numbers which for any nonzero vectors $\bar{k} = (k_1, \ldots, k_p) \in \mathbb{Z}^p$ and $\bar{l} = (l_1, \ldots, l_q) \in \mathbb{Z}^q$ satisfy the inequalities

$$|k_1 a_1 + \cdots + k_p a_p| > \exp(-\gamma |\bar{k}| \ln |\bar{k}|), \qquad |\bar{k}| = \max_{1 \leq i \leq p} |k_i|,$$

$$|l_1 b_1 + \cdots + l_q b_q| > \exp(-\gamma |\bar{l}| \ln |\bar{l}|), \qquad |\bar{l}| = \max_{1 \leq i \leq q} |l_i|,$$

where γ is a positive constant. In particular, it is well known that this condition is satisfied by algebraic numbers and by the logarithms of algebraic numbers which are linearly independent over $\mathbb{Q}$. Traditionally the three sets of numbers have been studied

1) $e^{a_i b_j}$,   2) $a_i, e^{a_i b_j}$,   3) $a_i, b_j, e^{a_i b_j}$, $1 \le i \le p, 1 \le j \le q$.

Many estimates for transcendence degree of these sets of numbers were proved after appearance of Gelfond's articles [9, 10]. We mention here papers by R. Tijdeman, A. Shmelev, W.D. Brownawell, G. Chudnovsky, M. Waldschmidt, E. Reyssat. P. Philippon, and others (see [40]). The best result now is the next assertion.

Theorem 5. *Among the numbers in the sets 1)–3) there are at least*

$$\left[\frac{pq}{p+q}\right], \qquad \left[\frac{pq+p}{p+q}\right], \qquad \frac{pq}{p+q},$$

respectively which are algebraically independent over $\mathbb{Q}$.

The lower estimate for tr deg K follows immediately from this theorem if in the set 2) we take $p = q = d$, $a_i = \beta^{i-1}$, $b_i = \beta^{i-1} \ln \alpha$, $i = 1, \ldots, d$.

P. Philippon in 1984 [31] proved a little weaker estimate for the sets 1), 2) and the estimate for the set 3) in Theorem 5. The main tools for obtaining these results were introduced by the author [20–23] and P. Philippon [32], and are connected with application of some classical ideas of "constructive" commutative algebra, developed in the works of K. Hentzelt, E. Noether, B.L. van der Waerden, W.L. Chow (see Section 5) to the theory of transcendental numbers. The estimates for the sets 1) and 2) in the Theorem 5 were established in 1987 by G. Diaz [6] who used a new construction of auxiliary function in the analytical part of the proof (for another variant see [24]).

For $d = 3$ the estimate for integral polynomials at the point $(\alpha^\beta, \alpha^{\beta^2}, \ldots, \alpha^{\beta^{d-1}})$ was established by Gelfond and Feldman in 1950. It was later improved by W.D. Brownawell [3] and G. Diaz [7]. In the general case $d \ge 4$ such an estimate is not proved yet. But it is possible to prove a lower estimate for the maximum of absolute values at this point of any set of polynomials which generate an ideal in $\mathbb{Z}[x_1, \ldots, x_{d-1}]$ with rank higher than $\frac{d-1}{2}$ (see [23]).

4. Elliptic Functions

Let $\wp(z)$ be the Weierstrass elliptic function with algebraic invariants g_2, g_3. This function satisfies an algebraic differential equation with algebraic coefficients and an addition theorem. That is why its transcendence behaviour is similar to the behaviour of e^z. First results about the transcendence of values of this function were proved by Schneider in 1934 [36]. The next assertion, the elliptic analogue of the Lindemann's theorem, was proved in 1983 by G. Wustholz [42] and P. Philippon [30] after the previous paper of G. Chudnovsky [5].

Theorem 6. *If* $p(z)$ *has the complex multiplication over the field* $k = \mathbb{Q}(\tau)$ *and* $\alpha_1, \ldots,$ α_m *are algebraic numbers, linearly independent over field* k, *then the numbers*

$$p(\alpha_1), \ldots, p(\alpha_m)$$

are algebraically independent over $\mathbb{Q}$.

Now it is possible to estimate a measure of the algebraic independence of these numbers (see [26]).

Theorem 7. *Let* $p(z)$, $\alpha_1, \ldots, \alpha_m$ *be the same as in Theorem 6*, H *and* d *be positive numbers such that*

$$\ln \ln H > c_5 * d^m * \ln(d + 1), \qquad d \geq 1.$$

Then the following inequality holds

$$\phi(p(\alpha_1), \ldots, p(\alpha_m); d, H) > H^{-c_6 d^m}. \tag{2}$$

Here c_5 *and* c_6 *are a positive constants, depending only on* $\alpha_1, \ldots, \alpha_m$, g_2 *and* g_3.

This inequality was established for $m = 1$ by G. Chudnovsky [5] in 1980. In the general form with exponent $m + \varepsilon$ instead of m in the right hand side of the inequality (2) it was proved by E.M. Jabbouri [11] as a consequence of a result for algebraic groups.

Now let $p(z)$ have algebraic invariants g_2, g_3 and possibly be without complex multiplication. Let $\alpha_1, \ldots, \alpha_m$ be algebraic numbers linearly independent over $\mathbb{Q}$. One can prove (see [26]) that, among the numbers $p(\alpha_1), \ldots, p(\alpha_m)$ at least there are $\left[\dfrac{m}{2}\right]$ algebraically independent over $\mathbb{Q}$ and there exists a quantitative result like in the exponential case.

In the general case (g_2 and g_3 may be transcendental numbers) if $a_1, \ldots, a_p$ and $b_1, \ldots, b_q$ are complex numbers with almost the same condition on linear forms in a_i and b_j there are results about the lower estimates for the transcendence degree of the fields $L_1 = \mathbb{Q}(g_2, g_3, p(a_1 b_1), \ldots, p(a_p b_q))$, $L_2 = \mathbb{Q}(g_2, g_3, a_1, \ldots, a_p, p(a_1 b_1),$ $\ldots, p(a_p b_q))$, $L_3 = \mathbb{Q}(g_2, g_3, a_1, \ldots, a_p, b_1, \ldots, b_q, p(a_1 b_1), \ldots, p(a_p b_q))$.

Small estimates were proved by W.D. Brownawell and K.K. Kubota, D.W. Masser and G. Wüstholz, R. Tubbs. In 1983 D.W. Masser and G. Wüstholz [19] under the assumption that g_2 and g_3 are algebraic, proved for $t = \operatorname{tr} \deg L_1$ the estimate

$$2^{t+2}(t + 8) \geq pq/(p + 2q).$$

Later in the papers of M. Waldschmidt [41] was published a sketch of the proof of the inequalities

$$\operatorname{tr} \deg L_1 \geq \frac{pq}{p + 2q} - 1, \qquad \operatorname{tr} \deg L_2 \geq \frac{pq + p}{p + 2q} - 1, \qquad \operatorname{tr} \deg L_3 \geq \frac{pq + 2q}{p + 2q} - 1.$$

The paper [37] of S. Shestakov contains the proof of the inequality $\operatorname{tr} \deg L_2 \geq \left[\dfrac{pq + p}{p + 2q}\right]$ from which follows "one half" of the elliptic variant of Gelfond's conjecture: *if* g_2, g_3 *are algebraic,* $\wp(z)$ *has complex multiplication over a quadratic field* k, β *is algebraic of degree* d *over* k *and* $u \neq 0$ *is such that* $\wp(u)$ *is algebraic, then among the numbers* $\wp(\beta u), \wp(\beta^2 u), \ldots, \wp(\beta^{d-1} u)$ *at least* $\left[\dfrac{d+1}{2}\right]$ *are algebraically independent over* $\mathbb{Q}$. In [29] a weaker estimate was proved.

The paper [41] contains estimates for fields, generated by the values of both exponential and elliptic function.

We should note here that recent articles of D.W. Masser and G. Wüstholz [18], P. Philippon [33] contain estimates for the number of zeros of polynomial defined on group varieties, which are very useful in the proofs of algebraic independence of numbers, especially of the values of elliptic and abelian functions.

5. Algebraic Base of Methods

Any homogeneous unmixed ideal $I \subset \mathbb{Z}[x_0, \ldots, x_m]$ may be characterized by numbers $\deg I$, $H(I)$, which are analogous to $\deg P$, $H(P)$ for polynomials $P \in \mathbb{Z}[x_0, \ldots, x_m]$. One may also define $|I(\bar{\omega})|$ for the point $\bar{\omega} = (\omega_0, \ldots, \omega_m) \in \mathbb{C}^{m+1}$, analogous to $|P(\bar{\omega})|$.

For exact definitions we need the notion of rank for ideals. The rank of the prime ideal $\mathfrak{p} \subset \mathbb{Z}[x_0, \ldots, x_m]$ is the maximal length of any increasing chain of prime ideals terminating with $\mathfrak{p}$. The rank of any ideal I is the minimal rank of prime ideals $\mathfrak{p}$, containing I. The rank of an ideal I will be denoted by $h(I)$. Recall that an ideal $I \subset \mathbb{Z}[x_0, \ldots, x_m]$ is called unmixed if all of the primary components of I have rank $h(I)$.

Let I be a homogeneous unmixed ideal of the ring $\mathbb{Z}[x_0, \ldots, x_m]$ and $r = m + 1 - h(I) \geq 1$. We introduce the linear forms

$$L_i(x) = \sum_{j=0}^{m} u_{ij} x_j, \qquad i = 1, 2, \ldots, r,$$

where u_{ij} are variables, $\bar{u}_i = (u_{i0}, \ldots, u_{im})$, and denote $\bar{I}(r)$ the ideal of $\mathbb{Z}[\bar{u}_1, \ldots, \bar{u}_m]$, consisting of the polynomials G for each of which there exists a natural number M such that $G x_i^M$ for all $i = 0, \ldots, m$ belong to the ideal generated in $\mathbb{Z}[x_0, \ldots, x_m, \bar{u}_1, \ldots, \bar{u}_r]$ by the linear forms $L_1, \ldots, L_m$ and the elements of I. One can prove that $\bar{I}(r) = (F)$ is a principal ideal. The polynomial F is a Chow form of I. We define (see [21]) $\deg I$ as the degree of F in the variables $\bar{u}_1$ and $H(I) = H(F)$. Now consider r skew-symmetric matrices $S^{(i)} = \|s_{jk}^i\|, 0 \leq j, k \leq m, 1 \leq i \leq r$, where we suppose that, except for the skew-symmetry $s_{jk}^i + s_{kj}^i = 0$, the variables s_{jk}^i are not connected by any algebraic relation over the ring $\mathbb{Z}[x_0, \ldots, x_m]$. If $\mathbf{æ}(F)$ is the polynomial in the variables $s_{jk}^i, j < k$, which is obtained from F by substituting the vectors $S^{(i)}\bar{\omega}, i = 1, \ldots, r$, for the variables $\bar{u}_i$, we define

$$|I(\bar{\omega})| = |\bar{\omega}|^{-r*\deg I} * H(\mathbf{æ}(F)).$$

Now the problem may be put as follows. For a given point $\bar{\omega} = (\omega_0, \ldots, \omega_m) \in \mathbb{C}^{m+1}$ we are searching for a lower estimate for $|I(\bar{\omega})|$ in terms of the $H(I)$ and $\deg I$. For the principal ideal $I = (P)$, its rank equals 1, and the quantities $|I(\bar{\omega})|$ and

$|P(\bar{\omega})|$, deg I and deg P, $H(I)$ and $H(P)$ are closely related, which gives us possibility to obtain the lower estimate for $|P(\bar{\omega})|$ in terms of the properties of P.

For example, the proof of the Theorem 7 is based on the next assertion.

Theorem 8. *Let D and H be positive numbers satisfying the inequalities*

$$\ln \ln H > c_7 * D^m * \ln(D+1), \qquad D \geq 1.$$

For every integer r, $1 \leq r \leq m$ there exist constants $\lambda_r > 0$ and $\mu_r \geq 0$ such that for all homogeneous unmixed ideals $I \subset \mathbb{Z}[x_0, \ldots, x_m]$ with conditions $h(I) = m + 1 - r$,

$$\deg I \leq \lambda_r D^{m+1-r}, \qquad \ln H(I) \leq \lambda_r D^{m-r} \ln H$$

the inequality holds $(\bar{\omega} = (1, \mathfrak{p}(\alpha_1), \ldots, \mathfrak{p}(\alpha_m)))$

$$\ln|I(\bar{\omega})| \geq -\mu_r(D * \ln H(I) + \deg I * \ln H)D^{r-1}.$$

The Theorem 8 may be proved by induction on r. The step of induction contains two stages.

i) *Reduction of the estimate of $|I(\bar{\omega})|$ to the analogous estimate for prime ideals.* It is based upon the fact that numbers deg I, $H(I)$ and $|I(\bar{\omega})|$ have almost linear behaviour under decomposition of the ideal into primary ideals.

Proposition 1 (see [22]). *Suppose that I is an unmixed homogeneous ideal in $\mathbb{Z}[x_0, \ldots, x_m]$, $h(I) \leq m$; there exist prime ideals $\mathfrak{p}_1, \ldots, \mathfrak{p}_s$ associated with I and natural numbers $k_1, \ldots, k_s$ such that*

1) $\sum_{l=1}^{s} k_l * \deg \mathfrak{p}_l \sim \deg I,$

2) $\sum_{l=1}^{s} k_l * \ln H(\mathfrak{p}_l) \leq \ln H(I) + m^2 \deg I,$

3) $\sum_{l=1}^{s} k_l * \ln |\mathfrak{p}_l(\bar{\omega})| \leq \ln |I(\bar{\omega})| + m^3 \deg I.$

ii) *Increasing of the rank of ideal.* The algebraic base of this stage is the next assertion.

Proposition 2 (see [24]). *Suppose that $\mathfrak{p}$ is a homogeneous prime ideal of $\mathbb{Z}[x_0, \ldots, x_m]$, $\mathfrak{p} \cap \mathbb{Z} = (0)$, $h(\mathfrak{p}) \leq m$ and Q is a homogeneous polynomial in $\mathbb{Z}[x_0, \ldots, x_m]$, $Q \notin \mathfrak{p}$. If $r = m + 1 - h(I) \geq 2$, then there exists an unmixed homogeneous ideal $J \subset \mathbb{Z}[x_0, \ldots, x_m]$ whose zeros coincide with the zeros of the ideal $(\mathfrak{p}, Q)$, for which $h(J) = m - r + 2$, and such that*

1) $\deg J \leq \deg \mathfrak{p} * \deg Q,$

2) $\ln H(J) \leq \deg Q * \ln H(\mathfrak{p}) + \deg \mathfrak{p} * \ln H(Q) + m(r+1) * \deg \mathfrak{p} * \deg Q,$

3) *If $\bar{\omega} \in \mathbb{C}^{m+1}$, $|Q(\bar{\omega})| < |\bar{\omega}|^{\deg Q}$ and ϱ is the distance between $\bar{\omega}$ and the set of zeros of the ideal $\mathfrak{p}$, then the following inequality holds*

$$\ln |J(\bar{\omega})| \leq \ln A + \deg Q * \ln H(\mathfrak{p}) + \deg \mathfrak{p} * \ln H(Q) + 11m^2 * \deg \mathfrak{p} * \deg Q,$$

where $A = |Q(\bar{\omega})| * |\bar{\omega}|^{-\deg Q}$, if $\varrho < |Q(\bar{\omega})| * |\bar{\omega}|^{-\deg Q}$, and $A = |\mathfrak{p}(\bar{\omega})|$ in opposite case.

If $r = 1$ the right side of the inequality in 3) is nonnegative.

If we can construct some polynomial Q, which is sufficiently small at the point $\bar{\omega}$ and does not belong to the ideal $\mathfrak{p}$, then the inequality from 3) and the induction assumption about the $|J(\bar{\omega})|$ give us a lower estimate for $|\mathfrak{p}(\bar{\omega})|$. The inequalities from 1) and 2) enable us to recompute this estimate in terms of $\deg \mathfrak{p}$ and $H(\mathfrak{p})$.

The constructions of the desired polynomials Q are based on the analytic properties of the functions considered and of course are different in different problems (for example in Theorems 4, 5, 7).

It is possible to axiomatize the situation and to prove a general assertion in which from the existence of a set of polynomials, sufficiently small at $\bar{\omega}$ and independent in some sense, follows a lower estimate for the number of algebraically independent coordinates of the vector $\bar{\omega}$. The first assertion of this kind for polynomials in one variable was proved by Gelfond [10] (see [40] for history). Recently P. Philippon [32] by means of the above ideas established a general algebraic independence criterion. One of the forms of it is

Theorem 9. *Suppose $\omega_1, \ldots, \omega_m$ are complex numbers algebraically independent over $\mathbb{Q}, \bar{\omega} = (\omega_1, \ldots, \omega_m)$; γ and η are positive numbers. Suppose that for any $N \geq N_0$ there exist polynomials $P_1, \ldots, P_M \in \mathbb{Z}[x_1, \ldots, x_m]$ with finite set of common zeros in the ball $B(\bar{\omega}, \exp(-3\gamma N^\eta))$ and with conditions*

$$\deg P_i + \ln H(P_i) \leq N, \qquad i = 1, \ldots, M,$$

$$\max_{1 \leq i \leq M} \{|P_i(\omega_1, \ldots, \omega_m)|\} < \exp(-\gamma N^\eta).$$

If γ is sufficiently large with respect of m, then the inequality $m > \eta - 1$ holds.

Jabbouri [12] obtained a quantitative form of this criterion which made it possible to estimate in this way the measure of algebraic independence of numbers.

References

1. Becker, P.G., Nishioka, K.: Measures for the algebraic independence of the values of Mahler type functions. C.R. Math. Rep. Sci. Canada **XI**, no. 3 (1989) 89–93
2. Beukers, F., Brownawell, W.D., Heckman, G.: Siegel normality. Ann. Math. **127** (1988) 279–308
3. Brownawell, W.D.: On the Gelfond-Feldman measure of algebraic independence. Comp. Math. **38**, no. 3 (1979) 355–368
4. Brownawell, W.D.: The effectivity of certain measures of algebraic independence. In: Seminaire de Theorie des Nombres, Paris 1984–1985. Birkhäuser, Basel 1986, pp. 41–50
5. Chudnovsky, G.V.: Algebraic independence of the values of elliptic functions at algebraic points; Elliptic analogue of the Lindemann-Weierstrass theorem. Invent. math. **61** (1980) 267–290
6. Diaz, G.: Grands degres de transcendance pour des familles d'exponentielles. C.R. Acad. Sci. Paris, Ser. 1 **305**, no. 5 (1987) 159–162

7. Diaz, G.: Une nouvelle mesure d'independance algebrique pour $(\alpha^\beta, \alpha^{\beta^2})$. Acta Arithmetica. **LVI** (1990) 25–32

8. Galochkin, A.I.: Estimate for the relative transcendence measure of the values of E-functions. Mat. Zametki **3**, no. 4 (1968) (Russian) 377–386

9. Gelfond, A.O.: On algebraic independence of algebraic powers of algebraic numbers. Dokl. Akad. Nauk USSR **64**, no. 3 (1949) (Russian) 277–280

10. Gelfond, A.O.: On the algebraic independence of transcendental numbers of certain classes. Usp. Mat. Nauk **4**, no. 5 (1949) (Russian) 14–48. [English transl.: Amer. Math. Soc. Transl. **2** (1962)]

11. Jabbouri, E.M.: Mesures d'independance algebrique de valeurs de fonctions elliptiques et abeliennes. C.R. Acad. Sci. Paris, Ser. I **303**, no. 9 (1986).

12. Jabbouri, E.M.: Sur un critere d'independance algebrique de P. Philippon. In: Seminaire d'Arithmetique de Saint-Etienne 1986–1987, IV.

13. Lang, S.: A transcendence measure for E-functions. Mathematika **9** (1962) 157–161

14. Loxton, J.H.: Automata and transcendence. In: Baker, A. (ed.) New advances in transcendence theory. Cambridge University Press, 1988, pp. 215–228

15. Loxton, J.H., van der Poorten, A.J.: Arithmetic properties of the solutions of a class of functional equations. J. Reine Angew. Math. **330** (1982) 159–172

16. Mahler, K.: Zur Approximation der Exponentialfunktion und des Logarithmus I. J. Reine Angew. Math. **166** (1932) 118–136

17. Mahler, K.: Arithmetische Eigenschaften einer Klasse transzendental-transcendenter Functionen. Math. Z. **32** (1930) 545–585

18. Masser, D.W., Wüstholz, G.: Zero estimates on group varieties I. Invent. math. **64** (1981) 489–516; II, Invent. math. **80** (1985) 233–267

19. Masser, D.W., Wüstholz, G.: Fields of large transcendence degree generated by values of elliptic functions. Invent. math. **72** (1983) 407–464

20. Nesterenko, Yu. V.: Estimates for the orders of zeros of functions of a certain class and applications in the theory of transcendental numbers. Izv. Akad. Nauk SSSR, Ser. Mat. **41**, no. 2 (1977) (Russian) 253–284. [English transl.: Math. USSR Izv. **11**, no. 2 (1977)]

21. Nesterenko, Yu. V.: On the algebraical independence of algebraic numbers to algebraic powers. In: Bertrand, D., Waldschmidt, M. (eds.) Approximations diophantiennes et nombres transcendants, Luminy 1982. (Progress in Mathematics, vol. 31). Birkhäuser, Basel 1983, pp. 199–220

22. Nesterenko, Yu. V.: On algebraic independence of algebraic powers of algebraic numbers. Mat. Sb. **123**(165), no. 4 (1984) (Russian) 435–459. [English. transl.: Math. USSR Sb. **51**, no. 2 (1985)]

23. Nesterenko, Yu. V.: On a measure of the algebraic independence of the values of certain functions. Mat. Sb. **128**(170), no. 4 (1985) (Russian) 545–568. [English transl.: Math. USSR Sb. **56**, no. 2 (1987)]

24. Nesterenko, Yu. V.: On a transcendence degree of certain fields generated by values of exponential function. Matem. Zametki **46**, no. 3 (1989) (Russian) 40–49

25. Nesterenko, Yu. V.: Effective bounds for the algebraic independence measure of the values of E-functions. Vestnik Moskov. Univ., Ser. I, no. 4 (1988) (Russian) 85–88

26. Nesterenko, Yu. V.: On a measure of the algebraic independence of the values of elliptic functions. In: Philippon, P. (ed) Approximations diophantiennes et nombres transcendants, Luminy 1990 (to appear)

27. Nishioka, K.: New approach in Mahler's method. J. Reine Angew. Math. **407** (1990) 202–219

28. Nishioka, K.: Algebraic independence measures of the values of Mahler functions. (To appear)

29. Philippon, P.: Varieties abeliennes et independance algebrique I. Invent. math. **70** (1983) 289–318

30. Philippon, P.: Varietes abeliennes et independance algebrique II. Invent. math. **72** (1983) 389–405
31. Philippon, P.: Criteres pour l'independance algebrique de familles de nombres. Math. Repts. Acad. Sci. Can. **6**, no. 5 (1984) 285–290
32. Philippon, P.: Criteres pour l'independance algebrique. Inst. Hautes Etudes Sci. Publ. Math. **64** (1986) 5–52
33. Philippon, P.: Lemmes de zeros dans les groupes algebriques commutatifs. Bull. Soc. Math. France **114** (1986) 355–383
34. Salichov, W.: Formal solutions of linear differential equations and their application in transcendental number theory. Trudy Moskov. Mat. Obshch. **51** (1988)
35. Salichov, W.: On algebraic independence of the values of hypergeometric E-functions. Dokl. Akad. Nauk SSSR **307**, no. 2 (1988) 284–287
36. Schneider, Th.: Transzendenzuntersuchungen periodischer Funktionen II; Transzendenzeigenschaften elliptischer Funktionen. J. Reine Angew. Math. **172** (1934) 70–74
37. Shestakov, S.: On transcendence degree of the fields generated by the values of elliptic function. Vestn. Moscow State Univ. (To appear)
38. Shidlovskij, A.B.: Transcendental Numbers. Nauka, Moskow, 1987 (Russian). [English transl.: Transcendental numbers. W. de Gruyter, Berlin New York 1989]
39. Siegel, C.L.: Über einige Anwendungen diophantischer Approximationen. Abh. Preuss. Akad. Wiss. Berlin, Phil.-Math. Kl., no. 1 (1929–1930) 1–70
40. Waldschmidt, M.: Algebraic independence of transcendental numbers. Gelfond's method and its developments. In: Jager, W., Moser, J., Remmert, R. (eds.) Perspectives in Mathematics, Anniversary of Oberwolfach. Birkhäuser, Basel 1984, pp. 551–571
41. Waldschmidt, M.: Algebraic independence of values of exponential and elliptic functions. J. Indian Math. Soc. **48** (1985) 215–228
42. Wüstholz, G.: Über das abelsche Analogon des Lindemannschen Satzes. Invent. math. **72** (1983) 363–388
43. Wüstholz, G.: Multiplicity estimates on group varieties. Ann. Math. **129** (1989) 501–517

Diophantine Problems and Linear Groups

Peter C. Sarnak

Department of Mathematics, Stanford University, Stanford, CA 94305, USA

1. Group Varieties

In this lecture we describe and exploit the relation between analytic Diophantine problems on homogeneous varieties and harmonic analysis on the corresponding groups. For the case $G = SL(2)$ this relation has been well studied and striking applications to analytic number theory have been found, especially by Iwaniec and his collaborators [I1]. Our focus here will be on general G where this aspect of the theory is still in a primitive state.

Let $V \subset A^n$ be a variety in affine n space and we assume that V is defined over $\mathbf{Q}$ (or more generally a number field $E/\mathbf{Q}$). Let $V(\mathbf{Z})$ denote the integral points in V and define the counting functions N by

$$N(T, V) = |\{m \in V(\mathbf{Z}) \mid \|m\| \leq T\}| \tag{1.1}$$

$$N(T, V, \xi, q) = |\{m \in V(\mathbf{Z}) \mid \|m\| \leq T, \ m \equiv \xi \ (\mathrm{mod} \ q)\}|, \tag{1.2}$$

where $T \geq 1, q \in \mathbf{N}$, $\xi \in V(\mathbf{Z})$, and $\|\cdot\|$ is some suitably chosen Euclidean norm on A^n.

The basic problem is to determine the asymptotic behavior of N as $T \to \infty$. In certain special cases where there are sufficiently many additive terms in the equations defining V, the Hardy–Littlewood method may be applied, see Schmidt [Sc]. One obtains an asymptotic in the form

$$N(T, v) \sim c \, T^{\beta}, \qquad \beta \in \mathbf{N}. \tag{1.3}$$

The constant c is given by a product of local densities [Sc]. Another case, due to Moroz [M], that can be successfully handled by algebraic number theory is algebraic varieties of the form $f_1(x_1) = f_2(x_2) = \ldots = f_r(x_r)$, where $f_j(x_j)$ is a norm form of an order in a number field. Besides these cases[1] very little is known even in terms of upper bounds for N, see S.D. Cohen's bound described in Serre [SER]. However, if we assume that V is also a homogeneous variety for a linear algebraic group action, then much more can be said. In particular if $V \simeq H \backslash G$ with $H(\mathbf{R}) \backslash G(\mathbf{R})$ an affine symmetric space then a reasonably complete

---

[1] Also see [F-M-T] for rational points on homogeneous projective varieties.

Proceedings of the International Congress
of Mathematicians, Kyoto, Japan, 1990
© The Mathematical Society of Japan, 1991

theory can be developed, see [D-R-S]. We give some concrete examples of the latter which go beyond the well studied case of quadratic forms.

Example 1. (a) For $k \neq 0$ let

$$V_k = \{x_{ij}, \ i = 1,2,3, \ j = 1,2,3 \mid \det(x_{ij}) = k\} \subset A^9. \tag{1.4}$$

V_1 is the group variety $G = SL(3)$, while for the general k, G acts on V_k by multiplication. In this way V_k decomposes into finitely many homogeneous spaces for G. We have [D-R-S]

$$N(T, V_k) = c_k \, T^6 + O(T^{23/4}) \qquad \text{as} \quad T \to \infty.$$

Here and elsewhere the constant c_k is explicitly computable and is given in terms of products of local densities.

(b) Let

$$\varphi(x_1, x_2, x_3, x_4, x_5, x_6) = x_1 x_4 x_6 + 2 x_2 x_3 x_5 - x_1 x_5^2 - x_6 x_2^2 - x_4 x_3^2$$

$$= \det \begin{pmatrix} x_1 & x_2 & x_3 \\ x_2 & x_4 & x_5 \\ x_3 & x_5 & x_6 \end{pmatrix} = \det X. \tag{1.5}$$

Let

$$W_k = \{x \in A^6 \mid \varphi(x) = k\}, \qquad k \neq 0.$$

$G = SL(3)$ acts on W_k by

$$X \mapsto {}^t g X g.$$

Again W_k decomposes into finitely many G orbits. We have

$$N(T, W_k) \sim d_k \, T^3 \qquad \text{as } T \to \infty.$$

(c) Let

$$D(a, b, c, d) = 18abcd + b^2 c^2 - 4ac^3 - 4b^3 d - 27a^2 d^2$$

be the homogeneous form of degree four in four variables which is the discriminant of binary cubics. Let

$$U_k = \{(a, b, c, d) \mid D(a, b, c, d) = k\}.$$

$SL(2)$ acts on binary cubics and hence on U, which is again a homogeneous space. Relative to the norm

$$\|(a, b, c, d)\|^2 = a^2 + \frac{b^2}{3} + \frac{c^2}{3} + d^2$$

we have for $k \neq 0$

$$N(T, U_k) \sim C_k \, T^{2/3} \qquad \text{as } T \to \infty.$$

For the purposes of this report we specialize now to the case that V is a group variety G. We assume that $G(\mathbf{R})$ is semisimple and that we have a realization

$$G \subset GL(m) \subset A^{m^2}. \tag{1.6}$$

Without loss of generality we assume that

$$K = G(\mathbf{R}) \cap O(m, \mathbf{R}) \tag{1.7}$$

is a maximal compact subgroup of $G(\mathbf{R})$. Using (1.6) we get from the Euclidean norm on A^{m^2} a norm, $\| \cdot \|$, on $G(\mathbf{R})$ and hence on V. Let

$$\alpha(G) = \lim_{T \to \infty} \frac{\log \int_{\|g\| \leq T} dg}{\log T}. \tag{1.8}$$

$\alpha(G)$ depends on the realization (1.6) but does not depend on the Haar measure used in definition (1.8). As usual let $\Gamma(q)$ be the principal congruence subgroup of $\Gamma(1) = G(\mathbf{Z}) = V(\mathbf{Z})$, that is

$$\Gamma(q) = \{\gamma \in G(\mathbf{Z}) \mid \gamma \equiv I \pmod{q}\}. \tag{1.9}$$

Denote by $W(q)$ the index $[\Gamma(1) : \Gamma(q)]$.

Theorem 1.1 [D-R-S]. *For $\xi \in G(\mathbf{Z})$*

$$N(T, V, \xi, q) \sim \frac{c_V}{W(q)} T^\alpha \qquad as \ T \to \infty.$$

Remarks 1.1. (a) A special case of this is well known. That is in hyperbolic spaces $(G(\mathbf{R}) = SO(n, 1))$ where the problem reduces to counting lattice points in a non Euclidean ball. This goes back to Delsarte [DE], Selberg [SEL2], Huber [HUB], Patterson [P], and Lax-Phillips [L-P]. Also in this case, Elstrodt-Mennicke-Grunewald [E-G-M] have investigated the arithmetic aspects of the asymptotics. Bartels [BA] obtains the asymptotics of lattice points in a non-Euclidean ball (which in general is quite different from the above counting problem)[2] for general Riemannian symmetric spaces, when $\Gamma \backslash G(\mathbf{R})$ is compact.

(b) The constant c_V is essentially $\mathrm{Vol}(G(\mathbf{Z}) \backslash G(\mathbf{R}))^{-1}$ which is non-zero by the Borel-Harish Chandra theorem [B-HC]. Moreover, by the theory of Tamagawa numbers, see [KNE], c_V is given by a product of local densities which at almost all places is the Hardy-Littlewood local density.

(c) If G is not semisimple some care must be taken. For example, if $V = \{(x, y) \mid x^2 - y^2 = 1\}$ then $V(\mathbf{Z})$ is finite, while for norm forms of number fields the asymptotics is of the form $c(\log T)^r$.

Theorem 1.1 can be improved to give remainder terms as well but these depend on the spectrum of $L^2(\Gamma(q) \backslash G(\mathbf{R})/K)$. More precisely, let $G(\mathbf{R}) = NAK$ be an

---

[2] See also Terras [TE], p. 248.

Iwasawa factorization of $G(\mathbf{R})$ and $W = \text{Weyl}(G(\mathbf{R}), A)$ the corresponding Weyl group. $S = G(\mathbf{R})/K$ is a globally Riemannian symmetric space. The algebra of invariant differential operators on S is isomorphic to the algebra of polynomials on $\mathfrak{a}_{\mathbf{C}}^*$ (where $\mathfrak{a} = \text{Lie}(A)$) which are W invariant [V]. Thus the spectrum of this ring of invariant operators acting on $L^2(\Gamma(q)\backslash G(\mathbf{R})/K)$ may be identified with a subset of $\mathfrak{a}_{\mathbf{C}}^*/W$ [V]. In fact it is a subset of $\hat{G}^1(\mathbf{R})$, the set of $\lambda \in \mathfrak{a}_{\mathbf{C}}^*/W$ which correspond to positive spherical functions. Note too that this spectral parameter space comes with a natural topology. We denote the above spectrum by $\sigma^1(\Gamma(q)\backslash G(\mathbf{R}))$, the super 1 denoting "class-1" representations [KN]. It consists of discrete eigenvalues as well as unitary Eisenstein series when $\Gamma(q)\backslash G$ is not compact [LS]. In our normalization of $\mathfrak{a}_{\mathbf{C}}^*$, $\varrho = 1/2$ sum of positive roots, corresponds to the spherical function 1 and $i\mathfrak{a}_{\mathbf{R}}^*$ to the tempered spectrum denoted $\hat{G}_{\text{temp}}^1(\mathbf{R})$. Note that the notion of spectrum $\sigma^1(\Gamma\backslash G(\mathbf{R}))$ makes sense for any discrete subgroup $\Gamma \leq G(\mathbf{R})$.

Definition 1.2. $\Gamma \leq G(\mathbf{R})$ (Γ discrete and not necessarily of finite volume) is called *tempered* if $\sigma^1(\Gamma\backslash G(\mathbf{R})) \subset \hat{G}_{\text{temp}}^1(\mathbf{R})$.

Theorem 1.3 [B-L-S]. *If $\Gamma < G(\mathbf{R})$ is tempered then*

$$N(T, \Gamma) = \sum_{\|\gamma\| \leq T} 1 = O_\varepsilon(T^{\alpha/2+\varepsilon}), \quad \text{for all } \varepsilon > 0.$$

One might (optimistically) hope that the "Ramanujan conjectures" of $G = SL(2)$ (see Selberg [SEL], Satake [SA]) would be true in this general setting. That is for all $q \geq 1$

$$\sigma^1(\Gamma(q)\backslash G(\mathbf{R})) \subset \hat{G}_{\text{temp}}^1(\mathbf{R}) \cup \{\varrho\}. \tag{1.13}$$

Such a bound would lead to optimally small remainder terms in Theorem 1.1, at least for smoothed versions thereof. Unfortunately (1.13) is not true in general. The first counterexamples to (1.13) (at least to its analogue at the finite places) were found in $SP(4)$, numerically by Kurokawa [KU] and using θ-liftings by Howe and Piatetski-Shapiro [H-PS]. Actually, we can see that (1.13) is not true in general for certain structural reasons that come from an examination of $N(T, \Gamma)$. Indeed the truth of (1.13) would imply the same for any $\Gamma(q) \subset \Gamma \subset \Gamma(1)$, i.e., any congruence subgroup. Now suppose that $\Delta = \cap_{i=1}^{\infty} \Gamma_i$ is an intersection of congruence groups. Δ need no longer be of finite index in $G(\mathbf{Z})$. It is easy to see [B-L-S2] that

$$\sigma^1(\Delta\backslash G(\mathbf{R})) \subset \bigcap_{j=1}^{\infty} \overline{\bigcup_{i=j}^{\infty} \sigma^1(\Gamma_i\backslash G(\mathbf{R}))}. \tag{1.14}$$

In particular if $\sigma^1(\Delta\backslash G(\mathbf{R}))$ is nontempered (and does not contain ϱ) then the same must be true of $\sigma^1(\Gamma\backslash G(\mathbf{R}))$ for $\Gamma = \cap_{i=1}^{\ell}\Gamma_i$ where ℓ is sufficiently large. That is (1.13) fails for suitable large q. To find Δ as above with $\sigma^1(\Delta\backslash G(\mathbf{R}))$ nontempered we note that if $H < G$ is a Q-subgroup then $\Delta = H(\mathbf{Z})$ is the intersection of congruence groups [B-L-S2]. Moreover we can compute $N(T, H(\mathbf{Z}))$ by virtue of

$H(\mathbf{Z})$ being (essentially) a lattice in $H(\mathbf{R})$. Hence using Theorem 1.3 we can show $H(\mathbf{Z}) = \varDelta$ is nontemprered by showing it has too many elements. We illustrate this with an example.

Let $G = SO(n+1, 1) \subset GL(n+2, \mathbf{R})$ be the orthogonal group of the form

$$F(x_1, \ldots, x_{n+2}) = x_1^2 + x_2^2 + \ldots + x_{n+1}^2 - x_{n+2}^2. \tag{1.15}$$

Then

$$\alpha(G) = n \qquad \text{for } n \geq 1. \tag{1.16}$$

If $H < G$ corresponds to the subgroup fixing $x_1, \ldots, x_m$ then $H \simeq SO(n+1-m, 1)$ and

$$\alpha(H(\mathbf{Z})) = n - m. \tag{1.17}$$

Hence H is nontempered if

$$1 \leq m < \frac{n}{2}. \tag{1.18}$$

Thus for $n \geq 3$, $G(\mathbf{Z})$ has congruence subgroups with exceptional spectrum, that is for which (1.13) fails. In fact, a more refined analysis shows that

$$\left\{ \varrho = \frac{n}{2}, \frac{n}{2} - 1, \frac{n}{2} - 2, \ldots, \frac{n}{2} - \left[\frac{n}{2}\right] \right\} \tag{1.19}$$

are limit points of exceptional eigenvalues of congruence subgroups of $G(\mathbf{Z})$. An explicit lower bound for the automorphic spectrum $\cup_q \sigma(\Gamma(q) \backslash G(\mathbf{R}))$ is developed in [B-L-S2] along these and related lines. The results show the important role played by all $\mathbf{Q}$-subgroups in the general Ramanujan conjectures. In fact one obtains a powerful and general method to construct automorphic forms and spectrum. This new method complements the only other general method for constructing automorphic forms viz. theta liftings [Ho].[3] In particular it leads to new results on nonvanishing of Betti numbers of arithmetic lattices [B-L-S2].

To end this section a word about the proofs of Theorem 1.1 and its generalizations. The problem reduces to one of counting points of $\Gamma = G(\mathbf{Z})$ (or subgroups thereof) lying in an expanding family of regions $R_T \subset G(\mathbf{R})$. One forms the Γ invariant function

$$F_T(g) = \sum_{\gamma \in \Gamma} \chi_{R_T}(\gamma g),$$

where χ_R is the characteristic function of R. Under suitable assumptions about R_T, F_T can be asymptotically spectrally analyzed. When successful, one deduces from the asymptotic behavior of certain eigenfunctions on $G(\mathbf{R})/K$, such as the spherical functions, that the main term in the asymptotics of $F_T(1)$ comes from the component of F_T along constants. The latter is $\text{Vol}(R_T)/\text{Vol}(G(\mathbf{Z}) \backslash G(\mathbf{R}))$. Various L^2 averaging arguments reduce the spectral analysis to a minimum so that a result like Theorem 1.3 may be deduced without any detailed knowledge of the nature of the spectrum (here $\text{Vol}(\Gamma \backslash G(\mathbf{R})) = \infty$).

---

[3] We add, however, that when lifting by the trace formula can be carried out successfully it typically yields the most complete picture known – see for example Arthur-Clozel [A-C].

2. Density Hypothesis

We have noted that the naive Ramanujan conjecture is false in most cases. Borrowing from the analytic theory of the zeta function we formulate certain weaker but still very useful hypotheses. If $v \in \hat{G}^1(\mathbf{R})$ is an eigenvalue occuring in $L^2(\Gamma(p)\backslash S)$, p a large prime, then we get a representation of the finite group $\Gamma(1)/\Gamma(p)$ on the eigenspace corresponding to v. Under suitable assumptions $\Gamma(1)/\Gamma(p) \simeq G(\mathbf{Z}/p\mathbf{Z})$ and we may use the lower bounds on the dimensions of the smallest nontrivial irreducible representations of $G(\mathbf{Z}/p\mathbf{Z})$ [H-H, LUS] to deduce a lower bound on the multiplicity $m(v, \Gamma(p))$ of v in $L^2(\Gamma(p)\backslash S)$

$$m(v, \Gamma(p)) \gg p^{\delta(G)}, \quad \delta(G) \geq 1. \tag{2.1}$$

In particular in such a case if an exceptional eigenvalue occurs it must do so with high multiplicity. The substitute for Ramanujan is a sharp upper bound for this multiplicity. Notice that we are sticking to the principal congruence subgroups $\Gamma(q)$. In fact this is forced on us in view of the Q-subgroup phenomenon discussed in the previous Section. We begin with a basic conjecture concerning the integer points on the group variety V_G. For what follows we assume that V_G is as in (1.6) and the norm is the Euclidean one on $\mathbf{R}^{m^2}$.

Main Conjecture 2.1 [S-X]. *For* $T, q \geq 1$

$$N(T, V, I, q) \ll_\varepsilon \frac{T^{\alpha+\varepsilon}}{W(q)} + T^{\alpha/2}, \quad \varepsilon > 0. \tag{2.2}$$

Remarks 2.1. (a) The point of this conjecture is the uniformity in T and q. The value of T for which it is most useful (and hardest to prove) is when the terms on the right hand side are equal. Theorem 1.1 implies (2.2) is true for fixed q.

(b) The Main Conjecture (MC) follows from the naive Ramanujan conjecture 1.13.

(c) MC has been established in two important cases:

(i) If Γ is any arithmetic lattice in $G(\mathbf{R}) = SL(2, \mathbf{R})$ or $SL(2, \mathbf{C})$ (see [BO] for a classification) and $\Gamma(q)$ are corresponding congruence subgroups then MC is true [S-X].

(ii) J. Katznelson [KA] has recently established MC for $G = SL(m)$, $m \geq 2$.

We consider the implications of MC. We begin with the density hypothesis and consider only the case $G(\mathbf{R})$ of (real) rank 1. The reason is that in this case there is a natural ordering of the spectrum. In the general case one must order by regions. If $G(\mathbf{R})$ is rank 1 then $\hat{G}^1(\mathbf{R})$ is contained in

$$i\mathfrak{a}_{\mathbf{R}}^* \cup (0, \varrho] \subset \mathfrak{a}_{\mathbf{C}}^*. \tag{2.3}$$

For $0 \leq v \leq \varrho$ let

$$M(v, q) = \sum_{v \leq \mu \leq \varrho} m(\mu, \Gamma(q)). \tag{2.4}$$

Thus $M(v,q)$ counts the number of exceptional eigenvalues which are more nontempered than v. Clearly

$$M(\varrho, q) = 1$$

$$M(0, q) = O(W(q)).$$

The Density Hypothesis is the linear interpolation of these

Density Hypothesis 2.2 [S-X]. *Assume* $G(\mathbf{R})$ *is rank 1, then*

$$M(v, q) = O_\varepsilon(W(q)^{1-v/\varrho+\varepsilon}), \quad for\ all\ \varepsilon > 0.$$

Proposition 2.3 [S-X]. MC $\Rightarrow$ *Density Hypothesis.*

Remarks 2.3. (a) Combining (2.3) with Remark 2.1 (c) we find that the Density Hypothesis is true for any arithmetic lattice Γ (relative to its principal congruence subgroups) in $SL(2, \mathbf{R})$ or $SL(2, \mathbf{C})$. For $\Gamma = SL(2, \mathbf{Z})$ this was first proved by H. Iwaniec and J. Szmidt [I-S] and M. Huxley [HU] using trace formulae. Iwaniec [I2] has obtained an even stronger bound for $\Gamma_0(q)$ in $SL(2, \mathbf{Z})$.

The Main Conjecture also implies bounds on the multiplicities of the $\pi \in \hat{G}(\mathbf{R})$, the unitary dual of $G(\mathbf{R})$, which appear discretely in the decomposition of the regular representation of $G(\mathbf{R})$ on $L^2(\Gamma(q)\backslash G(\mathbf{R}))$. Denote this multiplicity by $m(\pi, \Gamma(q))$. For π class 1, π corresponding to v as in the previous Section, we have $m(v, \Gamma(q)) = m(\pi, \Gamma(q))$ by the duality theorem [G-G-PS]. If $\pi \in \hat{G}(\mathbf{R})$ we let $p(\pi)$ be the infimum over all $p \geq 2$ such that all matrix coefficients of π are in $L^p(G)$ [KN]. For $\pi = \varrho$ the trivial representation, $p(\pi) = \infty$ and $m(\pi, \Gamma(q)) = 1$. For $\pi \in L^2(G)$, i.e., π discrete series, $p(\pi) = 2$ and $m(\pi, \Gamma(q))$ is of order $W(q)$, see deGeorge-Wallach [G-W]. Linearly interpolating leads to

Conjecture 2.3 [S-X]. *Let* $\pi \in \hat{G}(\mathbf{R})$

$$m(\pi, \Gamma(q)) = O_\varepsilon(W(q)^{2/p(\pi)+\varepsilon}).$$

We have

Proposition 2.4 [S-X]. *If* $G(\mathbf{R})$ *is of rank 1 then* MC $\Rightarrow$ *Conjecture 2.3.*

Remarks 2.4. (a) If we drop the assumption that $G(\mathbf{R})$ is of rank 1 then we can prove (in the case $G(\mathcal{O}_E)\backslash G(\mathbf{R})$ is compact) that MC implies a somewhat weaker bound for multiplicities.

(b) Conjecture (2.3) as well as the others are typically false for noncongruence families of $G(\mathbf{Z})$. See for example the Abelian covers in Section 3.

(c) Combining the bound in Conjecture 2.3 with Matsushima's formula [B-W] in the case $\Gamma\backslash G(\mathbf{R})$ is compact, gives rather sharp bounds for the Betti numbers of principal congruence subgroups of Γ.

Finally Conjecture 2.3 implies certain "arithmetic vanishing theorems" and in particular lower bounds on small eigenvalues. The lower bound (2.1) applies to any representation $\pi \in \hat{G}(\mathbf{R})$ and hence we conclude:

Under Conjecture 2.3, if $p(\pi) > 2/\delta(G)$ then $m(\pi, \Gamma(q)) = 0$ for all q. \hfill (2.5)

As an application of this where MC has been established we have the following [S-X]:

Let $\Gamma \leq SL(2, \mathbf{R})$ (or $SL(2, \mathbf{C})$) be arithmetic. Thus Γ is commensurable with $G(\mathcal{O}_E)$ where E is a number field over $\mathbf{Q}$ and G is the unit group of a quaternion algebra A over E which is ramified at all infinite places of E except one, call it v, and $A \otimes v \simeq M(2, \mathbf{R})$. Let $\Gamma(P) = G(P) \cap \Gamma$, where P is a prime ideal of $\mathcal{O}_E$. Denote by $\lambda_1(P)$ the smallest non zero eigenvalue of the Laplacian on $\Gamma(P) \backslash \mathbf{H}$, $\mathbf{H}$ being the hyperbolic plane. Clearly $\lambda_1(P) \leq \lambda_1(\Gamma)$ and the last may be arbitrarily small, see Randol [R].

Proposition 2.5 [S-X]. *For* $\mathrm{Norm}\,(P)$ *large*

$$\min\left(\lambda_1(\Gamma), \frac{5}{36}\right) \leq \lambda_1(\Gamma(P)) \leq \lambda_1(\Gamma).$$

The number 5/36 also appears in Huxley [HU] for essentially the same reason; viz. (2.5). This kind of vanishing theorem in (2.5) can also be derived for the graphs constructed in [L-P-S] to prove, in an entirely elementary fashion, that they are expanding families with good expansion coefficients.

Arithmetic vanishing results of the type of (2.5) have been established unconditionally for certain (in particular noncompact) congruence lattices in $SO(n, 1)$ by Elstrodt-Grunewald-Mennicke [E-G-M2] and Li-Piatetski Shapiro-Sarnak [L-PS-S]. The method used in these papers is quite different (and special) and goes via Kloosterman sums as in Selberg [SEL].

3. Variations

We have restricted our discussion to $G(\mathbf{R})$ and its spectrum. There are numerous Diophantine problems which are related to $G(\mathbf{Q}_p)$ for finite p. We illustrate this with certain Diophantine problems studied by Linnik [LIN, LIN2]. Let $F(x_1, \ldots, x_r)$ be a homogeneous form in r variables and of degree d. For $m \in \mathbf{N}$ and $x \in \mathbf{Z}^r$ for which

$$F(x) = m \tag{3.1}$$

we associate to x,

$$x' = \frac{x}{m^{1/d}}. \tag{3.2}$$

Thus $F(x') = 1$. With the use of his ergodic method Linnik was able, in certain cases, to study the distribution of these points x' on the surface $V_1 = \{F = 1\}$ as $m \to \infty$.

For example, let

$$F(x_1, x_2, \ldots, x_r) = x_1^2 + x_2^2 + \ldots + x_r^2, \qquad r \geq 3. \tag{3.3}$$

The basic problem is the equidistribution with respect to Lebesgue measure of the x' on the unit sphere S^{r-1}. Linnik [LIN] proved this for $r \geq 4$ and conditionally for $r = 3$ [LIN]. By the use of classical θ–functions [S1] this problem is immediately reduced to one of estimating Fourier coefficients of classical modular forms (of half integer weight if r is odd). For $r \geq 4$ standard estimates give the result. For $r = 3$ the estimates needed lie much deeper and were only proven recently by Iwaniec [I2]. The general case of this problem for quadratic forms in three variables requires similar estimates for Fourier coefficients of Maass forms as well. These were established by W. Duke [D] whose work resolves this instance of Linnik's problem completely.

Concerning forms of higher degree Linnik and Skubenko [L-S] considered the form

$$F(x_{ij}) = \det(x_{ij}), \quad i, j = 1, \ldots, n. \tag{3.4}$$

Using the ergodic method they show that if $\Omega \subset \mathbf{R}^{n^2}$ is a (nice) compact set and

$$R_\Omega(m) = |\{x \in \mathbf{Z}^{n^2} \mid F(x') \in \Omega\}| \tag{3.5}$$

then

$$R_\Omega(m) \sim \frac{\text{meas}(\Omega')}{\zeta(2)\zeta(3)\ldots\zeta(n)} \prod_{p_i^{k_i}} \frac{(p_i^{k_i+1} - 1)(p_i^{k_i+2} - 1)\ldots(p_i^{k_i+n-1} - 1)}{(p_i - 1)(p_i^2 - 1)\ldots(p_i^{n-1} - 1)}, \tag{3.6}$$

where $\Omega' = \Omega \cap SL(n, \mathbf{R})$, meas$(\Omega')$ its Euclidean content, and $m = p_1^{k_1} p_2^{k_2} \ldots p_v^{k_v} \to \infty$.

We outline a proof of (3.6) using harmonic analysis on $\hat{G}(\mathbf{Q}_p)$. Let $G(\mathbf{R}) = SL(n, \mathbf{R})$, $\Gamma = SL(n, \mathbf{Z})$. The usual Hecke operators [TE], T_p (or more generally T_m) are defined by

$$\left. \begin{aligned} T_p : L^2(\Gamma \backslash G(\mathbf{R})) &\to L^2(\Gamma \backslash G(\mathbf{R})) \\ T_p H(g) = \sum_{\substack{\det M_j = p \\ M_j \text{ integral} \\ M_j \neq M_k \pmod{\Gamma}}} & H\left(\frac{1}{p^{1/n}} M_j g\right) \end{aligned} \right\} \tag{3.7}$$

Let $f \in C_0^\infty(G(\mathbf{R}))$, then

$$\sum_{\substack{\det X = p \\ X, n \times n \text{ integral}}} f(x') = (T_p H)(I), \tag{3.8}$$

where

$$H(g) = \sum_{\gamma \in \Gamma} f(\gamma g). \tag{3.9}$$

Now

$$\int_{\Gamma \backslash G(\mathbf{R})} H(g) dg = \int_{G(\mathbf{R})} f(g) dg \tag{3.10}$$

(for a fixed Haar measure dg) from which it is easy to see that (3.6) follows from

$$\| T_m|_B \|_2 \ll m^{n-1-\delta} \tag{3.11}$$

for $\delta > 0$ $(\delta = \delta(n))$, where

$$B = \left\{ h \in L^2(\Gamma \backslash G(\mathbf{R})) \mid \int_{\Gamma \backslash G(\mathbf{R})} h(g) dg = 0 \right\}.$$

For $n \geq 3$, (3.11) follows from the classification of the spherical dual $\hat{G}(\mathbf{Q}_p)$ [TA], together with the standard relation between Hecke operators and representations of $G(\mathbf{Q}_p)$. For $n = 2$ we need well known, nontrivial bounds on the Fourier coefficients of holomorphic and Maass forms on $SL(2, \mathbf{Z}) \backslash \mathbf{H}$ to establish (3.11).

By examining more carefully the best δ that can be achieved in (3.11) (again this is the "Ramanujan Conjecture") or the best estimates for Fourier coefficients in the other cases, one can go well beyond the equidistribution of the points x'. In fact one can examine how well the points x' with $F(x) = m$ cover the set V_1. The harmonic analysis method above leads to sharp results [D-R-S]. For an elementary approach concerning the last problem, in the case $SL(n)$, see Harman [HA]. Determining the optimal covering exponent for these points (as $m \to \infty$) is an interesting and apparently difficult problem [D-R-S].

Finally we make comments related to Conjecture 2.3 but in connection with other families of coverings of $G(\mathbf{Z}) \backslash S$. Assume that $H_1(\Gamma, \mathbf{Z}) \simeq \mathbf{Z}^r$ with $r \geq 1$. Then there are uniquely defined $(\mathbf{Z}/q\mathbf{Z})^r$ Galois coverings of $\Gamma \backslash S$ given by $\Phi(q) \backslash S$ where

$$\Phi(q) = \{ \gamma \in \Gamma \mid p(\gamma) \equiv 0 \ (\text{mod } q) \}, \tag{3.12}$$

$p : \Gamma \to H_1(\Gamma, \mathbf{Z})$ being the canonical projection. An interesting example of this is the family $F_q : x^q + y^q = 1$ of Fermat curves. These are $(\mathbf{Z}/q\mathbf{Z})^2$ Galois coverings of $\Gamma(2) \backslash \mathbf{H}$ where $\mathbf{H} = \{ z \mid \text{Im } z > 0 \}$ and $\Gamma(2) < SL(2, \mathbf{Z})$ is the principal congruence subgroup of level 2, see [LA1]. Another example is the family of Hirzebruch surfaces [HI].

For such families the behavior of $m(\pi, \Phi(q))$, for $\pi \in \hat{G}(\mathbf{R})$ fixed, is closely tied to another Diophantine problem; one of counting the number of q–division points in an r dimensional torus which lie on a real analytic subvariety, see [S2]. For $\pi \in \hat{G}(\mathbf{R})$ which correspond to cohomology, the subvariety V in question is algebraic in the sense that it is a zero set (or intersection of such) of a trigonometric polynomial. For these there is a definite structure to the set of such torsion points lying on V. This was conjectured by Lang [LA2] and proved in Laurent [LAU]. See also [S2] for an effective solution. As a consequence one has

Theorem 3.1. [S2] *Let* $M(q) = \Phi(q) \backslash S$ *as above, then the j–th Betti numbers* $\beta^{(j)}(M(q))$ *are polynomial periodic in q, that is*

$$\beta^{(j)}(M(q)) = \sum_{\nu=0}^{r} a_\nu^{(j)}(q) q^\nu$$

with $a_\nu^{(j)}(q)$ *periodic in q.*

This is in sharp contrast to the congruence family. For the general $\pi \in \hat{G}(\mathbf{R})$ we cannot conclude such strong results. However, using some recent work of Bombieri-Pila [B-P] one can get asymptotic results for $m(\pi, \Phi(q))$ for the Fermat and related families [P-S]. Actually for the Fermat family we prove in [P-S] an exact formula for the number of exceptional eigenvectors coming from Eisenstein series residues. In the notation of (2.4) with obvious modifications

$$M^{\text{Eis}}(0, q) = 6 \left[\frac{q}{8} \right]. \tag{3.13}$$

What is striking about (3.13) is that the constant term of the Eisenstein series for the noncongruence groups $\Phi(q)$ corresponding to F_q, are not (as far as is known) computable in terms of classical zeta like functions.

We have touched on some aspects of the analysis of $G(\mathbf{Z}) \backslash G(\mathbf{R})$ which are directly related to analytic Diophantine problems. The aspects of the theory relating to L-functions associated to automorphic forms on general G, have been well developed at the hands of Langlands, Piatetski-Shapiro, and their coworkers.

Acknowledgements. I would like to thank M. Burger, W. Duke, Jian Shu Li, Z. Rudnick, and X. Xue. My various collaborations with them form the core of this report.

References

[A-C]   J. Arthur, L. Clozel: Simple algebras, base change, and the advanced theory of the trace formula. Ann. Studies, vol. 120, Princeton 1989

[BA]   H.J. Bartels: Nichteuklidische Gitterpunktprobleme und Gleichverteilung in linearen algebraischen Gruppen. Comment. Math. Helv. **57** (1982) 158–172

[BO]   A. Borel: Commensurability classes of Hyperbolic 3 manifolds. Ann. Sci. Norm. Sup. Pisa, Cl. Sci. **8** (1981) 1–33

[B-HC]   A. Borel, Harish–Chandra: Arithmetic subgroups of algebraic groups. Ann. Math. **75** (1962) 485–535

[B-W]   A. Borel, N. Wallach: Continuous cohomology, discrete subgroups, and representations of reductive groups. Ann. Math. Studies, vol. 94, Princeton 1980

[B-P]   E. Bombieri, J. Pila: The number of points on curves and ovals. Duke Math. J. **59** (1989) 337–358

[B-L-S1] M. Burger, Li Juan Shu, P. Sarnak: Ramanujan duals and automorphic spectrum. Preprint 1990

[B-L-S2] M. Burger, Li Juan Shu, P. Sarnak. In preparation

[DE]   J. Delsarte: Sur le Gitter Fuchsien. C.R. Acad. Sci. Paris **214** (1942) 147–149

[D]   W. Duke: Hyperbolic distribution problems and half integral weight Maass forms. Inv. math. **92** (1988) 73–90

[D-R-S] W. Duke, Z. Rudnick, P. Sarnak: The density of integral points on affine homogeneous varieties. Preprint 1990

[E-G-M1] J. Elstrodt, F. Grunewald, J. Mennicke: Arithmetic applications of the hyperbolic lattice point theorem. Proc. London Math. Soc. (3) **57** (1988) 239–283

[E-G-M2] J. Elstrodt, F. Grunewald, J. Mennicke: Poincaré series, Kloosterman sums and eigenvalues of the Laplacian for congruence groups acting on hyperbolic spaces. C.R. Acad. Sci. Paris **305** I (1987) 537–581

[F-M-T] J. Franke, Yu. Manin, Yu Tschinkel, Rational points of bounded height on Fano varieties. Inv. math. **95** (1989) 421–435

[G-G-PS] I. Gelfand, M. Graev, I. Piatetski-Shapiro: Generalized function, vol. 6. Saunders, Philadelphia 1969

[GW]   de George, N. Wallach: Limit formulas for multiplicities in $L^2(\Gamma \backslash G)$. Ann. Math. **107** (1978) 133–150

[HA]   G. Harman: Approximation of real matrices by integral matrices. J. Number Theory **34** (1990) 63–81

[H-H]   M. Harris, C. Hering: On the smallest degrees of projective representations of the group $PSL(n,q)$. Can. J. Math. **23** (1971) 90–102

[HO]   R. Howe: θ-series and invariant theory. Proc. Symp. A.M.S. XXXIII (1979) 315–322

[H-PS]   R. Howe, I. Piatetski–Shapiro: A counter example to the generalized Ramanujan conjecture for quasi split groups. Proc. Symp. A.M.S. XXXIII (1979) 315–322

[HUB]   H. Huber: Über eine neue Klasse automorpher Funktionen und Gitterpunkt Problem in der hyperbolischen Ebene. Comment. Mat. Helv. **30** (1956) 20–62

[Hu]   M. Huxley: Exceptional eigenvalues and congruence groups. AMS Contemp. Math. **53** (1985) 341–351

[I1]   H. Iwaniec: Spectral theory of automorphic functions and recent developments in analytic number theory. Proc. Int. Cong. 1986, Berkeley, pp. 444–456

[I2]   H. Iwaniec: Fourier coefficients of modular forms of half integral weight. Inv. math. **87** (1987) 385–401

[I3]   H. Iwaniec: Small eigenvalues for congruence groups. Acta Arithmetica, to appear.

[I-S]   H. Iwaniec, J. Szmidt: Density theorems for exceptional eigenvalues of the Laplacian for congruence groups. Banach Center Publ. **17** (1985) 317–331

[KN]   A. Knapp: Representation theory of semisimple groups. Princeton U. Press 1986

[KA]   J. Katznelson, Ph.D. Thesis, Stanford (to appear)

[KNE]   M. Knesser: Semisimple algebraic groups. In: Algebraic number theory, eds. Cassels and Frölich. Academic Press 1986, pp. 250–265

[KU]   N. Kurokawa: Examples of eigenvalues of Hecke operators on Siegel modular forms of degree two. Inv. math. **49** (1978) 149–165

[LA1]   S. Lang: Introduction to algebraic and Abelian functions. Springer, Berlin Heidelberg New York 1982

[LA2]   S. Lang: Fundamentals of Diophantine Geometry. Springer, Berlin Heidelberg New York 1983

[LAU]   M. Laurent: Equations Diophantiennes exponentielles. Inv. math. **78** (1984) 299–327

[L-P]   P. Lax, R. Phillips: The asymptotic distribution of lattice points in Euclidean and non Euclidean spaces. J. Funct. Anal. **46** (1982) 280–350

[LI]   Li Jian Shu: Kloosterman zeta functions on complex hyperbolic spaces. Preprint.

[L-PS-S] Li Jian Shu, I. Piatetski–Shapiro, P. Sarnak: Poincaré series for $SO(n,1)$. Proc. Ind. Acad. Sc. **97** (1987) 231–237

[LIN]   Y.V. Linnik: Ergodic properties of algebraic fields. Ergebnisse der Mathematik, vol. 45. Springer, Berlin Heidelberg New York 1967

[LIN2]   Y.V. Linnik: Additive problems and eigenvalues of the modular operators. Proc. Int. Cong. Stockholm 1962, pp. 270–284

[L-S]   Y.V. Linnik, B.F. Skubenko: Asymptotic distribution of integral matrices of third order. Vest. Leniner. Univ. Ser. Math. **13** (1964) 25–36

[L-P-S]   A. Lubotzky, R. Phillips, P. Sarnak: Ramanujan graphs. Combinatorica **8** (1988) 261–277

[LUS]   G. Lustig: Irreducible representations of finite classical groups. Inv. math. **43** (1977) 125–177

[M]   B. Moroz: On the numbe of integral points on a norm form variety in a cube like domain. J. Number Theory **27** (1987) 106–110

[P]   S.J. Patterson: A lattice point problem in hyperbolic space. Mathematica **22** (1975) 81–88

[P-S]   R. Phillips, P. Sarnak: The spectrum of Fermat curves. GAFA **2** (1991) 80–146

[R]   B. Randol: Small eigenvalues of the Laplace operator on compact Riemann surfaces. Bull. A.M.S. **80** (1974) 996–1000

[S1]   P. Sarnak: Some applications of modular forms. Cambridge University Press (to appear)

[S2]   P. Sarnak: Betti numbers of congruence subgroups. To appear in Israel Math. J.

[S-X]   P. Sarnak, X. Xue: Bounds for multiplicities of representations. To appear in Duke Math. J.

[SA]   Satake: Spherical functions and the Ramanujan conjecture. Proc. A.M.S. **9** (1967) 258–364

[SEL]   A. Selberg: On the estimation of Fourier coefficients of modular forms. Proc. Symp. Pure Math. **8** (1965) 1–15

[SEL2]   A. Selberg: Harmonic analysis and discontinuous groups in weakly symmetric Riemannian spaces with applications to Dirichlet series. J. Ind. Math. Soc. **20** (1956) 47–87

[SER]   J.P. Serre: Lectures on the Mordell–Weil theorem. Vieweg, Aspects Math. 1989

[SC]   W. Schmidt: The density of integer points on homogeneous varieties. Acta Math. **154** (1985) 243–296

[TA]   M. Tadic: Spherical unitary dual of the general linear group over a non Archimedian local field. Ann. Inst. Fourier **36** (1986) no. 2, 47–55

[TE]   A. Terras: Harmonic analysis on symmetric spaces and applications II. Springer, Berlin Heidelberg New York 1987

[V]   V. S. Varadarajan: The eigenvalue problem on negatively curved compact locally symmetric manifolds. A.M.S. Contemp. Math. **53** (1985) 449–463

Theory of Mordell-Weil Lattices

Tetsuji Shioda

Department of Mathematics, Faculty of Science, Rikkyo University
Nishi-Ikebukuro 3-chome, Tokyo 171, Japan

0. Introduction

The basic idea of Mordell-Weil lattices is to view the Mordell-Weil group (= the group of rational points on an elliptic curve or an abelian variety defined over some "global" field) as a Euclidean lattice by means of suitable inner product.

The necessary setup for this has been known for some time. First of all, the finite generation of such a group was established by Mordell and Weil in 1920s (thus named after them) in the case of elliptic curves over $\mathbb{Q}$ [Mo] or Jacobian varieties over a number field [W1]. After the general theory of abelian varieties was founded by Weil [W2], it was extended by Néron and Lang in 50s to a more general situation [L]. Second the notion of the canonical height on abelian varieties was developed by Néron, Tate and Manin in 60s [N2, T1, T3, M1; cf. L, Se2]. A more geometric method, based on the theory of elliptic surfaces [K], was used by the author [S1] and Cox-Zucker [CZ] in 70s.

Certainly the lattice-theoretic feature of Mordell-Weil groups has appeared in various works, e.g. in the statement of the Birch-Swinnerton-Dyer conjecture. But the idea to view them as lattices seems relatively new, and a systematic study of Mordell-Weil lattices has been done only very recently by the author [S3, S4, S5, S6, S7] and independently by N. Elkies [E], in the case of elliptic curves over function fields, or equivalently, in the case of elliptic surfaces. Actually the general scope of the theory of Mordell-Weil lattices should cover also the case of algebraic surfaces with higher genus fibration and the case of arithmetic surfaces as well (cf. [F, H] for the latter). But, in this talk, we shall restrict our attention to the case of elliptic surfaces.

In Part I, we review the definition and the basic results on Mordell-Weil lattices, and in Part II, we consider the Galois representations and algebraic equations arising from them, and discuss some applications. In the case of rational elliptic surfaces, we obtain some new insight into various problems related to E_6, E_7, E_8—a mysteriously rich subject in Mathematics. In particular, we can answer some questions raised by Weil [W3] and Manin [M2] concerning the Galois representation arising from the 27 lines on a cubic surface.

Proceedings of the International Congress
of Mathematicians, Kyoto, Japan, 1990
© The Mathematical Society of Japan, 1991

Acknowledgement. I would like to thank Professor J-P. Serre for many valuable comments on my work. I have had helpful discussions or correspondences with T. Ekedahl, N. Elkies, D. Gross, E. Horikawa, K. Oguiso, K. Saito, F. Sato, I. Shimada and P. Slodowy, and I thank them all.

Part I

1. Definition of Mordell-Weil Lattices

Let $K = k(C)$ be the function field of a smooth projective curve C over an algebraically closed field k. Let E be an elliptic curve defined over K, and let $E(K)$ denote the group of K-rational points of E, with the origin $O \in E(K)$. For what follows, the main reference is [S7].

Our basic tool is the associated elliptic surface $f : S \to C$ (the Kodaira-Néron model of E/K) and the intersection theory on S. Recall [K, N1, T2] that S is a smooth projective surface over k and f is a relatively minimal fibration with the generic fibre E. We can naturally identify the global sections of $f : S \to C$ with the K-rational points of E, and so $E(K)$ denotes the group of sections. For $P \in E(K)$, (P) will denote the image curve of $P : C \to S$.

We assume throughout that $(*)$ f is not smooth, i.e., there is at least one singular fibre. Then $E(K)$ is finitely generated by the Mordell-Weil theorem. On the other hand, let $N = NS(S)$ be the Néron-Severi group of S, i.e. the group of divisors modulo algebraic equivalence. It is a free module of finite rank under the assumption $(*)$, and it becomes an indefinite integral lattice with respect to the intersection pairing $(D \cdot D')$. Let T be the sublattice of N generated by the zero section (O), a fibre F and the irreducible components of fibres. Then T is a direct sum:

$$T = \langle (O), F \rangle_{\mathbb{Z}} \oplus V, \qquad V = \bigoplus_{v \in R} T_v \tag{1.1}$$

where $R = \{v \in C \,|\, F_v = f^{-1}(v) \text{ reducible}\}$ and T_v is generated by the irreducible components of F_v other than the identity component; each T_v is a root lattice of type A, D, E up to the sign. Let $L = T^{\perp}$ be the orthogonal complement of T in N. Then L is a negative-definite even integral lattice, as follows from the Hodge index theorem, the adjunction formula and the canonical bundle formula; see [S7, §7]. We call T or L the *trivial* or *essential* sublattice of $NS(S)$.

Now the map $P \to (P) \bmod T$ induces an isomorphism:

$$E(K) \simeq NS(S)/T \tag{1.2}$$

(cf. [S7, Th.1.3]). By splitting this isomorphism, we get a unique homomorphism ([S7, §8])

$$\varphi : E(K) \to NS(S) \otimes \mathbb{Q} \tag{1.3}$$

such that

$$\varphi(P) \equiv (P) \bmod T \otimes \mathbb{Q}, \qquad \mathrm{Im}(\varphi) \perp T \quad \text{and} \quad \mathrm{Ker}(\varphi) = E(K)_{\mathrm{tor}}. \qquad (1.4)$$

Theorem 1.1. *For* $P, P' \in E(K)$, *let*

$$\langle P, P' \rangle = -(\varphi(P) \cdot \varphi(P')). \qquad (1.5)$$

Then it defines the structure of a positive-definite lattice on $E(K)/E(K)_{\mathrm{tor}}$, *called the* Mordell-Weil lattice *of* E/K *or of* $f : S \to C$.

The explicit formula of the *height pairing* (1.5) is as follows:

$$\langle P, P' \rangle = \chi + (PO) + (P'O) - (PP') - \sum_{v \in R} \mathrm{contr}_v(P, P'). \qquad (1.6)$$

Here χ is the arithmetic genus of S (a positive integer under (*)). $(PQ) = ((P) \cdot (Q))$ is the intersection number of the curves (P) and (Q). The local contribution term $\mathrm{contr}_v(P, P')$ is a non-negative rational number, which is non-zero only if both P and P' pass through non-identity components of F_v; it is then determined by the type of F_v and the position of the components hit by P, P' ([S7, §8]).

On the other hand, let $E(K)^0$ be the subgroup of $E(K)$ consisting of those sections which pass through the identity component of every fibre. It is a torsion-free subgroup of finite index in $E(K)$.

Theorem 1.2. *With respect the height pairing,* $E(K)^0$ *is a positive-definite even integral lattice, which is isomorphic via the map* φ *to the opposite lattice* L^- *of the essential sublattice* L. *It will be called the* narrow Mordell-Weil lattice *of* E/K *or of* $f : S \to C$.

Observe that

$$\langle P, P' \rangle = \chi + (PO) + (P'O) - (PP') \in \mathbb{Z} \qquad \text{if } P \text{ or } P' \in E(K)^0, \qquad (1.7)$$

$$\langle P, P \rangle = 2\chi + 2(PO) \geq 2\chi \qquad \text{for any } P \in E(K)^0, P \neq 0. \qquad (1.8)$$

The former shows that $E(K)/E(K)_{\mathrm{tor}}$ is contained in the dual lattice M^* of $M = E(K)^0$.

Theorem 1.3. *Assume further that* $NS(S)$ *is unimodular. Then* $E(K)/E(K)_{\mathrm{tor}}$ *is equal to the dual lattice* M^* *of* $M = E(K)^0$. *Moreover we have* $[M^* : M] = \det M = (\det T)/|E(K)_{\mathrm{tor}}|^2$ *(see* [S7, Th.9.1]*).*

2. Basic Invariants of MWL

By the very definition, the Mordell-Weil lattices (abbreviated henceforth as MWL) are of Diophantine nature and have quite rich structure, as will be seen later. But, apart from that, they provide a purely algebraic method for constructing lattices of some interest.

In general, given a lattice L, one would like to know its basic invariants such as the rank $\mathrm{rk}(L)$, the determinant $\det(L)$ and, if L is positive-definite (which we assume now), the minimal norm $\mu(L)$; recall (cf. [CS] for what follows) that $\det(L) = |\det(\langle x_i, x_j \rangle)|$ for any $\mathbb{Z}$-basis $\{x_i\}$ of L, and $\mu(L) = \text{Min}(\langle x, x \rangle | x \in L - \{0\})$. Now a lattice L of rank r gives rise to a sphere packing in $L \otimes \mathbb{R} = \mathbb{R}^r$ by spheres of radius $\varrho_L = \frac{1}{2}\sqrt{\mu(L)}$. The density $\varDelta(L)$ is the ratio of the volume of one sphere to that of the fundamental domain of L, i.e. $\sqrt{\det(L)}$. The quantity $\delta(L) = \varDelta(L)/\text{vol(unit sphere)}$ is called the center density: we have $\delta(L) = (\sqrt{\mu(L)}/2)^r/\sqrt{\det(L)}$.

As for Mordell-Weil lattices, it follows from the results of §1 that their invariants can be expressed in terms of geometric data on the associated elliptic surface:

Theorem 2.1. *Let $M = E(K)^0$ be the narrow MWL of E/K. Then*

$$r = \text{rk}(M) = \varrho(S) - 2 - \sum_{v \in R} (m_v - 1), \tag{2.1}$$

where $\varrho(S) = \text{rk } NS(S)$ is the Picard number of S and m_v is the number of irreducible components of the singular fibre F_v. Further we have

$$\det M = \det NS(S) \cdot v^2/\det T, \quad v = [E(K):E(K)^0]. \tag{2.2}$$

$$\mu(M) = 2\chi + \text{Min}\{(PO)|P \in E(K)^0, P \neq 0\} \geq 2\chi. \tag{2.3}$$

It is not so easy in general to evaluate the rank and det, though we have a good lower bound 2χ for μ. In particular, it is a nontrivial problem to find E/K with large rank. A notable exception to this is the results of Shafarevich and Tate[ShT] and of the author [S2] that the rank can be arbitrarily large in case $p = \text{char}(K) > 0$; the proof uses supersingular Fermat curves [ShT] or supersingular Fermat surfaces [S2]. From the viewpoint of MWL, Elkies and the author have independently studied these cases and obtained, among others, the following (see [E] and [S4], cf. [G]):

Theorem 2.2. *Let E be the elliptic curve $y^2 = x^3 + t^{q+1} + 1$ over $K = k_1(t)$, k_1 any extension of the finite field of q^2 elements with $q = p^e \equiv -1 \pmod 6$. Then $E(K) = E(K)^0$ is an even integral lattice with the invariants:*

$$r = 2q - 2, \det = q^{2(\chi-1)}/b, \mu = 2\chi \quad (\chi = (q+1)/6), \tag{2.4}$$

where b is a positive integer (an even power of p) such that

$$b = 1 \; (e = 1), \quad b \geq p^{2(e-1)((p-5)/6)^e} \; (e > 1) \tag{2.5}$$

in which equality holds if $e = 3$. The integer b is equal to the order of the Shafarevich-Tate group of E/K for $k_1 = \mathbb{F}_{q^2}$ or of the Brauer group of S/k_1 (cf. [T1]). In particular, we have

$$\delta = \sqrt{b} \cdot ((q+1)/12)^{q-1}/q^{(q-5)/6}. \tag{2.6}$$

N.B. This gives denser sphere packings than previously known ones in certain dimensions; for instance, for $q = 41$, we have $r = 80$ and $\log_2 \delta = 40.14$, the previous record being 36 (cf. [CS, Table 1.3], [E], [S4, Ex.1.3]).

3. Rational Elliptic Surfaces

With the notation of § 1, suppose that S is a *rational* elliptic surface, i.e. birational to $\mathbb{P}^2$. Then we have $C = \mathbb{P}^1$, $K = k(t)$, $\chi = 1$, and $N = NS(S)$ is unimodular of rank $\varrho = 10$. Hence $r \leq 8$ by (2.1).

The structure of the MWL and MW group of a rational elliptic surface is completely determined by the reducible singular fibres. Namely, look at the embedding of lattices, deduced from (1.1):

$$V = \bigoplus_{v \in R} T_v \subset H = \langle (O), F \rangle^{\perp} \subset N. \tag{3.1}$$

Note that H is a negative-definite even unimodular lattice of rank 8, i.e. the root lattice E_8 up to the sign, and V is a direct sum of root sublattices (cf. [B, CS] for root lattices). Theorem 1.3 gives:

Theorem 3.1. *For a rational elliptic surface, the narrow MWL is isomorphic to the orthogonal complement of V in E_8, and the MWL is its dual lattice. The torsion subgroup of $E(K)$ is V'/V, where V' is the primitive closure of V in E_8, i.e. $V' = V \otimes \mathbb{Q} \cap E_8$. Thus we have*

$$E(K) \simeq (V^{\perp})^* \oplus (V'/V), \qquad E(K)^0 \simeq V^{\perp}, \tag{3.2}$$

where the isomorphism preserves the pairing. (See [S7, Th.10.3]).

The structure theorem in the case of rank $r = 8, 7$ or 6 is:

Theorem 3.2. *The narrow MWL $E(K)^0$ is the root lattice E_8, E_7 or E_6, according to whether $f : S \to \mathbb{P}^1$ has (i) no reducible fibres, (ii) only one reducible fibre of type I_2 or III ([K]), or (iii) only one reducible fibre of type I_3 or IV. Further $E(K)$ is torsion-free and isomorphic to E_8, E_7^* or E_6^* accordingly. (Cf.* [S7, Th.10.4]).

More generally, we have a complete structure theorem by using Dynkin's classification of $V \subset E_8$ (see [OS]). As a consequence, we obtain an effective result on the generators:

Theorem 3.3. *The Mordell-Weil group of any rational elliptic surface is generated by the sections P with $(PO) = 0$, hence by those P with $\langle P, P \rangle \leq 2$. If E/K is defined by the Weierstrass equation*

$$y^2 + a_1(t)xy + a_3(t)y = x^3 + a_2(t)x^2 + a_4(t)x + a_6(t), \tag{3.3}$$

with $a_i(t) \in k[t]$, $\deg a_i(t) \leq i$, then $E(K)$ is generated by rational points $P = (x, y)$ of the form:

$$x = gt^2 + at + b, \qquad y = ht^3 + ct^2 + dt + e \qquad (g, a, \ldots, e \in k). \tag{3.4}$$

There are at most 240 such rational points. (Cf. [S7, Th.10.10]).

Part II

4. Galois Representations Arising from MWL

From now on, we consider the following situation. Let k_0 be a perfect field and k its algebraic closure; let $G = \text{Gal}(k/k_0)$. Let C be an absolutely irreducible smooth projective curve defined over k_0 and $K = k(C)$ or $K_0 = k_0(C)$ the function field of C over k or k_0. Now let E denote an elliptic curve defined over K_0 and consider E/K. The associated elliptic surface $f : S \to C$ is now defined over k_0.

Obviously the Galois group G acts on $E(K)$, and $E(K_0)$ coincides with $E(K)^G$, the subgroup of G-invariants. First we note:

Lemma 4.1. (i) *The map φ defined by* (1.3) *is G-equivariant.* (ii) *The height pairing* (1.6) *on $E(K)$ is stable under G.* (*Cf.* [S7, Prop.8.13]).

Therefore we get a Galois representation on the MW group

$$\varrho : G = \text{Gal}(k/k_0) \to \text{Aut}(E(K), \langle \ , \ \rangle) = \text{a finite group} \qquad (4.1)$$

and its variant on the MWL or the narrow MWL of E/K, say M,

$$\varrho' : G \to \text{Aut}(M) \subset \text{GL}_r(\mathbb{Z}) \qquad (r = \text{rk}(M)). \qquad (4.2)$$

Let $\mathcal{K}/k_0$ be the extension corresponding to $\text{Ker}(\varrho)$; equivalently, $\mathcal{K}$ is the smallest extension of k_0 such that $E(\mathcal{K}(C)) = E(K)$. By definition, $\mathcal{K}/k_0$ is a finite Galois extension such that

$$\text{Gal}(\mathcal{K}/k_0) = \text{Im}(\varrho). \qquad (4.3)$$

The basic problem on the Galois representation (4.1) is this:

Problem 4.2. Determine the image of ϱ. In particular, we ask: (i) How big or (ii) how small can $\text{Im}(\varrho)$ be?

To study this will be a main theme in the subsequent sections.

Remark 4.3. The Galois representations ϱ or ϱ' arising from MWL are quite different from those arising from the torsion points of an elliptic curve or an abelian variety (e.g. the Tate modules), because we are dealing with points of *infinite order* here!

Next we note the connection to the Hasse zeta functions (cf. [M2, Se1]). We state the result in the simplest situation, which is related to the questions of Weil [W3] and Manin [M2]; see Theorem 8.4.

Proposition 4.4. *Let $k_0 = \mathbb{Q}$. Assume that S is a rational elliptic surface (over $\overline{\mathbb{Q}}$) and that the trivial lattice T is a trivial G-module. Then the Hasse zeta function of S over $\mathbb{Q}$ is given by*

$$\zeta(S/\mathbb{Q}, s) = \zeta(s)\zeta(s - 1)^{10 - r}\zeta(s - 2) \cdot L(s - 1, \varrho, \mathcal{K}/\mathbb{Q}), \qquad (4.4)$$

where $\zeta(s)$ is the Riemann zeta function and $L(s, \varrho, \mathcal{K}/\mathbb{Q})$ is the Artin L-function attached to the representation ϱ of $\mathrm{Gal}(\mathcal{K}/\mathbb{Q})$ on the MWL.

Proof. By [Se1], the Hasse zeta function is determined by the l-adic representation of G on $H^i_{\text{ét}}(S, \mathbb{Q}_l)$ $(0 \leq i \leq 4)$. The other cases being trivial, consider the case $i = 2$. For a rational surface, the cycle map induces a G-isomorphism

$$NS(S) \otimes \mathbb{Q}_l \simeq H^i_{\text{ét}}(S, \mathbb{Q}_l(1)). \tag{4.5}$$

On the other hand, the results in §1 imply

$$NS(S) \otimes \mathbb{Q} = L \otimes \mathbb{Q} \oplus T \otimes \mathbb{Q} \quad \text{and} \quad L \otimes \mathbb{Q} \simeq E(K) \otimes \mathbb{Q}, \tag{4.6}$$

where the latter is a G-isomorphism by Theorem 1.2 and Lemma 4.1. $\qquad\square$

(N.B. We have *exact* equality in (4.4), as pointed out by J-P. Serre.)

5. Algebraic Equations Arising from MWL

Given an elliptic curve E/K_0, take a singular fibre $F_v = f^{-1}(v)$ and a G-stable finite subset I of $E(K)$. Then we define

$$\Phi(X) = \Phi(E/K_0, v, I; X) = \prod_{P \in I} (X - \mathrm{sp}'_v(P)) \in k[X]. \tag{5.1}$$

Here, sp'_v is the *specialization map* at v, defined as follows. For any $P \in E(K)$, the section (P) intersects the fibre F_v at a unique smooth point of F_v; call it $\mathrm{sp}_v(P)$. The smooth part $F_v^{\#}$ of F_v is an algebraic group over k, which is the product of $\mathbb{G}_a$ or $\mathbb{G}_m$ by a finite group (at least if char $k \neq 3$; cf. [K, N1, T2]), and the map

$$\mathrm{sp}_v : E(K) \to F_v^{\#}(k) \tag{5.2}$$

is a G-equivariant homomorphism. We denote by $\mathrm{sp}'_v(P)$ the projection of $\mathrm{sp}_v(P)$ to the factor $\mathbb{G}_a$ or $\mathbb{G}_m$.

As a typical example of the set I, we have (by Lemma 4.1)

$$I_n = \{P \in E(K) | \langle P, P \rangle = n\} \qquad (n \geq 0). \tag{5.3}$$

Proposition 5.1. *Assume $v \in C(k_0)$. Then $\Phi(X)$ has coefficients in k_0, i.e. $\Phi(X) \in k_0[X]$. The splitting field of the algebraic equation $\Phi(X) = 0$ over k_0, say $\mathcal{K}'$, is contained in $\mathcal{K}$ defined in §4, and we have $\mathcal{K}' = \mathcal{K}$ if 1) sp'_v is injective and 2) I contains generators of $E(K)$.*

Proof. Immediate by Galois theory. $\qquad\square$

Remark 5.2. The algebraic equations (5.1) look somewhat analogous to the classical "division equations" arising from torsion points on an elliptic curve, whose study goes back to Abel and Galois ([ST, Ch.1]). As noted in Remark 4.3, we are concerned here with points of *infinite order* on an elliptic curve E, and in general the algebraic equations will lead to highly *non-abelian* extensions (see Theorems 6.1, 7.1).

6. Generic Galois Representation and Algebraic Equation of Type E_r

To illustrate the nature of the Galois representations and the algebraic equations arising from MWL, we consider some special cases. Let $E = E_\lambda$ be the elliptic curve, defined by one of the following equations, over $K_0 = k_0(t)$ where $k_0 = \mathbb{Q}(\lambda) = \mathbb{Q}(p_i, q_j)$.

$$(E_8) \qquad y^2 = x^3 + x\left(\sum_{i=0}^{3} p_i t^i\right) + \left(\sum_{i=0}^{3} q_i t^i + t^5\right)$$

$$\lambda = (p_0, p_1, p_2, p_3, q_0, q_1, q_2, q_3) \in \mathbb{A}^8$$

$$(E_7) \qquad y^2 = x^3 + x(p_0 + p_1 t + t^3) + \left(\sum_{i=0}^{4} q_i t^i\right)$$

(6.1)

$$\lambda = (p_0, p_1, q_0, q_1, q_2, q_3, q_4) \in \mathbb{A}^7$$

$$(E_6) \qquad y^2 = x^3 + x\left(\sum_{i=0}^{2} p_i t^i\right) + \left(\sum_{i=0}^{2} q_i t^i + t^4\right)$$

$$\lambda = (p_0, p_1, p_2, q_0, q_1, q_2) \in \mathbb{A}^6.$$

Let $f : S_\lambda \to \mathbb{P}^1$ be the associated elliptic surface. The fibre $f^{-1}(\infty)$ is an additive singular fibre of type II, III or IV according to the case (E_r) for $r = 8, 7$ or 6 and $f^{-1}(\infty)^\# \simeq \mathbb{G}_a \times \mathbb{Z}/(9 - r)$ [K, N1, T2]. Assume that λ satisfies the condition:

$$\text{every fibre of } f \text{ over } t \neq \infty \text{ is irreducible.} \tag{6.2}$$

Then the MWL $E(k(t))$ is isomorphic to E_r^* by Theorem 3.2, and we get the Galois representation

$$\varrho = \varrho_\lambda : \text{Gal}(k/k_0) \to \text{Aut}(E_r^*) = \text{Aut}(E_r).$$

Recall that $\text{Aut}(E_r) = W(E_r)$ for $r = 8$ or 7, and $W(E_r)\{\pm 1\}$ for $r = 6$, where $W(E_r)$ denotes the Weyl group of type E_r (cf. [B, CS]). In any case, we have $\text{Im}(\varrho_\lambda) \subset W(E_r)$. It should be noticed that $W(E_r)$ is almost a *simple* group: with the notation of simple groups in Atlas [C], $W(E_6)$ contains $U_4(2)$ (order $2^6 3^4 5$) as a subgroup of index 2, $W(E_7)/\{\pm 1\} \simeq S_6(2)$ ($= \text{Sp}_6(\mathbb{F}_2)$; order $2^9 3^4 5 \cdot 7$), and $W(E_8)$ has a subgroup H of index 2 such that $H/\{\pm 1\} \simeq O_8^+(2)$ (order $2^{12} 3^5 5^2 7$).

Now we assume that λ is *generic*, i.e. p_i, q_j are algebraically independent over $\mathbb{Q}$. Then (6.2) obviously holds.

Theorem 6.1. *Let λ be generic over $\mathbb{Q}$. Then (i) the image of the Galois representation ϱ_λ is the full Weyl group $W(E_r)$:*

$$\text{Im}(\varrho_\lambda) = W(E_r). \tag{6.3}$$

(ii) *Let $\mathscr{K}_\lambda/k_0$ be the Galois extension corresponding to $\text{Ker}(\varrho_\lambda)$, and let $\{P_1, \ldots, P_r\}$ be a basis of $E(k(t)) \simeq E_r^*$ consisting of minimal vectors. Further let $u_i = \text{sp}'_\infty(P_i) \in \mathscr{K}_\lambda \subset k$. Then $u_1, \ldots, u_r$ are algebraically independent over $\mathbb{Q}$, and we have*

$$\mathscr{K}_\lambda = k_0(u_1, \ldots, u_r) = \mathbb{Q}(u_1, \ldots, u_r) \tag{6.4}$$

$$\mathrm{Gal}(\mathbb{Q}(u_1, \ldots, u_r)/\mathbb{Q}(p_i, q_j)) = W(E_r). \tag{6.5}$$

(iii) $W(E_r)$ *acts on the vector space* $\sum \mathbb{Q}u_i$ *and hence on the polynomial ring* $\mathbb{Q}[u_1, \ldots, u_r]$, *and the ring of the invariants is:*

$$\mathbb{Q}[u_1, \ldots, u_r]^{W(E_r)} = \mathbb{Q}[p_i, q_j]. \tag{6.6}$$

In particular, p_i *and* q_j *form the fundamental invariants of the Weyl group* $W(E_r)$ *and we can explicitly write*

$$p_i \text{ or } q_j = J_d(u_1, \ldots, u_r), \tag{6.7}$$

where J_d *denotes a* $W(E_r)$-*invariant of degree* d, $d \in \{2, 5, 6, 8, 9, 12\}$, $\{2, 6, 8, 10, 12, 14, 18\}$ *or* $\{2, 8, 12, 14, 18, 20, 24, 30\}$ *for* $r = 6, 7$ *or* 8.

The *universal polynomial of type* E_r is defined as a special case of (5.1), and it has degree $N = 27, 56$ or 240 for $r = 6, 7$ or 8:

$$\Phi_{E_r}(X, \lambda) = \Phi(E_\lambda/\mathbb{Q}(\lambda)(t), \infty, I; X) = \prod_{P \in I} (X - \mathrm{sp}'_\infty(P)) \in \mathbb{Q}(\lambda)[X]. \tag{6.8}$$

Here $I = I_\mu$ is the set of minimal vectors for $r = 8$ or 7 (thus $\mu = 2$ or $3/2$, and $\#I = 240$ or 56 accordingly). For $r = 6$, we take $I = I_\mu^\pm$, either one of 2 orbits of $W(E_6)$ in I_μ ($\mu = 4/3$, $\#I = 27$), and modify the right hand side of (6.8) by $\prod(X - (-2)\,\mathrm{sp}'_\infty(P))$. Then we have

Theorem 6.2. *For* λ *generic,* $\Phi_{E_r}(X, \lambda)$ *is a monic irreducible polynomial in* X *with coefficients in the polynomial ring* $\mathbb{Z}[\lambda] = \mathbb{Z}[p_i, q_j]$, *and its splitting field over* $\mathbb{Q}(\lambda)$ *is equal to* $\mathscr{K}_\lambda$ *in Theorem 6.1.*

Considering the elliptic curve (6.1) over the field $\mathscr{K}_\lambda(t) = \mathbb{Q}(u_1, \ldots, u_r)(t)$, we have the following:

Theorem 6.3. *Suppose* $u_1, \ldots, u_r$ *are algebraically independent over* $\mathbb{Q}$, *and define* $\lambda = (p_i, q_j)$ *by (6.7) and the elliptic curve* E_λ *by (6.1). Then the Mordell-Weil group* $E(\mathbb{Q}(u_1, \ldots, u_r)(t)) \simeq E_r^*$ *is of rank* r, *and has a basis* $\{P_1, \ldots, P_r\}$ *such that* $\mathrm{sp}'_\infty(P_i) = u_i$. *More explicitly, we have* $P_i = (x, y)$ *where* x, y *are polynomials in* t *of the form (3.4) which are given as follows:*

(i) *If* $r = 8$, *then* $g = u_i^{-2}$, $h = u_i^{-3}$ *and*

$$a, \ldots, e \in \mathbb{Q}[u_1, \ldots, u_8][u_i^{-1}] \cap \mathbb{Q}(p_0, \ldots, q_3)(u_i).$$

(ii) *If* $r = 7$, *then* $c = u_i$ *and*

$$a, b, d, e \in \mathbb{Q}[u_1, \ldots, u_7] \cap \mathbb{Q}(p_0, \ldots, q_4)(u_i), \quad g = h = 0.$$

(iii) *If* $r = 6$, *then* $a = -2 \cdot u_i$, $c = 1$, *and*

$$b, d, e \in \mathbb{Q}[u_1, \ldots, u_6] \cap \mathbb{Q}(p_0, \ldots, q_2)(u_i), \quad g = h = 0.$$

Let us briefly sketch the proof of the above theorems (for details, we refer to [S6, §8]). By Theorem 3.2, we know the structure of the MWL $E_\lambda(k(t))$. For instance, if $r = 8$, it is E_8 and there exist 240 minimal vectors $P = (x, y)$ of the form (3.4). Substitute (3.4) into the equation (E_8), and we obtain polynomial relations among the coefficients $a, \ldots, g, h$. Then the successive elimination leads to a monic relation of $u = g/h = \mathrm{sp}_\infty(P)$ with coefficients in $\mathbb{Z}[\lambda]$, of degree 240, which must coincide with the universal polynomial of type E_8. This implies (6.6) and (6.7) by comparison of the coefficients of u^d, d being the degree of fundamental invariants. The rest of Theorems 6.1 and 6.2 follow immediately from this, and Theorem 6.3 from a closer look at the elimination process. □

Remark 6.4. Actually the proof shows that, in the above theorems, the ground field $\mathbb{Q}$ can be replaced by any field F, provided that its characteristic is different from a few primes which come into the denominators of the expression (6.7); they are $\{2, \ldots, 7\}$, $\{2, \ldots, 11, 29, 1229\}$ or $\{2, \ldots, 19, 41, 61, 199\}$ for $r = 6, 7$ or 8.

7. Galois Representations of Type E_r over $\mathbb{Q}$

We can compare the results of §6 with the classical theory of the generic algebraic equation of degree n. Let

$$F(X, \varepsilon) = X^n + \varepsilon_2 X^{n-2} + \cdots + (-1)^n \varepsilon_n \qquad (\varepsilon_i : \text{alg.indep.}/\mathbb{Q})$$

be such an equation, normalized so that the sum of the roots x_i is 0. Let $\mathscr{K}$ be the splitting field of $F(X, \varepsilon)$ over $k_0 = \mathbb{Q}(\varepsilon_2, \ldots, \varepsilon_n)$. Then

 (i) $\mathscr{K} = k_0(x_1, \ldots, x_n) = \mathbb{Q}(x_2, \ldots, x_n)$.
 (ii) $\mathrm{Gal}(\mathbb{Q}(x_2, \ldots, x_n)/\mathbb{Q}(\varepsilon_2, \ldots, \varepsilon_n)) = \mathfrak{S}_n = W(A_{n-1})$.
 (iii) $\mathbb{Q}[x_2, \ldots, x_n]^{W(A_{n-1})} = \mathbb{Q}[\varepsilon_2, \ldots, \varepsilon_n]$.
 (iv) $\varepsilon_i = $ the fundamental invariants of $\mathfrak{S}_n$.

Thus Theorems 6.1 and 6.2 (based on the theory of MWL) give a complete analogy for the exceptional type E_r ($r = 6, 7, 8$) of what the theory of generic equation does to the classical type A_{n-1}.

The latter theory has a standard application to number theory: (a) construction of Galois extensions of $\mathbb{Q}$ with Galois group $\mathfrak{S}_n$ (via specialization "downstairs" $(\varepsilon_i) \to (a_i) \in \mathbb{Q}^{n-1}$ and Hilbert's irreducibility theorem (cf. [L, Se2]), and (b) construction of an algebraic equation with the prescribed roots (via specialization "upstairs" $(x_i) \to (b_i) \in \mathbb{Q}^n$), which is indeed trivial.

In what follows, we describe similar application of our theory.

First consider the specialization "downstairs": $\lambda \to \lambda^0 \in \mathbb{Q}^r$. We keep the same notation as in §6, except that we write λ for λ^0 (to simplify printing) so that $\mathscr{K}_\lambda/\mathbb{Q}$ is now a Galois extension.

Theorem 7.1. Fix $r = 6, 7$ or 8, and let

$$\Lambda = \Lambda_r = \{\lambda = (p_i, q_j) \in \mathbb{Q}^r \,|\, \mathrm{Gal}(\mathscr{K}_\lambda/\mathbb{Q}) = W(E_r)\}. \tag{7.1}$$

Then Λ is a Zariski dense subset of $\mathbb{Q}^r$ (more precisely, it is the complement of a "thin set" in $\mathbb{Q}^r$ in the sense of [Se2]). There exist infinitely many $\{\lambda\} \subset \Lambda$ such that $\{\mathcal{K}_\lambda\}$ are linearly disjoint over $\mathbb{Q}$.

Proof. Immediate from Theorem 6.1 (esp. (6.5)) by applying Hilbert's irreducibility theorem [Se2, Ch.10].   □

We note that, for each $\lambda \in \Lambda$, we have an elliptic curve $E_\lambda/\mathbb{Q}(t)$ with the MWL $E_\lambda(\overline{\mathbb{Q}}(t)) = E_\lambda(\mathcal{K}_\lambda(t)) \simeq E_r^*$ and $E_\lambda(\mathbb{Q}(t)) = \{0\}$, and a rational elliptic surface S_λ defined over $\mathbb{Q}$, having the Hasse zeta function (4.4) with non-abelian Artin L-function $L(s, \varrho_\lambda, \mathcal{K}_\lambda/\mathbb{Q})$.

Theorem 7.2. *Every Galois extension of $\mathbb{Q}$ with Galois group $W(E_r)$ is obtained as $\mathcal{K}_\lambda$ for some $\lambda \in \Lambda$. In other words, every such extension arises from the MWL of the elliptic curve $E_\lambda/\mathbb{Q}(t)$ for some $\lambda \in \mathbb{Q}^r$.*

Proof. Let $K/\mathbb{Q}$ be any extension with Galois group $W(E_r)$. Viewed as a $W(E_r)$-module, K is equivalent to the regular representation, and hence it contains a subspace, say U, isomorphic to $E_r \otimes \mathbb{Q}$ (a natural irreducible representation of $W(E_r)$). Fixing such an isomorphism, choose a $\mathbb{Z}$-basis $\{u_i\}$ of $E_r^* \subset U$ and define $\lambda = (p_i, q_j)$ by (6.7). Then we have $\lambda \in \mathbb{Q}^r$, since p_i, q_j are $W(E_r)$-invariant. The splitting field $\mathcal{K}_\lambda$ of $\Phi_{E_r}(X, \lambda) \in \mathbb{Q}[X]$ is contained in K and contains U. We claim: $\mathcal{K}_\lambda = K$. Indeed, any $\sigma \in \mathrm{Gal}(K/\mathcal{K}_\lambda)$ acts trivially on $U \subset \mathcal{K}_\lambda$. But U is a faithful representation of $W(E_r)$. Hence $\sigma = 1$, which proves the claim. (Our original proof of the claim used the structure of $W(E_r)$ (cf. § 6). We owe the above simplification to T. Ekedahl).   □

We can also prove a formal analogue of Tate's conjecture for abelian varieties Let $\lambda = (p_i, q_j)$, $\lambda' = (p_i', q_j') \in \Lambda_r$ for $r = 6$ or 7.

Theorem 7.3. *The following conditions are equivalent:*
   (i) *The Galois representations ϱ_λ and $\varrho_{\lambda'}$ are equivalent.*
   (ii) *$\mathcal{K}_\lambda = \mathcal{K}_{\lambda'}$ and the roots $\{u_i\}$ and $\{u_i'\}$ of the polynomials $\Phi(X, \lambda)$ and $\Phi(X, \lambda')$ defined by (6.8) are related (up to ordering) by*

$$u_i' = A(u_i) \text{ (all } i) \text{ for some } A(X) = \sum a_v X^v \in \mathbb{Q}[X].$$

   (iii) *$\mathcal{K}_\lambda = \mathcal{K}_{\lambda'}$ and p_i', q_j' can be expressed by certain polynomials in p_i, q_j with $\mathbb{Q}$-coefficients.*

Proof. We consider the case $r = 6$. First assume (i). Then clearly $\mathcal{K}_\lambda = \mathcal{K}_{\lambda'}$, which we denote by K. There are exactly $n = 27$ minimal subfields M of K such that $[M : \mathbb{Q}] = n$, because $K/\mathbb{Q}$ has Galois group $W(E_r)$, which has so many maximal subgroups of index n (cf. [C]). Thus the n subfields $\mathbb{Q}(u_i)$ $(1 \le i \le n)$ of K give all such, and hence $\mathbb{Q}(u_i') = \mathbb{Q}(u_i)$ (all i) up to reordering. Let $u_1' = \sum_{v=0}^{n-1} a_v u_1^v$ $(a_v \in \mathbb{Q})$. Then by the action of $\sigma \in \mathrm{Gal}(K/\mathbb{Q})$ on both sides, we see (i) implies (ii). Conversely, if (ii) holds, then the permutations induced by σ on the sets $\{u_i\}$ and $\{u_i'\}$ are identical. Hence we have $\varrho_\lambda \sim \varrho_{\lambda'}$, i.e. (i). The equivalence of (ii) and (iii) follows from (6.6), (6.7).

The case $r = 7$ can be treated in the same way. (We omit the case $r = 8$ here, since more work is needed in addition to the information on the maximal subgroups in [C].) □

Our method allows us to find explicit examples.

Example 7.4 $(r = 6)$. Take $\lambda = (1, 0, 1; 1, 1, 1)$. The elliptic curve is:

$$E = E_\lambda : y^2 = x^3 + x(1 + t^2) + (1 + t + t^2 + t^4). \tag{7.2}$$

Then the MWL $E(\bar{\mathbb{Q}}(t)) \simeq E_6^*$. We claim: $\mathrm{Gal}(\mathcal{K}_\lambda/\mathbb{Q}) = W(E_6)$, i.e. $\lambda \in \Lambda_6$. Further $\mathcal{K}_\lambda$ is the splitting field of $F(X) = \Phi(X, \lambda)$ below:

$$F(X) = X^{27} + 12X^{25} + 60X^{23} + 264X^{21} + 1302X^{19} - 1344X^{18}$$
$$+ 3792X^{17} - 5568X^{16} + 22252X^{15} - 8832X^{14} + 57560X^{13}$$
$$+ 4224X^{12} + 39025X^{11} - 49728X^{10} + 88516X^9 - 50880X^8$$
$$- 95024X^7 - 150016X^6 - 35840X^5 - 16384X^4 - 104192X^3$$
$$+ 5888X - 4096.$$

This can be verified as follows. Look at the decomposition of $F(X)$ mod p into irreducible factors. For $p = 3$, it has 3 irreducible factors of degree 9 each, denoted symbolically by $(9)^3$. Similarly, for $p = 19$, it has $(2)(5)^3(10)$. Then the claim follows from:

Lemma 7.5 (Serre). *Let C_1 or C_2 be the conjugacy classes in $W(E_6)$ of the elements of order 9 or 10 with cycle type $(9)^3$ or $(2)(5)^3(10)$ in $\mathfrak{S}_{27}$ (cf. [Sw, p.57]). Suppose a subgroup H of $W(E_6)$ has the property that $H \cap C_i \neq \varnothing$ for $i = 1, 2$. Then $H = W(E_6)$.*

Proof. If not, take a maximal subgroup containing H. The possible orders of such are given in [C], but none of them are divisible by both 9 and 10 except one. But this last group does not contain any element of order 9. Hence a contradiction. □

(We can show further that the ramification occurs only at $p = 2, 137, 15784603$.)

Example 7.6 $(r = 7)$. Let $\lambda = (1, 1; 1, 1, 1, 0, 1)$ and

$$E = E_\lambda : y^2 = x^3 + x(1 + t + t^3) + (1 + t + t^2 + t^4). \tag{7.4}$$

Then the MWL $E(\bar{\mathbb{Q}}(t)) \simeq E_7^*$, and $\mathrm{Gal}(\mathcal{K}_\lambda/\mathbb{Q}) = W(E_7)$, i.e. $\lambda \in \Lambda_7$. The universal polynomial $F(X) = \Phi(X, \lambda)$ equals $f(Y)$ below $(Y = X^2)$, and the splitting field of $f(Y)$ over $\mathbb{Q}$ has the Galois group isomorphic to $W(E_7)/\{\pm 1\} \simeq S_6(2)$ (a simple group of order 1451520).

$$f(Y) = Y^{28} - 36Y^{27} + 594Y^{26} - 6084Y^{25} + 43935Y^{24} - 240192Y^{23}$$
$$+ 1039392Y^{22} - 3661764Y^{21} + 10681839Y^{20} - 26088660Y^{19}$$
$$+ 53894394Y^{18} - 95282532Y^{17} + 145821463Y^{16} - 194265660Y^{15}$$

$$+ 223728462 Y^{14} - 216948108 Y^{13} + 168475770 Y^{12} - 98229852 Y^{11}$$

$$+ 42796234 Y^{10} - 19590492 Y^9 + 14262444 Y^8 - 9949084 Y^7$$

$$+ 4609696 Y^6 - 1118808 Y^5 + 53521 Y^4 + 160 Y^3 + 4288 Y^2$$

$$- 5312 Y + 1024. \tag{7.5}$$

The proof is similar to Example 7.4. The decomposition type of $F(X)$ mod p is $(7)^8$ for $p = 47$, and $(3)^2(5)^4(15)^2$ for $p = 131$. This suffices to conclude $\mathrm{Gal}(\mathcal{K}_\lambda/\mathbb{Q}) = W(E_7)$ in view of [C]. □

In the above, we have taken $\mathbb{Q}$ as the ground field, but the same idea works in more general situation. Here is one such example.

Example 7.7 $(r = 6)$. Let $k_0 = \mathbb{F}_3(s)$ be the rational function field over the finite field $\mathbb{F}_3$. Then the following simple equation

$$F(X) = X^{27} + X^{12} - X^7 - X^2 - s^3 \in \mathbb{F}_3[s][X] \tag{7.6}$$

has the Galois group $W(E_6)$! It arises from the elliptic curve $E/k_0(t)$

$$y^2 = x^3 + xt + st + t^4 \tag{7.7}$$

having the MWL $E(\bar{k}_0(t)) = E_6^*$. This is proven by checking that $F(X)$ mod p has the conjugacy type C_1 or C_2 for the prime ideal $p = (s - 1)$ or $(s + 1)$ of $\mathbb{F}_3[s]$ and by applying Lemma 7.5.

8. Further Applications

We briefly mention other applications of Theorems 6.1, etc.

1) Construction of Elliptic Curves over $\mathbb{Q}(t)$ with Rank $r = 6, 7, 8$

We consider the analogy for E_r, mentioned before, of (b) writing down an algebraic equation with prescribed roots. By specializing the parameters "upstairs" (u_i) to some $(u_i^0) \in \mathbb{Q}^r$ in Theorems 6.1 or 6.3 in such a way that the condition (6.2) is satisfied, we obtain an elliptic curve $E = E_\lambda$ over $\mathbb{Q}(t)$, together with r rational points P_i. Observe that each P_i has the prescribed value u_i^0 as its essential parameter; e.g. $P_i = (-2u_i^0 t + \cdots, t^2 + \cdots)$ for $r = 6$. The condition (6.2) is equivalent to certain polynomial $\delta_0(u^0) \neq 0$ (see [S6]).

Theorem 8.1. *Assume $\delta_0(u^0) \neq 0$. Then the Mordell-Weil group $E(\mathbb{Q}(t)) \simeq E_r^*$ has rank $r\ (= 6, 7, 8)$, and $\{P_i\}$ forms a set of generators.*

See [S6] for the proof, as well as for explicit examples.

Corollary 8.2. *Given $E/\mathbb{Q}(t)$ as above, we have an infinite family $\{E^{(\tau)} | \tau \in \mathbb{Q} - \sum\}$ $(\#\sum < \infty)$ of elliptic curves over $\mathbb{Q}$ of rank $\geq r$, with r independent points $P_i^{(\tau)}$, by specializing $t \to \tau \in \mathbb{Q}$. Further*

$$\det(\langle P_i^{(\tau)}, P_j^{(\tau)} \rangle_{can}/h(\tau)) \to 1/2^r d \qquad \text{as } h(\tau) \to \infty, \tag{8.1}$$

where $\langle \ , \ \rangle_{can}$ is the canonical height and $h(\tau)$ the standard height (esp. $h(\tau) = \log|\tau|$ for $\tau \in \mathbb{Z}$), and $d = 1, 2$ or 3 for $r = 8, 7$ or 6.

This follows from a result of Néron, Silverman [Si], Tate [T3].

2) The 27 Lines on a Smooth Cubic Surface

In the equation (E_6) of (6.1), let $y = y' \pm t^2$ and $(x : y' : t : 1) = (X : Y : Z : W)$. Then we have a cubic surface $V = V_\lambda^\pm$ in $\mathbb{P}^3$:

$$Y^2 W \pm 2YZ^2 = X^3 + X(p_0 W^2 + p_1 ZW + p_2 Z^2) + q_0 W^3 + q_1 ZW^2 + q_2 Z^2 W. \tag{8.2}$$

It is smooth if and only if (6.2) holds. Under this assumption, the narrow MWL $E_\lambda(k(t))^0 \simeq E_6$ is isomorphic to the primitive part of $NS(V)$ ($= \langle \omega_V \rangle^\perp$ in the notation of [M2, Ch.4]). The 27 minimal sections of the form $P = (at + b, t^2 + dt + e)$ in the MWL $E_\lambda(k(t))$ are transformed into the 27 lines on $V = V_\lambda^+$ defined by the equation

$$X = aZ + bW, \qquad Y = dZ + eW. \tag{8.3}$$

(Similarly, the 27 sections $-P$ are mapped to the 27 lines on V_λ^-.)

Hence the universal polynomial $\Phi(X, \lambda)$ of type E_6 becomes the "algebraic equation of the 27 lines" on V_λ. As a consequence, we can easily deduce the following from our previous results.

Theorem 8.3. For $\lambda = (p_i, q_j)$ generic over any field F of char. $p \neq 2$, the Galois group of the 27 lines on the cubic surface V_λ over $F(\lambda)$ is equal to the full Weyl group $W(E_6)$.

Proof. This is immediate from Theorem 6.1 and Remark 6.4 if p is different from 3, 5, 7, in which case the assertion can be verified as in Example 7.7. (A similar result holds also for $p = 2$.) $\qquad\square$

Theorem 8.4. For $\lambda \in \Lambda_6 \subset \mathbb{Q}^6$, the Galois group of the 27 lines on V_λ over $\mathbb{Q}$ is $W(E_6)$, and the Hasse zeta function of $V_\lambda/\mathbb{Q}$ is given by

$$\zeta(V_\lambda/\mathbb{Q}, s) = \zeta(s)\zeta(s-1)\zeta(s-2)L(s-1, \varrho_\lambda, \mathcal{K}_\lambda/\mathbb{Q}). \tag{8.4}$$

Proof. Immediate from Theorem 7.1. $\qquad\square$

Example 8.5. We easily obtain an explicit example of a cubic surface over $\mathbb{Q}$ with the Galois group $W(E_6)$. For instance, Example 7.4 gives:

$$Y^2 W + 2YZ^2 = X^3 + XW^2 + XZ^2 + W^3 + ZW^2 + Z^2 W. \tag{8.5}$$

The 27 lines are defined over $\mathcal{K}_\lambda$ given by (7.3). Compare [Ek].

We believe that these results clarify some of Weil's statements at the end of his 1954 Congress address [W3]. Also we can essentially answer the problems raised by Manin [M2, Ch.4, 23.13] by translating Theorems 7.2 and 7.3 into the language of cubic surfaces.

We also have a systematic construction of cubic surfaces over $\mathbb{Q}$ such that all of the 27 lines (8.3) are defined over $\mathbb{Q}$.

Example 8.6. *Let V be the cubic surface*

$$Y^2W + 2YZ^2 = X^3 - X(78Z^2 + 59475W^2) + 18226Z^2W + 2848750W^3. \quad (8.6)$$

The 27 lines are given by

$$(X, Y) = (411W, -6916W), (-2Z + 575W, 74Z - 12600W),$$

$$(-4Z + 275W, 124Z - 2700W), (-6Z + 15W, 126Z + 1400W),$$

$$(-8Z - 85W, 56Z + 2700W), (-10Z - 49W, -110Z - 2376W), \quad (8.7)$$

etc.; the above 6 correspond to the generators of $E(\mathbb{Q}(t)) \simeq E_6^*$.

3) The 28 Double Tangents to a Plane Quartic Curve

This classical topic is closely connected with the case (E_7). In the same way as in 2) above, we can prove the following:

  (i) The Galois group of 28 double tangents on a generic quartic curve over any prime field is $W(E_7)/\{\pm 1\} \simeq S_6(2)$.

  (ii) Construction of a quartic over $\mathbb{Q}$ with the Galois group $\simeq S_6(2)$.

  (iii) Construction of such with all double tangents defined over $\mathbb{Q}$. As a by-product, we can also prove

  (iv) the rationality of the moduli space of plane quartic curves (genus 3) with a flex (over any base field of char $\neq 3$).

4) Deformation of E_r-Singularities

The equation (E_r) in (6.1) defines the universal deformation of the rational double point of type E_r; at $\lambda = 0$, we have $y^2 = x^3 + t^4$, $y^2 = x^3 + xt^3$ and $y^2 = x^3 + t^5$ for $r = 6, 7, 8$. By Theorem 6.1, we can easily reprove some results due to Brieskorn, Tjurina and others (cf. [Br, DP, Sl]). Indeed (6.6) implies immediately that the map

$$\pi : \mathbb{A}^r \to \mathbb{A}^r/W(E_r) \simeq \mathbb{A}^r, \qquad (u_i) \to \lambda = (p_i, q_j) \quad (8.8)$$

is a finite covering, ramified along the discriminant locus D ([B]), so that (in case $k = \mathbb{C}$) the monodromy map below is surjective:

$$\pi_1(\mathbb{C}^r - D) \to W(E_r). \quad (8.9)$$

Further we can describe the stratification of D according to the type of singularities (see [S5]); also it will be evident that our method applies to the case of char $p > 0$ as well.

On the other hand, let $\mathfrak{g}$ denote the simple Lie algebra of type E_r and $\mathfrak{h}$ the Cartan subalgebra, $\Delta = \{\alpha\}$ the root system in $\mathfrak{h}^*$, then the characteristic polynomial of the adjoint representation $\mathrm{ad}(H)$ on $\mathfrak{g}$ ($H \in \mathfrak{h} = \mathbb{A}^r$) is given by

$$\det(X - \mathrm{ad}(H)) = X^r \cdot \prod_{\alpha \in \varDelta} (X - \alpha(H))$$

$$= X^r \cdot \varPhi^*(X, \lambda) \qquad (\lambda = \pi(H)), \tag{8.10}$$

where $\varPhi^*(X, \lambda) = \varPhi(E_\lambda/k_0(t), \infty, I_2; X)$ is the special case of our algebraic equation (5.1) (cf. (6.8)), of degree 72, 126 or 240 for $r = 6, 7, 8$. According to K. Saito and Slodowy, the study of the roots of (8.10) led Killing [Ki] (about 100 years ago) to introduce the "root" systems in classifying Lie algebras.

Thus the theory of Mordell-Weil lattices has unexpectedly rich applications even in the simplest case of rational elliptic surfaces. We close this paper with the hope to see much more fruits in future coming from the study of the Mordell-Weil lattices.

References

[B]   Bourbaki, N.: Groupes et algèbres de Lie, Chap. 4, 5 et 6. Hermann, Paris 1968
[Br]   Brieskorn, E.: Singular elements of semisimple algebraic groups, Proc. ICM Nice 1970, II. Gauthier-Villars, Paris 1971, pp. 279–284
[C]   Conway, J. et al: Atlas of finite groups. Clarendon Press, Oxford 1985
[CS]   Conway, J., Sloane, N.: Sphere packings, lattices and groups. (Grundlehren der mathematischen Wissenchaften, vol. 290.) Springer, Berlin Heidelberg New York 1988
[CZ]   Cox, D., Zucker, S.: Intersection numbers of sections of elliptic surfaces. Invent. math. **53** (1979) 1–44
[DP]   Demazure, M., Pinkham, H., Teissier, B.: Séminaire sur les singularités des surfaces. (Lecture Notes in Mathematics, vol. 777.) Springer, Berlin Heidelberg New York 1980
[Ek]   Ekedahl, T.: An effective version of Hilbert's irreducibility theorem. In: Sém. théorie des nombres, Paris 1988–89, pp. 241–249, Birkhäuser, Boston Basel Berlin 1990
[E]   Elkies, N.: On Mordell-Weil lattices. Arbeitstagung Bonn, 1990
[F]   Faltings, G.: Calculus on arithmetic surfaces. Ann. Math. **119** (1984) 387–424
[G]   Gross, B.: Group representations and lattices. J. Amer. Math. Soc. **3** (1990) 929–960
[H]   Hriljac, P.: Heights and Arakelov's intersection theory. Amer. J. Math. **107** (1985) 23–38
[Ki]   Killing, W.: Die Zusammensetzung der stetigen endlichen Transformationsgruppen, II. Math. Ann. **33** (1889) 1–48
[K]   Kodaira, K.: On compact analytic surfaces II–III. Ann. Math. **77** (1963) 563–626; **78** (1963) 1–40; Collected Works, vol. III. Iwanami and Princeton Univ. Press, 1975, pp. 1269–1372
[L]   Lang, S.: Fundamentals of diophantine geometry. Springer, Berlin Heidelberg New York 1983
[M1]   Manin, Ju.: The Tate height of points on an Abelian variety, its variants and applications. Izv. Akad. Nauk SSSR, Ser. Mat. **28** (1964) 1363–1390; A.M.S. Transl. (2) **59** (1966) 82–110
[M2]   Manin, Ju.: Cubic forms, 2nd edn. North-Holland 1986
[Mo]   Mordell, L.: On the rational solutions of the indeterminate equation of the third and fourth degree. Proc. Camb. Phil. Soc. **21** (1922) 179–192
[N1]   Néron, A.: Modèles minimaux des variétés abéliennes sur les corps locaux et globaux. Publ. Math. I.H.E.S. **21** (1964)

[N2]   Néron, A.: Quasi-fonctions et hauteurs sur les variétés abéliennes. Ann. Math. **82** (1965) 249–331

[OS]   Oguiso, K., Shioda, T.: The Mordell-Weil lattice of a rational elliptic surface. Comment. Math. Univ. St. Pauli **40** (1991) 83–99

[Se1] Serre, J-P.: Facteurs locaux des fonctions zeta des variétés algébriques. Sém. DPP (1969/70); Oeuvres II, pp. 581–592. Springer, Berlin Heidelberg New York

[Se2] Serre, J-P.: Lectures on the Mordell-Weil theorem. Vieweg 1989.

[ShT] Shafarevich, I., Tate, J.: The rank of elliptic curves. Dokl. Akad. Nauk SSSR **175** 770–773; Sov. Math. Dokl. **8** (1967) 917–920

[ST]   Shimura, G., Taniyama, Y.: Modern number theory (in Japanese). Kyoritsu Publ., Tokyo 1957

[S1]   Shioda, T.: On elliptic modular surfaces. J. Math. Soc. Japan **24** (1972) 1–59

[S2]   Shioda, T.: An explicit algorithm for computing the Picard number of certain algebraic surfaces. Amer. J. Math. **108** (1986) 415–432

[S3]   Shioda, T.: Mordell-Weil lattices and Galois representation, I, II, III. Proc. Japan Acad. **65A** (1989) 267–271, 296–299, 300–303

[S4]   Shioda, T.: Mordell-Weil lattices and sphere packings. (To appear in Amer. J. Math.)

[S5]   Shioda, T.: Mordell-Weil lattices of type E_8 and deformation of singularities. (To appear)

[S6]   Shioda, T.: Construction of elliptic curves with high rank via the invariants of the Weyl groups. (To appear in J. Math. Soc. Japan)

[S7]   Shioda, T.: On the Mordell-Weil lattices. Comment. Math. Univ. St. Pauli **39** (1990) 211–240

[Si]   Silverman, J.: Heights and the specialization map for families of abelian varieties. J. Reine Angew. Math. **342** (1983) 197–211

[Sl]   Slodowy, P.: Simple singularities and simple algebraic groups. (Lecture Notes in Mathematics, vol. 815.) Springer, Berlin Heidelberg New York 1980

[Sw]   Swinnerton-Dyer, H.P.F.: The zeta function of a cubic surface over a finite field. Proc. Camb. Phil. Soc. **63** (1967) 55–71

[T1]   Tate, J.: On the conjectures of Birch and Swinnerton-Dyer and a geometric analog. Sém. Bourbaki 1965/66, no. 306

[T2]   Tate, J.: Algorithm for determining the type of a singular fiber in an elliptic pencil. (Lecture Notes in Mathematics, vol. 476.) Springer, Berlin Heidelberg New York 1975, pp. 33–52

[T3]   Tate, J.: Variation of the canonical height of a point depending on a parameter. Amer. J. Math. **105** (1983) 287–294

[W1]   Weil, A.: L'arithmetique sur les courbes algébriques. Acta Math. **52** (1928) 281–315

[W2]   Weil, A.: Variétés abéliennes et courbes algébriques, Hermann, Paris 1948/1973

[W3]   Weil, A.: Abstract versus classical algebraic geometry. Proc. ICM Amsterdam, vol. III (1954), pp. 550–558; Collected Works, vol. II, pp. 180–188. Springer, Berlin Heidelberg New York 1980

Collapsing Riemannian Manifolds and Its Applications

Kenji Fukaya

Department of Mathematics, Faculty of Science
University of Tokyo, Hongo, Bunkyo-ku, Tokyo 113, Japan

§1

One of the typical problems to which Hausdorff convergence and collapsing of Riemannian manifolds can be applied is the study of local structure of Riemannian manifolds.

Problem 1.1. Let $\mathscr{C}$ be a class of Riemannian manifolds. Can we find a positive number $\varepsilon_{\mathscr{C}}$ such that for each $p \in M \in \mathscr{C}$ there exists a neighborhood U_p of p with the following properties ?

(1.1.1) U_p contains $B_p(\varepsilon_{\mathscr{C}}, M)$, the metric ball of radius $\varepsilon_{\mathscr{C}}$ and centered at p.
(1.1.2)   We can control the topology of U_p.

The essential point here is that $\varepsilon_{\mathscr{C}}$ is independent of p and M.

In the case when $\mathscr{C}$ is the class of homogeneous spaces, this problem is solved in Zassenhauss [Z] and Kazdan-Margulis [KM]. In that case, U_p is controled by nilpotent groups. In the case of negatively curved manifolds, this problem is studied by Margulis [M], Heintze [H] and Gromov [G1].

We consider the following two classes.

$$\mathscr{M}_n = \{M \,|\, \dim M = n, |K_M| \leq 1\},$$
$$\mathscr{N}_n = \{M \,|\, \dim M = n, K_M \geq -1\}.$$

Here K_M stands for the sectional curvature. Problem 1.1 for class $\mathscr{M}_n$ is first studied by Gromov [G2].

§2

In §§2–6, we deal with the class $\mathscr{M}_n$. First let us take the following two extremal cases.

(1) When we consider the subclass $\mathscr{M}_n(i_0)$ of $\mathscr{M}_n$ consisting of manifolds with injectivity radius $> i_0$, we can choose $\varepsilon_{\mathscr{C}} = i_0$, since i_0-ball is diffeomorphic to Euclidean space.

Proceedings of the International Congress
of Mathematicians, Kyoto, Japan, 1990
© The Mathematical Society of Japan, 1991

(2) When we consider the subclass consisting of manifolds with small diameter, we can apply the following theorem. (In this case we choose $U_p = M$.)

Theorem 2.1 (Gromov [G2], Ruh [R]). *There exists $\varepsilon_n > 0$ such that if*

$$|K_M| \cdot \operatorname{Diam} M^2 < \varepsilon_n,$$

then M is diffeomorphic to N/Γ, where

(2.1.1) N *is a nilpotent Lie group,*

(2.1.2) Γ *is a discrete subgroup of the semidirect product $N \tilde{\times} \operatorname{Aut} N$, the group of affine diffeomorphisms of N,*

(2.1.3)   *The index, $[\Gamma : \Gamma \cap N]$ is finite.*

(1) is the case when no direction of M is degenerate. (2) is the case when every direction of M is degenerate. In general there are both degenerate and nondegenerate directions. Hence we have to separate those two directions. For this purpose we study the following:

Problem 2.2. For a small neighborhood U_p of p in a Riemannian manifold M^n, find a Riemannian manifold V^m such that

(2.2.1)   the injectivity radius of V is larger than a constant depending only on the class $\mathscr{C}$ containing M,

(2.2.2) V is Hausdorff close to U_p.

The Hausdorff closeness between spaces is defined as follows.

Definition 2.3 (Gromov [G3]). Let X and Y be compact metric spaces. A (not necessary continuous) map $f : X \to Y$ is said to be an ε-*Hausdorff approximation* if

(2.3.1)   the ε-neighborhood of $f(X)$ is Y,

(2.3.2) $|d(f(x), f(y)) - d(x, y)| < \varepsilon$, for each $x, y \in X$.

We say that the *Hausdorff distance*, $d_H(X, Y)$, between X and Y is smaller than ε if there exist ε-Hausdorff approximations from X to Y and from Y to X.

If V is as in Problem 2.2, a nondegenerate direction in U_p is one which is "parallel" to some direction in V.

§3

In fact the answer to Problem 2.2 itself is negative. But we can prove a bit modified statement. The construction of the space V in 2.2 is based on a limit argument. In other words, the space V is a limit of Riemannian manifolds contained in class $\mathscr{M}_n$. Hence to study Problem 2.2, we need to study the limit behavior of the sequence of Riemannian manifolds.

First we need to show that there exist sufficiently many convergent sequences. The following result ensures it.

Theorem 3.1 (Gromov [G3]). *Let $\mathscr{S}(D)$ be a set of all Riemannian manifolds with diameter $\leq D$ and Ricci curvature ≥ -1. Then $\mathscr{S}(D)$ is precompact in the set of all compact metric spaces, with respect to the Hausdorff distance.*

Next we need to study the following:

Problem 3.2. Let $M_i \in \mathscr{M}_n$ and X be a compact metric space. Suppose that M_i converges to X with respect to the Hausdorff distance. What kind of singularity can X have ?

We are mainly interested in the case when the dimension of X is smaller than that of M. In this case, we say that M_i collapses to X.

The first important example of collapsing Riemannian manifolds is discovered by M. Berger. In his example, a family of metrics on S^3 collapses to S^2. Here S^2 is regarded as the quotient of S^3 by an action of S^1.

In general, one can construct a collapsing family of metrics on manifolds M with torus action, provided each orbit is of positive dimension. The Hausdorff limit of this family is the quotient space M/T, which, in general, is singular. The other important example is a nilmanifold, which collapses to a point. (See [G2, BK].)

The systematic study of the collapsing phenomena for Riemannian manifolds was initiated by Cheeger-Gromov [CG1]. We discuss their results in §5.

Now we present results about Problem 2.2. In case when M_i is a $K(\pi, 1)$-space, we can prove the following:

Theorem 3.3 ([F4]). *Let M_i and X be as in Problem 3.2. Assume $\pi_k(M_i) = 1$, for $k > 1$. Then $X = Y/\Gamma$, where Y is a contractible manifold and Γ a discrete group of isometries.*

In fact we can prove the following:

(3.4.1) $\widetilde{M}_i$ converges to Z. Z is a manifold of the same dimension as M_i. The convergence is compact $C^{1,\alpha}$-convergence of metric tensors.

(3.4.2) Γ_i converges to a group G of isometries of Z.

(3.4.3) The connected component G_0 of G is a simply connected nilpotent Lie group.

(3.4.4) $Y = Z/G_0$, $\Gamma = G/G_0$.

Thus, roughly speaking, the Hausdorff convergence of $K(\pi; 1)$-manifolds is a generalization of the convergence of discrete subgroups of Lie group. Collapsing phenomenon occurs when discrete groups converge to a continuous group.

Before discussing the general case, we give applications of Theorem 3.3.

Theorem 3.5 ([F4]). *There exists a positive number ε_D such that if an n-dimensional Riemannian manifold M $(n > 1)$ satisfies*

(3.5.1) $\pi_k(M) = 1$, *for* $k > 1$,
(3.5.2) $|K_M| \le 1$, Diam $M \le D$,
(3.5.3) $\pi_1(M)$ *does not contain* $\mathbf{Z}^2$,

then

$$\text{Vol } M > \varepsilon_D \,.$$

Conjecture 3.6. *The assumption* Diam $M \le D$ *can be removed.*

In case $K_M \le 0$, this was proved by Heintze [H], Gromov [G1], Buyalo [Bu].
The following result is not a direct application of Theorem 3.3 but its proof uses an idea of the proof of Theorem 3.3.

Theorem 3.7 (Fukaya-Yamaguchi [FY1]). *There exists a positive number ε_D such that if*

$$\varepsilon_D > K_M > -1 \,,$$
$$\text{Diam } M < D \,,$$

then the universal covering space of M is diffeomorphic to $\mathbf{R}^n$.

Theorem 3.7 is an affirmative answer to a conjecture by Gromov [G4].

As we remarked before, the limit of manifolds in class $\mathcal{M}_n$ is not necessary a manifold. The reason is that a singular space can arise as the quotient space M/T of a manifold M by its torus action. Note, however, that the quotient space FM/T is always nonsingular. Here FM is the frame bundle of M and the action of T on FM is a lift of one on M. This observation leads us to the following:

Theorem 3.8 ([F3]). *Let $M_i \in \mathcal{M}$, FM_i be their frame bundles equipped with metrics induced by ones on M. Let X be a compact metric space. Suppose that FM_i converges to X with respect to Hausdorff convergence.*
Then X is a smooth manifold with $C^{1,\alpha}$-metric.
Furthermore, there exists a smooth $O(n)$ action on X such that M_i converges to $X/O(n)$ and that the connected components of isotropy groups are abelian.

Using Theorems 3.1 and 3.8, we can prove the following:
Let $M \in \mathcal{M}_n$, $p \in M$. Then there exists a manifold V such that

(3.9.1)   injectivity radius of V > const > 0,
(3.9.2) $d_H(FB_p(\varepsilon, M), V)$ is small.

§4

To apply Problem 2.2 to Problem 1.1, we need to study the relation between the topological structures of U_p and V_p in 2.2. Namely:

Problem 4.1. Let M_i and X be as in Problem 3.2. Find the relation between the topological structures of M_i and X.

We have the following two results:

Theorem 4.2 (Fukaya [F1,2], Yamaguchi [Y2]). *Let $M_i \in \mathcal{N}_n$ and N be a compact Riemannian manifold. Suppose that M_i converges to N with respect to the Hausdorff distance. Then for each suffciently large i, there exists a map $f_i : M_i \to N$, such that*

(4.2.1) *f_i is a locally trivial fibre bundle,*
(4.2.2) *f_i is an almost Riemannian submersion, namely*

$$e^{\varepsilon_i} > \frac{|f_*(V)|}{|V|} > e^{-\varepsilon_i}$$

holds for vectors $V \in TM$ perpendicular to the fibre,
(4.2.3) *the fibre of f_i is of almost nonnegative curvature. (See §7).*

If we assume $M_i \in \mathcal{M}_n$ in addition, then

(4.2.4) *the fibre is almost flat, namely it satisfies (2.1.1)–(2.1.3),*
(4.2.5) *the structure group of fibration is reduced to the group of affine diffeomorphisms.*

Theorem 4.3 ([F3]). *Let M_i and X be as in Theorem 3.8. Then, for each sufficiently large i, there exist $O(n)$ maps $f_i : FM_i \to X$ satisfying (4.2.1), (4.2.2), (4.2.4), (4.2.5).*

It follows from Theorem 4.3 that there exists a singular fibration: $M_i \to X/O(n)$. The singular fibre corresponds to the singular point of $X/O(n)$.

Using the results of §§2, 3, 4, we can prove the following:

Theorem 4.4 ([F5]). *There exists a positive number ε_n such that if $p \in M \in \mathcal{M}_n$ then there exists a neighborhood U_p of p with the following properties.*

(4.4.1) *U_p is diffeomorphic to a vector bundle over N_p/Γ_p, where N_p and Γ_p is as in (2.1.1)–(2.1.3).*
(4.4.2) *U_p contains $B_p(\varepsilon_n, M)$.*

Ghanaat, Min-no, Ruh [GMR] proved a closely related result.

§5

The nilpotent Lie group N_p in Theorem 4.4 depends on the point p. In other words the dimension of the collapsing directions changes from point to point. In fact, consider the metric

$$dt^2 + e^{t-2C}dx^2 + e^{-t-2C}dy^2,$$

on $[-C, C] \times T^2$. (Here x and y are coordinates of T^2.) Then, in a neighborhood of $\{\pm C\} \times T^2$, the dimension of collapsing direction is one, and in a neighborhood of $\{0\} \times T^2$ the dimension of collapsing direction is two.

In order to study the global behavior of the collapsing part of the manifolds, we need to study the above phenomenon. Cheeger and Gromov introduced the notion of the local action of the groups for this purpose.

Definition 5.1 (Cheeger-Gromov [CG1]). An F-structure on a manifold M is a subsheaf $\mathcal{F}$ of the sheaf of vector fields such that

(5.1.1) $\mathcal{F}$ is a sheaf of abelian Lie algebra,
(5.1.2)   for each $p \in M$ there exists a neighborhood U_p, its finite covering space $\tilde{U}_p$, an action of torus, T_p, on $\tilde{U}_p$, such that the lift of $\mathcal{F}$ to $\tilde{U}_p$ is induced by the action of T_p.

They proved the following two results:

Theorem 5.2 ([CG1]). *Let M be a manifold. Assume that there exists an F-structure on M such that the dimension of each orbit is nonzero. Then there exists a family of metrics g_ε on M such that*

(5.2.1) $(M, g_\varepsilon) \in \mathcal{M}_n$,
(5.2.2)   *the injectivity radius of (M, g_ε) is smaller than ε.*

Theorem 5.3 ([CG2]). *There exists a positive number ε_n with the following properties. Let $M \in \mathcal{M}_n$. Put*

$$M_{\text{thin}} = \{p \in M \mid \text{injectivity radius at } p < \varepsilon_n\}.$$

Then there exists an F-structure $\mathcal{F}$ on M_{thin} such that every orbit of it is of positive dimension.

Cheeger-Gromov applied these results to study Gauss-Bonnet type formula for noncompact manifolds. ([CG 3,4] etc.)

§6

In Theorem 4.4, we obtained a nilpotent structure but the result is local. In Sect. 5, we obtain more global structure but the structure is abelian and not nilpotent. To combine them, we need to introduce the action of sheaf of nilpotent Lie algebra.

Definition 6.1 (Cheeger-Fukaya-Gromov [CFG]). A *nilpotent structure* on a manifold M is a sheaf $\mathcal{N}$ of vector fields such that for each p there exists a neighborhood U_p and an action of group G_p on a Galois covering $\tilde{U}_p$ of U_p with the following properties.

(6.1.1) The connected component N_p of G_p is nilpotent.
(6.1.2) The deck transformation group Γ_p of the covering: $\tilde{U}_p \to U_p$ is contained in G_p.
(6.1.3) G_p is generated by N_p and Γ_p.
(6.1.4) $[\Gamma_p : \Gamma_p \cap N_p] = [G_p : N_p]$ is finite.
(6.1.5) The lift of $\mathcal{N}$ to $\tilde{U}_p$ is generated by the action N_p.

Theorem 6.2 ([CFG]). *There exist ε_n and k_n with the property that, for each $M \in \mathcal{M}_n$ there exists a nilpotent structure $\mathcal{N}$ such that the following holds in addition.*

(6.2.1) U_p *contains* $B_{\varepsilon_n}(p, M)$.
(6.2.2) *The injectivity radius of* $\tilde{U}_p$ *is larger than* ε_n.
(6.2.3) $[G_p : N_p] < k_n$.

Definition 6.3 ([CFG]). A Riemannian metric g on M is said to be (ε, k)-*round*, if there exists a nilpotent structure $\mathcal{N}$ on M such that it satisfies (6.2.1)–(6.2.3) in addition and that the section of $\mathcal{N}$ is a Killing vector field of the metric g. (In other words the action of N_p is isometry for induced metric.)

Theorem 6.4 ([CFG]). *For each δ there exist $\varepsilon = \varepsilon(\delta, n)$ and $k = k(\delta, n)$ such that for each Riemannian manifold (M, g) in the class $\mathcal{M}_n$ we can find a metric g_δ on M with the following properties.*

(6.4.1) (M, g_δ) *is* (ε, k)-*round*.
(6.4.2) $|g - g_\delta|_{C^{1,\alpha}} < \delta$.

We expect that there is a result similar to Theorem 5.2 for nilpotent structure. The proof of it is not yet complete.

§7

In this section we consider the class $\mathcal{N}_n$. To study the collapsing phenomena of Riemannian manifolds in class $\mathcal{M}_n$, the starting point was Theorem 2.1, which characterize the manifolds collapsing to a point. The corresponding problem for class $\mathcal{N}_n$ is the study of almost nonnegatively curved manifolds.

Problem 7.1. Find a characterization of manifolds M admitting a metric with

$$K_M \cdot \text{Diam } M^2 > -\varepsilon_n,$$

where ε_n is a positive small number depending only on dimension.

This problem should be very difficult, since it includes the characterization of nonnegatively curved manifolds. We describe some results concerning the topology of almost nonnegatively curved manifolds. (Some of them hold for manifolds of almost nonnegative Ricci curvature.)

Theorem 7.2 (Gromov-Gallot [G3, Ga]). *There exists a positive number C_n such that if M is an n-dimensional Riemannian manifold with*

$$\text{Ricci}_M \cdot \text{Diam } M^2 > -D,$$

then

$$b_1(M;\mathbf{Q}) = \text{rank } H_1(M;\mathbf{Q}) \le n - 1 + C_n^D.$$

In particular, if M is of almost nonnegative Ricci curvautre then the first Betti number does not exceed the dimension. This is a generalization of a classical result by Bochner [B].

Theorem 7.3 (Yamaguchi [Y2]). *There exists a positive number ε_n depending only on dimension such that if an n-dimensional Riemannian manifold M satisfies*

$$K_M \cdot \text{Diam } M^2 > -\varepsilon_n,$$

then a finite cover of M is a fibre bundle over $b_1(M;\mathbf{Q})$-dimensional torus.
If the first Betti number is equal to the dimension then M is diffeomorphic to a torus.

(Compare also [Y1].)

Theorem 7.4 (Fukaya-Yamaguchi [FY2]). *If M satisfies the assumption of Theorem 7.3 then the fundamental group of M contains a nilpotent subgroup of finite index.*

Theorem 7.4 is an affirmative answer to another conjecture in [G4].
In fact we can prove a bit more than Theorem 7.4. (The precise statement is omitted. See [FY2].) As a corollary we have:

Theorem 7.5 ([FY2]). *There exists c_n such that if M satisfies the assumption of Theorem 7.3 then we have*

$$b_1(M;\mathbf{Z}_p) \le n,$$

for each prime $p > c_n$.
In case when the equality holds, M is diffeomorphic to a torus.

Using Theorem 7.4 we can control the fundamental group of the small neighborhood. Namely we have the following:

Theorem 7.6 ([FY2]). *There exists a positive number ε_n depending only on dimension such that if $M \in \mathcal{N}_n$ and $p \in M$ then the image*

$$\mathrm{Im}\,[\pi_1(B_p(\varepsilon_n, M)) \to \pi_1(B_p(1, M))]$$

contains a nilpotent subgroup of finite index.

Conjecture 7.7. *Theorems 7.3–7.6 still hold if we replace sectional curvature by Ricci curvature.*

In the case when $M \in \mathcal{M}_n$ one can choose a neighborhood U_p which is $K(\pi; 1)$ and whose fundamental group is almost nilpotent. (See Theorem 4.4.) But we can not choose such U_p of uniform size for M in $\mathcal{N}_n$, since the nonnegatively curved manifold is not necessary a $K(\pi; 1)$-space. Hence Theorem 7.6 does not give enough information to determine the local topological structure of manifolds in the class $\mathcal{N}_n$. So far, our knowledge about higher homology or homotopy groups of (almost) nonnegatively curved manifolds is quite restricted. The best result seems to be Gromov's Betti number estimate, [G5].

As for Hausdorff convergence of manifolds in the class $\mathcal{N}_n$, we have Theorem 4.2 due to Yamaguchi. But we do not know so much about Problem 3.2 for this class. (Recently Burago and Gromov proved that the limit space is almost everywhere a manifold.)

On the other hand, Yamaguchi [Y2] constructed examples of collapsing family of metrics using an action of compact groups. Namely, on a manifold M on which a compact Lie group K acts, there exists a sequence of metrics g_ε such that $(M, g_\varepsilon) \in \mathcal{N}_n$ and that (M, g_ε) converges to M/K for Hausdorff distance. This construction is closely related to the fact that compact Lie group admits a metric of positive sectional curvature.

In noncollapsing situation, (namely the case we assume Vol $M >$ constant > 0, in addition,) Grove-Petersen-Wu [GPW] proved a strong result. They conjectured that the limit space is a topological manifold in this case.

References

[B]   S.Bochner: Vector fields and Ricci curvature. Bull. AMS **52** (1946) 776–797

[BK]   P.Buser, H.Karcher: Gromov's almost flat manifolds. Asterisque **81** (1981) 1–148

[Bu]   V.Buyalo: Volume and fundamental group of a manifold of nonpositive curvature. Math. USSR Sb. **50** (1985) 137–150

[CFG]   J.Cheeger, K.Fukaya, M.Gromov: Nilpotent structure and invariant round metric on collapsed Riemannian manifolds. Preprint

[CG1]   J.Cheeger, M.Gromov: Collpasing Riemannian manifolds while keeping their curvature bounded I. J. Diff. Geom. **23** (1986) 309–346

[CG2]   J.Cheeger, M.Gromov: Collpasing Riemannian manifolds while keeping their curvature bounded II. J. Diff. Geom. **32** (1990) 269–298

[CG3]   J.Cheeger, M.Gromov: On the characteristic numbers of complete manifolds of
        bounded curvature and finite volume. In: Differential Geometry and Complex
        Analysis. Springer, Berlin Heidelberg New York 1985
[CG4]   J.Cheeger: Chopping Riemannian manifolds. Preprint
[F1]   K.Fukaya: Collapsing Riemannian manifolds to ones of lower dimension. J. Diff.
        Geom. **25** (1987) 139–156
[F2]   K.Fukaya: Collapsing Riemannian manifolds to ones of lower dimension II. J.
        Math. Soc. Japan **41** (1989) 333–356
[F3]   K.Fukaya: A boundary of the set of the Riemannian manifolds with bounded
        curvatures and diameters. J. Diff. Geom. **28** (1988) 1–21
[F4]   K.Fukaya: A compactness of a set of aspherical Riemannain orbifolds. In: A
        Fete of Topology, ed. by Matsumoto, Mizutani, Morita. Academic press, Boston
        1988
[F5]   K.Fukaya: Hausdorff convergence of Riemannian manifolds and its applications.
        In: Recent topics in differential and analytic geometry, ed. by T.Ochiai. Advanced
        Studies in Pure Math. 18-I. Kinokuniya, Academic Press, Tokyo Boston 1990
[FY1]   K.Fukaya, T.Yamaguchi: Almost nonpositively curved manifolds. J. Diff. Geom.
        33 (1991) 67–90
[FY2]   K.Fukaya, T.Yamaguchi: The fundamental group of almost nonnegatively curved
        manifolds. Preprint
[Ga]   S.Gallot: A Sobolev inequality and some geometric applications. In: Spectra of
        Riemannian manifolds. Kaigai, Tokyo 1983
[GMR]   P.Ghanaat, Min-Oo, E.Ruh: Local structure of Riemannian manifolds. Preprint
[G1]   M.Gromov: Manifolds of negative curvature. J. Diff. Geom. **13** (1978) 223–230
[G2]   M.Gromov: Almost flat manifolds. J. Diff. Geom. **13** (1978) 231–241
[G3]   M.Gromov (J.Lafontaine and P. Pansu): Structure metrique pour les varietes
        riemanniennes. Cedic/Fernad Nathan, Paris 1981
[G4]   M.Gromov: Synthetic Geometry in Riemannian manifolds. Proceedings of ICM
        Helsinki 1978, vol. 1, pp. 415–419
[G5]   M.Gromov: Curvature, Diameter and Betti numbers. Comment. Math. Helv. **56**
        (1981) 179–195
[GPW]   K.Grove, P.Petersen, J.Wu: Geometry finiteness theorems via controlled topology.
        Invent. math. **99** (1990) 205–213
[H]   E.Heintze: Mannigfaltigkeiten negativer Krümmung. Habilitationsschrift, Univer-
        sität Bonn
[KM]   Kazdan, G.Margulis: A proof of Selberg's hypothesis. Math. USSR Sb. **75** (1968)
        162–168
[M]   G.Margulis: Discrete group of motions of manifolds of nonpositive curvature.
        AMS Trans. **109** (1977) 33–45
[R]   E.Ruh: Almost flat manifolds. J. Diff. Geom. **17** (1982) 1–14
[Y1]   T.Yamaguchi: Manifolds of almost nonnegative Ricci curvature. J. Diff. Geom.
        28 (1988) 157–167
[Y2]   T.Yamaguchi: Collapsing and Pinching in lower curvature bound. To appear in
        Ann. Math.
[Z]   H.Zassenhaus: Beweis eines Satzes über diskrete Gruppen. Abh. Math. Sem.
        Hansischen Univ. 12

Le Cercle à l'Infini des Surfaces à Courbure Négative

Etienne Ghys

Ecole Normale Supérieure de Lyon, 46, allée d'Italie F-69007 Lyon, France

1. Introduction

Soit S une surface compacte orientée munie d'une métrique riemannienne g de classe C^∞ à courbure variable strictement négative. Il est bien connu que le revêtement universel $\tilde{S}$ de S peut naturellement être compactifié par l'adjonction d'un bord à l'infini, noté $\partial\tilde{S}$ (voir par exemple [1]). Par définition, un point de $\partial\tilde{S}$ est représenté par un rayon géodésique, c'est-à-dire un plongement isométrique $r : [0, +\infty[\to \tilde{S}$ et deux tels rayons r_1 et r_2 définissent le même point du bord si la distance entre $r_1(t)$ et $r_2(t)$ est uniformément bornée. Si p est un point de $\tilde{S}$, on peut représenter chaque point de $\partial\tilde{S}$ par un unique rayon issu de p. Ainsi, le bord $\partial\tilde{S}$ s'identifie à l'espace des vecteurs unitaires de l'espace tangent à $\tilde{S}$ en p; il est donc homéomorphe à un cercle. Bien sûr, le groupe fondamental Γ de S opère sur $\tilde{S}$ et donc sur le cercle $\partial\tilde{S}$. Nous nous proposons ici de décrire quelques résultats relatifs à cette action. Notre but n'est pas de les démontrer mais d'essayer de les motiver et d'indiquer les liens qui les unissent.

2. L'aspect topologique et la cohomologie bornée

D'un point de vue topologique, il n'y a qu'une action à étudier. C'est un fait bien connu depuis longtemps.

Théorème. *Soient g_1 et g_2 deux métriques à courbure négative sur la même surface compacte S. Alors les actions du groupe fondamental de S sur les bords pour g_1 et g_2 du revêtement universel $\tilde{S}$ sont topologiquement conjuguées.*

Pour montrer ce théorème, on constate qu'un rayon r_1 dans $\tilde{S}$ pour la métrique g_1 n'est pas nécessairement un rayon pour g_2 mais que c'est un "quasi-rayon": la g_2-distance entre $r_1(t)$ et $r_1(t')$ est comprise entre $c^{-1}|t - t'|$ et $c|t - t'|$ pour une certaine constante $c > 0$. Ceci permet de montrer l'existence d'un rayon r_2 pour g_2 qui est à distance bornée de r_1. Les bords de $\tilde{S}$ pour g_1 et g_2 sont alors naturellement identifiés de manière Γ équivariante.

De manière plus conceptuelle, on peut associer un bord à un groupe de type fini G (voir [7] et [19]). Pour cela, on se fixe une partie génératrice finie et on considère

Proceedings of the International Congress
of Mathematicians, Kyoto, Japan, 1990
© The Mathematical Society of Japan, 1991

le graphe de Cayley de G correspondant à cette partie génératrice. Si une arête relie les éléments γ_1 et γ_2 de G, on équipe cette arête d'une métrique qui la rend isométrique à un intervalle de longueur $(n+1)^{-2}$ où $n = \inf(\|\gamma_1\|, \|\gamma_2\|)$ et $\|\gamma_i\|$ désigne la longueur de γ_i par rapport à la partie génératrice choisie. Le graphe de Cayley de G devient ainsi un espace métrique non complet (si G est infini). Le bord ∂G de G est alors défini comme l'espace qu'il faut ajouter à ce graphe pour le compléter. Il ne dépend pas du choix de la partie génératrice. Le groupe Aut(G) des automorphismes de G opère naturellement sur ∂G. En particulier, G opère sur ∂G (via les automorphismes internes). Ce bord est particulièrement utile lorsque G est un groupe hyperbolique au sens de M. Gromov [19]. Par exemple, si G est le groupe fondamental Γ d'une surface compacte S de genre supérieur ou égal à 2, le bord ∂G s'identifie à $\partial \tilde{S}$. Ainsi, *l'action que nous étudions n'est autre que celle du groupe Γ sur son bord.*

Examinons rapidement le cas particulier où la courbure de la métrique g est -1. Le revêtement universel $\tilde{S}$ s'identifie alors au disque de Poincaré D^2 et son groupe d'isométries directes à PSL$(2, \mathbb{R})$. Le bord de D^2 est le cercle, ici identifié à la droite projective réelle P^1. Dans ce cas, l'action étudiée de Γ sur P^1 est projective et provient d'un plongement de Γ dans PSL$(2, \mathbb{R})$ comme sous-groupe discret co-compact. L'espace de ces plongements (à conjugaison près) est l'espace de Teichmüller; il est de dimension finie. Bien que toutes ces actions projectives soient topologiquement conjuguées, il est facile de s'assurer qu'elles ne sont C^1-conjuguées que si elles sont projectivement conjuguées. D. Sullivan [31] montre même que si deux de ces actions sont conjuguées par une application mesurable qui respecte les ensembles négligeables au sens de Lebesgue, alors ces actions sont projectivement conjuguées.

Comment caractériser le type topologique de cette action? Si un groupe discret G opère sur le cercle en respectant l'orientation, on peut construire un fibré en cercles au dessus de l'espace d'Eilenberg-MacLane $K(G, 1)$. La classe d'Euler de ce fibré, élément de $H^2(G, \mathbb{Z})$, est évidemment un invariant de conjugaison topologique (respectant l'orientation). Cet invariant est cependant insuffisant. Si $G = \mathbb{Z}$ par exemple, ce second groupe de cohomologie est trivial. Dans [12], nous introduisons un invariant plus fin, élément de $H_b^2(G, \mathbb{Z})$, second groupe de cohomologie bornée à coefficients entiers [18]. Cette cohomologie est celle du sous-complexe du complexe d'Eilenberg-MacLane formé des cochaînes bornées (comme fonctions de G^n vers $\mathbb{Z}$). Voici comment on procède pour définir l'invariant en question.

Soient f_1, f_2, f_3 trois homéomorphismes directs du cercle S^1 et x un point base sur S^1. Posons $c(f_1, f_2, f_3) = 1$ si les points $f_1(x), f_2(x), f_3(x)$ sont distincts et placés dans un ordre cyclique rétrograde sur le cercle ou si $f_1(x) = f_3(x) \neq f_2(x)$ et posons $c(f_1, f_2, f_3) = 0$ dans les autres cas. Il est aisé de s'assurer que c est un cocycle. Si ϕ est une action du groupe discret G sur le cercle par homéomorphismes directs, on obtient ainsi, par image réciproque, une classe $\phi^*(c)$ dans $H_b^2(G, \mathbb{Z})$.

On vérifie facilement que l'image de cette classe dans la cohomologie usuelle est la classe d'Euler (voir aussi [25]). Nous appellerons cette classe la *classe d'Euler bornée* de l'action ϕ. Si $G = \mathbb{Z}$, on a $H_b^2(G, \mathbb{Z}) \simeq \mathbb{R}/\mathbb{Z}$ et la classe d'Euler bornée n'est autre que le nombre de rotation du générateur de l'action.

Théorème [12]. *Soient ϕ_1 et ϕ_2 deux actions du même groupe G sur le cercle, par homéomorphismes directs. On suppose que toutes les orbites de ϕ_1 et ϕ_2 sont denses dans le cercle. Alors, ϕ_1 et ϕ_2 sont topologiquement conjuguées, par un homéomorphisme direct, si et seulement si les classes d'Euler bornées $\phi_1^*(c)$ et $\phi_2^*(c)$ sont égales dans $H_b^2(G, \mathbb{Z})$.*

Lorsque les orbites ne sont pas supposées denses, on a un énoncé plus faible qui utilise une notion de semi-conjugaison (voir [12] et aussi [32]).

Revenons au cas où G est le groupe fondamental Γ de S. La classe d'Euler d'une action de Γ sur le cercle est un élément de $H^2(\Gamma, \mathbb{Z})$, c'est-à-dire un entier que nous noterons eu. Un théorème de Milnor-Wood [35], qui est d'ailleurs l'une des origines de la notion de cohomologie bornée, affirme que cet entier vérifie l'inégalité $|eu| \leq |\chi(S)|$ où $\chi(S)$ désigne la caractéristique d'Euler-Poincaré de S. Dans le cas de l'action de Γ sur le cercle $\partial\tilde{S}$ qui correspond à une métrique à courbure négative, on a en fait l'égalité $eu = \pm\chi(S)$.

Dans un joli article [28], S. Matsumoto montre que si Γ opère sur le cercle et si le nombre d'Euler est maximal, i.e. si $eu = \chi(S)$, alors la classe d'Euler *bornée* ne peut prendre qu'une valeur, à savoir celle correspondant à l'action de Γ sur son bord. Il peut alors en déduire la caractérisation topologique suivante qui répond positivement à une conjecture de W. Goldman [17].

Théorème [28]. *Considérons une surface compacte orientée S de genre supérieur ou égal à 2 et une action du groupe fondamental de S sur le cercle, par homéomorphismes directs. On suppose que toutes les orbites sont denses sur le cercle et que le nombre d'Euler vérifie $eu = \pm\chi(S)$. Alors, cette action est topologiquement conjuguée à l'action naturelle de ce même groupe sur le bord $\partial\tilde{S}$ du revêtement universel de S, pour n'importe quelle métrique à courbure négative sur S.*

L'hypothèse de densité des orbites est en fait inutile si l'on se restreint à des actions suffisamment différentiables.

Théorème [13]. *Considérons une surface compacte orientée S de genre supérieur ou égal à 2 et une action du groupe fondamental de S sur le cercle, par difféomorphismes directs de classe C^2 (resp. analytiques réels). On suppose que le nombre d'Euler du fibré associé vérifie $eu = \pm\chi(S)$ (resp. $eu \neq 0$). Alors, toutes les orbites sont denses dans le cercle.*

La démonstration se fonde sur les théorèmes de structure des feuilletages de codimension 1, à la Denjoy-Sacksteder.

Citons encore un résultat, obtenu avec J. Barge, et dont la démonstration utilise aussi la cohomologie bornée. Soit x un point base sur le bord $\partial\tilde{S}$ correspondant à une métrique g. Si γ_1, γ_2, γ_3 sont trois éléments de Γ, on peut considérer l'aire $a(\gamma_1, \gamma_2, \gamma_3)$ du triangle idéal de $\tilde{S}$ de sommets $\gamma_1(x)$, $\gamma_2(x)$, $\gamma_3(x)$. C'est un 2-cocycle borné sur Γ à valeurs réelles. Dans le cas de courbure -1, l'aire de ces triangles vaut $\pm\pi$ et la classe de cohomologie bornée de a est égale à 2π fois la classe d'Euler bornée (considérée comme classe réelle). La réciproque est plus délicate.

Théorème [2]. *Soit g une métrique riemannienne à courbure négative sur la surface compacte S. On suppose que tous les triangles idéaux du revêtement universel $\tilde{S}$ sont d'aire π. Alors, la courbure de g est constante, égale à -1.*

3. La différentiabilité du bord

Soient p et q deux points de $\tilde{S}$. Nous avons déjà observé que $\partial \tilde{S}$ est naturellement identifié aux cercles unités S_p^1 et S_q^1 dans les espaces tangents en p et q à $\tilde{S}$. Ainsi, il existe un homéomorphisme naturel π_{pq} entre S_p^1 et S_q^1. Quel est le degré de régularité de ces homéomorphismes? Un résultat classique de E. Hopf affirme qu'ils sont de classe C^1 [23] (voir aussi [22]). Un résultat plus fort a été obtenu par S. Hurder et A. Katok.

Théorème [24]. *Les homéomorphismes π_{pq} sont de classe $C^{2-\varepsilon}$ pour tout $\varepsilon > 0$.*

Ce résultat est à comparer au suivant sur lequel nous allons nous attarder un peu plus.

Théorème [15]. *Si les homéomorphismes π_{pq} sont de classe C^2, alors la courbure de la métrique considérée est constante.*

Ce théorème fait suite à un théorème local de S. Hurder et A. Katok [24] qui se fondait sur un résultat de [11] que nous mentionnerons plus loin. Il a donné lieu par la suite à un certain nombre de développements, tout spécialement en dimension supérieure. Puisque nous ne traitons ici que du cas des surfaces, nous ne décrirons pas ces développements qui aboutissent à une caractérisation analogue des espaces localement symétriques à courbure négative mais nous renvoyons à [3, 5, 6, 8, 20, 26]. Grâce à ces travaux, et tout particulièrement à ceux de M. Kanai, la démonstration du théorème précédent s'est significativement simplifiée. Nous allons esquisser ici une preuve assez élémentaire.

Soit $T_1 S$ le fibré unitaire tangent à S et ϕ_t le flot géodésique de S, agissant sur $T_1 S$. Soient $\mathscr{F}^s$ et $\mathscr{F}^u$ les feuilletages stables et instables faibles de ϕ_t. Les feuilles du relevé $\tilde{\mathscr{F}}^s$ de $\mathscr{F}^s$ au fibré unitaire $T_1 \tilde{S}$ sont constituées de tous les vecteurs unitaires qui définissent des rayons géodésiques asymptotes. L'espace des feuilles de $\tilde{\mathscr{F}}^s$ est donc homéomorphe au bord $\partial \tilde{S}$. Les fibres de la fibration en cercles de $T_1 \tilde{S}$ sont transverses à $\tilde{\mathscr{F}}^s$ et les applications d'holonomie induites sur ces fibres sont précisément les homéomorphismes π_{pq}. Ainsi, la différentiabilité des π_{pq} est équivalente à celle du feuilletage $\mathscr{F}^s$ (ou, d'ailleurs, de $\mathscr{F}^u$ puisque ces deux feuilletages sont conjugués par l'involution de $T_1 S$ envoyant un vecteur sur son opposé).

Plaçons nous donc dans l'hypothèse où ces feuilletages sont de classe C^2. La première étape consiste à montrer qu'ils sont en fait transversalement projectifs. Un disque D transverse à ϕ_t est muni d'une forme d'aire Ω (provenant de la forme de Liouville) et de deux champs de directions L^s et L^u, traces de $\mathscr{F}^s$ et $\mathscr{F}^u$. Ces structures permettent de définir une métrique pseudo-riemannienne q sur D: si v est un vecteur tangent à D et si v_u et v_s sont ses composantes sur L^s et L^u, on pose $q(v) = \Omega(v_u, v_s)$. C'est une métrique de classe C^2 si $\mathscr{F}^s$ est de classe C^2 et L^s et L^u sont les directions isotropes de q. On a ainsi construit une métrique pseudo-riemannienne sur le fibré normal à ϕ_t, évidemment invariante par ϕ_t. La courbure de q est donc une fonction continue constante sur les orbites de ϕ_t; elle est donc constante car ϕ_t est transitif.

En d'autres termes, tous les disques transverses sont localement isométriques à un même plan lorentzien de courbure constante. Il est très facile d'en conclure que $\mathscr{F}^s$ est muni d'une structure transverse projective (ou affine si la courbure de q était nulle). Chaque fibre de $T_1 S$, étant transverse à $\mathscr{F}^s$, est ainsi équipée d'une structure projective ou affine et les applications d'holonomie préservent cette structure. Puisque nous connaissons a priori la topologie de l'action de Γ sur le cercle, on montre que c'est en fait le cas projectif qui se présente et que l'action de Γ sur le cercle est C^2-conjuguée à celle d'un sous-groupe discret co-compact de PSL$(2, \mathbb{R})$.

Ainsi, si le feuilletage $\mathscr{F}^s$ est de classe C^2, il est C^2-conjugué au feuilletage stable du flot géodésique d'une métrique à courbure -1. Il faut encore en conclure que la métrique considérée est elle même à courbure constante. On dispose actuellement de deux méthodes pour y parvenir. La première consiste à développer encore des arguments analogues à ceux utilisés plus haut et à montrer qu'en fait ϕ_t est conjugué au flot géodésique d'une métrique à courbure constante. On peut alors conclure en appliquant le théorème de rigidité de A. Katok [27] qui caractérise la courbure constante en termes d'entropie. La seconde méthode est due à Y. Mitsumatsu et elle est reliée à l'invariant de Godbillon-Vey. Nous avons choisi de décrire ici cette méthode car cet invariant interviendra encore par la suite.

Rappelons la définition de cet invariant. Soit $\mathscr{F}$ un feuilletage de codimension 1, transversalement orientable et de classe C^2 sur une variété M et soit ω une 1-forme différentielle qui le définit. Il existe alors, d'après le théorème de Frobenius, une forme ω_1 telle que $d\omega = \omega \wedge \omega_1$. Il se trouve que la 3-forme $\omega_1 \wedge d\omega_1$ est fermée et que sa classe de cohomologie ne dépend que de $\mathscr{F}$; c'est la classe de Godbillon-Vey de $\mathscr{F}$, notée GV$(\mathscr{F})$. Lorsque M est une variété fermée orientée de dimension 3, l'évaluation de GV$(\mathscr{F})$ sur la classe fondamentale de M est un nombre réel; c'est l'invariant de Godbillon-Vey que nous noterons $gv(\mathscr{F})$. Par exemple, considérons le cas du feuilletage stable $\mathscr{F}^s$ du flot géodésique d'une métrique à courbure constante -1. Nous savons que le fibré $T_1 S$ s'identifie à un espace homogène PSL$(2, \mathbb{R})/\Gamma$. Sur PSL$(2, \mathbb{R})$, il existe une base de formes invariantes à droite ω, ω_1, ω_2 telle que:

$$d\omega = \omega \wedge \omega_1; \qquad d\omega_1 = \omega \wedge \omega_2; \qquad d\omega_2 = \omega_1 \wedge \omega_2.$$

La forme ω passe au quotient sur $T_1 S$ et définit le feuilletage $\mathscr{F}^s$. L'invariant de Godbillon-Vey est alors facile à calculer: c'est le volume de $T_1 S$. Le théorème de Gauss-Bonnet donne:

$$gv(\mathscr{F}^s) = 4\pi^2 \chi(S).$$

Dans [29], Y. Mitsumatsu eut l'idée de calculer l'invariant de Godbillon-Vey du feuilletage stable du flot géodésique d'une métrique à courbure négative variable, en supposant celui-ci de classe C^2. Voici le principe de son calcul. Sur $T_1 S$, la théorie du repère mobile permet de construire trois champs de vecteurs V, X_1, X_2 qui forment en chaque point un repère orthonormé et sont tels que:

i) V est le champ associé à l'action de $SO(2)$ sur T_1S,

ii) X_1 engendre le flot géodésique,

iii) $[V, X_1] = X_2; [V, X_2] = -X_1; [X_1, X_2] = -(k \circ p)V$ où k est la courbure de S et p est la projection de T_1S sur S.

Il existe une fonction F définie sur T_1S telle que le feuilletage stable $\mathscr{F}^s$ soit engendré par X_1 et $X_2 + FV$. La condition d'intégrabilité est facile à écrire; elle est équivalente à l'équation de Ricatti:

$$X_1(F) + F^2 + k \circ p = 0$$

où $X_1(F)$ désigne la dérivée de la fonction F dans la direction X_1. On dispose alors de toutes les données nécessaires à l'évaluation de l'invariant $gv(\mathscr{F}^s)$. Tous calculs faits, on trouve:

$$gv(\mathscr{F}^s) = 4\pi^2\chi(S) - 3 \int V(F)^2 \, d\text{vol}.$$

Revenons à notre situation. Si le feuilletage $\mathscr{F}^s$ est de classe C^2, nous avons vu qu'il est C^2-conjugué au feuilletage stable du flot géodésique d'une métrique à courbure -1. L'invariant $gv(\mathscr{F}^s)$ est donc égal à $4\pi^2\chi(S)$. La formule précédente montre alors que $V(F)$ est identiquement nul, c'est-à-dire que F est constant sur les fibres de p. Il est facile d'en conclure que la courbure est constante.

4. Le problème de l'invariance topologique de la classe de Godbillon-Vey

Soit $\mathscr{F}^s$ le feuilletage stable du flot géodésique d'une métrique de S à courbure -1. Nous avons déjà vu que l'invariant de Godbillon-Vey de $\mathscr{F}^s$ ne dépend pas du choix de cette métrique. Est-il possible de perturber $\mathscr{F}^s$ parmi les feuilletages de classe C^∞, hors de l'espace des feuilletages stables de flots géodésiques, de façon à faire varier la classe de Godbillon-Vey?

Nous savons que $\mathscr{F}^s$ peut être défini par une forme ω telle que $d\omega = \omega \wedge \omega_1$ avec $\omega_1 \wedge d\omega_1$ non singulière. Il est clair que cette même propriété est satisfaite pour tous les feuilletages $\mathscr{F}$, disons de classe C^∞, qui sont C^3-proches de $\mathscr{F}^s$. En dérivant la relation $d\omega = \omega \wedge \omega_1$, on constate qu'il existe une forme ω_2 telle que $d\omega_1 = \omega \wedge \omega_2$. Les formes $\omega, \omega_1, \omega_2$ sont alors linéairement indépendantes partout et on peut donc définir deux champs de vecteurs X et Y, tangents à $\mathscr{F}$ par:

$$\omega(X) = 0; \qquad \omega_1(X) = 1; \qquad \omega_2(X) = 0$$

$$\omega(Y) = 0; \qquad \omega_1(Y) = 0; \qquad \omega_2(Y) = 1$$

Des calculs extrêmement simples montrent alors que les champs X et Y préservent le volume $\omega \wedge \omega_1 \wedge \omega_2$ et que $[X, Y] = -Y$. Cette dernière relation décrit l'algèbre de Lie du groupe affine GA de la droite, i.e. du groupe des transformations $x \mapsto ax + b$ $(a > 0)$. On conclut donc que le feuilletage $\mathscr{F}$ est défini par une action localement libre de GA sur T_1S, préservant le volume $\omega \wedge \omega_1 \wedge \omega_2$. Ces actions sont complètement classées dans [11], ce qui permet alors d'obtenir le résultat de rigidité suivant:

Théorème [11]. *Tout feuilletage de classe C^∞ qui est suffisamment C^3-proche du feuilletage stable du flot géodésique d'une métrique à courbure -1, est C^∞-conjugué au feuilletage stable du flot géodésique d'une (autre) métrique à courbure -1.*

Résumons la situation. Toutes les actions de Γ de classe C^2 sur le cercle, de classe d'Euler maximale, sont topologiquement conjuguées. Deux actions projectives provenant de métriques à courbure -1 ne sont C^1-conjuguées que si les métriques correspondantes sont isométriques. Une action de classe C^∞, qui est C^3-proche d'une de ces actions projectives est C^∞-conjuguée à une autre action projective. Peut-on globaliser ce dernier énoncé?

Problème. *Soit Γ le groupe fondamental d'une surface compacte orientée S de genre supérieur ou égal à 2. Considérons une action de Γ sur le cercle par difféomorphismes directs de classe C^∞. Si le nombre d'Euler de l'action est maximal (i. e. égal à $\pm\chi(S)$), peut-on affirmer que l'action est C^∞-conjuguée à une action projective (associée à une métrique à courbure -1)?*

Ainsi, les déformations des feuilletages stables de flots géodésiques, dans l'espace des feuilletages de classe C^∞, gardent un type topologique constant mais ont aussi un invariant de Godbillon-Vey constant. C'est l'une des motivations de la question suivante.

Problème. *Soient $(M_1, \mathscr{F}_1)$ et $(M_2, \mathscr{F}_2)$ deux variétés compactes feuilletées, de codimension 1, transversalement orientables, et de classe C^2. On suppose qu'il existe un homéomorphisme h entre M_1 et M_2 qui envoie $\mathscr{F}_1$ sur $\mathscr{F}_2$. L'isomorphisme h^* entre les cohomologies de M_1 et M_2 envoie-t-il la classe de Godbillon-Vey de $\mathscr{F}_2$ sur celle de $\mathscr{F}_1$?*

Si l'on consent à un élargissement du domaine de définition de la classe de Godbillon-Vey, la réponse à la question précédente est négative. Nous allons mentionner rapidement deux tels élargissements.

Il n'est pas possible de définir un invariant pour les feuilletages de classe C^1. Ceci résulte de la contractibilité du classifiant des feuilletages transversalement orientables, de codimension 1 et de classe C^1, démontrée par T. Tsuboi [33]. La classe de Godbillon-Vey est cependant invariant par conjugaison de classe C^1 (voir [16] et [30]). D'autre part, dans [24], S. Hurder et A. Katok définissent un invariant pour les feuilletages de classe $C^{1+\varepsilon}$, avec $\varepsilon > 1/2$. En particulier, on peut calculer l'invariant du feuilletage stable du flot géodésique d'une métrique à courbure négative variable. Dans ce cas, la formule de Y. Mitsumatsu est encore valable et montre que l'invariant de Godbillon-Vey varie effectivement si la courbure varie. Cette même formule montre d'ailleurs que l'invariant prend sa valeur maximale précisément sur les métriques à courbure constante.

Dans une autre direction, on peut aussi définir l'invariant de Godbillon-Vey pour les feuilletages de classe C^2 par morceaux [10] et on peut aussi se poser le problème précédent dans ce contexte élargi. Nous y répondons négativement à travers le résultat suivant:

Théorème [14]. *L'action du groupe fondamental d'une surface à courbure négative sur le bord de son revêtement universel est topologiquement conjuguée à une action affine par morceaux.*

Ceci donne effectivement une réponse négative au problème car un feuilletage transversalement affine par morceaux a un invariant de Godbillon-Vey nul.

Pour démontrer ce dernier théorème, on utilise le fait que le flot géodésique ϕ_t possède une section de Birkhoff, c'est-à-dire qu'il existe une surface à bord Σ dans $T_1 S$ dont l'intérieur est transverse à ϕ_t et dont le bord est constitué d'orbites périodiques de ϕ_t. De plus, toute orbite de ϕ_t coupe Σ une infinité de fois et on peut donc définir une application de premier retour $r : \Sigma \to \Sigma$ (voir [4, 9]). Les feuilletages $\mathscr{F}^s$ et $\mathscr{F}^u$ tracent sur Σ deux feuilletages transverses ayant un comportement bien contrôlé au voisinage du bord. On montre alors que r est topologiquement conjugué à un difféomorphisme pseudo-Anosov qui agit de manière affine sur l'espace de ses feuilles stables. Par conséquent, $\mathscr{F}^s$ est topologiquement conjugué à un feuilletage transversalement affine par morceaux, les cassures correspondant aux feuilles contenant les courbes du bord de Σ. Une description explicite des homéomorphismes affines par morceaux ainsi obtenus a été récemment donnée par N. Hashiguchi [21].

D'une certaine façon, ce contre-exemple à l'invariance topologique de la classe de Godbillon-Vey est peu satisfaisant. On est tenté de penser que par cette conjugaison topologique, la classe de Godbillon-Vey s'est concentrée dans les singularités du feuilletage affine par morceaux. Il est effectivement possible de définir un "invariant discret" qui tient compte de ces singularités mais on montre que, même en ajoutant cette contribution, la classe n'est pas un invariant topologique.

Les deux extensions du domaine de définition de la classe de Godbillon-Vey que nous venons d'évoquer ont été récemment unifiées par T. Tsuboi [34]. Il définit un pseudogroupe d'homéomorphismes de la droite qui contient à la fois les difféomorphismes de classe $C^{1+\varepsilon}$ ($\varepsilon > 1/2$) et les homéomorphismes qui sont de classe C^2-par morceaux. Il montre ensuite comment définir une classe de Godbillon-Vey pour les feuilletages de codimension 1 dont la structure transverse est modelée sur ce pseudogroupe. Il semble bien que ceci soit "le" domaine maximal d'existence de l'invariant.

Le problème de l'invariance topologique dans le cadre des feuilletages de classe C^2 reste cependant ouvert.

References

1. Ballman, W., Gromov, M., Schroeder, V.: Manifolds of non-positive curvature. Progress in Mathematics **61** (1985)
2. Barge, J., Ghys, E.: Surfaces et cohomologie bornée. Invent. math. **92** (1988) 509–526
3. Benoist, Y., Foulon, P., Labourie, F.: Flots d'Anosov à distributions stables et instables différentiables. C.R. Acad. Sci. Paris 1990
4. Birkhoff, G.D.: Dynamical systems with two degrees of freedom. Trans. Amer. Math. Soc. **18** (1917) 199–300
5. Feres, R.: Geodesic flows on manifolds of negative curvature with smooth horospheric foliations. PhD Caltech. 1989

6. Feres, R., Katok, A.: Invariant tensor fields of dynamical systems with pinched Lyapunov exponents and rigidity of geodesic flows. Ergod. Theor. Dynam. Sys. **3** (1989) 427–433
7. Floyd, W.: Group completions and limit sets of Kleinian groups. Invent. math. **57** (1980) 205–218
8. Foulon, P., Labourie, F.: Flots d'Anosov à distributions de Liapounov différentiables. C.R. Acad. Sci. Paris **309** (1989) 255–260
9. Fried, D.: Transitive Anosov flows and pseudo-Anosov maps. Topology **22** (1983) 299–303
10. Fuchs, D.B., Gabrielov, A.M., Gelfand, I.M.: The Gauss-Bonnet theorem and the Atiyah-Patodi-Singer functionals for the characteristic classes of foliations. Topology **15** (1976) 165–188
11. Ghys, E.: Actions localement libres du groupe affine. Invent. math. **82** (1985) 479–526
12. Ghys, E.: Groupes d'homéomorphismes du cercle et cohomologie bornée. Contemp. Math. **58**, part III (1987) 81–105
13. Ghys, E.: Classe d'Euler et minimal exceptionnel. Topology **26** (1987) 93–105
14. Ghys, E.: Sur l'invariance topologique de la classe de Godbillon-Vey. Ann. Inst. Fourier **37** (1987) 59–76
15. Ghys, E.: Flots d'Anosov dont les feuilletages stables sont différentiables. Ann. Sci. Ec. Norm. Sup. **20** (1987) 251–270
16. Ghys, E., Tsuboi, T.: Différentiabilité des conjugaisons entre systèmes dynamiques de dimension 1. Ann. Inst. Fourier **38** (1988) 215–244
17. Goldman, W.: Discontinuous groups and the Euler class. PhD thesis, Dept. of Mathematics, University of California at Berkeley, 1980
18. Gromov, M.: Volume and bounded cohomology. Publ. Math. I.H.E.S. **56** (1982) 5–100
19. Gromov, M.: Hyperbolic groups. In: Essays in group theory (ed. S.M. Gersten). M.S.R.I. Pub. **8** (1987) 7–263
20. Hamenstadt, U.: A geometric characterization of negatively curved locally symmetric spaces. Preprint 1990
21. Hashiguchi, N.: PL-representations of the geodesic flows on closed surfaces. Preprint 1990
22. Hirsch, M., Pugh, C.: Smoothness of horocycle foliations. J. Diff. Geom. **10** (1975) 225–238
23. Hopf, E.: Statistik der geodätischen Linen in Mannigfaltigkeiten negativer Krümmung. Ber. Verh. Süchs. Akad. Wiss. Leipzig **91** (1939) 261–304
24. Hurder, S., Katok, A.: Differentiability, rigidity and Godbillon-Vey classes for Anosov flows. Preprint IHES, 1987
25. Jekel, S.: An elementary formula and a bound for the Euler class. Preprint
26. Kanai, M.: Geodesic flows of negatively curved manifolds with smooth stable and unstable foliations. Ergod. Theor. Dynam. Sys. **8** (1988) 215–239
27. Katok, A.: Entropy and closed geodesics. Ergod. Theor. Dynam. Sys. **2** (1982) 339–366
28. Matsumoto, S.: Some remarks on foliated S^1 bundles. Invent. math. **90** (1987) 343–358
29. Mitsumatsu, Y.: A relation between the topological invariance of the Godbillon-Vey invariant and the differentiability of Anosov foliations. Adv. Stud. Pure math. **5** (1985) 159–167.
30. Raby, G.: Invariance des classes de Godbillon-Vey par C^1-difféomorphismes. Ann. Inst. Fourier **38** (1988) 205–213
31. Sullivan, D.: Discrete conformal groups and measurable dynamics. Bull. Amer. Math. Soc. **6** (1982) 53–73
32. Takamura, M.: Semi-conjugacy and a theorem of Ghys. Preprint
33. Tsuboi, T.: On the foliated products of class C^1. Ann. Math. **130** (1989) 227–271
34. Tsuboi, T.: Area functional and Godbillon-Vey cocycles. Preprint
35. Wood, J.: Bundles with totally disconnected structure group. Comm. Math. Helv. **46** (1971) 257–273

Metric and Topological Measurements of Manifolds

Karsten Grove

Department of Mathematics, University of Maryland, College Park, MD 20742, USA

A basic question in riemannian geometry asks how properties detected by metric measurements of a riemannian manifold are reflected in properties detected by topological measurements. If (M, g) is riemannian manifold, a *metric measurement* can by definition be given in terms of its distance function. A *topological measurement* on the other hand is given in terms of algebraic topological invariants of M and/or of objects associated with M. Examples of invariants resulting from such measurements are e.g. *diameter, volume*, and *sectional curvature bounds* on the metric side, and *Betti numbers, characteristic numbers* etc. on the topological side.

Here we consider only metric invariants that assign to any (closed) riemannian manifold a number. These invariants can therefore be thought of as functions on the collection $\mathcal{M}$ of all isometry classes of (closed) riemannian manifolds, and as such they divide this class into smaller natural subclasses. Since topological properties are not affected by scaling of the metrics, one needs to normalize, or at least bound the *size* of (M, g). Quantities commonly chosen for this purpose are diameter or volume.

Ultimately one would like to give a complete topological description of a riemannian manifold making as few metric measurements as possible. As a first step one tries to bound the topology in terms of its geometry. Results of that type are usually referred to as *finiteness theorems*. In metrically extreme cases the goal is to determine all the possible topological types. Results of this kind are called *pinching theorems*.

As an illustration, suppose we normalize all closed, connected riemannian n-manifolds, M, $n \geq 2$, so as to have the same diameter as the unit n-sphere, i.e. diam $M = \pi$. The function *minsec*, that assigns to any such manifold the minimum of its sectional curvatures, is then bounded above by 1, i.e. minsec $M \leq 1$, according to the classical Bonnet-Myers theorem. For the superlevel sets, minsec $\geq k$ one has the following finiteness theorem.

0.1 Betti Number Theorem [G2]. *Let $n \geq 2$ be an integer and $k \leq 1$ a real number. There is a constant $C = C(n, k)$, such that for any field, F and any closed riemannian n-manifold, M with diam $M = \pi$ and sec $M \geq k$ one has* dim $H_*(M; F) \leq C$.

Proceedings of the International Congress
of Mathematicians, Kyoto, Japan, 1990
© The Mathematical Society of Japan, 1991

Near the extreme value 1 of minsec there is only one manifold topologically as stated in the following pinching theorem.

0.2 Diameter Sphere Theorem [GS]. *Any closed riemannian n-manifold M, $n \geq 2$, with* diam $M = \pi$ *and* sec $M > 1/4$ *is homeomorphic to the n-sphere.*

The bound sec $M > 1/4$ is optimal in this theorem as e.g. the real projective space of constant curvature $1/4$ shows. A corresponding *rigidity theorem* for manifolds with diam $M = \pi$ and sec $M \geq 1/4$ was proved in [GG1, 2].

In the general setting of (0.1), one cannot expect much more information (other than possibly improving the constant C [A]). To get more refined information it seems therefore natural to subdivide the superlevel sets minsec $\geq \ell$ in terms of other metric measurements.

The purpose here is to report on recent progress in this general area. This progress involves on the metric side the use of critical point theory for distance functions as conceived in [GS] (cf. [G2] and for a survey [Gr]), and the use of Hausdorff convergence of metric spaces as initiated in [G1] (cf. [P2] for a survey). On the topological side, powerful results from controlled topology as developed in [CF, F] and [Q1, 2, 3] play a crucial role.

1. Distance and Volume Comparison

Throughout we let M denote a closed, connected riemannian n-manifold with $n \geq 2$. Following Rinow [R], S_ℓ^n is the complete, simply connected n-dimensional space form of constant curvature ℓ. The distance function on M, S_ℓ^n, or any other metric space will be denoted by d. To distinguish point in S_ℓ^n from points in other metric spaces we use the notation $\bar{p}, \bar{q}, \ldots$, etc., rather than $p, q, \ldots$, etc.

For each $p \in M$ and $r > 0$, let $B(p, r)$, resp. $D(p, r)$ be the open, resp. closed r-ball in M centered at p. Moreover, let $\mathscr{D}(p) \subset T_p M$ be the region bounded by the tangent cut locus $\mathscr{C}(p)$, i.e. $v \in \mathscr{D}(p)$ if and only if the exponential map $\exp_p : T_p M \to M$ maps the segment, $[0, v]$ in $T_p M$ to a segment, i.e. minimal geodesic in M.

The *injectivity radius* of M, inj M, is by definition the largest $r > 0$ so that $\exp_p | B(0_p, r)$ is injective for all $p \in M$. The smallest $R > 0$ so that some $\exp_p | D(0_p, R)$ is surjective is easily seen to be the number $\min_p \max_q d(p, q)$, which makes sense in any compact metric space. This number is called the *radius* of M and is denoted by rad M.

Obviously rad $M \leq$ diam $M \leq 2$rad M. In particular, imposing a bound on the diameter is equivalent to imposing a bound on the radius.

Now suppose sec $M \geq \ell$. This condition can be viewed in the following way: For each $p \in M$ replace the euclidean metric on $B(0_p, r) \subset T_p M$ by a constant curvature ℓ metric via a radial conformal change $(r \leq \pi/\sqrt{\ell}$ if $\ell > 0)$. Then $\mathscr{D}(p)$ is a proper compact subset of S_ℓ^n for every $p \in M$, except when $\ell > 0$ and M isometric to S_ℓ^n. In the latter case we interpret $\mathscr{D}(p)$ as S_ℓ^n. The Toponogov triangle comparison theorem is then equivalent to

1.1 Distance Comparison Theorem. *The exponential map* $\exp_p : \mathcal{D}(p) \to M$ *is distance nonincreasing, when* $\mathcal{D}(p) \subset S_k^n$ *is given the induced metric from* S_k^n.

This is the basis also for volume comparison of metrically defined subsets of M with corresponding subsets in S_k^n. For clarity, we list three situations of obvious interest although they are all special cases of one general result [GP2] (cf. also [Du]).

1.2 Half Spaces. Fix $p \in M$ and a closed $Q \subset M$. Consider $\bar{p} = \exp_p^{-1}(p)$ and $\bar{Q} = \exp_p^{-1}(Q)$ in $\mathcal{D}(p) \subset S_k^n$. For the half spaces $H(p, Q) = \{x \in M | d(x, p) \le d(x, Q)\}$ and $H(\bar{p}, \bar{Q})$ in M and S_k^n respectively we have

$$\text{vol } H(p, Q) \le \text{vol } H(\bar{p}, \bar{Q}).$$

1.3. Swiss Cheeses. Define the swiss cheese K relative to $D(p, R)$ and $r : Q \to \mathbb{R}_+$ as the complement $K((Q, r); (p, R)) = D(p, R) - \bigcup_{q \in Q} B(q, r(q))$. Then

$$\text{vol } K((Q, r); (p, R) \le \text{vol } K((\bar{Q}, \bar{r}); (\bar{p}, R))$$

where $\bar{r} = r \circ \exp_p : \bar{Q} \to \mathbb{R}_+$.

1.4 Ball Collection. For the union $D(Q, r) = \bigcup_{q \in Q} D(q, r(q))$ one has $\text{vol } D(Q, r) \le \text{vol } D(I(Q), r \circ I^{-1})$, provided $I : Q \to I(Q) \subset S_k^n$ is an isometry.

The estimates (1.2)–(1.4) and (1.7), (1.8) below do not hold under the weaker curvature assumption $\text{Ric } M \ge (n - 1)k$. This, however, is sufficient for the following simple extension of the so-called Bishop-Gromov volume comparison theorem.

1.5 Relative Volume Comparison. *For finite* $Q \subset M$, *the function* $\text{vol } D(Q, R)/v_k^n(R)$ *is non-increasing in* R. *Here* $v_k^n(R) = \text{vol } D(\bar{p}, R)$ *in* S_k^n.

This theorem is the basis for various *packing arguments*. A subset $Q \subset M$ is called a *weak R-net* in M if $D(Q, R) = M$. Examples of such nets are provided by maximal sets Q for which $B(q, R/2)$, $q \in Q$ are mutually disjoint. For fixed $n \ge 2$, $k \in \mathbb{R}$ and $D > 0$ this together with (1.5) yields a *covering function*.

1.6 $N(\varepsilon) = C_1 \varepsilon^{-n}, \quad \varepsilon \in (0, D], \quad C_1 = C(n, k, D) > 0.$

For any closed riemannian n-manifold M with $\text{diam } M \le D$ and $\text{Ric } M \ge (n - 1)k$, i.e. for every $\varepsilon \in (0, D]$, any such M can be covered by $\le N(\varepsilon)$ closed ε-balls.

The relative volume comparison (1.5) can also be used to estimate the right hand sides of (1.2) and (1.3) in an important special case: (i) there is an $\bar{r} > 0$ such that $d(\bar{p}, \bar{q}) = \bar{r}$, $\bar{q} \in \bar{Q}$, and (ii) the set of directions of segments $\overline{pq}$, $\bar{q} \in \bar{Q}$ form a weak θ-net, $\theta \ge \pi/2$ in the unit sphere $S_1^{n-1} \subset T_{\bar{p}} S_k^n$. If furthermore $\bar{q}_1, \bar{q}_2 \in S_k^n$ are chosen so that $d(\bar{p}, \bar{q}_1) = d(\bar{p}, \bar{q}_2) = \bar{r}$ and $\text{ang}(\overline{pq}_1, \overline{pq}_2) = 2\pi - 2\theta$, then

1.7 $\text{vol } H(\bar{p}, \bar{Q}) \le \text{vol } H(\bar{p}, \{\bar{q}_1, \bar{q}_2\}),$ and

1.8 $\text{vol } K((\bar{Q}, r); (\bar{p}, R)) \le \text{vol } K(\{\bar{q}_1, \bar{q}_2\}, r); (\bar{p}, R)),$ r constant.

Now fix $v > 0$ in addition to N, k and D as above. Combining (1.2) and (1.7) with a *critical point theory argument* for $d: M \times M \to \mathbb{R}$ considered as the distance function in $M \times M$ from the diagonal $\Delta \subset M \times M$ gives constants $C_2 = C_2(n, k, D, v) \geq 1$ and $R = R(n, k, D, v) > 0$ such that

1.9
$$\varrho(\varepsilon) = C_2\varepsilon, \qquad \varepsilon \in (0, R]$$

is a *contractibility function* for any closed riemannian n-manifold M satisfying $\sec M \geq k$, $\operatorname{diam} M \leq D$ and $\operatorname{vol} M \geq v$, i.e. every ε-ball in such M is contractible in the concentric $C_2 \cdot \varepsilon$-ball.

The existence of covering and contractibility functions as in (1.6) and (1.9) are key ingredients in the finiteness and pinching theorems reported on here.

2. The Gromov-Hausdorff Topology

In this section we outline topological and geometric properties of Gromov-Hausdorff limit spaces (cf. [G1]) suited for our purposes.

Let $X, Y, Z, X_i, i = 1, 2, \ldots$ be compact metric spaces. If X and Y are isometrically embedded in Z, the classical *Hausdorff distance*, d_H^Z satisfies $d_H^Z(X, Y) < \varepsilon$ if and only if $Y \subset B(X, \varepsilon)$ and $X \subset B(Y, \varepsilon)$. For the *Gromov-Hausdorff distance* d_{GH} one has $d_{GH}(X, Y) < \varepsilon$ if and only if $d_H^Z(X, Y) < \varepsilon$ from some metric on $Z = X \amalg Y$ extending the ones on X and Y. Similarly $X = \lim X_i$ in the Gromov-Hausdorff topology if and only if $d_H^Z(X, X_i) \to 0$, $i \to \infty$ in some extended metric on $Z = X \amalg_i X_i$.

2.1 Precompactness Theorem [G1]. *A class χ of compact metric spaces is precompact if and only if there is a common covering function for all $X \in \chi$.*

Now let $\chi(N, \varrho)$ be the class of compact metric spaces which have N in (1.6) as a covering function and ϱ in (1.9) as a contractibility function. Since all spaces having (1.6) as covering function have dimension $\leq n$ by [PS], a result of Borsuk [Bo] combined with (2.1) implies (cf. also [P1]).

2.2 Compactness Theorem [GPW]. *The class $\chi(N, \varrho)$ is compact and $\dim X \leq n$ for any $X \in \chi(N, \varrho)$.*

Using the contractibility properties of $X \in \chi(N, \varrho)$ one can construct explicit homotopy equivalences between nearby spaces as in [GP1] (see [P1]). In particular

2.3 Homotopy Finiteness Theorem [P1]. *The class $\chi(N, \varrho)$ contains only finitely many (simple) homotopy types.*

The application of controlled topology is based on the important observation that the homotopy equivalences and corresponding homotopies alluded to above are controlled in size in terms of $d_{GH}(X, Y)$. In the context of manifolds, this together with the results of Begle [B] and Quinn [Q3] proves

2.4 Almost Compactness Theorem [GPW]. *Let $\mathcal{M}(N, \varrho)$ be the subset of closed topological n-manifolds in $\chi(N, \varrho)$. Any $X \in \overline{\mathcal{M}(N, \varrho)}$ is a generalized n-manifold and if $n \neq 3$ it admits a manifold resolution.*

Since $X \times S^1 \times S^1$ satisfies the disjoint disc property (cf. [D]) it follows from Edwards approximation theorem [E] that it is a manifold. Consequently, by results of Chapman and Ferry [CF, F] there are only finitely many homeomorphism types among $M \times S^1 \times S^1$, $M \in \mathcal{M}(N, \varrho)$. The controlled h-cobordism theorem [Q1, 2] is then used to complete the proof of

2.5 Topological Finiteness Theorem [GPW]. *The class of closed n-manifolds, $n \neq 3$, with metrics having (1.6) as covering functions and (1.9) as contractibility function contains at most finitely many homeomorphism types.*

We conclude this section with a brief discussion of geometric properties of limit spaces $X = \lim M_i$, where $\sec M_i \geq k$ and $\operatorname{diam} M \leq D$. First of all, X is an inner metric space, i.e. the distance between points in X is the infimum of lengths of curves joining them. In particular, one has the notion of geodesics in X. Moreover,

2.6 Exponential Maps [GP2]. *Let $p \in X = \lim M_i$, $\sec M_i \geq k$ and $\bar{p} \in S^n_k$. There is a compact subset $\mathscr{D}(p) \subset S^n_k$ and a distance nonincreasing surjective map $\exp_p : \mathscr{D}(p) \to X$. Moreover, $\exp_p$ maps segments from $\bar{p} \in \mathscr{D}(p) \subset S^n_k$ to segments in X from p, and any segment from p is the image of a segment from $\bar{p}$.*

It should be pointed out that the pairs $(\mathscr{D}(p), \exp_p)$ in (2.6) are by no means unique! Using the nontrivial fact that geodesics in X are limits of geodesics in M_i one gets

2.7 Curvature for Limits [GP2]. *$X = \lim M_i$, $\sec M_i \geq k$ has Toponogov curvature, $\sec X \geq k$, i.e. standard distance comparison holds for geodesic triangles in X.*

As an interesting consequence, X has everywhere constant dimension. This is of interest in situations of collapse (cf. also [Fu]).

3. The Volume Function

In this and the next section we consider sub- or superlevel sets of functions restricted to the classes of $\mathcal{M}^D_k(n)$, and $\mathcal{M}^R_k(n)$ of closed connected riemannian n-manifolds M, $n \geq 2$ with $\sec M \geq k$ and $\operatorname{diam} M \leq D$, $\operatorname{rad} M \leq R$ respectively.

The following finiteness theorem is an immediate corollary of (1.6), (1.9), (2.3), (2.5) and the fact that every closed topological n-manifold carries at most finitely many inequivalent smooth structures, when $n \neq 4$ (cf. [KS] for $n \geq 5$).

3.1 Diffeo/Top-Theorem [GP1, GPW]. *Fix $n \geq 2$, $k \in \mathbb{R}$ and D, $v > 0$. The class of closed riemannian n-manifolds M with* sec $M \geq k$, diam $M \leq D$ *and* vol $M \geq v$ *contains at most finitely many diffeomorphism-, homeomorphism-, and homotopy types when $n \neq 3, 4$, $n = 4$, and $n = 3$ respectively.*

We now turn to the corresponding metrically extreme cases of small and large volume.

Let $\{M_i\}$ be a sequence in the $\mathcal{M}_k^D(n)$ where vol $M_i \to 0$ as $i \to \infty$. Then a subsequence will converge in the Gromov-Hausdorff topology (cf. (1.6) and (2.1)) to a *lower dimensional* inner metric space X with diam $X \leq D$ and Toponogov curvature sec $X \geq k$. In the exceptional case where X is a riemannian manifold, a thorough investigation has been carried out in [Y2]: For large i, M_i fibers almost riemannian over X, and a finite cover of the fibers, F fiber almost riemannian over the $b_1(F)$-dimensional flat torus.

In the (weakly) extreme case of manifolds M with sec $M \geq k$, diam $M \leq D$ and vol $M \leq \varepsilon$, ε small, one would expect some similar kind of decomposition of M. This of course is related to the structure of general inner metric spaces X with sec $X \geq k$. For progress in this direction we refer to [BGP].

To treat the other extreme case of manifolds with large volume one first needs to have optimal bounds. Such bounds are not known for the class $\mathcal{M}_k^D(n)$. If on the other hand sec $M \geq k$ and rad $M \leq R$ we get from (1.4), (1.3) and (1.8) (cf. [GP4]) that

$$3.2 \qquad \text{vol } M \leq \begin{cases} v_k^n(R); & R \leq \pi/2\sqrt{k} \text{ if } k > 0 \\ \dfrac{R}{\pi/\sqrt{k}} \text{ vol } S_k^n & \text{if } k > 0 \text{ and } R \geq \pi/2\sqrt{k} \end{cases}$$

There are riemannian metrics on S^n (and $\mathbb{R}P^n$) showing that (3.2) is optimal (cf [GP4]). Moreover, using (1.3), (1.4), (1.8), (1.9), (2.6), (2.7) and controlled topology as in Sect. 2 one can prove the following pinching theorem in this (strongly) extreme case.

3.3 Large Volume Theorem [GP4]. *Given $n \geq 2$, $k \in \mathbb{R}$ and $R \in \mathbb{R}_+$ ($\leq \pi/2\sqrt{k}$ if $k > 0$). There is an $\varepsilon = \varepsilon(n, k, R) > 0$ such that any riemannian n-manifold M with* sec $M \geq k$, rad $M \leq R$ *and* vol $M \geq v_k^n(R) - \varepsilon$ *is topologically either S^n or $\mathbb{R}P^n$.*

In the case sec $M \geq k > 0$ and rad $M > \pi/2\sqrt{k}$, of course M is S^n topologically by (0.2). In all cases, however, there are (*singular*) *constant curvature k models* with equality in (3.2), and manifolds with volume close to this are metrically similar to these models (cf. [GP4]). In the special cases $k > 0$ and $R = \pi/2\sqrt{k}$, $\pi/\sqrt{k}$ corresponding to the only nonsingular models, it is possible to conclude diffeomorphism [OSY].

These pinching theorems may also be viewed as a solution to a generalized analog of Aleksandrov's area problem for convex surfaces in $\mathbb{R}^3$ [Al].

4. The Excess and Other Shape Invariants

The Theorems 0.1, 0.2, 3.1 and 3.3 are in essence exhaustive. Clearly therefore, in order to gain more information about individual manifolds $M \in \mathscr{M}_{kv}^D(n)$ where vol $M \geq v$ it is necessary to make more measurements.

Other than *maxsec*, *inj*, and to some extent, *FillRad* [G3], no such invariants seem to have been studied very extensively. This includes the *intermediate diameters*, diam $M = \text{diam}_0 M \geq \text{diam}_1 M \geq \cdots \geq \text{diam}_{n-1} M > \text{diam}_n M = 0$ considered by Urysohn [U] and more recently in [G3] (cf. also [P2]).

Here we discuss briefly a measurement based on the excess in the triangle inequality. Following [GP3] we define the *excess* of M by exc $M = \min_{(p,q)} \max_x (d(p, x) + d(x, q) - d(p, q))$. Obviously $0 \leq \text{exc } M \leq \text{diam } M$.

A smooth manifold M admits a Riemannian metric with exc $M = 0$ if and only if M is a twisted sphere (cf. [GP3]). On the other hand, any M admits a riemannian metric with exc $M/\text{diam } M$ arbitrarily small.

When restricted to $\mathscr{M}_{kv}^D(n)$, however, the sublevel sets exc $\leq \varepsilon$, ε small, contain only one topological type:

4.2 Exotic Sphere Theorem [GP3]. *Fix $n \geq 2$, $k \in \mathbb{R}$ and D, $v > 0$. There is an $\varepsilon = \varepsilon(n, k, D, v)$ such that any closed riemannian n-manifold M, with sec $M \geq k$, diam $M \leq D$, vol $M \geq v$ and exc $M \leq \varepsilon$, is a homotopy sphere.*

This result is also optimal in its context. Observe that in contrast to all other known sphere theorems in riemannian geometry, the metrically extreme case is represented not only by the unit sphere S_1^n but by any exotic sphere!

The other extreme exc $M = \text{diam } M$ occurs when M contains three points at maximal distance. This is related to the *triameter* of M, which is the second in an infinite sequence of invariants beginning with the diameter. That and another related infinite sequence beginning with twice the radius and the excess is currently under investigation in [GM].

5. Curvature Free Problems

We conclude in the generality we started, now scaling all closed riemannian n-manifolds M so that vol $M = \text{vol } S_1^n = v_1^n(\pi)$. In this generality we consider the superlevel sets inj $\geq i$.

5.1 Isoembolic Rigidity Theorem [Be]. *The injectivity radius of any closed riemannian n-manifold M so that vol $M = v_1^n(\pi)$ satisfies inj $M \leq \pi$, with equality if and only if M is isometric to S_1^n.*

5.2 Isoembolic Pinching Theorem [C2]. *For any $n = \mathbb{N}$ there is an $\varepsilon = \varepsilon(n)$, such that any closed riemannian n-manifold M with vol $M = v_1^n(\pi)$ and inj $M \geq \pi - \varepsilon$ is homeomorphic to S^n.*

5.3 Isoembolic Finiteness Theorem [GPW, Y1]. *Fix $n \in \mathbb{N}$ and $\varrho \in (0, \pi]$. There are at most finitely diffeomorphism-, homeomorphism-, or homotopy types of closed riemannian n-manifolds M satisfying* vol $M = v_1^n(\pi)$, inj $M \geq \varrho$ and $n \neq 3, 4$, $n = 4$, *or $n = 3$ respectively.*

The geodesic flow plays an essential role in the proof of these results. Here we only point out that a basic volume estimate from [C1] implies that the class of closed riemannian n-manifolds M with vol $M \leq v$ and inj $M \geq \varrho$ has a covering function as in (1.6). Since it has a contractibility function as in (1.9) with $C_2 = 1$ by assumption, (5.3) is a corollary of (2.3) and (2.5) just like (3.1) was.

The subclass consisting of closed riemannian n-manifolds M with Ric $M \geq (n - 1)\ell$, diam $M \leq D$ and inj $M \geq \varrho$ has recently been investigated in [AC]. In particular it is shown there, that this class contains at most finitely many diffeomorphism types also in dimensions 3 and 4.

References

[A]   U. Abresch: Lower curvature bounds, Toponogov's theorem, and bounded topology, I; II. Ann. Sci. Éc. Norm. Sup. **18**; **20** (1985/87) 563–633, 475–502

[Al]   A.D. Aleksandrov: Die innere Geometrie der konvexen Flächen. Academie-Verlag, Berlin 1955

[AC]   M.T. Anderson, J. Cheeger: C^α-Compactness for manifolds with Ricci curvature and injectivity radius bounded below. J. Diff. Geom. (to appear)

[B]   E.B. Begle: Regular convergence. Duke Math. J. **11** (1944) 441–450

[Be]   M. Berger: Une borne inférieure pour le volume d'une variétés riemannienne en function du rayon d'injectivité. Ann. Inst. Fourier **30** (1980) 259–265

[Bo]   K. Borsuk: On some metrizations of the hyper space of compact sets. Fund. Math. **41** (1955) 168–201

[BGP]   Y. Burago, M. Gromov, G. Perelman: A.D. Alexandrov's spaces with curvatures bounded from below I. Preprint

[CF]   T.A. Chapman, S. Ferry: Approximating homotopy equivalences by homeomorphisms. Amer. J. Math. **101** (1979) 583–607

[C1]   C.B. Croke: Some isoparimetric inequalities and eigenvalue estimates. Ann. Sci. Éc. Norm. Sup. **13** (1980) 419–435

[C2]   C.B. Croke: An isoembolic pinching theorem. Invent. math. **92** (1988) 385–387

[D]   R.J. Davermann: Detecting the disjoint disc property. Pac. J. Math. **93** (1981) 277–298

[Du]   O. Durumeric: Manifolds of almost half of the maximal volume. Proc. Amer. Math. Soc. **104** (1988) 277–283

[E]   R.D. Edwards: The topology of manifolds and cell-like maps. Proc. Int. Congress of Math. 1978. Acad. Sci. Fennica, Helsinki 1980, pp. 111–127

[F]   S. Ferry: Homotoping ε-maps to homeomorphisms. Amer. J. Math. **101** (1979) 567–582

[Fu]   K. Fukaya: These proceedings

[G1]   M. Gromov: Groups of polynomial growth and expanding maps. Publ. Math. IHES **53** (1981) 53–73

[G2]   M. Gromov: Curvature, diameter and Betti numbers. Comment. Math. Helv. **56** (1982) 179–195

[G3]   M. Gromov: Filling Riemannian manifolds. J. Diff. Geom. **18** (1983) 1–148

[GG1]   D. Gromoll, K. Grove: A generalization of Berger's rigidity theorem for positively curved manifolds. Ann. Sci. Éc. Norm. Sup. **20** (1987) 227–239

[GG2]   D. Gromoll, K. Grove: The low-dimensional metric foliations of euclidean spheres. J. Diff. Geom. **28** (1988) 143–156

[Gr]   K. Grove: Critical point theory for distance functions. Proc. A. M. S. Summer Inst., Los Angeles 1990

[GM]   K. Grove, S. Markvorsen: In preparation

[GP1]   K. Grove, P. Petersen. V: Bounding homotopy types by geometry. Ann. Math. **128** (1988) 195–206

[GP2]   K. Grove, P. Petersen, V: Manifolds near the boundary of existence. J. Diff. Geom. **33** (1991) 379–394

[GP3]   K. Grove, P. Petersen, V: A pinching theorem for homotopy spheres. J. Amer. Math. Soc. **3** (1990) 671–677

[GP4]   K. Grove, P. Petersen, V: Volume comparison à la Aleksandrov. Preprint

[GPW]   K. Grove, P. Petersen, V, J.-Y. Wu: Geometric finiteness theorems via controlled topology. Invent. math. **99** (1990) 205–213

[GS]   K. Grove, S. Shiohama: A generalized sphere theorem. Ann. Math. **106** (1977) 201–211

[KS]   P. Kirby, L. Siebenmann: Foundational essays on topological manifolds, smoothings and triangulations. Ann. Math. Stud., vol. 88. Princeton Univ. Press 1977

[OSY]   Y. Otsu, K. Shiohama, T. Yamaguchi: A new version of differentiable sphere theorem. Invent. math. **98** (1989) 219–228

[P1]   P. Petersen, V: A finiteness theorem for metric spaces. J. Diff. Geom. **31** (1990) 387–395

[P2]   P. Petersen, V: Gromov-Hausdorff convergence of metric spaces. Proc. A. M. S. Summer Inst. Los Angeles, 1990

[PS]   L. Pontrjagin, L. Schnirelmann: Sur une propriété métrique de la dimension. Ann. Math. **33** (1932) 156–162

[Q1]   F. Quinn: Ends of maps. J. Ann. Math. **110** (1979) 275–213

[Q2]   F. Quinn: Ends of maps, III. Dimensions 4 and 5. J. Diff. Geom. **17** (1982) 503–521

[Q3]   F. Quinn: An obstruction to the resolution of homology manifolds. Michigan Math. J. **34** (1987) 285–291

[R]   W. Rinow: Die innere Geometrie der metrischen Räume. Springer, Berlin Heidelberg 1961

[U]   P. Urysohn: Notes supplémentaires. Fund. Math. **8** (1926) 352–356

[Y1]   T. Yamaguchi: Homotopy type finiteness theorems for certain precompact families of Riemannian manifolds. Proc. Amer. Math. Soc. **102** (1988) 660–666

[Y2]   T. Yamaguchi: Collapsing and pinching under a lower curvature bound. Ann. Math. **133** (1991) 317–357

Symplectic Invariants

Helmut Hofer

Institut für Mathematik, Ruhr-Universität Bochum, Postbox 102148
W-4630 Bochum 1, Fed. Rep. of Germany

1. Introduction

In 1985 M. Gromov proved in his seminar paper *Pseudoholomorphic Curves in Symplectic Manifolds* [17] among other things a striking rigidity result, the so-called *squeezing theorem*.

Consider the vectorspace $\mathbb{C}^n$ equipped with its usual Hermitian inner product $(\cdot,\cdot)$. Denote by $\langle \cdot,\cdot \rangle = \mathrm{Re}(\cdot,\cdot)$ the associated real inner product and by $\sigma = -Im\,(\cdot,\cdot)$ the usual induced symplectic form. If $\mathbf{B}^{2n}(r)$ denotes the Euclidean r-ball and $\mathbf{Z}^{2n}(R) = \mathbf{B}^2(R) \times \mathbb{C}^{n-1}$ the symplectic cylinder of radius R, M. Gromov proved that $\mathbf{B}^{2n}(r)$ admits a symplectic embedding into $\mathbf{Z}^{2n}(R)$ iff $r \leq R$. This result exhibits a striking C°-rigidity phenomenon in the symplectic category. Hence already on the C°-level, volume-preserving and symplectic maps can be distinguished.

Coming from the variational theory of Hamiltonian dynamics, I. Ekeland and the author observed in [5, 6] that the study of periodic solutions of Hamiltonian systems can be effectively used to prove the squeezing theorem and – more important – gives even infinitely many new symplectic invariants.

Finally combining the key construction in [5, 6] with Floer's Instanton Homology [12–15] one is able to construct a new theory called Symplectic Homology [16].

2. Symplectic Capacities

Let us first motivate why periodic solutions of Hamiltonian systems occur as obstructions in a *symplectic rigidity theory*.

Given any hypersurface S in a symplectic manifold (M,ω) we see that S carries a distinguished 1-dimensional distribution $\mathscr{L}_S \to S$ which is defined by $\mathscr{L}_S = \ker(\omega\,|_S)$. We denote by $L_S(x)$ the leaf through $x \in S$. If S is a compact hypersurface, of particular interest are the closed characteristics, i.e. closed integral curves $L_S(x) \approx S^1$. We shall write $\mathscr{P}(S)$ for the set of periodic characteristics.

Now, suppose, coming back to Gromov's Theorem that we have an optimal symplectic embedding of some open subset $\mathscr{U}$ of $\mathbb{C}^n$ into $\mathbf{Z}^{2n}(r)$. Optimal here

Proceedings of the International Congress
of Mathematicians, Kyoto, Japan, 1990
© The Mathematical Society of Japan, 1991

means that there is no symplectic embedding for any $r' < r$. Of course, it is in general not clear, if there exists an optimal embedding. Identifying $\mathcal{U}$ with its image, we may assume without loss of generality that $\mathcal{U} \subset \mathbf{Z}^{2n}(r)$. Let us even suppose that $\partial\mathcal{U}$ is smooth. If we try to do, what is impossible by our assumption, namely to make $\mathcal{U}$ smaller, one likes to proceed as follows. It seems reasonable to try to push the set $(\partial\mathcal{U}) \cap (\partial\mathbf{Z}^{2n}(r))$ into $\mathbf{Z}^{2n}(r)$. Then the new set $\mathcal{U}'$, at least if $\bar{\mathcal{U}}'$ is compact, would strictly stay away from $\partial\mathbf{Z}^{2n}(r)$ and we would be able to replace $\mathbf{Z}^{2n}(r)$ by a smaller symplectic cylinder. The optimal way to carry out this procedure is, at least locally, to take a Hamiltonian $H : \mathbf{C}^n \to \mathbb{R}$, which is increasing on the parts of the characteristics of $\partial\mathcal{U}$ which touch $\partial\mathbf{Z}^{2n}(r)$. The associated Hamiltonian flow pushes (locally) $\partial\mathcal{U}$ into $\mathcal{U}$. However, there are obvious global obstructions against the existence of Hamiltonians increasing on leaves, namely characteristics in $\mathcal{P}(\partial\mathcal{U})$, i.e. periodic characteristics. There is another more diffuse obstruction, namely Poincare's recurrence: The characteristics properly parameterized define a volume preserving flow on $\partial\mathcal{U}$. So, there is a set of full measure consisting of points which return to every neighbourhood of themselves under the flow. Obviously this is a second obstruction.

The surprising fact is, that the above can be turned into a qualitative and quantitative theory of symplectic rigiditiy.

We start with the theory of symplectic capacities initiated in [5, 6], see also [20, 21, 23, 35]. Motivated by a break through results of C. Viterbo [34] concerning the existence of periodic solutions of Hamiltonian systems on a prescribed energy surface, a very simple, refined approach was given by E. Zehnder and the author in [24]. This is best described as follows. If E denotes the loop space of $\mathbf{C}^n$ and $\mathcal{U}$ is any bounded open set in $\mathbf{C}^n$ let $H : \mathbf{C}^n \to [0, +\infty)$ be a smooth Hamiltonian vanishing on $\mathcal{U}$ but growing sufficiently fast to infinity outside from $\mathcal{U}$. We define a map $\Phi_H : E \to \mathbb{R}$ by

$$\Phi_H(x) = \int x^*\lambda - \int H(x),$$

where the integrals are taken over $S^1 = \mathbb{R}/\mathbb{Z}$, and λ is a primitive of σ. Now, if H is in some sense "sharp" enough at $\mathcal{U}$ it turns out that every critical point of Φ_H with $\Phi_H(x) > 0$ is a 1-periodic solution of $\dot{x} = X_H(x)$, which is close to $\partial\mathcal{U}$. How close the periodic solutions are to $\partial\mathcal{U}$ depends solely on the growth of H outside of $\mathcal{U}$. This observation leads to an existence result of periodic solutions on almost all energy surfaces [24]. Precisely this abundance of periodic solutions is used in [5, 6]. One constructs a universal subset $\mathcal{F}$ of the power set of E and defines for every $H : \mathbf{C}^n \to [0, +\infty)$, which vanishes on some nonempty open set $\mathcal{U}$ a min-max-characterisation defining a number $c(H) \in [0, +\infty)$

$$c(H) := \sup_{F \in \mathcal{F}} \inf \Phi_H(F).$$

Obviously if $H \geq K$ then $c(K) \geq c(H)$. Finally one puts for a bounded subset $\mathcal{U}$ of $\mathbf{C}^n$

$$c(\mathcal{U}) := \inf c(H),$$

where the infimum is taken over all such H which vanish on $\mathcal{U}$. One can even show that for good sets $\mathcal{U}$ the number $c(\mathcal{U})$ is the action of a closed (perhaps iterated) characteristic on $\partial\mathcal{U}$. This number turns out to be a symplectic invariant. Moreover, different choices of $\mathcal{F}$ lead to infinitely many invariants.

These invariants satisfy a set of axioms, and are called symplectic capacities. To give the axioms, let us denote by $\Gamma = (0, +\infty]$ the extended strictly positive half line equipped with the obvious ordering "$\leq$". We view $(\Gamma, \leq)$ as a category. Let $\mathcal{S}$ denote the category of $(2n)$-dimensional symplectic manifolds together with the symplectic embeddings as the morphisms.

Definition 1. A symplectic capacity is a covariant functor $c : \mathcal{S} \to \Gamma$ such that
- $c(M, \alpha\omega) = |\alpha| \, c(M, \omega)$ for $\alpha \neq 0$,
- $c(\mathbf{Z}^{2n}(1), \sigma) < +\infty$,

where $\mathbf{Z}^{2n}(1)$ is equipped with the symplectic structure induced from $\mathbb{C}^n$.

We note that a convex combination of symplectic capacities as well as positive multiples are again capacities. So the collection of all capacities build a cone. One can show that this cone is infinite dimensional.

The first capacity constructed in [5] satisfies in addition the normalisation $c(\mathbf{B}^{2n}(1)) = c(\mathbf{Z}^{2n}(1)) = \pi$ and is therefore similar to Gromov's width [19]. But there are many other capacities [6], allowing for example to prove that the polydisk $\mathbf{B}^2(1) \times \ldots \times \mathbf{B}^2(1)$ can only be symplectically embedded into the round ball $\mathbf{B}^{2n}(R)$ iff $R \geq \sqrt{n}$.

Next we present some results which emphasize the importance of symplectic capacities.

Theorem 2. *A smooth embedding* $\Psi : \mathbf{B}^{2n}(\varepsilon) \to \mathbb{C}^n$ *preserving the capacity of open subsets of* $\mathbf{B}^{2n}(\varepsilon)$ *for some capacity c, is either symplectic or anti-symplectic.*

Even stronger, a homeomorphism $h : \mathbb{C}^n \to \mathbb{C}^n$ preserving locally capacity, has at every point x, where it is differentiable, either a symplectic or anti-symplectic linearization. If h and $h \times Id_\mathbb{C}$ preserve some capacities and h is smooth it is even symplectic. This motivates the following definition

Definition 3. A homeomorphism $h : \mathcal{U} \to V$ between two open sets $\mathcal{U}$ and $\mathcal{V}$ of $\mathbb{C}^n$ is said to be C°-symplectic if h and $h \times Id_\mathbb{C} : \mathcal{U} \times \mathbb{C} \to \mathcal{V} \times \mathbb{C}$ preserve every capacity of open subsets of their domains.

In view of the above results this is a natural C°-definition of "symplectic". It is not very practical yet since our knowledge of the cone of capacities is very limited. However, it is by now clear that the symplectic notion is a C°- rather than a C^1-concept. There are many known constructions for a symplectic capacity [5, 6, 16, 20, 21, 23, 35, 36]. A very interesting one is the following, which is only defined for subsets of $\mathbb{C}^n$ [20].

Let $\mathscr{C}$ be the vectorspace of all compactly supported smooth maps $H : [0, 1] \times \mathbb{C}^n \to \mathbb{R}$. We define a norm $\|H\|$ by

$$\|H\| := \int_0^1 \left[\sup_{x \in \mathbb{C}^n} H(t, x) - \inf_{x \in \mathbb{C}^n} H(t, x) \right] dt$$

Associated to $H \in \mathscr{C}$ is the time-1-map Ψ_H obtained by integrating the Hamiltonian system associated to H. The collection of all time-1-maps forms a group $\mathscr{D}$ under composition. We define a map $E : \mathscr{D} \to [0, +\infty)$, called the "Energy", by

$$E(\Psi) = \inf\{\|H\| \mid \Psi_H = \Psi\}.$$

One verifies easily that

$$E(\Psi^{-1}) = E(\Psi) = E(\Phi^{-1} \circ \Psi \circ \Phi) \qquad \text{and} \qquad E(\Psi \circ \Phi) \leq E(\Psi) + E(\Phi).$$

Of course the above would be true if $E \equiv 0$ (in fact that was everyone's favorite guess). However,

Theorem 4. $E(\Psi) = 0$ iff $\Psi = Id$. In particular $d : \mathscr{D} \times \mathscr{D} \to [0, +\infty)$ defined by $d(\Psi, \Phi) = E(\Psi^{-1}\Phi)$ defines a bi-invariant metric on $\mathscr{D}$. Moreover the map $\mathscr{C} \to \mathscr{D} : H \to \Psi_H$ is globally Lipschitz continuous with constant one.

Thus the infinite dimensional group $\mathscr{D}$ admits a bi-invariant metric. This distinguishes $\mathscr{D}$ from many other diffeomorphism groups. The d-topology on $\mathscr{D}$ is very strange. So, one has for example the following fact: Given any volume preserving diffeomorphism $\Psi : \mathbb{C}^n \to \mathbb{C}^n$ there exists a sequence $(\Psi_k) \subset \mathscr{D}$ such that $\Psi_k \to Id$ in $(\mathscr{D}, d)$ but

$$\int_K |\Psi_k(x) - \Psi(x)|^P \, dx \to 0 \text{ as } k \to +\infty$$

for every compact set K in $\mathbb{C}^n$ and every $1 \leq p < \infty$. Hence there seems to be no relation between the d-topology and the L^P_{loc}-topology, $1 \leq p < \infty$. However, [20], we have

Theorem 5. The inclusion map $j : (\mathscr{D}, d) \to C^\circ$ (which of course is not continuous) has closed graph. Here C° stands for the Frechet space of continuous maps $\mathbb{C}^n \to \mathbb{C}^n$, equipped with the metrizable compact open topology.

See also [36] for related results. The d-topology had been introduced in [20] in order to study symplectic fixed point problems.

As Y. Eliashberg pointed out recently [9] the d-topology occurs naturally in the study of the question to what extend does the interior of a symplectic manifold determine the symplectic topology on the boundary, see also [11].

Another important question, which one should mention is the following: *Does there exist a good model for the completion $\tilde{\mathscr{D}}$ of $\mathscr{D}$?* Note that $\tilde{\mathscr{D}}$ turns out to be a topological group! It is interesting to note that for every continuous Hamiltonian H which vanishes at infinity there exists a unique "time-1-map" in $\tilde{\mathscr{D}}$.

The d-topology on $\mathscr{D}$ and the symplectic phase space geometry in $\mathbb{C}^n$ are closely related. Let S be a compact, connected, smooth hypersurface in $\mathbb{C}^n$. Denote by B_S the bounded component of $\mathbb{C}^n \setminus S$. We write c_{EH} for the particular symplectic capacity introduced in [5]. We say the hypersuface S is of contact type if there exists a 1-form λ on S such that $d\lambda = \omega \mid S$ and λ does not vanish on the fibres of $\mathscr{L}_S \to S$. We say S is of restricted contact type if λ can be extended to

$\mathbb{C}^n$ satisfying globally $d\lambda = \omega$ (see [1] for examples). To get a quantitative idea about c_{EH} we just mention the following estimate

$$c_{EH}(B_S) \geq \sup\{\pi r^2 \mid \text{There exists a symplectic embedding } \mathbf{B}^{2n}(r) \to B_S\}.$$

Theorem 6. *Let S be a smooth, compact connected hypersurface of restricted contact type in $\mathbb{C}^n$. If $\Psi \in \mathcal{D}$ with $E(\Psi) \leq c_{EH}(S)$ then there exists $x \in S$ with $\Psi(x) \in L_S(x)$.*

This result has nice applications in a global Poincaré perturbation theory [28, 20]. As a consequence of 2.6 we can define a very intuitive capacity, called the displacement energy.

Definition 7. For a bounded subset $\mathcal{U}$ of $\mathbb{C}^n$ define $c(\mathcal{U}) = \inf\{E(\Psi) \mid \Psi(\bar{\mathcal{U}}) \cap \bar{\mathcal{U}} = \emptyset\}$. For an unbounded subset $T \subset \mathbb{C}^n$ put $c(T) = \sup\{c(S) \mid S \subset T, S \text{ bounded}\}$. We call c the displacement energy.

One can show that $c(\mathbf{B}^{2n}(1)) = c(Z^{2n}(1)) = \pi$. We recall the fact that a $\Psi \in \mathcal{D}$ with small energy could be close to a translation in L^p_{loc}, $1 \leq p < \infty$. Nevertheless 2.6 shows that small energy implies a lot of local recurrence.

3. Symplectic Homology

We have already explained the importance of closed characteristics in the theory of symplectic capacities. For the following they might be even more important. The starting point is similar to the construction of symplectic capacities. Namely we associate to suitable Hamiltonians H some invariant. Whereas it were numerical invariants in the contruction of capacities, we associate now to a Hamiltonian H a Floer-Homology group [16]. The contruction is well behaved with respect to the pointwise ordering on the Hamiltonians, in the sense that $H \leq K$ induces a unique morphism between the Floer groups, [16]. Thus instead of taking the infimum of numbers $c(H)$ over the set of Hamiltonians H vanishing on a set S in order to obtain some invariant for S, we just take the direct limit of the Floer groups. This construction explained now in more detail leads to symplectic homology.

The domain of the theory consists of all pairs (M, A), where M is a symplectic manifold of dimension $2n$ and A an open subset of $M \setminus \partial M$, whose closure $\bar{A}$ is contained in some open subset $\mathcal{U}$ of M, with $\bar{\mathcal{U}}$ being compact, such that $\partial \mathcal{U}$ is smooth and of contact type. Moreover $\pi_2(M) = 0$ (however, weaker hypotheses will do as well). A morphism $\Psi : (M, A) \to (N, B)$ is a symplectic embedding $\Psi : M \to N$ satisfying $\Psi(A) \subset B$. One studies now for each pair (M, A) the Hamiltonian systems associated to Hamiltonians $H : M \to \mathbb{R}$ which satisfy $H \mid \bar{A} < 0$, and $H(x) \equiv m(H) \in (0, +\infty)$, if $x \in M \setminus V$ for some open neighbourhood V of $\bar{A}$ contained in $M \setminus \partial M$. Moreover $H(x) \leq m(H)$ for all $x \in M$. Let us denote this class of Hamiltonians by $\mathcal{H}(M, A)$. One is interested in the 1-periodic solutions of the Hamiltonian system $\dot{x} = X_H(x)$. This problem is variational, i.e. the 1-periodic solutions are critical points of some functional Φ_H. Using an equivariant version of Floer's construction of a homology theory one can associate to Φ_H a cochain complex $C(H)$, which is essentially spanned by the 1-periodic solutions of $\dot{x} = X_H(x)$. As a new ingredient we have a real filtration of this cochain complex since the basis vectors carry a numerical invariant, namely their Φ_H-value. There is also a $\mathbb{Z}$-grading through some kind of Maslov-index. All homological algebra constructions like chain homotopy, chain map etc., are supposed to be compatible with the $\mathbb{R}$-filtration. We write

$$C = \bigcup_{\lambda \in \mathbb{R}} \left(\bigoplus_{k \in \mathbb{Z}} C_\lambda^k \right), \qquad C_\mu \subset C_\lambda \text{ for } \lambda \leq \mu.$$

Given two Hamiltonian $H \geq K$ one obtains a natural chain homotopy class (preserving the extra structure) $C(H) \rightarrow C(K)$, and for $\lambda \leq \mu$ induced maps $C_\lambda(H)/C_\mu(H) \rightarrow C_\lambda(K)/C_\mu(K)$. Applying the functor $\text{Hom}(\cdot, R)$ for some commutative ring we obtain for every pair (M, A) a directed system

$$(\text{Hom}(C_\lambda(H)/C_\mu(H)), R)_{H \in \mathcal{H}(M, A)}.$$

We pass to Homology and take the direct limit denoted by $I_\lambda^\mu(M, A)$. Symplectic embeddings $\Psi : (M, A) \rightarrow (N, B)$ induce group morphisms $\Psi^* : I_\lambda^\mu(N, B) \rightarrow I_\lambda^\mu(M, A)$. One has long exact sequences as well as isotopy invariance. For the details we refer the reader to [16] or [22]. We list now some applications. The first is a solution of Gromov's polydisk conjecture [22]. Let $a = (a_1, \ldots, a_n)$ be a n-tuplet of positive real numbers. We shall write $o(a)$ for the associated ordered n-tuplet (the a_i's written in increasing order). We write $\mathbf{D}^{2n}(a)$ for the open polydisk $\mathbf{D}^{2n}(a) = \mathbf{B}^2(a_1) \times \ldots \times \mathbf{B}^2(a_n)$.

Theorem 8. *Let a, b be ordered n-tuplets. Suppose $a_j \leq b_j$ for $j = 1, \ldots, n$ and $b_j^2 < a_1^2 + a_j^2$ for $j = 2, \ldots, n$. Assume $\mathbf{D}(r)$ is a third polydisk such that there exist symplectic embeddings $\Psi : \mathbf{D}(a) \rightarrow \mathbf{D}(r)$ and $\Phi : \mathbf{D}(r) \rightarrow \mathbf{D}(b)$ with $\Phi \circ \Psi$ being isotopic to the standard inclusion. Then we have for some permutation $\sigma(\tau)$ of τ*

$$a \leq \sigma(\tau) \leq b.$$

In particular if $a = b$ we have $a = o(r)$.
    Hence, if $\mathbf{D}(a) \simeq \mathbf{D}(b)$ we must have $o(a) = o(b)$, which is Gromov's conjecture.

It follows from the H-principle [18], that the space of symplectic embeddings $\mathbf{B}^2(2) \rightarrow \mathbf{D}^4(1, 1)$ is connected for the compact open topology. Take any such embedding, say $\hat{\Psi}$. Then $\hat{\Psi}(\mathbf{B}^2(2))$ has a symplectic tubular neighbourhood. Hence, we might slightly thicken $\mathbf{B}^2(2)$ to obtain a symplectic embedding $\Psi : \mathbf{D}^4(\varepsilon, 2) \rightarrow \mathbf{D}^4(1, 1)$ for some $0 < \varepsilon < 1$. If now $\Phi : \mathbf{D}^4(\varepsilon, \delta) \rightarrow \mathbf{D}^4(1, 1)$ is the composition of the standard inclusion of $\mathbf{D}^4(\varepsilon, \delta) \rightarrow \mathbf{D}^4(\varepsilon, 2)$ with Ψ for some $0 < \delta < 1$, we obtain the following result

Theorem 9. *Assuming $\varepsilon^2 + \delta^2 > 1$ the symplectic embedding Φ is not isotopic to the standard inclusion $j : \mathbf{D}^4(\varepsilon, \delta) \rightarrow \mathbf{D}^4(1, 1)$.*

This result is obtained by studying the maps induced by Φ and j in symplectic homology.

There is another application concerned with the question to which extent is the symplectic topology on the boundary determined by the symplectic topology of the interior [9, 11]. Let (M, ω) be a compact symplectic manifold with $\omega = d\lambda$ being an exact symplectic form. Suppose ∂M is of restricted contact type. We define the action spectrum $A(M)$ by

$$A(M) = \left\{ m \int (\lambda \mid P) \mid P \in \mathscr{P}(\partial M), m \in \mathbf{N}^* \right\}.$$

Theorem 10. *Let (M, ω) and (N, σ) be of the type described above. Suppose $M \setminus \partial M$ and $N \setminus \partial N$ are symplectomorphic by an exact symplectic map. Then $A(M) = A(N)$, provided the periodic trajectories on ∂M and ∂N are nondegenerate.*

One can even strengthen the statements by incorporating the Maslov-type-index giving the grading of the symplectic homology groups. The symplectic homology is so far not complete, but already it allows to prove some not so obvious results.

References

1. Benci, V., Hofer H., Rabinowitz, P.: A priori bounds for periodic solutions on hypersurfaces. In: Periodic solutions of Hamiltonian systems and related topics (eds. Rabinowitz et al.). Nato ASI, Ser. C 209. Reidel 1987
2. Ekeland, I.: Convexity methods in Hamiltonian mechanics. (Ergebnisse der Mathematik, 3. Folge, Band 19.) Springer, Berlin Heidelberg New York 1989
3. Ekeland, I.: Une theorie de Morse pour les systemes hamiltoniens convexes. Ann. IHP "Analyse non linéare" 1 (1984) 19–78
4. Ekeland, I., Hofer, H.: Convex Hamiltonian energy surfaces and their closed trajectories. Comm. Math. Phys. 113 (1987) 419–467
5. Ekeland, I., Hofer, H.: Symplectic topology and Hamiltonian dynamics. Math. Z. 200 (1990) 355–378
6. Ekeland, I., Hofer, H.: Symplectic topology and Hamiltonian dynamics II. Math. Z. 203 (1990) 553–567
7. Eliashberg, Y.: Three lectures in symplectic topology. Proceedings of the Conference on Differential Geometry in Cala Conone 1988, to appear
8. Eliashberg, Y.: Filling by holomorphic disks and its applications. Preprint 1989.
9. Eliashberg, Y.: Lecture at a Conference on Dynamical Systems, Lyon, July 1990.
10. Eliashberg, Y.: A theorem on the structure of wave fronts. Funct. Anal. Appl. 21 (1987) 65–72
11. Eliashberg, Y., Hofer, H.: Towards the definition of a symplectic boundary. In preparation
12. Floer, A.: Morse theory for Lagrangian intersections theory. J. Diff. Geom. 18 (1988) 513–517
13. Floer, A.: The unregularised gradient flow of the symplectic action. Comm. Pure Appl. Math. 41 (1988) 775–813
14. Floer, A.: Witten's complex and infinite dimensional Morse theory. J. Diff. Geom. 30 (1989) 207–221

15. Floer, A.: Symplectic fixed points and holomorphic spheres. Comm. Math. Physics **120** (1989) 576–611
16. Floer, A., Hofer, H.: Symplectic homology. In preparation.
17. Gromov, M.: Pseudoholomorphic curves in symplectic manifolds. Invent. math. **82** (1985) 307–347
18. Gromov, M.: Partial differential relations. Ergebnisse der Mathematik 1986.
19. Gromov, M.: Soft and hard symplectic geometry. Proceedings of the ICM at Berkeley 1986. AMS, 1987, pp. 81–89
20. Hofer, H.: On the topological properties of symplectic maps. Proc. Roy. Soc. Edinburgh **115A** (1990) 25–83
21. Hofer, H.: Symplectic capacities. Geometry of Low-dimensional Manifolds: 2. Edited by S. K. Donaldson and C. B. Thomas. London Mathematical Society Lecture Notes 151, pp. 15–34
22. Hofer, H., Wysocki, K.: Applications of symplectic homology. In preparation
23. Hofer, H., Zehnder, E.: A new capacity for symplectic manifolds. Analysis et cetera, edited by P. Rabinowitz and E. Zehnder. Academic Press 1990, pp. 405–428
24. Hofer, H., Zehnder, E.: Periodic solutions on hypersurfaces and a result by C. Viterbo. Inv. math. **90** (1987) 1–7
25. McDuff, D.: Examples of symplectic structures. Invent. math. **89** (1987) 13–36
26. McDuff, D.: Rational and ruled symplectic 4-manifolds. Preprint.
27. McDuff, D.: Elliptic methods in symplectic geometry. Progress in Math. Lecture, Boulder 1989
28. Moser, J.: A fixed point theorem in symplectic geometry. Acta mathematica **141** (1978) 17–34
29. Rabinowitz, P.: Periodic solutions of Hamiltonian systems. Comm. Pure Appl. Math. **31** (1978) 157–184
30. Rabinowitz, P.: Periodic solutions of a Hamiltonian system on a prescribed enery surface. J. Diff. Eq. **33** (1979) 336–352
31. Sikorav, J.C.: Rigidité symplectique dans le cotangent de T^n. Duke Math. J. **59** (1989) 227–231
32. Sikorav, J.C.: Systemes Hamiltoniens et topologie symplectique. Lecture notes Dipartimento Di Matematica Dell' Università Di Pisa, ETS Editrice
33. Viterbo, C.: New obstructions to embedding Lagrangian tori. Invent. math. **100** (1990) 301–320
34. Viterbo, C.: A proof of the Weinstein conjecture in $\mathbb{R}^{2n}$. Ann. Inst. Henri Poincare. Analyse non lineare **4** (1987) 337–356
35. Viterbo, C.: Capacités symplectiques et applications. Seminaire Bourbaki, June 1989, Asterisque, no. 695
36. Viterbo, C.: Symplectic topology as the geometry of generating functions. Preprint
37. Weinstein, A.: Lectures on symplectic manifolds. CBMS Reg. Conf. Series in Math. no 29, AMS, Providence, R.I. 1979
38. Weinstein, A.: On the hypotheses of Rabinowitz' periodic orbit theorems. J. Diff. Eq. **33** (1979) 353–388

Embedded Surfaces in 4-Manifolds

Peter B. Kronheimer

Merton College, Oxford OX1 4JD, UK

1. Introduction

The adjunction formula expresses the genus of a smooth curve C in a complex surface X in terms only of the homology class of C and the canonical class K of the surface:

$$2g - 2 = C \cdot C + K \cdot C.$$

An old and still elusive conjecture [Ki] asserts that, if Σ is any smoothly embedded 2-manifold in the same homology class as C, then the genus of Σ is at least as big as the genus of C, so

$$2g - 2 \geq \Sigma \cdot \Sigma + K \cdot \Sigma. \tag{1}$$

With Donaldson's introduction of gauge theory into 4-dimensional differential topology, there seems reason to hope that such a conjecture might well be approachable. However, almost the best result to date on questions of this sort is now twenty years old, proved independently by Rohlin [R] and Hsiang and Szczarba [HS]. Their result is not special to complex surfaces, but exploits a branched covering to prove a non-trivial lower bound on the genus whenever the homology class of Σ is divisible. Results due to Donaldson do allow an occasional improvement of this lower bound, but this still falls short of (1) by a factor of about 2, even in favourable cases.

A somewhat looser version of the above conjecture has been formulated by Morgan; we adapt it slightly here:

Conjecture A. *If X is a simply-connected, oriented 4-manifold for which Donaldson's polynomial invariants are defined and non-zero, and if Σ is any smoothly embedded, oriented 2-manifold with positive self-intersection, then the genus of Σ satisfies the inequality*

$$2g - 2 \geq \Sigma \cdot \Sigma.$$

To explain this conjecture we need to recall a few facts about the polynomial invariants [D2]. These are invariants of X which are defined whenever $b^+(X)$ (the dimension of a maximal positive subspace of the intersection form on the second homology) is odd and not less than 3. This is the case, for example, if X is a complex surface whose canonical bundle admits a non-zero holomorphic

Proceedings of the International Congress
of Mathematicians, Kyoto, Japan, 1990
© The Mathematical Society of Japan, 1991

section, and it is one of the two key results of [D2] that the invariants are indeed non-zero for such a surface. Thus simply-connected complex surfaces with $p_g > 0$ satisfy the hypotheses of the conjecture. (Here p_g, the geometric genus, is the dimension of the space of sections of K).

For a complex surface, the condition that K has a section is easily seen to imply that $K \cdot C$ is non-negative whenever C is a curve with positive self-intersection. So Conjecture A is consistent with the adjunction formula and the conjecture (1). It is weaker, in that the delicate term involving the canonical class is missing, and it says nothing about rational surfaces; on the other hand, it is a conjecture which is not restricted to algebraic cycles. An important case in which the two conjectures essentially coincide is when X is a $K3$ surface (a simply-connected surface with trivial canonical class). On account of this case, Conjecture A implies various local versions of (1). Thus it implies that if B^4 is a standard ball in $\mathbb{C}^2$ and C is a smooth, plane algebraic curve meeting ∂B^4 transversely, then the genus of $C \cap B^4$ is minimal among all 2-manifolds in B^4 with the same oriented boundary. This in turn implies, for example, that the unknotting number of the (p, q) torus knot is $\frac{1}{2}(p - 1)(q - 1)$; see [Mi].

Conjecture A is known to hold if the genus of Σ is 0 or 1; in other words, an embedded sphere or torus cannot have positive self-intersection if X has non-zero polynomial invariants. The fact that a sphere cannot have self-intersection 1 is a special case of the results of [D2] concerning connected sum decompositions; the other cases are due to Morgan, Mrowka and Ruberman [MMR].

The purpose of this article is to outline a proof of Conjecture A, but a proof with a hole. The argument uses gauge theory, but with a modification which makes it a little different from the usual set-up. We consider connections in some auxiliary $SU(2)$ or $SO(3)$ bundle over the manifold, but connections which are defined on the complement of the surface Σ and have non-trivial holonomy around the small linking circles of Σ. Such a set-up has been considered elsewhere, and the similar situation of a punctured 2-manifold is quite well-understood [Se, Si]. For the 4-dimensional case, some important analysis has been done by Sibner and Sibner [SS], but still some things are missing.

The basic properties of the moduli spaces of anti-self-dual connections associated to a compact 4-manifold are deduced from a simple package of results. This is the Kuranishi-type deformation theory, built on a particular Fredholm complex combined with the implicit function theorem. This package is missing for the "branched instantons" which we wish to exploit. In Section 2 we shall explain what one would expect of such a package were it to exist in our case. This will be our black box conjecture. Assuming the existence of a black box with the expected properties, we shall deduce Conjecture A.

This is not the place for a very detailed exposition. If the author, or anyone else, succeeds in making the necessary box, then full details will appear. If it turns out that such a box does not exist, then the reader will be glad that he or she did not read a longer paper.

The author wishes to thank Simon Donaldson for his generous help and advice. The key tunnelling trick described in Section 4 was suggested by him. The author must also thank Ed Witten for clearing up some of the author's early misconceptions and confusion concerning these branched instantons. Finally, thanks are due to Tom Mrowka and Carlos Simpson for their help at various times.

2. Connections with Holonomy

For some basic definitions and properties of anti-self-dual connections over 4-manifolds, we refer to [D2] and the references therein. We consider here an extension of the usual constructions which seems naturally suited to the problem at hand. Let Σ be a connected, oriented 2-manifold embedded in the oriented, Riemannian 4-manifold X, and let E be a hermitian vector bundle over $X \setminus \Sigma$. For the moment we shall take it that E has structure group $SU(2)$. Let $\mathcal{N}$ be the set of all connections A in E (compatible with the $SU(2)$ structure) which are anti-self-dual and have finite action; that is, the curvature $F(A)$ should satisfy

$$*F(A) = -F(A),$$

and

$$\int_{X \setminus \Sigma} |F(A)|^2 < \infty,$$

where $*$ is the Hodge star operator on 2-forms. Let M be the quotient of $\mathcal{N}$ by the action of the automorphisms of the bundle E. This M is the moduli space we would like to study. Note that the connections A are defined only on the complement of the surface Σ and are anti-self-dual with respect to an incomplete metric.

A local study of such connections was made by Sibner and Sibner in [SS]. Their first result is that the finite action condition alone implies that a connection on $X \setminus \Sigma$ has a well-defined holonomy on the small linking circles. To be precise, choose polar coordinates on a small element of plane normal to Σ at a point σ. Let h_r be the holonomy of the connection around the circle of radius r in this plane. Then the result is that, for almost all σ, the conjugacy class of h_r has a well-defined limit as r approaches zero, and this limit is independent of the choice of σ. Thus we can write

$$h_r \sim \exp 2\pi i \begin{pmatrix} \alpha & 0 \\ 0 & -\alpha \end{pmatrix},$$

for some real α in the interval $0 \le \alpha \le 1/2$. Let us write M_α for the subset of the moduli space M where the holonomy is α in this parametrization.

The first simple point is that the values 0 and 1/2 are very special values for the holonomy. When $\alpha = 0$, the holonomy matrix is the identity; so there is no geometrical obstruction to extending the connection. It follows from the technical result of [SS] that in this case, an anti-self dual connection A will extend smoothly to all of X, or to be more precise, there is a bundle E_0 over X with an anti-self-dual connection whose restriction to $X \setminus \Sigma$ is isomorphic to A. When α is $1/2$, the holonomy is -1. In this case, the connection will not necessarily extend as an $SU(2)$ connection, but in the associated $SO(3)$ bundle the holonomy is again trivial; so A extends to a smooth anti-self-dual connection in some $SO(3)$ bundle $E_{1/2}$ over X. In this sense then, M_0 and $M_{1/2}$ are both ordinary moduli spaces of anti-self-dual connections associated to X. (We shall generally write M_E for the usual moduli space of anti-self-dual connections on a bundle E.)

When $0 < \alpha < 1/2$, there are new phenomena. First, whereas an $SU(2)$ connection on a closed 4-manifold has only one characteristic number, namely the second Chern class k, a connection A with holonomy such as we are considering has two invariants. To defines these, first take a small tubular neighbourhood

N of Σ, and let Y be its boundary, a circle bundle over Σ. Since the holonomy around the circle fibres is non-trivial, its eigenspaces give a decomposition of $E|_Y$ as a sum of line bundles $L \oplus L^*$. We take L to be the eigen-bundle corresponding to the eigenvalue $e^{-2\pi i \alpha}$. On each circle fibre, we can choose a trivialization of E for which the θ component of the connection is approximately

$$A_\theta = \begin{pmatrix} \alpha & 0 \\ 0 & -\alpha \end{pmatrix} d\theta.$$

This preferred trivialization on the fibres allows us to extend $L \oplus L^*$, in an essentially unique way, as a topological bundle on N. Thus we get an extension $\bar{E}$ of E to all of X, and a decomposition of $\bar{E}|_\Sigma$ as a sum of line bundles. We define

$$k = c_2(\bar{E})[X],$$
$$l = -c_1(L)[\Sigma],$$

and call these the *instanton* and *monopole* number (the latter by analogy with the topology of monopole solutions on 3-space; see [A]).

Given suitable regularity of the connection near the surface, we can also define k and l by a formula of Chern-Weil type, which involves some correction terms due to the non-trivial topology of the neighbourhood. Near Σ, we should expect the curvature $F(A)$ to be approximately abelian, since the subbundle L on Y should be asymptotically parallel. If this is true to a sufficient degree, we can write

$$F(A) = \begin{pmatrix} \omega & 0 \\ 0 & -\omega \end{pmatrix},$$

with ω having a well-defined limit on Σ, in which case

$$k = -2\alpha l + \alpha^2 \Sigma \cdot \Sigma + \frac{1}{8\pi^2} \int_{X \setminus \Sigma} \mathrm{Tr}(F \wedge F), \tag{2}$$

and

$$l = \alpha \Sigma \cdot \Sigma - \frac{1}{2\pi i} \int_\Sigma \omega.$$

We shall write $M_{\alpha,k,l}$ for the moduli space of anti-self-dual connections with holonomy α, instanton number k and monopole number l. Our definition makes it, at present, just a set. We can topologize it, mimicking [T2] for example, but it is not clear what sort of a space would result. The rest of this article will be largely based on the assumption that, for a generic choice of Riemannian metric on X, the moduli space $M_{\alpha,k,l}$ will be (except at the reducible or flat connections) a smooth manifold whose dimension is given by the formula

$$d = 8k - 3(b^+ - b^1 + 1) + 4l - (2g - 2). \tag{3}$$

The first two terms here, involving the 4-dimensional characteristic number of the bundle and the topology of X, are familiar as the dimension of the ordinary moduli spaces associated to a closed 4-manifold. The remaining terms, $4l - (2g - 2)$, involve the surface. The formula is motivated by consideration of parabolic bundles on a complex surface, as is explained in the next section. For the moment, it might be helpful to note that in the case that Σ is S^2 contained in S^4 and k is zero, the formula gives $4l - 1$, in accordance with the dimension of the moduli space of hyperbolic monopoles; see [A] again.

The arguments we shall use later will assume a little more. We wish to consider the moduli spaces $M_{\alpha,k,l}$ as a family as α varies. Let us write $M_{*,k,l}$ for the union of $M_{\alpha,k,l}$ as α runs through $(0, \frac{1}{2})$. We cannot expect that, for a given metric, all the $M_{\alpha,k,l}$ will be smooth manifolds, but we can hope that the union $M_{*,k,l}$ is a manifold with a smooth map to to the half-interval:

Conjecture (Black Box). *For a generic choice of Riemannian metric on X, the space $M_{*,k,l}$ is, except at the reducible or flat connections, a smooth manifold of dimension $d + 1$, with d as shown in (3), and the map $\alpha : M_* \to (0, \frac{1}{2})$ is smooth. The usual transversality and weak compactness arguments, as applied to the usual moduli spaces of anti-self-dual connections, carry over to this case.*

It is on the basis of this conjecture, and one other described in Section 3, that we shall deduce Conjecture A. The conjecture is, necessarily, loosely phrased. Indeed, it would not be sufficient for our application if the conjecture were correct only by some accident. We have in mind that M_α should be generically a manifold *for the same reasons* that the usual moduli spaces M_E are, and that as α varies, M_α should behave rather as the ordinary moduli spaces do when the Riemannian metric varies in a 1-parameter family [D2]. The principal thing is that there should be a deformation theory describing the local structure of M_α, based on a Fredholm complex. Smoothness and compactness are secondary matters which should follow from standard techniques if the primary machinery is in place. In particular, the results of [SS] are what is needed for a weak compactness theorem along the lines of the essential theorem of Uhlenbeck [U]. Here there will be differences, though, since there will be a bubbling off phenomenon in which both the instanton and monopole number change in the weak limit. Detailed information concerning the possibilities is not important to our argument, but we do need to know that the limiting connection lives in a moduli space of smaller dimension.

3. Parabolic Bundles

Suppose X is a Kähler surface and $\Sigma \subset X$ is a smooth curve. By a *parabolic $SL(2, \mathbb{C})$-bundle* on X we shall mean a rank-2 holomorphic bundle $\mathscr{E} \to X$ with $\Lambda^2\mathscr{E}$ trivial, together with a holomorphic line sub-bundle $\mathscr{L} \subset \mathscr{E}|_\Sigma$. Pick a positive real number α and define the α-*degree* of a rank-1 subsheaf $\mathscr{F} \subset \mathscr{E}$ by the formula

$$\deg_\alpha(\mathscr{F}) = \langle c_1(\mathscr{F}), [\omega] \rangle \pm \alpha \langle [\Sigma], [\omega] \rangle,$$

where ω is the Kähler form. Here the sign is to be $+$ if $\mathscr{F}|_\Sigma$ is contained in $\mathscr{L}$ and is to be $-$ otherwise. Define $(\mathscr{E}, \mathscr{L})$ to be α-*stable* if $\deg_\alpha(\mathscr{F})$ is strictly negative for every such subsheaf.

Conjecture (Parabolic Bundles). *If Σ is a smooth curve in a Kähler surface X and $0 < \alpha < \frac{1}{2}$, then the set of irreducible connections $M_{\alpha,k,l}^{\mathrm{irr}} \subset M_{\alpha,k,l}$ is in one-to-one correspondence with the set of α-stable parabolic bundles $(\mathscr{E}, \mathscr{L})$ with $c_2(\mathscr{E}) = k$ and $c_1(\mathscr{L})[\Sigma] = -l$.*

This result is the expected analogue of the theorem of [D1] relating anti-self-dual connections to stable bundles on a complex surface. The terminology 'parabolic' is taken from [Se] which contains a corresponding result in complex dimension 1. An analytic proof of the result of [Se], together with a large part of what is necessary for the higher-dimensional case, is contained in [Si].

The above conjecture motivates everything else in this paper, but our later arguments are not logically dependent on this general result and will appeal to only one particular case, that of a ruled surface. To be precise, let $N \to \Sigma$ be a holomorphic line bundle of positive degree, and let X be the $\mathbb{C}P^1$ bundle obtained by adding a section at infinity. Let Σ be embedded in X as the zero section. We shall assume the truth of the parabolic bundles conjecture for this pair (X, Σ) in the case $k = 0$.

There are actually a couple of points which make this case technically a little easier. One is the circle symmetry in X, another is the local product structure near Σ. Exploiting these and the results of [Si], the author has no difficulty in constructing the map from parabolic bundles to $M_{\alpha,0,l}$. Surjectivity is still in question.

4. Outline of the Proof

Assuming the package of results in the black box, and assuming also the parabolic bundles conjecture in the case of the ruled surface described above, let us deduce Conjecture A.

For ease of exposition, we shall first treat a slightly simplified situation; the proof will occupy this section and the next. The necessary modifications for the general case are described in Section 6. For the simplified argument, we shall suppose that there is an $SO(3)$ bundle E over X, with $w_2(E)$ non-zero, such that the (ordinary) moduli space of anti-self-dual connections M_E is a finite set of points, cut out transversely by the anti-self-duality equations. The number of points in this moduli space, counted with sign according to their orientation, is a simple example of one of the polynomial invariants of X. We shall denote this number by q and we shall suppose it is non-zero.

Concerning Σ, we shall assume that its genus is *even* and that $w_2(E)$ is non-zero on Σ. This will simplify the excision argument in the following section. We assume also that there is an integer homology class η in $X \setminus \Sigma$ on which $w_2(E)$ is again non-zero. Finally, we shall suppose that $\Sigma \cdot \Sigma$ is square-free. The point of this condition is that it ensures that $\pi_1(X \setminus \Sigma)$ has no non-trivial representations in $SO(3)$, as can be deduced from the Chern-Weil formulae for k and l in the case that the curvature terms vanish.

The assumptions on Σ which we have just listed do not involve a loss of generality. If Σ_0 is any surface which satisfies the hypotheses of Conjecture A but violates the inequality which is the conclusion of that conjecture, then we can soon construct many other surfaces with the same property, and in particular, we can find surfaces Σ which satisfy the additional hypotheses of the previous paragraph. The trick is to take n nearby copies of Σ_0, intersecting transversely with positively-oriented intersection points, and then replace the intersection points with small handles. This gives a surface representing the class $n[\Sigma_0]$ which will violate the inequality of Conjecture A by a margin which is n times larger. We can then form a connected sum of this new surface with any other surface T disjoint from Σ_0 in X, to obtain a representative of $[T] + n[\Sigma_0]$ which still violates Conjecture A once n is large.

We shall take for granted the slight modification of the contents of the previous sections which is necessary to include the case of $SO(3)$ bundles. The difference from the $SU(2)$ case is the presence of the Stieffel Whitney class; we shall not build $w_2(E)$ into our notation for the moduli spaces, but shall take it that $M_{\alpha,k,l}$ denotes now the moduli spaces of connections with the same w_2 as E on $X \setminus \Sigma$. The instanton number now lives in $\mathbb{Z}/4$, and the monopole number l will be a half-integer if $w_2(\bar{E})$ is non-zero on Σ.

After these preliminaries, we can begin the proof. Let k be the instanton number of E and consider the corresponding moduli spaces $M_{\alpha,k,l}$. We shall consider a special value for the monopole number: we take

$$l = \frac{1}{4}(2g - 2). \tag{4}$$

The significance of this choice is that the dimension of $M_{\alpha,k,l}$ is now the same as the dimension of the ordinary moduli space M_E: the terms from Σ in the dimension formula (3) cancel out. Note that since g is even, l is a half-integer, as it should be when $w_2(E)$ is non-zero on Σ.

Since, the dimension of M_E is zero, so too is the dimension of $M_{\alpha,k,l}$. The latter will be generically an oriented 0-manifold to which we can attach an integer by counting the points with their signs (here we appeal to the compactness assertion in the black box). Further, the family $M_{*,k,l}$ gives a compact (again), oriented, 1-dimensional cobordism between any two of these 0-manifolds, so the integer we obtain is independent of α. Let us call this integer p. (It is important here that $b^+ \geq 3$, to avoid reducible connections in this one-dimensional family). To prove Conjecture A, we will calculate p in two different ways, using the limiting behaviour as α approaches first 0 and then 1/2.

The first is technically the harder of the two in our case. We shall argue, using small values of α, that $p = 2^g q$. So, roughly speaking, every point of the original moduli space M_E gives rise to 2^g points in M_α, counted algebraically, when α is small. The origin of the number 2^g is in the following result.

Proposition. *Let $\mathscr{E}$ be a stable rank-2 holomorphic bundle on a curve Σ. Assume that the degree of $\mathscr{E}$ is even if the genus of Σ is odd, and that the degree is odd otherwise. Let $\lambda = \frac{1}{2}\deg(\mathscr{E}) - \frac{1}{4}(2g - 2)$. Then generically the bundle has precisely 2^g holomorphic line sub-bundles of degree λ. Here the word generic refers to a Zariski open set in the set of pairs $(\mathscr{E}, \Sigma)$.*

For the proof, the Riemann-Roch theorem for families shows that 2^g is the right number: one applies the theorem to the family of bundles $\mathrm{Hom}(\mathscr{L}, \mathscr{E})$ as $\mathscr{L}$ runs through the Jacobian of line bundles of the given degree. It remains to show that the generically expected behaviour does occur, and for this one can consider a particularly degenerate curve, such as a sphere intersecting g tori, for which the problem is easy to analyse.

The idea here is that the above proposition tells us that $p = 2^g q$ in the case that Σ is a complex curve in a complex surface, provided we accept the conjecture of Section 3 concerning parabolic bundles. The proposition tells us that, when l is given by (4), the moduli space of parabolic bundles $(\mathscr{E}, \mathscr{L})$ generically covers the ordinary moduli space of bundles 2^g times (though there are some provisos here, to do with stability). Of course, the point is precisely that our 2-manifold Σ is not holomorphic; nevertheless, we can use an excision argument, based on the fact that a neighbourhood of Σ can always be given a complex structure, to show that the conclusion $p = 2^g q$ is valid generally. Some details of the excision argument are given in the next section.

We have not yet referred to the inequality $2g - 2 \geq \Sigma \cdot \Sigma$, but consider now what happens as α approaches $1/2$. As we said in Section 2, when α is $1/2$ we can identify the moduli space $M_{1/2,k,l}$ with an ordinary moduli space $M_{E_{1/2}}$ for some $SO(3)$ bundle $E_{1/2}$. This bundle is different from the E; it differs by a sort of twist along Σ, and its topology depends on l. The essential piece of information is the formula which relates the instanton number of $E_{1/2}$ to the old k and l: a short calculation gives

$$k(E_{\frac{1}{2}}) = k(E) + l - \tfrac{1}{4}\Sigma \cdot \Sigma .$$

(Compare this with (2) when $\alpha = 1/2$). If we recall that l is given by (4), we see that in our case the difference in the instanton numbers is a quarter of $(2g - 2) - \Sigma \cdot \Sigma$, which is precisely the number whose sign we are interested in.

Thus if the inequality of Conjecture A is violated, then $k(E_{1/2})$ is less than $k(E)$. Since the dimension of M_E is zero, this means that $M_{E_{1/2}}$ will have negative dimension and will therefore be empty. (Here again we use the fact that w_2 is non-zero, to rule out the possibility that the moduli space contains the trivial flat connection.) This would mean, however, that p must be zero. For otherwise the moduli spaces $M_{(1/2-\varepsilon),k,l}$ are non-empty for all $\varepsilon > 0$, and as ε goes to zero, Uhlenbeck's compactness theorem gives a limiting anti-self-dual connection which is either in $M_{E_{1/2}}$ (if there is no bubbling off) or in some moduli space of still more negative dimension (if some action is lost in the limit). Thus we have obtained a contradiction, since $p = 2^g q$, and q was supposed to be non-zero.

5. Excision

Excision arguments have become a flexible tool to analyse moduli spaces of anti-self-dual connections. Necessary ingredients are described and used in [T1, T2, Mr] and [MMR].

Consider deforming the Riemannian metric on X so that a collar of the boundary, Y, of the tubular neighbourhood $N \supset \Sigma$ is stretched out into a long cylinder $Y \times [-T, T]$. Let $M(T)$ be the corresponding moduli space of anti-self-dual connections in E. This contains q points, counted with sign. As T goes to infinity, connections in $M(T)$ converge weakly to give us connections on the disjoint union $X' \cup N'$, where these two are non-compact manifolds, with cylindrical ends isometric to $Y \times \mathbb{R}^+$. In our case, a simple dimension count shows that the limiting connection on N' must be flat, and the limiting connection on X' must have total action k (no action is lost). On the end of X', the connection will approach a flat connection on Y, and if we recall that $w_2(E)$ is non-zero on N, we see that this flat connection ϱ must be the pull-back of a flat $SO(3)$ connection on Σ with odd w_2. Write $\mathcal{R}(\Sigma)_{\text{odd}}$ for the space of these flat connections; it is a smooth manifold since all these connections are irreducible.

In the other direction, given an anti-self-dual connection in E over X' with action k, asymptotic to a flat connection from $\mathcal{R}(\Sigma)_{\text{odd}}$, we can "glue" it to the corresponding flat connection on N' to obtain a connection on X. The analysis of this gluing is simplified by the fact $\mathcal{R}(\Sigma)_{\text{odd}}$ is smooth. It can thus be shown that, when counted with signs, the number of such connections on X' is the same integer q. Let us write these connections as $A_1, \ldots, A_t$, and let $\varrho_1, \ldots, \varrho_t$ be the limiting elements of $\mathcal{R}(\Sigma)_{\text{odd}}$.

So far, this has been for the ordinary connections, without holonomy around Σ. Consider now the moduli space of connections on $N' \setminus \Sigma$ which have holonomy α and monopole number given by (4). There is a component of this moduli space which has the the the same dimension as $\mathcal{R}(\Sigma)_{\text{odd}}$ (which can be regarded as a moduli space of flat connections on N') and in which every connection is asymptotic to some element of $\mathcal{R}(\Sigma)_{\text{odd}}$ on the end. Calling this moduli space $M_\alpha(N')$, we have a smooth map $r : M_\alpha(N') \to \mathcal{R}(\Sigma)_{\text{odd}}$, given by this limiting connection [T2]. Although the former space is not compact due to possible bubbling off of monopoles, the weak compactness theorem will guarantee that r has a well-defined *degree*, say D, provided α is small. (For larger α, the degree may jump, on account of reducible connections.) Using Proposition 6.4 from [T2], we can ensure that the connections ϱ_i are regular values for r; so on N' there will be exactly D anti-self-dual connections with holonomy α asymptotic to each ϱ_i.

For small α, the analysis of the ordinary case goes through, and shows that the anti-self-dual connections with holonomy on E over X can be regarded as being obtained by gluing one of the connections A_i on X' to one of the connections of $M_\alpha(N')$. Whereas in the original set-up each A_i could be glued to just one flat connection on N', now there are D possibilities for each (counted algebraically). Thus we see that the total number of instantons with holonomy α (the number we called p) is equal to D times q.

The important thing we have gained now is the information that D depends only on the neighbourhood of Σ, not on the geometry of the rest of X; this is the excision principle. We can therefore calculate this covering degree using any convenient model. The simplest thing is the ruled surface described in Section 3, where the answer comes out as 2^g, for small α, by virtue of the proposition from Section 4 and the conjecture on parabolic bundles.

6. The General Case and Further Comments

For the general case we must start with an $SU(2)$ bundle, not an $SO(3)$ bundle. We therefore arrange that g is odd. Our original moduli space M_E will not be zero-dimensional, but as in [D2] we can cut down the moduli space by divisors V_j corresponding to 2-dimensional homology classes β_j in X. We choose the β_j so that they are orthogonal to $[\Sigma]$ in $H_2(X)$ and are therefore supported in $X \setminus \Sigma$. Provided we are in the stable range [D2], we can cut down the moduli space to a finite set of points whose number (counted algebraically) is an invariant of X and the chosen homology classes: this is the definition of the polynomial invariants.

Everything can be carried through for these cut-down moduli spaces. Being in the stable range means that simple dimension-counting arguments ensure the compactness which the non-zero Stieffel-Whitney class previously provided (see [D2] again). The only technical difference is in the excision step, where we are now forced to deal with $\mathcal{R}(\Sigma)_{\text{even}}$. Since this space contains reducible connections, we need to be sure that, after deformation, we can arrange for the flat connections ϱ_i involved in the gluing to be in the irreducible part. For this step, the positive sign of the self-intersection of Σ puts us in a good situation (see [T2], Proposition 12.2).

At the end of the day, the argument shows that the polynomial invariants vanish on the orthogonal complement of $[\Sigma]$ if the surface contradicts the inequality of the conjecture. But since infinitely many pairwise-independent homology classes can be represented by surfaces contradicting the inequality (see the previous section), it follows that the invariant must vanish identically.

In a more general setting, it seems that the ideas of this paper, if they can be carried through, would be useful for establishing relations between the polynomial invariant associated to a bundle E_0 and that associated to a corresponding twisted bundle $E_{1/2}$; each choice of l would give some relation. To calculate these, one first needs to know the analogue of the proposition of Section 4, for different values of the monopole number. For example, if μ is greater than the value λ given in that proposition, then only special rank-2 bundles $\mathscr{E}$ admit line subbundles of degree μ. This condition defines a subvariety of complex codimension $2(\mu - \lambda)$ in the moduli space of stable bundles over Σ, and one would need to know its dual cohomology class in terms of standard generators. Via the excision argument, this sort of information establishes the relationship between M_0 and M_α for small α, with similar considerations at $\alpha = 1/2$.

References

[A]   Atiyah, M.F.: Magnetic monopoles in hyperbolic space. In: Vector Bundles on Algebraic Varieties. Tata Institute of Fundamental Research, Bombay, 1968, pp. 1–33

[D1]   Donaldson, S.K.: Anti-self-dual Yang-Mills connections on complex algebraic surfaces and stable vector bundles. Proc. London Math. Soc. **50** (1985) 1–26

[D2]   Donaldson, S.K.: Polynomial invariants for smooth four-manifolds. Topology **29** (1990) 257–315

[HS]   Hsiang, W.C., Szczarba, R.H.: On embedding surfaces in four-manifolds. In: Algebraic Topology. (Proc. Sympos. Pure Math., vol. 22.) Amer. Math. Soc., Providence, RI, 1971, pp. 97–103

[Ki]   Kirby, R.C.: Problems in low-dimensional topology. In: Algebraic and Geometric Topology. (Proc. Sympos. Pure Math., vol. 22, pt. 2.) Amer. Math. Soc., Providence, RI, 1978, pp. 273–312

[Mi]   Milnor, J.: Singular Points of Complex Hypersurfaces. (Ann. Math. Studies, vol. 61.) Princeton University Press, 1968

[MMR]   Morgan, J.W., Mrowka, T., Ruberman, D.: Preprint (in preparation)

[Mr]   Mrowka, T.: A Mayer-Vietoris principle for Yang-Mills moduli spaces. Ph.D. thesis, Berkeley, 1989

[R]   Rohlin, V.A.: Two-dimensional submanifolds of four-dimensional manifolds. Funkt. Annal. Prilozen. **5** (1971) 48–60 (Russian). [English transl.: Funct. Anal. Appl. **5** (1971) 39–48]

[Se]   Seshadri, C.S.: Moduli of vector bundles on curves with parabolic structures. Bull. Amer. Math. Soc. **83** (1977) 124–126

[SS]   Sibner, L.M., Sibner, R.J.: Classification of singular Sobolev connections by their holonomy. Preprint, 1989

[Si]   Simpson, C.T.: Constructing variations of Hodge structure using Yang-Mills theory and applications to uniformization. J. Amer. Math. Soc. **1** (1988) 867–918

[T1]   Taubes, C.H.: Gauge theory on asymptotically periodic 4-manifolds. J. Diff. Geom. **25** (1987) 363–430

[T2]   Taubes, C.H.: L^2-moduli spaces on 4-manifolds with cylindrical ends, I. Preprint, 1990

[U]   Uhlenbeck, K.K.: Connections with L^p bounds on curvature. Commun. Math. Phys. **83** (1982) 31–42

Symplectic 4-Manifolds

Dusa McDuff

Department of Mathematics, State University of New York at Stony Brook
Stony Brook, NY 11794, USA

§ 1. Introduction

A symplectic structure on a $2n$-dimensional manifold V is a 2-form ω which is closed and non-degenerate, that is, $d\omega = 0$ and ω^n does not vanish. The main example is the standard form

$$\omega_0 = dx_1 \wedge dx_2 + dx_3 \wedge dx_4 + \cdots + dx_{2n-1} \wedge dx_{2n}$$

on $\mathbb{R}^{2n}$. Darboux showed that every symplectic form is locally diffeomorphic to ω_0. Thus symplectic manifolds, in contrast to Riemannian manifolds, are all locally isomorphic. We are concerned here with questions of their global structure.

When $n = 1$ a symplectic form is just an area form, and the corresponding global topology is well-understood. For example, the existence problem is trivial: clearly, if V is compact, orientable and connected, it has a symplectic structure in each non-zero class in $H^2(V; \mathbb{R})$. The uniqueness problem is almost as easy, since Moser [M] showed that an area form (or, in higher dimensions, a volume form) is determined up to isotopy by its cohomology class. (Two forms ω_0 and ω_1 on V are said to be *isotopic* if there is a family g_t of diffeomorphisms of V such that $g_0 = $ id. and $g_1^*(\omega_0) = \omega_1$. They are *symplectomorphic* if $g_1^*(\omega_0) = \omega_1$ for some diffeomorphism g_1.) However, in dimensions > 2, these questions are far from being understood.

This paper will describe selected topics in the 4-dimensional case, concentrating on my own work. As we shall see, there are situations in which a symplectic form is determined up to symplectomorphism by its cohomology class. For example, Gromov's celebrated uniqueness theorem says that this holds for symplectic forms on $\mathbb{C}P^2$ which are non-degenerate on some embedded 2-sphere in the homology class of $\mathbb{C}P^1$. However, this does not solve the problem of describing all symplectic structures on $\mathbb{C}P^2$ since it is unknown whether this condition holds for every symplectic form on $\mathbb{C}P^2$.

Almost all the results mentioned below use Gromov's technique of J-holomorphic spheres. They exploit the almost complex structures associated to a symplectic structure, and so emphasise the similarities between Kähler, complex and symplectic geometry. However, symplectic geometry cannot always be derived from Kähler (or even complex) geometry. For example, there are many non-Kähler

Proceedings of the International Congress
of Mathematicians, Kyoto, Japan, 1990
© The Mathematical Society of Japan, 1991

symplectic 4-manifolds. (There are even symplectic 4-manifolds which admit no complex structure: see [FGG].) There is also an example in [McD7] of a symplectic 4-manifold which has a disconnected "symplectically convex" boundary. By contrast, any complex manifold with pseudo-convex boundary must have connected boundary. So far, the geometric significance of a symplectic manifold being Kähler is not well understood. It would be very interesting to have some answers to the following questions:

1.1 *Is every simply-connected symplectic 4-manifold Kähler?*

This is not true in dimensions ≥ 10 by [McD1].

1.2 *Is every symplectic form on a 4-manifold determined up to symplectomorphism (or even isotopy) by its cohomology class?*

This is not true in dimension 8, and in dimension 6 there are examples of symplectomorphic but non-isotopic forms: see [McD2].

1.3 *Which 4-manifolds admit a symplectic structure?*

Very little is known about this unless V is a fibration or the form ω is invariant under a non-trivial S^1 action (see [Au, AH]).

§2. *J*-Holomorphic Curves in Almost Complex 4-Manifolds

Because the linear symplectic group deformation retracts onto the unitary group, there is a homotopy class of almost complex structures J associated to a symplectic form ω. (Recall that an almost complex structure J on V is an automorphism of the tangent bundle TV of V such that $J^2 = -Id$. Thus TV may be considered as a complex n-dimensional vector bundle, with J corresponding to multiplication by i.) When $n = 1$, every almost complex structure is integrable, that is, it comes from an underlying complex structure on V. This fact is one of the reasons why J-holomorphic curves have such nice properties.

Gromov realised that one could get a handle on the geometry of a symplectic manifold by considering properties of a more restricted family of almost complex structures on V, namely the set $\mathcal{J}(\omega)$ of all J such that $\omega(v, Jv) > 0$ for all non-zero $v \in TV$. Such J are said to be ω-*tame*. Recall from [G1] that the set $\mathcal{J}(\omega)$ is non-empty and contractible.

A map f from a Riemann surface (S, J_0) to (V, J) is said to be J-*holomorphic* if f satisfies the generalized Cauchy-Riemann equation $df \circ J_0 = J \circ df$. Because this equation is elliptic, the space $M_p(J, A)$, of all J-holomorphic maps f which have a fixed compact domain S and represent a fixed homology class A, is a finite dimensional manifold for generic J. Moreover, if J is ω-tame, the quotient of this space by the automorphism group $G = \text{Aut}(S)$ of (S, J_0) is either compact, or has a nice compactification. In fact, as the following proposition shows, homological conditions are often enough to guarantee compactness. For simplicity, we state a result about rational curves, i.e. curves $C = f(S)$ which are J-holomorphic images of spheres. The first Chern class of the complex vector bundle (TV, J) will be denoted by c. A proof may be found in [G1] or [McD4].

Theorem 2.1 [G1]. *Suppose that ω is integral (i.e. $[\omega] \in H^2(V, \mathbb{Z})$) and that $\omega(A) = 1$. Then, for generic ω-tame J, the moduli space $M_p(J, A)/G$ of rational A-curves is a compact manifold of dimension $2(c(A) + n - 3)$. Moreover, the bordism class of the evaluation map*

$$e(J) : M_p(J, A) \times_G S^2 \to V$$

given by $(f, z) \mapsto f(z)$ is independent of $J \in \mathcal{J}(\omega)$.

This result already has many interesting consequences: see [G1, 2] for example. In dimension 4 the theory is much sharper because, as the following results show, the geometric behaviour of the curves is governed by their homology classes.

Theorem 2.2. Positivity of Intersections ([G1]). *Let C and C' be distinct closed J-holomorphic curves in the almost complex 4-manifold V. Then C and C' have only a finite number of intersection points. Each such point x contributes a number $k_x \geq 1$ to the algebraic intersection number $C \cdot C'$. Moreover, $k_x = 1$ iff the curves C and C' intersect transversally at x.*

In particular, the curves C and C' are disjoint iff $C \cdot C' = 0$. This result is almost obvious if neither curve is singular at the point of intersection. The best way to prove the result in general is to perturb the curves so that their intersections avoid the singular points: see [McD8]. The next result gives a homological criterion for a curve to be embedded. By analogy with the integrable case, we define the virtual genus $g(C)$ of a closed curve C in an almost complex 4-manifold to be the number $g(C) = 1 + \frac{1}{2}(C \cdot C - c(C))$. If C is an embedded copy of a closed Riemann surface S, it is easy to check that the virtual genus $g(C)$ equals the genus g_0 of S. Conversely:

Theorem 2.3 [McD5]. *Let $C \subset V^4$ be the J-holomorphic image of a closed Riemann surface S of genus g_0. Then $g(C)$ is an integer which is greater than or equal to g_0, with equality if and only if C is embedded.*

The crucial point here is that one can define for each point $x \in C = \text{Im } f$ a local self-intersection number L. Int.(f, x) of Im f at x. One proves that this is always ≥ 0, and is strictly > 0 iff x is a singular point of f. In fact, it is not hard to see that one can choose local coordinates near $x \in V$ so that the lowest order terms in the Taylor expansion of f when regarded as a polynomial in z and $\bar{z}$ involve only z. More precisely, one can suppose that $f : D \to \mathbb{C}^2$ has the form

$$z \mapsto (z^k, z^m) + \text{terms in } z, \bar{z} \text{ of order} > m,$$

where $m > k \geq 1$ and k does not divide m. (Here, D is the unit disc in $\mathbb{C}$, and we identify a neighbourhood of x in V with a neighbourhood of $\{0\}$ in $\mathbb{C}^2$ in such a way that J corresponds to an almost complex structure on $\mathbb{C}^2$, which equals the standard almost complex structure J_0 at $\{0\}$.) If m and k are mutually prime, it is easy to see that L. Int.$(f, x) =$ L. Int.$(f_0, x) = (m - 1)(k - 1)$, where $f_0(z) = (z^k, z^m)$. In general, the proof of Theorem 2.3 involves perturbing f, using the techniques of Nijenhuis and Woolf. A more detailed analysis of the generalized Cauchy-Riemann equation shows that the jet of the J-holomorphic map at a critical point is dominated by its J_0-holomorphic part. More precisely,

Theorem 2.4 [McD8]. *Let* $f : (D, 0) \to (\mathbb{C}^2, 0)$ *be a J-holomorphic map, where J is an almost complex structure on $\mathbb{C}^2$ which equals the standard structure J_0 at $\{0\}$. Then there is a J_0-holomorphic map f_0 such that for sufficiently small $\varepsilon > 0$ the knots formed by intersecting $f(D)$ and $f_0(D)$ with the sphere $S^3(\varepsilon)$ of radius ε about $\{0\}$ are isotopic.*

§3. Applications to Symplectic 4-Manifolds; Uniqueness Results

In order to apply this theory one needs to know that the bordism class of the evaluation map $e(J)$ of Theorem 2.1 is non-trivial. For example, suppose that there is a symplectically embedded 2-sphere C in V^4. It is not hard to see that C can be parametrized in such a way that it is J-holomorphic for some ω-tame J. If in addition the self-intersection number $p = C \cdot C$ is ≥ -1 one can show that that the corresponding moduli space is generically non-empty. Thus V contains a $2(p + 1)$-dimensional family of (unparametrized) curves which may be used to analyse the structure of V.

The case $p = -1$ is rather special since then, by Theorem 2.2, there is at most one J-holomorphic curve in each homology class. Such an embedded 2-sphere is said to be *an exceptional sphere*. By analogy with the complex case, a symplectic 4-manifold which contains no exceptional spheres is said to be *minimal*.

The following results describe all minimal symplectic 4-manifolds which contain a symplectic 2-sphere with non-negative self-intersection.

Theorem 3.1 [McD6]. *Let* (V, ω) *be a closed symplectic 4-manifold (V, ω) which contains a symplectically embedded 2-sphere C with $C \cdot C = 0$. Suppose further that $(V - C, \omega)$ is minimal. Then (V, ω) is ruled, that is, there is a fibration $\pi : V \to M$ whose fibers are symplectically embedded 2-spheres, one of which we may assume to be C. Moreover the symplectic form ω is determined up to symplectomorphism by its cohomology class.*

Sketch of proof. Let A be the homology class of C. One first shows that, for any ω-tame J, the set of A-curves may be compactified by adding "A-cusp-curves". These must consist of two exceptional curves joined at one point and, because $A \cdot A = 0$, must be disjoint from all J-holomorphic A-curves by Theorem 2.2. Therefore, if J is chosen so that C is J-holomorphic, the minimality of $V - C$ implies that the moduli space of A-curves is compact. Using Theorem 2.2 again, one sees that there is at most one A-curve through each point. It follows that the evaluation map $e(J)$ has degree 1, and that there is exactly one curve through each point. Now observe that these curves are embedded by Theorem 2.3. With a little more work, one can show that they form the fibers of a fibration.

To prove the uniqueness statement, one first reduces to the case $V = S^2 \times S^2$ by some cutting and pasting. Next one constructs a symplectic section of V in the class $B = [\text{pt.} \times S^2]$. We are now in a situation considered by Gromov. Since $B \cdot B = 0$, the B-curves also fiber V. Thus there is a symplectomorphism from (V, ω) to $(S^2 \times S^2, \varrho)$ where ϱ is non-degenerate on all the slices pt. $\times S^2$ and $S^2 \times$ pt. The result now follows from:

Lemma 3.2 [K]. *Suppose that ω is a symplectic form on the product $M' \times M''$ of two compact 2-manifolds which is non-degenerate on all the slices $M' \times$ pt. and pt. $\times M''$. Then ω is isotopic to a product form.*

Proof. This is a straightforward calculation.

An explicit description of all possible symplectic ruled 4-manifolds is given in [McD6] §4. Basically, one gets all the Kähler ruled surfaces considered as symplectic manifolds, i.e. one forgets the complex structure, since this is not relevant. However, note that when V is a non-trivial bundle the cohomology classes $a = [\omega]$ which can be realised in this way satisfy the condition $(a(C))^2 < a^2(V)$, and so do not include every class a with $a^2(V) > 0$. Note also that (V, ω) is itself minimal unless V is the non-trivial bundle over S^2, in which case (V, ω) is $\mathbb{C}P^2$ blown up at one point with a standard Kähler form.

There is a corresponding and even simpler result when the sphere has positive self-intersection. It is the symplectic analogue of the fact that the only minimal complex surfaces which contain a rational curve of positive self-intersection are $\mathbb{C}P^2$ and $\mathbb{C}P^1 \times \mathbb{C}P^1$.

Theorem 3.3 [McD6]. *Let (V, ω) be a closed minimal symplectic 4-manifold (V, ω) which contains a symplectically embedded 2-sphere C with $p = C \cdot C > 0$. Then (V, ω) is symplectomorphic either to $\mathbb{C}P^2$ with its standard Kähler form or to $S^2 \times S^2$ with a product form. In the former case p must equal 1 or 4 and C is either $\mathbb{C}P^1$ or a quadric, and in the latter case p is even, and C is the graph of a holomorphic self-map of S^2.*

We will see in §4 below that, just as in the complex case, one can get rid of exceptional spheres by blowing them down. Thus these two theorems classify all symplectic 4-manifolds which contain a symplectically embedded 2-sphere C with $C \cdot C \geq 0$, modulo the question of the uniqueness of blow ups. They generalize and sharpen uniqueness results which Gromov obtained in [G1] for the manifolds $S^2 \times S^2$ and $\mathbb{C}P^2$. One expects that there are corresponding results for manifolds which contain a symplectically embedded 2-manifold M of higher genus, provided that $M \cdot M$ (or equivalently, $c(M)$) is sufficiently large to guarantee that the index (i.e. formal dimension of the corresponding moduli space) is positive. But it is unknown what happens if $M \cdot M$ is small. For example, if ω is a symplectic form on

$T^2 \times S^2$ which is non-degenerate on one slice $T^2 \times$ pt., it is not known whether ω must be symplectomorphic to a product form.

Note. These uniqueness results all derive from considering families of J-holomorphic spheres, and so apply only to 4-manifolds which contain such spheres. In some circumstances it is better to consider J-holomorphic discs whose boundary is constrained to lie on a 2-manifold M which is totally real, i.e. $J(T_x M) \cap T_x M = \{0\}$, except at isolated points. This approach yields the sharpest results when M is a 2-sphere which is contained in a J-convex hypersurface: see [E, B]. Note that M need not be symplectic.

§4. Blowing Up and Down, and Embeddings of Balls

In the category of complex manifolds, there is a well-known operation of blowing up a point in which one replaces a point x by the space of all complex lines through x. More formally, one cuts out a neighbourhood of x which is biholomorphic to $\mathbb{C}^n$ and glues back in a copy of the total space L of the canonical line bundle over $\mathbb{C}P^{n-1}$ by identifying the complement of the zero section in L with the deleted neighbourhood $\mathbb{C}^n - \{0\}$ of x. Thus the point x is replaced by an "exceptional divisor" which is a copy of $\mathbb{C}P^{n-1}$ with normal bundle L. When $n = 2$, this divisor is simply an embedded rational curve C with $C \cdot C = -1$. The converse process of "blowing down" replaces an exceptional divisor by a point.

In the symplectic case, when one blows down an exceptional divisor P one gets not a point but a ball $B(\lambda)$ whose radius λ is related to the cohomology class of ω by the formula $\pi\lambda^2 = \omega(A)$, where A is the positive generator of $H_2(P; \mathbb{Z})$. Conversely, in order to "blow up a point" of (V, ω) one chooses a symplectic embedding g of $B = B(\lambda)$ into V, cuts out the interior of the ball $g(B)$ and then forms the blow up $\tilde{V}$ by identifying the boundary sphere $g(\partial B)$ with $\mathbb{C}P^{n-1}$ via the Hopf map and smoothing the result: see [McD3]. It is not hard to check that ω induces a symplectic form $\tilde{\omega}$ on $\tilde{V}$. Note that $\tilde{V}$ is diffeomorphic to the manifold obtained by putting any integrable tame almost complex structure on a neighbourhood of some point $x \in V$, and then blowing V up at x. The point of the above description is that it gives the construction of $\tilde{\omega}$.

Clearly, by successive blowing down of a finite sequence of exceptional divisors one can make any manifold minimal, i.e. such that it does not contain any exceptional divisors. When $n = 2$, one can use the theory of J-holomorphic curves to control the exceptional divisors and hence can show:

Theorem 4.1 [McD3]. *Every symplectic 4-manifold (V, ω) covers a minimal symplectic manifold (V', ω') which may be obtained from V by blowing down a finite collection of disjoint exceptional curves. Moreover, (V', ω') is determined up to symplectomorphism by the homology classes of the blown down curves.*

Another important question is that of the uniqueness of blow ups. Not much is known about this except when one blows up one point of the standard $\mathbb{C}P^2$. In this case, the resulting manifold (X, ω) is the non-trivial S^2 bundle over S^2. Moreover, it is not hard to show that X contains a symplectically embedded 2-sphere C with $C \cdot C = 0$. Hence the theory of §3 applies, and one can show that all cohomologous forms are symplectomorphic. As a corollary, one finds:

Theorem 4.2 [McD3]. *For each $\lambda < 1$, the space of symplectic embeddings of the ball $B(\lambda)$ into the open unit ball in $\mathbb{C}^2$ is connected.*

The analogous statement is true if the target space is the standard $\mathbb{C}P^2$. However, this result is surprisingly delicate, and it is unclear if this holds for any other target space, even those as simple as $S^2 \times S^2$ or a convex subset of $\mathbb{C}^2$. This question is of interest because of the following result.

Proposition 4.3 [McD3]. *Let X_k be the complex surface obtained by blowing $\mathbb{C}P^2$ up at k points, and let S be a fixed copy of $\mathbb{C}P^1$ in X_k (disjoint from the blown up points). Then, there is a non-Kähler structure on X_k which is non-degenerate on S iff there are $\lambda_1, \ldots, \lambda_k$ such that the space of symplectic embeddings of the disjoint union $\coprod_i B(\lambda_i)$ into $B(1)$ is disconnected.*

References

[Au]   Audin, M.: Hamiltoniens périodiques sur les variétés symplectiques compactes de dimension 4. (Lecture Notes in Mathematics, vol. 1416.) Springer, Berlin Heidelberg New York 1990

[AH]   Ahara, K., Hattori, A.: 4-dimensional symplectic S^1-manifolds admitting a moment map. Preprint, Tokyo 1990

[B]   Bennequin, D.: Topologie symplectique, convexité holomorphe et structure de contacte. Séminaire Bourbaki, June 1990

[E]   Eliashberg, Ya.: Filling by holomorphic discs and its applications. In: Geometry of low dimensional manifolds, vol 2, ed. Donaldson and Thomas. CUP, Cambridge 1990

[FGG]   Fernandez, M., Gotay, M., Gray, A.: Four-dimensional parallelizable symplectic and complex manifolds. Proc. Amer. Math. Soc. **103** (1988) 1209–1212

[G1]   Gromov, M.: Pseudo-holomorphic curves in almost complex manifolds. Invent. math. **82** (1985) 307–347

[G2]   Gromov, M.: Soft and hard symplectic geometry. In: Proceedings of the ICM at Berkeley, 1986. Amer. Math. Soc., 1987, pp. 81–98

[K]   Kasper, B.: Examples of symplectic structures on fiber bundles. Ph.D. thesis, Stony Brook 1990

[McD1] McDuff, D.: Examples of simply connected non-Kählerian symplectic manifolds. J. Diff. Geom. **20** (1984) 267–277

[McD2] McDuff, D.: Examples of symplectic structures. Invent. math **89** (1987) 13–36

[McD3] McDuff, D.: Blowing up and symplectic embeddings in dimension 4. To appear in Topology 1991

[McD4] McDuff, D.: Elliptic methods in symplectic geometry. Bull. Amer. Math. Soc. **23** (1990) 311–358

[McD5] McDuff, D.: The local behaviour of holomorphic curves in almost complex 4-manifolds. To appear in J. Diff. Geom. 1991

[McD6] McDuff, D.: The structure of rational and ruled symplectic 4-manifolds. J. Amer.
        Math. Soc. **3** (1990) 679–712
[McD7] McDuff, D.: Symplectic manifolds with contact-type boundaries. Invent. math. **103**
        (1991) 651–671
[McD8] McDuff, D.: Singularities of J-holomorphic curves in almost complex 4-manifolds.
        Preprint, 1991
[M]   Moser, J.: On the volume elements on manifolds. Trans. Amer. Math. Soc. **120**
        (1965) 280–296

Rational Homotopy Theory and Deformation Problems from Algebraic Geometry

John J. Millson

Department of Mathematics, University of Maryland, College Park, MD 20742, USA

This paper is a description of research I have been doing over the last four years, applying some of the methods and ideas of rational homotopy theory as developed by Chen, Quillen and Sullivan, to deformations of flat and holomorphic bundles, complex manifolds and isolated singularities. My work is based on the fundamental observation of Pierre Deligne [D] that "in characteristic zero, a deformation problem is controlled by a differential graded Lie algebra, with quasi-isomorphic differential graded Lie algebras giving the same deformation theory." I would like to thank Pierre Deligne for providing me with this insight and Bill Goldman who was my collaborator in developing much of what follows. I would also like to thank Ragnar Buchweitz, Kevin Corlette, Steve Halperin, Jack Lee, Madhav Nori and Mike Schlessinger for helpful conversations. The interested reader will find details in [GM1, GM2, BM] and [M].

We begin by recalling that Sullivan showed how to recover *explicitly* the rational homotopy type of a simply-connected manifold M by replacing the de Rham algebra $\mathscr{A}^*(M)$ on M by a minimal free differential graded algebra quasi-isomorphic to it having finite dimensional cochain groups and decomposable differential. We recall that two differential graded algebras A and B are quasi-isomorphic if there is a chain of homomorphisms

$$A = A_0 \to A_1 \leftarrow A_2 \cdots \to A_n = B$$

all of which induce isomorphisms of cohomology. The rational homotopy type of M can be calculated from any differential graded algebra quasi-isomorphic to $\mathscr{A}^*(M)$.

We will make use of one concept from the now well-developed homotopy theory of differential graded algebras.

Definition. *A differential graded algebra A is* formal *if it is quasi-isomorphic to a differential graded algebra B with zero differential. The underlying algebra B is then necessarily isomorphic to the cohomology algebra of A.*

We now recall the celebrated theorem of [DGMS].

This work was supported in part by NSF grant DMS-85-01742.

Theorem. *The de Rham algebra $\mathscr{A}^*(M)$ of a compact Kähler manifold M is formal.*

The theorem is a consequence of the following considerations. We use the complex structure on M to decompose the exterior differential d on M in the usual way, $d = \partial + \bar{\partial}$. Then we have the following quasi-isomorphism

$$(\mathscr{A}^*(M), d) \leftarrow (\ker \partial, \bar{\partial}) \rightarrow \left(\frac{\ker \partial}{\operatorname{im} \partial}, 0\right)$$

The fact that $\bar{\partial}$ induces the zero differential on $\frac{\ker \partial}{\operatorname{im} \partial}$ is an immediate consequence of the "$\partial\bar{\partial}$-lemma" of Kähler geometry, [DGMS].

To carry Sullivan's ideas over to deformation theory, we start with a differential graded Lie algebra over a field $\mathbf{k}$ (we will consider only those algebras with finite dimensional first cohomology groups). Given such an algebra L we choose a complement $C^1(L)$ to the 1-coboundaries $B^1(L) \subset L^1$. We define a functor $A \to Y_L(A)$ on the category of Artin local $\mathbf{k}$-algebras by

$$Y_L(A) = \{\eta \in C^1(L) \otimes \mathfrak{m} : d\eta + \tfrac{1}{2}[\eta, \eta] = 0\}.$$

Here $\mathfrak{m}$ is the maximal ideal of the Artin local $\mathbf{k}$-algebra A. It is evident that the functor Y_L satisfies the hypotheses of Theorem 2.11 of [Sc1] and is consequently pro-representable by a complete local $\mathbf{k}$-algebra R_L (we will see later as a consequence of our main theorem that the isomorphism class of R_L does not depend on the complement $C^1(L)$).

One can apply the above construction to the following geometric situations:

(i) The twisted de Rham algebra with coefficients in the flat Lie algebra bundle ad P associated to a flat principal G-bundle P over a compact manifold M;
(ii) the Kodaira-Spencer algebra $(\oplus \mathscr{A}^{0,*}(M, T^{1,0}(M)), \bar{\partial})$ associated to a complex manifold M;
(iii) the tangent complex T associated to the germ (V, x) of an isolated singularity in $\mathbb{C}^n$.

It is then reasonably clear (and proved in [GM2]) from the above construction that in cases (i) and (ii) the algebra R_L is the completion of the analytic local ring $\mathfrak{O}_\mathscr{K}$ associated to the versal deformation space as constructed by Kuranishi of the given flat connection on P (resp. complex structure on M). In case (iii), it is proved in [BM] following ideas in [SS] that the ring R_T is isomorphic to the completion of the analytic local ring of the versal deformation space of the isolated singularity (V, x). We note two other important cases of differential graded Lie algebras L over a field $\mathbf{k}$ in which R_L is the completion of an analytic local $\mathbf{k}$-algebra.

In case L has zero differential, then R_L is the completion of the analytic local $\mathbf{k}$-algebra associated to the germ $(\mathscr{Q}_L, 0)$ where

$$\mathscr{Q}_L = \{\eta \in L^1 : [\eta, \eta] = 0\}.$$

We note that $\mathscr{Q}_L$ is a quadratic cone canonically associated to L. In case L^1 is finite dimensional, then R_L is the completion of the analytic local $\mathbf{k}$-algebra of the germ $(Y, 0)$ where

$$Y = \{\eta \in C^1 : d\eta + \tfrac{1}{2}[\eta, \eta] = 0\}.$$

Henceforth we will use $\mathscr{K}_L$ to denote the corresponding analytic germ.

Definition. *If (X, x) is an analytic germ which parameterizes a versal family for a deformation theory and L is a differential graded Lie algebra such that $R_L \approx \hat{\mathfrak{O}}_{X,x}$ then we will say that L is a* controlling *differential graded Lie algebra for that deformation theory.*

We have the following theorem (Theorem 4.1 of [GM2]).

The Comparison Theorem. *Suppose $f : L_1 \to L_2$ is a homomorphism of differential graded Lie algebras such that f induces an isomorphism on first cohomology and an injection on second cohomology. Then R_{L_1} and R_{L_2} are isomorphic.*

Corollary. *The isomorphism class of R_L does not depend on the choice of $C^1(L)$.*

Corollary. *If L_1 and L_2 are quasi-isomorphic, then R_{L_1} and R_{L_2} are isomorphic.*

Remarks. We do not require f to carry the complement $C^1(L_1)$ into $C^1(L_2)$. In case R_{L_1} and R_{L_2} are the complete local **k**-algebras associated to analytic germs (X_1, x_1) and (X_2, x_2), it follows from [A] that (X_1, x_1) and (X_2, x_2) are isomorphic.

If L is quasi-isomorphic to a differential graded Lie algebra with zero differential (one says L is formal), then it follows that

$$\mathscr{K}_L = \{\eta \in H^1(L) : [\eta, \eta] = 0\}.$$

Somewhat surprisingly this frequently happens. Carlos Simpson [S] has shown that the twisted de Rham algebra is formal if M is Kähler, the structure group G of the underlying principal bundle is linear and the monodromy representation $\varrho : \pi_1(M) \to G$ is completely reducible. Using Simpson's Theorem, the above general results and some standard results relating representations of $\pi_1(M)$ and flat connections we obtain the following theorem (recall that if G is a linear algebraic group over $\mathbb{R}$ (or $\mathbb{C}$) then the space of representations $\mathrm{Hom}(\pi_1(M), G)$ is an affine scheme over $\mathbb{R}$ (or $\mathbb{C}$)). The proof of this theorem is in [GM1].

Theorem. *Let G be a linear algebraic group over $\mathbb{R}$ (or $\mathbb{C}$) and M be a compact Kähler manifold. Let $\varrho : \pi_1(M) \to G$ be a completely reducible representation. Then the analytic local ring of $\mathrm{Hom}(\pi_1(M), G)$ at ϱ is isomorphic to that of the quadratic germ $(\mathcal{Q}, 0)$ where*

$$\mathcal{Q} = \{u \in Z^1(\pi_1(M), \mathfrak{g}) : [u, u] = 0 \quad in \quad H^2(\pi_1(M), \mathfrak{g})\}.$$

Here $Z^1(\pi_1(M), \mathfrak{g})$ is the space of Eilenberg-MacLane 1-cocycles with values in the Lie algebra $\mathfrak{g}$ of G and $[u, u]$ denotes the product obtained by combining the cup-product on group cochains with the bracket on $\mathfrak{g}$.

The proofs of the above results are too long to be given here; however, to emphasize the analogy with the theorem of [DGMS] referred to earlier, we give the original proof of [GM1] of the formality of the twisted de Rham algebra in case G is compact. We may decompose the exterior covariant differential d_∇ by type as $d_\nabla = \partial_\nabla + \bar{\partial}_\nabla$ (here ∇ is the covariant derivative operator on sections of ad $P \otimes \mathbb{C}$ associated to the flat connection on P). We then have the quasi-isomorphism

$$(\mathscr{A}^*(M, \text{ad }P_\mathbb{C}), d_\nabla) \leftarrow (\ker \partial_\nabla, \bar{\partial}_\nabla) \rightarrow \left(\frac{\ker \partial_\nabla}{\text{im }\partial_\nabla}, 0 \right).$$

Once again the induced differential on $\frac{\ker \partial_\nabla}{\text{im }\partial_\nabla}$ is zero by a $\partial_\nabla \bar{\partial}_\nabla$-lemma, see [GM1].
There is also an interesting formality result for the Kodaira-Spencer algebra.

Theorem. *Let M^n be a compact Kähler manifold admitting a nowhere zero top degree holomorphic form. Then the Kodaira-Spencer algebra of M is formal. Moreover, the cup-square from $H^1(M, T^{1,0}(M))$ to $H^2(M, T^{1,0}(M))$ is zero.*

Corollary. *The Kuranishi space (versal deformation space) of M is $H^1(M, T^{1,0}(M))$.*

The corollary is Bogomolov's Theorem. Our proof of the above theorem is a reinterpretation of proofs of Tian [Ti] and Todorov [To]. We let ω be the nowhere zero top degree holomorphic form on M. We then obtain an isomorphism of complexes $\Phi : (\mathscr{A}^{0,\bullet}(M, T^{1,0}(M)), \bar{\partial}) \rightarrow (\mathscr{A}^{n-1,\bullet}(M), \bar{\partial})$ given by

$$\Phi(\eta) = \iota_\eta \omega.$$

Here $\iota_\eta \omega$ denotes the contraction [FN] of the scalar form ω by the vector form η. We use Φ to transfer the graded Lie bracket from the Kodaira-Spencer algebra to $\bigoplus_{q=0}^n \mathscr{A}^{n-1,q}(M)$.

The Tian-Todorov Lemma. *Suppose $\eta_1 \in \mathscr{A}^{n-1,q_1}(M)$ and $\eta_2 \in \mathscr{A}^{n-1,q_2}(M)$ are both ∂-closed differential forms. Then the (transported) bracket $[\eta_1, \eta_2]$ is ∂-exact.*

Remark. This lemma is the analogue of the lemma in symplectic geometry that the bracket of two symplectic vector fields corresponds to an *exact* 1-form. It may be proved in the same way using the formalism of vector-valued forms of [FN].

Once the Tian-Todorov Lemma is proved, the formality of the Kodaira-Spencer algebra follows from the now familiar diagram

$$(\mathscr{A}^{n-1,\bullet}(M), \bar{\partial}) \leftarrow (\ker \partial, \bar{\partial}) \rightarrow \left(\frac{\ker \partial}{\text{im }\partial}, 0 \right).$$

Here $\ker \partial$ denotes $\ker(\partial : \mathscr{A}^{n-1,\bullet}(M) \rightarrow \mathscr{A}^{n,\bullet}(M))$ and $\text{im }\partial$ denotes $\text{im}(\partial : \mathscr{A}^{n-2,\bullet}(M) \rightarrow \mathscr{A}^{n-1,\bullet}(M))$. For details and the vanishing of the cup square the reader is referred to [GM2].

A class of examples which are easily analyzed is the class of compact complex parallelizable nilmanifolds. Let $M = \Gamma \backslash N$, N nilpotent complex with Lie algebra $\mathfrak{n}$ defined over $\mathbb{R}$ and Γ a cocompact lattice. Let L be the Kodaira-Spencer algebra and $\overline{L} \subset L$ the image of the left N-invariants. The inclusion $\overline{L} \to L$ is a quasi-isomorphism so $\mathscr{K}_L \cong \mathscr{K}_{\overline{L}}$. It is easy to see that

$$\mathscr{K}_{\overline{L}} = \mathrm{End}_{\mathrm{alg}}(\mathfrak{n})$$

the (germ at 0 of the) affine variety of Lie algebra endomorphisms of $\mathfrak{n}$. If we describe $\mathfrak{n}$ by generators and relations we can produce a very large number of germs that are Kuranishi spaces of complex manifolds. For example, let $\mathfrak{n}$ be the free Lie algebra on two generators X and Y subject to the relations

(i) all $(n+1)$-fold commutators $= 0$
(ii) $\mathrm{ad}^{n-1} X(Y) = \mathrm{ad}^{n-1} Y(X)$.

Then

$$\mathscr{K}_{\overline{L}} = \{ (X', Y') \in \mathfrak{n}^2 : X', Y' \text{ satisfy (ii) } \}$$

so the Kuranishi space of M is a homogeneous cone of degree n, see [GM2] for more details.

Remark. These examples provide realizations of all the obstructions to integrating an infinitesimal deformation of the complex structure of a complex manifold M.

We observe that $\overline{L}$ controls the deformation theory of locally left-invariant (i.e., descended from left-invariant complex structure on N) structures on M. Since $\overline{L} \to L$ is a quasi-isomorphism it follows that the two deformation theories are the same, that is, every complex structure on M sufficiently close to the locally bi-invariant one is locally left-invariant. The rest of this paper outlines deeper examples of such a comparison of two deformation theories.

We now describe applications of the above ideas to deformation of isolated singularities. We first summarize the fundamental results of Kuranishi in [K2]. Let V be an analytic subvariety of $\mathbb{C}^N$ with a normal isolated singularity at the origin. Let M be a link of V (the intersection of V with a small sphere centered at the origin). Then M has an induced CR-structure. Let $T^{1,0}(M)$ be the $(1,0)$ subspace of the complexified horizontal subspace $H \otimes \mathbb{C} \subset T(M) \otimes \mathbb{C}$. Choose a complement F to H in $T(M)$. The map τ of [K2], (8), gives

$$E = T^{1,0}(M) \oplus (F \otimes \mathbb{C})$$

the structure of a holomorphic vector bundle over M. Then $(\oplus \mathscr{A}^{0,*}(M, E), \overline{\partial}_b)$ is a complex, and Kuranishi used it to construct a finite-dimensional family $(\mathscr{K}_M, 0)$ of integrable CR-structures on M which is a versal deformation of the given CR-structure modulo a relation (coarser than isomorphism) designed to account for the above choice of sphere. However, he did not give $\mathscr{K}_M$ a complex analytic structure nor did he relate $\mathscr{K}_M$ to the versal deformation space of $(V, 0)$. The first problem was solved for the case dim $V \geq 4$ by Miyajima in [Mi1] completing earlier work of Akahori [Ak1]. In the rest of this paper we will show how the Comparison Theorem can be used to identify $\mathcal{K}_M$ with the versal deformation space of $(V, 0)$. The following result is proved in [BM]. It was proved independently by Miyajima in [Mi2] using results of Fujiki [F].

Theorem. *Suppose* $(V, 0)$ *is normal and satisfies*

(1) $\dim V \geq 4$;
(2) $\text{depth}_{\{0\}} V \geq 3$.

Then the base space of the versal deformation of $(V, 0)$ *is isomorphic to* $\mathcal{K}_M$.

Remarks. The assumption (2) is equivalent to the assumption that the Kohn-Rossi cohomology group $H^1(M, \mathfrak{O})$ vanishes by [Y]. If we do not assume (2), it can be shown that the base space for the versal deformation of $(V, 0)$ is isomorphic to a closed subgerm of $\mathcal{K}_M$. In [BM] we give a family of examples such that $(V, 0)$ is normal but the deformation space of $(V, 0)$ is a *proper* subgerm of $\mathcal{K}_M$.

We will now prove the above theorem by applying the Comparison Theorem many times.

Let T be the tangent complex associated to $(V, 0)$, see [B], [Sc3] or [P]. Then as stated above, T is a differential graded Lie algebra that controls the deformation theory of $(V, 0)$. We recall that this means that there is an isomorphism

$$R_T \approx \widehat{\mathfrak{O}}_{X,0}$$

where $(X, 0)$ is the analytic germ parametrizing the versal deformation of $(V, 0)$. Choose any Stein representative V of $(V, 0)$ and let L be the Kodaira-Spencer algebra of $U = V - \{0\}$. In [BM] we prove, following [Sc2], that under the assumption $\text{depth}_{\{0\}} V \geq 3$ we have

$$R_T \approx R_L.$$

It remains to compare the deformation theory of U with that of M.

We now consider the image of L under the restriction map from U to M. Unfortunately brackets in L do not behave well under this map and consequently it is necessary to replace L by a subalgebra L_{tan} of forms whose restrictions to M take values tangent to M. The definition of the algebra L_{tan} is somewhat involved. We may assume that r has no critical points on U and note that $\omega = \frac{i}{2} \partial \bar{\partial}(r^2)$ is the restriction of the Kähler form of $\mathbb{C}^n$. We consider the sub-graded vector space $D^{\bullet}$ of L defined by

$$D^0 = \{Z \in L^0 : \iota_{\bar{\partial}Z} \partial r|_M = 0\}$$

and for $i \geq 1$

$$D^i = \{\mu \in L^i : \iota_\mu \partial r|_M = 0, \quad \iota_\mu \omega|_M = 0\}.$$

It is proved in [Ak2] that $(D^{\bullet}, \bar{\partial})$ is a complex and it is easily seen that

$$D^+ = \bigoplus_{i \geq 1} D^i$$

is closed under the Frölicher-Nijenhuis bracket [FN]. Furthermore, it is a result of Akahori [Ak2] that the inclusion $D^\bullet \to L^\bullet$ is a quasi-isomorphism of complexes. Thus if we choose a complement C^1 to $\bar{\partial}D^0$ in D^1 and define

$$L_{\text{tan}} = C^1 \oplus \bigoplus_{i \geq 2} D^i,$$

we find that L_{tan} is a differential graded Lie algebra and that the inclusion $L_{\text{tan}} \to L$ induces an isomorphism of cohomology in degree greater than or equal to one. Thus by the Comparison Theorem we have (for any choice of C^1)

$$R_{L_{\text{tan}}} \approx R_L.$$

We now identify the image of j^*. To do this we need the CR-analogue of L_{tan} which was constructed by Akahori in [Ak1] and was the basis of the proof of the Akahori-Miyajima Theorem referred to above. We define $\overline{D}^\bullet$ to be the following sub-graded vector space of Kuranishi's complex $(\oplus \mathscr{A}^{0,*}(M,E), \bar{\partial}_b)$. Let θ be a contact form on M compatible with H. Then we define

$$\overline{D}^0 = \{Z \in \mathscr{A}^0(M,E) : \iota_{\bar{\partial}_b Z}\theta = 0\}$$

and for $i \geq 1$,

$$\overline{D}^i = \{\mu \in \mathscr{A}^{0,i}(M,E) : i_\mu\theta = 0, \quad \iota_\mu d\theta = 0\}.$$

We then choose a complement $\overline{C}^1$ to $\bar{\partial}_b \overline{D}^0$ in $\overline{D}^1$ and define

$$\overline{L} = \overline{C}^1 \oplus \bigoplus_{i \geq 2} \overline{D}^i.$$

We also define

$$\overline{D}^+ = \bigoplus_{i \geq 1} \overline{D}^i.$$

Now the restriction map j^* followed by the canonical map $\tau : T^{1,0}(U)|M \to E$ of [K2] gives a homomorphism of complexes

$$j^+ : D^+ \to \overline{D}^+.$$

Let $I^+ = \ker j^+$.

Lemma. I^+ is an ideal in D^+.

As a consequence of this lemma we find that $\overline{D}^+$ carries the structure of a differential graded Lie algebra such that j^+ is a homomorphism. Since j^* is surjective we may choose the complement C^1 so that it is carried into $\overline{C}^1$ by j^+. We obtain a homomorphism of differential graded Lie algebras

$$j^+ : L_{\text{tan}} \to \overline{L}.$$

In [Ak1], Akahori proved that the inclusion of $\overline{L}$ into Kuranishi's complex induced an isomorphism of cohomology of degree one or greater. In [Ak1] and

[Mi1], Akahori and Miyajima proved that Kuranishi's family $\mathscr{K}_M$ was isomorphic to the analytic subvariety $\mathscr{K}_{\overline{L}} \subset H^1(\overline{L})$ obtained by applying the well-known construction of [K1] to $\overline{L}$ with a suitable choice of complement $\overline{C}^1$, provided dim $V \geq 4$. In particular, it is immediate that

$$R_{\overline{L}} \approx \widehat{\mathfrak{D}}_{\mathscr{K}_M,0}\,.$$

Our theorem then follows by the Comparison Theorem from the fact that j^+ induces isomorphisms on cohomology in degrees 1 and 2 provided dim $V \geq 4$. To prove this latter fact we consider the following diagram of complexes:

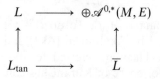

We have seen that the two vertical arrows are quasi-isomorphisms. But it follows from the discussion in [Y], pp. 81–82, that the top horizontal arrow induces an isomorphism on H^1 and H^2 in case dim $V \geq 4$.

The theorem of Buchweitz-Millson and Fujiki-Miyajima can be used to compute $\mathscr{K}_M$ if one has enough information about the ideal of the corresponding singularity. In [M] we use differential geometry to compute $\mathscr{K}_M$ directly for a certain class of CR-manifolds. Let $\mathscr{L}$ be a negative line bundle over a compact Kähler manifold N and let $U(\mathscr{L})$ be its unit circle bundle. In [M] we compute $\mathscr{K}_M$ for $M = U(\mathscr{L})$ in case $\mathscr{L}$ is sufficiently negative. Recall that if $\mathscr{L}$ is sufficiently negative, then $\mathscr{L}^{-1}$ gives a projective embedding of N into $\mathbb{CP}^m$ for some m. We then have the following theorem, [M], the CR-analogue of Theorem 2 of [Sc2].

Theorem. *Let $M = U(\mathscr{L})$ as above. If $\mathscr{L}$ is sufficiently negative then $\mathscr{K}_M$ is isomorphic to the parameter space for the versal projective deformation of N in $\mathbb{CP}^m$. Moreover, by Schlessinger's Theorem, this latter space is isomorphic to the deformation space of the singularity $(C_N, 0)$, where C_N is the affine cone over N.*

Example. Let N be a complex torus of dimension n and $\mathscr{L}$ *any* negative line bundle over N. Then by a direct computation one can show

$$\mathscr{K}_{U(\mathscr{L})} \approx \mathbb{C}^{\frac{n(n+1)}{2}}\,.$$

In case $\mathscr{L}$ is sufficiently negative, the above Theorem produces an isomorphism between $\mathscr{K}_{U(\mathscr{L})}$ and the parameter space of the versal deformation of $(C_N, 0)$ even though

$$\mathrm{depth}_{\{0\}}\, C_N = 2\,.$$

Remark. For any $\mathscr{L}$ as above one can obtain an analytic germ $(C_N, 0)$ by collapsing the zero section. It is reasonable to expect that the parameter space of the versal deformation of $(C_N, 0)$ is again $\mathbb{C}^{\frac{n(n+1)}{2}}$.

References

[A]   M. Artin: On solutions to analytic equations. Invent. math. **6** (1968) 277–291

[Ak1]   T. Akahori: The new estimate for the sub-bundles E_j and its application to the deformation of the boundaries of strongly pseudo-convex domains. Invent. math. **63** (1981) 311–344

[Ak2]   T. Akahori: The new Neumann operator associated with deformation of strongly pseudo-convex domains and its application to deformation theory. Invent. math. **68** (1982) 317–352

[AM]   T. Akahori, K. Miyajima: Complex analytic construction of the Kuranishi family on a normal strongly pseudo-convex manifold II. Publ. RIMS, Kyoto University **16** (1980) 811–834

[B]   R.O. Buchweitz: Contributions à la théorie des singularités. Ph.D. thesis, University of Paris

[BM]   R.O. Buchweitz, J.J. Millson: CR geometry and deformations of isolated singularities. In preparation

[D]   P. Deligne: Letter to J.J.Millson, April 24, 1986

[DGMS] P. Deligne, P.A. Griffiths, J.W. Morgan, D. Sullivan: Rational homotopy type of compact Kähler manifolds. Invent. math. **29** (1975) 245–274

[F]   A. Fujiki: Flat Stein completion of a flat (1,1)-convex concave map. Preprint

[FN]   A. Frölicher, A. Nijenhuis: Theory of vector-valued differential forms, Part I. Derivations in the graded ring of differential forms. Proc. Kon. Ned. Akad. Wet. Amsterdam **59** (1956) 338–359

[GM1]   W.M. Goldman, J.J. Millson: The deformation theory of representations of fundamental groups of compact Kähler manifolds. Publ. Math. IHES **67** (1988) 43–96

[GM2]   W.M. Goldman, J.J. Millson: The homotopy invariance of the Kuranishi space. Ill. J. Math. **34** (1990) 337–367

[K1]   M. Kuranishi: Deformations of compact complex manifolds. Les Presses de l'Université de Montréal 1971

[K2]   M. Kuranishi: Application of $\bar{\partial}_b$ to deformation of isolated singularities. Proceedings of Symposia in Pure Mathematics **30** (1977) 97–106

[M]   J.J. Millson: $\bar{\partial}_b$ and deformations of cones. Preprint

[Mi1]   K. Miyajima: Completion of Akahori's construction of the versal family of strongly pseudo-convex CR structures. Trans. A.M.S. **277** (1983) 163–172

[Mi2]   K. Miyajima: Deformations of a complex manifold near a strongly pseudo-convex real hypersurface and a realization of the Kuranishi family of strongly pseudo-convex CR-structures. Preprint

[P]   V.P. Palamodov: Deformations of complex spaces. In: Several Complex Variables IV, Encyclopedia of Mathematical Sciences, vol. 10 (S.G. Gindikin and G.M. Khenkin, eds.) Springer, Berlin Heidelberg New York, pp. 105–194

[S]   C. Simpson: Higgs bundles and local systems. Preprint

[Sc1]   M. Schlessinger: Functors of Artin rings. Trans. A.M.S. **130** (1968) 208–222

[Sc2]   M. Schlessinger: On rigid singularities. Conference on Complex Analysis, Rice University Studies 59 No. 1 (1972) 147–162

[Sc3]   M. Schlessinger: Infinitesimal deformation of singularities. Ph.D. thesis, Harvard University

[SS]   M. Schlessinger, J. Stasheff: Deformation theory and rational homotopy type. Preprint

[Ti]   G. Tian: Smoothness of the universal deformation space of compact Calabi-Yau
         manifolds and its Peterson-Weil metric. In: S.T. Tau (ed.) Math. Aspects of string
         theory. World Scientific Publ. 1987

[To]   A.N. Todorov: The Weil-Peterson geometry of the moduli space SU ($n \geq 3$)
         (Calabi-Yau) manifolds I. Preprint

[Y]   Stephen S.-T. Yau: Kohn-Rossi cohomology and its application to the complex
         Plateau problem I. Ann. Math. **113** (1981) 67–110

Geometry of Discriminant and Topology of Algebraic Curves

Eugenii I. Shustin

Department of Mathematics, Kuibyshev State University, ul. Acad. Pavlova 1
Kuibyshev 443011, USSR

Introduction

The subject of this lecture is some aspects of the theory of deformations of plane real and complex algebraic singular curves and its applications. Namely, in this large classical problem we shall consider (i) the independence of singular points deformations, (ii) connections between deformations of singularities and discriminant hypersurface in moduli space of plane curves of a given degree, (iii) applications of these results to classification of algebraic real and complex curves.

First results in this direction concerned the curves with nodes (in complex case – Severi [17], in real case – Brusotti [2]): any reduced curve with nodes can be deformed in the space of curves of a given degree so that arbitrary set of nodes is smoothed, and all other singularities are preserved.

The proof idea is that the germ of locus of curves with a given number of nodes is a transversal intersection of smooth germs of hypersurfaces corresponding to each singularity. The last assertion follows from nonspeciality of some linear system on the curve by means of Riemann-Roch theorem. It should be noted that Harris theorem on irreducibility of variety of given degree curves with a given genus [10] was proved, in fact, by the same ideas. We also point out that many methods of construction of real algebraic curves [8] are based on Brusotti theorem.

In this lecture we shall develop above results and approaches.

1. Notions and Notations

Further on we shall consider only plane algebraic projective real and complex curves (including irreduced and reducible curves). The polynomial defining a curve and the locus of curve points in $\mathbb{C}P^2$ we denote identically. The set of curves of degree n is parametrized by $\mathbb{P}^N$, $N = n(n + 3)/2$ (from now on $\mathbb{P}^N$ means $\mathbb{C}P^N$ and $\mathbb{R}P^N$ at the same time).

Let $F \in \mathbb{P}^N$ and $p \in F$ be an isolated singular point. *The singular point type*, or *singularity*, means the list of characteristic Puissex exponents of all local curve branches and the list of intersection numbers of branches. This is complete topo-

Proceedings of the International Congress
of Mathematicians, Kyoto, Japan, 1990
© The Mathematical Society of Japan, 1991

logical invariant of singular point. Symbols $\mu(p, F)$, $\delta(p, F)$ mean the Milnor number and δ-invariant of the point $p \in F$ respectively. By U_p we denote an open sufficiently small ball in $\mathbb{C}P^2$ centred at point p.

Remark 1. If $F \in \mathbb{R}P^N$, $p \in \mathbb{C}P^2 \setminus \mathbb{R}P^2$ then there is a conjugated singular point conj $p \in F$. In this case we call the pair $(p, \text{conj } p)$ by "singular point" understanding correspondingly all other terms.

The perturbation of $F \in \mathbb{P}^N$ is a sufficiently close to F curve $G \in \mathbb{P}^N \setminus \{F\}$. By *perturbation of the point* $p \in F$ we call any set $G \cap U_p$, where G is a perturbation of curve FL^m for some integer $m \geq 0$ and some fixed line $L \in (\mathbb{P}^2)^*$ non-crossing U_p. Nonsingular perturbations of the real point are called *smoothings* [25].

Two perturbations of $p \in F$ are called *equivalent*, if they can be transformed one into another by (equivariant in real case) diffeomorphism of U_p close to identity. Two perturbations of p are called *algebraic-topologically* (*AT*) *equivalent*, if they have the same sets of singularities, and (in real case) some neighbourhoods of their real parts can be transformed one into another by equivariant homeomorphism of U_p, close to identity.

By $V(n, S_1, \ldots, S_r)$ we'll denote the locus of degree n irreducible curves with r singular points of types $S_1, \ldots, S_r$ respectively. The germ $M(F) \subset \mathbb{P}^N$ of equisingular (ES) stratum (this is $\mu = $ const stratum) centered at reduced curve $F \in \mathbb{P}^N$ is called regular, if it is smooth and has codimension $\sum_{z \in F} c(z)$, where $c(z)$ is codimension of $\mu = $ const stratum in miniversal deformation base of singular point z.

2. Independence of Versal Deformations

Let p be an isolated singular point of curve $F \in \mathbb{P}^N$. Let Π, Π' denote two generic polar curves of F. For any branch Q of curve Π at point p we assign $\tau(Q) = s/r$, where s is the sum of orders of Q at point p and at the infinitely close to p points belonging to Π'. For a reduced curve F we assign $\tau = \max \tau(Q)$. It depends on different singularities of F, and is ever less than 1. Put

$$\lambda(F) = \begin{cases} 25/729, & \tau < 5\sqrt{2}/18 \\ 2\tau^2/9, & 5\sqrt{2}/18 \leq \tau < 5/9 \\ (1 - \tau + \tau^2 + (1 - 2\tau)\sqrt{1 + 2\tau - 2\tau^2})/9, & 5/9 \leq \tau \end{cases}$$

Theorem 1 [23, 24]. *If (i) F is reduced and satisfies*

$$\sum_{z \in F} \mu(z, F) \leq 4n - 5 \tag{1}$$

or (ii) F is reduced and irreducible and satisfies

$$\sum_{z \in F} (\mu(z, F) + \delta(z, \Pi)) \leq \lambda(F)n^2, \tag{2}$$

then some neighbourhood of F in $\mathbb{P}^N$ represents a joint versal deformation of all singular points of F.

This theorem fully describes properties of the discriminant hypersurface germ at $F \in \mathbb{P}^N$ under above conditions. The proof is based on the following: the intersection transversality of $\mathbb{P}^N$ and tangent spaces to orbits of Diff U_p-action for all $p \in \text{Sing } F$ can be concluded from nonspeciality of some linear system on suitable curve by means of Riemann-Roch theorem (in particular, polar curve Π can be taken as such a curve, and then one obtains condition (1)).

Corollary 1. *Under conditions of Theorem 1 a suitable perturbation of F realizes perturbations of any singular points, equivalent to a priori given perturbations of these points, while other singularities are preserved.*

This statement does not cover Severi and Brusotti theorems, but it does not contain restrictions on singular points types and their perturbations; also the right hand side of (2) is sharp with respect to exponent at n. For example, if F has only ordinary cusps as their singularities then (2) ensures the independent deformation of all cusps when

$$\#(\text{Sing } F) \le (7 - \sqrt{13})n^2/162,$$

since the other known estimates [4, 7] give

$$\#(\text{Sing } F) \le 3n - 1.$$

Corollary 2. *Under conditions of Theorem 1 $M(F)$ is regular.*

The above assertions do not contain restrictions on a position of singular points. Adding such restrictions one can weaken inequalities (1), (2) (in [3] there are those estimates for curves with ordinary singular points).

Compare (1) with known counterexamples. According to [14] varieties $V(9, A_{35})$, $V(11, A_{50})$ are not smooth. The Corollary 2 implies the smoothness of $V(9, A_k)$, $V(11, A_m)$ only for $k \le 31, m \le 39$.

3. Independence of Certain Perturbations Classes

Corollary 1 embraces every perturbation of arbitrary singularities, but it does not show what types of perturbations of each singular point exist. Here we'll use certain other approaches to our problem.

3.1 Perturbations of Quasiordinary Singular Points

Consider the following approach to problem of regularity of $M(F)$. Using Cremona transformations we can obtain that F had only ordinary singularities. Then regularity of $M(F)$ follows (see [3, 7]) from nonspeciality of some linear system on F. The exact assertions are as follows.

Definition 1. Let us consider the full resolution of singular point $p \in F$. Let $p_0 = p$, $p_1, \ldots, p_k$ be consequent infinitely close to p points of this resolution, which belong to branch Q of F centered at p. Define $\tilde{\beta}(Q, F)$ as sum of orders of branch Q at p_0, $\ldots, p_{k-1}$ (i.e. at those points, where proper inverse image of F has singularity, or intersection number of F with exceptional divisor exceeds 1). Let $\beta(Q, F)$ be equal to $\tilde{\beta}(Q, F) - 1$ (resp. $\tilde{\beta}(Q, F)$), if one of $p_1, \ldots, p_{k-1}$ is touch node, i.e. point of two smooth branches contact (resp. no touch node) of full inverse image of F. If order of F at p is 2, then we must subtract the order of Q at p from above defined value $\beta(Q, F)$. Put also

$$\tilde{\beta}(p, F) = \sum \tilde{\beta}(Q, F), \qquad \beta(p, F) = \sum \beta(Q, F).$$

It is easy to show that

$$c(p, F) - \delta(p, F) \le \beta(p, F) \le \mu(p, F) - 1$$

and that $\beta(p, F) = r$ for ordinary singular point of order r (this is an intersection point of r smooth branches with different tangents).

Definition 2. For any reduced curve F with irreducible components $F_1, \ldots, F_k$ and for a set $S \subset \operatorname{Sing} F$ we define vectors $d(F), b(F, S) \in \mathbb{Z}^k$ as follows:

$$[d(F)]_i = 3 \deg F_i - 1,$$

$$[b[F, S]]_i = \sum \beta(Q_1, F) + \sum \tilde{\beta}(Q_2, F),$$

where Q_2 (resp. Q_1) runs all branches of F_i centred at points $z \in S$ (resp. $z \in \operatorname{Sing} F \backslash S$). Put $\tilde{b}(F) = b(F, \operatorname{Sing} F)$, $b(F) = b(F, \varnothing)$.

Every inequality for vectors below means inequalities for each pair of corresponding coordinates of vectors.

Theorem 2 [18, 22, 24]. *If F is reduced and satisfying*

$$b(F) \le d(F) \tag{3}$$

then $M(F)$ is regular. If

$$\tilde{b}(F) \le d(F) \tag{4}$$

then the variety $V \subset \mathbb{C}P^N$ of curves, having singularities as F and disintegrating into components with the same singularities sets, is irreducible.

This theorem covers the Severi and Brusotti theorems and analogous results on curves with cusps and ordinary singularities [6, 7, 13, 26]. According to [27] varieties $V(6, 6A_2)$, $V(6, 3E_6)$ are reducible. Theorem 2 claims the irreducibility of $V(6, kA_2)$, $V(6, mE_6)$ for $k \le 4$, $m \le 2$.

Remark 2. The point $p \in F \backslash \operatorname{Sing} F$ is called *quasisingular* of order $k \ge 1$ with respect to fixed line f, smooth at p, if intersection number $(F \cdot f)_p = k$. We can take into account quasisingularities in Theorem 2 by adding to left hand sides of (3), (4) the values $\beta(p, F) = k - 1$, $\tilde{\beta}(p, F) = k$ respectively. Thus, using the supplemented

Theorem 2 one can prove Harris theorem [10] on irreducibility of $V(n, kA_1), k \geq 1$, Ran theorem [16] on irreducibility of $V(n, kA_1, S)$, where singularity S is transformed into set of quasisingularities and nodes by one blowing-up. Analogous approach gives

Theorem 3. *The variety* $V(n, kA_1, S_1, \ldots, S_r)$, *where* $k \geq 0, 0 \leq r \leq 9, S_1, \ldots, S_r$ *are ordinary singularities with* $\sum \beta(S_i) \leq 2n$, *is irreducible.*

In particular, this implies that the variety of curves on quadric $\mathbb{C}P^1 \times \mathbb{C}P^1$ with a given bidegree and a given genus is irreducible (see also [16]).

Remark 3. From proof of Theorem 2 it is possible to deduce that coefficients at monomials lieing under Newton diagrams of all singular points can be varied independently. Therefore perturbations determined by these monomials are independent too. Effective construction of those perturbations was suggested by Viro [25].

Definition 3. Singular point $p \in F \in \mathbb{C}P^N$ is called *quasiordinary* (*QO-point*), if (i) for some affine coordinates $p = (0; 0)$, and all sections of polynomial F on edges of the point p Newton diagram are nondegenerated (normal QO-point), or (ii) some embedding $\varphi : U_p \to \mathbb{C}^2$ transforms p into normal QO-point (in the case $p \in \mathbb{R}P^2$, $F \in \mathbb{R}P^N$ we assume $\varphi \circ \text{conj} = \text{conj} \circ \varphi$). Quasisingular points are QO-points in a natural sense.

This notion includes, for instance, zero-modal points, quasihomogeneous singular points. For normal QO-point $\tilde{\beta}(p, F) = k + m - r + 1$, where k, m are maximal absciss and ordinate of points on the Newton diagram, and r is a number of integer points on the Newton diagram.

Modifying the Viro construction [25] we'll define a set of *V-perturbations* of QO-point $p \in F$. Normalize the point p, and denote the Newton diagram of p by Γ and the Newton polygon of normalized curve F_0 by Δ_0. Take a set of polynomials $F_i(x, y) = \phi_i(x, y)x^t y^u$, where $x, y \not| \phi_i$, with Newton polygons $\Delta_i, 1 \leq i \leq k$, lieing under Γ. Assume that (i) $\Delta = \bigcup \Delta_i$ is convex, any nonempty intersection $\Delta_i \cap \Delta_j$ is a common edge of Δ_i, Δ_j or a common apex, there is also a continuous convex function $v : \Delta \to \mathbb{R}$, linear on each Δ_i and nonlinear on each union $\Delta_i \cup \Delta_j$, (ii) if $\Delta_i \cap \Delta_j = \gamma$ is a common edge, then sections of F_i, F_j on γ are identical square-free (except x, y) polynomials. Now V-perturbation is constructed by means of set $\{F_i\}$ according to [25]: (i) lift each curve φ_i upon toric manifold determined by Δ_i, (ii) glue these toric manifolds according to arrangement of $\Delta_0, \ldots, \Delta_k$, (iii) project the correspondingly glued curves $F_0, \varphi_1, \ldots, \varphi_k$ on the plane.

For a given V-perturbation π we'll define the set $T(\pi)$ of oriented graphs without oriented circles. Vertexes of graph $\Lambda \in T(\pi)$ correspond to polygons Δ_i, arcs of Λ correspond to common edges of $\Delta_i, 0 \leq i \leq k$. To curve $F_i, 1 \leq i \leq k$, and to graph $\Lambda \in T(\pi)$ we assign the vector $v(F_i, \Lambda)$. Its coordinates correspond to irreducible components of φ_i, and are equal to numbers of local branches of each component, defining by those edges of Δ_i, which correspond to arcs of Λ coming in Δ_i.

Theorem 4 [19, 22]. *Let* $F \in \mathbb{P}^N$ *be a reduced curve,* $\sigma \subset \text{Sing } F$ *be some set of zero-modal singular points,* $\tau \subset \text{Sing } F|\sigma$ *be some set of QO-points,* $\{\pi_z | z \in \tau\}$ *be a set of V-perturbations, defined by curves* $F_{zi}, z \in \tau, 1 \leq i \leq j_z$. *If there exist graphs* $\Lambda_z \in T(\pi_z), z \in \tau$, *for which*

$$b(F, S) + \sum_z v(F, \Lambda_z) \leq d(F), \tag{5}$$

$$b(F_{zi}, S_{zi}) + v(F_{zi}, \Lambda_z) \leq d(F_{zi}), \tag{6}$$

where $z \in \tau, 1 \leq i \leq j_z, S = \{z \in \tau | v(F, \Lambda_z) \neq 0\}, S_{zi} = \text{Sing } \phi_{zi} \cap \mathbb{C}^2 \backslash \{xy = 0\}$, *then a suitable perturbation of* F *realizes perturbations of* $z \in \tau$, *which are AT-equivalent to* $\pi_z, z \in \tau$, *and perturbations of points* $z \in \sigma$, *which are equivalent to their a priori given perturbations, while other singularities are preserved.*

Remark 4. If all curves $F_{zi}, z \in \tau, 1 \leq i \leq j_z$, are smooth in $\mathbb{C}^2 \backslash \{xy = 0\}$, or have there only nodes, then (5), (6) can be replaced by (3).

Remark 5. V-perturbations can be defined for arbitrary singular point as follows: the singular point should be resoluted and all infinitely close singular and quasi-singular points should be consequently perturbed.

3.2 Equigeneric and Equiclassical Perturbations

The independence of above perturbations is connected with properties of ES-stratum. Here we consider perturbations, which are connected with equigeneric (EG) stratum $g = \text{const}$ and equiclassical (EC) stratum $g = \text{const}$ and $c = \text{const}$; c is class of curve, i.e. degree of dual curve.

Theorem 5 (see [1, 15]). *For any reduced curve* $F \in \mathbb{C}P^N$ *there exists a perturbation that turns independently each singular point* $z \in F$ *into an arbitrary number of nodes from 0 to* $\delta(z, F)$.

This follows from that the tangent cone to EG-stratum at $F \in \mathbb{C}P^N$ is a linear system of adjoining curves of degree n, which is evidently nonspecial. It implies that curves with nodes form a dense subset of EG-stratum in $\mathbb{C}P^N$.

A perturbation from EC-stratum turns (see [5]) a singular point into $3\delta - \varkappa$ nodes and $\varkappa - 2\delta$ cusps, where $\varkappa$ is an intersection number of curve and its polar curve at this point. Namely (see [5]),

Theorem 6. *If a reduced irreducible curve* $F \in \mathbb{C}P^N$ *satisfies* $c > 2g - n$, *then the suitable perturbation of* F *turns independently each singular point* p *into* d *nodes and* k *cusps, where* d, k *are arbitrary satisfying*

$$0 \leq d \leq \delta, \qquad 0 \leq k \leq \min\{\varkappa - 2\delta, \delta - d\}.$$

According to [1, 4] EG-germ and EC-germ in a versal deformation base are irreducible. Therefore

Theorem 7. *If* $F \in \mathbb{C}P^N$ *satisfies* (1) *or* (2), *then EG-germ and EC-germ at F in* $\mathbb{C}P^N$ *are irreducible.*

3.3 Independence of Minimal Smoothings

Give one more example of unconditionaly independent perturbations – minimal smoothings (see [21]). *Minimal smoothing of singular point* $p \in \mathbb{R}P^2$ of order k, at which there are r real local branches of curve $F \in \mathbb{R}P^N$, is (i) a set of r disjunct intervals connecting the points of $F \cap U_p \cap \mathbb{R}P^2$ for $r > 0$, or (ii) a set of $s \leq k/2$ concentred circumferences in $U_p \cap \mathbb{R}P^2$ for $r = 0$.

Theorem 8 [21]. *A suitable perturbation of any reduced curve* $F \in \mathbb{R}P^N$ *realizes arbitrary a priori given minimal smoothings of all real singular points modulo AT-equivalence, while nodes can be preserved.*

Remark 6. Perturbations from germs of almost all linear families $F + tG$, $t \in \mathbb{R}$, $G \in \mathbb{R}P^N$, realize only minimal smoothings of all singular points.

4. Perturbations of Singular Points and Algebraic Curves

The approach used in Section 3.1 raises the following questions: 1) what part of all perturbations form V-perturbations? 2) what V-perturbations can be realized for a given singular point?

Consider semiquasihomogeneous singular point (that means Newton diagram lies on straight line). Then V-perturbations are constructed by gluing affine curves with complementary Newton triangle. On the other hand (see [23]) any perturbation can be obtained by formal gluing of those affine curves and replacing their singular points by suitable perturbations. If singular points of above affine curves have independent perturbations, then we have (i) *property A*: each perturbation of initial singular point can be obtained from affine curve with a given Newton triangle by means of (equivariant in real case) homeomorphism $\mathbb{C}^2 \to U_p$, or (ii) *property A'*: the same modulo AT-equivalence. If we can vary coefficients of a curve with a given Newton triangle at monomials on the edge of gluing, while a curve topology is preserved, then singular points of the same type have the same sets of V-perturbations modulo homeomorphisms of singular points neighbourhoods (*property B*), or modulo AT-equivalence (*property B'*).

Zero-modal singular points, ordinary 4th order point, point of quadratic contact of 3 smooth branches hold properties A and B. The ordinary 5th order point holds property A [23], and if all local branches are real then property B' is fulfilled for its smoothings [20]. Properties A', B' are fulfilled for smoothings of point of quadratic contact of 4 real branches [20]. Property A' is fulfilled for smoothings of ordinary 6th order point [12].

5. Applications to Study of Topology of Algebraic Curves

Firstly, above results on irreducibility mean that curves of a given kind are rigidly isotopic, i.e. they are connected by isotopy consisting of the same curves.

It might be interesting to construct curves with a given singularities. For example (see [19]), Theorem 4 implies existance of degree n complex curves with arbitrary cusps number from 0 to $7n^2/36$, and existence of degree n curves over $\mathbb{R}$ with arbitrary number of real cusps from 0 to $3n^2/16$.

All known types of nonsingular real curves were realized in the way of perturbation of singular curves. The majority of latest results about real curves [11, 12, 22, 25] were obtained by Viro method [25] through perturbation of singular points mentioned in Section 4. It should also be noted that there is M-curve of degree 8, whose construction does not satisfy conditions of Viro method and is based on Theorem 4 [22].

Another example of using above results is the complete description of discriminant in the space of plane real quartic curves and complete classification of inflexion points arrangements on these curves [9].

At last we'll mention the Hilbert-Rohn method (see [8, 20]), which allows to construct or to prohibit certain classes of real curves.

References

1. Arbarello, E., Cornalba, M.: A few remarks. Ann. Sci. Ec. Norm. Sup. **16** (1983) 467–488
2. Brusotti, L.: Sulla "piccola variazione" di una curva piana algebrica reali. Rend. Rom. Ac. Lincei (5) **30** (1921) 375–379
3. Catalisano, M.V.: Linear systems of plane curves through fixed "fat" points of $\mathbb{P}^2$. Queen's Pap. Pure Appl. Math. **83** (1989) D1–D34
4. Diaz, S.: Irreducibility of equiclassical stratum. J. Diff. Geom. **29** (1989) 489–498
5. Diaz, S., Harris, J.: Ideals associated to deformations of singular plane curves. Trans. Amer. Math. Soc. **309**, no. 2 (1988) 443–468
6. Gradolato, M.A., Mezzetti, E.: Curves with nodes, cusps and ordinary triple points. Ann. Univ. Ferrara, sez. 7 **31** (1985) 23–47
7. Gudkov, D.A.: On certain questions in topology of plane algebraic curves. Mat. Sb. **58**, no. 1 (1962) 95–127 (Russian)
8. Gudkov, D.A.: Topology of real projective algebraic varieties. Sov. Math. Surv. **29**, no. 4 (1974) 3–79
9. Gudkov, D.A.: Plane real projective quartic curves. (Lecture Notes in Mathematics, vol. 1346.) Springer, Berlin Heidelberg New York 1988, pp. 341–347
10. Harris J.: On the Severi problem. Inv. math. **84**, (1986) 445–461
11. Korchagin, A.B.: New M-curves of degree 8 and 9. Sov. Math. Dokl. **306**, no. 5 (1989) 1038–1041 (Russian)
12. Korchagin, A.B., Shustin, E.I.: Affine curves of degree 6 and smoothings of nondegenerated 6th order singular point. Math. USSR Izvestiya **33**, no. 3 (1989) 501–520
13. Lindner, M.: Über Mannigfaltigkeiten ebener Kurven mit Singularitäten. Arch. Math. **28** (1977) 603–610
14. Luengo, I.: The μ-constant stratum is not smooth. Inv. math. **90** (1987) 139–152
15. Nobile, A.: On specialisation of curves. Trans. Amer. Math. Soc. **282**, no. 2 (1984) 739–748
16. Ran, Z.: Families of plane curves and their limits: Enriques' conjecture and beyond. Ann. Math. **130**, no. 1 (1989) 121–157
17. Severi, F.: Vorlesungen über algebraische Geometrie. Anhang F. Teubner, Leipzig 1921
18. Shustin, E.I.: On varieties of singular algebraic curves. In: Methods of qualitative theory. Gorky Univ. Press, Gorky 1983, pp. 148–163 (Russian). [English translation: Shustin,

E.I.: On Manifolds of Singular Algebraic Curves. Selecta Math. Sov. **10**, no. 1 (1991) 27–37]

19. Shustin, E.I.: Gluing of singular algebraic curves. In: Methods of qualitative theory. Gorky Univ. Press, Gorky 1985, pp. 116–128 (Russian)
20. Shustin, E.I.: The Hilbert-Rohn method and smoothings of algebraic curves singular points. Sov. Math. Dokl. **31**, no. 2 (1985) 282–286
21. Shustin, E.I.: Hyperbolic and minimal smoothings of singular points. In: Methods of qualitative theory. Gorky Univ. Press, Gorky 1986, pp. 165–174 (Russian)
22. Shustin, E.I.: New M-curve of 8th degree. Math. Notices **42**, no. 6 (1987) 180–186 (Russian)
23. Shustin, E.I.: Versal deformations in the space of a fixed degree curves. Funct. Anal. Appl. **21**, no. 1 (1987) 90–91
24. Shustin, E.I.: Smoothness and irreducibility of varieties of singular algebraic curves. In: Arithmetic and geometry of algebraic varieties. Saratov Univ. Press/Kuibyshev branch, Kuibyshev 1989, pp. 102–117 (Russian)
25. Viro, O.J.: Gluing of algebraic hypersurfaces, smoothing of singularities and construction of curves. In: Proc. Leningrad Topological Internat. Conf., Nauka, Leningrad 1983, pp. 147–195 (Russian)
26. Wahl, J.: Deformations of plane curves with nodes and cusps. Amer. J. Math. **96** (1974) 529–577
27. Zariski, O.: Algebraic surfaces. Springer, New York Berlin Heidelberg 1971.

Applications of Hodge Theory to Singularities

Joseph H. M. Steenbrink

Mathematical Institute, University of Nijmegen, Toernooiveld
NL-6525 ED Nijmegen, The Netherlands

1. Introduction

We will describe some old and recent results in the theory of complex hypersurface singularities which have been obtained using methods from mixed Hodge theory. In the last 15 years, there has been a fruitful interaction between singularity theory and Hodge theory. The common interest is the study of degenerations. In the work of W. Schmid [21] the limit mixed Hodge structure of a one parameter polarized variation of Hodge structure was constructed. This has consequences for projective degenerating families, but does not apply to the case of a singularity. (However, Scherk's trick [19] made it apply to the isolated singularity case.) As P. Deligne observed in Metz 1974, the author's geometric approach [23] (filling in a divisor with normal crossings as a special fibre and analyzing the relative logarithmic de Rham complex) works also for the case of a singularity [24, 11].

In cases where a resolution of singularities is readily available, e.g. the Newton non-degenerate singularities, the resulting mixed Hodge structure on the vanishing cohomology has very computable discrete invariants (Hodge numbers, spectrum) which are closely related to interesting topological invariants like monodromy and intersection form. In the isolated singularity case, a description of the Hodge structure without resolution of singularities was first given by A. Varchenko [29] who used asymptotics of integrals over vanishing cycles. This work was the starting point of a description in the language of $\mathscr{D}$-modules, inspired by Pham's Gauss-Manin system [12] and leading to M. Saito's theory of mixed Hodge modules [13, 14, 15]. This formalism enables one to generalize many results to the case of non-isolated singularities.

2. Vanishing Cycles

Let X be a complex manifold of dimension $n + 1$ and let $f : X \to \mathbf{C}$ be a holomorphic function. Let $X_t = f^{-1}(t)$ for $t \in \mathbf{C}$ and let $i : X_0 \to X$ be the inclusion mapping. Define

$$X_\infty = \{(x, u) \in X \times \mathbf{C} \mid f(x) = \exp(2\pi i u)\}$$

and let $k : X_\infty \to X$ be given by $k(x, u) = x$. Choose an injective resolution $I^\bullet$ of $\mathbf{C}_{X_\infty}$ and define the *sheaf of nearby cycles* of f as

Proceedings of the International Congress
of Mathematicians, Kyoto, Japan, 1990
© The Mathematical Society of Japan, 1991

$$\psi_f(\mathbf{C}_X) = i^* k_* I^\bullet.$$

Let $x \in X_0$ and choose $\varepsilon, \eta > 0$ with $\eta \ll \varepsilon \ll 1$. By [10], the restriction of f to $\{z \in X \mid |z - x| < \varepsilon, \, 0 < |f(z)| < \eta\}$ is a C^∞ fibre bundle (the *Milnor fibration*). Let $X_{f,x}$ denote a Milnor fibre, i.e. a typical fibre of the Milnor fibration. Then

$$H^k(\psi_f(\mathbf{C}_X)_x) \simeq H^k(X_{f,x}, \mathbf{C}).$$

There is a natural morphism of complexes $\mathbf{C}_{X_0} \to \psi_f(\mathbf{C}_X)$ and one obtains the *sheaf of vanishing cycles* $\phi_f(\mathbf{C}_X)$ as the cone over this morphism. This gives a distinguished triangle

$$\mathbf{C}_{X_0} \longrightarrow \psi_f(\mathbf{C}_X) \xrightarrow{\text{can}} \phi_f(\mathbf{C}_X) \xrightarrow{+1}$$

and we have

$$H^k(\phi_f(\mathbf{C}_X)_x) \simeq \tilde{H}^k(X_{f,x}, \mathbf{C}).$$

So $\phi_f(\mathbf{C}_X)$ measures the difference in the local topology of the zero fibre and the nearby fibres of f. In general, we have $\tilde{H}^k(X_{f,x}, \mathbf{C}) = 0$ for $k > n$ and in the isolated singularity case, $\tilde{H}^k(X_{f,x}, \mathbf{C}) = 0$ for $k \neq n$.

Since [10], the study of the topology of the Milnor fibre of a holomorphic function germ has been an important issue in singularity theory. A first topic is the *monodromy*. Define $h : X_\infty \to X_\infty$ by $h(x, u) = (x, u + 1)$. Then we have an induced action $T = (h^*)^{-1}$ on $\psi_f(\mathbf{C}_X)$ and $\phi_f(\mathbf{C}_X)$ and their cohomology sheaves. The latter action is quasi-unipotent by the monodromy theorem [1, 7] and the Jordan blocks of T on $H^k(\phi_f(\mathbf{C}_X)_x)$ are of size at most $k + 1$.

3. Hodge Structure via Resolution

Let f be a holomorphic function on the complex manifold X and let $x \in f^{-1}(0)$. Then there exists a projective bimeromorphic mapping $\pi : \tilde{X} \to X$ such that

- $\tilde{X}$ is smooth;
- $E = (f \circ \pi)^{-1}(0)$ is a divisor with normal crossings on $\tilde{X}$;
- $E_x = \pi^{-1}(x)$ is a union of irreducible components of E.

Let e be a common multiple of all multiplicities of components of E. We map $\mathbf{C}$ to itself by $t \mapsto t^e$ and let $\tilde{Y}$ be the normalization of $\{(z, t) \in \tilde{X} \times \mathbf{C} \mid f(\pi(z)) = t^e\}$. The natural projections from $\tilde{Y}$ to $\tilde{X}$ and $\mathbf{C}$ are denoted by $\tilde{f}$ and ϱ respectively. By the Semi-stable Reduction Theorem [5], π and e can be chosen in such a way that $\tilde{Y}$ is smooth. Then $D = \tilde{F}^{-1}(0)$ is a *reduced* divisor with normal crossings and ϱ is a cyclic covering of degree e branched only along E. We put $D_x = \varrho^{-1}(E_x)$.

By [24] we have isomorphisms

$$H^k(\psi_f(\mathbf{C}_X)_x) \simeq \mathbf{H}^k(D_x, \Omega^\bullet_{\tilde{Y}/\mathbf{C}}(\log D) \otimes \mathcal{O}_{D_x})$$

$$\simeq \mathbf{H}^k(E_x, \varrho_*(\Omega^\bullet_{\tilde{Y}/\mathbf{C}}(\log D) \otimes \mathcal{O}_{D_x}))$$

where $\mathbf{H}$ denotes hypercohomology.

In [25], a quasi-isomorphism between $\Omega^\bullet_{\tilde{Y}/\mathbf{C}}(\log D) \otimes \mathcal{O}_{D_x}$ and a certain complex $K^\bullet$ has been constructed. The complex $K^\bullet$ carries an increasing *weight filtration*

L. and a decreasing *Hodge filtration* $F^{\bullet}$ which make it into a *cohomological mixed Hodge complex* in Deligne's sense (after a suitable rational structure has been added). In this way, $H^k(\psi_f(\mathbf{C}_X)_x)$ carries a mixed Hodge structure for each k.

The monodromy action T on $H^k(\psi_f(\mathbf{C}_X)_x)$ can be described in the following way. The semisimple part T_s is induced by the automorphism $(x, t) \mapsto (x, t \exp(2\pi i/e))$ of $\tilde{Y}$. Its nilpotent logarithm $N = -\log(T_u)/2\pi i$ is induced by an endomorphism v of the complex $K^{\bullet}$ which maps F^p to F^{p-1} and L_w to L_{w-2}. Hence N is a morphism of mixed Hodge structures of type $(-1, -1)$.

In the isolated singularity case, there is a nice description of the space $F^n H^n(X_{f,x}, \mathbf{C})$ due to Varchenko [29], in terms of the *geometrical weight* of holomorphic $(n + 1)$-forms on X. As a consequence of this description one concludes that, if $Gr_W^{2n} H^n(X_{f,x}) \neq 0$, then T_s has an eigenvalue 1 on it. This implies the following surprising supplement to the monodromy theorem in the isolated singularity case [2]: if T has a Jordan block of size $n + 1$, then for the eigenvalue 1 it has a Jordan block of size n.

4. Hodge Structure via $\mathscr{D}$-Modules

For each complex manifold X we have the coherent sheaf of rings $\mathscr{D}_X$ of germs of holomorphic differential operators on X. It is equipped with the increasing filtration $F.$ by the order of differential operators, such that $Gr^F \mathscr{D}_X \simeq \text{Sym}(\Theta_X)$ where Θ_X is the tangent sheaf of X.

Each $\mathscr{D}_X$-module M has its de Rham complex

$$DR(M) = \Omega_X^{\bullet} \otimes_{\mathscr{O}_X} M[\dim X].$$

This de Rham functor defines an equivalence of categories

$$DR : D_{rh}^b(\mathscr{D}_X) \xrightarrow{\sim} D_c^b(\mathbf{C}_X)$$

between the derived categories of bounded complexes of $\mathscr{D}_X$-modules with regular holonomic cohomology sheaves and of bounded complexes of $\mathbf{C}_X$-modules with constructible cohomology sheaves. The single regular holonomic $\mathscr{D}_X$-modules correspond in this way to the socalled *perverse* $\mathbf{C}_X$-complexes. These are characterized by the fact that for $K \in \text{Perv}(\mathbf{C}_X)$

$$\dim \text{supp} \, \mathscr{H}^i(K), \, \dim \text{supp} \, \mathscr{H}^i(\mathbf{D}K) \leq -i$$

where $\mathbf{D}$ denotes Verdier duality. See [9, 3].

It appears that for X a complex manifold of dimension $n + 1$ and $f : X \to \mathbf{C}$ a holomorphic function, $\psi_f(\mathbf{C}_X)[n]$ and $\phi_f(\mathbf{C}_X)[n]$ are perverse [8]. Hence they correspond via the de Rham functor to regular holonomic $\mathscr{D}_X$-modules $\psi_f(\mathscr{O}_X)$ and $\phi_f(\mathscr{O}_X)$. These are defined as follows. Consider the $\mathscr{D}_X[t, \partial_t]$-modules $\mathscr{O}_X[t, \partial_t]$ and $\mathscr{O}_X[t, \partial_t, \partial_t^{-1}]$ where the action is defined by

$$g(h\partial_t^k) = gh\partial_t^k$$

$$v(h\partial_t^k) = v(h)\partial_t^k - v(f)h\partial_t^{k+1}$$

$$t(h\partial_t^k) = fh\partial_t^k - kh\partial_t^{k-1}$$

$$\partial_t(h\partial_t^k) = h\partial_t^{k+1}$$

for g, h sections of $\mathcal{O}_X$ and v a section of Θ_X.

By [8], on $\mathcal{O}_X[\partial_t]$ and $\mathcal{O}_X[\partial_t, \partial_t^{-1}]$ there exist unique decreasing filtrations $V^\bullet$, discretely indexed by $\mathbf{Q}$, such that

1. $tV^\alpha \subset V^{\alpha+1}$ with equality for $\alpha > 0$
2. $\partial_t V^\alpha \subset V^{\alpha-1}$
3. each V^α is a finitely generated $\mathcal{D}_X$-module
4. $\partial_t t - \alpha$ is nilpotent on Gr_V^α.

One has

$$\psi_f(\mathcal{O}_X) = \bigoplus_{0<\alpha\leq 1} Gr_V^\alpha \mathcal{O}_X[\partial_t] = \bigoplus_{0<\alpha\leq 1} \psi_f^\alpha$$

$$\phi_f(\mathcal{O}_X) = \bigoplus_{0<\alpha\leq 1} Gr_V^\alpha \mathcal{O}_X[\partial_t, \partial_t^{-1}] = \bigoplus_{0<\alpha\leq 1} \phi_f^\alpha.$$

Observe that $can : \psi_f^\alpha \tilde{\to} \phi_f^\alpha$ for $\alpha \neq 1$.

The modules $\mathcal{O}_X[\partial_t]$ and $\mathcal{O}_X[\partial_t, \partial_t^{-1}]$ carry the increasing filtration $F_\bullet$ given by

$$F_p\mathcal{O}_X[\partial_t] = \bigoplus_{0\leq i\leq p-1} \mathcal{O}_X\partial_t^i$$

$$F_p\mathcal{O}_X[\partial_t, \partial_t^{-1}] = \bigoplus_{i\leq p-1} \mathcal{O}_X\partial_t^i.$$

These induce filtrations $F_\bullet$ on $\psi_f(\mathcal{O}_X)$, $\phi_f(\mathcal{O}_X)$ and their de Rham complexes. Moreover we have

$$T_s = \exp(-2\pi i\alpha), \quad N = -\partial_t t + \alpha \text{ on } Gr_V^\alpha.$$

The nilpotent endomorphism N of $\psi_f(\mathcal{O}_X)$ determines uniquely an increasing weight filtration L on it by $\mathcal{D}_X$-submodules in such a way that $N(L_j) \subset L_{j-2}$ and $N^k : Gr_L^{n+k} \tilde{\to} Gr_L^{n-k}$. The data of the bifiltered $\mathcal{D}_X$-module

$$(\psi_f(\mathcal{O}_X), F, L)$$

together with the rational structure

$$DR \psi_f(\mathcal{O}_X) \simeq \psi_f(\mathbf{C}_X)[n] \simeq \psi_f(\mathbf{Q}_X)[n] \otimes_{\mathbf{Q}_X} \mathbf{C}_X$$

constitute the ingredients of a *mixed Hodge module* on X [15]. We similarly have a mixed Hodge module structure on ϕ_f, with the same L on $\phi_f^\alpha \simeq \psi_f^\alpha$ for $\alpha \neq 1$ but with $N^k : Gr_L^{n+k+1} \tilde{\to} Gr_L^{n-k+1}$ on ϕ_f^1. The natural map $can : \psi_f \to \phi_f$ is a morphism of mixed Hodge modules.

If $f : (X, x) \to (\mathbf{C}, 0)$ has an isolated singularity, ϕ_f has support at x. Because mixed Hodge modules with support at a point correspond to mixed Hodge structures, we have a mixed Hodge structure on the vanishing cohomology again. This case has been studied in detail in [20]. In general, ϕ_f has support along the critical locus Σ of f, and one has to take the restriction to $\{x\}$ in the sense of mixed Hodge modules. This is a very delicate affair, even if Σ is a curve. One has to take iterated restrictions to divisors, and each of these steps involves taking the cone over the morphism *can* between nearby and vanishing cycles (for a defining function for the divisor). The resulting object $i_x^* \phi_f$ is no longer a single mixed Hodge structure, but a complex of mixed Hodge structures, satisfying $H^k(i_x^* \phi_f) \simeq H^{k+n}(X_{f,x}) = 0$ for $k \notin [-\dim \Sigma, 0]$.

If Σ has dimension one, there is a description of ϕ_f in terms of certain topological mappings and its restriction H to $\Sigma \setminus \{x\}$. Choose a holomorphic function germ g on (X, x) such that $\Sigma \cap g^{-1}(0) = \{x\}$. Observe that H is a variation of mixed Hodge structure (up to a shift) and that we have mixed Hodge structures $\psi_g(H)$ and $\phi_g(\phi_f)$. These have the following topological interpretation. At each point of $\Sigma \cap g^{-1}(\delta)$ for some small non-zero δ, take a slice transverse to Σ and let $F^\flat$ denote the union of the Milnor fibres of the restriction of f to all of these transverse slices. There is a natural way to embed $F^\flat$ into the Milnor fibre F of f. We have (taking $n \geq 2$ for convenience)

$$\psi_g(H) \simeq H^{n-1}(F^\flat)$$

$$\phi_g(\phi_f) \simeq H^n(F, F^\flat).$$

The structure of ϕ_f is now determined by H and a pair of mappings

$$c\tilde{a}n : \psi_g(H) \longrightarrow \phi_g(\phi_f)$$

$$v\tilde{a}r : \phi_g(\phi_f) \longrightarrow \psi_g(H).$$

Here *c̃an* is interpreted as the natural map $u : H^{n-1}(F^\flat) \to H^n(F, F^\flat)$ whereas *ṽar* is related to the map v on homology defined in the following way. Represent $\gamma \in H_{n-1}(F^\flat)$ by a cycle on $F^\flat$ and sweap it around x once following a characteristic homeomorphism of the monodromy of g the local system H. This gives a chain on F with boundary on $F^\flat$, representing the element $v(\gamma)$ of $H_n(F, F^\flat)$. One has $v\tilde{a}r = \sum_{i \geq 0} (-1)^i v^t (uv^t)^i / (i+1)$. One has the exact sequence of mixed Hodge structures

$$0 \longrightarrow H^{n-1}(F) \longrightarrow H^{n-1}(F, F^\flat) \xrightarrow{c\tilde{a}n} H^n(F) \longrightarrow 0.$$

Let $f : (X, x) \to (\mathbf{C}, 0)$ and $g : (Y, y) \to (\mathbf{C}, 0)$ be holomorphic function germs. Put

$$Z = X \times Y; \; z = (x, y); \; h = f + g : (Z, z) \longrightarrow (\mathbf{C}, 0).$$

(We will also denote h as $f \oplus g$). A result of Thom and Sebastiani [22] expresses the vanishing cohomology and monodromy of h in terms of the monodromies for f and g. Using the $\mathscr{D}$-module description above, one can show that

$$\phi_h(\mathcal{O}_Z) \simeq p_X^* \phi_f(\mathcal{O}_X) \otimes p_Y^* \phi_g(\mathcal{O}_Y)$$

and relate the Hodge filtrations at both sides. This is a weakened version of a general Thom-Sebastiani formula for mixed Hodge modules due to M. Saito and P. Deligne (unpublished).

5. Discrete Invariants

The main discrete invariants of a mixed Hodge structure (V, F, W) are its *Hodge numbers*

$$h^{pq}(V) = \dim Gr_F^p Gr_{p+q}^W V_{\mathbf{C}}.$$

If moreover an automorphism γ of finite order of V is given (like T_s in the case of the vanishing cohomology) it is convenient to split into eigenspaces of γ and to put

$$h_\lambda^{pq}(V, \gamma) = \dim \ker(\gamma - \lambda I \mid Gr_F^p Gr_{p+q}^W V_{\mathbf{C}}).$$

These invariants involve the weight filtration, which is in general quite hard to handle. Forgetting about the weight filtration leads one to the weaker invariant of the *spectrum*. For $f : (X, x) \to (\mathbf{C}, 0)$ a hypersurface singularity one defines

$$h_\lambda^p(f, x) = \sum_{j,q} (-1)^j h_\lambda^{pq}(H^j i_x^* \phi_f, T_s).$$

For each $\alpha \in \mathbf{Q}$ let $n_\alpha = h_\lambda^p(f, x)$ where $\lambda = \exp(-2\pi i \alpha)$ and $p \in \mathbf{Z}$ is determined by $n - p < \alpha \le n - p + 1$. Then

$$\mathrm{Sp}(f, x) := \sum_\alpha n_\alpha t^\alpha.$$

If $e \in \mathbf{N}$ is such that $T_s^e = I$, then $\mathrm{Sp}(f, x) \in \mathbf{Z}[t^{1/e}]$.

There are formulas for spectra of isolated quasi-homogeneous singularities in terms of the weights, and for Newton-nondegenerate isolated singularities in terms of the Newton diagram [16, 6]. A formula describing the behaviour of the spectrum in series of singularities was proven by M. Saito [17].

For spectra the Thom-Sebastiani formula takes the elegant form

$$\mathrm{Sp}(f \oplus g, (x, y)) = \mathrm{Sp}(f, x) \, \mathrm{Sp}(g, y)$$

(the right hand side is the product in the ring of fractional polynomials). This was conjectured in [24, 27], proved in the isolated singularity case in [29, 30] and in [18] in general.

For isolated singularities, the spectrum is semicontinuous under deformation of the singularity in a certain sense [30, 26]. The proof is based on a semicontinuity result for Hodge numbers and the Thom-Sebastiani formula. It appears that a similar semicontinuity result also holds for certain deformations of non-isolated singularities [28]. This gives e.g. necessary conditions for the existence of (globalizable) deformations between arbitrary normal surface singularities. These correspond to admissible (in the sense of [4]) deformations of weakly normal surface singularities, of which one considers the spectra. The proof uses the full strength of the formalism of mixed Hodge modules.

References

1. E. Brieskorn: Die Monodromie der isolierten Singularitäten von Hyperflächen. Man. Math. **2** (1970) 103–161
2. M.G.M. van Doorn, J.H.M. Steenbrink: A supplement to the monodromy theorem. Abh. Math. Sem. Univ. Hamburg **59** (1989) 225–233
3. M. Goresky, R. MacPherson: Intersection homology II. Invent. math. **71** (1983) 77–129
4. T. de Jong, D. van Straten: Deformations of non-isolated singularities. Chapter 3 of de Jong's thesis, University of Nijmegen 1988
5. G. Kempf, F. Knudsen, D. Mumford, B. Saint-Donat: Toroidal embeddings I. Lecture Notes in Mathematics, vol. 339. Springer, Berlin Heidelberg New York 1973
6. A.G. Khovanskii, A.N. Varchenko: Asymptotics of integrals over vanishing cycles and the Newton polyhedron. Soviet Math. Dokl. **32** (1985) 122–127
7. A. Landman: On the Picard-Lefschetz formula for algebraic manifolds acquiring general singularities. Thesis Berkeley 1967
8. B. Malgrange: Polynômes de Bernstein-Sato et cohomologie évanescente. Astérisque **101–102** (1983) 243–267
9. Z. Mebkhout: Une équivalence de catégories. Compos. Math. **51** (1984) 51–62; Une autre équivalence de catégories. Ibid, pp. 63–88
10. J. Milnor: Singular points on complex hypersurfaces. Ann. Math. Stud. 61. Princeton 1968
11. V. Navarro Aznar: Sur la théorie de Hodge-Deligne. Invent. math. **90** (1987) 11–76
12. F. Pham: Structures de Hodge mixtes associées à un germe de fonction à point critique isolé. Astérisque **101–102** (1983) 268–285
13. M. Saito: Gauss-Manin system and mixed Hodge structure. Proc. Japan Acad. **58**, Ser. A (1982) 29–32
14. M. Saito: Modules de Hodge polarisables. Publ. RIMS Kyoto Univ. **24** (1988) 849–995
15. M. Saito: Mixed Hodge modules. Publ. RIMS Kyoto Univ. 1990
16. M. Saito: Exponents and Newton polyhedra of isolated hypersurface singularities. Math. Ann. **281** (1988) 411–417
17. M. Saito: Preuve d'une conjecture de Steenbrink. Manuscript in preparation.
18. M. Saito, J.H.M. Steenbrink: Thom-Sebastiani formula for spectra of hypersurface singularities. Manuscript in preparation.
19. J. Scherk: On the monodromy theorem for isolated hypersurface singularities. Invent. math. **58** (1980) 289–301
20. J. Scherk, J.H.M. Steenbrink: On the mixed Hodge structure on the cohomology of the Milnor fibre. Math. Ann. **271** (1985) 641–665
21. W. Schmid: Variation of Hodge structure: the singularities of the period mapping. Invent. math. **22** (1973) 211–320
22. M. Sebastiani, R. Thom: Un résultat sur la monodromie. Invent. math. **13** (1971) 90–96
23. J.H.M. Steenbrink: Limits of Hodge structures. Invent. math. **31** (1976) 229–257
24. J.H.M. Steenbrink: Mixed Hodge structure on the vanishing cohomology. In: P. Holm (ed.): Real and complex Singularities. Oslo 1976. Sijthoff-Noordhoff, Alphen a/d Rijn 1977, pp. 525–563
25. J.H.M. Steenbrink: Mixed Hodge structures associated with isolated singularities. Proc. Symp. Pure Math. 40, Part **2** (1983) 513–536
26. J.H.M. Steenbrink: Semicontinuity of the singularity spectrum. Invent. math. **79** (1985) 557–565
27. J.H.M. Steenbrink: The spectrum of hypersurface singularities. Astérisque **179–180** (1989) 163–184

28. J.H.M. Steenbrink: Semicontinuity of the spectrum under certain deformations of non-isolated singularities. Manuscript in preparation
29. A.N. Varchenko: Asymptotic Hodge structure on the vanishing cohomology. Izv. Akad. Nauk SSSR, Ser. Mat. **45** (1981) 540–591 (in Russian) [English transl.: Math. USSR Izv. **18** (1982) 469–512]
30. A.N. Varchenko: On semicontinuity of the spectrum and an upper bound for the number of singular plints of projective hypersurfaces. Dokl. Akad. Nauk. **270** (1983) 1294–1297 (in Russian) [English transl.: Sov. Math. Dokl. **27** (1983) 735–739]

Trace Formulae in Spectral Geometry

Toshikazu Sunada and Manabu Nishio

Department of Mathematics, Faculty of Science, Nagoya University
Nagoya 464-01, Japan

A geometric analogue of the celebrated "Riemann hypothesis" is stated as

$$\lambda_1(M) \geq \lambda_0(\tilde{M}), \tag{1}$$

where $\lambda_1(M)$ is the first *positive* eigenvalue of the Laplacian Δ_M on a closed Riemannian manifold M, and $\lambda_0(\tilde{M})$ is the *bottom* of the spectrum of the Laplacian on the *universal* covering manifold $\tilde{M}$ (see [21]). For instance, if M is a Riemann surface with constant negative curvature, the inequality (1) turns out to be equivalent to the Riemann hypothesis for the Selberg zeta function defined by an Euler product over prime closed geodesics in M (A. Selberg [17]).

The view that we regard (1) as an analogue of the Riemann hypothesis is supported also by the fact that, under finite-fold covering maps, the eigenvalues behave like the non-trivial zeros of the zeta functions of number fields [19]. It is worthwhile, on the other hand, to note that prime closed geodesics play the same role as prime ideals, though, in general, there does not exist a well-suited zeta function connecting the eigenvalues directly with closed geodesics. We may establish, however, a weak relationship between those objects, which is embodied as an analogue of the Weil's explicit formula (cf. S. Lang [13]). This illustrates more intimate kinship between spectral geometry and number theory. To explain it, let $\varrho : \pi_1(M) \to U(N)$ be a unitary representation, and let $\lambda_0(\varrho) \leq \lambda_1(\varrho) \leq \cdots$ be the eigenvalues of the Laplacian Δ_ϱ acting on sections of the *flat* vector bundle associated with ϱ (which may be regarded as an analogue of the non-trivial zeros of an *L*-function). We define the (even) distribution $\Theta(\varrho) \in \mathscr{D}'(\mathbb{R})$ by

$$\langle \Theta(\varrho), f \rangle = \sum_{k=0}^{\infty} \hat{f}(\sqrt{\lambda_k(\varrho)}), \qquad f \in \mathscr{D}(\mathbb{R}),$$

where $\hat{f}(s) = \int_{-\infty}^{\infty} f(t) \cos(st)\, dt$. We also associate an (even) distribution Θ_α with each *free* homotopy class $\alpha \in [S^1, M]$ of closed curves by setting

$$\langle \Theta_\alpha, f \rangle = \int_{\Gamma_\sigma \backslash X} U_f(\sigma x, x)\, dx, \tag{2}$$

where $X = \tilde{M}$, $\Gamma = \pi_1(M)$, and we identify $[S^1, M]$ with the set $[\Gamma] = \{[\sigma]\}$ of

Proceedings of the International Congress
of Mathematicians, Kyoto, Japan, 1990
© The Mathematical Society of Japan, 1991

conjugacy classes in Γ, and write Γ_σ for the centralizer of σ. The function $U_f(x, y) \in C^\infty(X \times X)$ is the kernel function of the smooth operator

$$\hat{f}(\sqrt{\Delta_X}) = \int f(t) \cos(t\sqrt{\Delta_X}) \, dt.$$

We then obtain the following theorem which gives a refinement of the result by J. Chazarain [5] (see also [7, 9] and [25]).

Theorem I. (a) (trace formula) $\Theta(\varrho) = \sum_{[\sigma] \in [\Gamma]} \text{tr } \varrho(\sigma)\Theta_{[\sigma]}$.

(b) *The support of the distribution* $\Theta_{[\sigma]}$ *is contained in the set* $\{t \in \mathbb{R}; |t| \geq l_{[\sigma]}\}$, *where* $l_{[\sigma]}$ *is the length of the shortest closed geodesics in M whose homotopy class is* $[\sigma]$, *so that, when the right hand side of* (a) *is applied to a test function in* $\mathscr{D}(\mathbb{R})$, *the sum over* $[\Gamma]$ *reduces to a finite one.*

(c) *The singular support of* $\Theta_{[\sigma]}$ *is contained in the set* $\{\pm \text{length of closed geodesics in M with the homotopy class} [\sigma]\}$.

(d) *For the unit element* 1 *in* Γ, *the distribution* $\Theta_{[1]}$ *is temperate.*

(e) *The distribution* $\Theta_{[\sigma]}$ $(\sigma \neq 1)$ *is extended to a functional on the space of test functions f satisfying*

$$|f^{(k)}(t)| \leq C \exp(-(h + \delta)|t|), \qquad 0 \leq k < 1 + n/2$$

for some $\delta > 0$ *and* $C > 0$. *Here h is the exponential growth rate of volume of geodesic balls in X. Moreover, the sum* $\sum \text{tr } \varrho(\sigma) \langle \Theta_{[\sigma]}, f \rangle$ *converges absolutely and equals* $\langle \Theta(\varrho), f \rangle$. *If* Γ *is of polynomial growth, then* $\Theta_{[\sigma]}$ *is also a temperate distribution.*

(f) *For* $f_\tau(t) = (4\pi\tau)^{-1/2} \exp(-t^2/4\tau)$, *we have*

$$\langle \Theta(1), f_\tau \rangle = \mu_\tau(\Omega),$$

$$\langle \Theta_{[\sigma]}, f_\tau \rangle = \mu_\tau(\Omega_{[\sigma]}),$$

where 1 *denotes the trivial representation,* $\Omega_{[\sigma]}$ *is the connected component of the free loop space* $\Omega = C^0(S^1, M)$ *corresponding to the homotopy class* $[\sigma]$, *and* μ_τ *is the Wiener measure on* Ω.

We should point out that, by a slight modification, all the statements above are generalized to the case of normal covering manifolds. We shall give an outline of the proof.

Let X be a normal covering space of M with covering transformation group Γ. From the *finite propagation property* of the wave equation, it follows that, if $d(x, y) > R$ and $\text{supp}(f) \subset \{t \in \mathbb{R}; |t| < R\}$, then $U_f(x, y) = 0$. For each $R > 0$, if we put

$$H_R = \{x \in \Gamma_\sigma \backslash X; d(x, \sigma x) \leq R\},$$

then H_R is compact, so that $U_f(\sigma x, x)$ has compact support in $\Gamma_\sigma \backslash X$. Hence the distribution $\Theta_{[\sigma]}$ is well-defined (see (2)) and satisfies (b). Since $\lim_{R \to \infty} \sup R^{-1} \times$

$\log(\mathrm{vol}(H_R)) \le h$, applying the estimate established by J. Cheeger, M. Gromov and M. Taylor [6], we obtain (e).

Let $\mathbb{U}_{\varrho, f}(p, q) \in C^\infty(M \times M)$ be the kernel function of the operator

$$\hat{f}(\sqrt{\varDelta_\varrho}) = \int f(t) \cos(t\sqrt{\varDelta_\varrho})\, dt.$$

If we denote by ω the coveing map of X onto M, then

$$\mathbb{U}_{\varrho, f}(\omega(x), \omega(y)) = \sum_{\sigma \in \Gamma} \mathrm{tr}\, \varrho(\sigma) U_f(\sigma x, y).$$

Moreover, we find

$$\langle \Theta(\varrho), f \rangle = \mathrm{tr}\, \hat{f}(\sqrt{\varDelta_\varrho}) = \int_M \mathbb{U}_{\varrho, f}(p, p)\, dp.$$

The proof of (a) is now carried out in the same way as that of the Selberg trace formula (see [17, 18]). Actually, (a) is specialized down to the Selberg's formula when M is a constant negatively curved surface.

The assertion (c) is shown in a similar manner as [5], by analyzing the wave front set of $\Theta_{[\sigma]}$. In fact, one has the following expression for $\Theta_{[\sigma]}$ in terms of the distribution kernel $U_X(t, x, y)$ of the operator $\cos(t\sqrt{\varDelta_X})$:

$$\Theta_{[\sigma]} = (\pi_\sigma)_* (p_\sigma^*)^{-1}(D_\sigma^*) U_X,$$

where $D_\sigma : \mathbb{R} \times X \to \mathbb{R} \times X \times X$ is the map defined by $D_\sigma(t, x) = (t, \sigma x, x)$, $p_\sigma : \mathbb{R} \times X \to \mathbb{R} \times (\Gamma \backslash X)$ is the map induced by the covering projection $X \to \Gamma \backslash X$, and $\pi_\sigma : \mathbb{R} \times (\Gamma \backslash X) \to \mathbb{R}$ is the projection onto the first factor. We should note that $(D_\sigma^*) U_X$ is a Γ-invariant distribution on $\mathbb{R} \times X$, and that the pull-back p_σ^* yields an isomorphism of $\mathscr{D}'(\mathbb{R} \times (\Gamma \backslash X))$ onto the space of Γ-invariant distributions on $\mathbb{R} \times X$. We then make use of the fact that the map π_σ restricted to the support of $(p_\sigma^*)^{-1}(D_\sigma^*) U_X$ is *proper* (recall that H_R is compact), so that the push-forward $(\pi_\sigma)_*$ can be applied to this distribution. The rest of the proof is carried out by using the wave-front calculus.

The assertion (f) is a consequence of the observation that, for $f = f_\tau$, U_f is the fundamental solution $k(\tau, x, y)$ of the heat equation on X ([18]).

We now proceed to the proof of (d). We first recall that $\Theta(\varrho)$ is a temperate distribution. Indeed, this is deduced from the fact that the counting function $\varphi_\varrho(\lambda) = \#\{\lambda_k(\varrho) \le \lambda\}$ satisfies $\varphi_p(\lambda) \sim C_n \mathrm{vol}(M)\lambda^{n/2}$ as $\lambda \uparrow \infty$, and the equality $\langle \Theta(\varrho), f \rangle = \int \hat{f}(\sqrt{\lambda})\, d\varphi_\varrho(\lambda)$. We wish to establish a similar expression for $\Theta_{[1]}$, say

$$\langle \Theta_{[1]}, f \rangle = \int \hat{f}(\sqrt{\lambda})\, d\varphi(\lambda).$$

To define the increasing function $\varphi(\lambda)$, we first note that $\hat{f}(\sqrt{\varDelta_X})$ lies in the von Neumann algebra $\mathrm{End}_\Gamma(L^2(X))$ of Γ-equivariant endomorphisms of $L^2(X)$ and $\langle \Theta_{[1]}, f \rangle = \mathrm{tr}_\Gamma \hat{f}(\sqrt{\varDelta_X})$, where tr_Γ is a *trace* (called Γ-trace) in the sense of von Neumann algebras, and is equal, in this case, to

$$\int_\mathscr{D} U_f(x, x)\, dx,$$

$\mathscr{D}$ being a *fundamental domain* for the Γ-action (see M.F. Atiyah [1]). Let $\Delta_X = \int \lambda \, dE_\lambda$ be the *spectral resolution* of Δ_X, so that

$$\hat{f}(\sqrt{\Delta_X}) = \int \hat{f}(\sqrt{\lambda}) \, dE_\lambda.$$

We observe that E_λ is of Γ-trace class, and

$$\operatorname{tr}_\Gamma \hat{f}(\sqrt{\Delta_X}) = \int \hat{f}(\sqrt{\lambda}) \, d(\operatorname{tr}_\Gamma E_\lambda).$$

Hence we have only to set $\varphi(\lambda) = \operatorname{tr}_\Gamma E_\lambda$.

Consider the special case $f = f_\tau$. Then $\hat{f}_\tau(s) = e^{-\tau s^2}$, and

$$\int_{\mathscr{D}} k(\tau, x, x) \, dx = \operatorname{tr}_\Gamma \hat{f}_\tau(\sqrt{\Delta_X}) = \int e^{-\lambda \tau} \, d\varphi(\lambda).$$

The function $k(\tau, x, x)$ has the same asymptotic behavior as that of the heat kernel on M, and hence $k(\tau, x, x) \sim (4\pi\tau)^{-n/2}$ as $\tau \downarrow 0$, so that, applying the Tauberian theorem to φ, we get

$$\varphi(\lambda) \sim (4\pi)^{n/2} \operatorname{vol}(M) \lambda^{n/2} / \Gamma\left(\frac{n}{2} + 1\right). \tag{3}$$

This completes the proof of (d).   □

The function $\varphi(\lambda)$ includes much information of the spectrum of Δ_X. For instance, $\varphi(\lambda) = 0$ if $\lambda < \lambda_0(X)$, and $\varphi(\lambda) > 0$ if $\lambda > \lambda_0(X)$. The spectrum $\sigma(\Delta_X)$ coincides with the set of increasing points of φ. Let $\varphi(\lambda) = \varphi_{ac}(\lambda) + \varphi_s(\lambda) + \varphi_p(\lambda)$ be the (unique) decomposition into an *absolutely continuous* function φ_{ac}, a *singular continuous* function φ_s, and a *step* function φ_p. If

$$(\Delta_X, L^2(X)) = (\Delta_X, H_{ac}) \oplus (\Delta_X, H_s) \oplus (\Delta_X, H_p)$$

is the decomposition into the absolutely continuous, singular, and pure point parts of Δ_X, then $\varphi_{ac}(\lambda) = \operatorname{tr}_\Gamma E_\lambda P_{ac}$, $\varphi_s(\lambda) = \operatorname{tr}_\Gamma E_\lambda P_s$, and $\varphi_p(\lambda) = \operatorname{tr}_\Gamma E_\lambda P_p$, where P_{ac}, P_s, and P_p are the orthogonal projections onto H_{ac}, H_s, and H_p respectively. Therefore the sets of increasing points of the functions φ_{ac} and φ_s are just the absolutely continuous spectrum and the singular spectrum of Δ_X respectively. The point spectrum coincides with the set of jumping points of φ_p.

When $X = \tilde{M}$ and M has no *null*-homotopic closed geodesics except for point geodesics, the assertion (c) says that the singular support of $\Theta_{[1]}$ is $\{0\}$. It is an interesting problem to examine the decay of the function $\Theta_{[1]}$ at infinity. For instance, if $\Theta_{[1]}$ is in L^2 at infinity, it is shown that Δ_X has purely absolutely continuous spectrum. In fact, from the above argument, we have

$$\langle \Theta_{[1]}, f \rangle = \int \hat{f}(\lambda) \, d\psi(\lambda) = \langle \dot{\psi}, \hat{f} \rangle \qquad (\psi(\lambda) = \varphi(\lambda^2)),$$

so that $\hat{\dot{\psi}} = \Theta_{[1]} = T_1 + T_2$, where T_1 is a distribution with compact support and $T_2 \in L^2(\mathbb{R})$. In view of the Fourier inversion formula and the Parseval equality, we may write $\dot{\psi} = S_1 + S_2$, where S_1 is analytic and $S_2 \in L^2(\mathbb{R})$. Hence $\dot{\psi}$ is in L^2_{loc}, and φ is absolutely continuous. We conjecture that, if M is nonpositively curved, the function $\Theta_{[1]}$ decays with the order of $|t|^{-\alpha}$ at $t = \pm\infty$ with some $\alpha > 0$.

Example 1. When M is a flat torus and $X = \mathbb{R}^n$,

$$\varphi(\lambda) = (4\pi)^{-n/2} \operatorname{vol}(M)\lambda^{n/2}/\Gamma\left(\frac{n}{2} + 1\right) \qquad (\lambda \geq 0).$$

This follows from the Jacobi's inversion formula for the θ-series (a special case of (a) in Theorem I).

Example 2. Let M be a closed Riemann surface with constant negative curvature (-1), and let X be the universal covering. The Selberg trace formula leads to the identity

$$\operatorname{tr}_\Gamma \exp(-t\Delta_X) = (4\pi)^{-1} \operatorname{vol}(M) \int_0^\infty \exp\left(-t\left(\lambda + \frac{1}{4}\right)\right)\tanh \pi\sqrt{\lambda}\, d\lambda,$$

from which we easily deduce

$$\varphi(\lambda) = \begin{cases} (4\pi)^{-1} \operatorname{vol}(M) \displaystyle\int_0^{\lambda - 1/4} \tanh \pi\sqrt{\lambda}\, d\lambda & \text{for } \lambda \geq \dfrac{1}{4} \\[2ex] 0 & \text{for } \lambda < \dfrac{1}{4}. \end{cases}$$

Note that $\varphi(\lambda) \sim \frac{1}{6} \operatorname{vol}(M)(\lambda - \frac{1}{4})^{3/2}$ as $\lambda \downarrow \frac{1}{4}$.

Example 3. Let $X \to M$ be an infinite-fold covering with an *abelian* covering transformation group of rank b. Then $\varphi(\lambda)$ is smooth on an interval $(0, \delta)$ for some $\delta > 0$, and $\varphi(\lambda) \sim C\lambda^{b/2}$ as $\lambda \downarrow 0$. To see this, we make use of (e) in Theorem I:

$$\operatorname{tr} \exp(-\tau\Delta_\chi) = \sum_{[\sigma]} \operatorname{tr} \chi(\sigma)\langle\Theta_{[\sigma]}, f_\tau\rangle,$$

χ being a character of Γ. Integrating the both sides over the character group $\hat{\Gamma}$ with respect to the normalized Haar measure, we have

$$\operatorname{tr}_\Gamma \exp(-t\Delta_X) = \int_{\hat{\Gamma}} \operatorname{tr} \exp(-t\Delta_\chi)\, d\chi,$$

where we have applied the orthogonal relations of characters. Thus we obtain the equation

$$\int_{\hat{\Gamma}} d\chi \int e^{-t\lambda}\, d\varphi_\chi(\lambda) = \int e^{-t\lambda}\, d\varphi(\lambda),$$

which implies $\varphi(\lambda) = \int_{\hat{\Gamma}} \varphi_\chi(\lambda)\, d\chi$. We now recall the fact that $\lambda_0(1) = 0$ and $\lambda_0(\chi)$ is simple for χ in a small neighborhood of 1, from which it follows that when $\lambda \geq 0$ is small enough, $\varphi_\chi(\lambda) = 1$ for χ with $\lambda_0(\chi) \leq \lambda$, and $\varphi_\chi(\lambda) = 0$ otherwise. Hence we have

$$\varphi(\lambda) = \text{vol}\{\chi \in \hat{\varGamma}; \lambda_0(\chi) \le \lambda\}.$$

Since $\lambda_0(\chi)$ smoothly depends on χ around a neighborhood of 1, and the Hessian of $\lambda_0(\chi)$ at $\chi = 1$ is positive definite [11], the assertion follows easily.

For a normal *infinite-fold* covering $X \to M$, the bottom $\lambda_0(X)$ lies in the continuous spectrum of $\varDelta_X$ [21], that is, $\varphi(\lambda) \to 0$ as $\lambda \downarrow \lambda_0$. In view of the above exmaples, we conjecture that there exist positive constants c and C such that

$$\varphi(\lambda) \sim C(\lambda - \lambda_0)^c \qquad \text{as} \qquad \lambda \downarrow \lambda_0. \tag{4}$$

We now go back to the analogue of "Riemann hypothesis". The inequality (1) does *not* hold in general. Even for constant negatively curved surfaces, we have many counter examples (B. Randol [15], P. Buser [4], and [19]). H. Donnelly [8] established, however, a weak version of (1). To be exact, consider a set $\{M_i\}_{i=1}^\infty$ of finite-fold subcoverings of a normal covering $X \overset{\varGamma}{\to} M$, and let $\varGamma_i$ be the subgroup of finite index in $\varGamma$ corresponding to the covering map $X \to M_i$. Let us introduce the counting function $\varphi_{M_i}(\lambda) = \max\{k; \lambda_k(M_i) \le \lambda\}$, where $\lambda_1(M_i) \le \lambda_2(M_i) \le \cdots$ are the positive eigenvalues of the Laplacian on M_i (so that $\varphi_M = \varphi_1 - 1$). The "Riemann hypothesis" (1) for M_i is stated as $\varphi_{M_i}(\lambda) = 0$ for $\lambda < \lambda_0(\tilde{M})$. Suppose that $d(M_i) \to \infty$, where

$$d(M_i) = \inf_{\sigma \in \varGamma_i} \inf_{x \in X} d(x, \sigma x),$$

which, in the case that $X = \tilde{M}$, coincides with the infimum of length of non-null homotopic closed geodesics in M_i. Then, by [8], we have

$$\limsup_{i \to \infty} \text{vol}(M_i)^{-1}\varphi_{M_i}(\lambda) = 0 \qquad \text{if } \lambda < \lambda_0(X),$$

$$\liminf_{i \to \infty} \text{vol}(M_i)^{-1}\varphi_{M_i}(\lambda) > 0 \qquad \text{if } \lambda > \lambda_0(X).$$

In other words, the set of eigenvalues in the interval $[0, \lambda_0(X))$ is "thin" in the totality of eigenvalues on covering manifolds M_i.

We shall give a more precise statement.

Theorem II. *Let $X \overset{\varGamma}{\to} M$ be a normal covering, and let $\{\varrho_i\}_{i=1}^\infty$ be a sequence of finite-dimensional unitary representations of $\varGamma$. Suppose that*

$$\lim_{i \to \infty} (\dim \varrho_i)^{-1} \text{Re tr } \varrho_i(\sigma) = 0$$

for every $\sigma \ne 1$ in $\varGamma$. Then

$$\lim_{i \to \infty} (\dim \varrho_i)^{-1}\varphi_{\varrho_i}(\lambda) = \varphi(\lambda)$$

at every continuity point λ of φ.

Let $\{M_i\}$ be as above. As an example of $\{\varrho_i\}$, let us take the regular representation ϱ_i of $\varGamma$ on $L^2(\varGamma/\varGamma_i)$. Then

$$\text{tr } \varrho_i(\sigma) = \#\{\mu\Gamma \in \Gamma\backslash\Gamma_i; \mu^{-1}\sigma\mu \in \Gamma_i\}.$$

It is easily checked that $d(M_i) \to \infty$ if and only if, for $\sigma \neq 1$, $\text{tr } \varrho_i(\sigma) = 0$ for sufficiently large i. Furthermore we observe that $\varDelta_{M_i}$ is unitarily equivalent to $\varDelta_{\varrho_i}$. Thus we obtain

Corollary 1. *If* $d(M_i) \to \infty$, *then*

$$\lim_{i\to\infty} \text{vol}(M_i)^{-1}\varphi_{M_i}(\lambda) = \text{vol}(M)^{-1}\varphi(\lambda)$$

at all the continuity points of φ, *and hence for continuity points* $\alpha < \beta$ *of* φ, *we have*

$$\lim_{i\to\infty} \text{vol}(M_i)^{-1}\#\{\lambda_k(M_i); \alpha < \lambda_k(M_i) \leq \beta\} = \text{vol}(M)^{-1}(\varphi(\beta) - \varphi(\alpha)).$$

Especially, $\sigma(\varDelta_X) \subset$ *the closure of* $\bigcup_i \sigma(\varDelta_{M_i})$.

This corollary gives a partial generalization of a result by H. Huber [10] (see also [26]).

Corollary 2. *Under the same condition as above, if* Γ *is amenable, then* $\sigma(\varDelta_X) =$ *the closure of* $\bigcup_i \sigma(\varDelta_{M_i})$.

In fact, from amenability, it follows that $\sigma(\varDelta_{M_i}) \subset \sigma(\varDelta_X)$ for every M_i (see R. Brooks [2], and [19]).

It is interesting to observe a similarity of Corollary 1 to the *thermodynamical-limit-results* in statistical physics. The conjecture (4) on φ bears a resemblance to that on critical exponents for phase transitions.

To prove Theorem II, it suffices to show

$$\lim_{i\to\infty} (\dim \varrho_i)^{-1} \int_0^\infty e^{-\lambda t}\, d\varphi_{\varrho_i}(\lambda) = \int_0^\infty e^{-\lambda t}\, d\varphi(\lambda) \tag{5}$$

(cf. M.A. Shubin [16]). The left hand side of (5) is equal to

$$\lim_{i\to\infty} (\dim \varrho_i)^{-1} \text{tr} \exp(-t\varDelta_{\varrho_i})$$

$$= \text{tr}_\Gamma \exp(-t\varDelta_X) + \lim_{i\to\infty} \sum_{\substack{[\sigma]\\ \sigma\neq 1}} (\dim \varrho_i)^{-1}(\text{Re tr } \varrho_i(\sigma))\langle\Theta_{[\sigma]}, f_t\rangle$$

$$= \text{tr}_\Gamma \exp(-t\varDelta_X) = \int_0^\infty e^{-\lambda t}\, d\varphi(\lambda).$$

This completes the proof. $\qquad\qquad\square$

We are now concerned with a fine structure of the spectrum of $\varDelta_X$ on a normal infinite-fold covering manifold X. If the covering transformation group Γ is abelian, the spectrum has *band structure*, namely $\sigma(\varDelta_X)$ consists of a series of intervals without accumulations ([12]). We conjecture that this is true for general cases. An evidence supporting this conjecture will be explained below.

We identify $L^2(X)$ with $L^2(\Gamma, V) = L^2(\Gamma) \otimes V$ in a natural manner, where $V = L^2(\mathscr{D})$. We define the subalgebra $C_r^*(\Gamma, \mathscr{K})$ of $\mathrm{End}_\Gamma(L^2(X))$ to be the completion of $C_0(\Gamma, \mathscr{K}) = C_0(\Gamma) \otimes \mathscr{K}$ with respect to the operator norm, where $\mathscr{K}$ is the algebra of compact operators of $L^2(\mathscr{D})$, and $C_0(\Gamma) \otimes \mathscr{K}$ acts on $L^2(X)$ by

$$f \in L^2(\Gamma, V) \mapsto Af \in L^2(\Gamma, V)$$

$$(Af)(\sigma) = \sum_{\mu \in \Gamma} A(\mu^{-1}\sigma)f(\mu), \qquad A \in C_0(\Gamma, \mathscr{K}).$$

The algebra $C_r^*(\Gamma, \mathscr{K})$ is what we call the *reduced* group C^*-algebra with value in $\mathscr{K}$. From the finite propagation property for the wave equation, we deduce that $\hat{f}(\sqrt{\Delta_X}) \in C_0(\Gamma, \mathscr{K})$ for $f \in \mathscr{D}(\mathbb{R})$. Given a rapidly decreasing function $f \in \mathscr{S}(\mathbb{R})$, by approximating f by functions f_n in $\mathscr{D}(\mathbb{R})$ in the L^1-sense, we find that $\hat{f}(\sqrt{\Delta_X}) \in C_r^*(\Gamma, \mathscr{K})$. In fact,

$$\|\hat{f}(\sqrt{\Delta_X}) - \hat{f}_n(\sqrt{\Delta_X})\| \le \sup |\hat{f} - \hat{f}_n| \le \|f - f_n\|_{L^1}.$$

Especially $\exp(-t\Delta_X) \in C_r^*(\Gamma, \mathscr{K})$. Applying again an approximation-argument, we observe that if two real numbers a and b are *not* in the spectrum of Δ_X ($b > a$), then the projection $E_b - E_a$ is in $C_r^*(\Gamma, \mathscr{K})$. We thus have

Theorem III [22]. *Suppose that there exists a positive constant C such that* $\mathrm{tr}_\Gamma P \ge C$ *for every non-trivial orthogonal projection P in $C_r^*(\Gamma, \mathscr{K})$. Then the spectrum of Δ_X has band structure (possibly with degenerate intervals, say $[\lambda, \mu]$ with $\lambda = \mu$, corresponding to isolated eigenvalues). Moreover, the number of components of the spectrum which intersect with $(-\infty, \lambda]$ has the following asymptotic estimate*

$$\limsup_{\lambda \to \infty} \frac{C\Gamma(1 + n/2)N(\lambda)}{(4\pi)^{-n} \mathrm{vol}(M)\lambda^{n/2}} \le 1.$$

The asymptotic estimate is a consequence of the asymptotic formula (3) for $\varphi(\lambda)$.

A large class of discrete groups seems to satisfy the condition in the above theorem (cf. M. Pimsner [14]). For example, a free product of finite number of finite groups and infinite cyclic groups satisfies the condition with $C =$ the reciprocal of the least common multiple of the orders of the finite groups. The Kadison conjecture says that one may take 1 as C if Γ is torsion free.

We conclude the discussion with pointing out that the method employed throughout may be applied to a Schrödinger operator on X with a Γ-invariant potential, thereby leading to *non-abelian* Bloch's theory.

Acknowledgement. The authors are grateful to H. Donnelly for drawing their attention to the paper [8].

References

1. Atiyah, M.F.: Elliptic operators, discrete groups and von Neumann algebra. Astérisque **32–33** (1976) 43–72

2. Brooks, R.: The fundamental groups and the spectrum of the Laplacian. Comment. Math. Helv. **56** (1981) 581–598

3. Brooks, R.: The spectral geometry of tower of coverings. J. Diff. Geom. **23** (1986) 97–107

4. Buser, P: On Cheeger's inequality: $\lambda_1 \geq \frac{1}{4}h^2$. Am. Math. Soc. Proc. Symp. Pure Math. **36** (1980) 29–77

5. Chazarain, J.: Formule de Poisson pour les variétés riemanniennes. Invent. math. **24** (1974) 65–82

6. Cheeger, J. Gromov, M., Taylor, M.: Finite propagation speed, kernel estimates for functions of the Laplace operator, and the geometry of complete Riemannian manifolds. J. Diff. Geom. **17** (1982) 15–53

7. Donnelly, H.: On the wave equation asymptotics of a compact negatively curved surface. Invent. math. **45** (1978) 115–137

8. Donnelly, H.: On the spectrum of towers. Proc. A.M.S. **87** (1983) 322–329

9. Duistermaat, J.J., Guillemin, V.W.: The spectrum of positive elliptic operators and periodic bicharacteristics. Invent. math. **29** (1975) 39–79

10. Huber, H.: Ueber das spectrum des Laplace-operators auf kompakten Riemannschen flächen, Comment. Math. Helv. **57** (1982) 627–647

11. Katsuda, A., Sunada, T.: Homology and closed geodesics in a compact Riemann surface. Amer. J. Math. **109** (1987) 145–156

12. Kobayashi, T., Ono, K., Sunada, T.: Periodic Schrödinger operators on a manifold. Forum Math. **1** (1989) 69–79

13. Lang, S.: Algebraic number theory. Addison-Wesley, Massachusetts 1970

14. Pimsner, M.: KK-groups of crossed products by groups acting on trees. Invent. math. **86** (1986) 603–634

15. Randol, B.: Small eigenvalues of the Laplace operator on compact Riemann surfaces. Bull. AMS. **130** (1974) 996–1000

16. Shubin, M.A.: The spectral theory and the index of elliptic operators with almost periodic coefficients. Usp. Mat. Nauk **34** (1979) 95–135

17. Selberg, A.: Harmonic analysis and discontinuous groups in weakly symmetric spaces with application to Dirichlet series. J. Indian Math. Soc. **20** (1956) 47–87

18. Sunada, T.: Trace formula, Wiener integrals and asymptotics. Proc. Japan-France Seminor, "Spectra of Riemannian Manifolds", Kaigai Publ. Tokyo 1983, pp. 159–169

19. Sunada, T.: Riemannian coverings and isospectral manifolds, Ann. Math. **121** (1985) 169–186

20. Sunada, T.: Unitary representations of fundamental groups and the spectrum of twisted Laplacians. Topology **28** (1989) 125–132

21. Sunada, T.: Fundamental groups and Laplacians. Proc. Taniguchi Symp. "Geometry and Analysis on Manifolds" 1987. Lecture Notes in Mathematics, vol. 1339. Springer, Berlin Heidelberg New York 1988, pp. 248–277

22. Sunada, T.: Group C^*-algebras and the spectrum of a periodic Schrödinger operator on a manifold. Preprint 1989

23. Sunada, T.: Fundamental groups and Laplacians. Monograph in Japanese. Kinokuniya, Tokyo 1988

24. Sunada, T.: Trace formula for Hill's operators. Duke Math. J. **47** (1980) 529–546

25. Sunada, T.: Trace formula and heat equation asymptotics for a non positively curved manifold. Amer. J. Math. **104** (1982) 795–812

26. DeGeorge, D.L., Wallach, N.R.: Limit formulas for multiplicities in $L^2(\Gamma \backslash G)$. Ann. of Math. **107** (1978) 133–150

27. Efremov, D.V., Shubin, M.A.: Spectrum distribution function and variational principle for automorphic operators on hyperbolic space. Équations aux dérivées partielles, Séminaire 1988–1989 Ecole Polytechnique, Exposé n° VIII.

Kähler-Einstein Metrics on Algebraic Manifolds

*Gang Tian**

Department of Mathematics, State University of New York
Stony Brook, NY 11794, USA

1. Introduction

It is one of fundamental problems in differential geometry to find a distinguished metric on a smooth manifold. H. Poincaré's Uniformization theorem settles this problem for Riemann surfaces. That is, there is a unique metric with constant curvature in each Kähler class on a Riemann surface. Trying to generalize it to higher dimensions, E. Calabi conjectured in the 50s the existence of Kähler-Einstein metrics on a compact Kähler manifold with its first Chern class definite. A Kähler-Einstein metric is a Kähler metric with constant Ricci curvature.

In the middle of the 70s, this conjecture was solved by S. T. Yau in case the first Chern class is vanishing and Aubin and Yau, independently, in case the Chern class is negative (cf. [Y1]). The uniqueness in these two cases was done by E. Calabi himself in the 50s. Such Kähler-Einstein metrics were then applied to studying projective manifolds. For instance, Yau used these metrics to show the Miyaoka-Yau inequality on surface of general type, its generalized version in higher dimensions and the characterization of the quotients of the complex hyperbolic spaces (cf. [Y2]). We also refer readers to [CY, Ko, Ts, TY1] for the generalizations of these to quasi-projective manifolds.

However, this conjecture of Calabi still remains open in general in case the first Chern class is positive. In this paper, we will survey the recent progress on this part of Calabi's conjecture, including the uniqueness and the existence of Kähler-Einstein metrics with positive scalar curvature, the outline of the complete solution for Calabi's conjecture in case of complex dimension two, etc. Some related problems will also be discussed.

From now on, we always denote by M a compact Kähler manifold with positive first Chern class $C_1(M)$, that is, M is a smooth Fano variety. Then we can choose a Kähler metric g with its Kähler class ω_g representing $C_1(M)$. In local coordinates $(z_1, \cdots, z_n)$ of M with $\dim_C M = n$, if g is represented by positive hermitian metrices $\{g_{i\bar{j}}(z)\}_{1 \leq i,j, \leq n}$,

$$
\omega_g = \frac{\sqrt{-1}}{2\pi} \sum_{i,j=1}^{n} g_{i\bar{j}} dz_i \wedge d\bar{z}_j.
$$

* This work is partially supported by a grant from NSF.

Proceedings of the International Congress
of Mathematicians, Kyoto, Japan, 1990
© The Mathematical Society of Japan, 1991

It is well-known that the Ricci curvature form $Ric(g)$ also represents the first Chern class, therefore, there is a smooth function f in $C^\infty(M, R)$ such that

$$Ric(g) = \omega_g + \frac{\sqrt{-1}}{2\pi}\partial\bar\partial f$$

and

$$\int_M (e^f - 1)\omega_g^n = 0$$

where $\omega_g^n = \omega_g \wedge \cdots \wedge \omega_g$. In local coordinates, the Ricci curvature has the following expression

$$R_{i\bar j} = \partial_i\partial_{\bar j}\log\det(g_{k\bar l})$$

$$Ric(g) = \frac{\sqrt{-1}}{2\pi}\sum_{i,j=1}^n R_{i\bar j}dz_i \wedge d\bar z_j.$$

It follows from this that the existence of a Kähler-Einstein metric on M is equivalent to the solvability of the following complex Monge-Ampéré equations,

$$\begin{cases} \left(\omega_g + \frac{\sqrt{-1}}{2\pi}\partial\bar\partial\varphi\right)^n = e^{f-t\varphi}\omega_g^n & \text{on } M \\ \left(\omega_g + \frac{\sqrt{-1}}{2\pi}\partial\bar\partial\varphi\right) > 0 & \text{on } M \end{cases} \tag{1.1}_t$$

where $n = \dim_C M$.

In case $(1.1)_1$ has a solution φ, then $\omega_g + \frac{\sqrt{-1}}{2\pi}\partial\bar\partial\varphi$ gives the Kähler form of a Kähler-Einstein metric.

Theorem 1.1 (Bando and Mabuchi [BM]). *The solution of $(1.1)_1$ is unique modulo the connected component $\text{Aut}_0(M)$ of $\text{Aut}(M)$ containing the identity if it exists, where $\text{Aut}(M)$ denotes the group of all holomorphic automorphisms of M. In particular, it implies the uniqueness of the Kähler-Einstein metric on M if it exists.*

Since Yau's solution of Calabi's conjecture in case of vanishing first Chern class more than fifteen years ago, it has been known that the solvability of $(1.1)_t$ follows from an a priori C^0-estimate for the solutions, namely, there is a uniform constant C depending only on M and g such that for any solution φ of $(1.1)_t$,

$$\sup_M |\varphi| \le C. \tag{1.2}$$

However, such an estimate (1.2) does not exist in general due to the analytic obstructions discussed in the next section.

2. Some Obstructions

Let $\eta(M)$ be the Lie algebra of the automorphism group $\text{Aut}(M)$, that is, $\eta(M)$ consists of all holomorphic vector fileds on M. In [Fu], Futaki defined an analytic invariant $F : \eta(M) \to C$ as follows,

$$F(X) = \int_M X(f)\omega_g^n, \qquad X \in \eta(M).$$

Theorem 2.1 *Suppose that (M, g) admits a Kähler-Einstein metric with positive scalar curvature. Then*
  1. *(Matsushima, 1957 [Ma]) The Lie algebra $\eta(M)$ is reductive;*
  2. *(Futaki, 1983, [Fu]) The Futaki invariant F is identically zero on $\eta(M)$.*

Blowing up a one or two points in CP^2, we obtain a complex surface with positive first Chern class and non-reductive Lie algebra of holomorphic vector fields. Therefore, by the result of Matsushima, such a surface does not admit a Kähler-Einstein metric. In [Fu], Futaki constructed a Fano 3-fold with reductive Lie algebra of holomorphic vector fields, but nonvanishing Futaki invariant. These obstructions indicate that there is no a priori C^0-estimate for the solution of $(1.1)_t$ in general.

Both obstructions of Matsushima and Futaki come from holomorphic vector fields. One might expect that the absence of holomorphic vector field, as is the case with most Fano variaties, would exclude these obstructions. However, there is a very disturbing example constructed in [T1]. Before we give this example, we first remark that Calabi's conjecture can be generalized to compact Kähler orbifolds with positive Chern class and the previous discussions are still effective. Using Volume Comparison Theorem [Bi], one can show

Theorem 2.2 [T1]. *Let (M, g) be a n-dimensional Kähler-Einstein orbifold with positive first Chern class. Then for any singular point x in M,*

$$\sharp(G_x) < \frac{(2n - 1)^n \text{Vol}(S^{2n})}{C_1(M)^n} \tag{2.1}$$

where G_x is the local uniformization group of x in M, $C_1(M)^n$ is the volume of M with respect to any Kähler metric with the associated Kähler form in $C_1(M)$ and $\text{Vol}(S^{2n})$ is the volume of S^{2n} with respect to the standard metric. In particular, in case $n = 2$, (2.1) becomes

$$\sharp(G_x) < \frac{48}{C_1(M)^2}. \tag{2.2}$$

Example 2.1 [T1]. By blowing up CP^2 successively at $[0,0,1]$, $[0,1,0]$, $[1,0,0]$, $[0,1,1]$ seven times, one can obtain a rational surface $\tilde{M}$ without holomorphic vector fields and containing a Hirzebruch-Jung string $D = E_1 + E_2 + E_3 + E_4$ with E_i smooth rational curves and $E_1^2 = E_4^2 = -3$, $E_2^2 = E_3^2 = -2$, $E_i \cdot E_j = 1$ if $i - j = \pm 1$; $= 0$ otherwise.

Let M be the 2-dimensional Kähler orbifold obtained from $\tilde{M}$ by collapsing the Hirzebruch string D. Then M has positive first Chern class $C_1(M)$ with $C_1(M)^2 = 3$ and an orbifold singular point x of type C^2/σ, where $\sigma : (z_1, z_2) \to$

$(e(\frac{1}{16})z_1, e_2(\frac{7}{16})z_2)$ for the Euclidean coordinates z_1, z_2 on C^2, $e(\frac{l}{m}) = \exp(\frac{2\pi l\sqrt{-1}}{m})$.
In particular, $\sharp(G_x) = 16$ in equal to $48/C_1(M)^2$. Therefore, by Theorem 2.2, M
does not admit any Kähler-Einstein orbifold metric. On the other hand, M has
no holomorphic vector filed since $\tilde{M}$ does not.

This example indicates that holomorphic vector fields may not be only obstructions in solving Calabi's conjecture. The author believes that the existence
of Kähler-Einstein metric with positive scalar curvature should be closely related
to the geometry of pluri-anti-canonical divisors.

3. Some Existence Theorems

In this section, let (M, g) be a compact Kähler manifold with the Kähler form ω_g
in $C_1(M)$ as given in the first section. We will give some existence results. First of
all, let us mention two known results among others, for special Kähler manifolds
with positive first Chern class, in particular, those manifolds having a lot of
holomorphic automorphisms. In [Ma2], Matsushima showed that every simply-
connected homogeneous Kähler manifold admits a Kähler-Einstein metric. In
case M is a P^1-bundle of certain type with C^*-actions which restrict to C^*-actions
on the fibers P^1, Koiso and Sakane [KS] showed that M admits a Kähler-Einstein
metric iff the Futaki invariant is zero. The manifold M in all these examples has
very high homogeneity. In fact, until 1986, Siu [Si] and the author [T2] provided
the first examples of compact Kähler-Einstein manifolds with positive first Chern
class and without holomorphic vector fields. Precisely, Siu showed the existence
of Kähler-Einstein metrics on the 2-dimensional Fermat surface and the surface
obtained by blowing up CP^2 at three generic points, while the author showed
the existence of Kähler-Einstein metrics on n-dimensional Fermat hypersurfaces
of degree n or $n + 1$. Their proofs are completely different. In fact, the examples
of the author are just corollaries of a general existence theorem proved in the
same paper [T2]. Let us first describe this result.

Let G be any compact subgroup in $\text{Aut}(M)$. This group G preserves the first
Chern class $C_1(M)$, therefore we may take g, f to be G-invariant. Define

$$P_G(M, g) = \left\{ \phi \in C^\infty(M, g) \right.$$

$$\left. | \ \phi \text{ is } G\text{-invariant on } M, \omega_g + \frac{\sqrt{-1}}{2\pi} \partial\bar{\partial}\phi > 0, \ \sup_M \phi = 0 \right\}$$

$$\alpha(M) = \sup\{\alpha \mid \exists C_\alpha > 0 \text{ s.t. } \int_M e^{-\alpha\phi} dV_g \leq C_\alpha \text{ for all } \phi \in P_G(M, g)\}.$$

It is not hard to check that $\alpha(M)$ is independent of the particular choice of G, g,
that is, $\alpha(M)$ is a holomorphic invariant.

Theorem 3.1 [T2]. *Let M be a compact Kähler manifold with $C_1(M) > 0$ as
above. Then M admits a Kähler-Einstein metric whenever $\alpha(M) > n/n + 1$.*

Theorem 3.1 is proved by deriving an a priori C^0-estimate for the solutions
of $(1.1)_t$.

Remark. There is another proof of Theorem 3.1 given by W. Ding later in [Di]. His approach is interesting by itself and was based on an inequality proposed by T. Aubin.

Example 3.1. Let M be a smooth complete intersection in CP^{pq} defined by Fermat polynomials $\sum_{i=0}^{pq} a_{ji}z_i^p = 0$ for $1 \leq j \leq q$, where $[z_0, \cdots, z_{pq}]$ is the homogeneous coordinate of CP^{pq} and $p \geq 2, q \geq 1$ are positive integers. Then by Adjunction Formula (cf. [GH]), one can easily see that $C_1(M)$ is just the restriction of the hyperplane section on CP^{pq}. and the maximal compact subgroup G in $\text{Aut}(M)$ contains all transformations of form $\sigma_j : [z_0, \cdots, z_{pq}] \rightarrow [z_0, \cdots, e(\frac{1}{p})z_j, \cdots, z_{pq}]$ for $1 \leq j \leq pq$. The dimension n of M is $(p-1)q$. One can show (cf. [T3])

$$\alpha(M) \geq 1.$$

In particular, these complete intersections have Kähler-Einstein metrics with positive scalar curvature.

Besides those mentioned above, one can find in [TY2, T3, Na] etc. more examples of Kähler-Einstein manifolds with positive scalar curvature by using this theorem.

The same proof as that for Theorem 3.1 in [T2] also shows

Theorem 3.2 (cf. [TY2]). *Any compact Kähler manifold M with positive first Chern class admits a Kähler metric g_t with Kähler form ω_{g_t} in $C_1(M)$ and Ricci curvature $\geq t$ for any t between $\min\{1, \frac{n+1}{n}\alpha(M)\}$ and 1.*

By Bishop's Volume Comparison Theorem, one can derive from it

Corollary 3.1. *For any compact Kähler manifold M with $C_1(M) > 0$, we have*

$$C_1(M)^n \leq C_n \alpha(M)^{-n}$$

where C_n is a constant depending only on n.

In case $n = 2$, we have the following complete solution of Calabi conjecture.

Theorem 3.3 [T4]. *Let M be a complex surface with positive first Chern class, i.e., a Del-Pezzo surface. Then M admits a Kähler-Einstein metric iff $\eta(M)$ is reductive.*

We will outline the proof of this theorem in the next section. Next we briefly discuss the application of L^2-estimate for $\bar{\partial}$-operators to estimating $\alpha(M)$ from below.

Theorem 3.4 (Theorem 3.1 in [T2]). *Let $\{\phi_i\}$ be a sequence of functions in $P(M, g)$, λ be a positive number. Then there exists a subsequence $\{i_k\}$ of $\{i\}$ and a subvariety S_λ of M with $\dim_C S_\lambda \leq \dim_C M - 1$ such that*
1. $\forall z \in M - S_\lambda, \exists r > 0, C > 0, s.t.$

$$\int_{B_r(z)} e^{-\lambda \phi_{i_k}(w)} dV_g(w) \leq C$$

2. $\forall z \in S_\lambda$,

$$\lim_{k\to\infty} \int_{B_r(z)} e^{-\lambda\phi_{i_k}(w)} dV_g(w) = +\infty \quad \text{for all } r.$$

This theorem can be proved with help of Theorem 5.2.4 in Hörmander's book [Ho]. It implies that the solution φ_t of $(1.1)_t$ should behave well outside a subvariety. Note that the degree of this subvariety can be controlled in terms of the Chern numbers of M. Obviously, $\alpha(M) \geq \lambda_0$ is equivalent to $S_\lambda = \emptyset$ for $\lambda < \lambda_0$. Therefore, it is important to understand S_λ for $\lambda \leq 1$. Since there is no nonconstant holomorphic function on M, one can easily show that each S_λ is connected by the same arguments as in the proof of the Theorem 3.4 (cf. [T2]). In fact, recently, A.Nadel has obtained more information on such a subvariety S_λ for $\lambda < 1$ by Bochner identities for $(0, q)$-forms with $0 < q \leq n$. It can be explained as follows. Denote by I_λ the ideal sheaf of S_λ counting multiplicity, then for each open subset $U \subset M$, $I_\lambda(U)$ consists of those local holomorphic functions f such that $\int_U |f|^2 e^{-\lambda\varphi_{i_k}(w)} dV_g$ are uniformly bounded for all k. On the other hand, one can easily choose the metric g such that $Ric(g) + \frac{\sqrt{-1}}{2\pi}\partial\bar\partial\varphi_{i_k} > 0$ on M. Therefore $h^i(M, I_\lambda) = 0$ for all $i \geq 1$, $\lambda < 1$ by same argument as that in the proof of Kodaira's Vanishing Theorem (cf. [KM, Na]).

For more applications of L^2-estimates to bounding $\alpha(M)$ from below, the readers may refer [T5, TY2].

4. Solution of Calabi's Conjecture for Surfaces

Theorem 3.3 gives the complete solution to Calabi's conjecture for complex surfaces. In this section, we describe briefly the three major steps in the proof. Let M be a complex surface with positive first Chern class. In the first step, we deform the complex structure on M to obtain a smooth family of Kähler surfaces $\{M_t\}_{0 \leq t \leq 1}$ with positive first Chern class such that $M_1 = M$ and M_0 admits a Kähler-Einstein metric g_0. The existence of such a family follows from the main theorem in [TY2] and the classification theory of complex surfaces. Now we define $E = \{t \in [0,1] | M_{t'} \text{ has a Kähler-Einstein metric for any } t' \leq t\}$. Then E is nonempty. It is not hard to prove by the Implicit Function Theorem that E is open in $[0,1]$. Therefore, in order to show that $M = M_1$, has a Kähler-Einstein metric, it suffices to prove that E is closed. We should point out that the closedness is the most difficult part in the solution of Calabi's conjecture. We accomplish this in the following two steps. Without losing generality, we may assume that $E = [0, 1)$. Let g_t $(0 \leq t < 1)$ be the Kähler-Einstein metrics on M_t. We need to show that g_t converges to a metric g_1 on M in C^2-topology. Since $\{M_t\}_{0 \leq t \leq 1}$ is a smooth family of Kähler surfaces with positive first Chern class, there are Kähler metrics $\tilde{g}_t$ on M_t $(0 \leq t \leq 1)$ such that their Kähler forms $\omega_{\tilde{g}_t}$ are in $C_1(M_t) = C_1(M)$. Therefore, for each $t < 1$, $\omega_{g_t} = \omega_{\tilde{g}_t} + \frac{\sqrt{-1}}{2\pi}\partial\bar\partial\varphi_t$ for a unique function φ_t such that

$$\omega_{g_t}^2 = e^{\tilde{f}_t - \varphi_t} \omega_{\tilde{g}_t}^2 \quad \text{on } M_t \tag{4.1}_t$$

where $\tilde{f}_t$ are smooth functions determined by $\tilde{g}_t$ $(0 \leq t \leq 1)$, namely,

$$Ric(\tilde{g}_t) - \omega_{\tilde{g}_t} = \frac{\sqrt{-1}}{2\pi}\partial\bar{\partial}\tilde{f}_t \quad \text{on } M_t$$

and

$$\int_{M_t}(e^{\tilde{f}_t} - 1)\,\omega_{\tilde{g}_t}^2 = 0.$$

Now, in order to have g_t converge to a metric on M, it suffices to derive an a priori C^0-estimate of φ_t ($0 \leq t \leq 1$).

In the second step in the solution of Calabi's conjecture, we give a partial C^0-estimate for φ_t.

Theorem 4.1. *For any positive integer $m \equiv 0$ mod 6, there is a constant C_m independent of φ_t such that there are an orthonormal basis $\{S_i(t)\}_{0 \leq i \leq N_m}$ of $H^0(M, K_M^{-m})$ with respect to the inner product induced by $\tilde{g}_t$ and a sequence of positive numbers $0 < \lambda_0(t) \leq \lambda_1(t) \leq \cdots \leq \lambda_{N_m}(t) = 1$ satisfying:*

$$\left\| \varphi_t - \sup_M \varphi_t - \frac{1}{m}\log\left(\sum_{i=0}^{N_m}\lambda_i(t)\|S_i(t)\|_{\tilde{g}_t}^2\right)\right\|_{C^0(M_t)} \leq C_m \qquad (4.2)_t$$

where $\|\cdot\|_{\tilde{g}_t}$ are the norms of $K_{M_t}^{-m}$ induced by $\tilde{g}_t$.

Remark. It is interesting to know whether or not m has to be the multiple of 6 in $(4.1)_t$.

The estimate in $(4.1)_t$ in particular implies that each $\varphi_t - \sup_{M_t} \varphi_t$ is uniformly bounded in any compact subset outside the zero locus of $S_{N_m}(t)$. That is why $(4.2)_t$ is called a partial C^0-estimate of φ_t. Since there is no known local estimate for the solutions of complex Monge-Ampere equations, such a partial C^0-estimate is very precious. The key observation in the proof of $(4.2)_t$ is: by some computations and the maximum principle, one can show that $(4.2)_t$ is equivalent to

$$\log\left(\sum_{i=1}^{N_m}\|S_i'(t)\|_{g_t}^2\right) \geq C_m' \qquad (4.3)_t$$

where C_m' is a constant independent of t, $\|\cdot\|_{g_t}$ is the hermitian metric on $K_{M_t}^{-m}$ induced by g_t and $\{S_i'(t)\}_{0 \leq t \leq N_m}$ is an orthonormal basis of $H^0(M, K_M^{-m})$ with respect to the inner product induced by g_t. In [T4], $(4.3)_t$ is proved by using Cheeger-Gromov's compactness Theorem, Uhlenbeck's Yang-Mills estimates and L^2-estimates for $\bar{\partial}$-operators.

In the last step of our solution for Calabi's conjecture, we first prove a numerical criterion for the existence of a Kähler-Einstein metric on M. To describe it, let us denote by $P_{G,m,k}(M, \tilde{g}_1)$ the collection of all G-invariant functions of the form $\frac{1}{m}\log(\sum_{i=1}^{k}\|S_i\|_{\tilde{g}_1}^2)$, where $\{S_i\}_{0 \leq t \leq N_m}$ is an orthonormal basis of $H^0(M, K_M^{-m})$ with respect to the metric $\tilde{g}_1$, $0 \leq k \leq N_m$, $N_m = h^0(M, K_M^{-m}) - 1$. Define

$$\alpha_{m,k}(M) = \sup\left\{\alpha | \exists C_\alpha > 0, \text{ s.t. } \int_M e^{-\alpha\varphi}dV_g \leq C_\alpha, \text{ for any } \varphi \text{ in } P_{G,m,k}(M, \tilde{g}_1)\right\}.$$

Theorem 4.2. *Fix an integer m such that $(4.1)_t$ hold for all $t < 1$. If either $\alpha_{m,1}(M) > 2/3$ or $\alpha_{m,2}(M) > 2/3$ and $\alpha_{m,1}(M) > l_0/(l_0 + 1)$, where $l_0 = \max\{0, 2 - \frac{3\alpha_{m,2}(M)-2}{\alpha_{m,2}(M)}\}$, then M admits a Kähler-Einstein metric.*

Remark. In general, if we replace 3 by $n + 1$ and 2 by n in Theorem 4.2 and the partial C^0-estimate in Theorem 4.1 is valid for solutions of $(1.1)_t$, then the above conclusion is still true. It is just Theorem 6.1 in [T5].

In [T4], by direct computation, it is proved that $\alpha_{6,1}(M) \geq 2/3$, $\alpha_{6,2}(M) > 2/3$ for any complex surface M with positive first Chern class. Therefore, the proof for Theorem 3.3 is finished.

The lower bounds of these $\alpha_{m,k}(M)$ seem to be closly related to Mumford's stability of M in CP^{N_m} in the Chow variety. Therefore, it is quite possible that there is a connection between the existence of Kähler-Einstein metrics on M and Mumford's stability of M in the Chow variety.

5. Some Problems Related to Calabi's Conjecture in Higher Dimensions

According to the author's opinion, in order to solve Calabi's conjecture in higher dimensions, one has to understand further the behavior of the solutions of $(1.1)_t$ (cf. §1). The difficulty is the lack of some fundamental estimates for complex Monge-Ampere equations. To get around it, we posed and verified a partial C^0-estimate for the solutions of $(1.1)_1$, on complex surfaces. Such an estimate is crucial in the solution of Calabi's conjecture for complex surfaces. The auther believes such an estimate holds on n-dimensional complex Kähler manifold (M,g) with positive first Chern class. For reader's convenience, we state its stronger version.

Conjecture. *Given $t > 0$ and $\mu > 0$. There are constants $m_0 < m_1$, C, depending only on n, t, μ, satisfying: For any n-dimensional compact Kähler manifold (M,g) with $\mathrm{Ric}(g) \geq t\omega_g$ and $\sum_{i=1}^{2n} |b_i(M)| \leq \mu$, where $b_i(M)$ is the ith betti number of M, there is an integer m in $[m_0, m_1]$ such that*

$$\inf_M \log \left(\sum_{i=0}^N \|S_i\|_g^2 \right) \geq C \qquad (5.1)$$

where $\|\cdot\|_g$ is the hermitian metric on K_M^{-m} with $m\omega_g$ as its curvature form, $\{S_i\}_{1 \leq i \leq N}$ is any orthonormal basis of $H^0(M, K_M^{-m})$ with respect to the inner product induced by g and $\|\cdot\|_g$ and $N = \dim H^0(M, K_M^{-m}) - 1$.

The author is able to affirm this conjecture under two more assumptions: $\mathrm{Ricci}(g) \leq \mu\omega_g$ and $\int_M |\mathrm{Rm}|_g^n dV_g \leq \mu$, where Rm denotes the riemannian curvature tensor of g.

One may try to establish (5.2) by using Hörmander's L^2-estimate for $\bar{\partial}$-operator and constructing local nonvanishing sections of K_M^{-m} for m large, as we did in [T4] for complex surfaces. It boils down to understanding the local structure of a compact Kähler manifold with positive Ricci curvature. Naturally, it leads to the following problem.

Problem. Classify all complete Ricci-flat Kähler manifolds X with the following properties:

1) There is a point $x_0 \in X$ such that the geodesic balls $B_r(x_0)$ have Euclidean volume growth as r goes to infinity.

2) $\lim\limits_{r\to\infty} r^{-2n+4} \int_{B_r(x_0)} \|R(g)\|^2 dV_g = 0$ where $R(g)$ denotes the curvature tensor of g.

In case $n = 2$, (X, g) is clearly the Euclidean space C^2. In case $n \geq 3$, we can obtain such (X, g) as follows. Define $X = X_1 \times \cdots \times X_l$. Each X_i is a resolution of C^{n_i}/Γ_i with $n_i \geq 3$ and $\Gamma_i \subset SU(n_i)$ such that the push-down to $(C^{n_i}\backslash\{0\})/\Gamma_i$ of the standard holomorphic n_i form on C^{n_i} can be nonvanishingly extended across X_i. In [T6], by using the method in [TY3], it is proved that there exists a complete Ricci flat Kähler metric g_i on X_i with Euclidean volume growth and $\int_{X_i} \|R(g_i)\|^{n_i} dV_{g_i} < +\infty$. Take g to be the product metric of g_i ($1 \leq i \leq l$). Then (X, g) satisfies 1) and 2) above.

Are such (X, g) the only ones with 1) and 2)?

6. Degeneration of Kähler-Einstein Manifolds

In order to compactify the moduli space of Kähler-Einstein metrics with positive scalar curvature, we are bound to study Kähler-Einstein orbifold metrics as indicated by the work of M.Anderson [An] and Nakajima [Nk]. Such a metric is defined to be a Kähler orbifold metric on a normal Kähler orbifold such that its Ricci curvature form is a constant multiple of its Kähler form. On one hand, there are some analytic obstructions from holomorphic vector fields to the existence of Kähler-Einstein orbifold metrics analogous to those by Matsushima and Futaki in case of Kähler manifolds, on the other hand, the absence of holomorphic vector field does not assure the existence of a Kähler-Einstein orbifold metric as indicated by the example in § 2 (cf. § 2).

Problem 6.1. Determine when a Kähler orbifold $\tilde{M}$ with $C_1(\tilde{M}) > 0$ has a Kähler-Einstein metric.

It can be regarded as a generalized form of Calabi's conjecture. Of course, The discussions in previous sections, as well as those methods in [T2, TY2], etc., can be applied to this general case with slight modification. However, there are some additional difficulties due to the presence of singularities. For instance, the singularity may cause higher multiplicity of the anticanonical divisor, consequently, the invariant $\alpha(\tilde{M})$ gets smaller (cf. Section 3). Nevertheless, one can produce a lot of Kähler-Einstein orbifolds by means of the methods in [T2], etc. For example, we can prove that there is a Kähler-Einstein orbifold metric on the minimal model of the surface obtained by blowing up CP^2 at 7 or 8 points such that no six of them are on a common quadratic curve and exact three of them are collinear.

The simplest Kähler orbifold with positive first Chern class is the 2-dimensional projective variety with only rational double points as singularities. The minimal resolution of it is an almost Del-Pezzo surface, i.e., the rational surface with numerically positive anticanonical line bundle K^{-1} and $K^2 > 0$. It seems to be even nontrivial to solve Problem 6.1 in this particular case. The methods in [T2, T4] can be used for this purpose.

Not every Kähler-Einstein orbifold can be in the boundary of the moduli space of Kähler-Einstein metrics. Here the distance function on the moduli is defined in terms of some suitable limit, such as Hausdorff limit (cf. [Gr, T4]). In [T4], it is proved that the dimensions of all plurianticanonical divisors are preserved under Hausdorff limit for compact complex surfaces. In fact, the same proof also yields

Theorem 6.1. *Let $\{(M_i, g_i)\}$ be a sequence of Kähler-Einstein manifolds of complex dimension n. Suppose that (M_i, g_i) converge to a Kähler-Einstein orbifold (M_∞, g_∞) in the sense (cf. [Gr]): there are subsets V_i in M_i and V_∞ in M_∞ of real Hausdorff dimension less than $2n-2$ with respect to g_i and g_∞, respectively, such that for any $\varepsilon > 0$, there are diffeomorphisms ϕ_i from $M_\infty - B_{2\varepsilon}(V_\infty, g_\infty)$ into $M_i - B_\varepsilon(V_i, g_i)$ satisfying: (1) The image of ϕ_i contains the exterior of $B_{4\varepsilon}(V_i, g_i)$ for each i; (2) The pull-backs $\phi^* g_i$ converge to g_∞ in C^∞-topology. Then $H^0(M_i, K_{M_i}^{-m})$ converge to $H_0(M_\infty, K_{M_\infty}^{-m})$ for all m in Z. In particular, the dimensions of pluri-anticanonical or canonical sections stay same under the above convergence.*

Problem 6.2. Characterize the Kähler-Einstein orbifold in the boundary of the moduli space of Kähler-Einstein manifolds.

In fact, the author does not know an example of Kähler-Einstein orbifold with positive scalar curvature in the boundary of the moduli space such that its anticanonical ring is different from that of those Kähler-Einstein manifolds in the moduli. It is certainly an interesting question to be explored. We end this paper by a thereom on the degeneration of Kähler-Einstein surfaces (cf. Theorem 7.1 in [T4]).

Theorem 6.2. *Let $\{(M_i, g_i)\}$ be a sequence of Kähler-Einstein surfaces with $C_1(M_i)^2$ equal to $9-n$ ($5 \leq n \leq 8$) and positive first Chern class. By taking a subsequence, we may assume that they converge to a Kähler-Einstein orbifold (M_∞, g_∞) (cf. [An], [Nk]). Then*

    1. *if $n = 8$, then M_∞ has either only rational double points or two singular points of type $C^2/Z_{l,2}$ besides rational points($2 \leq l \leq 7$);*

    2. *if $n = 7$, then M_∞ has either only rational double points or two singular points of type $C^2/Z_{l,2}$ besides rational points;*

    3. *if $n = 5, 6$, M_∞ has at most two singular points of type $C^2/Z_{l,2}$ or $C^2/Z_{l,3}$ ($1 \leq l \leq 3$) besides rational double points. Moreover, in case M_∞ has two of such singular points, one of them must be of type $C^2/Z_{l,2}$, while the other is of type $C^2/Z_{l,3}$.*

Here $Z_{p,q}$ denotes the finite cyclic group in $U(2)$ generated by diag $\left(e(\frac{1}{pq^2}), \right.$ $\left. e(-\frac{1}{pq^2} + \frac{1}{q}) \right)$.

We in fact also have the explicit description of those M_∞ with singular points other than rational double points (cf. §6, §7 in [T4]). In fact, there are only a few of them.

References

[An] Anderson, M.: Ricci curvature bounds and Einstein metrics on compact manifolds. J. Amer. Math. (1989)

[Au] Aubin, T.: Réduction du cas positif de l'equation de Monge-Ampére sur les varietés Kählerinnes compactes á la démonstration dún intégalité. J. Funct. Anal. **57** (1984) 143–153

[Bi] Bishop, R.T., Crittenden, R.J.: Geometry of manifolds. Pure and Applied Math., vol. XV. Academic Press, New York London 1964

[BM] Bando, S., Mabuchi, T.: Uniqueness of Einstein Kähler metrics modulo connected group actions. (Adv. Stud. Pure Math., no. 10.) Kinokuniya, Tokyo 1987

[Ca] Calabi, E.: The space of Kähler metrics. Proc. Int. Congress Math., Amsterdam, vol. 2 (1954)

[CY] Cheng, S.Y., Yau, S.T.: Inequality between Chern numbers of singular Kähler surfaces and characterization of orbit space of discrete group of $SU(2,1)$. Contemp. Math. **49** (1986) 31–43

[Di] Ding, W.: Remarks on the existence problem of positive Kähler-Einstein metrics. Math. Ann. **282** (1988) 463–471

[Fu] Futaki, S.: An obstruction to the existence of Einstein-Kähler metrics. Invent. math. **73** (1983) 437–443

[GH] Griffiths, P., Harris, J.: Principles of algebraic geometry. Wiley, New York 1978

[Gr] Gromov, M., Lafontaine, J., Pansu, P.: Structure metrique pour les varietes Riemanniennes. Cedic/Fernand, Nathen 1981

[Ho] Hörmander, L.: An intrudction to complex analysis in several variables. Van Nostrand, Princeton 1973

[Ko] Kobayashi, R.: Einstein-Kähler metrics on open algebraic surfaces of general type. Tohoku Math. J. **37** (1985) 43–77

[KS] Koiso, N., Sakane, Y.: Non-homogeneous Kähler-Einstein metrics on compact complex manifolds. (Lecture Notes in Mathematics, vol. 1201.) Springer, Berlin Heidelberg New York 1986

[Ma1] Matsushima, Y.: Sur la structure du group d'homeomorphismes analytiques d'une certaine varietie Kaehlerinne. Nagoya Math. J. **11** (1957) 145–150

[Ma2] Matsushima, Y.: Sur les espaces homogériens Kählériens d'un group réductif. Nagoya Math. J. **11** (1957) 53–60

[Na] Nadel, A.: Multiplier ideal sheaves and existence of Kähler-Einstein metrics of positive scalar curvature. Proc. Natl. Acad. Sci. USA, vol. 86, no. 19 (1989)

[Nk] Nakajima, H.: Hausdorff convergence of Einstein 4-manifolds. J. Fac. Sci. Univ. Tokyo **35** (1988) 411–424

[Si] Siu, Y.T.: The existence of Kähler-Einstein metrics on manifolds with positive anticanonical line bundle and a suitable finite symmetry group. Ann. Math. (1988)

[T1] Tian, G.: Some Notes on Kähler-Einstein metrics with positive scalar curvature. Preprint

[T2] Tian, G.: On Kähler-Einstein metrics on certain Kähler manifolds with $C_1(M) > 0$. Invent. math. **89** (1987) 225–246

[T3] Tian, G.: Kähler metrics on algebraic manifolds. Thesis, Harvard University

[T4] Tian, G.: On Calabi's conjecture for complex surfaces with positive first Chern class. Invent. math. **101**, no. 1 (1990) 101–172

[T5] Tian, G.: On one of Calabi's problem. Proc. of AMS Summer Institute in Santa Cruz, 1989 (to appear)

[T6] Tian, G.: Complete Ricci-flat Kähler metrics on the resolution of the quotients of C^n by finite groups. In preparation

[Ts]   Tsuji, H.: An inequality of Chern numbers for open varieties. Math. Ann. **277**, no. 3 (1987) 483–487

[TY1]   Tian, G., Yau, S.T.: Existence of Kähler-Einstein metrics on complete Kähler manifolds and their applications to algebraic geometry. Math aspects of string theory (San Diego, California, 1986), pp. 574–628, Adv. Ser. Math. Phys., 1. World Sci. Publishing, Singapore 1987

[TY2]   Tian, G., Yau, S.T.: Kähler-Einstein metrics on complex surfaces with $C_1(M)$ positive. Comm. Math. Phys. **112** (1987)

[TY3]   Tian, G., Yau, S.T.: Complete Kähler manifolds with zero Ricci curvature, II. Submitted to Invent. math.

[Y1]   Yau, S.T.: On the Ricci curvature of a compact Kähler manifold and the complex Monge-Ampére equation, I^*. Comm. Pure Appl. Math. **31** (1978) 339–441

[Y2]   Yau, S.T.: On Calabi's conjecture and some new results in algebraic geometry. Proc. Nat. Acad. Sci. USA **74** (1977) 1798–1799

Ensembles Limites et Applications

Francis Bonahon

Department of Mathematics, University of Southern California
Los Angeles, CA 90089-1113, USA

Nous nous proposons de faire le point sur quelques progrès récents accomplis dans l'étude des ensembles limites des variétés à courbure négative, et surtout de montrer comment on peut utiliser les ensembles limites comme moyen d'attaque pour d'autres problèmes de géométrie et de topologie.

Les travaux de l'auteur ont été partiellement subventionnés par la N.S.F. et la Fondation A.P. Sloan. Cet article a été écrit en grande partie pendant que l'auteur visitait l'Université de Californie à Davis, qu'il souhaite vivement remercier de son hospitalité.

1. Ensembles Limites des Variétés de Courbure Négative

Considérons une variété riemannienne complète M à courbure majorée par une constante strictement négative. Dans son revêtement universel $\tilde{M}$, les rayons géodésiques issus d'un point base $\tilde{x}_0$ balaient $\tilde{M}$ et, puisque la courbure est négative, ne se recoupent pas deux à deux. On peut ainsi compactifier $\tilde{M}$ par sa *sphère à l'infini* $\tilde{M}_\infty$, en adjoignant un point au bout de chacun de ces rayons géodésiques. L'espace $\tilde{M} \cup \tilde{M}_\infty$ ainsi défini est homéomorphe à une boule fermée.

L'hypothèse sur la courbure permet de rendre cette construction indépendante du choix du point base $\tilde{x}_0 \in \tilde{M}$ (voir par exemple [EbON]). En effet, pour chaque rayon géodésique issu d'un autre point base $\tilde{x}_0'$, il existe un unique rayon géodésique issu de $\tilde{x}_0$ qui lui est asymptote. Ceci fournit une identification naturelle entre les rayons géodésiques issus de $\tilde{x}_0$ et ceux issus de $\tilde{x}_0'$, et montre que l'espace $\tilde{M} \cup \tilde{M}_\infty$ est en fait indépendant du choix du point base $\tilde{x}_0$. En particulier, l'action du groupe fondamental $\pi_1(M)$ sur $\tilde{M}$ s'étend continûment à $\tilde{M} \cup \tilde{M}_\infty$.

Un cas important est celui où la métrique de M a courbure constante -1. Le revêtement universel $\tilde{M}$ est alors isométrique à l'espace hyperbolique $\mathbb{H}^n$, défini comme la boule ouverte unité de $\mathbb{R}^n$ munie de la métrique qui est $2/(1 - \|x\|^2)$ fois la métrique euclidienne au point x. La compactification de $\mathbb{H}^n$ par sa sphère à l'infini $\mathbb{H}^n_\infty$ est simplement la compactification usuelle de la boule unité par la sphère unité dans $\mathbb{R}^n$. Dans ce cas, $\tilde{M} \cup \tilde{M}_\infty$ a une structure particulièrement riche. Par exemple, $\tilde{M} \cup \tilde{M}_\infty$ admet une structure différentiable qui est respectée par les isométries de $\tilde{M}$. Ceci n'est en général plus vrai si la courbure n'est pas constante (voir [Gh] et les références qui y sont mentionnées).

Par un argument simple, toutes les orbites de l'action de $\pi_1(M)$ sur $\tilde{M}$ ont le même ensemble Λ_M de points d'accumulation dans $\tilde{M} \cup \tilde{M}_\infty$. Ce sous-ensemble fermé de $\tilde{M}_\infty$ est l'*ensemble limite* de M. Pour une définition plus "physique", imaginons un observateur situé en un point x_0 de M. La sphère visuelle de cet observateur, formée de tous les rayons géodésiques issus de x_0, s'identifie à l'ensemble de tous les rayons géodésiques issus d'un point $\tilde{x}_0 \in \tilde{M}$ de la préimage de x_0, et donc à la sphère à l'infini $\tilde{M}_\infty$. Si l'on suppose que la lumière se propage le long de géodésiques, notre observateur (ponctuel) voit de nombreuses copies de lui-même, correspondant aux arcs géodésiques de M joignant x_0 à lui-même. L'ensemble limite est alors formé des points d'accumulation de ces images de l'observateur dans sa sphère de directions.

Quand M est compacte, l'ensemble limite Λ_M est la sphère à l'infini $\tilde{M}_\infty$ toute entière. Mais sinon, Λ_M peut avoir une structure particulièrement complexe dans $\tilde{M}_\infty$. Par exemple, considérons le cas classique où S est une surface fermée et où $M = S \times \mathbb{R}$ est munie d'une métrique de courbure constante -1 pour laquelle $S \times [-1, +1]$ est à bord convexe. Alors, l'ensemble limite Λ_M dans $\tilde{M}_\infty = \mathbb{H}^3_\infty = S^2 \subset \mathbb{R}^3$ est une courbe de Jordan qui, ou bien est un cercle, ou bien n'est nulle part différentiable et est de dimension de Hausdorff strictement comprise entre 1 et 2 (voir [Bow, Su1]). A ce propos, on peut signaler une caractéristique non négligeable des ensembles limites: ils donnent lieu à de jolis dessins (voir par exemple [Th2] pour quelques specimens). En effet, l'action de $\pi_1(M)$ sur la sphère à l'infini $\tilde{M}_\infty$ respecte l'ensemble limite Λ_M, et lui confère ainsi de multiples symétries fractales.

2. Propriétés de Mesure des Ensembles Limites

Pour simplifier, limitons-nous au cas où la métrique de M a courbure constante, par exemple égale à -1. Nous avons déjà indiqué que son revêtement universel $\tilde{M}$ est alors isométrique à l'espace hyperbolique $\mathbb{H}^n$, ce qui identifie la sphère à l'infini $\tilde{M}_\infty$ à la sphère unité $S^{n-1} = \mathbb{H}^n_\infty$ de $\mathbb{R}^n$. Cette identification $\tilde{M}_\infty \cong S^{n-1}$ est bien définie modulo composition par isométrie de $\mathbb{H}^n$, c'est à dire modulo composition par un élément du groupe de transformations de $\mathbb{R}^n \cup \infty$ engendré par les inversions par rapport aux sphères (rondes) de dimension $n-1$ qui sont orthogonales à S^{n-1}. Il s'ensuit facilement dans ce cas que l'on peut alors parler de propriétés de négligeabilité pour la mesure de Lebesgue, ainsi que de dimension de Hausdorff, pour l'ensemble limite Λ_M dans $\tilde{M}_\infty$.

L'un des premiers résultats dans cette direction est dû à L.V. Ahlfors [Ah2] qui a montré en dimension inférieure ou égale à 3 que, sous l'hypothèse que l'action de $\pi_1(M)$ sur $\tilde{M}$ admet un polyhèdre fondamental borné par un nombre fini de faces totalement géodésiques (on dit alors que M est *géométriquement finie*), l'ensemble limite Λ_M est ou bien de mesure de Lebesgue nulle dans la sphère à l'infini $\tilde{M}_\infty$ ou bien égal à $\tilde{M}_\infty$ tout entière. Ahlfors émit aussi la conjecture que cette propriété est vraie sous une hypothèse beaucoup plus faible [Ah1]:

Conjecture. Si M est une variété de dimension 3 à courbure négative constante dont le groupe fondamental $\pi_1(M)$ est de type fini, alors l'ensemble limite Λ_M est ou bien de mesure de Lebesgue nulle dans la sphère à l'infini $\tilde{M}_\infty$ ou bien égal à $\tilde{M}_\infty$ tout entière.

En dimension 2, une surface à courbure négative constante dont le groupe fondamental est de type fini est toujours géométriquement finie.

Ces résultats d'Ahlfors ont depuis été améliorés par D. Sullivan dans [Su3], où il montre que la dimension de Hausdorff de l'ensemble limite Λ_M d'une variété géométriquement finie M est strictement inférieure à la dimension de la sphère à l'infini $\tilde{M}_\infty$ quand Λ_M n'est pas $\tilde{M}_\infty$ tout entière. On pourra également considérer [Su1] pour une extension de ces résultats en dimension supérieure, ainsi que pour une relation entre la dimension de Hausdorff de Λ_M et le spectre du Laplacien sur M. D'un autre côté, Sullivan a aussi exhibé dans [Su2] des variétés (géométriquement infinies) M de dimension 3 pour lesquelles Λ_M est de dimension de Hausdorff 2 sans être égal à $\tilde{M}_\infty$, et pour lesquelles $\pi_1(M)$ est de type fini.

De gros progrès vers une solution complète de la conjecture d'Ahlfors ont été accomplis depuis une quinzaine d'années, grâce à un programme lancé vers le milieu des années 1970 par W.P. Thurston. En effet, Thurston a alors proposé un modèle conjecturel pour la géométrie des variétés de dimension 3 à courbure négative constante et dont le groupe fondamental est de type fini. Selon cette conjecture, une telle variété M est nécessairement difféomorphe à l'intérieur d'une variété compacte, de sorte que ses bouts sont topologiquement de la forme $S \times [0, \infty[$ où S est une surface compacte. De plus, pour un tel bout difféomorphe à $S \times [0, \infty[$, la surface S peut être décomposée en un nombre fini de sous-surfaces S_i telles que la géométrie de la métrique de M sur chacun de ces $S_i \times [0, \infty[$ est de deux types possibles: Le premier type est analogue à celui que l'on observe pour tous les bouts d'une variété géométriquement finie, et est tel que la métrique sur les sections $S_i \times t$ croît exponentiellement avec t. Le second type se comporte plutôt comme un tube de section bornée, et les tranches $S_i \times t$ avec t entier ont une aire bornée.

Par une généralisation d'un argument de L.V. Ahlfors pour les variétés géométriquement finies [Ah2], Thurston avait observé que toute variété satisfaisant cette conjecture satisfait aussi la conjecture d'Ahlfors, en ce sens que l'ensemble limite Λ_M dans la sphère à l'infini $\tilde{M}_\infty$ est ou bien de mesure de Lebesgue nulle ou bien égal à $\tilde{M}_\infty$.

Cette conjecture de Thurston est maintenant démontrée par les travaux successifs de Thurston [Th1], l'auteur [Bo1], et R. Canary [Ca] si l'on sait déjà que M est topologiquement sage, c'est à dire que M est difféomorphe à l'intérieur d'une variété compacte. Les résultats de [Bo1], améliorant ceux de [Th1], ne nécessitent pas l'hypothèse de sagesse topologique sur M (et l'entraînent en fait comme corollaire), mais réclament une condition supplémentaire sur le groupe fondamental $\pi_1(M)$, par exemple que celui-ci ne se décompose pas en un produit libre non-trivial. Les résultats récents de [Ca], qui requièrent la sagesse topologique

mais pas d'hypothèse sur le groupe fondamental, utilisent [Bo1] et [Th1] et sont
basés sur un astucieux argument de revêtement ramifié.

3. Topologie des Ensembles Limites

Que peut-on dire de l'espace topologique Λ_M, muni de son action de $\pi_1(M)$?

Un point de départ est le cas où M a un cœur convexe compact. Un *cœur
convexe* pour M est une sous-variété M_c à bord convexe telle que $M - \text{int}(M_c)$
est difféomorphe à $\partial M_c \times [0, \infty[$. En particulier, toute variété compacte M admet
un cœur convexe compact $M_c = M$.

Quand M admet un cœur convexe compact, il est alors connu depuis
longtemps que l'espace topologique Λ_M et son action de $\pi_1(M)$ peuvent être
décrits uniquement en termes de la structure algébrique du groupe $\pi_1(M)$ (voir
[Mo, Ni, Fl, Gr]). En effet, identifions $\pi_1(M)$ à l'orbite d'un point base $\tilde{x}_0 \in \tilde{M}$
situé au-dessus du cœur convexe M_c, ainsi qu'à l'ensemble des sommets du graphe
de Cayley de $\pi_1(M)$ associé à un choix de générateurs. Par convexité et com-
pacité de M_c, la fonction distance induite par la métrique de M sur cette orbite
est équivalente à celle provenant du graphe de Cayley. Par ailleurs, l'ensemble
limite Λ_M est l'espace des suites de l'orbite de $\tilde{x}_0$, où l'on identifie deux telles
suites quand elles sont asymptotes, c'est à dire quand l'angle sous lequel elles sont
vues depuis $\tilde{x}_0$ converge vers 0. Les références mentionnées plus haut développent
plusieurs méthodes pour traduire cette relation d'asymptoticité en termes de la
métrique induite sur l'orbite par la métrique de M. Ceci permet de décrire Λ_M de
manière purement combinatoire, en termes d'une certaine relation d'asymptoticité
sur les suites de sommets du graphe de Cayley de $\pi_1(M)$.

Par exemple, comme on l'a vu au Chap. 1 dans le cas d'une surface, si
M est une variété fermée à courbure négative, l'ensemble limite de n'importe
quelle métrique de courbure négative à cœur convexe compact sur $M \times \mathbb{R}^n$ est
homéomorphe à l'ensemble limite de M, et donc à la sphère unité de l'espace
euclidien de même dimension que M.

Quand la métrique de M n'admet pas de cœur convexe compact, on conjecture
que l'ensemble limite Λ_M est un certain quotient de l'*ensemble limite algébrique*
Λ_M^a associé au groupe $\pi_1(M)$ par la recette qui produirait l'ensemble limite si M
avait un cœur convexe compact. Cette conjecture est notamment vérifiée quand la
métrique de M est géométriquement finie (voir [Fl, Tu]), auquel cas Λ_M est obtenu
à partir de Λ_M^a en pinçant certaines paires de points associées aux sous-groupes
de $\pi_1(M)$ correspondant aux cusps de M.

Quand M n'est pas géométriquement finie, cette conjecture n'est guère
vérifiée que dans des cas très particuliers, dûs à J.W. Cannon et W.P. Thurston
[CaTh]. Dans certains de leurs exemples, l'ensemble limite algébrique Λ_M^a est
homéomorphe au cercle tandis que l'ensemble limite Λ_M est une sphère de dimension 2, ce qui fait que la projection $\Lambda_M^a \to \Lambda_M$ fournit une courbe de Peano
recouvrant toute la sphère!

4. Utilisations des Ensembles Limites

Après avoir parlé de la structure des ensembles limites, nous nous proposons maintenant de montrer comment on peut utiliser ceux-ci pour attaquer d'autres problèmes de toplogie et de géométrie. L'idée générale est, pour analyser des objets qui ne sont définis que modulo déformation, de considérer des structures sur l'ensemble limite qui sont associées à ces objets et qui sont définies sans aucune relation d'équivalence particulière.

Pour éviter trop de généralité, concentrons-nous sur un exemple. Dans [Bo1], on est amené à considérer l'espace Γ des classes d'homotopie libre de courbes fermées sur une surface fermée M de courbure négative, et on se trouve confronté au problème suivant: définir et étudier des directions asymptotiques dans Γ vis-à-vis de la forme d'intersection géométrique $i : \Gamma \times \Gamma \to \mathbb{N}$, où le *nombre d'intersection géométrique* $i(\alpha, \beta)$ de α et $\beta \in \Gamma$ est le minimum du nombre de points d'intersection (avec multiplicités) de courbes a et b représentant respectivement α et β.

Pour une approche formelle de ce problème, on peut considérer l'espace $\mathbb{R}_+^\Gamma$ des applications de Γ dans $\mathbb{R}_+$, muni de la topologie produit, et plonger Γ dans $\mathbb{R}_+^\Gamma$ en envoyant $\gamma \in \Gamma$ sur l'application $\alpha \mapsto i(\alpha, \gamma)$. Les directions asymptotiques que l'on veut étudier correspondent alors aux demi-droites de $\mathbb{R}_+^\Gamma$ qui sont dans l'adhérence des demi-droites $\mathbb{R}_+\gamma \subset \mathbb{R}_+^\Gamma$, $\gamma \in \Gamma$.

Cette approche formelle n'est malheureusement pas commode à manier. On peut toutefois raisonner de manière plus géométrique, de la façon suivante.

Le point de départ est que, à cause de l'hypothèse de courbure négative, chaque courbe fermée de M est homotope à un unique multiple d'une géodésique fermée. (Par convention, une géodésique fermée ne tourne pas plusieurs fois autour d'une autre géodésique, et représente donc un élément indivisible du groupe fondamental). Ceci établit une bijection entre les éléments de Γ et les géodésiques fermées de M munies d'un poids entier positif.

Relevons la situation dans le revêtement universel $\tilde{M}$. La préimage d'une géodésique fermée de M donne une famille de géodésiques de $\tilde{M}$, invariante par l'action du groupe fondamental $\pi_1(M)$. A chaque élément de Γ, on a ainsi associé un poids entier positif et un sous-ensemble $\pi_1(M)$-invariant de l'espace $G(\tilde{M})$ des géodésiques de $\tilde{M}$. On vérifie que ce sous-ensemble est fermé et discret. Maintenant, pour analyser un sous-ensemble fermé discret A muni d'un poids positif, un moyen naturel est de considérer la mesure de Dirac ainsi définie, pour laquelle la masse d'un sous-ensemble est le produit du poids par le nombre d'éléments de A contenus dans ce sous-ensemble. On associe ainsi à chaque élément de Γ une mesure de Dirac $\pi_1(M)$-invariante sur $G(\tilde{M})$.

Nous venons ainsi d'interpréter les éléments de Γ comme éléments de l'espace $\mathscr{C}(M)$ des mesures (de Radon positives) sur $G(\tilde{M})$ qui sont invariantes par l'action de $\pi_1(M)$. Pour des raisons expliquées dans [Bo1], les éléments de $\mathscr{C}(M)$ sont appelés *courants géodésiques*.

Munissons $\mathscr{C}(M)$ de la *topologie vague*, pour laquelle une suite de mesures α_n converge vers α si et seulement si les intégrales $\alpha_n(f)$ convergent vers $\alpha(f)$ pour toute fonction $f : G(M) \to \mathbb{R}$ à support compact. Notons que l'on peut additionner deux éléments de $\mathscr{C}(M)$, et multiplier un élément de $\mathscr{C}(M)$ par un nombre réel positif. Le résultat suivant est alors démontré dans [Bo3].

Proposition. *Le sous-espace* $\mathbb{R}_+\Gamma \subset \mathscr{C}(M)$, *formé des éléments de la forme* $\lambda\gamma$ *où* $\lambda \in \mathbb{R}_+$ *et* $\gamma \in \Gamma \subset \mathscr{C}(M)$, *est dense dans* $\mathscr{C}(M)$.

On peut donc considérer les courants géodésiques sur M comme des "classes d'homotopie diffusées".

Pour le moment, cette construction semble dépendre du choix d'une métrique à courbure négative sur M. Toutefois, chaque géodésique de $\tilde{M}$ est caractérisée par ses deux points limites distincts dans la sphère à l'infini $\tilde{M}_\infty$, laquelle est aussi égale à l'ensemble limite Λ_M puisque M est compacte. Par conséquent, $G\left(\tilde{M}\right)$ s'identifie à $\Lambda_M \times \Lambda_M - \Delta$, où Δ est la diagonale du produit. Ainsi qu'on l'a vu au Chap. 3, l'ensemble limite Λ_M peut être décrit en termes de la structure algébrique du groupe fondamental $\pi_1(M)$, et l'espace $\mathscr{C}(M)$ est par conséquent indépendant de la métrique de M et dépend uniquement du groupe $\pi_1(M)$. En fait, toute cette construction peut être developpée dans le cadre plus général des groupes à courbure négative (aussi appelés hyperboliques) au sens de Gromov [Gr]; ce point de vue est exposé dans [Bo3].

Notons que, pour la surface compacte M, l'espace $G\left(\tilde{M}\right)$ est particulièrement simple puisqu'il est homéomorphe à un anneau ouvert. Toutefois, l'action de $\pi_1(M)$ sur cet anneau est nettement plus complexe.

Revenons maintenant à la forme d'intersection géométrique $i : \Gamma \times \Gamma \to \mathbb{N}$. Dans [Bo1], on montre:

Proposition. *La forme d'intersection géométrique* $i : \Gamma \times \Gamma \to \mathbb{N}$ *s'étend en une (unique) application bilinéaire continue* $i : \mathscr{C}(M) \times \mathscr{C}(M) \to \mathbb{R}_+$.

En particulier, il existe une application continue $\mathscr{C}(M) \to \mathbb{R}_+^\Gamma$ qui associe $\gamma \mapsto i(\alpha, \gamma)$ à $\alpha \in \mathscr{C}(M)$, et on vérifie que cette application est propre. Un résultat de J.-P. Otal [Ot] affirme que, réciproquement, un courant géodésique α est complètement déterminé par ses nombres d'intersection avec les éléments de Γ (c'est non-trivial même dans le cas apparemment plus simple où α est un élément de Γ). Par densité des multiples réels d'éléments de Γ dans $\mathscr{C}(M)$, on en déduit:

Théorème. *L'application* $\mathscr{C}(M) \to \mathbb{R}_+^\Gamma$ *qui associe* $\gamma \mapsto i(\alpha, \gamma)$ *à* $\alpha \in \mathscr{C}(M)$ *induit un homéomorphisme de* $\mathscr{C}(M)$ *sur l'adhérence de* $\mathbb{R}_+\Gamma$ *dans* $\mathbb{R}_+^\Gamma$.

Ainsi, les directions asymptotiques de Γ que l'on voulait étudier au début de cette section correspondent exactement au demi-droites issues de l'origine dans $\mathscr{C}(M)$.

La définition de $\mathscr{C}(M)$ a été inspirée par une construction antérieure, due à W.P. Thurston [Th1, Th3], de "classes d'homotopie de courbes simples diffusées" sur les surfaces appelées *laminations géodésiques mesurées*. En fait, l'espace $\mathscr{ML}(M)$ des laminations géodésiques mesurées se plonge de manière naturelle dans $\mathscr{C}(M)$. De plus, de la même façon qu'un élément $\gamma \in \Gamma$ est représenté par une courbe simple si et seulement si son nombre d'auto-intersection géométrique $i(\gamma, \gamma)$ est égal à 0, $\mathcal{ML}(M)$ est exactement formé des $\alpha \in \mathcal{C}(M)$ tels que $i(\alpha, \alpha) = 0$ (voir [Bo1, Bo2]).

Les courants géodésiques peuvent également être utilisés pour étudier l'espace $\mathcal{M}(M)$ de toutes les classes d'isotopie de métriques à courbure négative sur une surface compacte M. Une difficulté ici est que ces métriques ne sont définies que modulo isotopie, ce qui fournit trop de degrés de liberté. La considération de l'ensemble limite Λ_M permet de "geler" ces isotopies

En effet, si m est une métrique à courbure négative sur la surface M, nous avons vu plus haut que l'espace topologique $G(\widetilde{M}) = \Lambda_M \times \Lambda_M - \Delta$ muni de son action de $\pi_1(M)$ est indépendant de la métrique m. Toutefois, m dépose une structure supplémentaire sur cet espace. En effet, $G(\widetilde{M})$ peut aussi être considéré comme l'espace des orbites du flot géodésique de la métrique m sur $\widetilde{M}$, agissant sur le fibré tangent unitaire $T_1(\widetilde{M})$. Ce flot géodésique respecte la métrique induite par m sur les orbites du flot et, par un résultat classique de Liouville, respecte aussi la forme volume de $T_1(\widetilde{M})$. Il s'ensuit que l'on peut factoriser la mesure volume sur $T_1(\widetilde{M})$ en le produit d'une mesure L_m sur $G(\widetilde{M})$ et de la mesure induite par m le long des orbites du flot. Cette mesure L_m est la *mesure de Liouville* associée à la métrique m. Elle est clairement invariante par l'action de $\pi_1(M)$ et fournit donc un élément de $\mathcal{C}(M)$ associé à la métrique m.

Le nombre d'intersection géométrique des courbes avec cette mesure de Liouville a une interprétation intéressante:

Proposition. *Le nombre d'intersection géométrique $i(\gamma, L_m)$ de $\gamma \in \Gamma$ avec L_m est égal à 4 fois la longueur du multiple de m-géodésique fermée représentant la classe d'homotopie γ.*

On vérifie que la mesure L_m sur $G(\widetilde{M})$ dépend uniquement de la classe d'isotopie de m. Un résultat récent de J.-P. Otal [Ot] affirme que, réciproquement, on peut reconstruire la métrique m à partir de sa mesure de Liouville L_m, modulo isotopie, en ce sens que:

Théorème. *L'application $\mathcal{M}(M) \to \mathcal{C}(M)$ qui à la métrique m associe sa mesure de Liouville L_m est injective, et est un homéomorphisme sur son image.*

Il serait intéressant d'avoir une caractérisation de cette image (voir [Bo2] pour une caractérisation des mesures de Liouville associées aux métriques de courbure constante).

On a ainsi transformé le problème de l'étude de classes d'isotopies de métriques en l'étude de mesures sur un espace donné.

Un sous-espace important de $\mathcal{M}(M)$ est l'*espace de Teichmüller* $\mathcal{T}(M)$ de la surface M, formé des métriques de courbure constante -1. Cet espace est aussi l'espace des classes d'isotopie de structures conformes sur M, et joue par conséquent un important rôle en analyse complexe.

Dans [Th1, Th3], Thurston a introduit une compactification de l'espace de Teichmüller $\mathscr{T}(M)$ en lui adjoignant à l'infini le projectivisé de l'espace $\mathscr{ML}(M)$ des laminations mesurées. L'espace $\mathscr{C}(M)$, qui contient les deux espaces $\mathscr{T}(M)$ et $\mathscr{ML}(M)$, fournit un cadre pour cette construction qui est sans doute plus naturel que celui originellement utilisé par Thurston. En effet, on démontre dans [Bo2] le résultat suivant :

Théorème. *L'union des demi-droites issues de l'origine dans $\mathscr{C}(M)$ qui sont asymptotes à $\mathscr{T}(M) \subset \mathscr{C}(M)$ est exactement égale à $\mathscr{ML}(M)$. La compactification de $\mathscr{T}(M)$ par l'espace de ces demi-droites que l'on obtient de cette façon est exactement la compactification de Thurston.*

De plus, par des travaux de W.P. Thurston et S. Wolpert [Wo], la restriction à $\mathscr{T}(M) \subset \mathscr{C}(M)$ de la forme d'intersection géométrique i se trouve reliée à la forme de Weil-Petersson sur $\mathscr{T}(M)$; voir [Bo2].

L'idée motivant ce programme est que, si l'on convertit des espaces d'objets définis seulement modulo isotopie en des espaces de mesures, on peut utiliser des méthodes d'analyse pour démontrer des propriétés de continuité et développer un calcul différentiel sur ces espaces. Par example, la continuité de la forme d'intersection sur $\mathscr{C}(M)$ fournit immédiatement des propriétés de continuité intéressantes sur la compactification de Thurston de $\mathscr{T}(M)$.

Le développement d'un calcul différentiel se heurte toutefois à une difficulté : En analyse classique, un vecteur tangent à un espace de mesures est une distribution. Mais, comme on l'a indiqué plus haut, la sphère à l'infini $\tilde{M}_\infty = \Lambda_M$ et donc l'espace $G(\tilde{M})$ ne possèdent pas de structure différentiable privilégiée. On ne peut donc pas parler de distributions sur $G(\tilde{M})$ de manière intrinsèque.

On résoud cette difficulté en observant que Λ_M possède une structure Hölder privilégiée, c'est à dire une fonction distance bien définie modulo la relation d'équivalence qui identifie deux distances d_1 et d_2 quand il existe des constantes $A > 0$ et $v > 0$ telles que $A^{-1} d_1(x,y)^{1/v} \leq d_2(x,y) \leq A d_1(x,y)^v$ pour tous x, y (voir par exemple [Gr]). On peut ainsi parler de fonctions Hölder $G(\tilde{M}) \to \mathbb{R}$. Définissons une *distribution Hölder* sur $G(\tilde{M})$ comme une forme linéaire continue sur l'espace des fonctions Hölder $G(\tilde{M}) \to \mathbb{R}$ à support compact. Ainsi, si l'on choisit sur $G(\tilde{M})$ une structure différentiable compatible avec la structure Hölder, une distribution Hölder est une distribution au sens usuel avec une forte propriété de régularité. L'espace $\mathscr{H}(M)$ des distributions Hölder sur $G(\tilde{M})$ qui sont invariantes par l'action de $\pi_1(M)$ semble être le bon cadre pour developper un calcul différentiel du type cherché.

Par exemple, considérons l'espace des laminations géodésiques mesurées $\mathscr{ML}(M) \subset \mathscr{C}(M) \subset \mathscr{H}(M)$. Thurston a construit une structure linéaire par morceaux naturelle sur $\mathscr{ML}(M)$, ce qui fournit une notion combinatoire de vecteur tangent à $\mathscr{ML}(M)$. Dans [Bo4], on établit une correspondance entre ces vecteurs tangents et les vecteurs de $\mathscr{H}(M)$ qui sont tangents à son sous-espace $\mathscr{ML}(M)$. Ceci fournit une description beaucoup plus géométrique des vecteurs tangents de $\mathscr{ML}(M)$, interprétant ceux-ci comme laminations géodésiques munies de distributions Hölder transverses invariantes. Réciproquement, il existe une classification combinatoire simple des distributions Hölder transverses invariantes dont on peut munir une lamination géodésique donnée, ainsi qu'un critère simple pour déterminer celles qui proviennent de vecteurs tangents à $\mathcal{ML}(M)$. Par ailleurs, un grand nombre d'objets définis sur $\mathcal{ML}(M)$ ont une extension naturelle au cadre Hölder. En particulier, ceci permet de calculer les dérivées de certaines fonctions longueur sur $\mathcal{ML}(M)$.

De manière similaire, l'espace de Teichmüller $\mathcal{T}(M) \subset \mathcal{C}(M) \subset \mathcal{H}(M)$ admet en chacun de ses points un espace linéaire tangent dans $\mathcal{H}(M)$, isomorphe à l'espace tangent défini par la structure différentiable naturelle de $\mathcal{T}(M)$.

Références

[Ah1]   Ahlfors, L.V.: Finetely generated Kleinian groups. Amer. J. Math. **86** (1964) 412–423

[Ah2]   Ahlfors, L.V.: Fundamental polyhedrons and limit sets of Kleinian groups. Proc. Nat. Acad. Sci. USA **55** (1966) 251–254

[Bo1]   Bonahon, F.: Bouts des variétés hyperboliques de dimension 3. Ann. Math. **124** (1986) 71–158

[Bo2]   Bonahon, F.: The geometry of Teichmüller space via geodesic currents. Invent. math. **92** (1988) 139–162

[Bo3]   Bonahon, F.: Geodesic currents on hyperbolic groups. In: Alperin, R. (ed.) Arboreal group theory. (Mathematical Sciences Research Institute Publications, vol. 19). Springer, Berlin Heidelberg New York 1991

[Bo4]   Bonahon, F.: Geodesic laminations with transverse distributions. Prépublication, 1990

[Bow]   Bowen, R.: Hausdorff dimension of quasi-circles. Publ. Math. IHES **50** (1979) 11–25

[Ca]   Canary, R.: Ends of Hyperbolic 3-manifolds. Prépublication. Stanford University, 1990

[CaTh] Cannon, J., Thurston, W.P.: Group invariant Peano curves. Prépublication, Brigham Young University and Princeton University, 1984–90

[EbON] Eberlein, P., O'Neill, B.: Visibility manifolds. Pac. J. Math. **46** (1973) 45–109

[Fl]   Floyd, W.J.: Group completions and limit sets of Kleinian groups. Invent. math. **57** (1980) 205–218

[Gh]   Ghys, E.: Le cercle à l'infini des surfaces de courbure négative. Ces Comptes-Rendus, p. 501

[Gr]   Gromov, M.: Hyperbolic groups. In: Gersten, S.M. (ed.) Essays in group theory. Springer, Berlin Heidelberg New York, 1987, pp. 75–263

[Mo]   Morse, H.M.: A one-to-one representation of geodesics on a surface of negative curvature. Amer. J. Math. **43** (1921) 33–51

[Ni]   Nielsen, J.: Untersuchungen zur Topologie der geschlossenen zweiseitigen Flächen I. Acta Math. **50** (1927) 189–358; II. Acta Math. **53** (1929) 1–76; III. Acta Math. **58** (1932) 87–167

[Ot]   Otal, J.-P.: Le spectre marqué des longueurs des surfaces à courbure négative. Ann. Math. **131** (1990) 151–162

[Su1]   Sullivan, D.: The density at infinity of a discrete group of hyperbolic motions. Publ. Math. IHES **50** (1979) 171–202

[Su2]   Sullivan, D.: Growth of positive harmonic functions and Kleinian group limit sets of zero planar measure and Hausdorff dimension 2. In: Looijenga, E., Siersma, D., Takens, F. (eds.) Geometry Symposium, Utrecht 1980 (Lecture Notes in Mathematics, vol. 894). Springer, New York Heidelberg Berlin 1981, pp. 127–144

[Su3]   Sullivan, D.: Hausdorff measures old and new, and limit sets of geometrically finite Kleinian groups. Acta Math. **153** (1984) 259–277

[Th1]   Thurston, W.P.: The topology and geometry of 3-manifolds. Notes de cours, Princeton University, 1976–79

[Th2]   Thurston, W.P.: Three-dimensional manifolds, Kleinian groups and hyperbolic geometry. Bull. Amer. Math. Soc. **6** (1982) 357–381

[Th3]   Thurston, W.P.: On the geometry and dynamics of diffeomorphisms of surfaces. Bull. Amer. Math. Soc. **19** (1988) 417–431

[Tu]   Tukia, P.: A remark on a paper of Floyd. In: Drasin, D., et al. (eds.) Holomorphic functions and moduli II. (Mathematical Sciences Research Institute Publications, vol. 11). Springer, Berlin Heidelberg New York 1988

[Wo]   Wolpert, S.: Thurston's Riemannian metric for Teichmüller space. J. Diff. Geom. **23** (1986) 143–174

Foliations and 3-Manifolds*

David Gabai

California Institute of Technology, Pasadena, CA 91125, USA

The goal of this talk is two fold. First in Sections 1 and 2 I survey some key results relating foliations and the topology of 3-manifolds. In particular in Section 2 I will follow the evolution of one enormously important idea "foliated height functions" as it has evolved over the last century. Second in Section 3 I discuss a recent development, the essential lamination.

1. 1-Dimensional Foliations

Since $\chi(M) = 0$ for all closed 3-manifolds M, it follows that M has a smooth vector field, hence a smooth codimension-2 foliation. By requiring a bit of structure on the foliation one imparts great structure to the 3-manifold. I briefly discuss one such structure.

A Seifert fibred space is a compact 3-manifold M which is almost a S^1 bundle over a compact surface, i.e. there exists a projection $\pi : M \to N$ such that for each $x \in N$ there exists a D^2 neighborhood of x such that $\pi^{-1}(D^2) = D^2 \times S^1$ and $\pi((r, \theta_1), (1, \theta_2)) = (r, p\theta_1 + q\theta_2)$ where $p \neq 0$ and p, q are relatively prime and depend on x and $\theta \in \mathbb{R}$ mod 2π.

Remark. Seifert fibred spaces were classified up to fibre preserving homeomorphism in 1928 by Seifert [S]. Their topological classification was obtained by Orlik, Vogt, and Zieschang [OVZ] in 1967. Seifert fibred spaces have one of six geometric structures as discussed in Scott's survey article [Sc1].

In 1972 Epstein [E] showed that M has a C^1 foliation by circles if and only if M is a Seifert fibred space.

Seifert fibred spaces play a central role in 3-dimensional topology. In fact Thurston's geometrization conjecture [Th3] for 3-manifolds asserts that a closed oriented 3-manifold which has no torus [JS], [Jo] or sphere decomposition [M], [K] is either a Seifert fibred space or a hyperbolic 3-manifold.

Very recently (fall 1990) Casson and the author [G1] have independently announced proofs of the conjecture that S^1 convergence groups [GM] are Fuchsian groups. This result implies the Seifert fibred space conjecture, i.e. if M is a closed

* Partially Supported by NSF grant DMS-8902343 and a Sloan foundation research fellowship.

orientable irreducible 3-manifold, then M is a Seifert fibred space if and only if $\pi_1(M)$ contains an infinite cyclic normal subgroup. [Me] using the work of [Sc2] and others reduced the Seifert fibred space conjecture to the convergence group conjecture.

Regarding the qualitative types of codimension-2 foliations on 3-manifolds I cite the following two remarkable results.

Theorem (Schweitzer [Sw] 1976). S^3 has a C^1 foliation by lines.

Theorem (Vogt [V] 1989). $\mathbb{R}^3$ has a C^1 foliation by circles.

2. 2-Dimensional Foliations

I will try to give a chronological account of codimension-1 foliations in 3-manifolds. For simplicity, unless otherwise stated all manifolds will be closed and orientable.

In 1863 Möbius proved the following result (see [St]).

Theorem. If S is a smooth closed surface in $\mathbb{R}^3$, then S is diffeomorphic to a surface of genus $\dot{g}$.

Idea of Proof. Find enough parallel planes to chop S up into discs, annuli, and spheres with 3-holes. Manipulate the pieces to get a standard form.   □

This is a foliations proof of a topological theorem. Möbius uses a foliation of $\mathbb{R}^3$ by parallel planes to decompose S, then rebuilds S in a recognizable way. This "idea" will be generalized many times over for increasingly sophisticated applications. Here is one such generalization.

Theorem (Alexander 1923 [A]). *A PL embedded S^2 in $\mathbb{R}^3$ bounds a 3-Ball. (The PL Shoenflies theorem).*

Idea of Proof. Consider a foliation of $\mathbb{R}^3$ by parallel planes. A generic such foliation induces a foliation on S^2 which has a finite number of critical points at isolated levels. Analyzing the foliation and the embedding one observes that saddles can be cancelled with centers after an isotopy of S^2, thus, after isotopy, one obtains an S^2 with exactly one maximum and one minimum. This S^2 evidently bounds a 3-cell, hence so does the original one.   □

The study of foliations did not begin as a recognized subject in its own right until the late 1940s with the work of Ehresmann and Reeb. (Perhaps as an offshoot of the study of fibre bundles which was a hot new topic in the 1940s.) The story goes that Ehresmann suggested the following thesis problem to his student Reeb. Show that there exists no codimension-1 foliation on S^3.

Ehresmann-1940s
– Introduced the idea of holonomy [Eh]. Advisor to Reeb and influenced Haefliger's thesis work (whose advisor was DeRham).

Reeb-1948 [R]
- Reeb foliation of S^3.
- Reeb Stability theorem - If F is a codimension-1 foliation on M and L is a compact leaf with trivial holonomy group, then L has a neighborhood in M homeomorphic to $L \times I$ and $F \mid L \times I$ is the product foliation.
- If F is transversely orientable and F contains a S^2 leaf, then $M = S^2 \times S^1$. (He more generally showed that the set of compact leaves with finite π_1 is closed.)

Remark. Later Lickorisch [L], Zieschang [N], and Wood [W] showed that all closed 3-manifolds have codimension-1 foliations.

Haefliger 1959 [H]
- Showed that if γ is a closed curve transverse to F, an analytic foliation of M, then $0 \neq \gamma \in \pi_1(M)$. In particular this implied that M has infinite fundamental group.
- Showed that the union of compact leaves of a codimension-1 foliation is closed in M.

Novikov 1962 [N]. If $M \neq S^2 \times S^1$, F is a C^2 foliation and has no Reeb components then
- $\pi_2(M) = 0$.
- π_1(leaves) inject into $\pi_1(M)$.
- γ transverse to F implies that $0 \neq \gamma \in \pi_1(M)$.

Remark. It was known to experts that C^2 was not crucial, being used only to push curves normally off of surfaces. Thus a transverse line field provides the needed structure in the C^0 case. See also [So] for a C^0 proof.

Sacksteder 1962 [Sa]. If F is a C^2 foliation on M then
- F has no holonomy implies that F is defined by a closed 1-form. In particular the lifted foliation on the universal cover $\tilde{M}$ is the product foliation $\mathbb{R}^2 \times \mathbb{R}$.
- exceptional minimal sets have resiliant leaves, i.e. leaves with holonomy elements which are contractions.

Remark. A C^0 "version" of Sacksteder's first theorem can be found in [I]. Tischler [Ti] showed that a foliation defined by a closed 1-form can be perturbed slightly to obtain a fibration over S^1.

Stallings 1962 [St]
- If M is irreducible, $\partial M \neq \emptyset$ and the inclusion map of $T \rightarrow M$ induces an isomorphism on π_1 where T is a component of boundary M, then $M = T \times I$.

Rosenberg 1968 [Ro]
- If F is a foliation without Reeb components, then M is irreducible.
- (with Sondow). If M is C^2 foliated by planes, then M is the 3-torus.

Remark. The Rosenberg-Sondow theorem uses C^2 in an essential way for it invokes Sacksteder's theorem. Although unstated there (see Sect. 3) a C^0 proof follows by [I].

The main result of Novikov's paper is that a Reeb component is the obstruction to isotoping a disc rel boundary into a leaf, where the induced foliation of the D^2 is the standard foliation by circles with one singular point at the origin. (Consider a meridianal disc transverse to a Reeb foliated $D^2 \times S^1$.)

Rosenberg's key observation in the proof of his first theorem is that Novikov's result implies that there is no obstruction to carrying out Alexander's argument to show that a smoothly imbedded S^2 in a foliated manifold without Reeb components bounds a B^3.

A generalization of the Möbius, Alexander, Rosenberg line is the following result of

Rousserie and Thurston 1972 [Rou, Th1]

- If T is a compact embedded π_1 injective surface in M and F is a foliation on M which has no Reeb components, then T can be isotoped to be transverse to F except at a finite number of saddle and circle tangencies.

Remark. A circle tangency looks exactly like the rim of a volcano. Rousserie further shows that the obstruction to isotoping away the circle tangencies in the case of T a torus is the existence of a cylindrical component, i.e. an annulus bundle over S^1 whose boundary components are leaves and whose interior leaves are annuli whose ends spiral "in the same direction" about the ends. More generally it is well known to experts that the obstruction to eliminating the circle tangencies is the existence of generalized Reeb components, i.e. of bundles over S^1 with fibre a compact surface S with boundary. The boundary tori are leaves and the interior leaves are homeomorphic to $\overset{\circ}{S}$, and nearly tangent to S except near the ends which spiral in the same direction about the boundary tori.

Thurston 1976 [Th2]

- Defined the Thurston norm on $H_2(M)$. $\|z\| = \{\min -\chi(S') \mid [S] = z \in H_2(M)$ where $S' = S - S^2$ components$\}$.
- compact leaves of foliations without Reeb components are Thurston norm minimizing, i.e. compact leaves are topologically minimal in their homology class.

Remark. Corresponding definitions and results were given for manifolds with boundary. Thurston's Pseudo-Anosov theory [FLP] is also of central importance.

Palmeira 1978 [P]

- If F is a foliation on M without Reeb components, then the universal covering of M is $\mathbb{R}^3$. Furthermore the induced foliation on $\mathbb{R}^3$ is topologically equivalent to a product of a foliation on $\mathbb{R}^2$ and $\mathbb{R}$.

Sullivan 1979 [Su]

- If F is C^2 and taut, (i.e. there exists a closed curve transverse to F which hits all the leaves), then there exists a Riemannian metric on M such that all the leaves are minimal, i.e. locally area minimizing.

Remark. Sullivan's result was inspired by a letter from Herman Gluck. This result was generalized by Harvey and Lawson [HL] who showed that F was in fact calibrated, i.e. there exists a metric so that compact portions of leaves are minimal area in their homology class. A corresponding topological statement can be made regarding the Thurston norm.

Gabai [G2, G3]

– 1981 – If T is a Thurston norm minimizing surface in an oriented irreducible manifold M, (∂M possibly nonempty) then there exists a finite depth taut transversely oriented foliation F on M such that T is a leaf.

– 1985 – If T is a minimal genus surface for a knot in S^3, then there exists a finite depth taut transversely oriented foliation F of $S^3 - \overset{\circ}{N}(k)$ such that T is a leaf of F and $F \mid \partial N(k)$ is a foliation by circles.

Remark. The difference between the 1985 and 1981 theorems is that the foliation $F \mid \partial N(k)$ given by the former would be a suspension of a homeomorphism of the circle with some fixed points. See [G3] for other constructions.

Applications

– If $z \in H_2(M, \partial M)$, then $g(z) = 2t(z)$ where g denotes the gromov norm and t denotes the Thurston norm. Viewed another way we obtain a positive solution to a conjecture of Thurston [Th2] that the norm on homology based on singular or immersed surfaces is equal to the "classical" norm based on embedded surfaces. In particular the immersed genus of a knot equals the embedded genus of a knot, which is the higher genus analogue to Dehn's lemma. [G2]

– Positive solutions to Property R and Poenaru conjectures. More generally we show that a homology $S^2 \times S^1$ manifold N obtained by surgery on a knot k in S^3 is irreducible or $S^2 \times S^1$. Furthermore the extension of a minimal genus surface for k becomes a Thurston norm minimizing surface representing the generator of N. One can view Property R as asserting that $\{S^3 \text{ knot theory}\} \cap \{S^2 \times S^1 \text{ knot theory}\} = 0$. [G3]

– Knots in $S^2 \times S^1$ or torus bundles over S^1, which are not contained in 3-cells, are determined by their complements. [G4]

– If T is a Thurston norm minimizing surface in M disjoint from a boundary component P, then T remains norm minimizing (and hence π_1 injective) in all but at most one of the manifolds obtained by Dehn filling on P. [G4]

– Sela showed by example that the bad surgeries corresponding to distinct norm minimizing surfaces may be distinct. [Se]

– Computing the genus of an oriented link in S^3. [G5], [G6]

– Property P for satellite knots. [G7]

– Superadditivity of knot genus under band connnect sum. [G8], [Sch]

– The classification of knots in solid tori such that non trivial surgery yields a solid torus. In particular there exists a unique knot in $D^2 \times S^1$ such that exactly two surgeries yield $D^2 \times S^1$. [G7] reduced the problem to 1-bridge braids in solid tori and Berge [B] solved the problem for 1-bridge case. See also [G9].

Key elements of the proofs of the theorems.
- Sutured manifold theory. Just as a Haken 3-manifold can be decomposed to a 3-cell via splitting along incompressible surfaces, the content of sutured manifold theory is that a 3-manifold M with nontrivial $H_2(M, \partial M)$ can be decomposed, with control, along taut sufaces to obtain a 3-cell. Dual to this decomposition is a finite depth foliation such that the cores of the finite depth leaves correspond to the splitting taut surfaces.
- Finite depth foliations "are like" compact surfaces, i.e. technically finite depth foliations can be treated more like incompressible surfaces rather than "thick" foliations.
- Height function arguments. A crucial moment in [G3] was the defining the notion of thin presentation of a knot in S^3. A thin presentation is essentially a bridge presentation where the local maxima are required to be as low as possible and the minima as high as possible, thus from the point of view of the horizontal planes in S^3 the knot is as thin as possible. With respect to a thin presentation a lamination can be put into a normal form. In particular there will be level planes where the induced lamination will have no inessential curves. These arguments should be viewed as sophisticated versions of Möbius, Alexander, Rosenberg, ...
- Combinatorial arguments.

I close this section with the following results, whose proofs push variants of the foliated height function arguments to high technical levels.

Gabai 1984 [G10]
- (Simple Loop Conjecture) $f : S \to T$ is a map of closed surfaces and $\emptyset \neq \ker f_\# : \pi_1(S) \to \pi_1(T)$, then there exists a simple loop in $\ker f_\#$.

Gabai Kazez 1985 [GK1]
- (Homotopy Classification of Maps of Surfaces) If $f, g : S \to T$ are maps of positive degree of closed oriented surfaces, then there exists a homeomorphism $h : S \to S$ such that f is homotopic to $g \circ h$ if and only if $f_\#(\pi_1(S)) = g_\#(\pi_1(S))$ and $\deg f = \deg g$.

Remark. The idea behind these results is to view f (or g) as a branched immersion in $T \times I$, then put the double curve locus into normal form. See [GK 2] for the version for maps of closed surfaces.

Gordon Luecke 1988 [GL]
- Knots in S^3 are determined by their complements.

Remark. The proof involved two steps.

i) If there exists two knots with distinct complements, then each knot complement has a foliation by planar surfaces coming from the foliation of S^3 by level S^2's. When these foliations are viewed in a single knot complement one finds two planar surfaces which intersect each other in an essential way. This step was done independently by myself and [GL] using thin presentations.
ii) There do not exist such planar surfaces in S^3. This step required a deep combinatorial topology argument.

3. Essential Laminations

Example. If $f : T \to T$ is a pseudo anosov homeomorphism of a closed surface T and Λ is the f-invariant (stable measured) foliation, then λ, the suspension of Λ in M, the T bundle over S^1 defined by f, is a singular foliation on M. One can get singular foliations in different manifolds by doing Dehn surgery to the singular circles. Fried was aware of this construction in the late 70's. Independently (at the same time) Ghys and Thurston knew how to create nonsingular foliations when the singular locus consisted of $2n$ prongs.

Remark.
  i) ([G11] 1981) λ is an example of a singular foliation with a finite number of singular circles. λ restricted to cross sections of neighborhoods of the circles look like index $n/2$ singularities of line fields where $n = 1$ or $n \le -1$. The author observed that Novikov's theorem extends to such singular foliations provided $n < 0$. (There exists a singular foliation of S^3 with a single singular circle of type $1/2$ and no compact leaves. Consider the foliation transverse to the fibration of the (-2,3,7) Pretzel knot.)
  ii) ([G11] 1986) Good singular foliations can be constructed transverse to finite depth foliations. In some sense most manifolds contain such foliations.
  iii) The notion of essential lamination [GO] developed with Oertel generalizes the above notion of "good" singular foliation. Recently Hatcher and Oertel have shown that essential laminations in non Haken manifolds can be viewed as singular foliations (of a possibly more general type).

Definition. A codimension-1 lamination λ of M^3 is a decomposition of a closed subset of M by surfaces (called leaves) such that M is covered by charts of the form $\mathbb{R}^2 \times \mathbb{R}$ and leaves of λ pass through in $\mathbb{R}^2 \times$ pts. λ is an essential lamination [GO] of M if
  i) $M - \lambda$ is irreducible.
  ii) No leaf of λ is a torus bounding a solid torus.
  iii) If V is the closure (in the path metric) of a complementary region of λ in M, then ∂V is incompressible and end incompressible.

Theorem [GO] 1987. *Let M be a closed oriented 3-manifold with an essential lamination λ, then*

  i) *M is irreducible.*
  ii) *π_1 (leaves) inject into $\pi_1(M)$.*
  iii) *π_1 (transverse efficient loops) inject into $\pi_1(M)$, transverse efficient arcs cannot be homotoped rel boundary into a leaf.*
  iv) *the universal cover of M is $\mathbb{R}^3$.*

Remark. Essential laminations generalize the notions of foliations without Reeb components and incompressible surfaces. Like finite depth foliations essential laminations can be manipulated as compact surfaces. While in some sense most 3-manifolds do not contain incompressible surfaces, most 3-manifolds do contain essential laminations.

Example. Essential laminations can be obtained by "blowing air" into the singular loci of the above good singular laminations. See [GO, Ha, O, Na, D] for other constructions.

Conjecture. *If M is a closed aspherical 3-manifold, then M is finitely covered by a manifold with an essential lamination.*

Conjecture. *If M is closed oriented and aspherical, then M has an essential lamination or is Seifert fibred.*

Remark. The second optimistic conjecture implies the first. The first conjecture is implied by Waldhausen's conjecture that every aspherical 3-manifold is finitely covered by a Haken manifold. There is also Thurston's conjecture that every hyperbolic manifold is finitely covered by a surface bundle over S^1. The following result combined with [M2, W2] (see [JN]) shows that there exist examples of aspherical Seifert fibred spaces with out essential laminations.

Theorem (Brittenham [B], Claus [C] 1990). *If λ is an essential lamination of a Seifert Fibred Space, then there exists a sublamination which is either horizontal or vertical (after isotopy).*

Remark. Vertical means that it is a union of Seifert fibres and horizontal means that it is transverse to the Seifert fibration.

The case for λ a compact surface was due to [Wa]. Thurston [Th1] obtained the result when λ was a C^2 foliation in a S^1 bundle. [EHN] generalized that result to C^2 foliations in Haken Seifert fibred spaces.

Theorem ([GK3] 1990). *If M has an essential lamination and $0 \neq \gamma \in \pi_1(M)$, then $\overset{\circ}{D}{}^2 \times S^1$ is the covering space of M with π_1 generated by γ.*

Theorem (Gabai-Kazez 1990). *If M has a pseudoanosov flow and $f, g : M \to M$ are homotopic homeomorphisms, then f is isotopic to g.*

Remark. It suffices to consider the case that $f = $ id. The hypotheses imply the existence of two transverse (stable and unstable) essential laminations. The proof proceeds by first showing that f is isotopic to f' which fixes a simple closed curve common to both laminations.

Homotopy implies isotopy for irreducible 3-manifolds is an old conjecture. Waldhausen [Wa] established the case for M Haken, a number of authors established the case for M a spherical space form or more generally a Seifert fibred space, see [Sc3] and [BO]. See also [HS].

Theorem. *If λ is an essential lamination of the closed 3-manifold M by planes, then M is the 3-torus.*

Proof. The closure of each complementary region of λ, with the path metric, is $\overset{\circ}{D}{}^2 \times I$, therefore λ extends to a foliation also called λ of M by planes. It follows from Theorem 3.1 [I], the lift $\tilde{\lambda}$ of λ to the universal covering of M has space of leaves $\mathbb{R}$. Since each leaf of λ is a plane, $\pi_1(M)$ acts freely and order preserving on $\mathbb{R}$, thus by a theorem of Holder $\pi_1(M)$ is archemedian and hence free abelian. This last argument is essentially Proposition 4.1 of [I]. Now argue as in [Ro] to conclude that $M = S^1 \times S^1 \times S^1$. ☐

Bibliography

[A]   Alexander, J. W.: On the subdivision of 3-space by a polyhedron. Proc. Nat. Acad. Sci. **10** (1924) 6–8

[B]   Berge, J.: The knots in $D^2 \times S^1$ with nontrivial Dehn surgeries yielding $D^2 \times S^1$. Topology Appl. (to appear)

[BO]   Boileau, M., Otal, J. P.: Groupe des diffeotopies de certains varietes de Seifert. Compt. Rend. Acad. Sci. **303** (1986) 19–22

[Br]   Brittenham, M.: Essential laminations in Seifert fibred spaces. Cornell University Thesis, 1990

[C]   Claus, E.: University of Texas Thesis. In preparation

[D]   Delman, C.: Cornell University thesis, 1991. In preparation

[E]   Epstein, D.: Periodic flows on 3-manifolds. Ann. Math. **95** (1972) 66–82

[Eh]   Ehresmann, C.: Sur la theorie des varietes feuilletees. Rend. Mat. Appl. Ser. 5, vol. 10, Rome, 1951

[EHN]   Eisenbud, D., Hirsch, U., Neumann, W.: Transverse foliations of Seifert bundles and self homeomorphism of the circle. Comm. Math. Helv. **56** (1981) 638–660

[FLP]   Fathi, A. Laudenbach, F. Poenaru V.: Travaux de Thurston sur les surfaces. Asterisque **66-67** (1979)

[G1]   Gabai, D.: Convergence groups are Fuchsian groups. Preprint

[G2]   Gabai, D.: Foliations and the topology of 3-manifolds. J. Diff. Geom. **18** (1983) 445–503

[G3]   Gabai, D.: Foliations and the topology of 3-manifolds III. J. Diff. Geom. **26** (1987) 479–536

[G4]   Gabai, D.: Foliations and the topology of 3-manifolds II. J. Diff. Geom. **26** (1987) 461–478

[G5]   Gabai, D.: Genera of the arborescent links. AMS Memoirs **339** (1986) 1–98

[G6]   Gabai, D.: Foliations and genera of links. Topology **23** (1984) 381–394

[G7]   Gabai, D.: Surgery on knots in solid tori. Topology **28** (1989) 1–6

[G8]   Gabai, D.: Genus is superadditive under band connected sum. Topology **26** (1987) 209–210

[G9]   Gabai, D.: 1-Bridge braids in solid tori. Topology Appl. (to appear)

[G10]   Gabai, D.: The simple loop conjecture. J. Diff. Geom. **22** (1985) 143–149

[G11]   Gabai, D.: Personal communication

[GK1]   Gabai, D., Kazez, W.H.: The classification of maps of surfaces. Invent. math. **90** (1987) 219–242

[GK2]   Gabai, D., Kazez, W.H.: The classification of maps of nonorientable surfaces. Math. Ann. **281** (1988) 687–702

[GK3]   Gabai, D., Kazez, W.H.: Manifolds with essential laminations are covered by solid tori. Preprint

[GL]   Gordon, C., Luecke, J.: Knots are determined by their complements. JAMS **2** (1989) 371–415

[GM]   Gehring F.W., Martin G.J.: Discrete quasiconformal groups I. Proc. London Math. Soc. (3) **55** (1987) 331–358

[GO]   Gabai, D., Oertel, U.: Essential laminations in 3-manifolds. Ann. Math. **130** (1989) 41–73

[H]   Haefliger, A.: Varietes feuilletes. Ann. Scuola Norm. Sup. Pisa (3) **16** (1962) 367–397

[Ha]   Hatcher, A.: Examples of essential laminations. Preprint

[HL]   Harvey, R., Lawson Jr., H. Blaine: Calibrated foliations. Amer. J. Math. **104** (1980) 607–633

[HS]   Hass, J., Scott, P.: Homotopy and isotopy in non Haken 3-manifolds. In preparation

[I]   Imanishi, H.: On the theorem of Denjoy-Sacksteder for codimension one foliations without holonomy. J. Math. Kyoto U. **14** (1974) 607–634

[J]   Johannson, K.: Homotopy equivalences of 3-manifolds with boundary. (Lecture Notes of Mathematics, vol. 761). Springer, Berlin Heidelberg New York 1979

[JN]   Jankins, M., Neumann, W.D.: Rotation numbers of products of circle homeomorphisms. Math. Ann. **271** (1985) 381–400

[JS]   Jaco, W., Shalen, P.: Seifert fibred spaces in 3-manifolds. Mem. Amer. Math. Soc. **220** 1979

[K]   Kneser, H.: Geschlossene Flächen in dreidimensionalen Mannigfaltigkeiten. Jahresber. Deutsch. Math. Verein. **38** (1929) 248–260

[L]   Lickorish, W.B.R.: A foliation for 3-manifolds. Ann. Math. (2) **82** (1965) 414–420

[M1]   Milnor, J.: A unique factorization theorem for 3-manifolds. Amer J. Math. **84** (1962) 1–7

[M2]   Milnor, J.: On the existence of a connection with curvature zero. Comm. Math. Helv. **32** (1957) 215–223

[Me]   Mess, G.: The Seifert conjecture and groups which are coarse quasiisometric to planes. Preprint

[N]   Novikov, S.P.: Topology of foliations. Trans. Moscow Math. Soc. (1965) 268–304

[Na]   Naimi, R.: Personal communication

[O]   Oertel, U.: Example of an affine lamination. Personal communication

[OVZ]   Orlik, P., Vogt, E., Zieschang, H.: Zur Topologie gefaserter dreidimensionaler Mannigfaltigkeiten. Topology **6** (1967) 49–64

[P]   Palmeira, C.F.B.: Open manifolds foliated by planes. Ann. Math. **107** (1978) 109–131

[R]   Reeb, G.: Sur certaines proprietes topologiques des varietes feuilletees. Actualites Sci. Indust. no 1183. Hermann, Paris 1952, pp. 91–158

[Ro]   Rosenberg, H.: Foliations by planes. Topology **6** (1967) 131–138

[Rou]   Rousserie, R.: Plongements dans les varietes feuilletees et classification de feuilletages sans holonomie. IHES **43** (1973) 101–142

[S]   Seifert, H.: Topologie dreidimensionaler gefaserter Raum. Acta Math. **60** (1933) 147–288. [English transl.: Topology of 3-dimensional fibred spaces. Pure Appl. Math. Ser. **89** (1980) 360–423

[Sa]   Sacksteder, R.: Foliations and pseudogroups. Amer. J. Math. **87** (1965) 79–102

[Sc1]   Scott, P.: The geometries of 3-manifolds. Bull. London Math. Soc. **15** (1983) 401–487

[Sc2]   Scott, P.: There are no fake Seifert fibred spaces with infinite π_1. Ann. Math. **117** (1983) 35–70

[Sc3]   Scott, P.: Homotopy implies isotopy for some Seifert fibred spaces. Topology **24** (1985) 341–351

[Sch]   Scharlemann, M.: Sutured manifolds and generalized Thurston norms. J. Diff. Geom. (to appear)

[Se]   Sela, Z.: Torus fillings that reduce Thurston norm. Preprint

[St]   Stillwell, J.: Classical topology and combinatorial group theory. (Graduate Texts in Mathematics, vol. 72). Springer, Berlin Heidelberg New York 1980

[Sta]   Stallings, J.: On fibering certain 3-manifolds. Topology of 3-manifolds. Prentice Hall, 1962, pp. 95–100

[Su]   Sullivan, D.: A homological characterization of foliations consisting of minimal surfaces. Comm. Math. Helv. **54** (1979) 218–223

[Sw]   Schweitzer, P.: Counterexample to the Seifert conjecture and opening closed leaves of foliations Ann. Math. **100** (1974) 386–400

[Th1]   Thurston, W.P.: Foliations of three-manifolds which are circle bundles. Berkeley Thesis 1972

[Th2]   Thurston, W.P.: A norm for the homology of 3-manifolds. AMS Memoirs **339** (1986) 99–130

[Th3]   Thurston, W.P.: Three dimensional manifolds, Kleinian groups and hyperbolic geometry. BAMS (N.S.) **6** (1982) 357–381

[Ti]   Tischler, D.: On fibering certain foliated manifolds over S^1. Topology **9** (1970) 153–154

[V]   Vogt, E.: Preprint 1989

[W1]   Wood, J.: Foliations on 3-manifolds. Ann. Math. (2) **89** (1969) 336–358

[W2]   Wood, J.: Bundles with totally disconnected structure group. Comm. Math. Helv. **46** (1971) 257–273

[Wa]   Waldhausen, F.: On 3-manifolds which are sufficiently large. Ann. Math. **87** (1968) 56–88

The Differential Calculus of Homotopy Functors

Thomas G. Goodwillie

Department of Mathematics, Brown University, Providence, R1 02912, USA

0. Introduction

My purpose here is to explain a method in homotopy theory. The following result is perhaps the best example to date of a statement that can be proved by this method:

Theorem 1. *For any 2-connected map of topological spaces $Y \to X$ the fiber of $A(Y) \to A(X)$ and the fiber of $TC(Y) \to TC(X)$ are weakly homotopy equivalent.*

Here A is Waldhausen's algebraic K-theory functor from spaces to spectra, and TC is another functor which I will discuss below. "Fiber" means homotopy fiber.

If we write $\tilde{A}$ for the reduced functor $\tilde{A}(Y) = \text{fiber}(A(Y) \to A(*))$ and similarly for TC, then in the case when X is a point we have the statement:

Corollary 2. *For any 1-connected topological space Y the spectra $\tilde{A}(Y)$ and $\tilde{T}C(Y)$ are weakly homotopy equivalent.*

I will not say much now about the other functor TC, except that $TC(X)$ is closely related to the free loop space ΛX (the space of all unbased maps from the circle to X) and is easier to study than $A(X)$ from the point of view of algebraic topology.

The theorem stated above represents the work of several people. In particular, the definition of the functor TC, and of a map $A \to TC$ which is crucial to the proof, uses work of Bökstedt-Hsiang-Madsen. A p-completed version of Theorem 1 (proved by the method outlined below) was the main result of [BCCGHM]. The theorem stated above is only a marginal advance over this, since a rational version ([G1], Corollary on p. 349) has been known for some time. (The final steps in the proof of Theorem 1 will appear in [G6].)

1. Summary of the Method

The proof of Theorem 1 uses a kind of deformation theory. The goal is to describe the change in $A(X)$ produced by a given (small) change in X. It turns out that to

Proceedings of the International Congress
of Mathematicians, Kyoto, Japan, 1990
© The Mathematical Society of Japan, 1991

achieve this it is enough to describe the *infinitesimal* change in $A(X)$ produced by an *infinitesimal* change in X. By this I mean: to give an approximate description of the change in $A(X)$ produced by a very small change in X. (A small change in X is a highly connected map $Y \to X$.)

In a little more detail, the method is this:

There is a natural map of spectra from $A(X)$ to $TC(X)$, called the cyclotomic trace map. Denote its homotopy fiber by $F(X)$. There is a constant c such that for any k-connected map of spaces $Y \to X$ the map of homotopy fibers

$$\text{fiber}(A(Y) \to A(X)) \to \text{fiber}(TC(Y) \to TC(X))$$

is $(c + 2k)$-connected. In other words, the map $F(Y) \to F(X)$ is about twice as highly connected as the map $Y \to X$ upon which it depends. By a certain general principle (Proposition 5 below), it follows that the map $F(Y) \to F(X)$ is in fact ∞-connected when the map $Y \to X$ is at least 2-connected. (In other words, up to weak equivalence $F(X)$ depends only on $\pi_1(X)$ if X is connected.) This yields the conclusion of Theorem 1.

The general principle used above is analogous to the following fact from differential calculus: If a function f (in a suitable domain, and satisfying suitable differentiability hypotheses) is such that $|f(x) - f(y)| < C|x - y|^2$, then f is locally constant. A more familiar statement of this fact is that if the derivative of f is identically zero then f is locally constant.

Section 2 explains the idea "derivative of a homotopy functor". Section 3 states the general principle mentioned above. Section 4 discusses what one needs to know about Waldhausen's functor A in order to apply the principle here. Section 5 describes the other functor TC. Section 6 discusses that part of the proof which involves the map from A to TC. Details may be found in [BCCGHM, G2, G3, G4, and G6].

2. Differentiation of a Functor

For a more detailed account of the ideas below, see [G2].

2.1 The Definition

The idea can be made quite general, but for concreteness let us suppose that F is a functor from spaces to spectra. We always assume that it is a *homotopy functor*, meaning that it takes equivalences to equivalences. (Throughout, an *equivalence* of spaces or spectra means a weak homotopy equivalence.)

In calculus the concept of derivative, or differential, of a function f at a point x is a way of systematically describing the quantity $f(y) - f(x)$ with an accuracy like $|y - x|^2$. In a similar way the next two definitions serve to describe the $2k$-homotopy type of the fiber of $F(Y) \to F(X)$ when the map $Y \to X$ is k-connected.

Definition 3. The *derivative* $\partial_x F(X)$ of F at the based space (X, x) is the homotopy colimit (as $k \to \infty$) of the spectra $\Omega^k \text{fiber}(F(X \vee S^k) \to F(X))$.

The maps in the limit system are (loosely speaking) induced by the diagrams

$$F(X \vee S^{k-1}) \longrightarrow F(X \vee D_+^k) \sim F(X)$$

$$F(X) \sim F(X \vee D_-^k) \longrightarrow F(X \vee S^k)$$

(Note that $F(X \vee D^k)$ is equivalent to $F(X)$.) Up to equivalence the derivative is determined by X and by knowing which component of X contains the point x. The spectrum $\partial_x F(X)$ is a functor of the based space (X, x), and any based map $(X, x) \to (Y, y)$ which is an equivalence induces an equivalence $\partial_x F(X) \to \partial_y F(Y)$.

There is a more general construction. If $f : Y \to X$ is a map of spaces, think of Y as a space over X, think of the mapping cylinder of f as the *fiberwise cone* of Y over X (another space over X), and denote it by $C_X Y$. Let $\Sigma_X Y$, the *fiberwise suspension* of Y over X, be the union along Y of two copies of $C_X Y$.

Definition 4. The *differential* of $(D_X F)(Y)$, defined for any map $Y \to X$, is the homotopy colimit of the spectra $\Omega^k \text{fiber}(F(\Sigma_X^k Y) \to F(X))$,

The maps in the limit system are defined using diagrams

$$F(\Sigma_X^{k-1} Y) \longrightarrow F(C_X \Sigma_X^{k-1} Y) \sim F(X)$$

$$F(X) \sim F(C_X \Sigma_X^{k-1} Y) \longrightarrow F(\Sigma_X^k Y)$$

For fixed X the differential $D_X F$ is a functor from spaces over X to spectra. It is a homotopy functor in the sense that it preserves equivalences, where a map of spaces over X is called an equivalence if as a map of spaces it is a (weak homotopy) equivalence. We have $(D_X F)(X) \sim *$ and $(D_X F)(X \vee S^0) \sim \partial_x F(X)$.

Note that there is a natural map

$$\text{fiber}(F(Y) \to F(X)) \to (D_X F)(Y)$$

The functor $D_X F$ is intended to be an excisive functor that approximates $Y \mapsto \text{fiber}(F(Y) \to F(X))$, much as in calculus the differential of a function f at a point x is a linear function that approximates $f(y) - f(x)$. To explain this I need some language.

2.2 Excision

A commutative diagram $\mathscr{X}$ of spaces (or spectra)

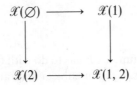

is a *cofiber square* if the canonical map to $\mathscr{X}(1, 2)$ from the homotopy pushout (union along $\mathscr{X}(\varnothing)$ of the mapping cylinders of $\mathscr{X}(\varnothing) \to \mathscr{X}(1)$ and $\mathscr{X}(\varnothing) \to \mathscr{X}(2)$) is an equivalence. It is a *fiber square* (resp. k-connected) if the canonical map from $\mathscr{X}(\varnothing)$ to the homotopy pullback (fiber product over $\mathscr{X}(1, 2)$ of the path fibrations of $\mathscr{X}(1) \to \mathscr{X}(1, 2)$ and $\mathscr{X}(2) \to \mathscr{X}(1, 2)$) is an equivalence (resp. k-connected). Equivalently, a square diagram $\mathscr{X}$ of spectra is k-connected if the *iterated fiber*, the homotopy fiber of the map

$$\text{fiber}(\mathscr{X}(\varnothing) \to \mathscr{X}(1)) \to \text{fiber}(\mathscr{X}(2) \to \mathscr{X}(1, 2))$$

of homotopy fibers, is $(k - 1)$-connected.

A functor F (say from spaces, or spaces over X, to spectra) is *excisive* if it takes cofiber squares to fiber squares. This is a very strong condition. Homotopy functors occurring in nature usually satisfy a much weaker, but useful, condition, called *stable excision*: there is a constant c_2 such that if the maps $\mathscr{X}(\varnothing) \to \mathscr{X}(1)$ and $\mathscr{X}(\varnothing) \to \mathscr{X}(2)$ in a cofiber square are respectively k_1- and k_2-connected, then the diagram $F(\mathscr{X})$:

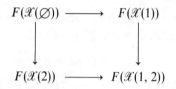

is $(k_1 + k_2 + c_2)$-connected.

If F satisfies stable excision then, for each X, $D_X F$ satisfies excision; we may think of $D_X F$ as a (reduced) homology theory on the category of spaces over X. Moreover, stable excision for F implies that the map from fiber$(F(Y) \to F(X))$ to $(D_X F)(Y)$ is approximately $2k$-connected for any k-connected map $Y \to X$.

2.3 The Principle

Theorem 1 is proved by applying the following principle with $F = \text{fiber}(A \to TC)$ and $\varrho = 1$. The term "ϱ-analytic" will be explained in Section 3.

Proposition 5. *If F is a ϱ-analytic functor from spaces to spectra such that $(D_X F)(Y)$ is trivial for all X and all $Y \to X$, then for every $(\varrho + 1)$-connected map $Y \to X$ of spaces the map $F(Y) \to F(X)$ is an equivalence.*

"Trivial" means equivalent to a point (all homotopy groups are trivial). If F satisfies a suitable limit axiom, so that up to equivalence it is determined by its behavior on finite CW complexes, then it is enough to assume that $\partial_x F(X)$ rather than $D_X F$ is trivial.

3. Analytic Functors

"Analyticity" of a homotopy functor F has to do with the behavior of F with respect to cubical diagrams of spaces. By an *n-cubical diagram* we mean a functor $\mathscr{X}$ from the partially ordered set of all subsets of $\{1, \ldots, n\}$ to the category of spaces. Analyticity of F involves one condition, stable $(n-1)$st order excision, for each n. Stable first-order excision is stable excision as defined in Sect. 2.2.

Stable $(n-1)$st order excision concerns certain n-cubical diagrams $\mathscr{X}$, namely the *strong cofiber cubes*. Call $\mathscr{X}$ a strong cofiber cube if, for each $1 \le i < j \le n$ and $S \subset \{1, \ldots, n\} - \{i, j\}$, the diagram

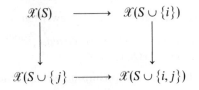

is a cofiber square. The condition is that there is a constant c_n such that, whenever $\mathscr{X}$ is a strong cofiber cube in which the map $\mathscr{X}(\varnothing) \to \mathscr{X}(i)$ is k_i-connected for all i, with $k_i > \varrho$, then $F(\mathscr{X})$ is $(c_n + \Sigma k_i)$-connected. (An n-cubical diagram of spectra is called k-connected if its iterated fiber – the spectrum obtained by taking homotopy fibers in each of the n directions in turn – is $(k-1)$-connected.) Note that c_n is allowed to be negative.

For $n = 1$ this simply says that there is a constant c_1 such that for any k-connected map $\mathscr{X}(\varnothing) \to \mathscr{X}(1)$ of spaces the map $F(\mathscr{X}(\varnothing)) \to F(\mathscr{X}(1))$ is $(k + c_1)$-connected, at least if $k > \varrho$.

Definition 6. The functor F is ϱ-analytic if it satisfies $(n-1)$st order as above for all $n \ge 1$, and if the numbers c_n are bounded below by $c - \varrho n$ for some constant c.

Most homotopy functors occurring in nature are ϱ-analytic for some ϱ, and in many cases it is a routine matter to verify this. The identity functor from spaces to spaces is 1-analytic, as is Waldhausen's functor A.

The proof of Proposition 5 uses an unusual inductive argument. It is not difficult, but I will not take time to explain it here; see [G3].

Proposition 5 expresses one of two main consequences of analyticity. The other, the existence of a "Taylor tower" for a functor in analogy with the Taylor series of a function, is not used in the proof of Theorem 1. It is explained in [G5].

4. The Derivative of K-Theory

In order to use Proposition 5 for proving Theorem 1, it is necessary first of all to know (up to natural equivalence) what the derivative of the functor A is. The answer turns out to be this:

Theorem 7. *For a based space (X, x) the spectrum $\partial_x A(X)$ is related by a chain of natural equivalences to $\Sigma^\infty(\Omega(X, x)_+)$.*

This is the unreduced suspension spectrum of the based loopspace of X. (The subscript "$+$" adds a disjoint basepoint.)

Theorem 7 (3.3 of [G2]) is proved indirectly; it is reduced to a corresponding statement (Theorem 7' below) about smooth manifolds, using a major theorem of Waldhausen:

Theorem 8 (Waldhausen [W1]). *There is a natural weak equivalence of spectra between $A(X)$ and the product $\Sigma^\infty(X_+) \times Wh^{\mathrm{Diff}}(X)$, where $Wh^{\mathrm{Diff}}(X)$ is a natural double delooping of the differentiable pseudoisotopy spectrum $\mathscr{P}^{\mathrm{Diff}}(X)$.*

In view of Theorem 8, Theorem 7 may be rewritten:

Theorem 7'. *For a based space (X, x) the spectrum $\partial_x \mathscr{P}^{\mathrm{Diff}}(X)$ is related by a chain of natural equivalences to $\Omega^2 \Sigma^\infty(\Omega(X, x)/X)$.*

It is notable that, while the relationship between K-theory and pseudoisotopy theory expressed in Theorem 8 is usually viewed as a way of reducing geometry to algebra, in this instance the flow of information is in the other direction. In this connection see also Sect. 5.3.

Recall that the underlying space of the spectrum $\mathscr{P}^{\mathrm{Diff}}(X)$ is essentially defined as a limit of spaces $P^{\mathrm{Diff}}(M)$ for manifolds M (compact, with boundary, of arbitrarily large dimension) of the homotopy type of X. The space $P^{\mathrm{Diff}}(M)$ is the simplicial group of all diffeomorphisms of $M \times I$ which are the identity along $(M \times 0) \cup (\partial M \times I)$.

Therefore, to "compute" $\partial_x \mathscr{P}^{\mathrm{Diff}}(X)$ is essentially to solve the following problem: For a smooth manifold M with an attached handle h of index $k \geq 3$, determine the $2k$-homotopy type of the fiber of $P^{\mathrm{Diff}}(M) \to P^{\mathrm{Diff}}(M \cup h)$. This is done in [G2] using Morlet's "disjunction lemma" and an old-fashioned differentiable general-position argument.

5. The Functor TC

I will now say something about the functor TC which occurs in the statement of Theorem 1. There are really two questions to address: How is it defined, and what does it turn out to be?

5.1 Definition of TC

I will not be very specific about this. TC is related to Bökstedt's "topological Hochschild homology" (THH). For details see [BCCGHM], [BHM], or [G4].

Recall that, according to one way of thinking about the K-theory of (based, connected) spaces, $A(BG)$ is the K-theory spectrum of the "ring up to homotopy" $\Omega^\infty \Sigma^\infty(|G|_+)$. The latter is to be thought of as the "group ring" $k[G]$ of the simplicial group G over the ground "ring" $k = QS^0$. Heuristically,

$$\text{connective spectrum} = \text{infinite loop space}$$

$$= \text{abelian group up to homotopy}$$

$$= k\text{-module}$$

and the group structure of G gives $k[G]$ a multiplication compatible with its additive structure. These ideas can be made precise by using a suitable notion of "ring up to homotopy", for example Bökstedt's notion of FSP (functor with smash product).

For such a "ring" R, Bökstedt defines a K-theory spectrum $K(R)$. Both the Quillen K-theory of rings and the Waldhausen K-theory of spaces are included as special cases (the cases of a discrete ring R and a group ring $k[G]$ respectively). He also defines a spectrum $THH(R)$; heuristically it is the simplicial object

$$R$$

$$\uparrow \downarrow \uparrow$$

$$R \otimes R$$

$$\uparrow \downarrow \uparrow \downarrow \uparrow$$

$$R \otimes R \otimes R$$

$$\cdots$$

with face and degeneracy maps given by the product and unit of R, respectively, as in the definition of the standard chain complex for Hochschild homology. The "tensor products" are meant to be over k and are really smash products of spectra.

Bökstedt defines a map of spectra $K(R) \to THH(R)$; it is modeled on the "trace map" from K-theory to Hochschild homology defined by Dennis for an ordinary ring R.

Very roughly speaking, TC is related to THH as cyclic homology is related to Hochschild homology. For any FSP there is a spectrum $TC(R)$ with a map $TC(R) \to THH(R)$. The trace $K(R) \to THH(R)$ lifts to a map $K(R) \to TC(R)$, called the *cyclotomic trace*. (After p-completion this is the same as the map of that same name constructed in [BHM]).

Let the simplicial group G be a loop group for the space X, and let R be $k[G]$. In this case we sometimes write $TC(X)$ instead of $TC(R)$. Thus in this case the cyclotomic trace is a map $A(X) \to TC(X)$. It is this which is used in the proof of the theorem.

5.2 Description of TC

From a computational point of view the main thing to know about $TC(X)$ is that it is related in a certain way to the free loop space $AX = \text{Map}(S^1, X)$. Again let G be a simplicial loop group for X.

First of all, it is fairly easy to see that $THH(k[G])$ is equivalent to $\Sigma^\infty(AX_+)$. This is essentially because AX is equivalent to the realization of the simplicial space

$$G$$

$$\uparrow \downarrow \uparrow$$

$$G \times G$$

$$\uparrow \downarrow \uparrow \downarrow \uparrow$$

$$G \times G \times G$$

$$\cdots$$

(the "cyclic bar construction" or "cyclic nerve" of G).

To describe $TC(X)$ we must consider some additional structure that the space AX has. Let the circle group S^1 act on AX in the usual way, and let $\Delta_p : AX \to AX$ be the pth power map (composition with the standard map $S^1 \to S^1$ of degree p).

It turns out that the functor $X \mapsto TC(X)$ is very closely related to the functor $X \mapsto B(X) = \Sigma^\infty \Sigma((ES^1 \times_{S^1} AX)_+)$, although to say exactly how they are related it is apparently necessary to consider separately the profinite homotopy type and the rational homotopy type.

Concerning the profinite type, the statement is that after p-completion (p a prime) the spectrum $TC(X)$ becomes part of a fiber square

$$
\begin{array}{ccc}
TC(X) & \longrightarrow & B(X) \\
\downarrow & & \downarrow{\scriptstyle \text{Trf}} \\
\Sigma^\infty AX_+ & \xrightarrow[1-\Delta_p]{} & \Sigma^\infty AX_+
\end{array}
$$

Here Trf is the S^1-transfer associated to the bundle

$$(AX \sim) ES^1 \times AX \to ES^1 \times_{S^1} AX$$

and $1 - \Delta_p$ is the difference between two stable maps, the identity and the map induced by Δ_p.

This, it turns out, has the consequence that for 1-connected spaces X there is a natural equivalence, after p-completion and passage to reduced functors, between $TC(X)$ and

$$\Sigma^\infty(X_+) \times \text{fiber}(e \circ \text{Trf} : B(X) \to \Sigma^\infty(X_+))$$

where the map is the composition of the transfer and the map induced by evaluation $AX \to X$ at a point in the circle.

Concerning the rational type, the statement is that for 2-connected maps $Y \to X$ there is a natural equivalence, after rationalization, between the fiber of $TC(Y) \to TC(X)$ and the fiber of $B(Y) \to B(X)$. (This is not, however, induced by a natural map $TC \to B$ or $B \to TC$.)

5.3 Generalizations

Theorem 1 can be generalized so as to apply to more than the K-theory of spaces. There is considerable evidence for the following:

Conjecture 9. *For any 1-connected map $R \to S$ of FSP's the resulting map of spectra from* fiber$(K(R) \to K(S))$ *to* fiber $(TC(R) \to TC(S)$ *is an equivalence.*

This can be deduced from Theorem 1 in some cases, namely those in which $\pi_0(R)$ $(= \pi_0(S))$ is an integral group ring $\mathbb{Z}[\pi]$. In particular, it is true for the map $QS^0 = k = R \to S = \mathbb{Z}$. Unfortunately, this does not yet amount to a computation of the fiber of $A(*) \to K(\mathbb{Z})$ in any real sense, because $TC(\mathbb{Z})$ is still a fairly mysterious object.

6. The Derivative of *TC*

After producing a map from A to TC, it remains to show that it induces an equivalence $\partial_x A(X) \to \partial_x TC(X)$. This is done in two steps.

The first step is to show that $\partial_x A(X)$ and $\partial_x TC(X)$ are abstractly equivalent, in the sense that these two functors from based spaces (X, x) to spectra are related by a chain of natural equivalences. I have already said that $\partial_x A(X)$ is abstractly equivalent to $\Sigma^\infty \Omega(X, x)_+$. The same is true of $\partial_x TC(X)$. Of course I cannot begin to explain why, since I have not even defined TC here, but to get the idea I invite the reader to work out the equivalences (see Section 2 of [G2]):

$$\partial_x \Sigma^\infty \Lambda(X)_+ \sim \mathrm{Map}(S^1, \Sigma^\infty \Omega(X, x)_+)$$

$$\partial_x B(X)_+ \sim \Sigma^\infty \Omega(X, x)_+.$$

The second step is to prove:

Lemma 10. *The cyclotomic trace $A \to TC$ induces an equivalence $\partial_x A(X) \to \partial_x TC(X)$.*

The trick in proving this is to begin with the case when X is the suspension ΣY of a connected space Y.

To see that this special case is enough, one classifies all the natural maps $\Sigma^\infty \Omega(X, x)_+ \to \Sigma^\infty \Omega(X, x)_+$ in the homotopy category of homotopy functors from based spaces to spectra. It turns out that the only maps which are equivalences when X is a simply-connected suspension are those which are equivalences for all X.

The argument which proves the lemma in the case $X = \Sigma Y$ is essentially the main argument of [CCGH]. It relies on a tool which is only available in the suspension case: the cyclotomic trace can be composed with another natural map as follows:

$$\coprod_{n \geq 1} D_n(Y) \to \Omega \tilde{A}(\Sigma Y) \to \Omega \tilde{T}C(\Sigma Y).$$

Here $D_n(Y)$ is the divided power $\Sigma^\infty(E(\mathbb{Z}/n)_+ \wedge_{\mathbb{Z}/n} Y^{[n]})$. (I am writing $Y^{[n]}$ for the smash product of n copies of Y.) The composed map above induces a map of derivatives

$$\partial_y \left(\coprod_{n \geq 1} D_n(Y) \right) \to \partial_y \Omega \tilde{T}C(\Sigma Y) \sim \Omega \Sigma^\infty \Omega \Sigma Y_+$$

which, more or less by direct examination, is seen to be an equivalence. It follows that the map

$$\Omega \Sigma^\infty \Omega \Sigma Y_+ \sim \partial_y \Omega \tilde{A}(\Sigma Y) \to \partial_y \Omega \tilde{T}C(\Sigma Y) \sim \Omega \Sigma^\infty \Omega \Sigma Y_+$$

induced by the cyclotomic trace is a split surjection, and from this one concludes without much trouble that it is an equivalence.

As a by-product this yields the main result of [CCGH], which can now be viewed as a special case of Corollary 2:

Theorem 11. *For connected spaces Y there is a natural equivalence of spectra*

$$\Omega A(\Sigma Y) \sim \Omega A(*) \times \coprod_{n \geq 1} D_n(Y).$$

References

[B]   Bökstedt, M.: Topological Hochschild homology. Preprint, Bielefeld
[BCCGHM]   Bökstedt, M., Carlsson, G., Cohen, R., Goodwillie, T., Hsiang, W.-c., Madsen,
            I.: The algebraic K-theory of simply-connected spaces (p-complete case).
            Preprint
[BHM]   Bokstedt, M., Hsiang, W.-c., Madsen, I.: The cyclotomic trace and algebraic
            K-theory of spaces. Preprint, Aarhus
[CCGH]   Carlsson, G., Cohen, R., Goodwillie, T., Hsiang, W.-c.: The free loop space and
            the algebraic K-theory of spaces. K-theory **1** (1987) 53–82
[G1]   Goodwillie, T.: Relative algebraic K-theory and cyclic homology. Ann. Math.
            124 (1986) 347–402
[G2]   Goodwillie, T.: Calculus I, The derivative of a homotopy functor. K-Theory
            4 (1990) 1–27
[G3]   Goodwillie, T.: Calculus II, analytic functors. To appear in K-theory
[G4]   Goodwillie, T.: Notes on the cyclotomic trace. In preparation
[G5]   Goodwillie, T.: Calculus III, The taylor series of functor. In preparation
[G6]   Goodwillie, T.: Calculus IV, Applications to K-theory. In preparation
[W1]   Waldhausen, F.: Algebraic K-theory, concordance, and stable homotopy. In:
            Ann. Math. Studies **113** (1987) 392–417

Dehn Surgery on Knots

Cameron McA. Gordon

Department of Mathematics, The University of Texas at Austin
Austin, TX 78712, USA

1. Introduction

In [D], Dehn considered the following method for constructing 3-manifolds: remove a solid torus neighborhood $N(K)$ of some knot K in the 3-sphere S^3 and sew it back differently. In particular, he showed that, taking K to be the trefoil, one could obtain infinitely many non-simply-connected homology spheres in this way. Let $M_K = S^3 - \overset{\circ}{N}(K)$. Then the different resewings are parametrized by the isotopy class r of the simple closed curve on the torus ∂M_K that bounds a meridional disk in the re-attached solid torus. We denote the resulting closed oriented 3-manifold by $M_K(r)$, and say that it is obtained by r-*Dehn surgery on* K.

More generally, one can consider the manifolds $M_L(\mathbf{r})$ obtained by $\mathbf{r}$-Dehn surgery on a k-component link $L = K_1 \cup \cdots \cup K_k$ in S^3, where $\mathbf{r} = (r_1, \ldots, r_k)$. It turns out that *every* closed oriented 3-manifold can be constructed in this way [Wal, Lic]. Thus a good understanding of Dehn surgery might lead to progress on general questions about the structure of 3-manifolds.

Starting with the case of knots, it is natural to extend the context a little and consider the manifolds $M(r)$ obtained by attaching a solid torus V to an arbitrary compact, oriented, irreducible (every 2-sphere bounds a 3-ball) 3-manifold M with ∂M an incompressible torus, where r is the isotopy class (*slope*) on ∂M of the boundary of a meridional disk of V. We say that $M(r)$ is the result of r-*Dehn filling on* M.

An observed feature of this construction is that

generically, the topology of M persists in $M(r)$.

We shall illustrate this slogan by stating some results that give restrictions on the "exceptional" slopes r for which $M(r)$ represents some sort of degeneration of M. Specifically, we shall consider the following questions:

(1) When is an essential surface destroyed by Dehn filling?
(2) When is an essential surface created by Dehn filling?
(3) When is $M(r)$ "small"?

The results we shall state show that the present state of knowledge on these questions is quite good.

Proceedings of the International Congress
of Mathematicians, Kyoto, Japan, 1990
© The Mathematical Society of Japan, 1991

First notice that, roughly speaking, anything can happen under a *single* Dehn filling. For if we are interested in when $M(r)$ has a certain property, then we can simply start with a closed 3-manifold Q with that property, and let $M = Q - \overset{\circ}{N}(K)$ for some suitable knot K in Q. Then there is a Dehn filling on M that yields Q. This puts a limitation on the kind of results one expects to obtain. However, the theorems we shall state will usually assert that (for any M) the set of exceptional slopes (for the particular kind of degeneration in question) is small. In fact, if $\Delta(r,s)$ denotes the minimal geometric intersection number of two slopes r, s on ∂M, then the theorems will often be of the form that give an upper bound on $\Delta(r,s)$ for any pair of exceptional slopes r and s.

We conclude this introduction with some conventions that we shall adopt throughout the paper. All 3-manifolds and surfaces will be assumed to be compact and orientable, and M will always denote (as above) an irreducible 3-manifold with ∂M an incompressible torus. A knot K will always be assumed to be a (non-trivial) knot in S^3 unless otherwise stated, and we reserve the notation M_K $(= S^3 - \overset{\circ}{N}(K))$ for this case. Slopes on ∂M_K will be parametrized by $\mathbb{Q} \cup \{\infty\}$ in the usual way, using a meridian-longitude basis $\{\mu, \lambda\}$ for $H_1(\partial M_K)$. Thus $r \leftrightarrow a/b$ if and only if $[r] = a\mu + b\lambda$ in $H_1(\partial M_K)$. Here, ∞ denotes $1/0$ (so $M_K(\infty) \cong S^3$).

Finally, since the condition $\Delta(r,s) = 1$ occurs frequently in the sequel, we remark that any set of slopes, such that $\Delta(r,s) = 1$ for every pair of distinct slopes in the set, has at most 3 elements. Also, for slopes on ∂M_K, $\Delta(r, \infty) = 1$ if and only if r is an integer.

2. When Is an Essential Surface Destroyed by Dehn Filling?

A properly embedded surface F in a 3-manifold Q is *essential* if either F is incompressible, not parallel into ∂Q, and not a 2-sphere, or F is a 2-sphere which does not bound a 3-ball in Q.

Essential surfaces play an important role in the theory of 3-manifolds. For instance there is the Kneser-Milnor prime factorization theorem, which involves cutting the manifold up into canonical pieces along essential 2-spheres. Essential tori feature in a similar way in the torus decomposition theory of Jaco-Shalen and Johannson. And of course there is the Haken-Waldhausen theory of irreducible 3-manifolds that contain essential surfaces (*Haken manifolds*), based on the idea of successively cutting the manifold along essential surfaces to eventually obtain a disjoint union of 3-balls.

Coming to the title of this section, let us take the case of a *closed* essential surface in M.

As examples, first consider the case of a (p,q)-*cable* K in a solid torus $S^1 \times D^2$. In other words, take a simple closed curve on $\partial(S^1 \times D^2)$ that winds around p times meridionally and q times longitudinally ($q \geq 2$) and push it into $\text{int}(S^1 \times D^2)$. Then it turns out that there are infinitely many Dehn surgeries on K that yield a solid torus again. In particular, there are infinitely many Dehn surgeries on K under which $\partial(S^1 \times D^2)$ compresses. To put this in the setting of our question, we simply take a knot K_0 in a closed 3-manifold Q (such that $M_0 = Q - \overset{\circ}{N}(K_0)$ is irreducible and ∂M_0 is incompressible), and identify $N(K_0)$ with $S^1 \times D^2$ by some homeomorphism, so that K becomes a knot in Q. (We say that K is a (p,q)-*cable of* K_0.) Then the torus $S = \partial N(K_0)$ is essential in $M = Q - \overset{\circ}{N}(K)$, but compresses in $M(r)$ for infinitely many slopes r.

Analogous examples with surfaces S of higher genus can be constructed by starting with a curve C in the boundary of a handlebody X of genus ≥ 2 such that $\partial X - C$ is incompressible in X, pushing C into $\overset{\circ}{X}$ to get K, say, taking a suitable embedding of X in some closed 3-manifold Q, and letting $M = Q - \overset{\circ}{N}(K)$ and $S = \partial X$.

An example of a different kind is the knot K in $S^1 \times D^2$ illustrated in Fig. 1.

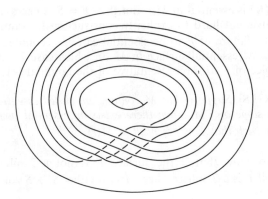

Fig. 1

Let $W = S^1 \times D^2 - \overset{\circ}{N}(K)$. Then, parametrizing slopes on $\partial N(K)$ using the meridian and longitude that come from the embedding of K in S^3 defined by Fig. 1, we have

$$W(\infty) \cong W(18) \cong W(19) \cong S^1 \times D^2,$$

so $\partial(S^1 \times D^2)$ compresses in $W(r)$ for 3 distinct slopes r. (In fact K is the unique non-cable curve in $S^1 \times D^2$ (which is not a core and does not lie in a 3-ball) with this property [Ga2, B1, Sc].) Note that

$$\Delta(18, 19) = \Delta(18, \infty) = \Delta(19, \infty) = 1.$$

Similar examples in handlebodies of genus ≥ 2 have been constructed by Berge (unpublished). Again, as above, these can be used to construct essential closed surfaces S in irreducible 3-manifolds M with torus boundary such that S compresses under 3 distinct Dehn fillings on M.

With these examples in mind we have the following result of Wu.

Theorem 2.1 [Wu2]. *Let S be a closed essential surface in M which compresses in $M(r)$ and $M(s)$. Then either $\Delta(r, s) = 1$, or the core K of the solid torus V in $M(s) = M \cup V$ can be isotoped onto S and S compresses in $M(r)$ for infinitely many r.*

In particular, if S is a torus then either $\Delta(r,s) = 1$ or K is a cable. In this case the theorem was originally proved in [CGLS].

Let us now consider when an essential surface *with boundary* in M is destroyed by Dehn filling. By this we mean the following. Let F be an essential surface in M with non-empty boundary, so each boundary component of F has slope r, say. (We say that r is a *boundary-slope*.) Then F gives rise to a closed surface $\widehat{F}$ in $M(r) = M \cup V$ by capping off the boundary components of F with meridian disks of V, and we are interested in the question: is $\widehat{F}$ essential in $M(r)$?

It is clear that the answer is not always "yes." For instance, we can start with a closed surface S in S^3, and take a knot K which punctures S in such a way that $F = S - \overset{\circ}{N}(K)$ is essential in M_K. But then $\widehat{F} = S$ is necessarily inessential in $M_K(\infty) = S^3$. (It is not hard to explicitly construct such examples.) However, it turns out that the failure of $\widehat{F}$ to be essential can be accounted for either by the presence of a *closed* essential surface in M (in which case Theorem 2.1 applies), or by a bad choice of F as a representative of the boundary-slope r.

Theorem 2.2 [CGLS]. *Suppose that M contains no closed essential surface, and let r be a boundary-slope on ∂M. Then there exists an essential surface F in M with boundary-slope r such that $\widehat{F}$ is essential in $M(r)$.*

One can also show that if the surface $\widehat{F}$ is not a 2-sphere, then $M(r)$ is irreducible, and if $\widehat{F}$ is a 2-sphere, then $M(r)$ is either $S^1 \times S^2$ or a connected sum of two lens spaces.

3. When Is an Essential Surface Created by Dehn Filling?

Suppose that $M(r) = M \cup V$ contains an essential surface S. It is straightforward to show that if S is moved (by an isotopy if $M(r)$ is irreducible, or in general, by a sequence of disk-swappings) so as to minimize the number of components of $S \cap V$, then either S lies in M or $F = S - \overset{\circ}{V}$ is an essential surface in M with boundary-slope r. The title of this section refers to the second possibility. In this context we have the following basic result of Hatcher.

Theorem 3.1 [Ha]. *M has only finitely many boundary-slopes.*

Corollary 3.2. *If M does not contain a closed essential surface then $M(r)$ contains a closed essential surface for only finitely many r.*

Since essential spheres and tori play a special role in 3-manifold theory, it is of interest to consider when they are created by Dehn filling.

In the case of essential spheres, the cabling construction described in Sect. 2 provides some examples. For if K is a (p,q)-cable of a knot K_0 in a 3-manifold Q, then $M = Q - \overset{\circ}{N}(K)$ contains an essential annulus with boundary-slope r where $\Delta(r,s) = 1$, s being the meridian of K. This annulus becomes a 2-sphere in $M(r)$, and in fact it splits $M(r)$ as a connected sum $M_0(r_0) \# L(q,p)$, where $M_0 = Q - \overset{\circ}{N}(K_0)$ and r_0 is some slope on ∂M_0. If we now choose Q to contain an essential sphere (and K_0 to be any knot in Q such that M_0 is irreducible), then

M is irreducible, whilst $M(r)$ and $M(s) = Q$ contain essential spheres. (It is not obvious, but in fact follows from [GLu1], that, here, $M_0(r_0)$ cannot be S^3.)

Another example where $M(r)$ and $M(s)$ contain essential spheres with $\Delta(r,s) = 1$, which does not come from cabling, is given in [GLi].

The next theorem says that this is the worst that can happen.

Theorem 3.3 [GLu3]. *If $M(r)$ and $M(s)$ each contain an essential sphere then $\Delta(r,s) = 1$.*

Turning to essential tori, a key example here is the exterior M_K of the figure eight knot K (see Fig. 2). M_K does not contain an essential torus, but $M_K(4)$, which is homeomorphic to $M_K(-4)$ since K is amphicheiral, does. Note that $\Delta(4,-4) = 8$.

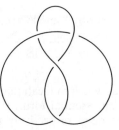

Fig. 2

Theorem 3.4 [Go]. *If M does not contain an essential torus, but $M(r)$ and $M(s)$ do, then $\Delta(r,s) \le 8$.*

In fact one can do a little better. The figure eight knot exterior belongs to the family of manifolds obtained by Dehn surgery on one component of the Whitehead link L (see Fig. 3). More precisely, let $W = S^3 - \overset{\circ}{N}(L)$ and let $W(p/q)$ be the manifold with torus boundary obtained by p/q-Dehn filling on one boundary component of W (where we use the obvious meridian-longitude parametrization). Then $W(1)$ is homeomorphic to the exterior of the figure eight knot. It also turns out that for $M = W(-5)$, $W(-5/2)$, and $W(2)$, M is irreducible and contains no essential torus, whilst there are r,s such that $M(r)$ and $M(s)$ contain essential tori with $\Delta(r,s) = 8,7$, and 6 respectively. One can show that these examples are the only ones with $\Delta(r,s) > 5$.

Addendum 3.5 [Go]. *In the setting of Theorem 3.4,*

if $\Delta(r,s) = 8$, then M is homeomorphic to $W(1)$ or $W(-5)$;
if $\Delta(r,s) = 7$, then M is homeomorphic to $W(-5/2)$;
if $\Delta(r,s) = 6$, then M is homeomorphic to $W(2)$.

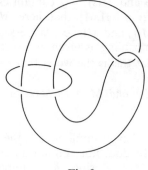

Fig. 3

4. When Is $M(r)$ Small?

Examples of what we might mean by "small" are:

When is $\pi_1(M(r))$ trivial? finite? cyclic?

When is $M(r)$ homeomorphic to S^3? a lens space? $S^1 \times S^2$? a Seifert fibred space?

This is a convenient point at which to recall that by [T], M is either hyperbolic, or Seifert fibred, or contains an essential torus. (For instance, if $M = M_K$ then the last two possibilities correspond respectively to K being a torus knot or a satellite knot.) As it is straightforward to describe $M(r)$ when M is Seifert fibred [He], and as the case when M contains an essential torus is to a large extent covered by Theorem 2.1, the hyperbolic case is the most important to understand. Here there is the following result of Thurston.

Theorem 4.1 [T]. *If M is hyperbolic then $M(r)$ is hyperbolic for all but finitely many r.*

A hyperbolic manifold is one whose interior has a complete Riemannian metric of constant negative (sectional) curvature. If we are willing to sacrifice "constant" in the conclusion of Theorem 4.1, then we have the following theorem, due to Gromov and Thurston, and improved in [BH] using a result of Adams.

Theorem 4.2. *If M is hyperbolic then $M(r)$ has a Riemannian metric of negative curvature for all but at most 24 values of r.*

It is conjectured that a 3-manifold with negative curvature is in fact hyperbolic. Even in the absence of a proof of this conjecture it is known that such a manifold must have infinite fundamental group, cannot contain an essential sphere or torus, cannot be Seifert fibred, etc.

Let us turn to the more specific question: When is $\pi_1(M(r))$ cyclic?

As a first set of examples, let K be a torus knot. Then $M_K(r)$ is a lens space for infinitely many r [Mo]. Note that in this case M_K is Seifert fibred.

A set of examples of a different flavor comes from the knot K in $S^1 \times D^2$ depicted in Fig. 1 and discussed in Sect. 2. Embedding $S^1 \times D^2$ in S^3 by applying k meridional twists to the embedding shown in Fig. 1, we obtain an infinite family of knots K_k, $-\infty < k < \infty$, (which are in fact hyperbolic). The two non-trivial Dehn surgeries on K that yield $S^1 \times D^2$ determine two Dehn surgeries on K_k, with slopes r_k and s_k, say, that yield lens spaces, where $\Delta(r_k, s_k) = \Delta(r_k, \infty) = \Delta(s_k, \infty) = 1$. For instance K_0 is the $(-2, 3, 7)$ pretzel knot shown in Fig. 4 (this example was discovered earlier by Fintushel and Stern [FS]), for which

$$M_{K_0}(18) \cong L(18, 5) , \quad M_{K_0}(19) \cong L(19, 7) , \quad M_{K_0}(\infty) \cong S^3 .$$

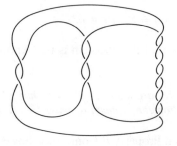

Fig. 4

The following Cyclic Surgery Theorem shows that these examples are extremal.

Theorem 4.3 [CGLS]. *Suppose that M is not Seifert fibred. If $\pi_1(M(r))$ and $\pi_1(M(s))$ are cyclic then $\Delta(r, s) = 1$.*

Less is known about when $\pi_1(M(r))$ is finite. An interesting example here is the manifold $W(-5)$ defined in Sect. 3. It is shown in [We] that there exist r, s with $\Delta(r, s) = 3$ such that $\pi_1(W(-5)(r))$ and $\pi_1(W(-5)(s))$ are finite.

Question 4.4. *If M is hyperbolic and $\pi_1(M(r))$ and $\pi_1(M(s))$ are finite, is $\Delta(r, s) \leq 3$?*

The case where $M = M_K$ for a satellite knot K (so M contains an essential torus) is discussed in [BH], where it is shown that $\Delta(r, s) \leq 5$, and that this bound is attained.

Regarding the question of when $\pi_1(M(r))$ is trivial, the following conjecture is still open.

Conjecture 4.5. $\pi_1(M(r))$ *is trivial for at most one slope r.*

5. Dehn Surgery on Knots in S^3

Continuing the theme of the previous section, let us ask: when can we have two Dehn fillings on M that give small manifolds, one of which is S^3? Clearly there is a Dehn filling on M that gives S^3 if and only if M is homeomorphic to M_K for some knot K in S^3, with the slope that determines the filling corresponding to the meridian of K. So our question becomes: given a knot K in S^3, when does non-trivial Dehn surgery on K yield a small manifold?

First we have the result that S^3 can never be obtained by non-trivial Dehn surgery.

Theorem 5.1 [GLu2]. $M_K(r) \not\cong S^3$ if $r \neq \infty$.

An immediate corollary of this theorem is that knots are determined by their complements.

Corollary 5.2. If K_1 and K_2 are knots in S^3 such that $S^3 - K_1$ is homeomorphic to $S^3 - K_2$, then there is a homeomorphism of pairs $(S^3, K_1) \cong (S^3, K_2)$.

The following so-called Property P Conjecture, however, is still open, although it is known to be true for many classes of knots.

Conjecture 5.3. $\pi_1(M_K(r)) \neq 1$ if $r \neq \infty$.

Of course, Conjectures 4.5, 5.3, and Theorem 5.1 are closely related, and are equivalent if the Poincaré Conjecture is true.

After S^3 (the unique 3-manifold of Heegaard genus zero), we might consider when Dehn surgery on a knot K yields a manifold of Heegaard genus one, *i.e.*, $S^1 \times S^2$ or a lens space.

For homological reasons, the only Dehn surgery that could possibly give $S^1 \times S^2$ is 0-Dehn surgery. However, a result of Gabai implies that this never happens.

Theorem 5.4 [Ga1]. $M_K(0)$ is irreducible.

Regarding lens spaces, Berge [B2] gives an explicit construction which yields several infinite classes of knots with a Dehn surgery yielding a lens space. (These include the infinite family described in Sect. 4 with *two* such Dehn surgeries.) It appears to be not entirely outside the bounds of possibility that every such knot can be constructed in this way.

Question 5.5. *Does every knot K such that $M_K(r)$ is a lens space for some r appear in Berge's list?*

An affirmative answer to this question would imply, in particular, the truth of the following conjecture.

Conjecture 5.6. $M_K(r) \not\cong RP^3 \, (= L(2, 1))$, $L(3, 1)$, or $L(4, 1)$.

Finally, we mention that for satellite knots, the question of when $\pi_1(M_K(r))$ is cyclic is completely solved.

Theorem 5.7 [Wan, Wu1, BL]. *Let K be a satellite knot. Then $\pi_1(M_K(r))$ is cyclic if and only if K is the $(2pq \pm 1, 2)$-cable of a (p,q)-torus knot and $r = 4pq \pm 1$, in which case $M_K(r) \cong L(4pq \pm 1, 4q^2)$.*

We now turn to the creation of essential spheres and tori by Dehn surgery on a knot K.

As far as spheres are concerned, recall the discussion of cables in Sect. 3. In our present setting of knots in S^3, if K is the (p,q)-cable of K_0 then the boundary-slope of the essential annulus in M_K is pq, and

$$M_K(pq) \cong M_{K_0}(p/q) \# L(q,p).$$

(It is convenient here to allow K_0 to be unknotted, in which case K is the (p,q)-torus knot and $M_K(pq) \cong L(p,q) \# L(q,p)$. Also, if K_0 is non-trivial then $\pi_1(M_{K_0}(p/q)) \neq 1$ by [CGLS, Corollary 2].) The following Cabling Conjecture of González-Acuña and Short asserts that these are the only examples where essential spheres arise.

Conjecture 5.8 [G-AS]. *$M_K(r)$ contains an essential sphere if and only if K is a (p,q)-cable and $r = pq$.*

Although this conjecture is still open, several partial results exist. For instance, there is Theorem 5.4 above. Also, it is known to be true if K is a satellite knot [Sc], an alternating knot [MT], a strongly invertible knot [E-M], and others [G-AS]. It is also known that if $M_K(r)$ contains an essential sphere then r is an integer [GLu1] and $M_K(r)$ has a lens space summand [GLu2]. (Incidentally, it is an immediate consequence of this last result that no non-prime homology sphere can be obtained by Dehn surgery on a knot in S^3. There remains the interesting question of whether there are *prime* homology spheres that cannot be so obtained.)

Regarding essential tori, some examples are known, but the general picture is not yet clear. However, the following conjecture seems reasonable. (Recall that M_K contains an essential torus if and only if K is a satellite knot.)

Conjecture 5.9. *If K is not a satellite knot and $M_K(r)$ contains an essential torus then $\Delta(r, \infty) \leq 2$.*

(There are examples with $\Delta(r, \infty) = 2$.)

6. The General Knot Complement Problem

Theorem 5.1 above asserts that if $M(r) \cong M(s) \cong S^3$, then $r = s$. More generally, one can ask: when is $M(r) \cong M(s)$? One has to be a little careful with orientations here, for M may have an orientation-reversing automorphism. For instance, if K is an amphicheiral knot (such as the figure eight knot), then $M_K(r) \cong M_K(-r)$. However, if we orient M, then this determines an orientation of $M(r)$, and we can ask: when is $M(r) \cong M(s)$ by an orientation-preserving homeomorphism? No counterexamples are known to the following conjecture.

Conjecture 6.1. *If $M(r) \cong M(s)$ by an orientation-preserving homeomorphism then $r = s$.*

This would imply the truth of the following Oriented Knot Complement Conjecture.

Conjecture 6.2. *If K_1 and K_2 are knots in a closed, oriented 3-manifold Q such that $Q - K_1$ and $Q - K_2$ are homeomorphic by an orientation-preserving homeomorphism, then there exists an orientation-preserving homeomorphism $h : Q \to Q$ such that $h(K_1) = K_2$.*

Mathieu [Ma] has given examples where $Q - K_1$ and $Q - K_2$ are homeomorphic (by an orientation-reversing homeomorphism), but there is *no* homeomorphism of pairs $(Q, K_1) \cong (Q, K_2)$.

7. Intersections of Surfaces

The proofs of the theorems stated above use several techniques from modern 3-dimensional topology. See [Sh] for instance, for a nice account of some of the ideas that went into the proof of the Cyclic Surgery Theorem (Theorem 4.3). However, here we shall focus on just one technique, namely, the analysis of intersections of properly embedded surfaces. The point is that many of the particular forms of what we have called the "degeneration" of the topology of M in $M(r)$ imply the existence of a (useful) surface F in M with non-empty boundary such that each boundary component has slope r on ∂M. For instance, as was pointed out in Sect. 3, if $M(r)$ contains an essential surface S which cannot be moved into M, then M contains an essential surface F with boundary-slope r such that $\widehat{F} \cong S$. Hence if $M(r_1)$ and $M(r_2)$ contain such surfaces S_1, S_2 respectively, then we get corresponding essential surfaces F_1, F_2 in M with boundary-slopes r_1, r_2. Putting F_1 and F_2 in mutual general position, $F_1 \cap F_2$ will consist of circles and arcs properly embedded in F_1 and F_2. Note that the pattern of intersections on the boundary, that is, the triple $(\partial M; \partial F_1, \partial F_2)$ is completely standard: all the components of ∂F_i are parallel, $i = 1, 2$, and each component of ∂F_1 meets each component of ∂F_2 in $\Delta(r_1, r_2)$ points. A final important point is that we may assume that no arc of $F_1 \cap F_2$ is isotopic in F_i (keeping its endpoints fixed) into ∂F_i, $i = 1, 2$. This easily follows from the fact that F_1 and F_2 are essential.

The idea is to then analyze the pattern of arcs of $F_1 \cap F_2$ as they lie on F_1 and F_2. We record the intersections on the boundary by numbering the boundary components of F_i in order as they appear on ∂M, $i = 1, 2$, and labelling the points of $\partial F_1 \cap \partial F_2$ on F_1 with the number of the corresponding boundary component of F_2 (and vice versa). It is convenient to regard the disks $\widehat{F}_i - \overset{\circ}{F}_i$ as "fat" vertices, and the arcs of $F_1 \cap F_2$ as edges, so that we get graphs Γ_1, Γ_2 in $\widehat{F}_1, \widehat{F}_2$ respectively, whose vertices and edge-endpoints are labelled as just described. Typically, one now proceeds to show that if $\Delta(r_1, r_2)$ is greater than some Δ_0, then the pair of graphs Γ_1, Γ_2 must contain certain configurations of faces which have topological implications for M (or $M(r_1)$ or $M(r_2)$) inconsistent with the hypotheses. This type of argument was first used by Litherland [Lit].

Theorems 3.3 and 3.4 are proved in this way, the surfaces S_i being spheres and tori respectively. The proof of Theorem 2.1 is similar in spirit, only here the surfaces S_i are disks, whose boundaries lie on the closed surface S, realizing the compression of S in $M(r_i)$, $i = 1, 2$. (This set-up is also an ingredient of the proof of Theorem 4.3.) In Theorem 5.1, the hypothesis is that $M(r_1) \cong M(r_2) \cong S^3$, which of course does not contain any essential surface. Nevertheless, using an idea of Gabai [Ga1], 2-spheres S_i in $M(r_i)$, $i = 1, 2$, can be chosen so that the corresponding punctured spheres F_i in M still satisfy the key condition that no arc of $F_1 \cap F_2$ can be isotoped into the boundary in either F_1 or F_2. This allows one to apply the same philosophy.

8. Conclusion

As mentioned in the Introduction, the importance of Dehn surgery lies in the fact that every closed oriented 3-manifold can be obtained by Dehn surgery on some link in S^3. Thus one might regard a Dehn surgery presentation of a 3-manifold as being analogous to a Heegaard splitting, with the study of Dehn surgery on knots corresponding to the study of Heegaard splittings of genus one. However, I think this analogy is misleading, because, unlike Heegaard splittings of genus one, Dehn surgery on knots already seems to contain much of the complexity of the general case. This, together with the progress that has been made in the case of a single component, makes it hopeful that Dehn surgery might be a useful way to approach general problems about 3-manifolds. At any rate, the next step should be to find appropriate generalizations of the results discussed above to the case where solid tori are attached along the boundary components of an irreducible 3-manifold M whose boundary is a disjoint union of an arbitrary number of tori.

References

[B1]   Berge, J.: The knots in $D^2 \times S^1$ which have nontrivial Dehn surgeries that yield $D^2 \times S^1$. Topology and its Applications **38** (1991) 1–19

[B2]   Berge, J.: Obtaining lens spaces by surgery on knots. To appear in Proceedings of International Conference on Knots 90. de Gruyter

[BH]   Bleiler, S.A., Hodgson, C.G.: Spherical space forms and Dehn surgery. To appear

[BL]   Bleiler, S.A, Litherland, R.A.: Lens spaces and Dehn surgery. Proc. Amer. Math. Soc. **107** (1989) 1127–1131

[CGLS] Culler, M., Gordon, C.McA., Luecke, J., Shalen, P.B.: Dehn surgery on knots. Ann. Math. (2) **125** (1987) 237–300

[D]   Dehn, M.: Über die Topologie des dreidimensionalen Raumes. Math. Ann. **69** (1910) 137–168

[E-M]   Eudave-Muñoz, M.: Band sum of links which yield composite links. To appear

[FS]   Fintushel, R., Stern, R.: Constructing lens spaces by surgery on knots. Math. Z. **175** (1980) 33–51

[Ga1]   Gabai, D.: Foliations and the topology of 3-manifolds. III. J. Diff. Geom. **26** (1987) 479–536

[Ga2]   Gabai, D.: Surgery on knots in solid tori. Topology **28** (1989) 1–6

[G-AS] González-Acuña, F., Short, H.: Knot surgery and primeness. Math. Proc. Camb.
        Phil. Soc. **99** (1986) 89–102
[Go]   Gordon, C.McA.: Boundary slopes of punctured tori in 3-manifolds. To appear
[GLi]   Gordon, C.McA., Litherland, R.A.: Incompressible planar surfaces in 3-mani-
        folds. Topology and its Applications **18** (1984) 121–144
[GLu1] Gordon, C.McA., Luecke, J.: Only integral Dehn surgeries can yield reducible
        manifolds. Math. Proc. Camb. Phil. Soc. **102** (1987) 97–101
[GLu2] Gordon, C.McA., Luecke, J.: Knots are determined by their complements. J.
        Amer. Math. Soc. **2** (1989) 371–415
[GLu3] Gordon, C.McA., Luecke, J.: Reducible manifolds and Dehn surgery. To appear
[Ha]   Hatcher, A.E.: On the boundary curves of incompressible surfaces. Pacific J.
        Math. **99** (1982) 373–377
[He]   Heil, W.: Elementary surgery on Seifert fiber spaces. Yokohama Math. J. **22**
        (1974) 135–139
[Lic]   Lickorish, W.B.R.: A representation of orientable combinatorial 3-manifolds.
        Ann. Math. **76** (1962) 531–540
[Lit]   Litherland, R.A.: Surgery on knots in solid tori, II. J. London Math. Soc. (2) **22**
        (1980) 559–569
[Ma]   Mathieu, Y.: Sur des noeuds qui ne sont pas déterminés par leur complément et
        problèmes de chirurgie dans les variétés de dimension 3. Thèse, L'Université de
        Provence, 1990
[MT]   Menasco, W.W., Thistlethwaite, M.B.: Surfaces with boundary in alternating knot
        exteriors. To appear
[Mo]   Moser, L.: Elementary surgery along a torus knot. Pacific J. Math. **38** (1971)
        734–745
[Sc]   Scharlemann, M.: Producing reducible 3-manifolds by surgery on a knot. Topol-
        ogy **29** (1990) 481–500
[Sh]   Shalen, P.B.: Representations of 3-manifold groups and applications in topology.
        Proceedings ICM Berkeley, 1986, pp. 607–614
[T]   Thurston, W.: Three dimensional manifolds, Kleinian groups and hyperbolic
        geometry. Bull. Amer. Math. Soc. **6** (1982) 357–381
[Wal]   Wallace, A.H.: Modifications and cobounding manifolds. Can. J. Math. **12** (1960)
        503–528
[Wan]   Wang, S.: Cyclic Surgery on knots. Proc. Amer. Math. Soc. **107** (1989) 1091–1094
[We]   Weeks, J.R.: Hyberbolic structures on three-manifolds. Ph.D. thesis, Princeton
        University, 1985
[Wu1]   Wu, Y-Q.: Cyclic surgery and satellite knots. Topology and its Applications **36**
        (1990) 205–208
[Wu2]   Wu, Y-Q.: Incompressibility of surfaces in surgered 3-manifolds. To appear in
        Topology

Parametrized Morse Theory and Its Applications

*Kiyoshi Igusa**

Department of Mathematics, Brandeis University, Waltham, MA 02254, USA

We will discuss the basic concepts of parametrized Morse theory and show how they are related to the stability theorem, Waldhausen's $A(X)$ and higher Franz-Reidemeister torsion.

§ 1. Standard Morse Theory

In standard Morse theory one usually takes a compact smooth (C^∞) manifold M with boundary the union of three manifolds $\partial_0 M \cup \partial_1 M \cup D \times I$ where $I = [0, 1]$, $D \times 0 = \partial\partial_0 M$ and $D \times 1 = \partial\partial_1 M$. However for the purpose of these notes we will stick mainly to the case when D is empty. We will also write $X = \partial_0 M$. Thus $\partial M = X \coprod \partial_1 M$.

Any generic smooth function $f : M \to I$ with $f^{-1}(i) = \partial_i M$ for $i = 1, 2$ will be a Morse function with a finite number of nondegenerate singularities $x_1, \ldots, x_m$. Each critical point x_i has an index $\mathrm{ind}(x_i) \geq 0$. We can construct a finite relative CW complex
$$Y = X \cup e_1 \cup e_2 \cup \cdots \cup e_m$$
with one cell e_i for each critical point x_i and $\dim(e_i) = \mathrm{ind}(x_i)$. From this relative cell complex we can construct the cellular chain complex $C_*(Y, X)$ with coefficients in a ring R or with twisted coefficients given by a locally constant sheaf $\mathcal{F}$ over M. If this chain complex is acyclic then we get a torsion invariant $\tau(Y) = \tau(M)$ which is an element of some quotient group of $K_1 R$ independent of the choice of the function f (assuming $K_0 Z \to K_0 R$ is a monomorphism).

Example 1.1. $M = S^1 \times I$, $X = \partial_0 M = \emptyset$, $\partial_1 M = S^1 \times \{0, 1\}$ and f is the "height" in the following drawing.

In Example 1.1 there are two critical points x_0, x_1 of indices 0, 1 and $Y = e^0 \cup e^1$ is a cricle. There is no torsion unless we take twisted coefficients in which case $\tau(Y) = \tau(M) = \pm(1 - h^{\pm 1})$ where h is the holonomy of the coefficient sheaf around the circle $S^1 \times 1/2$ in M and the sign ambiguities are due partially to the fact that the 1-cell e^1 is unoriented.

---

* Supported by NSF Grant No. DMS9002512.

Proceedings of the International Congress
of Mathematicians, Kyoto, Japan, 1990
© The Mathematical Society of Japan, 1991

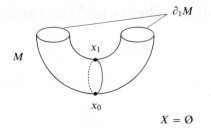

$$X = \emptyset$$

§2. Parametrized Morse Theory

The study of parametrized Morse theory began with J. Cerf who used 1 and 2 parameter families of functions on $M = X \times I$ to prove the following theorem with definitions given below.

Theorem 2.1 [C2]. *If X is a simply connected with dimension ≥ 5 then $\mathscr{C}(X)$ is connected.*

Definition 2.2. If X is a compact smooth manifold then the *concordance space* of X is defined to be the space $\mathscr{C}(X) = \mathrm{Diff}(X \times I \text{ rel } X \times 0 \cup \partial X \times I)$ where $\mathrm{Diff}(M \text{ rel } A)$ denotes the space of all self diffeomorphisms of M which are the identity on A with the strong C^∞ topology [C1].

Instead of taking families of functions on a single manifold we will consider functions on "families of manifolds." This is an equivalent point of view. We take a smooth bundle (i.e. submersion) $p : E \to B$ with fiber M so that E contains the trivial subbundle $X \times B$. If $B = S^{k+1}$, the isomorphism classes of such bundles are in one-to-one correspondence with the elements of $\pi_i \mathrm{Diff}(M \text{ rel } X)$. We are interested in calculating these groups and of finding computable invariants.

Given a smooth bundle $M \to E \to B$ as above we will construct maps:

$$\theta : B \to |\mathscr{E}.(X)| = \{\text{cell complexes and characteristic maps}\}$$

$$\lambda\theta : B \to |\mathscr{K}.(Z, F)| = \{\text{filtered chain complexes}\}$$

The second map is constructed from the first by composition with a map $\lambda : |\mathscr{E}.(X)| \to |\mathscr{K}.(Z, F)|$. Taking $B = S^{k+1}$ we will get homomorphisms

$$\theta_+ : \pi_k \mathrm{Diff}(M \text{ rel } X) \to \pi_{k+1} |\mathscr{E}.(X)|$$

$$\lambda_*\theta_* : \pi_k \mathrm{Diff}(M \text{ rel } X) \to \pi_{k+1} |\mathscr{K}.(Z, F)|.$$

It turns out that θ_* is an isomorphism in many cases. Its purpose it to compute the group $\pi_k \mathrm{Diff}(M \text{ rel } X)$. The map $\lambda_*\theta_*$ is used to obtain algebraic invariants to detect particular elements of $\pi_k \mathrm{Diff}(M \text{ rel } X)$.

The idea behind θ and λ is as follows. We take a "good" function $f: E \to I$ and consider it as a family of functions on the fibers $M_t = p^{-1}(t)$ of $p: E \to B$. This will give us a "family of cell complexes" Y_t which we interpret as a map from B to the "space of all cell complexes." By a "linearization" process we get a map to the "space of all filtered chain complexes." We will explain what these terms mean.

§3. Waldhausen's Expansion Category

We use a model for the "space of all cell complexes" due to Waldhausen [W1]. Our nomenclature is however different: what we call the "expansion space" is the loop space of Waldhausen's expansion space. Our *expansion space* $|\mathscr{E}.(X)|$ is the geometric realization of a simplicial category $\mathscr{E}.(X)$. The objects of $\mathscr{E}_0(X)$, which form the vertices of $|\mathscr{E}.(X)|$, are given by pairs $(Y, \{\psi_e\})$ where Y is a finite relative cell complex $Y = X \cup e_1 \cup \cdots \cup e_m$ and $\psi_{e_i}: D^{n_i} \to Y$ are *characteristic maps* (i.e. parametrizations) for the cells. The objects of $\mathscr{E}_k(X)$, which form the k-simplices of $|\mathscr{E}.(X)|$, are given by pairs $(Y, \{\psi_e\})$ where Y is a k parameter family of finite relative cell complex $Y = X \times \varDelta^k \cup e_1 \times \varDelta^k \cup \cdots \cup e_m \times \varDelta^k$ and each $\psi_{e_i}: D^{n_i} \times \varDelta^k \to Y$ is a k parameter family of characteristic maps. To avoid set theoretic difficulties we assume that Y is a subspace of a fixed infinite dimensional contractible space, e.g., $\mathbb{R}^\infty$. This also clarifies our assertion that morphism are inclusion maps. Figure 3.1 gives an example for $k = 1$.

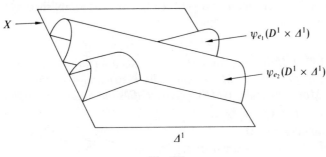

Fig. 3.1

The morphisms of $\mathscr{E}_k(X)$, which may be viewed as gluing maps for the k-simplices of $|\mathscr{E}.(X)|$, are given by expansions. *Expansions* are compositions of *elementary expansions* which are given by inclusion maps: $Y \to Y \cup D^n \times \varDelta^k$ where the k parameter family of disks $D^n \times \varDelta^k$ is attached to the k parameter family of cell complexes Y along the southern hemisphere $S^{n-1} \times \varDelta^k$. Figure 3.2 shows a path in $|\mathscr{E}.(X)|$ which is obtained by gluing two edges $A, B \in \mathscr{E}_k(X)$ along a morphism $f: \partial_1 A \to \partial_0 B$ from an endpoint of A to an endpoint of B.

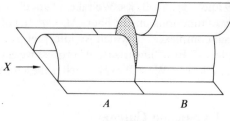

A B

Fig. 3.2

An old unpublished result of Waldhausen gives a computation of the homotopy type of some of the components of the expansion space in terms of the Waldhausen K-theory of X. A proof of this is being prepared for publication in [IW].

Theorem 3.3 (Waldhausen). $A(X) \simeq Q(X_+) \times B|\mathscr{E}.^h(X)|$ where $A(X) = $ Waldhausen K-theory of X [W2], $Q(X_+) = \Omega^\infty \Sigma^\infty (X \coprod \text{pt.})$ $B|\mathscr{E}.^h(X)|$ is a nonconnected delooping of $|\mathscr{E}.^h(X)|$, $\mathscr{E}.^h(X)$ is the simplicial full subcategory of $\mathscr{E}.(X)$ consisting of pairs (Y, ψ) where $Y \simeq X$.

§4. The Framed Function Theorem

Suppose we have a smooth bundle $M \to E \to B$ as in Section 2 above. Then we would like to construct a map $\theta : B \to |\mathscr{E}.(X)|$. This is accomplished by the following two theorems.

Theorem 4.1 [I1]. *If* dim $B \leq$ dim M *there exists a smooth function* $f : E \to I$ *so that for each* $t \in B$, *the map* $f_t : M_t \to I$ *given by restricting* f *to the fiber over* t *is a generalized Morse function* (GMF), *i.e. it satisfies the following conditions.*

a) $f_t^{-1}(0) = X$, $f_t^{-1}(1) = \partial_1 M_t$,
b) f_t *is nonsingular on* ∂M_t,
c) f_t *has only* A_1 *and* A_2 *singularities.*

Since $f : E \to I$ is being considered as a family of functions $f_t : M_t \to I$ we consider the *fiberwise singular set* $\sum(f)$ of f which is defined to be the set of all x in E so that x is a singularity of $f_t : M_t \to I$ where $t = p(x)$ is the image of x in B.

Definition 4.2. $y \in M$ is an A_k singularity of $g : M \to \mathscr{R}$ if g can be written in local coordinates as

$$g = \pm x_0^{k+1} + \sum_{i=1}^{n} \pm x_i^2 + C.$$

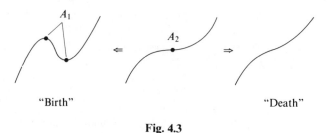

"Birth" "Death"

Fig. 4.3

By the Morse lemma A_1 singularities are the same as nondegenerate singularities. A_2 singularities are also called *birth-death* singularities. This is because an A_2 singularity can be perturbed to give either two A_1 singularities in "cancelling position" or to give no singularities. This is illustrated in Fig. 4.3. (As before f is the "height" function given by the distance from the bottom of the page.)

Theorem 4.1 is not sufficient for our purposes. Given a smooth bundle $M \to E \to B$ and a smooth function $f: E \to I$ as described in Theorem 4.1 we can do Morse theory on the family of generalized Morse functions $f_t: M_t \to I$ and we get a family of cell complexes parametrized by B. However we do not get a family of characteristic maps $\{\psi_e\}$ as required in the definition of the expansion category. Therefore we do not get a map from B into its geometric realization. To remedy this situation we have another theorem.

Theorem 4.4 [I2] (Framed Function Theorem). *In Theorem* 4.1 *the smooth map* $f: E \to I$ *can be chosen so that it admits a "framing." This consists of a Riemannian metric for E and a function ξ on the fiberwise singular set of f of the following form.*

a) *For each critical x of $f_t: M_t \to I$, $\xi(x) = \xi_t(x) = (\xi^1, \ldots, \xi^i)$ is an orthonormal framing for the nonpositive eigenspace of $D^2 f_t(x)$, (i.e. the sum of all eigenspaces of $D^2 f_t(x)$ corresponding to nonpositive eigenvalues).*

b) *For each j, $\xi_t^i(x)$ is a continuous function on its domain.*

c) *At each birth-death singularity the last vector ξ^i lies in $\ker D^2 f_t(x)$ and the intrinsically defined third derivative $D^3 f_t(x)(\xi^i, \xi^i, \xi^i)$ is positive.*

Furthermore, if $\dim B < \dim M$ *then* (f, ξ) *is unique up to framed homotopy.*

We will call the pair (f, ξ) a *fiberwise framed function* on E.

Figure 4.5 illustrates the definition of a framing. In Fig. 4.5.1 there are two A_1 singularities of indices 0, 1. Thus a framing associates a unit tangent vector ξ^1 to the A_1 point of index 1. This goes continuously to the indicated framed A_2 singularity. The unit vector ξ^1 must point to the right at the A_2 singularity by condition (c) in Theorem 4.4. In Fig. 4.5.2 the same A_1 singularities with the opposite framing cannot be cancelled in the same way, although there is a more complicated deformation which eliminates these two framed singularities. In Fig. 4.5.3 the framing vector ξ^2 plays the role of ξ^1 in the first two examples.

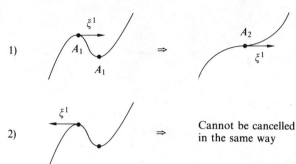

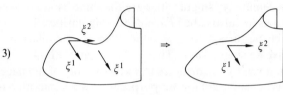

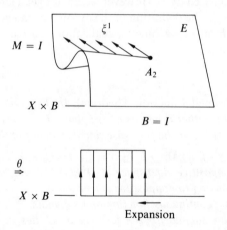

Fig. 4.5

Fig. 4.6

Now suppose that $M \to E \to B$ is a smooth bundle and $(f, \xi) : E \to I$ is a fiberwise framed function. Then for each t in B we get a GMF $f_t : E \to I$ and a framing ξ_t. The GMF f_t tells us now to construct a finite relative cell complex Y_t and the framing ξ_t is exactly what is needed to construct the characteristic maps $\{\psi_t\}$ for Y_t. Consequently we get a map $\theta : B \to |\mathscr{E}.(X)|$. The uniqueness of (f, ξ) up to framed homotopy implies that the map θ is unique up to homotopy. We note that the dimension condition $\dim B < \dim M$ can be circumvented by "stabilization," i.e. by taking the product with a large disk D^N to get $M \times D^N \to E \times D^N \to B$.

Figure 4.6 gives an example of this construction.

§5. Application to Pseudoisotopy

Suppose that $M = X \times 1$. Then smooth bundles $M \to E \to S^{k+1}$ are classified by the homotopy group $\pi_k \mathscr{C}(X)$ and the framed function theorem gives us a homomorphism

$$\theta_* : \pi_k \mathscr{C}(X) \to \pi_{k+1} |\mathscr{E}.(X)|$$

This map is an isomorphism if $3k \leq \dim X - 9$ basically because a smooth bundle can be reconstructed from a family of cell complexes by embedding it in Euclidean space and taking a neighborhood. Another way to say this is that we get a highly connected map

$$\theta : B\mathscr{C}(X) \to |\mathscr{E}.^h(X)|$$

where $B\mathscr{C}(X)$ is a nonconnected delooping of $\mathscr{C}(X)$. This leads to the following stability theorem. The details can be found in [13].

Theorem 5.1 (Stability Theorem). *The suspension map* $\sigma : \mathscr{C}(X) \to \mathscr{C}(X \times I)$ *is* $\sim \dim X/3$-*connected.*

Using Theorem 3.1 we also get another proof of the following famous theorem of Waldhausen [W2]:

Theorem 5.2 (Waldhausen). $A(X) \simeq Q(X_+) \times B^2 \mathscr{P}(X)$ *where* $\mathscr{P}(X)$ *is the direct limit of concordance spaces* $\mathscr{C}(X) \to \mathscr{C}(X \times I) \to \cdots \to \mathscr{C}(X \times I^n)$ *with respect to suspension and* $B^2 \mathscr{P}(X)$ *denotes a two-fold nonsimply-connected delooping of* $\mathscr{P}(X)$.

The space $\mathscr{P}(X)$ is called the *stable pseudoisotopy* (or *concordance*) space of X.

§6. The Space of Filtered Chain Complexes

This section is a report on recent joint work with John Klein.

One way to obtain K-theory invariants for smooth bundles is to triangulate the base space B and associate to each simplex the total singular complex of its inverse image. This gives a functor from the category of simplices of B to a category of chain complexes with additional structure. If this category is carefully constructed it will have homotopy groups closely related to algebraic K-groups. One suitable construction is the simplicial category of "filtered chain complexes."

Suppose that Z is a space, R is an associative ring with 1 and $\mathscr{F}$ is a locally constant sheaf of nonzero finitely generated (f.g.) free R-modules over Z. Then we construct the "space of filtered chain complexes." This is the geometric realization of a simplicial category $\mathscr{K}.(Z, \mathscr{F})$ whose objects in every degree are quadruples $(S, C, \{S^A\}, \{l_{A,e}\})$ where S is a free R-complex, C is a finite poset over Z. $\{S^A\}$ is a family of subcomplexes of S indexed by the order ideals in C and $\{l_{A,e}\}$ is a family of cohomology classes indexed by pairs (A, e) where A is an order ideal in C and e is a minimal element of the complement $C \backslash A$ of A in C. The precise definition is rather technical so we will explain it with two examples. We recall that an *order ideal* in a poset C is a subset A of C so that A contains any element of C which is less than any element of A.

Example 6.1. Let $f: M \to I$ be a Morse function as explained in Section 1, let ξ be a framing for f and let $\mathscr{F}$ be any locally constant sheaf of f.g. free R-modules over M. Then an object $(S, C, \{S^A\}, \{l_{A,e}\})$ of $\mathscr{K}_0(M, \mathscr{F})$ is given as follows.

a) $S = C_*(M, X; \mathscr{F})$ is the relative singular complex of (M, X) with coefficients in $\mathscr{F}$.
b) $C = \sum(f)$ is the singular set of f ordered by critical value. Thus $a < b$ iff $f(a) < f(b)$.
   i) $\mathrm{ind}: C \to N$ is given by $\mathrm{ind}(x) = $ *index of f at x.*
   ii) $p: C \to M$ is the inclusion map.
c) $S^A = C_*(f^{-1}[0, f(e)] \backslash (C \backslash A), X; \mathscr{F})$ for any order ideal A in C where e is a minimal element in $C \backslash A$. (If no such e exists then $A = C$ and $S^C = S$.)
d) If A is an order ideal in C and $e \in A$ is maximal with $\mathrm{ind}(e) = i$ then let

$$l_{A,e}: H_i(S^{A \cup e}, S^A) \overset{\sim}{\to} \mathscr{F}_{p(e)}$$

be the isomorphism specified by the framing $\xi(e)$ where $\mathscr{F}_{p(e)}$ is the stalk of $\mathscr{F}$ over $p(e)$.

Example 6.2. Let $\pi = \pi_1 X$ and let $B\pi$ be an open $K(\pi, 1)$ neighborhood of X in $\mathbb{R}^\infty$. Let $\mathscr{F}$ be any locally constant sheaf of f.g. free R-modules over $B\pi$. Let $\mathscr{E}.^\pi(X)$ denote the simplicial full subcategory of $\mathscr{E}.(X)$ whose objects are pairs (Y, ψ) so that Y is a subspace of $B\pi$. Then there is a simplicial functor $\lambda: \mathscr{E}.^\pi(X) \to \mathscr{K}.(B\pi, \mathscr{F})$ given in degree zero by $\lambda(Y, \psi) = (S, C, \{S^A\}, \{l_{A,e}\})$ where $S, C, \{S^A\}, (l_{A,e}$ are given as follows.

a) $S = C_*(Y, X; \mathscr{F}_Y)$ where $\mathscr{F}_Y$ is the restriction of $\mathscr{F}$ to Y.
b) C is the set of cells of Y. (The indexed poset C is actually part of the structure of (Y, ψ).)
   i) $\mathrm{ind}: C \to N$ is given by $\mathrm{ind}(x) = \dim(x)$.
   ii) $p: C \to B\pi$ is given by $p(x) = \psi_x(0)$.
c) $S^A = C_*(Y^A, X; \mathscr{F}_{Y^A})$ where Y^A is the closed subcomplex of Y corresponding to A.
d) If A is an order ideal in C and $e \in A$ is maximal with $\mathrm{ind}(e) = i$ then let

$$l_{A,e}: H_i(S^{A \cup e}, S^A) \overset{\sim}{\to} \mathscr{F}_{p(e)}$$

be the isomorphism specified by the characteristic map ψ_e.

Let $\mathscr{K}.^h(Z, \mathscr{F})$ denote the simplicial full subcategory of $\mathscr{K}.(Z, \mathscr{F})$ whose objects are those quadruples $(S, C, \{S^A\}, \{l_{A,e}\})$ so that S is acyclic. Assume that Z is connected and that the natural map $K_0 Z \to K_0 R$ is a monomorphism. Then using ideas from the proof of Theorem 3.3 we obtain the following.

Theorem 6.3 [IK]. *There is a homotopy fiber sequence:*

$$|\mathscr{K}.^h(Z, \mathscr{F})| \to Q(Z_+) \to \mathbb{Z} \times B\,\mathrm{Gl}(R)^+.$$

Corollary 6.4 (Higher Franz-Reidemeister Torsion). *Suppose that $\pi_1 M = \pi_1 X = \pi$ is a finite group, $R = \mathbb{C}$ and $\mathscr{F}$ has unitary holonomy, i.e. $\mathscr{F}$ is given by a unitary representation of π. Let $M \to E \to S^{2k}$ be a smooth bundle as in Section 2 so that the map in homology $H_*(M, X; \mathscr{F}) \to H_*(E, X \times S^{2k}; \mathscr{F})$ induced by the inclusion of M in E is a monomorphism. Then there is a naturally defined torsion invariant $\tau(E) \in \mathbb{R}$. Furthermore these invariants detect the rational homotopy groups of $\mathscr{P}(*)$ and are thus nontrivial for all even k.*

Remark. Higher Franz-Reidemeister torsion was first constructed by Wagoner [W] but only for bundles over circles. Klein [K] was the first to construct higher Franz-Reidemeister torsion invariants for bundles over higher dimensional spheres but he assumed that $H_*(M, X; \mathscr{F}) = 0$.

References

[C1] Cerf, J.: Topologie de certain espaces de plongements. Bull. Soc. Math. France **89** (1961) 227–380

[C2] Cerf, J.: La stratification naturelle des espaces de fonctions différentiables réelles et le théorème de la pseudo-isotopie. I.H.E.S. Publ. Math. **39** (1970) 5–173

[I1] Igusa, K.: On the homotopy type of the space of generalized Morse functions. Topology **23** (1984) 245–256

[I2] Igusa, K.: The space of framed functions. Transactions of the A.M.S. **301** (1987) 431–477

[I3] Igusa, K.: The stability theorem for smooth pseudoisotopies. K-Theory **2** (1988) 1–355

[IK] Igusa, K., Klein, J.: Filtered chain complexes and higher Franz-Reidemeister torsion. In preparation

[IW] Igusa, K., Waldhausen, F.: The expansion space and its application to pseudoisotopy theory. In preparation

[K] Klein, J.: The cell complex construction and higher R-torsion for bundles with framed Morse function. Ph.D. Thesis, Brandeis University, 1989

[W] Wagoner, J.: Diffeomorphisms, K_2, and analytic torsion. Proc. Symp. Pure Math., vol. 32, part I, AMS (1978) 23–33

[W1] Waldhausen, F.: Algebraic K-theory of topological spaces I. Proc. Symp. Pure Math. vol. 32, part I, AMS (1978) 35–60

[W2] Waldhausen, F.: Algebraic K-theory of spaces, Lecture Notes in Mathematics, vol. 1126. Springer, Berlin Heidelberg New York 1985, pp. 318–419

Rigidity in Geometry and Topology

F. Thomas Farrell[1] and Lowell E. Jones[2]

[1] SUNY, Binghamton, NY 13902, USA
[2] SUNY, Stony Brook, NY 11794, USA

Here is an outline of the paper.

In §1 we consider whether the class of closed Riemannian manifolds (closed = connected, compact, with no boundary) with sectional curvature satisfying $K \leq 0$ everywhere is smoothly rigid or is topologically rigid. In other words if M, N are two such manifolds with isomorphic fundamental groups, then are M, N diffeomorphic to one another or homeomorphic to one another? We review some of the history of the problem beginning with the rigidity results of Bieberbach and Mostow. Our two main results state that this class of manifolds is not smoothly rigid, but it is topologically rigid in dimensions greater than four.

In §2 we analyze the space $\text{TOP}(M)$ of all self-homeomorphisms of a closed Riemannian manifold M whose sectional curvature satisfies $K \leq 0$ everywhere. In this setting rigidity expresses itself through the simplicity of the homotopy type of $\text{TOP}(M)$, e.g. the rational homotopy groups of $\text{TOP}(M)$ vanish through a stable range of dimensions greater than one. We review results which reduce this problem to the study of the space $\mathscr{P}(M)$ of stable topological pseudoisotopies of M. Our main result in this section states that up to homotopy the space $\mathscr{P}(M)$ can be constructed in a simple way from the stable pseudoisotopy spectrum $\mathscr{P}_*(S^1)$ of the circle.

In §3 we formulate a general conjecture whose truth would certainly imply the main topological results of §1, §2, and would also imply many other well known conjectures in algebraic K-theory and algebraic topology (e.g. the Novikov Conjectures, the Borel Conjecture in dimensions ≥ 5, and Conjectures 1.8, 1.9, 2.8 of this paper). Roughly speaking our conjecture states that for any connected CW complex X the stable topological pseudoisotopy Ω-spectrum $\mathscr{P}_*(X)$ can be constructed in a simple way from the Ω-spectra $\{\mathscr{P}_*(X_H) : H \in \mathscr{G}(X)\}$, where $\mathscr{G}(X)$ denotes the collection of all cyclic by finite subgroups of $\pi_1 X$ and where $X_H \to X$ denotes the covering projection corresponding to any $H \in \mathscr{G}(X)$. This same conjecture is also made for the Ω-spectra valued functors associated to surgery L-theory, algebraic K-theory, and to smooth pseudoisotopy theory. Based on our recent computations of the surgery L-groups and the stable pseudoisotopy spectra of cocompact discrete subgroups of virtually connected Lie groups (cf. 3.4), we believe it is only a matter of time before these conjectures are verified when $\pi_1 X$ is a discrete subgroup of a virtually connected Lie group. For more general classes of fundamental groups the authors regard these conjectures as only estimates which best fit the known data at

Proceedings of the International Congress
of Mathematicians, Kyoto, Japan, 1990
© The Mathematical Society of Japan, 1991

this time. If these estimates prove incorrect for a larger class of groups, then one should look for the minimal class of subgroups $\mathscr{G}(X)$ of each $\pi_1 X$ which makes the preceeding conjecture true ($\mathscr{G}(X)$ is now forced to be larger then the class of cyclic by finite subgroups of $\pi_1 X$).

§1. Rigidity

We begin with a statement of the rigidity results of Bieberbach [5] and of Mostow [26, 27].

1.1 Bieberbach's Rigidity Theorem. *Let $f: N \to M$ be a homotopy equivalence between closed flat (i.e. sectional curvature satisfies $K = 0$ everywhere) Riemannian manifolds. Then f is homotopic to an affine diffeomorphism.*

1.2 Mostow's Rigidity Theorem. *Let $f: N \to M$ be a homotopy equivalence between closed locally symmetric spaces with sectional curvature satisfying $K \leq 0$ everywhere. Suppose that M has no closed one or two-dimensional totally geodesic subspaces which are local factors of M. Then f is homotopic to a diffeomorphism, which becomes an isometry after adjusting the normalizing factors of M.*

These two results suggest the following problem.

1.3 Problem. Let $f: N \to M$ denote a homotopy equivalence between two closed Riemannian manifolds which both have sectional curvature satisfying $K \leq 0$ everywhere. Is f homotopic to a diffeomorphism or to a homeomorphism?

1.4 History of Problem. In addition to the Rigidity Theorems of Bieberbach and Mostow, there are the following results which give partial answers to Problem 1.3. Eells and Sampson [9] showed that f is homotopic to a harmonic map $\bar{f}: N \to M$, and Al'ber [1] and Hartmann [18] showed that $\bar{f}$ is uniquely determined by f if $K < 0$ holds everywhere for M. Lawson and Yau then suggested that $\bar{f}: N \to M$ should be a diffeomorphism whenever $K < 0$ holds everywhere for both N, M. This was proven by Schoen and Yau [33] when dim $M = 2$, and follows from Mostow's Rigidity Theorem when N, M are locally symmetric spaces. The next theorem (cf. 1.5) gives counterexamples to the Lawson-Yau problem in dimensions greater than six, but the problem remains open in dimensions three through six. If $K = 0$ holds everywhere for M in 1.3 then it follows from work of Gromoll and Wolf [17], and from work of Yau [39], that $K = 0$ also holds everywhere for N; thus f must be homotopic to an affine diffeomorphism by Bieberbach's Rigidity Theorem. If M is an irreducible locally symmetric space of rank ≥ 2 Gromov [3] has shown that after rescaling the metric on M f will be homotopic to an isometry. Eberlein [7, 8] independently proved the same result under the hypothesis that the universal cover of M is reducible. Cheeger showed in the mid-1970s that the bundles of orthonormal two-frames $V_2(N)$, $V_2(M)$ are homeomorphic provided $K < 0$ holds everywhere for both N, M; and then, under the same hypothesis, Gromov showed that the unit sphere bundles $S(N)$, $S(M)$ are homeomorphic, via a homeomorphism which preserves the orbits of the geodesic flows. Mishchenko [25] showed that f pulls the rational Pointrjagin classes of M back to those of N; and Farrell and Hsiang [10] showed that $f \times \mathrm{id} : N \times \mathbf{R}^3 \to M \times \mathbf{R}^3$ is properly homotopic to a homeomorphism, where $\mathrm{id} : \mathbf{R}^3 \to \mathbf{R}^3$ denotes the identity map.

The authors [11, 15] have proven the following two theorems which more or less settle Problem 1.3. In the first of these theorems we let M denote any closed real hyperbolic manifold with sectional curvature satisfying $K = -1$ everywhere, and we let $\Sigma_1, \Sigma_2, \ldots, \Sigma_n$ denote a complete list of nondiffeomorphic exotic spheres of dimension equal dim $M > 4$.

1.5 Theorem. *Given any number $\delta > 0$ there is a finite sheeted covering space projection $\hat{M} \to M$ which satisfies the following properties.*

(a) No two of the manifolds $\hat{M}$, $\hat{M} \# \Sigma_1, \ldots, \hat{M} \# \Sigma_n$ are diffeomorphic, but they are all homeomorphic.

(b) Each $\hat{M} \# \Sigma_i$ supports a Riemannian metric all of whose sectional curvature values lie in the interval $(-1 - \delta, -1 + \delta)$.

1.6 Theorem. *Let $f : N \to M$ denote a homotopy equivalence between two closed manifolds, where M is a closed Riemannian manifold with sectional curvature satisfying $K \leq 0$ everywhere. Then f is homotopic to a homeomorphism, provided dim $M \geq 5$.*

1.6.1 Remark. There is the following more general version of 1.6: any homotopy equivalence $h : (N, \partial N) \to (M \times I^k, \partial(M \times I^k))$ of compact manifold pairs, which is a homeomorphism of the boundaries, is homotopic rel ∂ to a homeomorphism, provided $k + \dim M \geq 5$, where $k \geq 0$ is an integer and I^k denotes the k-fold Cartesian product of the unit interval with itself.

That the conclusion to Theorem 1.6 should remain true for any closed aspherical manifold M is known as the *Borel Conjecture*. In the remainder of this section we formulate two conjectures in terms of generalized homology theory and the Whitehead group, which taken together are even more general than Borel's conjecture.

1.7 The Assembly Map and Generalized Homology Theory. We remind the reader that Ω-spectra are the "coefficients" of generalized homology (and cohomology) theories. Let S_* denote an Ω-spectrum, that is S_* is a collection of spaces $\{S_i : i \in \mathbf{Z}\}$ together with given homotopy equivalences $S_i \cong \Omega S_{i+1}$ and base points $s_i \in S_i$ for all $i \in \mathbf{Z}$. For any topological space X and for each integer j define a space $\mathbb{H}_j(X, S_*)$ to be the direct limit space $\lim_{i \to \infty} \Omega^i(X \times S_{j+i}/X \times s_{j+i})$. Note that the collection of spaces $\{\mathbb{H}_j(X, S_*) : j \in \mathbf{Z}\}$ is a Ω-spectrum, which we denote by $\mathbb{H}_*(X, S_*)$ and call the *homology spectrum* for X with coefficients in S_*. Recall that the *homology groups* $H_j(X, S_*)$ for X with coefficients in S_* are defined to be the homotopy groups $\pi_j(\mathbb{H}_*(X, S_*))$ of the Ω-spectrum $\mathbb{H}_*(X, S_*)$.

Let $\mathcal{L}_*^h(\ )$ denote the Ω-spectra valued homotopy functor whose value on a connected CW complex X is the equal to the Ω-spectrum of surgery classifying spaces with four-fold periodicity (cf. [30]) for the group $\pi_1 X$ with orientation data given by the zero homomorphism $w : \pi_1 X \to \mathbb{Z}_2$, where the homotopy groups of $\mathscr{L}_*^h(X)$ are the surgery groups $L_*^h(\pi_1 X, w)$ defined by Wall [37]. Let "pt." denote the space with one point. Note that for each integer j we get a map $\phi_j : X \times \mathscr{L}_j^h(\text{pt.}) \to \mathscr{L}_j^h(X)$ by identifying each $y \times \mathscr{L}_j^h(\text{pt.})$, $y \in X$, with the subset $\mathscr{L}_j^h(y) \subset \mathscr{L}_j^h(X)$. Define $A_j : \mathbb{H}_j(X, \mathscr{L}_*^h(\text{pt.})) \to \mathscr{L}_j^h(X)$ to be the direct limit of the composition of maps

$$\Omega^i(X \times \mathscr{L}_{j+i}^h(\text{pt.})/X \times s_{j+i}) \xrightarrow{\Omega^i(\phi_{j+i})} \Omega^i(\mathscr{L}_{j+i}^h(X)) \cong \mathscr{L}_j^h(X),$$

where s_{j+i} is the base point of $\mathscr{L}_{j+i}^h(\text{pt.})$. Note that the collection of all such maps $A_* : \mathbb{H}_*(X, \mathscr{L}_*^h(\text{pt.})) \to \mathscr{L}_*^h(X)$ is a mapping of Ω-spectra, which is called the *assembly map* (cf. [29, 38]).

1.8 Conjecture. *Let X denote any connected aspherical CW complex with torsion free $\pi_1 X$. Then the assembly map $A_* : \mathbb{H}_*(X, \mathscr{L}_*^h(\text{pt.})) \to \mathscr{L}_*^h(X)$ is an equivalence of Ω-spectra.*

1.9 Conjecture. *The Whitehead group $Wh(\Gamma)$ of any torsion free group Γ is zero.*

1.10 Remark. There is a more general version of Conjecture 1.8 (which we do not write down because of space limitations) that covers the case of surgery when the surgery orientation data $w : \pi_1 X \to \mathbb{Z}_2$ is not assumed to be the zero homomorphism. Conjecture 1.8 (in its dual cohomology form) and Conjecture 1.9 are possibilities that many experts have been aware of for some time (cf. [20, 28, 31]). It has also been known for some time (cf. [24, 37]) that if X is a connected aspherical compact manifold (possibly with boundary) satisfying $\dim(X) \geq 5$ then Conjectures 1.8, 1.9 (with $\Gamma = \pi_1 X$ in 1.9) are together equivalent to the following strong topological rigidity property for X: any homotopy equivalence $h : (N, \partial N) \to (X \times I^k, \partial(X \times I^k))$ of compact manifold pairs, which is a homeomorphism of the boundaries is homotopic rel ∂ to a homeomorphism, where $k \geq 0$ is an integer and I^k is the k-fold Cartesian product of the unit interval with itself. Thus it follows from 1.6.1 that Conjectures 1.8, 1.9 are true for X equal any closed Riemannian manifold with sectional curvature satisfying $K \leq 0$ everywhere. Note that $X \times T^n$ also is a closed Riemannian manifold having $K \leq 0$ everywhere, where T^n is the flat n-torus. So by 1.9 we have that $Wh(\pi_1 X \oplus \mathbb{Z}^n) = 0$ holds for all integers $n \geq 0$. Now this last equality together with the Bass-Heller-Swan formula [4] implies that $\tilde{K}_0(Z\pi_1 X) = 0$ and $K_{-i}(Z\pi_1 X) = 0$ for all integers $i \geq 1$. We have just derived the following theorem.

1.11 Theorem. *Let X denote a closed Riemannian manifold with sectional curvature satisfying $K \leq 0$ everywhere. Then we have that $Wh(\pi_1 X) = 0$, $\tilde{K}_0(Z\pi_1 X) = 0$, and $K_{-i}(Z\pi_1 X) = 0$ for all integers $i \geq 1$. Moreover Conjecture 1.8 is also true for X.*

§2. Spaces of Self-homeomorphisms

Let $\text{TOP}(M)$ denote the space of all self-homeomorphisms of the closed manifold M. Topologists study $\text{TOP}(M)$ by analyzing the three associated spaces $G(M)$,

$\overline{\text{TOP}}(M)$, $P(M)$ of self-homotopy equivalences of M, blocked homeomorphisms of M, and pseudoisotopies of M. We have the inclusions $\text{TOP}(M) \subset \overline{\text{TOP}}(M) \subset G(M)$ since we may think of all of these spaces as being semi-simplicial spaces in which a typical k-simplex Δ consists of a selfmap $f : \Delta^k \times M \to \Delta^k \times M$ of the Cartesian product of the standard k-simplex with M which satisfies the following properties: if $\Delta \in \text{TOP}(M)$ then for each $p \in \Delta^k$ the restricted map $f : p \times M \to p \times M$ is a homeomorphism; if $\Delta \in \overline{\text{TOP}}(M)$ then for each face $\Delta \in \Delta^k$ the restricted map $f : \Delta \times M \to \Delta \times M$ is a homeomorphism; if $\Delta \in G(M)$ then for each face $\Delta \in \Delta^k$ the restricted map $f : \Delta \times M \to \Delta \times M$ is a homotopy equivalence. Recall that a pseudoisotopy of M is a homeomorphism $h : M \times [0, 1] \to M \times [0, 1]$ such that $h | M \times 0$ is equal the inclusion $M \times 0 \subset M \times [0, 1]$.

The following lemma, which is due to Hatcher [19], shows the role of $\overline{\text{TOP}}(M)$ and $P(M)$ in analyzing $\text{TOP}(M)$.

2.1 Lemma. *There is a spectral sequence with $E_{p,q}^1 = \pi_q(P(M \times I^p))$ which converges to $\pi_{q+p+1}(\overline{\text{TOP}}(M)/\text{TOP}(M))$.*

For any aspherical closed manifold M it is well known that there is a homotopy equivalence $G(M) \cong \text{Out}(\pi_1 M) \times K(\text{center}(\pi_1 M), 1)$, where $\text{Out}(\pi_1 M)$ denotes the outer automorphism group and where $K(\Gamma, 1)$ denotes the aspherical space having $\pi_1(K(\Gamma, 1)) = \Gamma$. If M is a closed aspherical manifold for which Conjectures 1.8, 1.9 hold then we have (by 1.10) that the inclusion $\overline{\text{TOP}}(M) \subset G(M)$ is a homotopy equivalence provided $\dim M \geq 5$. Thus we have the following lemma.

2.2 Lemma. *Let M denote a closed aspherical manifold with $\dim M \geq 5$ for which Conjectures 1.8, 1.9 hold (e.g. M is a Riemannian manifold with $K \leq 0$ everywhere). Then there is a homotopy equivalence $\overline{\text{TOP}}(M) \cong \text{Out}(\pi_1 X) \times K(\text{center}(\pi_1 X), 1)$.*

The effect of Lemmas 2.1, 2.2 is to reduce the study of $\text{TOP}(M)$ to the study of the spaces of pseudoisotopies $P(M \times I^p)$, $p = 0, 1, 2, 3, \ldots$. Note that there is an "inclusion" map $P(M \times I^p) \to P(M \times I^{p+1})$ gotten by forming the product of any pseudoisotopy $h : (M \times I^p) \times [0, 1] \to (M \times I^p) \times [0, 1]$ with the identity map $\text{id} : I \to I$. Thus we may form the direct limit space $\mathscr{P}(M) = \lim_{p \to \infty} P(M \times I^p)$, which is more amenable to homotopy analysis than its approximating spaces $P(M \times I^p)$. One of the benefits of this stabilization process is that $\mathscr{P}(M)$ now fits in as the zero'th space of an Ω-spectrum $\mathscr{P}_*(M)$ (cf. [19]). Moreover $\mathscr{P}_*(\ )$ is an Ω-spectra valued continuous functor defined on the category of all topological spaces; the value of $\mathscr{P}_i(\ )$, $i \geq 0$, on the space X is the semisimplicial space of all stable pseudoisotopies over $X \times \mathbf{R}^i$ which have compact support in the factor X and are bounded in the factor $\mathbf{R}^i$. Note that any homotopy information obtained about $\mathscr{P}(M)$ can be translated into information about the spaces $P(M \times I^p)$ by using the following lemma.

2.3 Lemma. *For any closed smooth manifold M the inclusion map $P(M \times I^p) \subset \mathscr{P}(M)$ induces an isomorphism on homotopy groups through dimension $d = (\dim M + p - 7)/3$ provided $(\dim M + p) > 10$ holds.*

This last lemma is proven by Igusa [22] for spaces of smooth psdeudoisotopies. Results of Burghelea and Lashof [6], and of Goodwillie (unpublished), then imply that the lemma also holds for spaces of topological pseudoisotopies.

We will now precisely formulate a result which states roughly that $\mathscr{P}_*(M)$ can be constructed in a simple way from $\mathscr{P}_*(S^1)$, where S^1 is the circle. Towards this end we need the terminology and notation of the following two subsections.

2.4 Homology Theory with Coefficients in Stratified and Twisted Ω-Spectra and the Assembly Map.

A mapping $f: E \to X$ is called a *generalized Siefert fibration* if there is a triangulation T for X, and for each simplex $\varDelta \in T$ there is a continuous map $g_\varDelta: Y_\varDelta \times \varDelta \to f^{-1}(\varDelta)$, such that the following hold: $f \circ g_\varDelta$ is equal projection onto the second factor of $Y_\varDelta \times \varDelta$, and $Y_\varDelta$ is connected; the restricted map $g_\varDelta | Y_\varDelta \times (\varDelta - \partial\varDelta)$ is a homeomorphism onto $f^{-1}(\varDelta - \partial\varDelta)$; and the restricted map $g_\varDelta | Y_\varDelta \times (\sigma - \partial\sigma)$ is a covering space projection onto $f^{-1}(\sigma - \partial\sigma)$ for each face $\sigma \in \varDelta$.

Let $\mathscr{S}_*(\ )$ denote a given Ω-spectra valued continuous functor from the category of path connected topological spaces, and let $f: E \to X$ denote a given generalized Siefert fibration. Note that the collection of Ω-spectra $\{\mathscr{S}_*(f^{-1}(p)): p \in X\}$ is a system of stratified and twisted coefficients over X with respect to which we can (roughly) define the *homology spectrum* for X, denoted by $\mathbb{H}_*(X, \mathscr{S}_*(f))$, by

$$\mathbb{H}_j(X, \mathscr{S}_*(f)) = \lim_{i \to \infty} \Omega^i\left(\left(\bigcup_{p \in X} \mathscr{S}_{i+j}(f^{-1}(p))\right) \Big/ \left(\bigcup_{p \in X} s_{i+j,p}\right)\right)$$

where $s_{i+j,p} \in \mathscr{S}_{i+j}(f^{-1}(p))$ are the base points. The *k-th homology group* of X with respect to the system of coefficients $\{\mathscr{S}_*(f^{-1}(p)): p \in X\}$, denoted by $H_k(X, \mathscr{S}_*(f))$, is defined to be $\pi_k(\mathbb{H}_*(X, \mathscr{S}_*(f)))$. There is also an *assembly map* $A_*: \mathbb{H}_*(X, \mathscr{S}_*(f)) \to \mathscr{S}_*(E)$ induced as in 1.7 from the inclusions $\mathscr{S}_*(f^{-1}(p)) \to \mathscr{S}_*(E)$. For more details about the homology spectrum $\mathbb{H}_*(X, \mathscr{S}_*(f))$ and the assembly map A_* the reader is referred to Quinn [29], Yamasaki [38], and to Farrell and Jones [14].

2.5 The Generalized Siefert Fibration $p: E \to G$ and the Map $f: E \to M$.

Let S denote the space of all maps $\{S^1 \to S^1\}$ which have constant non-zero speed. Note that S is a semi-group with respect to the composition operation. Let F denote the space of all essential closed geodesics $\{g: S^1 \to M\}$ in the closed Riemannian manifold M with sectional curvature satisfying $K \leq 0$. Note that S "acts" on the right of F by $(g, h) \to g \circ h$ for $g \in F, h \in S$, and S "acts" on the right of $F \times S^1$ by $(g, x, h) \to (g \circ h, h^{-1}(x))$ for $g \in F, x \in S^1, h \in S$. We define $p: E \to G$ to be the quotient of the first factor projection $F \times S^1 \to F$ under the right actions of S. Note that the map $F \times S^1 \to M$ defined by $(g, x) \to g(x)$ induces a map $f: E \to M$. We note that G parametrizes the collection of (unparametrized) essential *t-simple* closed geodesics $g: S^1 \to M$ (where "*t-simple*" means that there is a no rotation $r: S^1 \to S^1$ other than the identity which satisfies $g = g \circ r$), in fact any such geodesic is equal to the restricted map $f: p^{-1}(y) \to M$ for some $y \in G$. The authors have proven that $p: E \to G$ is a generalized Siefert fibration.

The following two results have been proven by the authors [15].

2.6. Theorem. *Let M be a closed Riemannian manifold with sectional curvature satisfying $K \leq 0$ everywhere. Then the composite map*

$$\mathbb{H}_*(G, \mathscr{P}_*(p)) \xrightarrow{\mathscr{P}_*(f) \circ A_*} \mathscr{P}_*(M)$$

is an equivalence of Ω-spectra, where A_* is the assembly map for the generalized Siefert fibration $p : E \to G$ (cf. 2.4, 2.5) and where $\mathscr{P}_*(f)$ is the image of the map $f : E \to M$ under the functor $\mathscr{P}_*(\ )$.

2.6.1 Remark. Theorem 2.6 reduces the study of $\mathscr{P}_*(M)$ to the Ω-spectrum $\mathscr{P}_*(S^1)$ about which there is a good deal known. For example Waldhausen [35, 36] shows that $\pi_i(\mathscr{P}_*(S^1)) \otimes \mathbf{Q} = 0$ for $i \geq 0$, and Anderson and Hsiang [2] show that $\pi_i(\mathscr{P}_*(S^1)) = 0$ for $i < 0$; results of Igusa [23] imply that $\pi_0(\mathscr{P}_*(S^1)) = \mathbf{Z}_2^\infty$, where $\mathbf{Z}_2^\infty$ denotes the direct sum of a countably infinite number of copies of $\mathbf{Z}_2$. By using Theorem 2.6 and a spectral sequence argument we get these same equalities for $\pi_i(\mathscr{P}_*(M))$.

2.7 Corollary. *Let M be as in 2.6 with* $\dim M > 10$ *and let i be any integer satisfying* $1 \leq i \leq (\dim M - 7)/3$. *Then*

$$\pi_i(\mathrm{TOP}(M)) \otimes \mathbf{Q} = \left\{ \begin{array}{c} \mathrm{center}(\pi_1 M) \otimes \mathbf{Q}, \text{if } i = 1 \\ 0, \text{if } i \neq 1 \end{array} \right\}.$$

Moreover the forgetful map $\pi_0(\mathrm{TOP}(M)) \to \mathrm{Out}(\pi_1 X)$ *is surjective with kernal equal* $\mathbf{Z}_2^\infty$.

The proof of 2.7 consists of combining Lemmas 2.1, 2.2, 2.3 with Remarks 1.6.1, 2.6.1.

We end this section by recalling a conjecture made by the authors in [14]. To precisely formulate this conjecture we associate to any space X a generalized Siefert fibration $\bar{p} : \bar{E} \to \bar{G}$ as follows. Let $\bar{F}$ denote the space of all essential continuous maps $\{ g : S^1 \to X \}$ and let S be the semi-group of 2.5. Note S acts on the right of $\bar{F} \times S^1$ and of $\bar{F}$ as described in 2.5. Our first candidate for $\bar{p} : \bar{E} \to \bar{G}$ is gotten by imitating the construction of $p : E \to G$ in 2.5, i.e. taking the quotient of the projection map $\bar{F} \times S^1 \to \bar{F}$ under the right actions of S. However with this definition it is not clear that $\bar{p} : \bar{E} \to \bar{G}$ is a generalized Siefert fibration. So instead we define $\bar{p} : \bar{E} \to \bar{G}$ to be a semisimplicial version of our first candidate: replace $\bar{F}$ and S by their associated semisimplicial objects and proceed as before. By imitating the construction of $f : E \to M$ in 2.5 we get a map $\bar{f} : \bar{E} \to X$. (The reader is referred to [14; § 4] for more details).

2.8. Conjecture. *Let X denote any path connected aspherical CW complex with torsion free $\pi_1 X$. Then the composite map*

$$\mathbb{H}_*(\bar{G}, \mathscr{P}_*(\bar{p})) \xrightarrow{\mathscr{P}_*(\bar{f}) \circ A_*} \mathscr{P}_*(X)$$

is an equivalence of Ω-spectra, where A_* is the assembly map for the generalized Siefert fibration $\bar{p}: \bar{E} \to \bar{G}$ and where $\mathscr{P}_*(\bar{f})$ is the image of the map $\bar{f}: \bar{E} \to X$ under the functor $\mathscr{P}_*(\ )$.

2.8.1 Remark. Unlike Conjectures 1.8, 1.9, which have been around for some time, Conjecture 2.8 was first formulated in 1989 by the authors [14] as an attempt to generalize Theorem 2.6. It requires some argument (cf. [14; §4]) to see that 2.6 and 2.8 give the same calculation for $\mathscr{P}_*(M)$ when M is as in 2.6.

§3. The Isomorphism Conjectures

The starting point of this section consists of Conjectures 1.8, 2.8. Simply stated these conjectures offer recipes for computing the values of the functors $\mathscr{L}_*^h(\ )$ and $\mathscr{P}_*(\ )$ on a space X in terms of their values on a collection of much simpler spaces (a point or a circle respectively). In this section we formulate a more general conjecture along these lines, where we drop the hypothesis of 1.8, 2.8 that X be aspherical with torsion free $\pi_1 X$, and we add to our list of Ω-spectra valued continuous functors $\mathscr{S}_*(\ )$ which we wish to analyze by such a conjectured recipe.

3.1 The Functors $\mathscr{P}_*(\ )$, $\mathscr{P}_*^{\text{diff}}(\ )$, $\mathscr{L}_*^\infty(\ )$, $\mathscr{K}_*(\ )$. We have already discussed $\mathscr{P}_*(\ )$ in §2; $\mathscr{P}_*^{\text{diff}}(\ )$ is the functor whose value on a space X is the Ω-spectrum defined by Hatcher [19] of stable smooth pseudoisotopies which have compact support in X; $\mathscr{L}_*^\infty(\ )$ is the functor whose value on a path connected space X is the Ω-spectrum (with four-fold periodicity) of surgery classifying spaces for the group $\pi_1 X$ with orientation data given by the zero homomorphism $w: \pi_1 X \to \mathbf{Z}_2$, having the surgery groups $L_*^\infty(\pi_1 X, w)$ defined by Ranicki [32] for homotopy groups; and $\mathscr{K}_*(\ )$ is the functor whose value on the path connected space X is the algebraic K-theory Ω-spectrum for the group ring $\mathbf{Z}\pi_1 X$ defined by Gersten [16] and by Wagoner [34].

3.2 The Generalized Siefert Fibration $\varrho(X): \mathscr{E}(X) \to \mathscr{B}(X)$ and the Map $f(X): \mathscr{E}(X) \to X$. Recall that a group H is *cyclic by finite* (also called *virtually cyclic*) if either H is finite or if there is a short exact sequence of groups $\mathbf{Z} \to H \to F$ where F is finite and $\mathbf{Z}$ is the infinite cyclic group. Let X denote a path connected CW complex, and let $\mathscr{G}(X)$ denote the collection of all cyclic by finite subgroups of $\pi_1(X, x_0)$ where x_0 is a given base point of X. Define a category $\mathscr{C}(X)$ as follows. The objects of $\mathscr{C}(X)$ are equivalence classes of based connected covering space projections $p: (Y, y_0) \to (X, x_0)$ such that $\text{Image}(\pi_1 p) \in \mathscr{G}(X)$, where two such covering projections $p: (Y, y_0) \to (X, x_0)$ and $p': (Y', y_0') \to (X, x_0)$ are equivalent if there is a homeomorphism $h: (Y, y_0) \to (Y', y_0')$ satisfying $p = p' \circ h$. A map is $\mathscr{C}(X)$ from $p: (Y, y_0) \to (X, x_0)$ to $p': (Y', y_0') \to (X, x_0)$ consists of an (unbased) covering projection $q: Y \to Y'$ satisfying $p = p' \circ q$. We define $\mathscr{B}(X)$ to be the nerve of the category $\mathscr{C}(X)$. In more detail the vertices of $\mathscr{B}(X)$ are the objects of $\mathscr{C}(X)$, and a k-simplex in $\mathscr{B}(X)$ consists of a k-tuple $\langle p_1, p_2, \ldots, p_k \rangle$ of maps p_i in $\mathscr{C}(X)$ such that $\text{domain}(p_i) = \text{range}(p_{i-1})$ holds for all $1 < i \le k$. The face operators are defined by $\partial_0 \langle p_1, \ldots, p_k \rangle = \langle p_2, \ldots, p_k \rangle$, $\partial_k \langle p_1, \ldots, p_k \rangle = \langle p_1, \ldots, p_{k-1} \rangle$, $\partial_i \langle p_1, \ldots, p_k \rangle =$ $\langle p_1, \ldots, p_{i-1}, p_{i+1} \circ p_i, p_{i+2}, \ldots, p_k \rangle$ for all $0 < i < k$. We get $\varrho(X) : \mathcal{E}(X) \to \mathcal{B}(X)$ as follows. For each k-simplex $\Delta = \langle p_1, \ldots, p_k \rangle$ in $\mathcal{B}(X)$ we set

$$\varrho(X) : \varrho(X)^{-1}(\Delta - \partial_0 \Delta) \to \Delta - \partial_0 \Delta$$

equal to the standard projection

$$(\Delta - \partial_0 \Delta) \times Y \to \Delta - \partial_0 \Delta,$$

where Y is the domain of p_1 (or more precisely where the covering projection $(Y, y_0) \to (X, x_0)$ is the domain of p_1). We note that the union of all the composite maps

$$(\Delta - \partial_0 \Delta) \times Y \xrightarrow{\text{proj.}} Y \xrightarrow{p_1} X$$

gives a well defined map $f(X) : \mathcal{E}(X) \to X$.

3.3 Conjecture. *Let X denote any path connected CW complex and let $\mathscr{S}_*(\ )$ denote any of the Ω-spectra valued continuous functors $\mathscr{P}_*(\ ), \mathscr{P}_*^{\mathrm{diff}}(\ ), \mathscr{L}_*^\infty(\ ), \mathscr{K}_*(\ )$. Then the composite map*

$$\mathbb{H}_*(\mathcal{B}(X), \mathscr{S}_*(\varrho(X))) \xrightarrow{\mathscr{S}_*(f(X)) \circ A_*} \mathscr{S}_*(X)$$

is an equivalence of Ω-spectra, where A_* denotes the assembly map for the generalized Siefert fibration $\varrho(X) : \mathcal{E}(X) \to \mathcal{B}(X)$ (cf. 2.4, 3.2) and where $\mathscr{S}_*(f(X))$ is the image of the map $f(X) : \mathcal{E}(X) \to X$ under the functor $\mathscr{S}_*(\ )$.

There is a more general version of Conjecture 3.3 for Ω-spectra associated to surgery theory in which the orientation data $w : \pi_1 X \to Z_2$ is not assumed to be the zero homomorphism. Space limitations prevent us from stating it here.

We can verify Conjecture 3.3 for many spaces X for the functor $\mathscr{L}_*^\infty(\ )$, but for these same spaces our proofs often break down for the functor $\mathscr{L}_*^h(\ )$ (or for the functor $\mathscr{L}_*^s(\ )$) if $\pi_1 X$ contains some elements of finite order. So at this time we do not know if 3.3 is a reasonable conjecture for $\mathscr{L}_*^h(\ )$ (or for $\mathscr{L}_*^s(\ )$). Is it possible that the truth of our 3.3 for $\mathscr{L}_*^h(\ )$ (or for $\mathscr{L}_*^s(\ )$) is just a formal consequence of the truth of 3.3 for both $\mathscr{L}_*^\infty(\ )$ and $\mathscr{K}_*(\ )$?

In the following remarks we consider some of the consequences of the truth of Conjecture 3.3 for each of the Ω-spectra valued functors in 3.1.

3.3.1 $\pi_1 X$ Torsion Free and X Aspherical. Conjecture 1.8 is equivalent to Conjecture 3.3 for $\mathscr{S}_*(\ ) = \mathscr{L}_*^h(\ )$ in 3.3. Conjecture 2.8 is equivalent to Conjecture 3.3 for $\mathscr{S}_*(\ ) = \mathscr{P}_*(\ )$ in 3.3, and each of these conjectures implies Conjecture 1.9. Conjecture 3.3 for $\mathscr{K}_*(\ )$ is equivalent to the usual assembly map $A_* : \mathbb{H}_*(X, \mathscr{K}_*(\text{pt.})) \to \mathscr{K}_*(X)$ being an equivalence of Ω-spectra, which in turn is equivalent to $\mathrm{Wh}_i(\pi_1 X) = 0$ and $K_{-i}(\mathbf{Z}\pi_1 X) = 0$ for all $i \geq 1$ and $\bar{K}_0(\mathbf{Z}\pi_1 X) = 0$.

3.3.2 The Novikov Conjectures. In general the truth of Conjecture 3.3 for the functor $\mathscr{L}_*^\infty(\ )$ implies the truth of Novikov's original conjecture, which stated in more modern terms claims that the assembly map $A_* : \mathbb{H}_*(X, \mathscr{L}_*^\infty(\text{pt.})) \to \mathscr{L}_*^\infty(X)$

is rationally injective on the homotopy group level. The truth of Conjecture 3.3 for either $\mathscr{K}_*(\ )$, $\mathscr{P}_*(\ )$, or for $\mathscr{P}_*^{\mathrm{diff}}(\ )$ implies the truth of the K-theoretic Novikov conjecture, which claims that the assembly map $A_* : \mathbb{H}_*(X, \mathscr{K}_*(\mathrm{pt.})) \to \mathscr{K}_*(X)$ is rationally injective on the homotopy group level.

3.3.3. On the Role of Finite Subgroups of $\pi_1 X$. Many authors (Quinn [31] and Yamasaki [38] were among the first) have recognized the role that finite subgroups should play in any calculation of the Ω-valued functors $\mathscr{S}_*(\ )$ of 3.1. However that the class of cyclic by finite subgroups of $\pi_1 X$ might be enough to detect all of $\mathscr{S}_*(X)$ first occured as a possibility in the authors' Theorem 2.6 and in their Conjecture 2.8.

The authors obtain the following partial verification of Conjecture 3.3 in an ongoing research project.

3.4 Theorem. *Suppose that in 3.3 $\pi_1 X$ is isomorphic to a cocompact discrete subgroup of a virtually connected Lie group. Then the map of 3.3 is an equivalence for $\mathscr{S}_* = \mathscr{P}_*$, $\mathscr{P}_*^{\mathrm{diff}}$, or $\mathscr{L}_*^{\infty}$, and is a rational equivalence for $\mathscr{S}_* = \mathscr{K}_*$.*

References

1. Al'ber, S.I.: Spaces of mappings into manifolds of negative curvature. Dokl. Akad. Nauk. SSSR **178** (1968) 13–16
2. Anderson, D.R., Hsiang, W.-c.: The functors K_{-i} and pseudoisotopies of polyhedra. Ann. Math. **105** (1977) 201–223
3. Ballman, W., Gromov, M., Schroeder, V.: Manifolds of nonpositive curvature. Birkhäuser, 1985
4. Bass, H., Heller, A., Swan, R.: The Whitehead group of a polynomial extension. Inst. Hautes Etudes Sci. Publ. Math. **22** (1964) 64–79
5. Bieberbach, L.: Über die Bewegungsgruppen der Euklidischen Räume II: Die gruppen mit einem endlichen Fundamentalbereich. Math. Ann. **72** (1912) 400–412
6. Burghlea, D., Lashof, R.: Stability of concordances and the suspension homomorphism. Ann. Math. **105** (1977) 449–472
7. Eberlein, P.: Rigidity of lattices of nonpositive curvature. Erg. Theor. Dyn. Syst. **3** (1983) 47–85
8. Eberlein, P.: Euclidean de Rham factor of a lattice of nonpositive curvature. J. Diff. Geom. **18** (1983) 209–220
9. Eells, J., Sampson, J.H.: Harmonic mappings of Riemannian manifolds. Amer. J. Math. **86** (1964) 109–160
10. Farrell, F.T., Hsiang, W.-c.: On Novikov's conjecture for nonpositively curved manifolds I. Ann. Math. **113** (1981) 199–209
11. Farrell, F.T., Jones, L.E.: Negatively curved manifolds with exotic smooth structures. J. Amer. Math. Soc. **2** (1989) 899–908
12. Farrell, F.T., Jones, L.E.: A topological analogue of Mostow's rigidity theorem. J. Amer. Math. Soc. **2** (1989) 257–370
13. Farrell, F.T., Jones, L.E.: Classical aspherical manifolds. CBMS Regional Conference Series in Mathematics, vol. 75. Amer. Math. Soc. 1990
14. Farrell, F.T., Jones, L.E.: Computations of stable pseudoisotopy spaces of aspherical manifolds. Proc. of Intern. Conf. on Algebraic Topology, held in Poznan, Poland in June 1989 (to appear)

15. Farrell, F.T., Jones, L.E.: Rigidity and other topological aspects of compact nonpositively curved manifolds. Bull. Amer. Math. Soc. **22** (1990) 59–64
16. Gersten, S.: On spectrum of algebraic K-theory. Bull. Amer. Math. Soc. **78** (1972) 216–219
17. Gromoll, D., Wolf, J.: Some relations between the metric structure and the algebraic structure of the fundamental group in manifolds of nonpositive curvature. Bull. Amer. Math. Soc. **77** (1971) 545–552
18. Hartman, P.: On homotopic harmonic maps. Canad. J. Math. **19** (1967) 673–687
19. Hatcher, A.: Concordance spaces, higher simple homotopy theory, and applications. Proc. Symp. Pure Math. **32** (1978) 3–21
20. Hsiang, W.-c.: Geometric applications of algebraic K-theory. Proc. Int. Congress of Math., Warsaw 1983, pp. 99–118
21. Hu, B.: A PL geometric study of algebraic K-theory. Trans. Amer. Math. Soc., in press.
22. Igusa, K.: The stability theorem for pseudoisotopies. K-theory **2** (1988) 1–355
23. Igusa, K.: On the algebraic K-theory of A_∞-ring spaces. (Lecture Notes in Mathematics, vol. 967.) Springer, Berlin Heidelberg New York 1982, pp. 146–194
24. Kirby, R.C., Siebenmann, L.C.: Fundational essays on topological manifolds, smoothings, and triangulations. (Ann. Math. Stud., vol. 88.) Princeton Univ. Press, 1977
25. Mishchenko, A.S.: Infinite dimensional representations of discrete groups and higher signatures. Izv. Akad. Nauk. SSSR Ser. Mat. **38** (1974) 81–106
26. Mostow, G.D.: Quasi-conformal mappings in n-space and the rigidity of hyperbolic space forms. Inst. Hautes Etudes Sci. Publ. Math. **34** (1967) 53–104
27. Mostow, G.D.: Strong rigidity of locally symmetric spaces. Princeton Univ. Press, Princeton 1973
28. Nicas, A.: On higher Whitehead groups of a Bieberbach group. Trans. Amer. Math. Soc. **287** (1985) 853–859
29. Quinn, F.: Ends of maps II. Invent. math. **68** (1982) 353–424
30. Quinn, F.: A geometric formulation of surgery. Topology of manifolds. Proc. Georgia Topology Conf., Markham Press 1970. pp. 500–511
31. Quinn, F.: Applications of topology with control. Proc. Int. Congress of Math. Berkeley 1986, pp. 598–606
32. Ranicki, A.: Algebraic L-theory II. Proc. London Math. Soc. **27** (1973) 126–158
33. Schoen, R., Yau, S.T.: On univalent harmonic maps between surfaces. Invent. math. **44** (1978) 265–278
34. Wagoner, J.: Delooping classifying spaces in algebraic K-theory. Topology **11** (1972) 349–370
35. Waldhausen, F.: Algebraic K-theory of topological spaces I. Proc. Symp. Pure Math. **32** (1978) 35–60
36. Waldhausen, F.: Algebraic K-theory of generalized free products. Ann. Math. **108** (1978) 135–256
37. Wall, C.T.C.: Surgery on compact manifolds. Academic Press 1971
38. Yamasaki, M.: L-groups of crystallographic groups. Invent. math. **88** (1987) 571–602
39. Yau, S.T.: On the fundamental group of compact manifolds of nonpositive curvature. Ann. Math. **93** (1971) 579–585

Mapping Class Groups of Surfaces and Three-Dimensional Manifolds

Shigeyuki Morita

Department of Mathematics, Faculty of Science, Tokyo Institute of Technology
Ohokayama Tokyo 152, Japan

1. Introduction

Let Σ_g be a closed oriented surface of genus g and let $\mathcal{M}_g$ be its mapping class group. Namely it is the group of all isotopy classes of orientation preserving diffeomorphisms of Σ_g. It is also called the Teichmüller modular group because it acts on the Teichmüller space $\mathcal{T}_g$ properly discontinuously with the quotient space $\mathbf{M}_g$: the Riemann moduli space of compact Riemann surfaces of genus g. A classical theorem of Nielsen asserts that $\mathcal{M}_g$ can be naturally identified with the proper outer automorphism group of $\pi_1(\Sigma_g)$. Thus the mapping class group appears in diverse branches of mathematics and have been investigated from various points of view. In this article we would like to describe some of the recent progress in the topological aspects of the theory of the mapping class group. More precisely we will concern ourselves with the following three topics which are all of cohomological nature and mutually closely related: a brief review of some of the known results about the cohomology of $\mathcal{M}_g$, in particular various properties of some canonical cohomology classes of $\mathcal{M}_g$, called the characteristic classes of surface bundles (§2), the study of $\mathcal{M}_g$ via its action on the lower central series of $\pi_1(\Sigma_g)$ (§3) and the interplay between the structure of $\mathcal{M}_g$ and topological invariants of three dimensional manifolds (§4).

There has been a great deal of results which are relevant to the subjects of this article and our coverage here is necessarily limited. The reader is referred to survey articles [3, 10, 13, 14, 40] and their bibliographies for more informations.

2. The Cohomology of $\mathcal{M}_g$

By the cohomology of $\mathcal{M}_g$ we mean the Eilenberg-MacLane cohomology of the group $\mathcal{M}_g$ whose definition is given purely algebraically. However it has also geometrical meanings at least from the following two viewpoints. One is that, by virtue of the result of Earle-Eells [6], the classifying space for oriented Σ_g-bundles has the homotopy type of an Eilenberg-MacLane space $K(\mathcal{M}_g, 1)$ for $g \geq 2$ so that elements of the cohomology group of $\mathcal{M}_g$ can be naturally considered as *characteristic classes of surface bundles*. The other is the fact that the rational

Proceedings of the International Congress
of Mathematicians, Kyoto, Japan, 1990
© The Mathematical Society of Japan, 1991

cohomology of $\mathcal{M}_g$ is canonically isomorphic to that of the moduli space $\mathbf{M}_g$. For both of technical and theoretical reasons it is convenient to introduce the mapping class group $\mathcal{M}_{g,q}^p$ of a general compact oriented surface $\Sigma_{g,q}^p$ of genus g with p distinguished points and q boundary components (or equivalently q embedded discs). Here $\mathcal{M}_{g,q}^p$ is defined to be the group of path components of diffeomorphisms of $\Sigma_{g,q}^p$ which fix the p points and restrict to the identity on the boundary.

One of the most useful methods to compute (co)homology of a group is to find a space or a cell complex on which it acts naturally and then analyse the action. As for the mapping class group, besides the Teichmüller space it is natural to consider various cell complexes which are constructed out of isotopy invariants of surfaces, e.g. simple closed curves or arcs. By an extensive use of this method Harer obtained several fundamental results concerning the homology of the mapping class group (see [10] for a survey of his results). One of his results is about the stability of the (co)homology with respect to the genus and it may be stated roughly as

Theorem (Harer [8]). *The cohomology* $H^k(\mathcal{M}_{g,q}^p; \mathbf{Z})$ *is independent of* g *and* q *for* $g \geq 3k + 1$ *and the same is true for the rational cohomology for* $g \geq 3k$.

In the case where $q \neq 0$ Ivanov has improved the stability range (see [14]). By virtue of the above theorem we can speak of the *stable cohomology* of $\mathcal{M}_g$.

Now we recall the definition of some canonical classes in the stable cohomology of $\mathcal{M}_{g,q}^p$. Thus let $\pi : E \to X$ be an oriented $\Sigma_{g,q}^p$-bundle. This means that it is an oriented differentiable fibre bundle with fibre Σ_g and there are given $(p+q)$ cross sections $s_i : X \to E$ $(i = 1, \cdots, p + q)$ such that their images are disjoint and the normal bundles of the last q sections are trivialized. Let ξ be the tangent bundle along the fibres, namely it is an oriented plane bundle over E consisting of all tangent vectors of E which are tangent to the fibres. We write $e \in H^2(E; \mathbf{Z})$ for the Euler class of ξ and consider the element $e_i = \pi_*(e^{i+1}) \in H^{2i}(X; \mathbf{Z})$ where $\pi_* : H^{2i+2}(E; \mathbf{Z}) \to H^{2i}(X; \mathbf{Z})$ is the Gysin homomorphism. We also set $\sigma_i = s_i^*(e) \in H^2(X; \mathbf{Z})$ $(i = 1, \cdots, p)$. These definitions are functorial in the obvious sense so that we have well defined cohomology classes $e_i \in H^{2i}(\mathcal{M}_{g,q}^p; \mathbf{Z})$ and $\sigma_i \in H^2(\mathcal{M}_{g,q}^p; \mathbf{Z})$. We may call e_i the i-th characteristic class of surface bundles. Mumford [31] defined, in the context of algebraic geometry, canonical classes κ_i in the Chow ring of the moduli space $\mathbf{M}_g$ (or in fact its compactification $\overline{\mathbf{M}}_g$) and it turned out that the class e_i is the topological version of κ_i (up to signs) although our work has been done independently. Miller [23] also considered the same class. According to a recent paper of Witten [39], these characteristic classes have physical meanings.

Theorem. *The homomorphism*

$$\Phi : \mathbf{Q}[e_1, e_2, \cdots, \sigma_1, \cdots, \sigma_p] \to H^*(\mathcal{M}_{g,q}^p; \mathbf{Q})$$

is injective up to degree $\frac{1}{3}g$.

The above theorem for the case $p = 0$ is due to Miller [23] and the author [24] independently and was proved by generalizing an earlier work by Atiyah [1]. Miller also observed that the stable cohomology $\lim H^*(\mathcal{M}_{g,1}; \mathbf{Q})$,which exists by the stability theorem of Harer, is the tensor product of a polynomial algebra on even generators with an exterior algebra on odd generators. The general cases of the above theorem were proved by the author [24]. As a corollary to the above theorem, we have the following result which was one of our original motivation for the present work.

Corollary [24]. *The natural homomorphism* $\mathrm{Diff}_+\Sigma_g \to \mathcal{M}_g$ *does not have a right inverse for* $g \geq 18$.

We mention that the original Nielsen realization problem was solved affirmatively by Kerckhoff [20] and also that it is still unknown whether the homomorphism $\mathrm{Homeo}_+\Sigma_g \to \mathcal{M}_g$ has a right inverse or not. As Mumford says in [31] for the case $p = 0$ one of the main questions concerning the cohomology of the mapping class groups (or the moduli spaces of compact Riemann surfaces) is whether the stable cohomology $\lim H^*(\mathcal{M}_{g,q}^p; \mathbf{Q})$ is isomorphic to $\mathbf{Q}[e_1, e_2, \cdots, \sigma_1, \cdots, \sigma_p]$ or not. In low dimensions it is known that $\mathcal{M}_{g,q}^p$ is perfect except for finite cases ([7, 33]) and also we have

Theorem (Harer [7]). $H^2(\mathcal{M}_{g,q}^p; \mathbf{Z}) \cong \mathbf{Z}^{p+1}$ *with the generators* $\frac{1}{12}e_1, \sigma_1, \cdots, \sigma_p$ *for* $g \geq 5$.

On the other hand for a fixed g it is known that the homomorphism Φ : $\mathbf{Q}[e_1, e_2, \cdots, \sigma_1, \cdots, \sigma_p] \to H^*(\mathcal{M}_{g,q}^p; \mathbf{Q})$ is far from being injective nor surjective. The latter statement is due to Harer-Zagier and is a consequence of the following result.

Theorem (Harer-Zagier [11], see also Penner [32]). *The orbifold Euler characteristic of* $\mathcal{M}_g^1$ *is equal to* $\zeta(1 - 2g)$ *where* ζ *is the Riemann zeta function.*

By making use of the above result, they have shown that the true Euler characteristic of $\mathcal{M}_g$ grows more than exponentially with respect to g and even takes negative values infinitely often.

Finally we consider the kernel of the homomorphism Φ, namely relations between the characteristic classes of surface bundles. It has certainly a big kernel simply because the rational cohomology of the mapping class groups are finite dimensional. More precisely Harer [9] has determined the virtual cohomological dimensions of the mapping class groups and in particular $H^*(\mathcal{M}_g; \mathbf{Q}) = 0$ for $k \geq 4g - 4$. There are several known relations between the characteristic classes. For example Mumford proved that for a fixed g, the classes e_i are polynomials on $e_1, \cdots, e_{g-2}$ for all $i \geq g - 1$ (see [24, 31] for the proof as well as other relations). Here we describe one particular relation which was proved by Harris [12] and the author [26] independently and by completely different methods. The following is along the lines of our papers [25, 26]. First of all it was proved that $H^1(\mathcal{M}_g^1; H)$, where H stands for the first homology of Σ_g, is an infinite cyclic group and a characterization was given for a crossed homomorphism $k : \mathcal{M}_g^1 \to H$ to represent its generator. It is amusing to point out here that there have been known various ways of describing such a crossed homomorphism reflecting the many-sided features of the mapping class group; complex analytic construction due to Earle [5], two definitions in the context of combinatorial group theory [25, 26], determinant of certain matrix representation [27], a very simple geometrical one due to Furuta (unpublished, see [30]) and a closely related one due to Trapp [36] in terms of the winding number and finally Wagoner's definition [37], a particular example of his pseudo isotopy invariants using algebraic K-theory, should be completed to give yet another one. By making use of the above crossed homomorphism, we proved in [26] a topological version of Earle's embedding theorem [5] which embedds any surface bundle into its associated family of Jacobian manifolds explicitly. We also proved that the crossed homomorphism k gives rise to a canonical cocycle for the first characteristic class $e_1 \in H^2(\mathcal{M}_{g,1}; \mathbf{Z})$. Namely the 2-cochain c of $\mathcal{M}_{g,1}$ defined by $c(\varphi, \psi) = k(\varphi) \cdot k(\psi^{-1})$ $(\varphi, \psi \in \mathcal{M}_{g,1})$ is a cocycle representing the class e_1, where $\cdot$ denotes the intersection number. Then by pulling back a canonical cohomology class of the total space of the family of Jacobian manifolds to that of the given surface bundle, we obtain

Theorem (Harris [12], Morita [26]). *Let $e \in H^2(\mathcal{M}_g^1; \mathbf{Q})$ be the Euler class of the tangent bundle along the fibres of the universal Σ_g-bundle $\mathcal{M}_g^1 \to \mathcal{M}_g$ and let $e_1 \in H^2(\mathcal{M}_g^1; \mathbf{Q})$ be the first characteristic class. Then we have $\{2g(2-2g)e - e_1\}^{g+1} = 0$ in $H^{2g+2}(\mathcal{M}_g^1; \mathbf{Q})$.*

3. Representations of $\mathcal{M}_g$ via Its Action on $\pi_1(\Sigma_g)$

As was mentioned in the introduction, the mapping class group $\mathcal{M}_g$ is naturally isomorphic to the proper outer automorphism group of the fundamental group $\pi_1(\Sigma_g)$ and similarly $\mathcal{M}_{g,1}$ can be identified with a certain subgroup of the automorphism group of $\pi_1(\Sigma_{g,1})$ which is a free group of rank $2g$. These isomorphisms provide us with an algebraic method of investigating the structure of the mapping class groups in terms of their actions on the fundamental groups of surfaces. Apart from the general theory due to Magnus and his school on the structure of the (full) automorphism groups of free groups, it was Johnson who has first employed the above idea systematically in the study of the structure of (various subgroups of) the mapping class groups. By this method he obtained several fundamental results concerning the structure of the Torelli group $\mathcal{I}_g$ which is the subgroup of $\mathcal{M}_g$ acting trivially on the homology of Σ_g (see [18]). In this section we summarize the method and results of Johnson and their extensions by the author.

We denote Γ_1 for $\pi_1(\Sigma_{g,1})$ and inductively define $\Gamma_k = [\Gamma_{k-1}, \Gamma_1]$. We may call the quotient group $N_k = \Gamma_1/\Gamma_k$ the k-th nilpotent quotient of Γ_1. We simply write H for $N_2 \cong H_1(\Sigma_g; \mathbf{Z})$ and let $\mathcal{L} = \oplus_{k \geq 1} \mathcal{L}_k$ be the free graded Lie algebra over $\mathbf{Z}$ generated by the elements of H. It is a classical result that we can identify $\mathcal{L}_k$

with Γ_k/Γ_{k+1} so that there exists a central extension $0 \to \mathcal{L}_k \to N_{k+1} \to N_k \to 1$. Now the action of $\mathcal{M}_{g,1}$ on Γ_1 induces that on N_k and we have a representation

$$\varrho_k : \mathcal{M}_{g,1} \longrightarrow \mathrm{Aut}N_k.$$

The image of the representation ϱ_2 is equal to the subgroup of $\mathrm{Aut}N_2$ which preserves the skew symmetric pairing on $N_2 = H$ defined by the intersection number so that ϱ_2 is essentially equal to the classical representation of $\mathcal{M}_g$ onto the Siegel modular group $\mathrm{Sp}(2g; \mathbf{Z})$. Now let us write $\mathcal{M}(k)$ for $\mathrm{Ker}\varrho_k$ so that in particular $\mathcal{M}(1) = \mathcal{M}_{g,1}$ and $\mathcal{M}(2)$ is the Torelli group $\mathcal{I}_{g,1}$. $\mathcal{M}(k)$ is the subgroup of $\mathcal{M}_{g,1}$ acting trivially on N_k. Now it can be shown that we have a central extension

$$0 \longrightarrow \mathrm{Hom}(H, \mathcal{L}_k) \longrightarrow \mathrm{Aut}N_{k+1} \longrightarrow \mathrm{Aut}N_k \longrightarrow 1$$

where the abelian group $\mathrm{Hom}(H, \mathcal{L}_k)$ acts on N_{k+1} by the formula $f(\gamma) = \gamma f(\bar{\gamma})(f \in \mathrm{Hom}(H, \mathcal{L}_k), \gamma \in N_{k+1}$ and $\bar{\gamma} \in H$ is the image of γ in $H_1(N_{k+1}) = H)$. Hence we obtain a homomorphism

$$\tau_k : \mathcal{M}(k) \longrightarrow \mathrm{Hom}(H, \mathcal{L}_k)$$

which is the restriction of the homomorphism $\varrho_{k+1} : \mathcal{M}_{g,1} \to \mathrm{Aut}N_{k+1}$ to the subgroup $\mathcal{M}(k)$. The canonical isomorphism $H^* \cong H$ makes it possible to write $\tau_k : \mathcal{M}(k) \to \mathcal{L}_k \otimes H$ and $\mathrm{Ker}\tau_k = \mathcal{M}(k+1)$. The homomorphisms τ_k were introduced by Johnson [16, 18] and we call them Johnson's homomorphisms.

Before going further, here we review Johnson's results on the structure of the Torelli group. There are two natural types of mapping classes which belong to the Torelli group. One is the Dehn twist along a bounding simple closed curve and is called a BSCC map. The other is the composition $\varphi\psi^{-1}$ of two (say right handed) Dehn twists φ and ψ along mutually homologous simple closed curves on Σ_g (or $\Sigma_{g,1}$) and is called a BP (bounding pair) map. At the time when Johnson began his study on the Torelli group, it was known that the Torelli group is generated by these two types of mapping classes (Powell [33]) and also it had been asked whether the subgroup $\mathcal{K}_g$ of $\mathcal{I}_g$ (resp. $\mathcal{K}_{g,1}$ of $\mathcal{I}_{g,1}$) generated by all BSCC maps has a finite index or not.

Theorem (Johnson [17]). *The Torelli group is generated by a finite number of BP maps for $g \geq 3$.*

Theorem (Johnson [16, 18]). *The image of the homomorphism $\tau_2 : \mathcal{I}_{g,1} \to \Lambda^2 H \otimes H$ is equal to $\Lambda^3 H \subset \Lambda^2 H \otimes H$ and $\mathrm{Ker}\tau_2 = \mathcal{M}(3)$ coincides with $\mathcal{K}_{g,1}$ so that we have a short exact sequence*

$$1 \longrightarrow \mathcal{K}_{g,1} \longrightarrow \mathcal{I}_{g,1} \overset{\tau_2}{\longrightarrow} \Lambda^3 H \longrightarrow 1.$$

In particular $\mathcal{K}_{g,1}$ has an infinite index in $\mathcal{I}_{g,1}$.

There are other abelian quotients of the Torelli group than the homomorphism τ_2. Namely Birman and Craggs [4] have produced many $\mathbf{Z}/2$-valued homomorphisms on the Torelli group by making use of the Rohlin invariant for homology 3-spheres. Combining these two types of abelian quotients, Johnson [19] succeeded to determine the abelianization of the Torelli group completely.

Now we go back to the general theory. We have Johnson's homomorphism $\tau_k : \mathcal{M}(k) \to \mathrm{Hom}(H, \mathcal{L}_k) = \mathcal{L}_k \otimes H$. Let us define a submodule $\mathcal{H}_k$ of $\mathcal{L}_k \otimes H$ to be the kernel of the natural surjection $\mathcal{L}_k \otimes H \to \mathcal{L}_{k+1}$ defined as $\xi \otimes u \mapsto [\xi, u]$ ($\xi \in \mathcal{L}_k, u \in H$). Then one can show that $\mathrm{Im}\tau_k \subset \mathcal{H}_k$ (see [28]). For $k = 2$, $\mathcal{H}_2 = \Lambda^3 H$ so that $\mathrm{Im}\tau_2 = \mathcal{H}_2$ and in [27] we have shown that $\mathrm{Im}\tau_3$ is a submodule of $\mathcal{H}_3$ of index a power of two. Now there is a natural structure of a Lie algebra over $\mathbf{Z}$ on the graded module $\oplus_{k\geq 2}\mathrm{Hom}(H, \mathcal{L}_k)$ such that the graded submodule $\oplus_{k\geq 2}\mathcal{H}_k$ is a Lie subalgebra (the degree should be shifted down by one). The graded module $\oplus_{k\geq 2}\mathcal{M}(k)/\mathcal{M}(k+1)$ also admits a similar structure and Johnson's homomorphisms $\{\tau_k\}_{k\geq 2}$ induce an injection

$$\oplus_{k\geq 2}\mathcal{M}(k)/\mathcal{M}(k+1) \longrightarrow \oplus_{k\geq 2}\mathcal{H}_k$$

of graded Lie algebras. It is a very important problem to identify the image of this homomorphism. Computations so far show that the rank of $\mathrm{Im}\tau_k$ is much smaller than that of $\mathcal{H}_k$ for even $k \geq 4$. Now $\mathrm{Aut}N_k$ is an extension of $\mathrm{GL}(2g; \mathbf{Z})$ by a nilpotent group whose associated graded module is $\oplus_{i=2}^{k-1}\mathrm{Hom}(H, \mathcal{L}_i)$ and the representation $\varrho_k : \mathcal{M}_{g,1} \to \mathrm{Aut}N_k$ induces an identification of the quotient group $\mathcal{M}_{g,1}/\mathcal{M}(k)$ as an extension of $\mathrm{Sp}(2g; \mathbf{Z})$ by a nilpotent group whose graded module is $\oplus_{i=2}^{k-1}\mathrm{Im}\tau_i$. Now we apply Sullivan's general theory [35]. Thus let $N_k \otimes \mathbf{Q}$ be the Malcev completion of the nilpotent group N_k. Then any automorphism of N_k extends uniquely to that of $N_k \otimes \mathbf{Q}$ so that there is an embedding $\mathrm{Aut}N_k \to \mathrm{Aut}N_k \otimes \mathbf{Q}$, where the latter group is an extension of $\mathrm{GL}(2g; \mathbf{Q})$ by a nilpotent Lie group $\mathcal{G}_k$ over $\mathbf{Q}$ whose associated graded vectorspace is $\oplus_{i=2}^{k-1}\mathrm{Hom}(H, \mathcal{L}_i) \otimes \mathbf{Q}$. Now it is easy to see that $\mathrm{Aut}N_k \otimes \mathbf{Q}$ is a linear algebraic group over $\mathbf{Q}$ and its maximal normal unipotent subgroup is precisely equal to $\mathcal{G}_k$. Hence, by virtue of the Levi-Chevalley decomposition theorem, we can conclude that $\mathrm{Aut}N_k \otimes \mathbf{Q}$ is isomorphic to a *split* extension $\mathcal{G}_k \rtimes \mathrm{GL}(2g; \mathbf{Q})$. Although there is no canonical splitting, we can make explicit calculations involving choices of $\mathrm{GL}(2g; \mathbf{Q})$-invariant 2-cocycles for the successive extension classes of the nilpotent group N_k for small k and we obtain

Theorem. (*i*) *The representation* $\varrho_3 : \mathcal{M}_{g,1} \to \mathrm{Aut}N_3$ *induces an embedding of* $\mathcal{M}_{g,1}/\mathcal{K}_{g,1}$ *into the split extension* $\Lambda^3 H \rtimes \mathrm{Sp}(2g; \mathbf{Z})$ *as a subgroup of finite index such that the associated crossed homomorphism* $\tilde{k} : \mathcal{M}_{g,1} \to \Lambda^3 H$ *has the property that its restriction to the Torelli group* $\mathcal{I}_{g,1}$ *is equal to* $2\tau_2$.

(*ii*) *The representation* $\varrho_4 : \mathcal{M}_{g,1} \to \mathrm{Aut}N_4$ *induces an embedding of* $\mathcal{M}_{g,1}/\mathcal{M}(4)$ *into* $(\mathcal{H}_3 \tilde{\times} \Lambda^3 H) \rtimes \mathrm{Sp}(2g; \mathbf{Z})$ *as a subgroup of finite index where* $\mathcal{H}_3 \tilde{\times} \Lambda^3 H$ *denotes the central extension of* $\Lambda^3 H$ *by* $\mathcal{H}_3$ *defined by the bracket operation* $\Lambda^3 H \otimes \Lambda^3 H \to \mathcal{H}_3$ *and* $\mathrm{Sp}(2g; \mathbf{Z})$ *acts on it naturally.*

In view of the fact that the crossed homomorphism $\tilde{k}$ in the above theorem followed by the contraction $\Lambda^3 H \to H$ is nothing but the crossed homomorphism $k : \mathcal{M}_{g,1} \to H$ of §2, it seems to be worthwhile to propose

Problem. Give a definition of the crossed homomorphism $\tilde{k} : \mathcal{M}_{g,1} \to \Lambda^3 H$ in the context of complex analysis and/or algebraic K-theory.

4. Algebraic Structures of $\mathcal{M}_g$ and Topological Invariants of 3-Manifolds

Let us recall the classical theorem on the existence and stable uniqueness of the Heegaard splittings of 3-manifolds. First of all it says that any closed oriented 3-manifold can be expressed as $M_\varphi = H_g \cup_\varphi -H_g$ for some $\varphi \in \mathcal{M}_{g,1}$ where H_g is an oriented handlebody of genus g with $\partial H_g = \Sigma_g$ and M_φ is the manifold obtained from the disjoint union of H_g and $-H_g$ by identifying the boundaries by the map φ. Secondly two such manifolds $M_\varphi(\varphi \in \mathcal{M}_{g,1})$ and $M_\psi(\psi \in \mathcal{M}_{g,1})$ are orientation preserving diffeomorphic each other if and only if the two elements φ and ψ are equivalent in the disjoint union $\amalg_g \mathcal{M}_{g,1}$ of the mapping class groups $\mathcal{M}_{g,1}$ generated by the following two types of moves. One is to replace $\varphi \in \mathcal{M}_{g,1}$ by any other element in its double coset with respect to the subgroup $\mathcal{N}_{g,1}$ consisting of all elements which extend to diffeomorphisms of H_g. The other is to replace $\varphi \in \mathcal{M}_{g,1}$ by $i_*(\varphi)\iota \in \mathcal{M}_{g+1,1}$ where $i : \mathcal{M}_{g,1} \to \mathcal{M}_{g+1,1}$ is a natural injection and ι is the right angled rotation of the "last handle" of Σ_{g+1}. Thus to define a topological invariant of 3-manifolds is equivalent to give a function defined on the mapping class groups $\mathcal{M}_{g,1}$ which is invariant under the above two moves. Although the situation here is much more subtle than that in the case of oriented links in S^3 and the braid groups, it seems to be still reasonable to expect that there should exist a deep connection between the algebraic structure of the mapping class group and topological invariants of 3-manifolds. As a supporting evidence for that, extending and completing earlier works by Birman-Craggs [4] and Johnson [15], we will describe below how the Casson invariant of homology 3-spheres influences the structure of the subgroup $\mathcal{K}_{g,1}$ of $\mathcal{M}_{g,1}$.

Let $\iota_g \in \mathcal{M}_{g,1}$ be the product of right angled rotation of each handle of (an appropriate model of) Σ_g so that M_{ι_g} is the 3-sphere. For each element $\varphi \in \mathcal{K}_{g,1}$ we consider the manifold $W_\varphi = H_g \cup_{\iota_g\varphi} -H_g$ which is easily seen to be an oriented homology 3-sphere. It was proved in [27] that any homology 3-sphere appears like this and two such manifolds $W_\varphi(\varphi \in \mathcal{K}_{g,1})$ and $W_\psi(\psi \in \mathcal{K}_{g,1})$ are orientation preserving diffeomorphic if and only if φ and ψ are equivalent in $\lim_{g\to\infty} \mathcal{K}_{g,1}$ generated by the move which replaces a given element $\varphi \in \mathcal{K}_{g,1}$ by $\iota_g^{-1}\eta\iota_g\varphi\eta' \in \mathcal{K}_{g,1}$ for any $\eta, \eta' \in \mathcal{N}_{g,1}$. Thus to define a topological invariant for oriented homology 3-spheres is equivalent to give a function on $\lim_{g\to\infty} \mathcal{K}_{g,1}$ which is invariant under the above move.

Now let $\lambda^* : \mathcal{K}_{g,1} \to \mathbf{Z}$ be the mapping defined by $\lambda^*(\varphi) = \lambda(W_\varphi)(\varphi \in \mathcal{K}_{g,1})$ where $\lambda(W_\varphi)$ is the Casson invariant of the homology 3-sphere W_φ. Recall that $\mathcal{K}_{g,1}$ is the subgroup of $\mathcal{M}_{g,1}$ generated by all BSCC maps. Suppose $\varphi \in \mathcal{K}_{g,1}$ is such a Dehn twist along a simple closed curve ω on $\Sigma_{g,1}$. Then since the surface Σ_g is now considered as a particular Heegaard surface of $S^3 = H_g \cup_{\iota_g} -H_g$, ω is naturally a knot in S^3. Moreover it is easy to see that the homology sphere W_φ is nothing but the one obtained by performing -1 surgery on S^3 along this knot. Hence $\lambda(W_\varphi)$ is equal to the negative of the Casson invariant of the knot ω and this in turn can be calculated at the level of the surface Σ_g (and not in S^3) because the compact surface which ω bounds on $\Sigma_{g,1}$ is naturally a Seifert surface of ω. Combining these considerations, it can be shown that the mapping $\lambda^* : \mathcal{K}_{g,1} \to \mathbf{Z}$ is a *homomorphism*. Here we recall from §3 that we have Johnson's homomorphism $\tau_3 : \mathcal{K}_{g,1} \to \mathcal{H}_3$ which gives a large abelian quotient of $\mathcal{K}_{g,1}$. However it turns out that the homomorphism λ^* cannot be covered by τ_3 and we need a deeper additive invariant for $\mathcal{K}_{g,1}$. Such an invariant can be obtained by making use of the first characteristic class $e_1 \in H^2(\mathcal{M}_{g,1}; \mathbf{Z})$ of surface bundles introduced in §2 as follows. As was described in §2, the cohomology class e_1 can be represented by a cocycle $c \in Z^2(\mathcal{M}_{g,1}; \mathbf{Z})$ defined as $c(\varphi, \psi) = k(\varphi) \cdot k(\psi^{-1})$ where $k : \mathcal{M}_{g,1} \to H$ is a crossed homomorphism representing a generator of $H^1(\mathcal{M}_{g,1}; H)$. On the other hand Meyer [22] defined a 2-cocycle τ of the Siegel modular group $\mathrm{Sp}(2g; \mathbf{Z})$ which represents the signature class $\in H^2(\mathrm{Sp}(2g; \mathbf{Z}); \mathbf{Z})$ and its pull back under the representation $\mathcal{M}_{g,1} \to \mathrm{Sp}(2g; \mathbf{Z})$ is $-\frac{1}{3}e_1$. Hence there exists a mapping $d : \mathcal{M}_{g,1} \to \mathbf{Z}$ such that $\delta d = c + 3\tau$. Although the mapping d does depend on the choice of the crossed homomorphism k, its restriction to the subgroup $\mathcal{K}_{g,1}$ does not and moreover it is a homomorphism there. This is the invariant we seek for. Namely there exists a certain homomorphism $q : \mathcal{H}_3 \to \mathbf{Q}$ and we have

Theorem [27]. *The function* $\frac{1}{24}d + q \circ \tau_3$ *defined on* $\mathcal{K}_{g,1}$ *is invariant under the equivalence relation on* $\lim_{g \to \infty} \mathcal{K}_{g,1}$ *so that it defines an invariant for oriented homology 3-spheres and this coincides with the Casson invariant.*

Remarks. 1. At present there are only a few known informations about the structure of the groups $\mathcal{K}_g$ and $\mathcal{K}_{g,1}$. For example it is not known whether they are finitely generated or not for $g \geq 3$. Also in view of the results in this section, it would be worthwhile to determine the abelianizations of them.

2. Recently the author obtained an interpretation of the invariant d as the signature defect of certain framed 3-manifolds [30]. It would be desirable to have further geometrical or analytical meaning of it, perhaps along the lines of Atiyah's paper [2] which treats the case of genus 1.

3. In some sense the invariant d can be understood to be the effect of the Casson invariant for knots on the structure of the group $\mathcal{K}_{g,1}$ (see [29]). It would be interesting to investigate how other invariants for knots influence the structure of $\mathcal{K}_{g,1}$.

4. It seems to be reasonable to conjecture that we can generalize the definition of the Casson invariant to possibly arbitrary 3-manifold by making use of the invariant d and the explicit description of the quotient $\mathcal{M}_{g,1}/\mathcal{M}(4)$ given in §3. In particular one can ask whether Walker's extension to rational homology 3-spheres [38] can be expressed like this or not.

5. Recently Reshetikhin and Turaev [34] have defined new topological invariants for 3-manifolds, motivated by Witten's ideas in quantum field theory, and there are certain representations of $\mathcal{M}_g$ associated to them. Kohno [21] has also made use of certain projective representations of $\mathcal{M}_g$, which arise naturally

in the conformal field theory, to derive topological invariants via the Heegaard splittings. There seems to be much work to be done here from the point of view of the structure of $\mathcal{M}_g$. For example we can ask whether there exists a natural number k such that the subgroup $\mathcal{M}(k)$ is contained in the kernel of these representations or not.

References

1. Atiyah, M.F.: The signature of fibre-bundles. In: Global analysis, papers in honor of K. Kodaira, Tokyo University Press, 1969, pp. 73–84
2. Atiyah, M.F.: The logarithm of the Dedekind η-function. Math. Ann. **278** (1987) 335–380
3. Birman, J.: Mapping class groups of surfaces. In: Braids, Contemp. Math. **78** (1988) 13–43
4. Birman, J., Craggs, R.: The μ-invariant of 3-manifolds and certain structural properties of the group of homeomorphisms of a closed oriented 2-manifold. Trans. Amer. Math. Soc. **237** (1978) 283–309
5. Earle, C.J.: Families of Riemann surfaces and Jacobi varieties. Ann. Math. **107** (1978) 255–286
6. Earle, C.J., Eells, J.: The diffeomorphism group of a compact Riemann surface. Bull. Amer. Math. Soc. **73** (1967) 557–559
7. Harer, J.: The second homology group of the mapping class group of an orientable surface. Invent. math. **72** (1983) 221–239
8. Harer, J.: Stability of the homology of the mapping class groups of orientable surfaces. Ann. Math. **121** (1985) 215–249
9. Harer, J.: The virtual cohomological dimension of the mapping class group of an orientable surface. Invent. math. **84** (1986) 157–176
10. Harer, J.: The cohomology of the moduli space of curves. In: Theory of moduli. (Lecture Notes in Mathematics, vol. 1337). Springer, Berlin Heidelberg New York 1988, pp. 138–221
11. Harer, J., Zagier, D.: The Euler characteristic of the moduli space of curves. Invent. math. **85** (1986) 457–485
12. Harris, J.: Families of smooth curves. Duke Math. J. **51** (1984) 409–419
13. Ivanov, N.V.: Algebraic properties of the mapping class groups of surfaces. Preprint, Leningrad 1985
14. Ivanov, N.V.: Complexes of curves and the Teichmüller modular group. Uspekhi Mat. Nauk **42** (1987) 49–91 [English transl.: Russ. Math. Surv. **42** (1987) 55–107]
15. Johnson, D.: Quadratic forms and the Birman-Craggs homomorphisms. Trans. Amer. Math. Soc. **261** (1980) 235–254
16. Johnson, D.: An abelian quotient of the mapping class group $\mathcal{I}_g$. Math. Ann. **249** (1980) 225–242
17. Johnson, D.: The structure of the Torelli group I: A finite set of generators for $\mathcal{I}_g$. Ann. Math. **118** (1983) 423–442
18. Johnson, D.: A survey of the Torelli group, Contemp. Math. **20** (1983) 165–179
19. Johnson, D.: The structure of the Torelli group II and III. Topology **24** (1985) 113–144
20. Kerckhoff, S.P.: The Nielsen realization problem. Ann. Math. **117** (1983) 235–265
21. Kohno, T.: Topological invariants for 3-manifolds using representations of mapping class groups I. Topology (to appear)
22. Meyer, W.: Die Signatur von Flächenbündeln. Math. Ann. **201** (1973) 239–264
23. Miller, E.Y.: The homology of the mapping class group. J. Diff. Geom. **24** (1986) 1–14
24. Morita, S.: Characteristic classes of surface bundles. Invent. math. **90** (1987) 551–577

25. Morita, S.: Families of Jacobian manifolds and characteristic classes of surface bundles
    I. Ann. Inst. Fourier **39** (1989) 777–810
26. Morita, S.: Families of Jacobian manifolds and characteristic classes of surface bundles
    II. Math. Proc. Camb. Phil. Soc. **105** (1989) 79–101
27. Morita, S.: Casson's invariant for homology 3-spheres and characteristic classes of
    surface bundles I. Topology **28** (1989) 305–323
28. Morita, S.: On the structure and the homology of the Torelli group. Proc. Japan Acad.
    65 (1989) 147–150
29. Morita, S.: On the structure of the Torelli group and the Casson invariant. Topology
    (to appear)
30. Morita, S.: Signature defects of certain framed 3-manifolds and the Casson invariant.
    In preparation
31. Mumford, D.: Towards an enumerative geometry of the moduli space of curves. In:
    Arithmetic and geometry. Progr. Math. **36** (1983) 271–328
32. Penner, R.C.: Perturbative series and the moduli space of Riemann surfaces. J. Diff.
    Geom. **27** (1988) 35–53
33. Powell, J.: Two theorems on the mapping class group of a surface. Proc. Amer. Math.
    Soc. **68** (1978) 347–350
34. Reshetikhin, N., Turaev, V.G.: Invariants of 3-manifolds via link polynomials and
    quantum groups. Invent. math (to appear)
35. Sullivan, D.: Infinitesimal computations in topology. Publ. Math. I.H.E.S. **47** (1977)
    269–331
36. Trapp, R.: A linear representation of the mapping class group $\mathcal{M}$ and the theory of
    winding numbers. Preprint
37. Wagoner, J.: K_2 and diffeomorphisms of two and three dimensional manifolds. In:
    Geometric topology. Academic Press 1979, pp. 557–577
38. Walker, K.: An extension of Casson's invariant to rational homology spheres. Preprint
39. Witten, E.: On the structure of the topological phase of two dimensional gravity. Nucl.
    Phys. **B 340** (1990) 281–332
40. Wolpert, S.: The topology and geometry of the moduli space of Riemann surfaces. In:
    Arbeitstagung Bonn 1984. (Lecture Notes in Mathematics, vol. 1111). Springer, Berlin
    Heidelberg New York 1985, pp. 431–451

Cyclic Cohomology and Invariants of Multiply Connected Manifolds

Henri Moscovici

Department of Mathematics, The Ohio State University
231 West 18th Ave., Columbus, OH 43210-1174, USA

To motivate the subject-matter of this report, let me begin by recalling two of the early applications of the Atiyah-Singer Index Theorem, concerning two important manifold invariants. The first result, which was in fact a precursor of the Index Theorem, is the Hirzebruch Signature Theorem. It asserts that the L-genus $\lambda(M) = L(M)[M]$ of a closed oriented smooth $4k$-dimensional manifold M is equal to the index of the signature operator $\mathcal{B}^{+}$ attached to a Riemannian metric, which in turn coincides with the signature of the manifold. The second application requires M to be in addition a Spin-manifold and equates its $\widehat{A}$-genus $\alpha(M) = \widehat{A}(M)[M]$ with the index of the Dirac operator $\mathcal{D}^{+}$ associated to a Spin-structure. These results not only explain the integrality of the above invariants (under the given conditions) but also have other, more subtle, implications. Thus, Hirzebruch's theorem implies that the L-genus is an invariant of the oriented homotopy type of M. On the other hand, in conjunction with Lichnerowicz's formula for the Dirac Laplacian, the index theorem for the Dirac operator implies that if $\alpha(M) \neq 0$, then M can admit no metric of (strictly) positive scalar curvature.

It is a remarkable fact that for simply connected manifolds both results are 'sharp'. That is, the L-genus is, up to a scalar multiple, the only rational oriented cobordism invariant which is also a homotopy type invariant. The parallel result for the $\widehat{A}$-genus, asserting that for a simply connected closed Spin-manifold M of dimension $4k$ $(k > 1)$ the vanishing of $\alpha(M)$ is the only obstruction to the existence on M of a metric of positive scalar curvature, was recently proved by S. Stolz.

The uniqueness part of the preceding statements ceases to hold in the non-simply connected case. Indeed, let T be a torus and let $f : M \to T$ be a continuous map. For each cohomology class $\eta \in H^{*}(T, \mathbb{Q})$, define the corresponding higher L (resp. $\widehat{A}$)-genus by

$$\lambda_{f,\eta}(M) = (L(M) \cup f^{*}(\eta))[M] \quad (\text{resp. } \alpha_{f,\eta}(M) = (\widehat{A}(M) \cup f^{*}(\eta))[M]).$$

Lusztig [Lu] found an ingenious index theoretical interpretation for the higher L-genera $\lambda_{f,\eta}(M)$, in terms of the index bundle of a family of twisted signature operators, and used it to prove that they are invariants of the oriented homotopy

Proceedings of the International Congress
of Mathematicians, Kyoto, Japan, 1990
© The Mathematical Society of Japan, 1991

type of M. The same method can be adapted to give the vanishing of the higher $\hat{A}$-genera $\alpha_{f,\eta}(M)$ in the presence of a metric of positive scalar curvature (cf. [GL1]).

More generally, the torus in the above definition may be replaced by any $K(\Gamma, 1)$ space, where Γ is a finitely presented group. According to a conjecture of Novikov [N], the numbers $\lambda_{f,\eta}(M)$, $\eta \in H^*(\Gamma, \mathbb{Q})$ ought to be homotopy invariants. There is a parallel conjecture for the $\hat{A}$-genera, made by Gromov and Lawson [GL2], implying that if the Spin-manifold M admits a metric of positive scalar curvature then $\alpha_{f,\eta}(M) = 0$, for all $\eta \in H^*(\Gamma, \mathbb{Q})$. For both conjectures one may assume without loss of generality that $\Gamma = \pi_1(M)$ and f is the classifying map of the universal cover $\widetilde{M} \to M$.

The preceding discussion was meant to underscore the following points. First, for non-simply connected manifolds, there are 'higher' versions of the classical genera, which are involved in interesting open problems. Secondly, an index theoretical interpretation of the higher genera provides an effective tool for handling such problems. It was also noted above that the Families Index Theorem supplies the desired interpretation in the case when Γ is free abelian, essentially because its dual, the torus $\widehat{\Gamma}$, is a nice space. In general, though, the 'dual object' of Γ is no longer an ordinary space and therefore beyond the scope of standard index theory. It is, however, a 'non-commutative space' in the sense of A. Connes [C1]. Moreover, Connes' theory of cyclic cohomology and its pairing with K-theory [C1, C2] offers precisely the right context for developing the appropriate generalization of index theory. The first two sections of this report will survey the 'higher' index theorems obtained along these lines in [CM2] (cf. also [CM1]) and [CGM], as well as their applications to the conjectures of Novikov and Gromov-Lawson.

Like the higher genera, the higher indices of elliptic operators are manufactured out of group cohomology classes, viewed as classes in the cyclic cohomology of the group ring $\mathbb{C}\Gamma$. Somewhat surprisingly, the unstable component of the cyclic cohomology of $\mathbb{C}\Gamma$ appears to be intimately related to non-local invariants when $K(\Gamma, 1)$ is a manifold. This point of view will be put forth in the last section, where I will interpret in this light the results in [MS1] and [MS2], relating η-invariants and analytic torsion to periodic geodesics.

1. Group Cocycles and Higher Indices

1.1 Covering Indices of Elliptic Operators

Let M be a closed smooth manifold, Γ a finitely presented group, $f : M \to B\Gamma$ a continuous map to the classifying space of Γ and $\widetilde{M}$ the pull-back of the universal cover $E\Gamma \to B\Gamma$. Let $C_r^*(\Gamma)$ denote the C^*-algebra obtained as the completion of the group ring $\mathbb{C}\Gamma$ acting on the Hilbert space $\ell^2(\Gamma)$. Then $\widetilde{M} \times_\Gamma C_r^*(\Gamma)$ is a $C_r^*(\Gamma)$-bundle over M. Using it, Mishchenko [Ms2] and Kasparov [K1] have associated to an elliptic pseudodifferential operator D on M a K-theory class $Index_{f,C_r^*(\Gamma)}D \in K_0(C_r^*(\Gamma))$. In [CM2] this assignment was refined to produce an element $Index_{f,\Gamma} D \in K_0(\mathbb{C}\Gamma \otimes \mathfrak{R})$, where $\mathfrak{R}$ stands for the ring of infinite complex matrices $(a_{ij})_{1 \leq i,j < \infty}$ satisfying the rapid decay condition $\sup((i+j)^k |a_{ij}|) < \infty$, for all $k \in \mathbb{R}$. We sketch the construction of this covering index.

Let $\Psi(M, E)$ denote the algebra of pseudodifferential operators of classical type and integral order acting on the space of smooth sections of a vector bundle E over M. Let $\Psi^{-\infty}(M, E)$ denote the ideal formed by all smoothing operators. The quotient algebra $CS(M, E)$ is referred to as the algebra of complete symbols with coefficients in E. When E is the trivial line bundle, we simplify the notation by omitting E and by setting $\mathfrak{A} = \Psi(M)$ and $\mathfrak{B} = \Psi^{-\infty}(M)$. Upstairs, on the covering manifold $\widetilde{M}$, we let $\widetilde{\mathfrak{A}}$ be the subalgebra of $\Psi(\widetilde{M})$ consisting of all Γ-invariant, properly supported pseudodifferential operators and let $\widetilde{\mathfrak{B}}$ be the subalgebra of $\widetilde{\mathfrak{A}}$ formed of smoothing operators. Clearly $\widetilde{\mathfrak{A}}/\widetilde{\mathfrak{B}}$ is isomorphic to $CS(M) = \mathfrak{A}/\mathfrak{B}$.

For E a $\mathbb{Z}_2$-graded vector bundle on M, an odd elliptic pseudodifferential operator $D \in \Psi(M, E)$ defines an $\mathfrak{A}$-linear map D^+ between two finitely generated projective right $\mathfrak{A}$-modules, which becomes an isomorphism, $CS(D^+)$, on tensoring with $CS(M) = \mathfrak{A}/\mathfrak{B}$. Note that D^+ can also be viewed as a quasi-isomorphism over $\mathfrak{B}$. Now identify $CS(M)$ with $\widetilde{\mathfrak{A}}/\widetilde{\mathfrak{B}}$. A well-known construction in algebraic K-theory, employed in the definition of the connecting homomorphism $\partial : K_1(\widetilde{\mathfrak{A}}/\widetilde{\mathfrak{B}}) \to K_0(\widetilde{\mathfrak{B}})$, associates to $CS(D^+)$ an element $\partial(D^+)$ in $K_0(\widetilde{\mathfrak{B}})$. We remark parenthetically that the same construction applied to $\mathfrak{A}/\mathfrak{B} = CS(M)$ gives rise to the ordinary index of D^+, on identifying $K_0(\mathfrak{B})$ with $\mathbb{Z}$.

The algebra $\widetilde{\mathfrak{B}}$ can also be regarded as the convolution algebra of smooth compactly supported functions on the differentiable groupoid $\widetilde{M} \times_\Gamma \widetilde{M}$. As such, it can be homomorphically mapped to $\mathbb{C}\Gamma \otimes \mathfrak{B}$ and further, after identifying $\mathfrak{B} = \Psi^{-\infty}(M)$ with $\mathfrak{R}$, to $\mathbb{C}\Gamma \otimes \mathfrak{R}$. Although these homomorphisms are non-canonical they induce however a well-defined map, independent of all choices made,

$$\theta : K_0(\widetilde{\mathfrak{B}}) \to K_0(\mathbb{C}\Gamma \otimes \mathfrak{R}).$$

Definition 1. The *index of D^+ associated to the covering $\widetilde{M}$* is defined as

$$Index_{f,\Gamma} D^+ = \theta(\partial(D^+)) \in K_0(\mathbb{C}\Gamma \otimes \mathfrak{R}).$$

Remarks. 1. The ring $\mathbb{C}\Gamma \otimes \mathfrak{R}$ is very likely the smallest group ring associated to Γ with the property that its K_0-group is nontrivial. Indeed, for Γ torsion-free, the reduced K_0-group of the group ring $\mathbb{C}\Gamma$ is believed to be 0. By contrast, the group $K_0(\mathbb{C}\Gamma \otimes \mathfrak{R})$ appears to be, in general, uncountable.

2. Under the map from $K_0(\mathbb{C}\Gamma \otimes \mathfrak{R})$ to $K_0(C_r^*(\Gamma) \otimes \mathfrak{K}) = K_0(C_r^*(\Gamma))$, induced by the inclusion of $\mathbb{C}\Gamma$ into $C_r^*(\Gamma)$ and of $\mathfrak{R}$ into the ring $\mathfrak{K}$ of compact operators on the infinite dimensional separable Hilbert space, the image of $Index_{f,\Gamma} D^+$ coincides with $Index_{f,C_r^*(\Gamma)} D^+$. However, our covering index seems to be a finer invariant, which depends on the *complete,* not just the principal, symbol of the given operator.

1.2 Higher Analytic Indices

Next, we will use group cocycles to produce numerical invariants out of the covering index. Let $c \in Z^k(\Gamma, \mathbb{C})$ be a homogeneous group cocycle; without loss of generality, we may assume that c is alternating. One associates to c a cyclic cocycle τ_c on the algebra $\mathbb{C}\Gamma \otimes \mathfrak{R}$ by

$$\tau_c(a_0 \otimes A_0, \ldots, a_k \otimes A_k) = \mathrm{Tr}(A_0 \ldots A_k) \sum_{g_0 g_1 g_2 \cdots g_k = 1} a_0(g_0) a_1(g_1) \ldots a_k(g_k)$$
$$\cdot \, c(1, g_1, g_1 g_2, \ldots, g_1 g_2 \ldots g_k) \, ;$$

the cyclic cohomology class $\tau_{[c]} = [\tau_c] \in HC^k(\mathbb{C}\Gamma \otimes \mathfrak{R})$ depends only on the group cohomology class $[c] \in H^k(\Gamma, \mathbb{C})$.

One employs now Connes' pairing [C1] between cyclic cohomology and K-theory to define the *covering higher indices* of the elliptic operator D^+, as follows:

$$Index_{f, \eta} D^+ = \langle \tau_\eta, Index_{f, \Gamma} D^+ \rangle, \quad \eta = [c] \in H^k(\Gamma, \mathbb{C}), \ k \text{ even}.$$

When $\eta = [1] \in H^0(\Gamma, \mathbb{C})$, the above index coincides with the Γ-index of Atiyah and Singer [A]. Their Index Theorem for Covering Spaces asserts that

$$Index_{f, [1]} D^+ = Index \, D^+,$$

the right hand side being the ordinary index of D^+.

In the same spirit, relying on the fact that the differentiable groupoids $\widetilde{M} \times_\Gamma \widetilde{M}$ and $M \times M$ are locally isomorphic, one can show that the pairing $\langle \tau_\eta, Index_{f, \Gamma} D^+ \rangle$ depends only on the Alexander-Spanier cohomology class $f^*(\eta) \in H^k(M, \mathbb{C})$ and the K-homology class $[D^+] \in K_0(M)$. More precisely, one has the equality

$$\langle \tau_\eta, Index_{f, \Gamma} D^+ \rangle = \langle f^*(\eta), [D^+] \rangle,$$

where the right hand side is a certain natural pairing between $H^{\mathrm{ev}}(M, \mathbb{C})$, viewed as Alexander-Spanier cohomology, and the K-homology group $K_0(M)$. This pairing can also be used to attach *intrinsic higher indices* to the elliptic operator D^+ by setting

$$Index_\xi D^+ = \langle \xi, [D^+] \rangle,$$

for each Alexander-Spanier cohomology class $\xi \in H^{\mathrm{ev}}(M, \mathbb{C})$. When $\xi = [1] \in H^0(M, \mathbb{C})$ one recovers the ordinary index of D^+. The above discussion can thus be summarized as follows.

Proposition 1 ([CM2], §5). $Index_{f, \eta} D^+ = Index_{f^*(\eta)} D^+, \ \forall \eta \in H^{\mathrm{ev}}(B\Gamma, \mathbb{C})$.

1.3 Cohomological Formulae

To relate the higher indices introduced above to higher genera, one needs to express them, just like in the classical case, in purely cohomological terms. The basic result in this direction is the following generalization of the Atiyah-Singer Index Theorem.

Theorem 2 ([CM2], §3). *Let $\xi \in H^{2q}(M, \mathbb{C})$. With $\sigma(D^+) = $ principal symbol of D^+ and $\Im(M) = $ the index class of M^n, one has:*

$$Index_\xi D^+ = \frac{1}{(2\pi i)^q} \frac{q!}{(2q)!} (-1)^n (ch\,\sigma(D^+) \cup \Im(M) \cup \xi)[T^*M].$$

Remarks. 3. In fact we obtained a more refined result, at the level of differential forms not just cohomology classes, which extends the Local Index Theorem. Its proof, based on the heat equation method, uses the full force of Getzler's symbolic calculus [G].

4. The above result, in a non-compact variant due to J. Roe [R], was also used, in a rather striking way, by S. Weinberger [W] to give a direct analytic proof of Novikov's celebrated theorem on the topological invariance of rational Pontrjagin classes.

Combining Proposition 1 and Theorem 2, one obtains the desired Higher Index Formula for Covering Spaces:

Theorem 3 [CM2], §5). *Let $\eta \in H^{2q}(B\Gamma, \mathbb{C})$. Then*

$$Index_{f,\eta} D^+ = \frac{1}{(2\pi i)^q} \frac{q!}{(2q)!} (-1)^n (ch\,\sigma(D^+) \cup \Im(M) \cup f^*(\eta))[T^*M].$$

1.4 Applications to the Novikov and Gromov-Lawson Conjectures

Theorem 3, applied to the signature operator β^+ on an even-dimensional manifold M^{2k}, gives the identity

$$\lambda_{f,\eta}(M) = C_{k,q} \langle \tau_\eta, Index_{f,\Gamma} \beta^+ \rangle, \quad \forall \eta \in H^{2q}(B\Gamma, \mathbb{C}),$$

where $C_{k,q}$ is an explicit (nonzero) constant depending only on the dimension of M and the degree of η. Thus, the Novikov conjecture would follow if one could prove that $Index_{f,\Gamma} \beta^+ \in K_0(\mathbb{C}\Gamma \otimes \mathfrak{R})$ is an invariant of the homotopy type of M. In view of Remark 2, this is unlikely to happen. However, its image $Index_{f,C_r^*(\Gamma)} \beta^+ \in K_0(C_r^*(\Gamma))$ *is a homotopy invariant.* Indeed, this follows from Mishchenko's fundamental construction of an a priori homotopy invariant $\sigma[M, f]$ in the Witt group $L(\mathbb{C}\Gamma)$ [Ms1], together with the fact that $Index_{f,C_r^*(\Gamma)} \beta^+$ coincides with the image of $\sigma[M, f]$ under the map sending $L(\mathbb{C}\Gamma)$ to $L(C_r^*(\Gamma)) = K_0(C_r^*(\Gamma))$ (cf. [MsS], [K2]).

The problem which emerges is therefore to extend the pairing of $H^{ev}(\Gamma, \mathbb{C})$ with $K_0(\mathbb{C}\Gamma \otimes \mathfrak{R})$ to a pairing with $K_0(C_r^*(\Gamma))$. There is a strategy for that, devised by Connes in a similar context [C3]. It consists of finding an intermediate subalgebra, which is closed under holomorphic functional calculus and therefore has the same K_0-group, to which cyclic cocycles of the form τ_c extend by continuity. In our situation, the natural candidate for this subalgebra is $\mathfrak{C}\Gamma =$ the smallest subalgebra of $C_r^*(\Gamma) \otimes \mathfrak{R}$ (C^*-tensor product) which contains $\mathbb{C}\Gamma \otimes \mathfrak{R}$ and is closed under holomorphic functional calculus. The following two conditions, involving the choice of a word length function $g \mapsto |g|$, are sufficient for the successful application of the above strategy to Γ:

Polynomial Cohomology: $\forall \eta \in H^{2q}(\Gamma, \mathbb{C})$, $\exists c \in Z^{2q}(\Gamma, \mathbb{C})$ of polynomial growth, with $\eta = [c]$;

Rapid Decay (cf. [J]): $\exists s \in \mathbb{N}$ and $C > 0$ such that $\forall a \in \mathbb{C}\Gamma$

$$\|a\|_{C_r^*(\Gamma)} \le C\left(\sum (1 + |g|)^{2s}|a(g)|^2\right)^{1/2}.$$

Theorem 4 ([CM2], §6). *Any finitely presented group Γ with the above two properties satisfies both the Novikov and the Gromov-Lawson conjectures.*

Indeed, for such a group one has, $\forall \eta \in H^{2q}(B\Gamma, \mathbb{C})$,

$$\lambda_{f,\eta}(M) = C_{k,q}\langle \tau_\eta, \; Index_{f,C_r^*(\Gamma)}\rlap{/}{D}^+ \rangle,$$

which proves the homotopy invariance of the higher L-genera. On the other hand, applying Theorem 3 to the Dirac operator $\rlap{/}{D}^+$ gives

$$\alpha_{f,\eta}(M) = (2\pi i)^q \frac{(2q)!}{q!}\langle \tau_\eta, \; Index_{f,C_r^*(\Gamma)}\rlap{/}{D}^+ \rangle.$$

By an extension of Lichnerowicz' argument, Rosenberg [Ro] has shown that if M admits a metric of positive scalar curvature then $Index_{f,C_r^*(\Gamma)}\rlap{/}{D}^+ = 0$. Therefore, all higher $\widehat{A}$-genera also vanish under this hypothesis.

Remark. 5. The conclusion of the above theorem can be strengthened to assert that the assembly map (see [K2], [K3]) $\mu : K_*(B\Gamma) \to K_*(C_r^*(\Gamma))$ is rationally injective.

The two properties stated above are satisfied by a large class of finitely presented groups, including the remarkable class of Gromov's hyperbolic groups [Gr]. The fact that a hyperbolic group satisfies the Polynomial Cohomology property is due to Gromov, while the Rapid Decay property was proved by P. de la Harpe [Ha], building on prior results by Haagerup [H] and Jolissaint [J].

Corollary 5 ([CM2], §6). *The hyperbolic groups satisfy the Novikov and the Gromov-Lawson conjectures.*

2. Asymptotic Cocycles and Higher Indices

We now turn to a different procedure for constructing numerical indices, based on Connes' pairing for entire cohomology [C2].

2.1 Almost Flat Bundles and Almost Representations

Let M be a closed connected Riemannian manifold, E a Hermitian vector bundle over M and ∇ a metric preserving connection on E. With $\Theta = \nabla^2$ being the curvature of ∇, we set

$$\|\Theta\| = \sup\{\|\Theta_x(X, Y)\| \; ; \; \|X \wedge Y\| \leqq 1, \; X, Y \in T_x M, \; x \in M\}.$$

Definition 2. (i) Given $\varepsilon > 0$, the connection ∇ is called ε-*flat* if $\|\Theta\| < \varepsilon$.

(ii) A vector bundle E is called ε-*flat* if it admits a metric preserving connection ∇ with $\|\Theta\| < \varepsilon$.

(iii) A class $\xi \in K^0(M)$ is called *almost flat* if for each $\varepsilon > 0$, there exist ε-flat bundles E^+, E^- such that $[E^+] - [E^-] = \xi$.

Note that the last notion is independent of the Riemannian structure on M. We denote by $K^0_{\mathrm{af}}(M)$ the subgroup of $K^0(M)$ formed of almost flat classes. If $h : N \to M$ is a continuous map, then $h^* : K^0(M) \to K^0(N)$ maps $K^0_{\mathrm{af}}(M)$ to $K^0_{\mathrm{af}}(N)$.

Remark. 6. If M is simply connected, there exists $\varepsilon_M > 0$ such that any ε_M-flat bundle is flat and therefore, trivial. On the other hand $K(\Gamma, 1)$-manifolds with Γ 'large' tend to have 'many' almost flat bundles (see 2.4).

Let $\mathfrak{B}$ be a Banach algebra. Given a linear map ϱ of $\mathfrak{B}$ to the algebra $\mathfrak{L}(V)$ of all linear operators on a finite-dimensional Hilbert space V, we introduce the 'curvature' operators $\theta(a, b) = \varrho(ab) - \varrho(a)\varrho(b)$ and set, for any finite subset $F \subset \mathfrak{B}$,

$$\|\theta\|_F = \sup\{\|\theta(a, b)\| \; ; \; a, b \in F\}.$$

Definition 3. Given $\varepsilon > 0$ and a finite subset $F \subset \mathfrak{B}$, an (ε, F)-*representation* of $\mathfrak{B}$ is a linear map $\varrho : \mathfrak{B} \to \mathfrak{L}(V)$ such that: (i) $\|\varrho(a)\| \leq \|a\|$, $\forall a \in \mathfrak{B}$; (ii) $\varrho(1) = 1$; (iii) $\|\theta\|_F < \varepsilon$.

Let $p \in M$ be a fixed base point and let $\Gamma = \pi_1(M, p)$. We denote by $\Omega(M, p)$ the space of loops based at p. Choose a cross-section $\gamma : \Gamma \to \Omega(M, p)$ to the canonical projection $\pi : \Omega(M, p) \to \Gamma$ such that: $\gamma(1) = 1$ and $\forall g \in \Gamma$ with $g^2 \neq 1$, $\gamma(g^{-1}) = \gamma(g)^{-1}$. Given $F \subset \ell^1(\Gamma)$, one can associate to any ε-flat bundle E an $(\varepsilon c_F, F)$-representation $\varrho_E : \ell^1(\Gamma) \to \mathfrak{L}(E_p)$, by assigning to an element $g \in \Gamma$ the parallel transport operator $\varrho_E(g)$ along the loop $\gamma(g)$; if $g^2 = 1$, $\varrho_E(g)$ is replaced by $\frac{1}{2}(\varrho_E(g) + \varrho_E(g)^*)$. Indeed, since the effect of the parallel transport along loops can be estimated in terms of the curvature, one has

$$\|\theta_E\|_F \leq c_F \|\Theta_E\|,$$

where Θ_E denotes the curvature of E, θ_E is the curvature of ϱ_E and c_F is a constant depending on F.

2.2 Pairing with K-Theory and an Index Formula

Let $\varrho : \mathfrak{B} \to \mathfrak{L}(V)$ be an (ε, F)-representation of a Banach algebra $\mathfrak{B}$. Its character $\phi = (\phi_{2k})$ is a discrete version of the asymptotic cocycles introduced in [CM2, §4], and is given by the following formula:

$$\phi_{2k}(a_0, a_1, \ldots, a_{2k}) = (-1)^k \frac{(2k)!}{k!} Tr(\varrho(a_0)\theta(a_1, a_2) \ldots \theta(a_{2k-1}, a_{2k})), \quad a_j \in \mathfrak{B}.$$

Recall that asymptotic cocycles pair with K-theory, by means of the pairing formula for entire cohomology ([C2], [GS]):

$$\langle \phi, e \rangle = \sum_{k=0}^{\infty} \frac{(-1)^k}{k!} \phi_{2k}(e - \frac{1}{2}, e, \ldots, e), \qquad e^2 = e \in \mathscr{M}_N(\mathfrak{B}).$$

Proposition 6 ([CGM], Prop. 9). *Let $e^2 = e \in \mathscr{M}_N(\mathfrak{B})$. There exists a finite set $F \subset \mathfrak{B}$ and $\delta > 0$, such that for any (ε, F)-representation ϱ of $\mathfrak{B}$ with $0 < \varepsilon < \delta$, the above series converges. Moreover, with ϕ denoting the character of ϱ one has*

$$\langle \phi, e \rangle = Tr(\Pi - \frac{1}{2}),$$

where $\Pi \in \mathscr{M}_N(\mathfrak{L}(V)) = \mathfrak{L}(V \otimes \mathbb{C}^N)$ is the projection defined by

$$\Pi = \frac{1}{2\pi i} \int_{|z-1|=\frac{1}{2}} ((\varrho \otimes 1)(e) - z)^{-1} dz.$$

In the remainder of this section $\widetilde{M}$ will denote the universal cover of M^n and $f : M \to B\Gamma$ will be its classifying map. Consider an elliptic operator D^+ on M. From the preceding discussion, it follows that there exists a $\delta > 0$ such that, for any ε-flat bundle E on M with $0 < \varepsilon < \delta$, the pairing

$$Index_{E,\Gamma} D^+ := \langle \phi_E, Index_{f,\Gamma} D^+ \rangle,$$

between the character ϕ_E of the associated almost representation $\varrho_E : \ell^1(\Gamma) \to \mathfrak{L}(E_p)$ and the covering index $Index_{f,\Gamma} D^+ \in K_0(\mathbb{C}\Gamma \otimes \mathfrak{R})$ is well-defined.

Theorem 7 ([CGM], Thm. 10). *With the above assumptions, one has*

$$Index_{E,\Gamma} D^+ = (-1)^n (ch\, \sigma(D^+) \cup \mathfrak{I}(M) \cup ch\, E)[T^*M] = Index\, D_E^+.$$

The proof is similar in spirit with that of Theorem 3 and involves two stages. The first, corresponding to Proposition 1, converts the above pairing into one between the 'asymptotic' Alexander-Spanier cocycle associated to the character ϕ_E and the K-homology class of D^+. The second consists in establishing the analogue of the index formula of Theorem 2.

2.3 Pairing with L-Theory and a Signature Formula

The extra virtue of the type of cocycles considered above is that they can be paired directly with the L-theory. Indeed, $\ell^1(\Gamma)$ is not merely a Banach algebra, but one equipped with a compatible involution. So, it is natural to work in the context of Banach $*$-algebras. An obvious modification of Definition 3 gives the notion of an $(\varepsilon, F) - *$-representation for a Banach $*$-algebra $\mathfrak{B}$.

Proposition 8 ([CGM], Prop. 5). *Let $\mathfrak{B}$ be a Banach $*$-algebra and let $H = H^*$ be a self-adjoint invertible element in $\mathcal{M}_N(\mathfrak{B})$.*

(i) If $\varrho : \mathfrak{B} \to \mathfrak{L}(V)$ is an $(\varepsilon, F) - *$-representation, with $F =$ the set of entries of H and H^{-1} and $0 < \varepsilon < 1/2N^2$, then $(\varrho \otimes 1)(H) \in \mathcal{M}_N(\mathfrak{L}(V)) = \mathfrak{L}(V \otimes \mathbb{C}^N)$ is self-adjoint and invertible.

(ii) If H' is another representative of the Witt class $[H] \in L(\mathfrak{B})$, there exists a finite set $F' \subset \mathfrak{B}$ and $\delta > 0$, such that for any (ε, F')-representation ϱ of $\mathfrak{B}$ with $0 < \varepsilon < \delta$ one has:

$$\text{Signature}\,(\varrho \otimes 1)(H) = \text{Signature}\,(\varrho \otimes 1)(H').$$

Remark that the almost representation associated to an almost flat bundle is automatically a $*$-representation, therefore the above construction applies. From Proposition 6 and Theorem 7, one obtains:

Theorem 9 ([CGM], Thm. 6). *Let $H \in \mathcal{M}_N(\ell^1(\Gamma))$ be a representative of the Mishchenko class $\sigma[M, f] \in L(\ell^1(\Gamma))$. There exists a $\delta > 0$ such that, for any ε-flat bundle E on M^{2k} with $0 < \varepsilon < \delta$, one has:*

$$\text{Signature}\,(\varrho_E \otimes 1)(H) = 2^k(\mathscr{L}(M) \cup ch\,E)[M],$$

where $\mathscr{L}(M)$ denotes the stable L-class of M.

Corollary 10 [CGM]. *Let M, N be two closed oriented manifolds and let $h : N \to M$ be an oriented homotopy equivalence. Then, for any class $\xi \in K_{af}^0(M)$, one has $(\mathscr{L}(N) \cup ch\,h^*\xi)[N] = (\mathscr{L}(M) \cup ch\,\xi)[M]$.*

Remark. 7. According to Theorem 7, the higher indices constructed in this section, unlike the higher indices of Section 1, are genuine indices of elliptic operators with suitable coefficients. Thus, it makes sense to try to obtain the above result by proving directly the homotopy invariance of the index of the signature operator with coefficients in an almost flat bundle. Such a direct proof was recently achieved by Hilsum and Skandalis [HS].

2.4 Application to the Novikov Conjecture

The notion of an almost flat bundle can be extended to the case when the base space is a simplicial complex and the bundle is infinite-dimensional (but endowed with a 'superconnection', in a suitable sense). Moreover, Theorem 9 continues to function in this extended context. With these modifications, one can show that $K^0(B\Gamma) = K_{\mathrm{af}}^{i0}(B\Gamma)$, and therefore that the Novikov conjecture holds, for a 'very large' class of finitely presented groups Γ. This class includes all the cases for which the conjecture was previously established, in particular the following:

(1) $\Gamma = \pi_1(N)$ where N is a complete Riemannian manifold of non-positive sectional curvature (cf. [K3]) or, more generally, $\widetilde{N}$ is hyperspherical (see [GL2]);

(2) Γ is a discrete subgroup of a finitely connected Lie group (cf. [K3]) or of an algebraic group over a local field (cf. [KS]);

(3) Γ is hyperbolic in Gromov's sense (cf. [CM2]).

3. Unstable Cyclic Cohomology and Non-Local Invariants

3.1 Cyclic Cohomology of Group Rings

Let Γ be a finitely generated group. For simplicity, we will assume that it is torsion-free and of finite cohomological dimension (over $\mathbb{Q}$) n. This implies that the cyclic cohomology of $\mathbb{C}\Gamma$ stabilizes above $n : HC^{2k+n}(\mathbb{C}\Gamma) = HC^n(\mathbb{C}\Gamma)$ and $HC^{2k+1+n}(\mathbb{C}\Gamma) = HC^{n+1}(\mathbb{C}\Gamma)$, for $k = 0, 1, 2, \ldots$.

Burghelea [B] showed that $HC^*(\mathbb{C}\Gamma)$ decomposes into a direct sum parametrized by the conjugacy classes $[\gamma]$ in Γ:

$$HC^*(\mathbb{C}\Gamma) \cong H^*(B\Gamma, \mathbb{C}) \otimes HC^*(\mathbb{C}) \oplus \sum_{[\gamma] \neq 1} H^*(BN_\gamma, \mathbb{C}),$$

where $N_\gamma = \Gamma_\gamma / C_\gamma$, $\Gamma_\gamma =$ the centralizer of γ and $C_\gamma =$ the (free) group generated by γ. On the other hand, for certain classes of groups Γ, including discrete subgroups of general linear groups over fields of characteristic 0, Eckmann [E] showed that in the stable range, i.e. for $k \geq n$, $H^k(BN_\gamma, \mathbb{C}) = 0$, $\forall [\gamma] \neq 1$.

Thus, it was precisely the stable component of $HC^*(\mathbb{C}\Gamma)$ which was used in the preceding sections to construct higher indices of elliptic operators. The aim of this section is to suggest that the unstable component of $HC^*(\mathbb{C}\Gamma)$ plays a similar, albeit more subtle, rôle with respect to non-local invariants such as analytic torsion and η-invariants.

3.2 η-Invariants and Analytic Torsion via Cyclic Homology

Let M^{4k-1} be a closed connected Riemannian manifold. The η-invariant η of M (see [APS]), is meant to measure the 'signature' of the signature operator $\mathcal{B}^{\mathrm{ev}}$ acting on the space $\Omega^{\mathrm{ev}} = \Omega^{\mathrm{ev}}(M)$ of exterior differential forms of even degree on M. We recall that η is defined as the value at $s = 0$ of the analytic continuation of the function

$$\eta(s) = \sum (\mathrm{sign}\lambda)|\lambda|^{-s}, \qquad \mathrm{Re}(s) \gg 0,$$

where λ runs over the set of non-zero eigenvalues of $\mathcal{B}^{\mathrm{ev}}$, each eigenvalue being repeated according to its multiplicity. It can be easily shown that η involves only the 'middle' degree forms, more exactly than $\eta = \eta^{2k}(0)$, where

$$\eta^{2k}(s) = \sum (\text{sign } \lambda) |\lambda|^{-s}, \qquad \text{Re}(s) \gg 0,$$

with λ running this time over the set of non-zero eigenvalues of the operator $*d$ acting on $d^*\Omega^{2k}$.

This definition can be reformulated in terms of the cyclic homology of the algebra $\mathcal{A} = C^\infty(M)$. We recall ([C1]) that:

(i) $HH_*(\mathcal{A})$ is canonically isomorphic to $\Omega^*(M)$; under this isomorphism, the exterior derivative $d : \Omega^n(M) \to \Omega^{n+1}(M)$ corresponds to the operator $\beta = B \circ I :$ $HH_n(\mathcal{A}) \to HH_{n+1}(\mathcal{A})$;

(ii) $HC_n(\mathcal{A}) \cong \Omega^n(M)/d\Omega^{n-1}(M) \oplus H^{n-2}(M) \oplus H^{n-4}(M) \oplus \cdots$.

By transport of structure, both $HH_*(\mathcal{A})$ and $HC_*(\mathcal{A})$ become (graded) pre-Hilbert spaces.

Thus, ignoring the zero modes, η can be viewed as the *regularized signature* of $*\beta : HC_{2k-1}(\mathcal{A}) \to HC_{2k-1}(\mathcal{A})$.

There is a similar interpretation for the analytic torsion. Namely, if τ denotes the analytic torsion of M (see [RS]), with the above notation one has:

$$\log \tau = \sum_{q \geq 0} (-1)^q \log \det_\zeta (\beta^* \beta : HC_q(\mathcal{A}) \to HC_q(\mathcal{A})).$$

Here $\deg_\zeta$ denotes the ζ-determinant of the corresponding positive elliptic operator, with the zero modes removed. Regarding $\log \deg_\zeta (\beta^* \beta : HC_q(\mathcal{A}) \to HC_q(\mathcal{A}))$ as a regularized dimension of $HC_q(\mathcal{A})$, the above formula identifies $\log \tau$ as the *regularized Euler characteristic* of $HC_*(\mathcal{A})$.

Let $\mathcal{E}_\varrho = \widetilde{M} \times_\Gamma E$ be a flat bundle associated to a unitary representation ϱ of $\Gamma = \pi_1(M)$ on E. Twisted invariants τ_ϱ and η_ϱ can be defined in similar terms. To this end, one needs a twisted version of the cyclic homology of the algebra $\mathcal{A} = C^\infty(M)$. This is obtained by replacing the entries $C(\mathcal{A})_{p,q} = \mathcal{A}^{\otimes q-p+1}$, if $q \geq p$ and 0 otherwise, in the (b, B)-bicomplex $C(\mathcal{A})$ of the algebra $\mathcal{A}$, with $C(\mathcal{A}, \mathcal{E}_\varrho)_{p,q} = (E \otimes \widetilde{\mathcal{A}}^{\otimes q-p+1})^\Gamma$, if $q \geq p$ and 0 otherwise. Here $\widetilde{\mathcal{A}} = C^\infty(\widetilde{M})$ and $(E \otimes \widetilde{\mathcal{A}}^{\otimes n})^\Gamma = $ the subspace of Γ-invariant elements of $E \otimes \widetilde{\mathcal{A}}^{\otimes n}$. Note that $1 \otimes b$ and $1 \otimes B$ commute with the (tensor product) action of Γ and therefore induce boundary operators for the bicomplex $C(\mathcal{A}, \mathcal{E}_\varrho)$, which we will continue to denote b and B. One can thus define the *cyclic homology of $\mathcal{A}$ with coefficients in $\mathcal{E}_\varrho$* as $HC_*(\mathcal{A}, \mathcal{E}_\varrho) = $ the homology of the total complex associated to $C(\mathcal{A}, \mathcal{E}_\varrho)$ and the *Hochschild homology of $\mathcal{A}$ with coefficients in $\mathcal{E}_\varrho$* as $HH_*(\mathcal{A}, \mathcal{E}_\varrho) = $ the homology of the 'vertical column complex' $\{(E \otimes \widetilde{\mathcal{A}}^{\otimes n})^\Gamma, b\}$.

3.3 Regularized Index Formulae

In [MS1] and [MS2], the η-invariants and the torsion invariants of locally symmetric manifolds of non-positive sectional curvature were expressed as special values of zeta-functions constructed out of the periodic geodesics. Our purpose here is to explain how to understand these results as some kind of 'regularized index formulae'.

Let $M = \Gamma \backslash \widetilde{M}$ with $\widetilde{M}$ a globally symmetric space of non-compact type and Γ a discrete, torsion-free, co-compact subgroup of orientation-preserving isometries. M inherits a (locally symmetric) Riemannian metric g, of non-positive sectional curvature. The connected components of the periodic set of the geodesic flow Φ, acting on the unit tangent bundle SM, are parametrized by the non-trivial conjugacy classes $[\gamma]$ in $\Gamma = \pi_1(M)$. Each connected component M_γ is itself a closed locally symmetric manifold of non-positive sectional curvature. Φ restricts to a periodic flow on each M_γ and the quotient $\widehat{M}_\gamma = M_\gamma / \Phi$ is an orbifold; we denote by $\chi(\widehat{M}_\gamma)$ its Euler characteristic. All orbits of Φ in M_γ have the same length ℓ_γ and m_γ will denote the multiplicity of a generic orbit.

Let ϱ be an *acyclic* unitary representation of Γ and denote by τ_ϱ the analytic torsion of M with coefficients in $\mathcal{E}_\varrho$. The main result in [MS2] reads as follows:

$$\log \tau_\varrho = - \sum_{[\gamma] \neq 1} \chi(\widehat{M}_\gamma) \mathrm{Tr}\varrho(\gamma) \frac{e^{-s\ell_\gamma}}{m_\gamma} \Bigg|_{s=0} ,$$

where the evaluation at 0 is made after analytic continuation.

We remark that, as a topological space, $\widehat{M}_\gamma = BN_\gamma$ (see 3.1). Moreover, the orientation of $\widehat{M}_\gamma$ defines a fundamental class $[\widehat{M}_\gamma]$ in the top-degree rational homology of BN_γ and one has:

$$\chi(\widehat{M}_\gamma) = e(\widehat{M}_\gamma)[\widehat{M}_\gamma] ,$$

where $e(\widehat{M}_\gamma)$ denotes the Euler class of the orbifold. The collection of classes $e(M, g) = \{ e(\widehat{M}_\gamma) \in H^*(BN_\gamma, \mathbb{Q}); [\gamma] \neq 1 \}$ can be viewed as an element of a suitable completion $\widehat{HC}^*(\mathbb{Q}\Gamma)$ of $HC^*(\mathbb{Q}\Gamma)$, while the collection $[M, g] = \{ [\widehat{M}_\gamma] \in H_*(BN_\gamma, \mathbb{Q}); [\gamma] \neq 1 \}$ defines an element in a similar completion of $HC_*(\mathbb{Q}\Gamma)$. Therefore, the right hand side of the above formula for torsion can be interpreted as a regularized pairing $\langle e(M, g), [M, g] \rangle_\varrho$, depending on the representation ϱ, the regularization being given by the geodesic spectrum. We recall that $\log \tau_\varrho$ is also defined by a regularization procedure, but involving the 'dual' data, namely the spectrum of the Laplace operator associated to the metric.

A similar interpretation can be given to the η-invariant formula proved in [MS1].

We conclude by observing that all the above expressions make sense for any closed $K(\Gamma, 1)$-manifold which admits a metric of non-positive sectional curvature. Examples of such formulae, for the torsion of certain manifolds which are not locally symmetric, can be found in the work of D. Fried (cf. [F] and references therein). It is plausible that such formulae continue to hold on more general non-positively curved closed $K(\Gamma, 1)$-manifolds.

References

[A]   M. F. Atiyah: Elliptic operators, discrete groups and von Neumann algebras. Asterisque **32–33** (1976) 43–72

[APS]   M. F. Atiyah, V. K. Patodi and I. M. Singer: Spectral asymmetry and Riemannian geometry. I, Math. Proc. Camb. Phil. Soc. **77** (1975) 43–69; II, **78** (1975) 405–432; III, **79** (1976) 71–99

[B]   D. Burghelea: The cyclic homology of group rings. Comment. Math. Helv. **60** (1985) 354–365

[C1]   A. Connes: Non-commutative differential geometry. Publ. Math. IHES **62** (1985) 41–144

[C2]   A. Connes: Entire cohomology of Banach algebras and characters of θ-summable Fredholm modules. K-theory **1** (1988) 519–548

[C3]   A. Connes: Cyclic cohomology and the transverse fundamental class of a foliation. In: H. Araki, E. G. Effros (eds.) Geometric Methods in Operator Algebras. Pitman, Research Notes in Mathematics Series **123** (1986) 52–144

[CGM]   A. Connes, M. Gromov and H. Moscovici: Novikov's conjecture and almost flat vector bundles. C. R. Acad. Sci. Paris, Ser. I **310** (1990) 273–277

[CM1]   A. Connes and H. Moscovici: Novikov's conjecture and hyperbolic groups. C. R. Acad. Sci. Paris, Ser. I **307** (1988) 475–480

[CM2]   A. Connes and H. Moscovici: Cyclic cohomology, the Novikov conjecture and hyperbolic groups. Topology **29** (1990) 345–388

[E]   B. Eckmann: Cyclic homology of groups and the Bass conjecture. Comment. Math. Helv. **61** (1986) 193–202

[F]   D. Fried: Counting circles. In: Dynamical systems. (Lecture Notes in Mathematics, vol. 1342). Springer, Berlin Heidelberg New York, pp. 196–215

[G]   E. Getzler: Pseudodifferential operators on supermanifolds and the Atiyah-Singer index theorem. Commun. Math. Phys. **92** (1983) 163–176

[GS]   E. Getzler and A. Szenes: On the Chern character of a theta summable Fredholm module. J. Funct. Analysis **84** (1989) 343–357

[Gr]   M. Gromov: Hyperbolic groups. In: S.M. Gersten (ed.) Essays in Group Theory. MSRI Publ. **8** (1987) 75–264

[GL1]   M. Gromov and H. B. Lawson: Spin and scalar curvature in the presence of a fundamental group. Ann. Math. **111** (1980) 209–230

[GL2]   M. Gromov and H. B. Lawson: Positive scalar curvature and the Dirac operator on complete Riemannian manifolds. Publ. Math. IHES **58** (1983) 83–196

[H]   U. Haagerup: An example of a non-nuclear C^*-algebra which has the metric approximation property. Invent. math. **50** (1979) 279–293

[Ha]   P. de la Harpe: Groupes hyperboliques, algèbres d'opérateurs et un théorème de Jolissaint. C. R. Acad. Sci. Paris, Ser. I **307** (1988) 771–774

[HS]   M. Hilsum and G. Skandalis: Invariance par homotopie de la signature à coefficients dans un fibré presque plat. Preprint, Collège de France, 1990

[J]   P. Jolissaint: Rapidly decreasing functions in reduced C^*-algebras of groups. Trans. Amer. Math. Soc. **317** (1990) 167–196

[K1]   G. G. Kasparov: Topological invariants of elliptic operators, I: K-homology. Math. USSR Izv. **9** (1975) 751–792

[K2]   G. G. Kasparov: K-theory, group C^*-algebras and higher signatures. Conspectus, Parts I, II. Chernogolovka Preprint 1981

[K3]   G. G. Kasparov: Equivariant K-theory and the Novikov conjecture. Invent. math. **91** (1988) 147–201

[KS]   G. G. Kasparov and G. Skandalis: Groupes agissant sur des immeubles de Bruhat-
         Tits, K-théorie opératorielle et conjecture de Novikov. C. R. Acad. Sci. Paris, Ser.
         I, **310** (1990) 171–174
[Lu]   G. Lusztig: Novikov's higher signature and families of elliptic operators. J. Diff.
         Geom. **7** (1971) 229–256
[Ms1]   A. S. Mishchenko: Homotopy invariants of non-simply connected manifolds. I,
         Math. USSR Izv. **4** (1970) 509–519
[Ms2]   A. S. Mishchenko: C*-algebras and K-theory. (Lecture Notes in Mathematics,
         vol. 763). Springer, Berlin Heidelberg New York, pp. 262–274
[MsS]   A. S. Mishchenko and Yu. P. Solov'ev: Representations of Banach algebras and
         Hirzebruch type formulae. Math. USSR Sbornik **39** (1981) 189–205
[MS1]   H. Moscovici and R. J. Stanton: Eta invariants of Dirac operators on locally
         symmetric manifolds. Invent. math. **95** (1989) 629–666
[MS2]   H. Moscovici and R. J. Stanton: R-torsion and zeta functions for locally symmetric
         manifolds. Invent. math. (in press)
[N]   S. P. Novikov: Analogues hermitiens de la K-théorie. Actes, Congrès Intern. Math.
         t. **2**, (1970) 39–45
[RS]   D. B. Ray and I. M. Singer: R-torsion and the Laplacian on Riemannian mani-
         folds. Adv. Math. **7** (1971) 145–210
[R]   J. Roe: Exotic cohomology and index theory for complete Riemannian manifolds.
         Preprint 1990
[Ro]   J. Rosenberg: C*-algebras, positive scalar curvature and the Novikov conjecture.
         Publ. Math. IHES **58** (1983) 409–424
[W]   S. Weinberger: An analytic proof of the topological invariance of rational Pontr-
         jagin classes. Preprint 1990

State Sum Models in Low-Dimensional Topology

Vladimir G. Turaev

LOMI, Fontanka 27, Leningrad 191011, USSR and Louis Pasteur University
7, rue René Descartes, F-67084 Strasbourg, France

1. Introduction

In recent years certain fundamental ideas of statistical mechanics and quantum physics have penetrated through psychological barriers between physics and topology. The original impetus to this development was given in 1984 by the V. Jones's discovery of a new polynomial invariant of links in the 3-sphere S^3. His construction uses braids, Hecke algebras and von Neumann algebras. The later progress has involved 2-dimensional conformal field theory, theory of quantum groups and other subjects, creating a unified overwhelmingly rich area of study. Of especial importance have been the methods of statistical mechanics first applied by L. Kauffman to reproduce the Jones polynomial, and the ideas of quantum field theory brought into the subject by E. Witten.

From the viewpoint of topology the main new notion attained in the course of this development is the notion of topological quantum field theory (TQFT; see [At1]). There are several complementary approaches to constructing TQFTs in dimension 3. Here I present an approach based on the theory of quantum groups (see [Dr1, FRT1, Ji1]).

The state sum models appear in this context as a technical tool to define invariants of geometric objects. I give here an abstract description of the models; examples are abundant below. On the geometric part the models use certain "images" of the objects and "moves" relating these images. The moves should be local transformations of images following certain patterns. Any images of topologically equivalent objects should be related by a finite sequence of moves. One also distinguishes two sorts of local geometric blocks of the images called atoms and interactions. (This terminology is due to G. Kuperberg). Each interaction should give rise to a tuple of atoms, "involved in this interaction". On the statistical mechanics side one uses the notions of "algebraic initial data", "state", and "state sum". The initial data (in its simplest form) consists of a finite set I of "colors" and a weight function which is a numerical function on certain tuples of colors. A state on an image is an I-valued function on the set of atoms. Having such a state one associates with each interaction the tuple of colors of atoms involved in this interaction. One forms the product of weights of these tuples over all interactions and sums up these products over all states. To ensure topological invariance of the sum it suffices to check invariance under the moves. Their local nature usually enables one to reformulate the invariance

Proceedings of the International Congress
of Mathematicians, Kyoto, Japan, 1990
© The Mathematical Society of Japan, 1991

as a collection of algebraic equations on the weights. Solving these equations is fundamental to justify the model.

This scheme admits variations and extends to relative situations, producing linear operators (rather then numbers) forming a TQFT.

It seems that in the last decades of the 20th century characterized by the Bourbaki style in mathematics one may reject the methods outlined above as old-fashioned and non-elegant. A good excuse is that due mainly to E. Witten we have a global physical ideology behind the constructions. To justify this mathematically non-rigorous ideology we apply the down-to-earth technique of state sum models, moves et cetera. The models play here the role similar to that of simplicial chain complexes in homology theory.

The theory of state sum models has revived certain geometric ideas of K. Reidemeister, M. Newman, J. Alexander developed in the 1920–1930s; it involved more recent results of J. Cerf, R. Kirby, R. Fenn and C. Rourke, as well as the theory of special spines of 3-manifolds due to B. Casler, S. Matveev, and R. Piergallini, and the theory of categories of tangles due to D. Yetter and the author. Bringing these topological results in a row and relating them to the methods coming from physics and representation theory seem to be one of the most attractive features of the theory.

I will describe here 3 types of models: vertex models on link diagrams in $\mathbb{R}^2$, face models on link diagrams and link shadows, simplicial models on triangulations of 3-manifolds.

For an earlier discussion of state sum models in knot theory see [Jo2]. For other approaches see [Ja1, Ku1].

2. Vertex Models on Link Diagrams

The 2-dimensional vertex models of statistical mechanics are concerned with 4-valent graphs in $\mathbb{R}^2$, the edges and vertexes being respectively the atoms and interactions (see [Ba1]). These models have been adjusted to the topological setting of link diagrams in $\mathbb{R}^2$ where one has to take into account the over/undercrossing information and to treat the local extremums of the height function $(x, y) \mapsto y : \mathbb{R}^2 \to \mathbb{R}$ (reduced to the diagram at hand) as interactions. This produces isotopy invariants of colored framed oriented links in S^3 (see [AW1, Jo1, Tu1, Re1, RT1]).

Fix a set I of left modules over a commutative ring with unit K. Assume that each module $V \in I$ is equipped with a finite K-basis J_V.

By a colored link in S^3 we mean a finite collection of disjoint imbedded circles in S^3, each circle being equipped with a "color" $V \in I$. A link $L \subset S^3$ is framed (resp. oriented) if L is equipped with a non-singular normal vector field (resp. with orientation of L). Link diagrams are the images of links under pushing off into $\mathbb{R}^3 = S^3 \backslash \{\infty\}$ and projecting to $\mathbb{R}^2$, with an account taken of over/undercrossings. (To construct a diagram of a framed link L one first deforms L to make its framing orthogonal to $\mathbb{R}^2$).

We consider only the diagrams whose self-crossings are distinct from local extremums (of the height function). The classical Reidemeister moves relate any two diagrams of a link modulo ambient isotopy of the diagrams in $\mathbb{R}^2$. After a small modification the moves work also for diagrams of framed links. D. Yetter and independently the author gave a list of moves relating ambiently isotopic diagrams (see [Ye1, Tu2, Tu3]); the simplest of these moves introduces 1 point of local maxima and 1 point of local minima on a small vertical segment of a diagram.

The initial algebraic data of the model consists of 6 weight functions

$$R, \overline{R} : \coprod_{V,W \in I} J_V^2 \times J_W^2 \to K ; \quad \alpha, \beta, \mu, \nu : \coprod_{V \in I} J_V^2 \to K . \tag{1}$$

Let $\mathscr{D} \subset \mathbb{R}^2$ be a diagram of a colored framed oriented link L. The self-crossings of $\mathscr{D}$ and the local extremums of the height function split $\mathscr{D}$ into a finite number of arcs. A state s on $\mathscr{D}$ is a mapping, associating with each such arc e of $\mathscr{D}$ an element of the set $J_{V(e)}$ where $V(e) \in I$ is the color of the component of L projecting to the loop of $\mathscr{D}$ which contains e. For a state s on $\mathscr{D}$ and a positive (resp. negative) self-crossing a of $\mathscr{D}$ we put $\langle a \rangle_s = R(i, \ell, j, k)$ (resp. $\langle a \rangle_s = \overline{R}(i, \ell, j, k)$) where i, ℓ, j, k are the values of s on the 4 arcs incident to a; see Fig. 1. For a local extremum b, as in Fig. 2, we put $\langle b \rangle_s = f(i, \ell)$ where respectively $f = \alpha, \beta, \mu, \nu$. Put

$$\langle \mathscr{D} \rangle_s = \prod_a \langle a \rangle_s \prod_b \langle b \rangle_s \text{ and } \langle L \rangle = \langle \mathscr{D} \rangle = \sum_s \langle \mathscr{D} \rangle_s$$

where a, b, s run respectively over crossings, local extremums and states of $\mathscr{D}$.

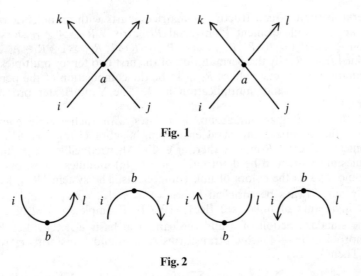

Fig. 1

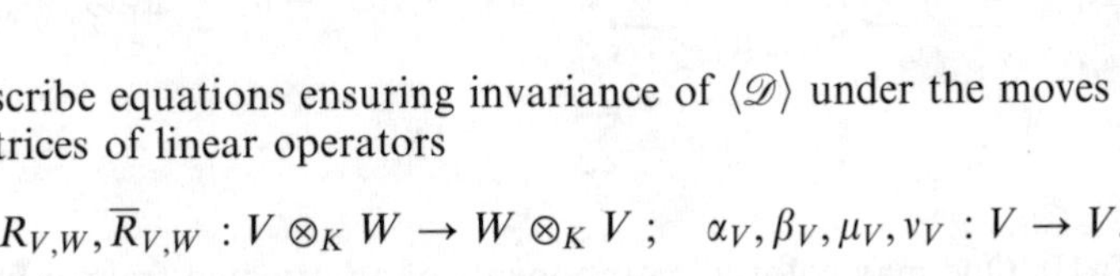

Fig. 2

To describe equations ensuring invariance of $\langle \mathscr{D} \rangle$ under the moves one treats (1) as matrices of linear operators

$$R_{V,W}, \overline{R}_{V,W} : V \otimes_K W \to W \otimes_K V ; \quad \alpha_V, \beta_V, \mu_V, \nu_V : V \to V . \tag{2}$$

The state sum $\langle \mathscr{D} \rangle$ depends solely on $\mathscr{D}$ and these operators. Here are equations corresponding to certain moves: $\beta_V = \alpha_V^{-1}$, $\nu_V = \mu_V^{-1}$, $\overline{R}_{V,W} = R_{W,V}^{-1}$,

$$(R_{V,W} \otimes 1_U) \circ (1_V \otimes R_{U,W}) \circ (R_{U,V} \otimes 1_W) \qquad (3)$$

$$= (1_W \otimes R_{U,V}) \circ (R_{U,W} \otimes 1_V) \circ (1_U \otimes R_{V,W}).$$

The latter equation is the Yang-Baxter equation ; the operators satisfying the equation are called R-matrices. This equation ensures invariance of $\langle \mathscr{D} \rangle$ under the third Reidemeister move, known also as the braid relation. There are also a few equations relating R, α, μ (see [Tu3]). Note that without loss of generality one may assume that $\alpha_V = \beta_V = 1_V$ for all $V \in I$. Note also that one usually imposes an additional condition that a positive $360°$-twist of the framing along a link component of color V leads to multiplication of the link invariant by an invertible constant depending only on V. This condition is fullfilled in the examples below.

According to [Re2], any system of invertible R-matrices $\{R_{V,W} : V \otimes W \to W \otimes V\}_{V,W \in I}$ satisfying a non-degeneracy condition may be enriched by $\{\alpha, \beta, \mu, \nu\}_V$ to satisfy all equations ensuring isotopy invariance of $\langle L \rangle$.

Using the operators (2) one may extend the invariant $L \mapsto \langle L \rangle$ to operator invariants of I-colored framed oriented tangles in $\mathbb{R}^2 \times [0, 1]$ (see [Tu2, Tu3, Re2, RT1]). This produces a linear tensor representation of the category of tangles in the category of K-modules.

3. Quantum Groups as a Source of Invariants

A general method of constructing R-matrices starts with an algebra with unity A and an invertible element ("universal R-matrix") $R \in A \otimes A$ satisfying the equation $R_{12}R_{13}R_{23} = R_{23}R_{13}R_{12}$, where $R_{12} = R \otimes 1$, $R_{23} = 1 \otimes R \in A^{\otimes 3}$, and R_{13} is obtained from R_{23} by the permutation of the first two tensor multiples. For any left A-modules V, W one defines $R_{V,W}$ to be the composition of the permutation $V \otimes W \to W \otimes V$ and multiplication by R. The Yang-Baxter property of R implies (3).

The theory of quantum groups associates with each simple complex Lie algebra $\mathfrak{g}$ the quantized universal enveloping algebra $U_q(\mathfrak{g})$ over $\mathbb{C}$ possessing a canonical universal R-matrix (here $q \in \mathbb{C}$). All irreducible finite-dimensional $\mathfrak{g}$-modules are known to be deformable into $U_q(\mathfrak{g})$-modules. Thus, one may use these $\mathfrak{g}$-modules as the colors of link components. The system $\{R_{V,W}\}$ arising in this way may always be enriched by $\alpha_V = \beta_V = 1_V$, $\mu_V, \nu_V = \mu_V^{-1}$ to produce isotopy invariants of links (see [Re1], [We1]). Example: $\mathfrak{g} = sl_n(\mathbb{C})$, $V = \mathbb{C}^n$ with the standard action of $\mathfrak{g}$ and the canonical basis $e_1, \ldots, e_n$. Let E_i^j be the homomorphism $V \to V$ which transforms e_i in e_j and transforms e_k with $k \neq i$ into 0. Then

$$R_{V,V} = -q \sum_i E_i^i \otimes E_i^i + \sum_{i \neq j} E_i^j \otimes E_j^i + (q^{-1} - q) \sum_{i<j} E_i^i \otimes E_j^j ;$$

$$\mu_V = \sum_i q^{2i-m-1} E_i^i$$

(see [Ji1, Tu1]). One may color all components of an oriented framed link $L \subset S^3$ with this V and get an invariant $\langle L \rangle$ which is a Laurent polynomial over q. Multiplying $\langle L \rangle$ by a power of $-q^{-n}$ one makes it independent of the framing.

The resulting polynomial is (up to reparametrization) a reduction $P_L(q^n, q - q^{-1})$ of the Homfly polynomial $P_L(x, y)$ (see [Jo1, Tu1, Li1]). For $n = 2$ this reduction is the Jones polynomial. In the case $n = 2$ one may reformulate the model to produce the Kauffman [Ka1] bracket model for the Jones polynomial. Similarly, the fundamental representations of the Lie algebras of series B, C, D give rise to a series of 1-variable reductions of the 2-variable Kauffman polynomial (see [Tu1]). For invariants corresponding to $\mathfrak{g} = G_2$ see [Re1, Ku2].

For any m-component link in S^3 one may define a universal invariant sitting in $U_q(\mathfrak{g})^{\otimes m}$ and comprising all invariants corresponding to various colorings of components by $\mathfrak{g}$-modules (see [La1]).

4. Invariants of 3-Manifolds

The invariants of links in S^3 discussed above may be combined to produce topological invariants of closed oriented 3-manifolds and links in these manifolds. The relevant construction was introduced and applied to link invariants derived from $sl_2(\mathbb{C})$ in [RT2]. For the case of other classical semisimple Lie algebras see [TW1]. The construction is based on the classical theorem of V. Rochlin and W. Lickorish presenting any closed oriented (connected) 3-manifold as the result of surgery on S^3 along a framed link. R. Kirby introduced in 1978 moves on links which relate any presentations of homeomorphic manifolds. An equivalent system of local moves was introduced by R. Fenn and C. Rourke.

On the algebraic part the construction uses a modular Hopf algebra (see [RT2]). This is a quasitriangular Hopf algebra A with an R-matrix $R \in A \otimes A$ equipped with a finite collection $\{V_i\}_{i \in I}$ of irreducible finite dimensional A-modules satisfying certain axioms. Let, now, a closed oriented 3-manifold N be the result of surgery on a framed link $L \subset S^3$ with m components $L_1, \ldots, L_m$. Provide L with an arbitrary orientation. For each sequence $j = (i_1, \ldots, i_m) \in I^m$ denote by L^j the link L whose components $L_1, \ldots, L_m$ are colored respectively with $V_{i_1}, \ldots, V_{i_m}$. Here is the invariant of N:

$$\tau_A(N) = b^m c^{\sigma(L)} \sum_{j=(i_1,\ldots,i_m) \in I^m} \left(\langle L^j \rangle \prod_{r=1}^m \langle 0_{i_r} \rangle \right)$$

where: b, c are normalizing constants, associated with $A, R, \{V_i\}_i$; 0_i is the trivial knot in S^3 with zero framing and the color V_i; $\sigma(L)$ is the signature of the oriented 4-manifold W_L bounded by N and obtained by adding 2-handles to the 4-ball B along $L \subset \partial B$. This definition slightly differs from the one in [RT2] (cf. also [KM1]) ; equivalence of definitions follows in the case $A = U_q(sl_n)$ from computations of [RT2, TW1] and in general case from results of K. Walker (private communication). If L' is a colored framed oriented link in N then one pushes L' in $S^3 \setminus L$ and defines $\tau_A(N, L')$ using $L^j \coprod L'$ instead of L^j.

With a primitive complex root of unity q of degree $r \geq 3$ one associates a finite-dimensional modular Hopf algebra $U = U_q$ which is a quotient of $U_q(sl_2\mathbb{C})$ (see [RT2]). Here $I = \{0, 1/2, 1, \ldots, (r-2)/2\}$ and V_i is the q-deformation of the irreducible $sl_2\mathbb{C}$-module of dimension $2i + 1$. Thus with the pair (N, L') we may associate the function $q \mapsto \tau_{U_q}(N, L')$ which is a topological invariant of this pair. For colored links in S^3 this function coincides with the value of the colored Jones polynomial in q. In general it is unknown if this function may be extended in a topologically invariant manner to the unit circle or, what is even more intriguing, to the open unit disc in $\mathbb{C}$. For a further study of $\tau_U(N)$ see [KM1, Li2].

For even r one may refine $\tau_U(N)$ to an invariant $\tau_U(N;\theta)$ where θ is an arbitrary spin structure on N, if $r = 0$ (mod 4), and $\theta \in H^1(N;\mathbb{Z}/2\mathbb{Z})$, if $r = 2$ (mod 4). Such θ defines a sublink $L(\theta)$ of the link $L \subset S^3$ presenting N: it consists of components of L such that θ does not extend across the corresponding 2-handles of W_L. One defines $\tau_U(N;\theta)$ in the same way as $\tau_U(N)$ but using only integral colors $0, 1, \ldots, [(r-2)/2]$ for the components of $L(\theta)$. Using [RT2, §7] and [KM1, Theorem 4.20] one may show that $\tau_U(N;\theta)$ is a topological invariant of $(N;\theta)$ and $\tau_U(N) = \sum_\theta \tau_U(N;\theta)$. The same refinement applies to (N, L'). In particular, for $q = \sqrt{-1}$, $r = 4$ the invariant $\tau_U(N;\theta)$ is equivalent to the Rochlin invariant $\mu(N;\theta) \in \mathbb{Z}/16\mathbb{Z}$: for any spin structure θ on N one has $\tau_U(N,\theta) = \exp(-3\pi\sqrt{-1}\mu(N;\theta)/8)$.

5. The Jones-Witten TQFT

Witten [Wi1] conjectured existence of a 3-dimensional TQFT extending the Jones polynomial of links in S^3 (or, more exactly, extending the value of this polynomial in a fixed root of unity q) to oriented 3-dimensional cobordisms and to 3-cobordisms with tangles sitting inside. It was shown in [RT2] that each modular Hopf algebra $(A, R, \{V_i\}_{i\in I})$ gives rise to a 3-dimensional TQFT. For $A = U_q$ one gets the Jones-Witten TQFT. (Though [RT2] treats only cobordisms with links, one may apply the same methods to tangles in cobordisms). The construction uses presentations of 3-cobordisms by tangles in $\mathbb{R}^3$ and operator invariants of tangles. One of the advantages of this approach is the relative simplicity of the definition of the linear space associated with a surface. For the closed oriented surface Σ_g of genus g this linear space is

$$\Psi_g = \bigoplus_{i_1,\ldots,i_g \in I} \left(V_{i_1} \otimes V_{i_1}^* \otimes \cdots \otimes V_{i_g} \otimes V_{i_g}^* \right)^0,$$

where for an A-module V one denotes by V^* the dual module and by V^0 the maximal splitting submodule of V with trivial action of A. (In [RT2] we involved a projector acting in Ψ_g which was later proven by K. Walker and the author, and independently N. Reshetikhin to be the identity). The Teichmüller modular group of Σ_g projectively acts in Ψ_g; the corresponding 2-cocycle is closely related to the 2-cocycle of the projective Shale-Weil representation of the symplectic group, see [CLM1], [Tu5]. For $A = U_q$, q being a primitive root of even degree, one may refine the TQFT along the lines of Sect. 4.

6. Face Models on Link Shadows

Face models differ from vertex models mainly in that the role of atoms is played by 2-faces. A face model on link diagrams in $\mathbb{R}^2$ producing the value in $q \in \mathbb{C}$ ($q^r = 1$) of the Jones polynomial and its versions for colored links in S^3 was introduced in [KR1]. In [Tu4] this model was reformulated avoiding the use of coordinates, and generalized to diagrams on oriented surfaces. The natural topological setting for the model is the one of shadows (see [Tu4]). A shadow on an oriented closed surface Σ is a finite collection s of closed curves on Σ lying in general position (possibly with intersections) such that each component of $\Sigma \backslash s$ is equipped with a number, called the gleam of the component. Each link diagram $\mathscr{D}$ on Σ (presenting a link in $\Sigma \times \mathbb{R}$) canonically produces a shadow $s(\mathscr{D})$ on Σ; it has the same underlying loops as $\mathscr{D}$ whereas the gleam of a component X of $\Sigma \backslash s(\mathscr{D})$ is defined as the number of jumps up along ∂X diminished by the number of jumps down along ∂X and divided by 2 (the orientation in ∂X is specified by that of $X \subset \Sigma$). One defines Reidemeister-type moves on shadows and calls a class of shadows related by these moves a (framed) shadow link on Σ. If M is a closed oriented 3-manifold fibering over Σ with the fiber S^1 then each (framed) link in M considered up to isotopy canonically produces a (framed) shadow link on Σ. In particular, each link $L \subset S^3$ produces a shadow link on S^2 via the Hopf mapping of the diagrams. One may reconstruct L looking at its shadow on S^2.

The algebraic data used in the model consist of a finite set I, a set adm $\subset I^3$ invariant under permutations, two functions $i \mapsto u_i$, $i \mapsto v_i : I \to \mathbb{C} \backslash 0$ and a function $I^6 \to \mathbb{C}$, called "the symbol".

A coloring of a shadow s is an I-valued function on the set of underlying loops of s. Let s be a colored shadow on Σ. An area-coloring η of s is an I-valued function on the set of components of $\Sigma \backslash s$ such that if 2 components X, Y of $\Sigma \backslash s$ meet along an arc of s of color i then $(i, \eta(X), \eta(Y)) \in$ adm. For each crossing point a of s one defines $|a|_\eta$ to be the symbol of the 6-tuple formed by the properly ordered colors of 2 loops of s traversing a and η-colors of 4 components of $\Sigma \backslash s$ incident to a. The symbol should be symmetric enough to make $|a|_\eta$ well-defined. For a component X of $\Sigma \backslash s$ with η-color i put $|X|_\eta = \exp(2u_i x) v_i^\chi$ where x and χ are the gleam and Euler characteristic of X. One forms the product of $|a|_\eta, |X|_\eta$ over all crossings and all components of $\Sigma \backslash s$ and sums up the products over all η. Under proper algebraic conditions, this model yields invariants of colored shadow links on Σ and colored framed links in M.

The model extends to colored shadow 3-valent graphs on Σ (and the graphs in M; a graph is colored if its edges are equipped with elements of I). The new ingredients are the additional multiples $|f|_\eta$ which are symbols of 6-tuples associated in the obvious way with all 3-valent vertices f of the graph. In fact, the setting of 3-valent graphs in 3-manifolds rather then links is natural for all models in this paper (cf. [RT1]).

The algebra $U_q(sl_2\mathbb{C})$ with $q = \exp(2\pi\sqrt{-1}h/r)$ gives rise to a suitable algebraic data. Here: $r \geq 3$; $(h,r) = 1$; $I = \{0, 1/2, 1, \ldots, (r-2)/2\}$; the symbol is the slightly normalized q-deformation of the Racah-Wigner $6j$-symbol (see [KR1]);

$$u_i = \pi\sqrt{-1}(i - i(i+1)hr^{-1}), v_i = (-1)^{2i}(q_0^{2i+1} - q_0^{-2i-1})(q_0 - q_0^{-1})^{-1},$$

where $q_0 = \exp(\pi\sqrt{-1}h/r)$; a triple (i,j,k) belongs to adm iff $r - 2 \geq i+j+k \in \mathbb{Z}$ and $|i - j| \leq k \leq i + j$. For colored links in S^3 the corresponding face model on S^2 produces the values in $\exp(\pi\sqrt{-1}h/2r)$ of the invariants considered in Sect. 3, the case $A = U_q(sl_2\mathbb{C})$.

Using shadow links on S^2 with complex gleams, and using the shadow version of the Kirby-Fenn-Rourke moves one may define shadow 3-manifolds (over $\mathbb{C}$) and the corresponding shadow version of the Jones-Witten TQFT. A remarkable property of shadow links and 3-manifolds over $\mathbb{C}$ is that one may continuously deform their topological types.

7. Simplicial Models

By a simplicial model I mean a statistical model on triangulations of polyhedra
producing a topological invariant. Such a model for 3-manifolds was introduced in
[TV1]. The geometric result underlying the model is the theorem of J. Alexander
applicable in all dimensions: any two triangulations of a compact PL-manifold
may be related by star subdivisions and their inverses. (Note that each 3-manifold
admits a essentially unique PL-structure.) The algebraic data for the model is
the same as in Sect. 6. The model gives rise to a 3-dimensional TQFT defined
for non-oriented (possibly, non-orientable) 3-cobordisms (see [TV1]), and 3-
cobordisms with links sitting inside (see [Tu6]). In particular for links in S^3 one
gets a simplicial model on the link exteriors producing the values of the Jones
polynomial in the roots of unity.

A state η on a triangulated compact 3-manifold N is an I-valued function
on the set of edges of N. For a 3-simplex T of N denote by $|T|_\eta$ the symbol of
the 6-tuple consisting of the η-colors of edges of a 2-face of T followed by the
η-colors of the opposite edges of T. Put

$$|N, \partial N|_\eta = w^{-2\alpha} \prod_e v_{\eta(e)} \prod_T |T|_\eta$$

where e, T run respectively over all edges and 3-simplexes of N; α is the number
of vertexes of N; w is a constant computed from the algebraic data. For a
closed N the state sum $|N| = \Sigma_\eta |N|_\eta$ is a topological invariant of N if the alge-
braic data satisfy simple conditions which mimic the orthogonality relation and
Biedenharn-Eliot identity for $6j$-symbols. Under the same algebraic conditions
one similarly constructs for each 3-cobordism (N, Σ_1, Σ_2) a topologically invariant linear operator $Q(\Sigma_1) \to Q(\Sigma_2)$ where $Q(\Sigma)$ is a finite-dimensional Euclidean
space associated with any closed surface Σ (see [TV1]). Elements of $Q(\Sigma)$ are
geometrically represented by I-colored 3-valent graphs imbedded in Σ.

The invariant $N \mapsto |N|$ extends to I-colored framed links in N with oriented
neighborhoods. Let $L = L_1 \cup \ldots \cup L_m \subset N$ be such a link with m components
(here $\partial N = \emptyset$). Let U be a closed regular neighborhood of L. Put $M = N \backslash U$ and
triangulate M. Put $\Sigma = \partial M = m(S^1 \times S^1)$ and equip Σ with the triangulation
induced by that of M and the orientation induced by that of U. Let $\psi_1, \ldots, \psi_m \subset \Sigma$
be meridians of $L_1, \ldots, L_m$.

For a sequence $j = (i_1, \ldots, i_m) \in I^m$ equip each ψ_r with the color i_r and denote
by ψ_j the union of these m colored loops on Σ. Denote by L' the family of m
colored loops on Σ obtained by shifting L along its framing. Denote by γ the
3-valent graph on Σ formed by the segments $\{e^*\}$ dual to edges $\{e\}$ of Σ. Here e^*
crosses e once and connects the barycenters of the two 2-simplexes of Σ meeting
along e. For a state η of M denote by γ^η the graph γ with the coloring $e^* \mapsto \eta(e)$.
Put

$$|N, L| = w^{2-2m} \sum_{j=(i_1,\ldots,i_m)} \left(\prod_{r=1}^m v_{i_r} \sum_\eta |M, \partial M|_\eta \langle \gamma^\eta |L'|\psi_j \rangle \right)$$

where the last multiple is the state sum invariant of the shadow graph $s =
\gamma^\eta \cup L' \cup \psi_j$ on Σ. Here γ is assumed to lie in general position with respect to
$L' \cup \psi_j$; the gleam of a component X of $\Sigma \backslash s$ is defined as follows: looking onto
∂X compute the number of passages from ψ_j to L' and from L' to γ, subtract the

number of opposite passages, and divide by 2 (the orientation in ∂X is specified by that of $X \subset \Sigma$). Topological invariance of $|N, L|$ is ensured by the conditions mentioned above and the Racah identity for the symbol.

One may generalize this model to models derived from modular Hopf algebras or modular tensor categories. The definition of $|N|$, $|N, L|$ may be reformulated in terms of simple spines of 3-manifolds and link diagrams on the spines which makes computations more accessible.

The initial data associated with $q = \exp(2\pi\sqrt{-1}h/r)$ in Sect. 6 satisfy all necessary algebraic equations. The relationships between the corresponding invariants $|N| \in \mathbb{R}$ and $\tau_U(N) \in \mathbb{C}$ (defined for oriented N) are the following: $|N|$ equals $\tau_U(N)\overline{\tau_U(N)}$ (up to normalization). For links in S^3 the approaches of Sect. 2–3, Sect. 6, and Sect. 7 are equivalent. In particular, if all components of a framed link $L \subset S^3$ are provided with the color $1/2 \in I$ and the right-hand orientation of neighborhoods then $|S^3, L|$ is equal to the value of the Kauffman bracket polynomial $\langle L \rangle$ in $\exp(\pi\sqrt{-1}h/2r)$.

The invariant $|N|$ corresponding to q may be refined to an invariant $|N, \Delta|$ where $\Delta \in H^1(N; \mathbb{Z}/2)$ (see [TV1]). It is equal to the sum of $|N|_\eta$ over η such that the 1-cocycle $e \mapsto 2\eta(e) \bmod 2$ presents Δ. Clearly, $|N| = \Sigma_\Delta |N, \Delta|$. For oriented closed N and even r one has $|N, \Delta| = \sum_\theta \tau_U(N; \theta)\overline{\tau_U(N; \theta + \Delta)}$.

Both $\tau_U(N)$ and $|N|$ have "physical" meaning: they are mathematical versions of invariants which come up respectively in the 3-dimensional quantum field theory with non-abelian Chern-Simons action (see [Wi1]) and (in the limit $q \to 1$) in 3-dimensional quantum gravity (see e.g. [HP1]).

References

[AW1] Akutsu, Y., Wadati, M.: Knot invariants and the critical statistical systems. J. Phys. Soc. Japan **56** (1987) 839–842

[At1] Atiyah, M.: Topological quantum field theories. Publ. Math. IHES **68** (1989) 175–186

[Ba1] Baxter, R.J.: Exactly solved models in statistical mechanics. Academic Press, London 1982

[CLM1] Cappell, S.E., Lee, R., Miller, E.Y.: Invariants of 3-manifolds from conformal field theory. Preprint 1990

[Dr1] Drinfeld, V.G.: Quantum groups. Proc. Int. Congress of Mathematicians. Academic Press, Berkeley, Cal., vol. 1, 1986, pp. 798–820

[FRT1] Faddeev, L.D., Reshetikhin, N.Y., Takhtajan, L.A.: Quantization of Lie groups and Lie algebras. Algebra i Analysis **1** (1989) 178–207 (Russian)

[HP1] Hasslacher, B., Perry, M.: Phys. Lett. **103B** (1981) 21

[Ja1] Jaeger, F.: Composition products and models for the Homfly polynomial. Preprint 1987

[Ji1] Jimbo, M.: Quantum R-matrix for the generalized Toda system. Commun. Math. Phys. **102** (1986) 537–547

[Jo1] Jones, V.F.R.: Notes on a talk in Atiyah's seminar, 1986

[Jo2] Jones, V.F.R.: On knot invariants related to some statistical mechanical models. Pacif. J. Math. **137** (1989) 311–334

[Ka1] Kauffmann, L.M.: State models and the Jones polynomial. Topology **26** (1987) 395–407

[Ku1] Kuperberg, G.: Involutory Hopf algebras and 3-manifold invariants. Preprint 1990

[Ku2]   Kuperberg, G.: The quantum G_2 link invariant. Preprint 1990

[KM1]   Kirby, R., Melvin, P.: On the 3-manifold invariants of Witten and Reshetikhin-Turaev for $sl(2,\mathbb{C})$. Preprint 1990

[KR1]   Kirillov, A.N., Reshetikhin, N.Y.: Representations of the algebra $U_q(sl_2)$, q-orthogonal polynomials and invariants of links. In: Kac, V.G. (ed.), Infinite dimensional Lie algebras and groups. (Adv. Ser. In Math. Phys., vol. 7). World Scientific, Singapore 1988, pp. 285–339

[La1]   Lawrence, R.J.: A universal link invariant using quantum groups. In: Differential geometric methods in theoretical physics, XVII International conference. World Scientific, Singapore 1988, pp. 55–63

[Li1]   Lickorish, W.B.R.: Polynomials for links. Bull. London Math. Soc. **20** (1988) 558–588

[Li2]   Lickorish, W.B.R.: Three-manifolds and the Temperley-Lieb algebra. Preprint 1990

[Re1]   Reshetikhin, N.Y.: Quantized universal enveloping algebras, the Yang-Baxter equation and invariants of links, I, II. LOMI preprints E-4-87, E-17-87, Leningrad 1988

[Re2]   Reshetikhin, N.Y.: Quasitriangular Hopf algebras and invariants of tangles. Algebra i Analysis **1** (1989) 169–188 (Russian)

[RT1]   Reshetikhin, N.Y., Turaev, V.G.: Ribbon graphs and their invariants derived from quantum groups. Comm. Math. Phys. **127** (1990) 1–26

[RT2]   Reshetikhin, N.Y., Turaev, V.G.: Invariants of 3-manifolds via link polynomials and quantum groups. Invent. math. **103** (1991) 547–597

[Tu1]   Turaev, V.G.: The Yang-Baxter equations and invariants of links. Invent. math. **92** (1988) 527–553

[Tu2]   Turaev, V.G.: The Conway and Kauffman modules of the solid torus with an appendix on the operator invariants of tangles. LOMI preprint E-6-88, Leningrad, 1988

[Tu3]   Turaev, V.G.: Operator invariants of tangles and R-matrices. Izv. Akad. Nauk SSSR **53** (1989) 1073–1107 (Russian) [English transl.: Math. USSR-Izv. **35** (1990) 411–444]

[Tu4]   Turaev, V.G.: Shadow links and IRF-models of statistical mechanics. Publ. Inst. Rech. Math. Av., Strasbourg (1990)

[Tu5]   Turaev, V.G.: Quantum representations of modular groups and Maslov indices. To appear

[Tu6]   Turaev, V.G.: Quantum invariants of links and 3-valent graphs in 3-manifolds. Preprint 1990

[TV1]   Turaev, V.G., Viro, O.Y.: State sum invariants of 3-manifolds and quantum $6j$-symbols. Preprint 1990

[TW1]   Turaev, V.G., Wenzl, H.: Quantum invariants of 3-manifolds associated with classical simple Lie algebras. Preprint 1991

[We1]   Wenzl, H.: Representations of braid groups and the quantum Yang-Baxter equation. Pacif. J. Math. **145** (1990) 153–180

[Wi1]   Witten, E.: Quantum field theory and Jones polynomial. Comm. Math Phys. **121** (1989) 351–399

[Ye1]   Yetter, D.N.: Markov algebras. In:"Braids", AMS Contemporary Mathematics, vol. 78. AMS, Providence, RI, 1988

Canonical and Minimal Models of Algebraic Varieties

Yujiro Kawamata

Department of Mathematics, University of Tokyo, Hongo, Tokyo 113, Japan

In this talk we shall explain a minimal model program for arbitrary dimensional algebraic varieties over a field k of characteristic zero. Starting with the working hypothesis on the finite generatedness of the canonical ring, we shall discuss the process to obtain minimal and canonical models (we refer the reader to [KMM] for more general argument, or [Sh2] for more brief survey from the same point of view). We shall also review some recent results on the boundedness of indices of singularities in dimension 3.

We shall put emphasis on the idea of logarithmic formalism, which provides natural setup for the statements and proofs of new theorems such as vanishing or cone-contraction theorems, rather than just generalizations of previously known results. Many results will be obtained simultaneously for the log and non-log cases. There will also be applications to the reverse direction from the log case to the non-log case.

§1. Log-canonical Divisors

1.1 Let X be a nonsingular projective variety over k and K_X the canonical divisor, i.e., $\mathcal{O}_X(K_X) = \det \Omega^1_X$. The *canonical ring* $R(X) = \sum_{m=0}^{\infty} H^0(X, mK_X)$ is the fundamental birational invariant of X, and the *Kodaira dimension* $\kappa(X)$ is defined to be trans. $\deg_k R(X) - 1$ if $R(X) \neq k$ and $-\infty$ otherwise. Thus $\kappa(X) \leqq \dim X$, and if the equality holds, then X is said to be of *general type*. The basic question is the following:

Question 1. Is $R(X)$ a finitely generated graded k-algebra?

The answer to this question is known to be "yes" if $\dim X = 2$ by Zariski and Mumford [Z, Mu], and if $\dim X = 3$ by the combination of recent results of Mori, Kawamata, Benveniste, Shokurov and Fujita (for the case $\kappa(X) = 3$, the construction of the minimal model is by [Ka5, Ka8, Mo4, Sh1] and the rest by [Ka4] or [B], whereas the case $\kappa(X) < 3$ is by [Fu3]). The proof in dimension 2 works in

Proceedings of the International Congress
of Mathematicians, Kyoto, Japan, 1990

arbitrary characteristic, but our result in dimension 3 needs the assumption that the characteristic is zero because we use the vanishing theorem of Kodaira type.

1.2 The key word of our investigation is the $\mathbb{Q}$-*divisor*. A $\mathbb{Q}$-divisor $D = \sum_{j=1}^{t} d_j D_j$ is a formal sum of distinct prime divisors D_j on a normal variety X with rational coefficients d_j. The *integral part* $[D] = \sum [d_j] D_j$ is a usual Weil divisor. D is said to be $\mathbb{Q}$-*Cartier* if mD is a Cartier divisor for some positive integer m. X is called $\mathbb{Q}$-*factorial* if an arbitrary prime divisor on X is $\mathbb{Q}$-Cartier.

The *intersection number* $(D \cdot C) \in \mathbb{Q}$ of a $\mathbb{Q}$-Cartier $\mathbb{Q}$-divisor D and a complete curve C on X is defined by $(D \cdot C) = (mD \cdot C)/m$. D is said to be *nef* (or *numerically semipositive*) if $(D \cdot C) \geq 0$ for all C.

1.3 Let us define the notion of the *log-canonical divisor*. Let X be a nonsingular variety and $B = \sum_{j=1}^{t} D_j$ a normal crossing divisor on X. The sheaf of logarithmic 1-forms $\Omega_X^1(\log B)$ is a locally free sheaf with a local basis $dx_1/x_1, \ldots, dx_r/x_r, dx_{r+1}$, $\ldots, dx_n$ if B is defined by $\prod_{j=1}^{r} x_j = 0$ for a local parameter system $\{x_1, \ldots, x_n\}$. Corresponding to its determinant sheaf, we define the log-canonical divisor to be $K_X + B$, which is related to questions concerning the open variety $X - B$.

But more important is its $\mathbb{Q}$-divisor version $K_X + D$ with $D = \sum_{j=1}^{t} d_j D_j$ for some rational numbers d_j which satisfy the inequalities $0 < d_j < 1$, i.e., D is effective and $[D] = 0$. Here, the experience shows that the conditions on the d_j are necessary to avoid counterexamples. This $\mathbb{Q}$-divisor version of the log-canonical divisor is related to the covering spaces. For example, if $f: X \to Y$ is a morphism of affine lines $X = Y = \mathbb{A}^1$ given by $f^*(y) = x^e$, then $K_X = f^*(K_Y + (e-1)/e \cdot B)$ for $B = \mathrm{div}(y)$.

In this paper a *smooth pair* (X, D) is meant to be a pair consisting of a nonsingular projective variety X and an effective $\mathbb{Q}$-divisor on it such that $[D] = 0$ and $\mathrm{Supp}(D)$ has only normal crossings. We can extend our question to the *log-canonical ring* of a smooth pair (X, D) defined by $R(X, D) = \sum_{m=0}^{\infty} H^0(X, [m(K_X + D)])$. (X, D) is said to be of *log-general type* if trans. $\deg_k R(X, D) = \dim X + 1$.

Question 2. Is $R(X, D)$ finitely generated?

The answer is known to be "yes" only in dimension 2 [Ka1, Fu2].

1.4 We shall explain Fujita's reduction ([Fu3], also [Mo3], [N]) which reduces Question 1 (*non-log-case*) for a variety X with $0 \leq \kappa(X) < \dim X$ (or even Question 2 for a smooth pair (X, C), see [Mw]) to Question 2 (*log-case*) for another smooth pair (Y, D) of log-general type such that $\dim Y = \kappa(X)$.

After replacing X by a suitable birational model, one can find a proper surjective morphism $f: X \to Y$ with connected fibers to a nonsingular projective variety Y, a

$\mathbb{Q}$-divisor $D = D' + D''$ on Y, and a positive integer d which satisfy the following conditions:

(a) D' is nef,
(b) (Y, D'') is a smooth pair and f is smooth over $Y - \text{Supp}(D'')$,
(c) dim $Y = \kappa(X)$ and (Y, D) is of log-general type,
(d) $f^* : H^0(Y, [d \cdot m(K_Y + D)]) \simeq H^0(X, d \cdot m \, K_X)$ for all $m \in \mathbb{N}$.

The $\mathbb{Q}$-divisor D' comes from the moduli of fibers of f. The property (a) is proved by using the theory of variation of Hodge structures ([Fu1] if dim $Y = 1$, and [Ka2] in general by using [G]). The degenerations of the fibers of f give rise to D''. The contribution of semistable degenerations to D'' is zero. So by the semistable reduction in codimension one, we understand that the coefficients of D'' are dominated by those of coverings, hence less than one. Then by using Kodaira's lemma (cf. [Ka3]), one can reduce Question 1 for X to Question 2 for some lower dimensional smooth pair.

§2. Log-Canonical Model

2.1 In higher dimensional algebraic geometry, varieties must be allowed to have certain mild singularities. Let X be a normal variety and $D = \sum_{j=1}^{t} d_j D_j$ an effective $\mathbb{Q}$-divisor such that $[D] = 0$. The pair (X, D) is said to have only *log-terminal singularities*, or simply (X, D) is called *log-terminal*, if the following conditions are satisfied:

(a) $K_X + D$ is $\mathbb{Q}$-Cartier,
(b) for a resolution of singularities ([H]) $\mu : Y \to X$ with a normal crossing divisor $\sum_{i=1}^{t+s} E_i$ on Y such that the E_i $(1 \leq i \leq t)$ are the strict transforms of the D_i and the E_i $(t + 1 \leq i \leq t + s)$ are all the prime divisors such that $\text{codim}_X \mu(E_i) \geq 2$, if we write

$$K_Y + \sum_{i=1}^{t+s} E_i = \mu^*(K_X + D) + \sum_{i=1}^{t+s} a_i E_i \qquad (*)$$

then we have $a_i > 0$ for all i.

The rational numbers a_i are called *log-discrepancy coefficients*. Note that $a_i = 1 - d_i$ for $1 \leq i \leq t$. In the case $D = 0$, considering the formula

$$K_Y = \mu^* K_X + \sum_{i=t+1}^{t+s} b_i E_i, \qquad (**)$$

we say that X has only *terminal* (resp. *canonical*) *singularities*, if $b_i > 0$ (resp. $b_i \geq 0$) hold for all i. Since $b_i = a_i - 1$, we have the implications

$$X : \text{terminal} \Rightarrow X : \text{canonical} \Rightarrow (X, 0) : \text{log-terminal}.$$

If (X, D) is log-terminal and $D^\#$ is an effective Cartier divisor on X, then the pair $(X, D + \varepsilon D^\#)$ is also log-terminal for a sufficiently small positive rational number ε.

For example, if dim $X = 2$, then

$$X : \text{terminal} \Leftrightarrow \text{nonsingular}$$

$$X : \text{canonical} \Leftrightarrow \text{rational double points } \mathbb{C}^2/\Gamma, \ \Gamma \subset \text{SL}(2, \mathbb{C})$$

$$(X, 0) : \text{log-terminal} \Leftrightarrow \text{quotient singularities } \mathbb{C}^2/\Gamma, \ \Gamma \subset \text{GL}(2, \mathbb{C}).$$

Though the above definitions are simple, it is hard to understand what these singularities really are. All quotient singularities of any dimension are log-terminal (with $D = 0$), but there are a lot more of them. In dimension 3, a detailed classification of terminal singularities are obtained by [MS], [D] and especially [Mo2]. The results in [Mo2] were fundamental for the success of Mori's flip theorem in dimension 3 ([Mo4]). But it seems too difficult to classify canonical or log-terminal singularities even in dimension 3. A structure theorem of three-dimensional canonical singularities is proved in [Ka8], and it was one of the key steps toward the flip theorem in [Mo4]. In general, we know only that if (X, D) is a log-terminal pair, then X has only rational singularities (Elkik [E], also [Fu4]).

2.2 Since $R(X)$ is determined by the function field $k(X)$, one might say that Question 1 is a purely algebraic one. But the solutions so far are geometric. In fact, the affirmative answer to Question 2 for a smooth pair (X, D) of log-general type is equivalent to the existence of the *log-canonical model* of (X, D); a pair (X', D') of a normal projective variety and a $\mathbb{Q}$-divisor is said to be the log-canonical model of (X, D) if the following conditions are satisfied:

(a) (X', D') has only log-terminal singularities,

(b) $K_{X'} + D'$ is an ample $\mathbb{Q}$-Cartier divisor,

(c) there exists a birational map $\gamma : X \dashrightarrow X'$ which is surjective in codimension one (i.e., every prime divisor on X' has a strict transform on X),

(d) $\gamma_* D = D'$ and $K_X + D \geq \gamma^*(K_{X'} + D')$, where γ_* is the direct image homomorphism of codimension one cycles and $\gamma^* = \alpha_* \beta^*$ if $\beta = \gamma\alpha$ for birational morphisms α and β.

If $R(X, D)$ is finitely generated, then we set $X' = \text{Proj}(R(X, D))$ (cf. [R1]). Conversely, the condition (d) implies that $R(X, D)$ is isomorphic to the log-canonical ring $R(X', D')$, which is similarly defined for the log-terminal pair. Hence the existence of the log-canonical model implies the finite generatedness.

§ 3. Minimal Model Program

3.1 We want to construct the log-canonical model via log-minimal models. A *log-minimal model* (X', D') of a smooth pair (X, D) is a pair which satisfies (a), (c), (d) of (2.2) and (b') $K_{X'} + D'$ is nef.

A log-minimal model is defined just by a numerical property that $K + D$ is nef. But it has a good geometric property at least if it is of log-general type:

Theorem (Kawamata, Benveniste, Shokurov). *If (X, D) is a log-minimal model of log-general type, then there exists a positive integer m such that the linear system $|m(K_X + D)|$ is free. Therefore, the locus on X where $K_X + D$ is numerically trivial is contractible, and $R(X, D)$ is finitely generated.*

3.2 Starting from an arbitrary smooth pair (X, D) of log-general type, we should reduce the locus where the log-canonical divisor is *numerically negative* in order to obtain a log-minimal model. The key idea of the minimal model program is to look at the contraction morphisms associated with *extremal rays*, or the *extremal contractions* ([Mo1]). In the following, we combine cone and contraction theorems; in fact, they are proved at the same time ([Ka6]).

Cone-Contraction Theorem (Mori, Kawamata, Reid, Shokurov). *Let X be a normal projective $\mathbb{Q}$-factorial variety and D a $\mathbb{Q}$-divisor such that the pair (X, D) has only log-terminal singularities. Suppose that $K_X + D$ is not nef. Then there exists a projective surjective morphism $\varphi : X \to Y$ with connected fibers to a normal projective variety Y such that $(1) - (K_X + D)$ is φ-ample, and (2) $\varrho(X/Y) = 1$ (i.e., φ is not an isomorphism, and for any two curves C_i $(i = 1, 2)$ with $\varphi(C_i)$ points, there exists a rational number q such that $(B \cdot C_1) = q(B \cdot C_2)$ for all prime divisors B on X).*

In the course of the proof of the above two theorems, the log-discrepancy coefficients a_i play a guiding role. The proof is by induction, and we need the relative version of the cone-contraction theorem over a base space S in order to have Property (2) (cf. [Ka6]).

3.3 φ is called an *extremal contraction*. We define its *degenerate locus* $E = \mathrm{Exc}(\varphi) = \{x \in X; \varphi \text{ is not an isomorphism at } x\}$. φ is a birational morphism if and only if $E \neq X$. There are three types of φ as follows.

(a) *Fibering type*: $\dim Y < \dim X$. Then $R(X, D) = k$.

(b) *Divisorial contraction*: E is a prime divisor. Then Y is again $\mathbb{Q}$-factorial and $(Y, \varphi_* D)$ is log-terminal. So we have a log-minimal model or apply the cone-contraction theorem again.

(c) *Small contraction*: $\mathrm{codim}_X E \geq 2$. The *log-flip conjecture* asserts that there exists another projective birational morphism $\varphi^+ : X^+ \to Y$ such that $\mathrm{codim}_{X^+} \mathrm{Exc}(\varphi^+) \geq 2$ and $K_{X^+} + D^+$ is φ^+-ample, where D^+ is the strict transform of D. Then (X^+, D^+) is again $\mathbb{Q}$-factorial and log-terminal. The birational transformation $X \dashrightarrow X^+$ is called the *log-flip* of φ.

Given a smooth pair, if the log-flip conjecture is true, and if there does not exist an infinite chain of successive log-flips (*termination conjecture*), then we obtain a log-minimal model after a finite number of birational transformations of divisorial contractions and log-flips.

Note that the log-flip conjecture is a special case of Question 2, because X^+ should coincide with $\mathrm{Proj}(\mathscr{R}(X/Y, D))$, where the *relative log-canonical ring*

$$\mathscr{R}(X/Y, D) = \sum_{m=0}^{\infty} \varphi_* \mathcal{O}_X([m(K_X + D)])$$ is a graded $\mathcal{O}_Y$-algebra. In particular, the conjecture is local. It is easy to see that an affirmative answer to Question 2 implies the finite generatedness of $\mathscr{R}(X/Y, D)$.

704   Yujiro Kawamata

A *flip* is a log-flip in the case where $D = 0$ and X has only terminal singularities. The existence of flips in dimension 3 is proved in [Mo4]. If X is a nonsingular and of dimension 4, the flip exists by [Ka9]. Shokurov has some idea toward the log-flip conjecture in dimension 3.

One can prove that $\mathrm{Exc}(\varphi)$ is covered by a family of rational curves ([Ka11, appendix] by using the vanishing theorem explained below and [MM]). As an application, Wilson [W] proved that there exists a rational curve on any Calabi-Yau 3-fold X whose Picard number is greater than 19. In fact, he proved the existence of an effective Cartier divisor D which is not nef. Since $(X, \varepsilon D)$ is log-terminal for $0 < \varepsilon \ll 1$, the cone-contraction theorem yields the result.

3.4 The following generalization [Ka3, V] of the Kodaira vanishing theorem is essential in the proof of the above theorems.

Vanishing Theorem (Kawamata, Viehweg). *Let (X, D) be a smooth pair and L a Cartier divisor on X. If $L - (K_X + D)$ is ample, then $H^i(X, L) = 0$ for $i > 0$.*

The idea of the proof is as follows. One constructs a $\mathbb{Q}$-divisor D' on X and a Galois covering $\pi : Y \to X$ from a nonsingular projective variety Y such that π^*D is a usual divisor, (X, D') is a smooth pair, $K_Y = \pi^*(K_X + D')$ and $D \leq D'$ ([Ka2]). Then by [Kod], $H^i(Y, \pi^*(L + D' - D)) = 0$ for $i > 0$. Taking the $\mathrm{Gal}(Y/X)$-invariants, we obtain the theorem.

One can extend the theorem to the case where the pair (X, D) is log-terminal ([KMM]). As an application, if $L - K_X$ is nef and big (i.e., $(L - K_X)^{\dim X} > 0$), then $L - (K_X + \varepsilon B)$ is ample for some effective divisor B and $0 < \varepsilon \ll 1$ by Kodaira's lemma, hence the vanishing (cf. [Ka3]).

§4. Flip to Flop

4.1 In the notation of (3.3), a *flop* is a special kind of log-flip where X has only canonical singularities and K_X is numerically trivial along the fibers of φ. We can take D to be a small $\mathbb{Q}$-divisor such that (X, D) is log-terminal and $-D$ is φ-ample, since φ is an isomorphism in codimension one.

Given $\varphi : X \to Y$, the existence of flops in dimension 3 is proved by [R2] in the case where X has only terminal singularities and by [Ka8] in general. By [Ka8] and [Ko1], if X and X' are birationally equivalent minimal 3-folds (i.e., $\mathbb{Q}$-factorial projective 3-folds with terminal singularities and nef canonical divisors), there exists a sequence of flops which connects X and X'. In particular, a 3-fold of general type has only a finite number of minimal models ([KM]). Flops of 3-folds with terminal singularities are geometrically symmetric and preserves the analytic types of singularities ([Ko1]). But they are not symmetric in general even in dimension 3. In dimension > 3, the dimensions of $\mathrm{Exc}(\varphi)$ and $\mathrm{Exc}(\varphi^+)$ may also be different.

4.2 There are two approaches to make flips from flops. The first method uses double coverings. Given a small contraction $\varphi : X \to Y$ with φ-ample $-K_X$ as in

(3.2), we replace Y by its affine open subset, since flip is local. We take double covers $\tilde{X} \to X$ and $\tilde{Y} \to Y$ which ramify along a general element $B_X \in |-2K_X|$ and its image $B_Y \in |-2K_Y|$, respectively. The morphism φ induces a small contraction $\tilde{\varphi}: \tilde{X} \to \tilde{Y}$ such that $K_{\tilde{X}}$ is numerically trivial for $\tilde{\varphi}$. So if one can prove that $\tilde{X}$ has only canonical singularities, then the existence of flop $\tilde{X} \dashrightarrow \tilde{X}^+$ produces a flip $X \dashrightarrow X^+$ as the $\mathbb{Z}/(2)$-quotient. This approach is first found in [Ka8] to prove the existence of a flip for semistable degeneration of surfaces, and then applied to general 3-folds in [Mo4].

The second approach is the cone construction. Given $\varphi: X \to Y$ as before, we take the "total spaces" of K; we let $CX = \mathrm{Spec}\left(\sum\limits_{m=0}^{\infty} \mathcal{O}_X(-mK_X)\right)$ and $CY = \mathrm{Spec}\left(\sum\limits_{m=0}^{\infty} \mathcal{O}_Y(-mK_Y)\right)$. Here, both graded algebras are finitely generated, since $-K_X$ is a φ-ample $\mathbb{Q}$-Cartier divisor. One can prove that CX has only canonical singularities and K_{CX} is numerically trivial for the induced morphism $C\varphi: CX \to CY$. Thus if the flop $CX \dashrightarrow (CX)^+$ exists, the strict transform X^+ of X gives the flip in one dimension less.

§5. Bounding the Indices

5.1 The *index* r of a terminal singularity is the smallest positive integer such that rK is a Cartier divisor. For example, the cyclic quotient singularity $\mathbb{C}^3/\langle(\zeta^a, \zeta^{-a}, \zeta)\rangle$ for $\zeta = \exp(2\pi\sqrt{-1}/r)$ and $(r, a) = 1$ has index r. By [Mo2], an arbitrary terminal singularity of dimension 3 has a small deformation with only quotient singularities of the above type. Note that these quotient singularities are rigid [Sch].

We have to deal with singular varieties in dimension 3, and so results on nonsingular varieties must be modified. Let X be a projective 3-fold with only terminal singularities. Then we have

$$\chi(\mathcal{O}_X) = \frac{1}{24}\left\{c_1 c_2 + \sum_P (r_P - 1/r_P)\right\},$$

where $c_1 c_2 = -(\mu^* K_X \cdot c_2(Y))$ for a resolution $\mu: Y \to X$, the summation is taken over all virtual quotient singularities of X, where singular points of X are replaced by its generic small deformations, and r_P are their indices ([Ka7, R3]). More generally, Fletcher calculated the correction terms for $\chi(mK_X)$ ([Fl]).

As applications of the above formula, we have some boundedness results. By [Ka7] and [Ms], if K_X is numerically trivial, then $12K_X \sim 0$ if $\chi(\mathcal{O}_X) \geq 2$ and $120K_X \sim 0$ if $\chi(\mathcal{O}_X) = 1$.

By [Ka10], there exist universal constants $m_1 \in \mathbb{N}$ and $m_2 \in \mathbb{Q}$ such that if X is a $\mathbb{Q}$-Fano 3-fold (a projective 3-fold having only terminal singularities and ample anti-canonical divisor $-K_X$) with Picard number $\varrho = 1$, then $m_1 K_X$ is a Cartier divisor and $(-K_X)^3 \leq m_2$. Hence all the $\mathbb{Q}$-Fano 3-folds with $\varrho = 1$ form a bounded family.

Let S be a minimal nonsingular projective surface with $\kappa(S) = 1$, $f: X \to \Delta = \{t \in \mathbb{C}; |t| < 1\}$ a relative minimal model of a semistable degeneration of surfaces such that $f^{-1}(t)$ $(t \neq 0)$ are deformations of S, and m_i $(i = 1, \ldots, e)$ the multiplicities of multiple fibers of the elliptic fibration of S. If $m = $ l.c.m.$\{12, m_i\}$, then mK_X is a Cartier divisor by [Ka11]. We note that a similar boundedness result for the case $\kappa(S) = 2$ would yield a geometric compactification of the moduli space of such surfaces (cf. [KSB]).

References

[B]   Benveniste, X: Sur l'anneau canonique de certaines variétés de dimension 3. Invent. math. **73** (1983) 157–164

[D]'   Danilov, V.I.: Birational geometry of toric 3-folds. Math. USSR Izv. **21** (1983) 269–279

[E]   Elkik, R.: Rationalité des singularités canoniques. Invent. math. **64** (1981) 1–6

[F1]   Fletcher, A.R.: Contributions to Riemann-Roch on projective 3-folds with only canonical singularities and applications. Proc. Symp. Pure Math. **46**, no. 1 (1987) 221–231

[Fu1]   Fujita, T.: On Kähler fiber spaces over curves. J. Math. Soc. Japan **30** (1978) 779–794

[Fu2]   Fujita, T.: Fractionally logarithmic canonical rings of algebraic varieties. J. Fac. Sci. Univ. Tokyo IA-**30** (1984) 685–696

[Fu3]   Fujita, T.: Zariski decomposition and canonical rings of elliptic threefolds. J. Math. Soc. Japan **38** (1986) 19–37

[G]   Griffiths, P.: Periods of integrals on algebraic manifolds III. Publ. Math. IHES. **38** (1978) 779–794

[H]   Hironaka, H.: Resolution of singularities of an algebraic variety over a field of characteristic zero. Ann. Math. **79** (1964) 109–326

[Ka1]   Kawamata, Y.: On the classification of non-complete algebraic surfaces. (Lecture Notes in Mathematics, vol. 732.) Springer, Berlin Heidelberg New York 1979, pp. 251–232

[Ka2]   Kawamata, Y.: Characterization of abelian varieties. Comp. Math. **43** (1981) 253–276

[Ka3]   Kawamata, Y.: A generalization of Kodaira-Ramanujam's vanishing theorem. Math. Ann. **261** (1982) 43–46

[Ka4]   Kawamata, Y.: On the finiteness of generators of a pluri-canonical ring for a 3-fold of general type. Amer. J. Math. **106** (1984) 1503–1512

[Ka5]   Kawamata, Y.: Elementary contractions of algebraic 3-folds. Ann. Math. **119** (1984) 95–110

[Ka6]   Kawamata, Y.: The cone of curves of algebraic varieties. Ann. Math. **119** (1984) 603–633

[Ka7]   Kawamata, Y.: On the plurigenera of minimal algebraic 3-folds with $K \equiv 0$. Math. Ann. **275** (1986) 539–546

[Ka8]   Kawamata, Y.: The crepant blowing-up of 3-dimensional canonical singularities and its application to degenerations of surfaces. Ann. Math. **127** (1988) 93–163

[Ka9]   Kawamata, Y.: Small contractions of four dimensional algebraic manifolds. Math. Ann. **284** (1989) 595–600

[Ka10]   Kawamata, Y.: Boundedness of $\mathbb{Q}$-Fano threefolds. To appear in Proc. Symp. Novosibirsk 1989

[Ka11]   Kawamata, Y.: Moderate degenerations of algebraic surfaces. To appear in Proc. Symp. Bayreuth 1990

[KMM]   Kawamata, Y., Matsuda, K., Matsuki, K.: Introduction to the minimal model problem. Adv. Stud. Pure Math. **10** (1987) 283–360

[KM]   Kawamata, Y., Matsuki, K.: The number of minimal models for a 3-fold of general type is finite. Math. Ann. **276** (1987) 595–598

[Kod]   Kodaira, K.: On a differential method in the theory of analytic stacks. Proc. Natl. Acad. Sci. USA **39** (1953) 1268–1273

[Ko1]   Kollár, J.: Flops. Nagoya Math. J. **113** (1989) 15–36

[KSB]   Kollár, J., Shepherd-Barron, N.I.: Threefolds and deformations of surface singularities. Invent. math. **91** (1988) 299–338

[MM]   Miyaoka, Y., Mori, S.: A numerical criterion of uniruledness. Ann. Math. **124** (1986) 65–69

[Mo1]   Mori, S.: Threefolds whose canonical bundles are not numerically effective. Ann. Math. **116** (1982) 133–176

[Mo2]   Mori, S.: On 3-dimensional terminal singularities. Nagoya Math. J. **98** (1985) 43–66

[Mo3]   Mori, S.: Classification of higher-dimensional varieties. Proc. Symp. Pure Math. **46**, no. 1 (1987) 269–331

[Mo4]   Mori, S.: Flip theorem and the existence of minimal models for 3-folds. J. Amer. Math. Soc. **1** (1988) 117–253

[Mw]   Moriwaki, A.: Semi-ampleness of the numerically effective part of Zariski decomposition. J. Math. Kyoto Univ. **26** (1986) 465–481

[Ms]   Morrison, D.: A remark on Kawamata's paper [Ka7]. Math. Ann. **275** (1986) 547–553

[MS]   Morrison, D., Stevens, G.: Terminal quotient singularities in dimensions three and four. Proc. Amer. Math. Soc. **90** (1984) 15–20

[Mu]   Mumford, D.: The canonical ring of an algebraic surface (Appendix to [Z]). Ann. Math. **76** (1962) 612–615

[N]   Nakayama, N.: The singularity of the canonical model of compact Kähler manifolds. Math. Ann. **280** (1988) 509–512

[R1]   Reid, M.: Canonical 3-folds. In: Géométrie algébriques Angers 1979, pp. 273–310

[R2]   Reid, M.: Minimal models of canonical 3-folds. Adv. Stud. Pure Math. **1** (1983) 131–180

[R3]   Reid, M.: Young person's guide to canonical singularities. Proc. Symp. Pure Math. **46**, no. 1 (1987) 345–414

[Sch]   Schlessinger, M.: Rigidity of quotient singularities. Invent. math. **14** (1971) 17–26

[Sh1]   Shokurov, V.V.: The nonvanishing theorem. Math. USSR Izv. **26** (1986) 591–604

[Sh2]   Shokurov, V.V.: Numerical geometry of algebraic varieties. Proc. Intl. Congress of Math. Berkeley 1986, pp. 672–681

[V]   Viehweg, E.: Vanishing theorems. J. Reine Angew. Math. **335** (1982) 1–8

[W]   Wilson, P.M.H.: Calabi-Yau manifolds with large Picard number. Invent. math. **98** (1989) 139–155

[Z]   Zariski, O.: The theorem of Riemann-Roch for high multiples of an effective divisor on an algebraic surface. Ann. Math. **76** (1962) 560–615

Flip and Flop

János Kollár

Department of Mathematics, University of Utah, Salt Lake City, UT 84112, USA

Dedicated to my teacher
Teruhisa Matsusaka

The study of smooth projective varieties naturally breaks into two parts. One part is to study the relationship between birationally equivalent varieties and the other part is to study the different birational equivalence classes. The aim of this talk is to review some interesting features of birational equivalence for three dimensional varieties over $\mathbb{C}$.

For smooth projective curves birational equivalence implies isomorphism, hence the first part of the problem is not interesting.

For smooth projective surfaces the simplest birational transformation is the blowing up. One removes a point of a surface S and replaces it with a copy of $\mathbb{P}^1$ corresponding to the local complex directions at the removed point. Any birational equivalence is a composite of blowing ups and their inverses. There are no other kinds of birational transformations.

In dimension three one can blow up points and smooth curves. It is conjectured that any birational transformation is a composite of blowing ups and their inverses, though at the moment no one seems to know how to prove this. However it became clear in the last decade that even if the above factorisation is possible it may not be the best way of factoring birational transformations. A large theory has been developped by several people, I refer to the talk of (Mori 1990) in the same volume for a general overview. My aim in this talk is to concentrate on certain special birational transformations between threefolds, called flip and flop.

For technical reasons it is natural to investigate these transformations not only for smooth threefolds but also for certain singular ones. The smallest class of singularities that allows the theory to work well is given by the following definition:

Definition 1. Let X be a normal variety and let $f : X' \to X$ be a resolution of singularities with exceptional divisors $E_i \subset X'$. Assume that $\mathcal{O}(mK_X)$ is locally free for some $m > 0$. The smallest such m will be called the index of X. One can write

$$\mathcal{O}(mK_{X'}) \cong f^* \mathcal{O}(mK_X) \otimes \mathcal{O}(\sum a_i E_i),$$

for some integers a_i. We say that X has terminal (resp. canonical) singularities if $a_i > 0$ (resp. $a_i \geq 0$) for every i. The survey paper (Reid 1987) is the best source of information concerning terminal (and canonical) singularities.

The informal definition of flip or flop is the following. Let X be a threefold and let $C \subset X$ be a compact curve. Remove C from X and try to replace it with another compact curve C^+ in such a way that $C^+ \cup (X - C)$ becomes a threefold X^+. To avoid the unexciting possibility of $C = C^+$ we require that X and X^+ be nonisomorphic. Such an X^+ need not exist at all and it also need not be unique. It turns out that the nature of the transformation $X \; --> X^+$ depends on the sign of $C \cdot K_X$. The $C \cdot K_X = 0$ case is called flop, the $C \cdot K_X < 0$ case called flip. Very little is known about the case $C \cdot K_X > 0$ and I will not consider it at all.

The formal definitions are the following:

Definition 2. (2.1) A three dimensional curve neighborhood is a pair $C \subset X$ where C is a proper connected curve and X is the germ of a normal threefold along C. One can think of X as an analytic representative of the germ.

(2.2) A three dimensional curve neighborhood $C \subset X$ is called contractible if there is a morphism $f : C \subset X \to P \in Y$ satisfying the following properties:

(i) Y is the germ of a normal singularity around the point P;

(ii) $f(C) = P$;

(iii) $f : X - C \to Y - P$ is an isomorphism.

f and Y are uniquely determined by $C \subset X$. f is called the contraction morphism of $C \subset X$.

(2.3) Two three dimensional curve neighborhoods $C_i \subset X_i$ are called bimeromorphic if there is an isomorphism $(X_1 - C_1) \cong (X_2 - C_2)$.

(2.4) Let $C_1 \subset X_1$ be a three dimensional curve neighborhood. Let H_1 be a line bundle such that $H_1^{-1}|C_1$ is ample. A three dimensional curve neighborhood $C_2 \subset X_2$ bimeromorphic to $C_1 \subset X_1$ is called the opposite of $C_1 \subset X_1$ with respect to H_1 if there is a line bundle H_2 on X_2 such that $H_2|C_2$ is ample and

$$(X_2 - C_2, H_2^m | X_2 - C_2) \cong (X_1 - C_1, H_1^n | X_1 - C_1),$$

for some positive integers m, n. By (Kollár 1991, 2.1.6) the opposite is unique, though it need not exist.

Definition 3. Let $f : C \subset X \to P \in Y$ be a three dimensional contractible curve neighborhood. Assume that X has terminal singularities and that $K_X|C = 0$. Let H be a line bundle on X such that $H^{-1}|C$ is ample.

The opposite of $C \subset X$ (if it exists) will be called the flop of $C \subset X$. (It turns out that the flop is independent of the choice of H.)

Definition 4. Let $f : C \subset X \to P \in Y$ be a three dimensional contractible curve neighborhood. Assume that X has terminal singularities and that $-K_X|C$ is ample.

The opposite of $C \subset X$ with respect to K_X (if it exists) will be called the flip of $C \subset X$.

The simplest example of flops is the following:

Example 5. Let V be the total space of the line bundle $\mathcal{O}(-1, -1)$ over $\mathbb{P}^1 \times \mathbb{P}^1$. Both of the projections $\pi_i : \mathbb{P}^1 \times \mathbb{P}^1 \to \mathbb{P}^1$ can be extended to morphisms

$$p_i : (\mathbb{P}^1 \times \mathbb{P}^1 \subset V) \to (C_i \cong \mathbb{P}^1 \subset X_i).$$

It is easy to see that X_i is smooth and the normal bundle of $C_i \subset X_i$ is $\mathcal{O}(-1) + \mathcal{O}(-1)$. $p_2 \circ p_1^{-1} : X_1 \; -\,-> X_2$ is bimeromorphic but it is not an isomorphism.

This example was first noticed by (Atiyah 1958). It was used in a systematic way by (Kulikov 1977) to study the birational transformations of threefolds that have a basepoint-free pencil of K3 surfaces.

Examples of flips are much more difficult to present since X cannot be smooth. A fairly exhaustive list is given in (Kollár-Mori 1991).

From the logical point of view (which is the reverse of the historical order) the theory of flips and flops has three large parts.

I. Prove that flops and flips exist.
II. Describe $C^+ \subset X^+$ in terms of $C \subset X$ as precisely as possible.
III. Applications.

The answer to part I is very simple to state:

Theorem 6 (Reid 1983). *Flops exist.*

Theorem 7 (Mori 1988). *Flips exist.*

The existence of flips is one of the most difficult results in three dimensional geometry; see (Kollár 1990a) for an introduction. Previously an important special case was settled by (Tsunoda 1987; Shokurov 1985; Mori 1985; Kawamata 1988) using different methods. (The order corresponds to the order in which the proofs were announced. The only complete published version is the last one.)

It is interesting to note that the above proofs provide very little information beyond the existence of X^+. Thus the answer to part II proceeds along very different lines.

For flops there is a detailed structure theory. This is based on the following observation:

Theorem 8 (Mori 1987). *Let $f : C \subset X \to P \in Y$ be a three dimensional contractible curve neighborhood such that $K_X|C = 0$. Assume in addition that X is smooth (or that it has only index one terminal singularities). Then Y embeds into $\mathbb{C}^4$ and in suitable local coordinates its equation can be written as $x^2 + h(y, z, t)$*

*for some power series h. Let $\tau : Y \to Y$ be the involution $(x,y,z,t) \mapsto (-x,y,z,t)$.
Then*

$$\left(f^+ : X^+ \to Y\right) \cong (\tau \circ f : X \to Y) .$$

In particular, X^+ is smooth iff X is smooth.

If X has higher index points then the symmetry breaks down in some cases. However in these cases another explicit description is possible, see (Kollár 1989; Kollár 1991, 2.2.2). There are always many similarities between X and X^+:

Theorem 9. *Let X_1 and X_2 be projective threefolds with terminal singularities only. Assume that X_2 is obtained from X_1 by a sequence of flops. The following objects do not depend on j (up to isomorphism):*

(9.1) The intersection homology groups $IH^i(X_j, \mathbb{C})$ together with their Hodge structures (Kollár 1989, 4.12).

(9.2) The collection of analytic singularities of X_j (Kollár 1989, 4.11).

(9.3) The miniversal deformation space $\operatorname{Def} X_j$ (Kollár-Mori 1991, 12.6).

(9.4) The integral cohomology groups $H^i(X_j, \mathbb{Z})$ (Kollár 1991, 3.2.2).

(9.5) $\operatorname{Pic} X_j \subset \operatorname{Weil} X_j$ (Kollár 1991, 3.2.2).

(9.6) $h^0(X_j, \mathcal{O}(D))$ for every Weil divisor D.

(9.7) $h^i(X, \mathcal{O}(mK_{X_j}))$ for every i and m.

The isomorphisms are canonical except possibly for (9.2).

Describing $C^+ \subset X^+$ in terms of $C \subset X$ is harder for flips. The first breakthrough was a result of (Shokurov 1985) which showed that the singularities of X^+ are less "difficult" than the singularities of X in a complicated technical sense. For a given singularity the "difficulty" is not easy to compute but the definition is very well suited to inductive proofs.

It turns out that the situation is very complicated. The problem is that the singularities of X^+ depend not only on the singularities of X but also on the global structure of X. If X is given by coordinate patches and transition functions along $C \cong \mathbb{P}^1$ then the singularities of X^+ do not depend continuously on the transition functions. This makes any description very subtle. Several special cases are worked out in (Kollár-Mori 1991); many more – but not all – cases are treated in unpublished works of Mori. I would like to mention one patholgy that seems interesting:

Proposition 10 (Kollár-Mori 1991, 13.5, 13.7). *Let $C^+ \subset X^+$ be the flip of $C \subset X$. Then the number of irreducible components of C^+ is less than or equal to the number of irreducible components of C. For every $m > 0$ there are examples where C^+ is irreducible and C has m irreducible components.*

Let us now turn to some of the applications. The most important application of flops is the following result which shows that birational maps between certain threefolds can be factored into a sequence of flops. This factorisation is much more useful than a factorisation into blowing ups and downs would be. Its importance is clear only in light of the minimal model theory of threefolds (see e.g. (Mori 1990)).

Theorem 11 (Reid 1983; Kawamata 1988; Kollár 1989). *Let X_1 and X_2 be normal projective threefolds with terminal singularities only. Assume that they have $\mathbb{Q}$-factorial singularities. Assume furthermore that K_{X_i} has nonnegative intersection number with any curve $C \subset X_i$ for $i = 1, 2$. Then any birational map $f : X_1 \dashrightarrow X_2$ can be obtained as a composite of flops.*

Another application is to compact nonprojective threefolds. Under certain conditions there is a very economical way of finding a birationally equivalent model which is projective.

Theorem 12 (Kollár 1991, 5.2.3). *Let X be a proper algebraic threefold with $\mathbb{Q}$-factorial terminal singularities. Assume that K_X has nonnegative intersection number with any curve $C \subset X$. Then after finitely many flops one obtains a proper algebraic threefold X^+ with $\mathbb{Q}$-factorial terminal singularities such that X^+ admits a birational morphism $g : X^+ \to Y$ onto a projective variety Y. Y has terminal singularities and g contracts only finitely many curves.*

Flips are one step of Mori's program (also called minimal model program) for threefolds, in fact the most difficult step. Their main importance is thus derived from the importance and applications of the program. See (Kollár 1991) or (Mori 1990) for details.

There are many problems concerning flips even in dimension three. The most interesting open question about flips is the following.

Conjecture 13. Reid's Conjecture on General Elephants (Reid 1987; Kollár-Mori 1991). *The contraction map provides a one-to-one correspondence between the following two sets:*
  Extremal neighborhoods:

$$EN := \left\{ \begin{array}{l} \textit{Three dimensional contractible curve neighborhoods } C \subset X \textit{ such that} \\ X \textit{ has canonical singularities and } -K_X|C \textit{ is ample.} \end{array} \right\}$$

  Flipping singularities:

$$FS := \left\{ \begin{array}{l} \textit{Three dimensional normal singularities } P \in Y \textit{ such that } K_Y \textit{ is not} \\ \mathbb{Q}\textit{-Cartier and some } D \in |-K_Y| \textit{ has a Du Val singularity at } P. \end{array} \right\}$$

Reid's original hope was that this equivalence can be used to obtain a proof of the existence of flips. To do this one needs to produce a member of $|-K_Y|$ with Du Val singularity and then to use this member to construct X^+. It is still to be seen whether either of these steps can be done in the spirit envisaged by Reid. A proof along these lines would considerably enhance our understanding of flips.

(Kollár-Mori 1991, 3.1) shows that for every flipping singularity there is a corresponding extremal neighborhood. The opposite direction is also proved for extremal neighborhoods with terminal singularities and irreducible C in (Kollár-Mori 1991, 1.7). The general case probably requires different methods. There is also the possibility that the correspondence is more complicated. The conjecture implies that every non-Gorenstein singularity on an extremal neighborhood is pseudo-terminal. I do not see any a priori reason why this should be so. On the other hand, even certain terminal singularities cannot occur on extremal neighborhoods, thus unexpected restrictions are possible.

The higher dimensional problems are discussed in the talk of (Kawamata 1990).

Acknowledgement. Partial financial support was provided by the NSF under grant numbers DMS-8707320 and DMS-8946082 and by an A. P. Sloan Research Fellowship.

References

Atiyah, M. (1958): On analytic surfaces with double points. Proc. Roy. Soc. **247**, 237–244

Kawamata, Y. (1988): The crepant blowing-up of 3-dimensional canonical singularities and its application to the degeneration of surfaces. Ann. Math **127**, 93–163

Kawamata, Y. (1990): Canonical singularities and minimal models of algebraic varieties. These Proceedings, p. 699

Kollár, J. (1989): Flops. Nagoya Math. J. **113**, 14–36

Kollár, J. (1990): Minimal models of algebraic threefolds: Mori's Program. Asterisque **177–178**, 303–326

Kollár, J. (1991): Flips, flops, minimal models etc. J. Diff. Geom. (to appear)

Kollár, J., Mori, S. (1991): Soon to be written up

Kulikov, V. (1977): Degenerations of K3 surfaces and Enriques surfaces. Math. USSR Izv. **11**, 957–989

Mori, S. (1985): Minimal models for semistable degenerations of surfaces. Lectures at Columbia University. Unpublished

Mori, S. (1987): Personal communication

Mori, S. (1988): Flip theorem and the existence of minimal models for 3-folds. Journal AMS **1**, 117–253

Mori, S. (1990): Birational Classification of algebraic threefolds. These Proceedings, p. 235

Reid, M. (1983): Minimal models of canonical threefolds, Algebraic Varieties and Analytic Varieties. Adv. Stud. Pure Math. vol. 1 (Iitaka, S., ed.). Kinokuniya and North-Holland, pp. 131–180

Reid, M. (1987): Young person's guide to canonical singularities. Algebraic Geometry Bowdoin 1985, Proc. Symp. Pure Math. vol. 46, 345–416

Shokurov, V. (1985a): The nonvanishing theorem. Izv. Akad. Nauk SSSR Ser. Mat. **49**, 635–651

Shokurov, V. (1985b): Letter to M. Reid

Tsunoda, S. (1987): Degenerations of surfaces. Algebraic Geometry, Sendai, Adv. Stud. Pure Math. vol. 10 (Oda, T., ed.). Kinokuniya and North-Holland, pp. 755–764

Linear Series on Algebraic Varieties

*Robert K. Lazarsfeld**

Department of Mathematics, University of California, Los Angeles, CA 90024, USA

In recent years, ideas and methods from the theory of vector bundles have been applied to study some classical sorts of questions concerning linear series on curves, surfaces and other algebraic varieties. Our purpose here is to survey some of the problems to which these and related techniques have been applied.

Many of the results we describe in one way or another involve the equations defining projective varieties. There are several motivations for studying questions along these lines. The first is historical—the equations defining varieties have been of interest to geometers at least since Noether, and it is natural to try to clarify and extend classical results as much as possible. Secondly, there is a "practical" motivation. It has become common for geometers to use computers to analyze explicit varieties, and for the most part the only way to describe a variety to a computer is by giving its equations. Finally, it seems likely that vector bundle methods will play an increasingly important role in the study of linear series, and the classical questions we consider here have proven to be good testing grounds for these techniques.

Limitations of space preclude more than a cursory discussion of methods of proof. We refer to [L3] for an overview of some of these. However we have attempted to survey some of the many interesting open problems in this area. We work throughout over the complex numbers $\mathbb{C}$.

§ 1. Castelnuovo-Mumford Regularity

Let $X \subset \mathbb{P}^r$ be an irreducible variety of dimension n, and denote by $\mathscr{I}_X$ the ideal sheaf of X in $\mathbb{P}^r$. Recall that one says that X is *k-regular* if $H^i(\mathbb{P}^r, \mathscr{I}_X(k - i)) = 0$ for $i \geq 1$ ([M2], [M3]); the *regularity* reg(X) of $X \subset \mathbb{P}^r$ is the least such k. The interest in this concept stems partly from the fact that reg(X) governs the complexity of computing the syzygies and other invariants of X ([BS1, BS2]). For example, a theorem of Mumford (cf. [EG]) states that X is k-regular if and only if for every $p \geq 0$, the minimal generators of the kth module of syzygies of the homogeneous

* Partially supported by N.S.F. grant DMS 89-02551.

ideal I_X occur in degrees $\leq k + p$. This has led to a substantial body of work aimed at bounding the regularity of X in terms of its invariants or (more recently) the degrees of its defining equations. In each case there is a fascinating tension between the situation for arbitrary schemes $X \subset \mathbb{P}^r$ — for which results and examples show that the regularity can be very large — and the geometric situation in which X is smooth or at least reduced, where one expects much stronger statements.

We start with the *Castelnuovo problem* of bounding the regularity of X in terms of its invariants. For an arbitrary scheme $X \subset \mathbb{P}^r$, Gotzmann [Gotz, G3] has given a best-possible bound in terms of the Hilbert polynomial of X. For instance, if $X \subset \mathbb{P}^r$ is a curve of degree d and arithmetic genus g, his result states that $\mathrm{reg}(X) \leq \binom{d-1}{2} + d - g$. The question now arises if one can do better assuming X is smooth or reduced.

In practice the main difficulty is to control $H^1(\mathbb{P}^r, \mathscr{I}_X(k))$, which is the classical question of whether hypersurfaces of degree k trace out a complete linear series on X. A theorem of Castelnuovo asserts that when $X \subset \mathbb{P}^r$ ($r \geq 3$) is a smooth curve of degree d, then $H^1(\mathbb{P}^r, \mathscr{I}_X(k)) = 0$ for $k \geq d - 2$, and consequently X is $(d-1)$-regular. This is optimal for curves in $\mathbb{P}^3$, but not in general. In fact, Castelnuovo's result was completed in [GLP], where vector bundle techniques are used to show that if $X \subset \mathbb{P}^r$ is a reduced irreducible non-degenerate curve of degree d, then X is $(d + 2 - r)$-regular. Furthermore the borderline examples are classified: roughly speaking, X fails to be $(d + 1 - r)$-regular if and only if it has a $(d + 1 + r)$-secant line. This possibility of giving geometric explanations for extremal algebraic behavior is one of the interesting themes in this circle of ideas; another example appears in [G3].

These results for curves have led various people to hope that if $X \subset \mathbb{P}^r$ is a smooth non-degenerate complex projective variety of degree d and dimension n, then X is $(d + n + 1 - r)$-regular [EG]. At least when $r \geq 2n + 1$, this would be the best possible linear inequality. It is proved for surfaces in [P] and [L1], and for 3-folds in [R] when $r \geq 9$. In higher dimensions, the situation is much less clear. There is a rather weak bound of Mumford's (cf. [BEL, (2.1)]), but new ideas are apparently needed to treat the problem in general. It would already be very interesting to show that X is d-regular. There is also no clear understanding of what to expect if X is singular (but reduced and irreducible). For an arbitrary scheme $X \subset \mathbb{P}^r$, Ravi [Rav] asks whether one has the inequality $\mathrm{reg}(X_{\mathrm{red}}) \leq \mathrm{reg}(X)$; this may be too much to wish for, but it would be nice to settle the matter.

As we have noted, attempts to bound the regularity of a projective variety $X \subset \mathbb{P}^r$ are motivated in part by the desire to bound the complexity of computing the syzygies and related invariants of X. Bayer has remarked that it is then natural to ask for bounds in terms of the degrees of the defining equations of X – which are usually primary data when one sets out to make an actual computation – rather than the degree of X, which may not be so obvious in a given situation. For an arbitrary scheme, this regularity can again be horrendously large: there are examples due to Mayr-Meyer-Bayer-Stillman [BS1] of schemes $X \subset \mathbb{P}^r$ defined by hypersurfaces of degree d with $\mathrm{reg}(X) \geq (d - 2)^{2^{(r/10)}}$. However when X is smooth, the regularity grows much more slowly with d and r. In fact, it comes as a pleasant surprise to note that elementary arguments using the Kodaira vanishing theorem lead to the optimal bound in this case:

Theorem 1.1 [BEL]. *Assume that $X \subset \mathbb{P}^r$ is a smooth variety of dimension n and codimension e defined by hypersurfaces of degrees $d_1 \geq d_2 \geq \cdots \geq d_m$. Then X is $(d_1 + \cdots + d_e - e + 1)$-regular, and X fails to be $(d_1 + \cdots + d_e - e)$-regular if and only if it is the complete intersection of hypersurfaces of degrees $d_1, d_2, \ldots, d_e$.*

Note that only the degrees of the first $e = \mathrm{codim}(X, \mathbb{P}^r)$ defining equations come into play here. It would be very interesting to know to what extent one can allow X to be singular.

§ 2. Syzygies of Algebraic Varieties

It is useful to think of the results described in the previous section as being *extrinsic* in nature, in the sense that they refer to a given projective embedding $X \subset \mathbb{P}^r$, with the main difficulties coming from the fact that X might not be linearly normal. But one can also take a more *intrinsic* point of view, where one deals with the embedding $X \subset \mathbb{P} = \mathbb{P}(H^0(L))$ defined by the complete linear system associated to a very ample line bundle L on X. Here it is usually not so hard to compute $\mathrm{reg}(X)$, so one can ask for more precise information. A very interesting line of inquiry – inaugurated by Green in [G1] – is to study the syzygies of X.

We start with some notation. Let L be a very ample line bundle on a projective variety X, defining an embedding $X \subset \mathbb{P} = \mathbb{P}(H^0(L))$. Denote by $S = \mathrm{Sym}^\bullet H^0(L)$ the homogeneous coordinate ring of the projective space $\mathbb{P}$, and consider the graded S-module $R = R(L) = \oplus H^0(X, L^d)$. Let $E_\bullet = E_\bullet(L)$ be a minimal graded free resolution of R:

$$\cdots \longrightarrow \oplus S(-a_{2,j}) \longrightarrow \oplus S(-a_{1,j}) \longrightarrow \begin{array}{c} S \\ \oplus \\ \oplus S(-a_{0,j}) \end{array} \longrightarrow R \longrightarrow 0.$$

$$\begin{array}{ccc} \| & \| & \| \\ E_2 & E_1 & E_0 \end{array}$$

We have indicated here the fact that R has a canonical generator in degree zero. It is easy to see that all $a_{0,j} \geq 2$ and all $a_{i,j} \geq i + 1$ when $i \geq 1$. To extend classical results on normal generation and presentation we ask when the first few E_i are as simple as possible:

Definition 2.1. The line bundle L saisfies *property (N_p)* if

$$E_0 = S \quad \text{when } p \geq 0$$

and

$$E_i = \oplus S(-i-1) \text{ [i.e. all } a_{i,j} = i+1] \qquad \text{for} \qquad 1 \leq i \leq p.$$

Note that if $E_0 = S$, then $E.$ determines a resolution of the homogeneous ideal $I = I_{X/\mathbb{P}}$ of X in $\mathbb{P}(H^0(L))$. Thus the definition may be summarized very concretely as follows:

L satisfies (N_0) $\Leftrightarrow$ X embeds in $\mathbb{P}(H^0(L))$ as a projectively normal variety;

L satisfies (N_1) $\Leftrightarrow$ (N_0) holds for L, and the homogeneous ideal I_X of X is generated by quadrics;

L satisfies (N_2) $\Leftrightarrow$ (N_0) and (N_1) hold for L, and the module of syzygies among quadratic generators $Q_i \in I$ is spanned by relations of the form

$$\sum L_i Q_i = 0,$$

where the L_i are *linear* polynomials;

and so on. Properties (N_0) and (N_1) are what Mumford [M3] calls respectively *normal generation* and *normal presentation*.

A classical theorem of Castelnuovo, Mattuck and Mumford states that if X is a smooth curve of genus g, and if $\deg(L) \geq 2g + 1$, then L is normally generated. Fujita and St. Donat proved that if $\deg(L) \geq 2g + 2$, then L satisfies (N_1). These results were extended and clarified by Green [G1], who showed that if $\deg(L) \geq 2g + 1 + p$, then L satisfies (N_p). Green's theorem was recovered as a consequence of an analogous statement of finite sets in [GL3], where it was also proved that if $\deg(L) = 2g + p$, then L fails to satisfy (N_p) if and only if either X is hyperelliptic or Φ_L embeds X with a $(p + 2)$-secant p-plane.

The result just quoted gives a first indication of the fact that (at least conjecturally) the syzygies of a smooth curve X are intimately connected with its geometry. The crucial invariant here is the *Clifford index* $\mathrm{Cliff}(X)$ of X. Referring for instance to [GL1] for the precise definition, suffice it to say that $\mathrm{Cliff}(X)$ is a non-negative integer which measures from the point of view of special divisors how general X is in moduli. For instance $\mathrm{Cliff}(X) = 0$ if and only if X is hyperelliptic, and $\mathrm{Cliff}(X) = 1$ if and only if X is trigonal or a smooth plane quintic. One has the inequality $0 \leq \mathrm{Cliff}(X) \leq [(g - 1)/2]$, and $\mathrm{Cliff}(X) = [(g - 1)/2]$ when X is a general curve of genus g.

The hope is that uniform results on the syzygies of a curve X can be strengthened to take into account its Clifford index. The clearest expression of this philosophy which has actually been proved to date is a theorem of [GL1] to the effect that if L is a very ample line bundle on X with $\deg(L) \geq 2g + 1 - 2 \cdot h^1(L) - \mathrm{Cliff}(X)$, then L is normally generated. This includes the result of Castelnuovo et al. mentioned above, and for example taking L to be the canonical bundle Ω it yields Noether's celebrated theorem that Ω is normally generated unless X is hyperelliptic. One conjectures [GL1] that if L is a very ample line bundle with $\deg(L) \geq 2g + 1 + p - 2 \cdot h^1(L) - \mathrm{Cliff}(X)$, then L satisfies (N_p) provided that Φ_L does not embed X with a $(p + 2)$-secant p-plane. This would imply all the known results on syzygies of curves. But its most interesting consequence would be a beautiful conjecture of Green's on the syzygies of canonical curves:

Conjecture 2.2 [G1]. $\mathrm{Cliff}(X)$ *is the least integer p for which* (N_p) *fails for the canonical bundle* Ω.

Green's conjecture has already sparked a considerable amount of work. For instance, drawing on some ideas of Chang and Ran as well as some computer computations by Bayer and Stillman, Ein [E] proves (2.2) on a general curve for $p \leq 3$. The most striking progress to date is due to Voisin [V1] and Schreyer [S], who verify the first non-classical case $p = 2$. Schreyer also gives examples to show that (2.2) fails in characteristic 2. A possible approach to this conjecture is suggested in [PR]. In a somewhat different direction, one has

Conjecture 2.3 ([GL1], Conj. 3.7). *One can read off the "gonality" of a curve X from the grading of the resolution $E_. = E_.(L)$ of any one line bundle of sufficiently large degree.*

In fact, one expects the gonality to be determined by the tail end of $E_.$. One hopes that (2.3) should be easier to establish than (2.2).

Results on normal generation and presentation of curves have traditionally gone hand in hand with analogous results for abelian varieties. Suppose then that X is an abelian variety of dimension n, and let L be an ample line bundle on X. A theorem of Mumford [M1, M3], Koizumi [K] and Sekiguchi [Sek] states that $L^{\otimes k}$ is normally generated provided that $k \geq 3$, and Mumford [M3] proved that X is scheme-theoretically cut out by quadrics under the embedding defined by $L^{\otimes k}$ provided that $k \geq 4$. By analogy with the situation on curves, it is then natural to conjecture [L3] that if $k \geq p + 3$, then $L^{\otimes k}$ satisfies (N_p). Kempf [Kmf] has recently made considerable progress on this conjecture. Specifically, he proves if $p \geq 1$ then (N_p) holds for $L^{\otimes k}$ so long as $k \geq 2p + 2$. It would be wonderful to prove the conjecture in general.

Finally, what can one say in higher dimensions? Green [G2] proved that on an arbitrary smooth variety X of dimension n, any sufficiently positive line bundle L satisfies (N_p). But naturally one would like an explicit result. Mukai observed that the known theorems deal with embeddings defined by bundles of the type $K_X \otimes P$, where P is an explicit multiple of a suitably positive bundle. He suggested that in general one should aim for statements having this shape. In this direction one has

Theorem 2.4 [EL]. *Let A be a very ample line bundle on X, and let B be a numerically effective bundle on X. If $k \geq n + 1 + p$, then $K_X \otimes A^{\otimes k} \otimes B$ satisfies (N_p).*

So for example, $K_X \otimes A^{\otimes n+1}$ is normally generated (which has a quick proof: cf. [BEL], [AS] and [ABS]), and in the embedding defined by $K_X \otimes A^{\otimes n+2}$ the homogeneous ideal I_X is generated by quadrics. It would be very interesting to prove analogous statements assuming only that A is ample. Butler [But] shows that if X is a projective bundle over a curve, and if A is an ample line bundle on X, then $K_X \otimes A^{\otimes 2n+1}$ is normally generated. A rash optimist might be tempted to speculate that $K_X \otimes A^{\otimes n+2+p}$ satisfies (N_p) whenever A is an ample line bundle on an arbitrary smooth variety X. However even when $n = 3$ it is unknown whether $K_X \otimes A^{\otimes n+2}$ is very ample (this is a celebrated conjecture of Fujita), so that the moment it seems premature to hope for statements of this type for syzygies. However as Mukai points out, it might be reasonable to ask for sharp theorems for surfaces. When $X = \mathbb{P}^n$, results on syzygies and related issues amount to statements about multiplicative properties of subspaces of the polynomial ring. Green [G2] has obtained some general theorems along these lines, for which he has given interesting applications to the Hodge theory of hypersurfaces. We refer the reader to [G4] for a survey.

§3. Linear Series and Vector Bundles on Surfaces

As we remarked in the Introduction, most of the results described above are proved using vector bundle techniques. For questions involving regularity and syzygies, the arguments are largely cohomological in nature. However on surfaces, bundles have been used in a more geometrical fashion to study linear series. In this final section we survey two such applications. Henceforth X denotes a smooth complex projective surface.

Consider a curve C on X, and a line bundle A on C which is generated by its global sections. Thinking of A as a coherent sheaf on X, there is a natural surjective evaluation map $e_{C,A}: H^0(A) \otimes_{\mathbb{C}} \mathcal{O}_X \to A$. Set $F = F_{C,A} = \ker e_{C,A}$; thus F is a vector bundle of rank $h^0(C, A)$ on X. Philosophically, F encodes information about the pair (C, A) into a geometric object that lives globally on X. When X is a K3 surface, an analysis of these bundles together with a theorem of Mukai's [Mk1, Mk2] on the smoothness of the moduli space of simple bundles on X leads to the following

Theorem 3.1 [L2]. *Let X be a K3 surface, and let $C_0 \subset X$ be a smooth curve having the property that every curve in the linear series $|C_0|$ is reduced and irreducible. Then the general member $C \in |C_0|$ behaves generically in the sense of Brill-Noether theory, i.e. the varieties $W_d^r(C)$ of special linear series on C have the postulated dimensions and $W_d^r(C)$ is smooth away from $W_d^{r+1}(C)$.*

We refer for instance to [ACGH] for a fuller discussion of the Brill-Noether-Petri package. The hypothesis is satisfied for instance when $\text{Pic}(X) = \mathbb{Z} \cdot [C_0]$, and the theorem then leads to a very quick proof of an important result of Gieseker's [Gies] concerning the behavior of the varieties $W_d^r(C)$ on a general curve of genus g. The bundles $F_{C,A}$ were studied on an arbitrary surface by Tjurin [T]. They also lead to the proof [GL2] of a conjecture of Harris, Mumford and Green to the effect that all curves in a given linear series on a K3 surface have the same Clifford index. Related results appear in [DM] and [Par]. Reid [Reid] has some intriguing ideas about how bundles of this type might be relevant to the study of canonical surfaces. More recently, Voisin [V2] has used an infinitesimal analogue of the $F_{C,A}$ to prove a very beautiful result about the Wahl map $\gamma_\Omega: \bigwedge^2 H^0(C, \Omega) \to H^0(C, \Omega^3)$ on a smooth curve C of genus g.

Finally, no survey of vector bundles and linear series would be complete without mentioning the very interesting work of Igor Reider [Rdr] on linear series on a smooth projective surface X. Suppose that L is a line bundle on X such that $K_X \otimes L$ fails to be globally generated at a point $x \in X$. Then $H^1(X, \mathcal{I}_x \otimes K_X \otimes L) \neq 0$, and so by Grothendieck duality and a well-known construction of Serre's one can construct a non-split extension of the form $0 \to \mathcal{O}_X \to E \to L \to 0$. Reider's idea — which has its antecedents in a proof by Mumford of the Ramanujam vanishing theorem — is that suitable numerical conditions on L will force E to be Bogomolov-unstable. Analyzing this instability geometrically then leads to the following

Theorem 3.2 [Rdr]. *Assume that L is a nef line bundle on X such that $K_X \otimes L$ is not globally generated. If $c_1(L^2) \geq 5$, then there is an effective divisor $D \subset X$ such that either $c_1(L) \cdot D = 0$ and $D^2 = -1$, or $c_1(L) \cdot D = 1$ and $D^2 = 0$.*

Reider proves an analogous statement for the failure of $K_X \otimes L$ to be very ample. This theorem has a surprising range of applications, among them a quick proof of Bombieri's results on pluricanonical maps of surfaces. Reider's theorem has attracted a lot of attention, and it has been extended in several directions. We refer to [Kot] and the conference proceedings [Som] for a sampling of some of the work in this direction.

References

[ABS]   M. Andreatta, E. Ballico, A. Sommese: On the projective normality of the adjunction bundles II. Preprint

[AS]   M. Andreatta, A. Sommese: On the projective normality of the adjunction bundles. Preprint

[ACGH]   M. Arbarello, M. Cornalba, P. Griffiths, J. Harris: Geometry of algebraic curves. Springer, Berlin Heidelberg New York 1985

[BS1]   D. Bayer, M. Stillman, On the complexity of computing syzygies. In: L. Robbianno (ed) Computational aspects of commutative algebra. Academic Press, 1989, pp. 1–13

[BS2]   D. Bayer, M. Stillman: A criterion for detecting m-regularity. Inv. math. **87** (1987) 1–11

[BEL]   A. Bertram, L. Ein, R. Lazarsfeld: Vanishing theorems, a theorem of Severi, and the equations defining projective varieties. To appear in J. of the AMS

[But]   D. Butler: Tensor products of global sections for vector bundles over a curve with applications to linear series. To appear

[DM]   R. Donagi, D. Morrison: Linear systems on K3 sections. J. Diff. Geom. **29** (1988) 49–64

[E]   L. Ein: Some remarks on the syzygies of general canonical curves, J. Diff. Geom. **26** (1987) 361–366

[EL]   L. Ein, R. Lazarsfeld: A theorem on the syzygies of smooth projective varieties of arbitrary dimension. To appear

[EG]   D. Eisenbud, S. Goto: Linear free resolutions and minimal multiplicity. J. Alg. **88** (1984) 89–133

[Gies]   D. Gieseker: Stable curves and special divisors: Petri's conjecture. Inv. math. **66** (1982) 251–275

[Gotz]   G. Gotzmann: Eine Bedingung für die Flachheit und das Hilbertpolynom eines graduierten Ring. Math. Z. **158** (1978) 61–70

[G1, G2]   M. Green, Koszul cohomology and the geometry of projective varieties I and II. J. Diff. Geom. **19** (1984) 125–171 and **20** (1984) 279–289

[G3]   M. Green: Restrictions of linear series to hyperplanes, and some results of Macau-
         lay and Gotzmann. In: E. Ballico, C. Ciliberto (eds), Algebraic curves and projec-
         tive varieties (Lect. Notes in Mathematics, vol. 1389.) Springer, Berlin Heidelberg
         New York 1989, pp. 76–86

[G4]   M. Green: Koszul cohomology and geometry. In: M. Cornalba et al. (eds) Lectures
         on riemann surfaces. World Scientific Press, 1989, pp. 177–200

[GL1]   M. Green, R. Lazarsfeld: On the projective normality of complete linear series on
         an algebraic curve. Inv. math. **83** (1986) 73–90

[GL2]   M. Green, R. Lazarsfeld: Special Divisors on curves on a K3 surface. Inv. math.
         89 (1987) 357–370

[GL3]   M. Green, R. Lazarsfeld: Some results on the syzygies of finite sets and algebraic
         curves. Comp. Math. **67** (1988) 301–314

[GLP]   L. Gruson, R. Lazarsfeld, C. Peskine: On a theorem of Castelnuovo and the
         equations defining space curves. Inv. math. **72** (1983) 491–506

[Kmf]   G. Kempf: Projective coordinate rings of abelian varieties. In: J.I. Igusa (ed.),
         Algebraic analysis geometry and number theory. Johns Hopkins Press 1989,
         pp. 423–438

[K]   S. Koizumi: Theta relations and projective normality of abelian varieties. Am. J.
         Math. **98** (1976) 865–889

[Kot]   D. Kotschick: Stable and unstable bundles on algebraic surfaces. To appear

[L1]   R. Lazarsfeld: A sharp Castelnuovo bound for smooth surfaces. Duke Math. J. **55**
         (1987) 423–238

[L2]   R. Lazarsfeld: Brill-Noether-Petri without degenerations. J. Diff. Geom. **23** (1986)
         299–307

[L3]   R. Lazarsfeld: A sampling of vector bundle techniques in the study of linear series.
         In: M. Cornalba et al. (eds) Lectures on Riemann surfaces. World Scientific Press,
         1989, pp. 500–559

[Mk1]   S. Mukai: Symplectic structure of the moduli space of sheaves on an abelian or
         K3 surface. Inv. math. **77** (1984) 101–116

[Mk2]   S. Mukai, On the moduli space of bundles on K3 surfaces, I. In: Vector bundles
         on algebraic varieties. Oxford Univ. Press, 1987, pp. 341–413

[M1]   D. Mumford: On the equations defining abelian varieties. Inv. math. **1** (1966)
         287–354

[M2]   D. Mumford: Lectures on curves on an algebraic surface. Ann. Math. Stud, vol.
         59 (1966)

[M3]   D. Mumford: Varieties defined by quadratic equations. Corso CIME 1969. In:
         Questions on algebraic varieties. Rome 1970, pp. 30–100

[P]   H. Pinkham: A Castelnuovo bound for smooth surfaces. Inv. math. **83** (1906)
         321–332

[Par]   G. Pareschi: Exceptional linear systems on curves on del Pezzo surfaces. To appear

[PR]   K. Paranjape, S. Ramanan: On the canonical ring of a curve. In: Algebraic geome-
         try and commutative algebra. Kinokuniya, 1988

[R]   Z. Ran: Local differential geometry and generic projections of threefolds. J. Diff.
         Geom. To appear

[Rav]   M.S. Ravi: Regularity of ideals and their radicals. Manuscr. Math. **68** (1990) 77–87

[Reid]   M. Reid: Quadrics through a canonical surface. In: A. Sommese et al. (eds)
         Algebraic geometry. (Lect. Notes in Mathematics, vol. 1417.) Springer, Berlin
         Heidelberg New York 1990, pp. 191–213

[Rdr]   I. Reider: Vector bundles of rank 2 and linear systems on algebraic surfaces. Ann.
         Math. **127** (1988) 309–316

[S]   F. Schreyer: A standard basis approach to the syzygies of canonical curves. To
         appear

[Sek]   T. Sekiguchi: On the normal generation by a line bundle on an abelian variety. Proc. Jap. Acad. **54** (1978) 185–188

[Som]   A. Sommese et al. (eds) Algebraic geometry. Proceedings, L'Aquila 1988. (Lecture Notes in Mathematics, vol. 1417.) Springer, Berlin Heidelberg New York 1990

[T]   A. Tyurin: Cycles, curves and vector bundles on an algebraic surface. Duke Math. J. **55** (1987)

[V1]   C. Voisin: Courbes tetragonales et cohomologie de Koszul. J. Reine Angew. Math. **387** (1988) 111–121

[V2]   C. Voisin: Sur l'application de Wahl des courbes satisfaisant la condition de Brill-Noether-Petri. To appear

Mixed Hodge Modules and Applications

Morihiko Saito

Research Institute for Mathematical Sciences, Kyoto University, Kyoto 606, Japan

1. Mixed Hodge Module

Since the theory of mixed perverse sheaves was established in char $p > 0$ by Beilinson-Bernstein-Deligne-Gabber [3], it has been conjectured that there would exist objects in char 0 corresponding philosophically to mixed perverse sheaves. This conjecture was solved recently by introducing the notion of *mixed Hodge Modules* [25–28].

For a complex algebraic variety X, we have an abelian category $\mathrm{MHM}(X)$ of mixed Hodge Modules on X with a forgetful functor rat : $\mathrm{MHM}(X) \to \mathrm{Perv}(\mathbb{Q}_X)$ which assigns the underlying perverse sheaf over $\mathbb{Q}$. By [2, 3], rat is naturally extended to a morphism of the bounded derived category $D^b \mathrm{MHM}(X)$ to $D_c^b(\mathbb{Q}_X)$, which is also denoted by rat. They satisfy the formalism of mixed sheaves, i.e.

Theorem 1.1. *The categories $D^b \mathrm{MHM}(X)$ are stable by the standard functors like f_*, $f_!$, f^*, $f^!$, ψ_g, φ_g, $\mathbb{D}$, $\boxtimes$, $\otimes$, $\mathscr{H}om$, and these functors are compatible with the corresponding functors on the underlying complexes of $\mathbb{Q}$-vector spaces via the functor* rat.

Proposition 1.2. *Every mixed Hodge Module $\mathscr{M}$ has a functorial increasing filtration W in $\mathrm{MHM}(X)$, called the weight filtration of $\mathscr{M}$, such that $\mathscr{M} \to \mathrm{Gr}_k^W \mathscr{M}$ is an exact functor and $\mathrm{Gr}_k^W \mathscr{M}$ is semisimple.*

If X is smooth, $\mathrm{MHM}(X)$ is a full subcategory of the category of (M, F, K, W) where M is a holonomic $\mathscr{D}_X$-Module with a good filtration F [14] and with rational structure given by an isomorphism $\alpha : \mathrm{DR}(M) \simeq \mathbb{C} \otimes K$ in $\mathrm{Perv}(\mathbb{C}_X)$ for a perverse sheaf K defined over $\mathbb{Q}$ [3], and W is a pair of filtrations on M and K compatible with α and gives the weight filtration in Proposition 1.2. Here DR is the de Rham functor (shifted to the left by $\dim X$), and $\mathrm{DR}(M) \in \mathrm{Perv}(\mathbb{C}_X)$ by [14]. (We assume that an analytic holonomic $\mathscr{D}$-Module on a smooth algebraic variety has an algebraic stratification. We can use also algebraic (filtered) $\mathscr{D}$-Modules [5] by GAGA and the extendability of mixed Hodge Modules.) For $X = \mathrm{pt}$, we have

Theorem 1.3. $\mathrm{MHM}(\mathrm{pt})$ *is the category of polarizable mixed $\mathbb{Q}$-Hodge structures* [8].

Proceedings of the International Congress
of Mathematicians, Kyoto, Japan, 1990
© The Mathematical Society of Japan, 1991

Here a mixed Hodge structure is called polarizable if its graded pieces are. This result can be generalized to the case of admissible variation of mixed Hodge structure in the sense of Steenbrink-Zucker [41] (in the one-dimensional case) and Kashiwara [16], cf. Theorem 1.5.

We say that a mixed Hodge Module $\mathcal{M}$ is *pure of weight n*, if $\mathrm{Gr}_k^W \mathcal{M} = 0$ for $k \neq n$. A pure Hodge Module is semi-simple, and a mixed Hodge Module is obtained by successive extensions of pure Hodge Modules. We say that a pure Hodge Module $\mathcal{M}$ has *strict support* if its support is irreducible and $\mathcal{M}$ has no sub or quotient with smaller support. By semisimplicity, a pure Hodge Module is a direct sum of pure Hodge Modules with different strict supports, called the *decomposition by strict support* (which is unique). A pure Hodge Module is also called a *polarizable Hodge Module* in [26], and we denote by $\mathrm{MH}_Z(X,n)^p$ the category of pure Hodge Modules of weight n with strict support Z. Then $\mathrm{MH}_Z(X,n)^p$ depends only on Z, and $\mathcal{M} \in \mathrm{MH}_Z(X,n)^p$ is generically a polarizable variation of Hodge structure in the sense of Griffiths [13]. An important fact is that its converse is true: any polarizable variation of Hodge structure can be extended uniquely to a pure Hodge Module, i.e.

Theorem 1.4. *We have an equivalence of categories*

$$\mathrm{MH}_Z(X,n)^p \simeq \mathrm{VHS}_{\mathrm{gen}}(Z, n - \dim Z)^p.$$

Here the right hand side is the category of polarizable variations of Hodge structures of weight $n - \dim Z$ defined on (nonempty) smooth subvarieties of Z. This result may be viewed as analogue of Brylinski's conjecture [6], cf. (5). Actually, the notion of polarizable Hodge Module is weaker than that of pure Hodge Module. But they coincide by Theorem 1.5 below, because a polarizable variation of Hodge structure is an admissible variation of Hodge structure by Schmid [37], and pure Hodge Modules are stable by intermediate direct images in the sense of [3] by Theorem 1.1 and the estimate of weight for direct images, cf. [27, 2.26].

Theorem 1.5. *Admissible variations of Hodge structures are mixed Hodge Modules,* cf. [27, 3.27].

Since the condition for mixed Hodge Modules is Zariski local, we can construct mixed Hodge Modules by induction on dimension, using Theorem 1.5 and the following:

Proposition 1.6. *Let g be a function on an algebraic variety X. Put $Y = g^{-1}(0)$, $U = X \setminus Y$. Let $\mathrm{MHM}(U,Y)_{\mathrm{ex}}$ be the category whose objects are pairs of mixed Hodge Modules $\mathcal{M}'$ on U and $\mathcal{M}''$ on Y endowed with morphisms $u : \psi_{g,1}\mathcal{M}' \to \mathcal{M}''$, $v : \mathcal{M}'' \to \psi_{g,1}\mathcal{M}'(-1)$ such that $vu = N$, where $\psi_{g,1}$ is the unipotent monodromy part of ψ_g, and N is the logarithm of the unipotent part of the monodromy, tensored by $(2\pi i)^{-1}\mathbb{Z}$. Then we have an equivalence of categories*

$$\mathrm{MHM}(X) \simeq \mathrm{MHM}(U,Y)_{\mathrm{ex}}, \quad \text{cf. } [27, 2.28].$$

We have also a generalization of the Kodaira vanishing:

Theorem 1.7. *Let Z be a projective variety with an ample line bundle L, and $\mathscr{M} \in \mathrm{MHM}(Z)$. Let X be a smooth variety containing Z so that $\mathscr{M}$ is represented by $(M, F, K, W) \in \mathrm{MHM}(X)$. Then $\mathrm{Gr}_p^F \mathrm{DR}\, \mathscr{M} := \mathrm{Gr}_p^F \mathrm{DR}\, M \in D(\mathcal{O}_Z)$ depends only on $\mathscr{M}$, and we have*

$$H^i(Z, \mathrm{Gr}_p^F \mathrm{DR}\, \mathscr{M} \otimes L) = 0 \quad \text{for} \quad i > 0,$$
$$H^i(Z, \mathrm{Gr}_p^F \mathrm{DR}\, \mathscr{M} \otimes L^{-1}) = 0 \quad \text{for} \quad i < 0.$$

Convention 1.8. In this paper we use left $\mathscr{D}$-Modules. In [26, 27], right Modules are used, because they are more convenient for the theory of direct image and duality. For the correspondence between left and right *filtered* $\mathscr{D}$-Modules, we use $\otimes_{\mathcal{O}_X}(\Omega_X^{\dim X}, F)$ with $\mathrm{Gr}_i^F \Omega_X^{\dim X} = 0$ for $i \neq -\dim X$, see [28, 34].

2. Idea of Construction

2.1 One-Dimensional Case [37]

Let S be an open disc with coordinate t, and $S^* = S \setminus \{0\}$. Let $\mathbb{H}$ be a polarized variation of Hodge structure of weight n on S^* with L the underlying local system defined over $\mathbb{Q}$. For simplicity, we assume the monodromy is *unipotent*. We denote by $\mathcal{O}_{S^*}(L)$ the corresponding locally free $\mathcal{O}_{S^*}$-Module with integrable connection, and F the Hodge filtration on it. Let $\mathscr{L}$ be Deligne's canonical extension [9] of $\mathcal{O}_{S^*}(L)$. By Schmid's nilpotent orbit theorem [37], the filtration F on $\mathcal{O}_{S^*}(L)$ is naturally extended to a filtration F of $\mathscr{L}$ whose graduation is free. Let $\mathscr{L}(0) = \mathscr{L}/t\mathscr{L}$ the fiber of the vector bundle $\mathscr{L}$ at the origin. Let $\pi : \tilde{S}^* \to S^*$ be a universal covering, and $L_\infty = \Gamma(\tilde{S}^*, \pi^* L)$ the multivalued sections of L. Then we have a canonical isomorphism

$$\mathscr{L}(0) \simeq \mathbb{C} \otimes L_\infty. \tag{1}$$

See [9]. The left hand side of (1) is equipped with the induced filtration F, called the limit Hodge filtration, and the right hand side with rational structure. The theorem of Schmid [37] (conjectured by Deligne) asserts that they form a mixed Hodge structure, called the limiting mixed Hodge structure, where the weight filtration is given by the monodromy filtration shifted by n. Here the monodromy filtration W is a finite increasing filtration which is uniquely characterized by the properties: $NW_j \subset W_{j-2}$ and $N^i : \mathrm{Gr}_i^W \xrightarrow{\sim} \mathrm{Gr}_{-i}^W (i > 0)$, cf. [37, 11], and N is the logarithm of the unipotent part of the monodromy. The primitive part $P\,\mathrm{Gr}_{n+i}^W := \mathrm{Ker}\{N^{i+1} : \mathrm{Gr}_{n+i}^W \to \mathrm{Gr}_{n-i-2}^W\}$ is naturally polarized by the polarization induced by that of the variation of Hodge structure (combined with the isomorphism N^i). Since the above construction may be viewed as the definition of the nearby cycle functor ψ_t [10] for variations of Hodge structures, the result of Schmid can be interpreted as the stability theorem of polarized variations of Hodge structures under $P\,\mathrm{Gr}^W \psi_t$ in the one-dimensional case.

2.2 Geometric Case [38]

Assume there is a projective morphism $f : X \to S$ such that f is smooth over S^*, $f^{-1}(0)$ is a reduced divisor with normal crossings, and $L = R^n f_* \mathbb{Q}_X|_{S^*}$. Steenbrink [38] constructed a theory which gives a geometric proof of Schmid's theorem (see [39] for the nonreduced case). Although the notion of perverse sheaf and monodromy filtration on it was not yet known at that time, he constructed the weight filtration on the nearby cycle sheaf $\psi_f \mathbb{Q}_X$ [10], which represents the monodromy filtration in the abelian category of perverse sheaves (up to a shift of complex, cf. [3]). Moreover, he showed that the primitive part $P \operatorname{Gr}_i^W \psi_f \mathbb{Q}_X$ is a direct sum of constant sheaves supported on the intersections of $i + 1$ irreducible components of the singular fiber (up to Tate twist and shift of complex), i.e. the stability of constant sheaves by $P \operatorname{Gr}_i^W \psi_f$ in the reduced normal crossing case. Using E_2-degeneration of the monodromy weight spectral sequence, this stability implies Schmid's theorem: the stability of the variations of Hodge structures by $P \operatorname{Gr}_{n+i}^W \psi_t$, because the direct image of the monodromy filtration on $\psi_f \mathbb{Q}_X$ induces the monodromy filtration on $\psi_t L$ (cf. also [26, 35] for further information).

The above construction can be extended to the vanishing cycle sheaf $\varphi_f \mathbb{Q}_X$ [10]. Using these, we can show the decomposition theorem for the direct image of filtered $\mathscr{D}$-Module $\int_f (\mathscr{O}_X, F)$ which implies that of $\mathbb{R} f_* \mathbb{Q}_X$, cf. [24, II]. The proof uses the V-filtration on the direct image, which was inspired by Varchenko's theory of asymptotic Hodge filtration [42], see also [24, I]. (Note that the definition of the direct image of filtered $\mathscr{D}$-Modules in [24] differs from the standard one by a shift of complex and filtration.)

2.3 Generalization

The first step to generalize the above proof of the decomposition theorem is the use of the filtration of Kashiwara [17] and Malgrange [22]. Let g be a holomorphic function on X, and $i_g : X \to X \times \mathbb{C}$ the embedding by graph. For a filtered holonomic $\mathscr{D}_X$-Module (M, F) (i.e. M is holonomic and F is a good filtration), we define $(\tilde{M}, F) = \int_{i_g} (M, F)$ (i.e. $\tilde{M} = M[\partial_t]$, cf. [28, (1.3)] for F). By [15], $\tilde{M}$ has the filtration of Kashiwara and Malgrange, denoted by V. We assume M regular and quasi-unipotent (i.e. V is indexed by $\mathbb{Q}$, and $t\partial_t - \alpha$ is nilpotent on $\operatorname{Gr}_V^\alpha$ locally on X) and

$$
\begin{aligned}
t : F_p V^\alpha \tilde{M} &\xrightarrow{\sim} F_p V^{\alpha+1} \tilde{M} \quad \text{for} \quad \alpha > -1, \\
\partial_t : F_p \operatorname{Gr}_V^\alpha \tilde{M} &\xrightarrow{\sim} F_{p+1} \operatorname{Gr}_V^{\alpha-1} \tilde{M} \quad \text{for} \quad \alpha < 0.
\end{aligned}
\tag{2}
$$

We define

$$
\begin{aligned}
\psi_g(M, F) &= \oplus_{-1 < \alpha \leq 0} \operatorname{Gr}_V^\alpha (\tilde{M}, F), \\
\varphi_g(M, F) &= \oplus_{-1 < \alpha < 0} \operatorname{Gr}_V^\alpha (\tilde{M}, F) \oplus \operatorname{Gr}_V^{-1} (\tilde{M}, F[-1]).
\end{aligned}
\tag{3}
$$

(Same for $\psi_g M, \varphi_g M$, forgetting F.) Then we have canonical isomorphisms

$$
\operatorname{DR} \psi_g M = \psi_g \operatorname{DR} M[-1], \quad \operatorname{DR} \varphi_g M = \varphi_g \operatorname{DR} M[-1],
\tag{4}
$$

generalizing (1). This was obtained by Malgrange [22] in the case $M = \mathscr{O}_X$, and the general regular holonomic case is due to Kashiwara [17]. In the case when the Hodge filtrations F of $\psi_g M, \varphi_g M$ are good and (2) is satisfied, these isomorphisms were also constructed in [26, 3.4.12] generalizing the construction of Malgrange (see also [33, 2.4] for further information), and used in the proof of the main theorems in an essential way.

For a smooth variety X, we define $\mathrm{MH}(X, n)$, the category of Hodge Modules of weight n, to be the largest full subcategory of the category of filtered holonomic $\mathscr{D}$-Modules with rational structure as in 1, whose objects admit decomposition by strict support and are stable by: (i) restrictions to Zariski-open subsets, (ii) strict support decomposition, (iii) $\mathrm{Gr}^W \psi_g$, $\mathrm{Gr}^W \varphi_g$ for any function g, where W is the monodromy filtration shifted by $n - 1$ or n. Here (2) is always assumed when we take ψ_g, φ_g. We assume also that, if an object has point support, it is isomorphic to the direct image of Deligne's $\mathbb{Q}$-Hodge structure of weight n [8], cf. [26, 5.1.6]. The conditions are well-defined by induction on the dimension of the support. In fact, if $g^{-1}(0) \supset \mathrm{supp}\, M$, we have $\psi_g M = 0$, and $\varphi_g(M, F) = (M, F)$ is equivalent to $gF_p M \subset F_{p-1} M$ and follows from (2), cf. [26, 3.2.6]. So, in the stability by (iii), it is enough to consider the case when (M, F, K) has strict support and $g^{-1}(0) \not\supset \mathrm{supp}\, M$, using the stability by (ii). Moreover, in this case, it is enough to assume the stability by $\mathrm{Gr}^W \psi_g$ with condition (2) and the surjectivity of ∂_t in (2) for $\alpha = 0$, cf. [26, 5.1.15]. Note that the last conditions are equivalent to

$$F_p \tilde{M} = \sum_i \partial_t^i (V^{>-1} \tilde{M} \cap j_* j^{-1} F_{p-i} \tilde{M}) \tag{5}$$

where $j : X \times \mathbb{C}^* \to X \times \mathbb{C}$. This means that the Hodge filtration F is uniquely determined by its restriction to the complement of $g^{-1}(0)$ combined with the filtration V associated with g. The inverse functor of the equivalence of categories in Theorem 1.4 is given in this way, where we use also the regularity of the underlying $\mathscr{D}$-Modules [5, 18, 23]. (In the case of analytic $\mathscr{D}$-Modules, the regularity follows from the above inductive condition of Hodge Module, cf. [26, 5.1.18], but we have to assume it at infinity for algebraic $\mathscr{D}$-Modules. Note that the regularity [5, 18, 23] is used only for the uniqueness of the underlying $\mathscr{D}$-Modules of open direct images.) So Theorem 1.4 may be viewed as analogue of Brylinski conjecture [6], where Kashiwara-Kawai's order filtration is replaced by the filtration V. Here it is enough to take one g locally on X. But, for the stability by direct images, it is necessary to consider all g, because we have to deal with all f and the pull-backs of functions by f.

A polarization of a Hodge Module $\mathscr{M} = (M, F, K) \in \mathrm{MH}(X, n)$ is a duality $\mathbb{D}\mathscr{M} = \mathscr{M}(n)$, whose condition is defined inductively by the stability under $P\,\mathrm{Gr}^W \psi_g$, cf. [26, 5.2.10]. Here (n) is the Tate twist, which is defined essentially by the shift of filtration by n, cf. [27, (2.17.7)], and the dual $\mathbb{D}\mathscr{M} = (\mathbb{D}(M, F), \mathbb{D}K)$ is defined using the theory of dual of filtered induced $\mathscr{D}$-Modules and filtered differential complexes (e.g. its compatibility with de Rham functor), cf. [26, 2.4-5]. Note that $\mathbb{D}(M, F)$ is well-defined (e.g. $\mathbb{D}^2 = \mathrm{id}$ holds), because $\mathrm{Gr}^F M$ is Cohen-Macaulay over $\mathrm{Gr}^F \mathscr{D}_X$, cf. [26, 5.1.13], so that the filtration of the dual complex $\mathbb{D}(M, F)$ is strict and $\mathbb{D}(M, F)$ is filtered quasi-isomorphic to a filtered $\mathscr{D}_X$-Module. The category $\mathrm{MH}(X, n)^p$ is defined to be the category of polarizable Hodge Modules.

The merit of the inductive definition using nearby and vanishing cycles is that we can show by induction on dimension the stability of polarizable Hodge Modules by the cohomological direct image under a projective morphism and also the relative hard Lefschetz property. In fact, generalizing the proof of the decomposition theorem in 2.2, we can prove inductively the stability of the stability condition in the definition of polarizable Hodge Modules by the direct image, where the direct image of filtered $\mathscr{D}$-Modules is defined using a canonical resolution by filtered induced $\mathscr{D}$-Modules [26] (this definition coincides with that of Brylinski and Laumon using a factorization of a morphism f). A key point of the proof is the strictness of the Hodge filtration F on the direct image, which allows us to deal with filtered Modules instead of filtered complexes. When the image of the morphism is a point, we use Lefschetz pencil, and the assertion is reduced to the case of the projection of $\mathbb{P}^1$ to a point. Then we can apply Zucker's Hodge theory [44]. For the induced polarization on the direct image, we use also the duality of filtered $\mathscr{D}$-Modules for a proper morphism and its compatibility with Verdier duality [43], see [26, 2.5]. It should be noted that the definition of Hodge Module is obtained by axiomatizing some arguments which are repeatedly used in the proof of the stability by projective direct images. Combining with 1.4, we get a canonical pure Hodge structure on the intersection cohomology of a polarizable variation of Hodge structure. This Hodge structure coincides with that of Cattani-Kaplan-Schmid [7] and Kashiwara-Kawai [19, 20] where the singularity of the variation is a divisor with normal crossings.

2.4 Mixed Case

Once the pure objects are constructed, it is not difficult to define the mixed objects. We denote by $\mathrm{MHW}(X)^p$ the category of (M, F, K, W) as in 1 such that $\mathrm{Gr}_n^W(M, F, K) \in \mathrm{MH}(X, n)^p$. We can show by induction on dimension that $\mathrm{MHW}(X)^p$ is an abelian category such that every morphism is strictly compatible with the filtrations F and W in a strong sense, cf. [26, 5.1.14].

By definition, an object of $\mathrm{MHW}(X)^p$ is obtained by successive extension of pure Hodge Modules. Here arbitrary extensions are not allowed for mixed Hodge Modules, and we control the extensions by using mainly the stability by nearby cycle and vanishing cycle functors. The imposed conditions are natural generalizations of those of Steenbrink-Zucker [41]: the existence of the relative monodromy filtration and some condition on the filtrations F and W (i.e. the compatibility [26, 1.1.13] of the three filtrations F, W and V on $\tilde{M}$), cf. [27, 2.2-3]. We define the full subcategory of mixed Hodge Modules $\mathrm{MHM}(X)$ in $\mathrm{MHW}(X)^p$ using the stability by nearby cycle and vanishing cycle functors, smooth pullbacks and the existence of (and the stability by) the direct images by open embeddings whose complements are locally principal divisors, cf. [27, 4.2]. For singular X, $\mathrm{MHM}(X)$ is defined by using locally defined closed embeddings of X into smooth varieties. We can show the stability of the mixed Hodge Modules by the cohomological direct image under a projective morphism, generalizing the arguments of Steenbrink-Zucker [41] in the one-dimensional base space case. Using this, we can construct the functors in Theorem 1.1, where we use also Theorem 1.5 and Kashiwara's result [16] for the stability by external products $\boxtimes$, cf. [27]. Combining with 1.5, we get a canonical mixed Hodge structure on the cohomology of an admissible variation of mixed Hodge structure. We have also the decomposition theorem for the direct image of a pure Hodge Module by a proper morphism, using the semisimplicity of pure Hodge Modules, cf. [27, 4.5]. (See [30] for the last two functors in 1.1.)

3. Applications

3.1 Kollár's Conjecture [21, 32]

Let X be an irreducible algebraic variety, and $\mathbb{H}$ a polarizable variation of Hodge structure of weight $n - \dim X$ on a smooth open subvariety U of X. By Theorem 1.4, there exists uniquely a polarizable Hodge Module $\mathcal{M}$ on X extending $\mathbb{H}$. If X is a closed subvariety of smooth X', $\mathcal{M}$ is represented by $(M, F, K) \in \mathrm{MH}_X(X', n))^p$. Let $\mathscr{L}$ be the underlying $\mathcal{O}_U$-Module of $\mathbb{H}$ with Hodge filtration F. Put $q(\mathbb{H}) = \max\{p : F^p \mathscr{L} \neq 0\}$ and

$$S_X(\mathbb{H}) = \Omega_{X'}^{\dim X'} \otimes_{\mathcal{O}_{X'}} F_p M \quad \text{for} \quad p = \min\{p : F_p M \neq 0\}.$$

Then, $S_X(\mathbb{H})$ is an extension of $\Omega_U^{\dim U} \otimes_{\mathcal{O}_U} F^{q(\mathbb{H})} \mathscr{L}$, which is expressed using the filtration V, cf. (5). (It is more natural to use right $\mathscr{D}$-Modules, cf. 1.8.)

Let $f : X \to Y$ be a proper surjective morphism of irreducible algebraic varieties with relative dimension d. Then for a generic point $y \in Y$, the intersection cohomology $IH^i(X_y, \mathbb{H}|U_y)$ has a canonical pure Hodge structure, cf. 2.2, where $X_y = f^{-1}(y)$ and U_y is the smooth points of $U \cap X_y$. They form a variation of Hodge structure on a smooth open subvariety of Y, which we denote by $\mathbb{H}^i$. As a corollary of the decomposition theorem for the direct image of pure Hodge Modules, we get

$$\mathbf{R}f_* S_X(\mathbb{H}) \simeq \bigoplus_{q(\mathbb{H}^i)=q(\mathbb{H})+d} S_Y(\mathbb{H}^i)[-i]. \tag{6}$$

3.2 Illusie's Conjecture [35]

Let $f : X \to S$ be a projective morphism of a complex manifold to an open disc such that $X_0 := f^{-1}(0)$ is a divisor with normal crossings. Consider the (second) spectral sequence

$$E_2^{p,q} = H^p(X_0, \mathscr{H}^q \psi_f \mathbb{Q}_X) \Rightarrow H^{p+q}(X_0, \psi_f \mathbb{Q}_X)(\simeq H^{p+q}(X_t, \mathbb{Q})). \tag{7}$$

Here $X_t := f^{-1}(t)$ for $t \neq 0$ sufficiently near 0, and the last isomorphism depends on the lifting of t to the universal covering of $S \setminus \{0\}$. The spectral sequence degenerates at E_3, and the induced filtration coincides with the kernel filtration [41, 45]. This can be generalized to the theory of kernel spectral sequence, cf. [35].

3.3 Steenbrink's Conjecture [40, 29]

Let $f : (X, x) \to (\mathbb{C}, 0)$ be a germ of holomorphic function on a complex manifold. We define the *spectrum* $\mathrm{Sp}(f, x)$ using the Milnor monodromy and the mixed Hodge structure on the vanishing cohomology [40], (cf. also [39, 42]). Here we express the spectrum as a fractional Laurent polynomial (i.e., an element of $\mathbb{C}[t^{1/n}, t^{-1/n}]$ for some integer n), and multiply t on that of [40] to simplify the multiplication of spectra. Assume $\dim \mathrm{Sing} f = 1$, and let $\{Z_k\}$ be the irreducible components of $\mathrm{Sing} f$. Let $g : (X, x) \to (\mathbb{C}, 0)$ be a linear form (i.e., $dg(x) \neq 0$) such that $\{f = g = 0\}$ has isolated singularity. Let m_k be the mapping degree of the restriction of g to Z_k. Then, for $r \in \mathbb{N}$ sufficiently large, we have

$$\mathrm{Sp}(f + g^r, x) - \mathrm{Sp}(f, x) = \sum_{k,j} t^{\alpha_{k,j} + \beta_{k,j}/m_k r} (t - 1)/(t^{1/m_k r} - 1). \qquad (8)$$

Here $\alpha_{k,j}$ are the exponents of the spectrum of the restriction of f to $g^{-1}(t)(t \neq 0)$ near Z_k, and $\beta_{k,j} \in (0, 1]$ are the logarithm (divided by $-2\pi i$) of the eigenvalues of the monodromy along Z_k. The range of r is determined by the discriminant of (f, g).

3.4 Intersection Cohomology of Link [12]

Let X be a complex algebraic variety, and Z a closed subvariety of X. Then the *link* of Z in X is defined to be a level set of a suitable real analytic function on X which vanishes exactly on Z. We can define a canonical mixed Hodge structure on the intersection cohomology of the link, and get estimates of weight, e.g. $IH_c^i(L)$ has weights $\leq i$ for $i < \dim X - \dim Z$. Using these, we can show, for example, that a (real) torus cannot be a link of a subvariety of codimension > 1.

3.5 Thom-Sebastiani Type Theorems

Let f and g be holomorphic functions on X and Y respectively. For $x \in X$ and $y \in Y$, put $z = (x, y) \in X \times Y$. Then

$$\mathrm{Sp}(f + g, z) = \mathrm{Sp}(f, x) \, \mathrm{Sp}(g, y). \qquad (9)$$

(See [36, 42] for the isolated singularity case.) This follows from the Thom-Sebastiani type theorems for regular holonomic filtered $\mathscr{D}$-Modules (with Steenbrink) and for mixed Hodge Modules (with Deligne).

3.6 Algebraic Cycles [31]

If the Hodge conjecture is true, the mixed motives over $\mathbb{C}$ and with $\mathbb{Q}$-coefficients, whose existence is still conjectural [1], might be quite close to the mixed Hodge structures of *geometric origin* (using mixed Hodge Modules instead of mixed perverse sheaves in [3]). In fact, let $\mathbb{Q}^H$ be the trivial Hodge structure of rank 1 and type $(0,0)$, and $\mathbb{Q}_X^H = a_X^* \mathbb{Q}^H$, $\mathbf{R}\Gamma(X, \mathbb{Q}^H) = (a_X)_* \mathbb{Q}_X^H$ with $a_X : X \to \mathrm{pt}$. Then we can define a cycle class map

$$\mathrm{cl}^{\mathrm{MH}} : CH^p(X)_{\mathbb{Q}} \to \mathrm{Ext}^{2p}_{D^b \mathrm{MHM}(X)^{\mathrm{go}}}(\mathbb{Q}_X^H, \mathbb{Q}_X^H(p)))$$
$$= \mathrm{Ext}^{2p}_{\mathrm{MHS}(\mathbb{Q})^{\mathrm{go}}}(\mathbb{Q}^H, \mathbf{R}\Gamma(X, \mathbb{Q}^H)(p)) \qquad (10)$$

where $CH^p(X)_{\mathbb{Q}}$ is the Chow group tensored by $\mathbb{Q}$, $\mathrm{MHM}(X)^{\mathrm{go}}$ is the category of mixed Hodge Modules of geometric origin, and $\mathrm{MHS}(\mathbb{Q})^{\mathrm{go}} = \mathrm{MHM}(\mathrm{pt})^{\mathrm{go}}$. By the canonical filtration on $\mathbf{R}\Gamma(X, \mathbb{Q}^H)$, we get a filtration L on the target of the morphism such that

$$\mathrm{Gr}_L^i = \mathrm{Ext}^i_{\mathrm{MHS}(\mathbb{Q})^{\mathrm{go}}}(\mathbb{Q}^H, H^{2p-i}(X, \mathbb{Q})(p)), \qquad (11)$$

and it induces a filtration L on $CH^p(X)_{\mathbb{Q}}$ via $\mathrm{cl}^{\mathrm{MH}}$. The cycle map is surjective if and only if the Hodge conjecture is true for any variety. In this case $\mathrm{Gr}_L^i \, \mathrm{cl}^{\mathrm{MH}}$

are also surjective, and we get a filtration L on $CH^p(X)_{\mathbb{Q}}$ such that (11) holds. Compare [4]. Moreover $Gr^i_L CH^p(X)_{\mathbb{Q}} = 0$ for $i > p$, and this filtration is separated if and only if cl^{MH} is injective. The injectivity of cl^{MH} seems to be related with the surjectivity of the cycle map of Bloch's higher Chow group $CH^p(X, 1)_{\mathbb{Q}}$ to $Ext^{2p-1}_{D^b MHM(X)^{go}}(\mathbb{Q}^H_X, \mathbb{Q}^H_X(p))$.

References

1. Beilinson, A.: Height pairing between algebraic cycles. (Lecture Notes in Mathematics, vol. 1289.) Springer, Berlin Heidelberg New York 1987, pp. 1–26
2. Beilinson, A.: On the derived category of perverse sheaves. Ibid, pp. 27–41
3. Beilinson, A., Bernstein, J., Deligne, P.: Faisceaux pervers. Astérisque **100** (1982)
4. Bloch, S.: Lectures on algebraic cycles. Duke Univ. Math. Series 4, Durham 1980
5. Borel, A.: Algebraic $\mathscr{D}$-Modules. Academic Press, Boston 1987
6. Brylinski, J.-L.: Modules holonomes à singularités régulières et filtrations de Hodge II. Astérisque **101-102** (1983) 75-117
7. Cattani, E., Kaplan, A., Schmid, W.: L^2 and intersection cohomologies for a polarizable variation of Hodge structure. Inv. math. **87** (1987) 217–252
8. Deligne, P.: Théorie de Hodge I, II, III. Actes Congrès Intern. Math. (1970) 425–430; Publ. Math. IHES **40** (1971) 5–58; ibid. **44** (1974) 5–77
9. Deligne, P.: Equation différentielle à points singuliers réguliers. (Lecture Notes in Mathematics, vol. 163.) Springer, Berlin Heidelberg New York 1970
10. Deligne, P.: Le formalisme des cycles évanescents. In: SGA7 XIII and XIV. (Lecture Notes in Mathematics, vol. 340.) Springer, Berlin Heidelberg New York 1973, pp. 82–115 and 116–164
11. Deligne, P.: Conjecture de Weil II. Publ. Math. IHES **52** (1980) 137–252
12. Durfee, A., Saito, M.: Mixed Hodge structure on the intersection cohomology of links. Comp. Math. **76** (1990) 49–67
13. Griffiths, P.: Periods of integrals on algebraic manifolds I, II, III. Amer. J. Math. **90** (1968) 568–626; ibid. 805–865; Publ. Math. IHES **38** (1970) 125–180
14. Kashiwara, M.: On the maximally overdetermined system of linear differential equations I. Publ. RIMS, Kyoto Univ. **10** (1975) 563–579
15. Kashiwara, M.: On the holonomic systems of differential equations II. Inv. math. **49** (1978) 121–135
16. Kashiwara, M.: A study of variation of mixed Hodge structure. Publ. RIMS, Kyoto Univ. **22** (1986) 991–1024
17. Kashiwara, M.: Vanishing cycle sheaves and holonomic systems of differential equations. (Lecture Notes in Mathematics, vol. 1016.) Springer, Berlin Heidelberg New York 1983, pp. 136–142
18. Kashiwara, M., Kawai, T.: On the holonomic system of microdifferential equations III. Publ. RIMS, Kyoto Univ. **17** (1981) 813–979
19. Kashiwara, M., Kawai, T.: The Poincaré lemma for variations of polarized Hodge structures. Publ. RIMS, Kyoto Univ. **23** (1987) 345–407
20. Kashiwara, M., Kawai, T.: Hodge structure and holonomic systems. Proc. Japan Acad. **62** (1986) 1–4
21. Kollár, J.: Higher direct images of dualizing sheaves I, II. Ann. Math. **123** (1986) 11–42; **124** (1986) 171–202
22. Malgrange, B.: Polynôme de Bernstein-Sato et cohomologie évanescente, Astérisque **101-102** (1983) 243–267
23. Mebkhout, Z.: Une autre équivalence de catégories. Comp. Math. **51** (1984) 63–88

24. Saito, M.: Hodge filtration on Gauss-Manin systems I, II. J. Fac. Sci. Univ. Tokyo, Sect. I A **30** (1984) 489-498; Proc. Japan Acad., Ser. A **59** (1983) 37–40
25. Saito, M.: Hodge structure via filtered $\mathscr{D}$-Modules. Astérisque **130** (1985) 342-351
26. Saito, M.: Modules de Hodge polarisables. Publ. RIMS, Kyoto Univ. **24** (1988) 849–995
27. Saito, M.: Mixed Hodge modules. Publ. RIMS, Kyoto Univ. **26** (1990) 221–333
28. Saito, M.: Decomposition theorem for proper Kähler morphisms. Tohoku Math. J. **42** (1990) 127–148
29. Saito, M.: Vanishing cycles and mixed Hodge modules. Preprint IHES 1988
30. Saito, M.: Extension of mixed Hodge modules. Comp. Math. **74** (1990) 209–234
31. Saito, M.: Hodge conjecture and mixed motives I, II. Preprint RIMS-691 and 692, 1990
32. Saito, M.: On Kollár's conjecture. Preprint RIMS-693, 1990
33. Saito, M.: Duality for vanishing cycle functors. Publ. RIMS, Kyoto Univ. **25** (1989) 889–921
34. Saito, M.: Introduction to mixed Hodge Modules. Astérisque **179-180** (1989) 145–162
35. Saito, M., Zucker, S.: The kernel spectral sequence of vanishing cycles. Duke Math. J. **61** (1990) 329–339
36. Scherk, J., Steenbrink, J.: On the mixed Hodge structure of the Milnor fiber. Math. Ann. **271** (1985) 631–665
37. Schmid, W.: Variation of Hodge structure: the singularity of period mapping. Inv. math. **22** (1973) 211–319
38. Steenbrink, J.: Limits of Hodge structures. Inv. math. **31** (1976) 229-257
39. Steenbrink, J.: Mixed Hodge structure on the vanishing cohomology. In: Real and complex singularities, Oslo 1976. Sijthoff-Noordhoff, Alphen a/d Rijn 1977, pp. 525–563
40. Steenbrink, J.: The spectrum of hypersurface singularity. Astérisque **179-180** (1989) 163-184
41. Steenbrink, J., Zucker, S.: Variation of mixed Hodge structure I. Inv. math. **80** (1985) 489–542
42. Varchenko, A.: Asymptotic Hodge structure in the vanishing cohomology. Math. USSR Izv. **18** (1982) 469-512
43. Verdier, J.-L.: Dualité dans les espaces localement compacts. Séminaire Bourbaki no. 300 (1965/66)
44. Zucker, S.: Hodge theory with degenerating coefficients, L_2-cohomology in the Poincaré metric. Ann. Math. **109** (1979) 415–476
45. Zucker, S.: Variation of mixed Hodge structure II. Inv. math. **80** (1985) 543–565

L_2-Cohomology of Algebraic Varieties

*Leslie Saper**

Department of Mathematics, Duke University, Durham, NC 27706, USA

In this article I shall sketch the theory of L_2-cohomology and its applications in algebraic geometry. The basic principle is that for appropriate Riemannian metrics on a smooth, Zariski open subset of a complex projective variety X, the L_2-cohomology represents invariants of the full variety X, for example, the intersection cohomology and its Hodge structure. The primary obstacle to proving such results is to relate L_2-growth conditions to a suitable theory of "weights" for which one can establish a semi-purity theorem. Two examples will be presented whose proofs, although rather different, exhibit this pattern: the case of Kähler varieties with isolated singularities [37–39], and the case of locally symmetric varieties (Zucker's conjecture) [40, 41].

1. L_2-Cohomology

1.1 Definition and Basic Properties

References are [10] and [45]. Let M be a Riemannian manifold (assumed oriented for simplicity) and $\mathbb{E}$ a flat complex vector bundle with a (not necessarily flat) Hermitian metric. Let $L_2^{\bullet}(M; \mathbb{E})$ denote the Hilbert space of measurable $\mathbb{E}$-valued forms ϕ which satisfy the L_2-growth condition

$$\int_M |\phi|^2 dV < \infty.$$

Note that the dependence of this condition on the metric of M arises in two opposing fashions: in the metric on the exterior powers of the cotangent bundle, and in the volume form.

Definition 1. The L_2-*cohomology* $H_{(2)}^{\bullet}(M; \mathbb{E})$ is the cohomology of the complex

$$\mathrm{Dom}(d) = \{ \phi \in L_2^{\bullet}(M; \mathbb{E}) \mid d\phi \in L_2^{\bullet+1}(M, \mathbb{E}) \}.$$

Here $d\phi$ is taken in the sense of currents.

---

* The author is an Alfred P. Sloan Research Fellow and is supported in part by NSF Grants DMS-8705849 and DMS-8957216.

Proceedings of the International Congress
of Mathematicians, Kyoto, Japan, 1990
© The Mathematical Society of Japan, 1991

Similarly we can define the L_2-$\bar{\partial}$-cohomology $H^{p,\bullet}_{(2),\bar{\partial}}(M;E)$ if M is a Hermitian manifold and E is a Hermitian holomorphic vector bundle. By a regularization result of Cheeger [10], we obtain the same cohomology in either case if we require ϕ above to be smooth.

Quasi-isometry Invariance. Although in general the L_2-cohomology depends on the metrics involved, it is an invariant of their quasi-isometry class (two metrics g_1, g_2 are *quasi-isometric* if $Cg_2 \leq g_1 \leq C'g_2$ for constants C, $C' > 0$).

Hodge Theory. Clearly we have

$$H^{\bullet}_{(2)}(M;\mathbb{E}) = \mathrm{Ker}(d) \big/ \overline{\mathrm{Range}(d)} \bigoplus \overline{\mathrm{Range}(d)} \big/ \mathrm{Range}(d)$$

$$= \mathrm{Ker}(d) \cap \mathrm{Ker}(d^*) \bigoplus \overline{\mathrm{Range}(d)} \big/ \mathrm{Range}(d), \tag{1}$$

where d^* is the Hilbert space adjoint. In general $\mathrm{Dom}(d^*)$ is smaller than $\mathrm{Dom}(\delta) = \{\phi \in L^{\bullet}_2(M;\mathbb{E}) \mid \delta\phi \in L^{\bullet-1}_2(M;\mathbb{E})\}$, where δ is the formal adjoint to d; by a result of Gaffney [15], they are equal if M is complete.

Since the second term of (1) is either 0 or infinite dimensional, we have the following result:

Theorem 2. *Say* $\mathrm{Range}(d)$ *is closed (e.g., if* $\dim H^{\bullet}_{(2)}(M;\mathbb{E}) < \infty$*). Then*
  1) Hodge Theorem:

$$H^{\bullet}_{(2)}(M;\mathbb{E}) = \mathrm{Ker}(d) \cap \mathrm{Ker}(d^*) = \mathrm{Ker}(\Delta),$$

where $\Delta = d^*d + dd^*$ *has domain in the sense of unbounded operators.*
  2) Poincaré Duality: If furthermore, $d^* = \delta$ *(e.g., if* M *is complete), then the Hodge star operator induces conjugate linear isomorphisms*

$$H^i_{(2)}(M;\mathbb{E}) \xrightarrow{\sim} H^{m-i}_{(2)}(M;\mathbb{E}^*),$$

where $m = \dim M$.

Kähler Package. Assume now that M is a complete Kähler manifold and $\mathbb{E} = \mathbb{C}$. In this case the formal identity $\Delta = 2\square$ between the d- and $\bar{\partial}$-Laplacians holds in the sense of unbounded operators [47]. Consequently we have the

Theorem 3. *Let* M *be complete Kähler of dimension* n *and say* $\mathrm{Range}(d)$ *is closed. Then*
  1) Hodge Decomposition:

$$H^i_{(2)}(M;\mathbb{C}) = \bigoplus_{p+q=i} H^{p,q}_{(2),\bar{\partial}}(M;\mathbb{C}).$$

  2) Hard Lefschetz: If L *denotes the operation of wedging with the Kähler class, there are isomorphisms*

$$L^i : H^{n-i}_{(2)}(M;\mathbb{C}) \xrightarrow{\sim} H^{n+i}_{(2)}(M;\mathbb{C}).$$

Analogous statements hold when $\mathbb{E}$ underlies a polarizable variation of Hodge structure (see [44]).

1.2 Example: Conical Metrics

Let X be a closed triangulated pseudomanifold of dimension n, that is, a compact simplicial complex which equals the union of its n-simplices and in which every $(n-1)$-simplex is a boundary face of exactly two n-simplices. For example, it is well-known that any compact complex analytic variety admits such a structure. If $\Sigma \subset X$ is the codimension 2 skeleton, then $X^{\text{reg}} = X \setminus \Sigma$ is naturally a smooth manifold, dense in X.

Definition 4. A Riemannian metric on X^{reg} is called *conical* if it is quasi-isometric to a piecewise flat metric with respect to the triangulation of X.

Although the metric on X^{reg} is incomplete, Cheeger [10] has proved the

Theorem 5. *If X is an admissible triangulated pseudomanifold, the L_2-cohomology $H^{\bullet}_{(2)}(X^{\text{reg}};\mathbb{C})$ of a conical metric is finite dimensional and invariant under subdivision. It satisfies the Hodge theorem and (if X^{reg} is oriented) Poincaré duality.*

Here admissibility is an inductively defined condition on the links which is always satisfied if X admits a stratification with even codimensional strata (e.g., a complex analytic variety).

The proof of Theorem 5 is based on the following local calculation [10]:

Proposition 6. *Let N be a Riemannian manifold of dimension $m-1$; if m is even assume* Range(d) *is closed in $L_2^{m/2-1}(N)$. Let $c(N)$ be the closed cone on N and let its regular part $c(N)^{\text{reg}}$ have the conical metric $dx^2 + x^2 ds_N^2$. Then*

$$H^i_{(2)}(c(N)^{\text{reg}};\mathbb{C}) = \begin{cases} H^i_{(2)}(N;\mathbb{C}) & i < m/2, \\ 0 & i \geq m/2. \end{cases} \tag{2}$$

The truncation at middle degree arises from the interplay between the volume form and the metrics on the bundles of forms; see Sect. 4.1 for a further discussion in the case of a complete Kähler metric.

2. Intersection Cohomology

Independently of the preceding developments, and also partly motivated by the desire to find a homology theory on singular spaces satisfying Poincaré duality, Goresky and MacPherson had also encountered a local calculation such as (2), but in the topological context of their intersection homology. References are [5, 16, 17]; see also the historical survey [20].

2.1 Definition and Basic Properties

Let $X = X^0 \supset \Sigma = X^1 \supset X^2 \supset \cdots \supset X^n$ be a stratified topological pseudomanifold. In other words, the strata $S^k = X^k \setminus X^{k+1}$ are manifolds of even codimension $2k$ and S^0 is dense. (The restriction to even codimensional strata is not essential.) Furthermore, a point in S^k has a cofinal system of local neighborhoods of the form $B \times c(L)$, where B is a ball in S^k and L is a stratified topological pseudomanifold of dimension $2k - 1$.

Let $\mathbb{E}$ be a flat complex vector bundle on $X^{\text{reg}} = X \setminus \Sigma$. The *(middle perversity)* *intersection cohomology* $IH^{\bullet}(X;\mathbb{E})$ was originally defined by imposing restrictions on the dimension of intersection of a geometric chain with the singular strata; it may also be defined as the hypercohomology $\mathbf{H}^{\bullet}(X;\mathscr{IC}^{\bullet}(\mathbb{E}))$, where

$$\mathscr{IC}^{\bullet}(\mathbb{E}) = \tau_{<n} Ri_{n*} \ldots \tau_{<1} Ri_{1*} \mathbb{E} \tag{3}$$

is Deligne's sheaf. Here $i_k : X \setminus X^k \hookrightarrow X \setminus X^{k+1}$ is the inclusion and $\tau_{<k}$ truncates sheaf cohomology in degrees $\geq k$.

Intersection cohomology is independent of the choice of stratification; when X is smooth and $\mathbb{E}$ is everywhere defined, it reduces to ordinary cohomology. True to its name, there is an intersection pairing effecting Poincaré duality. And when X is a projective algebraic variety and $\mathbb{E}$ underlies a polarizable variation of Hodge structure, Saito [35, 36] has given $IH^{\bullet}(X;\mathbb{E})$ a natural Hodge structure.

It is clear from (3) that intersection cohomology admits a local characterization. Namely, for $x \in S^k$, let $U = B \times c(L)$ be a local neighborhood of x as above. Then the local calculation

$$IH^i(U;\mathbb{E}) \cong IH^i(c(L);\mathbb{E}) \cong \begin{cases} IH^i(L;\mathbb{E}) & i < k, \\ 0 & i \geq k, \end{cases} \tag{4}$$

essentially characterizes intersection cohomology; here the isomorphism for $i < k$ is given by the natural "attaching" map.

2.2 Relations with L_2-Cohomology

In view of (2) and (4), Cheeger [10] proved:

Theorem 8. *Let* X^{reg} *have a conical metric. Then* $H^{\bullet}_{(2)}(X^{\text{reg}};\mathbb{C}) \cong IH^{\bullet}(X;\mathbb{C})$.

One is thus led to try extending the classical Hodge-de Rham theory to singular spaces by using metrics on X^{reg} which somehow reflect the singular compactification. However, for complex projective varieties, it would be more fruitful to have a Kähler metric. It is rare that there exist Kähler conical metrics, but there is a natural quasi-isometry class of Kähler metrics – namely, those induced from the Fubini-Study metric by a projective embedding. Cheeger, Goresky, and MacPherson have conjectured [12]:

Conjecture 9. *For the metric induced on X^{reg} by the Fubini-Study metric of any projective embedding, $H_{(2)}^{\bullet}(X^{\text{reg}}; \mathbb{C}) \cong IH^{\bullet}(X; \mathbb{C})$. Furthermore, this induces a natural Hodge decomposition.*

Cheeger's work [10, 11] covers the case of varieties with analytically conical singularities, while Hsiang and Pati [18] and Nagase [26] have established the first part of the conjecture for surfaces with isolated singularities. Recently Ohsawa [31] has proven the first part for varieties with isolated singularities, using a result of Saper [39] (see Theorem 11 below). Since the metric is incomplete, the second part of the conjecture does not follow from Theorem 3. Nagase [25] has obtained partial results for surfaces; Pardon [32] has shown that a Hodge decomposition *cannot* in general be given by the L_2-$\bar{\partial}$-cohomology groups without conceivably imposing boundary conditions.

3. Complete Kähler Metrics

Let X be a projective algebraic variety of complex dimension n. In view of Theorem 3, we turn to *complete* Kähler metrics on X^{reg} in order to construct Hodge decompositions on intersection cohomology.

3.1 Degenerating Coefficients

Assume X is smooth and let $\mathbb{E}$ underlie a polarizable variation of Hodge structure on $X^{\text{reg}} = X \setminus D$, where D is a divisor with normal crossings. Endow X^{reg} with a complete Kähler metric which has Poincaré singularities normal to each component of D; in local coordinates, if $D = (z_1 \cdots z_k)$, the metric is quasi-isometric to

$$\sum_{i=1}^{k} \frac{dz_i d\bar{z}_i}{|z_i|^2 (\log |z_i|^2)^2} + \sum_{i=k+1}^{n} dz_i d\bar{z}_i. \tag{5}$$

The isomorphism $H_{(2)}^{\bullet}(X^{\text{reg}}; \mathbb{E}) \cong IH^{\bullet}(X; \mathbb{E})$ was proved by Zucker [44] when $n = 1$; the general case was proved independently by Cattani, Kaplan, and Schmid [9] and Kashiwara and Kawai [19]. Note that when $\mathbb{E}$ is trivial, the theorem is in fact quite simple; the main point here is the degeneration of coefficients.

3.2 Locally Symmetric Varieties

The prime examples of complete Kähler metrics naturally associated to *singular* varieties are those on locally symmetric varieties.

Let D be a bounded symmetric domain endowed with the Bergman metric, or equivalently, a Hermitian symmetric space of noncompact type G/K equipped with a group invariant metric. For Γ an arithmetically defined group of automorphisms of D, the quotient $\Gamma \backslash D$ is called a *locally symmetric variety*. It may have finite quotient singularities, but this is not important here; the main point is that it may be noncompact and that there is a natural, though singular, compactification $\Gamma \backslash D^*$ (due to Baily and Borel [1] and Satake [42]) as a normal projective variety. Briefly, $\Gamma \backslash D^*$ is obtained by adjoining arithmetic quotients $\Gamma_F \backslash F$ of rational boundary components F of D, which are themselves locally symmetric varieties of lower rank and dimension.

For example, when D is the complex n-ball, $\Gamma \backslash D^*$ is formed by adding singular cusp points. After resolving the singularities with a smooth exceptional divisor, the metric is quasi-isometric to (compare (5))

$$\frac{dz_1 d\bar{z}_1}{|z_1|^2 (\log |z_1|^2)^2} + \frac{1}{|\log |z_1|^2|} \sum_{i=2}^{n} dz_i \, d\bar{z}_i. \tag{6}$$

The following theorem was conjectured by Zucker after verifying some cases with G having $\mathbb{Q}$-rank 1 [45]. It was established independently by Saper and Stern [40, 41], and by Looijenga [22]; Looijenga and Rapoport have also given a third proof [23]. Previously special cases were proven by Borel [4], Borel and Casselman [6] (see also [8]), and Zucker [45, 46]. An indication of Saper and Stern's proof will be given in Sect. 4.3.

Theorem 10. *Let E be a finite dimensional complex representation space of G with admissible metric, and let $\mathbb{E}$ be the corresponding metrized local system on $\Gamma \backslash D$. Then we have a natural isomorphism*

$$H_{(2)}^{\bullet}(\Gamma \backslash D; \mathbb{E}) \cong IH^{\bullet}(\Gamma \backslash D^*; \mathbb{E}).$$

3.3 Kähler Varieties

For a general Kähler variety X, one would like a complete Kähler metric on X^{reg} analogous to those on locally symmetric varieties. In particular, one would like a generalization of Theorem 10.

Note that one could always take a resolution $\tilde{X} \to X$ with exceptional divisor D having normal crossings, and consider a Poincaré metric on $\tilde{X} \backslash D$ as in Sect. 3.1. However this is not the metric we want, since its L_2-cohomology is strictly larger than $IH^{\bullet}(X; \mathbb{C})$, namely $H^{\bullet}(\tilde{X}; \mathbb{C})$. Instead, Saper [37–39] constructed a metric which, locally on D, is quasi-isometric (modulo a perturbation) to the metrics on toroidal resolutions of locally symmetric models. Using this metric (which depends on $\tilde{X}$), he established the following theorem; a brief indication of the proof will be given in Sect. 4.2. The case of isolated analytically conical singularities was also proven later with a different metric by Ohsawa [30] (see also [28, 29]).

Theorem 11. *Let X be a Kähler variety with isolated singularities. There exists an infinite discrete set of complete Kähler metrics on X^{reg} for which we have*

$$H_{(2)}^{\bullet}(X^{\mathrm{reg}}; \mathbb{C}) \cong IH^{\bullet}(X; \mathbb{C}) \tag{7}$$

and

$$H_{(2), \bar{\partial}}^{0, \bullet}(X^{\mathrm{reg}}; \mathbb{C}) \cong H_{\bar{\partial}}^{0, \bullet}(\tilde{X}; \mathbb{C}). \tag{8}$$

Here $\tilde{X}$ is any resolution of X.

A priori, the different metrics referred to in the theorem could induce different Hodge decompositions on $IH^\bullet(X;\mathbb{C})$ via (7); Zucker [47] has shown in fact that they all agree with Saito's Hodge structure. This is used in the proof of (8), which asserts that the L_2-$\bar\partial$-Betti numbers $h^{0,q}_{(2),\bar\partial}$ are birational invariants. The corresponding assertion for the L_2-$\bar\partial$-index $\sum_q(-1)^q h^{0,q}_{(2),\bar\partial}$ was conjectured in the contexts of Conjecture 9 and Theorem 10 by MacPherson [24]. Recently Pardon and Stern [33] have given a proof of MacPherson's conjecture (and (8)) for the metric induced from the Fubini-Study metric provided one uses L_2-$\bar\partial$-cohomology with Dirichlet boundary conditions.

4. Outline of Proofs

Let X be a projective variety of dimension n. To establish the isomorphism $H^\bullet_{(2)}(X^{\mathrm{reg}};\mathbb{E}) \cong IH^\bullet(X;\mathbb{E})$ for some metric on X^{reg}, the main point is to show that L_2-cohomology has the correct local calculation (see (4)). That is, for a neighborhood U of a point on a complex codimension k stratum with link L, we need

$$H^i_{(2)}(U^{\mathrm{reg}};\mathbb{E}) \cong H^i_{(2)}(c(L)^{\mathrm{reg}};\mathbb{E}) \cong \begin{cases} H^i_{(2)}(L^{\mathrm{reg}};\mathbb{E}) & i < k, \\ 0 & i \geq k. \end{cases} \tag{9}$$

In practise, proving just the vanishing condition is usually sufficient.

4.1 Heuristic Arguments

In order to develop a heuristic device for understanding why the vanishing result might hold, applicable to both Theorem 10 and Theorem 11, we will consider in this subsection the basic example of a space X with a single singular stratum of codimension k. For simplicity we will only treat constant coefficients, however this is not necessary.

Write $c(L)^{\mathrm{reg}} = \mathbb{R}^{\geq 0} \times L$ with r the coordinate on the first factor ($r \to \infty$ near the vertex) and π the projection on the second. Analogously to Borel's formula [3] for the metric on a cusp of a locally symmetric variety, consider the metric

$$dr^2 + \sum_\gamma e^{-\gamma r}\pi^* ds_\gamma^2,$$

on $c(L)^{\mathrm{reg}}$, where $TL = \bigoplus_\gamma TL_\gamma$ is some decomposition into subbundles indexed by *weights* $\gamma \in \mathbb{R}^{\geq 0}$ and ds_γ^2 is a metric on TL_γ. Then

$$dV = e^{-\delta r}dr\,dV_L \tag{10}$$

where $\delta = (1/2)\sum_\gamma(\dim TL_\gamma)\gamma$ is the middle weight, while for a pure wedge product $\psi = \psi_1 \wedge \cdots \wedge \psi_i$ ($\psi_j \in \Gamma(TL^*_{\gamma_j})$) with weight $\mu = \sum \gamma_j$,

$$|\pi^*\psi|^2 = e^{\mu r}|\psi|_L^2. \tag{11}$$

By separation of variables, it is reasonable to assume that an L_2-cohomology class can be represented either as $\pi^*\psi$ or $f(r)dr \wedge \pi^*\psi$, where ψ is a closed form on L. From (10) and (11) we immediately see that $\pi^*\psi$ is L_2 if and only if $\mu < \delta$. On the other hand, if $f(r)dr \wedge \pi^*\psi$ is L_2, it may be expressed as a cocycle via $(\int_0^r f(t)dt)\pi^*\psi$ or $(-\int_r^\infty f(t)dt)\pi^*\psi$, which are L_2 precisely when $\mu < \delta$ or $\mu > \delta$ respectively [45]. This suggests that if the above notion of weight actually descends to a decomposition (or at least a filtration) of $H^i(L;\mathbb{C})$ and $H^i_{(2)}(c(L)^{\text{reg}};\mathbb{C})$, then

$$H^i_{(2)}(c(L)^{\text{reg}};\mathbb{C})_\mu = \begin{cases} H^i(L;\mathbb{C})_\mu & \mu < \delta, \\ H^1_{(2)}(\mathbb{R}^{\geq 0}) \otimes H^{i-1}(L;\mathbb{C})_\mu & \mu = \delta, \\ 0 & \mu > \delta, \end{cases} \qquad (12)$$

where the subscript μ denotes the part of weight μ. The group $H^1_{(2)}(\mathbb{R}^{\geq 0})$ is infinite dimensional because, as is well known, Range(d) on $\mathbb{R}^{\geq 0}$ is not closed.

Even ignoring the infinite dimensional contribution, (12) does not achieve the desired truncation (9) at and above middle degree; instead we encounter a truncation at and above middle weight. We are thus forced to prove a semi-purity result, which that asserts these two truncations are the same, namely

$$H^i(L;\mathbb{C})_\mu = 0 \quad \text{for} \quad \begin{cases} i \geq k \text{ and } \mu \leq \delta, \text{ or} \\ i < k \text{ and } \mu \geq \delta . \end{cases} \qquad (13)$$

4.2 Kähler Varieties with Isolated Singularities

Assume X has isolated singularities and let U be a contractible neighborhood of Sing X. We will realize the above heuristic argument in the proof of Theorem 11 by relating L_2-growth weights in U^{reg} to the weights of the mixed Hodge structure on $H^\bullet(U^{\text{reg}};\mathbb{C})$.

Let $\tilde{X} \to X$ be a resolution of X with normal crossing exceptional divisor D having smooth components. Let $D^{(p)}$ denote the disjoint union of p-fold intersections of components of D. Recall [13] that the q-th row of the E_1 term of the weight spectral sequence converging to $H^\bullet(U^{\text{reg}};\mathbb{C})$ is

$$\cdots \longrightarrow H^{q-4}(D^{(2)}) \longrightarrow H^{q-2}(D^{(1)}) \longrightarrow H^q(D^{(1)}) \longrightarrow H^q(D^{(2)}) \longrightarrow \cdots,$$

where the differentials are formed from alternating sums of Gysin maps and of restriction maps.

On the other hand, a principal technical result of [39] is the construction of an analogous spectral sequence $\{E_{(2),r}^{-p,q}\}$ converging to $H^\bullet_{(2)}(U^{\text{reg}};\mathbb{C})$. In this L_2 spectral sequence, the q-th row of $E_{(2),1}$ may be represented by forms with L_2-growth weight q, or by $dr\wedge$ forms with L_2-growth weight $q-1$. Consequently we see that

$$E_{(2),1}^{-p,q} = \begin{cases} E_1^{-p,q} & q < n, \\ H^1_{(2)}(\mathbb{R}^{\geq 0}) \otimes E_1^{-p,n} & q = n+1, \\ 0 & \text{otherwise}, \end{cases}$$

which is the rigorous version of (12) (note that $\delta = n$ here).

The required semi-purity (13) follows from the following stronger result. It was deduced as a consequence of the decomposition theorem [2, 35, 36] by Steenbrink [43]; an independent proof was given by Navarro Aznar [27].

Proposition 12. *The weight filtration W of the mixed Hodge structure on $H^{\bullet}(U^{\mathrm{reg}};\mathbb{C})$ satisfies*

$$\mathrm{Gr}_{\mu}^{W} H^{i}(U^{\mathrm{reg}};\mathbb{C}) = 0 \quad for \quad \begin{cases} i \geq n \text{ and } \mu \leq i, \text{ or} \\ i < n \text{ and } \mu > i. \end{cases}$$

Consequently, $E_2^{-p,q} = 0$ for $q - p \geq n$, $p \leq 0$ or $q - p < n$, $p > 0$.

For the intersection cohomology of the link of nonisolated singularities, the proposition has been generalized by Durfee and Saito [14].

4.3 Locally Symmetric Varieties

Let X now be the Baily-Borel-Satake compactification of a locally symmetric variety; we present some of the highlights in Saper and Stern's proof of Theorem 10, Zucker's conjecture. The technical approach used is to establish on $c(L)^{\mathrm{reg}}$ the estimate

$$\|d\phi\|^2 + \|d^*\phi\|^2 \geq c\|\phi\|^2 \qquad (\phi \in \mathrm{Dom}(d) \cap \mathrm{Dom}(d^*), \deg \phi \geq k), \qquad (14)$$

where $c > 0$; it is well-known that this implies the desired vanishing of L_2-cohomology (9). The argument follows the heuristic reasoning of Sect. 4.1 by exploiting the fairly immediate connection between L_2-growth weights and Lie-theoretic weights. In contrast, the proof of Looijenga [22] (and even more so, Looijenga and Rapoport [23]) passes from L_2-growth weights to mixed Hodge weights via a local Hecke operator.

When H is a Lie subgroup of G, we will denote the Lie algebra of H by $\mathfrak{h}$, and when appropriate we will use the notation Γ_H (resp. K_H) to denote the arithmetic subgroup (resp. maximal compact subgroup) induced in H by Γ (resp. K).

For simplicity, we begin by assuming we are in the $\mathbb{Q}$-rank 1 case, with a single boundary component F. Let $Z = A \ltimes {}^{\circ}Z$ be the centralizer of F, where $A \cong \mathbb{R}^+$ is a one-dimensional $\mathbb{Q}$-split torus representing the dilations transverse to F, and ${}^{\circ}Z$ is a certain complement, representing movement in the link directions. The link L is $\Gamma_Z \backslash {}^{\circ}Z / K_Z$; thus since A commutes with K_Z, there is an adjoint action of A on the tangent bundle TL. We decompose TL into weight spaces $\bigoplus TL_{\gamma}$ under this action; after we have fixed a coordinate r on $\mathfrak{a}$ (with $r \to \infty$ near F), we may by abuse of notation consider the weights γ as nonnegative real numbers. Then Borel's formula for the metric [3] is

$$dr^2 + \sum_{\gamma} e^{-\gamma r} ds_{\gamma}^2,$$

where ds_{γ}^2 is a metric on TL_{γ}. In fact, after suitably rescaling r, we have $\gamma = 0$,

1, or 2. Thus we see the connection between L_2-growth weights and Lie-theoretic weights.

It turns out that it suffices to prove (14) assuming that ϕ is a weight vector for the induced action of $\mathfrak{a}$, say of weight μ. It also suffices to assume that ϕ is an eigenvector for $dr \wedge \iota_{\partial/\partial r}$, say with eigenvalue j; note that $j = 1$ (resp. 0) if ϕ belongs to (resp. is orthogonal to) the ideal generated by dr. In realizing the previous heuristic reasoning of Sect. 4.1, we never actually pass to cohomology; we merely seek to establish the estimate (14) for all possible values of μ and j. Thus the first step is accomplished by proving the estimate

$$\|d\phi\|^2 + \|d^*\phi\|^2 \geq (\mu - \delta)^2 \|\phi\|^2 + (\mu - \delta)(1 - j) \int_{\partial c(L)} |\phi|^2 \, dV_L, \qquad (15)$$

where δ is the middle weight (the weight of $\mathfrak{a}$ on $\left(\bigwedge^{\dim \mathfrak{z}} \mathfrak{z}\right)^{1/2}$). This proves (14) whenever

$$\mu \neq \delta \text{ and } j = 1, \quad \text{or}$$
$$\mu > \delta,$$

which corresponds to (12) of the heuristic argument.

For the second step, we concentrate on the link. Decompose $^\circ Z = M \ltimes N$, where M is reductive and N is the unipotent radical. Thus the link fibers over $\Gamma_M \backslash M / K_Z$ with nilmanifold fiber $\Gamma_N \backslash N$. By a result of Zucker [45], we may assume our forms are harmonic in the fibers. The Koszul Laplacian $\Delta_{\mathfrak{m},0}$ on $\bigwedge^\bullet (\mathfrak{m}/\mathfrak{k}_Z)^* \otimes H^\bullet(\mathfrak{n}; E) \otimes \delta$ extends to an action on forms and it suffices to assume that ϕ is also an eigenvector of $\Delta_{\mathfrak{m},0}$, say with eigenvalue ν. Similarly to (15), we may prove the estimate

$$\|d\phi\|^2 + \|d^*\phi\|^2 \geq \nu \|\phi\|^2. \qquad (16)$$

The following proposition tells us when (16) yields a nontrivial estimate; it is the analogue of the heuristic semi-purity (13).

Proposition 13. *In the above context,*

$$\nu > 0 \quad \text{for} \quad \begin{cases} \deg \phi - j \geq k \text{ and } \mu \leq \delta, \text{ or} \\ \deg \phi - j < k \text{ and } \mu \geq \delta. \end{cases} \qquad (17)$$

The first line of (17) follows from [41, Prop. 10.2, Prop. 11.1, §12]; the second line follows by a dual argument. The proof uses a result of Raghunathan [34], valid for any coefficient sytem over $\Gamma_M \backslash M / K_Z$, to obtain a condition equivalent to $\nu = 0$; a Weyl group argument using Kostant's theorem [21] allows us to analyze this for the coefficients $H^\bullet(\mathfrak{n}; E) \otimes \delta$. In the $\mathbb{R}$-rank 1 case, the proposition reduces to a result of Casselman [7].

In the case of $\mathbb{Q}$-rank > 1, we form a distinguished cover of $c(L)^{\text{reg}}$, with each open set consisting of the points that are near a particular flag of boundary components. The above argument generalizes to prove the estimate for forms compactly supported in each such open set; one replaces Z by the $\mathbb{Q}$-parabolic subgroup Q which is the normalizer of the corresponding flag. In order to combine the separate estimates into one for all of $c(L)^{reg}$, we use a partition of unity $\{\eta_Q\}$; terms involving $|d\eta_Q|$ arise which can be controlled by choosing the elements of the distinguished cover to have sufficiently large overlap.

References

1. Baily W., Borel, W.: Compactification of arithmetic quotients of bounded symmetric domains. Ann. Math. **84** (1966) 442–528
2. Beilinson, A., Bernstein, J., Deligne, P.: Faisceaux pervers, Proceedings of the Conference "Analyse et topologie sur les espaces singuliers", Juillet 1981. Astérisque **100** (1982)
3. Borel, A.: Some properties of arithmetic quotients and an extension theorem. J. Diff. Geom. **6** (1972) 543–560
4. Borel, A.: L_2-cohomology and intersection cohomology of certain arithmetic varieties. In: Sally, J., Srinivasan, B. (eds.) Emmy Noether in Bryn Mawr. Springer, Berlin Heidelberg New York 1983
5. Borel, A., et. al.: Intersection Cohomology. Birkhäuser, Boston 1984
6. Borel, A., Casselman, W.: Cohomologie d'intersection et L^2-cohomologie de variétés arithmétiques de rang rationnel 2. C. R. Acad. Sci. Paris **301** (1985) 369–373
7. Casselman, W.: L^2-cohomology for groups of real rank one. In: Trombi, P. (ed.) Representation theory of reductive groups. Birkhäuser, Boston 1983
8. Casselman, W.: Introduction to the L^2-cohomology of arithmetic quotients of bounded symmetric domains. In: Suwa, T., Wagreich, P. (eds.) Complex analytic singularities. (Advanced Studies in Pure Mathematics, vol. 8.) North-Holland, Amsterdam 1986
9. Cattani, E., Kaplan, A., Schmid, W.: L^2 and intersection cohomologies for a polarizable variation of Hodge structure. Invent. math. **87** (1987) 217–252
10. Cheeger, J.: On the Hodge theory of Riemannian pseudomanifolds. In: Geometry of the Laplace operator. (Proc. Symposia Pure Math., vol. 36.) Amer. Math. Soc., Providence 1980
11. Cheeger, J.: Hodge theory of complex cones. Astérisque **101–102** (1983) 118–134
12. Cheeger, J., Goresky, M., MacPherson, R.: L^2-cohomology and intersection homology of singular algebraic varieties. In: Yau, S.-T. (ed.) Seminar on Differential Geometry. (Annals of Mathematics Studies, no. 102.) Princeton Univ. Press, Princeton 1982
13. Durfee, A.: Mixed Hodge structures on punctured neighborhoods. Duke Math. J. **50** (1983) 1017–1040
14. Durfee, A., Saito, M.: Mixed Hodge structures on the intersection cohomology of links. Comp. Math. To appear
15. Gaffney, M.: A special Stokes' theorem for complete Riemannian manifolds. Ann. Math. **60** (1954) 140–145
16. Goresky, M., MacPherson, R.: Intersection homology theory. Topology **19** (1980) 135–162
17. Goresky, M., MacPherson, R.: Intersection homology II. Invent. math. **72** (1983) 77–130
18. Hsiang, W.-C., Pati, V.: L^2-cohomology of normal algebraic surfaces I. Invent. math. **81** (1985) 395–412
19. Kashiwara, M., Kawai, T.: The Poincaré lemma for variations of polarized Hodge structure. Publ. R.I.M.S., Kyoto Univ. **23** (1987) 345–407
20. Kleiman, S.: The development of intersection homology theory. In: Duren, P. (ed.) A century of mathematics in America, part II. Amer. Math. Soc., Providence 1989
21. Kostant, B.: Lie algebra cohomology and the generalized Borel-Weil theorem. Ann. Math. **74** (1961) 329–387

22. Looijenga, E.: L^2-cohomology of locally symmetric varieties. Comp. Math. **67** (1988) 3–20

23 Looijenga, E., Rapoport, M.: Weights in the local cohomology of a Baily-Borel compactification. Preprint, January 1990

24. MacPherson, R.: Global questions in the topology of singular spaces. In: Proceedings of the International Congress of Mathematicians, Warsaw, 1983. Elsevier, Amsterdam 1984

25. Nagase, M.: Pure Hodge structure of the harmonic L^2-forms on singular algebraic surfaces. Publ. R.I.M.S., Kyoto Univ. **24** (1988) 1005–1023.

26. Nagase, M.: Remarks on the L^2-cohomology of singular algebraic surfaces. J. Math. Soc. Japan **41** (1989) 97–116

27. Navarro Aznar, V.: Sur la théorie de Hodge des variétés algébriques à singularités isolées. Astérisque **130** (1985) 272–307

28. Ohsawa, T.: Hodge spectral sequence on compact Kähler spaces. Publ. R.I.M.S., Kyoto Univ. **23** (1987) 265–274. Also, Supplement to "Hodge spectral sequence on compact Kähler spaces" Preprint, August 1990

29. Ohsawa, T.: Hodge spectral sequence and symmetry on compact Kähler spaces. Publ. R.I.M.S., Kyoto Univ. **23** (1987) 613–625

30. Ohsawa, T.: An extension of Hodge theory to Kähler spaces with isolated singularities of restricted type. Publ. R.I.M.S., Kyoto Univ. **24** (1988) 253–263

31. Ohsawa, T.: Cheeger-Goresky-MacPherson's conjecture for the varieties with isolated singularities. Preprint, July 1990

32. Pardon, W.: The L_2-$\bar{\partial}$-cohomology of an algebraic surface. Topology **28** (1989) 171–195

33. Pardon, W., Stern, M.: L^2-$\bar{\partial}$-cohomology of complex projective varieties. Preprint, August 1990

34. Raghunathan, M.S.: Vanishing theorems for cohomology groups associated to discrete subgroups of semisimple Lie groups. Osaka J. Math. **3** (1966) 243–256. Also, Corrections to "Vanishing theorems ..." Osaka J. Math. **16** (1979) 295–299

35. Saito, M.: Modules de Hodge polarisables. Publ. R.I.M.S., Kyoto Univ. **24** (1988) 849–995

36. Saito, M.: Mixed Hodge modules. Publ. R.I.M.S., Kyoto Univ. **26** (1990) 221–333

37. Saper, L.: L_2-cohomology and intersection cohomology of certain varieties with isolated singularities. Invent. math. **82** (1985) 207–255

38. Saper, L.: L_2-cohomology of isolated singularities. Preprint, November 1985

39. Saper, L.: L_2-cohomology of Kähler varieties with isolated singularities. Preprint, July 1990

40. Saper, L., Stern, M.: L_2-cohomology of arithmetic varieties. Proc. Natl. Acad. Sci. USA **84** (August 1987) 5516–5519

41. Saper, L., Stern, M.: L_2-cohomology of arithmetic varieties. Ann. Math. **132** (1990) 1–69

42. Satake, I.: On compactifications of the quotient spaces for arithmetically defined discontinuous groups. Ann. Math. **72** (1960) 555–580

43. Steenbrink, J.H.M.: Mixed Hodge structures associated with isolated singularities. In: Singularities. (Proc. Symposia Pure Math., vol. 40, part 2.) Amer. Math. Soc., Providence 1983

44. Zucker, S.: Hodge theory with degenerating coefficients: L_2 cohomology in the Poincaré metric. Ann. Math. **109** (1979) 415–476

45. Zucker, S.: L_2 cohomology of warped products and arithmetic groups. Invent. math. **70** (1982) 169–218

46. Zucker, S.: L_2-cohomology and intersection homology of locally symmetric varieties, II. Comp. Math. **59** (1986) 339–398

47. Zucker, S.: The Hodge structures on the intersection homology of varieties with isolated singularities. Duke Math. J. **55** (1987) 603–616

Nonabelian Hodge Theory

Carlos T. Simpson

Laboratoire de Topologie et Geometrie, U.F.R.M.I.G.
Université Paul Sabatier, 118, route de Narbonne, F-31062 Toulouse-Cedex, France

Introduction

Classically, Hodge theory and related constructions provided extra structure to abelian topological invariants of the usual topological spaces associated to algebraic varieties over $\mathbb{C}$. I would like to explain how, in analogy with these abelian constructions, certain nonabelian topological invariants of complex algebraic varieties have extra structures.

We will be concerned with two related invariants, the space R of framed n-dimensional representations of $\pi_1(X, x)$, and M, its universal categorical quotient by the action of $G\ell(n)$. These are related to the nonabelian cohomology $H^1(X, G\ell(n))$, which is the set of isomorphism classes of representations, and which has a structure of non-Hausdorff space. The moduli space M is the set of Jordan equivalence classes of representations, the universal Hausdorff space to which the cohomology space maps. The representation space R is the cohomology of X relative to a choice of base point x. We will interpret R and M as nonabelian analogues of the abelian cohomology $H^1(X, \mathbb{C})$.

The Nonabelian Hodge Theorem

Suppose X is smooth and projective over $\mathbb{C}$. A *harmonic bundle* on X is a C^∞ vector bundle E with differential operators ∂ and $\bar{\partial}$, and algebraic operators θ and $\bar{\theta}$ (operators from E to one-forms with coefficients in E), such that the following hold. There exists a metric K so that $\partial + \bar{\partial}$ is a unitary connection and $\theta + \bar{\theta}$ is self adjoint. And if we set $D = \partial + \bar{\partial} + \theta + \bar{\theta}$ and $D'' = \bar{\partial} + \theta$, then $D^2 = 0$ and $(D'')^2 = 0$. With these conditions, (E, D) is a vector bundle with flat connection, and $(E, \bar{\partial}, \theta)$ is a *Higgs bundle*: a holomorphic vector bundle with holomorphic section $\theta \in H^0(\text{End}(E) \otimes \Omega^1_X)$ such that $\theta \wedge \theta = 0$. Fix a class of Kähler metric for X. A Higgs bundle is *stable* (resp. *semistable*) if, for any coherent subsheaf $F \subset E$ preserved by θ with $0 < \text{rk}(F) < \text{rk}(E)$, we have

$$\deg(F)/\text{rk}(F) < (\text{resp.} \leq) \deg(E)/\text{rk}(E).$$

Proceedings of the International Congress
of Mathematicians, Kyoto, Japan, 1990
© The Mathematical Society of Japan, 1991

Theorem 1. *There is a natural equivalence between the categories of harmonic bundles on X and semisimple flat bundles (or representations of $\pi_1(X)$). There is a natural equivalence between the categories of harmonic bundles and direct sums of stable Higgs bundles with vanishing Chern classes. The resulting correspondence between representations and Higgs bundles can be extended to an equivalence between the categories of all representations of $\pi_1(X)$, and all semistable Higgs bundles with vanishing Chern classes.*

The first part of the theorem, relating harmonic bundles and representations, is due to Corlette [2] and Donaldson [6]. It is a generalization of the theorem of Eells and Sampson – a harmonic bundle is equivalent to a representation of the fundamental group in $G\ell(n)$ together with an equivariant harmonic map from the universal cover of X to $G\ell(n)/U(n)$. This equivalence uses the Bochner formula of Siu [18]. The part relating harmonic bundles and Higgs bundles is a generalization of the theorem of Narasimhan and Seshadri [11] (when $\theta = 0$ a Higgs bundle is just a vector bundle, and the corresponding representation is unitary). Part of this generalization involves the higher dimensional Hermitian-Yang-Mills theory for stable vector bundles developed by Donaldson [5] and Uhlenbeck, Yau [19]. Hitchin [9] gave the definition of stable Higgs bundles, and the proof of their correspondence with irreducible flat bundles, when X is a curve. These two directions are combined for the general statement in [14]. The last statement of the theorem comes from a formality result for the collection of complexes which control extensions [17].

A generalization to the case of noncompact curves is carried out in [16]. The noncompact higher dimensional case is still open (Corlette has some recent work on nonabelian L^2 cohomology in the higher dimensional case). For the remainder of the present discussion we will stick with the assumption that X is compact.

The set of flat bundles is analogous to the abelian de Rham cohomology, while the set of Higgs bundles (E, θ) is analogous to the abelian Dolbeault cohomology, $H^1(\mathcal{O}_X) \oplus H^0(\Omega_X^1)$. The first two parts of Theorem 1 may be interpreted as giving harmonic representatives for certain nonabelian de Rham and Dolbeault cohomology classes. The fact that the notion of harmonic representative (harmonic bundle) is the same in both cases, is the analogue of the abelian Kähler identity $\Delta_d = 2\Delta_{\bar{\partial}}$.

Variations of Hodge Structure. There is a natural $\mathbb{C}^*$ action on the category of semistable Higgs bundles with vanishing Chern classes, defined by $t : (E, \theta) \mapsto (E, t\theta)$.

Lemma 2. *The semisimple representations which are fixed by this action of $\mathbb{C}^*$ are exactly those which underly complex variations of Hodge structure.*

Proof. If (E, θ) is fixed, then there is an action of $\mathbb{C}^*$ on E, corresponding to a decomposition $E = \bigoplus E^p$ with $\theta : E^p \rightarrow E^{p-1} \otimes \Omega_X^1$. The spaces E^p will be mutually orthogonal with respect to the harmonic metric [14]. A harmonic bundle with orthogonal decomposition which is preserved by $\bar{\partial}$ and shifted once by θ is exactly a complex variation of Hodge structure. $\square$

Rigid Representations. A representation ϱ of $\pi_1(X)$ is *rigid* if any nearby representation is conjugate to it. The correspondence given by Theorem 1 is continuous on the moduli space of semisimple representations (see below). Thus, if a semisimple representation is rigid, then it must be fixed by $\mathbb{C}^*$, and Lemma 2 implies that it comes from a complex variation of Hodge structure. This can be strengthened:

Proposition 3. *Suppose ϱ is a rigid semisimple representation of $\pi_1(X)$. Then there is a $\mathbb{Q}$-variation of Hodge structure $V_{\mathbb{Q}}$ such that ϱ is a direct factor of the monodromy representation of $V_{\mathbb{Q}} \otimes \overline{\mathbb{Q}}$ (that monodromy is a sum of conjugates of ϱ). If ϱ is two dimensional, then the variation of Hodge structure $V_{\mathbb{Q}}$ can be assumed to have Hodge types $(1,0)$ and $(0,1)$ only.*

For an ℓ-adic analogue, suppose X is defined over $K \subset \mathbb{C}$. If a representation ϱ is rigid, and integral at a prime ℓ, then the resulting representation into $PG\ell(n, \overline{\mathbb{Q}}_\ell)$ descends to a representation of the algebraic fundamental group $\pi_1^{\text{alg}}(X_{K'})$ for some finite extension K'/K. These results suggest the following conjectures (which are actually special cases of the conjectures in [15]).

Conjecture 4 (rigid $\Rightarrow$ motivic). *Suppose ϱ is a rigid semisimple representation of $\pi_1(X)$. Then ϱ is a direct factor in the monodromy representation of a motive (i.e. family of varieties) over X.*

Conjecture 5 (rigid $\Rightarrow$ integral). *Suppose ϱ is a rigid semisimple representation of $\pi_1(X)$. Then ϱ is isomorphic to a representation with coefficients which are algebraic integers.*

Note that the first would imply the second. By Proposition 3, a rigid integral two dimensional representation comes from a family of abelian varieties, so in this case the second statement would imply the first. In the case of two dimensional representations, one can show that a Zariski dense representation into $S\ell(2)$ is either rigid, or else it is equal in $PS\ell(2)$ to a representation pulled back (via an algebraic map) from a curve with orbifold structure. So Conjecture 5 would imply that any such two dimensional representation is pulled back, either from a curve with orbifold structure, or a subspace of the moduli space of abelian varieties.

Restrictions on Fundamental Groups. The nonabelian Hodge theorem, and related considerations, have led to restrictions on which groups can be fundamental groups of smooth projective varieties (or really, of any compact Kähler manifolds). This subject begins with Siu's rigidity theorem [18]. Siu's method has been used in [1], [2], and [13]. A typical example is:

Theorem 6 (Carlson, Toledo). *Suppose Γ is a discrete cocompact subgroup of $SO(n,1)$ $(n > 2)$. Then Γ is not the fundamental group of a compact Kähler manifold.*

The monodromy group of a variation of Hodge structure must be a *Hodge group* [17]. For connected simple groups, the Hodge groups are all except the complex groups, and those isogenous to $S\ell(n, \mathbb{R})$ $(n \geq 3)$, $SU^*(2n)$ $(n \geq 3)$, $SO(p, q)$ $(p, q$ odd$)$, $E_{6(6)}$, or $E_{6(-26)}$. By work of many people, almost all lattices in real groups are known to be rigid, so some further nonexistence results can be derived from Proposition 3 (again, indirectly, from Siu's Bochner formula).

Theorem 7. *Suppose G is a real group which is not a Hodge group. Suppose $\Gamma \subset G$ is a rigid lattice. Then Γ is not the fundamental group of a compact Kähler manifold. If Γ is a group of the above form, and if Γ' is a semi-direct product of Γ with any other group (in which Γ is the quotient), then Γ' is not the fundamental group of a smooth projective variety.*

Siu's technique has been used to obtain other topological information. In a related direction, Corlette has used harmonic maps to prove some rigidity results about lattices in real groups. With a generalization of the Bochner technique applying to spaces with quaternionic structures, he proves that any cocompact lattice Γ in $Sp(n, 1)$ $(n \geq 2)$, or F_4^{-20}, is superrigid over archimedean fields [3].

Moduli Spaces

Fix a number n. The *Betti* versions of our topological invariants are easy to define. $R_B(X, x)$ is the space of representations of the fundamental group $\pi_1(X, x)$ on the fixed space $\mathbb{C}^n$. It is an affine variety defined by the generators and relations of the fundamental group. The group $G\ell(n, \mathbb{C})$ acts. There is a universal categorical quotient $M_B(X)$, sometimes called the character variety. When X is a smooth projective variety, R and M have *algebraic de Rham* and *Dolbeault* realizations, as provided by the following theorem.

Theorem 8. *Suppose X is a smooth projective variety over $\mathbb{C}$, with a point $x \in X$. There exist algebraic varieties $R_{DR}(X, x)$, $R_{Dol}(X, x)$, $M_{DR}(X)$, and $M_{Dol}(X)$, quasiprojective over $\mathbb{C}$, which are moduli spaces as follows. R_{DR} is a fine moduli space for the set of (E, ∇, β) where (E, ∇) is a vector bundle with integrable algebraic connection and $\beta : E_x \cong \mathbb{C}^n$ is a frame. R_{Dol} is a fine moduli space for the set of (E, θ, β) where (E, θ) is a semistable Higgs bundle with vanishing Chern classes, and β is a frame for E_x. The group $G\ell(n, \mathbb{C})$ acts on R_{DR} and R_{Dol}, and M_{DR} and M_{Dol} are the universal categorical quotients. Their points parametrize semisimple objects.*

The construction of the moduli space of Higgs bundles was treated in Hitchin's paper [9], and later by Nitsure [12], in the case when X is a curve. The higher dimensional case is treated in a preprint by the author (containing some errors – particularly with regard to the next proposition – which have subsequently been corrected).

Proposition 9. *The Riemann-Hilbert correspondence between bundles with integrable connection and representations provides isomorphisms of the associated complex analytic spaces, $(R_{DR})^{\text{an}} \cong (R_B)^{\text{an}}$ and $(M_{DR})^{\text{an}} \cong (M_B)^{\text{an}}$.*

The Riemann-Hilbert correspondence is not algebraic. It provides an example of two different algebraic varieties whose associated complex analytic spaces are isomorphic, generalizing the example of Serre.

Proposition 10. *The correspondence of Theorem 1 provides a homeomorphism between the usual topological spaces underlying M_{Dol} and M_B. However, the extended correspondence between the points of R_{Dol} and R_B provided by the last part of Theorem 1, is not continuous.*

This proposition was treated by Hitchin in his case [9]. Hitchin also discussed several important properties of the moduli space M_{Dol}, all of which carry over to the case of higher dimensional X:

Theorem 11. *There is a natural algebraic action of $\mathbb{C}^*$ on the space M_{Dol}, representing the action defined previously. There is a proper map from M_{Dol} to a vector space, compatible with an action of $\mathbb{C}^*$ on the vector space having only positive weights. Thus if p is any point in M_{Dol}, then the limit $\lim_{t \to 0} tp$ exists in M_{Dol}.*

A corollary of this theorem, the continuity of Proposition 10, and the characterization of Lemma 2, is that any representation of $\pi_1(X)$ can be deformed to a complex variation of Hodge structure. This provides the mechanism for proving the restrictions stated in the second half of Theorem 7.

The Gauss-Manin Connection. Suppose $X \to S$ is a smooth projective family over a base scheme S, with section of base points $x : S \to X$. Then the above constructions give families of moduli spaces $R_{DR}(X/S, x)$ and $M_{DR}(X/S)$ over S. There is an analogue of the Gauss-Manin connection.

Theorem 12. *Suppose X/S is smooth and projective with base point $x(S)$. Let $(S \times S)^\wedge$ denote the formal neighborhood of the diagonal in $S \times S$, similarly in $S \times S \times S$. There is a canonical isomorphism*

$$\varphi : p_1^*(R_{DR}(X/S))|_{(S \times S)^\wedge} \xrightarrow{\cong} p_2^*(R_{DR}(X/S))|_{(S \times S)^\wedge},$$

such that on $(S \times S \times S)^\wedge$ the cocycle condition $p_{23}^*(\varphi)p_{12}^*(\varphi) = p_{13}^*(\varphi)$ is satisfied. The same holds for the universal categorical quotient $M_{DR}(X/S)$. These isomorphisms are compatible with the local analytic trivializations which arise from the Riemann-Hilbert correspondence.

Proof. The category of vector bundles with connection on X/S can be interpreted as a category of crystals on X/S. If $S_0 \subset S$ is a closed subscheme defined by a nilpotent ideal, and X_0 is the inverse image of S_0, then the category of crystals on X/S is equivalent to the category of crystals on X_0/S_0. Hence the scheme $R_{DR}(X/S)$ over S depends functorially only on the family $X_0 \to S_0$ and the inclusion $S_0 \subset S$, although the existence of $R_{DR}(X/S)$ seems to depend on the existence of a family $X \to S$. This gives a crystal of schemes over the stratifying site of S, which is equivalent to the structure described in the theorem. $\square$

At points where S and $R_{DR}(X/S)$ are smooth, the structure provided by this theorem may be described more simply as a splitting $\pi^* TS \to T(R_{DR}(X/S))$ of the quotient of tangent bundles, satisfying an integrability condition. If S is an artinian scheme, it gives a natural isomorphism $R_{DR}(X/S) \cong R_{DR}(X_0) \times S$.

The Hyperkähler Structure. The identification given by Theorem 1 is not complex analytic. Thus, the same space is given several complex structures.

Theorem 13 (Hitchin). *Denote by M^s, the set of smooth points of the moduli space. The isomorphism $M_{Dol}^s \cong M_B^s$ is smooth. Let I and J denote the complex structures on M^s obtained from M_{Dol}^s and M_B^s respectively. Then $K = IJ$ is a complex structure, and this triple gives a quaternionic structure in the tangent space at any point. The resulting quaternionic manifold M^s has a natural hyperkähler structure.*

This was pointed out by Hitchin [9] in the case when X is a curve; the general case has been treated in [4] and recently, in detail, by Fujiki [7]. We refer to Hitchin's paper for the definition of hyperkähler manifold.

The Space of λ-Connections. Here is a definition due to Deligne – the contents of this subsection were all described or outlined by him in correspondence [4]. Suppose $\lambda \in \mathbb{C}$. A λ-*connection* on a vector bundle E is an operator $\nabla : E \to E \otimes \Omega_X^1$, such that $\nabla(ae) = \lambda d(a)e + a\nabla(e)$. If $\lambda \neq 0$ then a λ-connection ∇ corresponds to a connection $\lambda^{-1}\nabla$. On the other hand, if $\lambda = 0$ then a bundle with integrable λ-connection is just a Higgs bundle (E, θ), $\theta = \nabla$. So this definition provides a deformation from the notion of connection to the notion of Higgs bundle. Its importance comes from the resulting moduli space.

Theorem 14. *Suppose X is smooth and projective over $\mathbb{C}$, with base point x. There exist quasiprojective algebraic varieties $W^+ \to \mathbf{A}^1$ and $Z^+ \to \mathbf{A}^1$ which are moduli spaces for λ-connections as follows. W^+ is a fine moduli space for $(\lambda, E, \nabla, \beta)$ where $\lambda \in \mathbf{A}^1$, (E, ∇) is a vector bundle with integrable λ-connection (semistable with vanishing Chern classes in the case $\lambda = 0$), and β is a frame for E_x. The group $G\ell(n, \mathbb{C})$ acts on W^+ with universal categorical quotient Z^+.*

The fibers W_1^+, W_0^+, Z_1^+, and Z_0^+ are respectively the moduli spaces R_{DR}, R_{Dol}, M_{DR}, and M_{Dol}. There are natural actions of $\mathbb{C}^*$ on W^+ and Z^+, defined by $t : (\lambda, E, \nabla, \beta) \mapsto (t\lambda, E, t\nabla, \beta)$, covering the standard action on $\mathbf{A}^1$.

Deligne thought of the space of λ-connections as a way of approaching the notion of quaternionic structure for the singular space M. There are complex analytic spaces W and Z over $\mathbb{P}^1$ characterized by the following properties. There are actions of $\mathbb{C}^*$ covering the standard action on $\mathbb{P}^1$, and antilinear involutions σ compatible with the action of $\mathbb{C}^*$ (and the map to $\mathbb{P}^1$) relative the involution $t \mapsto (\bar{t})^{-1}$. There are identifications

$$W|_{\mathbf{A}^1} \cong W^+ \qquad Z|_{\mathbf{A}^1} \cong Z^+$$

compatible with the action of $\mathbb{C}^*$. On the fibers $W_1 \cong R_B$ and $Z_1 \cong M_B$, the involution σ takes a representation to the complex conjugate representation.

A harmonic bundle E yields a family of holomorphic bundles $(E, \bar{\partial} + \lambda \bar{\theta})$ with λ-connections $\nabla = \lambda \partial + \theta$, indexed by $\lambda \in \mathbf{A}^1$. The resulting section $\mathbf{A}^1 \to W^+$ is holomorphic and extends to a section $\mathbb{P}^1 \to W$. Refer to these sections (and their projections in Z) as *preferred sections*. The involution σ maps a preferred section to the preferred section corresponding to the complex conjugate representation. We get a trivialization $Z \cong M \times \mathbb{P}^1$, where M denotes the moduli space of harmonic bundles.

Proposition 15 (Deligne). *The trivialization $Z \cong M \times \mathbb{P}^1$ is a homeomorphism, and on the set of smooth points it identifies Z^s with the twistor space for the quaternionic structure of M^s.*

Proof. There is a $\mathbb{P}^1$ of complex structures q in the quaternions. The twistor space is $M^s \times \mathbb{P}^1$ endowed with a complex structure which makes the horizontal projective lines holomorphic, and which endows each fiber $M^s \times \{q\}$ with the complex structure q deduced from the quaternionic structure of M^s. One has to check that the identification $Z_q \cong M^s$ takes the complex structure of Z_q to the complex structure q of M^s. $\qquad\square$

Deligne proposes a way of defining the notion of quaternionic structure for a singular space such as M by some axioms for a twistor space such as Z.

Compactifications. The space of λ-connections suggested by Deligne turns out to be very useful for compactifying the moduli spaces.

Lemma 16. *Suppose z is a point of Z^+. Then the limit $\lim_{t \to 0} tz$ exists in Z^+. Also, the set of fixed points of $\mathbb{C}^*$ in Z^+ is compact.*

Proposition 17. *Suppose Z^+ is any algebraic variety with an algebraic action of $\mathbb{C}^*$, such that the limits $\lim_{t \to 0} tz$ exist, and such that the set of fixed points is compact. Let $S^+ \subset Z^+$ denote the subset of points s such that $\lim_{t \to \infty} ts$ exists in Z^+. Then there is a universal geometric quotient $P = (Z^+ - S^+)/\mathbb{C}^*$, and P is a proper scheme (not necessarily projective).*

We can apply this construction to the moduli space of λ-connections Z^+, to obtain a compactification of M_{DR}.

Corollary 18. *Let Z^+ be the space of λ-connections. Let $P_{DR} = (Z^+ - S^+)/\mathbb{C}^*$ be the proper scheme defined above. Then $M_{DR} \subset P_{DR}$ is an open subset, providing a compactification of the space of bundles with integrable connection.*

Proof. Note that the subset $S^+ \subset Z^+$ is contained in the fiber over $0 \in \mathbf{A}^1$. Let $D = (Z_0^+ - S^+)/\mathbb{C}^* \subset P_{DR}$. It is a closed subset, and the complement is equal to $(Z^+ - Z_0^+)/\mathbb{C}^* \cong M_{DR}$. $\qquad\square$

There is a similar compactification $M_{Dol} \subset P_{Dol}$ where the complement is the same divisor D. In fact, these fit together into a relative compactification of Z^+ over $\mathbf{A}^1$.

The Hodge Filtration. Here is another use for the space of λ-connections. It can be thought of as the nonabelian analogue of the Hodge filtration on the de Rham moduli space. Suppose V is a finite dimensional vector space with an exhaustive decreasing filtration F^p. Define a locally free sheaf $\xi(V, F)$ over $\mathbf{A}^1$ as the submodule of $V \otimes \mathbf{C}[t, t^{-1}]$ generated by the $t^{-p} F^p$.

Lemma 19. *The construction ξ provides an equivalence between the category of vector spaces with exhaustive decreasing filtrations, and the category of locally free sheaves on $\mathbf{A}^1$ with action of $\mathbf{C}^*$ covering the standard action on $\mathbf{A}^1$.*

The vector space V is recovered as the fiber of $\xi(V, F)$ at 1, whereas the fiber at 0 is canonically the associated graded $\mathrm{Gr}_F(V)$. Given a real structure, one can glue together $\xi(V, F)$ and $\xi(V, \overline{F})$ using the involution $t \mapsto (\bar{t})^{-1}$, to obtain a locally free sheaf $\xi(V, F, \overline{F})$ on $\mathbf{P}^1$ with action of $\mathbf{C}^*$ and antilinear involution.

In view of this lemma, we can define a *nonabelian filtration* to be a scheme over $\mathbf{A}^1$ with action of $\mathbf{C}^*$. The spaces W^+ and Z^+ over $\mathbf{A}^1$, together with their $\mathbf{C}^*$ actions, may then be interpreted as the *nonabelian Hodge filtrations* of the fibers $R_{\mathrm{DR}} = W_1^+$ and $M_{\mathrm{DR}} = Z_1^+$. The spaces $\sigma(W^+)$ and $\sigma(Z^+)$ (which are the restrictions of W and Z to the affine neighborhood of infinity) are the complex conjugates of the Hodge filtrations. The Dolbeault spaces R_{Dol} and M_{Dol} become the associated graded spaces of the nonabelian Hodge filtrations (just as for abelian cohomology). Lemma 16 is analogous to saying that the Hodge filtration is concentrated in positive degrees.

These definitions are compatible with the mixed Hodge structure defined by Morgan [10] and Hain [8] on the nilpotent completion of the fundamental group. Let $A = (\mathbf{C}\pi_1)^\wedge$ be the completion of the group algebra at the augmentation ideal. They show that this has a Hodge filtration F giving a mixed Hodge structure (the real structure being the completion of the real group algebra, and the weight filtration given by powers of the augmentation ideal). By Lemma 19, this data is equivalent to the sheaf of algebras $\xi(A, F, \overline{F})$ together with its $\mathbf{C}^*$ action and involution.

Given a complete augmented algebra such as A, one can define a formal scheme $R(A)$, the space of representations infinitesimally close to the trivial one. For $A = (\mathbf{C}\pi_1)^\wedge$ as above, the space $R(A)$ is naturally isomorphic to the completion of R_{B} at the point corresponding to the trivial representation.

Theorem 20. *The family of algebras $\xi(A, F, \overline{F})$ over $\mathbf{P}^1$ gives rise to a family of formal schemes $R(\xi(A, F, \overline{F}))$ over $\mathbf{P}^1$. This family of formal schemes (with $\mathbf{C}^*$ action and involution) is naturally isomorphic to the formal neighborhood of the trivial preferred section in W. The isomorphism is compatible with the one described above in the fiber over 1.*

Conversely, Tannaka duality allows the mixed Hodge structure $\xi(A, F, \overline{F})$ to be recovered from the data of the formal neighborhoods of the spaces W, together with some information on how they fit together for different ranks.

Speculations

It is natural to wonder how the results outlined above might fit into a notion of *nonabelian Hodge structure*. The action of $\mathbb{C}^*$ on the space of representations gives, by Tannaka duality, an action of $\mathbb{C}^*$ on the *proalgebraic completion* $\varpi_1(X)$ (the inverse limit of the Zariski closures of the images, over all finite dimensional representations of $\pi_1(X)$). The action of $S^1 \subset \mathbb{C}^*$ on the proreductive quotient ϖ_1^{red} satisfies some axioms which might be taken as a preliminary definition of pure Hodge structure on a group [17]. The map $S^1 \times \pi_1(X) \to \varpi_1^{red}$ is continuous when the right side is given the inverse limit of analytic topologies. And the action of $-1 \in S^1$ is a Cartan involution of the real form $(\varpi_1^{red})_{\mathbb{R}}$ (inverse limit of real Zariski closures). The restrictions of the first half of Theorem 7 may be derived from these axioms (one would like to interpret all those kinds of topological restrictions as following from the nonabelian Hodge structure).

This definition refers to the continuous structure of the space of representations only in the crudest way. Preferable would be a notion of nonabelian Hodge structure for the space of representations. We have defined the nonabelian Hodge filtration. What about weight filtrations and polarizations? Probably the hyperkähler metric will play a role. The mixed Hodge structure on $(\mathbb{C}\pi_1)^{\wedge}$ will provide the power series for the nonabelian Hodge structure at the trivial representation (Theorem 20).

With or without a precise definition, we are afforded the pleasant opportunity of revisiting all the old glories of Hodge theory, and trying to provide their nonabelian analogues. The subject started with Griffiths' transversality result. Preliminary heuristic calculations seem to show that an analogous transversality relates the Gauss-Manin connection defined in Theorem 12, to the Hodge filtration interpreted as the space Z^+. The lifts of vector fields from the base S, to the restriction of $Z^+(X/S)$ over $\mathbb{A}^1 - \{0\}$, will have poles of order ≤ 1 along Z_0^+ (and hence also along the divisor at infinity in the compactificaton).

This raises the question of whether there is a good notion of *variation of nonabelian Hodge structure* over S. One could try to replicate the works of Schmid and others about degenerating variations. Is there an appropriate sense in which the Gauss-Manin connection has *regular singularities*? And is there some type of limiting nonabelian Hodge structure? This type of theory would be in the service of devissage, trying to obtain information about the topology of a space X by fibering $X \to Y$ and studying the family of fundamental groups of the fibers. For example, the moduli space $M(X)$ (with all of its structures) becomes a constant subspace of the varying family $M(X_y)$. The wide array of strutures on $M(X_y)$ seems to make this an unlikely event.

Let me close with some wild speculation on how this picture might fit into an overall view of Hodge theory. We have treated two topological invariants, the space R of framed representations, and the quotient M. For which topological invariants will there be such a theory? Suppose $F : \text{Top} \to \text{Sch}$ is a functor from the category of topological spaces to the category of schemes. Are there some simple conditions which would guarantee that, for a smooth projective variety X, $F(X)$ has a "Hodge structure"? Perhaps (to say something entirely unsubstantiated) there might be an action of $\mathbb{C}^*$ on the set of such functors F, and the functors which are fixed are of Hodge type.

References

1. Carlson, J.A., Toledo, D.: Harmonic mappings of Kähler manifolds to locally symmetric spaces. Publ. Math. I.H.E.S. **69** (1989) 173–201
2. Corlette, K.: Flat G-bundles with canonical metrics. J. Diff. Geom. **28** (1988) 361–382
3. Corlette, K.: Archimedean superrigidity and hyperbolic geometry. Preprint
4. Deligne, P: Letter, March 20 (1989)
5. Donaldson, S.K.: Anti self dual Yang-Mills connections over complex algebraic surfaces and stable vector bundles. Proc. London Math. Soc. **50** (1985) 1–26
6. Donaldson, S.K.: Twisted harmonic maps and self-duality equations. Proc. London Math. Soc. **55** (1987) 127–131
7. Fujiki, A.: Hyperkähler structure on the moduli space of flat bundles. Preprint
8. Hain, R.: The de Rham homotopy theory of complex algebraic varieties, I. K-Theory **1** (1987) 271–324
9. Hitchin, N.J.: The self-duality equations on a Riemann surface. Proc. London Math. Soc. **55** (1987) 59–126
10. Morgan, J.: The algebraic topology of smooth algebraic varieties. Publ. Math. I.H.E.S. **48** (1978) 137–204
11. Narasimhan, M.S., Seshadri, C.S.: Stable and unitary bundles on a compact Riemann surface. Ann. Math. **82** (1965) 540–564
12. Nitsure, N.: Moduli space of semistable pairs on a curve. Proc. London Math. Soc. (3) **62** (1991) 275–300
13. Sampson, J.H.: Applications of harmonic maps to Kähler geometry. Contemp. Math. **49** (1986) 125–133
14. Simpson, C.T.: Constructing variations of Hodge structure using Yang-Mills theory and applications to uniformization. J.A.M.S. **1** (1988) 867–918
15. Simpson, C.T.: Transcendental aspects of the Riemann-Hilbert correspondence. Illinois J. Math. **34** (1990) 368–391
16. Simpson, C.T.: Harmonic bundles on noncompact curves. J.A.M.S. **3** (1990) 713–770
17. Simpson, C.T.: Higgs bundles and local systems. Publ. Math. I.H.E.S. (to appear)
18. Siu, Y.T.: Complex analyticity of harmonic maps and strong rigidity of complex Kähler manifolds. Ann. Math. **112** (1980) 73–110
19. Uhlenbeck, K.K., Yau, S.T.: On the existence of Hermitian-Yang-Mills connections in stable vector bundles. Comm. Pure and Appl. Math. **39-S** (1986) 257–293

Arithmetic and Hyperbolic Geometry

Paul Vojta

Department of Mathematics, University of California, Berkeley, CA 94720, USA

We begin by recalling Faltings' theorem (née Mordell's conjecture):

Theorem 0.1 (Faltings, 1983). *Let C be a curve of genus > 1 defined over a number field k. Then the set $C(k)$ of k-rational points on C is finite.*

Compact Riemann surfaces of genus > 1 are also characterized by the property that they are hyperbolic, in any of the senses discussed in the next section. Recent work has suggested that a similar link exists, formally at least, between arithmetic properties of algebraic varieties (of any dimension) and complex analytic properties of the corresponding complex manifold. The goal of this talk is to discuss generalizations of Theorem 0.1 which have been motivated by this analogy.

1. Hyperbolicity

We begin by defining several notions of hyperbolicity. Throughout this section let X be a connected complex manifold, not necessarily compact.

Definition 1.1. The manifold X is *negatively curved* if there exists a $(1, 1)$-form ω on X and a constant $\kappa > 0$ such that all holomorphic sectional curvatures of ω are $< -\kappa$.

Definition 1.2 (Kobayashi, 1970). The Poincaré distance d_{hyp} on the open unit disc $\mathbb{D} \subseteq \mathbb{C}$ is the distance given infinitesimally by the form

$$\frac{dz \wedge d\bar{z}}{(1 - |z|^2)^2} .$$

Then X is said to be *Kobayashi hyperbolic* if there exists a distance d_X on X (with $d_X(x, y) > 0$ whenever $x \neq y$) such that for all holomorphic maps $f : \mathbb{D} \to X$, $f^* d_X \leq d_{\text{hyp}}$.

Definition 1.3. The manifold X is *Brody hyperbolic* if all holomorphic maps from $\mathbb{C}$ to X are constant.

Proceedings of the International Congress
of Mathematicians, Kyoto, Japan, 1990
© The Mathematical Society of Japan, 1991

Actually, the latter two definitions are valid when X is a complex space. We have

$$\text{negatively curved} \implies \text{Kobayashi hyperbolic} \implies \text{Brody hyperbolic}.$$

The converse of the second arrow holds if X is compact, but not in general.

A general reference on hyperbolicity is Lang (1987); see also Kobayashi (1970).

2. Conjectures in Number Theory

The first conjecture relating complex analytic and arithmetic properties is the following.

Conjecture 2.1 (Lang, 1974). *A complete variety X/k has only finitely many rational points if the corresponding complex space is Kobayashi hyperbolic.*

More recently the above statement, which is qualitative in nature, has been replaced by a more quantitative version. To describe it, we first state a theorem.

Theorem 2.2. *Let D be an effective divisor on a complete irreducible nonsingular curve C, and assume that D has no multiple points. Let K be a canonical divisor on C, let A be an ample divisor on C, and fix $\varepsilon > 0$. Then for almost all symbols "?,"*

$$m(D, ?) + T_K(?) \leq \varepsilon\, T_A(?) + O(1). \tag{1}$$

Supplied with one set of definitions, this theorem is Nevanlinna's Second Main Theorem. Indeed, let C be a compact Riemann surface, and let $f : \mathbb{C} \to C$ be a holomorphic function, regarded as an infinite collection of maps $f_r : \overline{\mathbb{D}}_r \to C$ (here $\overline{\mathbb{D}}_r$ is the closed disc of radius r, $r > 0$). Replace the symbol "?" with r, and interpret "almost all" to mean all r outside a set of finite Lebesgue measure. Choose some distance function $\text{dist}(P, Q)$ on C, and extend it to divisors by the formula

$$\text{dist}(P, \Sigma n_Q \cdot Q) = \prod \text{dist}(P, Q)^{n_Q}.$$

Then, for any divisor D on C let

$$m(D, r) = \int_0^{2\pi} -\log \text{dist}(f(re^{i\theta}), D)\, \frac{d\theta}{2\pi};$$

$$N(D, r) = \sum_{w \in \mathbb{D}_r} \text{ord}_{z=w} f^* D \log \frac{r}{|w|};$$

$$T_D(r) = m(D, r) + N(D, r).$$

This gives a weak form of the Second Main Theorem for curves.

With a different set of definitions, we obtain an arithmetic result which goes by several names, depending on the genus of the curve. Let C be a curve defined over a number field k, and assume also that D is defined over k. We now replace "?" by points $P \in C(k) \setminus \operatorname{Supp} D$, and interpret "almost all" to mean all but finitely many points P. Fix a finite set S of places of k, and for all places v of k let $\operatorname{dist}_v()$ denote some fixed distance on C in the v-adic topology, extended to divisors as before. It is assumed that these are chosen uniformly in some sense, as in Lang (1983). Then for any divisor D on C defined over k, let

$$m(D,P) = \frac{1}{[k : \mathbb{Q}]} \sum_{v \in S} -\log \operatorname{dist}_v(P,D) \, ;$$

$$N(D,P) = \frac{1}{[k : \mathbb{Q}]} \sum_{v \notin S} -\log \operatorname{dist}_v(P,D) \, ;$$

$$T_D(P) = m(D,P) + N(D,P)$$

$$= \frac{1}{[k : \mathbb{Q}]} \sum_{v} -\log \operatorname{dist}_v(P,D) \, .$$

The quantity T_D is thus the Weil height of P relative to D; in particular if D is ample then there are only finitely many points $P \in C(k)$ of bounded height. In the arithmetic case we will also write $h_D(P)$ for the height.

Proof of Theorem 2.2 (Algebraic Variant). Let g denote the genus of C.

Case 1. If $C(k)$ is finite then the inequality must hold; by Faltings' theorem 0.1 this must hold if $g > 1$. Conversely, if $g > 1$ then K is ample, so taking $D = 0$, $A = K$, and $\varepsilon < 1$, we see that Theorem 2.2 implies a bound on the height of rational points; hence it implies Mordell's conjecture.

Case 2. If $g = 0$ then Theorem 2.2 is equivalent to Roth's theorem:

Theorem (Roth, 1955). *For each $v \in S$ fix $\alpha_v \in \overline{\mathbb{Q}}$. Then for all $x \in k \setminus \{\alpha_v\}$,*

$$\frac{1}{[k : \mathbb{Q}]} \sum_{v \in S} -\log \min(\|x - \alpha_v\|_v, 1) < (2 + \varepsilon) h(x) \, . \tag{2}$$

Here $h(x) = h_A(x)$ with $A = \mathcal{O}(1)$. For details, see Vojta (1987), Sect. 3.2.

Case 3. If $g = 1$, then Theorem 2.2 is equivalent to an approximation statement on elliptic curves, derived by Lang from Roth's theorem using methods of Siegel. See Lang (1960b) for details. $\qquad\square$

Thus we find that the Second Main Theorem of Nevanlinna theory translates into the number field case, giving several known theorems. Generally speaking, any such theorem in Nevanlinna theory should translate into a true statement for number fields. Such statements often agree with diophantine conjectures previously made. See also Sect. 8, as well as Vojta (1987), Chap. 4.

Again, this framework is related to hyperbolicity because proofs of Nevanlinna theory exist using notions of curvature.

In Nevanlinna theory, the analogues of Roth's theorem and Mordell's conjecture are both proved by a common proof. This suggested that Mordell's conjecture should be provable by methods similar to those used in the proof of Roth's theorem. In fact, such a proof was found in Vojta (1989) and Vojta (1990).

3. Sketch of Roth's Proof

Roth's proof proceeds by contradiction. Assume that there are infinitely many $x \in k$ not satisfying (2). Then we choose a finite set with certain properties. These are used to construct an auxiliary polynomial, the properties of which lead to a contradiction.

To start, let $\alpha_1, \ldots, \alpha_m$ be all the conjugates of all α_v, $v \in S$. We construct a polynomial $P(x_1, \ldots, x_n)$, of degree d_i in x_i, with a specified type of zero at all points $(\alpha_i, \ldots, \alpha_i)$, $1 \leq i \leq m$. This is viewed as a linear algebra problem, with the coefficients of the polynomial as variables. The vanishing conditions are set up so as to use up almost all of the degrees of freedom in choosing P; by the Dirichlet Box Principle, we can find such a polynomial in $R[x_1, \ldots, x_n]$ with bounded coefficients, where R is the ring of integers of k. This step is often referred to as Siegel's lemma.

Next we choose a set of counterexamples to (2) such that

$$1 \ll h(x_1) \ll \cdots \ll h(x_n)$$

and such that, for all $v \in S$, the numbers

$$\frac{-\log \min(\|x_i - \alpha_v\|_v, 1)}{h(x_i)}, \qquad 1 \leq i \leq n$$

are close to each other. This involves a sphere packing argument.

Finally, for each $v \in S$, using the Taylor expansion of P at $(\alpha_v, \ldots, \alpha_v)$, the type of zero of P at that point, and the bound on the coefficients of P, we obtain a bound for $\|P(x_1, \ldots, x_n)\|_v$ and therefore for its norm. If the norm is too small, we obtain a contradiction unless $P(x_1, \ldots, x_n) = 0$. A similar argument applied to certain derivatives of P implies that they vanish as well. Hence P has a certain type of zero at $(x_1, \ldots, x_n)$. But this contradicts Roth's lemma: the type of zero of P is bounded from above, depending on the size of the coefficients of P and the heights of the x_i.

4. Vojta's Proof

A key difficulty in extending Roth's proof to the case of rational points is dealing with the absence of the α's, which are needed to (a) provide vanishing conditions which use up the degrees of freedom in choosing P, and (b) prove the vanishing of $P(x_1, \ldots, x_n)$, via the assumption that (2) fails to hold.

This difficulty is overcome by means of intersection theory; in particular, we make use of the fact that the diagonal divisor Δ on $C \times C$ is a divisor with negative self-intersection. Also let F_1 (resp. F_2) be a divisor on $C \times C$ coming from a divisor F of degree 1 on the first (resp. second) factor. Then let

$$\Delta' = \Delta - F_1 - F_2 \,;$$
$$Y_r = \Delta' + a_1 F_1 + a_2 F_2 \,,$$

where $a_i \in \mathbb{Q}$ are chosen so that $a_1/a_2 = r$ and Y_r^2 is positive but small. Then, by Riemann-Roch, high multiples of Y_r will have relatively few global sections; this is how we use up our degrees of freedom in this case. (Here r is a large rational number.)

The proof replaces many of the classical arguments involving polynomials and absolute values with the language of arithmetic intersection theory, as developed by Arakelov, Gillet, and Soulé. Therefore, we work on an arithmetic scheme W of dimension 3 corresponding to $C \times C$, and replace the Siegel's lemma argument with a use of the Riemann-Roch theorem of Gillet and Soulé, to construct a section y of a multiple of Y_r with certain arithmetical properties.

Next we choose points, only two this time, such that

$$1 \ll h(P_1) \ll h(P_2)$$

and satisfying a certain sphere packing condition again – this time in the Mordell-Weil group of the Jacobian of C. Let X be the arithmetic surface corresponding to C; for $i = 1, 2$ let E_i be the (Arakelov) divisor on X corresponding to P_i, and let E be the curve on W corresponding to (P_1, P_2). Then this sphere packing condition implies that $(E_1 . E_2)$ on X will be small, and therefore $(E . \Delta)$ on W will be small. Then $(E . Y_r)$ will be small and in fact negative. Thus y must be zero on E; similarly certain derivatives must vanish, giving a lower bound for the type of zero as before. Again, this produces a contradiction.

As is the case with Roth's theorem, this proof – as well as all proofs derived from it – is ineffective; i.e., it can give a bound on the *number* of rational points but not on their heights. Thus it is often true that one cannot prove that a given set of rational points is the set of *all* rational points.

5. Extensions Due to Faltings

In Faltings (1990), Vojta's methods were generalized, giving two new theorems.

Theorem 5.1 (Conjectured in Lang (1960a)). *Let X be a complete subvariety of an abelian variety A, and assume that X does not contain any translates of any nontrivial abelian subvarieties of A. Let k be a number field over which X and A are defined. Then $X(k)$ is finite.*

Theorem 5.2. *Let A be an abelian variety, and let $Y \subseteq A$ be a subvariety. Let k be a number field over which A and Y are defined, let v be a place of k, and let $\mathrm{dist}_v(x, Y)$ be a v-adic distance on A. Fix $\varepsilon > 0$, and fix a height function $h(x)$ on A relative to some ample divisor. Then the set of $x \in A(k) \setminus Y$ such that*

$$-\log \operatorname{dist}_v(x, Y) > \varepsilon h(x)$$

is finite.

Corollary 5.3 (Conjectured in Lang (1960a)). *Let H be an ample divisor on A. Then the set of integral points on $A(k)$ relative to H is finite.*

In proving these results, Faltings adds the following ingredients to Vojta's work.

1. Use of the Poincaré divisor on $A \times A$ in place of Δ'.

2. Use of the product of A with itself n times instead of just two. The divisor he considers is again a combination of divisors from each factor, together with Poincaré divisors associated to the i-th and $(i+1)$st factors, $1 \le i < n$.

3. A new zero estimate (the "product theorem") is needed to deal with technical issues concerning the use of n factors.

4. Instead of proving separately an upper bound on the type of zero of the section y at $(P_1, \ldots, P_n)$, he builds this upper bound into the construction of y, using Minkowski's theorem on successive minima. Also, the use of the Gillet-Soulé Riemann-Roch theorem is replaced by more elementary manipulations.

6. Recent Work of Bombieri

A few months ago Bombieri reworked Vojta's proof, obtaining a more elementary proof using only the classical theory of heights, Siegel's lemma, and a few necessary results from algebraic geometry over a field; see Bombieri (1990). In particular, he also eliminated the use of the Gillet-Soulé Riemann-Roch theorem, but in a manner independent of the extensions due to Faltings.

Bombieri uses the same divisor Y as in Sect. 4, but he expresses it as a difference of two very ample divisors, as follows.

First choose a positive integer s such that

$$B := sF_1 + sF_2 - \Delta'$$

is very ample; write

$$\phi_B : C \times C \to \mathbb{P}^m$$

for the corresponding embedding. Likewise choose a divisor F of degree 1 on C and a positive integer N such that NF is very ample, giving an embedding

$$\phi_{NF} : C \to \mathbb{P}^n.$$

This gives another embedding

$$\psi : C \times C \to \mathbb{P}^n \times \mathbb{P}^n.$$

It then follows from the Enriques-Severi-Zariski lemma (Zariski 1952) that for all sufficiently large integers δ_1, δ_2, the map

$$\Gamma\left(\mathbb{P}^n \times \mathbb{P}^n, \mathcal{O}(\delta_1, \delta_2)\right) \to \Gamma\left(C \times C, \delta_1 NF_1 + \delta_2 NF_2\right)$$

is surjective. Thus, writing

$$dY = d_1 NF_1 + d_2 NF_2 - dB,$$

we find that for any sections s of dY and s' of dB, the product ss' is the restriction to $C \times C$ of a global section on $\mathbb{P}^n \times \mathbb{P}^n$. Thus, it is possible to revert to use of polynomials in this proof.

Since he is working with polynomials over a field, the arguments in Vojta's proof regarding analytic torsion and cohomology over finite fields are unnecessary.

Also, to obtain the upper bound on the type of zero of the section at (P_1, P_2), Bombieri uses suitably chosen projections from C to $\mathbb{P}^1$; thus he is able to use the same lemma as Roth uses.

This latter simplification, together with the ideas of Faltings concerning use of Siegel's lemma in place of the Gillet-Soulé Riemann-Roch theorem, remove the obstacles to combining this proof with Roth's original proof. Thus we have a unified proof of the number field version of Theorem 2.2.

Of course, combining the arithmetic proof with the proof in the case of Nevanlinna theory remains a distant goal.

7. Algebraic Points of Bounded Degree

The unified proof mentioned above can be further extended to give a generalization of Theorem 2.2 which contains Wirsing's generalization, Wirsing (1971), of Roth's theorem to algebraic points of bounded degree.

Before stating the theorem, however, let X be an arithmetic surface corresponding to C, and for all points $P \in C(\bar{k})$ let E_P denote the corresponding Arakelov divisor on X. Then let

$$d_{\mathrm{a}}(P) = \frac{(E \cdot E + \omega_{X/B})}{[k(P) : \mathbb{Q}]},$$

where $\omega_{X/B}$ is the relative dualizing sheaf.

Theorem 7.1. *Let C, k, D, K, A, and ε be as in Theorem 2.2. Also fix $v \in \mathbb{Z}$, $v > 0$. Then for all points $P \in C(\bar{k})$ with $[k(P) : k] \leq v$, we have*

$$m(D, P) + h_K(P) \leq d_{\mathrm{a}}(P) + \varepsilon h_A(P) + O(1).$$

This is a weak form of the following conjecture.

Conjecture 7.2. *Under the same conditions as Theorem 7.1, we have*

$$m(D, P) + h_K(P) \leq d(P) + \varepsilon h_A(P) + O(1),$$

where

$$d(P) := \frac{\log |D_{k(P)/\mathbb{Q}}|}{[k(P) : \mathbb{Q}]}.$$

This conjecture is very strong; in particular it implies the *abc* conjecture:

Conjecture 7.3 (Masser-Oesterlé). *Given $\varepsilon > 0$ there exists a constant $C = C(\varepsilon) > 0$ such that for all relatively prime integers a, b, and c with $a + b + c = 0$, we have*

$$\max(|a|, |b|, |c|) < C \cdot \prod_{p|abc} p^{1+\varepsilon}.$$

This conjecture, in turn, implies the "asymptotic Fermat conjecture:"

Conjecture 7.4. *For all sufficiently large integers n, the only rational solutions to the equation*

$$X^n + Y^n + Z^n = 0$$

are the trivial ones, i.e., $XYZ = 0$.

8. Open Problems

In addition to the conjectures mentioned just above, the following problems are still unresolved at this time.

1. (Lang, 1965) Replace the error term $\varepsilon h_A(P)$ in the arithmetic variant of Theorem 2.2 with something sharper, e.g., $(1 + \varepsilon) \log h_A(P)$.

2. (Vojta, 1987) Prove a formula similar to (1) for rational points on varieties of higher dimension. In that case, however, one would first have to exclude a proper Zariski-closed subset Z from the set of points P. See also Schmidt (1975) and Vojta (1987), Example 3.5.1 for a discussion of a partial answer to this question, where the variety is projective space and the divisor D is a collection of hyperplanes.

3. (Lang, 1960b, p. 29) Generalize Theorem 5.1 to the case where X may contain the translate of an abelian subvariety of A. In this case the translated abelian subvarieties may contain infinitely many rational points, but it is conjectured that the rational points are not Zariski dense, unless X itself is a translated abelian subvariety of A. In Nevanlinna theory the analogue of this conjecture is called Bloch's conjecture; it was proved independently by Kawamata (1980) and Green and Griffiths (1980), both using work of Ochiai (1977). (In loc. cit., Lang actually makes the stronger conjecture that the rational points are contained in finitely many translated sub-abelian varieties contained in X; this conjecture is a consequence of the above conjecture, by the Kawamata structure theorem.)

References

Bombieri, E. (1990): The Mordell conjecture revisited. (To appear)

Faltings, G. (1983): Endlichkeitssätze für abelsche Varietäten über Zahlkörpern. Invent. math. **73**, 349–366; corrigendum Invent. math. **75**, 381

Faltings, G. (1990): Diophantine approximation on abelian varieties. Ann. Math. (to appear)

Green, M., Griffiths, P. (1980): Two applications of algebraic geometry to entire holomorphic mappings. The Chern Symposium 1979 (Proceedings of the International Symposium on Differential Geometry in Honor of S.-S. Chern, held in Berkeley, California, June, 1979). Springer, New York Heidelberg Berlin, pp. 41–74

Kawamata, Y. (1980): On Bloch's conjecture. Invent. math. **57**, 97–100

Kobayashi, S. (1970): Hyperbolic manifolds and holomorphic mappings. Marcel Dekker, New York

Lang, S. (1960a): Some theorems and conjectures in diophantine equations. Bull. AMS **66**, 240–249

Lang, S. (1960b): Integral points on curves. Publ. Math. IHES **6**, 27–43

Lang, S. (1965): Report on diophantine approximations. Bull. Soc. Math. France **93**, 177–192

Lang, S. (1974): Higher dimensional diophantine problems. Bull. AMS **80**, 779–787

Lang, S. (1983): Fundamentals of diophantine geometry. Springer, New York

Lang, S. (1987): Introduction to complex hyperbolic spaces. Springer, New York

Ochiai, T. (1977): On holomorphic curves in algebraic varieties with ample irregularity. Invent. math. **43**, 83–96

Roth, K. F. (1955): Rational approximations to algebraic numbers. Mathematika **2**, 1–20

Schmidt, W. M. (1975): Application of Thue's method in various branches of number theory. Proceedings of the International Congress of Mathematicians, R. P. James, ed., Canadian Mathematical Congress, Vancouver, pp. 177–186

Vojta, P. (1987): Diophantine approximations and value distribution theory. (Lecture Notes in Mathematics, vol. 1239). Springer, Berlin Heidelberg New York

Vojta, P. (1989): Mordell's conjecture over function fields. Invent. math. **98**, 115–138

Vojta, P. (1990): Siegel's theorem in the compact case. Ann. Math. (to appear)

Wirsing, E. (1971): On approximation of algebraic numbers by algebraic numbers of bounded degree. Proc. of Symp. in Pure Math. **XX** (1969 Number Theory Institute), pp. 213–247

Zariski, O. (1952): Complete linear systems on normal varieties and a generalization of a lemma of Enriques-Severi. Ann. Math. **55**, 552–592

Author Index

I indicates Volume I while II indicates Volume II.